INITIATIVES OF PRECISION ENGINEERING AT THE BEGINNING OF A MILLENNIUM

INITIATIVES OF PRECISION ENGINEERING AT THE BEGINNING OF A MILLENNIUM

10th International Conference on Precision Engineering (ICPE)
July 18–20, 2001, Yokohama, Japan

Sponsored by:
JSPE (Japan Society for Precision Engineering)

Co-sponsored by:
ASPE (American Society for Precision Engineering)
eu**spen** (European Society for Precision Engineering and Nanotechnology)

Edited by

Ichiro Inasaki
Keio University
Yokohama, Japan

KLUWER ACADEMIC PUBLISHERS

BOSTON / DORDRECHT / LONDON

Distributors for North, Central and South America:
Kluwer Academic Publishers
101 Philip Drive
Assinippi Park
Norwell, Massachusetts 02061 USA
Telephone (781) 871-6600
Fax (781) 871-6528
E-Mail <kluwer@wkap.com>

Distributors for all other countries:
Kluwer Academic Publishers Group
Distribution Centre
Post Office Box 322
3300 AH Dordrecht, THE NETHERLANDS
Telephone 31 78 6392 392
Fax 31 78 6546 474
E-Mail <services@wkap.nl>

Electronic Services <http://www.wkap.nl>

Library of Congress Cataloging-in-Publication Data

A C.I.P. Catalogue record for this book is available from the Library of Congress.

Printed on acid-free paper.

Printed in the United States of America.

Contents

Preface   xxiii

**Organizing Committee, International Advisory
Committee and Conference Executive Committee**   xv

Keynote Paper

Machining of Precision Parts and Microstructures
  E.Brinksmeier, O.Riemer, R.Stern   3

Roles of Quantum Nanostructures in Advanced Electronics
  H.Sakaki   12

Design of New Precision Machine Elements
  A.H.Slocum   18

Part I Cutting / Special machining

The Concept of Active Deflection Compensation and its Application
in Precision Forging
  E.Doege, J.Baumgarten, T.Neumaier   27

Coining of Thin Plates to Produce Micro Channel Structures
  G.Hirt, B.Rattay   32

Three-dimensional Micro-Forming Process of Thin Film Metallic Glass
in the Supercooled Liquid Region
  S.Hata, Y,Liu, T,Kato, A.Shimokohbe   37

Precision Cold Forging - Methods for Reduction of Working Pressure
  K.K.Tong, T.Muramatsu, C.M.Choy, S.X.Zhang, M.Enggalhardjo   42

Burr Formation in Micro-Machining Aluminum, 6061-T6
  K.Lee, B.Stirn, D.A.Dornfeld   47

Micro Structuring of High Aspect Ratio and Array by Means of Mechanical Machining
   K.Sawada, T.Kawai, Y.Takeuchi ... 52

Influence of Micro Machining on Strength Degradation of a Silicon Nitride Ceramic
   M.Wakuda, Y.Yamauchi, S.Kanzaki ... 57

Micro Material Processing Using UV Laser and Femtosecond Laser
   H.K.Tönshoff, A.Ostendorf, K.Koerber, C.Kulik, G.Kamlage ... 62

Subnanometer Fabrication of Optics by Plasma Chemical Vaporization Machining
   H.Takino, T.Kobayashi, T.Yamamoto, N.Shibata, Y.Gomei, K.Sugisaki ... 67

Micro Fabrication Using EDM Deposition
   S.Hayakawa , R.I.Ori, F.Itoigawa, T.Nakamura, T.Matsubara ... 72

Manufacture of Aspherical Fresnel Lens with Ideal Cross-sectional Profile with Oxygen-free Copper and Acrylic Resin
   N.Sornsuwit, Y.Takeuchi, T.Kawai, K.Sawada, T.Sata ... 77

Non-circular Machining Using Two Convex Milling Cutters
   Y.Nakao ... 82

A Method to Machine Three-Dimensional Thin Parts
   H.Obara, T.Watanabe, T.Ohsumi, E.Ninomiya, M.Hatano ... 87

Control of Surface Pattern of Mold Generated by Ball-End Milling
   A.Saito, X.Zhao, M.Tsutsumi ... 92

High-Speed Cutting - Fundamentals and Machine Tool Development
   H.K.Toenshoff, T.Friemuth, P.Andrae, C.Lapp ... 97

Dry Cutting of Stainless Steel Using Indexable Inserts Having Self-heat Absorbing Capability
   M.Murakawa, M.Jin ... 102

Measurement of Lubricant Applying Effect Influence on Lubrication by Oil-Submerged Cutting
   T.Kaneeda ... 107

Development of Inclined Ultrasonic-Vibration Cutting Tool Capable
of Mounting Commercial Indexable Inserts
　M.Jin, M.Murakawa　112

The Greentape Laser Sintering Method and its Applications
　K.Maekawa, T.Nishii, S.Iida, T.Suzuki　117

Conceptual Data Model for Advanced Rapid Prototyping
　K.G.Kobayashi, M.Fujii , F.B.Prinz　122

Machine Parts Manufacturing by Sheet Steel Lamination Technology
　J.Shinozuka, M.Yoshino, T.Obikawa　127

Textured Surface Produced by Anisotropic Etching of Silicon and its
Frictional Properties
　N.Moronuki, D.Nishi, K.Uchiyama　132

A Mechanical Vibration Assisted Plasma Etcher for Etch Rate
Improvement
　T.Hatsuzawa, M.Hirosawa, M.Hayase, T.Oguchi　137

Micromolding of Three-Dimensional Components
　W.N.P.Hung, Y.Ngothai, S.Yuan, C.W.Lee, M.Y.Ali　142

Surface Roughness of FIB Sputtered Silicon
　M.Y.Ali, N.P.Hung, S.Yuan　147

Micromachining in the Third Dimension
　F.Klocke, O.Ruebenach, T.Noethe　152

Optimization of Deep Hole Drilling Processes with Smallest
Drilling Diameters
　U.Heisel, M.Stortchak, R.Eisseler　157

Prediction of High Frequency Vibration in Fine Boring by Extended
Chatter Model
　E.Edhi, T.Hoshi　164

Thrust Force Analysis of Drilling Burr Formation Using Finite
Element Method
　S.Min, J.Kim, D.A.Dornfeld　169
Basic Studies on the Wear Behaviour of Modified and Coated

Diamond Tools for Precision Machining of Ferrous Materials
E.Brinksmeier, R.Gläbe ... 174

Difference in Wear Patterns of Diamond Cutting Tool Depending on Work Materials
H.Tanaka, S.Shimada, N.Ikawa, M.Higuchi, K.Obata ... 179

Ductile Mode Cutting of Single-crystal Silicon by Ultrasonic Vibration
S.Koshimizu, J.Otsuka ... 184

Regenerative Chatter Vibration in Ball End Milling of Curved Surfaces
B.W.Ikua, H.Tanaka, F.Obata, S.Sakamoto ... 189

Ultra-High Speed Discharge Control for Micro Electric Discharge Machining
S.Hara, N.Nishioki ... 194

Study of Contouring Micro EDM Characteristics
Z.Yu, K.P.Rajurkar, P.D.Prabhuram ... 199

Development of Desk-top Micro Electric Discharge Machine
S.Koga, N.Nishioki, S.Hara ... 204

Improving Process Characteristics of DRY-WEDM
C.Furudate, M.Kunieda, Y.Z.Bo, H.Yamada ... 209

U-shaped Curved Hole Creation by Means of Electrical Discharge Machining
T.Ishida, Y.Takeuchi ... 214

High Performance Slicing Method of Monocrystalline Silicon Ingot by Wire EDM
Y.Uno, A.Okada, Y.Okamoto, T.Hirano ... 219

Laser Ablation of Sapphire with a Pulsed Ultra-Violet Laser Beam
S.Tamura, H.Horisawa, S.Yamaguchi, N.Yasunaga ... 224

Metal Deposition on Glass by Laser Irradiation from Metal Powders
H.Tokura, H.Hidai ... 229

Ultra-Precision Cutting of Difficult-to-cut Metals
S.Sakamoto, H.Yasui, M.Kawada, S.Kondo   234

A Study on Metal Cutting under High Hydrostatic Pressure
M.Liu, J.Takagi   239

A New Interpolation Algorithm for Ultra-precision CNC Machining
S.Li, M.Zhang, Y.Dai, X.Xie   244

Six-Axis Control Character Line Finishing Using Ultrasonic
Vibrational Cutting Tool
K.Morishige, Y.Takeuchi   249

A New Diamond Turning Method for Fabrication of Convex
Aspheric Surface on Hard Brittle Material
J.Yan, K.Syoji, T.Kuriyagawa   254

Fracture Phenomena of Aramid Fiber During Oblique Cutting
E.Nakanishi, K.Isogimi   259

Development of a Practical Laser-Guided Deep-Hole Boring Tool:
Working Characteristics of Its Components
A.Katsuki, H.Onikura, T.Sajima, Y.Yuge, H.Murakami, T.Katayama   264

Oil Film on Water Fog Metalworking Fluid
T.Nakamura, T.Matsubara, F.Itoigawa, K.Kawata   269

On the On-set of Chip Formation and the Process Stability in Cutting
at Atomic Level
R.Rentsch   274

Analysis of Tool Temperature in High-Speed Milling
H.Sasahara, T.Nitta, K.Nishi   279

Research on Precise Cutting Performance of CVD Diamond
H.Zhang, Y.Yao, Z.Yuan   284

Machinability of TiAl Intermetallic Compounds
T.Furusawa, A.Ichikawa, H.Hino, S.Tsuji, S.Koroyasu   289

MDS Study on the Effect of Cutting Edge Radius of Diamond Tools
in Nanometric Cutting Process for Brittle Materials
   D.Li, S.Dong, Y.Liang , X.Luo, K. Cheng   294

An Attempt to Develop a Short Lasting Machinability Test for Steels
   J.C.Hamann, F.Meslin, Mactest   299

Micro Surface Damage Detecting and Repairing System by Precise
Micro Robots
   H.Aoyama, S.Miyamoto, R.Watanabe   304

Effect of Minimal Quantities of Lubricant in Micro Milling
   JRS.Prakash, M.Rahman, A.S.Kumar, SC.Lim   309

Development of MEMS IC Probe Card Utilizing Fritting Contact
   T.Itoh, K.Kataoka, T.Suga   314

Simulation of WEDM Using Discharge Location Searching Algorithm
   F.Han, M.Kunieda, T.Sendai, Y.Imai   319

Comparative Study of HSM and EDM in Injection Mould
Manufacturing
   M.R.Alam, K.S.Lee, M.Rahman, Y.F.Zhang, Y.D.Li, K.S.Sankaran   324

Net-Shaping via Aluminium Foaming
   Y.P.Kathuria   329

Development of Warm Forming Technique to Produce Thin Wall
Magnesium Components
   M.S.Yong, B.H.Hu, M.Enggalhardjo, C.M.Choy   334

Gas Flow Characteristics in a Laser Cut Kerf
   H.Horisawa   339

A New Method for Reduction of Residual Stress of Welded Joint
Using Ultrasonic Vibrational Load
   S.Aoki, S.Hirai, T.Nishimura, T.Hiroi   344

Three-dimensional Microcasting
   H.Noguchi, M.Murakawa   349

Improvement of Properties of SiC Mirror Surface Epitaxially
Grown on Silicon Substrate
    A.Kakuta, K.Hashimoto, N.Moronuki, Y.Furukawa   354

Part II Grinding / polishing / ultra-precision machining

Precision Grinding of Micro Aspherical Surface
    H.Suzuki, O.Horiuchi, H.Shibutani, T.Higuchi   361

Mirror Grinding of Silicon Wafer with Silica EPD Pellets
    T.Fukazawa, N.Fuwa, J.Ikeno, H.Shibutani, O.Horiuchi, H.Suzuki   366

New Fabrication Process of Deep Paraboloidal Mirror by Combination
of ELID-Grinding and Electro-forming Techniques for Laman
Spectroscopy Instrument
    H.Ohmori, J.Guo, W.Lin, T.Matsuzawa, S.Morita, H.Sato, H.Tashiro   371

A Study on the Surface Integrity of Single Crystal Silicon Ground by
CIFB-diamond Wheels (ELID) and Resin-bonded Diamond Wheels
    C.L.Chao, K.J.Ma, D.S.Liu, S.C.Sheu, Y.S.Lun, H.Y.Lin   376

A Study on Scratch Reduction Using the Sonic Dispersion of CMP Slurry
    S.H.Cho, H.J.Kim, H.Y.Kim, K.J.Kim, H.D.Jeong   381

A Study on the Chemical Mechanical Micro Machining (C3M) of Silicon
    S.C.Jeong, J.M.Park, H.D.Jeong   386

Development of a Lapping Film Utilizing Agglomerative Superfine
Silica Abrasives for Edge Finishing of a Silicon Wafer
    T.Enomoto, Y.Tani, K.Orii   391

Cryogenic Polishing Method of Optical Materials
    F.Zhang, R.Han, Y.Liu, S.Pei   396

Characteristics of Small Rotary Tool in Polishing of Fused Silica
    C.Liu, T.Kasai, H.Ohmori, W.Lin   401

Development of Electrostrictive Polyurethane(PU) Films and its
Characterization
    E-S Jeong, H-S Park, H-D Jeong, N-J Jo   406

Precision Machining of Future Silicon Wafers
   F.Klocke, D.Pähler   411

High Speed Grinding Performance and Material Removal Mechanism
of Silicon Nitride
   H.Huang, L.Yin   416

Precision Manufacturing of Micro Moulding Tools of Glass and
Sintered Carbide
   H.W.Hoffmeister, A.Wenda   421

An Experimental Study on Bore Honing Operation - Influence of
Unequal Over-run on Geometrical Error -
   J.Akbari   426

Investigations Regarding the Operating Range of Ultra Thin
Grinding Wheels on AlTiC
   H.H.Gatzen, C.Morsbach, G.M.Jones   431

Errors Analysis of Ball-Headed Wheel Discharge Dressing and Study
on Grinding Experiments in Ultra-precision Aspheric Grinding System
   M.Chen, S.Dong, F.Zhang, D.Li   436

Simulation on High Integrity Surface Generation of ϕ300 Si Wafer
with Fixed-Abrasive Solution
   K.Sagawa, H.Eda, L.Zhou, J.Shimizu   441

Effect of Wheel Revolutional Speed on Striped Pattern on Surfaces
Finished by High-Speed Reciprocation Grinding
   N.Yoshihara, T.Kuriyagawa, K.Syoji   446

Mechanisms for the Formation of Glossy Workpiece Surface in
Grinding and Polishing of Natural Stones
   H.Huang, X.Xu, H.Xu   451

Mechanical Properties and Grinding Characteristics of High
Purity Polycrystalline CBN Grits
   Y.Ichida, Y.Morimoto, R.Sato, M.Saijo, T.Koinuma   456

Influence of Cutting Edge Wear on Ductile-mode Grinding of Fine
Ceramics with Coarse Grain Size Diamond Wheel
   H.Yasui, Y.Hiraki, M.Sakata, M.Tsurusaki, Y.Murayama, S.Sakamoto 461

Study on Internal Grinding of Small Bore - Influence of
Bearing Characteristics on Grinding Accuracy -
　K.Yamauchi, J.Takagi　466

A Basic Study on Profile Grinding with ELID
　H.Shindo, H.Ohmori, T.Kasai　471

Grinding Characteristics of Cemented Carbide Concave Mirror by
Desk-top Type 4-Axes Machine "TRIDER-X" with ELID System
　Y.Uehara, H.Ohmori, Y.Yamagata, S.Moriyasu, S.Morita,
　K.Yoshikawa, M.Asami, T.Miura　476

Finish Surface Grinding of Titanium Alloys
　Z.Yuan, B.Zhu, Z.Lu, F.Zhang　481

Flattening of Micro-Functional Parts by ELID Lap Grinding
　N.Itoh, H.Ohmori, Y.Yamagata, S.Moriyasu, T.Kasai　486

Development of Small Tool by Micro Fabrication System Applying
ELID Grinding Technique
　Y.Uehara, H.Ohmori, Y.Yamagata, S.Moriyasu, W.Lin,
　K-I Kumakura, S-Y Morita, T.Shimizu, T.Sasaki　491

Effect of Polishing Pads in Finishing of Large Optical Elements
　W.Lin, H.Ohmori, Y.Yamagata, S.Moriyasu, C.Liu,, T.Kasai　496

Effects of Changes in Fluid Composition on Magnetorheological
Finishing (MRF) of Glasses and Crystals
　S.D.Jacobs, S.R.Arrasmith, I.A.Kozhinova, S.R.Gorodkin, L.L.Gregg,
　H.J.Romanofsky, T.D.BishopII, A.B.Shorey, W.I.Kordonski　501

Study of Magnetic Abrasive Finishing Process Using Pulsed
Finishing Pressure and Its Processing Characteristics
　S.Yin, T.Shinmura　506

Chain Forming Process of Magnetic Brush Development System Used
in Laser Printer
　N.Nakayama, H.Kawamoto, M.Yamaguchi, J.Wiphut　511

Part III Machine / element / measurement

Invited paper:
Precision Machine Tools
M.Weck, R.Hilbing, C.Peschke — 519

Effects of Bearing Surface Geometries on the Inclination Stiffness
of Aerostatic Thrust Bearings
S.Ohishi, K.Tanaka, T.Masada — 524

Development of a Three Axes Travelling Column Ultraprecision
Milling Machine
A.Herrero, R.Bueno — 529

Development of Hydrostatic Bearings with Groove Structures
M.Weck, J.Hennig, M.Winterschladen — 534

Assessment of Thermophysical Properties at Design Stage of Machine
Tool Structure with Thermal Symmetricity
M.Okabe, H.Sakamoto, S.Shimizu — 539

Concrete-Based Constrained Layer Damping
E.Bamberg, A.H.Slocum — 544

Effects of Manufacturing Errors on the Accuracy for TRR-XY
Hybrid PKM
T-H Chang, S-L Chen, M-H Hsei — 549

Effect of Static Stiffness of Grinding Systems on a Generating
Mechanism of Workpiece Geometrical Accuracy
H-S Lee, Y.Uchida — 554

Development of a New Reseatable Mechanism with High
Accurate Positioning Reproducibility
S.Koga, N.Nishioki — 559

Compensation of Periodic Errors by Gain Tuning in Multi-Probe
Sensors
I.Godler, T.Ninomiya — 564

Precision Positioning of a Surface Motor-Driven XY θ z Stage Using
a Surface Encoder
   W.Gao, T.Nakada, S.Kiyono   569

Feedforward Tracking Controller Design Based on a Limited
Bandwidth Desired Model
   L.Wang, F.Zhang, B.Su   574

Fuzzy-PID Hybrid Vibration Control of Slide Carriage of
Ultraprecision Machine Tool
   D.Shen, W.Jiachun, L.Dan   579

Invited paper:
Nanometrology - The Frontier of Precision
   R.J.Hocken   584

Development of Ultraprecision 5-Axis Machine Tool Equipped
with Elliptical Vibration Cutting Device
   T.Moriwaki, E.Shamoto, K.Tanaka, M.Matsuo, M.Osada   589

Development of On-machine Profile Measuring System with
Contact-type Probe
   S.Moriyasu, S.Morita, Y.Yamagata, H.Ohmori, W.Lin,
   J.Kato, I.Yamaguchi   594

An Instrument to Measure the Stiffness of MEMS Mechanisms
   J.Qiu, J.Sihler, J.Li, V.Sturgeon, M.Smith, A.Slocum   599

Development of X-ray Stepper with High Overlay Accuracy for 100-nm
LSI Lithography
   M.Fukuda, H.Morita, T.Haga, M.Suzuki, H.Tsuyuzaki,
   A.Shibayama, S.Ishihara, H.Aoyama, S.Mitsui, T.Taguchi, Y.Matsui 604

Wide Range Capacitive Gap Sensor with Reproducible Nano-meter-
order Resolution
   N.Ishikawa, N.Nishioki   609

Design and Development of Multi-Function Sensor Using a Group of
Micro Cantilevers
   K.Uchiyama, N.Moronuki   614

A Tool to Estimate the Adhesive Forces between Microcomponents Among
each other and between Microcomponents and Grippers or Magazines
J.Hesselbach, C.Graf   619

High Precision Collimation Using Talbot Interferometry
S.Haramaki, S.Yokozeki, A.Hayashi, H.Suzuki   624

A Novel Approach for Simultaneous Measurement of the Linear Guide
way Errors of a Machine Tool with a Volumetric Optical Encoder
U.Mueller, Y.Kagawa, K.Yamazaki, J.Braasch   629

Evaluation of Stages of Nano-CMM
M.Fujiwara, A.Yamaguchi, K.Takamasu, S.Ozono   634

Evaluation Method of Thermal Displacement of Machine Tools
S.Shimizu, N.Imai   639

Taut Wire Straightedge Reversal Artifact
J.G.Salsbury, R.J.Hocken   644

Automated System for Three-Dimensionl Roughness Testing
J.Rudzitis, J.Krizbergs, M.Skurba   649

Development of Micro-roughness Measuring Probe Using
Longitudinal Tapping Mode by Ultrasonic Vibration Sensor
A.Saitoh, K.Hidaka, T.Yamagiwa, K.Nishimura   654

Effects of Surface Roughness of Si Substrate and Pt-C Multilayer
Coated Film on X-Ray Reflectivity
Y.Namba, K.Zushi, T.Tanaka, K.Yamashita, Y.Tawara, T.Okajima   659

Detection of Surface Defects on Steel Ball in Bearing Production
Process Using a Capacitive Sensor : Performace of Prototype System
T.Matsuda, M.Sato, T.Yata, A.Kakimoto   664

Advanced 3D-Measuring Techniques for Quality Control of
Cutting Inserts
J.Leopold, I.Inasaki   669

Autonomous Profiling Measurement of Two Dimensional Section
FormsUsing Small-sized Ultrasonic Probe
K.Matsuki, K.Hidaka   674

Interferometry with Null Optics for Testing Aspherical Surface at
1nm Accuracy
  T.Gemma, S.Nakayama, H.Ichikawa, T.Yamamoto, Y.Fukuda,
  T.Onuki, T.Umeda   679

Analysis of Defects on SiO$_2$ Filmed Wafer -Evaluation of CMP
Defect Detection Schemes Using Computer Simulation(BEM)-
  T.Ha, T.Miyoshi, Y.Takaya, S.Takahashi   684

Palmtop Pantograph Mechanism with Large-Deflective Hinges for
Miniature Surface Mount Systems
  M.Horie, T.Uchida, D.Kamiya   689

Dynamics of Pin Electrode in Pin-To-Plate Gas Discharge System Used
for Ozone-less Charger in Laser Printer
  H.Kawamoto, K.Takasaki, H.Yasuda, N.Kumagai   694

Effects of Part Flexibility on Dynamic Behaviors of Machine Systems
with Clearances in the Joints
  Y.Wuyong, J.Linhong, J.Dewen   699

Analysis of Feed Drive System of X-Y Table Considering the Friction
  Q.Sun, X.Mei, M.Tsutsumi   704

Rotational Accuracy and Positioning Resolution of an Air-bearing
Spindle with Active Inherent Restrictors
  H.Mizumoto, S.Arii, M.Yabuya   709

Precision Control of Radial Magnetic Bearing
  X.Zhang, T.Shinshi, L.Li, K-B Choi, A.Shimokohbe   714

The Effects of Joint Clearance on the Motion Precision of the Machines
  J.Xiaohong, J.Linhong, J.Dewen, Z.Jichuan   719

Tool Trajectory Error for Table-tilting Machines
  Y-R Hwang, M-C Ho   724

Automated Calibration for Assembly Device Installation Based on Plug
& Produce Concept
  H.Kikuchi, Y.Maeda, M.Sugi, T.Arai   729

Calibration of 2-DOF Parallel Mechanism
  O.Sato, M.Hiraki, K.Takamasu, S.Ozono   734

Development of a Calibration Equipment Using Laser
Interferometer Combining a Variable Length Vacuum Cell - Structure
and Measurement -
  H.Masuda, Y.Kuriyama, H.Sakai, M.Ueda   739

Development of Calibration Equipment Using Laser
Interferometer Combining a Variable Length Vacuum Cell - Control
and Application -
  H.Oozeki, S.Kiyotani, M.Ogihara   744

A Measurement Force Estimation Method of Tapping Stylus
  K.Takahashi, M.Hayase, T.Hatsuzawa   749

Investigation of the Plastic Deformation and Fracture of Carbon
Nanotubes
  K.J.Ma, C.L.Chao, C.W.Tang, K.H.Chen, L.C.Chen   754

Molecular Dynamics Simulation on Dependence of Atomic-Scale Stick-
Slip Phenomenon upon Probe Tip Shape
  J.Shimizu, H.Eda, L.Zhou   759

Tool Wear Monitoring in Milling Process with Laser Scan Micrometer
  T.Matsumura, T.Murayama, E.Usui   764

Study on a Modified Concave Filter for Multi-Oscillating
Modalities Compensation in Ultra-Precision Machine Tool
  B.Su, L.Wang, S.Dong   769

New Practical Control of Point-To-Point Positioning Systems :
Robustness Evaluation
  Wahyudi, K.Sato, A.Shimokohbe   774

The Research of an Intelligent Open CNC System
  J.Zhang, L.Wang, S.Li, F.Zhang   779

Robot Intelligence Augmented by Information Technology
  S.Sakakibara   784

Standard API for Open-Architecture CNC and its Application to HMI
and Operation Monitoring
    S.Ueno, M.Mitsuishi, K.Muto, S.Takata   789

Development of Large Ultraprecision Mirror Surface Grinding System
with ELID
    K.Katahira, H.Ohmori, M.Anzai, Y.Yamagata, A.Makinouchi,
    S.Moriyasu, W.Lin   794

Development of a New Measurement System for Straightness Error by
a Heterodyne Interferometer with a Grating
    S.Asano, T.Goto, H.Tanimura, K.Mitsui   799

Development of Measuring System Equipped with Wire Probe to
Measure Fine Contour Shape of Penetrated Specimen
    H.Obara, M.Tsuji, T. Ohsumi, M.Hatano   804

Development of Ultra Precision Long End Standard Installing
Vacuumed Optical Path Laser Interferometer
    Y.Kuriyama, J.Ishikawa, M.Ueda, Y.Yokoyama   809

Flatness Measurement System Using Fizeau Type Solid Interferometer
with High Speed Acquiring Multi-phase Interferograms and Highly
Effective Error Decreasing Algorithm
    K.Kawasaki, Yasushi Ueshima, Naoki Mitsutani, Hiroshi Haino   814

Proposal of Absolute Length Measuring Machine by Combining
Crystalline Lattice Scale and Laser Interferometry
    P.Rerkkumsup, M.Aketagawa, K.Takada, T.Takagi,
    T.Watanabe, S.Sadakata   819

Evaluation of Acceleration and Deceleration Profile of Velocity and a
New Design Method Using Gaussian Distribution Curve
    M.Ogihara, Y.Zhang, H.Oozeki, Y.Kuriyama   824

Thermal and Mechanical FEA (Finite Element Analysis) of Wafer
Heating for EB Stepper
    S.Takahashi, Y.Miki, K.Morita, N.Hirayanagi, T.Fujiwara,
    M.Tateishi, K.Maeno   829

Part IV Manufacturing system / CAD / CAM

Modeling of Machining Process Planning based on Inverse Form-
Shaping Function
    F.Tanaka, N.Toyama, T.Kishinami   837

Easy Construction of 3D Model from 2D Information
    Y.Urabe, H.Aoyama   842

Virtual Reality for NC-programming
    H.K.Tönshoff, V.Böß, N.Rackow   847

Development and Estimation of Case Retrieval Model For
Machining Knowledge Support System on Machine Design
    Y.Fukushima, S.Nagasawa   852

Integrated Process Planning and Production Control -A Flexible
Approach Using Co-operative Agent Systems-
    H.K.Toenshoff, P.O.Woelk, O.Herzog, I.J.Timm   857

Dynamic Job Shop Scheduling Using a Neural Network: Multi-
stage Training and Selection of Input Information
    T.Eguchi, F.Oba, S.Toyooka   862

An Organizational Concept for Manufacturing of High Precision Products
    W.Eversheim, I.Fricker, M.Koschig, N.Michalas   867

Considering Feature Interactions for the Development of a Process
Planning System
    H. Muljadi, K.Ando, M.Ogawa, T.Sakurai   872

Strategic Product Development Decision-making Based on
Optimization Incorporating the QFD Method
    M.Yoshimura, K.Izui   877

Development of a Logistic Operating Curve for an Entire Manufacturing
Department -Logistic Process Operating Curve (LPOC)-
    H-P Wiendahl, M.Schneider   882

Fixturing Feature for Determining Optimum Clamping Points: Concept
of Fixturing Feature and Its Application to 2-D and 3-D Workpiece Models
    I.Zaitsu, H.Aoyama, T.Aoyama   887

Feature-Based Fixture Planning for Workpieces
   T.Inoue, I.Inasaki   892

Simulation of Manufacturing System Design for Uncertain Market Conditions
   H.Fujimoto, A.Ahmed   897

Construction of Master Model from 3D Scanned Data Using Recursive Subdivision Scheme
   B-Y Ren, I.Hagiwara, Q-X Meng   902

A Study on Knowledge Engineering for Equipment Design and Production
   S.Takeo, N.Kurochi, S.Tsuchiya, H.Nakamura   907

Life Cycle Support of Mechanical Products Using Network Agent
   M.Suzuki, K.Sakaguchi, H.Hiraoka   912

Aesthetic Design Based on "Kansei Language"
   M.Ota, H.Aoyama   917

A Design Review and Redesign System for NURBS Models with a Haptic Device
   H.Takahashi   922

Free-form Surface Manufacture for TRR-XY Hybrid Parallel Link Machine Tool
   T.H.Chang, S.L.Chen, C.A.Kang, K.W.Chen   927

Manufacturing Process Control Based on the Delta Operator
   T.Aoki   932

Assembly Reliability Evaluation Method (AREM)
   T.Suzuki, T.Ohashi, M.Asano   937

Maintenance Support System Equipped with Monitoring Function
   T.Tateno, M.Igoshi   942

Development of Software Platform for Realizing the Dynamic Reconstruction of Software Applications of Manufacturing Systems
   T.Satake, A.Hayashi, G.Zhang   947

Mechatronic Service and Investigation on the Example of a Forming and Milling Machine
J.Hamann, H-P Tröndle   952

Gaussian-Based Free Form Deformation and its Application to Fashion Design
N.Yoshida, M.Usui, K.Kitajima   957

A Proposal of Assembly Model Framework Specialized for Unified Parametrics
S.Sanami, N.Yoshida, K.Kitajima   962

Precision Machining of Thin-Walled Workpiece by NC Compensating Deflection
N.He, K.Wu, Z.Wang , C.Jiang, D.Zuo   967

Index of Contributors   973

Preface

Years of rapid development in manufacturing have created an environment that relentlessly searches for innovation and technological breakthroughs. Faced with ever-increasing market demand, manufacturing industries are striving to convey that momentum of advancement into the 21st century. From the viewpoint of global technology and engineering, it is crucial that the future aim of the industry is well perceived and that pertinent course of action is cooperatively taken. The purpose of the 10th International Conference on Precision Engineering (ICPE) is to promote research in various areas of production and precision engineering as well as to support the discussion of topics in advanced technologies of related fields.

On behalf of the Organizing Committee, I am pleased to inform you that the 10th ICPE was successfully held in Yokohama, Japan from July 18 to 20, 2001. This book presents the proceedings of that conference. The 10th ICPE holds a special meaning as it was realized by the collaboration of the American Society for Precision Engineering (ASPE), European Society for Precision Engineering and Nanotechnology (euspen), and the Japan Society for Precision Engineering (JSPE). These three societies have collaborated in various practical aspects of society activities for the past three years, and they jointly publish the journal "Precision Engineering". The 10th ICPE is their second collaboration. Similar regular joint international conferences are planned for the future.

The proceedings of the conference include 190 contributed papers from 10 countries. Two referees from the Program Committee reviewed each contributed paper in terms of originality and quality. It is our hope that this book will stimulate further research and development in the production and precision engineering.

I would like to express my sincere appreciation to the Organizing Committee, the International Advisory Committee, the Conference Executive Committee and all contributors for their considerable efforts in making this book available in its present form. I also wish to thank the Amada Foundation for Metal Work Technology, Electro-Mechanical Technology Advancing Foundation, FANUC FA and Robot Foundation, MAZAK Foundation, and Yokohama Convention Subsidy for their financial support. It should be mentioned that the conference was co-sponsored by the International Institution for Production Engineering Research (CIRP).

Ichiro Inasaki
Chairman, Organizing Committee
10th International Conference on Precision Engineering

Organizing Committee:
Chairman:
I.Inasaki (Keio Univ./Japan)
Executive secretary:
T.Aoyama (Keio Univ./Japan)
H.Tokura (Tokyo Institute of Technology /Japan)
Committee members:
S.Enomoto (Chiba Institute of Technology /Japan)
C.Evans(NIST /USA)
Y.Furukawa (Tokyo Metropolitan Univ. /Japan)
S.Ishihara (NTT Corp./Japan)
F.Kimura (Univ. of Tokyo /Japan)
S.Kiyono (Tohoku Univ./Japan)
T.Masuzawa (Univ. of Tokyo /Japan)
P.McKeown (Cranfield Univ./UK)
K.Mori (MEL /Japan)
T.Moriwaki (Kobe Univ./Japan)
T.Nishiguchi (Hitachi Corp./Japan)
H.Ohmori (RIKEN /Japan)
Y.Saito (Tokyo Inst.of Tech ./Japan)
S.Shimizu (Sophia Univ./Japan)
S.Takata (Waseda Univ./Japan)
Y.Uda (Nikon Corp./Japan)

International Advisory Committee :
T.Altan (The Ohio State Univ./USA)
Y.Altintas (The Univ. of British Columbia /Canada)
G.Arndt (Univ.of Wollongong /Australia)
M.Bonis (Compiegne /France)
C.D.Bouzakis (Aristoteles Univ./Greece)
E.Brinksmeier (Univ.Bremen /Germany)
R.Bueno (TEKNIKER /Spain)
G.Byrne (Univ. College Dublin /Ireland)
T.H.C.Childs (University of Leeds /UK)
T.E.Clayton (NIST /USA)
J.Corbett (Cranfield Univ./UK)
D.A.Dornfeld (Univ. of California /USA)
U.Heisel (Univ. Stuttgart /Germany)
R.J.Hocken (Univ. of North Carolina-Charlotte /USA)
R.Jablonski (Warsaw Univ. of Technology /Poland)
S.D.Jacobs (Univ. of Rochester /USA)
K.Jemielniak (Warsaw Univ. of Technology /Poland)
H.J.J.Kals (Univ. of Twente /Netherlands)

F.Klocke (T.H. Aachen /Germany)
H.Kunzmann (PTB-Braunschweig /Germany)
J.Lee (Univ. of Wisconsin-Milwaukee /USA)
J.M.Lee (Seoul National Univ./Korea)
K.I.Lee (Seoul National Univ./Korea)
L.C.Lee (Gintic Institute of Manufacturing Technology /Singapore)
F.LeMaitre (Ecole Centrale de Nantes /France)
D.A.Lucca (Oklahoma State Univ./USA)
L.Monostori (Hungarian Academy of Sciences /Hungary)
G.Peggs (National Physical Lab./UK)
M.Rahman (National Univ. of Singapore /Singapore)
S.Sartori (Italian National Research Council /Italy)
A.H.Slocum (MIT /USA)
H.K.Toenshoff (Univ. Hanover /Germany)
B.S.Y.Tsai (Industrial Technology Research Institute /Taiwan)
E.Uhlmann (TU Berlin /Germany)
H. Van Brussel (K.U.Leuven /Belgium)
M.Weck (T.H. Aachen /Germany)
R.Wertheim (ISCAR /Israel)
E.Westkaemper (Univ. Stuttgart /Germany)
S.Z.A.Zahwi (National Institute for Standards /Egypt)
G. X .Zhang (Tianjin Univ./China)

Conference Executive Committee :
Chairman
S.Inaba （FANUC LTD.）
Committee members:
AMADA FOUNDATION FOR METAL WORK TECHNOLOGY
ELECTRO-MECHANIC TECHNOLOGY ADVANCING FOUNDATION
FANUC FA and Robot Foundation
MAZAK Foundation
Yokohama Convention Subsidy

FANUC LTD.
NIKON CORPORATION
SEIKO EPSON CORPORATION
KURODA PRECISION INDUSTRIES
TOKYO SEIMITSU CO., LTD.

CANON INC.
DENSO CORPRATION
DISCO CORPORATION
Hitachi, Ltd. Production Engineering Research Laboratory

Japan Machine Tool Builders' Association
Kistler Japan Co., Ltd.
KONICA CORPORATION
Kosaka Laboratory Ltd. Measuring Instruments Dept.
Matsuura Machinery Corporation
MITSUBISHI HEAVY INDUSTRIES, LTD. HIROSHIMA RESEARCH &
DEVELOPMENT CENTER
Mitsubishi Electric Corporation Nagoya Works
Mitutoyo Corporation
MORI SEIKI, CO, LTD.
NACHI-FUJIKOSHI CORP.
NAGASE INTEGREX Co., Ltd.
NEC Corporation
NGK INSULATORS, LTD.
Nippon Lever K.K.
NIPPON TELEGRAPH AND TELEPHONE CORPORATION
NSK Ltd.
NTN corporation
RICOH COMPANY, LTD.
RIKEN CORUNDUM Co., LTD.
SAN SEIMITSU KAKO LAB., LTD.
SHIGIYA MACHINERY WORKS LTD.
Sumitomo Electric Industries, Ltd.
Taylor Hobson K.K.
THK CO., LTD.
Yamatake Corporation

ASAHI DIAMOND INDUSTRIAL CO., LTD. RESEACH AND
    DEVELOPMENT CENTER
ELIONIX. INC
Fuji Die Co., Ltd.
Ikegami precision tooling co., Ltd.
Maruto Instrument Co., Ltd.
MIKUNI ADEC CORPORATION
The NEXSYS Corp. (RIKEN Venture)
NORITAKE DIAMOND INDUSTRIES CO.,LTD
TAGA ELECTRIC CO., LTD.
TOSEI ENGINEERING CORP.
TOSHIBA MACHINE CO., LTD.
TZILLION CCEDSS CO., LTD.

Keynote paper

MACHINING OF PRECISION PARTS AND MICROSTRUCTURES

Ekkard Brinksmeier, Oltmann Riemer, Roland Stern

Laboratory for Precision Machining (LFM)
University of Bremen, Germany

Abstract

Many qualified technologies have been established for the manufacture of precision parts and microstructured surfaces in the field of MEMS, e.g. UV-lithography, silicon-micromachining, LIGA and in the range of energy assisted processes like Laser Beam Machining, Focused Ion Beam Machining or Electron Beam Machining. Nevertheless, mechanical processes, e.g. diamond machining, engraving, forming and molding, play a significant role for the generation of microstructured surfaces and precision parts.

An overview of recent micromachining processes will be considered. The thought will be given also to examples of applications, the selection of materials, machine tools, as well as handling- and mounting techniques.

Keywords

precision machining, microstructures, machine tools, materials, handling

1. INTRODUCTION

Although the products of micromanufacturing technologies have become a part of everyday life, people are frequently not aware of them. Microsystem technologies and also micro-engineering technologies have gained central importance in modern technology. Components and systems with microstructured surfaces reflect or transport light, connect surfaces without adhesives, shape sound waves or transport liquids. The design of these surfaces is often derived from what we find in nature. Examples are the leaves of a Lotus flower [BaNe97], moth eyes [Bletal99,Goetal98] or the skin of a shark [MaMT91]. The structured surface of the tropical plant allows to easily wash away dirt particles. The eye of the moth minimizes reflections by its pin-shaped surface to avoid attracting potential predators. The riblet structure of a shark's skin reduces the flow resistance significantly [fast00]. All these microstructures found in nature can be imitated and utilized in new products [WeDF00].

This paper provides a classification of commonly employed technologies in the machining of precision parts and microstructures. Some of the most frequently used methods and suited handling and mounting techniques are discussed. Selected applications in the field of micro-engineering technologies are of particular interest.

2. CLASSIFICATION

The machining of precision parts and microstructures can be subdivided into two general types of technologies (cf. Figure 1): Microsystem technologies (MST) and micro-engineering technologies (MET). MST are qualified for the manufacture of products of Micro Electro Mechanical Systems (MEMS) and Micro Opto Electro Mechanical Systems (MOEMS) including UV-lithography, silicon-micromachining and LIGA .

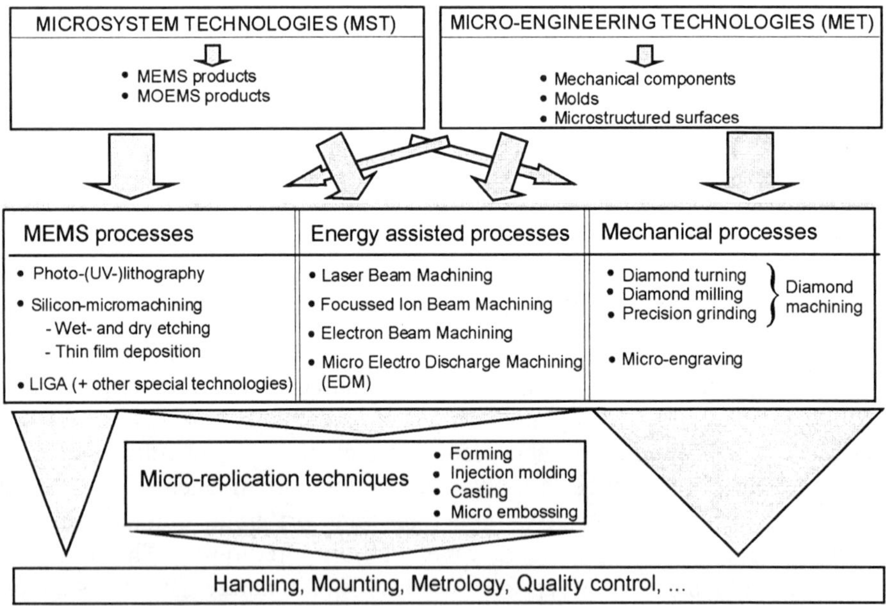

Figure 1. Process technologies for machining of precision parts and microstructures. (006)

Micro-engineering technologies (MET), on the other hand, comprise the production of highly precise mechanical components, molds and microstructured surfaces. Typical mechanical processes employed in this field are diamond machining (diamond turning, diamond milling and precision grinding), as well as micro engraving.

Furthermore, manufacturing processes like Laser Beam Machining, Focused Ion Beam Machining, Electron Beam Machining and Micro Electro Discharge Machining are classified as energy assisted processes.

Figure 1 shows these three groups of microstructuring processes. The size of the arrows indicates how frequently the three groups of processes are employed in the two base technologies MST and MET; the figure suggests that there can also be an overlap between the categories. The products can either be manufactured individually or as mass products using molds in micro-replication techniques like forming, injection molding and casting.

2.1 MEMS processes

Micro Electro Mechanical Systems (MEMS) include sensors, actuators as well as elements of data processing and interfaces to exchange information and energy with the environment. MEMS, manufactured by MST, can be found in a wide range of products. They are particularly applied in telecommunication systems, in the automotive industry, in chemical and medical applications and, furthermore, in the environmental and food industry. The most important MEMS processes are UV-lithography, silicon-micromachining and LIGA [mems01].

UV-lithography is the base process of all MST processes and therefore important in the production of microchips by the semiconductor industry.

Silicon-micromachining can be subdivided into wet- and dry etching technologies and thin film deposition technologies. Wet- and dry etching are used to generate structures into silicon work pieces (cf. Figure 2).

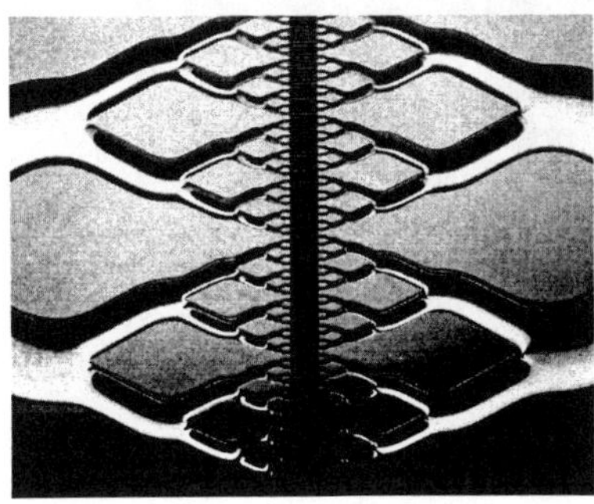

Figure 2. Glass silicon microfluidic system (micromixer), produced by a dry etching process: Deep-Reactive Ion Etching (DRIE) and anodic bonding. The size of the clipping is approximate 4 square millimeters (courtesy of Protron Mikrotechnik, Bremen).

Thin film deposition signifies technologies which are utilized to deposit functional layers, e.g. isolation layers or sensitive layers [dbank01].

The LIGA technology comprises the processes of X-ray lithography, electroforming and casting. LIGA enables the manufacture of micro-components made of non-silicon materials like plastics, metals and ceramics with almost any kind of lateral geometry and very high aspect ratios. For LIGA, in most cases, PMMA is used as resist material. Compared to optical or UV-lithography, almost parallel high energy synchrotron rays used in X-ray-lithography enables the manufacture of very deep structures with almost vertical and very smooth side walls. When these structures are produced in polymers, the exposed structured areas can be filled by electroplating with different metals like nickel, gold, copper or certain alloys. Once the PMMA is dissolved, metallic microstructures are left [MeB93,Sc00].

2.2   Energy assisted processes

Energy assisted processes comprise Laser Beam Machining, Focused Ion Beam Machining, Electron Beam Machining and Micro Electro Discharge Machining (Micro EDM).

Laser beams are used to join components and to manufacture microstructures. Different types of lasers have been established for machining of diverse materials (cf. Figure 3). Nd:YAG-lasers (λ = 1.06 µm) work well with metals and diamonds. Excimer lasers (λ = 193 nm – 350 nm) can be used to produce extremely small structures. For example, six lines of

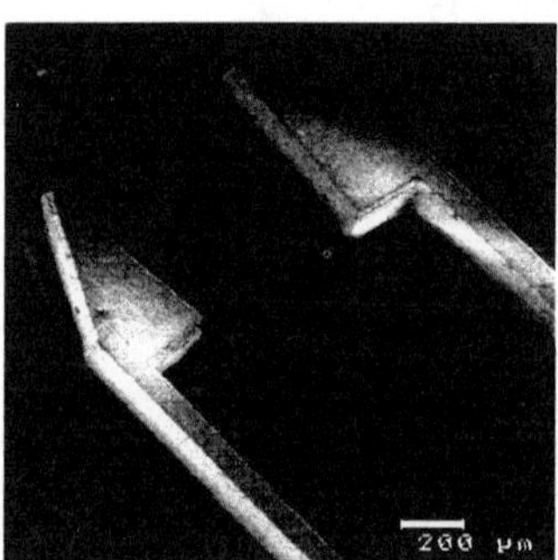

Figure 3. Left: Micro-gripper, machined by laser-assisted thermochemical etching using a cw-Nd:YAG laser (λ = 1064 nm). Thickness of the NiTi foil t = 200 µm. Right: Part of a large-area microstructure in PMMI, machined by laser-assisted photochemical etching using an excimer laser (λ = 248 nm). Depth of the structure d = 120 µm. (Courtesy of BIAS, Bremen).

written text can be miniaturized down to the size of a human hair. Nowadays, it is also easily possible, to produce a precise hole into a human hair [Vo01,ivam01]. Laser technologies are also used in rapid-prototyping to produce solid 3D microcomponents [Wo00].

Focused Ion Beam Machining is an alternative way of machining fine structures and allows to obtain extremely fine details. The removal rates, however, are very low, of the order of some $\mu m^3/s$. While gallium ions are typically used, the introduction of secondary gases into the system can either enhance sputter rates or cause selective deposition [RuGS99].

Alternatively, Electron Beam Machining can be employed yielding an even smaller spot size. The electron beam is used to write on an electron-sensitive film. The basic techniques are highly developed for IC mask production and particularly useful for the production of structured surfaces such as binary optics [Ev99].

Micro Electro Discharge Machining (Micro EDM) is employed in the field of micro-mold making and used for the production of micro valves, micro nozzles and MST components for medical and chemical process technologies. In the wire EDM process, thin wires with diameters down to 0.02 mm are used as electrodes. In the process of cavity sinking by Micro EDM, graphite or copper electrodes are manufactured by mechanical machining, LIGA-technology or by eroding itself. Micro EDM is applied to produce grooves and channels, bore holes, linear profiles, columns and even complex formed three dimensional structures [Wo00].

2.3 Mechanical processes

Mechanical processes are mainly employed for the direct manufacture of small numbers of precision parts. These processes comprise diamond turning and -milling, precision grinding and micro-engraving.

Diamond machining was originally developed for the production of optical mirrors and lenses with continuous and discontinuous surfaces. The generation of very smooth surfaces with optical quality can be achieved by diamond machining of non-ferrous metals, such as OFHC-copper, aluminum or electroless nickel. Diamond machining is employed for the manufacture of contact lenses, aspheric lenses, polygon mirrors, micro lens arrays and also the generation of microstructures like pyramids, cubes, gratings or channels with micrometer dimensions. Applications of diamond machined microstructured surfaces include diffractive optical elements (DOEs) [Her97,TiPR99], 2D and 3D roughness standards with sinusoidal, acute or triangular profiles and fusion targets for nuclear fusion research [BrPS97, BrPR94]. Some of these applications are depicted in Figure 4.

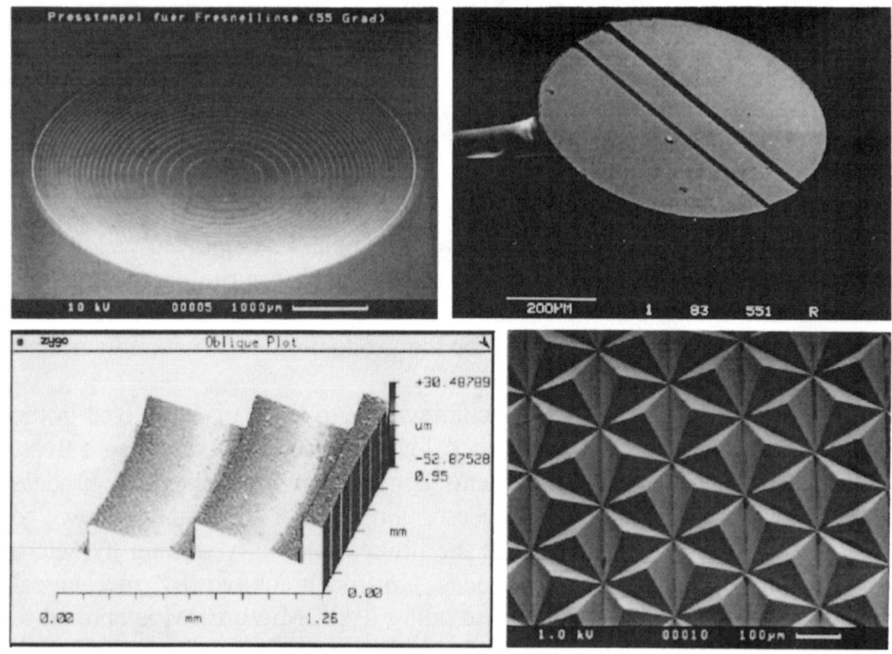

Figure 4. Examples of precision parts and microstructures: Bifocal intraocular lens with hyperbolic fresnel grooves, depth of grooves 4 μm (left). Aluminum target foil with a double step, pyramid height 25 μm, 640 μm in diameter (top right). Mold for retroreflective tapes; length of base 120 μm, height 8.4 μm (bottom right).

Precision grinding is applied for the manufacture of precision parts made of hard materials, e.g. glass lenses. It can be used for the fabrication of pins, grooves and microcavities with dimensions in the micrometer range. For precision grinding, submicron grains are essential [Ma00]. Differently shaped cylindrical grinding pencils with diameters from 1 mm down to 100 μm were developed by Gäbler et al. A CVD-diamond layer with a thickness of several micrometers was well suited as a cutting material for these machining processes with undefined cutting edges. Microstructures below 1 mm were produced in hard and brittle work piece materials like silicon, ceramic and glass [GäSWH99].

Micro-engraving with pyramidal diamond tools is used to generate microscopic marks, scales, lines etc. which can be found on injection molded adjustment screens for cameras and microscopes. This technique enables the manufacture of interrupted grooves which may be linear or curved. Engraving is performed at low cutting speeds less than 100 mm/min [BrPS97]. Another example of engraving is the production of diffraction gratings by ruling [MoW70].

2.4   Replication techniques

The economic mass production of microparts and microstructures is achieved by replication techniques like hot embossing, injection molding and casting.

Precision machined masters are used in the replication process of hot roller embossing. This technique is employed in the reproduction of planar surface relief microstructures in polymer films. As it is difficult to manufacture deep microstructures, hot stamper embossing is used [Ev99].

Traditionally, polymer parts with delicate geometries are precisely reproduced by injection molding techniques. Plastic lenses with microscopic optical elements are being mass-produced for a large variety of consumer goods and medical applications, ranging from bifocal intraocular lenses to projection screens and condensor lenses for pocket lights. Today, injection molding is also applied to shape microscopic green parts from metal and ceramic powders, which are later sintered to their final geometry. Figure 5 shows a microscopic gearwheel on a pencil lead. The gearwheel was produced by injection molding and sintering of stainless-steel powder [BrP99].

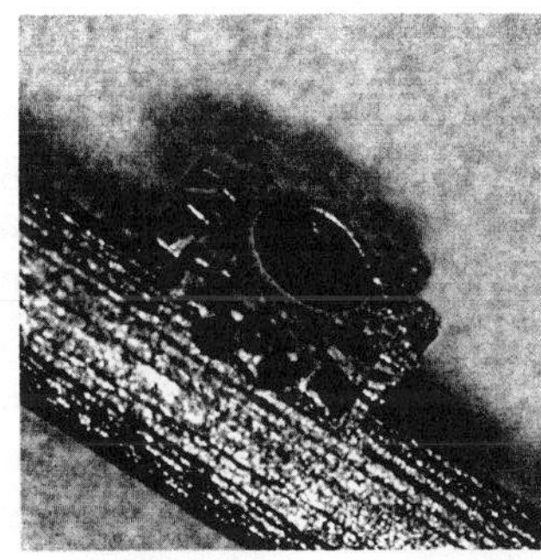

Figure 5. Microscopic gearwheel on a pencil lead (courtesy of FhG IFAM, Bremen).

2.5   Handling and mounting

Handling and mounting of MET-components is crucial to their application and currently makes up to 80 % of the total costs. Grippers (cf. Figure 3) generally need to be stiff but extremely sensitive and able to perform dynamic motions. They are typically built in a modular way to allow for multiple functions which is essential to move from the macro- to the microscopic scale.

Handling, mounting and quality control need to be performed subsequently in an automatic process. Nature sets an example: Insects like mites are acting with high precision "micro-hydraulic" grippers. Micro-grippers using sensors, electrostatic or magnetic forces for gripping are corresponding technical solutions. In MST only those techniques which allow mounting of a number of different elements in a few steps and which are suitable for mass production will be successful [Wo00].

3. SUMMARY

Machining of precision parts and microstructures is generally regarded as a key technology of the 21st Century. In all the examples mentioned above, different miniature components performing various functions are combined to form an intelligent system. Often, a combination of several microtechnologies are applied in order to realize single components. Terms like "microsystem technologies" ("microsystems"), "micromachining" or "Micro (Opto) Electro Mechanical Systems" (M(O)EMS) have been introduced internationally to approach tasks in this field. They all emphasize the innovation of this technology in contrast to microelectronics although they may also overlap. The more basic the level, e.g. a single microtechnology, the more difficult it is to distinguish "microsystem technologies" from conventional components and techniques. Several categories are necessary to describe the different ways of manufacturing precision parts and microstructures [vdi00]. This paper employs a categorization: The two main sets of technologies are microsystem technologies (MST) and micro-engineering technologies (MET). MEMS processes, energy assisted processes and mechanical processes can be assigned to these two main technologies. Suitable examples served to emphasize the possibilities of these technologies and the required handling and mounting techniques were discussed.

4. REFERENCES

[BaNe97]   Barthlott, W.; Neinhuis, C. Self-cleaning biological surfaces. The purity of the sacred lotus or escape from contamination in biological surfaces. Planta 1997; 202:1-8

[Bletal99] Bläsi B., Boerner V., Doll W., Dreibholz W., Gombert A., Heinzel A., Kubler V., Lobmann P., Manns P., Rose, K., Sporn, D., Wittwer, V. Periodic surface relief structures on large areas for optical applications. Proceedings of the 1st euspen Conf. Bremen, Germany. 1:522-525, 1999.

[BrPR94]   Brinksmeier, E., Preuß, W., Riemer, O. Manufacture of Shock-Wave Target-Foils for Nuclear Fusion Research. Proceedings UME 3, Aachen, Germany. 1:401-404, 1994.

[BrPS97]   Brinksmeier, E., Preuß, W., Schmütz, J. Manufacture of Microstructures by Diamond Machining. Proceedings of the 9th IPES / UME 4 int. Conf. Braunschweig, Germany. 503-507, 1997.

[BrP99]   Brinksmeier, E., Preuß, W. Fabrication of Precision Molds. Proceedings of the 1st euspen Conf. Bremen, Germany. 1:187-190, 1999.

[dbank01]   Danny Banks. Microengineering. MEMS, Micromachines, MST. http://www.dbanks.demon.co.uk/ueng/

[Ev99]   Evans, C.J., Bryan, J.B. "Structured", "Textured" or "Engineered" Surfaces. Annals of the CIRP Vol. 48/2/1999.

[fast01]   Fast Skin Speedo International. A new revolution in swimwear design. http://www.fastskin.com/index2.html

[GäSWH99]   Gäbler, J., L. Schäfer, A. Wenda, H.-W. Hoffmeister. Development and application of CVD diamond micro tools for milling and grinding. Proceedings of the 1st euspen Conf. Bremen, Germany. 1:434-437, 1999.

[Goetal98]   Gombert A., Rose K., Heinzel A., Horbelt W., Zanke Ch., Bläsi B., Wittwer V. Antireflective sub-micrometer surface relief gratings for solar applications, Solar Energy Materials and Solar Cells 1998; 54:333-340

[Her97]   Herzig, Hans Peter. *Micro-optics: Elements, systems and applications.* London, 1997.

[ivam01]   http://www.ivamnrw.com

[MaMT91]   Marentic, F. J.; Morris; L. Terry. Drag reduction article. United States Patent: 4986496, 1991.

[Ma00]   Masuzawa, T. State of the Art of Micromachining. Annals of the CIRP Vol. 49/2/2000.

[mems01]   MEMS View & Strategy. http://mems.kaist.ac.kr/tex_memsview.html

[MeB93]   Menz, W., Bley, P. *Mikrosystemtechnik für Ingenieure.* Weinheim, New York, Basel, Cambridge 1993.

[MoW70]   Moore, Wayne R., *Foundations of mechanical accuracy.* The MOORE special tool company. Bridgeport, Connecticut, 1970.

[RuGS99]   Russel P. E., Griffis D.P., and Stark, T.J. Focused Ion Beam Machining of molds. Proceedings of the ASPE Spring Topical Meeting Vol. 19, Chapell Hill, NC, 1999.

[Sc00]   Schäfer, L. Mikromotoren bieten die Grundlage für neue Produktentwicklungen. Maschinenmarkt 2000; 45:87-89

[TiPR99]   Tiziani, H., Pahlke, M. and Rocktaschel, M. Manufacture and application of diffractive elements for laser-beamshaping and diagnostics. Proceedings of the 1st euspen Conf. Bremen, Germany. 2:32-35, 1999.

[vdi00]   Berger, J., Büttgenbach, S., Karthe, W., Kergel, H., Lehr, H., Reichl, H. Microsystem Technology. http://www.vdivde-it.de/mst/papers/gmm00.

[Vo01]   Vollrath, K. Mikrotechnologie: Filigrane Kunststoffteile bringen neues Kostensenkungspotential in die Medizin. VDI nachrichten 2001; 8:22

[WeDF00]   Weck, M., Day, M., Fischer, S. Mikrozerspanung erfordert angepaßte Maschinenkonzepte. Maschinenmarkt 2000; 36:42

[Wo00]   Wohlgenannt, M. Funktionalität auf kleinstem Raum. F&M 2000; 108:20-23

ROLES OF QUANTUM NANOSTRUCTURES IN ADVANCED ELECTRONICS

Hiroyuki SAKAKI

University of Tokyo (IIS),

4-6-1 Komaba, Meguro-ku, Tokyo 153-8505

Abstract

In today's electronics, core layers in key semiconductor devices, such as transistors and lasers, are all set to be 10nm or less.   There one must not only control their layer thickness and flatness but also manipulate quantum or wave natures of electrons.   Trends in such nano-layer technology are examined to clarify its roles both in improving core devices and opening a new field of quantum-effect devices.   The reduction of in-plane dimensions of transistors to 100nm or less has been a main thrust for advancing LSI and high-speed communication.   We describe trends in such efforts and also recent progress in manipulating electron waves with 10nm-scale quantum wire or box/dot structures for advanced electronics applications.

Keywords

Nanostructures, Quantum Effects, Quantum Wells, Quantum Wires, Quantum Dots, Advanced Electronics, Transistors, Lasers, LSI

1. ROLES OF NANO-METER SCALE LAYERS IN ELECTRONICS

Remarkable developments of semiconductor electronics have been achieved by the persistent improvement of existing devices, such as transistors and lasers and also by the introduction of new devices. In most of these endeavors, ultrathin layers of semiconductors and insulators with the thickness of 10 to 1 nanometers (nm) play key roles. We discuss first the importance and potentials of such nano-layer technology [1-3].

1.1 Needs of nm-Scale Layers on Electronics and Optoelectronics

10nm-scale semiconductor layers are widely used in electronics as the core element of semiconductor lasers and field-effect transistors (FETs). This is because, in semiconductor lasers, the current to achieve laser oscillations can be reduced almost in proportion to the active layer thickness. Also, in field-effect transistors (FETs), the gate voltage to induce sufficient conductance modulation in FET channels can be lowered in proportion to the spacing d between the gate and the channel. In particular, in SiMOS FETs, which is the core device of LSIs, both the channel thickness d_1 and the gate insulator thickness d_2 are reduced as small as possible and recently set about at 5nm and 2nm, respectively. We discuss such trends and the importance of precisely controlling both the thickness and flatness of such layers. We also examine both prospects and limitations of this trend, which are imposed by the electron tunneling and other wave aspects of electrons

1.2 Quantum Effects in nm-Scale Layers and Their Use for New Electronics and Photonic Devices

According to quantum mechanics, electrons in semiconductors exhibit their wave natures, when confined in nano-meter (nm) acale structures. For example, electrons confined in 10nm scale GaAs films with the thickness L_z maintain their free-particle motion only along the (x, y) axis of the film plane and their motion along the z-axis (normal to the layer) is quantized to a series of standing wave states E_1, E_2, ..., E_n. As the wavelength λ_n and eigen energy E_n of such state E_n are respectively given by $(2L_z/n)$ and $(h^2/2m)(n\pi/L_z)^2 = 50\text{meV} \times n^2$, the majority of electrons in such films are accommodated in their ground level E_1. Two-dimensional electrons and holes confined in such films (quantum wells (QWs)) are now widely used in a variety of new electronic and photonic devices. Moreover, layered superlattices (SLs) of Esaki and Tsu, or a periodic stack of quantum well (QWs) and tunneling barriers, have provided a fertile field for electronics, where their unique cross-barriers transport and inter-subband (level) transition processes lead to the Bloch oscillation and new infrared detectors and lasers (quantum cascade lasers), opening a gateway to set of quantum devices. We describe the current state of these exploratory devices, where quantum phenomena in 10nm-scale films are cleverly use [1-3].

2. NEEDS AND SEEDS IN DOWN-SCALING OF IN-PLANE DIMENTIONS OF SEMICONDUCTOR DEVICES AND STRUCTURES

The reduction of in-plane dimensions of transistors has been a powerful and traditional recipe in upgrading the performance of LSIs and high-speed transistors indispensable for advanced information and communication systems. In addition, unique properties of electrons confined in 10nm-scale wire and dot structures offer new possibilities in creating semiconductor devices with unprecedented functions [2, 4-8]. We describe the current state of such efforts.

2.1. Down Scaling of Transistors: Needs and Prospects

Over the last three decades, the gate length (L_g) of FETs has been continuously reduced (1) to lower the operating voltage and power of FETs and (2) to improve their switching speed and high-frequency performances.

Moreover, the simultaneous reduction of the channel length and width has led to an increase in the packing density of FETs and the cost reduction in LSI chip production.   At present, the typical gate length for mass-produced FETs is getting close to 100nm and the gate lengths in test FETs newly developed in exploratory studies are getting as small as 35~10nm.   We discuss this scaling principle and point out its problems and potentials of this trend, as a variety of new phenomena are emerging as L_g plunges into the scale of 10nm.

2.2.   10nm-Scale Quantum Wire and Dot: Their Synthesis, Physics and Applications

As an attempt to expand the forefront of semiconductor physics and electronics, the author proposed in 1975 plenar arrays of quantum dots (QDs) and wires (QWRs) [4], where the free electron motion is totally forbidden in QDs and allowed only along the length direction of wires axis. In such systems, one can adjust the inter-dot or inter-wire coupling so that a series of minibands separated by real minigaps with zero density of states (DOS) can be formed in QD arrays.   In such systems, the gate-voltage control of their electron population will lead to unique transport phenomena and offer novel device functions.   He also pointed out in 1980 that 10nm-scale mono-mode QWRs are attractive materials for FETs as the elastic scattering of 1D electrons may be suppressed there [5].   Moreover, Arakawa and Sakaki proposed in 1982 the use of QDs and QWRs as new laser media as their peaked DOS may reduce the temperature sensitivity of laser performances [6].

Despite these attractive proposals, experimental studies of QDs and QWRs were quite slow in the early stage because of the difficulty of their fabrication.   In particular, the fabrication of 10nm scale dots and wires with large quantum level separations ($\geq$ 30meV) was considered to be almost hopeless until 1990. Following the lithographic approach which allowed the formation of 100nm-acale wires and dots [9, 10] several new methods have been developed to allow the formation of 10nm-scale dots and wires. The self-organized growth of dots and wires [10-17], the facet selective epitaxy on patterned substrates, the step-controlled epitaxy on tilted substrates, and the overgrowth onto the edge of QWs are some of such examples.

By employing such QDs and QWRs, numerous attempts have been made to fabricate FETs, memory devices [11, 12], lasers [13], photo-detectors [12, 14] and so on. We discuss first the present state of research on QD-based memories and QD-based photo-detectors, which make use of self-organized QDs embedded near the 2D or 1D electron channel of FET-like structures. The presence or absence of electrons in such QDs can be controlled either by the gate electric fields or by inter-band (near infrared) or inter-sublevel (mid-infrared) photons. Such changes of electron population in a single QD or ensemble of QDs can be detected by monitoring the channel conductance of a field-effect transistor (FET) placed in the vicinity of QDs. Both potentials and problems of using such phenomena for new FET memories, and near-, mid-, and far-infrared detectors will be described. We also report on the current state of research on QD lasers [13], discussing not only their performances but also important insights gained on unique aspects of zero-dimensional electrons and excitons in such dot structures [15-17]. If time allows, we briefly review the current state of research on 10nm-scale quantum wire structures and discuss their significance in relation to QD structures.

References

(1)   C. Weisbuch and B. Vinter, "Quantum Semiconductor Structures" Academic Press.

(2)   H. Sakaki, Solid State Comm. 92 (1994) 119, Physica E4 (1999) 56, and Phys. Status Solidi (b) 215 (1999) 291.

(3)   L. Esaki and R. Tsu, IBM J. Res. And Develop. 14 (1970) 61.

(4)   H. Sakaki et al, Thin Solid Fims 36 (1976) 497.

(5)   H. Sakaki, Jpn. J. Appl. Phys. 19 (1980) L735.

(6)   Y. Arakawa, H. Sakaki, Appl. Phys. Lett. 40 (1982) 939.

(7)   D.A.B. Miller et al, Appl. Phys. Lett. 52 (1988) 2154.

(8)   H. Sakaki et al, Appl. Phys. Lett. 57 (1990) 2800.

(9)   B.J. van Wees et al, Phys. Rev. Lett. 60 (1988) 848.

(10)   S. Tarucha et al, Phys. Rev. Lett. 77 (1996) 3613.

(11)   G. Yusa, H. Sakaki, Electron Lett. 32 (1996) 491.

(12)   G. Yusa, H. Sakaki, Appl. Phys. Lett. 70 (1997) 345.

(13)   For a recent review, see D. Bimberg, Solid State Electron (1998) (Proc. Int. Workshop on Nano Physics and Electronics, Tokyo, 1997) and V.M. Ustinov et al, J. Cryst. Growth 175 (1997) 689.

(14)   See for example, S.W. Lee, K. Hirakawa, and Y. Shimada, Appl. Phys. Lett. (1999) and Physica E7 (2000) 499.

(15)   T. Inoshita and H. Sakaki, Phys. Rev. B56 (1997) 4355 and X.Q. Li et al, Phys. Rev. B59 (1999) 5069.

(16)   S. Hameau el al, Phys. Rev. Lett. 83 (1999) 4152 and O. Verzelen el al, Phys. Rev. B62 (2000) R4809.

(17)   R. Ferreira and G. Bastard, Appl. Phys. Lett. 74 (1999) 2818.

DESIGN OF NEW PRECISION MACHINE ELEMENTS

Alexander H. Slocum

Professor of Mechanical Engineering
Massachusetts Institute of Technology
77 Massachusetts Avenue, Room 3-445
Cambridge, MA 02139

Abstract

History motivates the creation of lower cost and higher precision machine elements. This paper introduces a new low cost design for magnetically preloaded air bearings and self-compensating hydrostatic bearings. In addition, a new design concept for a low cost precision cylindrical grinding machine is introduced where the axis of motion carriages act as integral structural elements of the supporting frame to create a very small and rigid structural loop.

Keywords

Precision, machine element, hydrostatic bearings, grinding, self help

1. INTRODUCTION

History has shown that all good things turn into commodities, and so it is with precision machines and elements. Often it is the creation of a low-cost easy-to-use precision machine element that enables new types of precision machine tools to be developed, which in turn enable new precision commodities to be manufactured such as semiconductors and computer hard disks (Taniguchi).

1.1 Active or Passive Precision & Disruptive Technologies

In many designs, the limits of precision seem to be reached, and thus many researchers turn to a coarse-fine approach. Many early wafer steppers used a course position stage with a fine positioning stage on top [Stone]. Other researchers have used piezoelectric elements to fine position cutting tools or bearings [Patterson and Magrab]. Even air bearings have been servo-positioned by piezoelectric actuators to correct for error motions [Horikawa]. Indeed, modern disk drives use the tracks on the drives themselves as an end-point-feedback mechanism. However, in all these systems, it is important to get the performance as good as possible for as low cost as possible using conventional simple means, before more advanced strategies are applied. This is illustrated qualitatively in Figure 1.

Figure 1 also presents other qualitative relations between cost and performance that should be considered by precision machine designers [Slocum]. Of particular interest are disruptive technologies [Christensen] that one fears, after a design has already been committed to, or one searches for, before a design is found. Similarly, the coagulated edge product that for very little extra cost could yield a significant increase in performance is just asking to be displaced. The bleeding edge product is very expensive, or has very high margins, yet it yields only a small performance increase is also asking for a disruptive technology to displace it. Leading edge products, preferably disruptive to others, are most desirable. The question is: how are these technologies best identified?

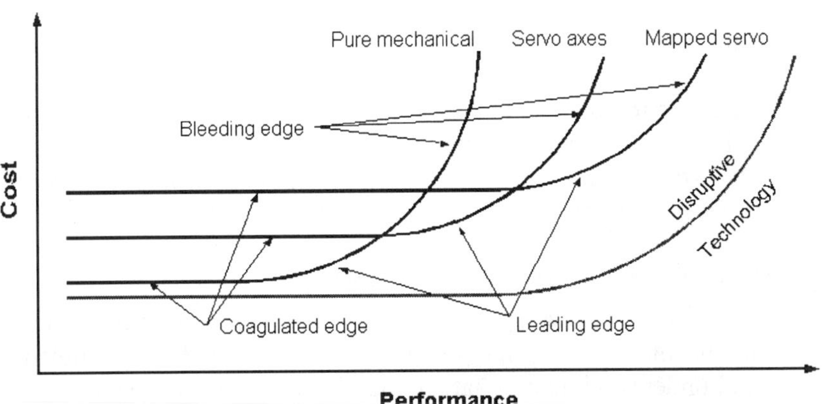

Figure 1. Cost performance conventional wisdom tradeoff metrics are often disrupted by new technologies.

2. DOMINANT SENSITIVE PARAMETERS

In order to identify potential disruptive technologies for precision machines, it often helps to consider what did design engineers do when they did not have high precision CNC production equipment on which to rely? The answer was they used self-checking designs such as double vee ways and square surface plates [Moore]. Kinematic or elastically averaged designs were often also employed generally by instrument makers or machine tool designers respectively [Blanding, Evans, Jones]. The overall sensitive parameter, however, appears to be preload, for it is preload that ensures repeatability.

It has been shown that repeatability is a most powerful system performance parameter [Hocken, Donaldson]. When designing a precision machine, all the elements must be individually optimized, but their performance as a system must also be kept in mind, for the chain is only as accurate as its weakest link. This requires the machine to be designed with a careful error budget that includes stiffness, thermal expansion elements as well as geometric terms [Slocum]. The author has noted that in many years of reviewing many different technologies and machines, that compliant structural loops and poor preloading practice are the dominant causes of poor performance.

Compliant structural loops occur when designers do not properly respect Abbe's or Saint Venant's principles. Designing for static stiffness while neglecting dynamic stiffness, damping, is also a common problem.

Poor preloading practice is often traceable to over or under constraint of bearings or actuators. In the former case this leads to increased loads, varying frictional loads, and premature wear. In the latter case, this leads to backlash, which is usually an obvious condition.

It is interesting to note that just as fundamental principles can be used to identify sensitive parameters in machine design, in a sense another fundamental principle, Maxwell's Reciprocity, can be applied philosophically to lead us to use the fundamental principle of self-help to create new designs

3. SELF-HELP MACHINE ELEMENTS

A key catalyst in the development of new low cost precision machine elements and machines is the principle of self-help [Pahl & Beitz] coupled with a fundamental understanding of what are the sensitive, and hence costly, parameters in a precision design. Indeed, applying the principle of self-help can often lead to the creation of disruptive technologies.

3.1 Preload

Preload is a key to repeatability. There are two fundamental means by which preload can be obtained, opposed geometry systems and opposed force systems. An opposed geometry system uses geometry to direct forces to oppose (preload) each other. An example is a back-to-back or face-to-face arrangement of ball bearings. An opposed force system, uses an applied force to maintain load on a bearing that would otherwise only bear a load in one direction. An

example is using the weight of a machine tool axis to preload a vee-and-flat
bearing rail. Note that the former requires both sides of a precision element to
be precision machined, and this is expensive. In the latter, complex shapes need
to be machined with great precision. Is there a simpler way?

3.1.1 Example: Vectored Magnetically Preloaded Air Bearings

If we start with the goal of maximizing accuracy and minimizing cost as key
functional requirements in creating a preloaded linear aerostatic bearing system,
then it forces us to first draw the simplest precision system from which to start:
two perpendicular planes. This leads to a machine axis, bearing rail that is
nominally rectangular, where two of the rectangle's perpendicular surfaces are
precision machined. If we next assume that there is an L-shaped carriage with
air bearing pads that can be placed over the bearing rail, we must ask ourselves
how can we apply preload? Vacuum regions are a possibility, but cannot
provide the high preload required for bi-directional load capacity. Alternatively,
we can add magnets to each face and obtain a robust modest cost system. But
can we do better?

By applying Occam's razor and Maudslay's Maxims (Keep It Super Simple)
we can indeed reduce complexity and cost (Slocum). All one has to do is draw
the required preload force vectors for each of the bearing pads, and then sum
them up. The single resulting vector is where the preload force must be applied.
Next, we ask what elements in the system can be used to apply the force? Figure
2 shows how we can start with the concept of a simple system, and add the next
element, a permanent magnet linear electric motor, to the system. The gap
between the magnets and the coils can be provided by milled features as opposed
to the high precision ground surfaces for the air bearings. Hence the attractive
force between the magnet track and the motor core can be harnessed to preload
the bearings. Since this force is typically 6 times the motor's axial force, the
system actually achieves a properly balanced force capability. The web site
http://pergatory.mit.edu/rcortesi/portf/axtrusion/index.html provides detail on
experimental results from a prototype system.

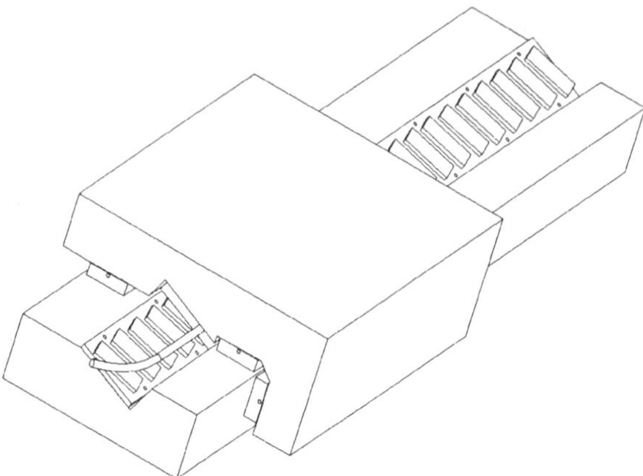

Figure 2. The Axtrusion™ configuration uses linear motor magnet track placed at an angle to equally preload aerostatic bearings.

3.2 Hydrostatic Compensation

Hydrostatic bearings are known to provide the ultimate in dynamic stiffness and accuracy; however, they have both high variable and fixed costs. The variable cost is associated with the cost of the pump system (power, filters, replacement) and it can be minimized if the flow rate is minimized and the system is made less sensitive to dirt which can plug flow restrictors and cause catastrophic failure.

The fixed cost can be reduced if we consider what are the most sensitive parameters in the system. The most sensitive fixed cost parameters in a hydrostatic bearing system are the bearing gap and the restrictor sizing, because the flow rates vary with gaps to the third power and diameters to the fourth power. Hence any small manufacturing error is greatly amplified, altering performance, or the size of the pump required. Thus, virtually all hydrostatic bearings systems must be hand tuned when the machine is assembled.

Zollern Corp. in Europe, http://www.zollern.de , has for many years sold modular self compensating hydrostatic bearings, which can be traced to a patent in the early 1940's and improved upon by many inventors and companies since then including their use in water hydrostatic bearing systems (Slocum). However, the self-compensating bearing systems themselves still require complex porting to restrict and collect flow on one side of a bearing, and then direct it to a hydrostatic pocket on the opposite side of the bearing.

Once again, Occam's razor and Maudslay's Maxims lead to the creation of a design where the compensation means and hydrostatic bearing pockets are connected directly, leading to the formation of surface self-compensated hydrostatic bearings for linear motion (Kane) and rotary motion (Wasson, Kotilainen) as shown in Figures 3 and 4 and described in detail on the website http://pergatory.mit.edu. The linear system shown in Figure 3 requires precision manufacturing capability commensurate with the manufacture of rolling element linear motion guides. The rotary motion system shown in Figure 4, on the other hand, can be cast or injection molded, and thus holds great promise as a modular precision machine element of the future.

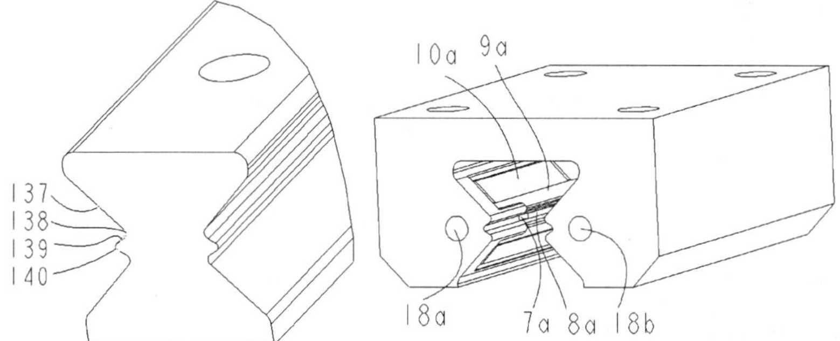

Figure 3. The HydroRail™ surface self compensated hydrostatic linear motion bearing truck: 18a & 18b supply, 7a supply entrance to compensation region, 8a compensation surface, 9a fluid path to pocket, 10a pocket, 137 load surface, 138 & 139 compensation surfaces, 140 supply surface.

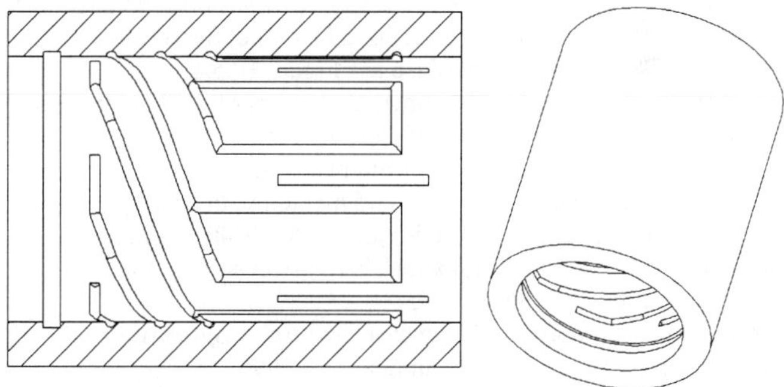

Figure 4. The HydroBushing™ surface self compensated hydrostatic rotary motion bearing.

4. CONCLUSIONS

All new machine elements will eventually be developed, the question is by whom and when. Fundamental principles of mechanics must always be obeyed, but business issues and philosophical principles can be invaluable catalysts. Just as this pattern can lead to new machine elements, application of this pattern can lead to the development of fundamental new machines that use these new machine elements.

5. REFERENCES

Blanding, Douglas, *Exact Constraint: Machine Design Using Kinematic Principles*, New York, ASME Press, 1999

Christensen, Clayton, *The Innovator's Dilemma*, Harvard Business School Press, 1997

Donaldson, Robert. The Deterministic Approach to Machining Accuracy. SME Fabricat. Technol. Symp., Golden, Colorado, 1972

Evans, C., *Precision Engineering: An Evolutionary View*, Cranfield, UK, Cranfield Press, 1992

Furukawa, Y. et al.,. Development of Ultra Precision Machine Tool Made of Ceramics. Ann. CIRP, Vol. 35, No. 1, 1986, pp. 279-282.

Hocken, Robert, Technology of Machine Tools, Vol. 5, Machine Tool Accuracy, Robert J. Hocken (ed.), Machine Tool Task Force. U.S. Dept. of Commerce National Technical Information Service Report UCRL-52960-5

Horikawa, O. et al. Vibration, Position, and Stiffness Control of an Air Journal Bearing. 1989 Int. Precis. Eng. Symp., Monterey CA, pp. 321-332.

Kane, N., Slocum, A. "Modular Hydrostatic Bearing with Carriage Form-Fit to Profile Rail", US Patent #5,971,614, Oct. 1999

Moore, W, *Foundations of Mechanical Accuracy*, Bridgeport, CT , Moore Special Tool Co., 1970.

Jones, R. *Instruments And Experiences*, New York, John Wiley & Sons, 1988.

Kotilainen, S., "Surface Self Compensated Hydrostatic bearings", MIT ME Ph.D. thesis 1999.

Kouno, E., McKeown, P. A Fast Response Piezoelectric Actuator for Servo Correction of Systematic Errors in Precision Engineering. Ann. CIRP, Vol. 33, 1984, pp. 369-373.

Pahl, G., Beitz, W., *Engineering Design, A Systematic Approach*, NY, Springer-Verlag, 1988

Patterson, S., Magrab, E. Design and Testing of a Fast Tool Servo for Diamond Turning. Precis. Eng., Vol. 7, No. 3, 1985, pp. 123-128. Also see Figure 10.5.1.

Slocum, A., Scagnetti, P., Kane, N., Brünnner, C. Design of Self Compensated Water-Hydrostatic Bearings. Precis. Eng., Vol. 17, No. 3, 1995, pp 173-185.

Slocum, A., *Precision Machine Design*, Detroit, MI, SME, 1995

Slocum, A., "Linear motion carriage system and method with bearings preloaded by inclined linear motor with high attractive force", US Patent #6150740, Nov., 2000

Stone, Stanley. "Design Case Study: Flexure Thermal Sensitivity." In *Precision Machine Design*, Alexander Slocum, Detroit, MI, SME, 1995

Taniguchi, Norio, *Nanotechnology : Integrated Processing Systems for Ultra-Precision and Ultra-Fine Products*, Oxford, Oxford Science Publications, 1996

Wasson, K., Slocum, A., "Integrated Shaft Self-Compensating Hydrostatic Bearing", US Patent #5,700,092, Dec. 23 1997

Part I

Cutting / special machining

THE CONCEPT OF ACTIVE DEFLECTION COMPENSATION AND ITS APPLICATION IN PRECISION FORGING

**Prof. Dr.-Ing. E. Doege, Dipl.-Ing. J. Baumgarten,
Dipl.-Ing. T. Neumaier**

Institute for Metal Forming and Metal Forming Machine Tools (IFUM),
University of Hanover, Germany

Abstract

Today's forging industry is subjected to an ever increasing pressure to manufacture complex net shape parts at minimum unit costs. One way of meeting this challenge is to combine a precision forging with a subsequent calibrating process. The Institute for Metal Forming and Metal Forming Machine Tools (IFUM) has developed an innovative tooling concept for calibrating processes which helps to raise the dimensional quality of the finished parts up to IT class 6 and to improve its surface quality as well as the process reliability. Moreover, IFUM has made initial investigations on applying this concept – named active deflection compensation – directly in precision forging processes with the aim of achieving further advances in accuracy.

Keywords

elastic die deflection, innovative die concept, precision forging

1 INTRODUCTION

The precision forging of geared parts frequently requires a subsequent grinding or calibrating operation – with calibrating standing for the finish forging of a part by a final forming operation to achieve a desired form or tolerance [JÜTTE 87, KÖNIG 96] – in order to achieve the required high degree of accuracy. In general, the accuracy of the formed part can be improved by 2 IT-classes employing a calibrating process which is, in most cases, more cost-effective than grinding operations.

A major problem, not only concerning calibrating but also any other cold, warm and hot bulk metal forming operation, is raised by the high compressive stresses that arise in the interior of the workpiece. These stresses act as pressure loads on the inner die walls and cause the walls to deflect elastically. This gives rise to a non-negligible deviation of the die geometry from its unloaded state (*Figure 1a,b*) which, in turn, might produce finished parts out of tolerance. Thereby, conventional calibrating dies are limited to achieving a dimensional quality of up to class IT7 which is not enough for certain applications. For example, running gears require class IT6 which – in case of conventional calibrating dies – can therefore only be achieved with the help of an additional costly grinding operation [JÜTTE 87].

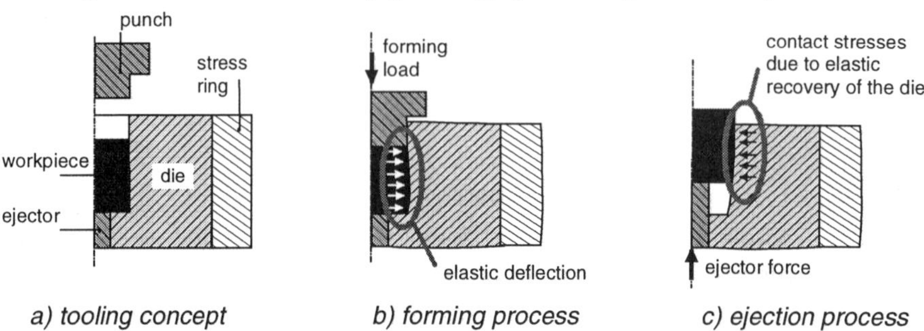

a) tooling concept   b) forming process   c) ejection process

Figure 1: Elastic deflection and elastic recovery of a conventional die

Another negative effect of the elastic die deflection is the jamming of the workpiece in the die occurring during the ejection process due to the die's elastic recovery (*Figure 1c*). Thus, the ejector forces necessary to eject the workpiece from the die after the forming operation can be very high – up to 20% of the maximum forming load in case of complex parts [DOHMANN 94, LAUFER 91]. Due to the elastic recovery of the die, high contact pressures act on the inner die walls and on the workpiece surface. These high contact pressures in conjunction with the extensive relative sliding motions during the ejection process can cause surface damage of both die and workpiece and – in the worst case – die failure [LENNARTZ 95].

All the specified drawbacks associated with the use of conventional dies call for the development of new tooling systems for calibrating operations which prevent the occurrence of elastic die deflection. Moreover, other bulk forming operations like e. g. cold forging or precision forging can be expected to benefit from these tooling concepts facilitating the inhibition of elastic die deflection, too.

2   ACTIVE DEFLECTION COMPENSATION

The key idea underlying the concept of active deflection compensation is to counterbalance the pressure loads on the inner die walls – which cause the die to deflect elastically – with the help of a counter pressure generated by an elastomer ring embedded in the lower die (*Figure 2a*). During the closing of the dies at the beginning of the calibrating / forming process, a compressive counter stress p_i is generated in the elastomer ring by the downward movement of the stopper ring attached to the upper die (*Fig-*

ure 2b). Being of equal magnitude, both the pressure arising in the work-piece and the counter pressure generated in the elastomer ring compensate each other at the inner die walls and thus, the elastic deflection of the lower die is inhibited. Consequently, the die geometry does not deviate from its unloaded state when loaded.

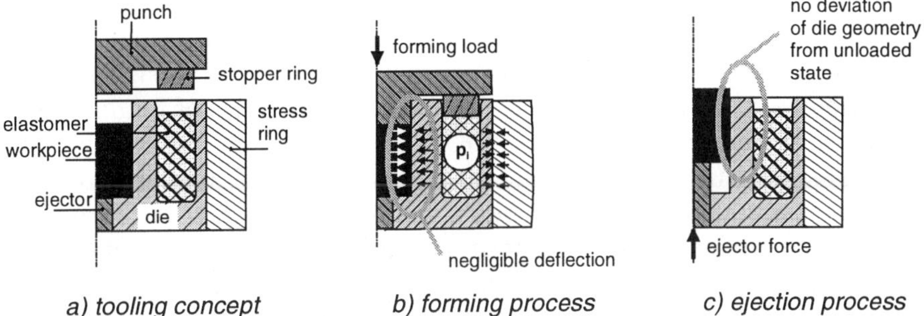

Figure 2: Innovative die concept with active deflection compensation

At the end of the forming process, i.e. during unloading and the opening of the dies, both the pressure in the workpiece and the counter pressure in the elastomer ring decline and the stresses in the die are relieved. For this reason and because of the die geometry not deviating from its unloaded state, the ejector force required when using active deflection compensation is far lower than when using conventional dies (*Figure 2c*).

In the course of previous investigations, Finite Element Analyses have given prove of the feasibility – i. e. its ability to inhibit elastic die deflection successfully – and the potential of the proposed tooling concept [DOEGE 00]. One expected benefit of the proposed concept is a substantial increase of the dimensional quality of parts calibrated using active deflection compensation compared to parts calibrated using conventional dies. Furthermore, due to the necessary ejector force being reduced considerably through active deflection compensation, even most complex parts can be ejected without difficulty. As another benefit, tool wear during the ejection process is reduced substantially. Finally, the process reliability and the workpiece surface quality are enhanced.

3 APPLICATION IN PRECISION FORGING

In order to determine whether the proposed concept is suitable for application in precision forging processes, a backward can extrusion process at hot working temperatures – the billet temperature has been set to 1100°C – involving both a conventional die and a die featuring active deflection compensation has been modelled and simulated using the FEA-packages MSC.Marc Version K7 and MSC.Marc AutoForge 3.1.

In order to investigate the tooling system's thermal behaviour in continuous operation, a set of 170 forging cycles has been simulated. Each forging cycle comprises the stages, warming of the billet, transport of the billet to the die, lying of the billet in the die before the forging operation, forging operation, remaining of the finished part in the die before its ejection, ejection process and application of coolant on the die. Due to the high computational cost, it is not appropriate to simulate every single forging cycle including the computationally expensive forging operation. Therefore, a method has been developed at IFUM to simulate a series of forging cycles based on the results the first cycle [DOEGE 97]. The final cooling stage has been modelled using additional heat fluxes applied as thermal boundary conditions on the die walls and the die bottom. The magnitude of said heat fluxes is determined by the composition and the amount of coolant (a water and lubricant emulsion) sprayed on the die. In order to model the process start phase, the coolant has not been applied during the first 20 cycles. Both the conventional die and the die featuring active deflection compensation have been cooled using the quantity of 6 ml coolant on the die wall and the die bottom.

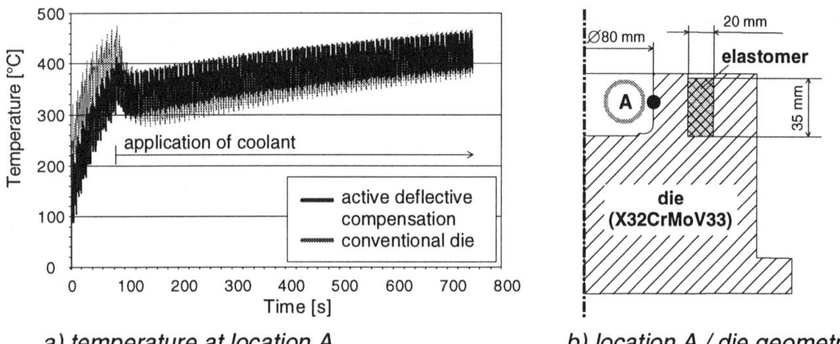

a) temperature at location A   b) location A / die geometry

Figure 3: Computed temperature at location A on inner die wall

In summary, the simulation results indicate that an application of the concept of active deflection compensation in precision forging is practicable. It could be determined that the elastomer ring embedded in the lower die does not obstruct the heat flux away from the inner die wall into the die body such that the temperature at the inner die wall exceeds the austenitizing temperature of common hot working steels during within the simulated number of forging cycles. *Figure 3* illustrates this fact showing that the temperature at location A on the inner die wall stays well below the austenitizing temperature of 600°C of common hot working steels for both the conventional die and the active deflection compensation.

Likewise, the results indicate that the thermal stability of the elastomers intended for use in the tooling system is sufficient for an application at hot working temperatures provided the cooling of the die is adequate. The results show that the temperature in the interior of the elastomer remains below 230°C during the simulated 170 forging cycles.

4   CONCLUSIONS AND OUTLOOK

The concept of active deflection compensation represents a practical way to inhibit the occurrence of elastic die deflection in calibrating operations and, therefore, permits the minimization of the problems associated with the use of conventional dies. The feasibility of the proposed concept, its benefits and its transferability to precision forging operations have been proven using Finite Element Analysis.

At present, IFUM is working on demonstrating the benefits of the proposed concept by realising a precision forging process with a subsequent calibrating operation involving a calibrating die featuring the concept of active deflection compensation. Additional effort is put into research concerning the utilization of the proposed concept in the field of cold forging and precision forging.

5   ACKNOWLEDGEMENTS

This project (AIF 12388N) is funded by the German Federal Ministry of Economics and Technology (BMWi) through the German Federation of Industrial Cooperative Research Associations "Otto von Guericke" e.V. (AiF) and the Industrial Cooperative Research Associations of Steel Forming e.V. (FSV). After completion of the project, a final report is going to be available at FSV, Goldene Pforte 1, D-58093 Hagen.

6   REFERENCES

DOEGE, E; AWISZUS, B.: *Prozeßführung beim Gesenkschmieden zum Ausgleich thermischer Einflüsse auf das Werkzeug*; DFG-report Aw 6/1-1 and Do 190/104-1; University of Hanover (1997)

DOEGE, E.; BAUMGARTEN, J.; NEUMAIER, T.: *An active deflection compensation for calibrating and cold forging dies.* Proc.: 10th Internat. Cold Forging Congress 2000, Fellbach, 13.-15. Sep, 2000, VDI-Berichte, Band 1555 (2000) page 103-117

DOHMANN, F.: *Untersuchung der Prozeßkette Umformen – Härten – Hartnachbearbeiten zur Herstellung einbaufertiger Laufverzahnungen.* Universität Gesamthochschule Paderborn (1994), report BMFT 02FT49210

JÜTTE, F.: *Kalibrieren von Stirnrädern durch Fließpressen.* VDI-Z, Band 129, Nr. 2 (1987), Düsseldorf: VDI-Verlag

KÖNIG, W.; KLOCKE, F.: *Fertigungsverfahren Band 4.* VDI-Verlag (1996)

LAUFER, M.: *Untersuchungen über das Kaltfließpressen gerad- und schrägverzahnter Stirnräder.* Fortschritt-Berichte VDI, Reihe 2: Fertigungstechnik (1991)

LENNARTZ, J.: *Kaltfließpressen von gerad- und schrägverzahnten Getriebewellen.* Fortschritt-Berichte VDI, Reihe 2: Fertigungstechnik (1995)

COINING OF THIN PLATES TO PRODUCE MICRO CHANNEL STRUCTURES

G. Hirt, B. Rattay
Saarland University, Institute of Materials Technology/Precision Forming, Saarbruecken, Germany

Abstract

Increasing miniaturization of production systems and products demands an expanding availability of metal forming technologies towards very small scale products and microscopic geometric details. A typical example are thin plates (thickness < 1 mm) with microscopic channel structures, which are required for micro heat exchangers, chemical micro reactors or other applications. In coining experiments sheet specimens from 0.5 to 0.8 mm thickness have been used to produce rib type channel structures with 0.4 mm width of the ribs as well as of the channels. The remaining thickness at the channels is about 0.3 - 0.6 mm, depending on the initial sheet thickness. The experimental results are compared to finite element simulations.

Keywords

micro metal coining, finite element analysis

1 INTRODUCTION

In many areas of industrial manufacturing an increasing trend to reduce component size can be observed. This is not only the case for electronic or mechatronic applications, but also in the field of general engineering like the micro heat exchangers or micro reactors. A typical feature of many of these applications is, that the functional structures of the component are very small and require very high precision.

These components have so far been produced by micro cutting, micro EDM or other processes which are mainly suited for small lot production. For mass production the coining process would be a very competitive alternative offering high precision as well as high productivity. However, coining of thin sheets to produce channel structures involves high plastic strains in the thickness direction, which are similar to bulk metal forming. Accordingly high forming loads are required and elastic die deflections may be almost of the same order of magnitude as the desired product structure. Fundamental questions concerning the manufacturing of metallic micro components by metal forming include especially size effects on flow stress, friction and metal flow. The coining process as a method to produce very fine metallic structures has been studied under various conditions [e.g. 1, 2].

2 MAIN GOALS OF THE PRESENT STUDY

Compared to classic coining processes where the coining depth is small compared to the product thickness, this study is intended to evaluate process opportunities for cases where the coining depth is of the same order of magnitude as the component thickness. This leads to large plastic strains as usually observed in bulk metal forming. Channel type structures have been chosen as demonstration geometry because they are directly related to a large group of potential applications (see Fig. 1).

The main goals of the study are:
- evaluation of the die loads to be expected
- influence of elastic die deflections
- evaluation of FEM for die- and process layout

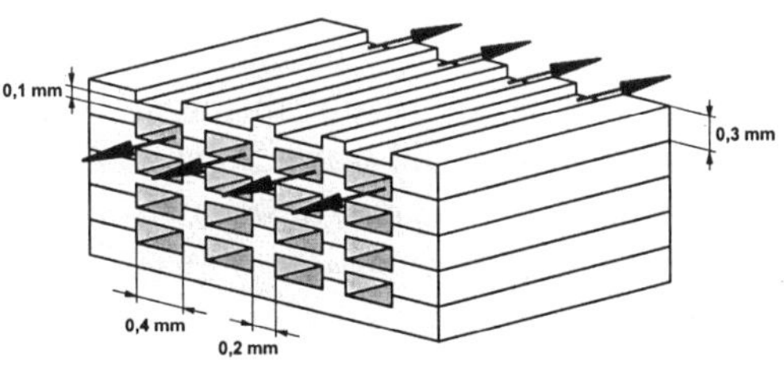

Figure 1: *Schematic view of a micro heat exchanger*

3 FEM-SIMULATION OF THE COINING PROCESS

In preparation of the experimental study numerical process simulation (FEM) was used primarily to evaluate the die design and to estimate the expected die loads and die deflections. In combination with the experimental results (see Chapter 4) the FE model has then been evaluated and qualified as a reliable process development tool.

3.1 Geometry and material data

The product geometry chosen for the coining evaluation was the rib type structure shown in Fig. 2. The simulation of the process was performed with a two dimensional model which included not only the sheet specimen and the die insert, but in some cases also the die frame in order to accurately model the die deflections (Fig. 2). The material data (sheets from Al99,8, AlMg3, CuZn37, X5CrNi1810; die frame: ARNE; die insert: VANADIS 6) were taken from literature which was found to be accurate enough for the purpose of estimating die loads, die deflections and metal flow.

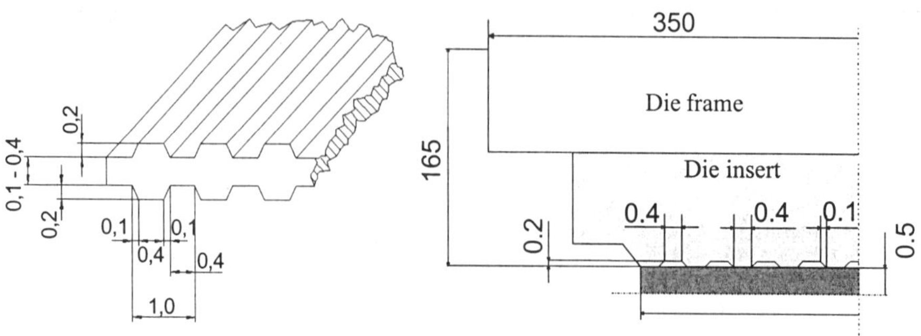

Figure 2: *Rib type structure (left) and simulation model (right)*

3.2 Typical simulation results

Even though the mechanical loads of the die frame are low (Fig. 3, left) there is a significant difference between the vertical displacement of upper edge of the die frame (u_z = -0.3 mm) and the actual displacement of the coining tool surface (u_z = -0.1 mm; Fig. 3, right) which means, that the amount of average elastic compression of the die is twice as large as the coining depth.

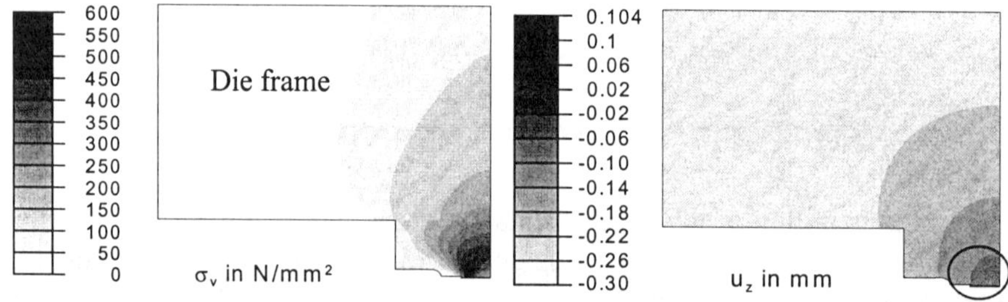

Die insert; see detail Fig. 2

Figure 3: *v. Mises equivalent stress (left) and vertical displacement (right) for the complete die in coining of aluminium*

Looking in detail at the die loads close to the coining surface it can be seen, that the expected equivalent stresses in the die are more than twice as high as the average flow stress or approximately 5 times the yield stress of these alloys (Fig. 4 left). This must be taken into account when choosing tool materials in order to avoid plastic tool deformation.

With respect to the precision of micro parts produced by coining it is also important to minimize thickness deviations resulting from inhomogenious die deflections in the vertical direction (Fig. 4 right).

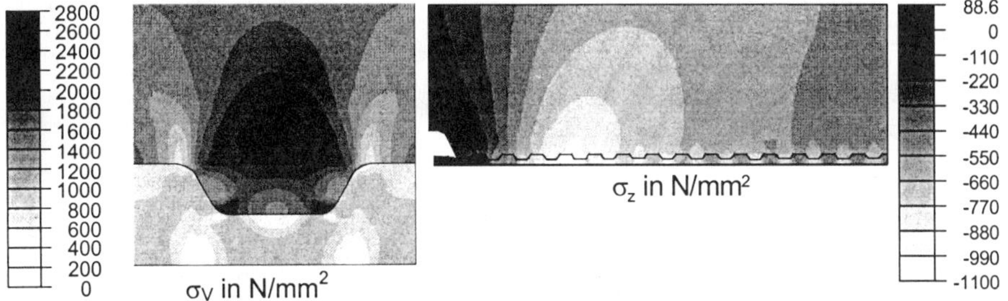

Figure 4: *v. Mises equivalent stress in coining of X10CrNiTi189 (left) and normal stress distribution in coining of aluminium (right); initial sheet thickness $h_0 = 0.6$ mm*

4 EXPERIMENTAL INVESTIGATION

4.1 Process conditions and procedure

The coining experiments were performed using a hydraulic press with $v_{max} = 56$ mm/s punch velocity under force control. The temperature was room temperature in all cases and all experiments were performed without lubrication. A series of experiments with different upper force limits was performed for each alloy and the achieved geometry was measured to derive a load stroke relationship.

4.2 Experimental results and FE evaluation

<u>Plastic strain distribution</u>

In the CuZn37 alloy the metallographic preparation clearly indicates a microstructure variation (dark x-lines) which corresponds very well with the plastic strain distribution achieved by numerical simulation (Fig. 5).

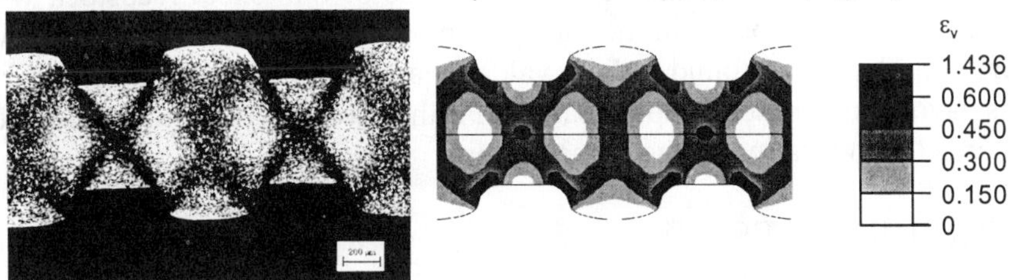

Figure 5: *Microstructure and plastic strain distribution (FEM) in coining CuZn37 channel structures; initial sheet thickness $h_0 = 0.8$ mm*

<u>Load versus coining depth and thickness distribution</u>

The measured part geometry and the coining depth as a function of force level show a good correlation with the FEM results (Fig. 6).

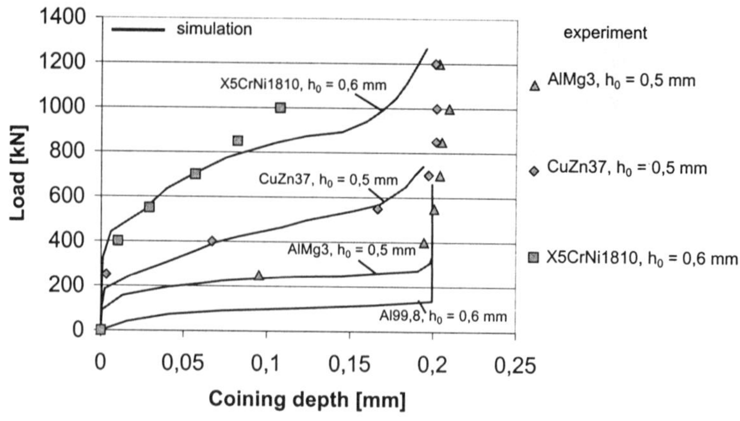

Figure 6: Comparison of simulation and experiment for different materials

5 DISTORTION AFTER UNLOADING

In some coining examples springback and distortion after unloading has been observed. This is especially to be expected, if the final structure is very thin and consequently has only little stiffness (Fig. 7).

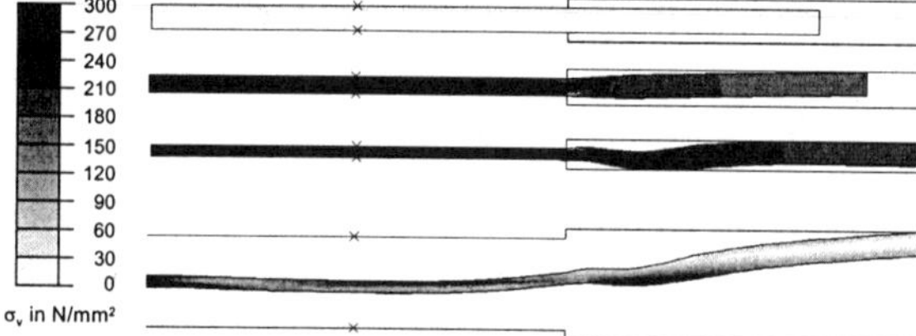

Figure 7: Simulation of distortion after unloading

6 SUMMARY AND OUTLOOK

Coining of sheet metal has been found to be capable of producing precise micro channel structures even in the case of thin sheet feedstock with the sheet thickness being in the range of the structure thickness. Also FEM-simulation has been found to be a valuable and fairly accurate tool for process layout and process optimisation. Further work will now be directed towards the evaluation of the process limits.

7 REFERENCES

[1] H. Ike, M. Plancak, *Controlling Metal Flow Of Surface Microgeometry In Coining Process*, Advanced Technology of Plasticity, Vol.II, Proc. of the 6th ICTP, Springer Verlag, Berlin, pp. 907-912, September 1999

[2] R. Neugebauer, A. Schubert, J. Kadner, T. Burkhardt, *High Precision Embossing Of Microparts*, Advanced Technology of Plasticity, Vol.II, Proc. of the 6th ICTP, Springer Verlag, Berlin, pp. 921-926, September 1999

THREE-DIMENSIONAL MICRO-FORMING PROCESS OF THIN FILM METALLIC GLASS IN THE SUPERCOOLED LIQUID REGION

Seiichi Hata*, Yongdong Liu*, Tomokazu Kato** and Akira Shimokohbe*

*Precision and Intelligence Laboratory, Tokyo Institute of Technology

**Graduate school of Tokyo Institute of Technology

Abstract

This paper introduces micro-forming for the three-dimensional micro structures of thin film metallic glass (TFMG). TFMG is a kind of amorphous thin film having suitable characteristics for MEMS because they are isotropic and homogeneous as well as free from defects originating from the crystal structure. Moreover TFMG soften in a certain temperature range called the supercooled liquid region, which makes it easily be formed into three-dimensional shapes. In this paper, new three-dimensional micro-forming process of TFMG in the supercooled liquid region and its application for micro actuator are reported.

Keywords

MEMS, Process, Micro-forming, Thin film, Amorphous, Metallic glass

1. INTRODUCTION

Metallic glasses are new amorphous alloys which have large glass-forming ability and significant supercooled liquid region before crystallization temperature. Accordingly, metallic glasses have attracted a great deal of research interests during the last decade. However, upon now, works on metallic glasses have focused mainly on bulk materials.

The fabrication method for Zr based TFMG had been reported as well as, micro-forming of a micro beam of TFMG using its bimetal effect in the supercooled liquid region [Hata et al., 1999]. To reduce the internal stress of the micro beam, annealing at the supercooled liquid region was applied. Moreover, Pd based TFMG could also be fabricated and its physical properties were reported [Liu et al., 2001].

This paper reports realization of a new three-dimensional micro structure of TFMG by using the large deformation capability of its supercooled liquid region. Although, a large plastic deformation of polysilicon had been reported [Yang et al, 1998], this paper can be characterized by 1) a new material (TFMG) for MEMS, 2) Free-inner and/or residual-stress micro-forming of TFMG at supercooled liquid state and 3) real three-dimensional deformation.

2. FABRICATION AND PROPERTIES OF TFMG

In this study, Zr based TFMG ($Zr_{75}Cu_{19}Al_6$, atomic %) was used. A RF magnetron sputtering equipment was applied to fabrication of the TFMG. Table 1 shows the sputtering condition and composition of the TFMG. Table 2 shows properties of the TFMG and a comparison with other materials. These TFMG exhibit both tensile strength and elastic limit further than poly-silicon or stainless steel.

Moreover, the TFMG have 70 K and stable supercooled liquid region. Figure 1 shows Time-Temperature-Transition (T.T.T) diagram of the TFMG. From this graphs the TFMG does not crystallize in the supercooled liquid region even if it is heated for one hundred to three thousand seconds.

<table>
<tr><td colspan="3">Table 1   Sputtering condition</td></tr>
<tr><td>RF power</td><td>150 W</td></tr>
<tr><td>Pre-sputtering time</td><td>30 min</td></tr>
<tr><td>Sputtering rate</td><td>22 nm/min</td></tr>
<tr><td>Argon gas flow rate</td><td>1.25 SCCM</td></tr>
<tr><td>Argon gas pressure</td><td>0.4 Pa</td></tr>
<tr><td>Target composition</td><td>$Zr_{75}Cu_{19}Al_6$</td></tr>
<tr><td>Thin film composition</td><td>$Zr_{75}Cu_{19}Al_6$</td></tr>
</table>

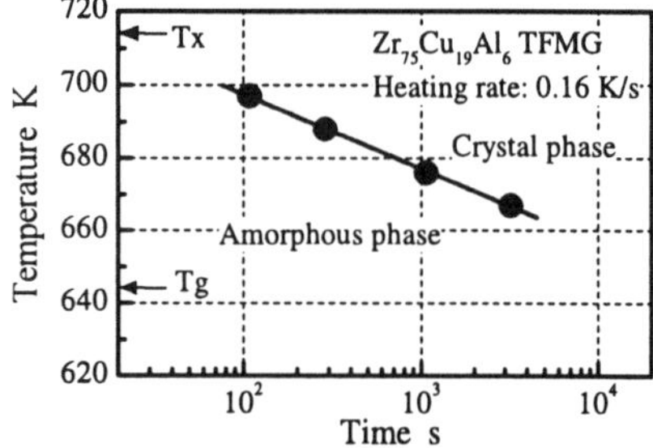

Figure 1   T.T.T. diagram

Table 2   Properties of TFMG

	TFMG ($Zr_{75}Cu_{19}Al_6$)	Polysilicon	SUS 304
Tg, Tx, ΔT	643 K, 713 K, 70 K	NA	NA
Young's modulus	By tensile test: 83.4 GPa By bending test: 93.5 GPa	170 GPa	197 GPa
Tensile strength	1.61 GPa	1.21 GPa	0.5 GPa
Elastic limit	1.97 %	0.71 %	Less than 0.2 %
Hardness	Hv 751.3	NA	Hv 540
Resistivity	160 μΩ·cm	NA	10-20 μΩ·cm
Density	5.51×10^3 kg/m^3	2.33×10^3 kg/m^3	8.0×10^3 kg/m^3

3. THREE DEMENTIONAL FORMING PROCESS
3.1 MEMS Process of TFMG

Figure 2 shows the lift-off process, which fabricates a planar spiral shape structure (spiral beam) of the TFMG. A SEM photograph of the fabricated spiral beam is shown in Figure 3. The spiral beam of the TFMG can be heated without be crystallizing to the supercooled liquid state. The TFMG spiral beam was softened and micro-formed by heating it to the supercooled liquid region. In this micro-forming process, two different methods using a micro jig were applied. One uses the viscous flow at supercooled liquid state, while the other uses annealing effect.

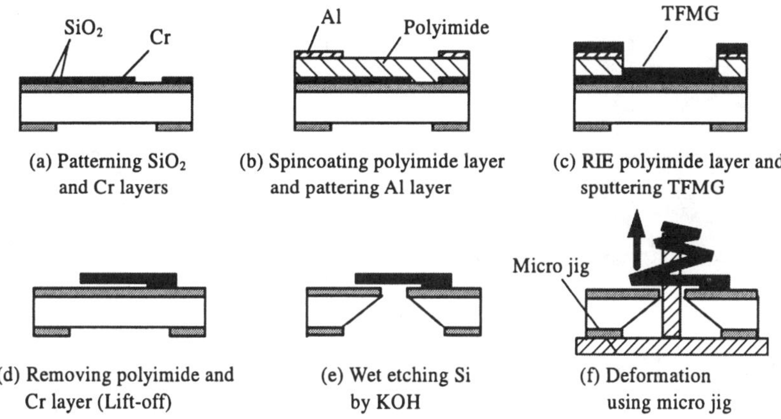

(a) Patterning SiO$_2$ and Cr layers

(b) Spincoating polyimide layer and pattering Al layer

(c) RIE polyimide layer and sputtering TFMG

(d) Removing polyimide and Cr layer (Lift-off)

(e) Wet etching Si by KOH

(f) Deformation using micro jig

Figure 2   Micro machining process of TFMG

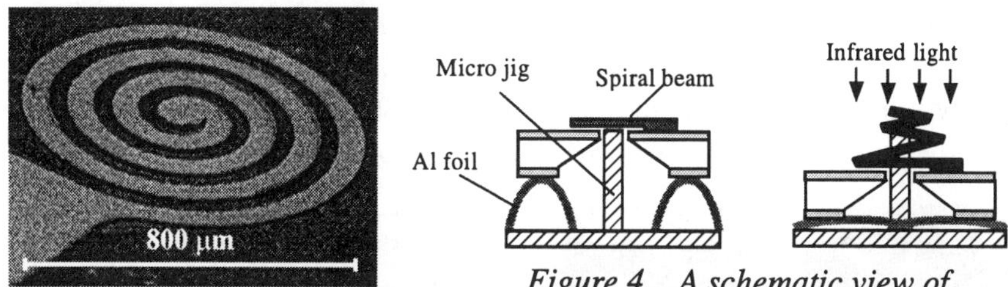

Figure 3   A micro spiral beam

Figure 4   A schematic view of method using the viscous flow

3.2 Micro Forming using Viscous Flow

In Figure 4, a schematic view of method using the viscous flow is shown. Arched aluminum foils were placed between the substrate and a micro jig. Using an infrared heating device, the substrate was heated to 678 K at a rate of 10 K/min in the vacuum chamber. The substrate temperature was held at 678 K for 300 seconds.

At this temperature, the foils were softened and deformed by substrate's weight. The micro jig pushed up the center of spiral beam through a hole of the substrate. Then the spiral beam of TFMG exhibiting the viscous flow in the supercooled liquid state was micro-formed into three-dimensional conical spring. Figure 5 shows a SEM photograph of the micro conical spring and section profiles of the spring before and after forming. After forming, a conical spring which has 350 μm height was micro-formed. In Figure 5, deformation of the spiral beam before forming is caused by residual stress at sputtering process.

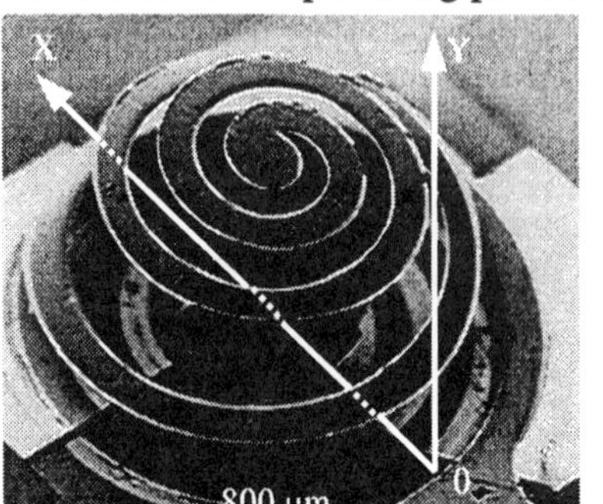
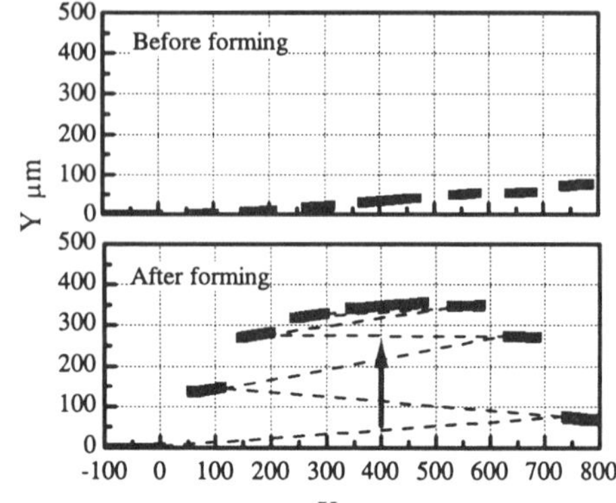

Figure 5   A micro conical spring by micro-forming using viscous flow

3.3 Micro Forming using Annealing Effect

Another micro conical spring was formed by using annealing effect. First, the spiral beam was pushed up using the micro jig and then the elastic stress in the beam was relaxed by heating it to the supercooled liquid region. The annealing temperature was 678 K at a rate of 10 K/min in the vacuum chamber. The annealing temperature was held at 678 K for 30 seconds. Finally, the beam was cooled at a rate of 10 K/min and the micro conical spring was formed. Figure 6 shows the results of this micro-forming.

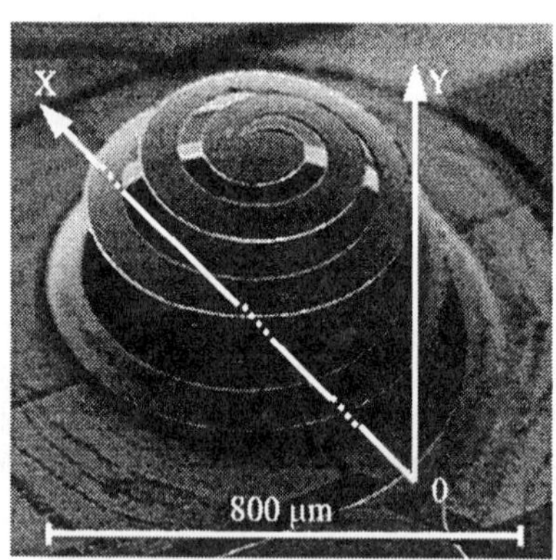
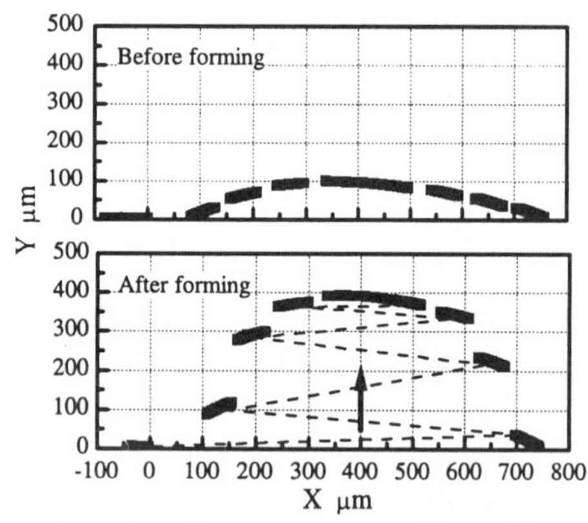

Figure 6   A micro conical spring by micro-forming using annealing effect

4. APPLICATION

An electrostatic micro actuator was fabricated as an application of the micro-forming process of TFMG. This micro actuator caused the conical spring to be a moving electrode. The insulating layer and substrate electrode were fabricated under the moving electrode by conventional MEMS process.

Figure 7 shows a driving micro actuator in SEM and change of height in the moving electrode center versus applied voltage. This actuator could be driven over 100 μm by 100 V applied voltage.   However, the height of this actuator did not change under 200μm even when applied voltage was increased up to 200 V and did not return to its original height when the voltage was decreased. This may be due to the fact that the moving electrode was adsorbed to the substrate electrode.

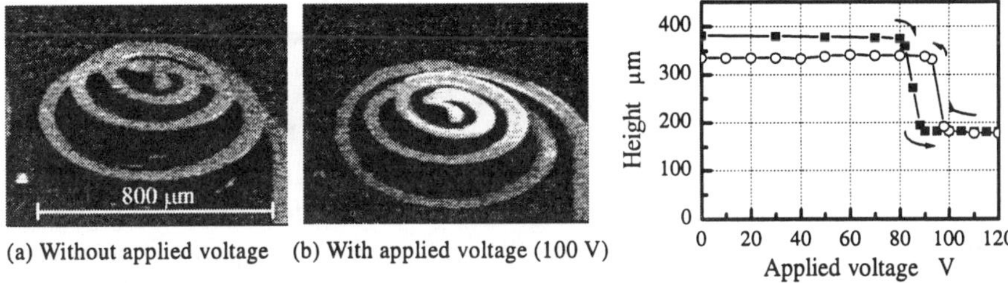

(a) Without applied voltage   (b) With applied voltage (100 V)

Figure 7   A micro actuator of the conical spring by the micro-forming

5. CONCLUTIONS

To realize a real three-dimensional micro structure, a new micro-forming process of TFMG in the supercooled liquid region was examined. By two kinds of micro-forming process, it was possible that fabrication of conical spring which an 800 μm diameter and over 350 μm height. This conical spring was applied to an electrostatic micro actuator which can move out of substrate. Further researches will be realizations of full-stroke and stepwise motion of the micro actuator.

This research was supported in part by a grant from JSPS Research for Future Program "Complex Integrated Machines" and Grant-in-Aid for Scientific Research (A) (2), 12355007, 2000. The experiment was carried out at the Vacuum Machining Laboratory in Tokyo Institute of Technology.

REFERENCES

S. Hata, K. Sato and A. Shimokohbe. Fabrication of Thin Film Metallic Glass and its Application to Microactuator. Proceedings of SPIE Conference on Device and Process Technologies for MEMS and Microelectronics; 1999 October 27-29; Queensland, Australia.

Y. Liu, S. Hata, K. Wada and A. Shimokohbe. Thin Film Metallic Glasses: Fabrication and Property Test, Proceedings of the 14th IEEE International Conference on Micro Electro and Mechanical Systems; 2001 January 21-25; Interlaken, Switzerland.

E.H.Yang and H. Fujita. Determination of the modification of Young's modulus due to Joule heating of polysilicon microstructures using U-shaped beams, Sensors and Actuators. 1998; A70: 185-190

PRECISION COLD FORGING – METHODS FOR REDUCTION OF WORKING PRESSURE

K.K. Tong, T. Muramatsu, C.M. Choy, S.X. Zhang, M. Enggalhardjo

Gintic Institute of Manufacturing Technology
71 Nanyang Drive
Singapore 638075.
Tel: (65) 7938478 Fax: (65) 7925362
E-mail: steven@gintic.gov.sg

Abstract

Reduction of forming pressure is inevitable to realise precision forging. This will result in improved dimensional accuracy for complex geometry components and increased tool life. The purpose of this paper is to review 3 proposed methods of reducing working pressure during cold forging. These methods were studied by conducting tests using force profiles measurement system, finite element study and metallurgical evaluation.

Keywords

Working pressure, forging, grain flow, load measurement, finite element.

1. INTRODUCTION

One of the most important subjects of research and development in advanced countries on forging is precision forging where high accuracy, complex and net shape components can be produced [1]. Cold forging has high potential to reduce manufacturing cost. If the work material could be completely filled up into the die cavity, desired accuracy of the product can be achieved, hence high productivity can be envisaged. However, the complete filling up of material into the die cavity is quite difficult because of high working pressure [2]. The process often involves uni/multi-axial loading, large deformation and substantial work hardening of the work material in order to achieve the required shape. Hence, punch and die used in cold forging often need to withstand forming stress up to 1500 N/mm^2.

2. METHODOLOGY

2.1 Process Parameters

The experiments were conducted using a 630-ton cold forging press (AIDA K1-6130). The three methods to reduce working pressure (loads) were evaluated [3] and are shown schematically in Fig. 1. The experiment conditions are tabulated in Table 1. The pre-chamfering method utilises radial changes of the blank before forging. Spread extrusion [4] and relief axis [5] methods were reported by Kunogi and Sawabe respectively. Detailed studies on the working loads and finite element simulation were conducted to further understand the proposed methods. The working material used in the experiments was AISI 1017; a medium carbon steel commonly utilised in cold forging components.

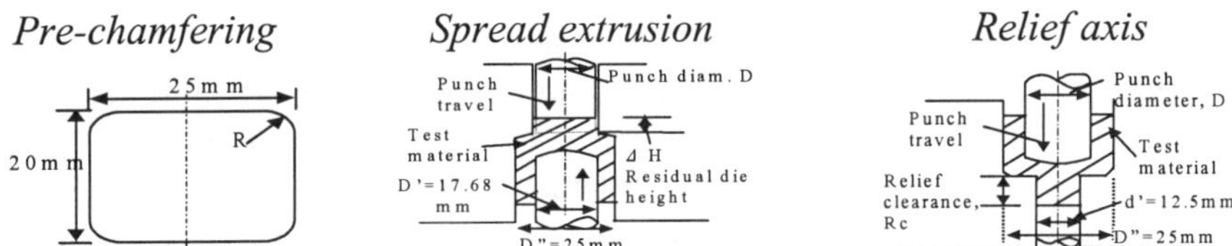

Fig. 1: Schematic diagram showing the construction of the three methods

Table 1: Testing conditions for the experiments (see Fig. 1 for reference)

Pre-chamfering	Spread extrusion				Relief axis	
Radius of curvature, R	Punch diameter, D mm				D mm	
	22.53	20.63	19.03	17.66	19.36	17.68
0 mm	⊿H=0.0	⊿H=0.0	⊿H=0.0	⊿H=0.5	Rc=0	Rc=0
4 mm	⊿H=1.5	⊿H=1.5	⊿H=1.5	⊿H=1.5	Rc=5	Rc=5
6 mm	⊿H=4.0	⊿H=4.0	⊿H=4.0	⊿H=4.0	Rc=10	Rc=10
10 mm	-	⊿H=6.5	-	-	Rc=15	Rc=15

2.2 Working Load Measurements

The experiments to measure the working loads (punch force) were obtained through a complex construction of data acquisition system and load sensors.

2.3 Finite Element Simulation

LS-DYNA version 950 was used for simulation on the pre-chamfering method and the results were compared with the actual experiment [6].

2.4 Grain Flow Studies

The final formed samples were sectioned, mounted, polished and etched to reveal the grain flow structures. The metallurgical samples were examined under the optical light microscope to study the material flow behaviour.

3. RESULTS AND DISCUSSION

3.1 Pre-Chamfering Method

3.1.1   Effect of Radii R on Material Deformation and Working Load

Fig. 2 shows the cold forging of the pre-form billet (with different radii) into the final axis symmetrical component. The gradual increase in effective stress at different punch displacement is shown in Fig. 3. It can be observed that substantial plastic deformation occurred at the reduction zone and will eventually result in the highest stress concentration area. The comparison of the deformation profile and the predicted stress/strain distributions with the actual macrograph showed that the FEA results conformed to the actual experimental results (Fig. 4). The full cycle of the forging process was recorded by the working load–displacement autographs for profile radius R of 0, 4, 6 and 10 mm. The process signatures of the actual experimental and finite element (FEA) results are shown in Fig.5.

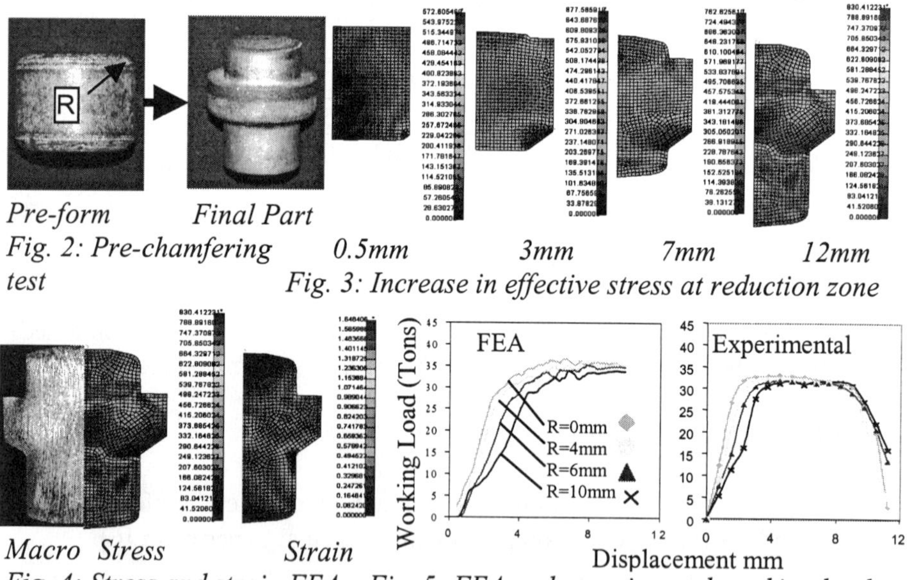

Fig. 2: Pre-chamfering test

Fig. 3: Increase in effective stress at reduction zone

Fig. 4: Stress and strain FEA   Fig. 5: FEA and experimental working load

The FE predicted pattern of the reduction of forming force was consistent with the experimental results. Both FE and experimental results show that with the use of profile radius up to R=10, the maximum forming load required was reduced approximately 4 to 5 ton. The load-displacement autographs showed that the gradient of the force increment reduces when profile radius was used. It is suggested that the reduction of the working load was attributed to the lower stress concentration at the reduction zone. It is believed that the use of profile radius promotes easy flow and deformation of the billet. In addition, with the lower stress concentration, strain hardening and forming force, longer tool life and a stable process can be envisaged.

3.2 Spread Extrusion Method

The working load of the punch was measured with respect to variations of the punch diameter (D=22.53mm, 20.63mm, 19.03mm and 17.66mm) and the residual die height ΔH (see Fig. 1 and Table 1 for reference). The billet was formed to its final shape and metallurgical analysis of the grain flow was conducted (Fig. 6). The grain flow pattern indicated the degree of work hardening at selected regions where redundant work was envisaged. The working load was reduced as the punch diameter decreased from 22.53mm to 17.66mm, as shown in Fig. 7. A smaller billet will reduce working pressure and a homogeneous material flow, resulting in higher tool life. In addition, the lower working load allows the die cavity to be filled completely by the material without reaching the die or punch critical breaking load, which will result in a more dimensionally accurate product.

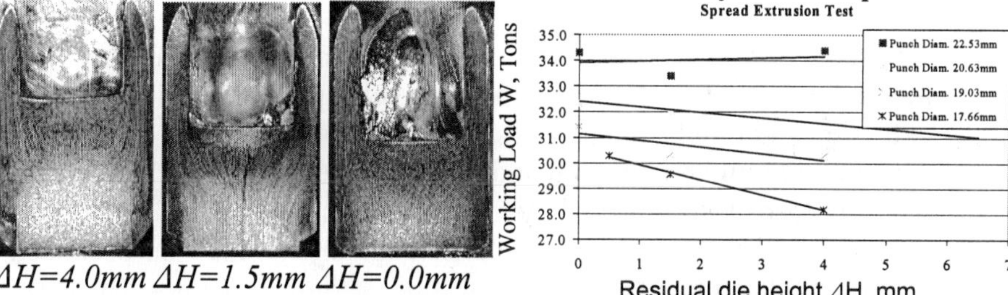

ΔH=4.0mm ΔH=1.5mm ΔH=0.0mm
Fig. 6: Cross-sections showing the grain flow during forming

Fig. 7: Working load W against residual die height ΔH

3.3 Relief-Axis Method

3.3.1 Effect of Relief Clearance Rc on Working Load

The working load of the punch was reduced as the relief clearance Rc increases from 0mm, 5mm, 10mm and 15mm (Fig. 8). The working load was at its lowest when a relief clearance of 15mm was utilised and the grain flow behaviour exhibited a consistent pattern not restricted by any significant dead zones. The reduction of the forming load was measured to be 11% and 8% for punch diameters 17.68mm and 19.36mm respectively. This method aims at simultaneous completion of die cavity filling at all contour portions of the products. This is achieved by allowing (non-restriction) the natural direction of the plastic flow of the material into relief section of the part, which can be removed easily by machining. This will result in desirable homogeneous deformation and accordingly the reduction of resistance for redundant work. In addition, it will prevent die breakage due to excessive working load during cavity filling of the part.

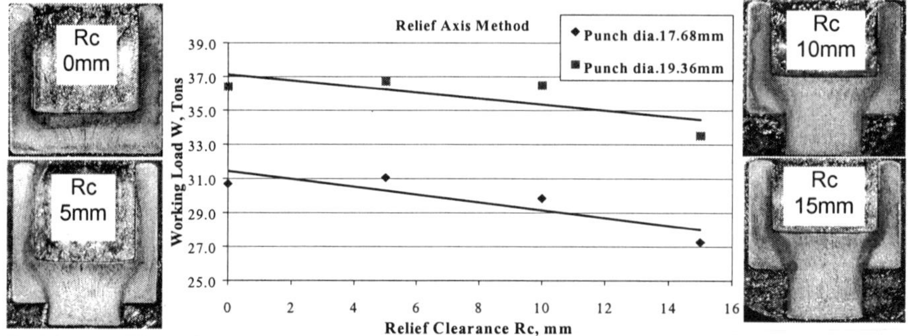

Fig. 8: Reduction of working load (W) with increase of relief clearance Rc. In addition, a smaller punch diameter exhibited lower working load.

4 CONCLUSIONS

1. The reduction of working pressure (load) can be envisaged by three methods discussed in this paper, namely pre-chamfering, spread extrusion and relief axis methods.
2. The die life will be improved and die breakages reduced when critical working loads were lowered by these techniques.
3. Dimensionally accurate and complex components may be produced when these techniques were applied.

REFERENCES:
[1] "Reduction of forming pressure by improved tool design of cold forging process for a small net shape component," Gintic's in-house report, C99-P-072A.
[2] "Improvement of product accuracy in cold forging die," K Kondo, Professor Emeritius, Faculty of Engineering, Nagoya University, Japan. (Internal report)
[3] "Precision cold forging and new shearing process," K Kondo, Professor Emeritius, Technical seminar, 9 Nov 2000, Gintic, Singapore.
[4] "Spread extrusion method," M. Kunogi, Report of Science Research Inst., 50(1956), 215.
[5] "Relief axis method," H. Sawabe, Basic and Application on Cold Forging, Sanpo(1968).
[6] "Verification of the reduction of forming force for injection extrusion using ANSYS/LSDYNA with 2D axis-symmetrical element," CM Choy, KK Tong, T Muramatsu, 3rd ASEAN ANSYS conference, Nov 9 –15, 2000, Singapore.

BURR FORMATION IN MICRO-MACHINING ALUMINUM, 6061-T6

Kiha Lee[1], Boris Stirn[2], David A. Dornfeld[1]

1. Department of Mechanical Engineering
University of California at Berkeley
Berkeley, California 94720, USA
2. Laboratory for Machine Tools and Production Engineering (WZL)
University of Technology at Aachen (RWTH)
52056 Aachen, Germany

Abstract

Experimental studies on micro-milling and micro-drilling in aluminum 6061-T6 have been carried out. A range of different cutting speeds, chip loads and depths of cut using a tool of 127μm diameter were considered. Drilling experiments with different feed to diameter ratios, cutting speeds and two drill diameters were conducted. The influence of the cutting parameters on burr size and burr type was observed. A comparison to burr formation in conventional machining is presented.

Keywords

Burr formation, Micro-Machining, Drilling, Milling

1. INTRODUCTION

A phenomenon similar to the formation of chips is the formation of burrs at the end of a cut. Burrs are undesirable because they present a hazard in handling machined parts and can interfere with subsequent assembly operations. Thus, they must be removed in subsequent deburring processes to allow the part to meet specified tolerances.

A number of burr removal processes exist for conventional machining and except for the cost, can usually be conveniently applied. In the micro-machining process, however, the burr is very difficult to remove and, more importantly, the burr removal can seriously damage the workpiece. Conventional deburring operations cannot easily apply to micro-burrs.

Micro-machining here refers to machining with miniaturized end mills and drills. Micro-machining in this work is defined as using tools as small as 50μm. The miniaturized end mills and drills have similar cutter geometries as conventional machine tools as shown in Figure 1.

Research on burr formation in micro-milling of stainless steel, brass, aluminum and cast iron (Damazo, *et al* 1999; Schaller, *et al* 1999) and of micro-drilling burr formation in Fe, Ag and SUS (Sugawara and Inagaki 1982) has been reported. Micro burr formation, however, with respect to cutting conditions has not been studied so far. The fundamental mechanisms are not well understood. In this paper, the size and type of burr created in aluminum, 6061-T6, are studied.

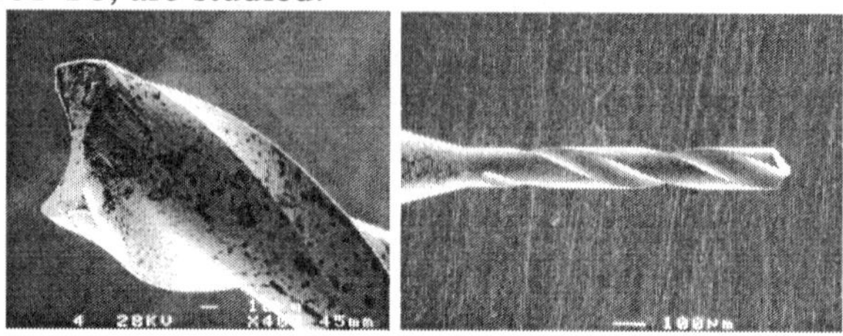

Figure 1. Micro end mill (127μm, left) and drill (130μm)

2. BURR FORMATION IN MICRO-MILLING

Chu (Chu 2000) conducted a set of conventional slot and step milling tests to analyze burr formation in milling aluminum, 6061-T6. Chu reported that the cutting speed has an insignificant influence on burr formation and larger feed per tooth induces large burrs in the feed direction.

A series of experiments (Table 1) was performed to determine the differences between conventional burr formation and micro burr formation in milling. A micro end mill (Robbjack Corp.) was used with either a Mori Seiki CNC Drilling Center TV-30 or an Air Turbine Tools spindle. Micro-drop coolant was used. Figure 2 shows the micro slots machined by a Mori Seiki CNC Drilling Center.

Various burrs were created at different locations of a workpiece, including side burr, top burr and entrance burr in the feed direction. Compared to conventional burr formation, important observation were made:

- Exit side burrs were always created regardless of cutting speed and feed. As seen in Figure 3, it is a "flag-type" burr.
- Curl-type entrance side burr was formed (Figure 3). With higher feedrate, the burr size increases. With higher cutting speed, the burr size decreases.
- Two different top burrs were created. Wavy-type burr is created when the tool enters the top surface of the workpiece. A ruptured-type burr results when the tool exits (Figure 3). These burrs are difficult to remove in micro slot milling. Conventional deburring operations cannot easily remove these burrs. Cutting speed has an insignificant effect on top burrs. Large feed per tooth induces large burrs.

On the basis of the results, burrs are barely evident for conditions using 40,000 rpm, 63.5 μm depth of cut and 0.826μm/tooth feed.

Cutting speed [rpm]	7500 rpm (Mori Seiki), 40,000 rpm (Air Turbine)
Feedrate [mm/min]	12.7, 38.1, 63.5 (Mori Seiki), 66.04, 198.12, 330.2 (Air Turbine)
Depth of cut [μm]	63.5, 127.0

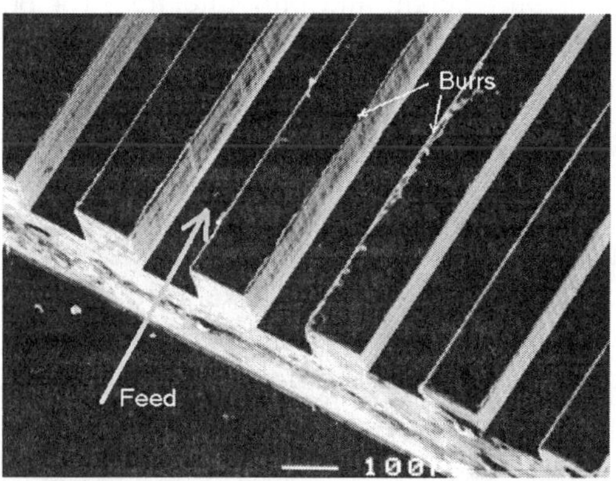

Figure 2. Slots (127μm end mills) at 7500 rpm, various feedrates and depths of cuts.

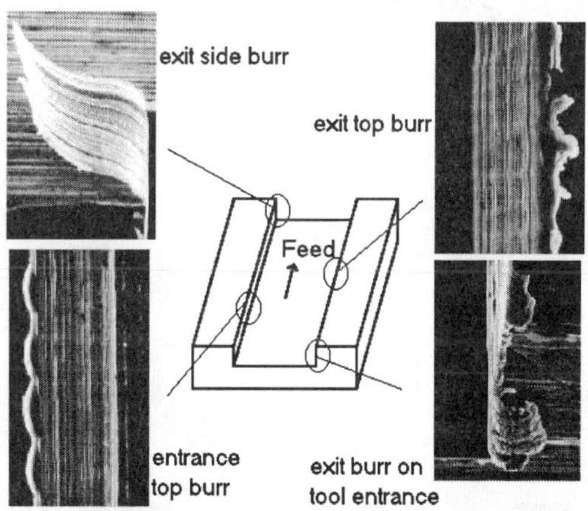

Figure 3. Burr formation in the micro end milling (the tool rotates clockwise)

3. BURR FORMATION IN MICRO-DRILLING

Previous research (Kim 2000) had established basic burr types in drilling, Figure 4. To investigate burrs in micro-drilling, experiments using a wide range of cutting parameters with two different drill diameters were carried out. A precision machine tool attached to a Mori Seiki CNC Drilling Center was used. The cutting conditions, which are based on the recommendation of the tool manufacturer Titex Tools, are given in Table 2.

Drill diameters	130μm, 250μm
Feed/diameter	0.00625, 0.0125, 0.025, 0.05, 0.1
Cutting speed	2.5-4.1 m/min (130μm), 6.7-11 m/min (250μm)
Coolant	Water-soluble emulsion

The values of burr height and burr thickness for both drill diameters are shown in Figure 5. Crown burr is represented by dark bars and uniform burr by white bars. The gray bars represent transient burr. Transient burr is believed to be the type of burr formed in the transition stage between a uniform burr and a crown burr as it has characteristics of both types (Kim, 2000).

The measurements were taken from SEM pictures. Five holes of each set were measured and the values were averaged. The analysis of the exit burrs shows that the effect of feed on the burr height, thickness and type is similar to conventional drilling. Important results are as follows:

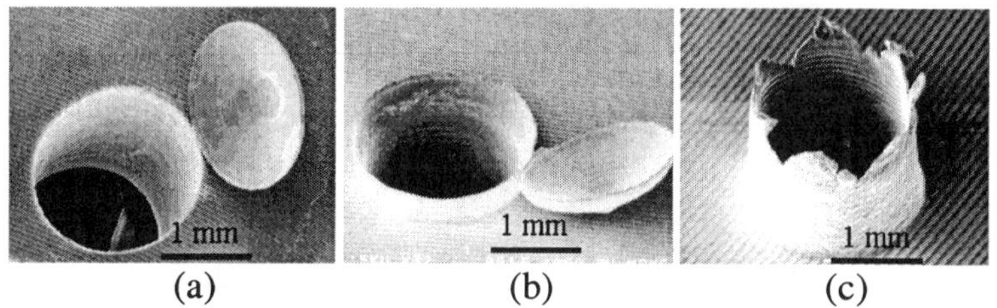

Figure 4. Drilling burr types; (a)(b) uniform burr, (c) crown burr (Kim 2000)

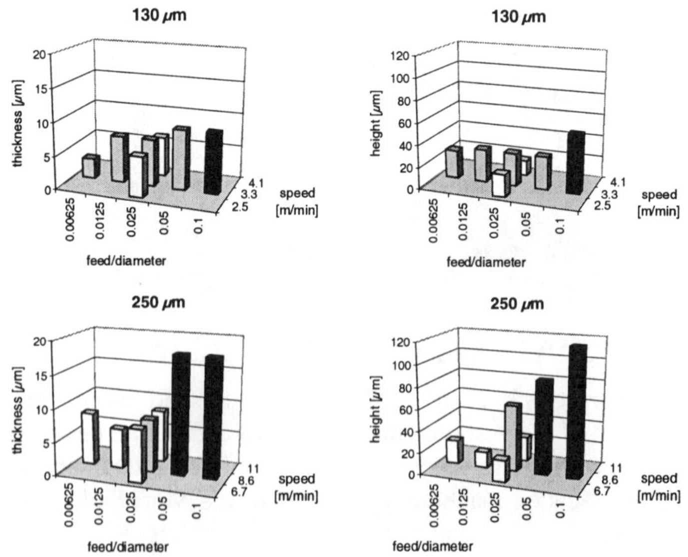

Figure 5. Burr height and thickness

- Burr height increases as the feed increases. With the 250μm drill the slow feed induces a uniform burr and the high feed induces crown burr formation, as in conventional drilling. No uniform burr was created by the 130μm drill. Here the slow feed induces a transient burr, Figure 6.
- The values of burr thickness correspond to the values of height. Increasing feed has the effect of increasing burr thickness as shown in Figure 5.
- No consistent effect of speed was determined, possibly because of the small variations of +/- 25%, due to the limits of the machine tool. The lower values of burr height at high and low speed might be the result of more or less vibrations induced by the machine or different sizes of build-up edges. Further experiments with a wider variation of cutting speed are necessary.

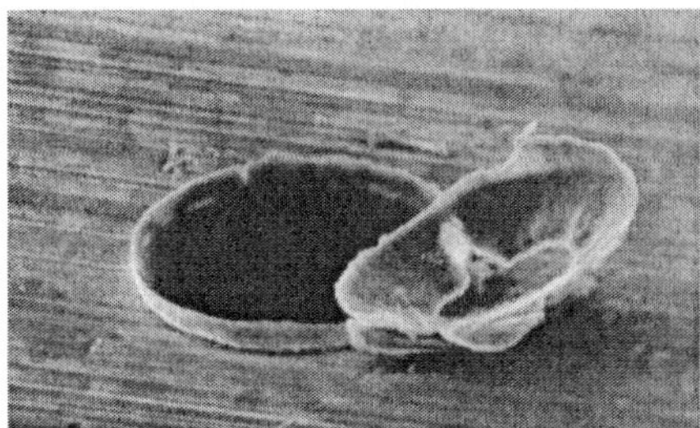

Figure 6. Exit burr at feed/diameter 0.00625, diameter 250μm (left) and 130 μm (right)

ACKNOWLEDGEMENTS

This work was supported by the Consortium on Deburring and Edge Finishing (CODEF) at University of California at Berkeley. The authors thank Air Turbine Technology and Robbjack Corp. for their support.

REFERENCES

Sugawara, A., and Inagaki, K., Effect of workpiece structure on burr formation in micro-drilling. Precision Engineering 1982, Vol. 4, No. 1, 9-14

Chu, C.H., Integrated Edge Precision Machining, Ph.D. dissertation, Department of Mechanical Engineering, University of California at Berkeley, 2000

Damazo, B. N., Davies, M. A., Dutterer, B. S., Kennedy, M. D., 1999, A summary of micro-milling studies, 1st International Conference of the European Society for Precision Engineering and Nanotechnology, Bremen, Germany, 31 May - 4 June, 1999. 322-324

Kim, J., Optimization and control of drilling burr formation in metals, Ph.D. dissertation, Department of Mechanical Engineering, University of California at Berkeley, 2000

Schaller, T., Bohn, L., Mayer, J., Schubert, K., Microstructure grooves with a width of less than 50μm cut with ground hard metal micro end mills, Precision Engineering 1999, Vol. 23, 229-235.

MICRO STRUCTURING OF HIGH ASPECT RATIO AND ARRAY BY MEANS OF MECHANICAL MACHINING

Kiyoshi Sawada, Tomohiko Kawai*, Yoshimi Takeuchi***

** FANUC Ltd., 3580, Shibokusa, Oshino, Yamanashi 401-0597, Japan*

*** The University of Erectro-Communications, 1-5-1 Choufugaoka, Choufu, Tokyo 182-8585*

Abstract

We introduce the ultra-precision micro structuring of high aspect ratio and array by means of mechanical machining with ultra-precision machine tool and single-crystal diamond tool.

Keywords

Ultra-precision machining, Micro machining, Diamond cutting

1.INTRODUCTION

In the medical field, micro needles have been studied for a painless hypodermic injection. Such needles should have small diameter and sharp tip not to hurt skin tissue. Recently, array structure of needles is investigated not to break the needles under the skin. For volume production of such precise parts, micro plastic molding is suitable and it needs a precise mold. Generally, silicon-based technology such as dry etching and EDM technology is not suitable using for it, because it is difficult to obtain pricise form accuracy and smooth surface roughness. On the other hand, mechanical machining has several advantages that it can produce free shape with smooth surface and shape edge. In this paper we introduce the machining of a micro-needle-array as our first report. We established the method of micro machining with high aspect ratio using our main technology of ultra-precision machine tool and single-crystal diamond end-mill.

2.MACHINING SYSTEM (FANUC ROBOnano Ui)

For nano-meter-positioning, very small stick-slip-motion due to friction is the biggest obstacle. The main feature of ROBOnanoUi is that full aerostatic air bearing structure is adopted in all moving section (slides, feed screws, nuts, and all motor units), and the solid friction is completely eliminated. This machine is a five-axis machining center with straightline three axes (X, Z, and Y) and rotational two axes (B, C). The resolution of CNC X, Y, Z-axes are 1nm, the feedback unit is 1/3nm and that of B, and C axes are 1/100000 degrees. And the machine tool has an air turbine spindle for micro milling. Also, the air bearing is adopted in this spindle.

3.MANUFACTURING OF MICRO INJECTOR ARRAY

Figure.1 shows a view of micro injector array, which we consider the target shape. A needle is 1mm in height, and $30\sim100\,\mu$ m in diameter, with the angle of point in 15 degrees. This injector array consists of 81(9x9) needles on 2.5mm square. A hole about $10\,\mu$ m diameter is bored in each needle for an injection. *Figure.2* shows a method of structuring the micro

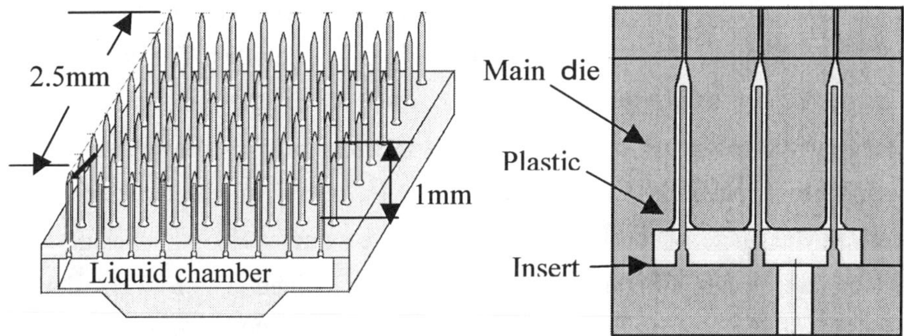

Figure.1 Micro injector array   Figure.2 Injection molding

injector array by plastic injection molding. The mold is consists of a main mold and an insert. The main mold, which is constructed with a hole array for injection molding of needle shape, is structure by mechanical machining a mold master and reversing it by electric casting. The insert is structure by mechanical machining. *Figure.3* shows a view of machining a micro injector array. A spindle is fixed on the C-table, and a workpiece is fixed on the B-table. For machining orthgonal three axes (X, Y, Z) are used. Y-axis is used for depth of cut. X, Z axes are used round tracking motion by harmony actions. On spindle, a single crystal diamond end-mill that has two edges (one is parallel to the spindle axes, the another is right angle) is installed.

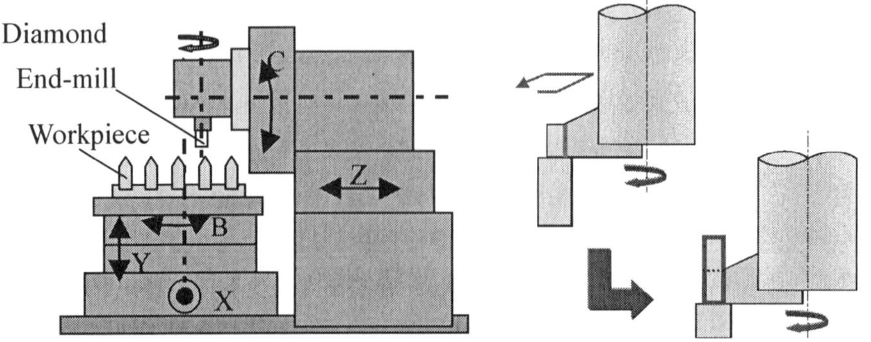

Figure.3 Machining construction   Figure.4 Machining method

3.EXPERIMENT OF MICRO HIGH ASPECT RATIO SINGLE STRUCTURE

As a test of machining of micro high aspect ratio structure, we machined a single pillar shape of 25 μ m in square, 1mm in height, and the top of the square pillar has sharpened angle of 15 degrees. The workpiece material is brass. *Figure.4* shows the method of the machining. The rotational tool move around and machine surroundings of the piller. When the machining of one cycle ends, the tool cuts axially of piller only by the width of the blade of the byte, and repeats the same cycle. As a result, the surface that has already been machined is not machined. The tool rotational speed is 50000r/min. In rough machining, feed rate is 100mm/min and depth of cut is 5 μ m until the work size became 50 μ m square. After that, feed rate is 10mm/min and depth of cut is 0.5 μ m as finish machining. The pickfeed is 100 μ m in corespondent to the width of tool blade. As the machining flued, kerosene was intermittently spraied on the workpiece. *Figure.5* is a machining result of the square piller, *Figure.6* shows an expansion picture in a top part of it, and in the edge part (*Figure.7*). The pillar did not break even by such long shape. Moreover, it could be machined without producing any burr and vibration at piller edge. The surface roughness of sidewall is 3 nm PV in both feed and pickfeed direction as shown in *Figure.8*.

54

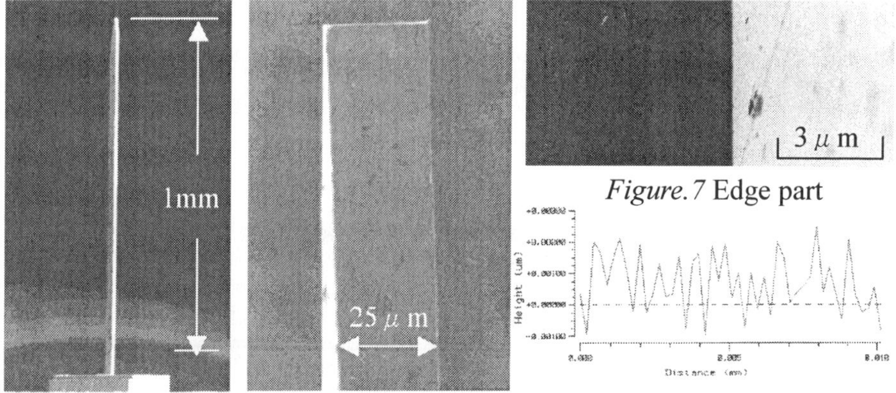

Figure.7 Edge part

Figure.5 Piller   Figure.6 Top part   Figure.8 Surface roughness

5.EXPERIMENT OF MICRO NEEDLE ARRAY

Next, As a test of machining of micro array structure, we machined a needle array shape of 30 μ m in diameter, 200 μ m in height, and the top of the needle has sharpened angle of 15 degrees. The workpiece material is brass. The pitch of each needles depend on the size diamond endmill, because, it is difficult to make the long and slender tool, which inserts through the needles in micro size. *Figure.9* shows the shape of diamond endmill. Therefore, we decided 250 μ m in the pitch of each needles. The needle array was machined one by one with the tool by using similar method as described above. The corn shape at the top of needle was machined by tracing an edge part of the tool blade along a ridgeline. The tool rotational

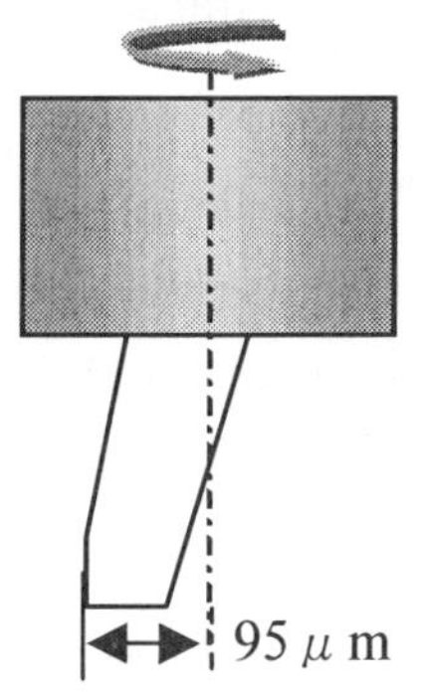

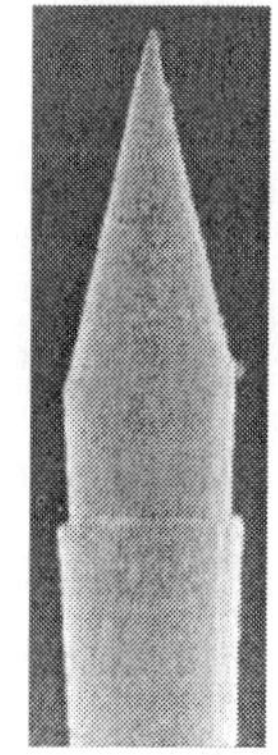

Figure.9 Tool shape   Figure.10 Needle array   Figure.11

Top part

55

speed is 50000r/min. In machining of corn shape, feed rate is 3mm/min and depth of cut is $0.5\,\mu$ m. In machining of cylinder shape, feed rate is 1mm/min and depth of cut is $45\,\mu$ m. Therefore, the total machining time is 1 hour for each needle. *Figure.10* shows a machining result of the needle array, *Figure.11* is an expansion picture in a top part of the needle array, which is some of scope for improvement at the shape of sidewall.

6.CONCLUSION

It was considerd that machining the shape of the high aspect ratio and array, like micro needle array, by means of mechanical machining was difficult. Because, it is difficult to make the high aspect ratio and micro diamond tool, which inserts through such shapes in micro size. Recently, we have come to obtain such tool, because of development of the grinding and lapping technology. Therefore, it has become to machine such shape. We picked up the mold of the micro injectior array as an example of a micro structure of the high aspect ratio. And we were obtained the excellent results, especially surface roughness and sharp edge, that were machined the micro square piller and the micro needle array as the preliminary examination by means of milling with a single-crystall diamond end-mill. In the future, we plan to be going to examine the plastic injection molding using the mold with a basal condition, smooth surface and sharp edge, in the good injection molding. However, in the present, there are a lot of problems shoud develop in the micro injection molding, for example, the problem of transcription accuracy, micro bubble, deformation, aged deterioration, etc.

References

Sawada, K., Kawai, T., Takeuchi, Y., and Sata, T., : Development of Ultraprecision Micro Grooving(Manufacturing of V-Shaped Groove), Trans.Jpn.Soc.Mech.Eng.InternationalJournal, Vol.43 NO.1, C (2000), p.170-176.

Acknowledgment

This work was performed by FANUC LTD and the laboratory of Prof. Takeuchi of the University of Erectro-Communications. A Part of this work was performed by FANUC LTD under the management of the Micromachine Center as the ISTF of MITI supported by NEDO.

INFLUENCE OF MICRO MACHINING ON STRENGTH DEGRADATION OF A SILICON NITRIDE CERAMIC

Manabu Wakuda[1], Yukihiko Yamauchi[2], Shuzo Kanzaki[2]

[1] *Synergy Ceramics Laboratory, FCRA*
[2] *Synergy Materials Center, AIST*

Abstract

This paper investigates two kinds of machining methods capable of dimpling a ceramic surface: abrasive jet machining (AJM) and laser beam machining (LBM). A discussion of the geometrical features of the dimples machined onto an Si_3N_4 surface is presented first, followed by the main topic, the effect of micro machining on the strength degradation. Dimpling by LBM can not be prevented from forming a sharp edge around the dimple, which may degrade the strength of the ceramic. The AJM dimple has a smooth-faced rounded shape, and only negligible deterioration of the strength takes place.

Keywords

Abrasive jet machining, laser beam machining, silicon nitride ceramic, flexural strength, micro machining

1. INTRODUCTION

Research work on ceramic materials has identified that surfaces with controlled porosity can exhibit good performance in tribological applications [Divakar, 1994 and MacBeth & Less, 1994]. This is attributed to the fact that pores on the surface act as fluid reservoirs and help to promote the retention of a fluid film at the sliding contact interface. In addition to this, particularly in instances where there are frequent start/stop operations, residual lubricant in the pores can play a role in avoiding dry running conditions.

Similar surface morphologies can be produced through micro machining processes. One of the most effective micro machining methods is laser beam machining (LBM). Although it is often pointed out that the technique causes thermal alterations in the subsurface characteristics of the material, excimer laser attracts much attention as a novel laser [Tönshoff & Gedrat, 1989]. As a competitive machining method, this study deals with abrasive jet machining (AJM), a specialized form of shot blasting, in which the surface of hard and brittle materials is engraved by impacting fine abrasives [Herbert, 1997].

This paper describes an attempt to generate a precise shallow hole on a silicon nitride ceramic surface using the AJM process. The geometrical features of the machined dimple are compared with that obtained with excimer laser beam machining. An investigation of the effect on material properties, in particular the strength of micro-dimpled specimens, is also discussed.

2. MICRO DIMPLING OF CERAMIC SURFACE

The test material was a commercially available pressureless sintered Si_3N_4 ceramic. The sample was machined into the form of 3 by 4 by 40 mm by grinding, based on the regulation for the four-point flexure testing. The tensile face was then lapped, and micro dimpled with AJM and LBM respectively as shown in Figure 1. The dimension of the dimples was fixed, for convenience, aiming at a target of 100 μm in diameter and 10 μm in depth. The blind holes were produced along the entire face, with a pitch of 500 μm in both the lateral and longitudinal directions.

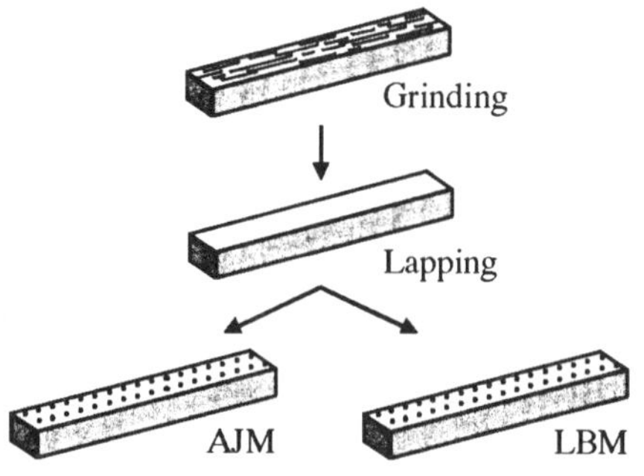

Figure 1. Flow of the machining procedure

In the AJM process, a micro-blaster (MB1–TR, Sintobrator Ltd.), capable of shooting a controlled mass of abrasives within a gas stream, was employed. The abrasive was silicon carbide of mesh size 800, corresponding to the size range 15–25 μm. A jet nozzle with an inner diameter of 8 mm was set at a distance of 100 mm above the workpiece surface. In order to mask the workpiece, a 0.1 mm thick laminate film was attached to the lapped surface, and the desired hole pattern was achieved by a photo resist method. The abrasive jet particles struck the workpiece through the open holes on the masking film, and thus engraved the surface.

Dimpling by LBM was performed using a KrF excimer laser, one of the most useful lasers for the application of micro machining of ceramics. In the process, four dimples were machined simultaneously using the mask projection technique [Ihlemann & Rubahn, 2000], and the operation was repeated in order to obtain the required dimple pattern on the entire face of the workpiece.

3. RESULTS AND DISCUSSION

3.1 Geometrical Features of Micro Dimples

The shape of the obtained dimples is first compared. Figure 2 shows photographs of the dimples machined by AJM and LBM. It is evident that the geometry depends greatly on the machining method. The AJM dimple had an inverted dome shape, whose sectional profile was composed of a very smooth curve. On the contrary, the base of the LBM dimple had a sharp edge, whose size was less than 5 μm, as a result of the hollow being deeper at the sides than in the center. This geometric feature is inevitably caused by the phenomenon that the reflected beam from the side wall is superimposed on the incident laser beam at the periphery of the base, which is a characteristic of micro machining of ceramic materials with excimer lasers [Miyamoto, 1990].

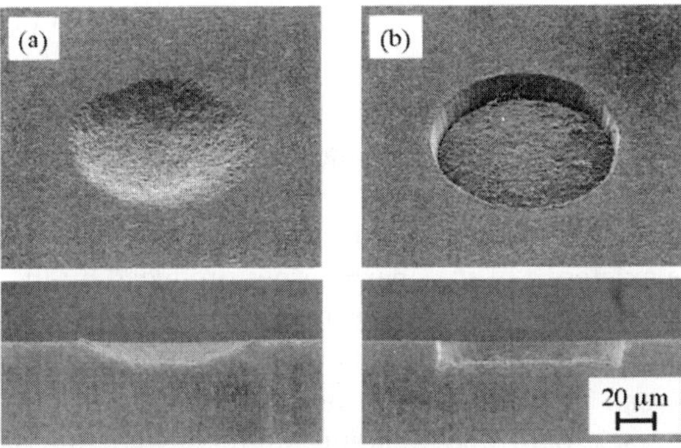

Figure 2. Micro dimples machined by (a) AJM and (b) LBM

In conclusion, the AJM and LBM processes can successfully generate the distributed dimples on the ceramic surface, although their geometry differs greatly depending on the machining method.

3.2 Strength of Micro Dimpled Ceramic

Flexural strength was evaluated by four-point bend tests on the lapped, AJM, and LBM workpieces. A total of ten samples was tested for each machining method. The mean strength is shown in Figure 3, with error bars showing maximum and minimum values. The lapped testpieces, i.e. non-dimpled ones, exhibited a flexural strength of approximately 1000 MPa, in agreement with the value quoted by the supplier. With regard to the AJM workpieces, there was no significant difference between the strength of these samples and the lapped ones. For the LBM specimens, however, the strength fell by about 20 % from that of the lapped workpiece due to the sharp-edged dimple shape.

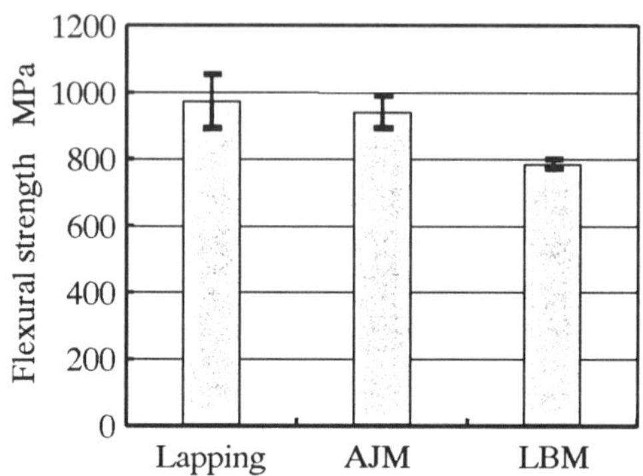

Figure 3. Flexural strength of the lapped, AJM, and LBM workpieces

Figure 4 shows the top view of typical AJM and LBM testpieces after the bending test. For the LBM specimens, the fracture path always passed through a single row of dimples. When micro machined by AJM, on the other hand, the fracture path did not necessarily go through the dimples. It is thought that the existence of the sharp corner produced by the LBM process enhanced the probability of fracture at the lower load. The fact that the deviation of the strength was surprisingly small in the case of LBM as shown in the figure 3, can be attributed to the same cause. In the AJM samples, since both the crack path and bending strength were the same as in the lapped samples, it is supposed that fracture occurred at pre-existing flaws within the material, without being affected by the existence of the dimples.

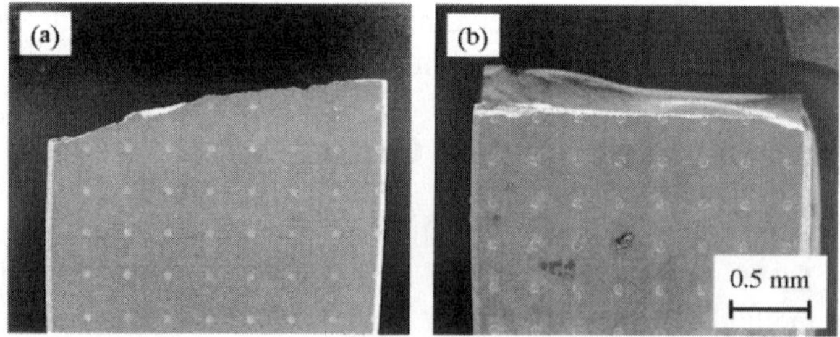

Figure 4. Bend test workpieces dimpled by (a) AJM and (b) LBM

4. CONCLUSIONS

Both the AJM and LBM processes were examined as micro dimpling methods for a silicon nitride ceramic. Through the machining tests, it was found that the geometry of the micro dimples is strongly dependent on the machining methods. An AJM dimple has a smooth-faced inverted dome shape, though an LBM dimple has an angular sectional profile. Additionally, the specimens micro-dimpled by LBM cause dramatic deterioration of the flexural strength, because of sharpened dimples. In the case of AJM, on the contrary, the strength is almost the same as the non-dimpled workpieces.

ACKNOWLEDGEMENTS

This work has been supported by METI, Japan, as part of the Synergy Ceramics Project. Part of the work has been supported by NEDO. The authors are members of the Joint Research Consortium of Synergy Ceramics.

REFERENCES

Divakar R. Sintered silicon carbides with controlled porosity for mechanical face seal applications. Journal of the Society of Tribologists and Lubrication Engineers 1994; 50(1):75–80.

Herbert D. Blast finishing. Metal Finishing 1997; 1:93–100.

Ihlemann J, Rubahn K. Excimer laser micro machining: fabrication and applications of dielectric masks. Applied Surface Science 2000; 154–155:587–92.

MacBeth J, Less S. Wear testing of silicon carbide mechanical seals in highly controlled diesel engine coolants. Proceedings of the 27th ISATA; 1994; 213–20.

Miyamoto I. Processing of ceramics by excimer lasers. Proceedings of the European Congress on Optics, SPIE 1990; 1279:66–76.

Tönshoff HK, Gedrat O. Removal process of ceramic materials with excimer laser radiation. Proceedings of the 2nd International Congress on Electro Optics and Science; 1989; 104–10.

MICRO MATERIAL PROCESSING USING UV LASER AND FEMTOSECOND LASER

H. K. Tönshoff, A. Ostendorf, K. Körber, C. Kulik, G. Kamlage

Laser Zentrum Hannover e.V., Hollerithallee 8, 30419 Hannover, Germany
(Tel.: +49-511/2788-0, Fax: +49-511/2788-100, E-mail: kb@lzh.de)

Abstract

System requirements for laser micro-machining strongly depends on the material to be processed. Since only one single laser type cannot serve to fulfill all requirements, each application asks for a well-considered system choice and a specific process optimization. This paper presents the possibilities and limitations concerning precision laser machining of different materials such as glass, ceramics, semiconductors and metals using 157 nm, 193 nm and 248 nm excimer as well as 780 nm Ti:sapphire femtosecond laser radiation.

Keywords

laser, excimer, VUV, femtosecond, micro-machining

1. INTRODUCTION

The manufacturing process of highly developed components in electronics, communication, information technology, medicine, and automotive industry require innovative machining technologies yielding greater accuracy, efficiency and flexibility. Since laser machining can fulfil most of these requirements there is a strong demand for the substitution of conventional techniques by innovative laser processes. Especially, excimer lasers have found numerous applications in micro-machining. For those lasers, emitting pulses in the nanosecond range, minimal thermal effects on ceramics and polymers are observed. Recent process developments using VUV laser beam sources also allow the precise machining of materials with extreme high band gaps, like flourinated polymers or fused silica. When processing materials with very high heat conductivities and low melting temperatures (e.g. metals) or other 'sensitive' materials such as organic tissues, typical pulsed UV lasers with ns pulses still can cause problems due to thermal effects on the bulk material. In this case shortening the pulse duration down to the femtosecond range is a very promising

This paper gives a survey of possibilities concerning precise laser machining of different materials (ceramics, glasses, semiconductors, metals) using excimer lasers (λ=157 nm, 193 nm, 248 nm) and Ti:sapphire femtosecond laser (λ=780 nm).

2. MATERIAL PROCESSING WITH UV AND VUV LASERS

For flexible structuring of a great variety of materials such as polymers, engineering ceramics, and glasses, excimer laser radiation has turned out to be a particularly appropriate tool [1-3]. The laser enables micro-drilling and micro-cutting in the μm range, e.g. the drilling of multichip-modules or ink jet printer nozzles and the manufacturing of micro-electro-mechanical systems, so called "MOEMS" [4].

Depending on the wavelength used, materials can be processed with a high degree of efficiency and surface quality: KrF laser radiation (λ=248 nm) is very powerful for the structuring of engineering ceramics (*figure 1a*) and a broad variety of polymers as well. The use of ArF lasers with a wavelength of λ=193 nm allows glass processing, with surface quality close to the requirements of optical components (*figure 1b*).

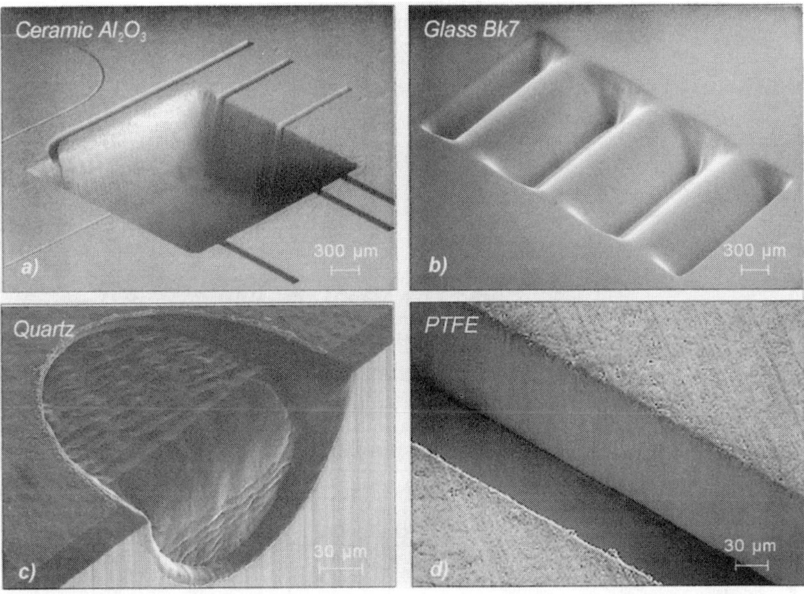

Figure 1a: 3-D structure (ceramic), b: micro-lens array (glass)
c: trimmed sensor (quartz), d: micro-cut (PTFE)

The use of the F_2 excimer laser (λ=157 nm), providing a photon energy of 7.9 eV, expands the group of materials that can be machined precisely by excimer laser radiation. Therefore, micro-structuring of quartz (*figure 1c*), applied for medicine, telecommunications, photonics and automotive industry, can be performed. Also polytetrafluorine (PTFE, Teflon®), known as being biocompatible, having outstanding tribological properties, and dielectric independency, can be processed with this innovative laser beam source (*figure 1d*).

3. FEMTOSECOND LASER MACHINING

Besides UV and VUV laser machining, ultrashort pulsed laser ablation is one of the most promising technologies among laser micro-machining. Significant advances in all-solid-state femtosecond lasers has led to a challenging phase of application research. This is because the ablation physics is quite different from that in conventional nanosecond and microsecond pulsed laser ablation. A low and well defined ablation threshold goes together with spatially and temporarily defined energy deposition into the material. Therefore, processing results are characterized by minimal thermal and mechanical damage, i.e. there is nearly no burr creation, melting, cracking, or any other type of changing materials properties [5-8].

In *figure 2*, some promising machining examples of different materials are shown. For instance, laser drilling of micro-holes (*figure 2a*) is a promising alternative to EDM processes for various injection or cooling components.

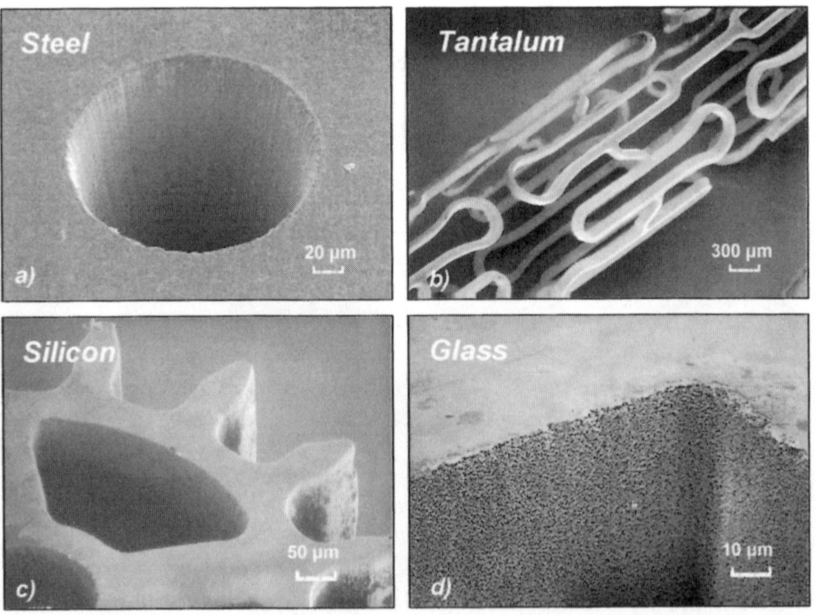

Figure 2a: micro-hole (steel), b: stent (tantalum)
c: micro-gear (silicon), d: micro-cut (glass)

Furthermore, precise cutting of tubes, so called "stents", with typical diameters of 1-2 mm (*figure 2b*), used as medical implants, can be produced by femtosecond lasers. The results show excellent quality that can hardly be achieved by any other technique. Different stent materials, metals as well as organic materials, can be processed without the necessity of post processing.

Femtosecond laser machining is also successful in cutting, drilling or structuring of semiconducting and dielectric materials. *Figures 2c* and *2d* show examples of cutting silicon and glass. These materials are known as quite difficult to machine by conventional techniques, due to the material's brittleness and thermal shock sensitivity.

4. COMPARISON OF THE LASER BEAM SOURCES

Depending on the target material and the laser process, different machining techniques are necessary to achieve economical fabrication. For instance, in excimer laser machining the use of flexible mask geometries is a very decisive factor. Process optimization by means of a flexible NC mask can lead to a reduction of the number of pulses. Less pulses means shorter processing times and cost reduction [4]. *Table 1* gives an overview about the typical classification of excimer and femtosecond laser processes.

Table 1: Characteristics of excimer and femtosecond laser processes

Classification	excimer-laser	femtosecond-laser
Beam profile	homogenous top head	gaussian
Typical repetition rate	200 – 400 Hz	1 –5 kHz
Process efficiency (ablation rate)	~ 0.7 µm/pulse (polymere) ~ 0.3 µm/pulse (ceramic, glass)	~ 1 µm/pulse
Typical size of ablation	2 x 2 to 500 x 500 µm²	∅ 5 to 200µm
Manufacturing techniques	drilling, cutting, structuring, annealing	drilling, cutting, structuring
Favorite materials	glass, polymere, ceramics	metals, semiconductors, organic materials
Running costs / pulse	~ 6 $\cdot 10^{-3}$ €	~ 0.9 $\cdot 10^{-3}$ €
Costs of laser source	~ 140,000 €	~ 200,000 €

5. CONCLUSION

Clearly, laser micro-machining offers a great variety of possible applications, as described in this paper. Any laser process requires a specific choice of laser parameters, i.e. essentially the fit of the wavelength, the pulse duration, the intensity and the pulse repetition rate. Therefore, a well-considered system choice is necessary.

However, the economical efficiency of laser micro-machining in general is limited by comparatively low ablation rates and high system costs. The excimer laser process is especially useful for brittle materials like ceramics and glasses. Compared to other machining techniques, femtosecond lasers have the highest potential for machining temperature sensitive materials like metals, semiconductors and organic materials.

6. REFERENCES

[1] M. Fiebig, J. Fair, M. Scaggs et al.: „Micromachining with 157 nm laser radiation", Conference Proceedings Euspen 99, Vol. 2, 31 May - 4 June 1999, Bremen, pp. 96-99

[2] H. K. Tönshoff, F. von Alvensleben, M. Rinke et al.: „Machining Concepts for Three-Dimensional Micro-Structuring with Pulsed UV-Lasers", Proceedings of MicroEngineering, 29 September –1 October, 1999, Stuttgart

[3] H. K. Tönshoff, F. von Alvensleben, C. Kulik: "Advanced 3D-CAD-Interface for Micromachining with Excimer Lasers", Proceedings of SPIE Vol. 3680, Design, Test and Microfabrication of MEMS and MOEMS, 1999, Paris, pp. 340-348

[4]   Gower, M.C.: „Laser applications in microelectronik and optoelektronik manufacturing IV. Excimer laser micromachining : a 15 year perspective". SPIE: Part of the SPIE Conference on Laser Applications in Microelectronik and Optoelectronik Manufacturing IV, Volume 3618, 25.-27.January 1999, San Jose, California. Washington: Society of Photo-Optical Instrumentation Engineers. – ISBN 0277-786X, pp. 251-260

[5] B.N. Chichkov, C. Momma, S. Nolte, F. von Alvensleben, A. Tünnermann: „Femtosecond, picosecond and nanosecond laser ablation of solids", Appl. Phys. A 63, 1996, pp. 109-115

[6] C. Momma, B.N. Chichkov, S. Nolte, F. von Alvensleben, A. Tünnermann, H. Welling, B. Wellegehausen: „Short-pulse laser ablation of solid targets", Opt. Commun.. 129, 1996, pp. 134-142

[7] H.K. Tönshoff, C. Momma, A. Ostendorf, S. Nolte, G. Kamlage: „Microdrilling of Metals with Ultrashort Laser Pulses", ICALEO 1998, Proceedings, Florida, p. A28

[8] F. Korte, S. Nolte, B. N. Chichkov, T. Bauer, G. Kamlage, T. Wagner, C. Fallnich, H. Welling: „Far-field and Near-field Material Processing with Femtosecond Laser Pulses", Appl. Phys. A 69,1999, pp. 7-11

SUBNANOMETER FABRICATION OF OPTICS BY PLASMA CHEMICAL VAPORIZATION MACHINING

Hideo Takino*, Teruki Kobayashi*, Takahiro Yamamoto*, Norio Shibata*,
Yoshio Gomei**, and Katsumi Sugisaki**

*Nikon Corporation
**Association of Super-Advanced Electronics Technologies (ASET)

Abstract

We discuss the fabrication of optics by plasma chemical vaporization machining (CVM) to obtain them with a shape accuracy below the nanometer level, that is, subnanometer accuracy. We used 88-mm-diameter spheric mirrors made of fused silica as workpieces, which were roughly polished to 9.24 nm RMS before plasma CVM. To achieve subnanometer accuracy, the plasma CVM conditions were adjusted. Under these conditions, we successfully fabricated the mirrors with the desired shape accuracy of 0.63 nm RMS. This demonstrated that plasma CVM is capable of fabricating optics with subnanometer shape accuracy.

Keywords

fabrication, optics, surface, plasma, atmosphere, chemical reaction, fused silica, extreme ultraviolet lithography

1. INTRODUCTION

In recent years, in the semiconductor industry, the establishment of technology for lithography using an extreme ultraviolet (EUV) wavelength of about 13 nm and related technologies has been strongly desired. Optics used in these technologies require a shape accuracy below the nanometer level, that is, subnanometer accuracy (Kinoshita *et al.*, 1999).

Plasma chemical vaporization machining (CVM) has been proposed by Mori for high-precision fabrication (Mori *et al.*, 1993, 2000a, 2000b). Plasma CVM is a chemical removal method, in which rf plasma generated in the proximity of an electrode at atmospheric pressure is used. Based on the plasma CVM principle, we developed a novel, computerized numerically controlled plasma CVM device for the fabrication of optics (Takino *et al.*, 1998a), and have shaped flat and aspherical surfaces with high accuracy using it (Takino *et al.*, 1998b, 1999). However, the shape accuracy of the fabricated optics was at the submicrometer level, which is insufficient for usage in EUV optics.

In the present study, we discuss the fabrication of optics with subnanometer shape accuracy by plasma CVM. The optics fabricated in this study were spherical mirrors made of fused silica. The desired shape accuracy of the mirrors was 0.95 nm RMS. To obtain such high accuracy efficiently, we used roughly polished surfaces as workpieces and corrected shape errors on these surfaces by plasma CVM.

2. FABRICATION DEVICE AND METHOD

Figure 1(a) shows the plasma CVM device used in this study (Takino *et al.*, 1998a, 2000). In this device, plasma is generated at the tip of the electrode to which rf power is supplied. The electrode is a conductive rod with a hemispherical tip. By placing the plasma in contact with the workpiece surface, the reactive radicals produced in the plasma react with the surface, resulting in the removal of the portion in contact with the plasma.

During the correction process, a workpiece is scanned relative to the electrode using a computerized numerical controller, in which the center of the curvature of the hemispherical tip is positioned on the normal of a targeted point on the workpiece surface, as shown in Fig. 1(b). In this arrangement, the gap distance between the workpiece surface and the electrode surface is minimum at the target point. Thus, plasma is concentrated at the target point to be removed, since the intensity of the electric field is high at a point where the gap distance is minimum.

To improve the shape accuracy of the workpieces, the shaping method described in reference (Takino *et al.*, 1999) was used. In short, in this method, the shape errors are approximated as a set of thin layers of thickness ε, as schematically shown in Fig. 2(a). By removing all the layers, the shape accuracy ε will be obtained, as shown in Fig. 2(b). The removal of each layer can be achieved by plasma scanning on the surface at a constant feed rate and a constant feed pitch, as shown in Fig. 2(c).

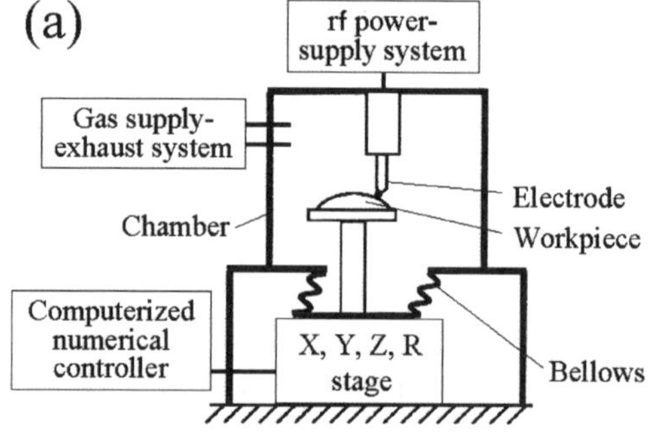

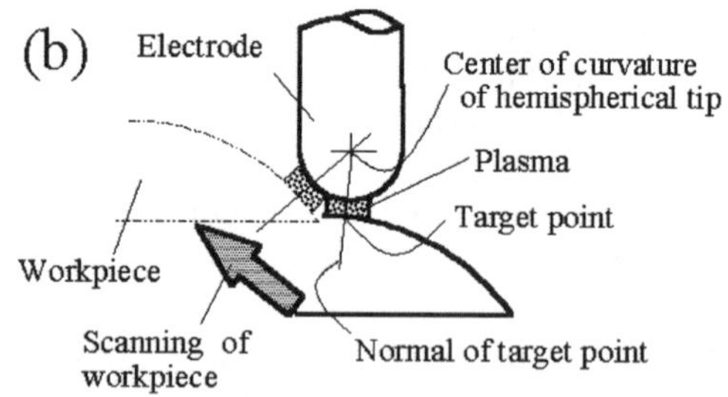

Figure 1. Schematic of plasma CVM device. (a) Outline of the device. (b) Detail of the electrode tip.

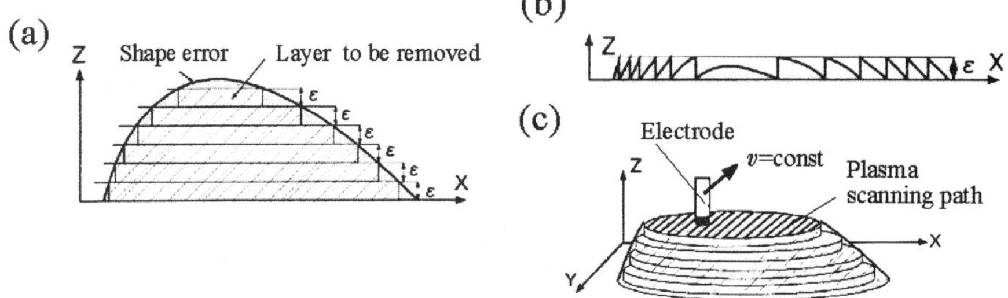

Figure 2. Schematic of shaping method. The cross sections of error shapes (a) before removal and (b) after removal. (c) Scanning path of the plasma for removing each layer.

3. REMOVAL CHARACTERISTICS

To realize a surface with high accuracy by the shaping method mentioned above, the thickness of the removed layer must be very thin. Thus, the process conditions for obtaining such thin layers were investigated experimentally. In the experiment, plasma was scanned on the workpiece to remove a rectangular area under various removal conditions, and then the removal depth (thickness of the layer) was measured with a commercial Fizeau-type laser interferometer. The removal was carried out using an electrode with a diameter of 4 mm. The process conditions are summarized in Table 1. As shown in Table 1, in order to reduce the removal rate so as to obtain a thin layer, the process conditions were set at lower rf power and lower concentration of SF_6 than

Table 1. Process conditions.

rf power	4.2 - 6.7 W
Intrachamber pressure	200 Torr
Flow rate ratio of SF$_6$ to He	He:SF$_6$ = 100:0.02
Gap distance	0.6 mm
Feed rate	13, 19 mm/s
Feed pitch	0.5 mm

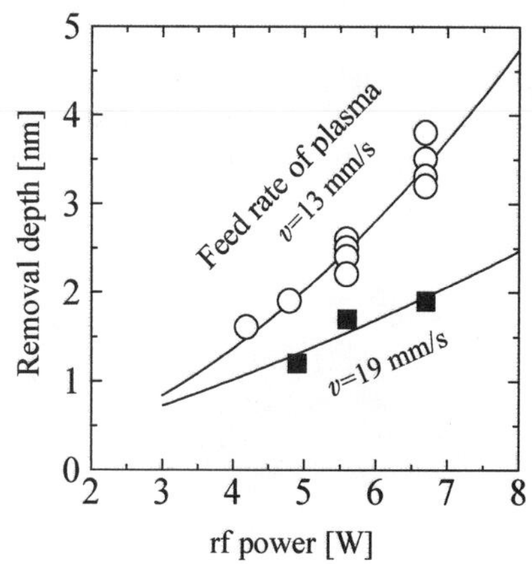

Figure 3. Relationship between the removal conditions and the removal depth.

in previous studies (Takino *et al.*, 1998a, 1998b, 1999, 2000). Moreover, to generate plasma stably even under such low rf power conditions, the intrachamber pressure was set lower than general plasma CVM conditions in which the pressure is 760 Torr (Mori *et al.*, 1993, 2000a, 2000b) and 600 Torr in our previous works.

Figure 3 shows the experimental results that reveal the relationship between the removal depth and the removal conditions. It is found that the removal depth can be controlled on the nanometer scale under these conditions.

4. FABRICATION OF OPTICS

Figure 4(a) shows the shape error of an 88-mm-diameter mirror obtained by roughly polishing before plasma CVM corrections, with a measured value of 9.24 nm RMS. The shape error was measured using the interferometer mentioned in section 3. This mirror was corrected to remove the layers with a thickness of 4 nm in the first step, and those of 2 nm in the second step followed by those of 1 nm in the third step. In each step, the correction process was performed three to five times.

Figure 4(b) shows the shape error after these corrections, demonstrating that the shape accuracy was improved to 0.63 nm RMS. The value of 0.63 nm RMS was obtained after an average of five measurements in which 2σ was 0.09 nm RMS. This indicates that the measuring was conducted with sufficient repeatability for this shape accuracy. In addition, with further processing to correct the error shown in Fig. 4(b), higher accuracy will be obtained, although we did not perform this because we attained the desired shape accuracy.

The surface roughness of the fabricated mirror was measured with a noncontact-type surface measuring instrument. The surface roughness of 0.22 nm RMS was obtained over the measuring length of about 250 μ m, demonstrating that the resultant surface is sufficiently smooth. Thus, the

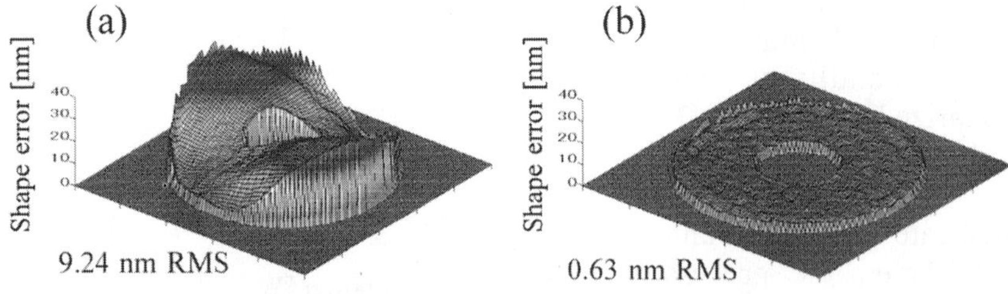

Figure 4. Fabrication of 88-mm-diameter mirror. Shape errors (a) before corrections and (b) after corrections.

shape accuracy and the roughness obtained here are sufficient for application in EUV optical systems.

5. CONCLUSIONS

We demonstrated that the removal depth can be controlled on the nanometer order by plasma CVM. Under the nanometer-order removal conditions achieved, we successfully fabricated optical surfaces with the desired shape accuracy at 0.63 nm RMS. Thus, we also demonstrated that plasma CVM is capable of fabricating optics with subnanometer accuracy. Moreover, the resultant surface has a smoothness of 0.22 nm RMS.

In addition, the mirrors shaped in this study will be used for the development of a point diffraction interferometer (PDI) (Sugisaki *et al.*, 2000), which is being carried out by the Association of Super-Advanced Electronics Technologies (ASET).

ACKNOWLEDGEMENTS

We gratefully acknowledge the support of this research by the New Energy and Industrial Technology Development Organization (NEDO).

REFERENCES

Kinoshita H., Watanabe T., Daniel J., Bajuk D., Platonov Y., Wood J. Development of 3-aspherical mirror optics for extreme ultraviolet lithography. Proc. the 9th ICPE; 1999 Aug. 29-Sept. 1; Osaka Japan: Jpn. Soc. Prec. Eng., 1999; 9-18.

Mori Y., Yamamura K., Yamauchi K., Yoshii K., Kataoka T., Endo K., Inagaki K., Kakiuchi H. Plasma cvm (chemical vaporization machining): an ultra precision machining technique using high-pressure reactive plasma. Nanotechnology 1993; 4: 225-229.

Mori Y., Yamamura K., Sano Y. The study of fabrication of the x-ray mirror by numerically controlled plasma chemical vaporization machining: development of the machine for the x-ray mirror fabrication. Rev. Sci. Instrum. 2000a; 71: 4620-4626.

Mori Y., Yamauchi K., Yamamura K., Sano Y. Development of plasma chemical vaporization machining. Rev. Sci. Instrum. 2000b; 71: 4627-4632.

Sugisaki K., Zhu Y., Gomei Y., Niibe M., Watanabe T., Kinoshita H. Present status of the ASET at-wavelength phase-shifting point diffraction interferometer. Proc. SPIE; 2000 Aug. 3-4; San Diego, USA: 4146: 47-53.

Takino H., Shibata N., Itoh H., Kobayashi T., Tanaka H., Ebi M., Yamamura K., Sano Y., Mori Y. Computer numerically controlled plasma chemical vaporization machining with a pipe electrode for optical fabrication. Appl. Opt. 1998a; 37: 5198-5210.

Takino H., Shibata N., Itoh H., Kobayashi T., Tanaka H., Ebi M., Taniguchi Y., Yamamura K., Sano Y., Mori Y. Fabrication of optical surfaces using plasma chemical vaporization machining (CVM) (2nd Rep.). Proc. Jpn. Soc. Prec. Eng. Spring Meeting; 1998 March 18-20; Kawasaki Japan: Jpn. Soc. Prec. Eng., 1998b; 513 (in Japanese).

Takino H., Shibata N., Itoh H., Kobayashi T., Tanaka H., Ebi M., Yamamura K., Sano Y., Mori Y. Fabrication of optical flat using plasma chemical vaporization machining with a pipe electrode. J. Jpn. Soc. Prec. Eng. 1999; 65: 1650-1651 (in Japanese).

Takino H., Ara K., Shibata N., Yamamura K., Sano Y., Mori Y. Optical fabrication using plasma chemical vaporization machining with a spherical-end rod electrode. Proc. Jpn. Soc. Prec. Eng. Spring Meeting; 2000 March 22-24; Tokyo Japan: Jpn. Soc. Prec. Eng., 2000; 191 (in Japanese).

MICRO FABRICATION USING EDM DEPOSITION

Shinya Hayakawa, Ricardo Itiro Ori, Fumihiro Itoigawa,
Takashi Nakamura and Tomio Matsubara
Nagoya Institute of Technology, Japan

Abstract

This paper describes a metal deposition process using micro electrical discharge machining (micro EDM) to fabricate microstructures. Steel is used for the tool electrode, and the EDM process is carried out in air. By feeding the tool electrode in the vertical direction, a micro rod with 0.14mm in diameter and 2.2mm in height is formed on the workpiece surface. A fillet-like line with uniform thickness is also drawn over the workpiece surface when the horizontal feed is applied independently of the servomechanism of the EDM machine and the horizontal feed rate is kept constant. It is considered from the observation of the discharging surface that the local gap distance at a discharge location becomes larger than that at other locations on the discharging surface and, therefore, the discharge points are not concentrated at a single point but distributed over the discharging surface.

Keywords

Electrical Discharge Machining, Deposition, Microstructure

1. INTRODUCTION

In recent years, micromachining and surface modification using electrical discharge machining (EDM) have shown significant progress. Since a micro tool electrode with 0.1mm or less in diameter can be easily obtained by the WEDG method (Masuzawa et al., 1985), micromachining is now used practically. Surface modification has been studied by Mohri et al. (1993) and Goto et al. (1997), and is practically employed in making an anti-wear coating with a hard material such as tungsten carbide, titanium carbide or some other metal carbides on the workpiece surface. In this process, a composite electrode (Mohri et al., 1993) or powder mixed working fluid (Satsuta et al., 1998) is usually used. Mohri et al. (1998) also attempted the deposition of tool electrode material using micro EDM, and succeeded in depositing tungsten on the workpiece. The process is carried out in EDM working oil, and a large discharge current is supplied. However, they have not succeeded in the deposition of metals commonly used for machine parts such as steel or aluminum alloy.

Table 1　Machining conditions of deposition process

Tool electrode	Mild steel (ϕ0.1mm)
Workpiece	Mild steel
Discharge current	2.5A
Discharge duration	$5\,\mu s$
Pulse interval	$450\,\mu s$
Polarity	Reversed
Working medium	Air

The authors also developed a metal deposition method using micro EDM aiming to fabricate a three-dimensional microstructure (Hayakawa et al., 2000). In this process, the tool electrode material, such as steel or aluminum alloy, is deposited on the workpiece when the tool electrode is used as the anode and the EDM process is carried out in gas. We also showed that this deposition process can easily be switched to a removal process simply by reversing the discharge polarity. In this study, the fabrication of a two-dimensional microstructure is attempted and the machining stability of this process is examined.

2.　EXPERIMENTAL METHOD

A scanning EDM machine is used for the experiment and a tool electrode with 0.1mm in diameter is prepared by the WEDG process. The experimental conditions are listed in Table 1. Mild steel is used for both the tool electrode and the workpiece, and the EDM process is carried out in air (Hayakawa et al., 2000). The discharge conditions, such as discharge current, discharge duration and polarity, are predicted from the transient temperature analysis of the tool electrode and the workpiece for a single pulse discharge (Hayakawa et al., 2000). In this analysis, we assume that the deposition process is realized when the temperature of the tool electrode exceeds the boiling point of the material and that of the workpiece is between the melting point and boiling point of the material.

3.　FABRICATION OF MICROSTRUCTURE

Initially, the EDM process is carried out by feeding the tool electrode only in the vertical direction. Figure 1 shows an overall view of the obtained object. The diameter of the rod is approximately 0.14mm, which is slightly thicker than that of the tool electrode, and it is approximately

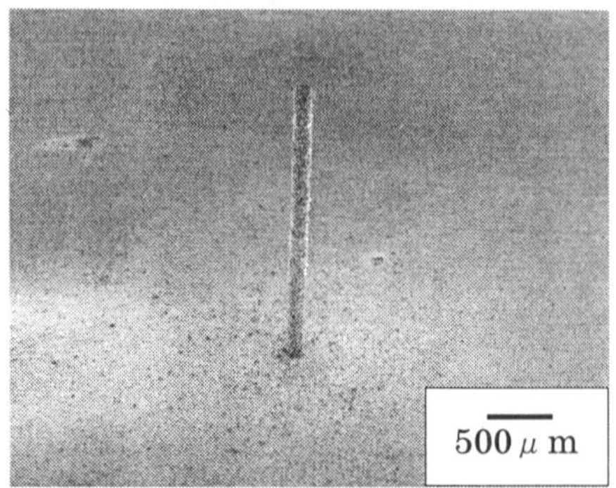

Figure 1 Overall view of the deposited rod

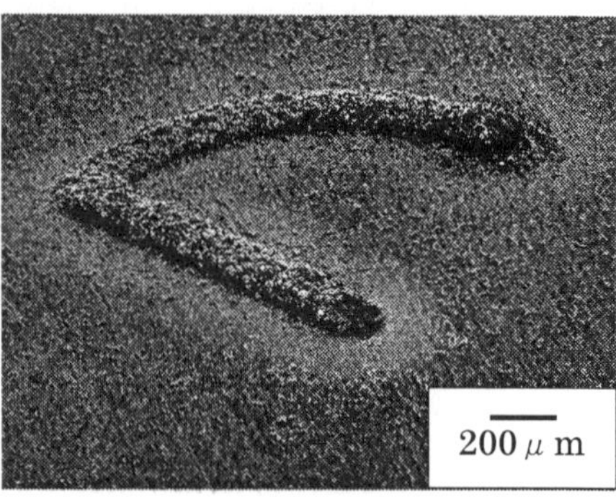

Figure 2 Fillet-like line including straight line, curved line and angle part

2.2mm in height. The time required to obtain this rod is approximately 6.3 hours.

Next, the horizontal feed is applied to the tool electrode in order to draw a fillet-like line over the workpiece surface. Figure 2 shows the deposited result including the straight line, curved line and angle part, which is obtained using a single tool path. It is found that machining stability during this process is achieved when the horizontal feed is applied independently of the servomechanism of the EDM machine, which maintains the appropriate gap distance in the vertical direction and applies the vertical feed due to the wear of the tool electrode. It is also found that a constant horizontal feed rate is needed in order to draw a line with uniform thickness.

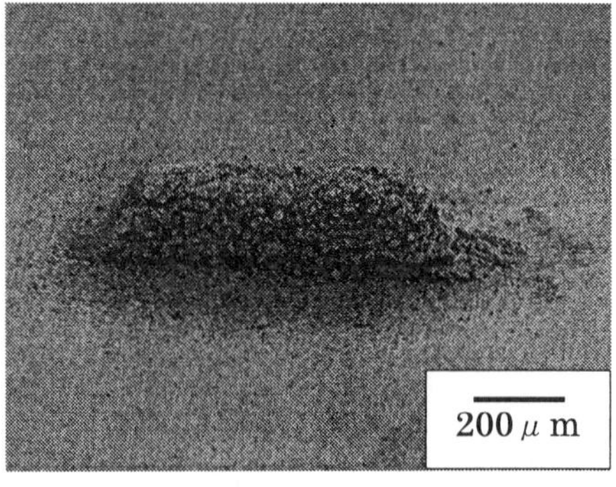

Figure 3 Deposited wall formed by laminating 8 layers

Then, a line with a large height, i.e., a wall, is achieved by repeating the process using the same tool path several times. Figure 3 shows the obtained wall which is formed by laminating eight layers. It can be observed that the height of the deposited wall is uniform. Since the horizontal feed rate was kept constant throughout the process, the thickness of each layer is uniform and the height of the deposited wall is proportional to the number of times the process was repeated.

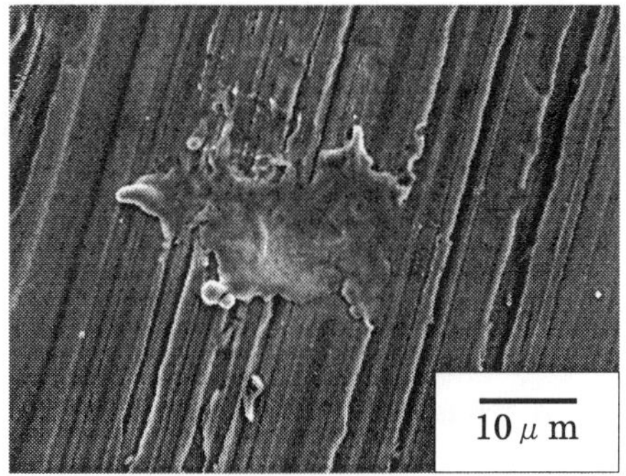

Figure 4 Material deposited on the workpiece surface by a single pulse discharge

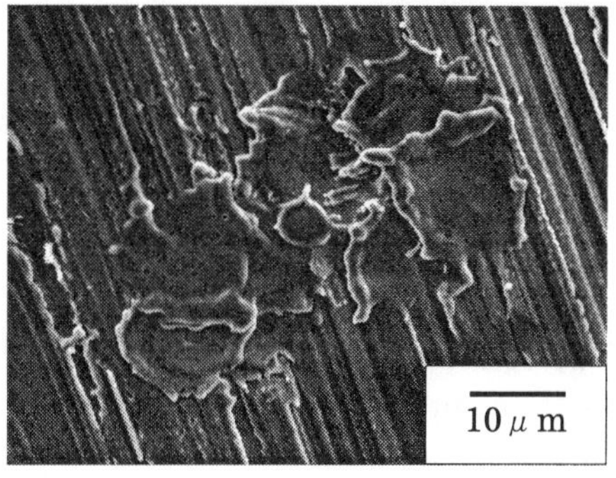

Figure 5 Discharging area after five discharges

4. OBSERVATION OF DEPOSITION PROCESS

In the processes mentioned in the previous section, the discharge points are distributed uniformly over the discharging surface; in other words, the deposition does not concentrate at the small protuberances that are produced by the previous discharges. In order to investigate the mechanism of machining stability mentioned above, the discharging surface is observed when the tool electrode is fed only in the vertical direction.

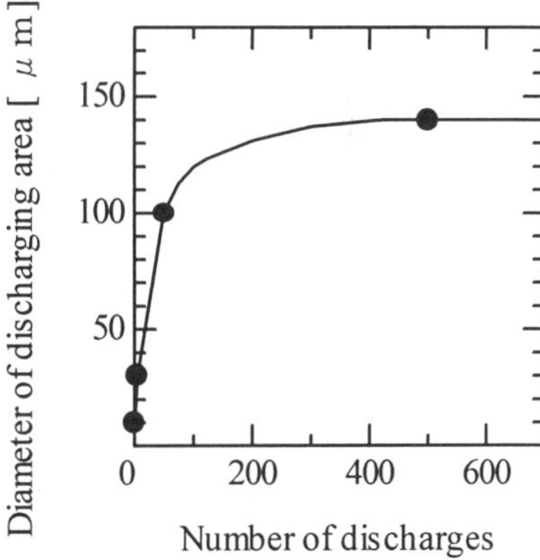

Figure 6 Relationship between number of discharges and diameter of discharging area

Figure 4 shows a material deposited on the workpiece surface by a single pulse discharge. As previously mentioned, the discharge conditions are predicted under the assumption that the temperature of the tool electrode exceeds the boiling point of the material. However, it can be seen that, when the material was deposited on the workpiece surface, it was not vaporized but melted.

Figure 5 shows the discharging area of the workpiece formed by five discharges, and the relationship between the number of discharges and the diameter of the discharging area is shown in Figure 6. It is found that some discharges following the first discharge occur near the first discharge, and the discharging area spreads with increasing number of discharges and saturates by the 500th discharge.

The measured ratio of the deposited volume to the electrode wear volume is approximately 90% or less, and the diameter of the deposited rod is approximately 1.4 times that of the tool electrode. These findings indicate that the thickness of the material deposited by each pulse discharge is less than the depth of the corresponding crater in the tool electrode. Therefore, it is considered that the local gap distance at the location becomes larger compared with other locations on the discharging surface, and the next discharge occurs at another location.

5. CONCLUSIONS

A metal deposition process using micro EDM to fabricate microstructures is developed and the machining stability of this process is discussed based on observation of the discharging surface. The following conclusions are experimentally obtained.

(1) A micro rod with 0.14mm in diameter and 2.2mm in height is fabricated by feeding the tool electrode in the vertical direction.

(2) A fillet-like line including a straight line, curved line and angle part is drawn with uniform thickness when the horizontal feed is applied independently of the servomechanism of the EDM machine and the horizontal feed rate is kept constant.

(3) The local gap distance at a discharge location is considered to become larger than that at other locations on the discharging surface, showing that the discharge points are distributed over the discharging surface.

ACKNOWLEDGEMENT

This study is funded by the Ministry of Education, Science and Culture of Japan (Grant-in-Aid for Scientific Research, Project No.12750096).

REFERENCES

Goto A., Magara T., Imai Y., Miyake H., Saito N. and Mohri N. Formation of hard layer on metallic material by EDM. J. JSEME 1997; 31(68): 26-31. (in Japanese)

Hayakawa S., Ori R.I., Itoigawa F., Nakamura T. and Matsubara T. Three-dimensional fabrication using micro EDM deposition. J. JSPE 2000; 66(12): 1943-1947. (in Japanese)

Masuzawa T., Fujino M. and Kobayashi K. Wire electro-discharge grinding for micro machining. Ann. CIRP 1985; 34(1): 431-434.

Mohri N., Saito N., Tsunekawa Y., Momiyama H. and Miyakawa A. Surface modification by electrical discharge machining. J. JSPE 1993; 59(4): 625-630. (in Japanese)

Mohri N. and Saito N. Surface modification by electrical discharge machining. J. JSPE 1998; 64(12): 1715-1718. (in Japanese)

Satsuta T., Hiraki K. and Ejiri K. Surface modification using EDM. Proc. Autumn Meeting of JSEME 1998; 51-52. (in Japanese)

Manufacture of Aspherical Fresnel Lens with Ideal Cross-sectional Profile with Oxygen-free Copper and Acrylic Resin

Nuttaphong SORNSUWIT[a], Yoshimi TAKEUCHI[a], Tomohiko KAWAI[b], Kiyoshi SAWADA[b] and Toshio SATA[c]

[a]Department of Mechanical Engineering and Intelligent Systems, The University of Electro-Communications, 1-5-1 Chofugaoka, Chofu, Tokyo 182-8585, JAPAN
[b]Basic Research Lab., FANUC Ltd., 3580 Oshinogusa, Oshino, Yamanashi 401-0597, JAPAN
[c]Toyota Technological Institute, 2-12-1 Hisakata, Tempaku-ku, Nagoya 468-8511, JAPAN

Abstract

Fresnel lens is typically manufactured by means of photolithography and mechanical cutting by lathe. The study deals with a new machining method of lens grooves by use of a side-edge of diamond tool to enhance the cutting efficiency and to acquire the ideal cross-sectional profile of lens groove which consists of a perpendicular plane, a sharp bottom and a curved slope. The machining experiment was conducted to cut lens grooves with a pitch of 300 μm on an oxygen-free copper and an acrylic resin plate. The result showed an ideal lens groove with the surface roughness of 8 nm(P-V). The accurate focus of 100 mm was also confirmed by the optical assessment with a manufactured acrylic fresnel lens.

Keywords

Ultraprecision machining, Aspherical fresnel lens,
Microgroove, Diamond tool

1.INTRODUCTION

Among optical lens in the market, spherical lens occupies the biggest share thanks to its easy manufacturing process. However, multiple lenses are required to compensate a spherical aberration, which prevents the optical components from reducing size and lightening weight or which even causes colour aberration and lower optical efficiency due to the deterioration in light transmissivity. Therefore, the study deals with the manufacture of aspherical Fresnel lens in order to cope with the problems.

Fresnel lens is typically manufactured by means of photolithography and mechanical cutting by lathe. Photolithography has problems of the limitation of material and the formation of a step-like cross-section of lens groove, while the method by conventional lathe has low productivity because the cutting is done by a sharp single point cutting edge. The manufacture by use of a rotating tool showed an excellent shape and a good surface roughness[1]. However, it takes long cutting time. The study adopted the method of lathe using a side cutting edge to create Fresnel lens with an ideal profile, an aspherical cross-sectional of lens grooves, as illustrated in *Figure 1*.

2. SHAPE OF IDEAL FRESNEL LENS

Fresnel lens consists of a lot of small rings arranged in a concentric circle and serves as a corresponding concave or convex lens, as shown in *Figure 1*. Each circular groove of Fresnel lens is a set of small pieces of the lens surface translated toward the plano-side of lens so as to keep the height constant, and also has the same lens center point. The adoption of Fermat's principle to modify the cross-section of lens groove enables the correction of lens focus which was out of alignment due to the translation to Fresnel lens. The specification of Fresnel lens designed in the study has the center lens groove of 1mm in radius and the others have a pitch of 300 μm. Index of refraction 1.49(Acrylic) is used to determined the profile of lens of 100 mm in focal length.

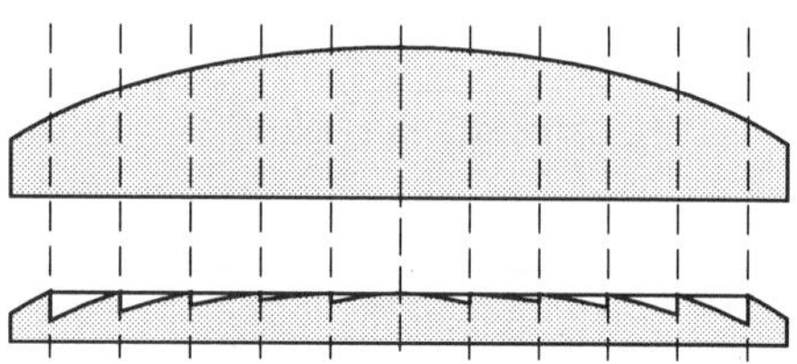

Fig. 1 Cross-section of Fresnel lens

According to Fermat's principle and *Figure 2*, all of light paths of ABC have to be equal in order to make light from point A condense at point C. In *Equation(1)*, which means that the light path along ABC is equal to that along the lens axis length OC, the profile of aspherical Fresnel lens can be obtained as *Equation(2)*. By the differentiation of *Equation(2)*, the position and its slope of 30 tiny divided parts in each lens groove is calculated to specify the shape of aspherical Fresnel lens for machining.

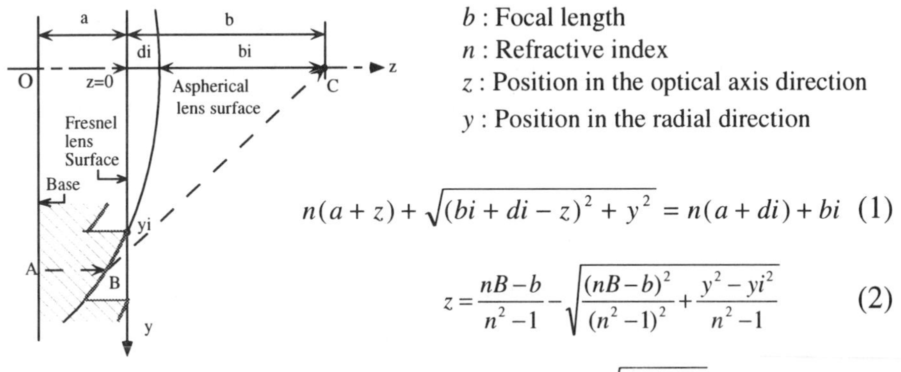

b : Focal length

n : Refractive index

z : Position in the optical axis direction

y : Position in the radial direction

$$n(a + z) + \sqrt{(bi + di - z)^2 + y^2} = n(a + di) + bi \quad (1)$$

$$z = \frac{nB - b}{n^2 - 1} - \sqrt{\frac{(nB - b)^2}{(n^2 - 1)^2} + \frac{y^2 - yi^2}{n^2 - 1}} \quad (2)$$

Fig. 2 Design of ideal shape

,where $B = \sqrt{b^2 + yi^2}$

3. CUTTING EXPERIMENT

An ultraprecision machining center with a resolution of 1 nm in linear axis and 0.00001 degree in rotational axis is used in the experiment in combination with an air turbine spindle having a single diamond tool with triangular cutting edge of 45 degree angle. The ideal lens groove shape should have a perpendicular plane, a sharp bottom and a curved slope with mirror surface. To achieve the ideal profile of lens, the diamond tool is fed to the bottom of groove and move along the designed aspherical lens groove. In this process, a sharp bottom and an aspherical profile of lens are machined using a simultaneous control of X,Z and B axis.

As is well known in turning, the low cutting speed near the center typically causes worse surface roughness. However, it is confirmed that the ultraprecision machining center used in the experiment could create a mirror surface at a slow cutting speed. The chip is expected to be a continuous one

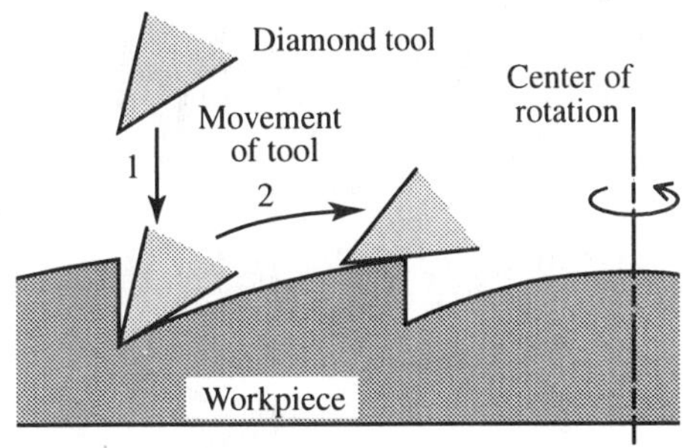

Fig. 3 Cutting method of lens surface

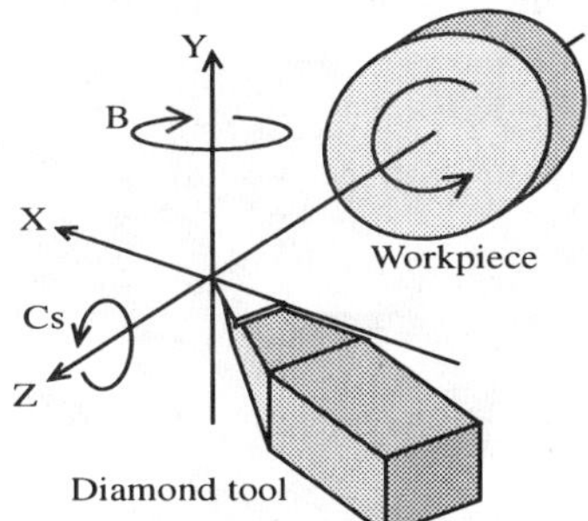

Fig. 4 Setup of cutting tool

and may convolve the tool and bring about undesirable defects, especially at the outer part of lens where the depth of groove increases and generates bigger chip. Therefore, rough cutting is interposed before finishing which machines side wall and curved lens groove in sequence. *Figure 3* shows the order of cutting to realize such a groove shape.

The experiment was conducted to cut grooves with a pitch of 300 μm on oxygen-free copper as the mold of Fresnel lens to evaluate the surface quality. The cutting condition is shown in *Table 1*.

The machined Fresnel lens, as shown in *figure 5-8*, on oxygen-free copper shows an excellent surface roughness of 8 nm(P-V) in both radial direction and circumferential direction. The cross-sectional shape, as shown in *Figure 6*, satisfied the required ideal lens groove one.

Table 1 Cutting condition

Spindle rotation		500 min⁻¹
Feed rate	Cutting direction	1.0 μm/rev.
	Feed direction	10 μm/rev.
Tool	Point angle	45°
	Material	Single crystal diamond
Workpiece	Size	φ 60 mm
	Material	Oxgen-free copper

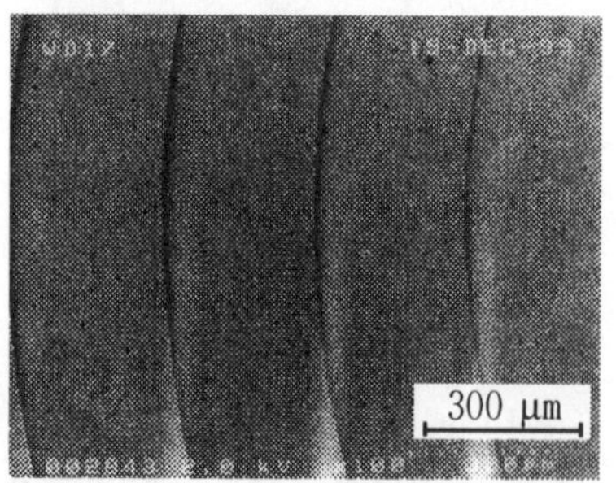

Fig. 5 Machined Fresnel lens

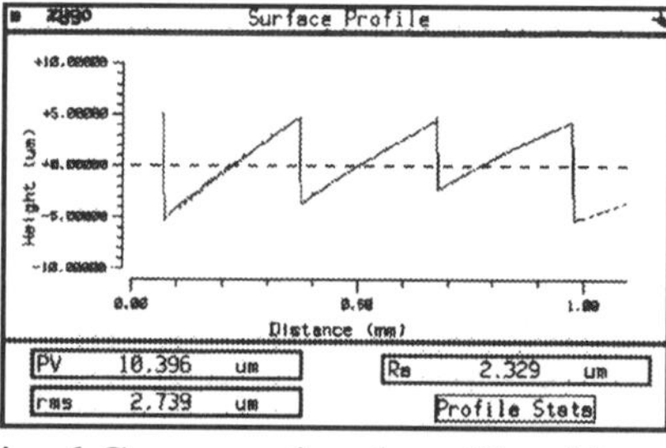

Fig. 6 Cross-sectional profile of lens

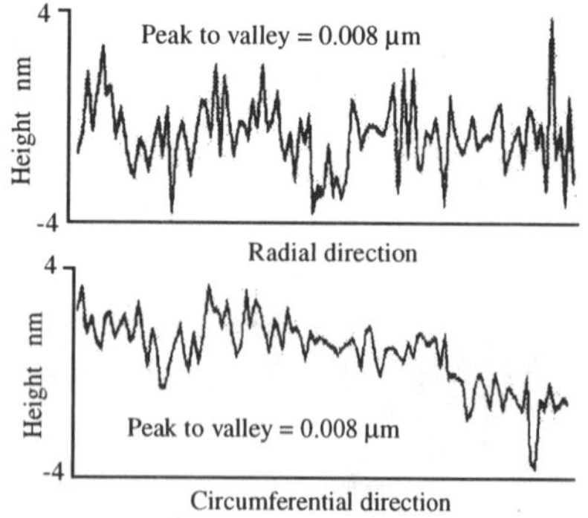

Fig. 7 SEM image of Fresnel lens   Fig.8 Surface roughness of lens groove

To make sure that a designed aspherical Fresnel lens has the proper function as an optical element, the cutting was also made on an acrylic resin plate with water poured as a cutting coolant, based on the results of manufactured oxygen-free copper Fresnel lens.

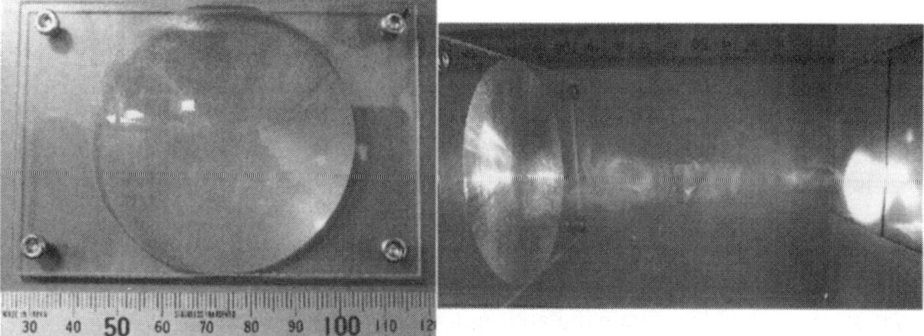

Fig.9 Fresnel lens on acrylic resin plate   Fig.10 Focal length inspection

The machined Fresnel lens on the acrylic resin plate, as shown in *Figure 9*, also reveals that a good lens shape was created without any burr or chatter. The focus of acrylic Fresnel lens is inspected and resulted a 100mm of focal length as designed, as shown in *Figure 10*. The optical property of the acrylic Fresnel lens confirmed the accurate groove shape of Fresnel lens manufactured by the method proposed in the study. The cutting time was also shorten due to the side edge cutting of diamond tool.

4. CONCLUSION

By employing the side edge cutting method, Fresnel lens consisting of a perpendicular side, a sharp bottom and a curved slope of lens groove were created with a surface roughness of 8 nm(P-V). The manufacture of Fresnel lens on an acrylic resin plate also realized an ideal shape and showed an accurate focus as designed.

5. REFERENCE

1)N. Sornsuwit, Y. Takeuchi, T. Kawai, K. Sawada and T. Sata, Manufacture of Fresnel Lens by Ultraprecision Machining Center Based on Friction-free Servo Mechanism, Proc. of 9th ICPE(Osaka), 1999, p159-164

2)N.Sornsuwit, Y. Takeuchi, K. Sawada and T. Sata, Metal Mold Manufacturing of Fresnel Lens by Use of Micro Grooving Technology, JSME International Journal, Series C, Vol. 43, No. 1, 2000

Non-circular machining using two convex milling cutters

Yohichi Nakao

Kanagawa University

Abstract

A non-circular machining method for producing parts with a large change of radius is presented. A machining operation based on a balance cut-type machining by two milling cutters is proposed for the non-circular profile parts, then a machine tool is developed. Two linear motors, moving on a guide rail, are used for the machine tool to actuate the cutting tools for cutting directions. An adaptive feedforward controller is used to maintain a good tracking performance. Machining results by the machine tool are presented. From machining results of circular cylinder parts, an advantage in machining time reduction by the balance cut-type machining is demonstrated. Effectiveness of the machining for non-circular parts is demonstrated by a machining of engine cam profile.

Keywords

Non-circular machining, Balance cut-type machining, Linear motor

1. INTRODUCTION

A non-circular machining is used for manufacturing of engine pistons, engine cams, mono pump shafts and other parts with non-circular cross sections. Polygon lathes are usually used to machine the products. In the non-circular machining by the polygon lathes, cutting angles, such as a rake angle and a relief angle, change periodically in every rotation of a workpiece spindle. Especially, a considerable change of the angles takes place in the machining of parts with a large change of radius of the cross sectional profile. Then the change results in deterioration in machinability. Consequently, the workpiece geometries machined by polygon lathes are strictly restricted to keep required cutting angles.

A different type of machine tool is therefore necessary to machine the products with a large change of radius of its cross section. In this paper, a machining method for non-circular cross section products will be first proposed. Applying a balance cut-type machining by two milling cutters for the non-circular machining is a feature of the machining method. In this method, there is no large change of the cutting angles even in the non-circular machining for the parts with a large change of radius. A linear motor drive system designed for the balance cut-type machining will be also presented. Structure of a designed machine tool and machining results will be described in the following sections.

2. MACHINING METHOD AND MACHINE TOOL
2.1 Machining Method

A machining method using two milling cutters is based on the balance cut-type machining and is illustrated schematically in Fig. 1. A workpiece fixed with a rotating spindle is placed between the two milling cutters. Then, a simultaneous machining by the two milling cutters from both sides of the workpiece can be attained. Each position for cutting motion, namely the motion along X-axis, of the milling cutters is controlled independently and synchronized with a rotational angle of the workpiece spindle. Non-circular profiles can then be machined by the control of C, X_1 and X_2 axis.

2.2 Designed Machine Tool

As illustrated in Fig. 1, the milling cutters and their drive systems are placed on a carriage moving for feed direction, namely Z-direction. AC servomotors actuate the workpiece spindle and the carriage.

In this study, two convex milling cutters are used for the machine tool. However, other types of milling cutters, such as ball end mill cutters, square end mill cutters and others, can be used for the proposed machining method. In order to avoid the interference between the cutters and the workpiece, the rotational axis of each cutting tools is tiled by 45 degree from the rotational axis of the workpiece.

Two milling cutters are fed for the cutting direction by linear motors moving on a straight guide rail, which is mounted on the carriage. Each linear motor can

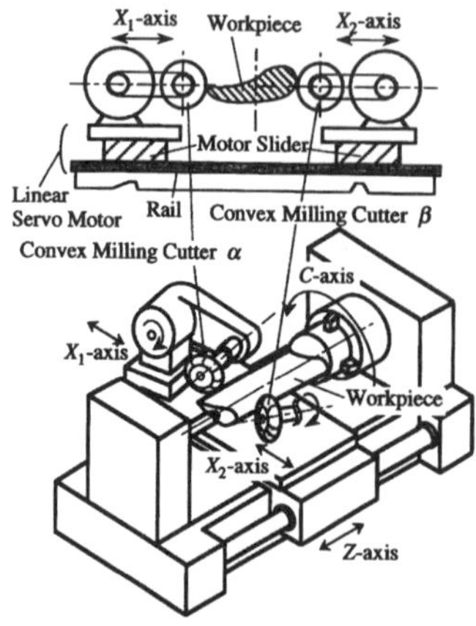

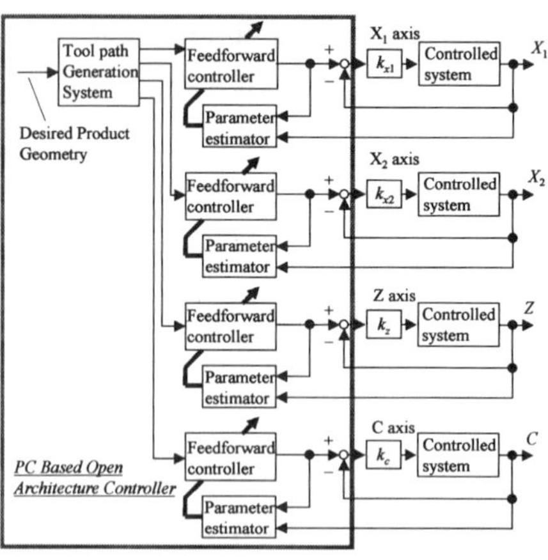

Fig. 1 Schematic of developed machine tool

Fig. 2 Block diagram of control system

be controlled independently, which is an important requirement for the non-circular machining.

2.3 Control System

A PC based open architecture controller controls the machine tool. A role of the machine tool controller for the non-circular machining is particularly important to keep a tracking error of the servo system for a cutting motion within a required level since the tracking error tends to increase with the increase of the workpiece spindle speed. As shown in Fig. 2, an adaptive feedforward controller[1] is integrated into the open architecture machine tool controller for each control axis to fulfill the requirement. The feedforward controllers can then maintain small tracking error for each servo drive system. Since on-line parameter estimators are also integrated into the feedforward controllers, any pre-adjustments for the controllers are not required. Moreover, it can adjust optimum parameters for the feedforward controllers to reduce tracking error, whenever cutting or operational conditions, such as feed, depth of cut etc., change during machining. In the PC based open architecture controller, a tool path generation system for convex milling cutter is also installed.

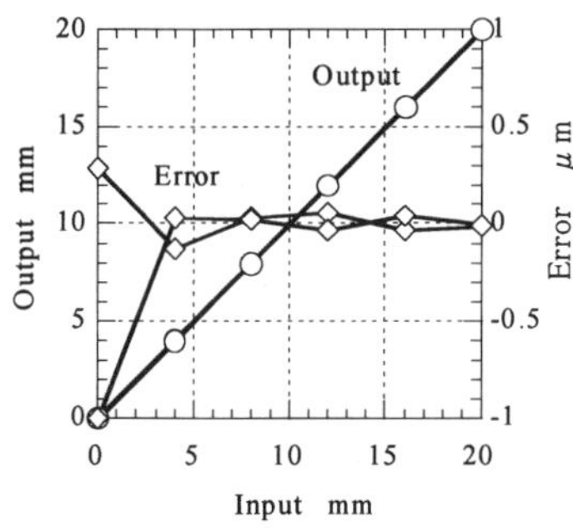

Fig. 3 Linear motor static characteristic

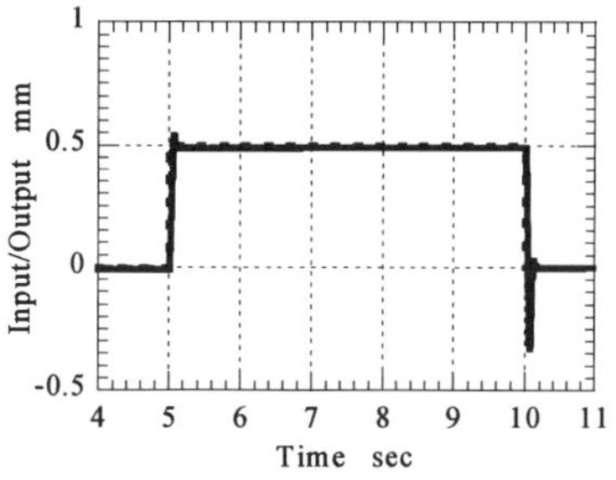

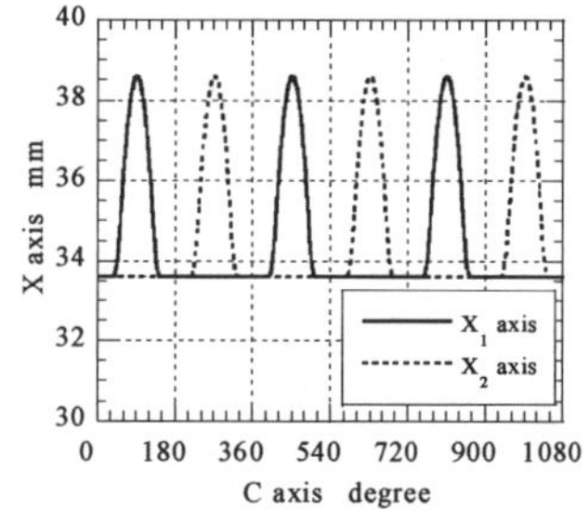

Fig. 4 Linear motor step response

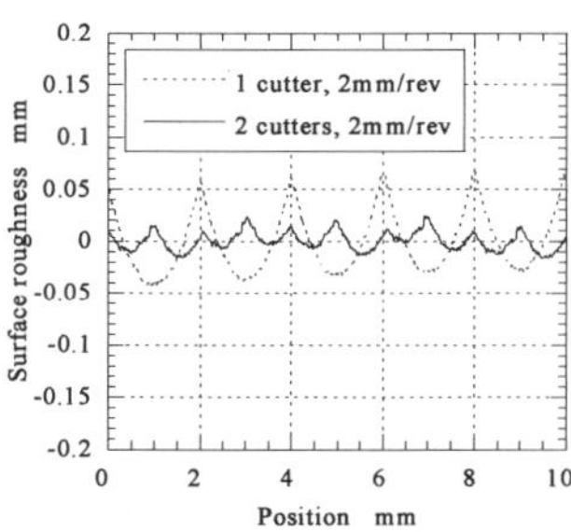

Fig. 5 Surface roughness

Fig. 6 Generated tool path

3. EXPERIMENTS

A static characteristic and a step response of the linear motor of X_1 axis are given in Fig. 3 and Fig. 4, respectively. A characteristic of linear motor of X_2 axis was similar to that of X_1 axis. Since the positioning error of the servo system of X_1 axis is within 1 μm, the linear motors have an adequate characteristic for machining applications. Figure 5 shows effects of the number of cutter used for machining on surface roughness. In these experiments, the machined parts were circular cylinders 50 mm in diameter. The feed rate of the carriage was set at 2 mm/rev, and the convex milling cutters of 5 mm in radius was used. A broken line in Fig. 5 represents a surface roughness by a single cutter machining. The magnitude of this surface roughness is larger than that of balance cut-type machining represented by a solid line because an actual feed in the balance cut-type machining is a half of that in the single cutter machining. A theoretical surface roughness for each case is 100 μm and 25 μm, respectively. The results are in a good agreement with obtained experimental results. These results show

Fig. 7 Machined part Fig. 8 Machined part

the proposed machining is effective for the reduction in machining time by setting double length feed rate.

Non-circular machining tests using the developed machine tool were conducted. An engine cam profile was taken as a machining example. A generated tool path for the engine cam is presented in Fig. 6. Machining conditions were 1 mm/rev in feed rate, 6 rpm in spindle speed and the workpiece material of polyvinyl chlroride resin. The machined part and another machined example are given in Fig.7 and Fig.8. The developed machine tool machined successfully these parts with large radius distribution without any troubles, which verify the effectiveness of proposed machining method for the non-circular parts.

4. CONCLUSIONS

A balance cut-type machining method by two milling cutters for non-circular parts, especially for parts with a large change of radius, has been presented. Based on the machining, a machine tool was developed and its structure has been described. Machining results of cylinder parts using the machine tool have verified that the machining method can reduce machining time by setting longer feed rate, keeping surface roughness quality. Machining results of non-circular parts have verified that the machining is effective in the non-circular parts with the large change of radius.

REFERENCE

[1] Nakao, Y, et al., Development of CNC Machine Tool with a Cylindrical Tool (2nd. Report), Trans. Jpn. Soc. Mech. Eng. (C) 62-604 (1996), 4702 (in Japanese).

A METHOD TO MACHINE THREE-DIMENSIONAL THIN PARTS

Haruki OBARA*, Takahiro WATANABE*, Tsuyoshi OHSUMI*,
Eiji NINOMIYA* and Masatoshi HATANO*
*Toyama University, Japan

Abstract

It is difficult to machine a comparatively wide but very thin three-dimensional part, because it is easily deformed by the clamping force or machining force used during machining. The developed method uses a low melting alloy (LMA) to machine three-dimensional thin parts without deformation. The parts are firmly supported and fixed on the LMA by adhesion or suction force. The method and some results are described.

Keywords

Low melting alloy, three-dimensional thin parts, adhesion, suction force

1. INTRODUCTION

It is not difficult to machine an even thin plate if it is placed on a flat table. However, sometimes a thin blank is already deformed during the production process, and is easily deformed by the clamping force or machining force, as shown in Figures 1 and 2. Consequently it is difficult to machine three-dimensional thin parts. A jig is required for the machining, but it is expensive, and thus not suitable for the production of a few parts. In this paper, a simple method using low melting alloy (LMA), whose melting point is below 100°C, to support the thin parts is proposed. This method is suitable for machining a few three-dimensional thin parts, or prototype parts for preliminary assessment before mass production.

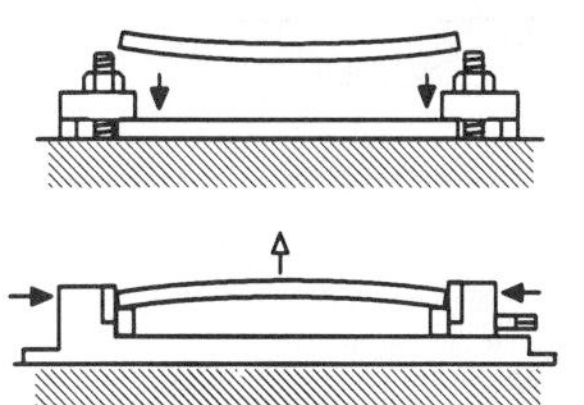

Fig.1 Deformation by clamping force

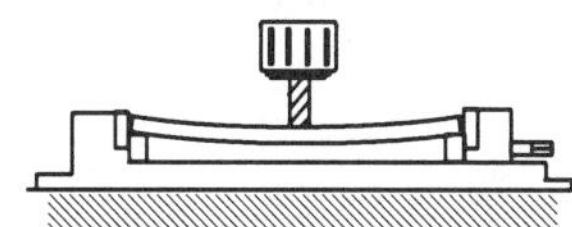

Fig.2 Deformation by machining force

2. LOW MELTING ALLOY

The low melting alloy we used is U-Alloy 70 [1]. Its melting point is 70℃ and it melts in boiling water. Additionally, this alloy has a remarkable characteristic in that it expands when solidifying. This characteristic enables the implantation and clamping of a blank in the LMA. After being used, the LMA is melted again and is reusable.

3. CASTING AND CLAMPING METHODS

The LMA is cast beneath the blank and supports it. This method is already used to machine small parts after implanting them in the LMA. When thin parts are required, the blank should be fixed firmly on the LMA; if not, the blank floats on the LMA during the machining and is damaged. We tested two methods to fix the blank on the LMA. The first method uses adhesive, and the other uses suction force. The first method is useful even if the LMA surface facing the blank is rough, because the gap between them is filled with the adhesive. The LMA and the adhesive are removed using boiled water after the machining. The method using suction force is available to make multiple parts because the melting of the LMA after each machining is not required. We tried the following three casting methods.

3.1 Upper Implanting Method

The blank is placed on the surface of the melted LMA in a holder. After coagulation of the LMA, the blank is picked up and replaced in the same position after applying the adhesive. The LMA is melted after the machining and the part is taken out. Then the adhesive is removed with boiled water or

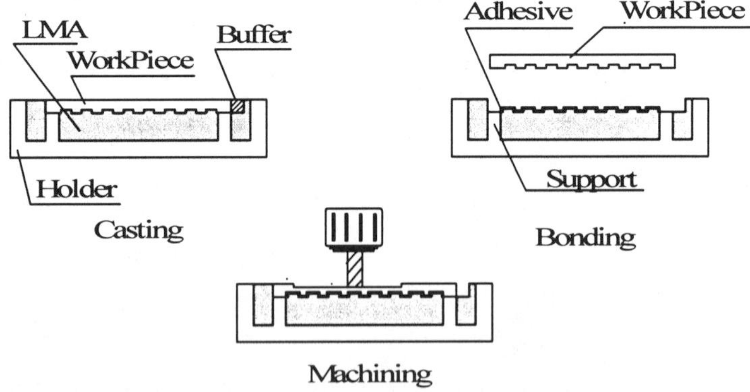

Fig.3 Upper implanting method

solvent. The buffer in the figure is used to enable easy pickup of the blank. Adhesives of cyanoacrylate or epoxy resin integral are used.

3.2 Lower Implanting Method

Adhesive is applied on the blank and the blank is covered with a holder, as shown in Figure 4. The melted LMA is poured into the holder. There is no need to pick up the blank before the machining. The parts are taken out after the machining, and the LMA and the adhesive are removed as described in section 3.1.

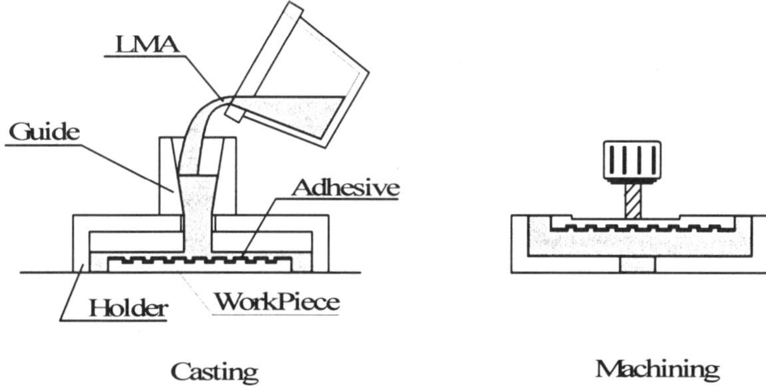

Fig.4 Lower implanting method

3.3 Suction Method

Many small pipes are placed on the blank in a holder, and the melted LMA is poured in the holder, as shown in Figure 5. After coagulation they are set on a base. The blank is vacuumed and machined. Clamping force is not strong. Huge machining force may nip out the blank.

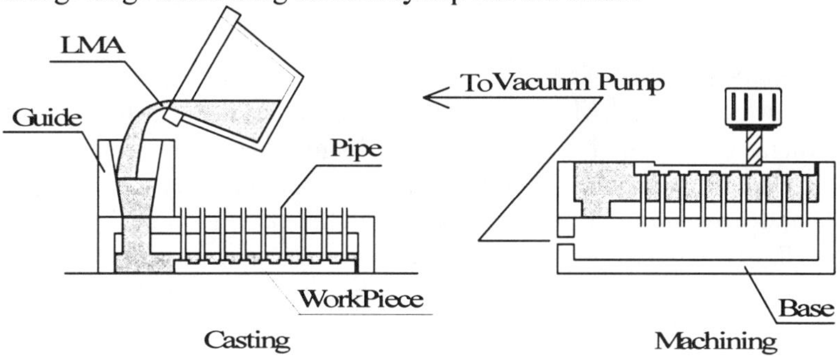

Fig.5   Suction method

3.4 Machining of Thin Walls

When a thin wall is required, a face of the wall is machined and the machined pocket is filled with LMA. Next another face of the wall is machined. It is necessary to use adhesive to machine such a thin wall. Figure 6 shows the procedure. Machining of honey-comb parts is one application. Figure 7 shows the procedure. Machining and pouring of LMA are repeated three times.

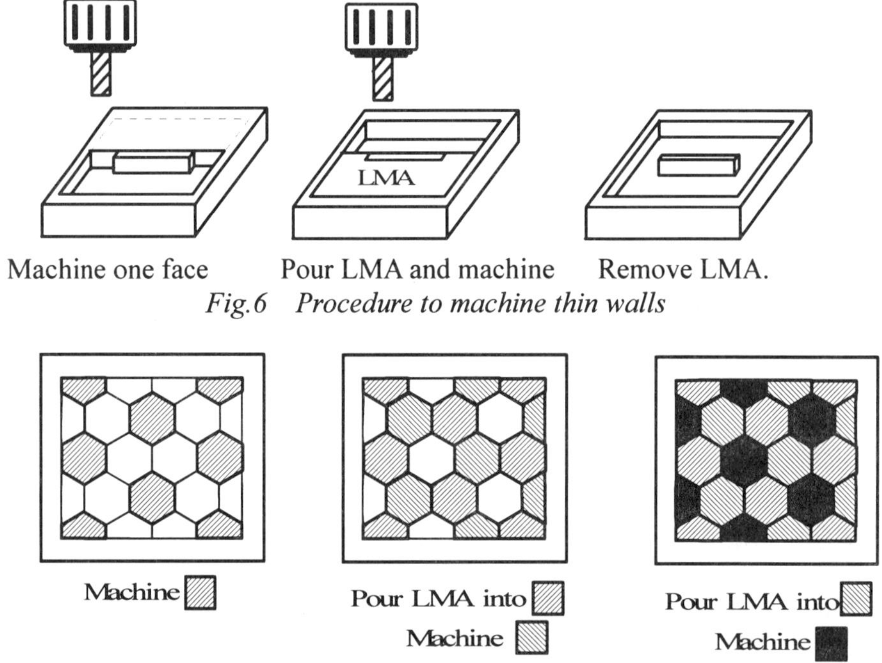

Fig.6   Procedure to machine thin walls

Fig.7   Procedure to machine honey-comb parts

4. THE RESULTS AND SAMPLES

The cast surface of the LMA should fit tightly on the blank. If not, the gap between them causes deformation of the blank during the machining. However we sometimes obtained a poor cast surface especially in the case of the upper and lower implanting methods. But the poor surface is covered with the adhesive. Consequently all methods were successively used for machining of 0.1 mm-thick parts. The lower implanting method is the most simple and preferable method.

Figure 8 shows a sample whose thickness is 0.05mm machined by a 40 mm square pocketing of a brass workpiece. The depth of cut per step was set less than 20% of the left thickness. Figure 9 shows a sample of honeycomb part of 0.1mm wall thickness. Figure 10 shows a sample of cross bellows of 0.1 mm thickness, and Figure 11 shows a 0.3 mm-thick propeller. Samples of Figures 8, 10 and 11 were made with the lower implanting method.

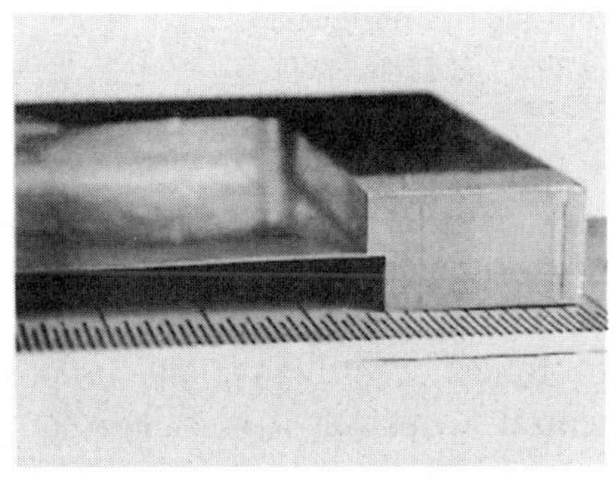

Fig.8 The cross section of t 0.05mm

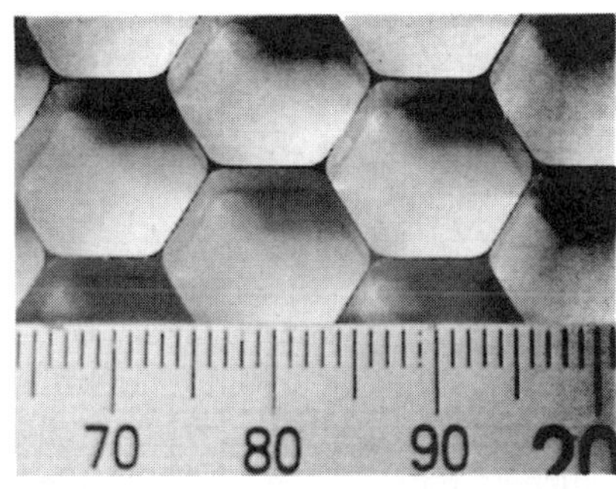

Fig.9 The honeycomb t 0.1mm

Fig.10 The cross bellows t 0.1mm

Fig.11 The propeller t 0.3mm

5. CONCLUSIONS

Methods to machine thin parts were developed where the blank is supported and fixed on a low melting alloy, with adhesive or suction force. Parts with a minimum thickness of 0.05mm and others were successively machined using these methods.

6. REFERENCE

[1] Catalog of low melting alloys: OSAKA ASAHI METAL MFG. CO. LTD., Japan

Control of Surface Pattern of Mold Generated by Ball-End Milling

Akinori SAITO, Xiaoming ZHAO and Masaomi TSUTSUMI

Graduate School of Bio-Applications and Systems Engineering,
Tokyo University of Agriculture and Technology

Abstract

In this research work, the generation mechanism of surface pattern in ball-end milling is analyzed theoretically and experimentally. In the experiment, the eccentricity of the tool axis and the angular position of the cutting edge of a ball-end mill are controlled according to the proposed method. From the experimental and theoretical results, it is found that the surface pattern generated on planes or cylindrical surfaces can be controlled by the proposed method.

Keywords

surface pattern, ball-end mill, surface finish, tool path, machining center

1. INTRODUCTION

Five-axis controlled machining centers are used for the production of molds with various curved surfaces. A lot of studies on the tool path generation for five-axis controlled machining centers have been reported [1]-[2]. However, the pattern of finished surface roughness has not been considered in their works. For reducing the hand-finishing process in the mold production, it is great important to control the pattern generated on the finished surface. If the surface pattern can be controlled, it is possible to reduce the lead time of the mold production and also to use the pattern for industrial design, etc.

This research paper describes the method for creating regular surface patterns on the finished surface of molds by adjusting the rotational axis of a ball-end mill using a boring head. In this study, experimentally and theoretically, the influence of the eccentricity of tool axis and the angular position of cutting edges on the surface pattern is investigated.

2. EXPERIMENT AND ANALYSIS

The experimental works has been mainly conducted using a three-axis controlled machining center and a solid ball-end mill with two cutting edges. The radius of the end mill is 5mm, and 70-30% brass is used as a work material for its good transcription. Cutting conditions are shown in Table 1 and the outline of experimental method is shown in Fig.1.

The tilting angle β of tool axis is given by tilting the cutting plane relative to X-Y plane as illustrated in Fig.1. A boring head is used as a tool holder to change the eccentricity of the ball end mill axis. The pattern of the finished surface is observed by a CCD camera, and the surface roughness is measured along the feed direction and the pick-feed one by a surface roughness measuring instrument. The feed rate is constant 1mm/rev throughout experiment.

In the simulation, the surface asperities are precisely calculated considering the relative motion of the tool to the machined surface as well as the angular position of cutting edge[3].

Table 1 Cutting conditions

Spindle speed n	240 min^{-1}
Feed speed f	240 mm/min
Pick feed f_p	0.5 mm
Depth of cut	0.1 mm
Tilting angle β	15°, 30°
Eccentricity e	1~40 μm

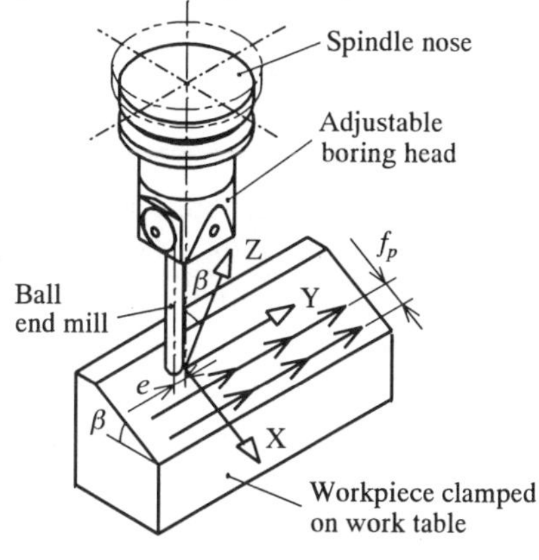

Fig.1 Outline of experiment

3. CONTROL OF SURFECE TEXTURE

Significant factors, which determine the surface pattern, are the eccentricity of tool axis and the angular position of cutting edge. However, it is not easy to synchronize the angular position of cutting edge with the movement of the linear motion axis in the experiments differing from the simulation.

In this study, it is assumed that the feed speed f and the spindle rotational speed n are kept constant throughout experiment. The angular position ω is calculated from the traverse time T of the tool along to the tool path. The traverse time T from point P_1 to point P_5 shown in Fig.2 is given by Eq.(1).

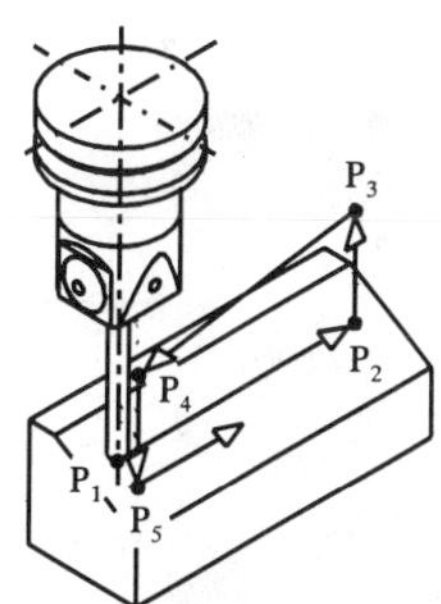

Fig.2 Tool path

$$T = \sum_{i=1}^{4} \frac{l_i}{f_i} \qquad (1)$$

Where, l_i is the distance from P_i to P_{i+1} (mm), and f_i is feed speed from P_i to P_{i+1} (mm/min). Thus, ω is expressed by Eq.(2).

$$\omega = 2 \pi T n \qquad (2)$$

Fig.3 shows the patterns of the machined surface and simulated one at different angular position ω of cutting edge. These values of ω are calculated from the tool path length. The pattern of the machined surface agrees well with the simulated one. Each angular position ω is controlled according to the proposed algorithm in which the traverse time T along the tool path is precisely calculated. Fig.4 shows a surface pattern generated by actual machining and its corresponding simulation. In the experiment as well as the simulation, the angular position ω is kept 0 degree. It can be seen in the figure that the machined surface pattern almost agrees with simulated one. It may be concluded from the figure that the surface pattern can be controlled by the proposed technique.

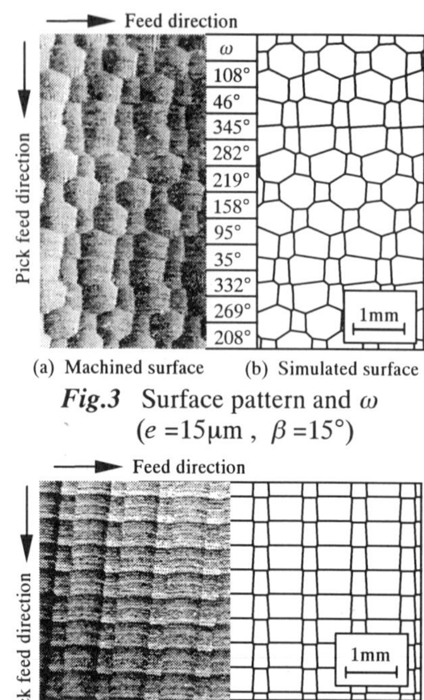

(a) Machined surface   (b) Simulated surface

Fig.3   Surface pattern and ω
(e =15μm , β =15°)

(a) Machined surface   (b) Simulated surface

Fig.4   Controlled surface pattern

4. EFFECT OF ECCENTRICITY ON FINISHED SURFACE
4.1 Effect of Eccentricity on Surface Pattern

Fig.5 shows various surface patterns corresponding to each eccentricity. In the experiment, the tilting angle β of tool axis is kept constant 30°. The pattern is composed of equal size hexagons when e=0.8μm. However, when the eccentricity becomes lager, the size and shape of

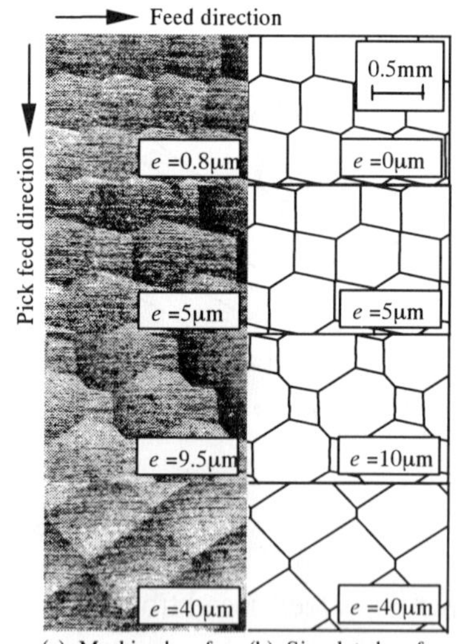

(a) Machined surface (b) Simulated surface

Fig.5   Influence of eccentricity on surface pattern (Plane, β =30°)

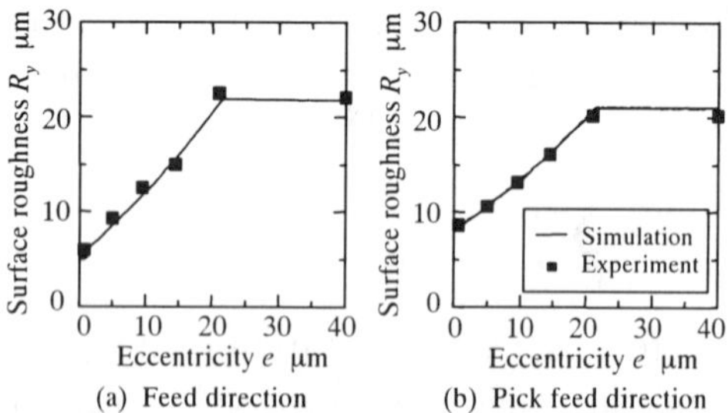

(a) Feed direction   (b) Pick feed direction

Fig.6   Relationship between eccentricity and surface roughness

adjoining two polygons are different from each other. In the case of e=40μm, only one cutting edge actually plays as an active cutting edge, so that the number of polygons is a half.

As shown in Fig.6, the surface roughness proportionally increases with the increasing eccentricity. However, the surface roughness is constant when the eccentricity is lager than 21μm. In this range, only one cutting edge works well to cut the work piece.

4.2 Effects of Eccentricity and Angular Position of Cutting Edges

Fig.7 shows the influence of the cutting edge angle ω on the surface pattern. As shown in the figure, the angle ω largely affects to the shape of polygon.

Fig.8 shows the relationship between the surface roughness and the angle ω. The surface roughness along the feed direction is independent of the angle ω and it takes a constant value. On the other hand, the surface roughness along the pick feed direction changes with the angle ω and it takes a minimum value when the angle is equal to 0 degree.

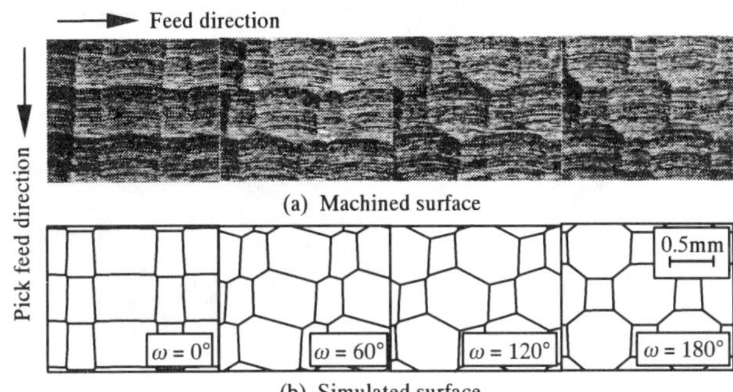

Fig.7 Influence of angler position ω of cutting edge on surface pattern (Plane, e =15μm, β =15°)

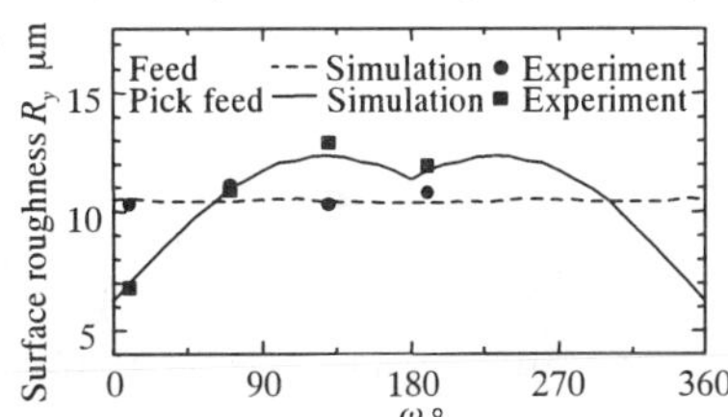

Fig.8 Relationship between ω and surface roughness

5. GENERATING PATTERN ON CYLINDRICAL SURFACE

If the surface pattern is generated on a cylindrical surface, the pattern can be applied to the functional surface such as a reflector. One of goals of this research work is to generate a regular pattern on a curved surface of plastic molds. In this experimental work, a five-axis controlled machining center has been used.

Feed speed at each block of NC program slightly changes if the NC data is created using a conventional CAD/CAM system. In addition, the length of each block is slightly different from each other. Such different feed speed and length of blocks affect to the surface pattern.

In this study, those feed speed and length of blocks are carefully controlled. Fig.9 shows an example of cylindrical surface. As shown in this figure, regular surface pattern is generated on the cylindrical surface. In this example, the angular position ω of cutting edge is kept 0 degree.

As mentioned above, the desired surface pattern can be created on a cylindrical surface if we can keep the actual feed speed and the length of each block constant.

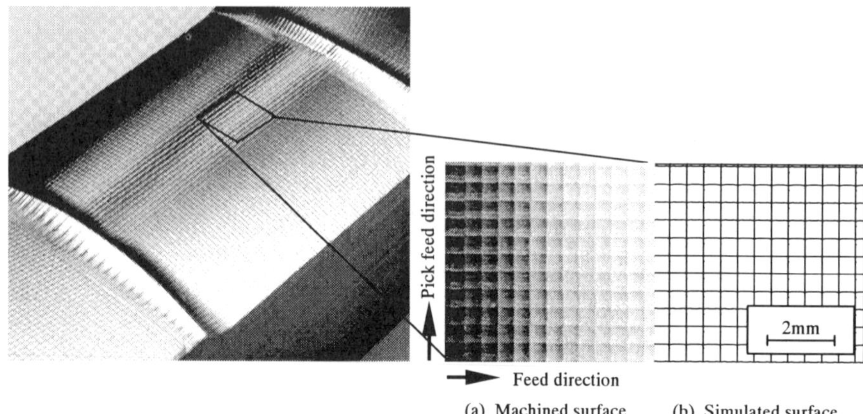

Fig.9 Surface pattern on cylindrical surface
(Cylindrical radius R=50mm, e =1μm, β =15°)

6. CONCULUSION

(1) The eccentricity affects to the surface roughness.
(2) The surface roughness along the pick feed direction is affected by the angular position of the cutting edge.
(3) The surface pattern on planes and cylindrical surfaces can be controlled by the proposed method.

REFERENCES

[1] D. Blackmore, M. C. Leu and K. K. Wang : Application of Flows and Envelopes to NC Machining, Ann. CIRP; 1992; 41: 493-496

[2] Rong-Shine Lin and Y. Koren : Efficient Tool-Path Planning for Machining Free-Form Surfaces, Trans. ASME, J. Manuf. Sci. Eng. ; 1996; 118: 20-28

[3] A. Saito, X. Zhao and M. Tsutsumi : Control of Surface Texture of Mold Generated by Ball-End Milling, J. JSPE ; 2000 ; 66, 2: 419-423 (in Japanese)

HIGH-SPEED CUTTING – FUNDAMENTALS AND MACHINE TOOL DEVELOPMENT

H.K. Tönshoff, T. Friemuth, P. Andrae, C. Lapp

Institute of Production Engineering and Machine Tools, University of Hannover

Abstract

The increasing demands for efficiency of machining processes have led to the development of new machining strategies like high-speed cutting (HSC) and high-performance cutting (HPC). At first the fundamentals concerning chip formation, temperatures and forces are described in this paper and the enhanced demands on involved system components like spindle systems cutting tools and machine tools are derived. Secondly, a new high-dynamic machine tool is presented. It is designed to meet the special requirements in high speed and precision cutting and now under construction at the IFW. Linear direct drives will be used in all axis in this machine. In one of the three axis a magnetic levitation guide will be used.

Keywords

high-speed cutting, machine tool, linear direct drives, magnetic levitation guide

1. INTRODUCTION

High-Speed cutting (HSC) is not really a new technology. First investigations have been performed by Salomon in the twenties /SAL25/. However, this investigations were only ballistic analysis. The improvement of machine tools and controls made HSC possible in machining operations in the seventies /KON77/. The work of Schulz in the eighties led to the introduction of the HSC in industrial applications /SCH81/. However, the fundamental effects in HSC are investigated in the last few years only /DAV98, TLU96, TOE99a/. The productivity of the HSC and newer technologies like high-performance cutting (HPC) is mostly limited by cutting tools and machine tools. At first, HSC was applied in machining tasks, where big workpieces with very long tool paths have to be machined. Now, the HSC will be applied in machining very small parts with high precision. This leads to new demands on the machine tools.

2. FUNDAMENTALS OF HIGH-SPEED CUTTING

The fundamental research of the mechanisms at high cutting speeds is forced in Germany with the research program of the German Research

Council, started in 1998. The results shown in the following, arise from the investigations performed in this program. The main effects of increasing cutting speeds on the process parameters in HSC are shown in figure 1.

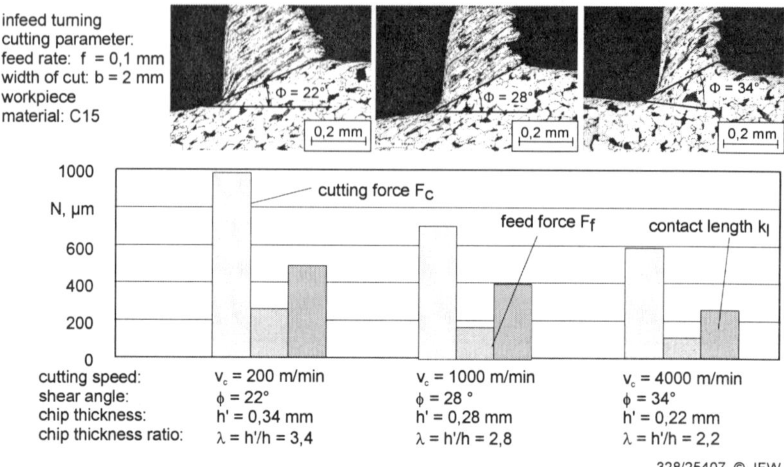

Figure 1. Process parameters in HSC

The cutting forces and the feed forces decrease with higher cutting speeds. Simultaneously, the chip formation changes. The shear angle increases and the chip thickness ratio as well as the contact length decrease. Other investigations show that the segmentation frequency of the chip rise significantly /TOE99a/. The knowledge of these mechanisms allows the definition of the load on the cutting tools in HSC. Figure 2 shows the strain towards the cutting tool.

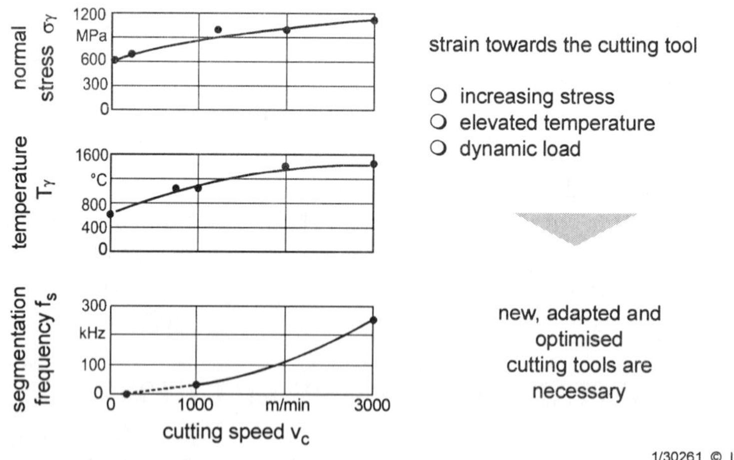

Figure 2. Strain towards the cutting tool

The temperature and the normal stresses are calculated. It is obvious that the mechanical load on the tool rises in spite of decreasing cutting forces due to a reduced contact length. The high stresses, the high temperature and the vibrations caused by the segmentation frequency have to be taken into account in the design of cutting tools for high speed machining. Further development of cutting tools and processes for HSC lead to rising demands on machine tools, also. Newer technologies like high performance cutting or circular milling require very high feed rates and high accuracy of the machine tools and controls. Additionally, the high speed cutting can be applied in machining of complex and small workpieces, if the acceleration of the axes allow high feed rates on short distances. The design of a machine tool, meeting the requirements, is presented in the next chapter.

3. NEW MACHINE TOOL FOR HIGH-SPEED CUTTING

The development of machine tools, feed drives and motion controls that are capable of high-speed precision motion is an important aim. In recent years, more and more linear direct drives have been used in high performance applications such as computer numerically controlled machine tools. To fulfil the high demands on machining accuracies, the complete system of the machine tool must be taken into account. A high-dynamic machine tool is now under construction. In this machine linear direct drives will be used. The main requirement for this new machine tool is a high amount of acceleration in all axes. This involves high mechanical loads on the structural components. These loads lead to displacements of mechanical components that disturb a precise machining process /TOE99b/. For this reason a completely new type of machine tool was designed, figure 3.

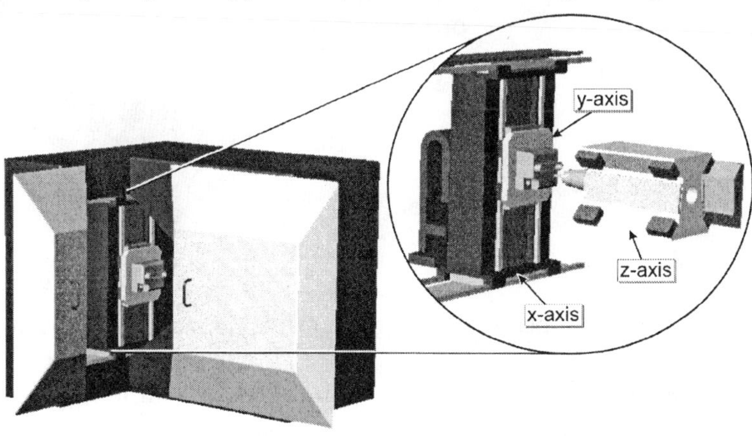

Figure 3. High-dynamic machine tool with magnetic guides

The technological new features are concentrated within the z-axis. Beneath two linear direct drives the guiding system is designed completely new. Here a magnetic levitation guide is used. Advantages of this non-contact system are freedom of wear and nearly no friction. The suspended axis is without any mechanical contact to the frame /POP99/. A multivariable control system, which makes use of a verified rigid body model of the structure, co-ordinates the action of the linear direct drives and the magnetically levitated guide system as an orthogonal actuator.

The shape of the z-axis body is found by use of a topology optimization process. This method assists the design engineer in finding a suitable shape of the design object already in an early stage of the formation phase. The goal of the topological optimization is to find the best use of material for a part that is subject either to a single load or to a multiple load distribution. The best use of material means in context with the topological optimization a "maximum stiffness" design. The objective function is predefined, the designer has only to define the structural problem and the percentage of material that has to be removed. The objective function of the topological optimization minimizes the energy of structural compliance while satisfying the constraints on the volume of the structure. Minimizing the compliance is equivalent to maximizing the global structural stiffness.

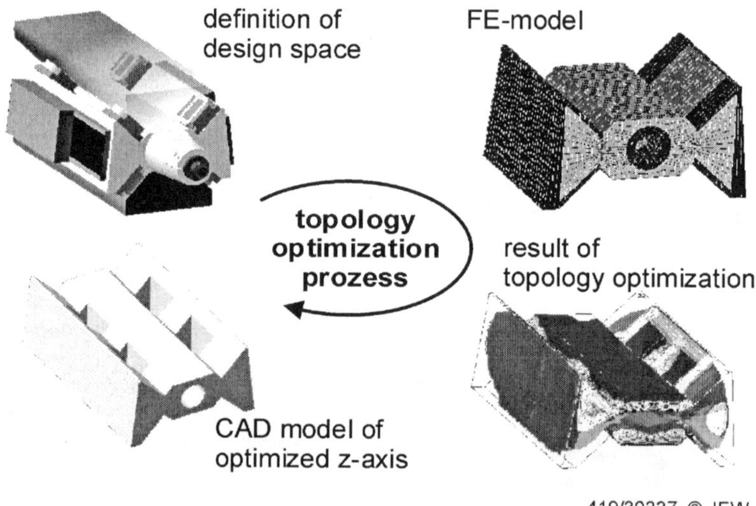

Figure 4. The process of the topological optimization of the z-axis

The process of the topological optimization, shown in figure 4, starts with the definition of the design space. The design space for a part is that particular space in the complete assembly, where the designer can freely arrange the material for that single part. That space can be derived from a

CAD assembly drawing. The next step is to generate a finite element model of the part that should be optimized. In the figure 4 a model of the z-axis is also shown. The dark colored elements are representing structure areas which are not taken into account for the topological optimization because of design reasons. The areas are needed to attach the linear motors, the magnetic bearings as well as the spindle to the structure. The bright colored elements are subject to the optimization. The loads coming into the structure from the linear motors, the bearings and the supposed machining forces are applied to the model. The different colors in the results plot of the topology optimization model are representing different pseudo densities. Dark colored areas represent high material density. In the last step a CAD model of the optimized part was generated.

Result of this design process is an optimized shape of the element. Higher stiffness at lower weight is the most important benefit. The low weight is required to achieve the aspired axis accelerations.

4. CONCLUSION

High-speed cutting leads to changed mechanisms in chip formation and to new demands on cutting tools and machine tools. High performance drives like linear direct drives and the achievable high accelerations require a very careful design of the machine tool structure. New drives and guides in combination with intelligent design procedures have the potential to improve the machine tool performance.

5. REFERENCES

DAV98 Davies M.A., Dutterer B., Pratt J.R., Schaut A.J. On the Dynamics of High-Speed Milling with long, Slender Endmills, Annals of the CIRP 47 (1998):55-60.

KON77 Koontz J.L. Ultra-high-speed machining, American Machines 1977, June:135-139.

POP99 Popp K., Ruskowski M., Tönshoff H. K., Kaak R., Lapp C. Auslegung einer kontaktlosen Werkzeugmaschinenachse, 4. Magdeburger Maschinenbautage, Magdeburg, 1999.

SAL25 Salomon C. Verfahren zur Bearbeitung von Metallen oder bei einer Bearbeitung durch schneidende Werkzeuge sich ähnlich verhaltender Werkstoffe, Deutsches Reichspatent 523 594, 27. November 1925.

SCH81 Schulz H., Arnold W., Scherer J. Hochgeschwindigkeits-Zerspanung: Neue Technologie oder Schlagwort? Werkstatt und Betrieb, 114 (1981) 8:527-31.

TLU96 Tlusty J., Smith S., Winfough W. Techniques for the Use of Long Slender End Mills in High-Speed Milling, Annals of the CIRP 45 (1996):393-96.

TOE99a Tönshoff H. K., Hollmann F. Spanen metallischer Werkstoffe mit hohen Geschwindigkeiten. Kolloquium des Schwerpunktprogramms der Deutschen Forschungsgemeinschaft, Bonn, 1999.

TOE99b Tönshoff H. K., Lapp C. Potential of Linear-Direct-Drives and Resulting Demands on Machine Tool Structures. Aerospace Manufacturing Technology Conference, Bellevue, USA 1999.

DRY CUTTING OF STAINLESS STEEL USING INDEXABLE INSERTS HAVING SELF-HEAT ABSORBING CAPABILITY

Masao MURAKAWA and Masahiko JIN

Department of Mechanical Engineering

Nippon Institute of Technology

Abstract

In this study, a dry cutting method that avoids the use of a cutting fluid in order to protect the environment is investigated.   As one means of achieving this purpose, a new cooling system in which heat generated during cutting operations is efficiently absorbed by the cutting tool is proposed.   Briefly, a heat absorption tool through which it is possible to run a coolant inside an indexable insert and an insert holder is fabricated, and a dry cutting method using this cutting tool is proposed.   As a result of cutting experiments using stainless-steel workpieces, in the case of the proposed dry cutting method using this insert, the tool life can be improved to more than twofold that in the case of the dry cutting using a conventional tool.

Key words: cutting, dry cutting, cutting heat, heat absorption, environmental problem

1. INTRODUCTION

Recently, it has been suggested that cutting fluid for cutting operations has harmful effects on the environment.   Furthermore, from the standpoint of production costs, eliminating the use of cutting fluid has the significant advantage of reducing production costs such as the costs of cutting fluid, coolant facilities, electricity consumption for supplying cutting fluid and the treatment of waste fluid from cutting operations.   Therefore, there is worldwide interest in reducing the use of cutting fluid as much as possible (Koening W. et al., (1994)).

In an effort to solve the problem, we have proposed a new cooling system in which the cutting heat is efficiently absorbed by the cutting tool, which differs from conventional cooling systems such as those using cutting fluid, oil mist or cold blast (Honma H. et al., (1996), Klocke F. et al., (1997), Heisel U. et. Al. (1994)).   Briefly, the system entails a dry cutting method involving the use of a cutting tool with high cooling efficiency, through which it is possible to run the coolant inside an indexable cutting insert and an insert holder (termed the heat absorption tool hereafter).

In this study, we investigate the effect of improving the tool life when cutting stainless-steel SUS304 using the heat absorption tool. Namely, in cutting stainless steel, a material with low thermal conductivity, heat is easily stored in the cutting tool. It is known that the machinability of this material is poor because problems in cutting often occur, such as a high tool wear rate and the strong adhesion property of stainless steel to the cutting tool. Thus, we investigate the tool life with respect to the dry cutting of stainless-steel SUS304 (JIS). As a result of the experiment, it is clarified that the proposed heat absorption tool is very effective for the dry cutting of stainless steel; furthermore, this method is proven to significantly improve tool life if the tool is used in combination with coated inserts.

2. PRINCIPLE OF THE DRY CUTTING METHOD USING THE HEAT ABSORPTION TOOL

The conceptual schema of the proposed dry cutting method using the heat absorption tool is shown in **Figure 1**, in comparison with that of the conventional method. The conventional cooling system functions in such a way that heat generated during cutting operations is decreased by circulating coolants, such as cutting fluid, oil mist or cold blast by convection, to the cutting point, as shown in Fig. 1 (a). In contrast, the cutting heat cooling system using the heat absorption tool involves method in which the cutting heat is decreased by absorbing the heat evolved in cutting operations in the cutting tool at high speed, as shown in Fig. 1 (b). The perfectly dry cutting conditions are such that there is no use of any cutting fluids or production of any noise from the blowing of cold air or oil mist, and that a pump for

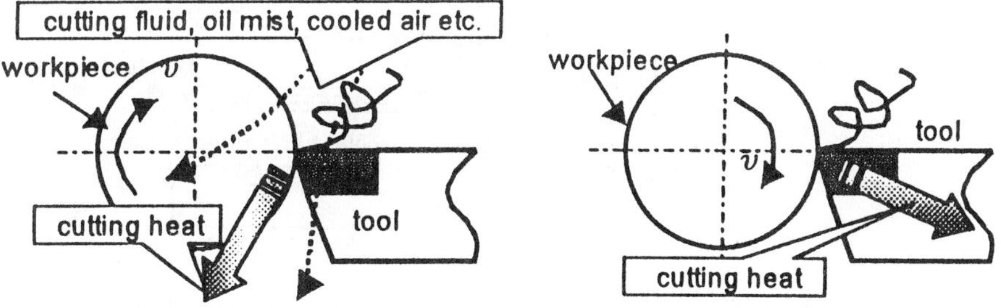

(a) Cooling by circulating coolants by convection   *(b) Cooling using the heat absorption tool*

Figure 1   Conceptual schema of the proposed dry cutting method using the heat absorption tool in comparison with that of the conventional method

supplying the cutting fluids and/or treatment of waste fluids is not needed. Although it appears that the 'cooling cutting method' in a former study (Bartle E. W., (1953), Chandiramani K. G., (1961)), (Okushima K., Kawashima Y., (1968)) has some resemblance to this technique, both techniques are different with respect to the following feature. The main purpose of the former study is to determine what cools the cutting tool to very low temperature; on the other hand, the main purpose of this study was to determine what increases the heat transfer or heat absorption rate.

3. EXPERIMENTAL

The heat absorption tool used in the experiment, composed of a heat absorption insert holder and a heat absorption insert, is shown in **Figure 2**. Holes for the water supply and drainage are drilled in the heat absorption insert holder. The heat absorption insert is prepared using a commercially available indexable insert in which a hole with a diameter of ϕ 1-2 mm is drilled by electrical discharge machining (EDM) to serve as a coolant aqueduct. The entrance and exit of the coolant aqueduct in the heat absorption insert are coupled to the coolant supply and drainage gate in the insert holder, respectively. Tap water with a temperature of 23 °C is used as a coolant and circulated at a flow rate of 800 cc/min.

The cutting conditions are listed in **Table 1**. The following four cutting methods were examined: dry cutting with the heat absorption tool using Al_2O_3-coated inserts (DCHA coating), wet cutting with the conventional tool using cemented carbide inserts (WCC), dry cutting with a conventional tool using Al_2O_3-coated inserts (DCC coating) and wet cutting with a conventional tool using Al_2O_3-coated inserts (WCC coating).

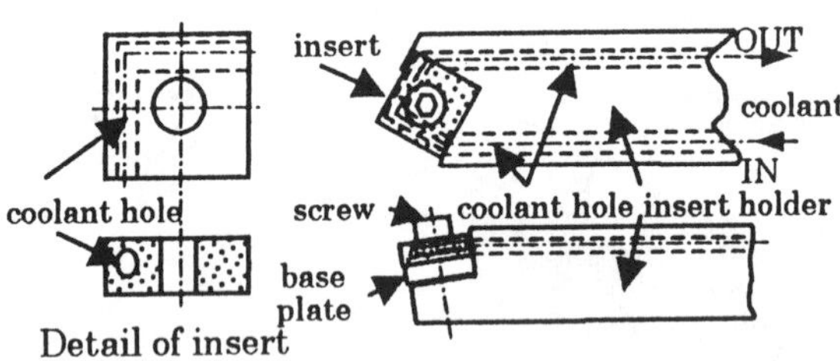

Figure 2   Heat absorption tool composed of a heat absorption insert holder and a heat absorption insert

4. EXPERIMENTAL RESULTS AND DISCUSSIONS

The relationship between flank wear width and cutting distance in each cutting method is shown in **Figure 3**. In the case of WCC, when cutting the stainless-steel SUS304 at the high cutting speed V of

Table 1   Experimental conditions for the cutting of stainless-steel SUS304

Tool／workpiece	Al_2O_3-coated K10 (SNMG120408) ／SUS304(JIS) or 304(AISI)
Dimensions of workpiece	ϕ 120 mm - L350 mm
Cutting speed V	200m/min
Depth of cut d／feed f	0.5 mm ／ 0.2 mm/rev
Cutting tool used & cooling conditions	A: Dry cutting with heat absorption tool using Al_2O_3-coated insert （DCHA coating） B: Wet cutting with conventional tool using cemented carbide insert （WCC） C: Dry cutting with conventional tool using Al_2O_3-coated insert （DCC coating） D: Wet cutting with conventional tool using Al_2O_3-coated insert （WCC coating）
Cutting fluid／flow rate (WCC or WCC coating)	Water-miscible fluid (2% dilution) ／20 l／min
Coolant／flow rate (DCHA coating)	Tap water (23℃)／800cc/min

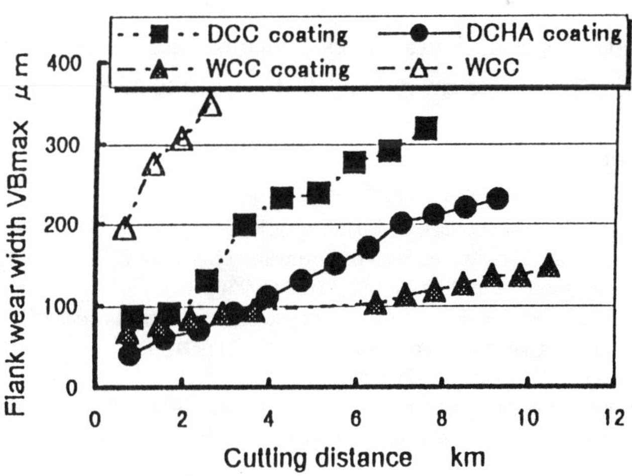

Figure 3   Relationship between flank wear width and cutting distance in each cutting method (Al_2O_3-coated K10, SUS304, v=200m/min, d=0.5mm, f=0.2mm)

200 m/min, the maximum flank wear width VB_{max}, VB_{max} =0.3 mm was reached at a cutting distance of only 2 km. On the other hand, in the case of the DCC coating, VB_{max} =0.3 mm at a cutting distance of 4 km; thus, the tool life was improved to approximately threefold that of WCC. Therefore, we confirmed that a coated tool is effective in the dry cutting of stainless steel. On the contrary, in the case of the DCHA coating, VB_{max} =0.3 mm at a cutting distance of 9 km; thus, the tool life was improved to approximately two- to threefold that of the DCC coating, although the improvement in tool life was not reached in the case of the WCC coating.

On the basis of this result, it is considered that the proposed heat absorption tool is very effective for the dry cutting of stainless steel; furthermore, this method is proven to significantly improve tool life if the tool is used in combination with coated inserts.

5. CONCLUSIONS

The results obtained in this study are as follows:

(1) As a dry cutting method that avoids the use of a cutting fluid, a cooling method whereby heat generated in cutting operations is effectively absorbed by the cutting tool is proposed to replace the conventional cooling method which utilizes fluid circulating by convection.

(2) In the cutting of stainless-steel SUS304 using an Al_2O_3-coated heat absorption insert, the tool life was improved to approximately sevenfold that in the case of wet cutting using noncoated carbide inserts and approximately twofold that in the case of dry cutting using an Al_2O_3-coated insert.

In conclusion, it is clarified that the dry cutting of stainless steel using the heat absorption tool is effective as an environmentally friendly method of cutting.

REFERENCES

Klocke F. and Eisenblatter G., Dry Cutting, Ann. CIRP, Vol. 46, No. 2 (1997), p. 1.

Koening W. and Rummenhoeiter S., Ecological Manufacturing, Prod. Eng., Vol. II, No. 1 (1994), p. 1.

Honma H., Yokogawa K. and Yokogawa M., Study on Environment Conscious CBN Cooling Air Grinding Technology (in Japanese), Journal of the JSPE, Vol. 62, No. 11 (1996), p. 1638.

Heisel U., Lutz M., Spath D., Wassmer R. and Walter U., Application of Minimum Quantity Cooling Lubrication Technology in Cutting Process, Prod. Eng., Vol. II, No. 1 (1994), p. 49.

Bartle E. W., Machinery (E), 83, 24, July (1953), p. 172.

Chandiramani K. G., Refrigeration in Machining, Int. Jnl. Prod. Res, July 14 (1961).

Okushima K, Kawashima Y, Study on Machining with Internally-Cooled Tool, (in Japanese), Journal of the JSPE, Vol. 34, No. 2 (1968), p. 97.

Maekawa K, Ohshima I, Murata R, Thermal Analysis of Internally Cooled Cutting Tools (in Japanese), Journal of the JSPE, Vol. 57, No. 11 (1991), p. 2011.

MEASUREMENT OF LUBRICANT APPLYING EFFECT INFLUENCE ON LUBRICATION BY OIL-SUBMERGED CUTTING

Toshiaki Kaneeda

Okayama University of Science, Okayama, Japan, 700-0005

Abstract

Applying material such as oleic acid or extreme pressure oil on the precut surface in ductile metal cuttings can dramatically improve the machinability, due to a reduction in friction between the lamella of the chip. Under optimal conditions, this lubricant applying effect can reduce the cutting forces by more than 90%. In the wet cutting operation at machine shops, cutting fluids can reach the precut surface. Therefore, it is reasonable to assume that lubrication by the cutting fluids involves the lubricant applying effect. We have investigated the extent of lubrication by cutting fluids using oil-submerged cutting experiments. The results show that the lubricant applying effect has a major role in lubrication under certain conditions, offering the possibility of semi-dry machining.

Keywords

Lubricant applying effect, oil-submerged cutting, lubrication, applied material, work hardening, semi-dry machining, oleic acid

1. INTRODUCTION

Applying material such as oleic acid or extreme pressure oil on the precut surface in ductile metal cuttings can dramatically improve the machinability, due to a reduction in friction between the lamella of the chip. This effect we define as the lubricant applying effect. [1] Under the optimal conditions the lubricant applying effect (called LAE hereafter) can reduce the cutting forces by more than 90%. It is also reasonable to assume that cutting fluids reach the precut surface in the wet cutting operation at machine shops. With a certain conbination of conditions such as the hardness distribution, the tool configuration and depth of cut, the effect will appear in the cutting operation.

Cutting fluids have two major roles in cutting. The first is lubrication. However, the extent of the lubrication has yet to be determined. The reasons are as follows: First, there are three directions from which the operator can supply the cutting fluids to the cutting process. Namely, (1) to the rake face along the chip flow direction, (2) to the rake face from the chip side, and (3) to the clearance face from the back side of the tool. As yet there is no agreement on which direction should be standard, and so the results of lubrication cannot easily be compared. Second, the lubrication is affected by the supply pressure, the amount of cutting fluid, the cutting conditions, and the operator. Therefore, it is very difficult to assess which of these factors affect the extent of the lubrication in the various conventional cutting methods. In order to investigate the extent of lubrication by cutting fluids, it is necessary to supply the cutting fluids to the tool and work material using a constant and steady supply method. Therefore, in this paper we have developed an oil-submerged cutting experiment.

The purpose of this paper is to measure the LAE on lubrication by cutting fluids. If the LAE has a major role in lubrication, utilizing the LAE can lead to semi-dry machining.

2. EXPERIMENTAL PROCEDURE

2.1 Experimental apparatus and work materials

Cutting experiments were conducted on an NC orthogonal precision cutting apparatus. A large-work hardening-capacity ductile metal, pure aluminum (A1050, 1/2), was used as the work material. The dimensions of the work material were $300 \times 35 \times 3$ mm. Cemented carbide (WC+Co) K10 was selected as the tool material. The rake angle was a constant $0°$. The depth of cut t_1 was widely varied from 5μm to 50μm and depth of cut in the last-pre cutting t_L was also varied between 10 μm to 100 μm. The latter depth t_L controls the work hardening in the zone to be cut, and is one of the major factors in LAE cutting. [2]

An oiliness agent, oleic acid, was selected as the applied material,

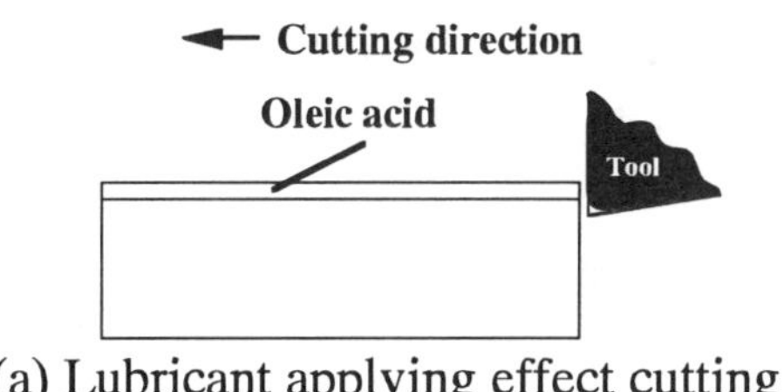

(a) Lubricant applying effect cutting

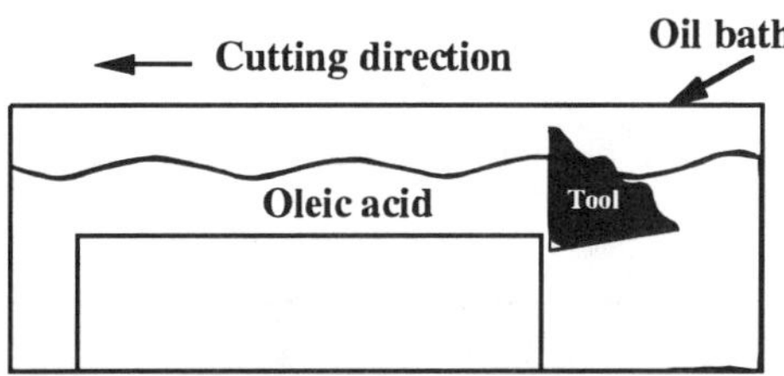

(b) Oil-submerged cutting

Figure 1. Cutting form

Table 1 Cutting conditions

Work material		Pure Al (1050 H24)
Tool material		Cemented carbide K10
Tool geometry	Rake angle α °	0
	Relief angle ε °	7
Cutting form		Orthogonal cutting
Cutting speeds V m/min		5.3, 25.7, 50.0
Depth of cut t_1 μm		5, 10, 20, 30, 40, 50
Depth of cut at last pre-cutting t_L μm		10, 20, 30, 40, 50, 70, 100
Applied materials		. Oleic acid

and in the case of oil-submerged cutting (called as OS hereafter) as the cutting fluid. This was chosen because of its high performance in lubrication.[3]

The applied material on the precut surface was less than 1 μm in thickness, which led to luster-less precutting surfaces.

In the OS, the work material and nose of the tool were submerged into the cutting fluid which was in an oil bath, as shown in Figure 1, and the work was machined under orthogonal cutting conditions. In this way, any deviation in supply of the cutting fluid could be eliminated.

A quartz force transducer (Kistler Co. type 9251A) was employed to measure the cutting forces. The cutting speeds were 5.0, 25.7 and 50.0 m/min. Table 1 summarizes the cutting conditions.

3. EXPERIMENTAL RESULTS AND DISCUSSION

3.1 Reduction in LAE and OS

Figure 2 shows the reduction in the tangential cutting force component

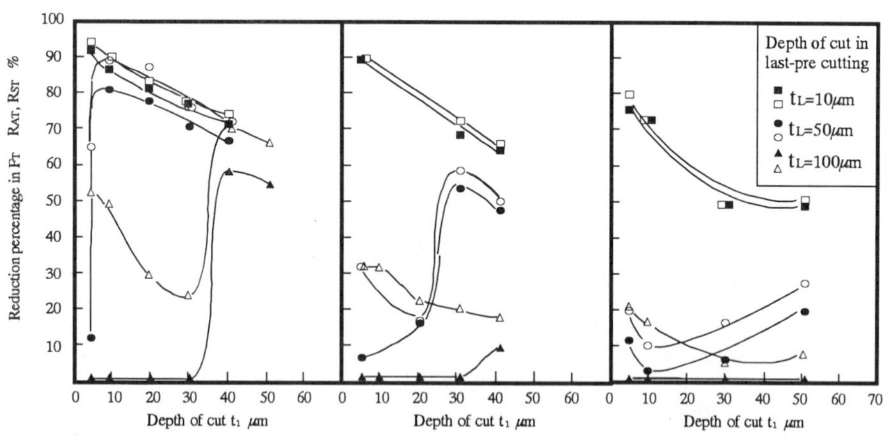

(a) V=5.0m/min   (b) V=25.7m/min   (c) V=50.0m/min

Solid symbols: Lubricant applying effect, Open symbols:Oil-submerged

Figure 2. Reduction in tangential cutting forces

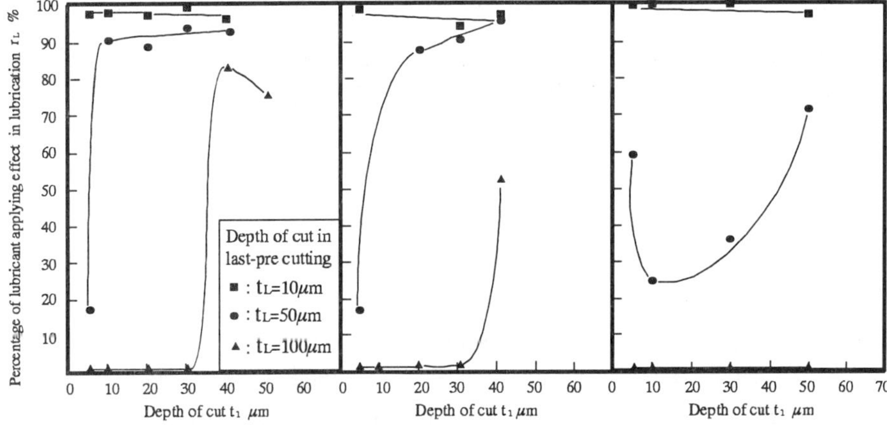

(a) V=5.0m/min   (b) V=25.7m/min   (c) V=50.0m/min

Figure 3. Percentage of LAE in lubrication

of the LAE and OS. The reduction can be given by subtracting the cutting forces in the LAE or the OS from those in the dry cutting. When t_1 is small and t_L is large, for example, V=5.0 m/min, t_1=5 μm, t_L =100 μm, and V=25.7 m/min, t_1=5 μm, t_L =100 μm, rather large differences in the reduction can be found between the LAE and OS.

Figure 3 shows the percentage of the LAE in the lubrication, that is, the ratio of the reduction in the LAE to in the OS. It is easy to see that the LAE can play a large role in the lubrication, especially when t_1=10 μm at all cutting speeds.

REFERENCES

[1] Kaneeda T, "Spreading effect in ductile metal cutting." *In Advancement of Intelligent Production*, Elsevier Science B.V. 1994 .

[2] Kaneeda T, Kohsaka H. Lubricant applying effect in ductile metal cutting (1st Report). J.of Japan Soc. Prec. Engng. 1995; 5:702-706.

[3] Kaneeda T, Noguchi K. Lubricant applying effect in ductile metal cutting (2nd Report). J.of Japan Soc. Prec. Engng. 2000; 1:74-79.

DEVELOPMENT OF INCLINED ULTRASONIC-VIBRATION CUTTING TOOL CAPABLE OF MOUNTING COMMERCIAL INDEXABLE INSERTS

Masahiko JIN and Masao MURAKAWA

Department of Mechanical Engineering

Nippon Institute of Technology

Abstract

In this study, a practical inclined ultrasonic-vibration cutting tool, on which commercially available indexable inserts of various shapes can be mounted, is investigated.   Such a tool can be achieved by using it in the longitudinal-vibration mode and optimizing the shape of the tool shank. Some cutting tools for turning operations such as cylindrical turning and facing are fabricated, and turning tests using indexable inserts are carried out. Consequently, it is found that the fabricated inclined ultrasonic-vibration cutting tool is as useful as a conventional cutting tool, and a high accuracy can be obtained when cutting workpieces of difficult-to-cut shapes.

Key words: cutting, ultrasonic-vibration cutting, cutting tool, indexable insert

1.   INTRODUCTION

In the ultrasonic-vibration cutting method, some advantages concerning the cutting ability, such as the reduction of cutting resistance and cutting burr, and the prevention of the built-up edge, can be achieved under the low-speed cutting condition (Murakawa, M. & Jin, M., (1998)).   Therefore, this cutting method has been expected to be applicable to cutting under conditions where a high cutting speed cannot be used, such as facing, microcutting and planing operations (Shamoto, E. & Moriwaki, T., (1999)).   However, this cutting method has the following two problems.   1) One is the problem that chipping of the tool edge and/or abnormal wear of the flank face easily occur when cutting hard materials.   2) Another is the problem that the methods of mounting the cutting chip are limited, such as brazing and fixing with a screw within the limits of small indexable inserts and a few limited insert shapes, because of the necessity of stabilizing the ultrasonic vibration of the cutting edge.   Therefore, the main problem has been that the practicability of the ultrasonic-vibration cutting method is low, since the various shapes, sizes and mounting attitudes of indexable inserts for various cutting operations cannot be applied, unlike the conventional cutting (CC hereafter) tool.   The authors have examined these problems. First, in the previous study (Jin, M. & Murakawa, M., (1999)), we investigated the former problem 1) and proposed the inclined ultrasonic-vibration cutting (IUVC hereafter) method as an effective countermeasure.

In this study, we investigate the latter problem 2). Namely, the IUVC tool system, in which indexable inserts of standard size and various shapes can be mounted in any fixed attitude, is developed.

2. DEVELOPMENT OF INCLINED ULTRASONIC-VIBRATION CUTTING TOOL WITH MOUNTABLE COMMERCIALLY AVAILABLE INDEXABLE INSERTS

In ultrasonic-vibration cutting, the tool edge must be continuously vibrated with fixed direction, frequency and amplitude. Hitherto, a bending-vibrational mode has been used for ultrasonic-vibration cutting. However, in that case, there arises the problem that the vibrational direction of the cutting edge inconveniently changes when the weight and/or mounting attitude of the insert is changed. This inferior vibrational mode causes the collision of the tool flank with the workpiece, which gives rise to further problems, such as abnormal wear of the flank and chipping of the cutting edge.

In an effort to solve the above-mentioned problems, first, we used a longitudinal-vibration mode of the tool shank because this vibrational mode has high stability. Next, we designed a tool in which the cross section of the upper part from the nodal point of vibration (part A in Fig. 1) of the tool shank is identical to the insert shape used, in order that the vibration direction will be fixed, even if we use inserts of different shapes or dimensions.

A schematic of the fabricated IUVC tool is shown in **Figure 1.** The tool shank ② includes both the body with a length of 240 mm as well as a length of 1 wavelength (λ) of the longitudinal-vibration mode, and the plates for mounting it to the tool holder ⑥. The plates are located on both sides of the body but are connected to the body at only two nodal points (A and B in Fig. 1) of the ultrasonic-vibration, as shown in Fig. 1. A cutting insert ① is mounted on the top face as well as the vibration loop of the tool shank, using a cap bolt ③. This tool is mounted on the tool post ⑧ of a lathe using the tool holder ⑥. The tool is mounted on the lathe with an angle of $10°$ toward both the depth of cut and feed directions of the lathe. The vibration is generated by a bolt-clamped ultrasonic transducer (BLT) ④ fixed to the bottom face of the tool shank using a set screw. This tool is ultrasonically vibrated at a resonant frequency of 20.8 kHz and an amplitude of approximately $7 \sim 10\,\mu$ m.

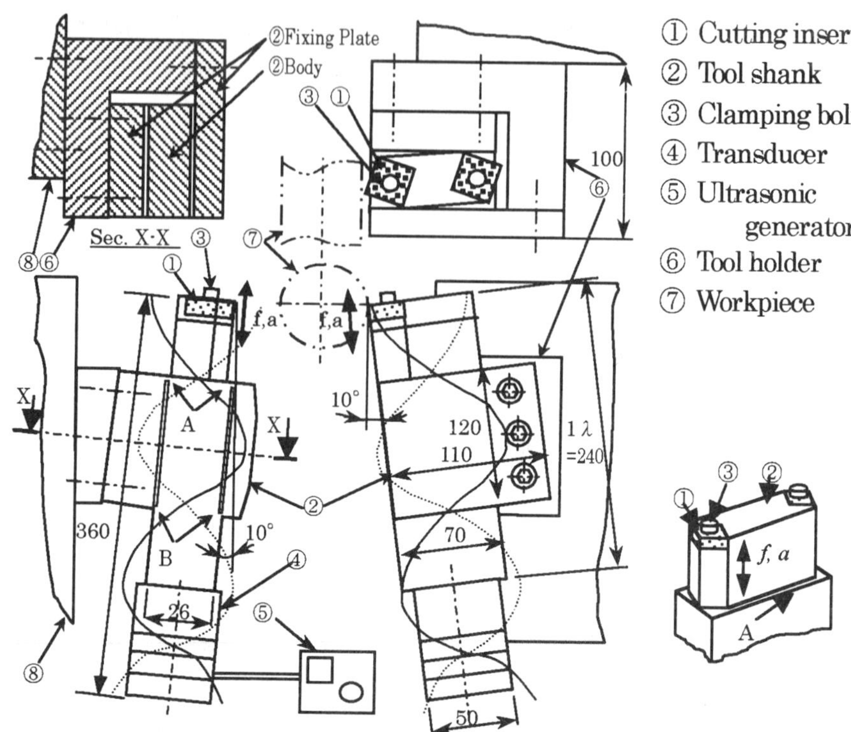

Figure 1   Schematic illustration of newly fabricated IUVC tool vibrated at 21kHz

3. EXPERIMENTAL

Turning experiments are carried out using the large lathe (swing over cross slide of ϕ450mm) mounted on the fabricated IUVC tool. The workpiece used in the experiment is that with three diameters, as shown in **Figure 2**. A quadrangular indexable insert for cylindrical turning (SNMA120404) and a triangular one for facing and corner turning (TNMG160404) were used in the experiment. The workpiece and insert material and cutting conditions are shown in **Table 1**.

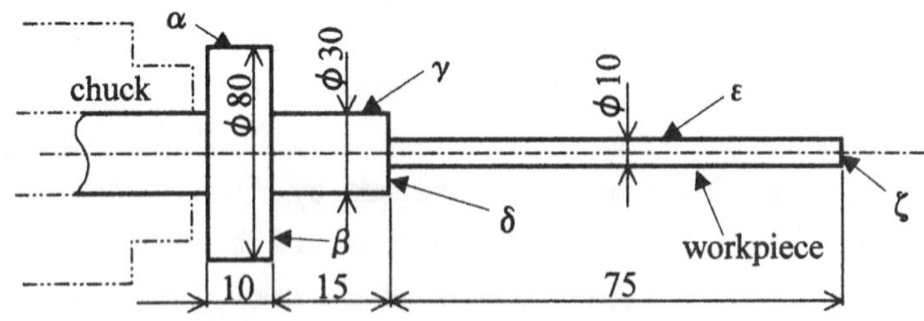

Figure 2   Configuration of workpiece

114

Table 1   Experimental conditions

Tool	Cemented carbide K10	Depth of cut d	0.1 mm
Workpiece	Carbon steel S45C	Feed rate S	0.1 mm/rev
(Hardness)	(27HRC)	Frequency f	20.8 kHz (IUVC)
Spindle speed N	740 min^{-1} (α : 186m/min	Amplitude a	10 μ m (IUVC)
(Cutting speed V)	γ :70m/min, ε : 23m/min)	Cutting fluid	None

4. EXPERIMENTAL RESULTS AND DISCUSSION

4.1 Turning conditions

As a result of the turning experiment, we recognized that the cutting condition using the fabricated IUVC tool was stable without looseness of the insert-clamping bolt and/or abnormal heat or vibration of the insert during the turning operation.

4.2 Accuracy of products

The appearance and the surface roughness of each part in the workpiece are shown in **Figure 3**.   In the case of turning the flange with the largest diameter of the workpiece, good appearance and surface roughness were obtained over both the cylindrical surface α and the end face β of the flange by both IUVC and CC. However, in the case of turning the shaft (ε , ζ) with the smallest diameter of the workpiece, a good cut surface could not be obtained by CC because the cutting speed was insufficient and the chatter arose due to insufficient stiffness of the shaft.   On the contrary, in the turning of that small shaft (ε , ζ), a good cut surface could be obtained without chatter by IUVC.   Therefore, the almost uniform and good surfaces were obtained over all parts of the workpiece by IUVC.

These experiment results clarified that the advantage of inclined ultrasonic vibration cutting, which is high accuracy, can be obtained when cutting the part of the workpiece with small diameter.   High accuracy cannot be obtained by conventional cutting because of insufficient cutting speed and stiffness of the workpiece.

5. CONCLUSIONS

The results obtained in this study are as follows.

(1) The inclined ultrasonic-vibration cutting tool system in which indexable inserts of standard sizes and various shapes can be mounted at any fixed attitude is developed.

(2) The experiment using the tool clarified that the advantage of inclined ultrasonic-vibration cutting, which is high accuracy, can be obtained when cutting the part of the workpiece with small diameter, which was not possible by

conventional cutting because of insufficient cutting speed and stiffness of the workpiece.

Finally, we would like to thank Mr. T. Kohinata and T. Tuchiya, graduate and former students at our institute, respectively, who assisted with the experiment.

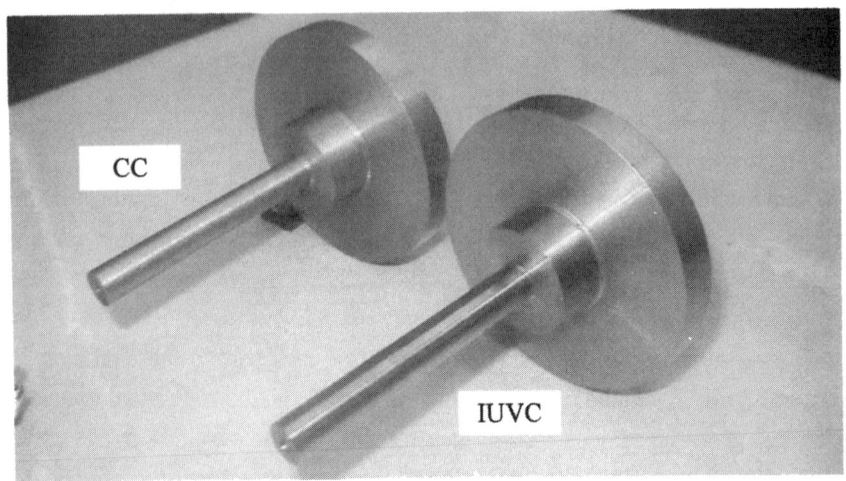

(a) Appearance of products

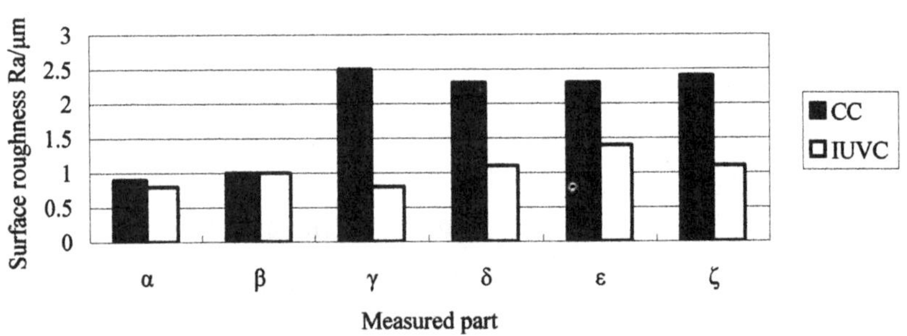

(b) Surface roughness

Figure 3   Comparison of cut surface between IUVC and CC using a triangular indexable insert (S45C, K10, N=740min⁻¹, d=0.1mm, S=0.1mm/rev, dry)

REFERENCES

Murakawa, M. & Jin, M.; Turning of Beta-Titanium Alloys by Means of Ultrasonic Vibration, Trans. NAMRI/SME, 26 (1998) 153.

Shamoto, E. & Moriwaki, T.; Ultraprecision Diamond Cutting of Hardened Steel by Applying Elliptical Vibration Cutting, Annals of the CIRP, 48, 1 (1999), 441.

Jin, M., Watanabe, T. & Murakawa, M.; Study on Prevention of Chipping of Tool Edge in Ultrasonic Vibration Cutting, Journal of the JSPE, 65, 12 (1999), 1814.

THE GREENTAPE LASER SINTERING METHOD AND ITS APPLICATIONS

Katsuhiro MAEKAWA and Tomohiro NISHII
The Research Center of Superplasticity, Ibaraki University
4-12-1 Nakanarusawa, Hitachi 316-8511, Japan

Shinichi IIDA and Tomoya SUZUKI
Ibaraki Plant, Kinzoku Giken Co., Ltd.
276-21 Motoishikawa, Mito 310-0843, Japan

Abstract

The present paper describes the greentape laser sintering (GTLS) method that allows us to fabricate metallic objects directly from powder, which is based on rapid prototyping technologies. The novelty lies in the use of a thin tape consisting of fine metal powder and organic binders. An automatic lamination forming system based on GTLS process has been developed, and a metallic sample is fabricated for its performance evaluation.

Keywords

Laser sintering, Direct metal fabrication, Rapid prototyping, Powder assembly

1. INTRODUCTION

Rapid prototyping technologies have made a shorter product development time possible and, in addition, imposed a major effect on other manufacturing processes such as injection molding and sand casting. Tooling and tooling inserts can be directly manufactured via rapid prototyping using new higher strength polymers. The main concern is that the parts could only offer a limited dimensional accuracy and surface finish, although metallic parts can be obtained quickly and inexpensively using rapid prototyping technologies.

From the latter point of view, the present paper reviews the greentape laser sintering (GTLS) method that allows us to fabricate metallic objects directly from powder (Maekawa, 1999). First, the GTLS method is outlined, then a newly developed laminate prototyping system is described together with a bladed-wheel prototyping experiment.

2. THE GREENTAPE LASER SINTERING SYSTEM

The GTLS technology is essentially a laminate prototyping system that utilizes a green tape which is a thin sheet formed from metal or ceramic powder by using organic binder. Laser sintering and lamination are repeated using the tape to directly produce a three-dimensional prototyping laminate. The doctor blade technique is applied to produce the green tape, which is a ceramic green sheet forming technique. With this technique, fine metal powder, organic binders, plasticizing agents, dispersion agents and solvents are mixed together and kneaded to form slurry. The slurry is then coated on a carrier sheet through medium of a fixation roll, and the solvent volatilized to solidify the mixture. The name "greentape" comes from the state of the objects before sintering.

Figure 1 schematically shows how to make a 3D object using the GTLS method. The carrier sheet is first removed and the greentape is placed on a substrate. When the powder is melted by the laser beam and solidified (the binder is evaporated), the volume of the sintered part is slightly decreased due to shrinkage or an increase in density. Before the second greentape is put on top of

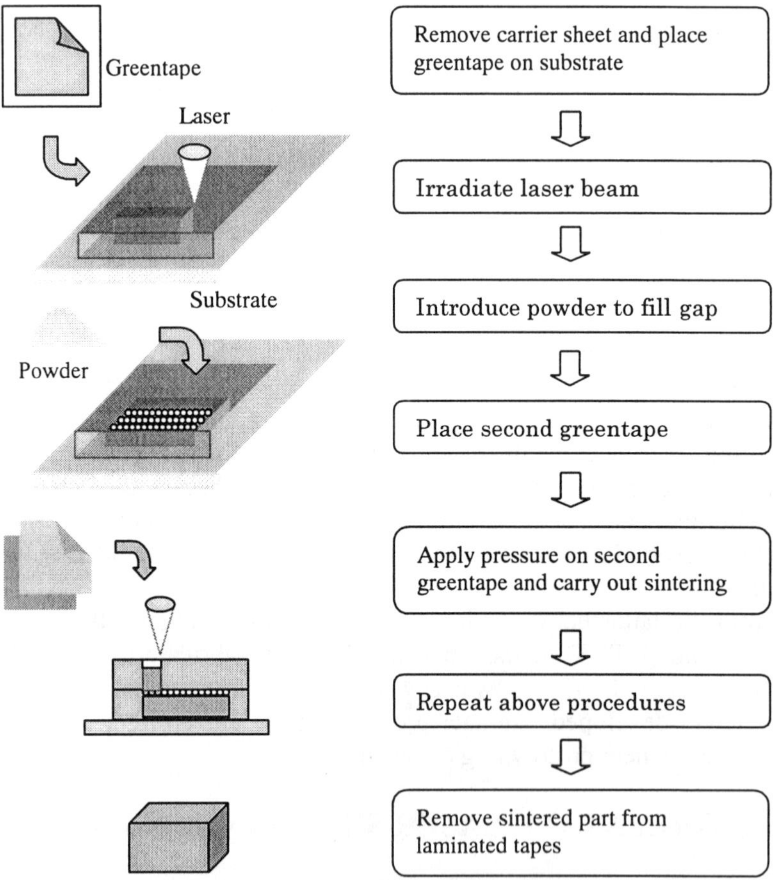

Figure 1. Principle of layering tapes using GTLS method

the first, more powder is introduced to fill the gap if necessary. Pressure is applied to the second tape and the powder in order to increase the density.

Nitrogen gas is overflowed to prevent the object from oxidizing. This process is repeated, using the sliced data that controls the sintering selectively. Finally, the sintered object comes out from the laminate when a solvent is used to dissolve the non-sintered part. Support is not necessary during layering.

Figure 2 shows a newly designed GTLS apparatus, which consists of a YAG laser with a wavelength of 1.06 μm and a nominal maximum power output of 150 W, a tape feeder, a gaseous chamber, a numerical control unit and a

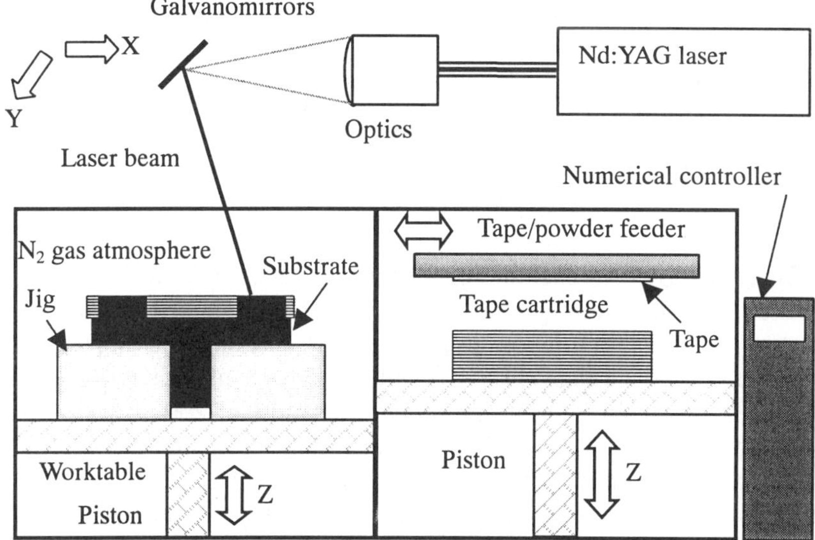

Figure 2. Schematic diagram of Greentape Laser Sintering (GTLS) system

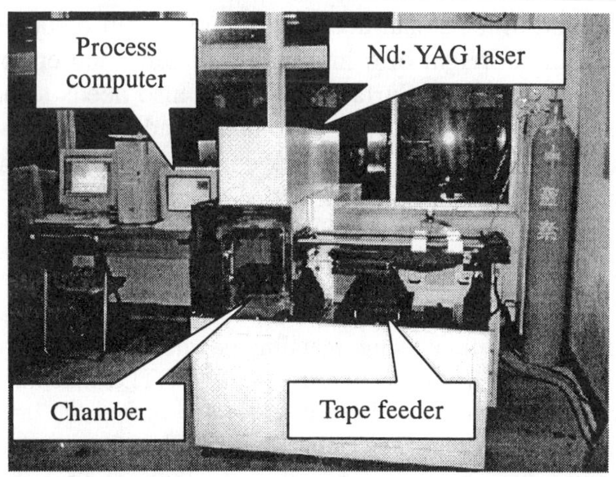

Figure 3. Appearance of GTLS system

119

computer for analysis and process control. The sintering is performed in a nitrogen atmosphere without preheating the tape and substrate. The laser irradiation area is 200×200 mm. The feeds of the sintering worktable and the tape feeding system are carried out by pulse motors and ball screws, the vertical feed resolution is 1 μm, and the stroke is 100 mm.

A DOS PC controls the laser beam, optics and galvanomirrors systems, whereas a Windows based PC generates NC data for the laser scanning in conjunction with the relevant process parameters such as beam power, scan speed and scan spacing. Synchronization between the tape feeding and sintering processes is done by the Windows based PC.

Figure 3 shows the realization of the design concept of GTLS. The technical novelty lies in the integration of GTLS processes, including tape feeding, sintering and layering in an inert gas atmosphere, with numerical control of the whole system. Unification of the tape feeding and the laser sintering unit leads to a fully automated sintering process.

3. SYSTEM PERFORMANCE

Figure 4 shows an example of a copper bladed wheel to be fabricated using the GTLS machine: (a) the CAD model, (b) the sliced model, (c) the simulation of laser irradiation paths, and (d) the sintered object. After reading the CAD data, the GTLS software starts to display the virtual object on the screen, then slices it into layers at a thickness of 200 μm. This is the thickness of the greentape too. The program further converts the sliced data into NC data that controls laser irradiation in conjunction with the process parameters. Before the NC data is sent to the sintering machine, radiation paths are checked using the simulation software.

Table 2 shows the laser fabrication conditions set by experience. These parameters produce a sintered object with the highest strength while maintaining a good dimensional accuracy. The sintered bladed wheel has a diameter of 30 mm, a height of 6 mm with a dimensional error of 0.5 mm, and a density of 4.46 g/cm^3. Surface roughness also needs to be improved. We plan to use finer powder as well as to add an additional powder feeding mechanism, and to incorporate a feedback control system into the worktable in order to improve the sintering accuracy.

4. SUMMARY

Direct laser fabrication using metallic powder tapes named Greentape Laser Sintering (GTLS) has been introduced. The GTLS machine architecture and the system performance have been disclosed. However the limitation of the dimensional accuracy and surface finish should be further improved in order to commercialize a high-performance automatic laminating system.

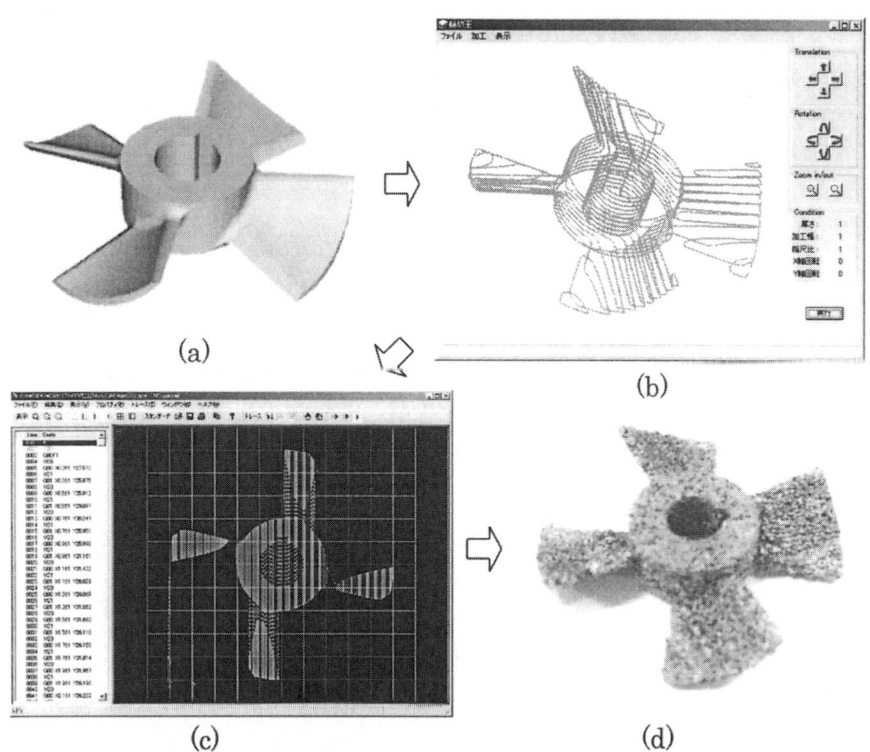

Figure 4. Bladed wheel fabricated using the GTLS system: (a) CAD model, (b) sliced model, (c) laser irradiation paths, and (d) sintered object

Table 2. Laser fabrication conditions

Powder	Cu (OFHC 99.99%, 40 μm diameter)
Tape composition	Cu, PVB, DOP, glycerine, toluene
Tape thickness	200 μm
Tape density	5.3 g/cm^3
Laser	Nd: YAG (1.06 μm wavelength)
Beam power	35 W
Beam diameter	0.25 mm
Scan speed	200 mm/min
Scan spacing	0.2 mm

REFERENCES

Maekawa, K., Kokura, S., Ohshima, I., Yokoyama, Y. Laser micro fabrication using thin powder tapes – process architecture and feasibility investigation. Manufacturing Systems, CIRP 1999; 29:131-135

CONCEPTUAL DATA MODEL FOR ADVANCED RAPID PROTOTYPING

K. G. Kobayashi, M. Fujii

Toyama Prefectural University, Dept. of Mechanical Systems Eng.
Kosugi, Toyama, 939-0398, JAPAN
E-mail: kobayasi@pu-toyama.ac.jp, miyu@cim.pu-toyama.ac.jp

F. B. Prinz

Stanford University, Rapid Prototyping Laboratory
Bldg. 530, Room 226, Stanford, CA 94305, U.S.A.
E-mail: fbp@cdr.stanford.edu

Abstract

Rapid prototyping (RP) has been developed as a key technology for mechanical manufacturing. STL format is used to transfer model data from CAD systems to RP hardwares, however it has insufficiency for stable and high-quality processing. Here, we analyze requirements for advanced RP applications and propose a conceptual data model. It includes representation of model and process data for manufacturing. These basic specifications could be formalized as a new AP in STEP.

Keywords

Rapid prototyping, STL, B-rep solid model, STEP.

1   INTRODUCTION

RP technique has been widely accepted not only for designing products but also for manufacturing a complicated part. Additive manufacturing or laminated production are the useful applications of RP. However, data communication sometimes cannot be done properly between CAD systems and RP hardwares.

The *de facto* standard to transfer 3D model data is STL (Standard Triangulated Language) format, which represents a shape by a set of triangles as a polygonal approximation of the original shape. STL format is quite simple and has the following problems;

- Incompleteness as solid model: A set of triangles is not constrained as 2-manifold to distinguish the material side from the outside.
- Approximation of free-form surfaces: To get a smooth object, the tessellation should be enough fine, which requires plenty of computational power.
- Lack of process support: Intermediate state during manufacturing process is required for verification.

For example, a stereo lithography system accepts STL data as an input and produces a tangible object as an output. A cross-section is calculated by slicing with a plane, and its pattern is painted out by laser to solidify resin. The system's capacity limits the amount of triangles and results the accuracy of object. If ill-formed STL data contains a gap, the sliced result has a problem, shown in Figure 1.

To improve these problems, B-rep solid model with free-form surfaces is applied to RP systems (Shu *et al* 1997). However, this direct method cannot be generalized by most of CAD and RP systems, because there are too many variations in model data formats to maintain their portability completely.

In this paper, we propose a conceptual data model for advanced RP and its applications instead of using STL format. We analyze requirements for advanced RP in section 2, give specifications of the data model in section 3, discuss its possible realization in section 4, and conclude in section 5.

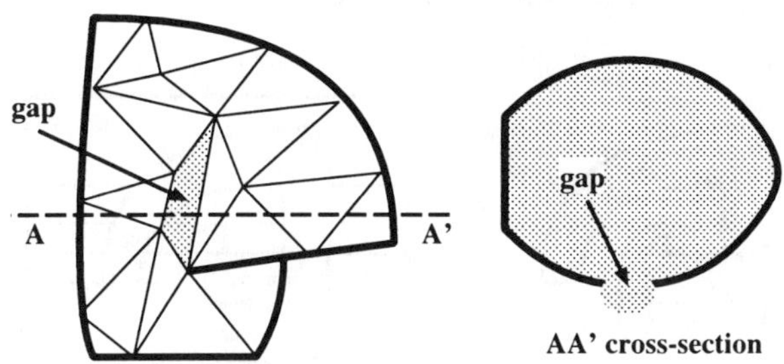

Figure 1   Gap in STL model.

2 ADVANCED RAPID PROTOTYPING

We define "advanced RP" as a technology in future stage to obtain fine/rough objects with scalable quality, to enable highly automated and stable processing, and, to operate seamlessly between different systems. These features are also true for the RP applications, such as layered manufacturing (Prinz *et al* 1997). The incompleteness of model shown in section 1 should be improved for the advanced RP.

Figure 2 shows an example object with heterogeneous materials, which can be built by layered manufacturing. The support materials do not remain in the final object, however their information has to be represented for handling the manufacturing process; such as, material addition, surface cutting, stress release, etc. This complicated sequence should be programmed in automatic manner.

The advanced RP is accomplished with the data model which satisfies the following requirements.

Facetted model: Polygon mesh is generated from free-form B-rep with arbitrary density. This is replaceable to STL.

Composite object: The object is composed of portions with different materials. They are geometrically related as a fixed connection in assembly model (Sugimura 2000).

Manufacturing process: Both of additive and removal machining processes are handled by referring the work piece model.

Portability: Data between CAD and RP systems is exchanged with a unified neutral format, which assures unlimited data usage by different environments.

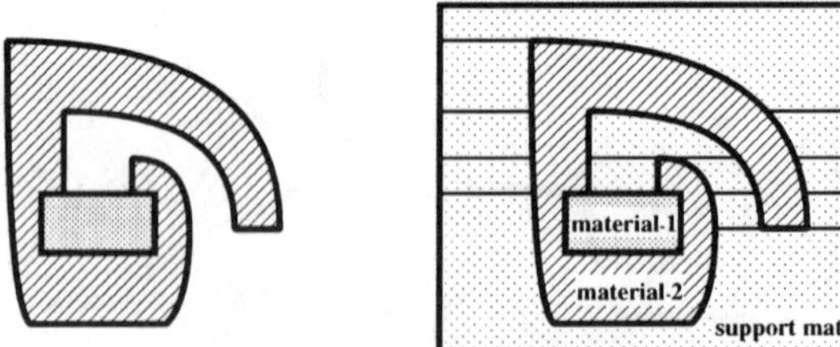

Figure 2   A composite part with heterogeneous materials.

3   SPECIFICATIONS OF DATA MODEL

Our proposal on data model for the advanced RP is shown as the following specifications, which satisfy the requirements in section 2.

B-rep model: A region filled by homogeneous material is represented by a B-rep model with free-form surfaces.

Dual representation: A facetted model is generated, if required. Each topological element (face, edge and vertex) has links to record the relationship between the two B-reps (Figure 3a).

Non-manifold extension: Different material regions are represented by different B-rep models, which are contacted by commonly used topological elements (Figure 3b). It is not necessary to adopt general non-manifold model (Masuda 1993).

Multi-state model: The work piece is represented by multi-state model. Each state of machining process is constructed from additional/removal volumes (Figure 4).

Process information: Material property, machining method and control parameters are represented.

Neutral format: The data format is unified as officially administrated or *de facto* standard. It avoids rupture of direct conversion softwares caused by combinational explosions.

4   POSSIBLE REALIZATION

When we realize the conceptual data model proposed in section 3, many software resources are required. That is, an activity model

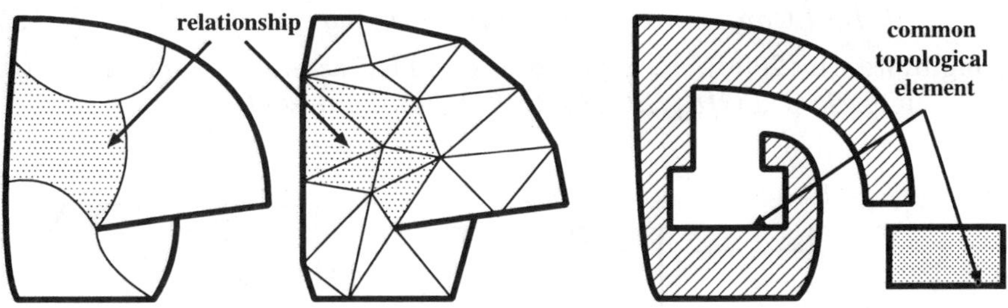

(a) free-form & facetted B-reps.   (b) commonly used elements.
Figure 3   Topological elements in the proposed model.

of RP, formal definitions of data structure, encoding rule to output data, conformance checks, geometric representations, tools to handle model data in each CAD & RP system, and so on.

ISO 10303, or STEP (Standard for the Exchange of Product Model Data), has been developed (Fowler 1995) and useful resources are accumulated. Domain-specific data exchange in STEP is defined by AP (Application Protocol). In ISO, an activity model of RP was suggested (Kishinami 1997), and a new AP development for advanced RP has started. We consider our proposal can be realized as new AP based on the existing resources in STEP.

5 CONCLUSIONS

Requirements for advanced RP in future stage were analyzed, to satisfy them a conceptual data model was proposed, and possible realization as AP in STEP was discussed.

6 REFERENCES

Fowler, J. (1995), STEP for data management, exchange and sharing. *Technology Appraisals Ltd., 1995.*

Kishinami, T. (1997), CAD/RP interface based on STEP. *ISO TC184/SC4/WG12 N165, Oct. 1997.*

Masuda, H. (1993), Topological operators and Boolean operations for complex-based non-manifold geometric models. *Computer Aided Design, Vol.25, No.2, pp.119–129, Feb. 1993.*

Prinz, F.B., Merz, R., Weiss, L.E. (1997), Building parts you could not build before. *Rapid Product Development (N.Ikawa, T.Kishinami, F.Kimura eds.), Chapman & Hall, pp.415–424, 1997.*

Shu, C., Mak, H. (1997), Tessellating STEP B-rep models. *Procs. of the 7th Int'l Conference on Rapid Prototyping, pp.70–78, March 1997.*

Sugimura, N. (2000), JNC proposal of STEP assembly model for products. *ISO TC184/SC4/WG12 N597, June 2000.*

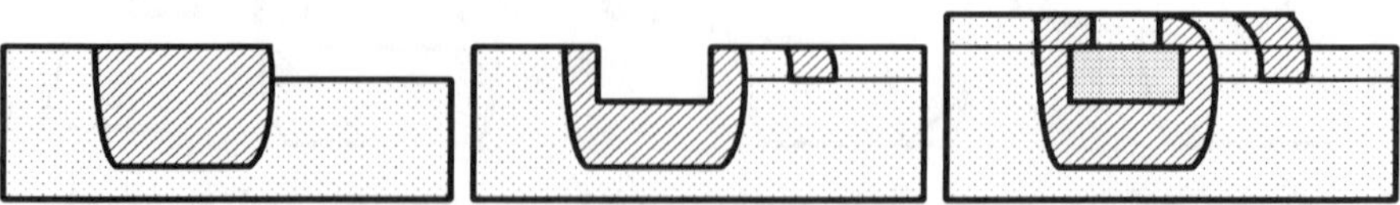

Figure 4   Multi-state model during machining process.

Machine Parts Manufacturing by Sheet Steel Lamination Technology

Jun Shinozuka, Masahiko Yoshino and Toshiyuki Obikawa

Department of Mechanical and Control Engineering
Tokyo Institute of Technology
2-12-1 O-Okayama, Meguro-ku, Tokyo 152-8552, Japan
TEL: +81-3-5734-2811, FAX: +81-3-5734-3982, E-mail: jshinozu@mes.titech.ac.jp

Abstract

An evolutional type of sheet steel lamination machine has been developed as an advanced smart manufacturing system. Since this machine is small and does not require complicated processes, steel machine parts can be manufactured in various places. This paper shows the automatic lamination processes in the developed machine and investigates the characteristics of the mechanical strength of manufactured parts.

Keywords

Manufacturing system, sheet steel lamination, rapid manufacturing system

1. INTRODUCTION

A rapid manufacturing system (RMS) was proposed as an advanced smart manufacturing system [Yoshino (1998), Obikawa (1998)]. The concept of RMS is that the various shapes of steel parts can be manufactured automatically and immediately on demand at various places without special cares.

The sheet steel lamination (SSL) technology using sheet blanks was adopted to realize RMS. The advantages of SSL are as follows. The processes consist of the repetitions of sheet steel lamination and cutting of a laminated sheet like RP processes. Based on the CAD data of the part, the tool path for cutting the sheet is calculated by a personal computer easily and immediately because the tool moves

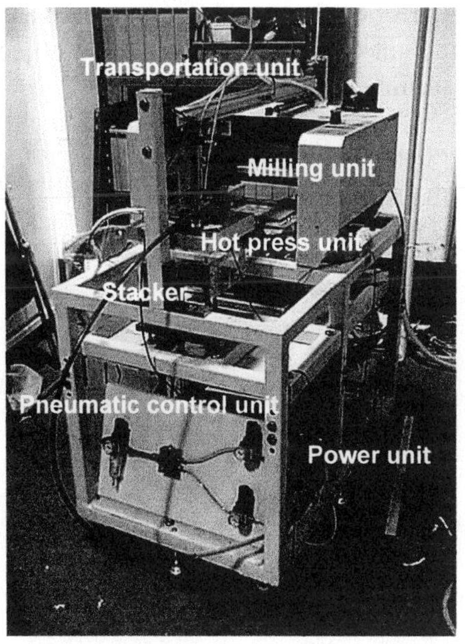

Figure 1 SSL machine developed

only on a plane. Neither a powerful spindle unit nor a machine tool with high rigidity are required because the axial depth of cut equals the thickness of the thin sheet and the diameter of a used end mill is small. No expert is needed to operate it. The strength of the products is expected to be close to that of the bulk steel products. SSL products are the most different from RP products in this point. Therefore, SSL may manufacture the steel parts automatically and immediately at every place in which CAD data of the products can be received through Internet or satellite circuit. From this point of view, SSL may be applied to not only the ubiquitous manufacturing, but also the urgent manufacturing of substitutes for failure parts during the restoration from disaster. A prototype machine of SSL developed was reported [Obikawa (1999), Yoshino (2000)].

In this paper, an evolution type of the SSL machine having automatic lamination process is shown, and the mechanical properties of product by SSL are investigated.

2. EVOLUTIONAL TYPE OF THE SSL MACHINE

Figure 1 shows an evolutional type of sheet steel laminating machine. This is composed of eight units: power unit, pneumatic control unit, stacker unit of cut sheets of steel, transportation unit, hot press unit, milling machine unit, air compressor and PC. The air compressor and PC are not shown in Fig. 1. The PC automatically controls the transportation of sheets of steel, the operations of hot press and end milling, and the measurement of the height of a partially completed object. It also calculates the tool path of end milling of each sheet.

The sheet steel 0.2mm thick coated with 0.04mm thick U alloy (50%Bi+27%Pb+13%Sn+10%Cd, melting point 70℃) on the both sides of the sheet was used for SSL.

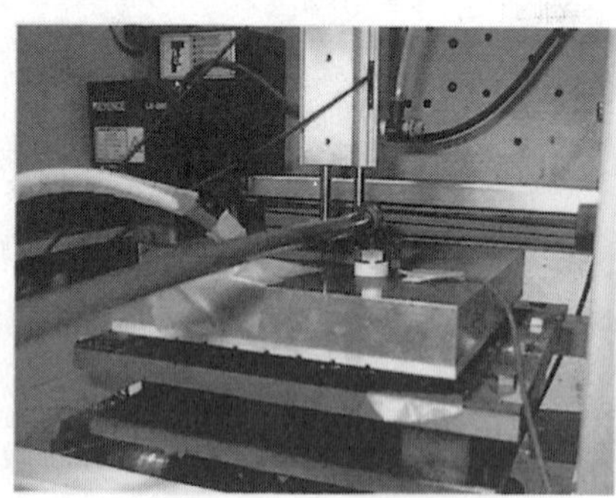

(a) hot press process   (b) end milling process

Figure 2 Processes of the sheet steel lamination

In a process of lamination, at first, a sheet of steel transported from the stacker was placed on a partially completed object. The sheet subject to a specified pressure by the hot press unit was heated rapidly to a temperature just above the melting point of the U alloy with an induction heater which was built in the hot press unit as shown in Fig. 2(a). Then it was cooled by flowing compressed air on its surface to weld sheets with U alloy completely. Next, the height of the object was measured with a laser displacement meter. This height determined the shape of the cross section of a designed part to be cut out from the just welded sheet by referring STL data. Finally end milling was conducted to cut out the shape along a tool path determined by considering the diameter of square end mill as shown in Fig. 2(b). The above lamination process was repeated until the object was completed.

The conditions of hot press in the lamination process significantly affect the welding strength. Hence the influences of the heating time and hot press pressure on the thickness of the laminating object were investigated. Five

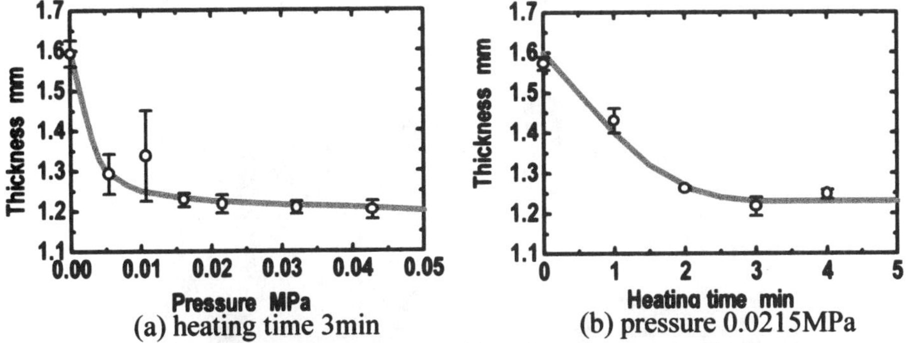

(a) heating time 3min   (b) pressure 0.0215MPa

Figure 3 Influence of the pressure and heating time on the thickness
of the laminated object

Figure 4 Synchronous pulleys manufactured by SSL method
Size; ϕ 55.0-67.0mm $\times$ 19.0mm, laminating time; about 50 hours

sheet steels laminated at a time was used for the investigation.

Figure 3 shows the results. The pressure was changed at a fixed heating time in Fig. 3(a), while the heating time was changed at a fixed pressure in Fig. 3(b). In Fig. 3(a), it is seen that the thickness approached to a value around 1.20mm when the press stress exceeded 0.02 MPa. In Fig. 3(b), the thickness became almost constant when the heating time exceeded 2 minutes. Form these results, hot press pressure 0.0215MPa and heating time 3 minutes were selected in the laminating process. Under these conditions the thickness of U alloy between layers is 0.045mm on the average.

Figure 4 shows synchronous pulleys manufactured by SSL. Two pulleys in this side are SSL products. Two pulleys in the back are factory products, which consist of three elements. In SSL, special care is not required when machining a keyway and teeth in a synchronous pulley. This is one of advantages in the manufacturing of machine parts by SSL.

3. MECHANICAL PROPERTIES OF SSL PRODUCTS

SSL products have an anisotropy in strength depending on the laminating direction. Figure 5 shows two types of damage of SSL products under

(a) parallel (b) orthogonal
Figure 5 Modes of damage

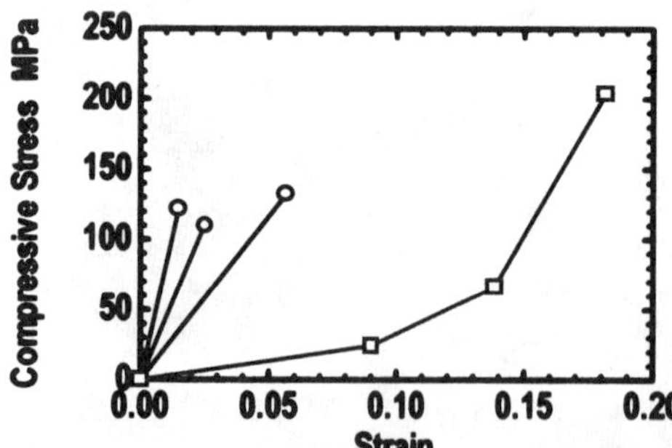

Figure 6 The influence of the lamination direction on the strength
○ lamination direction is parallel to the loading direction.
□ lamination direction is orthogonal to the loading direction.

compression. The test specimens were steel cubes manufactured by SSL. When the loading direction was parallel to sheet steels (mode A), the specimen was delaminated followed by buckling of sheet steels. When the loading direction is normal to sheet steels (mode B), some amount of U alloy were squeezed out. This leaded to the change in the size of the specimen.

Figure 6 shows the influence of the laminating direction upon the compressive strength of the specimen. For mode B compression, the laminated specimen had almost the same strength of about 200MPa as the sheet steel. On the other hand, for mode-A compression, the delaminating of sheet steels resulted in a decrease in strength by 80MPa. The strength of products may depend on the strength of U alloy directly. From the result, the laminating direction can affect strongly on the mechanical properties of the laminated products. Therefore, in the application of SSL to machine parts, it is necessary to determine the lamination direction based on stress analysis by finite element method.

4. CONCLUDING REMARKS

As described above, the manufacturing system developed adapted only very simple processes. Hence, it may be possible to manufacture machine parts in any place, for instance, in places far away from design offices, on shipboards or in space stations, to which it may be very difficult to supply parts immediately.

Acknowledgment

The authors would like to thank Mr. Koji Furusawa, Mr. Yuki Hanawa and Mr. Goki Yoshida, students in Tokyo Institute of Technology, for their cooperation in this study. They also would like to express their appreciation to NKK Corporation for supplying U-alloy coated sheet steel and NK-EXA Corporation for providing us a CAD system.

References

[1] Masahiko Yoshino, Toshiyuki Obikawa, Jun Shinozuka. Rapid Manufacturing System by Sheet Steel Laminating System. Transactions of the Japan Society of Mechanical Engineers 2000; 66 (C) (642): 313-318

[2] T. Obikawa, M. Yoshino, J. Shinozuka. Sheet steel lamination for rapid manufacturing. Journal of Materials Processing Technology 1999; 89-90: 171-176

[3] Toshiyuki Obikawa, et al. Rapid Manufacturing System by Sheet Steel Lamination. Proceedings of 14th International Conference on Computer Aided Production Engineering; 1998 Sept. 8-10: 265-270

[4] Masahiko Yoshino, et al. Mechanical Properties of Laminated Sheet Steel Products by Rapid Manufacturing System. Proceedings of 14th International Conference on Computer Aided Production Engineering; 1998 Sept. 8-10: 271-276

TEXTURED SURFACE PRODUCED BY ANISOTROPIC ETCHING OF SILICON AND ITS FRICTIONAL PROPERTIES

Nobuyuki MORONUKI, Daisuke NISHI, and Kenji UCHIYAMA

Tokyo Metropolitan University, Graduate School of Mechanical Engineering

1-1 Minami-ohsawa, Hachioji-shi, TOKYO, JAPAN, 192-0397

E-mail: moronuki-nobuyuki@c.metro-u.ac.jp

Abstract

"Textured" surface, defined here as a surface covered with periodic regular pattern, was produced by applying anisotropic etching of silicon. By choosing crystal orientation of the substrate, asymmetric sectional profile can be obtained, though the profile is limited to saw tooth only. Experimental results show that the increase of the friction force at the low contact pressure, which is generally known in the micro-mechanism research, is suppressed with texturing. In addition, when the textured silicon surface is mated with soft material, directionality of the friction, up to 2, was observed.

Keywords

Textured surface, Regularity, Anisotropic etching, Silicon, Friction

1. INTRODUCTION

"Textured surface" is defined as the surface that has a periodic regular shape on the surface and performs various functions such as optical, tribological and so on. Recently, textured surfaces are adopted in various applications (Evans et al., 1999). Anisotropic etching seems suitable for texturing since it produces regular shapes easily with simple facilities. The principle is that the final shapes consist of {111} crystal planes because of the difference of etching rates. On the other hand, in the field of micro-mechanism research, the reduction of friction is crucial because the friction at the bearing often constrains the motion. The surface force becomes dominant rather than the body force in the microscopic world. Thus, this paper aims to examine the applicability of the anisotropic etching to texturing, and evaluate effect of the texture on the frictional properties.

2. TEXTURING BY ANISOTROPIC ETCHING

Figure 1 shows the principle of the texturing process. When a periodic "line and space" mask pattern is fabricated on {nn1} (n: natural number) substrate aligned with [110] direction and anisotropically etched, V-shapes that consist of {111} crystal planes are obtained, because the etching rate against this plane is the slowest. The opening angle of the V-shapes is determined as 109.5 degrees from crystal structure. By choosing the substrate orientation, symmetric or asymmetric shapes can be obtained.

The substrate orientation and the spacing of the mask determine the V-angle and its depth. Thus the design parameter is only these two. Two photo-masks were prepared in the experiments: 4μm-line/8μm-space (pitch 12μm) and 3μm-line/3μm-space pattern (pitch 6μm) over 60x30m area. The patterns were transferred to SiO_2 mask (1μm thick) on φ100mm wafers with photolithography and then etched in KOH 35wt% solution (333K). The etching time can be estimated from the depth and the etching rate: about 15min in this case.

Figure 2 shows the SEM photo of the obtained textures. The (221) substrate produced slightly asymmetric saw tooth texture (left). The (111) four-degree-off substrate [equivalent to (1.16 1.16 1) plane] produced strongly asymmetric cross-sectional profile (right). The roughness of the etched surface as good as 20nm(Ra).

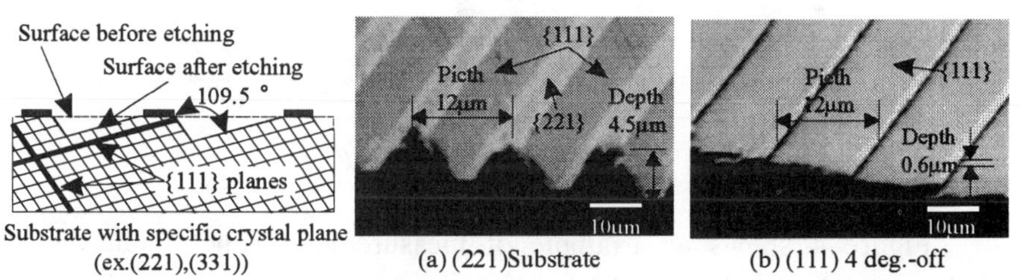

Fig.1 Principle   *Fig.2 SEM photo of the texture*

3. FRICTIONAL PROPERTIES
3.1 Experimental Setup

Figure 3 shows the setup for the friction measurement. Three high-resolution (0.5mN) force sensors construct the tripod structure and based on the geometric relation, the force components in the rectangular coordinates were calculated. The resolutions along each axis are also shown in the figure. The flexure hinges on both side of the sensor reduce the unnecessary moment that acts between mating surfaces to assure uniform contact over wide area. Beneath the measurement system, XYθ-stage is located to give the relative motion between the mating surfaces. The XY-motions are driven by linear motors with 0.1μm resolution feedback, and a PC acquires friction force during reciprocation repeatedly. The θ-motion is applied manually to adjust the orientation of the texture. The contact pressure was controlled by the weight set on the force measurement unit. Its position was controlled with a motor-driven linear stage and the experimental condition can be changed dynamically without human operation.

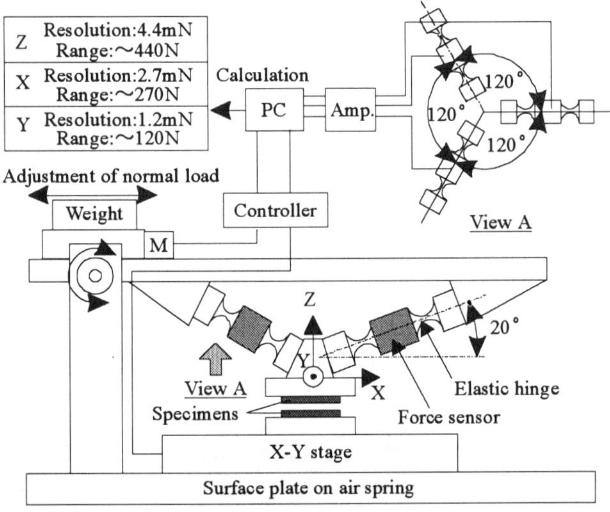

Fig.3 Experimental setup

Figure 4 shows an example of measurement, which indicates the relation between position and friction force during two reciprocations. The size of the texture is constant as 10x10mm through this study. The texture is similar to the one shown in Fig.2 (b), but 3μm-line/3μm-space pattern (pitch 6μm) was adopted here. The sliding speed was 30μm/s. It is found from the figure that the fluctuation between steady sliding is less than 11mN and the repeatability is good. The inclination at both turning points indicates the elastic stiffness of the measurement system. It is also found that the closed hysteresis loop has an offset in vertical direction, 42mN in this case, which is considered as the directionality of the friction.

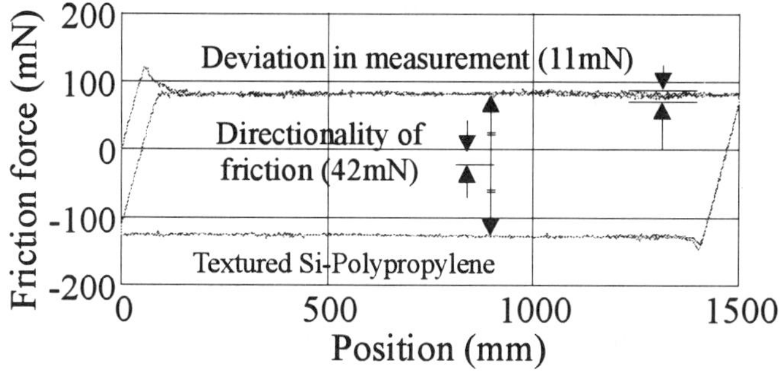

Fig.4 Frictional directionality (example)

3.2 Experimental Results And Discussion

Figure 5 shows the effect of contact pressure on the friction in case the textured silicon chips (Fig.2 (b)) were in contact. When the contact pressure is low, the coefficient of friction becomes large. This phenomenon is well known in the research field of micro-mechanism (Moronuki, 1999). The increase of friction is considered that the attractive force between the surfaces, such as meniscus, electrostatic force and so on, becomes dominant rather than the normal force externally applied by the mass.

The figure also shows the effect of the orientation of the texture. When the texture is aligned in same direction, the effect of the contact pressure is large. On the other hand when the texture is aligned in right angle, the effect of the contact pressure becomes small. The difference between these conditions is the real contact area, thus situation is well explained by the attractive force mentioned above. This result can be applicable to the reduction of friction in the micro-mechanisms.

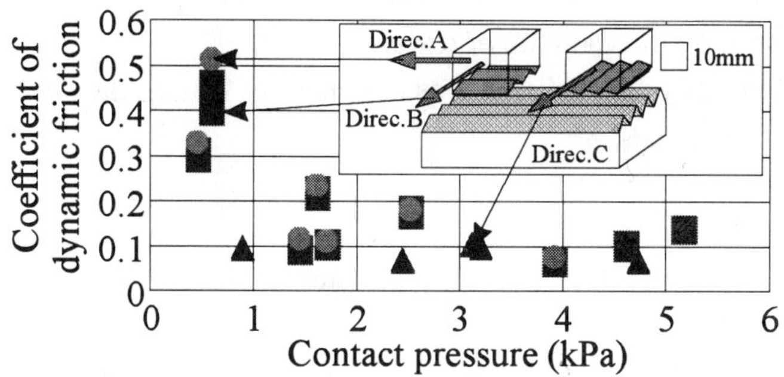

Fig.5 Effect of contact pressure and texture orientation

Figure 6 shows the effect of the combination of materials. In this experiment, textured silicon chip (same with Fig.4) and planer sample were made into contact and the directionality of the friction was evaluated. The horizontal axis denotes the contact pressure and vertical axis denotes the directionality (ratio of the friction in forward and reverse direction) , where the orientation of the texture is perpendicular to the motion. In case of Si-Si combination, no directionality was observed. However, in case of Si-polypropylene and Si-paper, directionality up to 2 was observed. It is considered that this directionality is caused by the elastic deformation of the softer material.

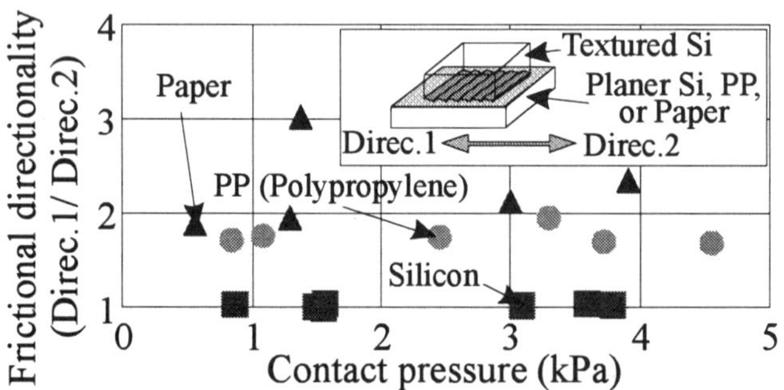

Fig.6 Material combination and friction directionality

4. CONCLUSIONS

Anisotropic etching of silicon is applied to the production of the texture and its frictional properties were examined. The results are summarized as follows:
- The design of the texture was discussed and demonstrated.
- The increase of friction force when the contact pressure is low can be suppressed by the adoption of texture.
- Directionality of the friction was obtained by mating the textured silicon with soft materials.

REFERENCES

Evans C. J. and Bryan J. B., "Structured", "Textured" or "Engineered" Surfaces, Annals of CIRP, 1999; 48, 2.

Moronuki N., Linear Motion Microsystem Fabricated on a Silicon Wafer, Int. J. Japan Soc. Prec. Eng., 1999; 33, 2.

A MECHANICAL VIBRATION ASSISTED PLASMA ETCHER FOR ETCH RATE IMPROVEMENT

HATSUZAWA Takeshi, HIROSAWA Minoru, HAYASE Masanori and OGUCHI Toshiaki

Precision and Intelligence Laboratory, Tokyo Institute of Technology
4259, Nagatsuta-cho, Yokohama, 226-8503   JAPAN
Tel/Fax : +81-45-924-5037, e-mail : hat@pi.titech.ac.jp

Abstract

A novel plasma etcher with a substrate vibration mechanism has been developed for etch rate improvement.   The etcher is based on an RIE(Reactive Ion Etching) system, in addition, it has an ultrasonic vibrator under the anode electrode.   Two vibration frequencies of 28kHz and 40kHz can be applied to the substrate by changing vibration units, and they have vibration amplitudes of 7 to 8 microns.. Some results of silicon etching are shown by SEM micrographs compared to those without vibration. The system has confirmed to have improvements in etch rate by 50% and anisotropy.

Keywords

Vibration assisted plasma etching, etch rate improvement, anisotropy

1.INTRODUCTION

Plasma etchers are one of the key equipments for LSI fabrication and micromachining.   So many improvements have been performed for etching selectivity, etch rate and damage reduction by using plasma generation technology such as RIE(Reactive Ion Etching), ECR(Electron Cyclotron Resonance), ICP(Inductive coupled plasma), pulsed-plasma etc.   However, no mechanical approach has applied for the improvement of plasma etchers. Against this backdrop, a mechanical vibration assisted plasma etcher has developed to improve etch rate and anisotropy.   The system has based on a RIE etcher together with a vibration stage.   The vertical vibration in the plasma causes a relative speed variation to plasma particles such as ions, neutral spices.   It increases the energy of collision and the angle of incident plasma particles.   The former effects improves the etch rate , and the latter enhances the anisotropic movement of the spices resulting in the etching anisotropy.

In this paper, the construction of the etcher, some results of silicon etching and considerations are presented.

2.PRINCIPLE OF VIBRATION ASSISTED   ETCHING

Plasma etching mechanisms are attributed to various types of reaction such as chemical etching and physical spattering shown in figure 1. Those mechanisms are mainly improved by plasma generation techniques. On the contrary, the principle of the vibration assisted plasma etching is interpreted by a model shown in figure 2, which is purely based on a mechanical vibration of the substrate against the bulk plasma. The species in plasma are moving at a velocity of V with a mean free path.   When the substrate is vibrated to plasma, the vibrating velocity of the substrate can be added to the vertical velocity component of the plasma spices resulting in the anisotropic movement.   This is particularly effective to neutral species, which are not accelerated by the electric field.   The energy amount Ez increased by the vibration is as follows,

$$\delta E_z = \frac{1}{T}\int_0^T \frac{1}{2} M_s u^2(t)\, dt = \frac{M_s U^2}{4} = M_s \pi^2 f^2 A^2 , \tag{1}$$

where T is a period of the vibration, A is an amplitude, M_s is the mass of a plasma species.   Some of the etching promotion effects by the vibration can be obtained as shown below,
1)   an enhanced spattering,
2)   an improved anisotropy, and
3)   a gas scrambling.

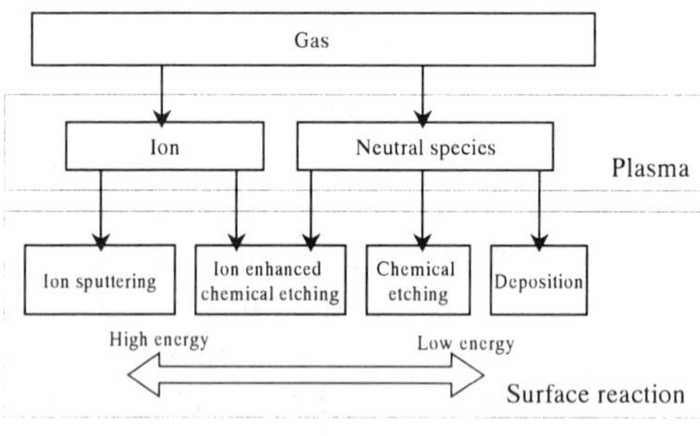

Figure 1. Models of etching.

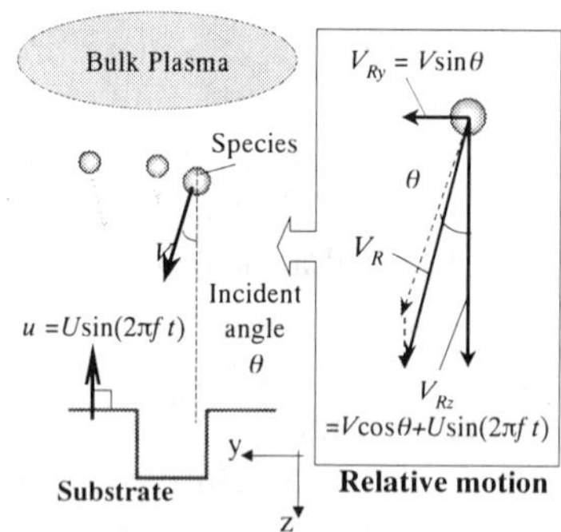

Figure 2. Effects of vibration.

3.CONSTRUCTION OF ETCHER

Figure 3 illustrates the schematic construction of the system.   It is based on a conventional RIE etcher for 4" wafers, however, the lower electrode has a vibration mechanism driven by a volted-Langevin exciter. Two exciter have chosen to switch the resonant frequency from 28kHz to 40kHz alternatively. To enhance amplitude, a step horn shown in figure 4 has been used for both exciters. Only central part of 18mm can be used as the vibration stage, while the outer part of the stage works as a non-vibrating electrode for plasma generation, which is shown in figure 5.

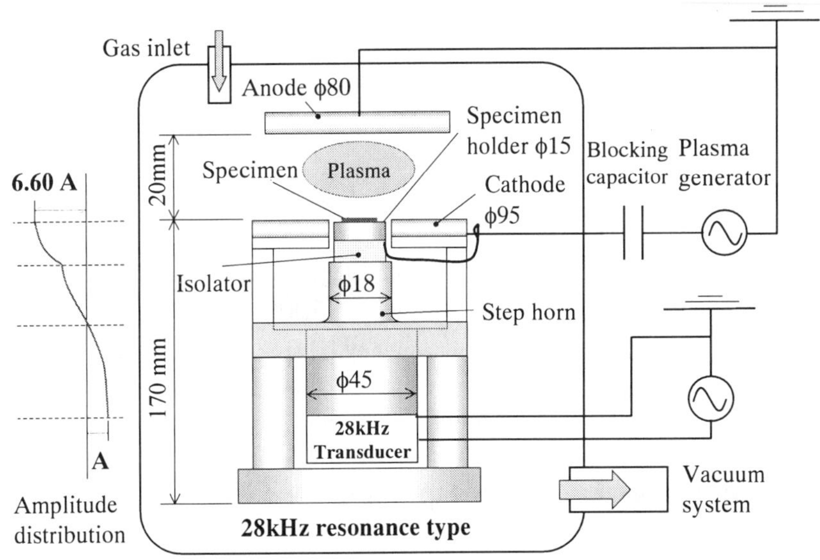

Figure 3. Schematic construction of the system.

Figure 4. Step horn.

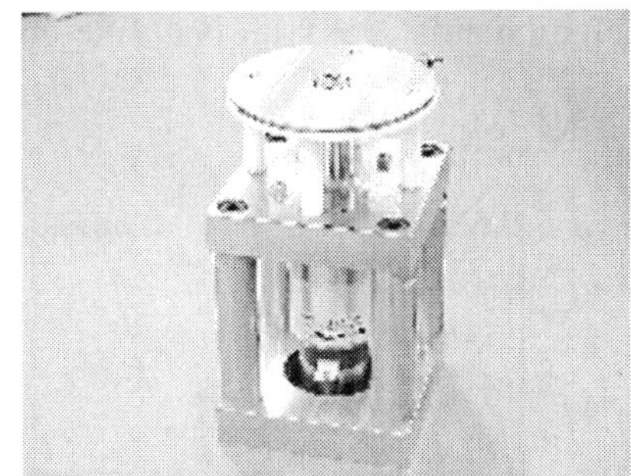

Figure 5. Lower electrode.

The frequency response of the vibration stage is shown in figure 6. Originally, the exciters have nominal resonant frequencies of 28kHz and 40kHz, however, additional mass of the horns change the frequencies slightly to 28.27kHz and 39.62kHz, respectively. Due to the horns, amplitudes are improved to 8μ m and 6μ m, which are 6.6 times from their original. Vibration energy of the horn, which is proportional to f^2A^2, is improved to 33 times for 28kHz horn, and 26 times for 40kHz.

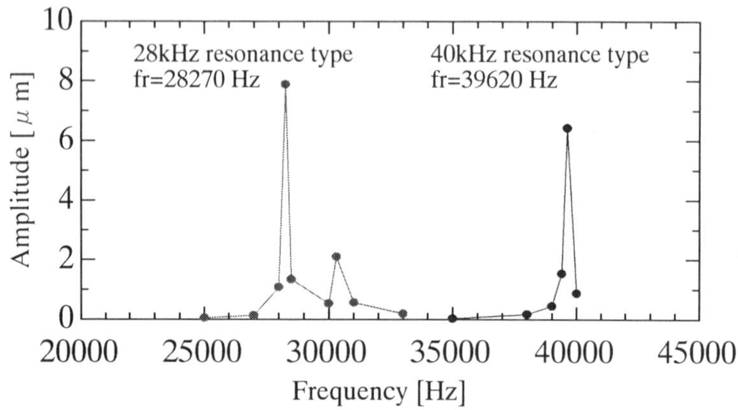

Figure 6. Frequency response of the horns.

4.EXPERIMENTAL RESULTS

An n-type (100) oriented silicon wafer with a 1μ m SiO_2 film is used for etching experiments. A simple line and space layout is used for the silicon etching mask by an EB lithography. After the mask etching process by buffered hydrogen fluoride, an SF6 gas is used for the silicon etching. The etching anisotropy *An* is defined by the geometry[1] shown in figure 7,

$$An = 1 - (S/d). \qquad (2)$$

Some of etching results are shown in figure 8. etch rate is improved by 16% at a frequency of 28kHz and an amplitude of 8.9 μ m, and 48% at 40kHz and 6 μ m. The dependence of etch rate and anisotropy on the specific energy *Ez/Ms* is

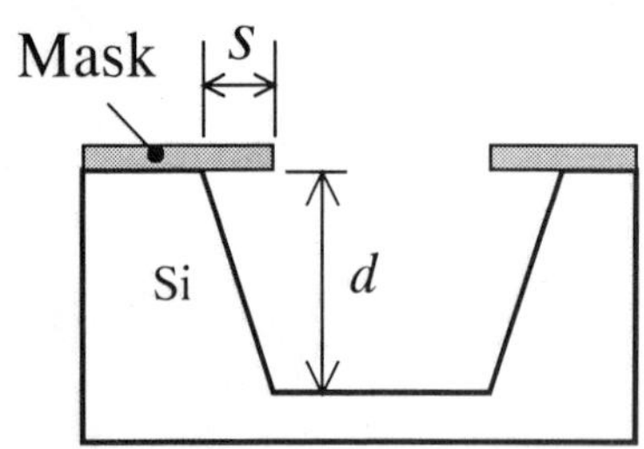

Figure 7. Definition of anisotropy.

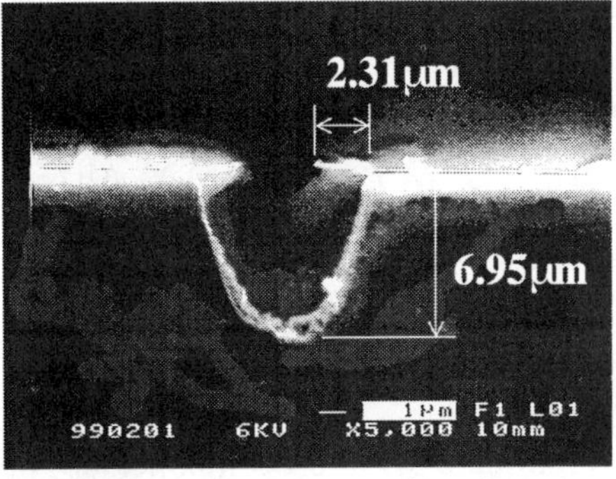

(a)15Pa and without vibration

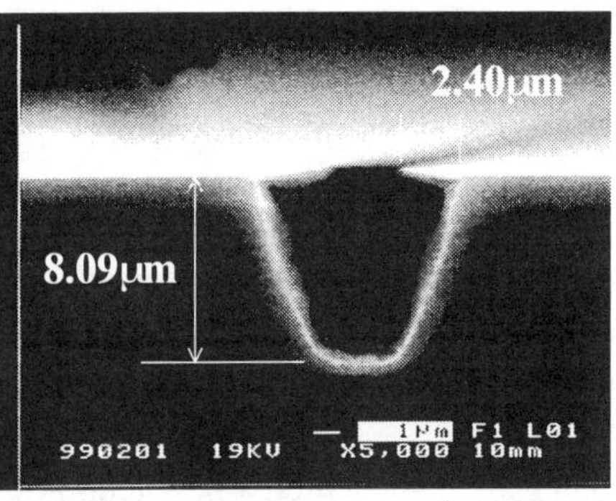

(b)15Pa with vibration（28kHz, amplitude 8.9μm）

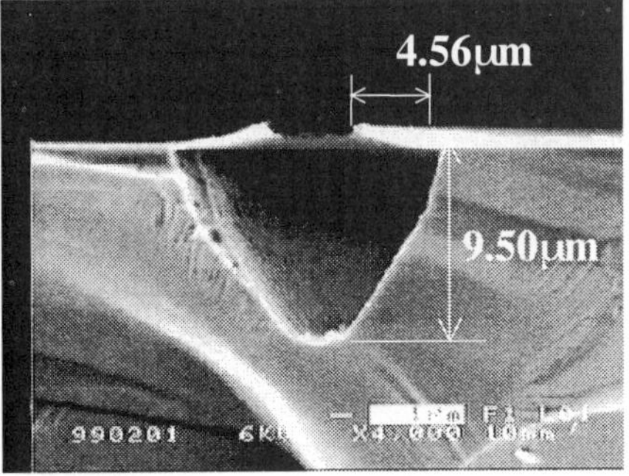

(c)40Pa and without vibration

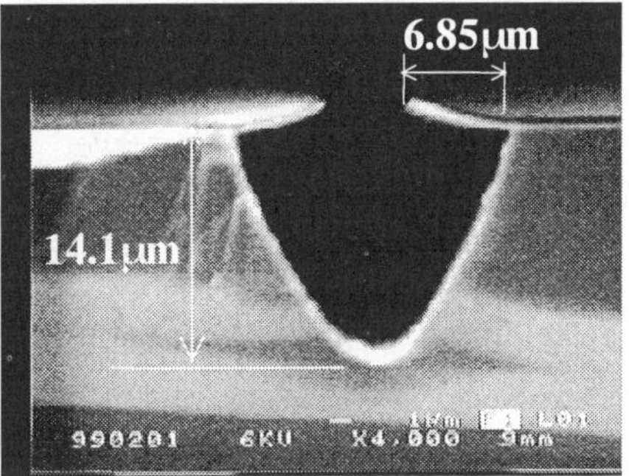

(b)40Pa with vibration（40kHz, amplitude 6.0μm）

Figure 8. Examples of etching with and without vibration.

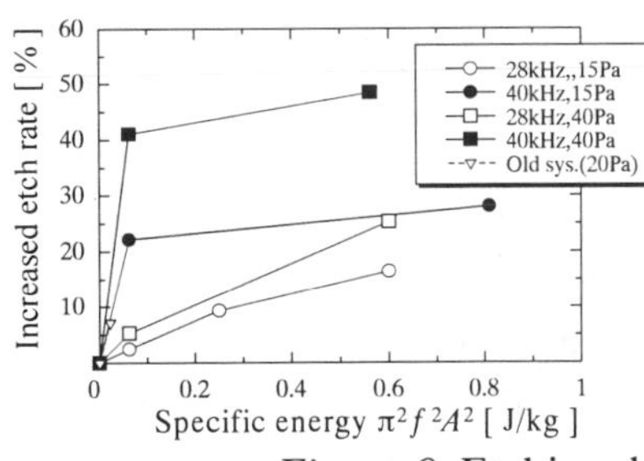

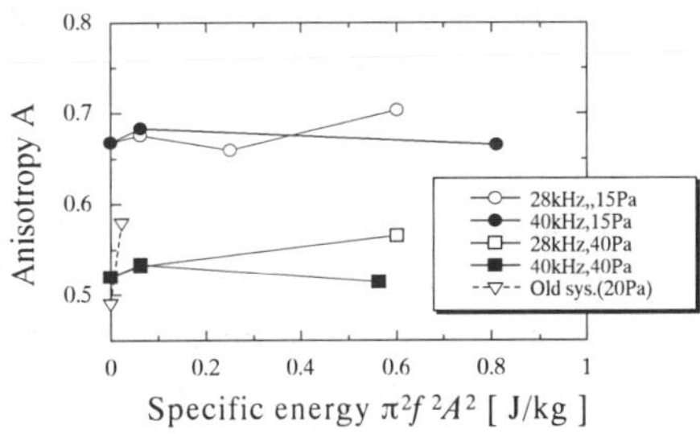

Figure 9. Etching dependence on specific energy.

shown in figure 9. Etch rate is improved as specific energy increase, however, anisotropy is dominated by gas pressure rather than vibration.

Reference

1. Rob Legtenberg et.al.: Anisotropic reactive ion etching of silicon using SF6/O2/CHF3 gas mixtures, J Electronchem. Soc.,142-6,2020(1995)

MICROMOLDING OF THREE-DIMENSIONAL COMPONENTS

W. N. P. Hung*, Y. Ngothai**, S. Yuan*, C. W. Lee*, and M.Y. Ali*

* *Precision Engineering & Nanotechnology Center, School of Mechanical & Production Engineering, Nanyang Technological University, Singapore*
** *Centre of Life Sciences & Chemical Technology, Ngee Ann Polytechnic, Singapore*

Abstract

This paper presented a new technique to fabricate 3D microcomponents. The simulation and molding of microgears of 70-1000μm diameters and 1:1 aspect ratio were completed. Micro-EDM was combined with focused ion beam sputtering to produce a microcavity for subsequent molding process. The simulation pinpointed locations of trapped gas, predicted filling time, stress distribution, and uniformity of a microgear. Subtle differences in simulation results were observed depending on the finite elements used. Limited success was found for predicting weld lines and surface finish. The molded microcomponents producing by this technique possessed features that were compatible with those fabricated by the LIGA technique, i.e., surface finish ~10 nm, accuracy < 1 μm, and aspect ratio >10.

Keywords

Micromolding, Simulation, Focused ion beam, 3D micro component.

1. INTRODUCTION

Fabrication of three-dimensional (3D) microcomponents is still in the developing stage. A promising method is to leverage from the matured injection molding process by filling a microcavity with plastics or metals. Although the LIGA technique [Rupercht et al, 1995] or lithography-based methods [Haisma et al, 1996] have been utilized to produce microcavities for molding, they are expensive and inflexible, not only because of the expensive masks, but also the strict control of the complex processes.

An alternative to produce 3D microcomponents and simulate the process is sought. The objectives of this paper are to (i) introduce a new technique to produce microcavities by combining micro-electrical discharged machining (EDM) and focused ion beam (FIB) sputtering, (ii) present results of computer simulation for the micromolding process, and (iii) compare the simulation results with data from actual molded microcomponents.

2. SIMULATION AND EXPERIMENT

High- and low-density polyethylene (HDPE, LDPE) and general-purposed polystyrene (GPPS) polymers were selected. The software MOLDFLOW (Release 1.1, Build 00055) was used for the simulation. Involute gears with 8 teeth, 25° pressure angle, 21µm core diameter, and 70-1000µm outside diameters were simulated and molded. The microgear was integrated onto a block for ease of handling. The mold temperature varied from 40°C to that above the "no-flow temperature (T_{nf})," suggested from the material data bank. Both 2.5D and 3D finite element modules were utilized for modeling. All the simulations assumed molding at atmospheric conditions.

Shear viscosity at different temperatures was measured with a controlled-stress rheometer (Carri-Med CSL500). A commercial plastic injection system was modified for micromolding. A microcavity on Ni-Be insert was formed by rough-eroding on a Micro-EDM (Panasonics MG-ED72W), and/or fine-sputtering with the focused Ga^+ ion beam (Micrion 9500EX). The FIB parameters were optimized for acceptable surface finish and sputtering rate. While scanning the sputtered area, the beam-shape produced a suitable draft angle for subsequent molding (Fig. 1). The mold temperature was controlled with a heating system using 4 cartridge heaters and 2 thermocouples for feedback. After molding, the packing pressure was maintained until the mold temperature dropped at least 20°C below T_{nf}. The mold was then water-cooled to room temperature. The molded microsamples were removed and observed in a scanning electron microscope to compare with the simulation results.

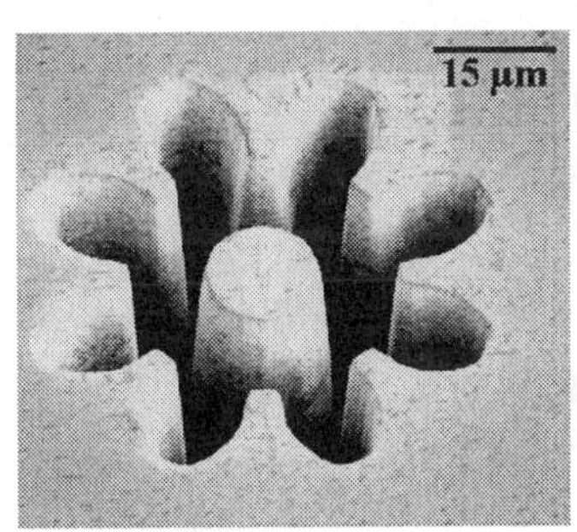

Fig. 1: A typical micro-cavity. Notice the inherent draft angle of the mold wall after FIB sputtering

Fig. 2: Shear viscosity of gerenal purposed polystyrene, low- and high-density polyethylene at different temperatures and shear rates.

3. RESULTS AND DISCUSSION

Figure 2 plots the steady state shear viscosity of the polymers versus shear rate at different temperatures. Notice that there was a strong shear-rate dependence of the viscosity for LDPE (120°C) and GPPS (170°C). The viscosity of the LDPE was high at temperature just above its glass transition temperatre T_g, but significantly reduced above the no-flow temperture T_{nf}. The T_{nf} were listed in the software data bank as 114, 124, and 130°C for LDPE, HDPE, and GPPS respectively. The T_{nf} was later shown experimently (by differential scanning calorimetry) to be the average temperature between the melting point T_m and the crystallized temperature T_c. The GPPS had higher T_{nf}, and required a higher molding temperature and pressure to flow through a 10μm dimameter opening (Fig. 3). This behavior was explained by comparing the chemical structures of the two mers; the bulky benzene ring in the PS mer would impede flowing of the material in restricted areas as compared to a more linear and compact structure in the PE mer.

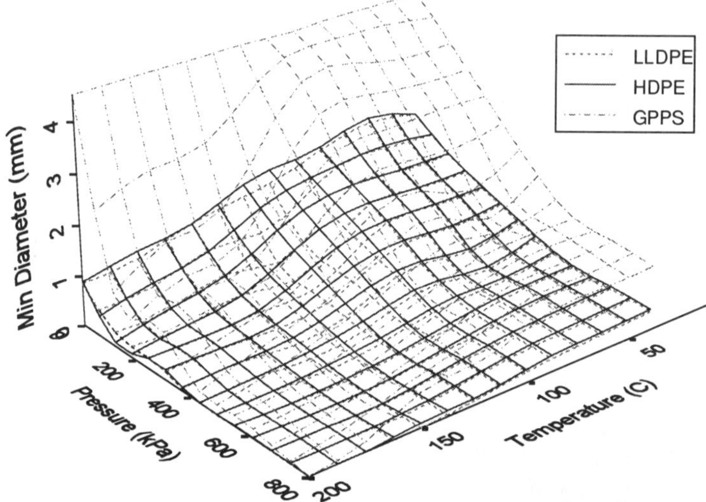

Fig. 3: Simulation results for minimal diameters in micromolding. Flow of polystyrene and polyethylene through a 3° taper cone with the smallest diameter of 10 μm. Mold temperature 40°C, 2.5D module.

Although modeling in the 2.5D (fusion) module was much simpler and used satisfactorily in macromolding, it had certain limitation in micromolding simulation [Weber et al, 1996]. The coarser elements in 2.5D module tend to distort the minute features of a component that has overall size less than 500 μm. In contrast, the finer elements in 3D module can preserve the geometrical features properly and were used mostly in this study. There was size limitation of a microcavity for successfully simulated by the MOLDFLOW software. The minimal gear diameter was found by trial and error to be 220μm unless it was integrated into a larger block. Similar size limitation was also reported when simulating with the C-MOLD software package. The minimum size for a cylinder was found to be 545μm in diameter and 120μm deep when integrated into a larger base [Ali et al, 2000].

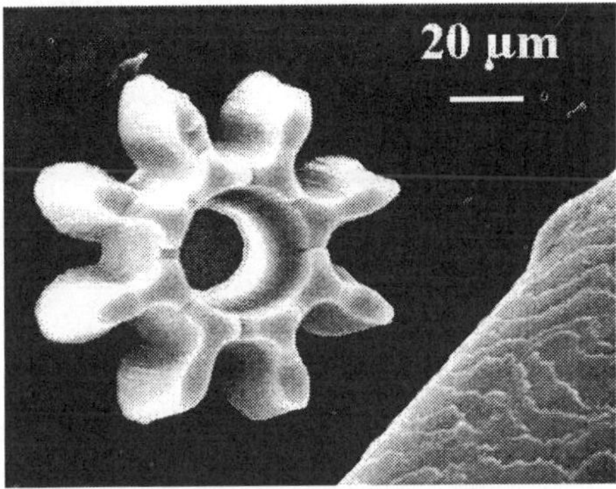

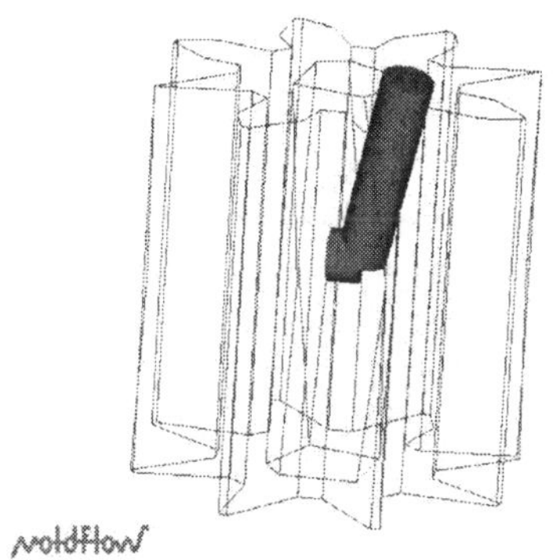

Fig. 4: A molded φ100μm microgear next to a human hair. HDPE-6801-YN, 210°C melt, 110°C mold, 15s injection and packing time, 691 kPa pressure.

Fig. 5: Simulated weld lines, away from the gate. 2.5D module, φ150μm diameter, 1:1 aspect ratio, HDPE-6750, 210°C melt, 124°C mold, 5s injection, 691 kPa pressure.

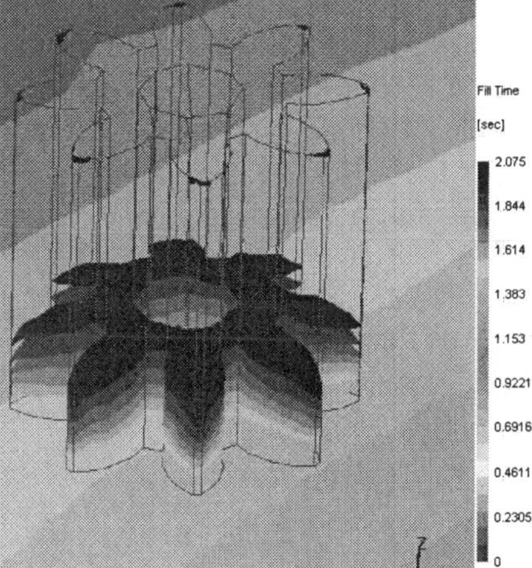

Fig. 6: Uneven pressure due to end effect. 3D module, φ500μm gear, 1:1 aspect ratio. HDPE-6801-YN, 210°C melt, 40°C mold, 15s injection and packing time, 691 kPa pressure.

Fig. 7: Delay in cavity filling. 3D module, φ150μm gear, 1:1 aspect ratio. HDPE-6750, 150°C mold, 2s injection, 691 kPa pressure, 15s packing.

Figure 4 shows a molded microgear next to a human hair. Prediction of weld lines was not completely successful since multiple weld lines are seen at the top half of the micro gear and between teeth (Fig. 4), but only one weld line is shown in the simulation (Fig. 5). Perhaps the incomplete weld line prediction could be improved when this option is available in 3D module. When a gear cavity was positioned near the gate, about one gear diameter or less, then non-uniformity of pressure, viscosity, flow velocity... was predicted by simulation (Fig. 6). Therefore, a microcavity in actual molding was positioned far away from the gate to avoid such end effect. The filling sequence was corrected simulated in 3D module (Fig. 7), i.e., the melt front traveled pass the microgear cavity to fill up the base before entering into the microcavity. Gas trap was predicted to be at the gear end when molding at atmospheric conditions. Parallel study of FIB sputtering showed it capability of producing small nanofeatures on Ni-Be at aspect ratio >10 and surface finish ~10 nm on nickel beryllium [Hung et al, 2000]. Using FIB to fabricate microcavities without a mask, therefore, was an advantage of this technique over those lithography-related processes.

4. CONCLUSIONS

Fabrication of microgears were completed. This study showed:

1. MOLDFLOW and C-MOLD softwares did not allow a feature smaller than 220μm, unless it is integrated into a larger block.
2. Subtle differences were found between simulated and experimental results due to discrepancies in material properties and experimental conditions. However, the simulation successfully predicted the filling of a microcavity, locations of trapped air pockets, and some weld lines.
3. Micro-EDM and FIB effectively replaced lithographic process for producing microcavities in molding.

REFERENCES

1. Ali, M.Y, Hung, N.P., and Yuan.S., Simulation of Micro-Injection Moulding. *Proceedings Int. Conf. on Precision Engineering*, Mar 2000, Singapore, pp. 535-540.
2. Haisma, J., Verheijen, M., Van den Heuvel, K., and Van den Berg, J., Mold-Assisted Nanolithography: A Process for Reliable Pattern Replication, *J. Vac. Sci. Technol. B*, Vol. 14(6), pp. 4124-4128, Nov/Dec 1996.
3. Hung, N.P. and Ali, M.Y., Simulation of Micro-Injection Moulding. *Proceedings Int. Conf. on Precision Engineering*, Mar 2000, Singapore, pp. 517-522.
4. Ruprecht, R., Bacher, W., Haußelt, J.H., Piotter, V., Injection Molding of LIGA and LIGA-Similar Microstructures Using Filled and Unfilled Thermoplastics. *SPIE Proceedings*, Vol. 2639, pp. 146-156, 1995.
5. Weber, L., Ehrfeld, W., Freimuth, H., Lacher, M., Lehr, H., Pech, B., Micromolding –A Powerful Tool for The Large Scale Production of Precise Microstructures, *SPIE Proceedings*, Vol. 2879, pp. 156-166, 1996.

SURFACE ROUGHNESS OF FIB SPUTTERED SILICON

M.Y. Ali, N. P. Hung, and S. Yuan
Precision Engineering and Nanotechnology Center
Nanyang Technological University, Singapore

Abstract

Mathematical models are developed to calculate the surface roughness of focused-ion-beam (FIB) sputtered surface. The surface roughness is the combination of the beam function and the material function. The beam function includes ion type, acceleration voltage, ion flux, intensity distribution, dwell time, etc; the material function includes the inherent material properties related to FIB micromachining. Surface of FIB sputtered (100) silicon was characterized using atomic force microscope. Reasonable agreement between the calculated and measured surface roughness was found.

Keywords

Focused ion beam, Beam profile, Sputtering, Surface roughness, Silicon.

1. INTRODUCTION

Challenges were identified when replicating three dimensional microcomponents with submicron accuracy, nano-leveled surface finish, and specific geometrical integrity (Ali, 2000; Vasile, 1997). LIGA was playing a strategic part in fabricating such microcomponents (Ehrfeld, 1995), but it was complex and expensive (Weber, 1996). Although fabrication using FIB was an alternative, the resulting geometrical integrity and surface finish were the major issues. Mathematical models for the geometrical integrity were developed (Nassar, 1998), but an estimation of sputtered surface finish is still not available. Preliminary work on sputtered surface of silicon and nickel beryllium were reported (Hung, 2000), this paper discusses the development and verification of a mathematical model.

2. MATHEMATICAL MODELS

The beam intensity distribution, tailing effects, and other beam parameters are grouped as beam function **B**. Selected material properties are combined in a separate group as material function **M**. These two functions are then combined to represent the surface roughness **R** in equation (1) where the operator "*" represents an appropriate functional relationship.

$$[R]=[M]*[B] \tag{1}$$

2.1 Beam Function

The function **B** is developed assuming circular beam, constant pixel spacing, insignificant effect of redeposited atoms on surface finish (effective removal of sputtered atoms), and the surface roughness in one dimensional and two dimensional space are the same. The sputtering time is also assumed to be long enough and transient surface finish is not an issue.

The beam profile is truncated at a level where the intensity fell to e^{-f} where the factor "f" is to be determined by experiment. The Gaussian beam intensity distribution is described in one dimensional space as (Sato, 1997):

$$\frac{J(x)}{J_0} = e^{\frac{-f\,x^2}{r^2}} \tag{2}$$

where $J(x)$ is the beam intensity at a distance x, J_0 is the peak intensity at $x=0$, and r is the beam radius. When the beam moves from one pixel to another, a part of the intensity overlaps. The cumulative intensity is split into a constant part φ_C and a variable part φ_V as:

$$\frac{J(x)}{J_0} = \sum_{i=0,1,2,\dots} e^{\frac{-f\,(x+iD)^2}{r^2}} = \varphi_C + \varphi_V \tag{3}$$

The function φ_V affects the surface finish while the function φ_C changes the base line of the surface. The total fluctuation of cumulative intensity "a" is expressed by:

$$a = \left| 1 + 2\times \sum_{i=1,2,3,\dots} e^{\frac{-f\,(iD)^2}{r^2}} - 2\times \sum_{i=1,2,3,\dots} e^{\frac{-f\,\{(2i-1)D\}^2}{4r^2}} \right| \tag{4}$$

The first term in equation (4) indicates the normalized intensity at the center of a beam (100%), the second term indicates the total contribution of intensity to that point, and the third term indicates the total intensity contribution to the neighboring point. When plotting the cumulative intensity curve for φ_V, the curve is similar to and can be simplified as:

$$\varphi_V = a\cos^2\left(\frac{\pi x}{D}\right) \tag{5}$$

Equation (4) determines the cumulative intensity fluctuation for unit flux and unit dwell time. So, for the ion flux ϕ and sputtering time T the beam function is:

$$\mathbf{B} = a\phi T\cos^2\left(\frac{\pi x}{D}\right) \tag{6}$$

2.2 Material Function

The material function represents the material behavior under the energetic focused ion beam. The basic form of this equation is expressed by:

$$M \quad = \quad K_1 K_2 \tag{7}$$

The function K_1, which is the contribution due to crystallographic structure, is normalized with respect to the (100) silicon, i.e., $K_1 = 1$. A correction factor is needed for sputtering along different crystalline orientations. The function, K_2 is defined as the volume of material removal per incident ion and determined by Sigmund's theory of sputtering (Sigmund, 1981). The sputtering yield was determined by both experimentally and theoretically to be 2.75 and 1.46 atoms/ion respectively. The latter was used for the calculation, $K_2 = 2.90 \times 10^{-29}$ m^3/ion for the atomic density of silicon of 5.08×10^{28} atoms/m^3.

2.3 Surface Roughness

The roughness model considers a surface texture formed by one raster. The function B just replicates its own functional shape if the material is "passive," i.e., $M=1$ in equation (1). Otherwise, the surface texture R is a modified form of B. Higher value of B indicates longer and more energetic collisions, while a higher value of M indicates more ion/material interaction. So the operator "*" is the multiplication and equation (1) becomes:

$$R \quad = \quad K_1 K_2 \, a \, \phi \, T \, \cos^2\left(\frac{\pi x}{D} \right) \tag{8}$$

Since the surface profile is now defined, its average finish R_a and peak-to-valley finish R_t are then calculated using standard procedure to be (Whitehouse, 1994):

$$R_a = \frac{K_1 K_2 a \phi T}{\pi}; \quad R_t = K_1 K_2 a \phi T = \pi R_a \tag{9}$$

3. MODEL VERIFICATION

A 50 keV Ga$^+$ FIB was used to sputter an (100) silicon wafer. Focused ion beam at factorial combination of beam parameters (Table 1) was directed to the four corners of 500×500 nm^2 windows to replicate the beam shape with no significant interference of neighboring beams. An AFM was used for measuring the sputtered profile. The profiles were well fitted with Gaussian curve to at least 3σ with average standard deviation of $\sigma = 140 \pm 4.4$ nm for the average diameter of 520 nm. A wrought estimation of beam size for sputtered volume calculation used f=1.0 [Sato, 1997] or f=0.7 [Prewett, 1991]. Such estimation was not suitable for the surface finish study since the beam tail was still energetic enough to sputter the surface. The average level of insignificant intensity was found at 11% ($\sim e^{-2.2}$) as illustrated in Figure 1. Knowing the beam diameter and pixel spacing, fluctuation of cumulative intensity was calculated using equation (4), and the surface finish was calculated using equation (9) (Table 2).

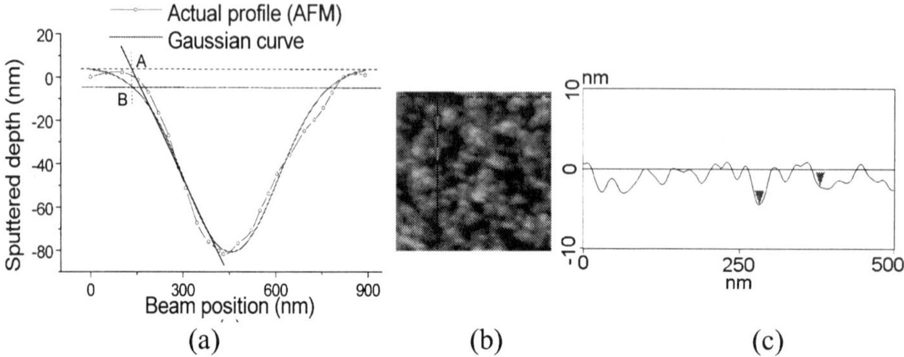

Fig. 1. (a) FIB profile where the intensity level AB (11%) was insignificant for sputtering, (b) AFM image of sputtered surface and (c) its cross section.

Table 1. The beam parameters used for the calculation of beam function.

#	Pixel D (nm)	Beam radius r (nm)	Amplitude "a"	Ion flux ϕ (ions/m^{2}s)	Sputtering time T (μs)	$a\phi T$ (ions/m^2)
1	25	260	0.1175	4.57×10^{24}	330	1.77×10^{20}
2	100	260	0.0785	4.57×10^{24}	864	3.10×10^{20}
3	25	350	0.1021	1.35×10^{25}	73	1.01×10^{20}
4	100	350	0.0764	1.35×10^{25}	215	2.22×10^{20}

Table 2. Surface roughness of FIB sputtered silicon.

#	$a\phi T$ (ions/m^2)	M (m^3/ion)	Surface finish (nm) Calculated		Surface finish (nm) Measured	
			R_a	R_t	R_a	R_t
1	1.8×10^{20}	2.9×10^{-29}	1.7	5.2	1.3	7.7
2	3.1×10^{20}	2.9×10^{-29}	2.8	9.0	3.1	13.5
3	1.0×10^{20}	2.9×10^{-29}	0.9	2.9	1.2	7.5
4	2.2×10^{20}	2.9×10^{-29}	2.0	6.4	2.9	11.3

Reasonable agreement of the calculated and measured R_a was seen from Table 2, but larger differences were seen for R_t. The deviation of surface finish (Figure 1c) could due to impurities in the wafer or near-perfect beam rastering mechanism.

4. CONCLUSIONS

Modeling and measurement of surface roughness of FIB sputtered silicon were made. This research showed:

1. The surface roughness function was combined with both the beam function and the material function.
2. For surface finish calculation, the effective beam diameter was defined where the intensity fell to $e^{-2.2}$ (11%) of the maximum.
3. Reasonable agreement of the calculated and measured surface finish was seen for the averaged roughness R_a, but larger deviation was found for the peak-to-valley roughness R_t.
4. Future work should include the effect of crystalline orientation of the substrate.

5. REFERENCES

1. Ali M. Y., Hung N. P., Yuan S. Simulation of Micro-Injection Molding, Proc. of the Int. Conf. on Precision Eng., Singapore, 2000; 535-40.
2. Ehrfeld W., Lehr H. Deep X-ray Lithography for the Production of Three-dimensional Microstructures from Metals, Polymers and Ceramics, Radiat. Phys. Chem 1995; 45 (3): 349-65.
3. Hung N. P., Ali M. Y., Yuan S. Producing LIGA-Competitive Microcomponents, Proc. of the Int. Conf. on Micromachining and Microfabrication Process Tech. VI, 2000; SPIE 4174: 49-57.
4. Nassar R., Vasile M. J., Zhang W. Mathematical Modeling of Focused Ion Beam Microfabrication, J. of Vac. Sc. and Tech. B 1998; 16 (1): 109-15.
5. Sato M. "Resolution." *Handbook of Charged Particles Optics*, J. Orloff, ed. NY: CRC Press, 1997.
6. Prewett P.D and Mair G.L.R. *Focused Ion Beam From Liquid Metal Ion Sources*, John Wiley, 1991.
7. Sigmund P. "Sputtering by Ion Bombardment: Theoretical Concepts." In *Sputtering by Particle Bombardment I*, R. Behrisch, ed. NY: Springer-Verlag, 1981.
8. Vasile M. J., Niu Z., Nassar R., Zhang W., Liu S. Focused Ion Beam Milling: Depth Control for Three-dimensional Microfabrication, J. of Vac. Sc. and Tech. B 1997; 15 (6): 2350-54.
9. Weber L., Ehrfeld W, Ferimuth H., Lacher M., Lehr H., Pech B. Micro Molding- A Powerful Tool for the Large Scale Production of Precise Microstructure, Proc. of the Int. Conf. on Micromachining and Microfabrication Process Tech. II, 1996, SPIE 2879: 156-67.
10. Whitehouse D. J. *Handbook of Surface Metrology*, Philadelphia, Institute of Physics Publishing, 1994.

MICROMACHINING IN THE THIRD DIMENSION

Prof. Dr.-Ing. Fritz Klocke[1,2]
Dipl.-Ing. Olaf Ruebenach[1]
Dipl.-Ing. Tobias Noethe[2]

[1]*Fraunhofer Institute for Production Technology IPT*
Steinbachstrasse 17, 52074 Aachen, Germany
[2] *Laboratory for Machine Tools and Production Engineering,*
Aachen Technical University
Steinbachstrasse 53B, 52074 Aachen, Germany

Abstract

Modified conventional processes show strong potential for the flexible machining of complex shaped micro parts. The current research activities are focussing of improving precision and enlarging the spectrum of machinable materials. Promising results have been achieved with ultrasonic assisted diamond turning and wire EDM cutting processes.

Keywords

Diamond Turning, Glass, Steel, Micromachining, Wire EDM, Process Development

1. INTRODUCTION

The manufacture of micro-systems products is characterised in many areas, by the silicon micro-mechanics technologies. The resultant restriction to silicon as the material of choice and to planar, lithographical machining technologies conflicts with the demands in terms of material and geometry, which modern micro-system products are required to fulfil. This applies particularly to small and medium sized serial manufacture of micro-engineering components. Modified conventional manufacturing techniques such as micro-Electro Discharge Machining and ultrasonic-assisted diamond machining, have the potential to manufacture high-quality parts with 3D geometries flexibly, economically and quickly. Part dimensions and surface structures in the range of a few micrometers can be achieved. The understanding of the principle mechanisms in material removal and surface generation are fundamental for the evaluation of the limitations and a successful development.

2. ULTRASONIC ASSISTED TURNING OF GLASS

2.1 Limitations of Diamond Turning

The mechanical and chemical interactions which take place in machining operations conducted on steel and glass, promote excessive diamond wear commonly referred to as "catastrophic tool wear" [CAS83, EVA91, PAU96]. Up until now, this has prevented the outstanding potential of ultraprecision machining with diamond tools from being applied to these groups of materials.

2.2 Principles of Ultrasonic-Assisted Cutting

As demonstrated in earlier publications [MOR91, MOR92, KLO00], the innovative technology of ultrasonic-assisted turning has already proven the potential to close this technological gap.

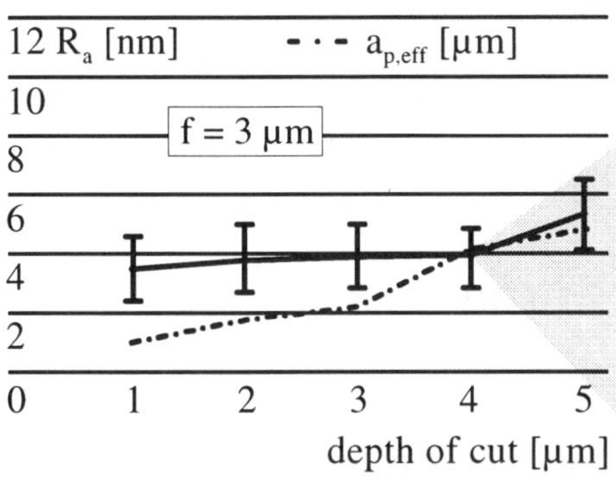

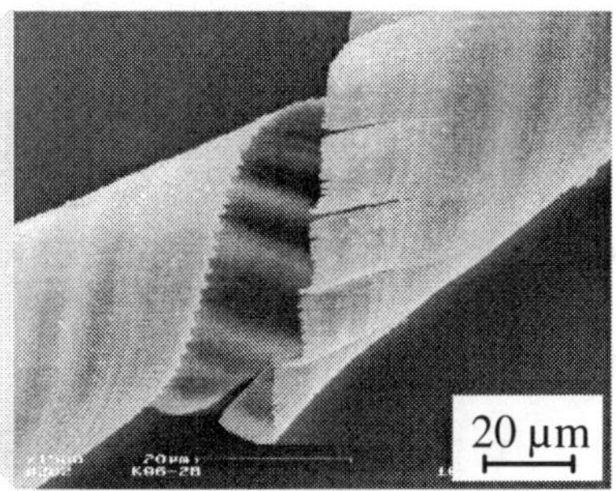

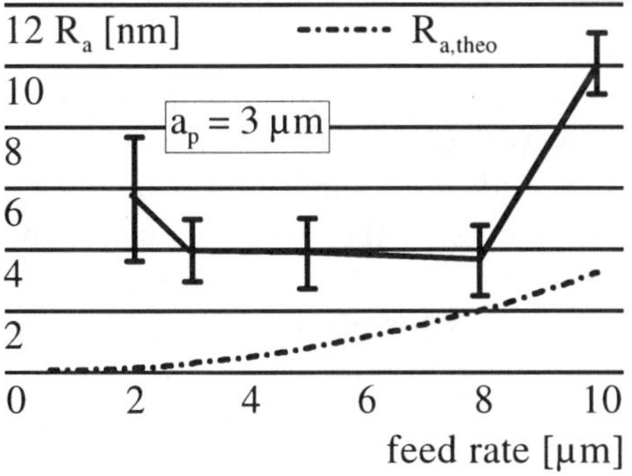

process parameter

cutting speed:	$v_{c\text{-rot}} < 6$ m/min
coolant:	spray mist
US-frequency:	$f = 38$ kHz
US-amplitude:	$\hat{A} = 3{,}5$ µm

Figure 2. Surface quality as a function of finishing parameters

In this hybrid process, the tool is excited during the turning operation to produce an vibration at ultrasonic frequency in the cutting direction. Thus, producing a high-dynamically interrupted cut, the actual tool contact time is reduced considerably, and with it, the load to which the diamond cutting edge is exposed. Consequently, tool wear decreases and the outstanding characteristics of the diamond cutting edge are retained, even when these critical materials are machined.

2.3. Research Activities at the Fraunhofer IPT

Now, extensive technological investigations have been performed to identify and evaluate the relevant variables and process parameters, which determine the material removal mechanisms. In this case the ductile machining of anorganic optical glasses with a defined cutting edge have been of special importance. In a first step the complex kinematics, which correspond to the combination of the constant spindle speed and the high-dynamic oscillation of the tool have been simulated. On the basis of these investigations the characterization of the material removal mechanisms has been realized by analyzing the structure of the chips, the surface topography and the surface integrity, respectively.

3. WIRE EDM CUTTING OF MICROSTRUCTURES

3.1 Situation

Wire EDM cutting operations have assumed what is virtually a monopoly position in the manufacture of cutting and pressing tools. This is due to the thermal material removal principle of this technique, which permits it to produce filigree and complex geometries with a high aspect ratio in even the hardest materials. However, the fact that only very low process forces occur, make this technique ideal for application in micro-engineering applications too. The requirements in the watch industry, the electronics and textile sectors have resulted in the use of thin erosion wires with diameters as small as 30 µm. Due to the extremely small cross-section of the wire, tungsten, which is highly resistant to thermal and mechanical load, is the material most frequently used. Compound wires, consisting of a high-strength steel core and a brass coating, have also established themselves.

3.2 Strategies in Process Development

It is vital, for the manufacture of structures in the micro-meter range, to reduce drastically the level of energy per discharge. Firstly, this prevents the wire from tearing. Secondly, the energy discharged, is the principal influence exerted on the surface and subsurface formation and on the wire vibrations. If the energy is too high, the dimensions of the discharge crater can impair the contour accuracy and narrow fins can bend as a result of the thermal induction of residual tensile stress. Static wire deviations caused by discharge forces, result in contour errors when corners and small radii are cut. Since very fine surfaces must be obtained in tool applications, trimcuts must be conducted after the contour generating main cut, with an even lower level of discharge energy.

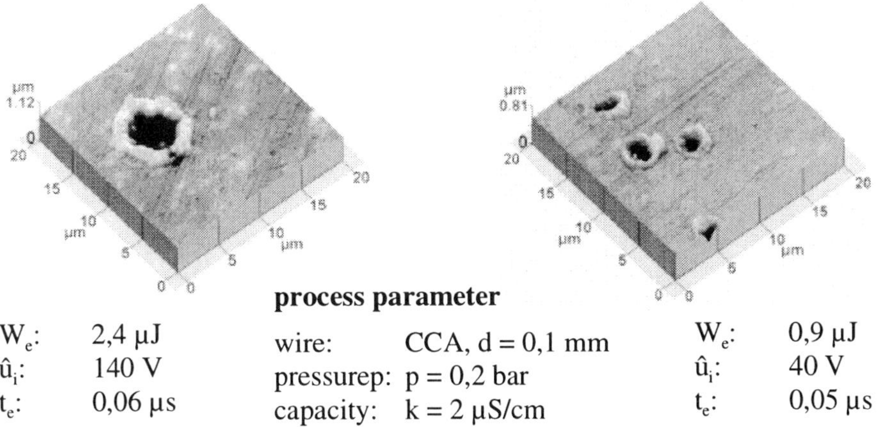

process parameter

W_e:	2,4 µJ	wire:	CCA, d = 0,1 mm	W_e:	0,9 µJ
$\hat{u}_i$:	140 V	pressurep:	p = 0,2 bar	$\hat{u}_i$:	40 V
t_e:	0,06 µs	capacity:	k = 2 µS/cm	t_e:	0,05 µs

Figure 3. Influence of discharge energy on crater formation

3.3 Research Activities at the WZL Aachen

Technology development for micro-wire EDM applications is still rather based on individual experience than on profound technological knowledge. Hence, the process design is very time consuming and the technological limits are not clear. The current research at the WZL is focussing on the main influences and limits on contour accuracy and surface quality in micro EDM using fine wires. Based on results from fundamental investigations fine wire cutting technologies for the manufacture of super-finished and accurate micro structures in steel and cemented carbide are derived.

References

CAS83 Casstevens, J.
Diamond Turning of Steel in Carbon-Saturated Atmospheres
Precision Engineering, Vol. 5, No 1(Jan 1983), S. 9-15

EVA91 Evans, C.
Cryogenic Diamond Turning of Stainless Steel
Annals of the CIRP, Vol. 40/1/1991, S. 571-575

PAU96 Paul, E.; Evans, C. J.; Mangamelli, A.; McGlauflin, M.L.; Polvani, R.S.
Chemical Aspects of Tool Wear in Single Point Diamond Turning
Precision Engineering 18: 4-19, 1996

MOR91 Moriwaki, T.; Shamoto, E.
Ultraprecision Diamond Turning of Stainless Steel by Applying Ultrasonic Vibration, Annals of the CIRP Vol. 40/1/1991, S. 559-562

MOR92 Moriwaki, T.; Shamoto, E.; Inoue, K.
Ultraprecision Ductile Cutting of Glass by Applying Ultrasonic Vibration
Annals of the CIRP Vol. 41/1/1992, S. 141-144

KLO00 Klocke, F.; Rübenach, O.
Ultrasonic Assisted Diamond Turning of Steel and Glass
International Seminar on Precision Engineering and Micro Technology, Aachen, July 19-20, 2000, Proceedings S. 179-190

OPTIMISATION OF DEEP HOLE DRILLING PROCESSES WITH SMALLEST DRILLING DIAMETERS

Prof. Dr.-Ing. Dr. h.c. mult. U. Heisel, Dr. Sc. Techn. M. Stortchak,
Dipl.-Ing. Dipl.-Gwl. R. Eisseler

*Institute for Machine Tools, University of Stuttgart, Holzgartenstr. 17, 70174 Stuttgart,
phone: ++49 711 121 3860; fax: ++49 711 121 3858, e-mail: heisel@po.uni-stuttgart.de*

Abstract

The machining of tough and high tensile materials with single-edge drilling tools of smallest diameter is only dissatifyingly mastered, because of the applied tools being very brittle and therefore fragile. The situation is made difficult by the opposed demands for higher productivity with respective cutting speed and rate of feed simultaniously to higher process security. At the moment, optimised cutting parameters for each individual application case have to be determined in expensive cutting tests. At the Institute for Machine Tools, University of Stuttgart, the cohesion between process parameters and their dependence on the control points are examined. The results enter a model of the machining process of single-edge drilling which in turn generates the basis for the optimisation model. By means of default conditions such as for example demanded machining quality, material to be machined, time of machining and so on, this leads to optimised values for the control points and moreover creates the precondition for the application of a process control system.

Keywords

Deep Hole Drilling, Process Optimization, Manufacturing Quality

1. INTRODUCTION

Single-edge drilling tools made of fully carbide with smallest diameters of 0,75 to 2 mm are frequently used to produce diesel injection systems and valves for the production of oil- and air ducts in hydraulic and pneumatic elements as well as deep drillings in bone nails, dental appliances and feeding stuff matrix. To also guarantee a reliable chip removal in deep drilling depths (L/D ≥ 200), cooling lubricant, washing outwards the arising chips through a bead, is brought to the working area through a cooling duct inside the tool. The required flow rate is assured by very high cooling lubricant pressures up to 250 bar as well as by optimised cross sections of cooling duct.With given tool parameters a very small wall thickness in the range of 0,1-0,2 mm result caused by the large cross sections of coolant ducts which makes the single-edge tool extremely fragile. This characteristic is particularly prominent during the production of components of the above mentioned diesel injection systems. With these components, injection pressures of approximately 2000 bar have to be controlled. The tendency of further increasing pressures simultaneously with smaller and lighter elements is met by tougher materials and less drilling parameters. Especially with tough and high tensile materials and high forward feed rates unfavourable chip forms develop in single-edge drilling, which tend to get stuck and therefore lead to an insufficient machining quality or even to the break of the drill. Within ever shorter intervals, such manufacturing tasks require time-intensive tests where the cutting data, the suitable material-tool combination as well as the cooling lubricant pressure is determined. Frequently it can be observed that users proceed empirical and rely on their own experiences. The present article introduces an approach, which allows the determination of suitable and adapted machining parameters with the help of an optimisation model.

2. INFORMATION SYSTEM

Such an optimisation model is the object of research work currently carried out at the Institute for Machine Tools. This system is based on a so called information model, built upon the following three models: model of the proceeding, process monitoring model and characteristics model of the drilling quality. The information model respectively the above mentioned models result in a characteristic overall process, where a functional cohesion between cutting speed as well as forward feed and drilling depth exist.

These functional processes can be pictured by a process diagram, showing the machining parameters of the drilling depth respectively the machining time. The process diagram is divided into the three essential parts of spot-drilling, actual drilling products as well as reboring. Against the background of an economical and safe overall process, each of these fields has to optimise itself according to set defaults, whereby the above mentioned models come into operation. They deliver the design parameters of the process diagram such as cutting speed and forward feed during spot-drilling and reboring as well as the parameter process between these procedures. The parameter and their process are determined with the help of characteristics and process control models in such a way that preferably high productivity results under consideration of default drilling quality. For the modelling, the functional dependencies of the parameters of the tool, the material and the machining course are determined in cutting tests. These interactive dependencies with their elements can be summarized in the set X_i.

$$X_i = \begin{cases} X_{HA} : \{v_c\} \\ X_{VA} : \{f\} \\ X_{WK} : \{\kappa_1, \kappa_2, \alpha, as, l_B\} \\ X_{WS} : \{d_t\} \\ X_{WF} : \{R_m, H, \varepsilon_B\} \\ X_{KS} : \{p_{KSS}\} \\ \vdots \end{cases} \tag{1}$$

Parameters:
HA: main drive, KS: lubricant, VA: feed drive, WF: material, WK: tool head, WS: tool shank

Elements:
as: center distance, d_t: shank diameter, f: forward feed, H: hardness, l_B: wear of cutting edge, p_{KSS}: lubrication pressure R_m: tensile strength, v_c: cutting speed, α: clearance angle, κ_1/κ_2: setting angles, ε_B: breaking elongation

With the knowledge of the cohesions within the parameter set X_i, a n-dimensional matrix of the functional dependencies can be generated, with whose help, the process course can be modeled. At the machining with maximised cutting parameters, unfavorable cutting forms respectively lengthes happen, especially with materials which are hard to cut. Left folded chips with a length of approximately 10 mm are presented in *Figure 1*.

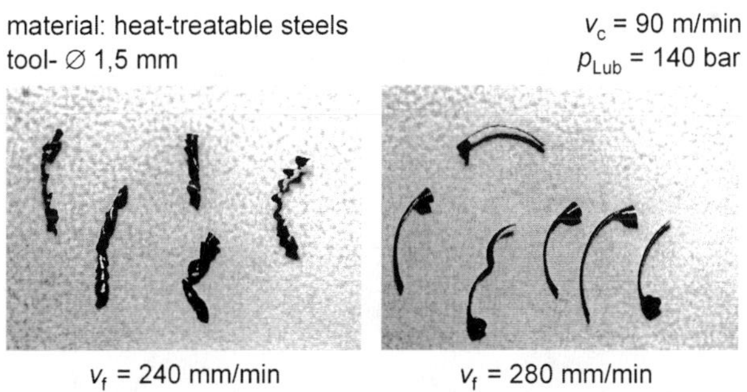

Figure 1: context between chip form and speed of feed

They develop during the machining of a tempering steel with a rate of feed minor than v_c = 240 mm/min and can be unproblematically ausgetragen from the bore hole also at higher drilling depths. At higher forward feed, the comma chips, presented on the right, occur, which are getting jammed between the tool and the bore wall and thus lead to tool failure and to damages at the bore wall. If the drilling moves to an instable process in such fringe ranges, the process control comes into operation.. With this, different process sizes are recorded and examined for significant characteristics indicating a failure. The process control reacts with machining parameters adapted to the failure until stabilisation is reached. Due to the limits arising with regard to a possibly high process security and machining quality simultaniously to a possibly short machining time, an decision range emerges out of the possible courses of process models, within which the control points have to lie. With the functional courses $\phi_{ti\text{-}j}$ for the upper bound (UL) and with $\xi_{ti\text{-}j}$ for the lower bound (LL), the following parametrical presentation results.

$$\begin{bmatrix} UL = \begin{cases} \phi_{t\,B-D}\,(X_{HA}, X_{VA}, X_{WK}, X_{WS}, X_{WF}, X_{DA}) \\ \phi_{t\,D-E}\,(X_{HA}, X_{VA}, X_{WK}, X_{WS}, X_{WF}, X_{DA}) \\ \phi_{t\,F-G}\,(X_{HA}, X_{VA}, X_{WK}, X_{WS}, X_{WF}, X_{DA}) \end{cases} \\ LL = \begin{cases} \xi_{t\,A-C}\,(X_{HA}, X_{VA}, X_{WK}, X_{WS}, X_{WF}, X_{DA}) \\ \xi_{t\,C-G}\,(X_{HA}, X_{VA}, X_{WK}, X_{WS}, X_{WF}, X_{DA}) \\ \xi_{t\,G-H}\,(X_{HA}, X_{VA}, X_{WK}, X_{WS}, X_{WF}, X_{DA}) \end{cases} \end{bmatrix} \qquad (2)$$

Figure 2 exemplary shows a process diagram with the three essential areas of spot-drilling, drilling and reboring, prepared for the machining of tempering steel with tools of 1,5 mm diameter.

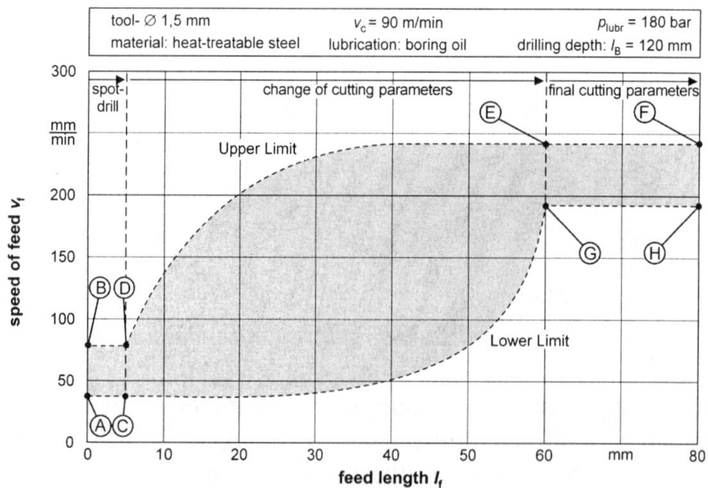

Figure 2: decision range for 1.5 mm-tools

The information model thereto delivers a lot of possible solutions regarding the machining parameters to be optimised out of a decision ranges, generated from predetermined quality criteria and an optimisation method which has to be determined.

3. OPTIMIZATION SYSTEM

To use the information system with regard to a computer supported optimisation, an optimisation system shall finally be created. With the help of algorithms, partly oppositional objective functions such as a short machining time simulaniously to a high machining quality and low costs are hereby transfered to a workable machining process. The basic tasks on developing an optimisation system, are therefore redefinition of its objects, the selection of the optimisation method as well as the creation of a method to convolute quality criteria and objective functions. With regard to the technological system 'single edge drilling' it follows convolution results:

$$\forall_R S \; \exists_R P : \{O_S \Rightarrow (\Phi_V \wedge F_V \wedge X_V \wedge Z_V)\} \qquad (3)$$

Therein, O_S are the objects of the optimisation system: Φ_V vector of the quality criteria, F_V vector of the objective functions, X_V vector of the project parameters and Z_V vector of the limits (here V-1, 2, ... , k, ...) /4, 5, 6/.

A multiparametrical optimisation results if local quality criteria, represented by the vector Φ_V exist. This vector is defined accordingly to the demands on the machining process. Thereby, the vector of the project parameters is changing to a n-dimensional vector, whereby n is indicating the number of the project parameters. Different project parameters implicate their own decision ranges resp. margins of error. Therewith the result of a working optimisation system is defined by project parameters ensuring an optimised process diagram with regard to the postulated criteria.

Figure 3 exemplary shows a corresponding optimised process diagram. Taking the criteria of shortest machining time and failure-free machining the optimised runs of forward feed and rotation speed were determined for drills with diameters between 1 and 2 mm.

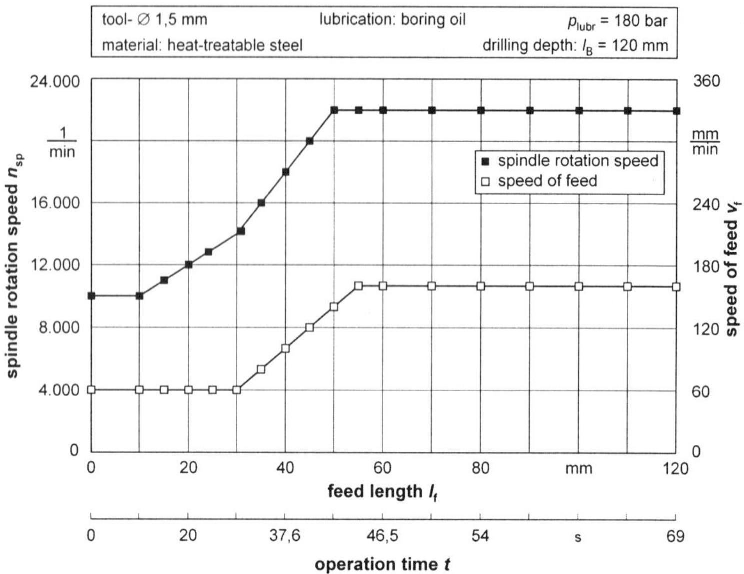

Figure 3: process diagram for 1.5 mm- tools

4. CONCLUSION

The machining of more and more heavier cutting materials and shorter intervals between type-changes in industrial scale manufacture mean high requests to the used machining methods. Particularly in the case of single edge drilling with smallest diameters it is nesessary to make time-consuming and expensive cutting tests to determine optimized cutting parameters. At the Institute for Machine Tools, University of Stuttgart, investigations to develop a so-called *optimisation system* are carried out. The precondition is an information system which includes results of cutting tests and a process control system. Having regard to criteria like quality, machining time and costs a computer supported method to determine optimized cutting parameters for single edge drilling can be created with the help of the information system.

5. REFERENCES

/1/   Heisel, U.; Eichler. R.: *Prozess-Integrity of Deep-Hole Drilling for Small Diameters.* In Annals of the German Academic Society for Produktion Engineering. 1, 1993, Nr. 1, S. 13 - 16

/2/   Hilbig, J.; Neyer, D.; Ermisch, N.: Der neue 1,2 l 3-Zylinder-
       Dieselmotor von Volkswagen. In VDI-Berichte, Band 1505,
       Technologien um das 3-Liter-Auto/VDI. Düsseldorf, VDI-Verlag,
       1999, S. 461-484.

/3/   Heisel, U.: *Technologie beim Bohren kleinster Durchmesser*. In:
       Begleitband zur Tagung "20 Jahre Zerspanungsforschung",
       Universität Dortmund, 1992, S. 4-64 - 4-71/4/ Stortchak, M., G.:
       Technological Systems for Finishing Gears. ISM, Kiev, 1994

/5/   Batischchev, D., I: *Methode der optimalen Projektierung*. Rundfunk
       und Kommunikation, Moskau, 1984

/6/   Nocedal, J.; Wright S., J.: *Numerical Optimisation*. Springer Verlag,
       New York, Berlin, 1999

PREDICTION OF HIGH FREQUENCY VIBRATION IN FINE BORING BY EXTENDED CHATTER MODEL

Evita EDHI, Tetsutaro HOSHI

Production Systems Engineering Department, Toyohashi University of Technology
Hibarigaoka, Tempaku Cho, Toyohashi City 441-8580, JAPAN
Tel. 081-532-47-0111, FAX. 081-532-45-7807
email: edhi@cherry.tutpse.tut.ac.jp, hoshi@cherry.tutpse.tut.ac.jp

Abstract

Based on previously proposed equation of the energy equilibrium derived from the theoretical model and another theory on the effect of two orthogonal radial orientations of bending natural vibration of the boring tool having non-circular cross section design, the study develops new equations for predicting y to x amplitude ratio of high frequency chatter vibration. Results of predictive calculations will be presented in comparison with the experimental results of cutting test to prove the validity of the new model.

Keywords

fine boring, high frequency chatter, extended chatter model.

NOMENCLATURE

a : (feed rate) x (depth of cut) (mm^2), b : Width of cut (mm).
k_d , k_{st} : Dynamic and static specific cutting force (N/mm^2).
x , y : Vibration displacement in X- and Y- directions.
β_d , β_{st} : Orientation angle of the dynamic and static cutting force (deg).
φ : Phase lag of inner modulation to outer modulation.
λ : Wave length of chatter mark (mm), μ : Overlap factor.

1. INTRODUCTION

High frequency chatter problem occurred in fine boring operation to machine inner surface of connecting rod of motor bike engine. The chatter frequency was identified from the chatter mark left on the machined surface to be as high as 10,000Hz [1]. It was also found that the cutting edge was vibrating in an oblong loop in X-Y plane normal to the boring tool axis.

Results of cutting experiment indicated that the mechanism of vibration can neither be explained by the existing theory of regenerative chatter nor by other known mechanisms of cutting vibration.

Based on those experimental observations, an extended chatter model was proposed that includes the cutting process and boring tool structure dynamics in X (opposite to depth of cut) and Y (cutting speed) directions to explain the onset of X-Y looping motion of the tool tip observed.

Equation for energy equilibrium was derived from the chatter model, as well as a theory on the effect of two orthogonal radial orientations of bending natural vibration of the boring tool with non-circular cross section.

The objective of the present study is to develop a method for predicting the amplitude ratio of X and Y vibration displacements that occurs at high frequency.

2. PREVIOUS FINDINGS ON THE HIGH FREQUENCY CHATTER
2.1. Boring Tools That Exhibit High Frequency Chatter

The boring tools are designed to finish 15mm diameter through hole as illustrated in Figure 1. Length to diameter ratio of the tools was from 1.2 to 1.5. Such small ratio value was originally considered to prevent chatter in its operation. However, tool **B** and **C** had chatter problem occurring at about 10,000Hz frequency. Tool **A** was prepared for comparison with extra mass designed at the front end. This tool was found to exhibit chatter at lower frequency about 5,000Hz.

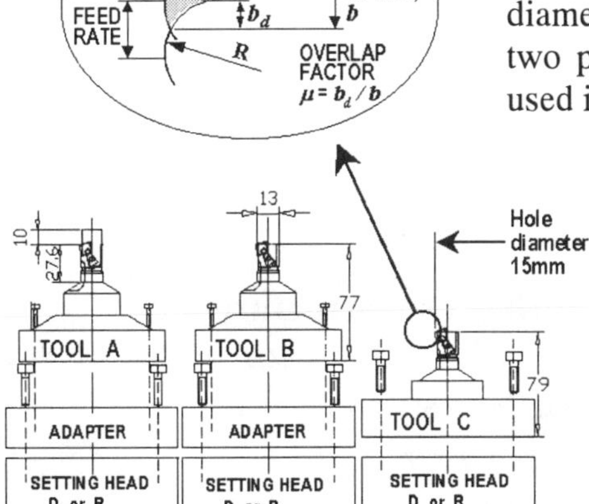

Either one of the tool was attached to the setting head whose function was to automatically adjust the finish bore diameter in production. Setting heads of two proprietary designs, D and R, were used in experiments.

Key parameters related to the chatter model are illustrated in the upper part of the figure, which are the width of cut b, overlap factor μ, and area of cut a.

Figure 1. Illustration of the boring tool used in study.

2.2. Fundamentals of the Extended Chatter Model

Cutting process dynamics due to regenerative effect and the imaginary part effect of inner modulation [2], often referred to as the cutting process damping, are presented in a stiffness polar diagram for X and Y directions by the lower and upper group of circles respectively, as shown in Figure 2.

Two solid curves are shown parallel to the real axis of the coordinate system. They represent the boring tool structure dynamics in X and Y directions, in terms of the stiffness frequency response function (FRF).

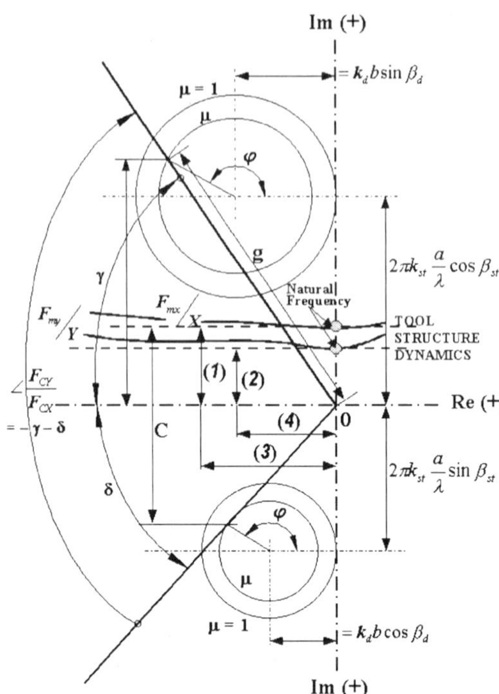

Figure 2. The extended chatter mode described in stiffness polar diagram.

2.3. Dynamic Process and Structure Models

The extended chatter model as represented in the figure is based on the following models of the cutting process and structure dynamics.

<u>Cutting Process Dynamics:</u>

$$F_{cx}/X = k_d\, b\, \cos\beta_d\,(\mu e^{j\varphi} - 1) - j\,2\pi k_{st}\frac{a}{\lambda}\sin\beta_{st}$$

$$F_{cy}/X = k_d\, b\, \sin\beta_d\,(\mu e^{j\varphi} - 1) + j\,2\pi k_{st}\frac{a}{\lambda}\cos\beta_{st}$$

$$(1)$$

<u>Structure Dynamics:</u>

The structural dynamics FRF of the boring tool in stiffness are approximated by those measured at natural frequency, which appear in the figure as two horizontal dashed lines, and are expressed by the following:

$$F_{mx}/X = \mathbf{(3)} + \mathrm{j}\,\mathbf{(1)}, \quad F_{my}/Y = \mathbf{(4)} + \mathrm{j}\,\mathbf{(2)} \tag{2}$$

2.4. Modal Orientations of the Boring Tool Structure

As marked in Figure 2, the expected time phase $\angle F_{CY}/F_{CX}$ between X and Y components of the dynamic cutting force is given by:

$$\angle F_{cy}/F_{cx} = -\delta - \gamma \tag{3}$$

As shown in Figure 3, there occur two orthogonal radial orientations om_1 and om_2 of the modes of bending natural vibration characteristic to the non-circular boring tool cross section.

The radial orientation of the vibration mode of the tool structure was found to affect the time phase $\angle F_{CY}/F_{CX}$ effective in the force components F_{m1} and F_{m2} to have 180° difference. Those time phases between force component $\angle F_{CY}/F_{CX}$ are further assumed to be equal to those between X and Y vibration displacement $\angle Y/X$, and they are given as follows depending on the mode orientations om_1 and om_2:

166

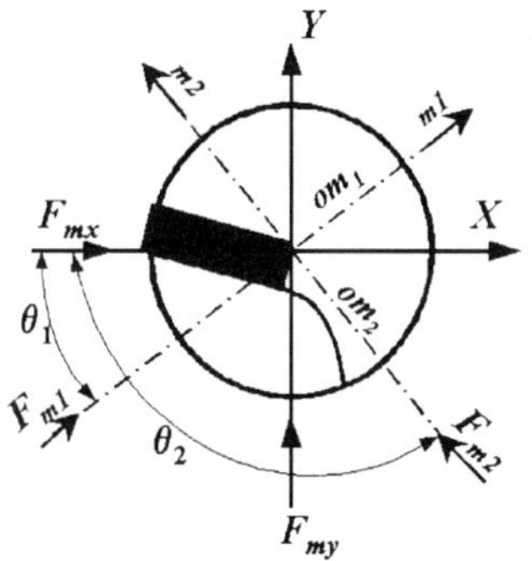

*Figure 3. Two orthogonal mode orientation **om₁** and **om₂**, characteristic to the non circular boring tool design.*

1. In the dynamic cutting force component exciting the boring tool at orientation om_1, which is in the direction close to the dynamic cutting force:

$$\angle Y/X = -\gamma - \delta \qquad (4\text{-}1)$$

2. In the dynamic cutting force component exciting the boring tool in the orthogonal orientation om_2 :

$$\angle Y/X = -\gamma - \delta + 180° \qquad (4\text{-}2)$$

2. 5. Energy Equilibrium

The energy equilibrium describing the supplied and dissipated energy in the system were derived as represented by the following equations:

Energy supplied by the X-Y looping:

$$ES1 = \pi k_d (\sin \beta_d) x\, y\, b\, (1 - \mu \cos\varphi)\sin \angle Y/X \qquad (5\text{-}1)$$

Energy supplied by regenerative effect :

$$ES2 = \pi k_d (\cos \beta_d) x^2 b\mu \sin \varphi + \pi k_d (\sin \beta_d) x\, y\, b\, \mu \sin \varphi \cos \angle Y/X \qquad (5\text{-}2)$$

Energy dissipated by the boring tool structure :

$$ED1 = \pi x^2 (\boldsymbol{1}) + \pi y^2 (\boldsymbol{2}) \qquad (5\text{-}3)$$

Energy dissipated by the cutting process damping :

$$ED2 = 2\pi^2 k_{st} \frac{a}{\lambda} \left(x^2 \sin \beta_{st} - x\, y \cos \beta_{st} \cos \angle Y/X \right) \qquad (5\text{-}4)$$

Equation of energy equilibrium : $ES1 + ES2 = ED1 + ED2 \qquad (5\text{-}5)$

When the energy equilibrium Eq. (5-5) is divided by πx^2, a following binomial equation of the amplitude ratio (y/x) is obtained:

$$(y/x)^2 (\boldsymbol{2}) - (y/x)\left[k_d b \sin \beta_d \left\{ (1 - \mu \cos\varphi)\sin \angle \frac{Y}{X} + \mu \sin \varphi \cos \angle \frac{Y}{X} \right\} + 2\pi k_{st} \frac{a}{\lambda} \cos \beta_{st} \cos \angle \frac{Y}{X} \right]$$

$$+ \left\{ (\boldsymbol{1}) + 2\pi k_{st} \frac{a}{\lambda} \sin \beta_{st} - k_d \mu b \sin \varphi \cos \beta_d \right\} = 0$$

$$(6)$$

Solution for (y/x) is given by: $\left. y/x = \pm \dfrac{g \sin \delta \pm \sqrt{g^2 \sin^2 \delta - 4(\boldsymbol{2})C}}{(\boldsymbol{2})} \right\} \qquad (7)$

(-)for mode om_1, and (+) for mode om_2 , where g, δ and C are as marked in Figure 2. Since the amplitude ratio (y/x) is real positive value by definition, among four possible values due to ± signs in Eq. (7), only the one with real positive value, and giving the minimum energy is to be accepted for prediction. In this case, φ is tentatively set to vary between 0° and 360° at 5°

intervals for calculation, and the result is obtained by finding the particular φ value that registers a real positive (y/x), and the minimum amount of the total energy supply.

4. RESULTS OF PREDICTIVE CALCULATION IN COMPARISON WITH CUTTING TEST

For variable combinations of the boring tool and cutting conditions, the amplitude ratio (y/x) has been predicted by computing Eq. (7), and the result is compared in Figure 4, with the amplitude ratio measured during cutting experiments in which chatter has occurred.

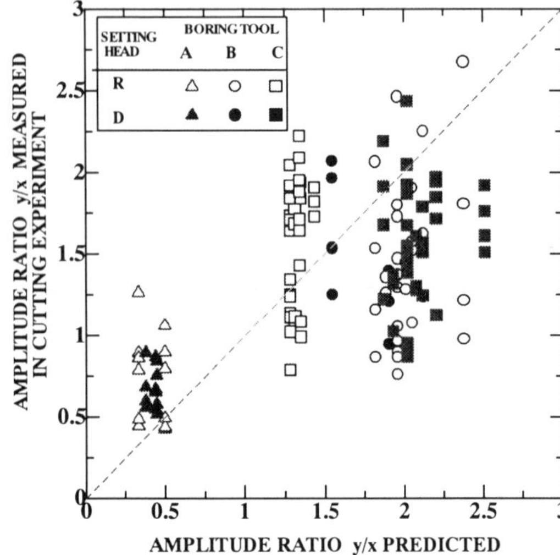

The predicted and experimental results are in fair correlation, suggesting that the prediction model is valid. For the real positive values to be obtained, it has been predicted also that chatter occurs in mode orientation om_1 with the tool **A**, and in om_2 with the tools **B** and **C**.

Figure 4. Comparison of predicted and experimentally measured amplitude ratio of X and Y vibration displacement.

5. CONCLUSION

Succeeding the previous research on theoretical model of the high frequency chatter involving the regenerative effect, cutting process damping and the tool tip X-Y looping, the present study has demonstrated that the energy equilibrium equation can be further developed into a model for predicting amplitude ratio of X- and Y-directional vibration displacements.

Results of the prediction have been confirmed to agree with experimental measurements obtained by cutting tests, therefore the validity of the original extended chatter model and the developed model for prediction has been confirmed in this study.

REFERENCE

[1] E.Edhi, T.Hoshi. A New Mechanism Explaining High Frequency Chatter Vibration Involving X-Y Looping in Fine Boring. DSC-Proc. of the 2000 ASME–IMECE, Orlando, FL.
[2] T. Hoshi: Cutting Dynamics Associated with Vibration Normal to Cut Surface, Annals of the CIRP, vol 21/1, 1972, p101-102.

THRUST FORCE ANALYSIS OF DRILLING BURR FORMATION USING FINITE ELEMENT METHOD

Sangkee Min, Jinsoo Kim, David A. Dornfeld
Laboratory for Manufacturing Automation
University of California at Berkeley, Berkeley, CA 94720, U.S.A.

Abstract

This study is focused on thrust force analysis of the drilling burr formation. A finite element model (FEM) for drilling burr formation was proposed with a stainless steel (AISI 304L) workpiece and a conventional twist drill. The thrust force variation during drilling process from FEM was obtained and compared with the experimental results. The estimation of burr size was briefly discussed.

Keywords

Finite element model, drilling burr, thrust force, stainless steel

1. INTRODUCTION

The burr is defined as material that is ragged and extended out of edges or surfaces of a workpiece after machining [Kim, 2000]. Since the burr is generated at the final stage of manufacturing, it damages the final precision integrity of parts. It requires an additional process, deburring which causes dimensional inaccuracy, changes surface integrity of machined workpiece, and sometimes results irrecoverable damages on the parts [Gillespie, 1975a]. Hence, it is very important to control burr formation at the design stage or process planning stage in precision manufacturing.

Many attempts have been made experimentally to understand drilling burr formation. However, it is very difficult to observe drilling burr formation in process because the drilling process is an enclosed process. Hence, experimental approaches have limitations of understanding the drilling burr formation. An analytical approach is also very difficult because of both the many parameters involved as well as the complexity of the problem [Guo, 2000]. An alternative way of studying the drilling burr formation is using finite element method (FEM). Once experimental validation is made, FEM gives in depth understanding of drilling burr formation.

Several drilling burr formation mechanisms have been proposed based on experimental observations [Gillespie, 1975b; Takazawa, 1985; Stein et. al., 1995]. In general, the drilling burr formation mechanism can be divided into five stages; steady-state cutting, initiation, development, initial fracture, and final burr. Kim measured thrust force throughout all the stages and observed the sudden drop of the thrust force at the burr initiation stage [Kim, 2000]. This is due to plastic deformation at the center of the drill on the exit surface of the workpiece, and is, therefore, closely related to burr type and size.

In this study, FE model of drilling burr formation is used to estimate thrust force at the steady-state cutting stage and the burr initiation point from the thrust force profile.

2. FINITE ELEMENT MODELING
2.1 Geometrical Finite Element Modeling

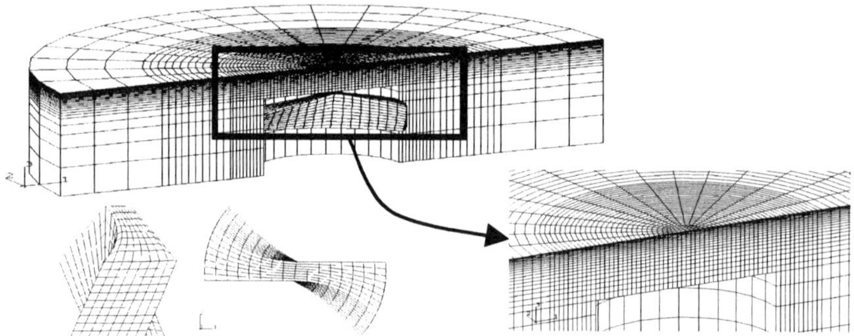

Figure 1. Initial configuration of finite element model of workpiece and drill

A stainless steel (AISI 304L) workpiece was used. The initial finite element mesh configuration and drill geometry are shown in Figure 1. To reduce the computational time, a blind hole was used. The thickness of the remaining part of the workpiece (1.5mm) ensures the steady-state cutting. The drill was assumed to be a rigid body and have a sharp edge. Tool wear was not considered. The point angle of the drill is 135°, helix angle 25°, and diameter 4 mm.

2.2 Material Modeling and Assumptions

Incremental plasticity using von Mises yield surface and associated flow rule were used to model the plastic behavior of the material. All the material properties were assumed to be isotropic. The strain rate dependency of material properties was modeled using the overstress power law because material properties, especially yield stress, vary at high strain rate (strain rate in drilling ranges from 10^{3} to 10^{5}).

Since a drilling process is an enclosed process that involves high strain rate, heat cannot be dissipated through the workpiece. Hence, an adiabatic thermal assumption was made. Built-up-edge and chip formation were not considered due to the complexity of the problem. Process parameters from experiments that generate a uniform burr and a crown burr were chosen, Table 1 [Kim, 2000]. Due to the highly non-linear dynamic character of the problem, the explicit code was used for FE simulation using ABAQUS/Explicit for computational efficiency.

Table 1. Process parameters for finite element simulations

Simulation	Feed		Speed ($\alpha=10^{-6}$)	
	f (mm/rev)	F=f/d	N (rpm)	S=α*N*d
F1S1	0.04	0.01	504	0.02
F1S2	0.04	0.01	1008	0.04
F2S1	0.08	0.02	504	0.02
F2S2	0.08	0.02	1008	0.04
F3S1	0.24	0.06	504	0.02

2.3 Failure Criterion

The essence of metal cutting in reality is removal of material from workpiece. How to model this concept is the biggest challenge in the FE modeling of metal cutting. In most cases, this concept is simulated by separation of elements. The chip separation criteria are related to the separation of elements that are mostly adopted in two-dimensional orthogonal cutting. Even though they used different measures for the chip separation criteria, the criteria were applied in the same way [Iwata et al., 1984; Carroll et al., 1988; Komvopoulos et al, 1991]. The parting line between the workpiece and the chip was predefined and the chip was formed when the element near the parting line meets a separation criterion.

In 3-dimensional finite element modeling of drilling, it is very difficult to define a parting line and arrange elements along this parting line because material in front of the drill deforms as the drill advances and this causes the parting line to be redefined instant by instant. Instead, elements close to the drill tip were removed when all the material points in an element meet a failure criterion.

Material failure was assumed to occur when the damage parameter, ω, the ratio of the incremental equivalent plastic strain to the equivalent plastic strain at failure exceeds one. Once an element satisfies the failure criterion, and then it becomes inactive in the remaining calculations [Hibbit, 1998].

$$\omega = \sum \left(\frac{\Delta \bar{\varepsilon}^{pl}}{\bar{\varepsilon}_f^{pl}} \right) \geq 1 \qquad (1)$$

3. RESULTS

The thrust force profiles from FE simulation and experiment are shown in Figure 2 along with the comparable burr formation states from FEM. The profile from experiment shows thrust force variation from tool engagement to the final burr formation while the profile from FE simulation shows it only from a part of steady-state cutting to the final burr formation. In the FE simulation, thrust force experiences large fluctuation right after the burr initiation point due to the removal of element and dynamic instability of finite element method. The burr initiation points, c and x in Figure 2, can be used to estimate the burr height and thickness using Kim's model [Kim, 2000]. (Results from this work will be presented in a future publication.)

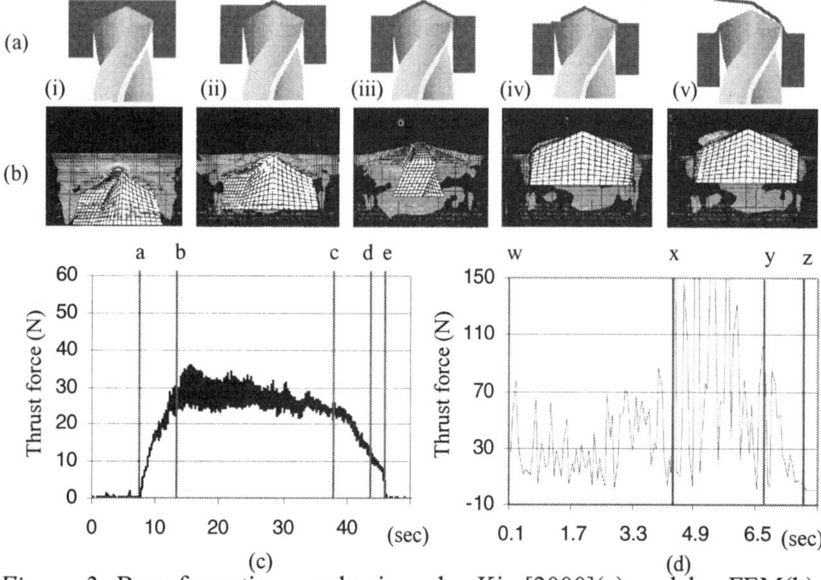

Figure 2. Burr formation mechanisms by Kim[2000](a) and by FEM(b). Thrust force profiles from experiment(c) and from FE simulation(d). The cutting condition is F1S1 in Table 1. (i)steady-state: b-c, w-x; (ii)initiation: c, x; (iii)development: c-d, x-y (iv)initial fracture: d, y; (v)final: e, z.

The thrust force for steady-state cutting was obtained from the average value during period b-c, Figure 2(c) and w-x, Figure 2(d). For the cutting conditions shown in Table 1, the average thrust forces from FE simulation are compared with experimental results in Figure 3. As feed parameter increases, the average thrust forces from experiments and FEM increase. Both FEM and experiments show high values of thrust force at higher speed parameter. The thrust forces were overestimated in FEM due to many factors that FEM may face such as the size of mesh and material properties. In general, the trend from FEM shows good agreement with experimental results.

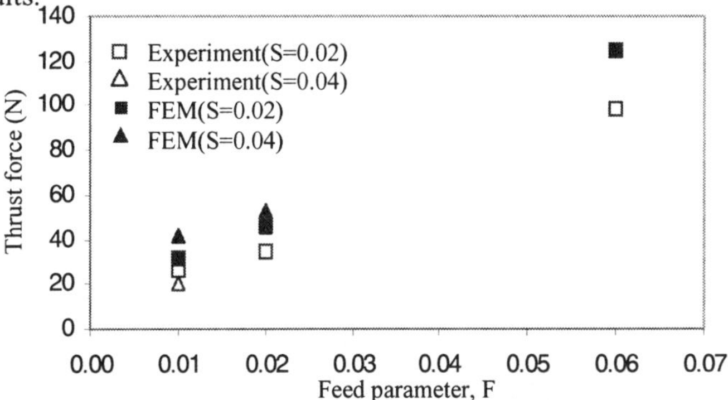

Figure 3. The average thrust forces in steady-state cutting

4. CONCLUSIONS

A three-dimensional finite element model of drilling burr formation was proposed. The average thrust forces from FEM showed good agreement with experimental results. A constant thrust force was shown during the steady-state drilling and the burr initiation point where an abrupt fluctuation in thrust force occurs was observed when the exit surface reaches its yield point.

REFERENCES

Carroll J.T., Strenkowski J.S., Finite Element Models of Orthogonal Cutting with Application to Single Point Diamond Turning, Int. J. Mech. Sci., 1988; 30:899-920.

Gillespie L.K., Hand Deburring of Precision Parts, Bendix Corporation Unclassified Topical Report, BDX-613-1443, 1975a.

Gillespie L.K., Burrs Produced by Drilling, Bendix Corporation Unclassified Topical Report, BDX-613-1248, 1975b.

Guo Y., Dornfeld D.A., Finite Element Modeling of Drilling Burr Formation Process in Drilling 304 Stainless Steel, Trans. ASME. J. Mfg. Sci. & Eng., 2000; 122:612-619.

Hibbit, Karlsson, and Sorenson, Inc., *ABAQUS/Explicit User's Manual*, Providence, 1998.

Iwata K., Osakada K., Terasaka Y., Process Modeling of Orthogonal Cutting by the Rigid-Plastic Finite Element Method, ASME Trans., J. of Eng. Mat. and Tech., 1984; 106:132-138.

Kim J., Optimization and Control of Drilling Burr Formation in Metals, Ph.D. Dissertation, University of California, Berkeley, 2000.

Komvopoulos K., Erpenbeck S.A., Finite Element Modeling of Orthogonal Metal Cutting, J. of Engineering for Industry, 1991; 113:253-267.

Stein, J., Dornfeld, D. A., An Analysis of Burs in Drilling Precision Miniature Holes Using a Fractional Factorial Design, ASME Symposium on Production Eng., Winter Annual Meeting, 1995.

Takazawa K., The Academic Challenge of Burr Technology in Japan, Japanese Society of Deburring and Surface Conditioning Technique, 1985.

BASIC STUDIES ON THE WEAR BEHAVIOUR OF MODIFIED AND COATED DIAMOND TOOLS FOR PRECISION MACHINING OF FERROUS MATERIALS

Ekkard Brinksmeier, Ralf Gläbe

Laboratory for Precision Machining (LFM)
University of Bremen, Germany

Abstract

As is well known, precision machining of steel with monocrystalline diamond tools is difficult to achieve due to catastrophic tool wear. The goal of this work is to explore the wear reducing effect: ion implanted diamond tools and of protective hard coatings. Moreover, the performance of nano grained alumina cutting tools were investigated.

Keywords

precision machining of steel, modified diamond tools, alumina tools

1. INTRODUCTION

Up to now, the responsible mechanisms for the catastrophic wear of diamond tools in precision machining of steel alloys are not fully understood. Simple correlations between tool wear and workpiece hardness or melting point do not seem to help predicting the wear mechanism. The most extensive discussion of diamond tool wear was presented by Paul and Evans [1].

In the past, several process modifications for reducing tool wear in diamond turning of steel have been proposed, i.e. cryogenic turning by cooling the tool-workpiece system with liquid nitrogen, turning in inert and other gas atmospheres, and ultrasonic vibration cutting [2].

Another approach to reduce tool wear is to modify the diamond tool or to change the tool material. Materials with a higher chemical stability against steel could be used. Some interesting tool modifications were presented in the literature. Knuefermann [3] has substituted the cutting material. He achieved a micro-roughness better than 40 nm Ra using a nano-grained CBN tool for precision machining of hardened steel. Masuda [4] observed a tool life of about 200 m with monocrystalline CBN tools turning hardened stainless steel (HRC52), whereas the polycrystalline CBN showed less resistance. Dautzenberg [5] found insufficient chemical stability of CBN tools when turning steel alloys. Alumina is chemically more inert than CBN and diamond, but has an insufficient hardness as reported by Dautzenberg. Krell [6] showed, that the use of micro-grained alumina substantially extends enlarges tool life.

In this work, modifications of diamond tools were investigated for maintaining the good mechanical properties of the tool material. The tool was coated with protective layers to build-up a diffusion barrier, or ion-implanted to change the chemical stability of the diamond lattice. It could be shown, that the hardness of polycrystalline alumina increases when the grain size decreases.

2. EXPERIMENTS

We have performed face turning experiments with coated diamond tools for introducing a barrier for diffusion, and with ion implanted tools for increasing the activation energy for chemical reactions. Moreover, the performance of aluminum oxide tools was compared to the performance of diamond tools.

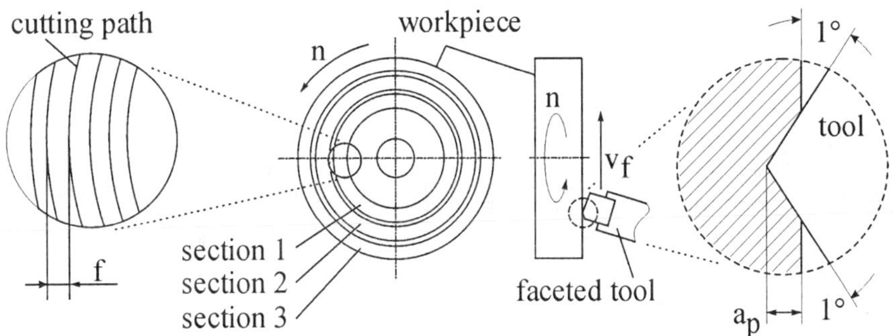

Figure 1. Experimental setup for experiments with modified tools.

All experiments were performed on a Hembrug Super-Mikroturn CNC precision lathe. MobilMet-423 mineral oil was used as a lubricant. The experiments were carried out with Ck01N carbon steel and faceted tools with a facet angle of 178°, 0° rake and 5° clearance angle. Due to a feed $f = 1300$ µm and a depth of cut $a_p = 3.6$ µm, the v-shaped cutting grooves did not overlap (cf. Figure 1). The cutting speed v_c was approximately 10 m/min. The cutting experiments were interrupted after cutting distances of 260 mm, 630 mm and 1130 mm (section 1, 2 and 3) for measuring the respective wear land widths. The observed shape of the wear land is outlined in Figure 2. The central wear land width VB_f and the average of the side wear land widths VB_1 and VB_2 were used for characterizing the flank wear.

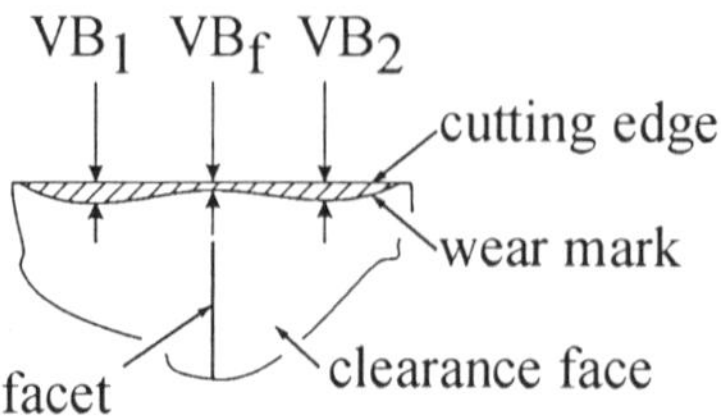

Figure 2. Wear criteria.

3. RESULTS AND DISCUSSION

3.1 Protective coatings

We have employed TiN coatings on diamond tools for the reduction of chemical wear in precision machining of steel. The protective TiN coating [7] was created by physical vapor deposition with a typical thickness of 10 nm and good adhesion to the diamond surface.

width of wear land

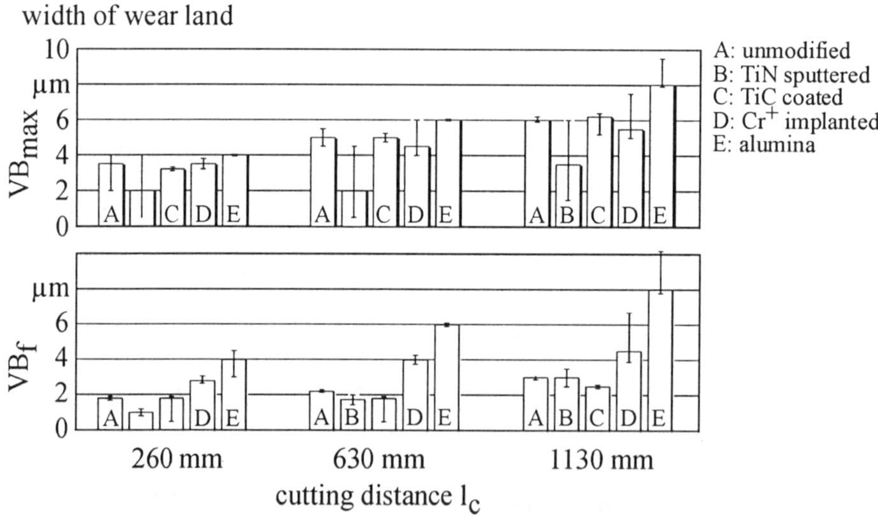

Figure 3. Wear of modified tools vs. cutting distance.

During the cutting process the TiN coating is worn abrasively, while the exposed diamond surface undergoes chemical wear. Flank wear is clearly reduced compared to the uncoated diamond (cf. Figure 3) unless the layer is removed.

In another experiment, we have investigated the wear behaviour of a TiC layer [8] which was created by glowing of a titanium sputtered diamond tool. The existence of the TiC interlayer was verified by an XPS analysis. Although, flank wear could be reduced by the TiC layer, the protective effect was less pronounced than with the TiN coating (cf. Figure 3).

3.2 Ion implantation

Ion implantation of diamond tools was motivated by the results of Zhang [9] who found an increase of the life of diamond dyes used for wire drawing after implanting nitrogen ions.

The crystal lattice of the faceted diamond tool was modified by implantation of Cr^+ ions with a kinetic energy between 120 keV and 180 keV yielding an ion concentration of 10^{16} - 10^{17} cm^{-2}. The lattice was partially healed by heating to 800°C during implantation. In order to achieve a high ion concentration close to the surface of the diamond, a sacrificial layer technique was used. The results of the turning experiments are shown in Figure 3. The observed widths of the wear lands were equal or larger than for unprotected tools.

3.3 Ceramic tools

We have employed a faceted sapphire tool and a sintered nano-grained alumina tool with extreme low porosity [10] for precision turning of Ck01N steel. The cutting edge radii of the tools were polished to approximately $r_\beta = 120$ nm. Interestingly, the sintered alumina tool performed slightly better than the sapphire. Flank wear near the facet was comparable to the wear observed in the reference experiments with unprotected diamond tools. The typical wear mark as outlined in Figure 2 was not found. Due to the polycrystalline structure of the tool material, the wear structure shows a combination of trans- and intercrystalline wear, the roughness of the machined surface was about 100 nm Ra.

4. CONCLUSIONS

We have performed different sets of experiments with geometrically defined cutting tools using diamond tool modifications in precision machining of steel.

- TiC and TiN coatings of diamond tools bear the potential of eliminating chemical wear but suffer from abrasive wear.
- Implantation of chromium ions in the diamond lattice did not improve tool wear.
- Sapphire and sintered alumina tools performed surprisingly well but did not surpass diamond tools.

Thus, we conclude that more research is needed on the chemistry of tool wear, the improvement of ultrasonic vibration cutting and coating technologies for diamond tools.

5. ACKNOWLEDGEMENTS

The authors like to thank the German Research Foundation (DFG) for the support of this work. We thank P. Mayr, A. Mehner, B. Felde, H.R. Stock, F. Seidel and J.-E. Doering of the Institute for Materials Science (IWT) for providing the coatings and the ion implanted diamond tools, and G. Grathwohl and D. Godlinski of the University of Bremen for providing the ceramic tools.

6. REFERENCES

[1] Evans, C.J.; Paul, E.; Mangamelli, A.; Mc Glauflin, M.L.: Chemical aspects of tool wear in single point diamond turning. Prec. Eng. 18(1996)1, p. 4-19.

[2] Brinksmeier, E.; Preuss, W.; Gläbe, R.: Single point diamond turning of steel. Proc. of the 1st International euspen Conference 1999; Bremen, Germany, p. 446 – 449

[3] Knuefermann, M.; Read, R.; Nunn, R.; Clark, I.; Fleming, A.: Ultrapräzisionsbearbeitung gehärteter Stahlbauteile mit Amborite DBN45. IDR; 34(2000)3; p. 222 – 230.

[4] Masuda, M.; Nishiguchi, T.: Mirror-like cutting of ferrous metals with CBN tools. Proc. o. t. 14th NAMRC Conf., May 28-30 1986, Univ. o. Minnesota, USA; Seite: 459 - 464

[5] Dautzenberg, J.H.; Taminiau, D.A.: High-Precision cutting of steel: Choosing tool material. Proceedings of the 3rd International Conference on Ultraprecision in Manufacturing Engineering; Aachen 2.-6. Mai 1994 (Hrsg.: Weck, Kunzmann); Riehm-Verlag; Seiten: 25 - 28

[6] Krell, A.; Blank, P.; Berger, L.; Richter, V.: Submicrometer Cutting Tools on the Basis of Al_2O_3 for Machining Alloyed hard cast iron and hardened steel. Proc. of the 23rd Annual Conf. on Composites, Advanced Ceramics, Materials, and Structures, January 25-29, 1999, Florida, USA

[7] Mayr, P.; Felde, B., Mehner, A.; Hoffmann, F.: Microstructure of alumina coatings deposition by sol-gel process. Proceedings of the EUROMAT 99 (Munich, Germany), Volume 11, p. 122-126.

[8] Mayr, P.; Felde, B.; Kohlscheen, J.; Mehner, A.; Hoffmann, F.: Diamond-Metal interface chemistry of thin titanium and chomium films on(100)-single crystal diamond surfaces. Diamond 2000: 11th European Conf. on Diamond and Diamond-Like Materials (Porto, Portugal)

[9] Zhang: Ion Implantation of Diamond Dies. Wire Industry; 52(1985); p. 314-317.

[10] Grathwohl, G.; Kunz, M.; Godlinski, D.: Development of ceramics micro tools for Precision maching. Proc. of the 1st International euspen Conference 1999, Bremen, Germany

DIFFERENCE IN WEAR PATTERNS OF DIAMOND CUTTING TOOL DEPENDING ON WORK MATERIALS

H. Tanaka[1], S. Shimada[2], N. Ikawa[1], M. Higuchi[3] and K. Obata[4]

[1]Faculty of Engineering, Osaka Electro-Communication University, 18-8 Hatsu-cho, Neyagawa, Osaka 572-8530 Japan
[2]Department of Precision Science and Technology, Osaka University, 2-1 Yamada-oka, Suita, Osaka 565-0871 Japan
[3]Faculty of Engineering, Kansai University, 3-3-35 Yamate-cho, Suita, Osaka 564-0073 Japan
[4]Allied Material Corporation, Kohtaka, Takino-cho, Katoh-gun, Hyogo 679-0221 Japan

Abstract

For profound understanding of wear mechanisms in diamond machining, thermodynamics analyses and an erosion test are carried out. The results show that there are three different mechanisms. First one is graphitization, the second oxidization-deoxidization reaction and the third carbide formation. The wear mechanisms depend strongly on the ambient oxygen and/or cleanliness of metal surface to be machined in contact with diamond. These results suggest how to extend applications of diamond tool and possible methods to suppress the tool wear.

Keywords

single point diamond turning, diamond tool, wear mechanism, erosion test, thermodynamics analysis

1. INTRODUCTION

Ultraprecision diamond turning has been fully established as a practical manufacturing technique of a variety of optical, electronic and mechanical components used in the field of advanced science and technology. However, it is well known that diamond tools show high wear rate in cutting of some workmaterials such as ferrous metals [1] and Si. Wear pattern of cutting tools strongly depends on workmaterials. These phenomena disturb the systematic understanding of wear mechanism and also may limit the application of diamond as a kind of an ideal material for ultraprecision cutting tool.

In this paper, aiming at the profound understanding of the differences in wear patterns of diamond tools in turning of iron (steel), copper and aluminum, thermodynamics analyses and erosion tests simulating the tool wear process are carried out.

2. DIFFERENCE IN WEAR PATTERN

Wear pattern of diamond cutting tools depends on workmaterials as shown in Fig. 1. For example, in turning of iron or steel, wear rate is very high in both of crater and flank wear together with grooving wear. In turning of copper, wear rate is very low and diamond cutting tool shows typical crater wear but little flank wear. In contrast, flank wear is dominant but crater wear cannot be observed in turning of aluminum.

The stresses at cutting edge during cutting are not so high as to break diamond even in turning of mild steel [2]. Diamond tool shows higher wear rate in turning of pure iron than steel the hardness of which is higher than pure iron. These results suggest that some thermo-chemical reaction between tool and workmaterial may cause the tool wear except in machining extremely hard materials.

In diamond turning, possible chemical reactions regarding tool wear mechanism on tool-work interface are oxidization of diamond and workmaterial, deoxidization of workmaterial oxide by diamond, carbonization of workmaterial and graphitization of diamond tool surfaces by catalytic reaction of workmaterial.

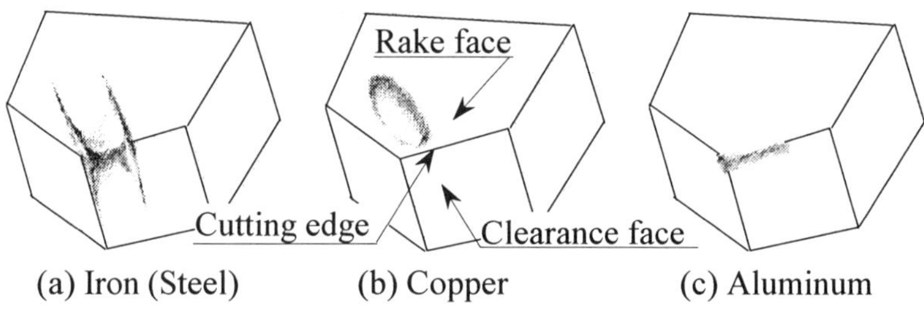

Figure 1. Difference in wear patterns of diamond cutting tool (iron: crater, flank and grooving wear, copper: crater wear, aluminum: flank wear)

3. THERMODYNAMICS OF DIAMOND WEAR

Based on thermodynamics analysis, the direction and stability of chemical reaction at a specified condition can be predicted by the difference between the Gibbs's free energy ΔG^0_T of the left and right sides of the reaction formula. Figure 2(a) shows ΔG^0_T in oxide formation of diamond and metals for 1 mol oxygen as a function of atmospheric temperature T. When $\Delta G^0_T < 0$, the reaction goes from the left to right sides of the reaction formula. Diamond and metals can be oxidized at the temperature higher than room temperature. Oxides which have larger negative value of ΔG^0_T in oxide formation are more stable than that with smaller one under the same temperature. Hence, when copper oxide is in contact with diamond, the latter can take oxygen away from copper oxide (deoxidization by diamond) and oxidizes for itself at the temperature higher than room temperature. Diamond can deoxidize iron oxide at the temperature higher than 900 K in the same manner as copper. Whereas, diamond cannot deoxidize aluminum oxide at the temperature lower than 2300 K. However, the energy levels of the transition state and consequently the amount of activation energy in the reaction process have not been estimated well so far. Hence, the reaction rate of the process cannot be predicted by the thermo-dynamics analysis alone.

On the other hand, aluminum can form carbide at any temperature and iron do higher than about 1000 K as shown in Fig. 2(b). Copper does not make carbide. The metals that can form carbide are presumed to have affinity with diamond. It is reported that the affinity of diamond for metals is related to the incompleteness in d orbital

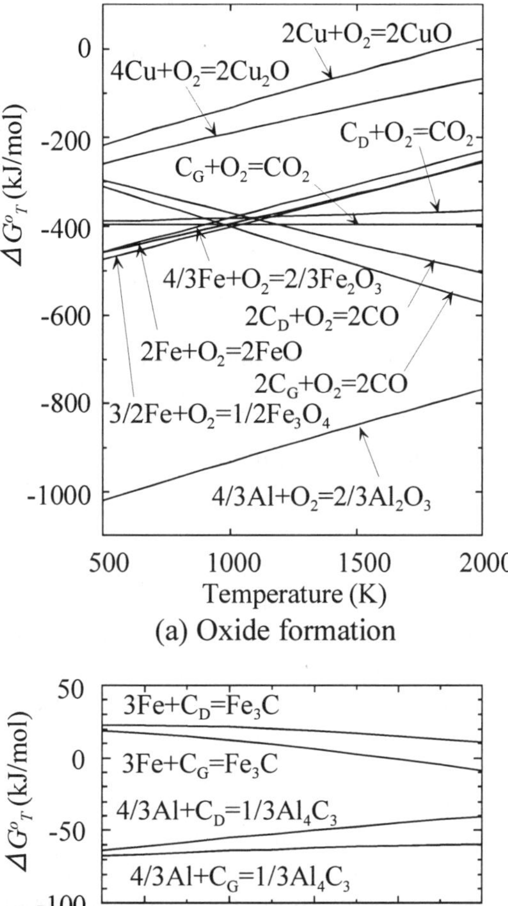

(a) Oxide formation

(b) Carbide formation

Figure 2. Change in Gibbs's free energy in oxide formation of copper, aluminum, iron and carbon for 1 mol oxygen and carbide formation of aluminum and iron for 1 mol carbon

electron configuration [3]. Accordingly, iron shows strong affinity for diamond due to the larger number of incomplete *3d* electrons. Therefore, graphitization of diamond can be performed on the interface with iron under elevated temperature.

181

4. EROSION TESTS

An erosion test that can simulate the wear process of diamond tool in turning is carried out. A diamond specimen in point contact with a curved metal wire of small diameter is heated in a vacuum under various temperatures. In the test, elliptical erosion pattern is generated on diamond surface due to thermo-chemical reaction with the metal wire. The volumetric wear of the erosion is calculated from the ellipse and arc, which approximate the pattern shape on the diamond surface and depth profiles measured by a profilometer.

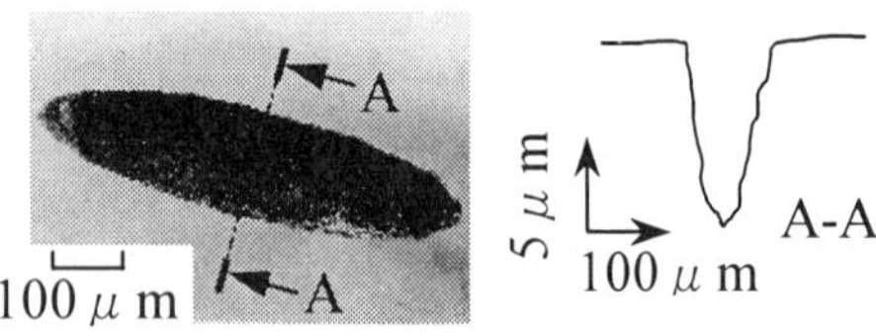

(a) Erosion pit   (b) Depth profile

Figure 3. Erosion pit and depth profile of wear pattern in contact with iron wire at 1173 K under the oxygen partial pressure of 1.3×10^{3} Pa for 30 minutes

Figure 3 shows the typical shape of erosion pattern tested with pure iron wire at 1173 K under 1.3×10^{-3} Pa. The surface of eroded pit turns black at the temperature above 973 K due to graphitization of diamond surface. Diffusion of carbon atoms can be observed in iron wire.

Figure 4 shows typical depth profiles of erosion pattern in contact with copper wire under different oxygen partial pressure. The volumetric wear increases as the oxygen partial pressure increases. The depth of erosion in the peripheral region of the contact area is larger than that in the central region. These results suggest that the existence of oxygen causes the increase of erosion and that easier arrival of ambient oxygen molecules to the interface between the diamond and copper wire causes larger erosion in the peripheral region than that in the central one. The similar wear pattern as is seen in diamond-copper combination can be observed in diamond-iron combination heated at 873 K. Neither graphitization on eroded surface nor diffusion of carbon atoms in iron wire cannot be observed in this case.

On the surface of diamond heated at the temperature above 873 K in contact with aluminum wire, no erosion can be observed. However, aluminum carbide is detected in contact area on diamond surface by Auger electron spectroscopy.

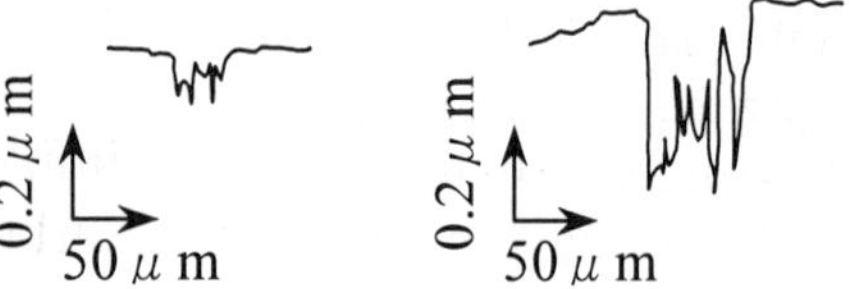

(a) $P_{O_2}=1.3\times10^{-3}$ Pa (b) $P_{O_2}=1.3\times10^{-2}$ Pa

Figure 4. Depth profile of wear pattern in contact with OFHC copper heated at 1173 K for 30 minutes under different oxygen partial pressure

5. POSSIBLE WEAR MECHANISMS

Based on the results of analyses and tests described above, proposal is made on three different mechanisms of diamond wear.

First one is graphitization of diamond surface due to thermally activated catalytic reaction of clean surface of the workmaterial to be machined. The carbon atoms the bonds of which are weakened comparing with those in bulk diamond are removed by abrasion of work surface or flowing chip or by diffusion into the workmaterial. In turning of iron or steel, this mechanism is dominant and wear rate is very high.

The second one is oxidization-deoxidization reaction. There are two steps in this mechanism. The surface of workmaterial is oxidized at first by ambient oxygen, and then the oxide of workmaterial is deoxidized by diamond. The existence of oxygen plays an important role in this kind of wear process. As rake face of cutting tool contacts loosely with chip, the ambient oxygen molecules easily penetrate into the interface between chip and rake face. The crater wear on rake face observed in turning of copper is attributed to this mechanism. If cutting temperature does not rise so high, similar wear pattern may be observed also in turning of iron.

The third one is carbide formation with workmaterial. The chemical reaction of this kind requires tight contact between reactants and temperature rise. Therefore, carbide formation occurs around the extreme cutting edge and flank face. The carbide is removed by abrasion of work surface or flowing chip. This type of wear can be observed in turning of aluminum.

6. CONCLUSIONS

Thermodynamics analyses and erosion tests well support the features of wear pattern of diamond tool and the wear mechanisms proposed.

Diamond turning under reduced oxygen atmosphere or using coolant without oxygen can be useful to suppress the tool wear in turning of copper. In turning of aluminum and steel, it is recommended to use some sort of "inhibitor" such as lubricant or surface layer formed to prevent diamond surface from contacting with clean surface of workmaterial at cutting edge and flank face, and to decrease cutting temperature for the suppression of tool wear.

REFERENCES

1. N. Ikawa, T. Tanaka, Thermal Aspects of Wear of Diamond Grain in Grinding, Annals of the CIRP 1971; **19**: 153.
2. N. Ikawa, S. Shimada, Microfracture of diamond as Fine Tool Material, Annals of the CIRP 1982; **31**, 1: 71.
3. K. Miyoshi, D. H. Buckley, Application of surface science 1980; **6**: 161.

DUCTILE MODE CUTTING OF SINGLE-CRYSTAL SILICON BY ULTRASONIC VIBRATION

Shigeomi KOSHIMIZU* and Jiro OTSUKA*

* :Department of Mechanical Engineering, Shizuoka Institute of Science and Technology,

2200-2 Toyosawa, Fukuroi-city, Shizuoka Prefecture, 437-8555 JAPAN

Abstract

Brittle materials have a critical depth of cut between ductile mode and brittle mode. In order to increase the critical depth of cut, ductile mode cutting of single crystal silicon was realized using ultrasonic vibration. It was clarified that during conventional cutting without ultrasonic vibration, the critical depth of cut was $dc = 0.15$ μm. however, during ultrasonic vibration cutting, the critical depth of cut was increased to $dc = 8.5$ μm.

Keywords

single-crystal silicon, ductile mode cutting, brittle mode cutting, ductile-to-brittle transition, critical depth of cut, ultrasonic vibration cutting

1. INTRODUCTION

The critical depth of cut in ductile-to-brittle transition is approximately 0.1 μm in the case of single-crystal silicon. In the actual cutting process, it is not easy to control the depth of cut less than 0.1 μm. It also can not realize high productivity. Therefore, it is purpose of this study to increase the critical depth of cut by applying ultrasonic vibration to the diamond tool in the cutting of single-crystal silicon.

2. EXPERIMENTAL EQUIPMENT AND PROCEDURE

2.1 Experimental equipment

Figure 1 shows the cutting test equipment. A workpiece of single-crystal silicon is fixed on an ultrafine motion table and is fed in the horizontal direction (Y direction). Since the fluctuation of the cutting direction (Z direction) during a cutting test can be controlled to less than 10 nm by the piezoelectric actuators, a precise cutting test can be carried out [1]. Ultrasonic vibration cutting tool is set on a column. The torsional vibration generated by the bolted Langevin-type oscillator is transmitted to the cutting edge of the diamond single-point tool. The oscillation frequency is 27 kHz and an amplitude of the tool edge is maintained at 30 μm during the cutting test.

2.2 Experimental procedure

The workpiece of mirror surface finished silicon wafer was fixed to a workpiece holder by an adhesive. The cutting surface was (100), and the cutting direction was <110>. The workpiece was inclined slightly from the horizontal. The ultrafine motion table was fed in the horizontal direction at feed speed $v = 2.5$ mm/s and a cutting test was performed. In this cutting test method, the depth of cut was continuously increased with table feed, which causes the mode of material removal to change from ductile to brittle. Subsequently, by observing the cutting groove using an atomic force microscope (AFM), the cutting surface quality and the transition of the cutting mode were studied, and the critical depth of cut dc in the ductile-to-brittle transition area was measured.

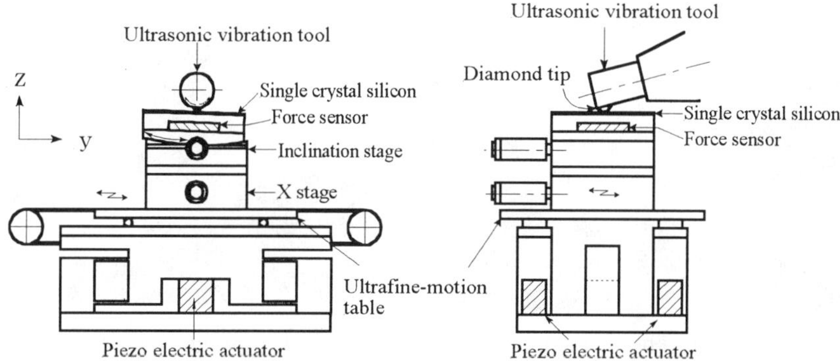

Figure 1   Experimental equipment

3. EXPERIMENTAL RESULTS

3.1 Effects of ultrasonic vibration

Hereafter, cutting without ultrasonic vibration is termed conventional cutting, while cutting with ultrasonic vibration is termed ultrasonic vibration cutting. The difference in cutting grooves between the conventional cutting and the ultrasonic vibration cutting is shown in *Figure 2*.   *Fig. 2(a)* shows an AFM image of a cutting groove obtained by conventional cutting, which indicates that chipping occurred soon after the initiation of cutting. The area in which brittle fracture was initially observed was determined to be the ductile-to-brittle transition area. The critical depth of cut dc in this area was measured and $dc = 0.15$ μm was obtained. On the other hand, an

AFM image of ultrasonic vibration cutting (*Fig. 2(b)*) shows that the depth of cut reaches 3.6 μm, and cracks and chippings are not observed around the cutting groove. Since the shape of the tool edge is also transcribed correctly, it is determined that the groove is cut in the ductile mode. The critical depth of cut was measured by laser microscope and $dc = 8.53$ μm was obtained.

Continuous flow-type chips, which are a feature of ductile mode cutting, were seen occasionally in scanning electron microscope (SEM) observations (*Figure 3*). On the surface of a chip, a line-formed pattern which is vertical to the cutting direction is observed. It indicates that repeated shearing occurred in the formation of chips. These phenomena are regarded as indicative of ductile mode cutting.

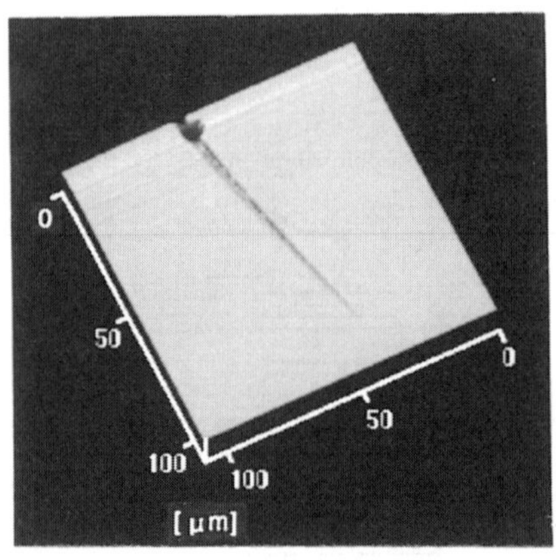

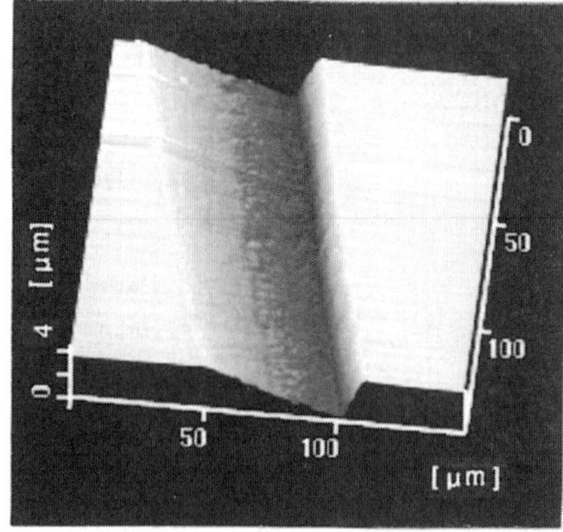

(a) Conventional cutting   (b) Ultrasonic vibration cutting

Figure 2   AFM image of cutting groove

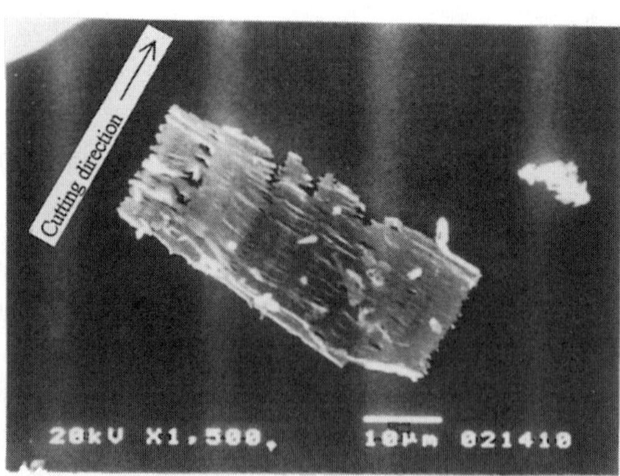

Figure 3   Chip in ultrasonic vibration cutting of single-crystal silicon

3.2   Subsurface damage

Figure 4 shows subsurface images of a cutting groove at approximately depth of cut $d = 5$ μm, observed by a transmission electron microscope (TEM). Since cracks are not observed under the machined surface, this cutting surface is determined to be ductile mode. In the topmost surface layer at approximately thickness of 1 μm, it can be judged that amorphous silicon was formed because an electron-beam diffraction image of this area showed a hollow pattern. Below the amorphous layer, a dislocation defect layer about 4 μm was observed. An electron-beam diffraction image of this area showed a diffraction pattern of silicon single crystal. In addition, the results of TEM observation of the ductile mode chips clarified that the chips were also transformed into amorphous silicon.

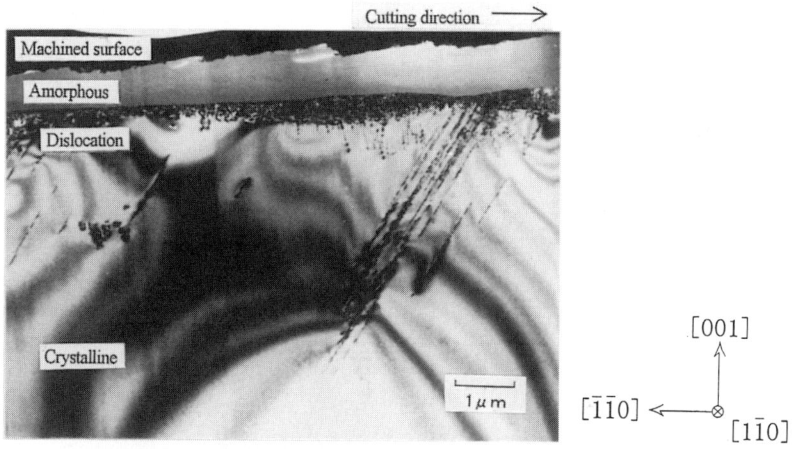

Figure 4   TEM observation of subsurface damage

4.   CONCLUSIONS

In this study, the ultrasonic vibration cutting of single-crystal silicon was conducted. The following conclusions were obtained:

1)   It was clarified that during ultrasonic vibration cutting, ductile chips were produced, the topmost surface layer in the cutting surface was changed to amorphous silicon, and cracks were not observed. Accordingly, it was determined that a ductile mode cutting was carried out.

2)   It was clarified that during conventional cutting without ultrasonic vibration, the critical depth of cut dc was 0.15 μm; however, during ultrasonic vibration cutting, the critical depth of cut dc was increased approximately 60 times to 8.5 μm.

ACKNOWLEDGMENTS

The authors express their sincere gratitude to Mr. H.Suzuki of Fuji Ultrasonic Engineering CO. LTD for his contribution of experimental apparatus.

REFERENCES

1) Shigeomi Koshimizu et al.: "An ultrafine motion table using piezoelectric actuator for ultraprecision grinding", Int. J. Japan Soc. Prec. Eng., Vol.30, No.4 (1996) 345-346.

REGENERATIVE CHATTER VIBRATION IN BALL END MILLING OF CURVED SURFACES

Ikua, B.W.*, Tanaka, H.**, Obata F.*, Sakamoto, S.**

*Graduate School of Tottori University
**Department of Mechanical Engineering, Tottori University
4-101 Minami, Koyama-cho, Tottori, Japan

Abstract

This paper presents an analysis of chatter vibration in ball end milling of curved surfaces using time domain approach. A model for dynamic cutting process, which takes into consideration the variation of helix angle of the ball end mill along the cutting edge, is developed. The vibration of the tool is calculated by using a lumped-parameter model with two degrees of freedom. The chatter stability limit is indicated by the critical nominal depth of cut. The results show that chatter stability is very low for low spindle speeds. Also, the stability is lower for low and high milling position angles, and higher for intermediate milling position angles.

Keywords

Ball end milling, curved surfaces, chatter vibration, stability limits

1. INTRODUCTION

Chatter vibration in machining operations usually has undesirable effects, such as accelerated tool wear, excessive noise, damage of the machine tool, poor surface finish and low dimensional accuracy of the machined part. A commonly used method for avoiding chatter vibrations in machining is to select low spindle speeds, and small depths of cut. However, using this method for chatter-free machining results in low productivity. Therefore, in order to maximize productivity, prediction of chatter vibration is essential.

The theory of chatter vibration for single point cutting tools has been discussed by several authors, for example, Tlusty [3] and Merrit [4]. This theory is applicable in operations such as turning where the directions of the cutting forces can be considered to be time invariant. It is, however, difficult to apply it to a milling process, due to the variation of uncut chip thickness and cutting force vector with spindle rotation. Nevertheless the theory can be used for rough estimation of stability limits in milling so that the apparently stable conditions can be identified prior to carrying out a more accurate numerical computation, thereby reducing the computation time [8].

Recently, various models for the prediction of chatter in end milling and face milling have been proposed [5,7]. An attempt to extend the application of these models to ball end milling, however, presents some difficulties. This is because in ball end milling, the cutting speed, helix angle and, consequently, the effective rake angle vary along the cutting edge. Little work has been reported on chatter vibration in ball end milling [1,2].

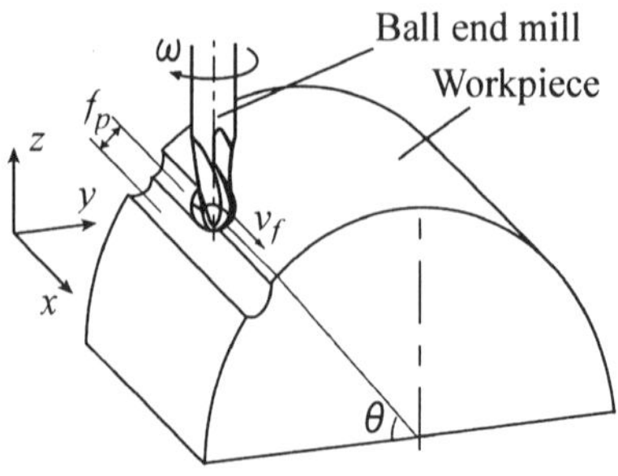

Figure 1. Workpiece geometry

In this paper, the chatter vibration in ball end milling of curved surfaces is investigated (see Fig. 1). The vibration of the ball end mill is approximated by a lumped-parameter model with two vibration modes, which are mutually perpendicular. The dynamic cutting forces are computed based on the tool geometry, the uncut chip thickness and the properties of the workpiece material.

Since the width of cut is dependent upon both the nominal depth of cut and the cross-feed in ball end milling, the critical nominal depth of cut was chosen as the criterion for the characterization of chatter stability limit.

2. MODEL FOR BALL END MILLING

Figure 2 shows the dynamic cutting model for ball end milling of a convex surface. In this figure, θ is the milling position angle, f_p is the cross-

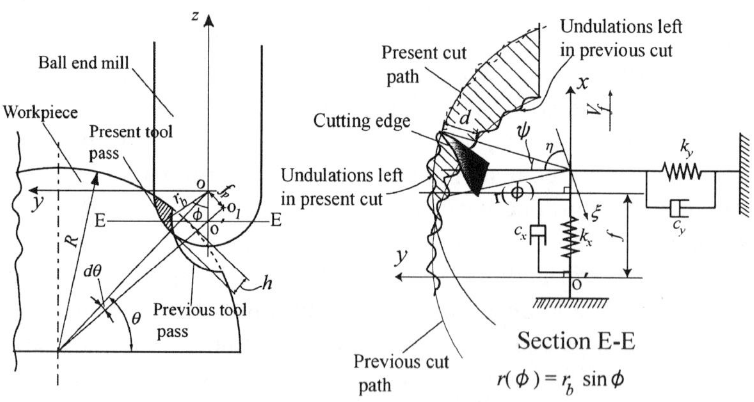

Figure 2. Dynamic cutting model

feed, h is the nominal depth of cut and f is the feed rate. ψ is the angle of rotation of a cutting edge element, measured clockwise from y-axis. k_i and c_i are the modal parameters of the tool-spindle system.

The instantaneous depth of cut $s(\psi,\phi)$ is defined in the radial direction from the ball center O, and it is evaluated in terms of h, f_p, f and θ.

The force components F_x and F_y are calculated from the instantaneous depth of cut, the tool geometry and the properties of the workpiece material [6]. These cutting forces excite vibrations according to the following equation,

$$
\left.
\begin{aligned}
\ddot{x} + 2\zeta_x \omega_{nx} \dot{x} + \omega_{nx}^2 x &= \frac{\omega_{nx}^2}{k_x} F_x \\[2em]
\ddot{y} + 2\zeta_y \omega_{ny} \dot{y} + \omega_{ny}^2 y &= \frac{\omega_{ny}^2}{k_y} F_y
\end{aligned}
\right\}
\tag{1}
$$

where ζ_i are the damping ratios and ω_{ni} are the natural angular frequencies associated with each mode. The tool displacements in the x- and y-directions are computed from Eq. (1), and the resultant tool displacement $\xi(t)$ is then obtained. The direction of the resultant displacement is defined by the angle η, which is given by $\eta = \tan^{-1}(x/y)$.

Since the cutting edges of the ball end mill are helical, the tool displacements have different effects on the instantaneous depth of cut at different points along the cutting edge. This complicates the computation of the uncut chip thickness along the cutting edge. This difficulty can, however, be overcome if we consider the apparent changes in f, h, f_p and θ due to the tool displacements, and modify the uncut chip thickness accordingly. The tool displacement $\xi(t)$ causes apparent changes in f, h, and f_p given by

$$
\left.
\begin{aligned}
\Delta f(t) &= \xi(t)\sin\eta(t) \\
\Delta h(t) &= \xi(t)\cos\eta(t)\cos\theta \\
\Delta f_p(t) &= \xi(t)\sin\eta(t)\sin\theta
\end{aligned}
\right\}
\tag{2}
$$

The new uncut chip thickness can therefore, be computed by incorporating these changes. For example, the apparent feed rate $f(t)$ at a certain instant in nth cut is given by

$$
f(t) = f + \xi(t)\sin\eta(t)\,\big|_{n-1} - \xi(t)\sin\eta(t)\,\big|_n
\tag{3}
$$

The cutting forces are calculated based on the modified chip geometry. The new displacements are stored and used in the next tooth pass, and so on. This process is repeated for a large number of spindle revolutions. If the amplitudes of the displacements grow with time, chatter is likely to occur, and the process is said to be unstable. If the vibrations decay with time, the process is stable. In this study, the nominal depths of cut are increased in steps of 0.05 mm, until when the vibrations are seen to increase.

3. RESULTS

Simulations have been carried out for ball end milling of a curved surface with radius of curvature of 40mm. The diameter of the ball end mill is 16 mm. The local helix angle γ was measured and found to vary with the position angle ϕ according to the following equation; $\gamma = 0.2904 + 0.10163\phi$.

The modal parameters of the tool-spindle system are k_x = 2.41MN/m, k_y = 3.08MN/m, F_{nx}=1850Hz, F_{ny}=1900Hz and $\zeta_x = \zeta_y = 0.02$. These parameters are determined experimentally. The feed rate is 0.05 mm/tooth, and the cross-feed is 0.5mm. The simulations were carried out in steps of 50 min^{-1}.

Figures 3(a) and (b) show computed x-displacements history at h=0.4mm and $\theta = 45°$, for spindle speeds of 1000 and 3000min^{-1}, respectively. These displacements are for the first 30 spindle revolutions. It can be seen that the amplitudes of the displacements increase with time for N = 1000min^{-1}, indicating instability, whereas the amplitudes do not change with time for N=3000min^{-1}, suggesting a stable condition.

Figure 4 shows the stability limits for ball end milling at $\theta = 45°$. It can be seen that the critical depth of cut is very low for low spindle speeds. For the range of speeds considered, higher stabilities are at spindle speeds of about 2500 and 4000min^{-1}.

The influence of milling position angle on the critical depth of cut at N=1500min^{-1} is shown in Fig. 5. It can be seen that the stability is low for low milling position angles ($\theta < 30°$) and higher for intermediate milling position angles ($\theta = 30 \sim 65°$).

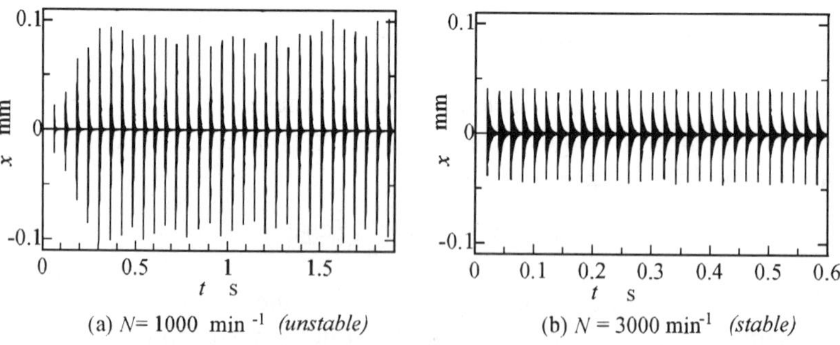

(a) N= 1000 min^{-1} *(unstable)*   (b) N = 3000 min^{-1} *(stable)*

Figure 3. Computed x-displacements (h = 0.4mm)

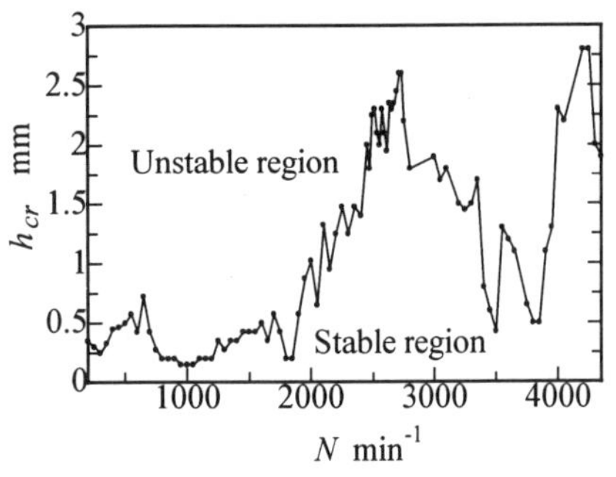

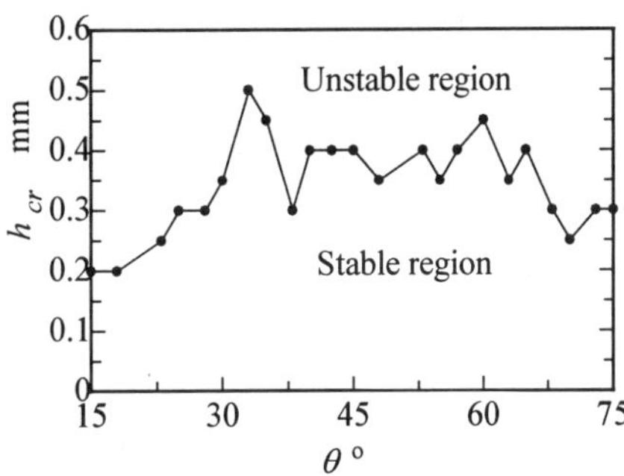

Figure 4. Computed stability limits ($\theta = 45°$)

Figure 5. Influence of milling position angle ($N = 1500$ min^{-1})

4. CONCLUSION

A time domain simulation model for the prediction of chatter vibration in ball end milling of curved surfaces has been presented. The model takes into consideration the variation of helix angle along the cutting edge of a ball end mill.

It was seen that chatter stability is very low for low spindle speeds. It was also seen that the stability is lower for low and high milling position angles, and high for intermediate milling position angles.

The work discussed in this paper has not been verified experimentally. This study is going on and the results of the experimental verification will be presented in near future.

References

[1] Abrari F., Elbestawi M.A., Spence A.D. On the dynamics of ball end milling: Cutting forces and stability analysis. Int J Mach Tools Manufact 1998; 38:215-37

[2] Altintas Y., Shamoto E., Lee P., Budak E. Analytical prediction of stability lobes in ball end milling. Trans ASME J Manufact Sci Eng 1999; 121:586-92

[3] Koenigsberger F., Tlusty J. "Theory of chatter stability analysis." In *Machine Tool Structures Vol. 1*. Pergamon Press, 1970.

[4] Merritt H.E. Theory of self-excited machine-tool chatter. Trans ASME J Eng Ind 1965; 87:447-54

[5] Smith S., Tlusty J. Efficient simulation programs for chatter in milling. Annals CIRP 1993; 42:463-6

[6] Tanaka H., Obata F., Ikua B.W., Sakamoto S., Ashimori M. Cutting forces and machining error in ball end milling of inclined flat surfaces. Int J Japan Soc Prec Eng 1999; 33:319-25

[7] Tlusty J., Ismail F. Basic Non-linearity in machining chatter. Annals CIRP 1981; 30:299-305

[8] Tsai M.D., Takata S., Inui M., Kimura F., Sata T. Prediction of chatter vibration by means of a model-based cutting simulation system. Annals CIRP 1990; 39:447-50

Ultra-high speed Discharge Control for Micro Electric Discharge Machining

Sotomitsu Hara, Nobuhisa Nishioki

Mitutoyo Corporation

Abstract

An ultra-high speed discharge control on micro electric discharge machine shows good performance, both in removal rate and surface roughness. A newly developed current-shutdown circuit that shuts down within 15nsec prevents crater damage. When a short circuit condition between the work and the electrode is detected, the logical judgment skips to the next processing pulse within just two clock cycles, so the discharge hit rate is not reduced.

Keywords

Electric Discharge Machine, Short detection, roughness, removal rate, gap control

1. Introduction

A Micro Electric Discharge Machine (mEDM) using WEDG[1] was developed to process small parts of a few millimeters in size with fine surface roughness. However, its processing speed is slow compared to conventional high power EDMs. It is desirable to have both fine surface roughness and a high processing speed.

In order to get small crater, a conventional mEDM pulsed power supply is made of a CR circuit (Capacitor and Resistor). The size of the discharge crater specifies the surface roughness. Therefore, it is important that this is small. However, in this system a long time is required to charge up the capacitor, so the discharge pulse rate is very low.

Electronically controlled pulsed power supplies are used practically. In this type, the average current is monitored, and the over current is decreased by increasing the gap between the electrode and the work. To increase the current, the gap must be narrowed. For this type, the response is slow, and a quick disturbance makes the system unstable.

Therefore, we have developed an alternative method of gap control and measured the discharge delay time. It is presumed that the normal discharge takes place a few nano-seconds after the voltage is supplied. If both electrodes are in contact and in the short circuit condition, the current increase at the same time after the voltage pulse. Thus, the two conditions are distinguishable, and they can be realized by an ultra high speed comparator using a high speed logic IC.

Generally speaking, a short circuit discharge is caused by a zeroing of the gap by lumps or wire diameter fluctuations. This newly developed high speed current sensing circuit provides just 15nsec time delay for the undesirable short circuit situation. After the delay time, the drive pulse voltage is extinguished perfectly and no craters were observed on the work surface.

2. Experimental Configuration

Fig.1 shows the system of gap control. This construction is the WEDG set up. BS wire runs slowly perpendicular to the plane of the paper. The workpiece is a tungsten rod; set on the Z axis table. A discharge spark removes a microscopic volume of the work, and the Z axis table moves forward compensating the gap expansion. In this way, the cutting process progresses continuously. The table position is controlled by the   NC apparatus pitched at 6.25 nm.

In order to stabilize the discharge condition, the gap should be sensed directly. However, due to several technical issues, this is not possible.

For the electronic circuits, two lines of FETs (Field Effect Transistor) are necessary to minimize the short-detect damage. FFT1 is for detecting short circuit current and is driven by V_{G1}. FFT2 is for processing, and is driven by V_{G2} . For a high speed response, a small packaged FET is preferable. These are cooled by water sinks because they are operated above the power consumption specification. V_{G1} pulse width : T_1 must be as narrow as possible; it can be trimmed to between 20nsec and

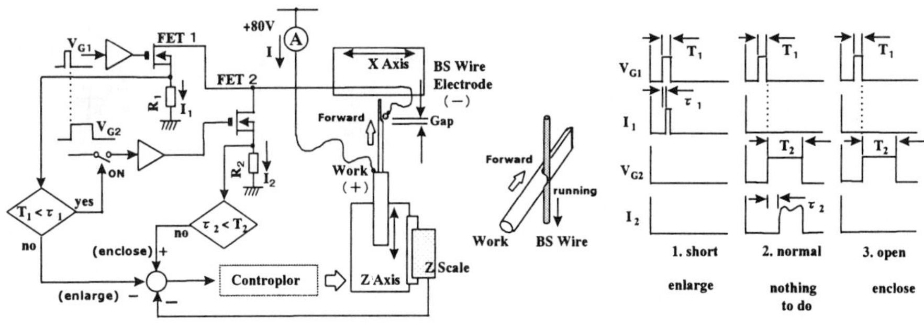

Figure 1.   The Electric Circuit and Algorithm

195

100nsec. V_{G2} pulse width : T_2 is set experimentally to a suitable value between 20nsec and 5000nsec according to the processing speed or roughness. V_{G2} is driven immediately after V_{G1} has fallen. The timing chart is drawn in the right part of Fig.1.

The distinction between normal discharge and the short circuit condition can be distinguished as follows. For normal discharge, the increasing discharge current requires several nano-seconds to ionize water molecules after the high voltage is supplied to the electrodes. The ionizing time is ($T_1 + \tau_2$) as shown in Fig.1.

For the short circuit condition, the current starts at the same time without ionization. By experiments, we confirmed that the ionization time is greater than 20nsec. Therefor if the current I_1 is detected within a period T_1, it is under the short circuit condition. τ_1 is the delay time of the current detection. Then, the next V_{G2} is shutdown, and a back command is sent to the Z axis. The gap is enlarged by repositioning the table. This is repeated until the short circuit condition is canceled.

In the third case, discharge does not occur, $T_2 < \tau_2$, and it is the open circuit condition. Then a forward command is sent to the Z axis to enclose the gap.
A resistor, R_1 (22 ohm) is connected to the FET source. The detection sensitivity is 0.18A. This is low enough so as not to cause damage to the work surface. The sensing current (voltage) is led to the high speed logic IC directly. No other device can be placed between them, since this would limit the pulse speed furthermore. The newly developed circuit with less than 15nsec response time makes it possible to prevent crater damage. The performance is shown in Fig.2. In (a), the V_{G1} and I_1 patterns are shown. The damage is so small that it can hardly be observed by an optical microscope as shown in (b). The minimum crater by single discharge pulse is clearly observed in (c).

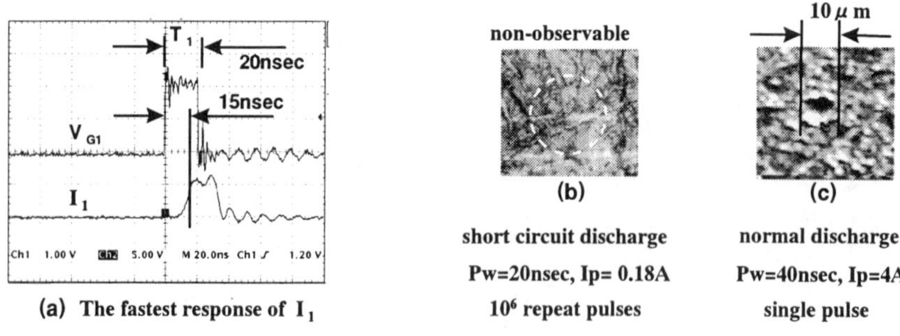

(a) The fastest response of I_1

Figure 2.   Short avoid performance

The cutting operation is examined practically in Fig.3 which shows four pulses for the three cases; the short circuit condition, normal discharge and the open circuit condition. For the short circuit condition, V_{G2} is kept to 0, and, of course, $I_2 = 0$. For normal discharge, I_1 is zero while V_{G1} is "High". Therefore V_{G2} is led to FET2, and the discharge current I_2 is observed. If I_2 is zero while V_{G2} is "High", it is in the open circuit condition.

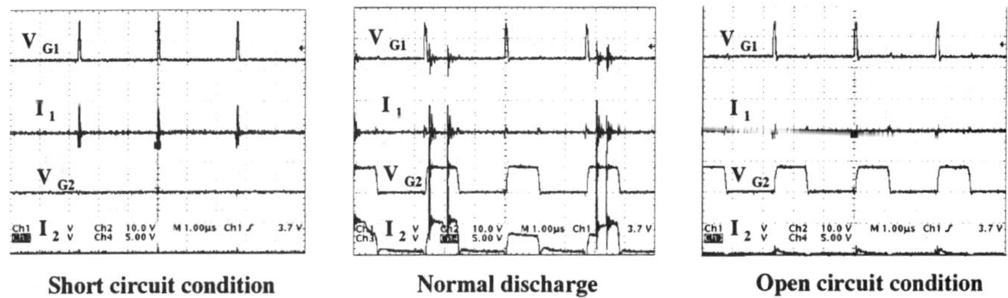

Figure 3.   The Pulse Patterns

The repeat frequency can be varied from 100kHz to 4 MHz. This pulse density is about ten times higher than conventional CR circuit. It is supposed the next values; C=4700pF - 470pF, R=1k, then the repeat frequency becomes 33k - 330kHz.

The processing speed is proportional to the three parameters product; I_2 and drive pulse duty and discharge hit probability. This is the average current itself, and can be set pretty wide range easily than CR circuit.

Results

Fig.4 shows the cut surface of tungsten. Three different resistors are used for 1 ohm, 10 ohm, and 100 ohm, with resulting currents I_2 of 4A, 0.4A, 0.04A respectively. The roughness Ra, Rz, and the removal rate are also listed. These are all cut in a single one pass, and the cutting depth is about 200 micrometer. The repeat rate is 400kpps. $I_2 = 4A$ gives high speed processing. This removal rate is about ten times higher than CR circuit [2]. $I_2 = 0.4A$ gives good performance for both the surface roughness and the removal rate. Long range waving is due to a side discharge of the BS wire, the fine ripples are dues to an outside disturbance or to fluctuations in the diameter of the BS wire.

An unexpectedly result is for $I_2 = 0.04A$. The roughness is comparable or better than

$I_2 = 0.4A$, though the removal rate is not much slower. At such a low current region, it has not until now been possible to realize such a steady gap control. The Z stage replacement is shown in Fig.5. It changes smoothly during the cutting process.

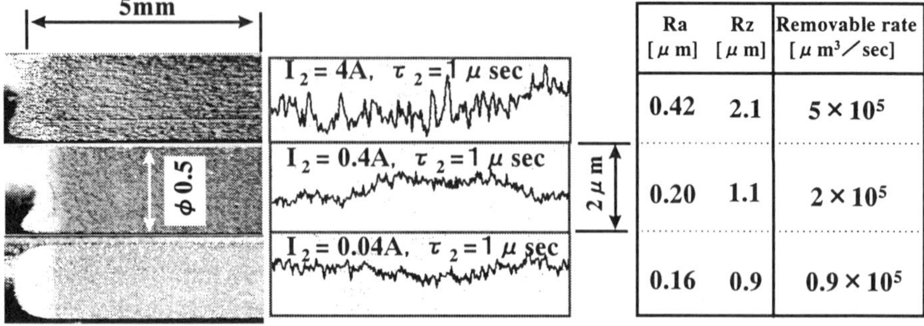

	Ra [μm]	Rz [μm]	Removable rate [μm³／sec]
	0.42	2.1	5×10^5
	0.20	1.1	2×10^5
	0.16	0.9	0.9×10^5

Figure 4. Cut Surface On Tungsten Rod

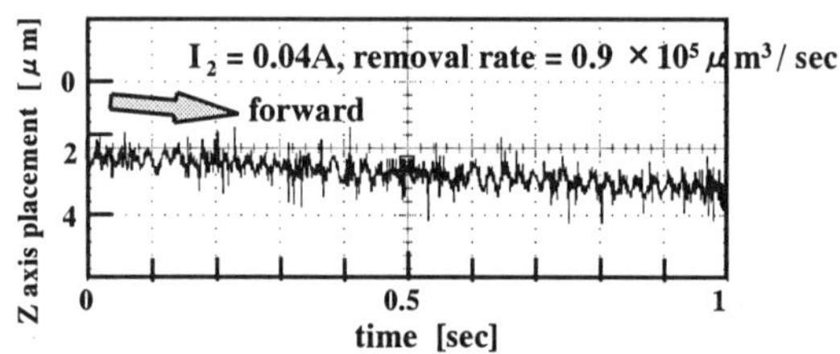

Figure 5. Z axis placement on cutting

4. Conclusion

1. A minimum 20nsec width and 10nsec pitched pulse generator was prepared.
2. A 15nsec protection response can prevent damage under the short circuit condition.
3. The discharge delay time control is stable even in the low current region.
4. Fine surface cutting of a tungsten rod was obtained.

References

1. T.Masuzawa : New Evolution of EDM Tech., J.J.S.P.E, Vol.64, No.12, p1713-1714
2. T.Masuzawa : Development of EDM System for Mass Production of microholes, SEISAN-KENKYU Monthly Journal of I.I.S., Unv.Tokyo, Vol.52, No.9, p34-37
3. Discharge Handbook No.2, P63, I.E.E.J., ISBN4-88686-308-6

STUDY OF CONTOURING MICRO EDM CHARACTERISTICS

Z. Yu, K. P. Rajurkar and P. D. Prabhuram

Center for Nontraditional Manufacturing Research
University of Nebraska-Lincoln
175 Nebraska Hall
Lincoln, NE 68588-0518, USA

Abstract

The integration of CAD/CAM system with micro EDM has simplified the procedure for generating complex 3D shapes, using simple shaped electrodes. Therefore, it is very important to understand the behavior of the contouring micro EDM process to achieve optimized machining results. This paper reports a comprehensive study on the influence of various parameters associated with the process.

Keywords

Micromachining, EDM, Electrode Wear, Material Removal Rate

1. INTRODUCTION

The recently developed uniform wear method has been proved to solve the electrode wear problem for the contouring 3D micro EDM by compensation during machining and has been integrated with commercial CAD/CAM software to generate tool paths for machining complex 3D shapes [Yu, Z., 1998, Rajurkar, K.P., 2000]. However, the process is time-consuming. In order to minimize machining time and enhance micro EDM productivity, it is essential to develop a database and use the most appropriate machining parameters. Although some results on the effect of frontal electrode gap and area on the process efficiency have been reported [Kruth, J.P., 1992], the comprehensive information about the influence of machining parameters on the process performance and the specific guidelines of selecting appropriate parameters are not available. This paper presents the results of an experimental investigation on the effect of machining parameters such as current, voltage, layer depth and feed on the material removal rate, electrode wear ratio and gap associated with the micro EDM process.

2. EXPERIMENTS

The experiments for the contouring micro EDM were conducted by machining slots in layers using the commercial Panasonic MG-ED72W micro EDM. The input parameters and experimental conditions listed in Table 1. The material removal rate (MRR) which is the rate of removal volume of workpiece to the machining time, electrode relative wear ratio (EWR) and discharge gap calculated from the difference between the slot width and the electrode diameter were measured. Additional experiments were carried out with micro hole drilling process with different capacitance values for comparison.

Table 1 Micro EDM Experimental Conditions.

Pulse generator	Relaxation type
Open circuit voltage	70, 80, 100 (V)
Capacitor	3300, 220, 100, 10 (pF)
Max. electrode feed	7, 14, 30 (μm/sec)
Layer depth	1, 2, 5 (μm)
Workpiece material	Stainless Steel AISI 304
Electrode material	Tungsten
Electrode diameter	50μm
Slot length	680μm
Designed slot depth	150μm

The length and width of machined slots were measured using a high resolution MicroVU model 441 video measuring system. The depth of the slots was measured by detecting several points along the bottom surface of the slot using an electrode of diameter smaller than the width of the slot. Using the electrode to detect the same reference point on the workpiece surface before and after machining the slot, the electrode wear length was calculated from the difference in the coordinate values. The electrode diameter was measured using the video measuring system.

3. ANALYSIS OF EXPERIMENTAL RESULTS

3.1 Influence on Material Removal Rate (MRR)

The MRR is a function of the discharge pulse energy. The pulse energy varies with either the voltage or capacitance. Increasing pulse energy results in the increase of MRR. Figure 1 shows the variation of MRR with respect to voltage. A similar trend is reflected in micro EDM hole drilling with the variation of capacitance as shown in Figure 2.

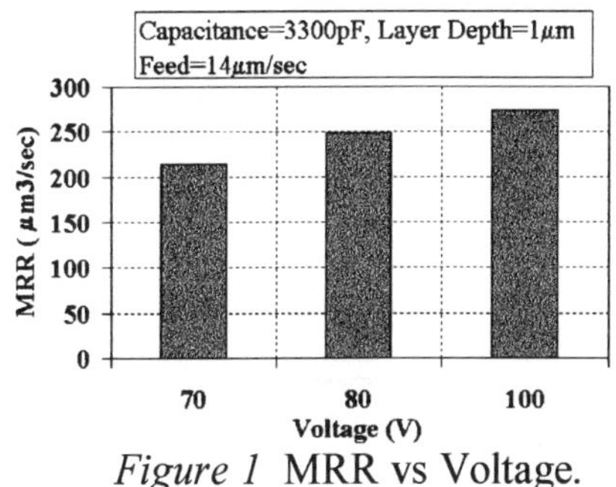

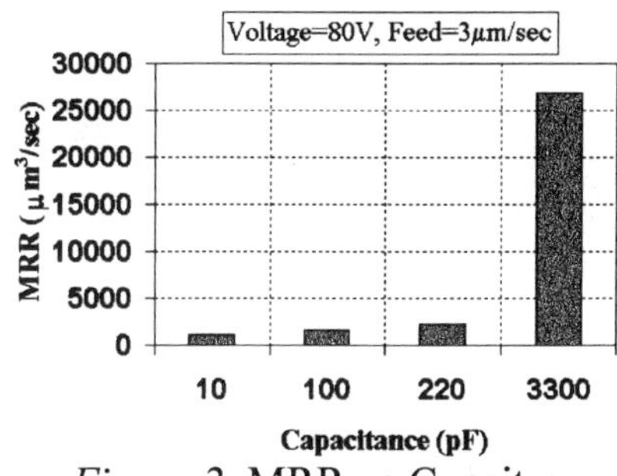

Figure 1 MRR vs Voltage.

Figure 2 MRR vs Capcitance
(For micro hole EDM drilling).

However, in micro contouring EDM, the MRR decreases with the increase in the capacitance i.e. discharge energy, as shown in Figure 3. This is because the applied feed rates to the tool electrode are so small that most of the time is spent on electrode movement without machining. Increasing the feed rate of the electrode leads to an increase in MRR (Figure 4). The increase of layer depth causes an increase in discharge area and hence results in higher MRR.

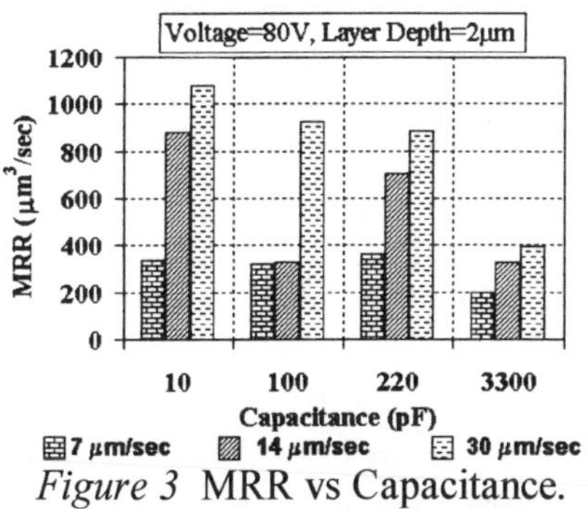

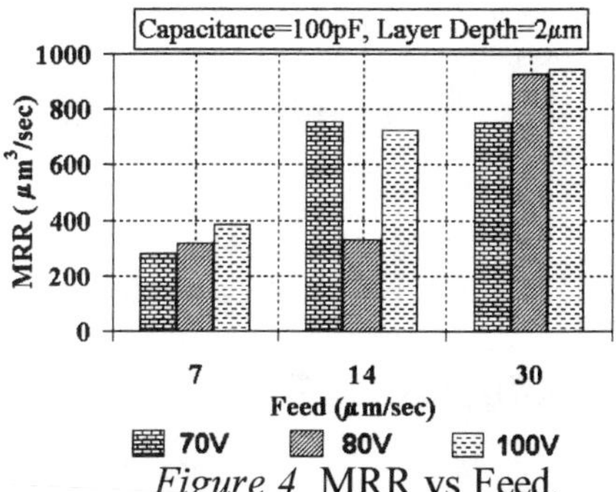

Figure 3 MRR vs Capacitance.

Figure 4 MRR vs Feed.

3.2 Influence on Electrode Wear Ratio (EWR)

The variation of EWR with voltage and capacitance are shown in Figures 5 and 6. It can be seen that the EWR increases with the increase in either voltage or capacitance. This variation in EWR is proportional to the discharge energy. There is a steep rise in the EWR with the increase in capacitance, particularly for the value of 3300pF. With the increase in the capacitance, the discharge energy increases. As a result of this, more heat is generated and the electrode consumption increases.

The EWR is also related to the discharge status. When the layer depth increases, the amount of workpiece material removed increases. When the layer depth increases for a particular feed, the time required for material

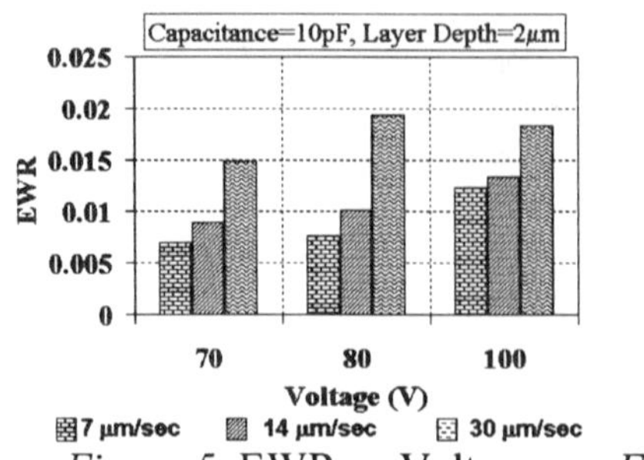

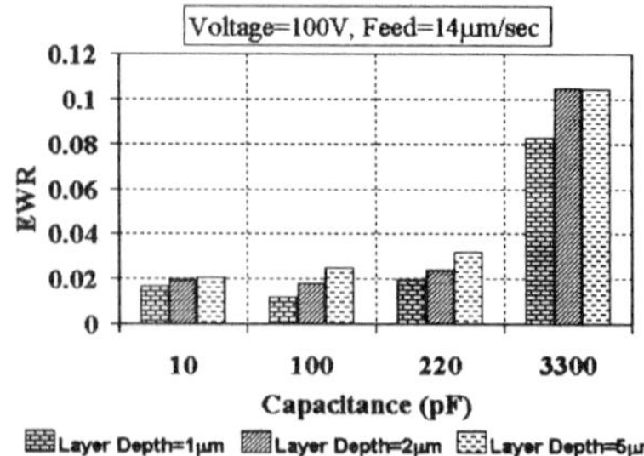

Figure 5 EWR vs Voltage.   *Figure 6* EWR vs Capacitance.

removal may be short and hence results in abnormal discharges leading to higher wear. The increase in EWR for higher layer depths is shown in Figure 7. When the electrode moving speed (feed) increases, the time available for machining the material is less and thereby results in abnormal sparking. It can be observed in Figure 8.

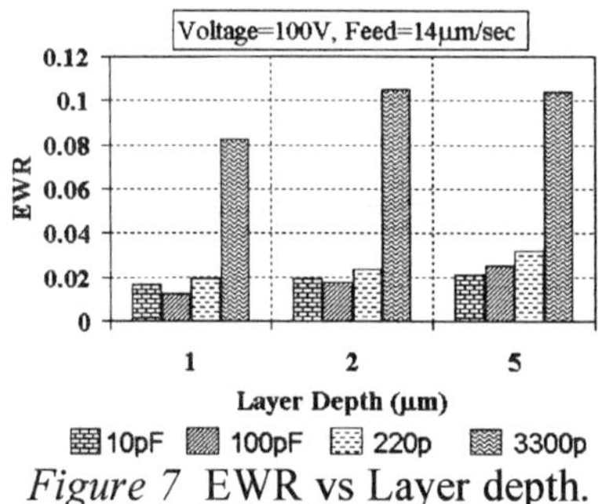

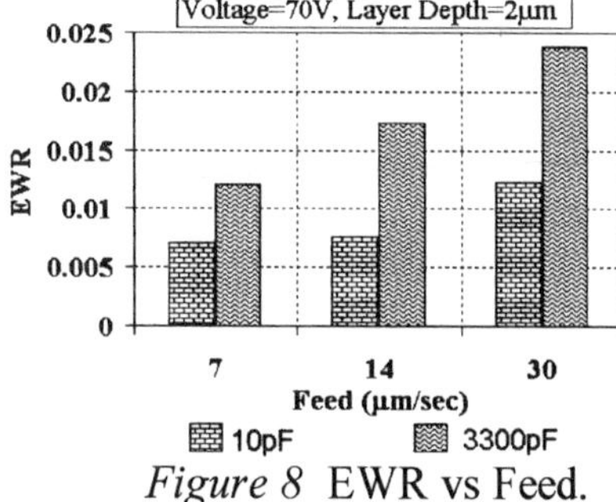

Figure 7 EWR vs Layer depth.   *Figure 8* EWR vs Feed.

3.3 Influence on Gap

The gap is a function of the discharge energy. The increase in capacitance or voltage results in larger discharge causing larger gaps (Figures 9 and 10).

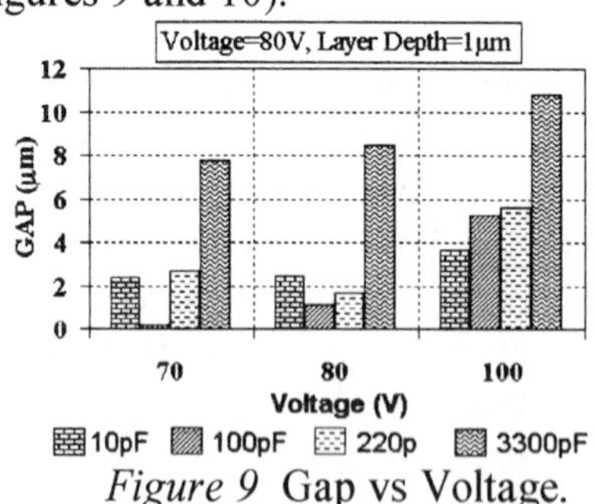

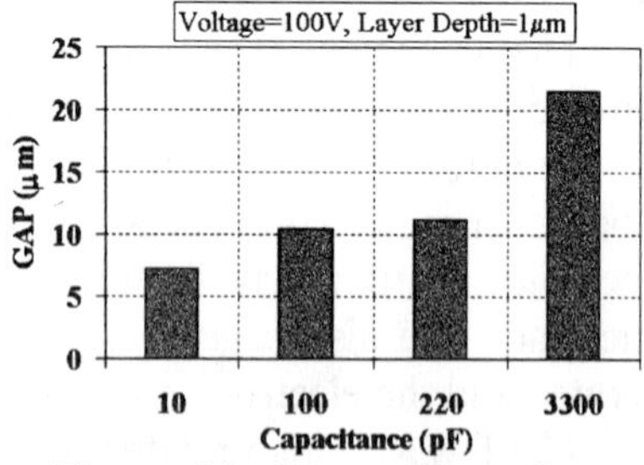

Figure 9 Gap vs Voltage.   *Figure 10* Gap vs Capacitance.

3.4 Profile of the bottom surface of machined slot

The profiles of bottom of the machined slots were measured by detecting several contact points along the bottom of the surface along the tool path. It was found (Figure 11) that the profile is within the layer depth except for some individual points, which were caused by the measurement error.

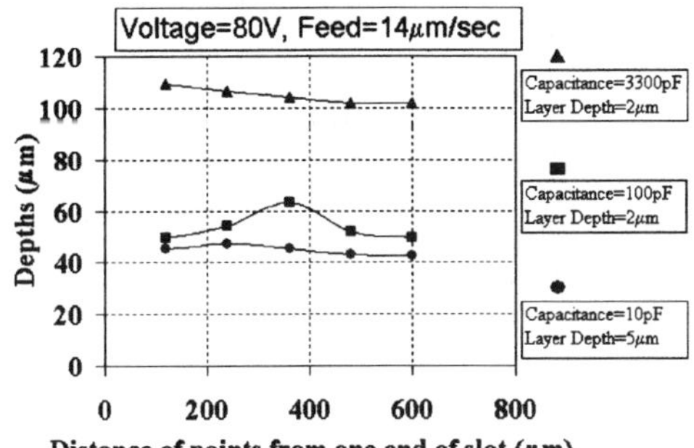

Figure 11 Bottom surface profile.

4. SUMMARY

This paper reports a study of the machining characteristics of contouring micro EDM. The experimental investigation provides some basic information about the influence of micro EDM parameters and may eventually lead to optimal parameters for machining 3D shapes. Moreover, the experimental results serve as a machining database for the current and future needs.

5. ACKNOWLEDGEMENTS

This study was conducted with the support from the Nebraska Research Initiative Fund (NRI) and the NSF grant #DMI-9908219. K.P. Rajurkar is currently on leave of absence at the NSF.

6. REFERENCES

1. Kruth, J.P., Lauwers, B., Clappaert, W. A Study of EDM Pocketing, Proceedings of the International Symposium For ElectroMachining X Conference; 1992; Magdeburg, Germany; 121-135.
2. Rajurkar, K.P., Yu, Z. 3D Micro-EDM Using CAD/CAM. Annals of CIRP 2000; 49/1: 127-130.
3. Yu, Z., Masuzawa, T., Fujino, M. 3D Micro-EDM with Simple Shape Electrode, Part 1: Machining of cavities with sharp corners and electrode wear compensation. International Journal of Electrical Machining 1998; 3:7-12.

DEVELOPMENT OF DESK-TOP MICRO ELECTRIC DISCHARGE MACHINE

Satoshi Koga,   Nobuhisa Nishioki,   Sotomitsu Hara

Mitutoyo Corporation

Abstract

In order to measure a size of fine parts, it is necessary to prepare a measuring instrument which is made up of more fine parts element. For that reason, we have developed the desk-top micro electric discharge machine to process fine parts. A stable processing requires well conditioned electrolyte and well controlled discharge gap. This paper mainly presents an mechanism outline of our machine. The moving stage is given a suitable performance of positioning by the application of voice-coil motor and nm linear-encoder.   And the wire feeder as an electrode is given a high stability by the double capstan method. The results from test processing obtained that our machine has a performance to process a fine parts.

Keyword

Micro-machining,   Electric discharge machining,   WEDG,
Precision positioning,   Linear Motor,   Wire feeding,   Double capstan,

1. INTRODUCTION

The desk-top micro electric discharge machine using WEDG[1] has been developed to process a mm sized fine works into parts of several 10µm sized microscopic mold patterns.   In the WEDG system at a first step, a tool which is suited for an object of a mold pattern is processed by using a wire electrode from material of a straight bar installed in the spindle.   And next, a work which become a new object of a mold pattern is processed by using the tool as an electrode without detaching from the spindle.   It is necessary to restrain the gap fluctuation between a work and a processing electrode within 100nm because of the needs to perform the electric discharge processing with high stability and quality for the demonstration of highly controlled electric discharge power supply (reported separately). Therefore, the means that makes fine feeding of the wire as electric discharge electrode and high positioning ability of a processing stage is indispensable. For the purpose, we have developed a highly precision positioning stage fitting for the part size and a highly accurate wire feeder (WEDG unit). The structure of the device is reported as follows.

2. THE MECHANISM OUTLINE

The appearance of a desk-top micro electric discharge processing machine is shown in *Figure 1*. The apparatus size is W400xD300xH200. In the spindle, to obtain non-contacted high accurate turn with highly controlled stiffness, an air bearing and a built-in spindle motor are adopted. By using a up-and-down Y stage of a fixed work installed in place of the WEDG unit on the X stage, 3 dimensional pattern machining at the next step has become possible as well.

The structure of the stage (X-axis) is shown in *Figure 2*. The cross roller guide, a high resolution hologram-scale detecting the stage position, and a voice coil motor is in use to drive

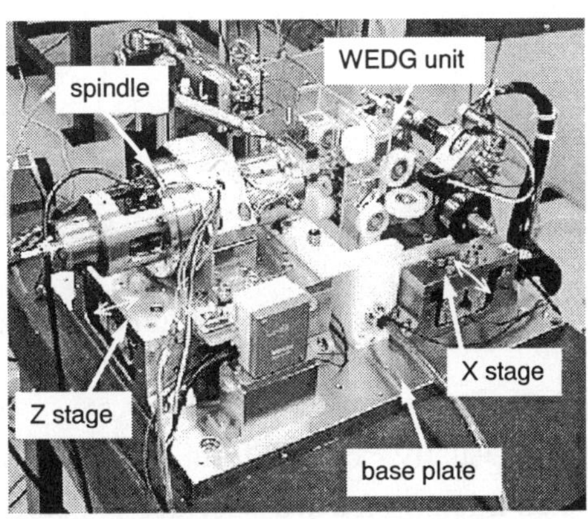

Figure 1. Desk-Top Micro Electric Discharge Machine

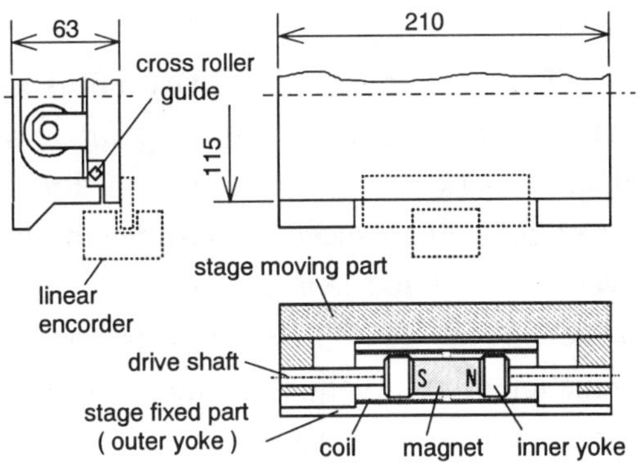

Figure 2. Stage schematic view

the stage. And *Figure 3* shows the block diagram of stage positioning controller. It is very effective to a high accurate positioning to drive the voice coil by direct current control because there are no element of elastic coupling like parts for coupling or joint and no thrust ripple. *Figure 4* shows a performance of the stage positioning which is sent 12.5nm steps and the measurement result of the stage positioning of when that is done. The positioning fluctuation is less than 6nmp-p, and there is no lost-motion during a return movement. The maximum static thrust of the stage is 30N at the time of 60W electric power consumption.

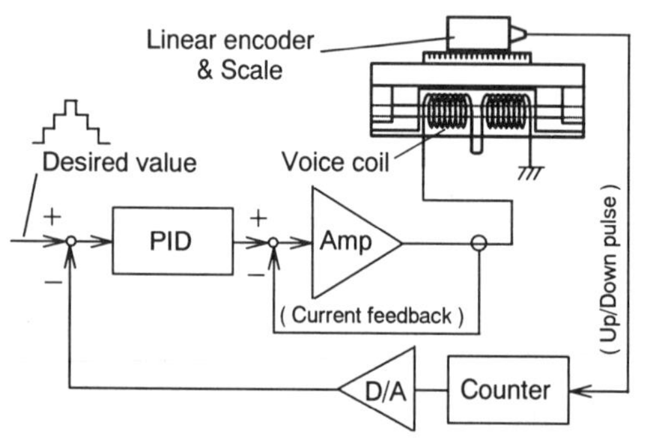

Figure 3. Block diagram of
stage positioning controller

Figure 4. Performance of stage
positioning

Usually, a wire form has local lumps and diametrical fluctuation. When a lump on a wire surface is detected, the damage of mis-discharge is avoided easily by turning the electric discharge from ON to OFF while the lump passing through the work area. But the diametrical fluctuation which can not be detected by the lump detector, must be controlled precisely to avoid mis-discharge.

The structure of the WEDG unit is shown in *Figure 5.* The double capstan method is done and used for a wire supplying and winding by independent two pared capstans with three pinch rollers respectively. By controlling the turn speed and also the torque of each capstan, specified wire traveling speed and tension are obtained. The result of the wire surface fluctuation in the WEDG unit is shown in *Figure 6.* Because the fluctuation of the rotation speed and wire tension between two capstans are restrained, it is able to make the wire surface fluctuation of a short cycle within 100nmp-p.

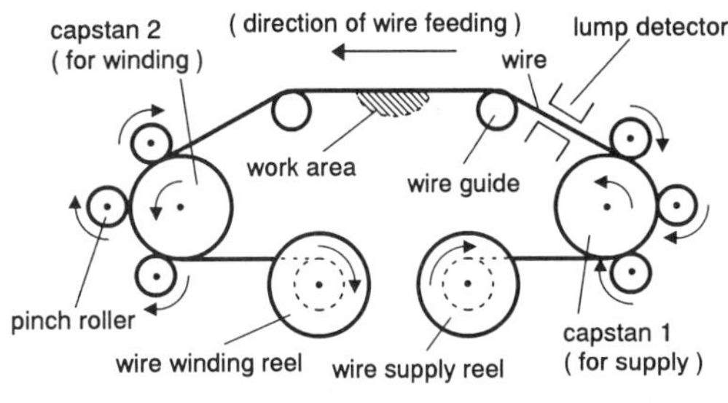

Figure 5. WEDG unit schematic view

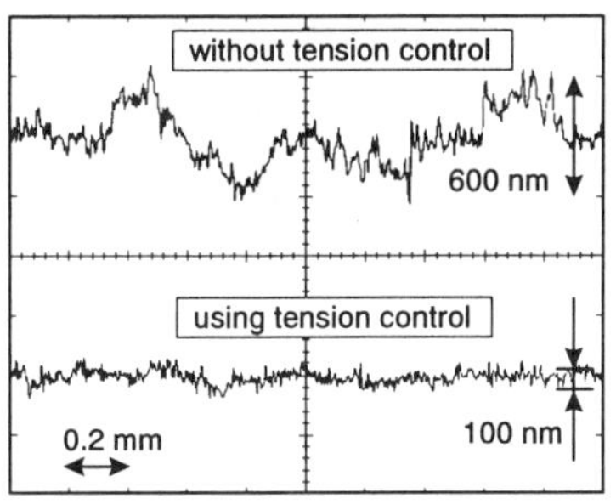

Figure 6. Wire pitching in feeding

3. THE DISCHARGE-CONTROLLER OUTLINE

The block diagram of discharge-controller for our machine is shown simply in *Figure 7*. The feature of this controller is to generate the discharge pulse at higher speed with using FET and a digital logic circuit. A waveform example of the discharge pulse which its width controlled by 10nsec step is shown in *Figure 8*. Our controller is able to control the time of the pulse width from 20nsec to 2500nsec, and to generate the high density pulse about hundred times as much as the conventional type using CR discharge. The gap between the work and the electrode is controlled to be not "too near" or "too far" for monitoring the waveform of discharge pulse train with time resolution of 10nsec order. These functions make it possible to process with high speed and quality.

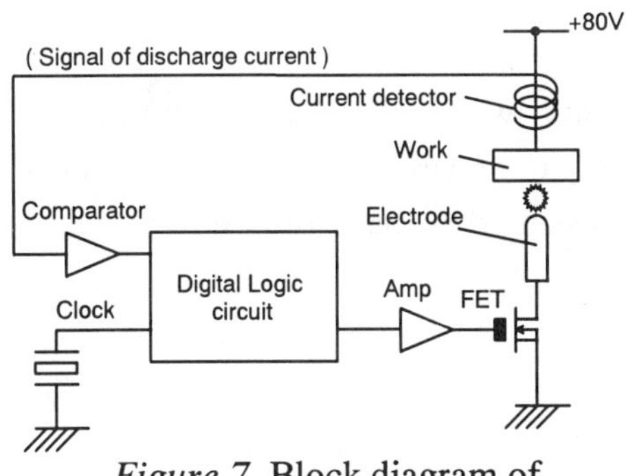

Figure 7. Block diagram of discharge controller

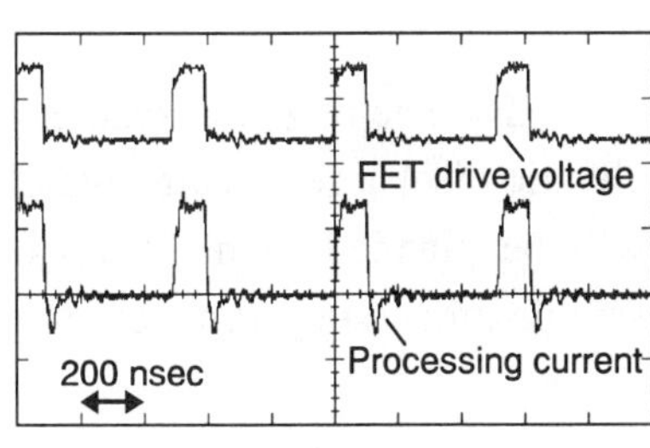

Figure 8. Waveform example of discharge pulse trains

4. RESULT OF TEST PROCESSING

Figure 9 shows the actual example of test processing. It is a micro ball-stylus for touch-detection sensor which is an application of measuring for the form of small hole inner or fine-pitch gear face. Moreover, the example of flat plate which is cut from cylindrical work and obtained by controlling the rotation angle of spindle axis is shown in *Figure 10*. These results from test processing show that our desk-top micro electric discharge machine has a performance to process a fine works. However, it is necessary to continue study for this machine, due to insufficient data of processing condition and accuracy.

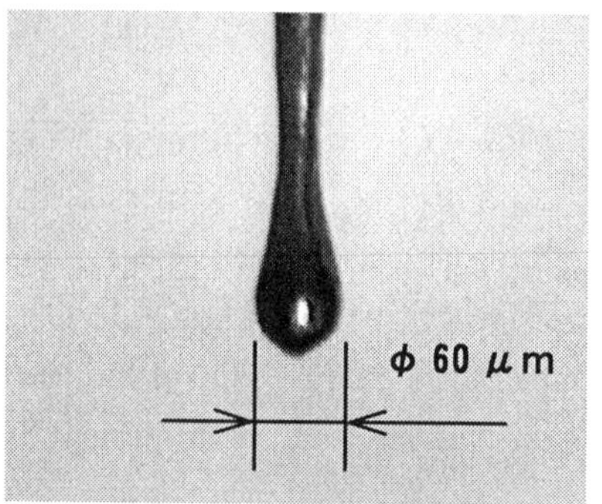

Figure 9. The processing example for a ball-stylus

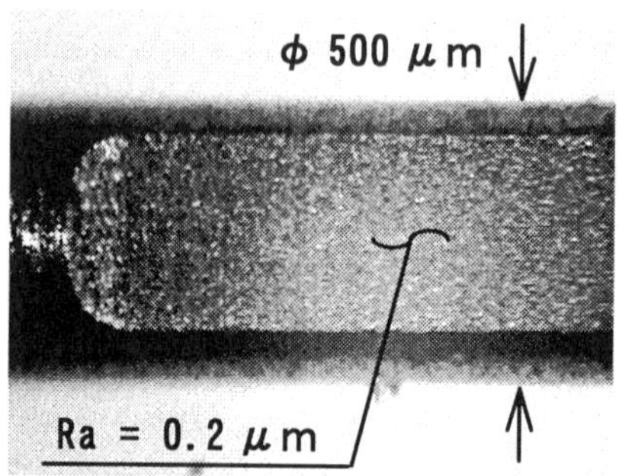

Figure 10. The example of processed surface

5. CONCLUSION

By keeping the system elements, we develop the desk-top micro electric discharge machine which is able to process fine parts accurately. It will be planned to make up a processing-data base through the various sample processing from now on.

Reference

(1) T Masuzawa & M Fujio : Wire Electro-Discharge Grinding for Micro-Machining, Ann. CIRP, 34, 1, 1985, pp431.

IMPROVING PROCESS CHARACTERISTICS OF DRY-WEDM

Chika Furudate*,   Masanori Kunieda*,   Yu Zhan Bo**,   and Hisanori Yamada**

Affiliation: * Tokyo University of Agriculture & Technology

Dept. of Mechanical Systems Engineering

** Sodick Co.,Ltd., Technical Engineering Dept.

Abstract

This paper describes the first attempt to conduct rough-cutting with dry wire electrical discharge machining (dry-WEDM). With the dry method, there is no corrosion, and edges can be cut more sharply than the conventional method. One drawback of the dry method however is the lower material removal rate compared to the conventional method. Moistened wire methods were thus also newly tested. It was found that the use of the least minimum dielectric liquid quantity required can improve the material removal rates of both dry-roughing and –finishing, and moreover the machining accuracy does not deteriorate to a great extent in dry-finishing.

Keywords

Dry-WEDM, rough-cut, finish-cut, machining accuracy, material removal rate, electrolytic corrosion

1.INTRODUCTION

The conventional electrical discharge machining (EDM) process is generally carried out using dielectric liquid. Hence, at the discharge spot, a bubble is generated due to the evaporation and dissociation of the dielectric liquid. The bubble expands explosively in the working gap, but rapid expansion is prevented by the influence of the inertia and viscosity of the dielectric liquid, resulting in extremely high pressure inside the bubble. Kunieda et al. [1] found that when the EDM gap is filled with dielectric liquid, a considerably large process reaction force is applied to the tool electrode at the moment dielectric breakdown occurs. On the other hand, dry-EDM is a new EDM process which is conducted in a gas atmosphere without using dielectric liquid. Dry-EDM was first attempted by Kunieda et al. [2] and applied to die-sinking process. They found that dry-EDM is characterized by extremely low tool electrode wear, shorter gap length, thinner white layer, lower residual stress, and ecologically clean process. Furthermore the process reaction force was found to be negligibly small when a discharge occurs in a gap filled with gas instead of liquid [1].

Since wire-EDM (WEDM) uses as tool electrode a thin and flexible wire, the wire electrode is subject to deformation and vibration due to the above mentioned reaction force and the electrostatic and electromagnetic disturbing forces [3] [4], resulting in unfavorable geometrical error of the machined surface. This fact motivated Furudate et al. [5] to perform a WEDM finish-cut in a gas atmosphere and found that this dry-WEDM produces excellent straightness of the finished surface because the process reaction force is negligibly small. Moreover, since there is no need to use water as dielectric liquid, corrosion of the workpiece can be prevented. However, dry-WEDM was found to have certain problems such as: [1] more susceptible to wire breakage due to localization of discharge spots, [2] lower material removal rate than the conventional WEDM, because of frequent occurrence of short circuiting resulting from the narrower discharge gap length. Hence, the present study aimed to propose some methods to resolve these problems.

So far dry-WEDM has been applied to finish-cutting only, because rough-cutting in dry condition was thought to be impossible due to the difficulty of removing debris particles out of the working gap in the absence of the dielectric liquid. However, if WEDM can be conducted from rough-cutting to finish-cutting consistently in dry conditions, water circulating, filtering and deionizing units or tank can be eliminated and the machine can be made compact. Furthermore, corrosion can absolutely be avoided and a higher machining accuracy achieved with fewer finish-cutting steps. Hence, the present work attempted to perform rough-cutting in dry conditions for the first time.

2. EXPERIMENTAL METHOD

To prevent wire breakage and to improve the material removal rate, the following two dry methods were newly tested, and their machining characteristics were compared with those of the conventional-WEDM method and normal dry-WEDM method. One is the moistened wire with water method, and the other is the moistened wire with oil method. With the conventional method, water, whose electric conductivity was adjusted moderately, was jetted from the upper and lower nozzles. With   the dry-method, no water was supplied and cutting was done in atmosphere. In the moistened wire methods, the wire surface was wiped using an absorbent cotton soaked with tap water or EDM oil just before the wire passes the upper wire guide. Another alternative for supplying minute amounts of dielectric liquid is by the use of mist, which was proposed by Tanimura et al. [6]. EDM using mist, however, does not produce machining accuracy comparable to dry-WEDM, which does not differ from the conventional WEDM to a considerable extent [5].

In finish-cut, the workpiece surface was preprocessed by surface grinding. In order to compare the machining characteristics, the actual depth of cut should be equal among the four methods tested: conventional, dry, moistened with water, and moistened with oil. Hence, different offsets were adopted according to the method to obtain equal actual depth of cut of about 9 μm. Cold tool steel (SKD11) 32mm in

Table 1. Working conditons

Wire electrode	ϕ 0.2, brass
Wire tension	10N
Wire winding speed	250mm/s
Workpiece	Cold tool steel (SKD11) 5mm in thickness (rough-cut) 32mm in thickness (finish-cut)
Discharge current	12A
Servo reference voltage	40V
Discharge period	$10\,\mu$s

Figure 1. Observation of rough-cut corners

	Conventional	Dry	Moistened with water	Moistened with oil
SEM ←→ 230 μm				
Optical microscope ←→ 50 μm				

thickness was used as the workpiece.

In rough-cut, a cold tool steel (SKD11) 3mm in thickness was used as the workpiece. It is well known that, when rough-cut of corner is carried out, vibration and distortion of the wire electrode cause over-cut of corner shapes and rounding off of sharp edges. Hence, a corner with right angle was cut to evaluate the cutting accuracies of the four methods.

The same working conditions that were used for finish-cut were used for rough-cut as shown in Table 1. This is because high energy pulses normally used for rough-cutting caused frequent wire breakage in rough-cutting under dry conditions.

3. APPLICATION OF DRY-WEDM TO ROUGH-CUT
3.1 Observation of WEDMed corner

The upper row in Figure 1 shows SEM images of rough-cut corners. The lower row shows optical microscopic photos of the workpiece surfaces which were etched after polished. With the conventional method, many pits caused by electrolytic corrosion was found on both the upper surface and rough-cut surface. The edge was also dulled. By contrast, in the dry-method, there was no corrosion, and the edge was extremely sharp. In the moistened with water method, the surfaces were damaged by electrolytic corrosion to a large extent, and significant rounding off of the edge was seen. Damage caused by electrolytic corrosion was greater than the conventional method, because tap water was used without deionization nature. In the moistened with oil method, a thick layer of resolidified materials was found adhered on the cut surface.

3.2 Gap length

Figure 2 shows a comparison of the gap length in rough-cutting. Half of the difference between the width of the cut groove and the diameter of wire (200 μm) was taken as the gap length. The measured gap length in the dry-method was narrower than that in the conventional method. In the moistened with oil method, the gap length was the smallest of all conditions. However, since the resolidified layer is removed by polishing for practical use, the resultant gap length after polishing will be 14 μm, which is the same value as that in the conventional WEDM.

3.3 Material removal rate

The material removal rates of rough-cutting are shown in Figure 3. The material removal rate in the dry-method was found to be one eighth of that in the conventional method. The material removal rate in the moistened with water method was almost equal to that in the conventional method, while that in the moistened with oil method was even higher than that in the conventional method.

4. IMPROVEMENT OF FINISH-CUTTING CHARACTERISTICS

4.1 Straightness

Figure 4 shows the difference in straightness of the finished surface measured parallel to the wire axis. The straightness in the moistened with water or oil method was worse than that in the dry-method, but considerably better than that in the conventional method. Under the conditions used in this experiment, all the surfaces finished were concave.

4.2 Surface roughness

Figure 5 shows a comparison of the surface roughness. The surface roughness was best in the dry method, and worst in the moistened with oil method.

4.3 Gap length

Figure 6 shows a comparison of the gap length. Gap lengths obtained from the dry and moistened wire methods were much narrower than that of the conventional method, suggesting that methods using less

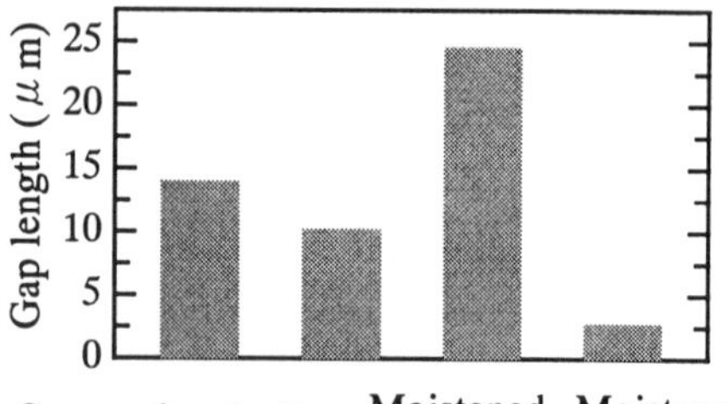

Figure 2. Comparison of gap length in rough-cut

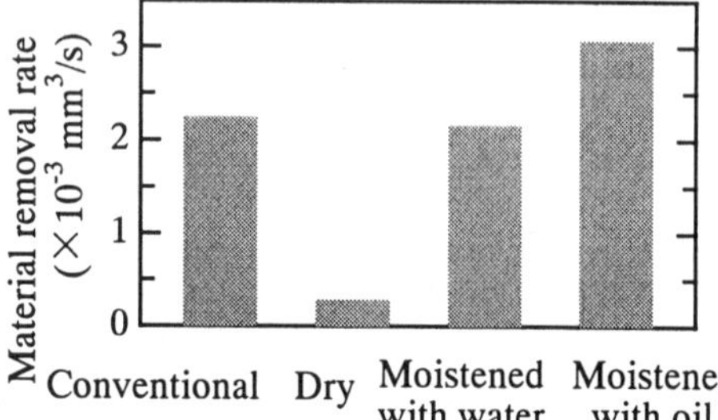

Figure 3. Comparison of material removal rate in rough-cut

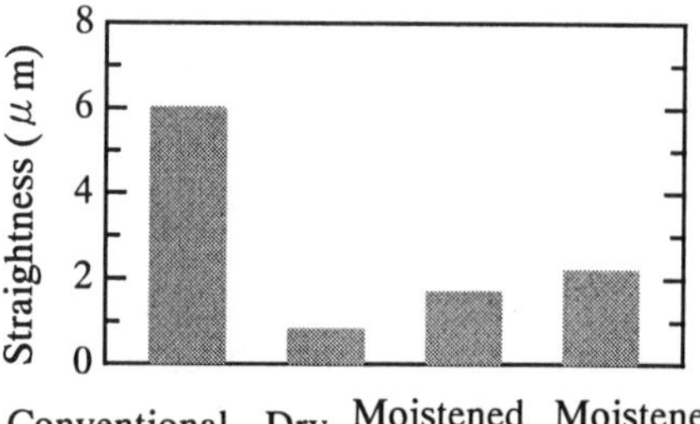

Figure 4. Comparison of straightness in finish-cut

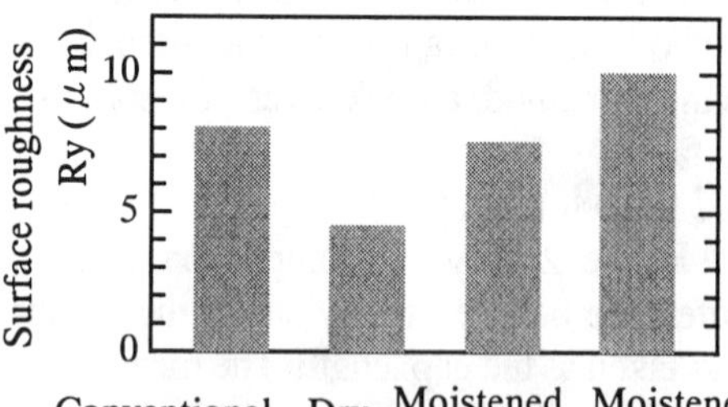

Figure 5. Comparison of surface roughness in finish-cut

212

dielectric liquid are more advantageous for finishing more complicated and finer contours.

4.4 Material removal rate

Figure 7 shows a comparison of the material removal rate between the four methods. Moistened wire methods are found to be very effective for increasing the material removal rate.

5.CONCLUSION

Rough-cutting was successfully performed for the first time in dry conditions, and the following conclusions were obtained:

(1) The accuracy of dry-rough cutting is better than that of the conventional method. However, material removal rate is much lower.

(2) With the moistened wire methods, the material removal rate can be increased to more or less the same as that in the conventional method. In the moistened with water method, however, surface quality is damaged by electrolytic corrosion. In the moistened with oil method, surface quality also drops due to the adherence of a thick layer of resolidified materials.

The moistened wire methods were also tested in order to improve the characteristics of dry-finish-cutting. It was found that least minimum dielectric liquid quantity required can improve the material removal rate without deteriorating the machining accuracy to a great extent.

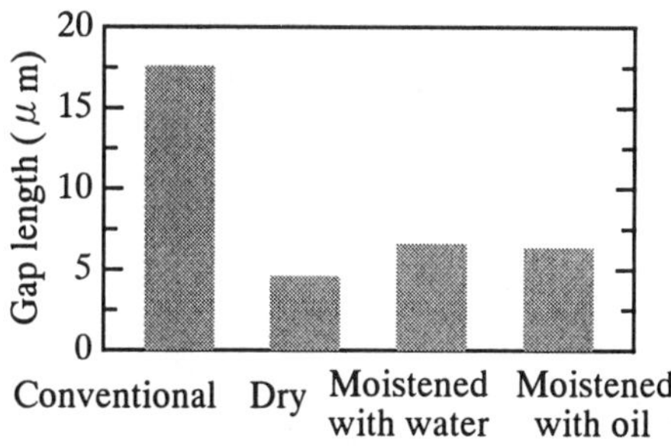

Figure 6. Comparison of gap length in finish-cut

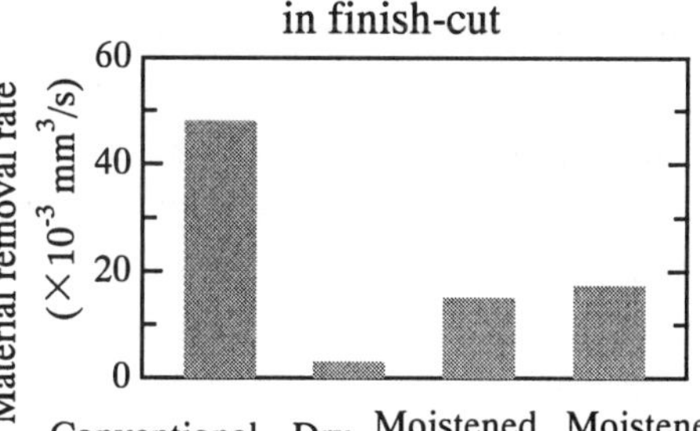

Figure 7. Comparison of material removal rate in finish-cut

ACKNOWLEDGEMENT

The authors wish to express their appreciation to the Die and Mold Technology Promotion Foundation for funding this study.

REFERENCES

1) Kunieda, M., Adachi, Y., and Yoshida, M. ; Study on Process Reaction Force Generated by Discharge in EDM, Proc.MMSS'2000, 2000, 313-324.

2) Kunieda, M. et al. ; Electrical Discharge Machining in Gas, Annals of the CIRP, 1977, 46, 1, 143-146.

3) Dekeyser, W. L. et al. ; Geometrical Accuracy of Wire-EDM, Proc. of ISEM 9, 226-232.

4) Obara, H. et al. ; Simulation of Wire EDM (3rd Report), JSEME, 2000, 34, 75, 30-37(in Japanese).

5) Furudate, C., and Kunieda, M., 2000, Study on Dry-WEDM, Proc, MMSS'2000,325-332.

6) Tanimura, T. et al. ; Development of EDM in the Mist, Proc. of ISEM 9, 1989, 313-316.

U-SHAPED CURVED HOLE CREATION BY MEANS OF ELECTRICAL DISCHARGE MACHINING

Tohru ISHIDA and Yoshimi TAKEUCHI

Dept. of Mechanical Engineering and Intelligent Systems, The University of Electro-Communications, 1-5-1, Chofugaoka, Chofu-shi, Tokyo, 182-8585, JAPAN
E-mail : ishidat@mce.uec.ac.jp, takeuchi@mce.uec.ac.jp

Abstract

This study deals with machining curved holes, aiming at application to water channels for cooling molds. In general, water channels have been manufactured by drilling operation. Therefore, they cannot help being polygonal-line-shaped pipe lines. This prevents the improvement of productivity. To solve the problem, we developed a new mechanism to machine curved holes, which is installed on an electrical discharge machine. The mechanism can make an electrode move along a curved trajectory simultaneously with performing electrical discharge machining. From experimental results, it is found that the mechanism is capable of machining U-shaped curved holes, difficult to create in a conventional manner.

Keywords

Electrical discharge machining, Water channel, U-shaped curved hole

1. INTRODUCTION

Most of mechanical engineers have taken it for granted that drilling is to machine a straight hole, since curved hole machining method is not industrially put into practical use and is not familiar for them at all. Therefore, straight holes have been used even in unsuitable cases. One example of the unsuitable cases is to employ drilling operation for manufacturing water channels. Water channels are pipe lines built in molds. Regulating flow rate and temperature of coolant running through water channels allows the thermal control of molds to obtain products without any defects occurring in molding process. Hence, the shape and arrangement of water channels are very important for achieving high productivity. However, water channels are formed as polygonal lines consisting of a series of straight holes, since they are conventionally machined by drilling. Accordingly, it is difficult to realize the appropriate shape and position of water channels, no matter how computers can design the ideal water channel system.

To solve this problem, it is strongly desired to establish a practical method of machining curved holes. With regard to curved hole machining, the mechanism using universal joints was devised [Fukui 1989], and the apparatus employing shape memory alloys was proposed [Ichiyasu 1997]. On the other hand, a new curved hole machining method without extra actuators and cumbersome control was devised [Ishida 2000], however, the devised mechanism could not easily change a curvature of machined curved hole.

To overcome the problem, it is intended to improve the device. The improved device chiefly consists of a helical compression spring, three wires, a shaft, three slider crank mechanisms and an electrode for electrical discharge machining, and is installed on an electrical discharge machine (EDM). From experimental results obtained under the various properties of the device, it is found that the device can create U-shaped curved holes by connecting two parallel straight holes previously machined.

2. STRUCTURE AND MOTION CONCEPT OF DEVICE

Figure 1 illustrates the structure of the device to make an U-shaped curved hole. An electrode is attached to the end of a helical compression spring, which is mounted on a shaft at the other end of the spring. The shaft is mounted on a head of an EDM. The EDM head is connected to three disks by three links respectively. Each disk is mounted on a fixture which rests on the bottom of a working tank of the EDM. Consequently, the links, the disks and the EDM head constitute three slider

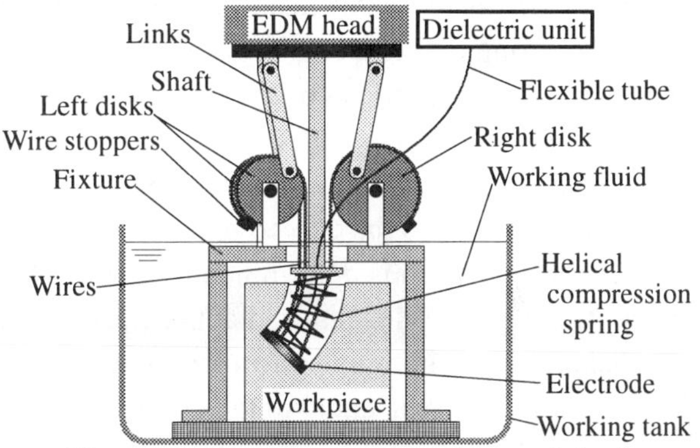

Figure 1. Structure of experimental device

crank mechanisms. Radius of two disks in left side is smaller than that of a disk in right. In addition, three wires are fastened in the equal angle of 120° on the end of the spring of the electrode side, and are guided to the disk respectively. Each wire is wrapped around the corresponding disk, and fastened on it. Chips are removed by supplying working fluid between the electrode and a workpiece through a flexible tube and a small hole of the electrode.

Let us explain the motion process of the device, using 2-D model illustrated in Figure 2. At the initial stage of the curved hole machining, the spring is deeply compressed to its solid length, as illustrated in Figure 2(a). When the EDM head is fed down, the spring naturally moves downward with the EDM

215

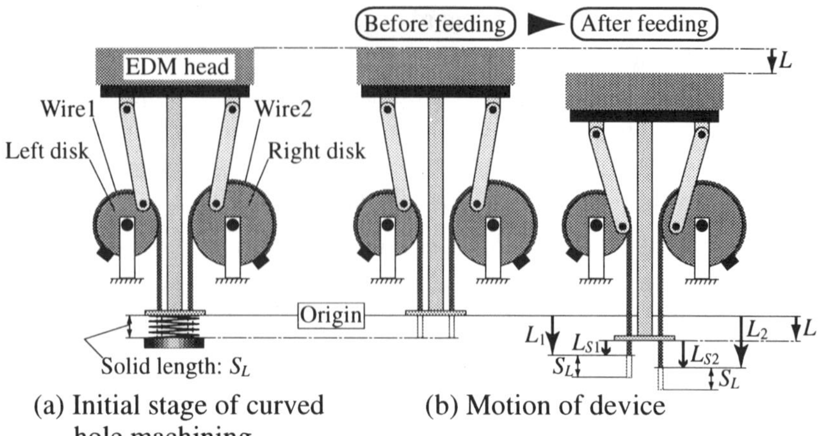

(a) Initial stage of curved
    hole machining

(b) Motion of device

Figure 2. Conceptual explanation of curved motion mechanism

head and stretches at the same time. In other words, the wires are drawn downward by the spring, as illustrated in Figure 2(b). Each length of the wires drawn down is different due to the difference of the disk radius. This leads to the fact that the spring stretches with the posture bent, simultaneously with its moving down, as the EDM head goes down. As a result, the electrode at the end of the spring moves along a curved trajectory. In addition, the curvature of the electrode trajectory can be changed by the combination of the disks with various radii.

That is to say, the devised mechanism converts the vertical movement of the EDM head into the curved motion of the electrode. Moreover, because the electrode position is determined by that of the EDM head, the discharge gap control with EDM is realized at the electrode. This means that the electrical discharge machining is performed on the electrode. Hence, the machined shape is expected to be equal to the curved trajectory of the electrode.

3. MOTION AND MACHINING EXPERIMENT

Before machining experiments, the spring motion and the electrode trajectory were subjected to test, changing the device properties. This yields the change in posture and position of the spring, and the change in position of the electrode, when the EDM head goes down, as shown in Figure 3. The properties of the slider crank mechanisms in the motion experiment are illustrated in Figure 4, and the specification of the spring is listed in Table 1. The electrode shape of oxygen-free copper is a cylinder of 20mm in diameter and 9.5mm in height. The total mass on the end of the spring is 22.7g. Figure 5 illustrates the electrode trajectory obtained from Figure 3. It is found that the electrode moves along a smooth curved trajectory, and that U-shaped curved hole can be created by connecting two smooth curved trajectories, as illustrated in Figure 6(a).

Two straight holes were in advance prepared by drilling a workpiece of aluminum alloy (A5052P). The prepared holes are 22mm in diameter and 32mm in depth. Then, the electrical discharge machining was twice done from the bottom of each straight hole, as shown in Figure 6(b). The machining condition is as follows; discharge current 20A, pulse duration 315μs, duty factor 32%, reverse polarity, working fluid of kerosene at flow rate 3cm³/s, and no jump.

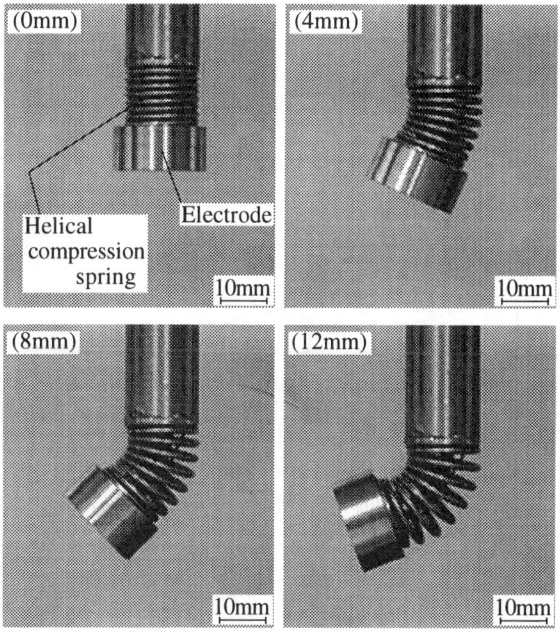

Figure 7 shows a sectional view of the holed workpiece. As seen from Figure 6(a) and 7, the shape of the electrode trajectory almost coincides with the machined shape of U-shaped curved hole. As a result, it is experimentally found that the device is valid to create curved holes. The machining time was 90 minutes for a curved hole and 180 minutes in total.

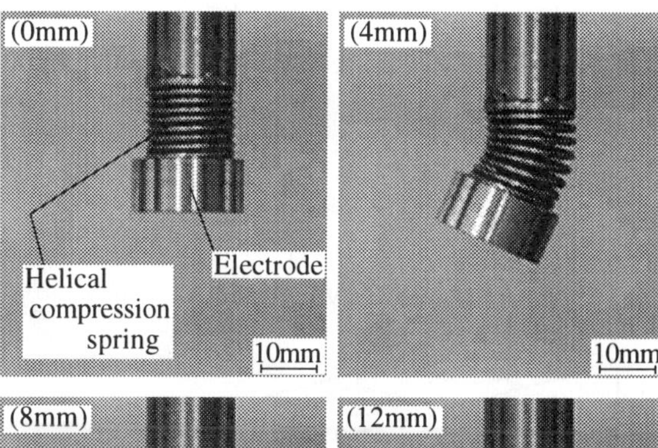

Figure 3. Actual motion of helical compression spring and electrode according to EDM head feed

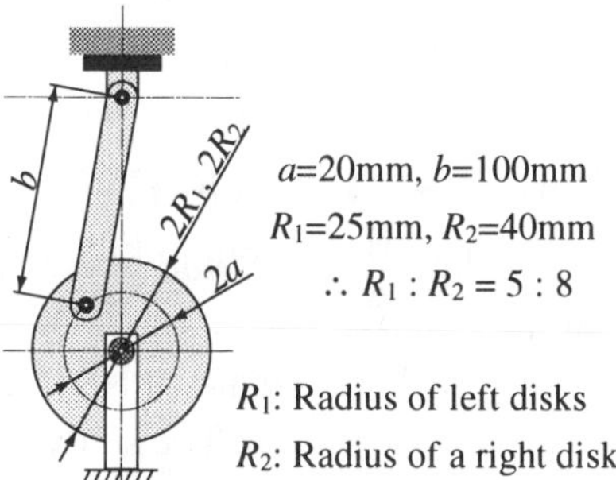

Figure 4. Properties of slider crank mechanisms

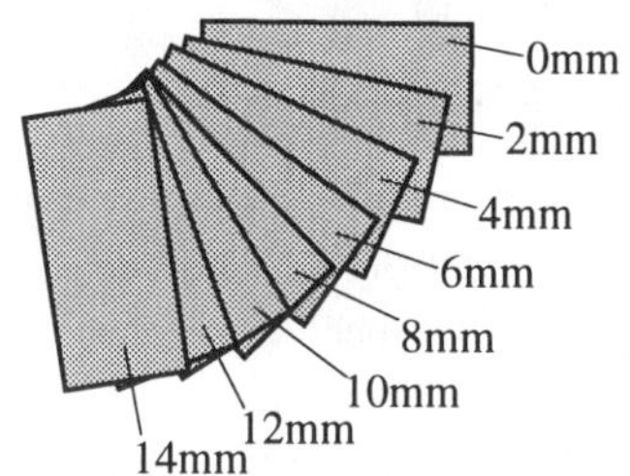

Figure 5. Change in electrode position

Table 1. Specification of helical compression spring

Material	SUS304-WPB
Outside diameter of coil	16.0 mm
Diameter of wire	1.40 mm
Free height	51.0 mm
Solid length	11.9 mm
Spring constant	1.51 N/mm

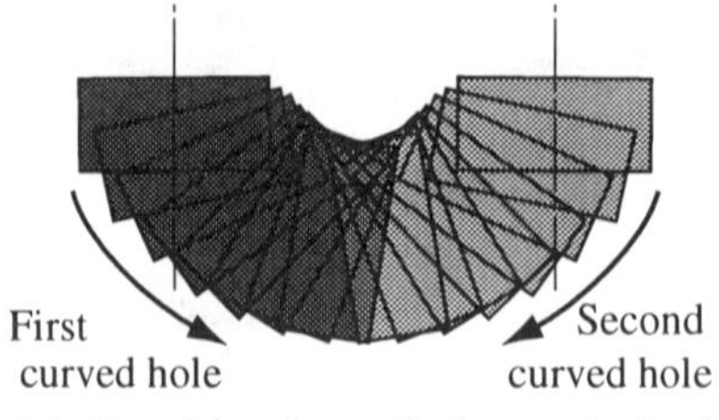

(a) Combination of electrode loci

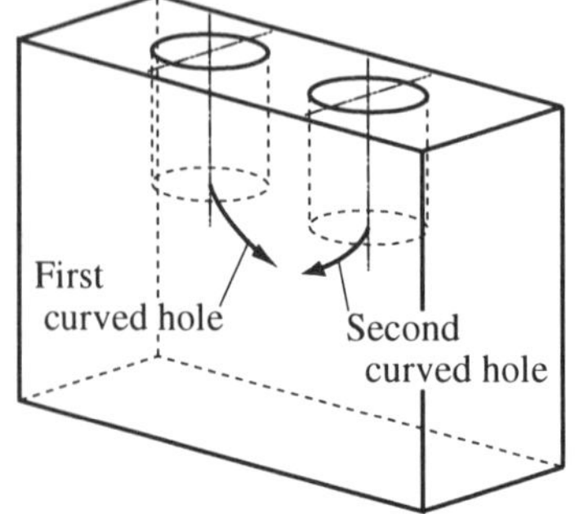

(b) Perspective view of workpiece

Figure 6. Machining manner of
U-shaped curved hole

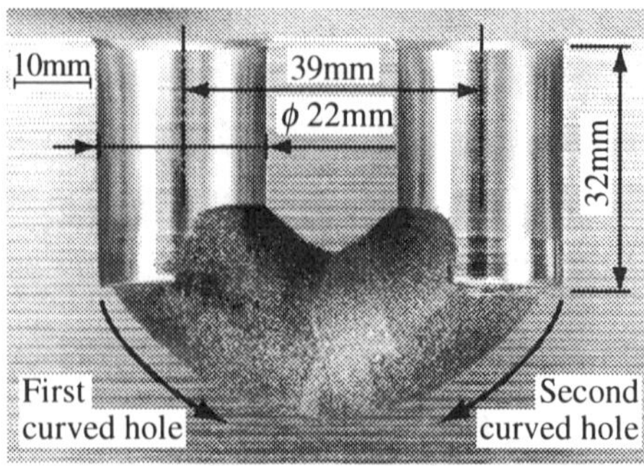

Figure 7. Sectional view of machined
U-shaped curved hole

4. CONCLUSION

In order to establish a new practical method for machining curved holes, we have proposed a new device mechanism, which is installed on a conventional EDM, and can convert the vertical motion of the EDM head into a curved motion of the electrode. In the study, slider crank mechanisms are introduced in the device. As a result, it is confirmed that the device have potential to machine U-shaped curved holes which are difficult to create with conventional machining methods.

5. ACKNOWLEDGEMENT

The authors would like to express their sincere appreciation to Mr. M. Sato, Executive Director and Director of R&D Center, Mr. K. Katsumata, Director, Mr. T. Tachibana at R&D Center of Makino Milling Machine Co., Ltd. for their cooperation. The study is partly supported by the Ministry of Education, Science, Sports and Culture, Grant-in-Aid for Encouragement of Young Scientists, 12750094, and by Mitutoyo Association for Science and Technology (MAST).

6. REFERENCES

Fukui M., Kinoshita N. Developing a 'mole' electric discharge digging machining. Annals of the CIRP 1989; 38, 1: 203-206.

Ichiyasu S., Takeuchi A., Watanabe K., Goto A., Magara T. Machining curved tunnel for coolant with mole EDM. Proc. of 4th Int. Conf. on Die & Mould Tech.; 1997 June 4-5, 224-230.

Ishida T., Takeuchi Y. Curved hole machining by means of electrical discharge phenomena and electrode feed mechanism. Proc.of 2000 Japan USA Symp. on Flexible Automation; 2000 July 23-26, CD-ROM No.13015, 1-6

HIGH PERFORMANCE SLICING METHOD OF MONOCRYSTALLINE SILICON INGOT BY WIRE EDM

Yoshiyuki UNO*, Akira OKADA*, Yasuhiro OKAMOTO*
and Tameyoshi HIRANO**

* Department of Mechanical Engineering, Okayama University
3-1-1, Tsushimanaka, Okayama 700-8530, Japan
** Toyo Advanced Technologies Co.,Ltd.
5-3-38, Ujina-higashi, Minamiku, Hiroshima 734-8501, Japan

Abstract

In this study, the slicing of silicon ingot by wire EDM is discussed, and the machining properties are experimentally investigated. Also, the accuracy of sliced wafer and the contamination on the machined surface are evaluated. Experimental results made it clear that the accuracy of wafer by this slicing method is almost the same as that by the conventional methods, and even higher slicing speeds are possible when multi type wire EDM slicing can be realized. Additionally, the contamination due to the adhesion and diffusion of wire electrode material into the sliced surface can be reduced by wire EDM under the condition of low current and long discharge duration. Therefore, wire EDM has a high potential as a new slicing method for monocrystalline silicon ingot.

Keywords

Slicing, Monocrystalline silicon ingot, Wire EDM

1. INTRODUCTION

In the slicing process of monocrystalline silicon ingot for manufacturing semiconductor, further improvements in machining efficiency and accuracy are strongly required, since the final flatness of wafer is significantly determined by this process. Conventionally, inner diameter blades have been used for slicing ingot[1]. However, some problems in efficiency still exist, like relatively large kerf loss and large cracks of about 30μm in depth on the sliced surfaces, resulted from mechanical machining. Moreover, this method is difficult to apply for large-scale wafers of 12 or 16 inches in diameter which are expected to be used in the near future. Therefore, multi wire saw slicing has been gradually applied[2]. In this method, thin piano wires are fed to the ingot with slurry which consists of abrasives and cutting oil. Multi wire saw slicing has many advantages such as having relatively small kerf loss, cutting large-scale wafers and multi wafers at the same time. However, there still remain the problems of slurry treatment and contamination of sliced surfaces.

From the above mentioned viewpoints, a new slicing method of silicon ingot by wire EDM (WEDM) is proposed. A silicon wafer used as a substrate for epitaxial film growth has low resistivity in the order of $0.01\,\Omega\cdot$cm, which makes it possible to slice silicon ingot by WEDM[3)-5)]. Also, it is expected that the length and the number of the cracks on the machined surface might be reduced, since the material removal is performed by repetition of micro craters and the machining force acting on the workpiece is extremely small. In this paper, the possibility of slicing silicon ingot by WEDM is discussed, and the machining properties such as slicing speed and surface roughness are experimentally investigated. Furthermore, the accuracy of sliced wafer and the contamination on the machined surface are evaluated.

2. EXPERIMENTAL PROCEDURE

Usually, high discharge current with short duration by condenser circuit has been applied in WEDM process because of higher cutting speeds and larger electrode wear allowed due to wire throwaway system. However, the wear of wire leads to adhesion or diffusion of wire material on the machined surface. This phenomenon is not suitable for the silicon process. Authors[4),5)] had done research on machining of monocrystalline silicon by EDM before, and it was made clear that the machining speed for silicon with transistor switching circuit was much larger than that for metal mold material, since Joule's heat generation in addition to heat conduction from arc column makes a great contribution to increase the removal rate of silicon[6)]. Therefore, slicing by WEDM with transistor switching circuit is discussed in this study. It is highly expected that the contamination of wire material on the machined surface can be reduced.

Figure1 shows the experimental apparatus. In this system, the discharge current pulse made by transistor switching circuit has a relatively long duration and a low current intensity. The used wire electrode is rewound around the wire winding drum and reused repeatedly, because of the high expectation of extremely low wear in the wire electrode. In this experiment, molybdenum wire

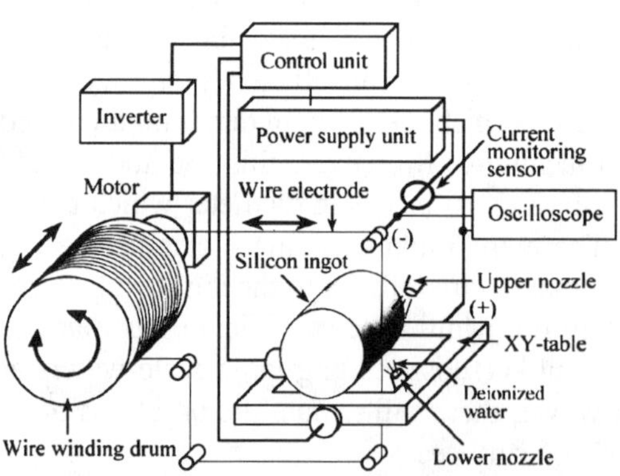

Figure 1. Schematic diagram of experimental apparatus

of 180µm in diameter is used as the electrode. P-type monocrystalline silicon ingot of 0.01 Ω·cm is used as the workpiece, and machining fluid is deionized water whose resistivity is about 10^6 Ω·cm.

3. MACHINING PROPERTIES

At first, the slicing speed and the surface roughness of the sliced surface using a silicon ingot of 40mm in thickness are discussed. The slicing speed increases with the increase of discharge current as shown in Figure 2, and it takes maximum at about 20µs under any discharge currents. It was also confirmed that the removal rate increases with the increase of wire feed rate because of the smooth exclusion of debris from the gap, which leads to stable machining performance. Additionally, Figure 3 shows the variations of surface roughness. It can be said that the surface roughness obtained by this method under short durations is almost the same as that in the conventional multi wire saw method.

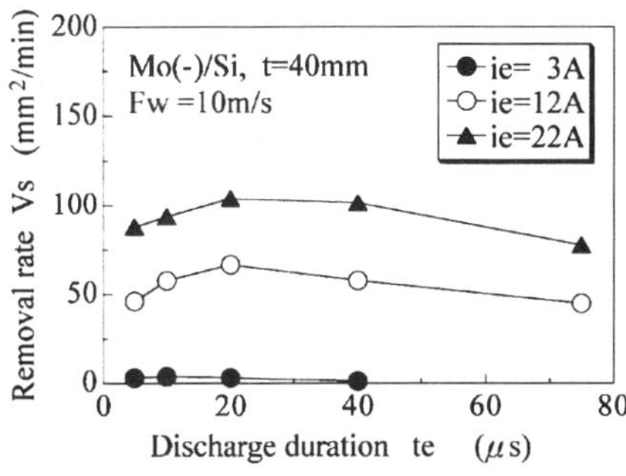

Figure 2. Removal rate

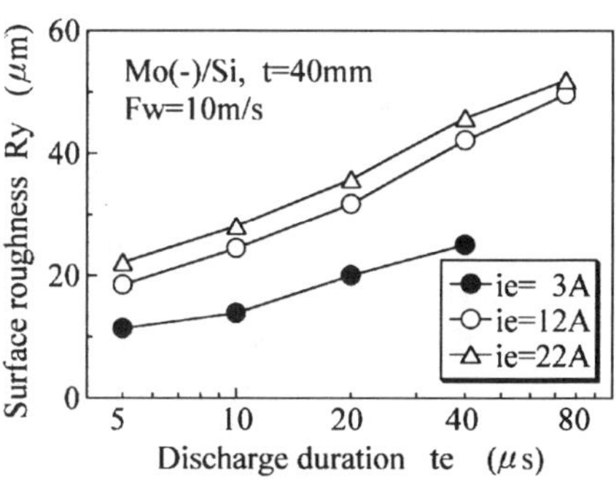

Figure 3. Surface roughness

4. SURFACE INTEGRITY

In order to investigate the contamination of sliced surface, XPS (X-ray Photoelectron Spectroscopy) analysis was carried out. *Figure4* shows the results of the analysis. As shown in the figure, oxygen exists on the sliced surface in both cases. It is considered that the machined surface is oxidized by electrical discharges in the deionized water. Of course, contamination of oxygen is not desirable. However, it is allowable to some extent, since some processes for removing oxygen are actually performed after the slicing process. Conventional WEDM is never applicable to slice silicon ingot, since copper and zinc from the wire material transfer to the silicon surface. On the other hand, in the case of the discussed method with the molybdenum wire, molybdenum hardly adheres to the machined surface. It is considered that adhesion or diffusion of wire material can be reduced by the discharge current waveform of the transistor switching circuit. Furthermore, by observing the cross section of sliced surfaces, the length of crack was measured. As a result, The length was about 20µm. On the other

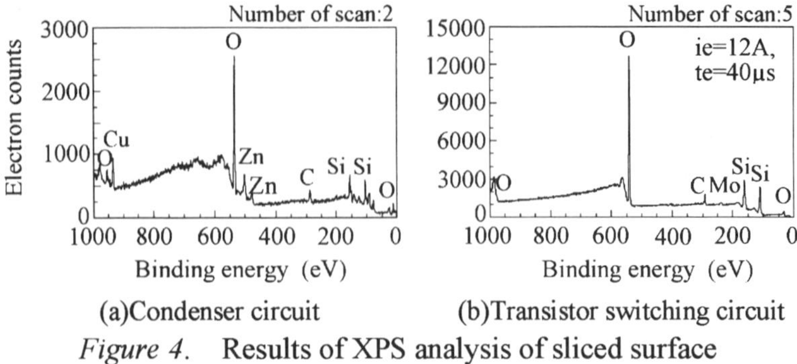

(a)Condenser circuit　　　(b)Transistor switching circuit

Figure 4.　Results of XPS analysis of sliced surface

hand, in the case of the conventional slicing method, the cracks are about 20-30μm in length. Therefore, this method is effective for high efficiency manufacturing of semiconductor, since a thinner crack layer leads to a shorter removing time for the layer.

5. SLICING OF 6 INCH SILICON INGOT

Next, slicing of a 6 inch ingot was carried out. *Figure5* shows the cutting speed and the removal rate with machining time. As shown in this figure, the cutting speed decreases with the increase of workpiece width, and the removal rate is almost constant at any time. It takes about 140min to slice a 6 inch ingot. That is, the average slicing speed is about 1.1mm/min. In the case of multi wire saw, it is 0.2-0.3mm/min. If multi slicing, in which many wafers are sliced at once can be realized like a wire saw slicing, this WEDM slicing method is applicable as a high efficient slicing method of silicon ingot.

Figure6 shows TTV (Total Thickness Variation) and Warp of the 6 inch wafer sliced by WEDM. Both are the most important dimensional parameters of wafer, that is, TTV is the difference between the maximum and minimum thickness, and Warp is defined as the difference between the maximum and

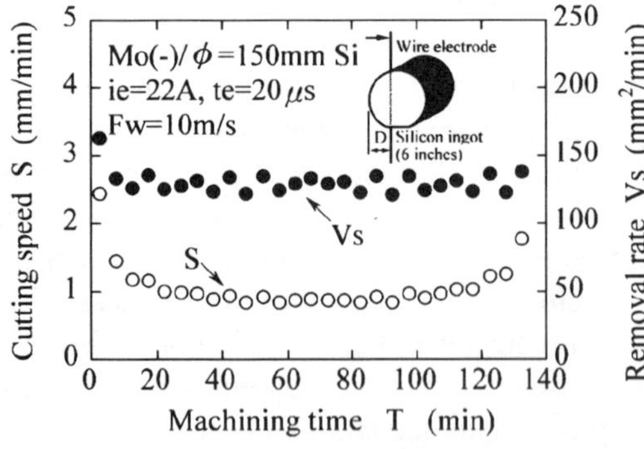

Figure 5.　Cutting speed and removal rate in slicing of 6 inch ingot

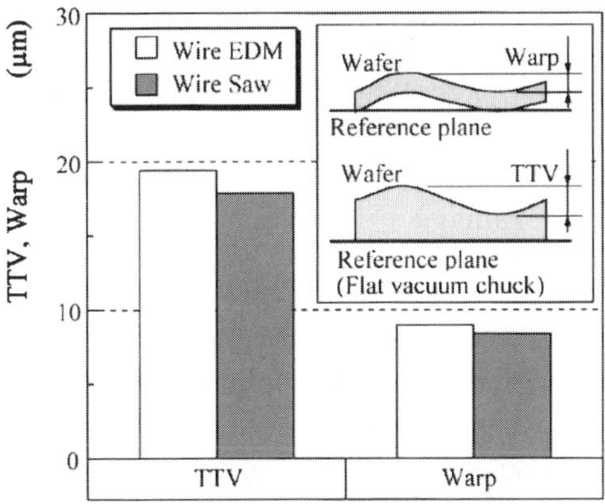

Figure 6.   TTV and Warp of wafer sliced by WEDM

minimum distances from a reference plane, as shown in the figure. Both values, in the case of WEDM, are almost the same as those in the case of multi wire saw. In addition, the kerf loss was about 250μm, which is also almost the same value as in multi wire saw.

6. CONCLUSIONS

A new slicing method by WEDM was discussed in this study. The experimental results clarified that WEDM has a high potential as a new slicing method for monocrystalline silicon ingot, since the surface roughness and the accuracy of wafer by this method are almost the same as that by multi wire saw method used conventionally. Moreover, the contamination due to adhesion and diffusion of wire electrode material to machined surface can be reduced by WEDM under the condition of low current and long discharge duration.

In the near future, high speed slicing is possible when multi type WEDM slicing can be realized.

7. ACKNOWLEDGEMENTS

This work was partly supported by Electric Technology Research Foundation of Chugoku. The authors would like to express their thanks to Sin-Etsu Handotai Co.,Ltd. and Sodick Co., Ltd. for their help through this research.

8. REFERENCES

1) F.Shimura. *Semiconductor Silicon Crystal Technology.* San Diego: Academic Press Inc., 1988.
2) K.Makino et al. Slicing by multi wire saw. J.of the Society of Grinding Engineers 1997; 41: 16-19.
3) D.Reynaerts et al. Microstructuring of silicon by EDM. Sensors and Actuators 1997; 60: 212.
4) Y.Uno et al. Fundamental study on EDM of single crystalline silicon. J.of JSPE 1997; 63: 1459.
5) Y.Uno et al. High efficiency fine boring of monocrystalline silicon ingot by electrical discharge machining. Precision Engineering (J.of ASPE) 1999; 23: 126-133.
6) T.Saeki et al. Transient workpiece temperature analysis in the EDM processes of high electric resistance materials considering Joule heating. J.of JSPE 1996; 62: 443.

LASER ABLATION OF SAPPHIRE WITH A PULSED ULTRA-VIOLET LASER BEAM

Shin Tamura, Hideyuki Horisawa,
Shigeru Yamaguchi, and Nobuo Yasunaga
Department of Precision Mechanics, School of Engineering, Tokai University

Abstract

A study on machining characteristics of sapphire using a short-pulse ultra-violet laser was conducted. As results, a control of the removal depth with the THG pulses was easier than those with SHG and fundamental pulses. Also, little redeposition of molten material around the craters was seen and a clean removal was found to be possible. Small pits with cracks were induced through a laser pulse irradiation at the spots of intrinsic lattice defects along the crystal orientation of the original polished-surface. The pits on the reverse side surface were also found at the points of lattice defects. Moreover, diamond-shaped pits or micro-plateaus with a flat bottom surface were formed near the beam center. Repetition of the pulse irradiation was found having similar effects to that of the chemical etching in cases of the THG pulse irradiation.

Keywords

Laser Machining, Nd:YAG laser, Ultra-Violet Laser Beam, THG wave, Sapphire

1. INTRODUCTION

Engineering ceramics such as alumina, or sapphire, silicon carbide, silicon nitride, etc., are attractive materials in their superior mechanical and corrosion properties, and are widely used in industry. Despite their remarkable properties, they present a considerable challenge in micromachining due to their extreme hardness, brittleness and corrosion-resistance.[1] On the other hand, for materials for the electronic devices, the pulsed laser machining with CO_2 laser, Excimar laser, or Nd:YAG laser is widely utilized in industry.[2] However, it has been inevitable to reduce redeposition of molten materials and thermal influences with those lasers. This must have been preventing the laser machining widely used for the micro-machining processes. Although it is expected that the high quality micro-machining with the reduced thermal influences is possible using a short-pulse ultra-violet laser beam, detailed characteristics of the processes and mechanism are not yet clear.[3~5] In order to investigate these points, the machining characteristics of sapphire using a short-pulse ultra-violet laser was conducted in this study. Detailed observations of laser-shot surfaces, or craters, were conducted. In addition, correlations of these surface conditions with pulse conditions, such as wavelength, duration, and fluence, were also investigated to explore the mechanism of the surface-photon interactions.

2. EXPERIMENTAL

In this experiment, an ultra-violet short pulse laser with the pulse duration of ~ 5 nsec was utilized to minimize the thermal influences of the laser pulse. As the ultra-violet laser beam oscillator, the third harmonic generation wave (THG: 355 nm) and the second harmonic generation wave (SHG: 532 nm) of a Q-switched Nd:YAG laser (HOYA Continuum, Surelite II -10) were used. The specification of the laser system is given in Table 1. The laser pulses were focused with a lens (f = 100 mm) and irradiated onto a workpiece surface in air under an atmospheric pressure. As the workpiece, a polished sapphire wafer (R [1012], t = 600 μm) was used. Detailed surface observations were conducted with a scanning electron microscope (SEM, JOEL JSM5410LV) and an atomic force microscope (AFM, SII SPA300).

Table 1. Specifications of a Q-sw Nd:YAG laser used in this study.

	Wavelength [nm]	Pulse width [nsec]	Pulse energy [mJ/pulse]
Fundamental	1064	6	250
SHG	532	4	130
THG	355	5	100

3. RESULTS AND DISCUSSION

3.1. Effects of Wavelength on Surface Removal Rate

SEM micrographs of a sapphire wafer surface irradiated with Nd:YAG pulses (λ = 1064 nm, 33 mJ/pulse) focused through an f = 100 mm lens are shown in Fig.1(a) and (b). As shown in Fig.1, cracks along the crystal orientations were seen. Also, edges of the crater became clearer with number of shots.

Figures.2(a) and (b) show SEM micrographs of a sapphire wafer surface irradiated with SHG of Nd:YAG pulses (λ = 532 nm, 33 mJ/pulse). Although cracks around craters were induced in Fig.2, a small amount of redeposition of molten material inside the crater was found. A small change in depth was found with the number of pulses in the range of 1 ~ 3 shots. Therefore, it must be difficult to control the depth of the craters with SHG pulses. Also, edges of the crater became clearer with shots of laser pulses.

SEM micrographs of the cases of THG of Nd:YAG pulses (λ = 355 nm, 33 mJ/pulse) are shown in Fig.3(a) and (b). In Fig.3 (a), small cracks occurring along the crystal orientations were seen. As number of pulse shots increased, small pits with cracks along the crystal orientation were induced and depth of the crater became larger, as in Fig.3 (b). From these results, it is clear that the control of the etching depth with the THG pulses is easier than those with the SHG and the fundamental waves. In addition, little redeposition of molten material around the craters were seen, and then the clean removal must be possible.

Therefore, with THG pulses, not only the reduction of thermal influences of a laser beam but the control of the surface removal rate in axial direction of the beam is possible, which used to be of substantial difficulty with the conventional laser processing. Moreover, from Figs. 1 ~ 3, it was found that the surface damage of the sapphire surface with SHG was more significant than that of THG and fundamental pulse cases. Although little difference of absorption rates of sapphire surface for fundamental wave ($\lambda = 1064$ nm), SHG wave ($\lambda = 532$ nm) and THG wave ($\lambda = 355$ nm) at room temperature, more energy must be converted into the heat on the surface in the case of focused SHG pulses. Also, pitches of the cracks for THG cases were smaller than those of SHG cases. This must be due to the smaller wavelength, or higher photon energy, of the THG pulses.

3.2. Effects of Pulse Energy on Surface Removal Process

Craters formed with different energy of THG pulses are shown in Fig.4 ((a)30 mJ and (b)35 mJ). With lower pulse energy (a), a number of small pits of ~ 1 μm in diameter occur. On the other hand, with higher pulse energy (b), almost whole area of irradiated surface was covered with straight cracks occurring along the crystal orientations. These cracks must be formed with a thermal shock through the sequential process of heating and cooling of a surface layer within a short duration along planes of cleavage, or cryatal orientations, of relatively small bonding energy.

SEM micrographs of the bottom surface, which is the reverse side of an irradiated surface, are shown in Fig.5 for THG pulses. Small pits were also found on the bottom surface. From the figures, size and numbers of those pits increase with the pulse energy and number of the shots.

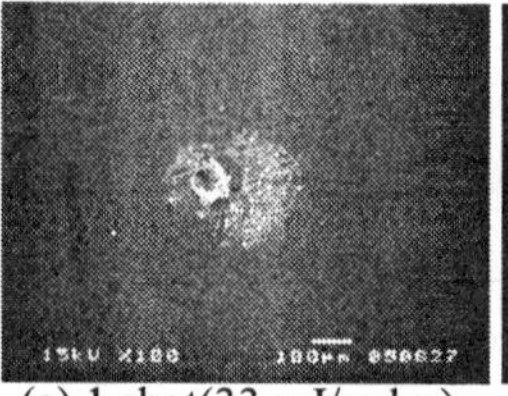 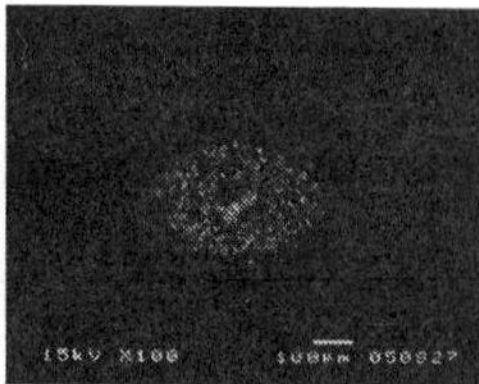

(a) 1 shot(33 mJ/pulse)   (b) 3 shots (33 mJ/pulse)
Figure 1. Craters with Nd:YAG pulses

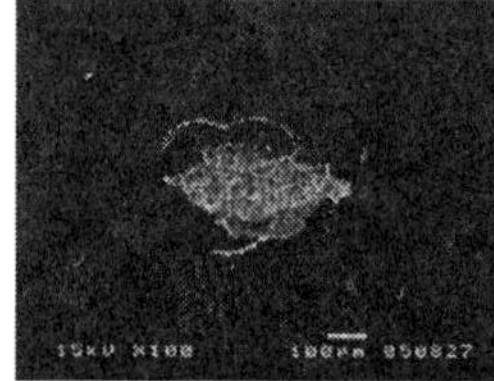 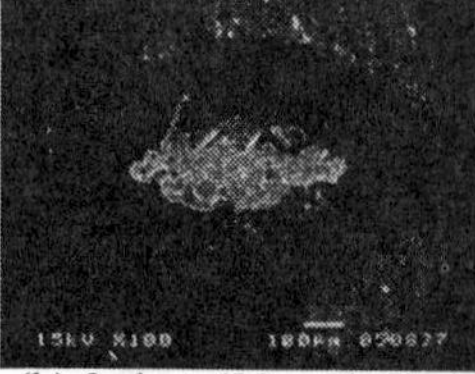

(a) 1 shot(33 mJ/pulse)   (b) 3 shots(33 mJ/pulse)
Figure 2.Craters with SHG pulses ($\lambda = 532$ nm) irradiated on sapphire surface.

(a) 1 shot(33 mJ/pulse)   (b) 3 shots(33 mJ/pulse
Figure 3. Craters with THG pulses ($\lambda = 355$ nm) irradiated on sapphire surface.

(a) 33 mJ/pulse (1 shot)   (b) 35 mJ/pulse (1 shot)   (a) 90 mJ/pulse (1 shot)   (b) 33 mJ/pulse (3 shots)

Figure 4. Craters with a THG pulse ($\lambda = 355$ nm)   *Figure 5*. Small pits with THG pulses ($\lambda = 355$ nm)
for different pulse energy.   on a reverse side of pulse irradiation.

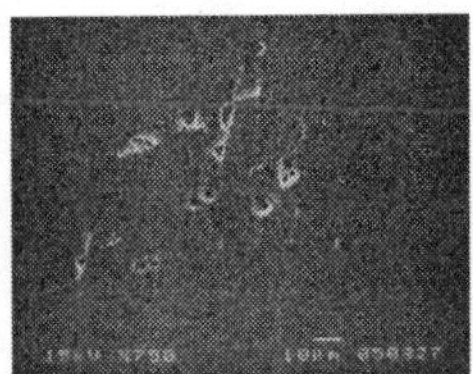 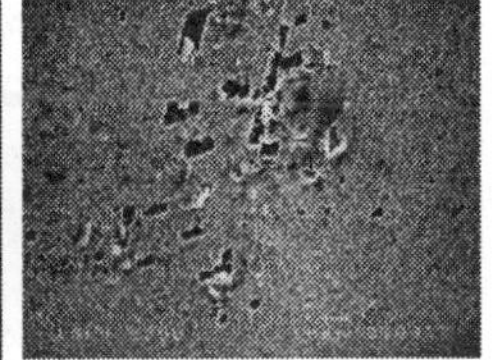 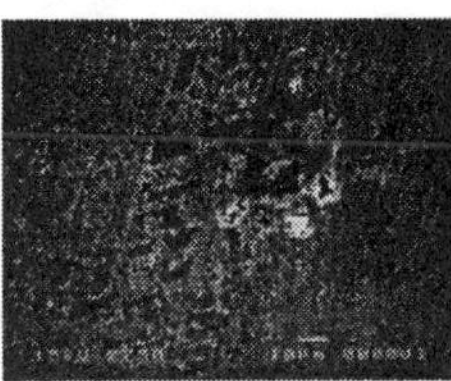 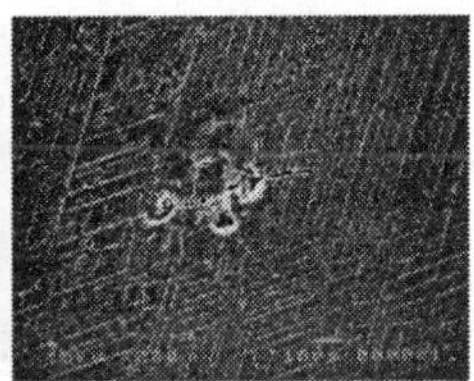

(a) Before chemical etching   (b) After 0.4μm-etching   (c) After 0.8μm-etching   (d) After 1.2μm-etching
(21 mJ/pulse, 1 shot)   at 473 K with 100   at 473 K with 100   at 473 K with 100
wt% H_3PO_4 wt% H_3PO_4 wt% H_3PO_4

Figure 6. Small pits with a THG pulse ($\lambda = 355$ nm) irradiated on sapphire surface.

3.3. Micro-structure of a Crater with THG Pulses

For the detailed diagnostics of craters formed with a THG pulse, the chemical etching with the hot phospheric acid (H_3PO_4) solution (100 wt%) was conducted with etching rate of 80 nm/min at 473 K. Sequential SEM images with the chemical etching duration are shown in Figs. 6 (b) ~ (d). As the surface layer of the irradiated point was removed, cracks connecting between the small pits appeared along the crystal orientations. Also, diamond-shaped pits of their width around 2~5 μm with a flat bottom surface appeared in Fig.6(b).

Figure 7. An AFM image of micro-plateaus.

From these results, it was found that these small pits with cracks are induced through the laser pulse irradiation at the spots of intrinsic lattice defects along the crystal orientation of the original polished-surface. With the increase of pulse energy and number of shots, cracks from these pits grew aligned along the crystal orientations as shown in Figs.4 (b) ~ 6(d). The pits on the reverse side surface were also found with the laser pulse irradiation at the points of lattice defects along the crystal orientations.

When the surface was chemically etched upto ~ 0.4 μm, as shown in Fig.6(b), small cracks appeared on the entire area of irradiated surface. From the AFM observations, it was found that pitches of these cracks were less than 100 nm along the crystal orientations and the depth of ~ 100 nm. It was also found that the pitch decreases with distance from a beam center. Near the beam center, some parts surrounded by cracks were removed forming diamond shaped small pits or plateaus of the width of 2~5 μm with a flat bottom surface. An AFM image of a typical micro-plateau is shown in Fig.7. The plateau had fairly flat surface with the flatness (roughness) of $R_y \sim 0.3$ nm and the height of ~ 100 nm.

3.4. Effects of Number of Pulse Shots on Surface Removal Rate with THG Pulses

Effects of the pulse repetition on the micro-structure of an irradiated point of the surface were also investigated. A SEM image of a crater formed with 3 shots of a THG pulse with 33 mJ energy is shown in Fig.3(b). Comparing with the chemically etched surface of Figs.6(a) ~ 6(d), the shapes and sizes of the small pits formed by the pulse repetition shown in Fig.3(b) were similar. Therefore, it is clear that the repetition of the pulse irradiation in cases of the THG pulse has similar effects to that of the chemical etching for the surface removal process.

4. CONCLUSION

A study on machining characteristics of sapphire using a short-pulse ultra-violet laser was conducted to investigate the detailed mechanism of the surface removal process. Following results were obtained.

(1) A control of the removal depth with the THG pulses was easier than those with SHG and fundamental waves. Also, little redeposition of molten material around the craters was seen and a clean removal was found to be possible.

(2) Small pits with cracks were induced through a laser pulse irradiation at the spots of intrinsic lattice defects along the crystal orientation of the original polished-surface. The pits on the reverse side surface were also found at the points of lattice defects.

(3) Diamond-shaped pits or micro-plateaus of their width around 2~5 μm with a flat bottom surface ($R_y \sim 0.3$ nm) were formed near the beam center.

(4) Repetition of the pulse irradiation has similar effects to that of the chemical etching in cases of the THG pulse irradiation.

REFERENCES

1) Yasunaga, N., Tarumi, N., Shinohara, K., and Mineta, S., "Fundamental Study of Laser Machining (3rd Report)", Proceeding of the 1980 Spring Symposium on Precision Engineering, 1980, pp.376-378 (in Japanese).

2) The Institute of Electrical Engineers of Japan, *Laser Ablation and Applications*, Corona Publishing (1999) (in Japanese).

3) Ohyanagi, T., Miyashita, A., Murakami, K., and Yoda, O., "Time-and Space Resolved X-Ray Absorption Spectroscopy of Laser-Ablated Si Particles", *Japan Journal of Applied Physics*, Vol.33, No.3, Part 1, No.5A, 1994, pp.2586-2592.

4) Kokai, F., and Koga, Y., "Time-of-Flight Mass Spectrometric Studies on the Plume Dynamics of Laser Ablation of Graphite", *Nuclear Instruments and Methods in Physics Research*, Vol.B121, 1997, pp.387-391.

5) Stoian, R., Ashkenasi, D., Rosenfeld, A., and Campbell, E. E., "Laser Ablation of Sapphire with Ultrashort Pulses," *Proceedings of SPIE* Vol.3885, pp.121 – 131 2000.

METAL DEPOSITION ON GLASS BY LASER IRRADIATION FROM METAL POWDERS

Hitoshi TOKURA and Hirofumi HIDAI

Department of Mechanical Sciences and Engineering, Tokyo Institute of Technology, 2-12-1 O-okayama, Meguro, Tokyo, 152-8552, JAPAN,

E-mail: htokura@ctrl.titech.ac.jp, hidai@ctrl.titech.ac.jp

Abstract

In this paper, a new, simple, high-speed method of selective metal deposition on glass substrates is proposed. Metal powders placed on a glass substrate were irradiated by an argon ion laser beam. Soda glass, Pyrex glass and silica glass were used, and aluminum and copper powders were chosen. Both metal powders could be deposited on all the glasses. Furthermore, silicon wafer, which placed on Pyrex glass substrate, could be bonded to glass by the similar method.

Keywords

Argon ion laser, Metal deposition, Laser bonding, Metal powder, Glass

1 INTRODUCTION

There is a growing need for the deposition of metal films on insulators. In particular, the use of focused lasers for direct writing or maskless patterned depositions have been extensively described in the literatures, such as LCVD (laser-induced chemical vapor deposition) [1,2] LIFT (laser-assisted forward transfer)[3,4], laser-assisted deposition from organo-metallic solutions[5-7], laser- enhanced electro- [8] or electroless [9, 10] plating, and photothermal decomposition of metal-doped polymer films[11,12]. Typically, the adherence of metal depositions to a smooth surface, such as glass or fused quartz, is low [13].

In this paper, a new, simple, high-speed method of selective metal deposition on glass substrates is proposed, in which metal powders placed on a glass substrate are irradiated by a laser through the glass from the other side. Furthermore, if silicon and glass could be bonded by the similar method, many applications are expected. Silicon-glass bonding is also examined by placing a silicon wafer on a glass instead of metal powders.

2 EXPERIMENTAL

Soda glass, Pyrex glass and silica glass were used as substrates, because they are popular materials and their thermal properties were varied. Aluminum and copper powders, with grain sizes of 7.0μm and

4.6μm, respectively, were chosen. The size of Si wafer was $16 \times 16 \times$ 0.5mm. In the experiments, the beam of an argon ion laser (Coherent, DBW20) was used at 488nm (for metal deposition), and 455- 529nm (for silicon-glass bonding) wavelength, because the glasses have high transparency of visible light. The laser beam was focused by a convex lens, with a focal length of 170mm. Glass substrates and metal powders were placed on a heater to control temperature, moreover they were fixed on an electronically controlled X-Y-Z stage, as shown in fig. 1.

Table 1 shows experimental conditions. The thickness of the metal powder layer was approximately 2mm and the powder was compressed with a roller. After irradiation, excessive or loosely adhered powder was

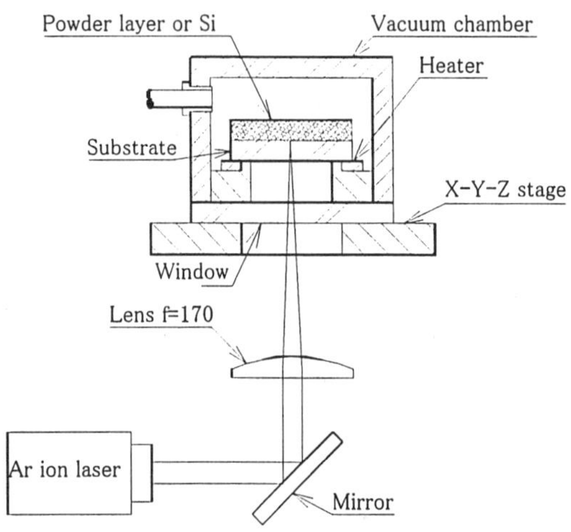

Fig.1　Illustration of the experimental setup

Table 1　Experimental conditions

	Metal depositon	Silicon-glass bonding
Laser	Ar ion laser	Ar ion laser
Wavelength	488 nm	455-529nm
Power	0 - 7.0 W	20 W
Beam mode	TEM$_{00}$	TEM$_{00}$
Scanning speed of moving stage	0.05 - 5.0mm/s	0.05mm/s
Focal length	170mm	170mm
Glass substrate	Soda, Pyrex, silica	Pyrex
Thickness of glass	2mm	2mm
Metal Powder (grain size)	Al(7μm), Cu(4.6μm)	-
Powder layer thickness	2mm	-
Size of silicon wafer	-	$16 \times 16 \times 0.5$ mm
Preheated temparature	20°C	20, 200, 400°C

brushed off, then the substrate was cleaned by ultra sonication.

3 DEPOSITION FROM POWDER

Glass substrates, metal powders, laser power and scanning speed were varied in deposition experiments. Both aluminum and copper were deposited on all the glasses under certain irradiation conditions. Fig. 2 shows SEM photos of the deposited metal lines at a laser power of 3W and scanning speed of 0.3mm/s. These micrographs show that both aluminum and copper powders are deposited well on soda glass. Each individual powder grain remains and cracks are observed on the glass due to rapid heating and cooling.

Aluminum deposited on Pyrex glass did not differ from that on soda glass. On the other hand, aluminum powder on silica glass was different from the others. In some parts of the deposited aluminum, grain shape could not be recognized, and no cracks were observed on the glass because silica glass has good thermal endurance.

4 SILICON-GLASS BONDING

Silicon-glass bonding was examined by placing silicon wafer instead of metal powder. Irradiation conditions were as shown in table 1. Firstly laser was irradiated at 20°C, but no changes were recognized. Secondly, silicon and glass were heated at 200°C and 400°C, then laser was irradiated, as a result, silicon and glass were bonded. Fig. 3 shows a

(a)   Aluminum on soda glass

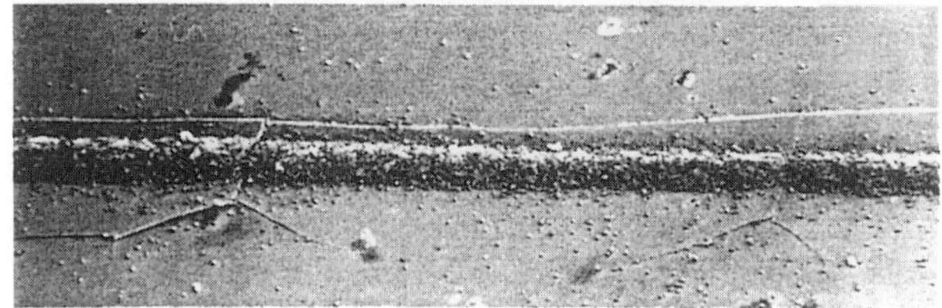

(b)   Copper on soda glass   0.5mm

Fig. 2   SEM micgrograph of deposited metals on soda glass

photo of bonded area. Fig. 4 is magnified micrograph of fig. 3. Bonded area is 200μn in width.

Figure 5 shows surface profiles of Si wafer bonded at 200°C and 400°C. The figure reveals that Si wafer are strained, and strain at 400°C is bigger than that at 200°C.

To clarify the bonded mechanism, silicon and Pyrex glass were etched by KOH and HF, respectively. Fig. 6 shows surface profiles. Glass surface is bulged, and Si becomes rough, however concave volume of Si is much small than convex volume of glass. It was inferred there was altered glass or Si, which could not etched KOH and HF.

5 CONCLUSIONS

A novel method of metal deposition on glass substrates is proposed. Aluminum and copper can be deposited on soda, Pyrex and silica glass by

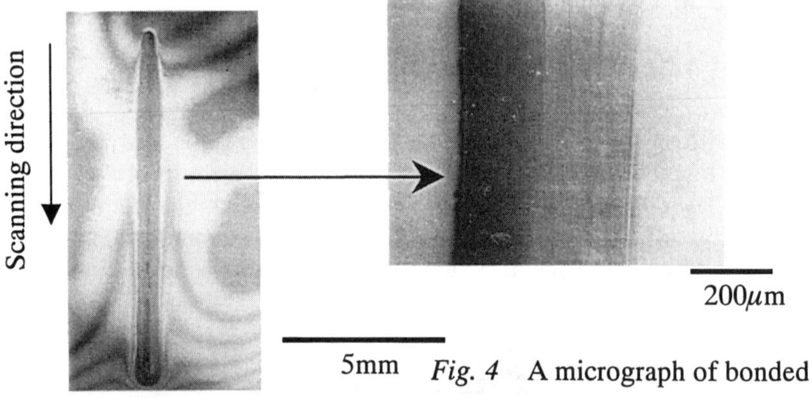

Fig. 3 A photo of bonded wafer and glass

Fig. 4 A micrograph of bonded wafer and glass

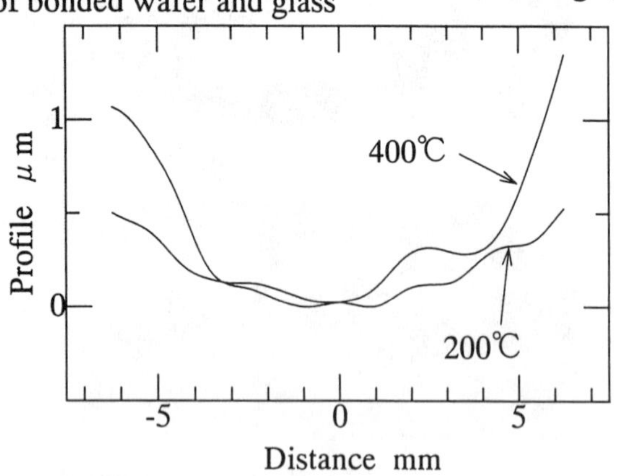

Fig.5 Surface profiles of bonded Si wafer

the argon ion laser irradiation. Furthermore, silicon can be bonded to glass
by placing silicon wafer on glass substrate instead of metal powder.

REFERENCE

[1] G. E. Blonder, G. S. Higashi and C. G. Fleming, Appl. Phys.
     Lett., 50, 12 (1987) 766.
[2] J. Y. Tsao and D. J. Ehrlich, Appl. Phys. Lett., 45, 6 (1984) 617.
[3] E. Pogarassy, C. Fuchs, F. Kerherve, G. Hauchecorne and J.
     Perriere, J Mater. Res., 4, 5 (1989) 1082.
[4] H. Esrom, J. Zhng, U. Kogelschatz and A. J. Pedraza, Appl.
     Surf. Sci., 86 (1995), 202
[5] Gupta and R. Jagannathan, Appl. Phys. Lett., 51, 26 (1987)
     2254.
[6] K. Bali, T. Szorenyi, M. R. Brook and G. A. Shafeev, Appl.
     Surf. Sci., 69 (1993) 75.
[7] T. Lin, H. Y. Lee and M. A. Souto, J. Mater. Res., 6, 4 (1991)
     760.
[8] R. J. von Gutfeld, E. E. Tynan, R. L. Melcher and S. E. Blum,
     Appl. Phys. Lett., 35, 9 (1079) 651.
[9] K. G. Mishra and R. K. Paramguru, J. Electrochem. Soc., 143, 2
     (1996) 510.
[10] G. Schrott, B. Braren, D. J. M. O'Sullivan, R. F. Saraf, P. Bailey
     and J. Roldan, J. Electrochem. Soc., 142, 3 (1995) 944.
[11] M. E. Gross, G. J. Fisanik, P. K. Gallagher, K. J. Schnoes and
     M. D. Fennell, Appl. Phys. Lett. 47, 9 (1985) 923.
[12] H. G. Muller, Appl. Phys. Lett., 56, 10 (1990) 904.
[13] G. A. Shafeev, L. Bellard, J. –M. Themlin, W. Marine and A.
     Cros, Appl. Surf. Sci., 86 (1995) 387.

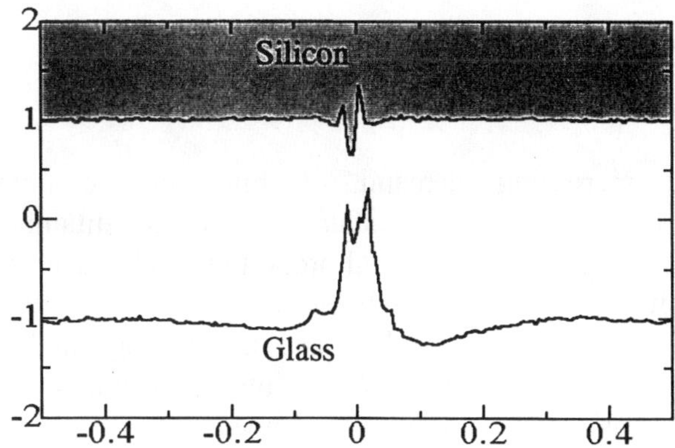

Fig. 6   Surface profiles of bonded Si wafer and glass surface

ULTRA-PRECISION CUTTING
OF DIFFICULT-TO-CUT METALS

Shigehiko SAKAMOTO, Heiji YASUI, Masaki KAWADA
and Sumihisa KONDO

Dept. of Mechanical Engineering & Materials Science,
Kumamoto University, Kumamoto 860-8555, Japan
Phone: 81-96-342-3758
Fax: 81-96-342-3764
E-mail: sak@mech.kumamoto-u.ac.jp

Abstract

Some attempts on ultra-precision cutting of difficult-to-cut metals with diamond bite are not sufficiently successful because the tool life is too short to produce the ultra-precision components. Special cutting method using high frequency vibration is only reported somewhat available[1]. Then in our previous researches[2],[3], a coated cemented carbide bite was tried to apply to ultra-precision cutting of titanium alloys. From the result, it is roughly found that the coated bite is available for ultra-precision cutting of the titanium alloys. This research describes on the possibilities of the ultra-precision cutting by the coated bite for stainless steels.

Keywords

Difficult-to-cut metal, Stainless steel, Ultra-precision cutting, Coated cemented carbide bite, Cutting fluid

1. INTRODUCTION

With increasing demand of high quality components using difficult-to-cut metals such as stainless steel and titanium alloys having excellent mechanical and chemical properties, high accuracy cutting technic for those difficult-to-metals has been increasingly requested in a wide field of industries. The diamond bite and the cutting fluid of kerosene are generally used to produce efficiently ultra-precision components of soft metals such as aluminum alloys, copper alloys and so on. For the difficult-to-cut metals, however, it is not successful to use the diamond bite, because the tool life is too short to produce the ultra-precision components[2].

Therefore there are little attempts on ultra-precision cutting of difficult-to-cut metals.

Then, in our previous research, coated cemented carbide bite was tried to apply to ultra-precision cutting of titanium alloys. From the result, it is roughly found that the coated bite is available for the titanium alloys. Using the coated bite with oil type cutting fluid, the surface roughness of α–β titanium alloy can be obtained below 60nm(P-V).

This research is one of a series of researches on ultra-precision cutting of difficult-to-cut metals with the coated bite. The difficult-to-cut metals used in this report are three kinds of stainless steels.

2. EXPERIMENTAL PROCEDURE

The experiments are carried out by face-turning the workpiece with the ultra-precision lathe. The overview of experimental setup is shown in Figure 1. The experimental conditions are summarized in Table 1. Workpiece materials used are three kinds of stainless steels of ϕ50mm in diameter and 10mm in thickness, which are SUS 304, SUS 316 and SUS 403 in JIS. Three kinds of bites which are cemented carbide bite, cermet bite and coated cemented carbide bite are used in this experiment. Cutting fluids used are oil type including extreme pressure additives which is found effective for ultra-precision cutting of titanium alloy[1],[2]. The observation and roughness measurement of the face-turned workpiece surfaces are done with Nomarski type of differential interference microscope and SEM, and stylus profilometer and surface interferometer (WYKO TOPO-3D).

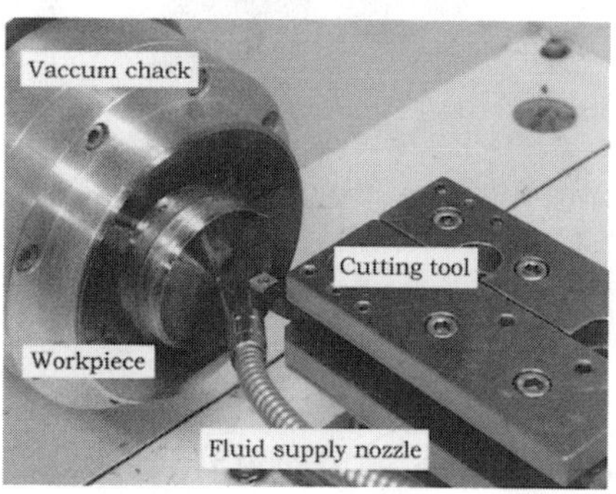

Figure 1

Overview of experimental setup

Table 1 Experimental conditions

Machine tool:	Ultra-precision Lathe ULC-100A, Toshiba Machine CO.,LTD.
Cutting tool:	Cemented carbide bite (Tool 1) Cermet bite (Tool 2) Coated cemented carbide bite (Tool 3) Nose radius: 0.4 mm Rake angle: 0 deg, Relief angle: 7 deg
Workpiece material:	SUS 304 (18Cr-8Ni) SUS 316 (18Cr-12Ni-2.5Mo) SUS 403 (13Cr-Si)
Cutting fluid:	Oil type cutting fluid
Cutting conditions	Cutting speed: V = 30~120 m/min Depth of cut: d = 5 μm Feed rate: f = 2, 4 μm/rev

3. RESULTS AND DISCUSSION

Figure 2 shows the microphotographs of face-turned SUS 316 surface. In figure, "Tool 1", "Tool 2" and "Tool 3" denote cemented carbide bite, cermet bite and coated cemented carbide bite, respectively. The 3D surface roughness for 256 μm square measured with WYKO TOPO-3D, however, is for coated cemented carbide bite. It is found from the figure that the SUS 316 surface face-turned by the coated cemented carbide bite "Tool 3" is smoothest and attains to the 3D surface roughness of about 83nm(P-V) as shown in Figure 3(d). On the other hand, the SUS 316 surfaces face-turned by cemented carbide bite "Tool 1" and cermet bite "Tool 2" have a lot of irregular flaws and are rougher than the surface face-turned by coated cemented carbide bite "Tool 3".

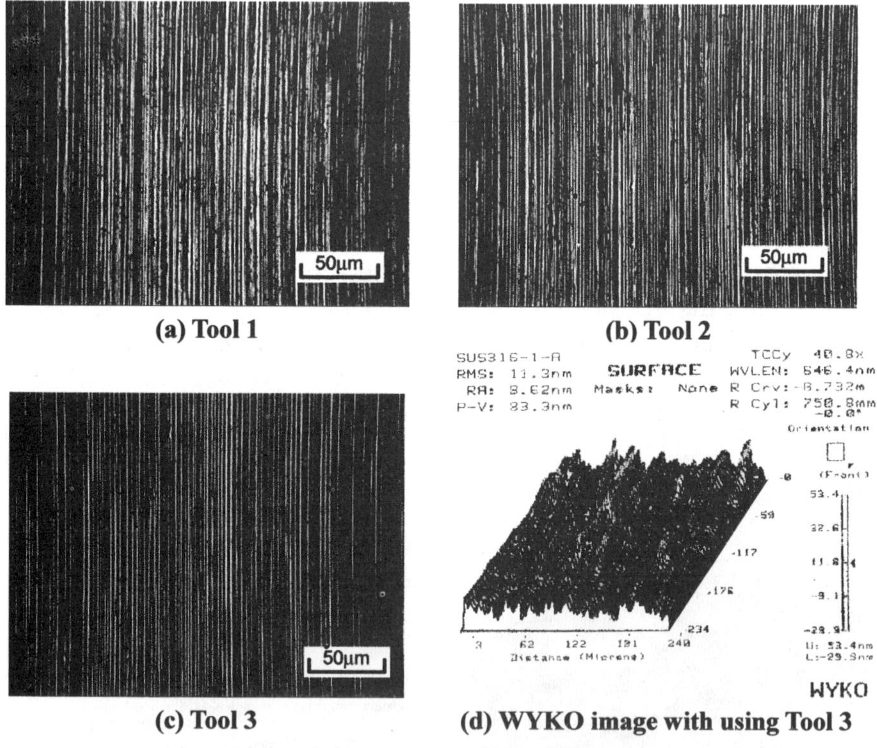

(a) Tool 1   (b) Tool 2

(c) Tool 3   (d) WYKO image with using Tool 3

Figure 2 Microphotographs and a WYKO 3D image of SUS 316 surface face-turned by three kinds of bites under using oil type cutting fluid (N=500rpm, V=60m/min, d=5μm, f=2μm/rev)

The 2D surface roughnesses of the three kinds of stainless steels face-turned are summarized for three kinds of bites in Table 2.   The surface roughness obtained using "Tool 1" and "Tool 2", however, is measured with stylus profilometer because those are too rough to measure with WYKO TOPO-3D.   It is clear that the surface roughness of all kinds of stainless steels face-turned by "Tool 1" and "Tool 2" are rougher than those by "Tool 3".

Table 2   Results of 2D surface roughness
(N=500rpm, V=60m/min, d=5μm, f=2μm/rev)

Unit: nm(P-V)

Cutting tool / Workpiece	Tool 1 Cemented carbide bite	Tool 2 Cermet bite	Tool 3 Coated cemented carbide bite
SUS 304	305	243	180
SUS 316	418	426	58.0
SUS 403	416	560	280
Measurement Method	Stylus profilometer		WYKO TOPO-3D

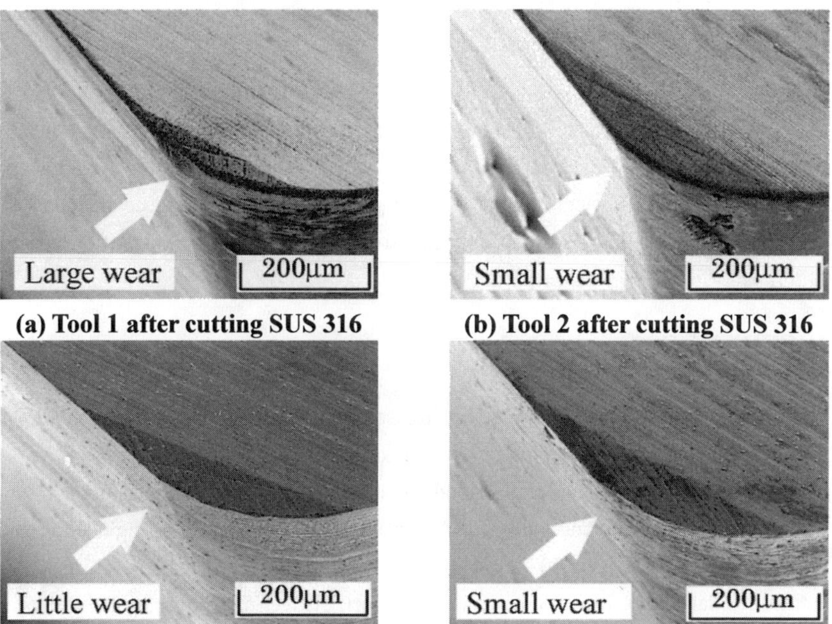

(a) Tool 1 after cutting SUS 316   (b) Tool 2 after cutting SUS 316

(c) Tool 3 after cutting SUS 316   (d) Tool 3 after cutting SUS 403

Figure 3   SEM photographs of cutting bite after face-turning stainless steel
(N=500rpm, V=60m/min, d=5μm, f=2μm/rev)

The smoothest surface obtained is about 58.0nm(P-V) for SUS 316 face-turned by "Tool 3". Figure 3 shows the representative SEM photographs of the cutting edges after cutting the stainless steels. From the figure 3, the wear of cutting edge after cutting the SUS 316 by "Tool 3" is observed remarkably little in comparison with other cases. On the other hand, the wear of the cutting edge of "Tool 1" and "Tool 2" after cutting SUS 316 become considerably large. However it is found that the coated cemented carbide bite does not wear so much after face-turning SUS 316 for length of 1,000m and the 2D surface roughness is kept about 100nm(P-V).

4. SUMMARY

With almost all the kind of coated bites, it is difficult to obtain the smoother surface than 100nm(P-V) for face-turned ultra-precision cutting of stainless steel. The special coated cemented carbide is found available for ultra-precision cutting of stainless steel (SUS316), because the bite wear is so small in the cutting length of about 1km and the finished surface roughness is smoother than about 100nm(P-V) in the 3D surface roughness for 256 μm square.

5. ACKNOWLEDGEMENT

The authors would like to thank Toshiba Machine Co., Ltd., Yushiro Chemical Industry Co., Ltd., UEX Co., Ltd, Daido Steel Co., Ltd, Toshiba Tungaloy Co., Ltd. and Osaka Diamond Industrial Co., Ltd. for providing cutting fluids, workpiece materials and cutting tools. A part of this research is supported by Grant-In-Aid for Scientific Research, 2000, in Japan.

6. REFERENCES

[1] E.Shamoto, T.Moriwaki: "Ultra-precise, ultrasonic vibration cutting work of mold steel using single crystal diamond tools", New Diamond, Vol.14, No.2, (1998), pp.26-27.

[2] H.Yasui, S.Sakamoto, M.Kawada, S.Kondo and A.Hosokawa: "Ultra-Precision Cutting of $\alpha-\beta$ Titanium Alloys", Proceedings of American Society of Precision Engineering 2000 Annual Meeting, (2000), pp.70-73.

[3] H.Yasui, S.Sakamoto, M.Kawada, and A.Fujimori: "Effect of Cutting Fluid on Surface Roughness in Ultra-Precision Cutting of Titanium Alloys", Proceedings of Spring Annual Meeting of JSPE, (2000), pp.231.

A STUDY ON METAL CUTTING UNDER HIGH HYDROSTATIC PRESSURE

Meng LIU*   Jun-ichiro TAKAGI*

*Department of Mechanical Engineering and Materials Science, Yokohama National University
79-5 Tokiwadai, Hodogaya-Ku, Yokohama, 240-8501 Japan

Abstract

The present paper deals with some experimental results of metal cutting under the hydrostatic pressure from 0MPa to 150MPa. The orthogonal cutting process was carried out in a developed cutting apparatus using hydraulic oil ISO VG 32 as hydraulic fluid. Main results obtained in this study are as followings: The hydrostatic pressure can improve the penetration of hydraulic oil into the chip-tool interface and tool-workpiece interface remarkably. The surface finish shows a tendency to become fine as the hydrostatic pressure increases for both aluminum and 0.45% steel.

Keywords

Orthogonal Metal Cutting, Hydrostatic Pressure, Lubrication, Surface Finish

1. INTRODUCTION

As for understanding the phenomena of cutting process under the special environment, it is expected to give the essential knowledge about the conventional cutting process and it also can give a possibility getting the hint to develop a new machining technology. Many contributions about metal cutting in the special environment, such as in vacuum and in argon gas etc. are reported [1-2]. To realize the ductile machining of silicon, the hydrostatic pressure was demonstrated experimentally [3]. The purpose of this research is in understanding the cutting phenomena under high hydrostatic pressure basically. The influence of the hydrostatic pressure on the cutting phenomena, especially the difference of chip formation and surface finish was discussed mainly.

2. EXPERIMENTAL APPARATUS AND PROCEDURES

To realize the cutting process under high hydrostatic pressure, a special cutting apparatus with a pressure vessel was developed. The hydrostatic pressure of maximum 200MPa can be given inside the pressure vessel. An orthogonal cutting apparatus was mounted in the pressure vessel. The orthogonal cutting process, which a round workpiece(ϕ 50mm(Diameter)$\times$2mm(Thickness)) can be turned and the depth of cut can be given in the radius direction of the workpiece, was carried out in this experimental apparatus. The size of the pressure vessel is ϕ 100mm(Diameter)$\times$150mm(Length). For the structure of the apparatus and the method of the sealing for the rotation shaft, the rotation speed of workpiece was designed only 4rpm. Cutting process was carried out at a very low cutting speed less than 0.628m/min. Hydraulic oil (ISO VG32) was used as the hydraulic fluid in this cutting apparatus. Two types cutting materials, mild carbon steel 0.45% steel (JIS S45C) and aluminum (JIS A1050P) were tested. To observe the side plastic flow of chips, the side face of workpieces was ground to obtain a fine surface before cutting test. Two kinds of cutting experiments, continuous cutting test and sudden stop cutting test were carried out in this study. In the continuous cutting test, chips were collected after the cutting test, then the free surface of chips was observed by SEM micrographs. The cutting ratio was evaluated by measuring the chip thickness and the depth of cut. In the sudden stop cutting test, the section of metal including partially formed chip was ground and polished, then etched to observe the shear strain. The etched surface was photographed through a microscope. Detail experimental conditions are listed in *Table1*.

Table1 Experimental conditions

Cutting apparatus	Hydrostatic pressure: 0～150MPa Pressure medium: Hydraulic oil ISO VG 32	
Cutting conditions	Tool	High-speed steel; Rake angle 10°, Relief angle 7°
	Workpiece	Aluminum (JIS A1050P) 0.45% carbon steel (JIS S45C) ϕ 50mm×2mm
	Cutting speed	0.6m/min
	Depth of cut	0.075mm/pass

3. EXPERIMENTAL RESULTS AND DISCUSS

Fig.1 shows the SEM micrographs of the free surface (side part and center

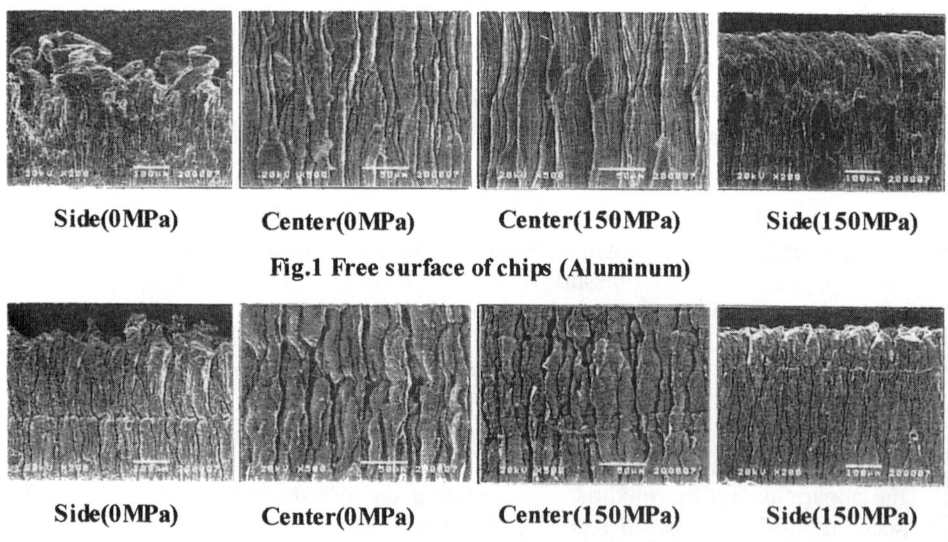

Fig.1 Free surface of chips (Aluminum)

Fig.2 Free surface of chips (0.45% steel)

part) of aluminum chips collected under the hydrostatic pressure of 0MPa and 150MPa. In this study, the cutting tests of hydrostatic pressure 0MPa were carried out in the oil environment. The collected chips were observed. Chips of aluminum were continuous under both the hydrostatic pressure 0MPa and 150MPa generally. From Fig.1, as the hydrostatic pressure increases form 0MPa to 150MPa, side flow of chip decreases remarkably and the chip shows a smooth side surface. In both conditions of the hydrostatic pressure, the free surface of chips (the back of the chips) is very wavy. It means cutting with inhomogeneous strain. But under high hydrostatic pressure, a tendency of regularity of the waviness and the closeness of spacing of the waviness can be observed.

Fig.2 shows the SEM micrographs of the free surface (side part and center part) of 0.45% steel chips under the hydrostatic pressure of 0MPa and 150MPa. About the collected chips, the chips of 0.45% steel were found to fracture periodically under the hydrostatic pressure of 0MPa, but chips were continuous under the hydrostatic pressure 150MPa. From Fig.2, the same results as aluminum, the closeness of spacing of the waviness can be observed, but it is not remarkable as aluminum. It may be assumed that the shear planes will be very closely spaced corresponding to the hydrostatic pressure.

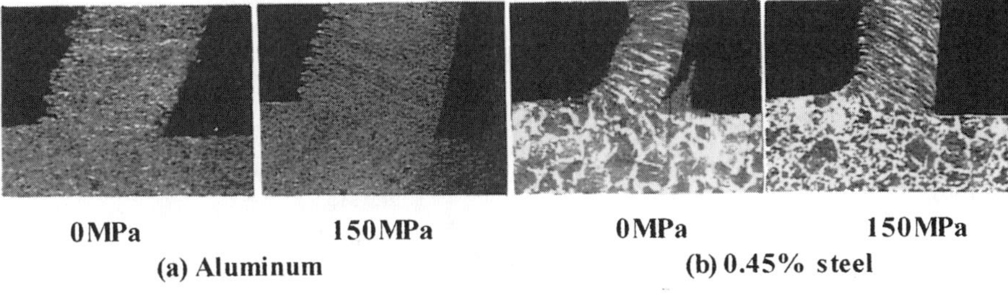

Fig.3 Photomicrographs of partially formed chip

Fig.3 shows photomicrographs of partially formed chip of aluminum and 0.45% steel under the hydrostatic pressure of 0MPa and 150MPa. As for 0.45% steel, as shown in the figure, as the hydrostatic pressure increases, the interaction which chips contact with the rake face of the tool increases. And a substantial shear stress sufficient to cause the secondary subsurface to shear can be guessed. As evident in *Fig.3(b)*, there exists the remarkable secondary shear. A result is also confirmed that there exists a built-up edge (BUE) under the hydrostatic pressure 0MPa and the BUE disappears under 150MPa for 0.45% steel. As for aluminum, the BUE was not observed clearly under the hydrostatic pressure of 0MPa and 150MPa.

Fig.4 shows the relation between the cutting ratio and the hydrostatic pressure. From the figure, in the conditions that the hydrostatic pressure changes from 0MPa to 150MPa and cutting speed is very low, aluminum and 0.45% steel show a directly opposite result about the cutting ratio. For aluminum, cutting ratio increases with the increase in the hydrostatic pressure. But for 0.45% steel, there exists a tendency that cutting ratio decreases with the increase in the hydrostatic pressure.

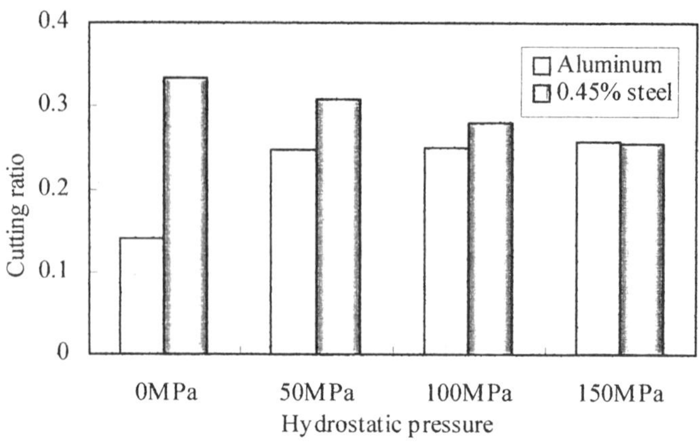

Fig.4 Relation between of cutting ratio and hydrostatic pressure

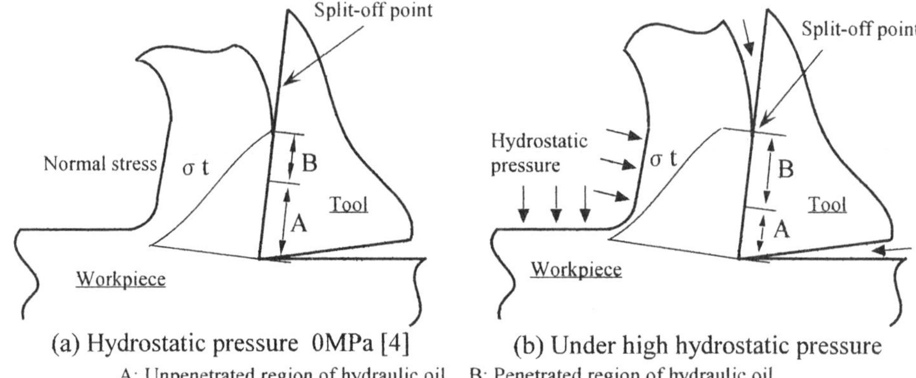

(a) Hydrostatic pressure 0MPa [4]   (b) Under high hydrostatic pressure

A: Unpenetrated region of hydraulic oil   B: Penetrated region of hydraulic oil

Fig.5 Cutting process under high hydrostatic pressure

From the above-mentioned results, It can be considered that the hydrostatic pressure affects the cutting process by improving the penetration of hydraulic oil into the interface between tool and chip, workpiece. This penetration of hydraulic oil can be guessed via three routes into the chip-tool interface and the tool-workpiece interface, from rake surface, from reliefe surface and both side face of the workpiece. *Fig.5* shows the cutting mechanism under hydrostatic pressure. The penetration of cutting fluid has the effect of reducing the friction between chip and tool [4-5]. As the hydrostatic pressure increases, the pressure medium hydraulic oil is easy to penetrate into the tool-chip interface and the tool-workpiece interface, and the effect of oil lubrication can be obtained. As the friction between chips and tool decreases, the shear angle increases and the cutting ratio increases. This can explain the result of the experiment of aluminum. But for 0.45% carbon steel, an opposite result, the decrease of cutting ratio was obtained. The result of 0.45% steel can be considered due to the change of the BUE. In the condition of hydrostatic pressure 0MPa, a BUE (as shown in Fig.3) was formed to give an increase in the effective rake angle. As

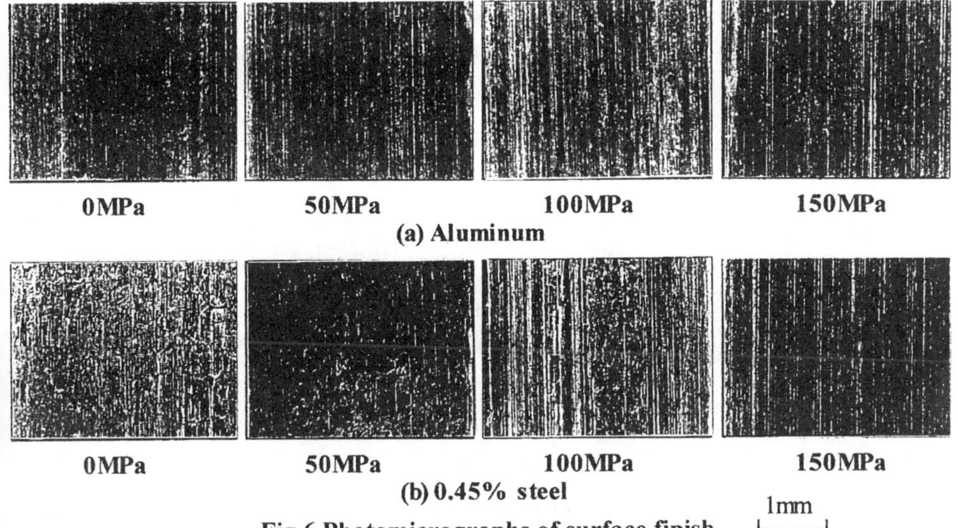

0MPa   50MPa   100MPa   150MPa

(a) Aluminum

0MPa   50MPa   100MPa   150MPa

(b) 0.45% steel

Fig.6 Photomicrographs of surface finish   1mm |———|

the hydrostatic pressure increases, the BUE disappears resulting a decrease in the effective rake angle. The change of BUE brought by the hydrostatic pressure can also explain the penetration of hydraulic oil into the chip-tool interface and tool-workpiece interface due to the hydrostatic pressure indirectly.

Fig.6 shows the photomicrographs of the finished surface of aluminum and 0.45% steel. As the hydrostatic pressure increases, finished surface shows a tendency to be fine for both aluminum and 0.45% steel. Especially for 0.45% steel, this result is very remarkable.

4. CONCLUSIONS

Main results obtain in this study are as followings:
1) The hydrostatic pressure can improve the penetration of hydraulic oil into the chip-tool interface and tool-workpiece interface remarkably.
2) The surface finish shows a tendency to became fine as the hydrostatic pressure increases for both aluminum and 0.45% steel.

REFERENCES

1. K. UEHARA et al: Cutting mechanism of carbon steel and titanium, Journal of the Japan Society Precision Engineering, 38, 4(1972)363-368.(In Japanese)
2. S. OGASAWARA et al: Effect of oxygen pressure on high speed cutting of metal materials, Journal of the Japan Society Precision Engineering, 63, 11(1997)1563-1568. (In Japanese)
3. J. YAN et al: On the ductile machining of silicon for micro electro-mechanical system(MEMS), opto-electronic and optical applications, Materials, Science & Engineering A, 297 (2001) 230-234.
4. K. MIZUHARA, E. USUI: Experimental evaluation of cutting fluid penetration into too-chip interface, Journal of the Japan Society Precision Engineering, 47, 3(1981)350-355.(In Japanese)
5. N. SHINOZAKI, H. YOSHIKAWA: The lubricating effect of cutting fluid (I), Journal of the Japan Society Precision Engineering, 24, 3(1958)140-145.(In Japanese)

A NEW INTERPOLATION ALGORITHM FOR ULTRA-PRECISION CNC MACHINING

Shengyi Li, Mingliang Zhang, Yifan Dai, Xuhui Xie

National university of Defense Technology

School of Mechtronical Engineering &Automation

Tel: 86-731-4573301

Abstract

In Optical design, aspheric lens are frequently adopted because optical system can be simplified and aberration of light can be reduced. Most conventional CNC machines support only the functions of straight line and circular interpolations and they can not machine aspheric surfaces. This paper introduces a new interpolation algorithm which is designed for ultra-precision machining. It can process general curves instead of pre-processing them out of CNC. Implementation of those algorithms is introduced in detail and some examples are given.

Keywords

CNC, real time, interpolation, aspheric, bi-arc approximation

1.INTRODUCTION

Ultra-precision machining technique plays an important role in manufacturing of high-tech products. For example, optical lens are typical ultra-precision parts. They are characterized with not only high accuracy but also special shape. Most conventional CNC machines support only the functions of straight line and circular interpolations and they can not machine aspheric curves directly.

In order to bridge this gap between application and limited capabilities offered by conventional CNC machines, several methods have been reported. The standard practice for machining an aspheric curve is to break the curve into a set of linear or circular segments out of CNC. Because those segments are generated out of practical machining environment, segmentation will affect the average tool feedrate and causes deterioration of the surface finish[1,2]. This method is not convenient and can not tap the machining potential of CNC machines. Thus this method is not adequate for the machining of ultra-precision parts. Several researchers have developed real time curve interpolators[3] which can be expressed as Figure 1(a). For general curves, calculation time of specific points in the curve is uncertain. If it is longer than interpolation period, those interpolators can not guarantee its real time characteristic.

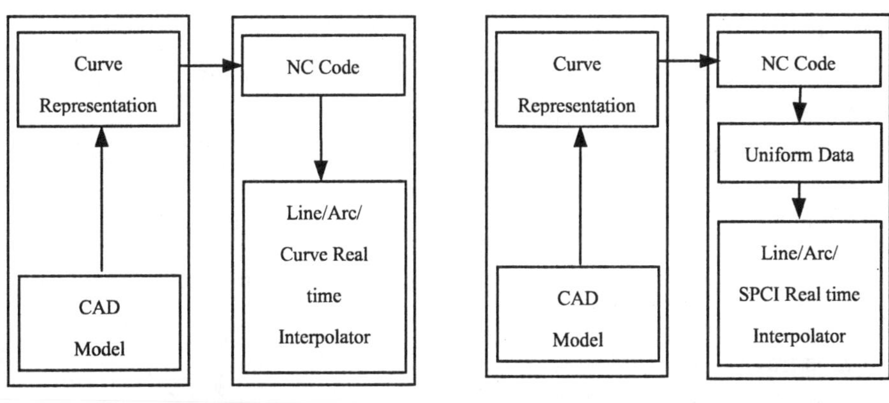

(a) Real Time Interpolator   (b) Quasi Real Time Interpolator.

Consequently, a new method needs to be developed (see Figure 1(b)). We have developed a quasi real time interpolator. The interpolator transfers general curves into data in a uniform format after machining command is sent out but before machining, and the uniform format data are processed in real time during machining. It has following advantages: (1) The machining parameters are obtained before transfer so that it can tap the machining potential of CNC machines; (2) Uniform format data is more compact than general NC codes; (3) Regardless of complexity of general curves, our interpolator can process them in real time. In order to achieve above goals, a method needs to be developed which can implement the transfer according to machining parameters in short time and generate minimum amount of data. Implementation of those algorithms is introduced and some examples are given.

2. APPROXIMATION OF GENERAL CURVES

Our method aims to bridge the gap between complex CAD model and limited capabilities offered by conventional CNC machines. Because uncertain complexity of general curves, we believe that approximation general curves by uniform curve is an ultimate method. For our quasi real time interpolator, an approximation algorithm must : (1) process general curve by a uniform means; (2) generate uniform data in short time; (3) generate appropriate volume of data. Although many papers have been published on the approximation of complex curves, few can meet the demands. We have explored below theorems (all proofs are omitted). Based on these theorems, a new approximation algorithm which divides general curves into spirals and then approximates each spiral by bi-arc is proposed.

A bi-arc[1,2,4] is a curve that is made by joining two arcs in a G^1 manner. A spiral[2] is a curve whose curvature is of one sign and is monotone-increasing or monotone-decreasing as the curve is traversed.

Theorem 1: The family of bi-arc of a spiral segment lies in the bi-arc region.

Bi-arc region is the region with boundary C_1 and C_2.

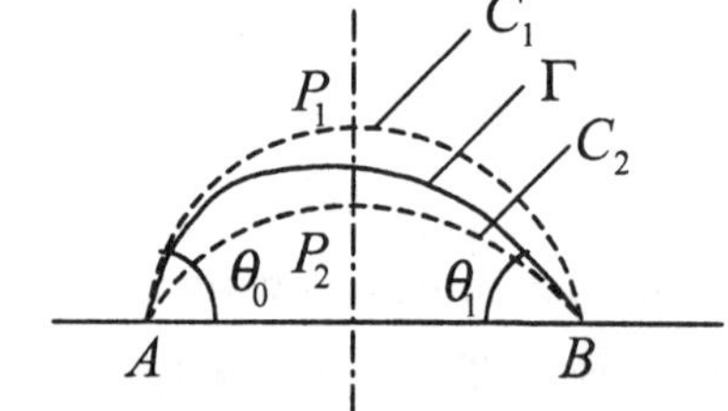

Figure 2. Spiral and its bi-arc region

Theorem 2: Optimum bi-arc approximation of a spiral segment satisfies below conditions: maximum approximation error is equal to minimum approximation error.

Theorem 3: Maximum error of bi-arc approximation of a spiral segment is

not greater than $\left| l \cdot (tg\dfrac{\max(\theta_0,\theta_1)}{2} - tg\dfrac{\min(\theta_0,\theta_1)}{2}) \right|$.

Based on Theorem 1 and Theorem 2, we have developed an optimal bi-arc approximation algorithm. For a trade-off of processing time and size of data and based on Theorem 3, a more efficient approximate optimal bi-arc approximation algorithm is developed.

3. REAL TIME INTERPOLATION

After approximation of general curves, we stored data in the memory of CNC. Because PC based CNC's memory is very cheap, so we need not worry the data volume. Resultant curve of bi-arc approximation is line and circle, so real time interpolation algorithm is the same. Because we approximate original curve after machining command is sent out, we can avoid acceleration or deceleration generated from segmentation by additional treatment according to machining parameters.

4.EXAMPLE

A polynomial $y = \displaystyle\sum_{n=0}^{N} x^n, x \in [1,2]$ is used to test our algorithm. The

algorithm is implemented by Matlab5.2，and run on a PC（PII233MHz，32M memory）

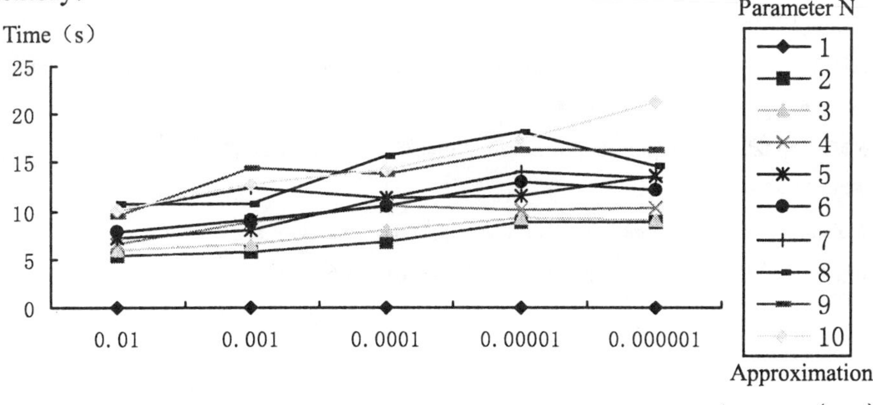

Figure 3.(a) Approximation Time of Our Algorithm

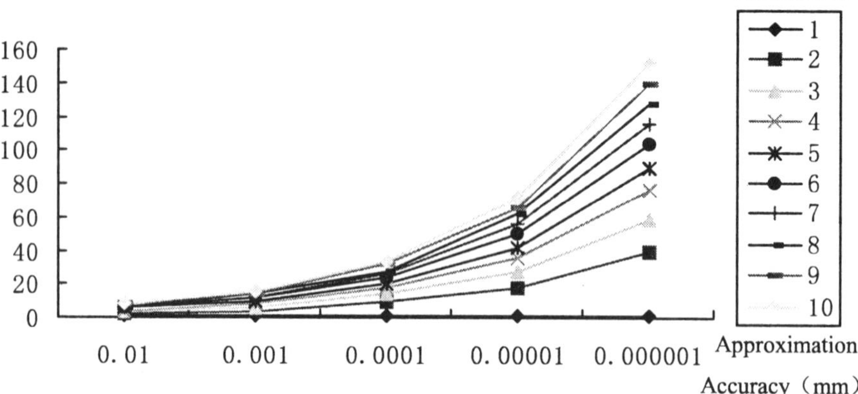

Figure 3.(b) Number of segments generated by our algorithm

From Figure 3, we can see that: (1) approximation time is so short that it can be ignored in a machining; (2) Number of segment is rather few so that we can store them in CNC.

5.CONCLUSION

Curves can be presented as either implicit or parameter form. Our algorithm can handle general curves, has no constraint, and is very stable. It can be transplanted into modern CNC to enhance their interpolation capability.

REFERENCES

1.D. S. Meek, D. J. Walton. Approximation of discret data by G^1 arc splines. Computer aided design 1992;6:301~306

2.D. S. Meek, D. J. Walton. Approximating quarlitic NURBS curves by arc splines. Computer aided design 1993;6:371~376

3.M. Shpitalni, Y. Koren, C. C. Lo. Realtime Curve interpolators, Computer aided design 1994;11:832~838

4.Young Joon Ahn,Hong Oh Kim,Kyoung,Yong Lee. G^1 arc spline approximation of quadratic Bezier curves. Computer aided design 1998;8:615~620

SIX-AXIS CONTROL CHARACTER LINE FINISHING USING ULTRASONIC VIBRATIONAL CUTTING TOOL

Koichi Morishige and **Yoshimi Takeuchi**

Dept. of Mechanical Engineering and Intelligent Systems

The University of Electro-Communications

1-5-1, Chofugaoka, Chofu-shi, Tokyo, 182-8585 JAPAN

m-shige@mce.uec.ac.jp, takeuchi@mce.uec.ac.jp

Abstract

The study deals with six-axis control character line finishing with the use of vibrational cutting tool. Six-axis control machining with non-rotational tools is a potential method to make a clear character line shape. However, the machining produces not good results due to the low cutting speed. In the study, the vibrational cutting is applied, which is effective in solving the problem such as low cutting speed. The cutter location data must be generated on the basis of the workpiece CAD data, considering the characteristic of the vibrational tool. The effectiveness of the devised method is experimentally confirmed.

Keywords

Six-axis control machining, Vibrational cutting, CAD/CAM

1. INTRODUCTION

Rotational cutting tools are usually used to increase cutting speed. However, in the case of machining with rotational cutting tools, the remains such as a circular arc-like part inevitably takes place at a corner, as shown in Fig.1(a). The character line, which is an intersection between surfaces, stands for the typical shape. The remains can be removed by EDM, which requires the preparation of a special electrode. However, changing from one process into another leads to longer setup time and the installation error.

In order to solve the above problems, six-axis control machining has been proposed with non-rotational cutting tools, as shown in Fig.1(b) [Takeuchi 96]. However, the practical use of six-axis control machining is difficult due to the low cutting speed of non-rotational tools. It is because the cutting speed is

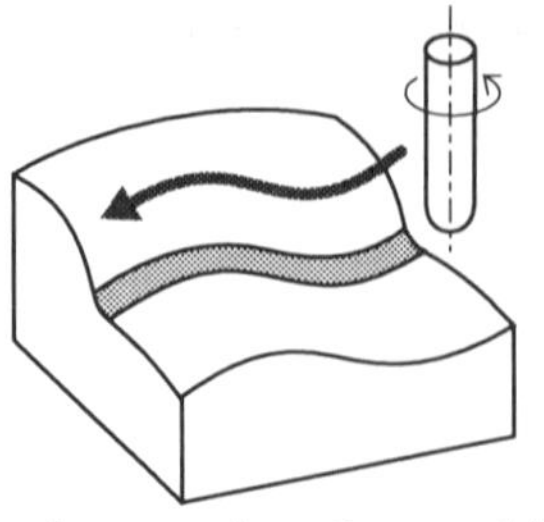 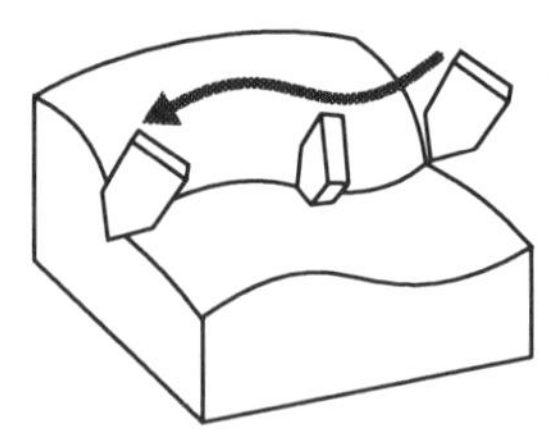

(a) Cutting remains after machining
with rotational cutting tool

(b) Character line finishing using six-axis
machining with non-rotational tool

Figure 1. Effectiveness of six-axis control character line finishing

equal to the feed rate. Likewise, there is a limitation in feed rate due to the capacity of the numerical control unit. Such low cutting speed causes the deterioration of surface quality, especially in the machining of malleable and soft materials like aluminum. With vibrational cutting, the problem on surface quality resulting from low cutting speed will be eliminated.

2. SIX-AXIS CONTROL CHARACTER LINE MACHINING

2.1 Vibrational Cutting Tool

The ultrasonic vibration is applied on the tip of a cutting tool during cutting. The cutting speed will be the feed rate plus the vibration speed. Using this process improves the roughness and appearance of the machined surface.

In this study, a commercially available vibrational cutting unit (SB-150: Taga Electric Co.) is used. This unit is composed of vibrational tools and a vibration element drive amplifier. The vibrational tool has a tip holder at the top of vibration part, in which the cutting tip is mounted. Figure 2 shows three kinds of tools that are prepared for the cutting unit. Each tool is attached into the oscillator that generates the ultrasonic vibration through the special holder that is designed to give an appropriate vibration to the cutting tip.

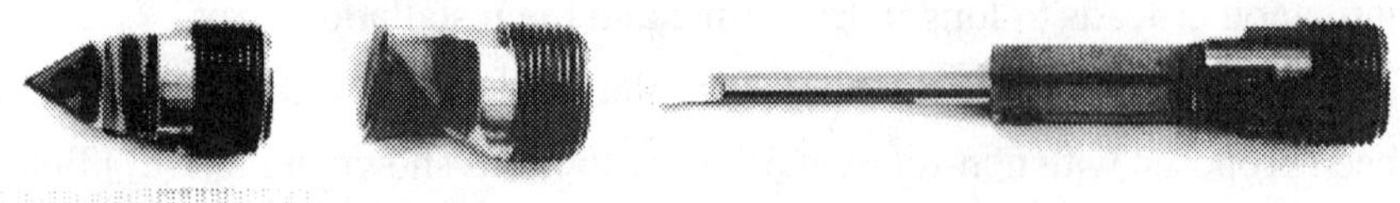

(a) Straight tool   (b) Knife tool   (c) Bore byte   10mm

Figure 2. Three types of tools for vibrational cutting

2.2 Tool Attitude Considering Vibrational Direction

In order to carry out an effective vibrational cutting, it is important for the cutting direction to be matched in the direction where the cutting tip vibrates. The cutting tip vibrates with a certain angle as seen from the side (Fig.3), thus the tool holder should be inclined by a pitching angle θ_p.

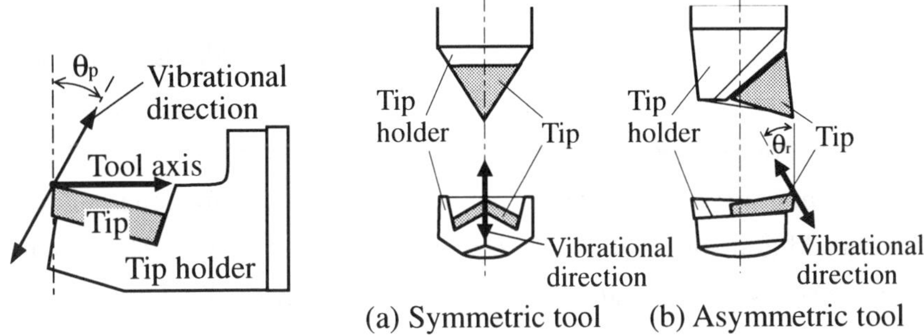

Figure 3. Pitching angle θ_p Figure 4. Rolling angle θ_r

Moreover, when the tool is seen from the down side, the vibrational direction of a symmetric point nose straight tool becomes parallel to the cutting direction, as shown in Fig.4(a). However, in case of an asymmetric knife tool and bore byte, the vibrational direction has a fixed angle, as shown in Fig.4(b). Therefore, the cutting direction should be matched with that of vibrational direction by rotating the rolling angle θ_r. Taking into account of the above-mentioned correction angles, the tool attitude can be easily calculated.

The tool attitude can be calculated as follows: consider a local orthogonal coordinate system at each cutting point P, as shown in Fig.5(a). The coordinate system has its origin at P, and is formed by the tool feed vector F and the initial

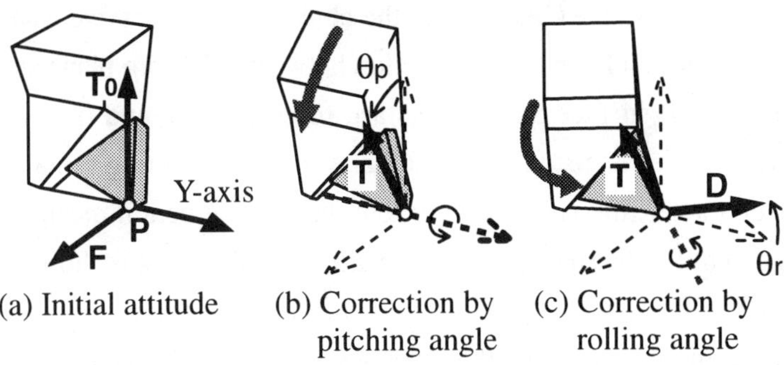

Figure 5. Tool attitude determination for six-axis control vibrational cutting

tool axis vector T_0, which corresponds to X-axis and Z-axis of the coordinate system, respectively. On the othor hand, the outer product of T_0 and F pertains to the Y-axis of the coordinate system. Figure 5(b) shows the tool axis T acquired in rotating T_0 by θ_p around the Y-axis. The tool direction D is also obtained through rotating Y-axis by θ_r around T, as shown in Fig.5(c).

2.3 Cutter Location for Character Line Machining

The CL data for the character line machining is calculated as follows. First, the target edge and the machining start point are selected, as illustrated in Fig.6. Then, one of two curved surfaces that make the target edge is chosen as the reference surface and cutting points P are sequentially generated on the edge. In addition, the tangent vector to the edge, F and the normal vector to the reference surface, T_0 are calculated at each P, respectively. T and D are obtained from the aforementioned process. If the tool interferes with the surfaces, the tool attitude is corrected by rotating vectors T and D around the vector F. Finally, the obtained sets of P, T and D are the output in the CL data for six-axis control character line machining.

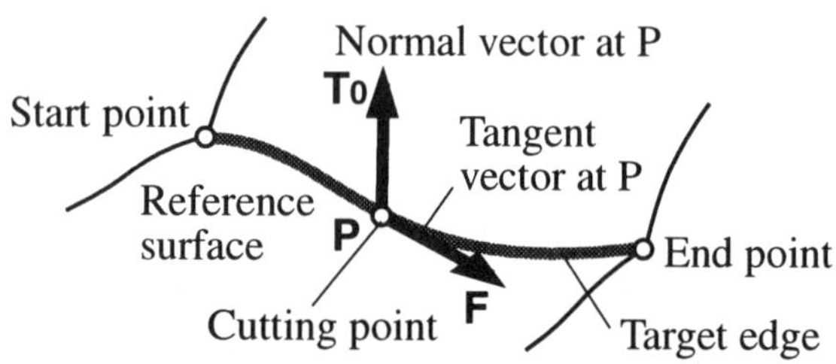

Figure 6. Generation of cutter location for character line machining

3. MACHINING EXPERIMENT

Figure 7 shows the CAD data of the object shape. The target shape is the character line formed from the intersection between a bottom plane surface S1 and a curved wall surface S2. At the start, rough cutting and finishing were done for the whole shape, using R5 ball end mill. The cutting remain formed along the character line is intensively removed by the pencil machining with R1.5 ball end mill. The shape appearance after pencil cutting is shown in Fig.8, after which it will be subjected to six-axis control character line finishing. Figure 9 shows machining with a bore byte tool. As a result, a clear character line was

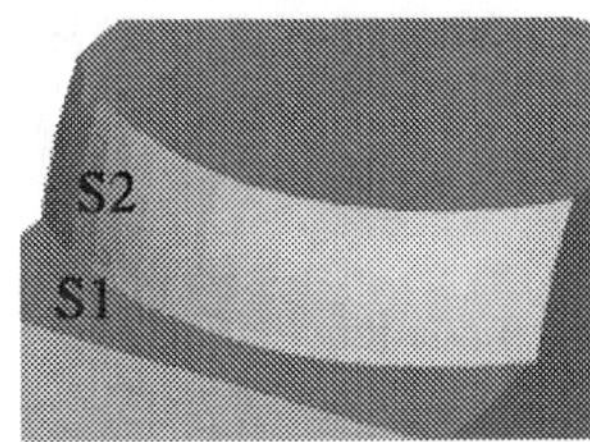

Figure 7. Shape CAD data

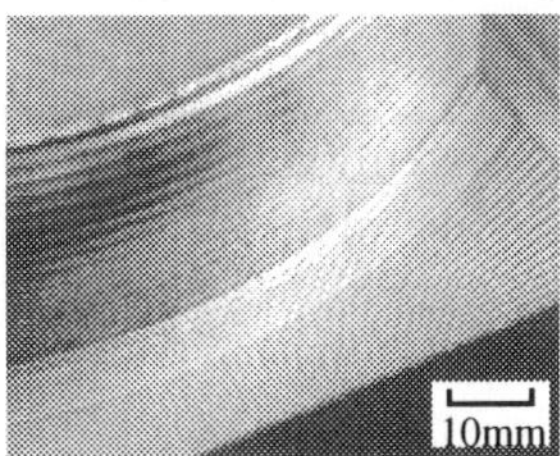

Figure 8. Machined shape after cutting with R5 ball end mill

Figure 9. Six-axis machining on MC

Figure 10. Finished character line

actually obtained by using such process (Fig.10).

4. CONCLUSION

Based on the experimental results, using the CL data with consideration on the special correction angles of vibrational cutting tool, it was found out that the deviced method allows a great improvement on the roughness and appearance of the machined surface, as well as on the possible depth of cut. Moreover, an excellent character line shape is obtained by using the CAM system specifically developed for the deviced method, and it was confirmed that the method is also applicable to the actual machining.

A part of the study is supported by Grant-in-Aid for Scientific Research ((B2)12450057) of the Ministry of Education, Science, Sports and Culture.

5. REFERENCES

Y Takeuchi, H Suzuki. Efficient and Accurate Manufacturing by Means of Multi-Axis Control Machine Tools. Japan/USA Symposium on Flexible Automation. ASME; 1996; 1: 343-347.

A NEW DIAMOND TURNING METHOD FOR FABRICATION OF CONVEX ASPHERIC SURFACE ON HARD BRITTLE MATERIAL

Jiwang Yan, Katsuo Syoji and Tsunemoto Kuriyagawa

Department of Mechatronics and Precision Engineering, Tohoku University
Aramaki-Aoba-01, Auba-ku, Sendai 980-8579, Japan

Abstract

The conventional method for diamond turning an aspheric surface is the arc-enveloping method. In the present study, a new method termed the straight-line enveloping method (SLEM) is proposed, in which the aspheric surface is generated using a straight-nosed diamond tool on a three-axis ultraprecision machine tool. This method is preferable to the conventional method for the machining of convex aspheric surfaces on hard brittle materials since it significantly improves efficiency and lowers tool wear. An experiment in which a large single crystal silicon aspheric lens is cut by applying the proposed method is described.

Keywords

Diamond turning, brittle material, aspheric surface, optical component

1. INTRODUCTION

Currently in the optics industry, aspheric surfaces are being utilized more and more in modern optical components such as lenses, mirrors and their molds. For most applications, aspheric surfaces are required to be manufactured on hard brittle materials such as single crystals, glasses and advanced ceramics. This paper presents a new diamond turning method for fabricating an aspheric surface on hard brittle material.

2. ASPHERIC SURFACE CUTTING METHOD

An aspheric surface is expressed as the difference between a sphere and an asphere at different heights above the optic axis[1]. An axis-symmetric aspheric surface can be generally described by:

$$z(x) = \frac{Cx^2}{1 + \sqrt{1 - (k+1)C^2 x^2}} + \sum_{i=1}^{m} a_i x^i \qquad (1)$$

where, $C=1/r$, in which r is the radius of curvature of the sphere surface; x is the distance from the optic axis (Z); k is the conic constant, a parameter representing the eccentricity of the conic surface; for even i, a_i are the aspheric deformation constants and for odd i, a_i are aspheric coefficients used to define other polynomial curves by setting $C=0$.

Conventionally, diamond turning of an axis-symmetric aspheric surface is carried out on a two-axis (X-Z) machine with a round-nosed tool by the arc-enveloping method[2]. When diamond turning with a round-nosed tool, undeformed chip thickness varies along the cutting edge. Therefore, when machining brittle material, a truly ductile response only occurs along the apex of the tool tip where the undeformed chip thickness is smaller than a critical value (critical depth d_c), while the upper material is fractured[3]. For a given d_c, tool feed must be kept smaller than a certain critical value $f_{\max}$ in order to obtain a ductile surface. The value of $f_{\max}$ can be determined by:

$$f_{\max} = d_c \sqrt{R} \cdot \sqrt{\frac{1}{2(d_c + y_c)}} \qquad (2)$$

where R is the tool radius and y_c is the crack penetration depth. In previous studies, $f_{\max}$ for silicon was reported to be approximately 1 μm/rev [4), 5)]. For a constant machining area, an extremely small tool feed corresponds to a long cutting distance which increases tool wear and lowers machining efficiency. In particular, when machining a large aspheric surface, tool wear becomes a serious problem.

To solve this problem, in the present study, a new method termed the straight-line enveloping method (SLEM) is proposed. A schematic representation of the method is shown in Fig.1. Here, a straight-nosed diamond tool (SNDT) is used instead of the conventional round-nosed tool. To generate an aspheric surface, the straight-nosed diamond tool is moved in the X and Z directions and rotated about a B-axis which is perpendicular to the X-Z plane.

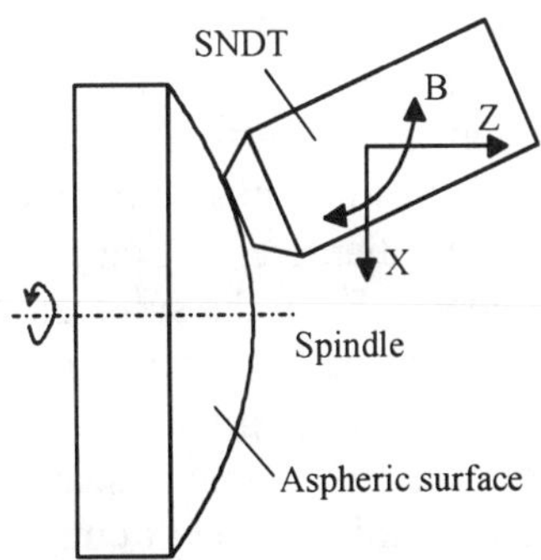

Fig.1 Schematic of the straight- line enveloping method (SLEM) for machining an aspheric surface

Thus, 3-axis (X-Z-B) simultaneous control is necessary and the aspheric surface is enveloped by the straight edge of the SNDT.

As noted in a previous paper by the authors[6], when the SNDT is used for cutting, undeformed chip thickness h is uniform along the entire width of the main cutting edge and h is determined by tool feed f and cutting edge angle κ. For a given d_c, the maximum tool feed f_{max} for ductile regime machining can be expressed by:

$$f_{max} = \frac{d_c}{\sin \kappa} \tag{3}$$

Therefore, by decreasing the cutting edge angle κ to a sufficiently small value, an extremely small undeformed chip thickness h can be obtained even at a large tool feed f. This enables ductile regime turning at a large tool feed and overcomes the problems of the arc-enveloping method.

3. EXPERIMENTAL PROCEDURE

Experiments were carried out on an ultraprecision lathe NACHI-ASP15. Figure 2 shows a photograph of part of the machine. This machine has an air-bearing spindle, two hydrostatic slide tables along the X-axis and Z-axis, respectively, and a hydrostatic rotation table about the B-axis. The X and Z slide tables can move at a revolution of 10nm and the B

Fig.2 Experimental setup

table can rotate down to an angular resolution of 0.001°, which enables fine adjustment of the cutting edge angle.

An infrared lens substrate made from single crystal silicon (100), 125mm in diameter, 15mm thick, was used as the specimen for the experiment. First, the substrate was contoured into a convex sphere surface having the same curvature radius as the objective aspheric surface by CG (curve generating) grinding. It was then bonded onto a diamond-turned aluminum blank using a heat-softened glue and vacuum chucked on the machine spindle. Rough cuts were performed for the purpose of obtaining an aspheric surface from the ground sphere surface and removing the ground-damaged layer.

A straight-nosed single crystal diamond tool having a 1.2 mm-long main cutting edge was used for the finishing cut. The rake angle was set to $-40°$ in order to achieve the maximum critical depth d_c[7], and the relief angle was set to 6°. A micro chamfer was fabricated on the rake face of the tool in order to obtain a short smoothing edge for a small surface roughness[6]. An included angle of 179.65° was formed and the cutting edge angle κ and the smoothing edge angle κ' were set to 0.24° and 0.11°, respectively.

Two cuts were performed under the following conditions: for the first cut, depth of cut a=10μm and tool feed rate f=50 μm/rev; and for the second cut, a=2μm and f=20μm/rev.   Therefore, the undeformed chip thickness h was 210nm and 84 nm, respectively.   The spindle rotation speed was set to 1000rpm and kerosene mist was used as the cooling fluid.

4. RESULTS AND DISCUSSION

Figure 3(a) shows a photograph of a single crystal silicon diamond-turned aspheric surface produced by the proposed method under the following conditions: a=10μm, f=50 μm/rev and h=210nm.   It is obvious that the cloudy regions form as a four-fold symmetric pattern spreading from the center of the specimen.   A detailed view of the center of the surface obtained using a Nomarski microscope is shown in Fig.3(b). The cloudy regions correspond to the severely damaged surface whereas the clear regions are less damaged.

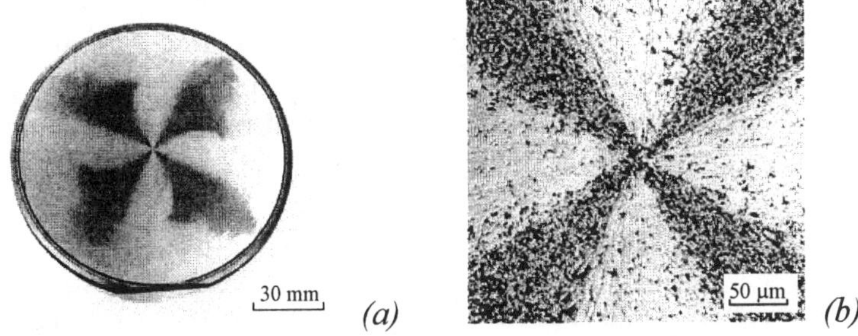

Fig.3   Silicon aspheric surface machined under f=50μm/rev, h=210nm

The anisotropy of the machined surface is due to the crystallographic effect of silicon.   In order to avoid the crystal anisotropy and to obtain homogeneous ductile surfaces, undeformed chip thickness h must be controlled to be less than a minimum critical depth d_{cmin} for all crystallographic orientations[6]. Here, the critical depths were measured using a method described previously[6] and it was found that d_{cmin} of silicon (100) was 185nm at a −40° rake angle. Next, cutting was performed under the following conditions: a=2μm, f=20 μm/rev and h=84nm.   Figure 4 shows a photograph of the resulting diamond-turned aspheric surface.   The entire surface has become clear without any cloudy region.   Figure 5(a) and (b) show the form error and surface roughness results, respectively.   The peak-to-valley value of the form error in the 90mm range was 1.36μm.   This form error is mainly caused by tool setting error and can be significantly reduced by using the on-machine measurement and compensation system.   The surface roughness was 78nmRmax.

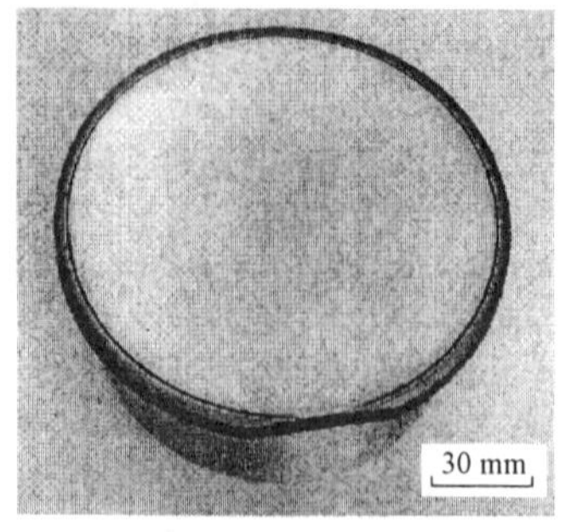

Fig.4 Silicon aspheric surface
(f=20μm/rev, h=84nm)

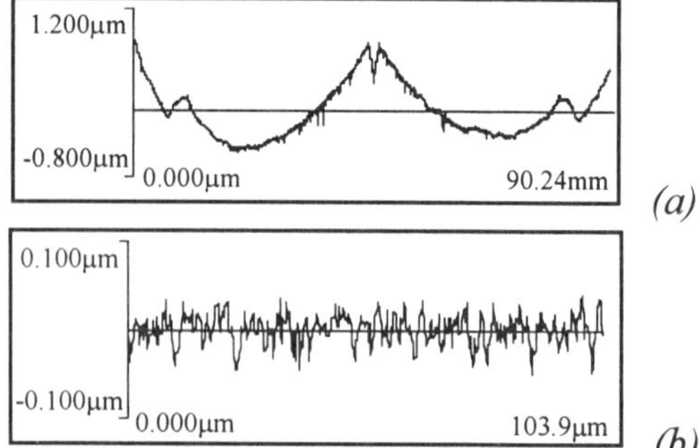

Fig.5 Form error (a) and surface roughness
(b) of the surface shown in Fig.4

5. CONCLUSIONS

A new method, termed the straight-line enveloping method (SLEM), for generating a convex aspheric surface on hard brittle materials has been proposed. This method uses a straight-nosed diamond tool on a three-axis ultraprecision machine tool and enables ductile regime turning of brittle material at large feed. An aspheric surface having 1.36μm form error and 78nmRmax surface roughness has been obtained on a single crystal silicon substrate at a tool feed rate of 20μm/rev.

ACKNOWLEDGEMENT

This research was supported by the Corning Research Grant 2000 as well as by grants from the Japan New Energy and Industrial Technology Development Organization (NEDO) RC-11H2003.

REFERENCES

1) D. C. O'Shea, *Element of Modern Optical Design*, Wiley, New York, 1985.
2) H. Suzuki, T. Kitajima and S. Okuyama: Study on Precision Cutting of Axis-symmetric Aspheric Surface, J. JSPE, 65, 3, (1999) 401.
3) W.S. Blackly and R.O. Scattergood: Ductile-Regime Machining Model for Diamond Turning of Brittle Materials, Prec. Eng., 13, 2 (1991) 95.
4) T. Nakasuji, S. Kodera, S. Hara, H. Matsunaga, N. Ikawa and S. Shimada: Diamond Turning of Brittle Materials for Optical Components, Ann. CIRP, 39, 1 (1990) 89.
5) T.P. Leung, W.B. Lee and X.M. Lu : Diamond Turning of Silicon Substrates in Ductile-regime, J. Mater. Proc. Tech., 73 (1998) 42.
6) J. Yan, K. Syoji, H. Suzuki and T. Kuriyagawa: Ductile Regime Turning of Single Crystal Silicon with a Straight-Nosed Diamond Tool, J. JSPE, 64, 9, (1998) 1345.
7) J. Yan, K. Syoji and T. Kuriyagawa: Ductile-Brittle Transition Under Large Negative Rake Angles, J. JSPE, 66, 7, (2000) 1130.

FRACTURE PHENOMENA OF ARAMID FIBER DURING OBLIQUE CUTTING

Eitoku NAKANISHI and Kiyoshi ISOGIMI

Dept. of Mechanical Eng., MIE University, Kamihama-cho 1515, Tsu, Mie, 514-8507 Japan.

Abstract

Due to the extreme difference of the mechanical properties between the matrix and fiber, the machining of the FRP, especially made by aramid fiber and polyester resin becomes very difficult. During machining, aramid fibers are difficult to be cut and are elongated easily. We tried to improve the surface appearance and machinability by employing the oblique cutting. The cutting phenomenon is investigated in comparison with the normal cutting, and the deformation of the aramid fiber and the creating mechanism of the fluffing are characterized. Moreover, the fluffs of the machined surface are observed by SEM in detail. We found that the fluffs of the aramid fibers are considerably affected by the inclination angle of the WC tool.

Keywords
Oblique cutting, Inclination angle, Aramid fiber, FRP, Fluffs

1. EXPERIMENTAL

We carried out the orthogonal and oblique cutting by using a lathe and the single point cutting tool of WC is used under 0.05 mm/rev in feed rate as shown in Fig.1.
A-FRP (Aramid Fiber Reinforced Plastic) is chosen as a specimen which contains only one bundle of aramid fibers and the matrix is polyester. The bundle of aramid fibers are placed in four kinds of orientations as shown in Fig.2. The cutting edge is inclined to cutting direction as shown in Fig.3.

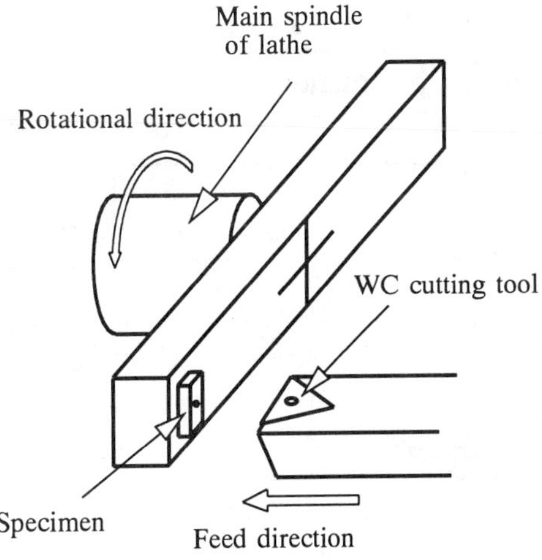

Fig.1 Appearance of the equipment for cutting

Effective rake angle α_e changes during oblique cutting, because it can be defined as follows[1]

$$\alpha_e = \arcsin(\sin^2 i + \cos^2 i \sin \alpha_n). \tag{1}$$

Where, i is inclined angle of the tool, α_n is normal rake angle of the tool. The value α_e increases with inclined angle i as shown in Fig.4. We measured the three components of the force F_p, F_r and F_t during machining, and we can get such a curve as shown in Fig.5. We measured two kinds of force in each component. One is the force required for machining the matrix (F_{pm}, F_{rm} and F_{tm}), and the other is required for cutting of fiber (F_{pf}, F_{rf} and F_{tf}).

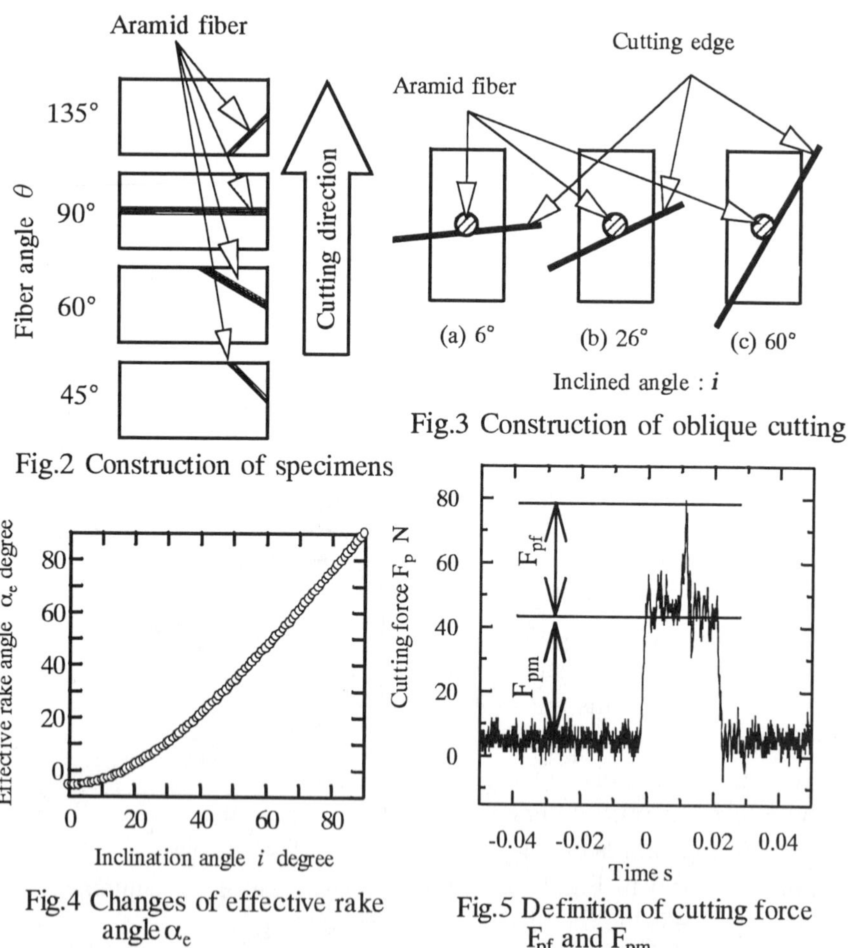

Fig.2 Construction of specimens

Fig.3 Construction of oblique cutting

Fig.4 Changes of effective rake angle α_e

Fig.5 Definition of cutting force F_{pf} and F_{pm}

2. EXPERIMENTAL RESULTS

Figure 6 shows the changes of force for cutting the fiber. All the components, especially F_{r_f} are considerably influenced by the i, and they increase with inclined angle i. Figure 7 shows the effect of inclined angle of cutting edge on the machined surface appearance under the condition of cutting speed $v = 23.4$ m/min. The aramid fibers are elongated and compressed strongly on the surface of the specimen, and long and wide fluffs appear on the machined surface. The appearances of aramid fibers on the machined surface are remarkably influenced by the fiber orientation. If angle θ is smaller than the right angle, at the time when the tool edge touches the fibers, the fibers will be lifted up by the tool edge from the machined surface of the matrix. The aramid fibers that have been lifted are not supported by the matrix, so it is difficult for the tool edge to cut them effectively during cutting. As a result, long fluffs of aramid fibers could be observed. On the contrary, when angle θ is greater than the right angle, for example, $\theta = 135$ degree, fibers will be compressed strongly between the tool edge and the matrix. Therefore aramid fibers are deformed by the forward cutting force, and it can be observed that the fibers resemble fish scales on the surface. During oblique cutting, the fluffs elongated obliquely due to the inclined cutting edge. And the fluffs become smaller and shorter.

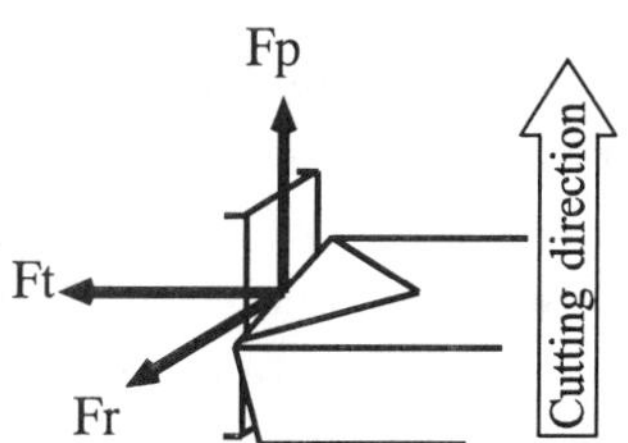

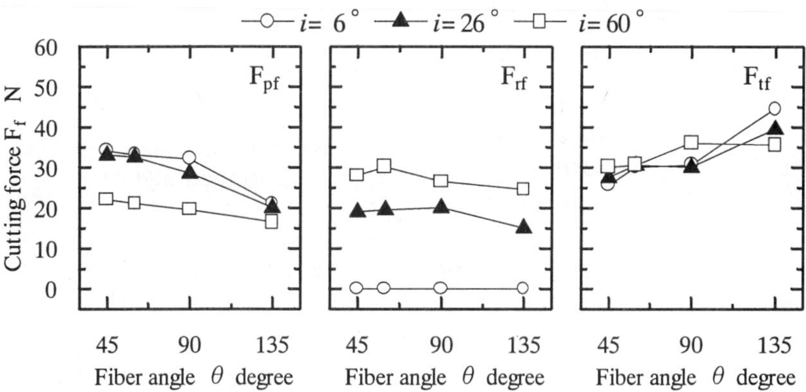

Fig.6 Effect of fiber angle θ on three components of cutting force for fiber

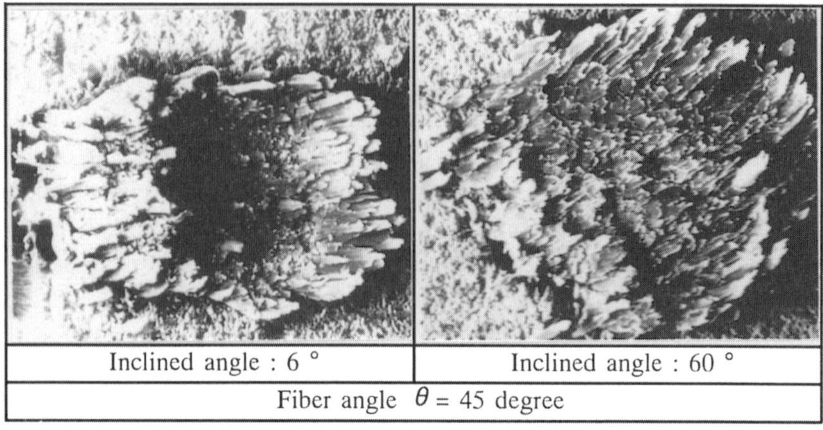

Inclined angle : 6 °	Inclined angle : 60 °
Fiber angle θ = 45 degree	

Fig.7 SEM photos of the fluffs on the machined surface
(V=23.4 m/min)

150 μ m

Maximum roughness of the machined surface Ry is measured to evaluate the surface in comparison with the normal type of cutting. Figure 8 shows that surface roughness Ry of the A-FRP specimen decreases with an increase in fiber angle θ. The value Ry for oblique cutting ($i = 60$ degree) is not affected by the fiber angle θ and when the fiber angle $\theta = 45$ degree, Ry becomes about 1/2 less than those in other cutting conditions. On the other hand, the value Ry for normal type of cutting (i=6 degree)is much influenced by the fiber angle θ.

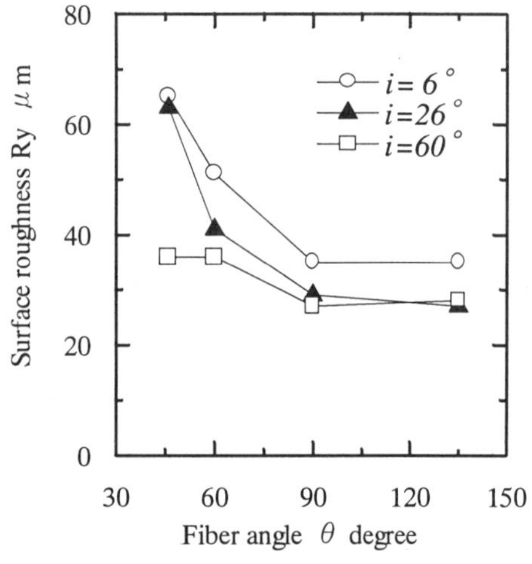

Fig.8 Effect of fiber angle θ on surface roughness Ry

When the oblique cutting, the tool edge is set obliquely to the cutting direction and the direction of the resultant force changes. Figure 9 shows the definition of ϕ which indicates the angle between the directions of fiber and the resultant forces. The white arrow indicates the direction of aramid fiber and shorter arrow indicates the resultant force. We measured this angle ϕ in each conditions.

Figure 10 shows the effect of angle ϕ on the surface roughness Ry. Ry is much affected by angle ϕ, and it decreases with angle ϕ. It is thought that if the angle ϕ becomes larger, the cutting force acts more effectively for cutting the aramid fibers and tool edges does not elongate the aramid fiber and Ry becomes small. When the inclined angle i is 60 degree, ϕ is always kept larger value than those in other conditions and the surface roughness Ry becomes small and the surface appearance is improved. On the other hand, the value Ry for normal type of cutting, when the fiber angle θ is 45 and 60 degree, the angle ϕ is smaller than those in other conditions and Ry becomes much larger.

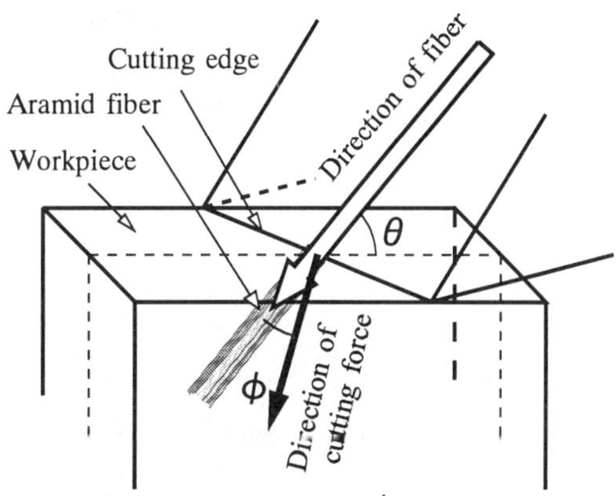

Fig.9 Definition of angle ϕ between the directions of fiber and cutting force

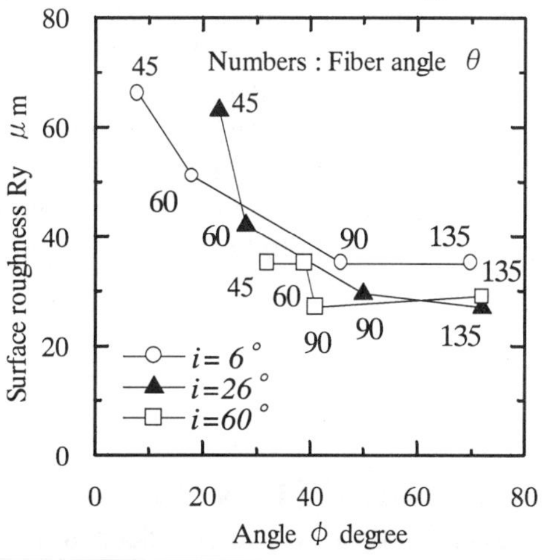

Fig.10 Effect of ϕ on surface roughness Ry

3. CONCLUSIONS

1. Fluffs creation of the machined surface is remarkably influenced by the inclined angle of cutting edge.

2. Surface appearance of machined surface looks to be improved by oblique cutting of $i = 60$ degree.

3. Surface roughness Ry for oblique cutting of $i = 60$ degree becomes about 1/2 less than that of normal cutting.

REFERENCES

[1] G.V. Stabler, "THE CHIP FLOW LAW AND ITS CONSEQUENCES", Proc. 5th Int. M.T.D.R. Conf., p.243, (1964).

DEVELOPMENT OF A PRACTICAL LASER-GUIDED DEEP-HOLE BORING TOOL:
WORKING CHARACTERISTICS OF ITS COMPONENTS

Akio Katsuki[1],   Hiromichi Onikura[1],   Takao Sajima[1]
Yoshikazu Yuge[2], Hiroshi Murakami[1], Takayuki Katayama[1]

[1] Department of Intelligent Machinery and Systems, Faculty of Engineering, Kyushu University 36, 6-10-1 Hakozaki, Higashi-ku, Fukuoka, 812-8581, JAPAN
TEL +81-92-642-3453, FAX +81-92-642-4120, E-mail kaetr@mech.kyushu-u.ac.jp
[2] Unitac Incorporated, 5, Shikashinmachi-3, Oomuta, 837-0901, JAPAN

Abstract

A practical laser-guided deep-hole boring tool with a diameter of 110 mm is produced. The tool consists of a counter-boring head, actuators for its attitude control, an active rotation stopper and a laser system for detecting its attitude. Performance of each component is examined. As a result, the stopper can move toward a rotational direction by an inchworm mechanism. The tool attitude can be controlled by the actuators and laser system. The trial experiments show that the practical laser guided deep hole boring can be performed by using a coupling.

Keywords

Deep hole boring, Adaptive control, Laser application

1. INTRODUCTION

Axial hole deviation in deep hole boring results in degradation of the quality and a decrease in the yield rate of products. A laser-guided deep-hole boring tool with a diameter of 110 mm was developed to prevent the hole deviation. Counter-boring of duralumin workpieces was performed with straightness of $\pm 10\mu m$ over 700mm [1].

On the basis of the experimental results of the original tool, a new deep-hole boring tool with a diameter of 110 mm is produced for practical use [2]. The tool is designed to increase performance of its manipulation, simplicity of detection of its attitude (position and inclination) and flexibility of machining method, e.g., oil supply and variation of cutting speed. Hole accuracy can be increased.

In the present research, performance of the active rotation stopper, correction method of detection errors of the tool position and rotation of the actuator holder, and the performance of the actuators for attitude control of the tool are examined.

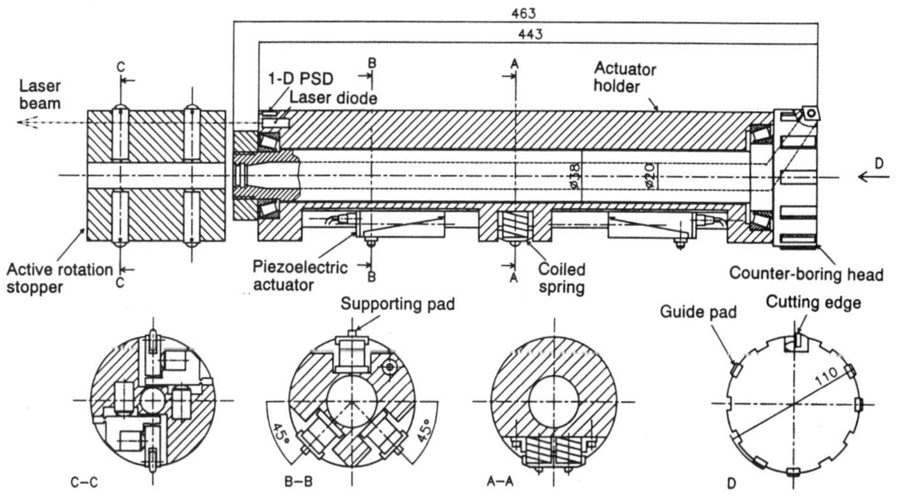

Figure 1 Practical high-performance laser-guided deep-hole boring tool

① Counter-boring head
② Actuator holder
③ Boring bar
④ Guide bush
⑤ Workpiece
⑥ Rotation stopper
⑦ Connecting plate
⑧ Ejector-type cleaner
⑨ Laser diode
⑩ Piezoelectric actuator
⑪ Interferometer
⑫ Corner cube prism
⑬ Industrial endscope
⑭ Mirror
⑮ Beam splitter
⑯ PSD δ
⑰ PSD i

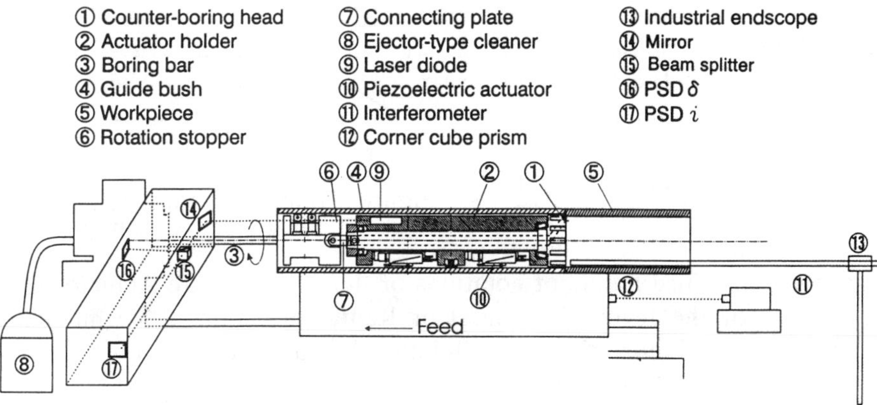

Figure 2 Experimental apparatus

2. STRUCTURE OF THE TOOL

The tool consists of a counter-boring head, the front and rear actuators mounted on an actuator holder, an active rotation stopper and a laser diode set in the back end of the holder (*Figure 1*). Tool attitude is detected by the laser diode and two PSDs (Position-Sensitive Detector) set behind it (*Figure 2*) [3].

2.1 Counter-Boring Head

Cutting speed is increased because the laser diode is set on the actuator holder under the condition of stationary state unlike the original tool [1]. Further data for tool attitude control is continuously acquired.

2.2 Rotation Stopper

The actuator holder rotates gradually during cutting toward opposite direction of rotation of the counter-boring head. For precise guidance of the tool, the rotation has to be restricted and corrected. The rotation is detected by a 1-D PSD next to the laser diode for tool attitude. A laser diode for detecting the rotation is set next to a corner cube prism (*Figure 2*).

Figure 3 shows the inchworm mechanism to rotate the stopper [4]. Piezoelectric actuators are used to move the levers. In *Figures 3* (c) and (f), the stopper rotates. *Figure 4* shows impressed voltage to each actuator (AFT, AFR, ART and ARR). First, second and third letters of notations of the actuators are Actuator, Front or Rear of the lever, Tangential or Radial direction of its movement, respectively. *Figure 5* shows a relationship between rotational angle of the stopper and time with moving cycle of each lever as a parameter.

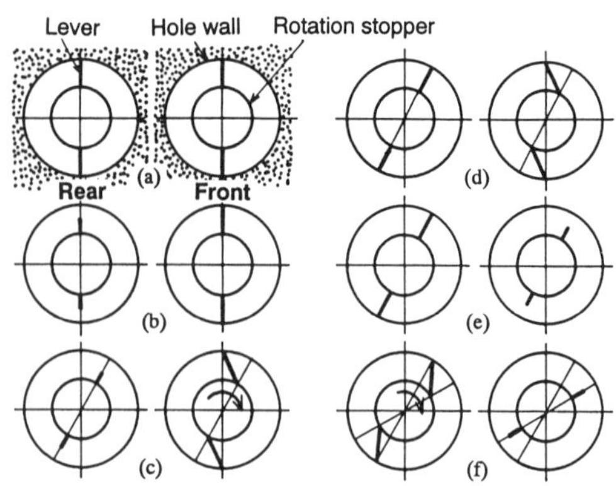

Figure 3 Rotation of the stopper by the inchworm mechanism

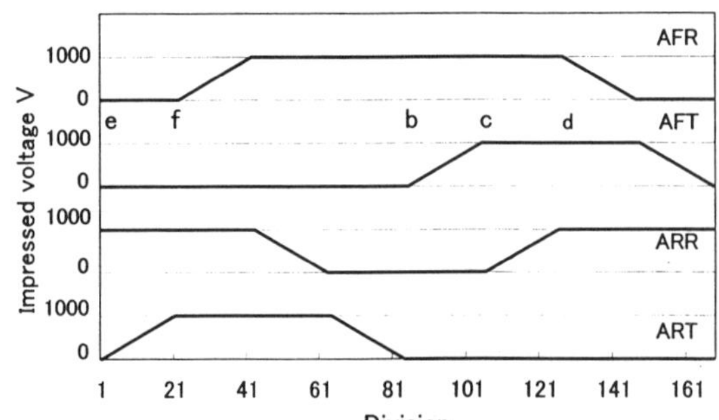

Figure 4 Impressed voltage to each piezoelectric actuator

3. TOOL GUIDING

The tool is set in the guide bush (*Figure 2*). Six actuators protrude equally to contact the hole wall of the guide bush. Then reference origins of the two PSDs for detecting tool attitude are decided. When the tool deviates from the reference axis during boring, the actuators are precisely manipulated to correct the tool attitude.

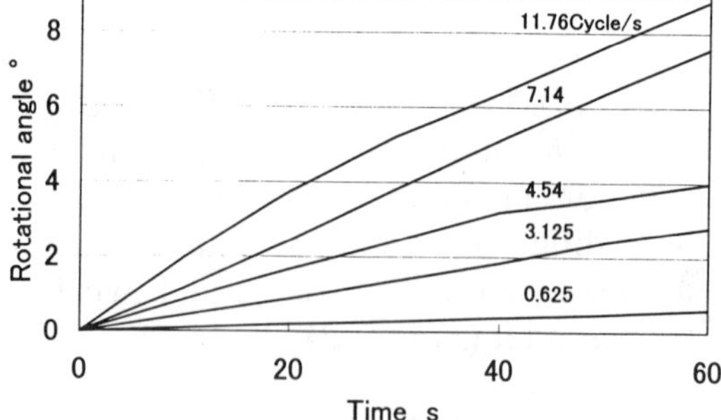

Figure 5 Rotational speed of the stopper

4. CORRECTION OF DETECTION ERRORS

Output of the PSD for detecting tool position changes due to rotation of an actuator holder. And also that for the rotation of the holder changes due to the displacement of the tool. Those interference has to be eliminated. *Figure 6* shows that the holder rotates by θ and displaces by x and y. S_{DI} and S_{RI} are the original positions of the laser diode for the tool position and the 1-D PSD, respectively. The follow-ing equations hold.

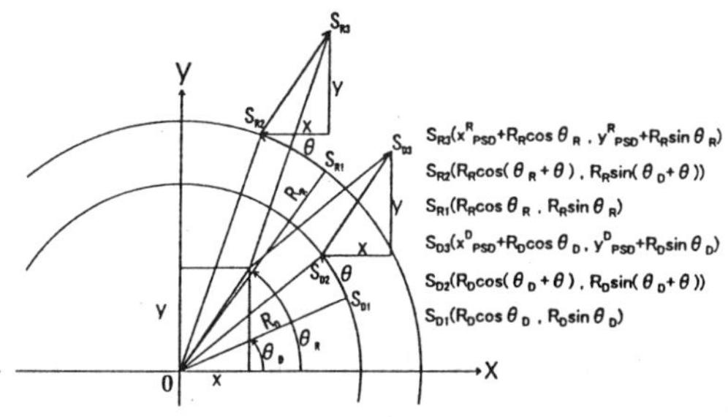

Figure 6 Displacement of the tool after rotation of the holder

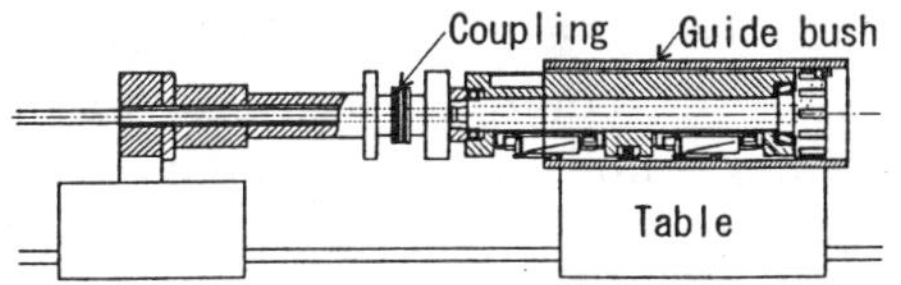

Figure7 Apparatus for simulation

$$x = x_{PSD}^{D} + R_D \cos\theta_D - R_D \cos(\theta_D + \theta) \quad (1)$$

$$y = y_{PSD}^{D} + R_D \sin\theta_D + R_D \sin(\theta_D + \theta) \quad (2)$$

If the 1-D PSD is set vertically to detect vertical movement, the y can be written as:

$$y = y_{PSD}^{R} + R_R \sin\theta_R - R_R \sin(\theta_R + \theta) \quad (3)$$

5. SIMULATION

A disk coupling is used instead of the stopper. Sinusoidal track shown by equation (4) is given to the tool to follow (*Figure 7*).

$$x = a \sin\left(2\pi f\left(\frac{N}{60}\right)t / w\right) \quad (4)$$

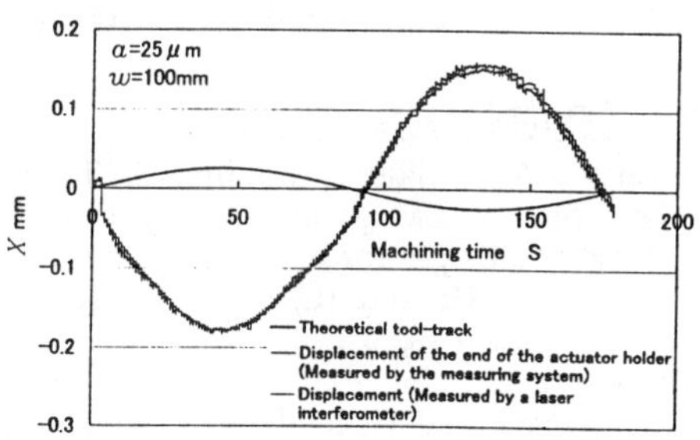

Figure 8 Simulation of tool guidance

where N is rotational speed and w wave length. *Figure 8* shows that three actuators on the rear side of the holder are controlled so that the tool follows the sinusoidal track.

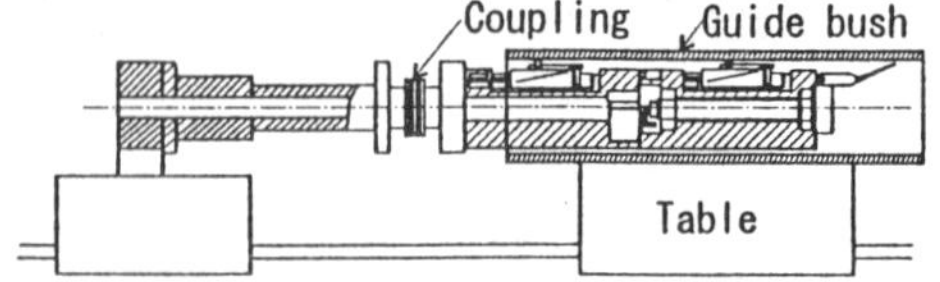

Figure 9 Measurement by the deep-hole evaluating probe

6. EXPERIMENTS

Performance of the stopper is not enough for precise control of rotation of the holder. The experiment is carried out using the coupling (*Figure 9*). The tool is guided following the sinusoidal track. Feed is 0.125 mm/rev and rotational speed 273rpm. Duralumin workpiece with a prebored 108-mm diameter hole is used. Hole accuracy is examined by the laser-guided deep hole evaluating probe (*Figure 9*) [3].

Each component for laser-guided deep hole boring fulfilled its function. Hole could be bored favorably. However hole accuracy for straightness could not be obtained due to unsatisfactory software for attitude control.

7. CONCLUSIONS

A practical laser-guided deep-hole boring tool with a diameter of 110 mm is produced. Performance of each component of the tool is examined through simulation and experiments. Under the condition that the coupling is used, the laser-guided tool can be used practically.

ACKNOWLEGEMENTS

Japan Science & Technology Corporation and the Ministry of Education, Science, Sports and Culture (Grand-in-Aid for Scientific Research, (C)(2))(Japan) which support this research are gratefully acknowledged. Mr. T. Moriyama and Mr. H. Murashita, undergraduate students are also appreciated for their cooperation

REFERENCES

[1] Katsuki,A., Onikura,H., Sajima,T., Takei, T., Thiele, D. Development of a High-Performance Laser-Guided Deep-Hole Boring Tool: Optimal Determination of Reference Origin for Precise Guiding. Precision Engineering 2000; 24; 1; 9-14.
[2] Katsuki,A., Onikura,H., Sajima,T., Thiele, D. Development of a High-Performance Laser-Guided Deep-Hole Boring Tool: A New Model for Application, and a Robot and a Probe for Extremely Deep Holes. Proceedings of the 1st International Conference and General Meeting of the euspen; 1999 May 31-June 4; Bremen, Germany; 254-257.
[3] Katsuki,A., Onikura,H., Sajima,T., Thiele, D. Development of a Laser-Guided Deep-Hole Evaluating Probe, Proceedings of the 2nd CIRP International Seminar on Intelligent Computation in Manufacturing Engineering(ICME 2000); 2000 June 21-June 23; Capri, Italy; 423-428.

OIL FILM ON WATER FOG METALWORKING FLUID

Takashi NAKAMURA*, Tomio MATSUBARA*,
Fumihiro ITOIGAWA* and Keiichi KAWATA**

* Nagoya Institute of Technology

** Aichi Industrial Research Institute

Abstract

Many types of ecological-friendly machining technologies have been developed in the last ten years. In this paper, a new type of metalworking fluid is introduced, where comparatively large water particles (fog) are covered with vegetable oil film and sprayed on a machining point with a soft air jet. Better machinability is obtained in the milling test for the oil film on water fog, compared with the conventional emulsion coolant. Machining performance is studied by changing the oil type and the supply condition.

Keywords

Ecology, Environmental concerns, Machining fluid, Minimum quantity lubricant, Oil film on water fog, Machinability

1. INTRODUCTION

Recently, environmental concerns have increased the cost of waste fluid treatment in traditional coolant systems. Thus, many types of ecological-friendly machining technologies have been developed, for example, MQL (Minimal Quantity Lubrication)[Heisel U. et al], Cooled air, and Dry machining. Each of these technologies has its advantages and disadvantages depending on the work materials, the process configurations, and the machining conditions. In the present paper, a new metalworking fluid supply system is proposed, and its property is examined with machining tests. The experimental results show that this system will be one of the effective techniques for ecological manufacturing.

2. MACHINING TESTS OF ALUMINUM MILLING

Figure 1 shows a concept of the oil film on water fog (OoW) metalworking fluid supply mechanism. Water particles (fog) are covered with thin film oil and spread on the tool or work surface by relatively gentle air stream. The water particles are expected to play three roles; carrying the minimal machining oil with them, spreading the oil on the tool and work surfaces effectively by their inertia, and chilling the tool and the work by their sensible heat and latent heat of evaporation. A nozzle for the OoW and an oil-water-air supply system have been designed and produced by cooperative research with a manufacturer. In the present study, a new nozzle (Fig. 2), which can widely change the spray conditions, is used for experiments. The oil used in OoW is a vegetable oil lubricant, which is harmless to the ecology and has a good ability to expand on a water surface.

Work materials of aluminum alloy are milled with a machining center as shown in Fig. 3 and machining forces are measured by a dynamometer.

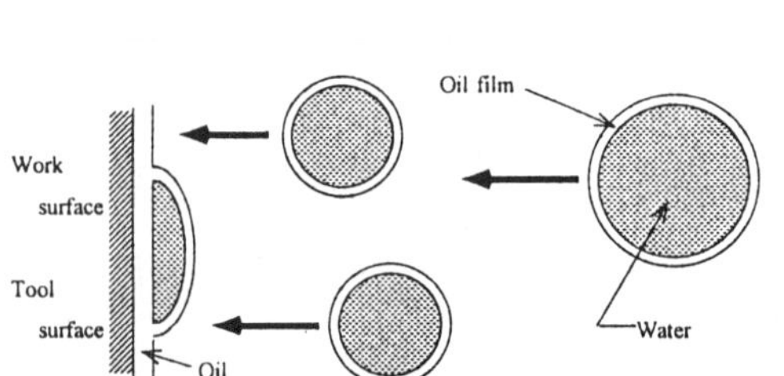

Figure 1. Concept of oil on water fog

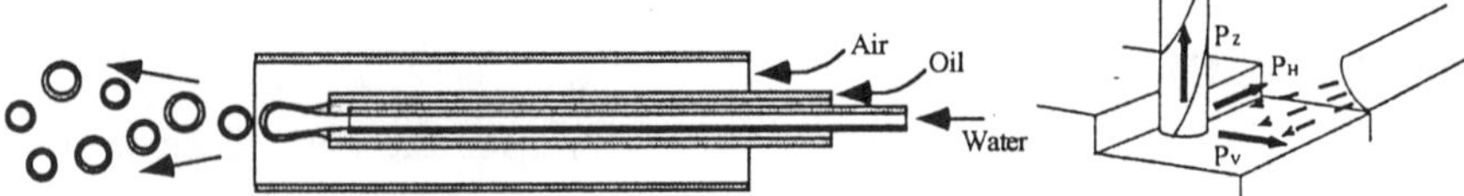

Figure 2. Nozzle for oil on water fog

Figure 3. Machining test

Table 1. Cutting conditions

Work material	A6063S
Tool	Square endmill, 2f
Tool material	Sintered carbide
Tool diameter	ϕ 12
Cutting direction	Up cutting
Depth of cut radial, axial	6 mm, 9 mm
Cutting speed	203 m/min
Feed rate	0.095mm/rev
Flood coolant supply rate	1200 l/h
Oil, water, air supply rate	0.009, 3, 2500 l/h

Table 2. Machining oils

1. Emulsion flood coolant
2. S-35 (kinetic viscosity 7.6 mm^2/s)
3. S-15 (kinetic viscosity 3.2 mm^2/s)
4. Pentaerithritoltetraester+2%TCP
5. Pentaerithritoltetraester+2%PS
6. Frying oil (A)
7. Frying oil (B)
8. Salad oil
9. Mineral oil (kinetic vis. 9.2 mm^2/s)
10. Mineral oil (ki. vis. 17.7 mm^2/s)

Machining conditions and the oils are given in Table 1 and Table 2, respectively, and the measured horizontal cutting forces are shown in Fig. 4. No.1 is the result of conventional emulsion flood coolant and the others are the OoW. The cutting forces of the OoW with ester oil (No. 2 to No. 8) show smaller values compared with those of the emulsion flood coolant. By contrast, the measured results of mineral oils (No. 9 and 10) show high cutting-force values. The reason for this seems to be that the mineral oils have no ability to expand over water surfaces.

The cutting forces are measured with changing the quantity of air supply. Experimental results are shown in Fig . 5, where the supply rate of water is 40 ml/min and the oil (s-35) is 9 ml/hour (volume ratio: 0.38%). The cutting forces increase as the quantity of air supply is decreased. When the air supply is decreased, the water particles have low velocity. Thus, the tool or work surfaces are insufficiently covered with the machining oil. The water particles have three roles; they carry the minimal machining oil with them, spread the oil on the tool and work surfaces effectively by their own inertia, and chill the tool and work by their sensible heat and the latent heat of evaporation. The air supply effects on the third role. The tool and work temperature increases when the air supply is decreased. The elevated temperature may diminish efficiencies of the machining oils. The minimum cutting forces are obtained where the air supply is about 100 l/mini, and the cutting forces increase as the air supply is increased. A total surface area of the water particles is increased as diameters of the water particles decrease. It seems that the thickness of oil film is not sufficient to cover the water

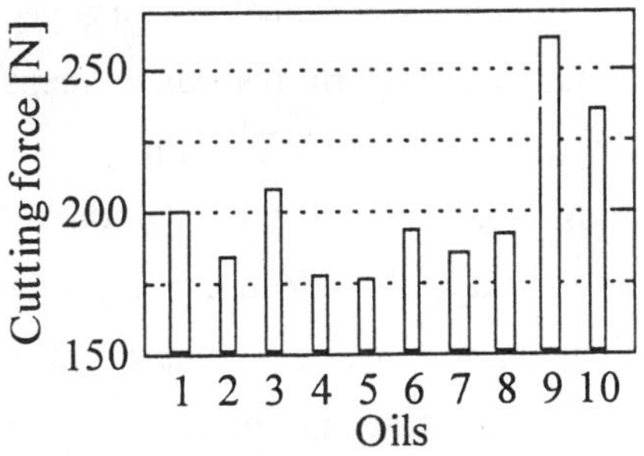

Figure 4. Measured cutting forces

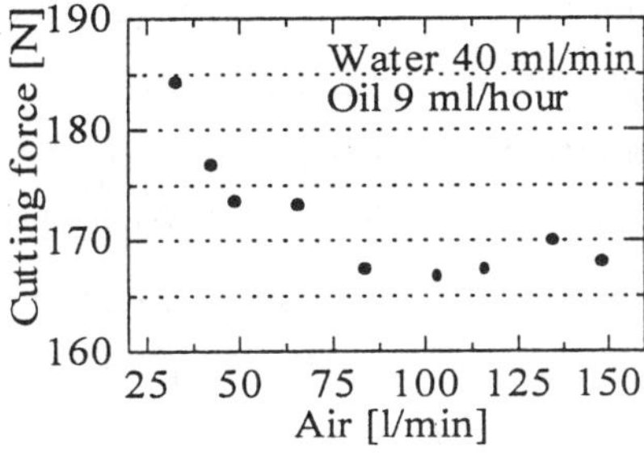

Figure 5. Effect of air supply rate

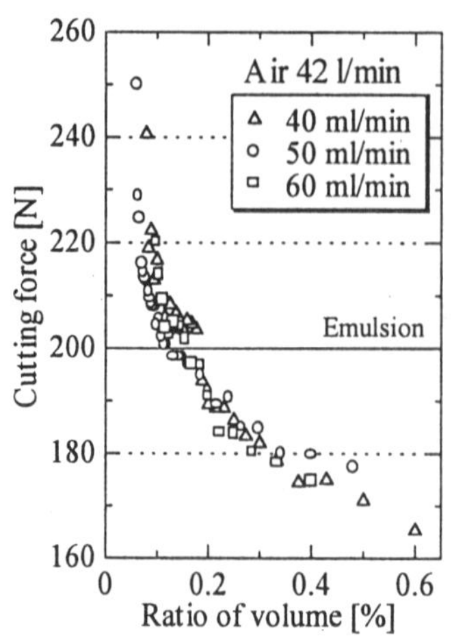

Figure 6. Effect of oil volume ratio

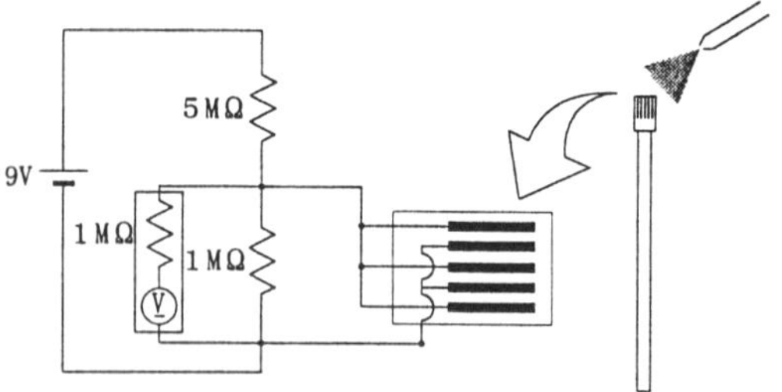

Figure 7. Measuring device for electric resistance of machining fuld

particles.

Effects of volume ratio of the machining oil (oil vol./ water vol.) on the machinability are examined and measured cutting forces are given in Fig. 6. The experimental results show that the cutting forces increase as the volume ratio of oil is decreased and that the quantity of OoW supply has small effect on the cutting forces.

3. PROPERTY OF OIL FILM ON WATER FOG

The property of the OoW is examined by a device designed to recognize an oil film on a surface on which the machining fluid is spread. Figure 7 shows the device by which electric resistance between gold printed circuits is recorded while the machining fluid is spread on it. Measured results are given in Fig. 8 where (a) water fog is spread, (b) oil-mist is spread, (c) OoW (S-35) is spread, (d) OoW (mineral oil) is spread on the printed circuits. When the water fog are spread with the air jet, the electric resistance decreases from an infinite value to a low level. The electric resistance keeps the infinite value when the oil-mist is spread. When the OoW with S-35 is spread, the decrease in resistance is slight and the resistance gradually increases to the infinite value. On the contrary, the electric resistance decreases to a low level and shows large fluctuations when a mineral oil is used for the OoW. As shown in Fig. 3, the measured results of mineral oils (No. 9 and 10) show high cutting-force values. The mineral oils have no

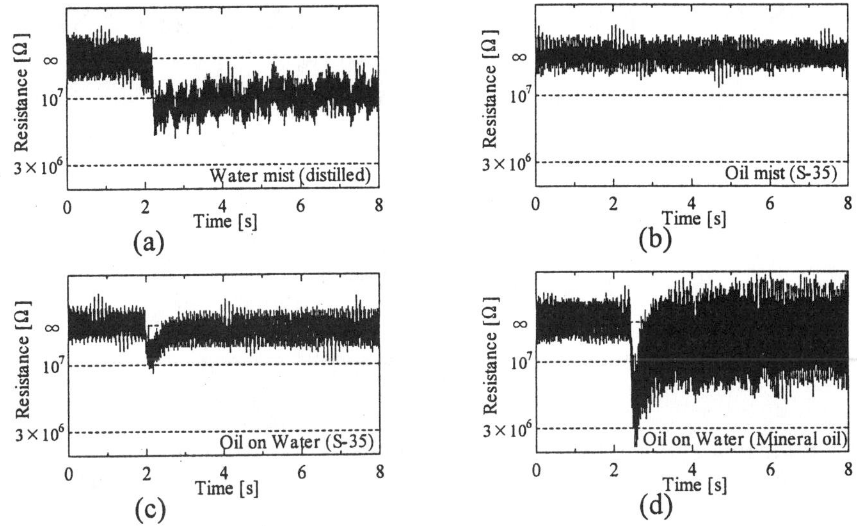

Figure 8. Measured electric resistance before and after mist spreading

ability to expand over water surfaces. The water particles are not covered with the oil and the water makes direct contact with the tool, work or printed circuits.

4. CONCLUSIONS

Conclusions obtained in this experimental study for the oil film on water fog are as follows.

(1) Vegetable oil-based lubricants are suitable for the oil film on water fog system, because they are harmless and have good abilities to expand on the surface of water particles.

(2) The water particles have three roles: they carry minimal oil, spread the oil on the tool and work, and chill the tool or work.

(3) A moderate air supply is necessary to obtain a superior performance of the oil film on water fog.

(4) The cutting forces in aluminum milling processes with the oil film on water fog indicate smaller values than those of flood emulsion coolant.

5. REFERENCE

Heisel U., Lutz M., Spath D. Wassmer R., Walter U. Application of Minimum Quantity Cooling Technology in Cutting Process, Production Engineering, II, 1, 1994, 49-54

ON THE ON-SET OF CHIP FORMATION AND THE PROCESS STABILITY IN CUTTING AT ATOMIC LEVEL

Dr.-Eng. Rüdiger Rentsch

Laboratory for Precision Machining,
University of Bremen, Bremen, Germany

Abstract

In cutting crystalline materials at extremely small depth of cut, the local crystalline orientation has a fundamental influence on the process as well as on the quality of the resulting surface. In this work results from molecular dynamics simulations of the cutting process at atomic level are presented. Specific orientations of the modeled fcc metal showed different chip formation behavior. In opposite to so-called 'soft orientations', 'hard orientations' did not show any chip formation until a minimum depth of cut was applied. Besides the influence of the crystalline orientation on the limits and the stability of chip formation, also the dependence of surface integrity, chip root and deformation zone on the cutting parameter are discussed.

Keywords

Molecular Dynamics, EAM, ultra precision cutting, micro cutting, surface finish, anisotropy, single-crystal, metal cutting, copper, chip formation

1. INTRODUCTION

The anisotropy describes orientation dependent properties of crystalline materials, which is at its maximum in single-crystalline structures. In ultra precision machining and in micro cutting so-called diamond turning and fly-cutting are applied as finishing operations to achieve high shape accuracy, extremely smooth surfaces and high surface integrity. Here the anisotropic crystalline properties determine the cutting process as the tool/work contact area is often much smaller than the grain size of the workpiece.

Although the nominal depth of cut is usually about a few microns, the surface is often generated at a fraction of that, reaching down to a few nanometer or atom layers only. Cutting at such conditions, the local crystalline orientation has a similarly fundamental influence on the specific chip formation process, the cutting forces and the surface roughness of the machined material. For a given tool and workpiece material not only the cutting parameters, such as cutting speed and depth of cut, play an important part in achieving atomically smooth surfaces, but also the local crystalline orientation of the machined material relative to the cutting direction, as seen in micro cutting of copper and silicon crystals [2,4]. In diamond turning of single-crystalline materials and in cutting polycrystalline materials, the local cutting conditions continuously change during operation and govern the process and the surface quality [4]. Knowing the limits and the mechanisms of chip formation for the specific crystalline orientations of a material forms the basis for optimizing the cutting process and its parameter in terms of shape accuracy, surface roughness and surface integrity.

2. MODELING AND SIMULATION

For the MD simulations an orthogonal cutting process set-up with different orientations of a single-crystalline fcc metal (copper) was chosen. The basics of molecular dynamics are described in great detail elsewhere [3]. For example, Fig.1 and 2 show sections of the employed process model.

The models contained on average about 15000 fully dynamic work and tool atoms, including thermostat boundary atoms to control the process temperature. Atoms continuously enter the model on the left side at cutting speed (here chosen to $v_c=100$ m/s) and move below the tip of the tool

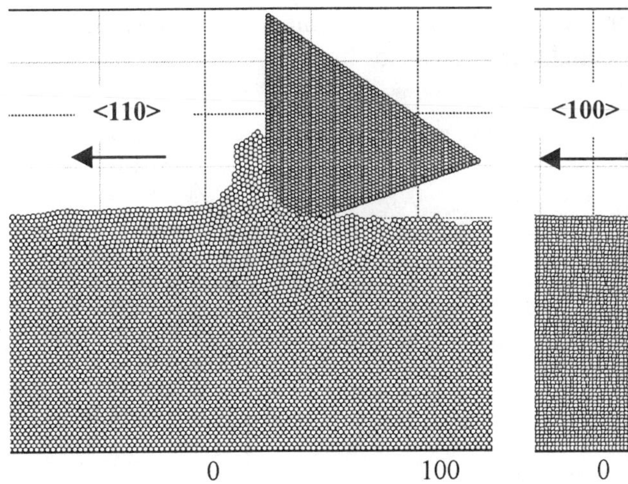

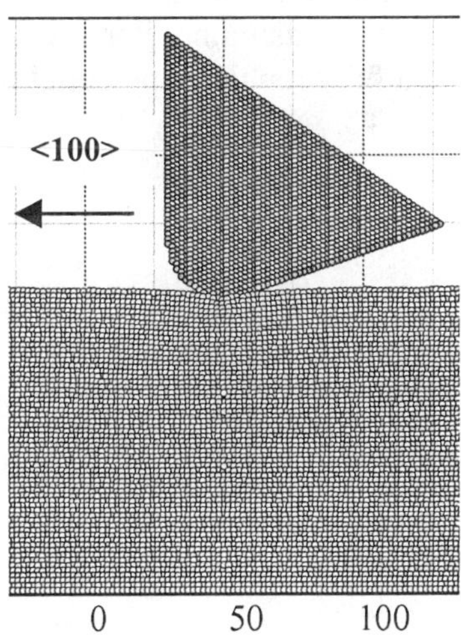

Figure 1. Chip formation at a_e=0.3 Å on {001} along <110> (L=60 nm)

Figure 2. No chip at a_e=3.0 Å on {001} along <100>

(radius 20 Å) to the right side, where they finally leave the simulation cell.

In order to model the three-dimensional bonding conditions of a crystalline material correctly, but still avoid modeling all atoms directly, so-called periodic boundary conditions (PBC) can be employed in combination with any periodic length of the crystal (see Fig. 3 and 4). Here, the simulation cell is surrounded by PBCs along the y-axis, so that the atoms in the simulation cell see shifted copies of themselves at the correct, three-dimensional lattice position in a virtual model space. Atoms crossing the PBC on one side will enter the simulation cell on the other, as it is common for PBCs. Usually PBCs require a minimum width of the simulation cell of at least the length of the cut-off radius, $r_{cut-off}$, which defines the maximum interaction distance between neighbor atoms. Since here the minimum width of the simulation cell, w_{cell} , is significantly smaller than the cut-off radius, multiple copies of the atoms in the simulation cell have to be considered within the force calculation.

This special model design allows to correctly model the three-dimensional crystal lattice, but restricts the motion of the atoms in a way, that only dislocations on planes perpendicular to the x/z-plane are active. With this set-up the material behavior for 2 crystal orientations was studied, a 'soft' orientation with active {111} shear planes (surface: {001}, cutting direction: <110>) and a 'hard' orientation ({001},<100>) with blocked {111} slip systems. For the Cu/Cu interactions a Finnes-Sinclair type EAM potential was employed and the C/C interaction in the diamond tool was approximated using a Lennard-Jones (LJ) pair potential with the bond strength of diamond. For the tool/work interface potential a weakly attractive and a fully repulsive potential were tested (for further details see [1, 3]).

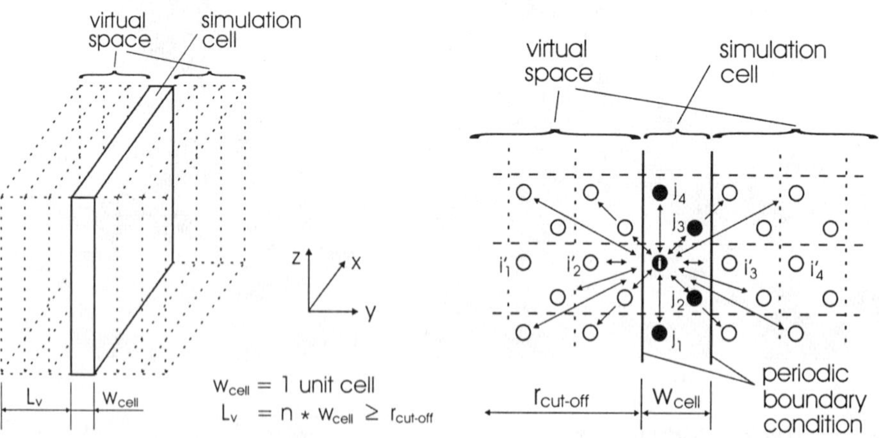

Figure 3. The simulation cell   Figure 4. Calculation of interactions

3. RESULTS

Fig. 1 shows the start of chip formation for the {001}/<110> crystal orientation (surface / cutting direction) with the tool at a depth of cut of a_e=0.3 Å. It was found, that the tool continuously generates a chip for this orientation, although it is not atomically sharp (cutting edge radius of 20 Å). The chip formation starts here almost from first contact and continues to larger depth of cut. Fig. 2 shows a similar set-up for the hard crystal orientation ({001}/<100>) with the tool at a depth of cut of a_e=3.0 Å and after moving the same distance. In opposite to the soft orientation, the hard orientation does not allow for a chip formation, although the applied depth of cut is 10 times larger. At this depth of cut, the tool causes mostly elastic deformation underneath its tip and few single dislocations, but no more significant plastic deformation. However, the chip formation can be initiated for the hard orientation by increasing the depth of cut beyond a_e=3.0 Å.

After activating the process at a depth of cut of a_e=10.0 Å, Fig. 5 shows chip formation for the hard crystal orientation, now at a_e=3.0 Å, for which no chip formation was seen before (same conditions as in Fig.1 and 2). Once a sufficient chip root is established, a stable chip formation is also possible at smaller depth of cut. Finally, at some point at even smaller depth of cut, the deformation by the tip of the tool is not sufficient anymore to support chip formation and the chip formation ends again.

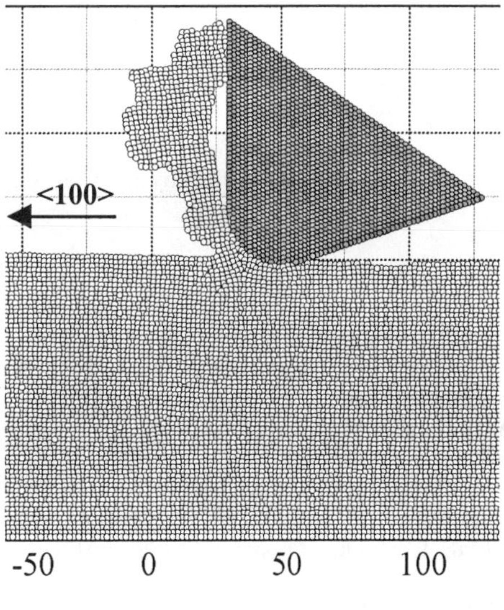

The soft orientation in Fig.1 shows a long pre-deformation zone ahead of the forming chip, that reaches also deep into the bulk, and a somewhat rougher surface than the hard orientation. The hard orientation seems to enforce localization of the deformation around the tip of the tool, leading to a specific chip formation process for the hard orientation and to a smaller chip root than for the soft orientation. Employing a fully repulsive tool/work interface potential has not changed the results on the whole.

Figure 5. Chip formation along <100> at a_e=3.0Å , after activation (surface: {001}; L=60 nm)

4. CONCLUSION

Cutting single crystalline material in two different directions can lead to significantly different chip formation behavior. It was found, that the minimum depth of cut for a stable chip formation is a function of the local crystalline orientation. While atoms can easily be removed from the surface in directions in which {111} planes can be activated (soft direction), atoms are stronger embedded in their surrounding structure along <100> direction ('hard'), which results in a strong localization of the deformation area. Presumably this strong localization causes the hysteresis like behavior in the chip formation for the hard orientation with respect to the depth of cut. Whenever the depth of cut reaches below the minimum depth of cut for one crystalline direction, the chip formation process will be interrupted. This additional criterion for process stability of chip formation is important in generating extreme smooth surfaces by a rotating tool or work, like in diamond turning and fly-cutting, as frequent interruptions deteriorate surface quality as well as roughness.

In real cutting experiments it is not possible to block some slip systems of a material. The fcc structure of the ductile copper will yield intensive dislocation activity and cross-slip between its easy dislocation slip systems, before less favorable slip systems determine the process. However, materials like bcc metals or semi-conductors cannot deform by activating the easy-slip systems of an fcc lattice. They deform on slip plane systems, that are harder to move, similar to the hard orientation discussed in this paper.

ACKNOWLEDGEMENTS

The author likes to acknowledge the funding of the related research project by the Deutsche Forschungsgemeinschaft DFG (German Research Council), Bonn, Germany.

5. REFERENCES

1. Rentsch, R., Influence of Crystal Orientation on the Nanometric Cutting Process, in *Series 'Berichte aus der Fertigungstechnik', precision engineering - nanotechnology*, proc. of the 1st. Int. conf. of Europ.Society f. Precision Eng. and Nanotechnology (euspen), Volume 1, Eds. McKeown, Corbett, et al., Shaker publishers, Aachen 1999, Germany, p. 250-253.

2. Rentsch, R., Inasaki, I., Brinksmeier, E., et al., Influence of Material Characteristics on the Micromachining Process, in *Materials Issues in Machining-III and The Physics of Machining Processes-III*, Eds. Stephenson and Stevenson, TMS, Cincinnati, Ohio, USA, 1996, p. 65-86.

3. Rentsch, R., Process modeling by means of molecular dynamics (MD), in *Bearbeitung neuer Werkstoffe, 2nd International Conference on Machining of Advanced Materials (MAM)*, reports of the German Society of Engineers VDI, 1276, Germany, 1996, p. 175-195.

4. Shibata, T., Fujii, Sh., Makino, E., Ikeda, M., Ductile-regime turning mechanism of single-crystal silicon, *Precision Engineering*, Elsevier Science, 4/5 1996, Vol.18 No.2/3, p. 129-137.

ANALYSIS OF TOOL TEMPERATURE
IN HIGH-SPEED MILLING

Hiroyuki Sasahara[1], Takehiko Nitta[2] , Kazufumi Nishi[1]

[1]*Department of Mechanical Systems Engineering, Tokyo University of Agriculture and Technology*

[2]*Tokyo NTT Data Communications Systems Corp.*

Abstract

In this study, temperature of cutting tool in high-speed milling was simulated using finite element method, and then the influences of cutting conditions and cutting edge geometry on tool temperature were studied. It was found out that a rapid temperature change is repeated in intermittent cutting, and the temperature around the cutting edge becomes very high within very short time. Combination of feed rate and radius depth of cut has a large influence on tool temperature. Thus it was suggested that maximum tool temperature becomes lower under the condition of high cutting speed and small radial depth of cut when the equal material removal rate is assumed. Also it was shown that the tool geometry such as rake angle and helix angle affect on tool temperature.

Keywords

High-speed milling, Tool temperature, Finite element modeling, Computer simulation

1. INTRODUCTION

In high-speed milling, it is important to avoid the problems such as short tool life, thermal expansion of tool and structure or the generation of an affected layer in machined surface, which are caused by heat generation during cutting. It is necessary to know the tool temperature in order to determine optimum cutting conditions and to obtain high productivity through the advantage of high-speed machining[Schulz, 1992]. Thermal behavior in milling process is modeled by finite element method and the effects of cutting conditions and tool geometry on tool temperature are studied in this paper.

2. MODELING

One of the thermal features of the milling process is to repeat the generation of heat during cutting term and cooling or thermal diffusion during uncutting term. In this study, the end mill is modeled by three-dimensional finite element method, and the heat flow is given intermittently, and the tool temperature is analyzed as three-dimensional transient analysis from the viewpoint of rapid temperature change of the tool.

Figure 1 shows a tool model with two straight cutting edges. Heat flux is given on tool-chip contact surface intermittently. And tool surface is heat transfer boundary to the air except upper end connects to shank. Thermal properties used in the simulation are shown in table 1.

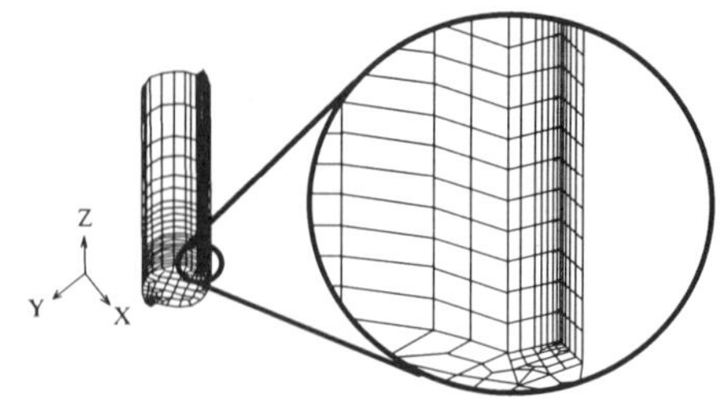

Fig.1 FEM model

Table 1 Analytical conditions

Thermal conductivity	W/mK	79.5
Density	kg/m^3	11200
Specific heat	J/kgK	402
Heat convection to air	W/m^{2}K	5.0
Heat convection to chuck	W/m^{2}K	2.855×10^6

The quantity of heat generation is supposed that all work done by each cutting edge changes to heat. Cutting force is estimated from the cutting removal cross section area and specific cutting force experimentally obtained. Then the work is derived as the product of the cutting force and cutting speed. The effect of the heat flow ratio into tool Rt will be discussed later in section 3.3. Quantity of heat flow given to one cutting edge is shown as

$$\dot{Q} = kd \cdot Rt \cdot V \cdot T \cdot A \quad [W] \tag{1}$$

kd is specific cutting force [N/mm^2], Rt is the heat flow ratio into tool, V is cutting speed [m/s], T is instantaneous uncut chip thickness [mm], A is axial depth of cut [mm]. kd is settled to 4302.5 N/mm^2 experimentally. This heat flow is distributed to the nodes on tool-chip contact surface. Tool-chip contact length is assumed three times of instantaneous uncut chip thickness.

3. EFFECT OF CUTTING CONDITIONS ON TOOL TEMPERATURE

3.1 Effect of feed rate and radial depth of cut

Table 2 Cutting conditions

Tool radius	mm	5.0
Spindle speed	min^{-1}	1000
Feed per tooth	mm/tooth	0.05, 0.1, 0.2
Axial depth of cut	mm	5.0
Radial depth of cut	mm	0.5, 1.0, 2.0, 3.0, 4.0, 5.0

Simulations are conducted under the conditions shown in table 2 with varying feed rate and radial depth of cut. Heat flow ratio into tool Rt is assumed 0.20 [Maekawa,1996]. Figure 2 shows one of the results. Although it shows the very short time after cutting starts, it becomes high

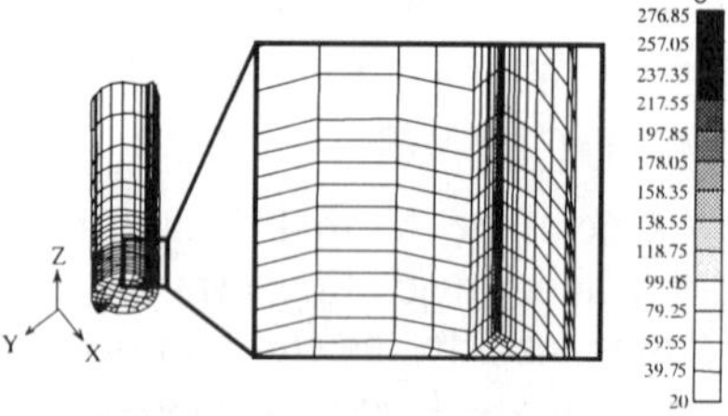

Fig.2 Temperature distribution
(S1000 min^{-1}, f=0.1mm/tooth, R.d=0.5mm)

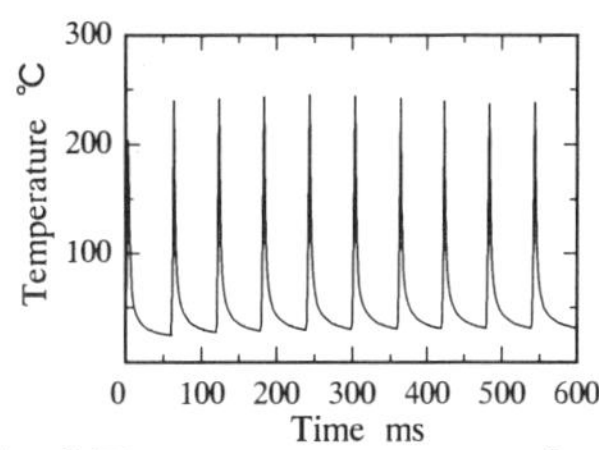

Fig.3 Temperature at cutting edge

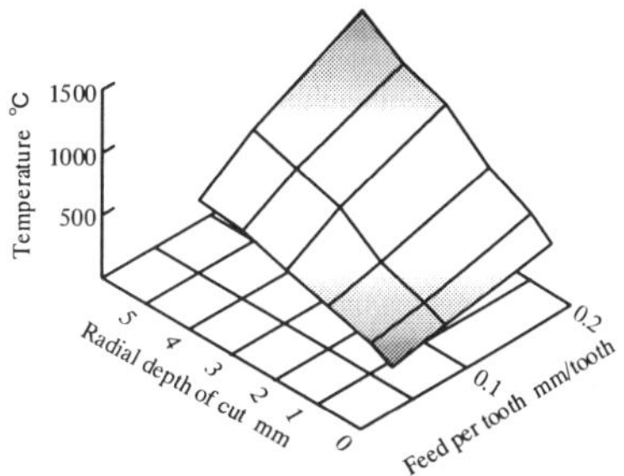

Fig.4 Effect of feed rate and radial depth of cut

temperature around the cutting edge . Figure 3 shows the transition of the temperature at cutting edge. Rapid rise and down of temperature are repeated.

Figure 4 shows the comparison of the maximum temperature under the combinations of feed rate and radial depth of cut. As feed rate and radial depth of cut become higher, the maximum temperature of the tool becomes high because cutting power increases. Tool temperature largely depends on feed rate and radial depth of cut.

3.2 In the case of constant MRR

It is desirable to get a long tool life with suppressing the tool temperature low and keeping high material removal rate (MRR). We employ the cutting conditions as shown in Fig.5 to make it

Material removal rate 2000 mm³/min

	Spindle speed min⁻¹	Feed per tooth mm/tooth	Radial depth of cut mm	Maximum tool temperature °C	
A	1000	0.05	4		520
B	1000	0.1	2		438
C	1000	0.2	1		511
D	2000	0.05	2		547
E	2000	0.1	1		430
F	2000	0.2	0.5		468

Fig.5 Maximum tool temperature under equal MRR

equal MRR 2000 mm³/min by setting the cutting conditions of spindle speed, feed rate and radial depth of cut. We assume the axial depth of cut is 5mm constant. Also Fig. 5 shows the maximum temperature under each condition. Tool temperature is lower in the case of B and E than other conditions. When radial depth of cut is large the temperature tends to be high even if feed rate is small. It seems small depth of cut and rather high feed rate would bring good results. But too large feed rate brings high temperature and poor machined surface quality. It seems that there is an optimum combination of feed rate and radial depth of cut, which needs more study.

3.3 High-speed machining and *Rt*

When cutting speed becomes high, heat generation increases. On the other hand, machining time would be short and cutting forces tend to decrease.

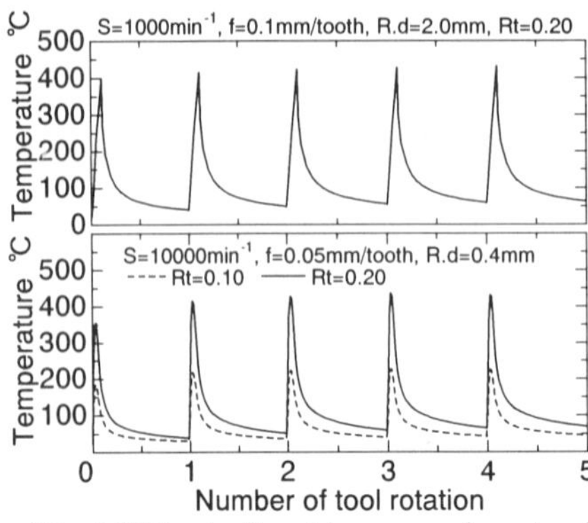

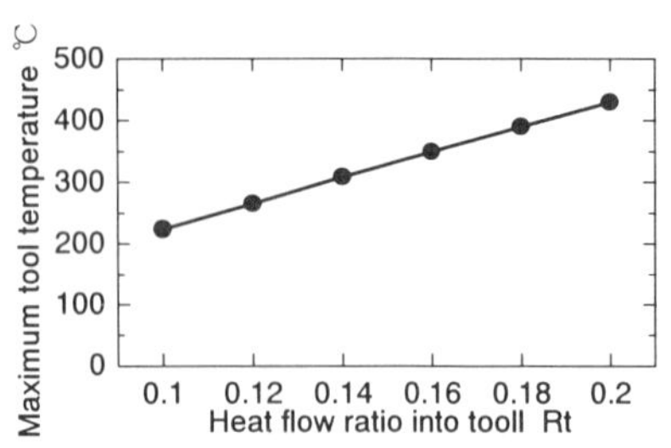

Fig.6 Effect of cutting speed on tool temperature under equal MRR

Fig.7 Effect of *Rt* on tool temperature

Also intermittent cycle affects on heating and cooling cycle, and the heat flow into ratio is considered to decrease at high cutting speed.

The equal MRR cutting conditions are set at S1000 and S10000. Also the effect of heat flow ratio into tool *Rt*, which is thought to lower at high cutting speed, is discussed. Figure 6 shows the analyzed results. Horizontal axis means the number of tool rotation. When *Rt* is 0.2, the maximum temperature indicate almost the same in both spindle speed. But the maximum temperature at S10000 lowers if we consider the decrease of *Rt* at high cutting speed as shown in Fig.7.

4. EFFECT OF TOOL GEOMETRY

4.1 Effect of rake angle

It can be clearly understood in the case of turning that the cutting temperature will become higher as the tool rake angle becomes low or negative because shear angle lows and heat generation will increase. But negative rake angle means large tool edge angle, which means higher heat capacity. It seems cutting temperature with negative rake angle tool doesn't always show higher temperature in milling process because the cycle of heating and cooling are repeated. To identify these, the effect of rake angle is discussed by using the models with rake angle 0 degree and −10 degree.

Figure 8 shows the result. Maximum temperature with 0 degree rake angle shows lower. It seems that an increase of heat generation caused by the increase of cutting force affects more largely than the increase of heat capac-

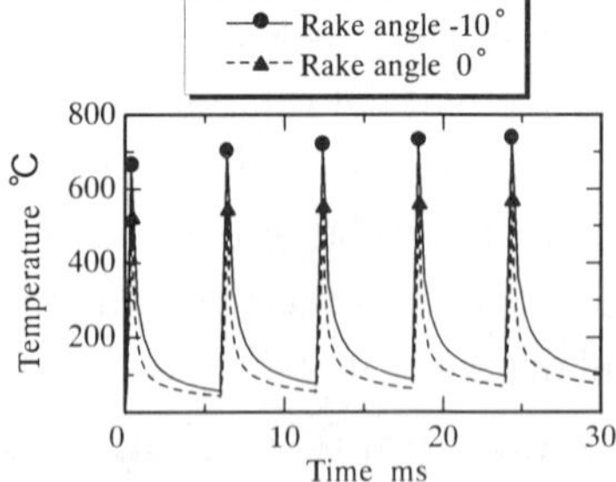

Fig.8 Effect of rake angle on tool temperature

ity.

4.2 Effect of Helix Angle

Strong helix angle tool is often used for difficult to machine materials or hard materials. The effect of helix angle is simulated here. Figure 9 shows the FEM model of end mill with 30 degree helical edges. Total quantity of heat given to tool model is equal on this model and the model with straight edge because we assume the equal tangential cutting force on both models. Heat source is moved along cutting edge from lower end to upward.

Figure 10 shows the transition of temperature on node a. The maximum temperature on helical edge shows lower than straight edge tool. It is because the length of the helical edge is longer than the straight edge, this means the quantity of heat per unit length of edge becomes smaller. On the case of straight edge tool, heat flux is given at the same time on the edge. But on helical edge, as heat source moves along cutting edge, the given heat can diffuse to cutter body more easily.

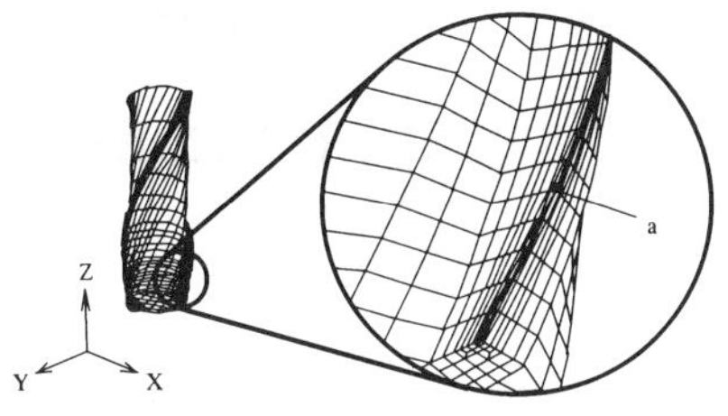

Fig.9 FEM model with two helical edges

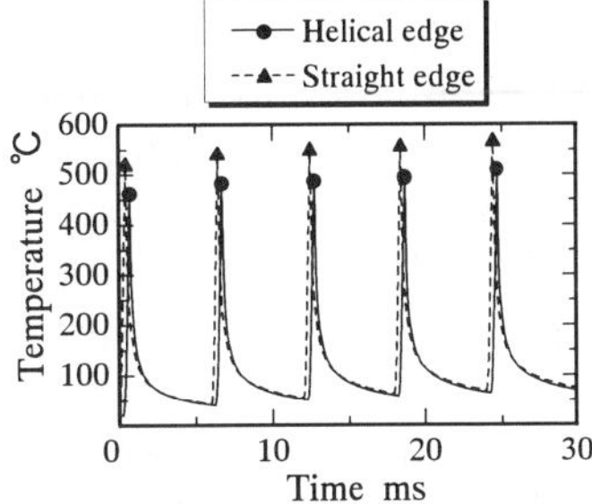

Fig.10 Transition of temperature on cutting edge

5. CONCLUSION

(1) Rapid temperature rise and drop are repeated around the cutting edge in milling.

(2) Under the equal MRR conditions, tool temperature tends to be lower when radius depth of cut is small, and there would be optimum combination of feed rate and radial depth of cut.

(3) Even if cutting speed becomes high, the equal MRR cutting conditions with small radial depth of cut doesn't always show temperature rise. Maximum tool temperature decreases depending on the decrease of heat flow ratio into the tool.

(4) Maximum tool temperature in milling decreases as rake angle become larger and helix angle become larger.

REFERENCES

(1) Maekawa K., Nakano Y., Kitagawa T., Finite Element Analysis of Thermal Behavior in Metal Machining, Journal of JSME(C),1996: 62-596, 1594.

(2) Schulz, H., Moriwaki, T., High-Speed Machining, Annals of the CIRP, 1992: 41-2,637.

RESEARCH ON PRECISE CUTTING PERFORMANCE OF CVD DIAMOND

Hongzhi ZHANG, Yingxue YAO, Zhejun YUAN

(Harbin Institute of Technology, P. R. China)

Abstract

The cutting edge quality of CVD diamond thick film tools has been analyzed by using Atom Force Microscope (AFM). The result shows that diamond crystallite size and hole defect between crystal grains are the main influence factors. In order to lower the crystal grains size and the hole defect, and to improve the cutting performance of tools, the nucleation density and compaction of diamond thick films have been increased in this paper. The crystal grain size has been fined, and cutting edge quality of thick film tools has improved. The cutting experiments with CVD diamond tools show that the surface roughness can reach Ra 0.024μm while cutting aluminum alloy, and Ra 0.048μm while cutting silicon carbide reinforced aluminum metal matrix composites (MMCs).

Keywords

Diamond film, Cutting tool, Crystal grain, Edge radius

1. INTRODUCTION

CVD diamond thick film can be widely used in the area of cutting tools [1]. The CVD diamond thick film is a kind of polycrystalline material, and it has the hole defect between crystal grains, so the texture of diamond thick film is one of the main factors which influence the cutting edge quality. The improvement of the texture of CVD diamond thick film and the edge sharpness of thick film tools are an important research content.

Professor YoshiKawa (Tokyo Institute of Technology , Japan) had made diamond thick film tools with edge radius less 0.1 μm by using the large size diamond particles as the nose of the tool [2]. But the synthesis technique was complex and the tool cost was increased. B.R.Stoner has synthesized highly oriented, textured diamond films via bias-enhanced nucleation and textured growth [3]. The method used by B.R.Stoner has decreased the hole defect extremely, but diamond films grow very slowly.

The author considers that an effective method of improving edge quality is to fine the crystal grain size, decrease the quantity and size of hole defects. To improve the nucleation density and to promote the secondary nucleation growth of crystallites can benefit to fine the crystal grain size and decrease the hole defect.

2. THE EXPERIMENTS OF IMPROVING NUCLEATION

The nucleation density has an important effect on crystal size, uniformity and compaction of diamond thick film. The intermediate layer of diamond-like amorphous carbon on the substrate has been deposited in this paper. The nucleation density has been improved in the experiments. The experiment devices show in reference [4]. The substrate material is tungsten, prior to deposition the substrate was scratched with 5μm diamond paste. For the comparison with traditional techniques, the experiments are divided into two groups, A and B. The experiment parameters are shown in table 1. The experiments results (in Figure 1) show that nucleation density has been increased 3~4 times by using the varying parameters deposit technique.

Table 1.Experiment parameters and experiment methods

Group	Experiment parameters			
	Q(L/min)	T (℃)	R_f	t (min)
A			0.98	5
B	4~6	800	0.70~0.75	2
			0.98	3

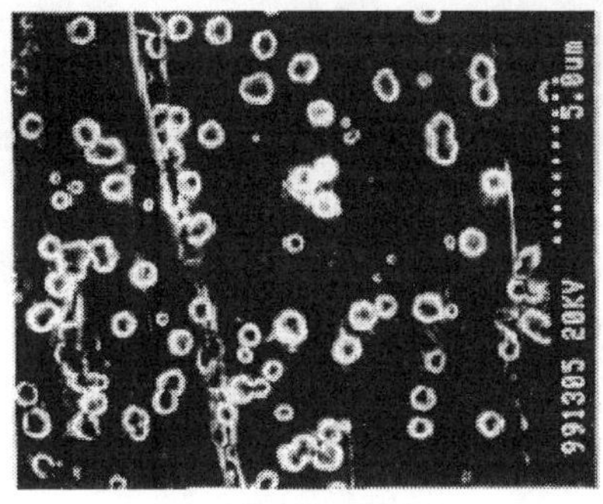
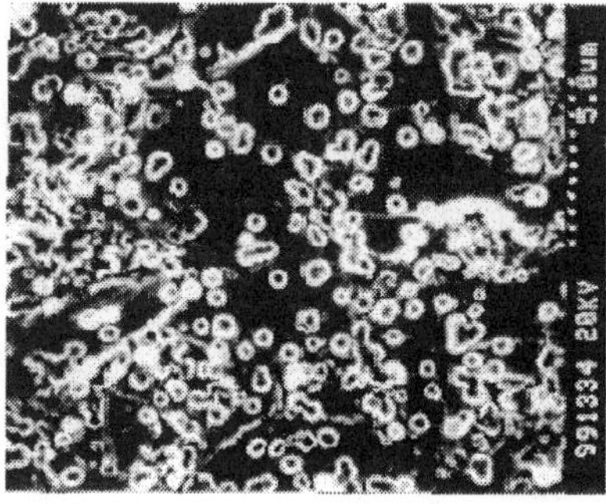

Figure 1. The contrast experiment results of nucleation

3. THE EXPERIMENTS OF FINING CRYSTAL SIZE

In the grown process of diamond thick film, alternative changing the flux ratio R_f of oxygen gas to acetylene, can promote the secondary nucleation on the surface of crystal grains, fine the crystal grain size, and decrease the hole defect size. The experiment parameters are shown in Table 2. Figure 2 shows the SEM micrographs of diamond film growth surface. The average crystal grain size in the varying parameters deposit technique has been decreased to half of original size.

Table 2. Experiment parameters and experiment methods

Technique	Steps	Experiment parameters			
		Q (L/min)	R_f	T (℃)	time (min)
Traditional			0.95		720
Varying Parameters	First step	5~6	0.95	700	55
	Second step		0.85		5
	Repeat first and second step		0.95/0.85		660

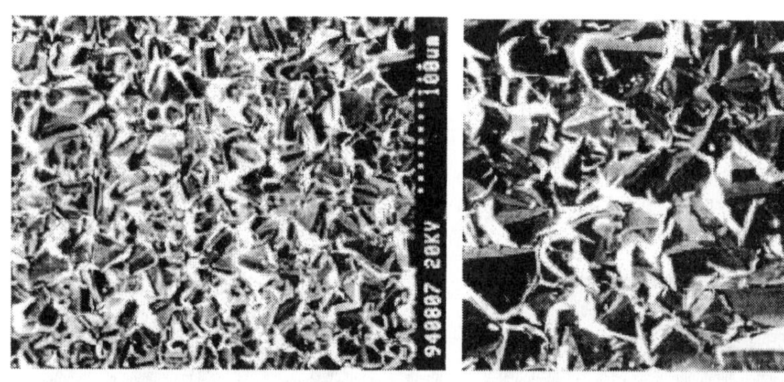

a) *Varying Parameters technique*   b) *Traditional technique*

Figure 2. The SEM micrographs of the diamond films surface

4. THE AFM ANALYSIS OF TOOL CUTTING EDGE

Figure 3 shows the three dimensions morphology of cutting edge. The cutting edge shown in figure 3 is composed of many crystal grains. In the process of tool grinding, with the decreasing of edge radius, the crystal grains in thick film also decrease. The crystal grains in cutting edge will fall off, and leave little scallops on the edge. So the sharpness of cutting edge will decrease. Figure 4 shows the cross section contour of the cutting edge. The edge radius is 2~3µm.

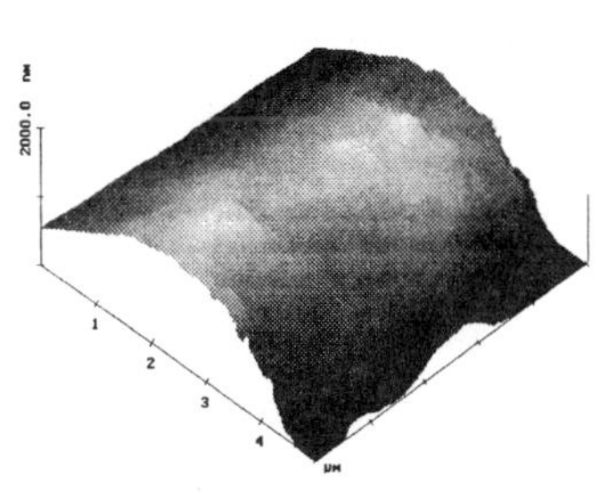

Figure 3.The three dimensions morphology of cutting edge

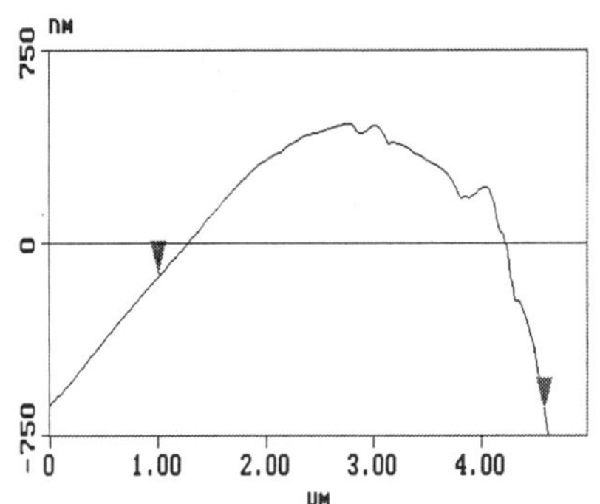

Figure 4.The contour curve of cutting edge cross section

5. THE MACHINING EXPERIMENTS WITH DIAMOND THICK FILM TOOLS

The experiment machine is an ultra-precision lathe made by Precision Engineering Research Institute in Harbin Institute of Technology. The tool geometry parameters: $\gamma_0=0°$, $K_r=K_r'=45°$, $\alpha_0=\alpha_0'=5°$. The cutting parameters are shown in Table 4. Figure 5 shows the workpiece surface roughness. The surface roughness of aluminum alloy is Ra 0.024µm. The surface roughness of silicon carbide reinforced aluminum metal composites is Ra 0.048µm.

Table 4 Experimental conditions and cutting parameter

Workpiece	size	V (m/min)	f (mm/r)	a_p (mm)	Ra (µm)	Machining surface
LY12	Φ155×120	700			0.024	Circular surface
Aluminum MMC	Φ80×20	300	0.001	0.003	0.048	End surface

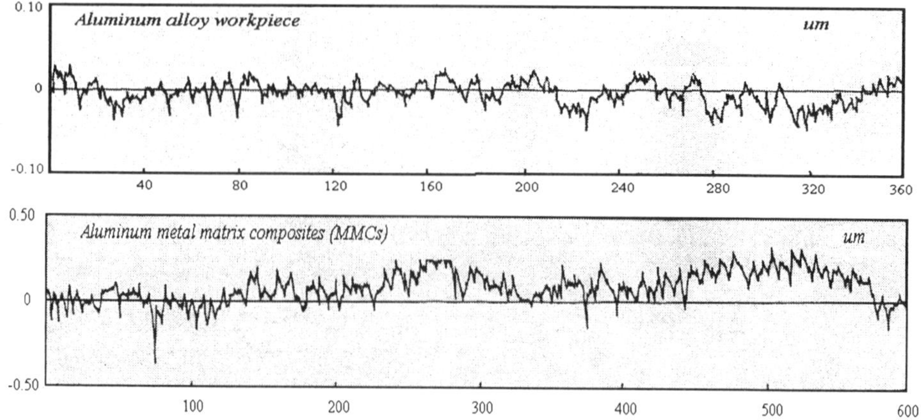

Figure 5. Measurement charts of workpiece surface roughness

6. CONCLUSIONS

1. The nucleation density is increased 3~4 times by using the technique of pre-depositing the intermediate layer of diamond-like amorphous carbon.

2. The average crystal grain size is decreased to half of original size by using the varying R_f deposit technique.

3. Crystal grains size influence tool edge radius, and the tool edge radius in this paper is 2~3μm.

4. The roughness of aluminum alloy workpiece is R_a=0.024μm, and that of aluminum metal matrix composite is R_a=0.048μm.

REFERENCES

[1] M.Murakawa, Mechanical application of thin and thick diamond film, Surface and coatings technology,49, 359-365(1991).

[2] M.YoshiKawa, Expanding the application of cutting tools made by chemical vapour deposition diamond, New Diamond, 11(1), 3-8(1995)

[3] B.R.Stoner, Highly oriented, textured diamond films on silicon via bias-enhanced nucleation and textured growth, J. Mater. Res.,8(6),1333-1340 (1993)

[4] Yoichi Hirose, The synthesis of high-quality diamond in combustion filmes, J. Appl. Phys.,68(12),6401-6405(1990)

MACHINABILITY OF TiAl INTERMETALLIC COMPOUNDS

Toshiaki Furusawa[*1], Atsushi Ichikawa[*2], Hiroshi Hino[*1],
Sinji Tsuji[*1] and Sadatoshi Koroyasu[*1]
*1 *School of Science and Engineering, Teikyo University*
*2 *Teikyo University of Engineering and Technology, Graduated*

Abstract

TiAl intermetallic compounds are cut in order to examine their machinabilty and to clarify the influence of machined defects on the mechanical strength. The main results are as follows: Diamond tools and cemented carbide tools exhibit streaky-abrasive-type damage. Surface and edge defects are formed along the lamellar layer direction. Bending strength is higher when the cutting direction is parallel to the edge. In the case of lower strength specimens, breakouts or cracks are observed at the edge. The average strength is improved by chamfering.

Keywords

TiAl intermetallic compounds, Cutting, Tool wear, Strength, Machinabilty

1. INTRODUCTION

Attempts[1] are being made to apply titanium aluminide intermetallic compounds in aerospace industries as novel materials because of their higher specific strength and higher toughness at elevated temperature compared to ceramics. Recently it has seen practical use in turbine blades of jet engines[2]. The application of this material has reached the stage of actual use. Since this material is difficult to shape by machining or plastic working because of a lack of ductility, methods such as casting and sintering are adopted for shaping under the currently existing conditions. However, since precise machining is required in the finishing process, the realization of suitable machining methods are urgently demanded. Moreover, it can be considered that the achievement of high-precision machining will bring about new applications of this material.

Because of the importance of clarifying the mechanical properties of this material in its final machined shape when applying it to precise products, we investigate the wear mechanisms[3] of a variety of tools during the cutting process, and examine the factors affected in the formation of the machined surface. In particular, we discuss the formation mechanisms of cracks on the finished surface and breakouts or cracks on the edges, and elucidate the influence of the initiation and propagation of these defects on the material strength[4].

2. EXPERIMENTAL EQUIPMENT AND METHODS

2.1 Work Material

The chemical composition and mechanical properties of the workpiece are shown in Table 1. This material has a lamellar structure of Ti_3Al layers termed the α_2 phase and TiAl layers termed the γ phase. The structure shows a rather large colony of which size is from 0.5 to 1.0 mm, as a result of using only casting with no heat treatment for shaping. Within one colony, the orientation of layers is almost homogeneous, but the orientation is random amongst colonies.

With respect to mechanical properties, this material is brittle because of small breaking strain. The hardness is almost the same value as that of the titanium alloy Ti-6Al-4V.

2.2 Machine Tool and Cutting Conditions

We used an ultra-precise lathe consisting of an air spindle and a granite bed.

The workpieces are cut to $20 \times 3.0 \times 3.0$mm for the strength tests and to $20 \times 15 \times 3.0$mm for cutting tests, using wire-cut electrical discharging machine (WEDM). Every specimen is adhered onto magnetic disk substrate which is then fixed to the spindle by a polyurethane vacuum chuck, and face cutting is finally performed.

For cutting tests, four kinds of diamond tools and three kinds of cemented carbide tools are prepared. The kinds of diamond are singlecrystalline and polycrystalline with average grain sizes of 50 μ m, 5 μ m and 0.5 μ m. We abbreviate these as SCD, PCD1, PCD2 and PCD3, respectively. K-grade cemented carbide tools are selected which are recommended for the cutting of titanium alloys and contained mainly tungsten carbides. These are materials of K10 rank, and K20 rank of the Japanese Industrial Standards (JIS) and ultrafine-particle material with average particle size of 0.7 μ m, abbreviated CC1, CC2 and CC3, respectively. Tool geometry and cutting conditions are shown in Table 2.

2.3 Strength Tests

Since the tensile tests are not reliable because of no fracture on the parallel portion of the testpiece for almost all workpieces, four-point bending tests, which are generally adopted for the strength tests of ceramics, are used for the strength tests. The test conditions are shown in Table 3. As machining specimens, four surfaces of testpieces are cut in the same direction.

Table 1   Chemical composition and mechanical properties of workpiece

Chemical composition(mass%)		
Al	Fe	Ti
33.93	0.08	Res.
Young's modulus	111GPa	
Bending strength	624MPa	
Breaking strain	2.3%	
Hardness(HV)	334	

Table 3   Conditions of four-point bending test

Test-piece		$2.7 \times 2.7 \times 20$mm
Span length	inner	6.5mm
	outer	16.5mm
Cross head speed		8.3×10^{-6}m/s

Table 2   Cutting tools and conditions

Tool geometry	
Cemented carbide	
$-5°,-6°,5°,6°,15°,15°,0.4$mm	
Diamond	
Rake angle	$0°$
Clearance angle	$5°$
Approach angle	$30°$
Rounded radius	0.4mm
Cutting conditions	
Spindle speeds	180,1000rpm
Cutting speeds	54-71,301-396m/min
Feed rate	0.03,0.05mm/rev
Depth of cut	5μ m
Cutting fluid	Synthetic oil

3. MACHINABILITY OF TiAl INTERMETALLIC COMPOUNDS

3.1 Tool Wear

Figure 1 shows flank wear of diamond tools. In the case of the singlecrystalline diamond tool, the amount of wear and the rate of increase of wear are rather large; this tendency is particularly marked at the beginning of cutting. In the case of polycrystalline diamond tools, there is almost the same tendency of wear between the three tools, but the cutting distance when flank wear reaches 30 μ m decreases with increasing size of diamond particles. PCD1 has the best antiwear property.

Figure 2 shows the results of cemented carbide tools. Similar to polycrystalline diamond tools, tools of smaller size of tungsten carbide particles have superior antiwear properties.

The scanning electron microscopy (SEM) photograph of a worn singlecrystalline diamond tool observed above the cutting edge is shown in Figure 3. Adhesives formed on rake and flank surfaces because of strong chemical affinity are removed using dissolved solution. Consequently, streaks and relatively large degrees of chipping are observed on rake and flank surfaces. It is considered that streaks are caused by abrasive wear due to the difference in hardness between the Ti_3Al layer and TiAl layer. In most cases, abrasive wear is a result of the accumulation of fine chipping, so that there is a tendency for large-size chipping to occur during diamond cutting. However, there seems to be heat wear[5] caused by the chemical reaction between diamond and titanium, as indicated by the formation of adhesive on the tool surface.

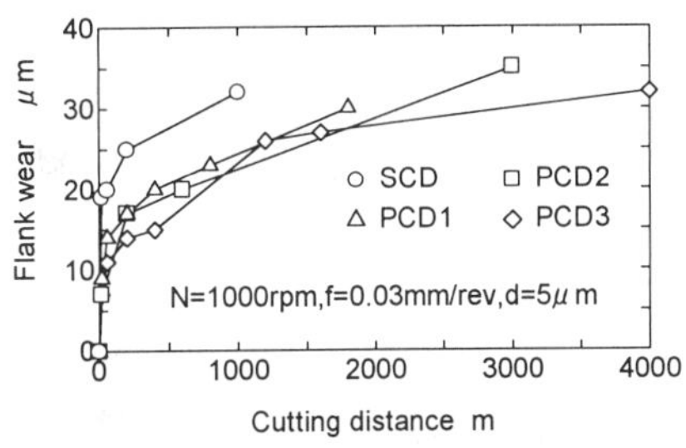

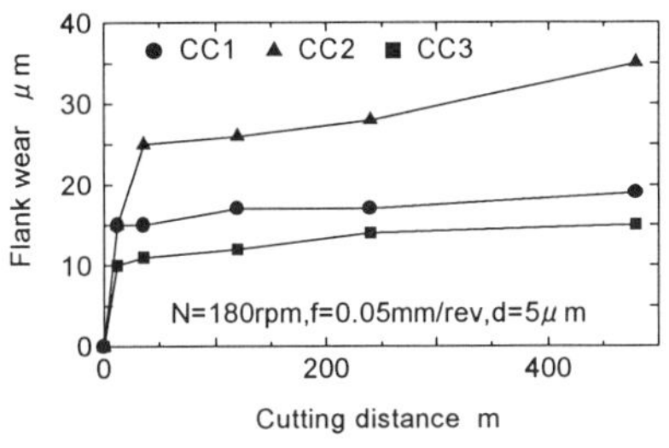

Figure 1   *Variation of flank wear machined by diamond tools*

Figure 2   *Variation of flank wear machined by carbide tools*

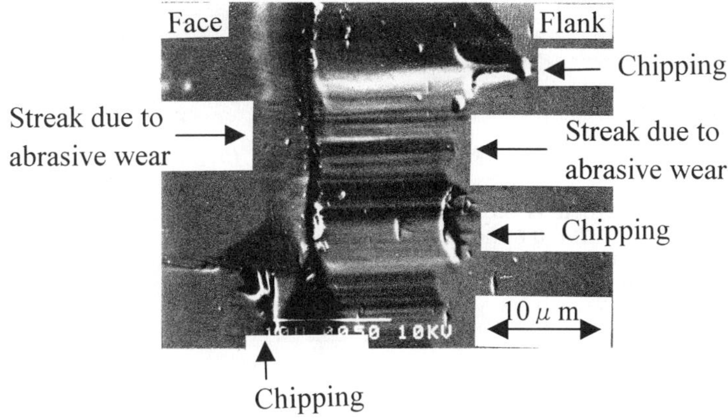

Figure 3   *Single-crystalline diamond tool with abrasive removed*

3.2 Formation of Finished Surface Defects and Edge Breakouts

The defect on a finished surface after cutting by the cemented carbide tool is shown in Figure 4. Defects with widths from 5 to 30 μ m and a long and narrow shape are formed at almost right angles to the cutting direction. Furthermore, these defects are observed in all cutting experiments with all tools and are observed with scattering states on the finished surface.

From the results of SEM investigation and the electron probe microscopic analyzer (EPMA) observation of the etched surface, it is ascertained that the defects are cleavage-type cracks. If the cutting direction is at right angles to the orientation of the lamellar structure composed of Ti_3Al and $TiAl$ layers, the cracks are formed with progress of the cutting edge.

Figure 5 shows the SEM photograph of breakout formed when the cutting edge detached from the workpiece edge. The large breakouts show a tendency to form when the cutting direction is at right angles to the lamellar orientation. Moreover, when the cutting direction is at right angles to the workpiece edge, breakout tends to be serve because the distance from the position of crack initiation to the workpiece edge is short.

4. INFLUENCE OF SURFACE DEFECTS AND EDGE BREAKOUTS ON STRENGTH

4.1 Influence of Defects Induced by Cutting

In order to clarify the influence of edge breakout on the mechanical strength, two types of specimens are prepared for four-point bending tests. That is, in one specimen, the cutting direction is at right angles to the edge, and in the other, the cutting direction is parallel to the edge. We call the former the R-type specimen and the latter the P-type specimen. The influence of the cutting direction on the bending strength is shown in Figure 6.

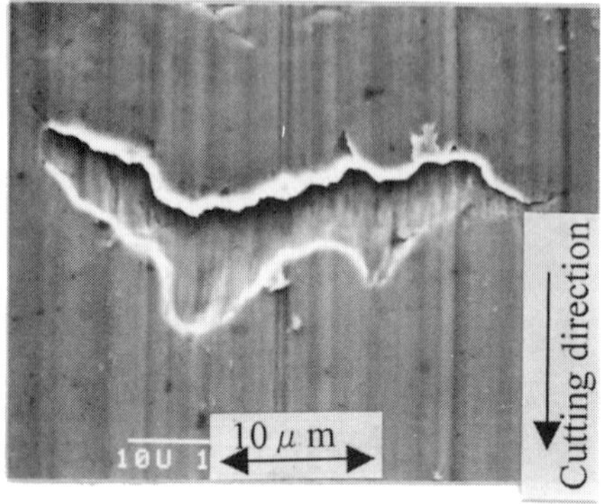

Figure 4   Defect on machined surface

Figure 5   Breakout at workpiece-edge

The P-type specimen is slightly superior to the R-type specimen in terms of average strength. From the optical microscope observation, it can be confirmed that regardless of specimen type, the main cracks propagate from edge breakouts or edge cracks in low-strength specimens.

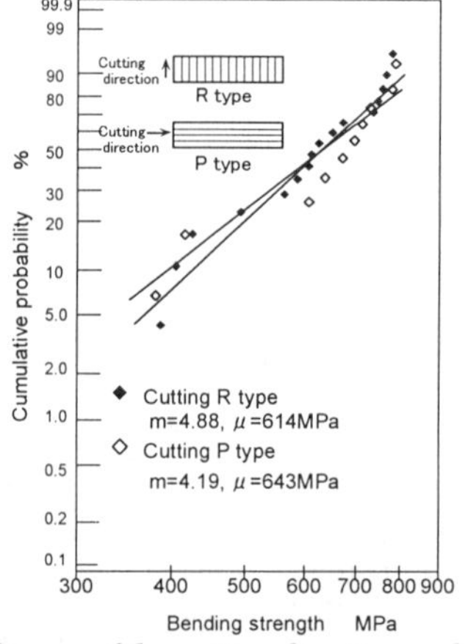

Figure 6   Influence of the cutting direction on bending strength

4.2 Influence of Chamfering

In order to examine the fracture mechanisms without forming defects at the workpiece edge, the bending strength is tested with chamfering for both types of specimens. The results are shown in Figure 7.

The values of strength increase for both types of chamfered specimens. The influence of chamfering is confirmed by the improvements of the strength and decrease of scatter. There are effective on R-type specimens because the R-type specimen, in which the cutting direction and the workpiece edge are at right angles, shows a tendency to form breakouts.

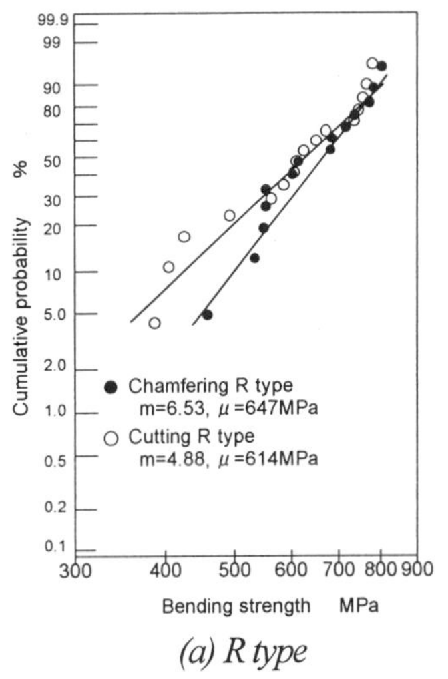

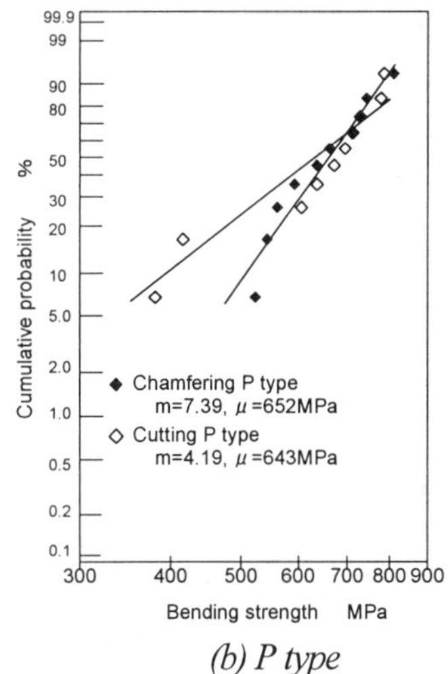

(a) R type   *(b) P type*

Figure 7   Influence of the cutting direction and chamfering on strength

5. CONCLUSION

Titanium aluminide intermetallic compounds were machined in order to examine their machinabilty and to clarify the influence of machined defects on the mechanical strength. The results obtained from this study are as follows.

(1) Diamond tools and cemented carbide tools exhibit streaky-abrasive-type damaged wear due to the lamellar structure of the workpiece composed of Ti_3Al and TiAl with different hardnesses. Polycrystalline diamond and cemented carbide made of large diamond or tungsten carbide particles have inferior antiwear properties.

(2) Long and narrow surface defects are formed along the layer direction of the lamellar structure due to crack propagation. Edge breakouts easily form when the cutting direction is at right angles to the edge of the workpiece. In particular, when the crack initiates near the edge caused by the lamellar structure of the workpiece, large breakouts are formed.

(3) Bending tests were conducted in which two types of specimen, with the cutting direction at right angles and parallel to the edge of the workpiece were prepared. The latter specimens had the higher strength. In the case of the lower strength specimens, breakouts or cracks were observed at the edge. As the distribution of strength shifts to the higher region, the average strength is improved by chamfering the two types of specimens.

6. ACKNOWLEDGMENTS

The authors wish to acknowledge contributions from The Light Metal Educational Foundation, Inc.

REFERENCES

1) H. Hino, T. Miyasita and T. Minakata, Application of TiAl-based Intermetallic Compounds, J. of Light Metals, 43-10(1993), 545.
2) M. Arai, R. Imamura, K. Matsuda, K. Nakagawa and T. Hosokawa, Development of Ti Al Blades for Large Pressure Turbine, Materia, 36-4(1997), 394.
3) T. Furusawa, H. Hino, S. Nakamura and S. Tsuji, High-Precision Cutting of Titanium Aluminide Intermetallic Compounds, Trans. of JSME , C, 64-624(1998), 3191.
4) T. Furusawa, A. Ichikawa, H. Hino, S. Tsuji and S. Koroyasu, Influence of Edge and Surface Defects on Mechanical Strength in Cutting of TiAl Intermetallic Compounds, Trans. of JSME , C, 66-651(2000), 3766.
5) N. Narutaki, Outline of Cutting Tool, J. of JSPE, 61-6(1995), 751.

MDS STUDY ON THE EFFECT OF CUTTING EDGE RADIUS OF DIAMOND TOOLS IN NANOMETRIC CUTTING PROCESS FOR BRITTLE MATERIALS

Dan Li   Shen Dong   Yingchun Liang   Xichun Luo
Precision Engineering Research Institute, Harbin Institute of Technology, China
Kai Cheng
School of Engineering, Leeds Metropolitan University, UK

Abstract

This paper presents an investigation of the effect of cutting edge radius of diamond tool in nanometic cutting process of brittle material based on molecular dynamics (MD) simulation. It shows that the specific cutting resistance increases drastically as the depth of cut is decreased. In general, the machined surface roughness decreased as the cutting edge radius decreased. But there is a minimum cutting edge radius for achieving a high accuracy surface, because the cutting tool is apt to wear and the machined surface is damaged when the cutting edge radius is smaller than it.

Keywords

Molecular dynamics simulation   Cutting edge radius Tool wear
Nanometric cutting process   Brittle materials

1. INTRODUCTION

Brittle materials have important application to fabricate intricate components with high quality in aviation, microelectronic, and optical industries. Though they are known for low machinability, by aid of a specially prepared fine diamond cutting tool on a highly reliable ultra-precision machine tool, it is possible to turn brittle material with good machined surface quality[1]. But the machining mechanism is still not fully understood. Some Pioneering works were initiated by Blackley, Scattergood[2] in the USA and Nakasuji[1], Shibata[3] in Japan, their studies focused on the mechanism of brittle-ductile transition, Blackley proposed a ductile-regime diamond cutting model of brittle materials; other studies about the effect of rake angle or crystalline orientation, etc in the ductile-regime machining were done by Kamimura[4], Kim[5], and Hung[6] etc. Applying Molecular Dynamics Simulation (MDS), Ikawa, Shimada and Inamura[7] in Japan began to investigate the mechanism of brittle-ductile transition in the micro-machining process of brittle materials from the atomic scale. Overall, a sound foundation has been laid for this study.

Unlike in conventional cutting where the depth of cut is significant compared to the edge radius (i.e.the edge radius is negligible), in nanometric cutting, the cutting edge radius of diamond tool plays an essential role in the chip formation when the depth of cut is only several nanometers and the measured radius of the diamond tool are in a range from 10nm to 45nm. Unfortunately, few studies of the effect of cutting edge radius have been published.

In this paper, Molecular Dynamics Simulation is applied to investigate the effect of cutting edge radius on nanometric cutting processes so that the processes can be better understood.

2. MDS ON ORTHOGONAL NANOMETRIC CUTTING

A 3-D MDS model of orthogonal cutting of a (001) plane of a single crystal of Ge by a diamond tool has been built (shown in Figure.1), which contains about 70,000 Newtonian work and tool atoms, and thermostat atoms to control the process temperature. In order to reduce the boundary

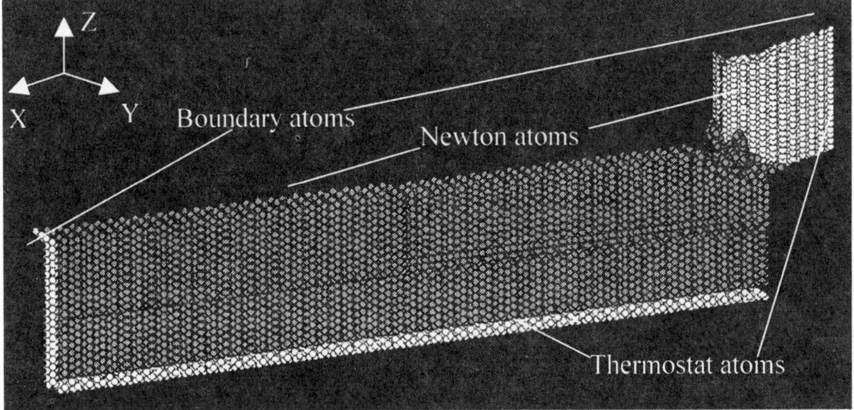

Figure 1. 3-D MDS nanometric cutting model

Effects, the hard boundary condition is adopted. The cutting speed of 100m/s is employed to reduce the computation time. Although this cutting speed is unrealistic, it was proved that there was little difference between the surface quality obtained under the cutting speed between at 20m/s and 200m/s by Shimada. The cutting direction is set at [010]. The width of cut, length of cut, the rake angle of the tool, clearance angle of the tool and the bulk temperature is 0.924nm, 7.35nm, 0°, 10°, 298K respectively. The depth of cut and tool edge radius vary from 0.49nm~2.45nm, 1.57nm~3.14nm, respectively.

The forces at the tool-work interface as well as the material themselves are calculated by difference of modified embedded-atom potential. The potential energy of the i-th atom is:

$$E_i = \frac{1}{Z_i} \sum_{j(\neq i)} E_s(r_{ij}) + [F_i(\frac{\rho_i}{Z_i}) - \frac{1}{Z_i} \sum_{j(\neq i)} F_i(\frac{\bar{\rho}_i(r_{ij})}{Z_i})]$$

$$(1)$$

Where E_s, F, Z_i, r_{ij} is the energy per atom of the reference structure, functional of electron density, the number of the nearest neighbors of i-th atom, the distance between i-th atom and j-th atom, respectively.

Theorem of equipartition of energy is used to calculate the equilibrium temperature of the system, and Kirchhoff's law is applied to rectify those parameters of equation (1) that relate to temperature, thus the interaction of thermal and force to the cutting tool can be studied in this MDS.

The initial positions of work and tool atoms are the sites of their crystal lattices, respectively. The initial velocities can be assigned from a Maxwell distribution at 298K. The computation time step is 10fs. The displacement and the velocity of atoms in the cutting region can be calculated by numerical difference.

3. RESUIT AND DISCUSSION

3.1 The Relationship Between Specific Cutting Resistance And The Depth Of Cut

Figure 2 shows the variation of specific cutting resistance with the depth of cut, it indicates that the specific cutting resistance increases drastically as the depths of cut is decreased. It can be explained that when depth of cut is close to the atomic dimension, it needs more specific energy to break atomic bond and initiate the new dislocations. It is different from that in conventional cutting, in which the deforms of work material happen between the crystallite, i.e. they come from the former dislocation.

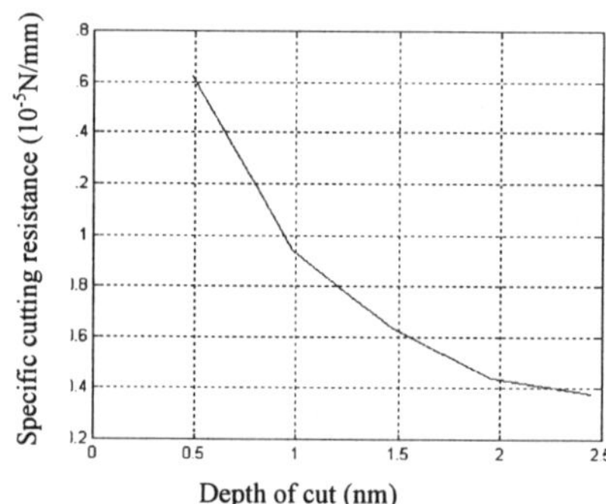

Figure 2. Relationship Between specific cutting resistance and the depth of cut

3.2 The Relationship Between The Machined Surface Roughness And The Cutting Edge Radius

Figure 3 shows the variation of machined surface roughness (Ra) with the cutting edge radius at the same depth of cut (1.47nm), it can be seen that the machined surface roughness decreases as the cutting edge radius decreases when the cutting edge radius is bigger than 2.2 nm. It can be explained that when the depth of cut is in several nanometers, the contact length between rake face and the chip is decreased, So in fact most of the

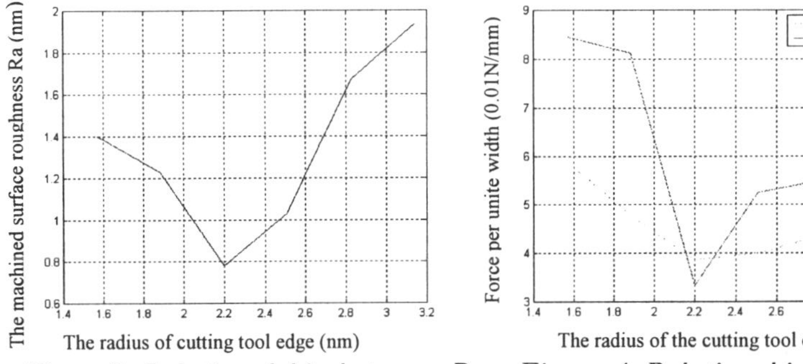

Figure3. Relationalship between Ra and cutting edge radius

Figure 4. Relationship between cutting force and cutting edge radius

work material is removed by the cutting edge. When the cutting edge radius is bigger than 2.2nm, the cutting force and thrust force increase with the increase of the cutting edge radius (as shown in Figure 4), and the drastically increasing thrust force can increase the vibration between the tool and the workpiece in the direction of depth of cut. It has been proved that the vibration in this direction has much more effect on the surface roughness[8].

3.3 The Minimum Cutting Edge Radius

It can be seen from Figure 3, when the cutting edge radius is smaller

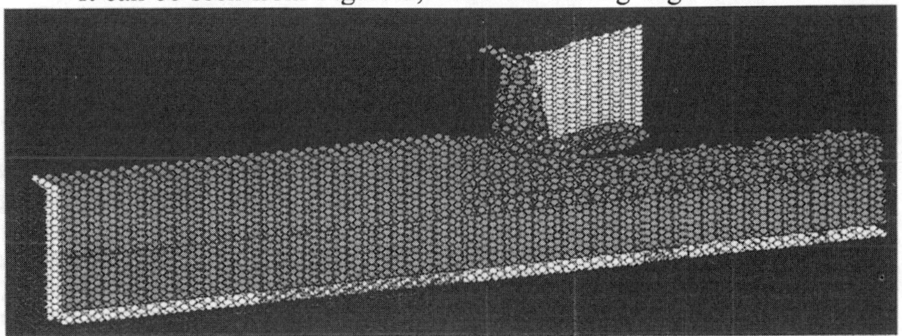

Figure 5. The tool wear in nanometric cutting process

than 2.2nm, with the decrease of cutting edge radius, the machined surface roughness is increased. Figure 5 is the snapshot when the length of cut is 7.35nm. It can be seen that there are distinct diffusions between work atoms and tool atoms. A few tool atoms at the tip of the tool are separated, and then move with the chip, only several of them adhere to the cutting edge and diffuse with work atoms on the rake face. That is to say the tip of the cutting edge of the tool begins to wear. By the first principle stress method and theorem of equipartition of energy, the stress and temperature are calculated in the MDS. The maximum tensile stress(38.5Gpa) and maximum temperature(546K) are all found at the tip of cutting edge. Due to the interaction of force and thermal, the cutting edge is apt to wear. The tool wear can result in the drastic vibration of the cutting force and damage the machined surface. To avoid the quick tool wear and achieve a high accuracy surface, the cutting edge radius should be bigger than the minimum cutting edge radius. As far as the nanometric cutting process of Ge, it is 2.2 nm.

4. CONCLUSIONS

Nanometric cutting processes of Germanium was studied by molecular dynamics simulation. It was found that the specific cutting resistance increases drastically as the depth of cut is decreased. In general, the roughness of the machined surface decreased as the cutting edge radius is decreased. But when the cutting edge radius is smaller than a critical value, the cutting tool is apt to wear and result in the damage of the machined surface. So there is a minimum cutting edge radius for achieving a high accuracy surface in nanometric cutting process.

5. ACKNOWLEDGEMENTS

The authors would like to thank the NSFC (National Natural Science Foundation of China) for the financial support for this project.

6. REFERENCES

1. T.Nakasuji, S.Kodera, S.Hara,H.Matsunaga, N.Ikawa, S.Shimada. Diamond turning of brittle materials for optical components. Journal of the CIRP 1990; 39: 89-92
2. W.S.Blackley,R.O.Scattergood. Ductile-regime machining model for diamond turning of brittle materials. Precision Engineering 1991; 13:95-102
3. Takayuki Shibata, Shigeru Fujii, Eiji Makino, and Masayuki Ikeda. Ductile-regime turning mechanism of single-crystal silicon, Precision Engineering 1996;18:129-137
4. Yasuyuki Kamimura, Hiomi Yamaguchi et al.. Ductile Regime Cutting of Brittle Materials Using a Flying Tool under Negative Pressure. Annals of the CIRP 1997; 46:451-454
5. Hyung-Sun Kim. Steve Roberts. Brittle-ductile Transition and Dislocation Mobility in Sapphire. Journal of American Ceramic Society. 1994; 77: 3099-3104
6. N.P.Hung and Y.Q.Fu. Effect of crystalline orientation in the ductile-regime machining of silicon. Int.J.Adv.Manuf Technol 2000;16:871-876
7. Toyoshiro Inamura, Shoichi Shimada,et al., Brittle/ductile transition phenomena observed in computer simulations of machining defect-free monocrystalline silicon. CIRP. 1997;46:31-34
8. S.C.Lin and M.F.Chang, A study on the effects of vibrations on the surface finish using a surface topography simulation model for turning. Int.J.Mach.Tools&Manu. 1998; 38:763-782

AN ATTEMPT TO DEVELOP A SHORT LASTING MACHINABILITY TEST FOR STEELS

J.C. Hamann*   F. Meslin*   MACTEST**

(*) Ecole Centrale de Nantes, BP92101, 44321 Nantes France
(**) Tekniker, Bosch, Ascometal, Fiat, Sidenor, CNR, ECN

Abstract

Machinability tends to remain a term which means 'all things to all men' [2]. We have investigate the possibility of quantifying machinability by an "easy to run" cutting test that can be adapted to end user requirements, based on a cutting force signal processing by a neural network. Results are encouraging but they need some improvement.

Keywords

Machinability control, cutting force, neural network

1. INTRODUCTION

Materials can exhibit machinability variations even if they fulfill severe requirements about their chemical composition, their mechanical properties and their micro-structure. The attempt here is to develop a short lasting and reliable machinability test as so as to implement it as an acceptance checking mean of material batches. We used our knowledge about the fundamentals of cutting in order to select specific force features from which we can derive information about the wearing of the tool. Since the link between both is complex, we used a nn in order to get this correlation by a learning process.

2. TEST PRINCIPLE

2.1 Related works

Figure 1-a presents a typical tool wear *vs* cutting time (t) curve. Three typical regions dissect the curve according to the following period of lifetime:

1. In the first region, $t < T_C$, wear increases rapidly. T_C is the point of inflection of the curve.

2. In the second region, $T_C < t < T_{M1}$, the wear propagation is more or less linearly dependent of time.

3. The third region, $T_{M1} < t < T_{M2}$, corresponds to the tool failure.

Previous studies [1] have shown that information about the tool wear evolution can be derived from the characteristics of the first region. Studies [6] have shown that such aspects (tool wear, surface quality ...) can be correlated with the cutting force by means of a neural network (nn). The leading idea is that a short lasting test (a few seconds) is enough to detect the salient features. Compared with other studies [3] where a nn was used to highlight the effect of wear on the cutting force, in our case, we used the nn to detect the tool loading features that will give rise to wear.

2.2 Experimental set up

The experimental set up is sketched out figure 2. 40 steel grades have been selected covering 8 families: free cutting, carbon, quench and tempered, case hardening, stainless, bearing, nitriding and micro-alloyed steels. The test consists in 4 successive 1 second lasting cuts with the same cutting edge in the following cutting conditions:

- SPUN P10 carbide insert with r_ϵ=0.8mm, κ_r=75°
- Vc=300m.min^{-1}, f=0.2mm.rev^{-1} and ap=2mm

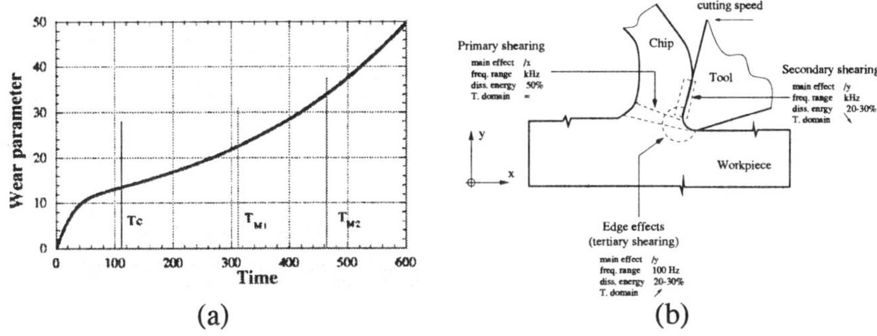

(a)   (b)

Figure 1: Conventional wear parameter evolution according to cutting time, and force features associated with the different tool-workmatrial interfaces.

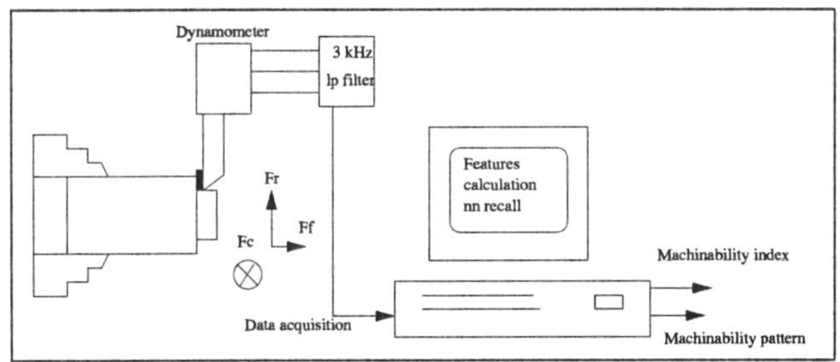

Figure 2: Sketch of the machinability test set up

3. DEFINITION OF A MACHINABILITY INDEX

We defined machinability for each grade by a "machinability pattern" which account for a machining behavior, and a "machinability index", i.e. a scalar parameter, which scores up the machinability. The classification of steels according to the machinability index has been validated by comparison with common classifiers (e.g. V_{20}) based upon the ISO 3685 standard.

3.1 Definition of machinability patterns and levels

The pattern can be viewed as a bar graph combining 3 machinability indicators, coded over 3 levels and selected according to expertise informa-

301

Area	Fc	Ff	Fr	spectral power
Primary shearing	x	x		too high
Secondary shearing		x		too high
edge effect		x	x	200-400 Hz

tion from steelmaker. From left to right: the sulfur content, the "machinability improvement" additions, such as oxyde modifiers, the wear kinetic of a reference tool in specific cutting conditions.

- $\begin{bmatrix} 1 & 0 & 0 \\ 1 & 1 & 1 \end{bmatrix}$: example of machinability of level 4.

3.2 Machinability level

The machinability level is simply defined by counting the number of '1' in the pattern. Following, 7 machinability levels can be defined corresponding to a total of 27 machinability patterns, or, let say, machining behaviors.

4. SELECTION OF FORCE FEATURES

The chip formation mechanism can roughly be split in 3 "interfaces": (i) the primary shearing plane, (ii) the secondary shearing, between the chip and the rake face, (iii) edge effects, e.g. rubbing, sometimes refers as "tertiary shearing". "Relevant force features" candidates were selected, according to figure 1-b and table 1, [4]. The final selection of force features was made using a recursive partitioning method [5] and the following set of features was found to discriminate all of the common machinability comparison cases:(i) the feed force (Ff), (ii) the (Ff/Fc) ratio, (iii) the (Fr/Fc) ratio, (iv) the 200 - 400Hz spectral power of the Ff component, (v) the ratio of the 200 - 400Hz spectral power of Ff and Fr. No time domain information hs been taken into account. This may be an explanation for the difficulties of pattern recall (see section 5).

5. RESULTS AND CONCLUSION

The nn has been trained for two cases : (i) in order to recall machinability patterns, (ii) in order to recall machinability level. Once trained, its reliability is measured in 3 ways: (i) with noisy inputs (up to 30 % of random noise), (ii) by testing its response with other steels, (iii) by checking the response with data from different laboratories. Good results were obtained in the case of machinability level determination with 98 % success [7], but very poor results for the determination of machinability patterns, although it is possible to predict the force features from the machinability pattern and the steel family. This can be explained by the fact that the machining behavior description is, at least, 3 dimensional (steel family, sulfur content, additive content) while the selected features are only 2 (spatial and frequency domain).

These are encouraging results but work is still necessary to improve the reliability of the test. This research was funded by EC during the MACTEST program.

References

[1] A. Ber, S.Kaldor, "The first seconds of cutting, wear behavior", Annals of the CIRP 31/1/1982, 13-17

[2] B. Mills and A.H. Redford,*Machinability of engineering materials*, App. Sci. Pub., 1983

[3] S. Rangwala and D.A. Dornfeld, "Integration of sensors via Neural Net ...",in Proc. of Winter Annual meeting of ASME, Vol 25, 109-120

[4] JC Hamann, F Le Maître et al.,"Selective transfer built up layer displacement in high speed machining ...", Annals of the CIRP 43/1/1994, 69-72

[5] D.M. Hawkins,"*FIRM Manual*", University of Minnesota, 1997

[6] X.P. Li, K. lyncaran, A.Y.C. Nee, "An hybrid machining simulator ...", in Proc. of the 1st CIRP int. workshop on modeling of mach. operation, Atlanta, May 1998,2D6-1-2D6-12

[7] C. Bertrand, S. Tudanca et al.,"Desarollo e implementatción de un nuevo ensayo de maquinabilidad ...", INVEMA Conference, San Sebastian, Spain, 25-27 October 2000

Micro Surface Damage Detecting and Repairing System by Precise Micro Robots

Hisayuki Aoyama, Satoshi Miyamoto and Ryoma Watanabe

Applied Micro Systems, Robotics Div.
Dept. of Mechanical Engineering & Intelligent Systems,
University of Electro-Communications
1-5-1, Chofu, Tokyo 182-8585, Japan

Abstract

In this report, an unique system in which many micro robots with micro tool can collaborate to detect the mechanical damages such scratched grooves and crack and to repair them automatically is described. One miniature robot can explore to scan over the surface with the stylus and propagate its acoustic signal to other small robots with micro repairing tool. Another small robot with microphones can distinguish the destination of signal source and approach to the location of damage to repair it automatically. The primary experiments that the small robots succeed in detecting mechanical damages and in repairing them by UV cement.

Keywords

micro fork, vibrating stylus, sound resonator, piezo film, surface damage repair, acoustic propagation, micro robot application.

1. INTRODUCTION

For these year, many types of micro robots have been developed as high precision positioning tools[1],[2],[3]. Also the micro robot has the potential performance for the application in the special environments where it is difficult for the conventional maintenance machines to extend their arms. Ocasionally some microscopic mechanical damages on the surface cause the serious problem in the plant such the power generator and the chemical reactor. So it is important to maintain such facilities to keep its good condition with low cost. It is, however, difficult to detect such small damages in the tube and behind the wall due to the lack of expandability of the conventional machines.

Our research group have developed many small robots with micro tools and sensors which are composed of piezo elements and electromagnets[4]. In the newly developed micro robots system, the robot with the special sensor can explore on the target surface to find out the microscopic crack. And then it can propagate the acoustic signal to navigate the other small robots with the repairing tool. When the robots close to the damage to be repaired, it can move to scan over the area again precisely and inject the cement into the crack by the solenoid driven capillary. The details and basic performances of this unique micro robots system will be discussed in the next sections.

2. MICRO ROBOT FOR SURFACE DAMAGE DETECTION

The micro robot which was developed for surface damage detection is shown in Fig.1. The size of this small robot is measured as less than 1 cubic inch. It has the piezo elements and electromagnets to move precisely based on an inch worm manner, as well as the micro fork with a thin stylus of 0.1mm diameter and the acoustic resonator. So the small robot can scan the tiny vibrating stylus of tungsten on the target surface. This commercially available U-shape micro fork of 1.5mm in width, 8mm in length and 0.5mm in thickness can be also excited

Fig.1 Micro robot with a vibrating stylus on a micro fork and an acoustic passive resonator

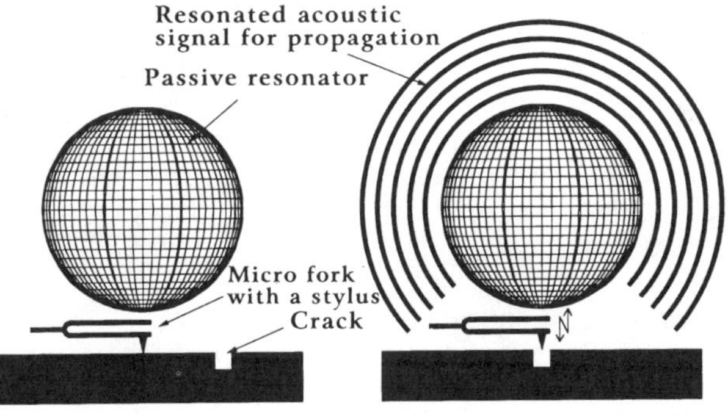

Fig.2 Principle of micro crack detection by a vibrating stylus and signal propagation by passive acoustic resonator

by the piezo film at the resonance frequency of 1000Hz. The amplitude of the stylus vibration can be varied from a few microns to several ten microns according to the input voltage. When the stylus traverses over the small crack, then the contact free micro fork can start to vibrate although the stylus vibration can be suppressed on the plain surface as shown in Fig.2. Thus the mechanical crack deeper than the specified amount of the stylus vibration as a threshold value can be detected easily by checking the acoustic signal. In order to amplify the faint acoustic signal and to propagate it around the robot with no electric device, the Helmhorzt passive resonator which is carefully fabricated is set close to the micro fork. This arrangement can provide the unique performance such accurate micro defect detection as well as signal propagation like an insect manner of a mosquito wing and a cicada's resonator.

3.MICRO ROBOT WITH MICROPHONES AND REPAIRING TOOL

After the location of the defect is identified by the small robot with a vibrating stylus and a sound resonator, the other small robot with a repairing tool have to move toward the destination automatically based on the acoustic signal. The piezo driven small robot as shown in Fig.3, is equipped with a pair of small microphones with the electronics for detecting the acoustic signal, a vibrating stylus on a micro fork for the damage detection and a solenoid driven pipette for injecting the cement into the damage. To prevent the system from being complicated, this small robot is required to maneuver itself by the reflective manner based on the simple control sequence.

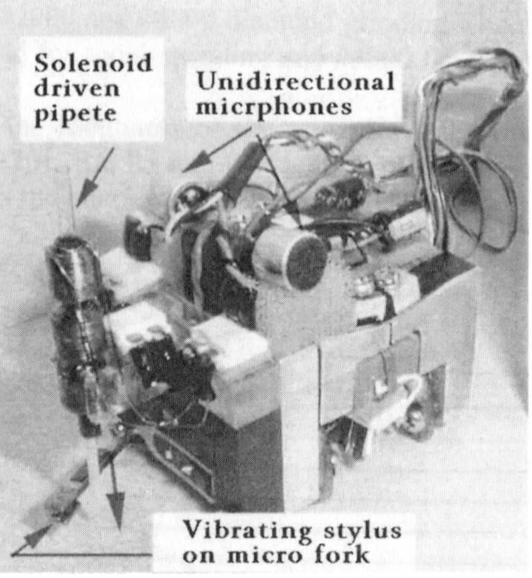

Fig.3 Micro robot with acoustic sensors for signal source detection and solenoid driven pipete for cement injection

A pair of tiny unidirectional microphones with the primary electronics are embedded on the robot to distinguish the direction of signal source which is generated by the vibrating stylus. The differential amount from two sensors can be used for switching either of two piezo elements to control the heading of the small robot. In order to accomplish the final approach to the defect and to repair it, another vibrating stylus and solenoid driven pipette are implemented on the robot. In the local area, this small robot can scan precisely on the target surface with the vibrating stylus and then UV reactive cement can be injected by the solenoid driven pipette as soon as it find out the damage.

4.CONTROL SEQUENCE

In Fig.4, the control sequence for the small robots is illustrated.
(1)The robot with a vibrating stylus can explore on the surface until it find out the crack deeper than the specified value.
(2)When it succeed in finding the damage, the small robot stops at the location since the contact free stylus begins to vibrate. The acoustic signal of 1000Hz can be propagated broadly as faint sound of micro fork vibration is amplified by the passive resonator. The other small robot with microphones, a stylus and the repairing tool can activate its piezo elements and electromagnets to move the signal source. Then the heading of the robot is controlled by switching the piezo elements based on the sensed acoustic signal.

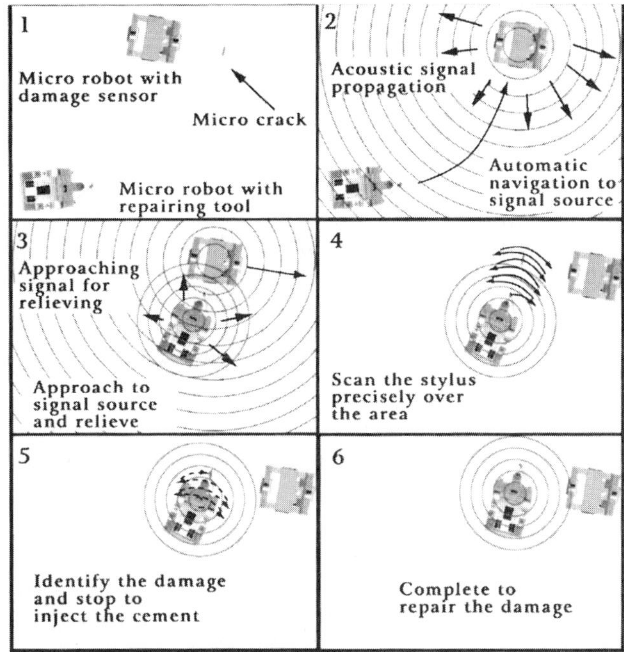

Fig.4 *Control sequence for small robots to collaborate for detecting,
navigating and repairing small surface defect*

(3) When the repairing robot can approach to the target area, it also gives off the sound signal of 2000Hz by using the micro buzzer. This signal can provide the trigger for the robot with the resonator to resume exploring again another area.

(4) After the robot moves away, the repairing small robot can search over the target area again by scanning its vibrating stylus.

(5) As soon as the stylus goes over the crack and begins to resonate, the small robot can stop at the position to inject the UV reactive cement by the solenoid driven pipette.

In the practical application, many small robots can be employed over the target surface to enhance their performance.

5. EXPERIMENTAL RESULTS

To check the performance of the system under the control sequence as mentioned in the previous section, several experiments such as microscopic damage gauging and acoustic signal propagating, automatic robot navigation based on the acoustic signal and the surface damage repairing were carried out.

5.1 Automatic navigation toward signal source

Fig.5 shows the typical experimental results that the small robot succeeds in exploring on the target surface to inspect the mechanical crack and then in navigating another small robot to approach to the signal source. The gauging performance from 30 to 80 micron was achieved and the range of signal detectable was approximately within 200mm. When the robot with repairing tool arrives at the point with the buzzer sound, then the small robot on the damage can move away to another destination.

5.2 Final damage detection and cement injection

At the final stage, the robot with tool can scan again around the surface defect by its vibrating stylus and stop to find out it. Then the small amount of liquid cement which is hardened by UV light can be injected. In Fig.6, the photographs show the damage before and after the operation as well as the small amount of cement can be injected from the pipette. It is found that the damage is fully covered by the UV cement.

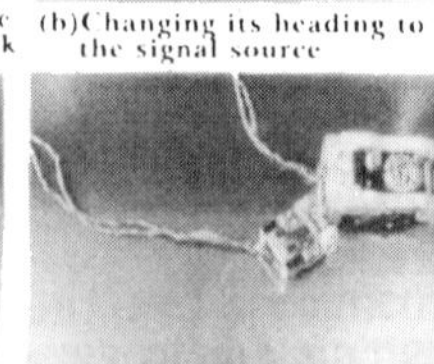

Fig.5 Automatic robot navigation based on acoustic signal from the detected damage

6.CONCLUSIONS

An unique system in which many micro robots with sensor and micro tool can collaborate to detect the cracks on the surface and to repair them automatically was introduced. In the several experiments, the small robots succeeded in detecting mechanical damages with the depth of several ten microns on the target surface, in navigating themselves to transport the micro repairing tool to the point with the help of acoustic signal propagation and in injecting the UV cement into them.

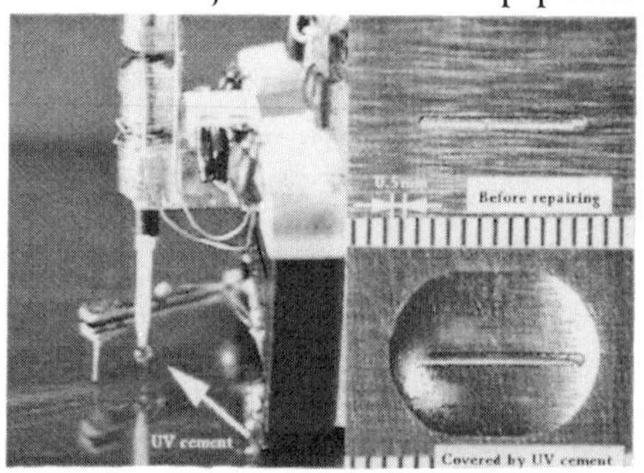

Fig.6 Solenoid driven pipete for UV cement injection and the repaired cracks

REFERENCES

1)S.Fatikow,U.Rembold and H.Worn; Design and control of flexible microrobots for an automated microassembly desktop-station, Proc. of SPIE, Vol.3202(1997)pp.66-78

2)S.Martel,K.Doyle,G.Martinez,I.Hunter and S.Lafontaine; Integrating a complex electronics system in a small scale autonomous instrumented robot:the Nano Walker Project, Proc. of SPIE, Vol.3834(1999)pp.63-74

3)C.Schmitt, J-.-M.Breguet, A.Bergander,R.Clavel; Micro Positioning Systems Using Piezo Technology: From Nanometer to Centimeter, Proc. of 2nd Int. Workshop on Microfactory,(2000)pp83-86

4)H.Aoyama and A.Hayashi; Multiple Micro Robots for Desktop Precison Production, Proc. of 1st Int. Conf. of EUSPEN(1999)pp.60-63

EFFECT OF MINIMAL QUANTITIES OF LUBRICANT IN MICRO MILLING

JRS.Prakash, M.Rahman, A.Senthil Kumar, and SC. Lim
Department of Mechanical Engineering, National University of Singapore,
10 Kent Ridge Crescent, Singapore 119260.

Abstract

Environmental, health, and waste disposal problems, which are associated with the conventional cooling technique, leads the interest switched to minimum quantities of lubricant (MQL) in machining. In this study the consumption rate of cutting oil is restricted to 1 mlh^{-1} against a benchmark flow rate of 42.0 lmin^{-1} in conventional cooling. The lubricating action of the cutting oil with extremely small amount of cutting fluid provides lower friction coefficient between the chip and tool. The effectiveness of this technique is investigated in terms of tool wear, chip shape, cutting speed and feed rate in micro milling.

Keywords

Micro machining, milling, MQL, mist coolant, tool life, tool wear.

1. INTRODUCTION

Micro machining process such as micro end milling has been developed in this decade to fabricate three dimensional miniature components. Unpredictable tool life and premature failure are major problems concerning micro milling [Tansel et al., 1998]. Cutting characteristics of micro end milling may be the same as those of the traditional end milling operations, but the wear and breakage mechanism are different [Rahman et al., 2000].

In machining, around 99% of work done is used up to heat the chip, the tool and workpiece material [Trent E.M, 1991]. The chip transports most of heat, remaining is conducted into the tool and workpiece material depending on the individual property of the materials. In this case the temperature of tool may rise up to 1000 °C [Kustas et al., 1997]. The rise in temperature at the tool chip interface may accelerate the wear rate. Machado and Wallbank revealed that the amount of lubricant required to provide effective lubrication layer in a typical machining operation is approximately around 0.1 mlh^{-1}. Extremely small amount of lubricant is blasted with adequate air, which mechanically break up into ultra fine droplets of up to 1 µm will be a useful technique to this problem. This novel approach is termed as minimal quantities of lubrication (MQL). Main benefits of MQL technique are tool, workpiece and chips remains dry, and thus obviating any need for further processing or finishing. So this technique is described as dry machining and such the drastic reduction in the consumption of MQL justifies a distinct cost advantage compared to flood cooling.

Klocke et al., investigated the machinability of the drilling process using MQL and concluded that it offers an alternative when 100% dry machining is not feasible. MQL system lowers friction coefficient between the chip and the tool than that in the case of dry cutting [Wakabayashi et al., 1998]. Since no significant work has been carried out using MQL in micro milling an investigation has been carried out in this study to evaluate the performance of the MQL technique in terms of cutting speed, feed rate, and axial depth of cut, chip shape, and tool life.

2. EXPERIMENTAL INVESTIGATION

The study is limited to micro end milling (down milling) operation. The experimental setup consists of a Makino V55 vertical machining center, and Ebara MQL system (Economist). Uncoated micro grain carbide end mill cutters of 1mm in diameter and helix angle 30 degrees are used to machine pure copper.

The data acquisition processes used in this experiment is shown in figure 1. Experiments are carried out under MQL method considering the following tool life criteria:

(i)   Average flank wear greater than or equal to 0.08 mm considering the nose radius and width of the cutting edge.

(ii)   Tool edge fracture or catastrophic failure.

The experimental setup is as shown in figure 2.

Experiments are carried over a cutting speed range of 30 to 90 m/min, feedrate of 175 to 350 mm/min and axial depth of cut 0.125 to 0.275 mm

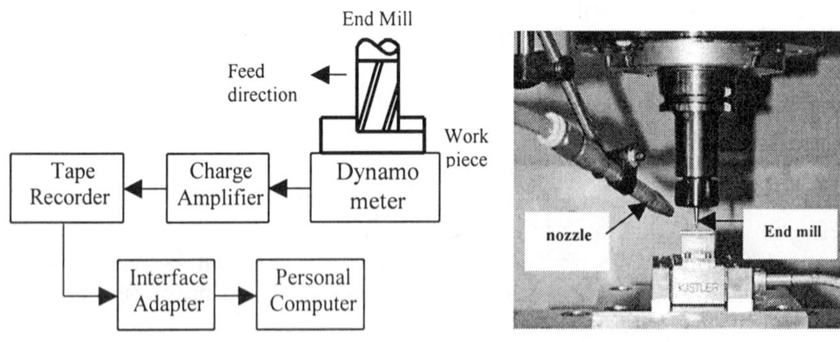

Figure 1. Schematic diagram   *Figure 2*. Experimental setup

while the radial depth of cut is kept at 30% immersion ratio. The content of vegetable based mist oil used in this experiment is Cr < 0.001w%, Fe < 0.001w%, Mo < 0.01w%, P <0.06w%, Zn < 0.01w%. The mist air is directly sprayed from the nozzle at 30-degree inclination towards the tool workpiece. The mist oil discharge rate is 1 cc/h and discharge pressure is 0.12 Mpa and carrier pressure is 0.22 Mpa. The Ebara MQL equipment is capable of producing fine mist of oil and air mixture at the required pressure. This is supplied through a nozzle at the machining region.

3. RESULTS AND DISCUSSIONS

The chips produced during MQL micro milling and dry cutting methods are shown in figures 3(a) and 3(b). The width of the chips produced by MQL is smaller in comparison with dry cutting. The MQL has a better chance of penetrating deep into the secondary deformation zone effectively reducing the area of tool chip contact length. This phenomenon is observed due to smaller chip length and width, this in turn reduces the friction.

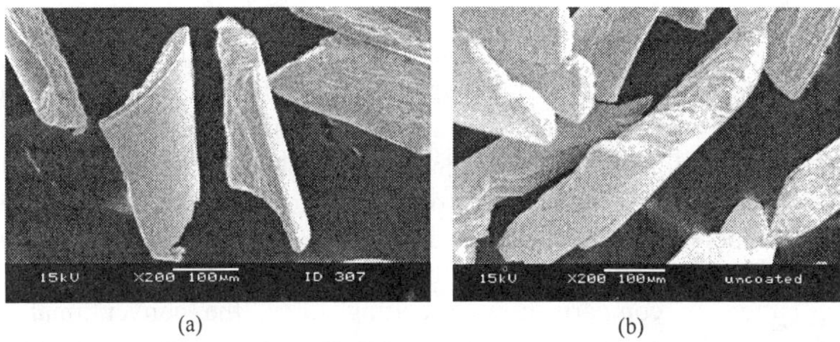

(a)   (b)

Figure 3.   Typical chip formation at cutting speed 59.7 m/min, feedrate 175mm/min and axial DOC 0.200 mm (a) MQL and (b) Dry cut

Effect of cutting speed for tool wear at axial depth of cut of 0.125 mm and feed 175 m/min is as shown in figure 4. From the experiment, it is observed that lower the cutting speed longer the tool life. The tiny oil droplets produced using the MQL technique penetrates deep into the chip tool interface at lower cutting speed thereby providing better lubrication and consequently lower rate of tool wear. In addition this will also aid in removing the heat generated during the cutting from there by increasing the tool life.

Feedrate has significant effect on tool wear while using mist coolant. From figure 5, with the given experimental condition it is observed that the tool life increases with the increase of feedrate. It can be clearly observed that at higher feedrate of 350 mm/min the life of cutter is prolonged. It is known that the tool wear is an abrasion process in the form of micro cutting coupled with the production of micro chips or wear particles. In end milling the abrasion of the tool is dependent on the sliding contact length between the tool and the workpiece and it reduces proportionally to the increase in feed. Therefore the tool life is expected to increase for certain feed range beyond which the tool may fail catastrophically.

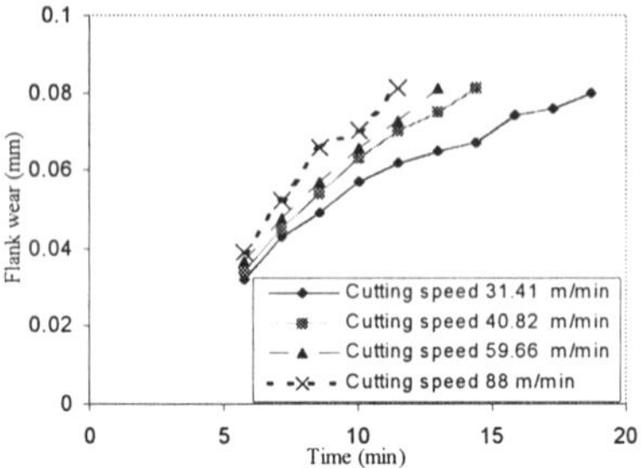

Figure 4. Effect of cutting speed on tool wear at axial DOC 0.125 mm and feed 175 mm/min

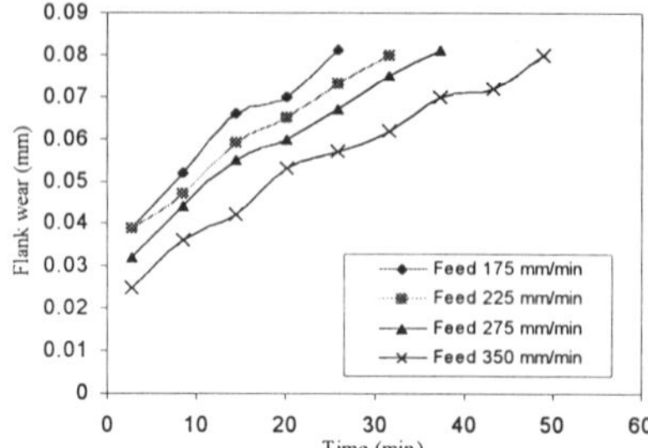

Figure 5. Effects of feedrate on tool wear at cutting speed 31.4 m/min and axial DOC 0.125 mm

Effect of depth of cut on tool wear is shown in figure 6. From the Table 1 it is shown that under the given conditions the total engagement angle for micro end milling cutter is 134.5° as compared with conventional end milling operation 70°. Larger engagement angle of cutter in contact with the work piece makes the cutting more stable in micro end milling. Here ratio of axial depth of cut to diameter of the cutter is not same because of the limitation of the machine tool to go for higher depth of cut in conventional milling cutter. In comparison with cutting force, the conventional end milling cutting force is more than the micro end milling operations, higher cutting force induce vibration to machine tool. The lower cutting force

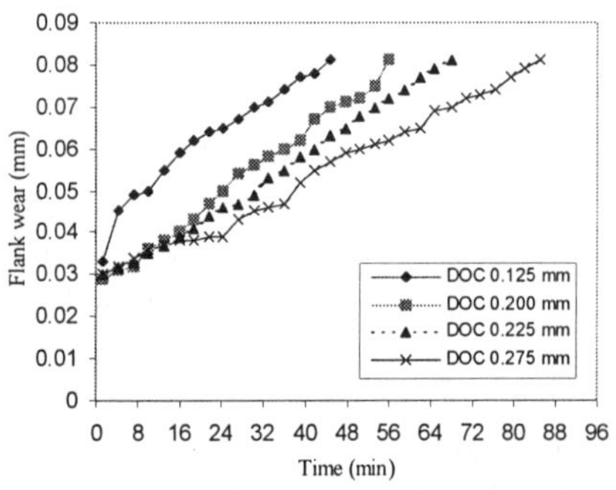

Figure 6. Effects of axial DOC under MQL on tool wear at feed 350 mm/min, and cutting speed 31.44 m/min

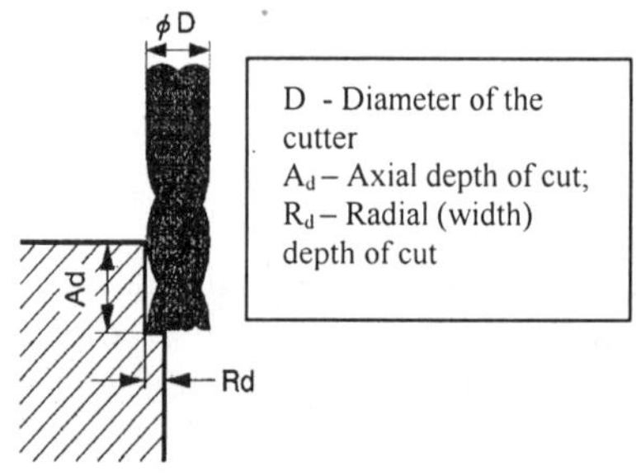

Figure 7. Schematic diagram of a milling process

makes the cutting more stable in case of micro end milling operations, which in turn increases the rigidity of the tool and this increase the tool life.

Table 1. Comparison of micro end milling with conventional end milling

Micro end milling	Conventional end milling
Diameter = 1 mm	Diameter = 16 mm
Helix angle = 30°	Helix angle = 30°
A_d = 0.275, R_d = 0.3D	Ad = 0.275, Rd = 0.3D
Total flute engagement = 134.5°	Total flute engagement = 70°
F_x = 4 N; F_y = 6 N; F_z = -3 N	F_x = 102 N; F_y = 250 N; F_z = -50 N

At a feedrate of 350 mm/min and cutting speed 31.4 m/min the tool life increased 1.5 times than that of lower depth of cuts. At axial depth of cut of 0.275 mm the observed tool life is 90 min.

4. CONCLUSIONS

Experiments are conducted to study the performance of MQL technique on micro milling of pure copper and following conclusions are drawn:

1. Within the range of experiments conducted it is clear that the tool life of micro grain carbide cutters increases with the increase of feedrate. Similar phenomenon is observed with the increase of depth of cut.
2. The size of cutting chips formed using MQL technique is found to be smaller compared to dry cutting. This phenomenon may be due to lower friction coefficient due to the presence of MQL.

5. REFERENCES

1. Klocke F., Elinesblatter G. Dry Cutting. Annals of the CIRP 1997; 46 n2:519-526.
2. Kustas F.M., Fehrehnbacher L.L., Komanduri R. Nanocoatings on cutting tool for dry machining. Annals of CIRP 1997; 46 n1: 39-42.
3. Machado A.R., Wallbank J. The effect of extremely low lubricant volumes in machining. Wear 1997; 210:76-82.
4. Rahman M, Senthil Kumar A, Prakash J.R.S. Micro milling of pure copper. Proceedings of the second international conference on advanced manufacturing technology (ICAMT –2000); 2000 August 16-17; Johor Bahru. Johor. Malaysia: 1:197-202.
5. Tansel I., Rodriguez O., Trujillo M.. Paz E., Li W. Micro-end-milling – I. Wear and breakage. International Journal of machine tool and manufacturing 1998; 38:1419-1436.
6. Trent E.M. Metal cutting. London: Butterworth-Heinmann Ltd, 1991.
7. Wakabayashi. T., Sato. H., Inasaki. I. Turning using extremely small amount of cutting fluids. JSME International Journal 1998; 7:143-148.

DEVELOPMENT OF MEMS IC PROBE CARD UTILIZING FRITTING CONTACT

Toshihiro Itoh, Kenichi Kataoka and Tadatomo Suga

Research Centar for Advanced Science and Technology (RCAST), The University of Tokyo

Abstract

In order to develop a micromachined probe card for IC testing, we investigated the relationship between contact forces and the fritting which is a contact make process with an ultra-small contact force, using force detection technique with an atomic force microscope (AFM) system. As a result, it has been clarified that a stable low-resistance contact to Al and Cu films can be obtained without applying contact forces larger than 10 μN, when the applied voltage is larger than 15 V. We also designed the fabrication process of electroplated-metal cantilever microprobes that are suitable for making contact with fritting.

Keywords

micromachined probe card, fritting contact, atomic force microscope

1. INTRODUCTION

MEMS probe cards made on Si substrates have some advantages over conventional needle ones. They can be applied to higher pad-density chips, and can be effective in high-frequency testing. Moreover, if a probe card consists of an array of actuator-integrated microprobes, it gains some further advantages, such as compensation of probe-pad distance deviation (Zhang et al. 1997), and direct switching (Itoh et al. 1999).

The critical problem of the micromachined probe cards is that each probe can not endure the force required to break the oxide on metal pad surface. According to Beiley et al. (1995), the required contact force is reduced by the

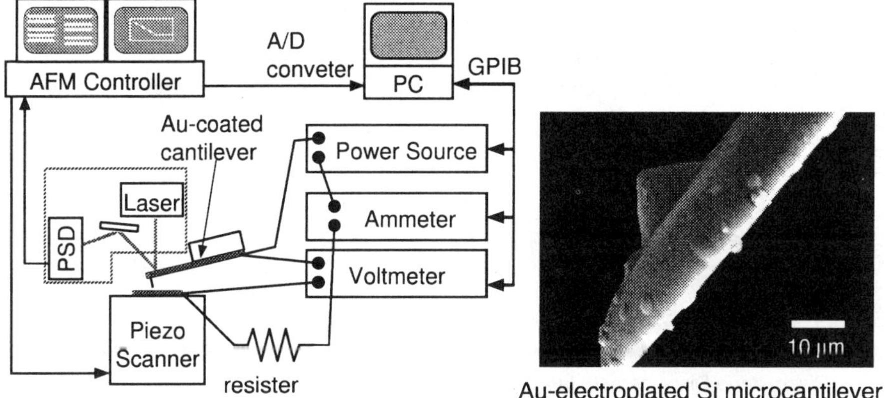

Figure 1. Experimental setup for contact resistance and force measurement using a Au-coated microcantilever, AFM photodetector and piezoelectric stage.

use of a kind of electrical breakdown, which is known as the fritting (Holm 1967). They showed that the fritting under the contact force of 0.7 mN guaranteed a low resistance of around 1.5 Ω between a tungsten probe and an Al pad. In the previous work, we have measured the characteristics of fritting contacts of electroplated Au and Ni bumps and showed that it is possible to get low resistance contact to Al pads without applying external force (Itoh et al. 2000).

In this paper, we clarify the characteristics of fritting contact between a micromachined Au cantilever and Al and Cu films using the AFM force detection techniques for more precise force measurement. Also, the preliminary probe fabrication for a MEMS probe card is reported.

2. AFM CHARACTERIZATION OF FRITTING CONTACT

Figure 1 illustrates the apparatus to measure the fritting characteristics between a microcantilever probe and pad metal films. The piezoelectric stage moves the sample to the probe, whose deflection is measured by the photo detector of a commercial AFM system . The probe was micromachined Si cantilever coated with 5-µm-thick electroplated Au film. Figure 2 shows how the contact resistance and probe deflection changes as the cantilever moves toward a sputtered Cu film. When a large voltage of 15 V was applied, fritting occurred at the same time as the cantilever touched the Cu film (Figure 2 (a)).

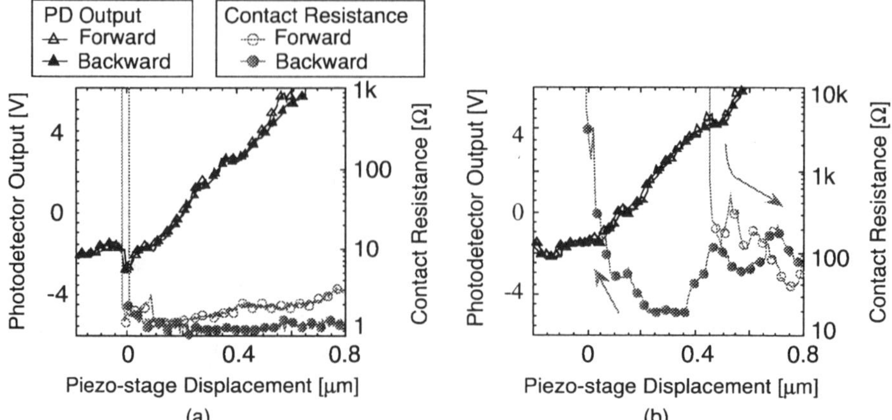

Figure 2. Cantilever deflection and contact resistance when driving a Cu film chip on the piezoelectric stage forward and backward to a Au-coated microcantilever probe. (a) 15 V and (b) 1.0 V were applied between the film and probe.

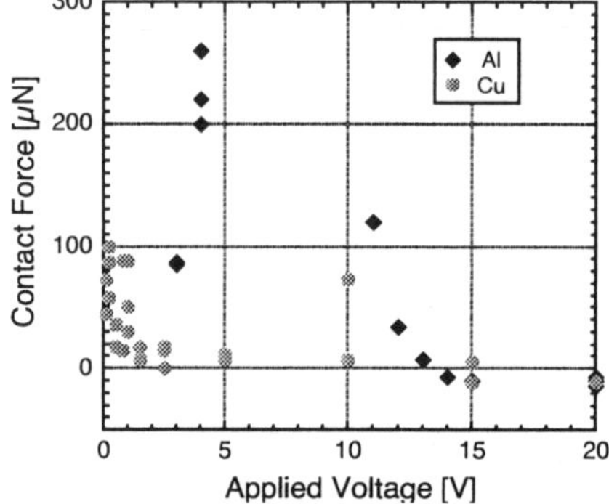

Figure 3. Relationship between the applied voltage and contact force necessary for the initiation of fritting on Al and Cu films.

Then the contact resistance fell down to a few ohms and it was stable. If the applied voltage was 1 V, fritting did not occur until 50 μN was applied (Figure 2 (b)). The contact resistance is larger than that of high-voltage contact and the value fluctuated as the contact force was changed by the piezo-stage movement. Figure 3 plots the contact force required to induce fritting at different voltage on Al and Cu films. When the applied voltage is larger than 15 V, fritting occurs at low contact force smaller than 10 μN. As the applied voltage decreases, the range of the force needed for fritting induction increases. The

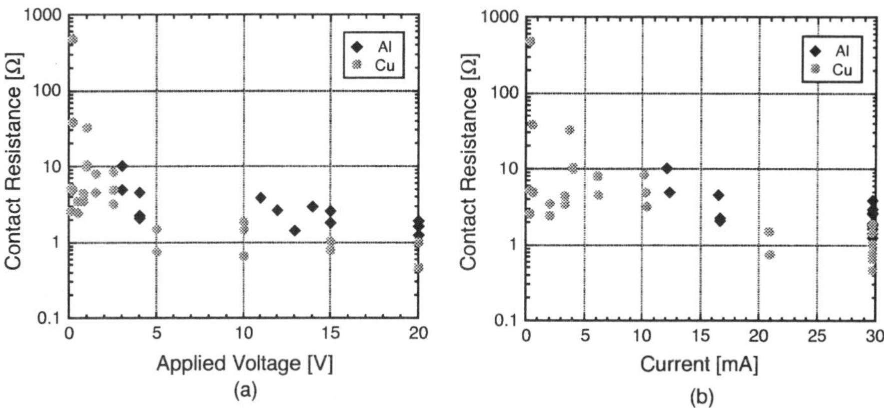

Figure 4. Dependence of the contact resistance on (a) applied voltage and (b) the fritting current.

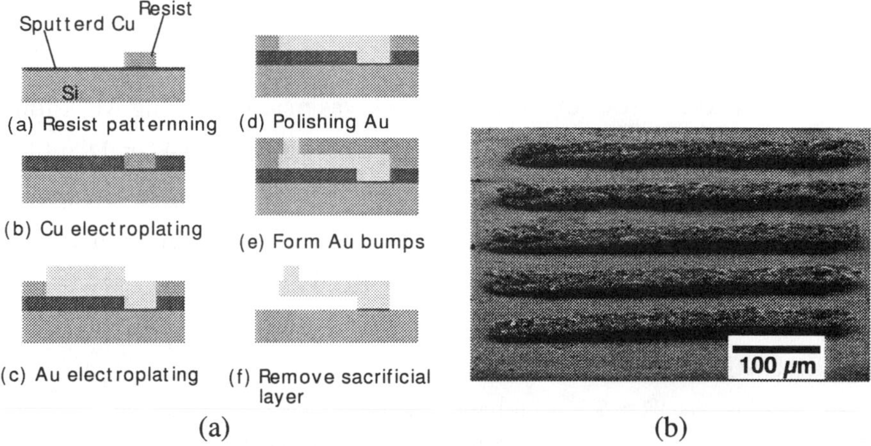

Figure 5. (a) Fabrication process for a Au microprobe and (b) SEM image of an array of fabricated probes.

fritting contact of Cu film is easily obtained with low contact force and low applied volatage in comparison with that of Al film. Furthermore, we have found that the contact resistance decreases with increasing current and applied voltage, as shown in Fig. 4. It can be considered that the applied voltage determines the maximum instantaneous current at initial fritting and the increase of current widens the diameter of current path formed by the initial fritting.

3. MICROPROBE FBRICATION

The fritting process using a Au-coated microprobe enables contact make to Al pad with a ultra-low contact force. Therefore, the microprobe constructed with only a electroplated Au film can be applied to the micromachined probe card, because the spring constant of the probe is not necessary to be large. Figure 5(a) shows the fabrication process for contact probe made on Si. The process includes electroplating of Cu as sacrifice layer, Au electroplating to make the cantilevers and bumps, and wet etching of sacrifice layer. The SEM image of the fabricated probes is shown in Fig. 5(b). Although it was shown that the proposed process is basically applicable to Au microprobe fabrication, the conditions of Au electroplating should be improved since the surface of probes are still very rough.

4. SUMMARY

The characteristics of fritting contact was measured using AFM force detection technique and it was clarified that stable contact to Al and Cu films were obtained without applying force larger than 10 μN, when the applied voltage is larger than 15 V. The basic fabrication process of micromachined probes made of electroplated gold was developed.

REFERENCES

Beiley M, Leung J and Wong S.S. :A micromachined Array Probe Card - Characterization, IEEE Trans. Components, Packaging, and Manufacturing Technology (CPMT) - Part B 1995; 18: 184-191.

Holm, Ragner, *Electrical Contacts - Theory and Application.* Berlin: Springer-Verlag, 1967

Itoh T, Suga T, Engelmann G, Wolf J, Ehrmann O and Reichl H. APPILICABILITY OF MEMS PROBE CARD TO WAFER-LEVEL TEST. Proceedings of the Pacific Rim/ASME International Intersociety Electronic & Photonic Packaging Conference; 1999 June 13-19; Maui. Hawaii: ASME: 123-130.

Itoh T, Suga T, Engelmann G, Wolf J, Ehrmann O and Reichl H. Characteristics of fritting contacts utilized for micromachined wafer probe cards, Rev. Sci. Instrum. 2000; 71: 2224-2226.

Zhang Y, Zhang Y, Worsham D, Morrow D and Marcus R.B. A NEW MEMS WAFER PROBE CARD. Proceedings of 10th International Workshop on Micro Electro Mechanical Systems; 1997 Jan. 26 - 30; Nagoya: IEEE: 395-399.

SIMULATION OF WEDM USING DISCHARGE LOCATION SEARCHING ALGORITHM

Fuzhu Han[*], Masanori Kunieda[*], Tomoko Sendai,[**] and Yoshihito Imai[**]

[*] *Tokyo University of Agriculture &Technology, Dept. of Mechanical Systems Engineering*

[**] *Mitsubishi Electric Corporation, FA Systems Department*

Abstract

This paper describes a WEDM simulation method. The simulation consists of the analysis of the wire vibration and search for discharge locations. It was found that using the probability method to determine discharge locations is more reasonable and realistic than using the deterministic method. The machined shape obtained from the simulation was consistent with those obtained from actual machining. It was also found that adaptive servo feed control easily results in the generation of streaks over the machined surface while constant feed rate control has no such tendency.

Key words

WEDM, simulation, wire vibration, discharge locations

1. INTRODUCTION

In WEDM, the explosive force generated at spark points and the electrostatic force applied between the wire electrode and workpiece results in wire vibration and flexure, which deteriorate the machining accuracy. In order to improve machining accuracy, it is thus necessary to understand the relationship between the spark points, explosive force, electrostatic force, wire vibration and flexure. In this research area, Yamada et. al.[1] measured the explosive force and acting time of impulse using the inverse-problem solving method and found that the explosive force in continuous pulse discharge differs from that in single pulse discharge. Kunieda et. al.[2] developed a discharge location searching algorithm which can simulate the distributions of the discharge locations, gap length, depth of workpiece removal and tool electrode wear, enabling the evaluation of EDMing accuracy. Obara et. al.[3] applied this algorithm to WEDM by taking into account the vibration of the wire electrode. With the algorithm developed by Kunieda et. al., the discharge location is deterministically determined by the gap distribution and the debris particle distribution. In actual machining however, the discharge location is probabilistically determined in the gap[4]. Hence, in order to reproduce the real phenomena in the gap, the

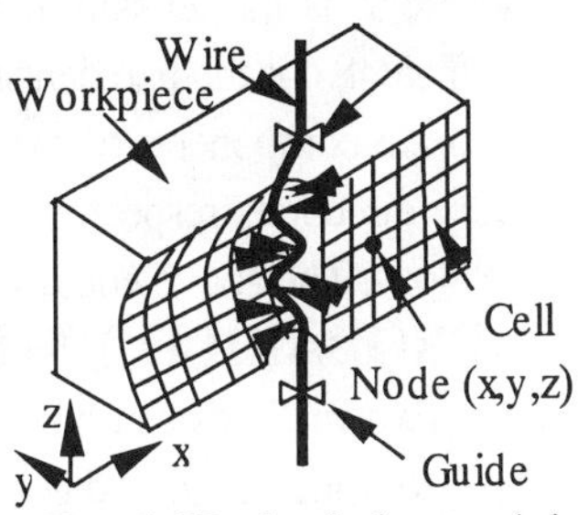

Fig. 1 3D simulation model

probability of discharge occurrence must be considered. In this study, WEDM simulation was conducted considering the probability of discharge occurrence, and the influence of the wire feed method on machining characteristics was investigated.

2. SIMULATION METHOD

The simulation is based on the actual discharge phenomena which is repeated for every pulse discharge. The simulation model is shown in *Fig. 1*. The simulation algorithm is composed of: 1) dividing the workpiece surface into cells, and calculating the gap distances of all cells, 2) distinguishing the gap state (open state, short state, discharge state) according to the gap distance, 3) if the gap distances of all cells are larger than the maximum distance within which dielectric breakdown can occur, the gap state is distinguished as an open state, the wire vibration due to the electrostatic force is analyzed, and the wire electrode is fed, 4) if there is at least one cell whose gap distance is shorter than zero, the gap state is distinguished as a short state, the wire vibration   is analyzed considering the collision of the wire with the workpiece, and the wire electrode is fed,   5) if the gap is neither in the open nor short state, it is distinguished as a discharge state in which the discharge location is first determined, the workpiece is then removed, after which the cells are rearranged at the discharge locations, the wire vibration due to the explosive force and electrostatic force is calculated, and lastly the wire electrode is fed. One pulse cycle consists of steps from 2) to 5), and the cycle is repeated until the preset machining time is over. Assuming the wire vibration to be a string vibration, the vibration equation of the wire in the x and y directions perpendicular to the

$$T\frac{\partial^2 y(z,t)}{\partial z^2} - c\frac{\partial y(z,t)}{\partial t} + f_y(z,t) = \rho\frac{\partial^2 y(z,t)}{\partial t^2} \qquad (1)$$

$$T\frac{\partial^2 x(z,t)}{\partial z^2} - c\frac{\partial x(z,t)}{\partial t} + f_x(z,t) = \rho\frac{\partial^2 x(z,t)}{\partial t^2} \qquad (2)$$

wire axis can be expressed as follows:

where T is the tension, c is the viscous damping coefficient, ρ is the line density of the wire, and $f_x(z,t)$, $f_y(z,t)$ are components of the explosive force in the x and y directions respectively. The explosive force is non-zero only at the discharge locations. The above equations are solved using the numerical calculation.

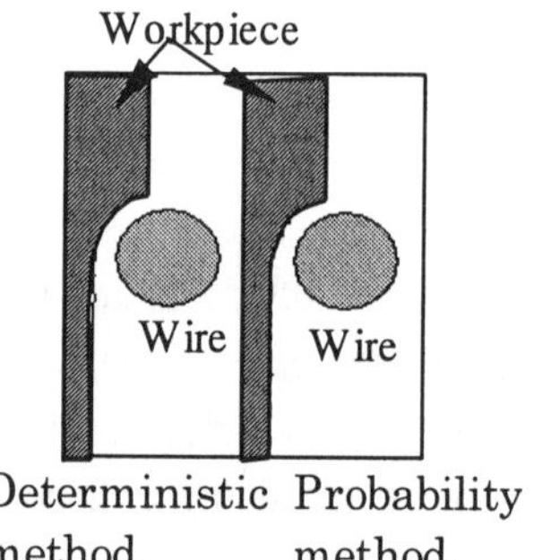

Fig. 2 Gap width distribution in x-y plane

3. PROBABILITY METHOD

Two methods can be considered for determining the discharge locations. One is the deterministic method in which the cell with the shortest gap width is selected as the discharge location. The other one is the probability method in which the discharge location is placed at the cell where the product of gap width and random number is smallest among all the cells. *Fig. 2* shows the simulation results of the gap width distribution in the x-y plane obtained without considering the wire vibration. The left part in *Fig.2* shows the results obtained from the deterministic method, and the right part shows that obtained from the probability method. It was found that the frontal gap is narrower than the lateral gap in the probability method while the gap width distribution along wire periphery is uniform in the deterministic method. The result of the gap width distribution obtained from the probability method is theoretically correct due to the formula proved by Kunieda et al.[5], and is consistent with the measurement result of the gap width distribution in actual machining. This clarifies that the method of determining discharge location using the probability method is essentially appropriate.

4. SIMULATION CONSIDERING WIRE VIBRATION

Fig.3 shows the simulated profile of the finished surface parallel to the wire axis when the wire vibration is considered. The simulation conditions are shown in *Table 1*. The zero offset value described in *Table 1* is the static distance between the wire side surface and workpiece in the direction perpendicular to the wire feeding direction before machining was preset at zero.

Table 1 Simulation conditions

Discharge duration t_e	$2\,\mu$ s
Discharge interval t_0	$20\,\mu$ s
Open voltage V_0	80V
Reference servo voltage V_g	20V
Length between two guides	60mm
Workpiece thickness	40mm
Offset value	$0\,\mu$ m

The left curve in *Fig.3* shows the simulated workpiece profile obtained with a wire feed speed of 1mm/min, and the right curve shows the profile obtained with a wire feed speed of 20mm/min, where position zero in the horizontal axis indicates the location of the preprocessed surface before finish cutting. It was found that convex shapes can be

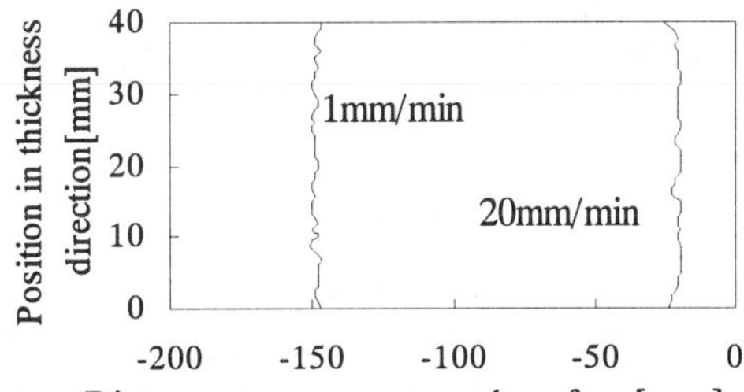

Fig. 3 Simulated profile of machined surface considering wire vibration

obtained at higher machining speeds, and concave shapes at lower machining speeds. These simulation results qualitatively agree with the results obtained from actual machining. The difference in shape results from the difference in the frequency of the impulse applied to the wire electrode due to the explosive force caused by the pulse discharge.

5. ADAPTIVE SERVO FEED CONTROL AND CONSTANT FEED RATE CONTROL

Generally there are two kinds of methods to feed the wire electrode in WEDM: the adaptive servo feed control and constant feed rate control. In order to keep the gap distance constant, the adaptive servo feed control is normally used, in which the wire feed rate is controlled so that the average working voltage in the gap can be kept constant. However, it is reported that several μm error is generated even with adaptive servo feed control when the depth of cut differs over the workpiece in the wire feed direction.[6] Hence, the simulation using these two kinds of wire feed methods was carried out for the finishing of a surface which has a 20 μm step over the preprocessed surface in the wire feed direction (see *Figs.4* and 5).

The simulation conditions are the same as those previously shown in *Table 1*. *Fig. 4* and *Fig. 5* show the simulation results with adaptive servo control and constant feed rate control, respectively. *Fig. 6* shows the change in the wire feed rate obtained from the simulation with adaptive servo feed control. With adaptive servo feed control, since the wire approaches the edge of the workpiece at the highest feed rate (*Fig. 6* "a"), a certain transient time (*Fig. 6* "b") is needed to achieve constant feed rate (*Fig. 6* "c"). This results in insufficient removal at the entrance, leaving a bulged shape (*Fig. 4* "a"). This

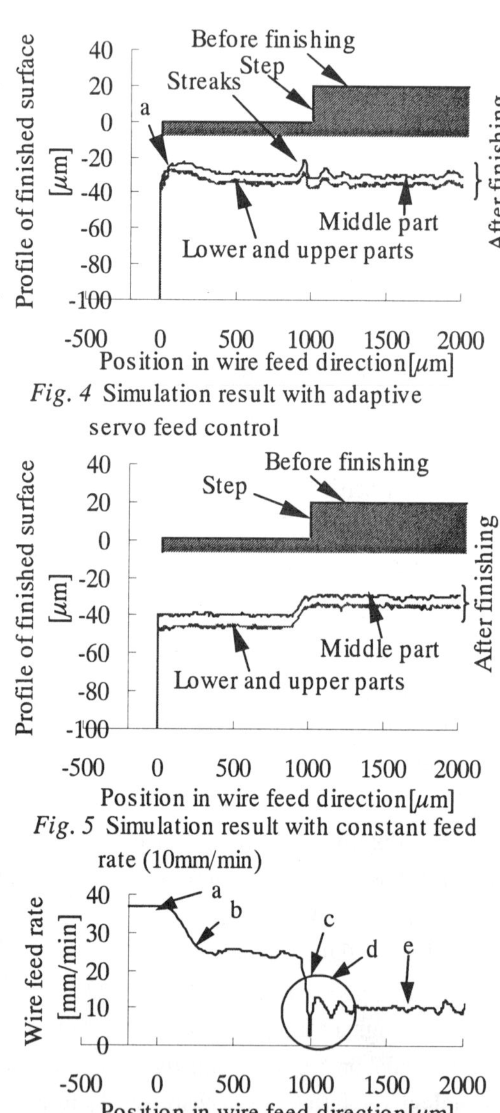

Fig. 4 Simulation result with adaptive servo feed control

Fig. 5 Simulation result with constant feed rate (10mm/min)

Fig. 6 Wire feed rate obtained from adaptive servo feed control

bulge can be reduced by shortening the response time of the adaptive control. When the wire electrode approaches the $20\,\mu$m step, the average working voltage becomes lower than the reference servo voltage due to the increase of the discharge area. To keep the average working voltage the same as the reference servo voltage, the wire feed rate is reduced by adaptive control. The rapid reduction, however, results in a transient vibration (*Fig. 6* "d") leading to generation of unfavorable streaks on the surface finished (*Fig. 4*). The streaks are formed parallel to the wire electrode, which results in a waviness of the finished surface in the wire feed direction. Here, it should be noted that the height of the step becomes reversed after finishing. This means that the height of the workpiece surface behind the step becomes lower by $2\,\mu$m than that in front of the step. These simulation results are relatively consistent with the experimental results obtained by Magara et. al.[6], indicating that the gap distance perpendicular to the wire feed direction is dependent on not only the average working voltage but also on the depth of cut over the workpiece. In contrast, in the case of constant feed rate control, although a significant error was generated at the step, no streaks were found at other places.

6. CONCLUSIONS

1) The gap width distribution obtained from the simulation using the probability method to determine discharge locations is consistent with the results measured in actual machining, which justifies the appropriateness of the simulation method proposed.

2) In the simulation considering the wire vibration, convex shapes were obtained at the higher machining speeds, and concave shapes were obtained at lower machining speeds. These results are qualitatively consistent with those obtained from the actual machining.

3) Adaptive servo feed control is likely to cause streaks over the machined surface, while constant feed rate control has no such tendency.

REFERENCES

1) Yamada H., Mohri N., Takezawa H., Furutani K., Magara T. Influence of Single Discharge Impulse on Vibration of Tool Electrode in Wire Electrical Discharge Machining, JSPE, 64, 2, 297 (1998)

2) Kunieda M., Kiyohara M. Simulation of Die-Sinking EDM by Discharge Location Searching Algorithm, IJEM, 3, 79 (1998)

3) Obara H., Ishizu T., Ohsumi T., Iwata Y. Simulation of Wire EDM, ISEM -12, 99 (1998)

4) Kunieda M., Nakashima T. Factors Determining Discharge Location in EDM, IJEM, 3, 53 (1998)

5) Kunieda M., Mori M. Simulation of Machining Accuracy in Die-Sinking EDM, JSEME, Vol. 28, No.59, 21 (1994)

6) Magara T., Yatomi T., Yamada H., Kobayashi K. Study on Machining Accuracy in WEDM, JSEME, Vol. 26, No. 52, 1(1992)

COMPARATIVE STUDY OF HSM AND EDM IN INJECTION MOULD MANUFACTURING

+M. R. Alam, +K. S. Lee, +M. Rahman, +Y. F. Zhang, *Y. D. Li and *K. S. Sankaran

+*Department of Mechanical Engineering, National University of Singapore*
Singapore 119260, Fax: (65) 874-5115
* *Makino Asia Pte Ltd, Singapore*

Abstract

As EDM is a rather slow process, mould makers are forced to look for alternative processes in order to stay competitive. With the emerging technology, high speed machining (HSM) is a much faster process compared to EDM. Recently, HSM is being considered as a replacement for EDM in the manufacturing of moulds. However, HSM may not be appropriate for machining all types of moulds. In this study, an algorithm has been proposed to select the most appropriate process whether HSM or EDM, or a combination of both for the manufacture of a particular mould. Several case studies have been conducted and the economics of different processes have been analysed to justify the suitability of the algorithm. From the case studies, it is concluded that, for the machining of a particular mould, HSM is preferred if found to be suitable. But the decision for the selection may depend on the set-up of the organization and constraints.

Keywords

High speed machining (HSM); EDM; mould manufacturing; economics.

1. INTRODUCTION

Process planning is an important task in a manufacturing organisation. Among the process planning activities, selection of appropriate processes to manufacture a part is the most important one. Timing is critical for any mould making company wishing to have the leading edge in today's world market. Many manufacturing companies are willing to pay high premium for a shorter delivery lead-time of a mould [1]. Machining of injection mould needs considerable amount of time and money. At the same time it needs high accuracy. The competition is fierce in this area too. Considering accuracy, surface finish and high rate of production for the competition, a lot of new technologies are emerging in recent years. Mould manufacturing companies are beginning to adopt advanced technologies and machinery to reduce the lead-time, cost and improve quality. Among the advanced technologies, high speed machining (HSM) is the most emerging one. It offers good surface finish, shortens the processing time and reduces cost [2]. Electrical Discharge

Machining (EDM) is an essential process in mould manufacturing. Some mould cavities cannot be manufactured without this process [3]. But EDM is a very slow process compared to HSM. As a result, delivery lead-time of mould cannot be reduced significantly due to the EDM process. So mould makers are forced to find alternative processes for EDM. However, HSM is the faster process compared to EDM, but it may not be appropriate for all types of moulds.

A few studies have been conducted to compare the economics between HSM and EDM. Aspinwall and Dewes [4] presented an analysis that compared the manufacture of a forging die by EDM with HSM. But the times and costs were based on the experience of the authors and not a real life example. Schumacher and Schulz [5] presented a comparison of EDM and HSM on the basis of typical applications, key features, process monitoring, equipment, work planning and staff knowledge of the process. But this comparison gives the general idea about two different processes. However, no algorithm has been proposed so far to select the most suitable method. In this study, an algorithm has been proposed to select the most appropriate processes for the manufacture of a particular mould and the economics of the processes have been analysed to justify the suitability of the algorithm.

2. PROCESS SELECTION ALGORITHM

Several processes such as milling, HSM, EDM and combination of HSM and EDM are frequently used in mould manufacturing. However, currently, there is no standard or algorithm for selection of the machining process and the decisions related to the selection are frequently made on the basis of the experience of the individuals. The proposed algorithm will give the rules of thumb to select the appropriate process for the manufacture of a particular mould.

2.1 Scope of the Work

In modern manufacturing approach, usually small and medium sized moulds with more ribs and complexities are manufactured in the mould shops. So the scope of this study is based on standard small to medium sized moulds having ribs and complexities. The ribs are usually 0.8-3 mm in width and their heights are 5-10 times of the widths. Such moulds are used for consumer electronics, printers, computer parts, telecommunication equipment, pagers, small cameras, signal lights for cars, electrical motors, health care items, etc.

2.2 Parameter selection and constraints

The main parameters for the selection of appropriate processes to manufacture moulds are tool length (effective length) to tool diameter ratio (L/D) and the hardness of the work piece material. Some other parameters and geometrical constraints of the features such as minimum radii of the corner and the bottom of the cavity, material removal rate, dimensions of the cavity, complexity of the mould, cost etc. are considered. Usually, if L/D is less than or equal to 3 and hardness is between 32 HRC and 55 HRC, generally HSM is a preferable choice and if L/D is greater than 3 and hardness is greater than 55 HRC, EDM is used. Figure 1 is drawn for low speed machining (LSM), HSM and EDM based on thumb rules applied for selection of appropriate machining processes.

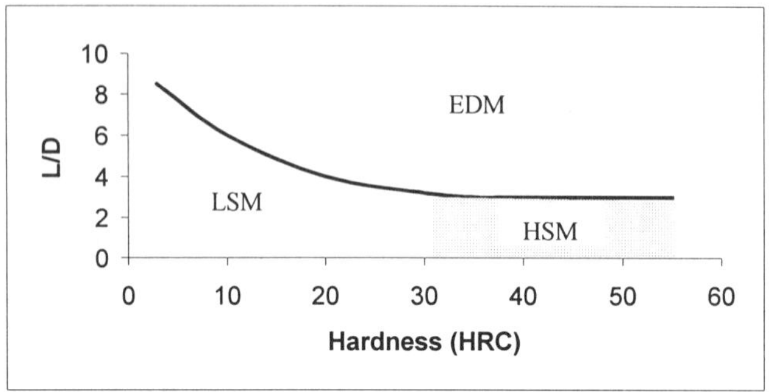

Figure 1: Appropriate zone for LSM, HSM and EDM

The use of low L/D ratio cutters is the best practice; because, high L/D ratio cutters reduce efficiency of machining. The diameter of the cutter must be less than the diameter of the blend for HSM. The effect of different L/D in the efficiency of machining is shown in Table 1 [6].

Table 1: Effect of different L/D in machining

L/D	Efficiency (%)
1	100
1.5	80
2	70
3	60
4	40
5	30

The high L/D ratio cutter increases run-out and a cutter with 0.01 mm run-out extends tool life up to 300% than one with 0.04 mm run-out [7]. On the other hand, run-out affects surface finish and contouring of the corners of the mould cavity.

3. ECONOMIC ANALYSIS FOR DIFFERENT PROCESSES

An economic module has been developed to compare the effectiveness of HSM, EDM and combination of HSM and EDM in mould manufacturing. The module calculates total processing time and cost to manufacture the moulds by using diffrent processes. Total processing time is calculated from the total set-up time, machining time, programming time and design time. All data are obtained by conducting different experiments. On the other hand, total processing cost is calculated by the addition of material cost, tool cost, machining cost, programming cost and design cost. Machining cost, design cost and programming cost are calculated from the multiplication of corresponding time and hourly rate of a particular company. In the following case study, economics for different processes are analysed.

4. CASE STUDY

Several case studies are tested with the algorithm. In this article, a cavity for an electronic component shown in Figure 2 is presented to verify the algorithm and compare the economics between HSM, EDM, and a combination of both. According to algorithm, HSM is appropriate for the machining of this cavity. Three different experiments are conducted to compare the economics and surface finishes. The first experiment is the machining of the cavity by HSM; the second experiment is the machining of cavity by EDM and third experiment is the machining using combination of both HSM and EDM.

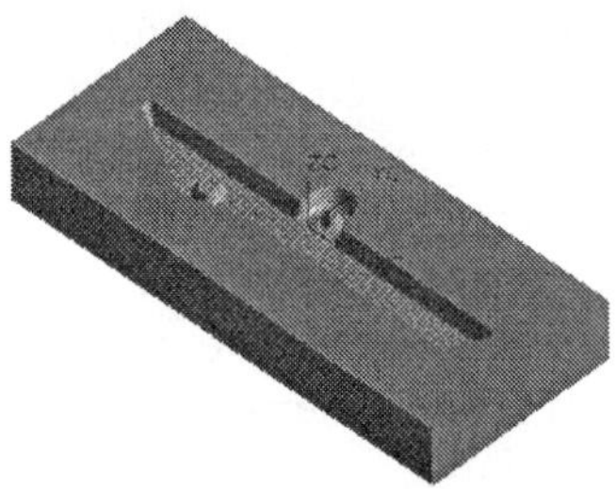

Figure 2: A cavity of an electronic component

The maximum length, width and height of the cavity shown in Fig. 2 are 117.77 mm, 28.30 mm and 5 mm respectively. There are two bosses having diameters 4.72 mm and 3.25 mm respectively in the cavity. Heights of the bosses are 3 mm. The blend radii in top and bottom of the boss are 1.55 mm and 0.75 mm respectively. The taper angle is 2 degrees. The blend radius all along the bottom edge is 2 mm.

The material used is ASAAB Stavx (AISI 420 MOD) and the hardness of the material is 51 HRC. Three ball nose end mill carbide cutters having diameter 6, 3 and 1.5 mm, coated with TiALN are used for the machining of cavity and copper electrodes used for the EDM experiments. The model of the electrode is shown in Figure 3 and comparative analysis of different experiments is shown in Table 2.

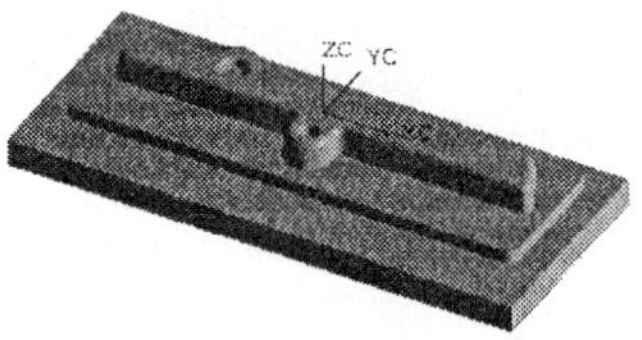

Figure 3: Electrode for EDM

Table 2: Comparative analysis of HSM, EDM and a combination of both

Process	Processing time (minutes)					Total cost (S$)						R_A (µm)
	Set-up	Machining	Program	Design	Total	Material	Tool	Machining	Program	Design	Total	
HSM	27	83.1	70		180.1	98	22	110.1	64.16		285.3	0.72
EDM	85	519.83	125	30	759.83	142	62.54	518.84	114.5	20	794.88	0.42
HSM + EDM	39.5	354.39	105	30	528.89	120	31.11	390.87	96.25	20	658.23	0.45

5. DISCUSSION AND CONCLUSION

In process selection algorithm (section 2.2), it was mentioned that there is already a division of effective zones for HSM and EDM based on L/D of the cutters and hardness of the material of the mould. But in a small area based on the estimated costs, both the HSM and EDM coincide. Using advanced CAM technology, advanced tooling and coating technology, machining of the mould of the coincident zone can be achieved by HSM. On the other hand, by using high-speed jump, graphite electrode with high-speed cutting, EDM can also be made appropriate for fabrication of the mould at that zone. In these circumstances, the decision of the selection of the processes depends on the nature of set-up of the organisations. If the organisations have more milling resources, they have to select HSM, otherwise EDM or a combination of both.

From the economic analysis of case study 1 (section 4), it is obvious that HSM requires minimum processing time and processing cost while EDM requires maximum processing time (321% more than HSM) and cost (200% more than HSM) to manufacture the cavity. It is concluded that, for the machining of a particular cavity, HSM is preferred if found to be suitable, because the cost and lead-time are the most important at the end (see Table 2). If HSM cannot be applied, a combination of HSM and EDM should be considered.

6. REFERENCES

1. Alam M. R., Lee K. S., Rahman M., Zhang Y. F., Automated process planning for the manufacture of sliders. Computers in Industry 2000; 43: 249-262.
2. Ikeda, T. et al., Ultra high speed milling of die steel with ball-nose end mill. Proceedings of 2nd International Conference on Die and Mould Technology (ICDMT), Singapore, 1992; 48-46.
3. Tanaka K. Precision die and mould manufacturing-a case study. Die and Mould Manufacturing International, April 1991; 47-50.
4. Schumacher B., Sculz, H., EDM competing HSC- tool manufacturing/hardened steel, CIRP-2000-WG E/C-2, January 2000; Paris, France.
5. Aspinwall, D. K., Dewes R. C., EPSRC case for support – ultra high speed machining of hardened ferrous alloys, school of manufacturing and mechanical engineering, University of Birmingham, 1994.
6. Technical guide- SECO Tools AB, Sweden.
7. Tooling systems- catalogue, MST Corporation, Japan 2000.

NET-SHAPING VIA ALUMINIUM FOAMING

Y. P. Kathuria
Laser X Co. Ltd. Chiryu-shi Aichi-ken 472 Japan
Email: ypkathuria@aimnet.ne.jp

Abstract

The present paper is focussed on the laser assisted aluminium foaming from a foamable precursor material and its possible applications in net shaping of 3D structures. Preliminary results suggest that a pore size gradient and a density gradient exist in the structure as the processing condition changes. The foam has large pores and lower density for slow processing speed, in contrast to the fast processing speed with small pore size but higher density. For pilot scale production of the foamed plates, sandwiches panel and localized foamed structure, specially intended for heat exchanger in the electronic industries, laser foaming technique may be the most appropriate.

Keywords

Laser foaming, net-shaping

1. INTRODUCTION

Recently aluminium foams[1-6] have evoked a special interest as a new alternative material due to their wide range of applications. It is a special class of porous material with cellular structure defined by randomly distributed air pores in metallic matrix. It posses high stiffness and low density. Additionally, the net-shaping of the aluminium foam has given a new dimension to the manufacturing technique. It results in a near net-shape semi-finished product for the light weight structures in the automotive and aerospace industries. For example, tailered blanks with foamable aluminium sandwich material offer the possibility of structural parts for car bodies. Therefore, its essential to introduce new technology, namely net-shaping via foaming and study its physical aspects with respect to production and quality.

2. FOAMING TECHNIQUES

To date several techniques[1] have been investigated for foaming and subsequently its various applications. We shall discuss below only the three techniques with emphasis on laser assisted foaming which can be used for the production of a near net-shape 3D porous structure.

2.1 Casting

The first method is the so-called, a conventional casting technique, in which the aluminium or aluminium alloy is melted in a graphite crucible. Ca and Mg can be added to increase the viscosity and decrease the surface tension of the melt. When the melts become suitable for foaming, the foaming agent titanium hydride is added to the melt. The melt is then rapidly stirred for dispersing the particles. The melt starts foaming and results in the formation of pores in the casting /mould. After the mould is cooled down, a semi-finished net-shape 3D-structure of the porous aluminium results.

2.2 Powder Metallurgy

In this method, aluminium alloy powder is mixed with a fraction of the foaming agent. The powder mixture is compressed to the extrusion billets by cold isostatic process (CiP), which is about 80 % in density. The CiP billets are then extruded to 100 % dense material. This results in a 3D-shaped precursor for foaming. To expand the precursor material, it is placed in an expansion mould. By heating up the mould and the precursor material, the foaming agent releases the gas and the precursor expands by forming the pores. After the mould is completely filled, it is cooled down thereby the pores gets stabilised. The results is a near net-shape semi finished products.

2.3 Laser Assisted Foaming

In this technique, which was introduced recently[5], foam is produced by mixing powdered material and a foaming agent and subsequently cold isostatic pressing the mixture to a foamable sandwich precursor material. The material is foamed by heating it up to its melting point by a high power laser beam irradiation. The unidirectional expansion of the foamable precursor material can be observed during the entire foaming process in the irradiation direction. The expansion in the other directions is relatively negligibly small. In the aluminium foaming process the laser radiation heats up the Al-alloy above its melting point of 660 ^{O}C and thereby the binding agent TiH_2 expands the material. It is believed that at high temperature, the foaming agent titanium hydride decomposes into titanium and hydrogen gas. Titanium released from the foaming agent remains in the aluminium and perhaps precipitates as $TiAl_3$, whereas the hydrogen gas tends to bubbles out, causing the melt to expand and form foam as a macroscale porosity. However, a fraction of the gas still remains trapped inside the solidified foam, due to its 20-times solubility difference in liquid and solid aluminium and thus accounts for the microscale porosity. The blowing or shield gas Ar is an additional help for the formation of the porosity may also become trapped inside the solidified foam.

The experimental details described elsewhere[5], were performed with a 5 kW cw CO_2 laser. The foamable but unfoamed Al-alloy sandwich sample was procured from Fraunhoffer Institute (IFAM) Bremen.[2] Its nominal chemical composition is AlSi7 with approximate 0.5 % by weight of TiH_2 as a foaming agent. A few of the test samples generated with this technique are shown in Figures 1 - 3 respectively.

3. RESULTS AND DISCUSSION

Figures 1a-1b, 2 and 3 show the side, pictorial and top view sections of the foam, which exhibits a closed cell structure. A strong variation in the cell size or structure can be seen depending upon the processing parameters and their shape varies from circular to irregular ellipsoidal or deformed circular structure. On carefully examination of the Figure 1, one find that initially the width of the unfoamed aluminium sandwich is 3.72 mm only. At higher processing speed, the width of foamed structure already becomes 5.03 mm (Figure 1a) and the average cell size is relatively fine ($\approx$ 0.2 mm), but still the base has not fully expanded. On the otherhand at lower processing speed the width of foamed structure is scaled up to 8.13 mm (Figure 1b) having a large size porosity and with a average cell size of about 3 mm. Besides that the cell shape is also partially deformed from its primitive shape due to the gravitational effects on foam, which is playing against the pressure of the blowing agent that is exerting an internal force on the cells that is equal in all directions.[4]

For further characterization, the density of the each aluminium foam block was calculated from the measured weight and the volume. The volume of each foamed specimen was calculated from its dimensions. The density of these Al-foamed blocks were measured to be $\rho^* =$ 1.84, 1.04, 0.93 and 0.88 gm/cm^3. Taking into account the density of the pure aluminium ($\rho_s =$ 2.6989 gm/cm^3), various other parameters for example relative density (ρ^*/ρ_s), porosity and densification strain etc can be calculated.[5, 6] To make an estimate of the cell wall yield strength, microhardness measurements were made on the cell edges that is at the cross-sectional area between the cells (Figure 3). The microhardness averaged to be 81 HV. A rough estimate of the cell wall yield strength was made from this microhardness measurement. This amounts to be $\sigma_{ys} \cong$ HV/3, that corresponds to 265 MPa which is slightly higher than (250 MPa) already reported.[3] This is possibly due to higher Si content.

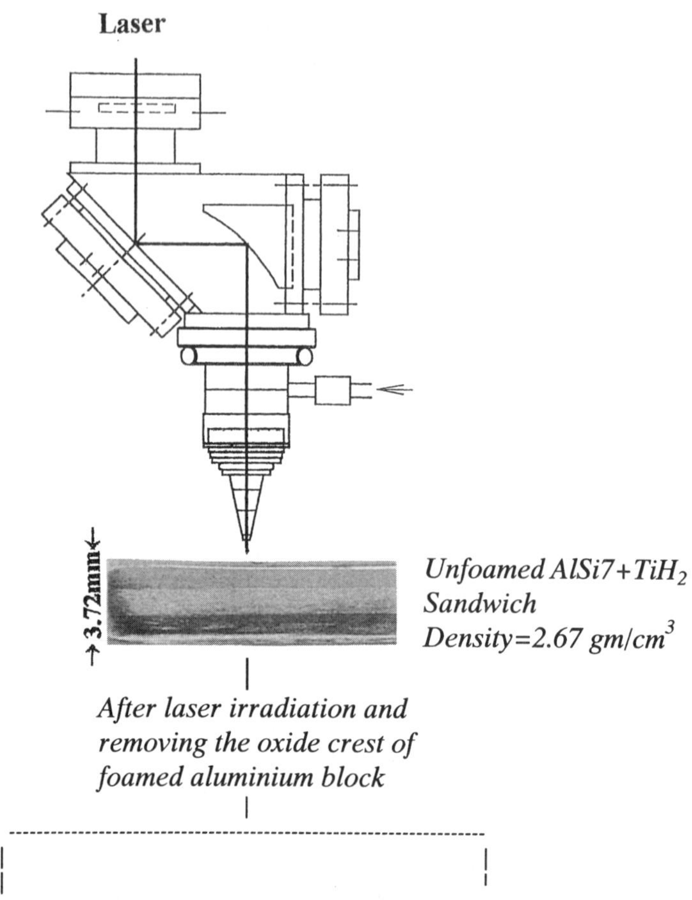

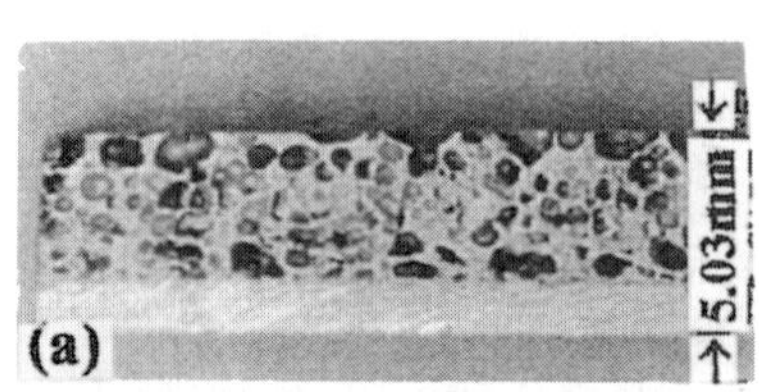

Density = 1.84 gm/cm³
Porosity = 32 %

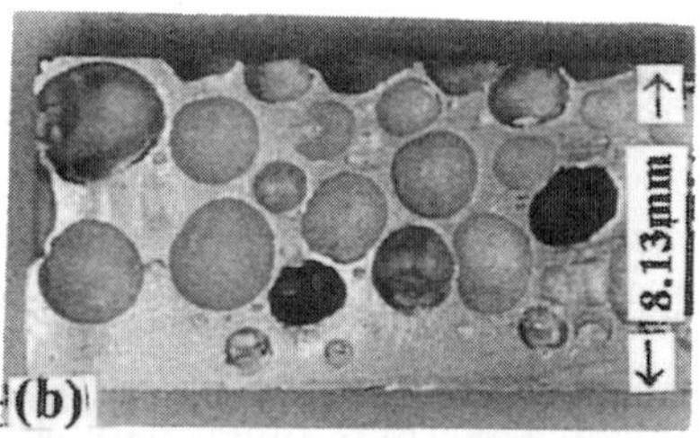

Density = 0.88 gm/cm³
Porosity = 67 %

Figure 1: Laser assisted AlSi7 foaming with experimental results. Laser: cw CO₂ ;P=5kW; Ar=30l/min.;Processing speed(V)=(a):2m/min.;(b):0.4m/min.

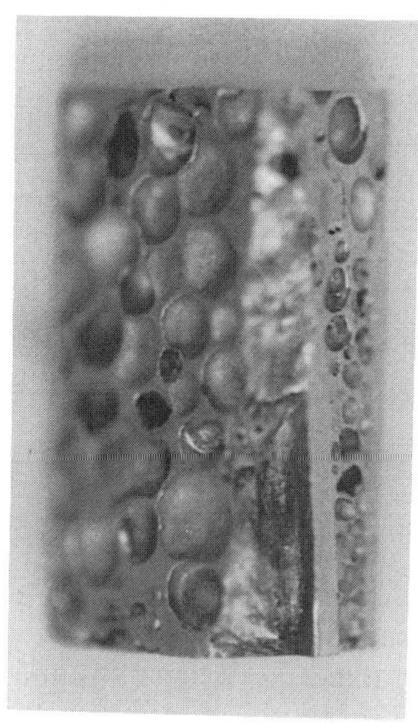

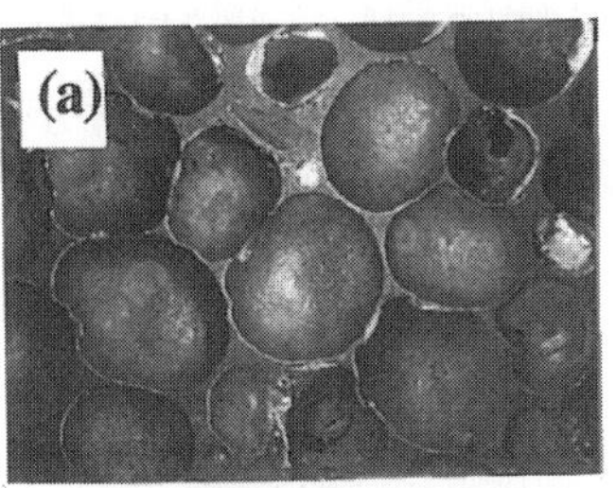

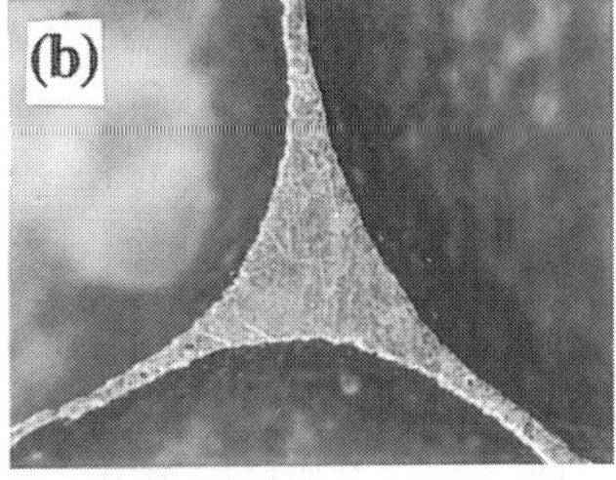

Figure 2: 3D rectangular near net-shape Al- foam structure.

Figure 3: Magnified view of the micro structure showing: (a) pores and (b) section between cell walls of Al-foam.

4. CONCLUSION

The paper demonstrate the feasibilty of laser assisted aluminium alloy foaming and net-shaping. The results yield a localized and unidirectional expansion of the foam in the direction of laser irradiation. Porous 3D rectangular structures with a relative density of 0.33 were created and a pore size gradient does exist with the slow and fast processing speed. There is a potential for substantial improvement in the process for net-shaping.

5. REFERENCES

1. G. J. Davies and Shu Zhen, Metallic foams: their production, properties and applications. J. Mater. Sci. 18 (1983) 1899-1911.
2. J. Banhart, J. Baumeister, Deformation characteristics of metal foam. J. Mater. Sci. 33 (1998) 1431-1440.
3. A. E. Simone and L. J. Gibson, Aluminium foams produced by liquid state processes. Acta Mater. 46 (1998) 3109-3123.
4. Samuel P. McManus, Production of polyurethane foams in space: gravitational and pressure effects on foam formation. Polymer Preprints 41 (2000) 1056-1057.
5. Y. P. Kathuria, Aluminium foaming using CO_2 - laser. J. Mater. Science Technol. Vol.17 No. 6 June (2001).
6. L. J. Gibson and M. F. Ashby (Ed.), *Cellular Solids* 2nd Edn., UK, 1997.

Development of Warm Forming Technique to Produce Thin Wall Magnesium Components

M.S.Yong, B.H.Hu, M.Enggalhardjo and C.M. Choy

Gintic Institute of Manufacturing Technology
71 Nanyang Drive
Singapore 638075
Tel: (65) 7938583   Fax: (65) 7925362
E-mail: msyong@gintic.gov.sg

Abstracts

This paper highlights the advantages of warm forming technique to produce thin-walled magnesium parts. The process parameters considered in the feasibility study were forming temperature, in the range of 28°C to 300°C and magnesium sheet (AZ31B-H24) thickness, in the range of 0.4 mm to 1 mm. Magnesium hand phone covers of a thickness down to 0.4 mm have been successfully produced. It was found that the optimum forming temperature is between 225°C and 275°C. Metallograhic examination shows a sound microstructure of the magnesium sheets formed at 250°C.

Keywords

Thin-Walled Magnesium, Sheet Forming, Warm Forming

1.   INTRODUCTION

Processing magnesium components has always been a challenge and that is due mainly to the chemical reactivity of magnesium, as magnesium in its molten form has a high affinity for oxygen. This causes particular problems in processing magnesium parts in the molten stage and extra precautions need to be taken. Typical process like hot chamber die casting has been the major process for manufacturing of thin wall magnesium components. The prerequisite of protective environment in processing the molten magnesium is always time consuming and tedious. Furthermore, the requirement of runner and gating system is another major drawback in processing magnesium parts by casting process. The inefficient use of raw material and issue on recycling of magnesium scrap will also incur an additional cost to the process. In addition, the non-metallic inclusions (NMIs) in magnesium arising from dross, sludge, slag and scale contaminants during melting, will lead to poor quality parts. Therefore, there is a need for development of a new or alternative technology to produce thin wall magnesium parts.

This paper explores the potential of using warm forming technique to produce thin wall magnesium components. The potential advantages of producing magnesium parts by warm forming process are: (i) the ability to produce thin wall magnesium components down to 0.4mm thickness; (ii) better surface finish (absense of blisters and flow marks); (iii) lower processing temperature compared to casting; (iv) lower energy consumption; (v) safer operation (due to the non-flammability of magnesium at low temperature); (vi) better quality parts (absents of non-metallic inclusions from melting); (vii) cheaper machinery and tooling cost for forming and stamping process; (viii) better utilisation of material (absents of runner and gating systems).

Despite the significant advantages of using warm forming on magnesium parts, up-to-date there has only been a handful of work done on the process [1][2]. Therefore, further studies on the process parameters, metallurgical analysis and deformation of material are required to establish the process for future production. In this study, the hand phone casing was used for the warm forming of thin wall magnesium parts to exploit the full potential of the process.

2.   EXPERIMENTAL PROCEDURE

The magnesium formed parts were produced using a 25 tonne mechanical press. The magnesium sheet used in this investigation was the commercial magnesium alloy AZ31B-H24. The processing parameters considered for warm forming of magnesium sheets: metal temperatures of 28±2°C, 150±5°C, 200±5°C, 225±5°C, 250±5°C, 275±5°C and 300±5°C; sheet thickness of 0.4 mm, 0.5 mm, 0.6 mm, 0.8 mm and 1 mm. Metallurgical analyses were conducted on the formed sample to evaluate the structure integrity. In the experiment, magnesium sheets were formed without the use of lubricant.

3.   EXPERIMENTAL RESULTS AND DISCUSSION

3.1   Effect of Temperature on the Formability of Mg Sheet

The effects of forming temperature on magnesium sheet are shown in Fig. 1. The pictures indicate that a minimum temperature of 225°C is required to produce crack free parts. This is expected as magnesium being a hexagonal close-packed (HCP) structure does not permit significant deformation at low temperature. On the other hand, the deformation of magnesium above 225°C will cause additional slip planes $\{10\bar{1}1\}$ to become operative [2][3]. This marked increase in plasticity of magnesium, due to an increased number of slip planes, allows magnesium parts to be formed without crack. Data by Winkler (Fig.2) show gradual increase of drawability from room temperature to 200°C for magnesium alloys AZ31, AM503 and AZ61 sheet [4]. The figure also indicated that increasing the temperature up to 300°C, in particular to AZ31 does not further improve the drawability. This argument may again be supported by the finding appeared in Fig. 1. The figure shows the forming of magnesium hand phone casing at a higher temperature do not further improve the form quality of the parts. Nevertheless, a minimum temperature is still required to produce crack free parts. Hence from the economical and feasible point of view, it can be concluded that the feasible range to form magnesium AZ31 lies between 225°C and 275°C.

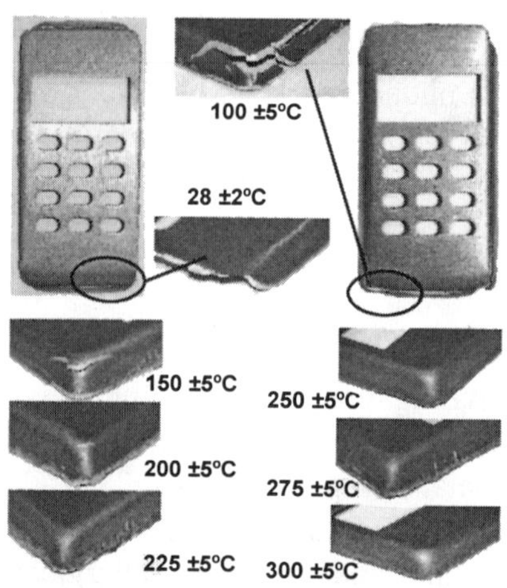

Fig. 1: Picture of the magnesium component of 0.8 mm thickness formed in various temperatures

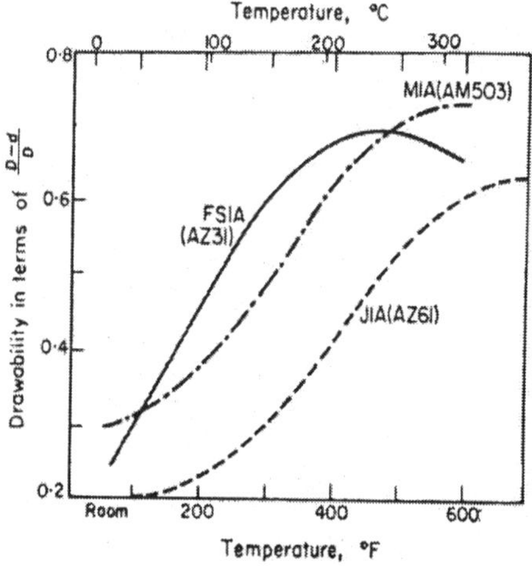

Fig. 2: Effect of temperature on drawability of magnesium alloy sheet [4].

3.2 Forming of Various Thickness Mg Hand Phone Casings

A series of experiments was conducted to evaluate the thinnest magnesium sheet which is formable using the forming process. Material thickness down to 0.4 mm has been successfully produced as shown in Fig. 3. The figure also shows that thinner materials (0.4 mm and 0.5 mm) are distorted after forming. The distortions may be attributed to excessive die clearance, as the die is designed to produce 0.8 mm thick parts. Therefore die clearance does play an importance role in ensuring the production of quality parts.

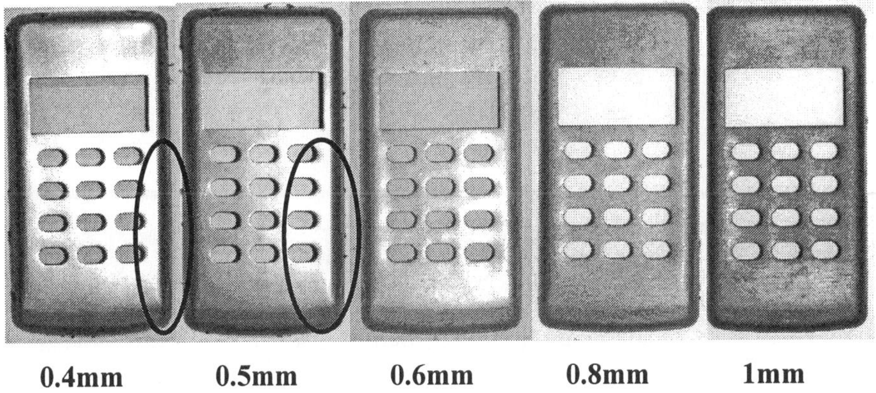

Fig. 3: Magnesium components with various thickness formed at temperature of 250°C. Note that the 0.4 mm and 0.5 mm are distorted.

3.3   Microstructural Evaluation on the Formed Casings

Metallographic examinations were conducted to evaluate the structural change in the magnesium formed parts. The microstructure taken along the bending section of a 0.8 mm thick magnesium hand phone cover produced at the forming temperature of 250°C, are shown in Fig. 4. Micrograph taken at location 4 of the structure shows significant increase in grain size at the outer surface. The increase in grain size may attribute to grain growth from the recrystallisation during forming and/or post forming of magnesium alloy AZ31 at elevated temperature. Despite the increase in grain size, a sound structure without fracture was observed throughout the bend.

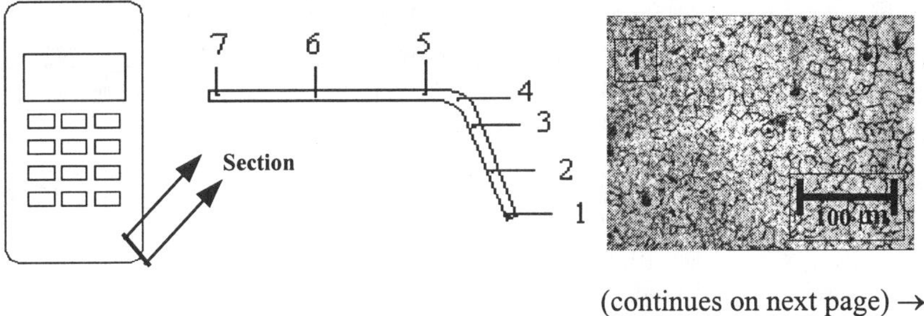

(continues on next page) →

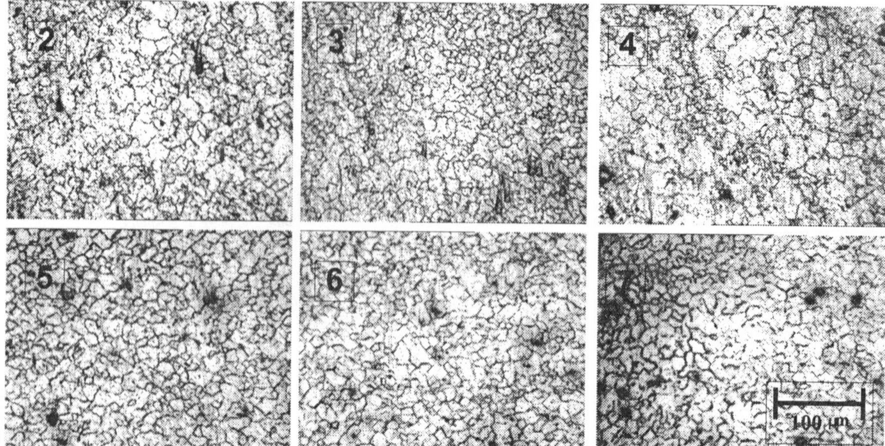

Fig. 4: Microstructures taken along the bending section of the magnesium hand phone cover

4.   CONCLUSIONS

1. Formability of magnesium greatly depends on the working temperature. It was determined that the feasible range to form magnesium AZ31B-H24 is between 225°C and 275°C.

2. Magnesium hand phone covers of thickness down to 0.4 mm have been successfully produced. It was also determined the excess die clearance will cause thinner magnesium parts to distort after form.

3. The microstructure examination of the magnesium parts formed at 250°C, shows sound structure.

5.   REFERENCES

[1]   Kumihiko, H., "Press Working of Magnesium Alloy Sheet - An Example of Electronics Part Panel", 10th Technological Salon References, The Japan Society for Technology of Plasticity, Sagamihara City, 2000, 26 July, pg 8 –11.

[2]   Emley, E.F., *Principle of Magnesium Technology*, Pergamon Press, 1968

[3]   Raynor, G.V., *The Physical Metallury of Magnesium and Its Alloys*, Pergamon Press, 1959.

[4]   Winkler, J.V., Trans. ASM, 1945, vol. 35, pg 225

GAS FLOW CHARACTERISTICS IN A LASER CUT KERF

Hideyuki Horisawa

Department of Precision Mechanics, Tokai University

Abstract

Characterization of a supersonic impinging jet in a laser cut kerf was conducted with a Schlieren method and a Computational Fluid Dynamics (CFD) analysis to investigate the gas-dynamic effects of the jet in a cutting process. A flow separation and an associated recirculation zone occurring at the bottom of a cutting front were found. Imbalanced flow separations and recirculation zones at both side walls, and flickers of the flow inside the kerf were also observed. From the cutting experiments, it was found that better cutting performances were achievable with deeper penetration of the flow field inside the kerf.

Keywords

Laser cutting, Supersonic jet, Impinging jet, Schlieren method, CFD analysis

1. INTRODUCTION

Laser cutting is a thermal process that results in higher quality and accuracy than other thermal processes. In the cutting process, a focused laser beam melts or evaporates the material, and a gas jet, coaxial to the beam, blows the molten and/or vaporized material out of a cut slit, or

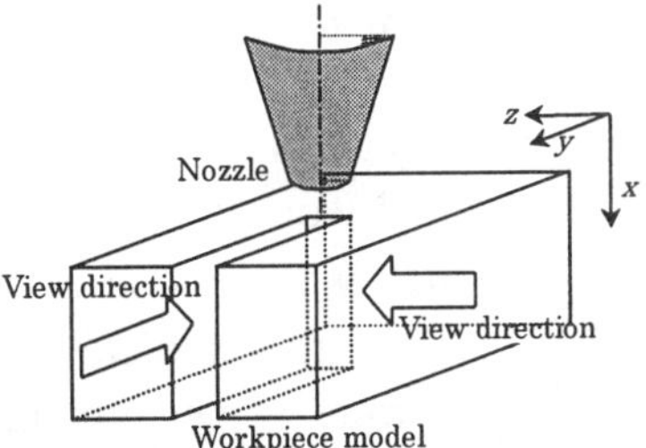

*Figure*1. Workpiece model for flow diagnostics and numerical analysis.

a cut kerf. Thus, not only the optical performance of a laser beam itself, but the characteristics of the gas jet, such as feed pressure, nozzle geometry, nozzle-workpiece separation, etc., also control the cutting performance. [1)-7)] Therefore, it is very important to characterize the flow field of the gas jet in the kerf for the evaluation of its contribution in cutting. In this study, in order to elucidate these effects, the flow visualization by a Schlieren method was conducted. In general cutting conditions, a supersonic underexpanded gas jet impinges on the workpiece surface and penetrates into a narrow cut kerf with its width of less than 1 mm forming complex flow structures. It is therefore very hard to diagnose the details of the flow in the kerf by only using probes or optical methods. In this study, as the method to overcome these difficulties, a Computational Fluid Dynamics (CFD) analysis was adopted to these complicated issues. Moreover, a correlation between the gas jet characteristics and cutting performances was also discussed.

2. FLOW DIAGNOSTICS AND CFD ANALYSES OF KERF INTERNAL FLOW

A schematic drawing of a workpiece model and a coordinate system used for flow diagnostics and numerical analyses of kerf internal flow are shown in Fig.1. The real cutting kerf geometry was approximated by in view direction transparent cutting kerf model made from polished clear quartz plates to investigate the flow field inside the kerf by a Schlieren method. Here, the simulated workpiece thickness tw = 10 mm and kerf width Wk = 0.6 mm. Nozzle diameter of dn = 2.0 mm was used, and

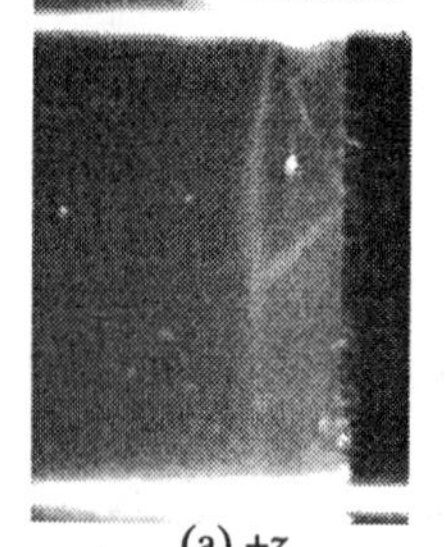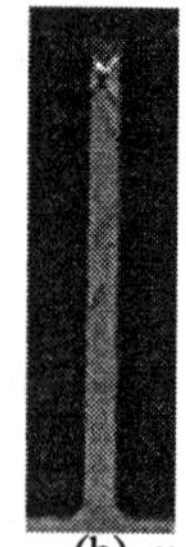

(a) +z   (b) -y

Figure 2. Schlieren photos of an impinging jet on kerf model ((a)Viewed from +z, (b) -y, for lnw: 0.5 mm, dn:2 mm, Ps: 0.5 MPa).

distance between the nozzle tip and workpiece surface was set lnw = 0.5 mm. Nitrogen gas was used for a coaxial gas jet, and its feed pressure, or stagnation pressure Ps, was changed up to ~ 1.3 MPa. An x-y-z coordinate system was defined as shown in Fig.1, taking its origin at a center edge of a cutting front on workpiece surface, + x-direction to a gas jet axis from a nozzle tip, y-direction along a kerf axis, and z-direction to kerf width. The Computational Fluid Dynamics (CFD) analyses were also conducted for a 3-Dimentional workpiece model of an $x \times y \times z$ = 10.6 mm$\times$8.0 mm$\times$7.0 mm space which was divided into meshes of $x \times y \times z$ = 108 grids $\times$ 110 grids $\times$ 72 grids in a typical case, and unsteady compressive Navier-Stokes equations were solved numerically using a finite volume method. For the analysis, a CFD code, CFD 2000 was utilized.

3. RESULTS AND DISCUSSION
3. 1. Flow Diagnostics of Kerf Internal Flow

Schlieren images of kerf internal flow are given in Fig.2. It can be seen in Fig.2(a) that width of the gas jet abruptly expands toward the kerf direction after passing through the kerf inlet. A flow (boundary layer) separation due to the expansion toward the + y-direction at a bottom part of a cutting front, an expansion fan associated with the expansion at the jet boundary and an oblique shock occurring from an edge of a kerf inlet at the cutting front can be seen. Images viewed from the − y-direction are given in Fig.2(b). A normal shock wave associated with the impingement of a supersonic jet on a flat plate is seen above the workpiece surface. Oblique shocks occurring from both edges of the kerf inlet, and flow separations from both kerf side walls toward a jet center, occurring unequally and flickering between the two side walls, were observed.

3. 2. CFD Analysis of Kerf Internal Flow

As shown in Fig.2, a flow field inside a kerf is extremely complicated, since the kerf is placed downstream of an underexpanded supersonic impinging jet, and width of a laser cut kerf is generally very narrow compared to its height. It is therefore very hard to diagnose the details of the flow in the kerf by only using optical methods or probes. In this study, a CFD analysis was adopted to overcome these complicated issues. Examples of the analytical results are shown in Figs.3 and 4, showing the kerf internal flow patterns, where dark colors for higher values and light colors for lower values. It is shown that a jet is abruptly compressed forming a normal shock wave above the workpiece surface, and penetrates inside a kerf, in which the flow is expanded and accelerated to both +x- and +y-directions. After the expansion, the flow is compressed again and is also decelerated. Moreover, after the second compression, the flow is expanded and is accelerated again to + x- and + y-directions. A flow separation occurs at the bottom part of a cutting front in Figs.3(a) and (c),

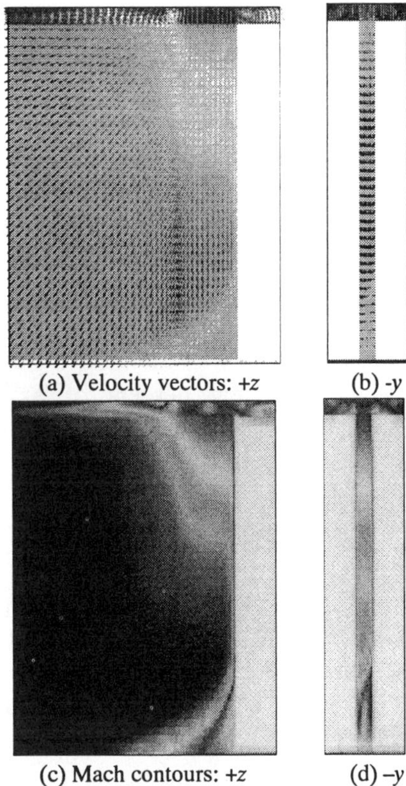

(a) Velocity vectors: +z (b) -y

(c) Mach contours: +z (d) –y

Figure 3. CFD results of an impinging jet on kerf model ((a)Velocity vector distribution viewed from +z, (b) Velocity vector distribution viewed from +-y at y = + 40 µm, (c)Mach contour distribution viewed from +z, (d) Mach contour distribution viewed from +-y at y = + 40 µm, for $l$nw: 0.5 mm, dn:2 mm, Ps: 0.5 MPa).

where the flow is expanded to + x- and + y-directions as well, and an associated recirculation zone is formed. In Figs.3 (b) and (d), imbalanced flow separations from both side walls downstream of the expansion zone, and flickers of the flow inside the kerf were also observed. Moreover, existence of recirculation zones at bottom parts of the kerf side walls associated with the flow separations were observed. A 3-D image of the flow field inside the kerf showing the 3-D structure of the flow separation and recirculation is shown in Fig.4. In addition, alternating formation/extinction of stagnation bubbles, or vortices,[8),9)] were also observed downstream of a normal shock on the workpiece surface, and these probably affect the normal shock vibration and also the flickers of the imbalanced flow inside the kerf.

In laser cutting using a nitrogen gas jet, in which oxidization of side walls of a kerf is not preferable, oxygen gas from atmosphere may flow into the kerf from the bottom of the workpiece through those recurculation zones, which may results in an undesirable oxidization of the cut surfaces. On the other hand, in laser cutting with an oxygen gas jet, in which a few percent of oxygen impurity at a cutting point results in the worse cutting performances, [10), 11)] nitrogen gas of atmosphere may inflow to the kerf from the bottom through those recirculation zones, resulting in the reduction of oxygen purity and hence in the worse cutting performances.

Since a shear force acting on a cutting front which drives the flow of molten layer is proportional to $\partial u_x / \partial y \sim \Delta u_x / \Delta y$, a relative shear force in an arbitrary unit is defined as $\Delta u_x / \Delta y$ and can be estimated using grid data from the CFD results. Here Δu_x is defined as the difference of x-component velocity between first- and second-grids from a cutting front surface in y-direction in a boundary layer, and Δy is the gap between those grids along a center line of the cutting front. The plots of the relative shear force $\Delta u_x / \Delta y$ along the grid points in x-direction, or depth, are

(a) 3-D image of a flow separation and a recirculation zone.

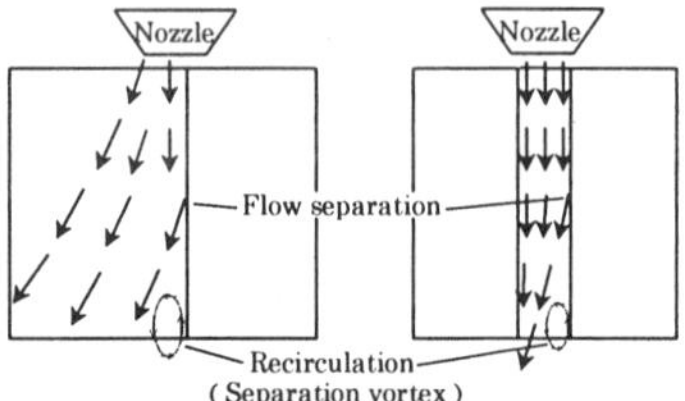

(b) Viewed from $+z$ (c) Viewed from $-y$

Figure 4. 3-D images of kerf internal flow pattern.

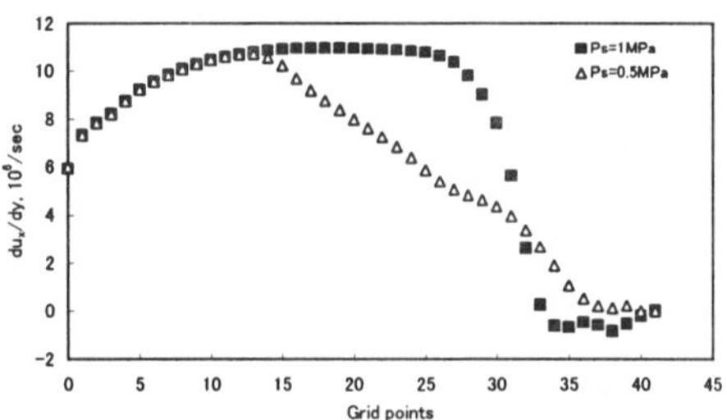

Figure 5. Estimated shear force in arbitrary unit acting on a center line of a cutting front for $Ps=1$ MPa and $Ps=0.5$ MPa.

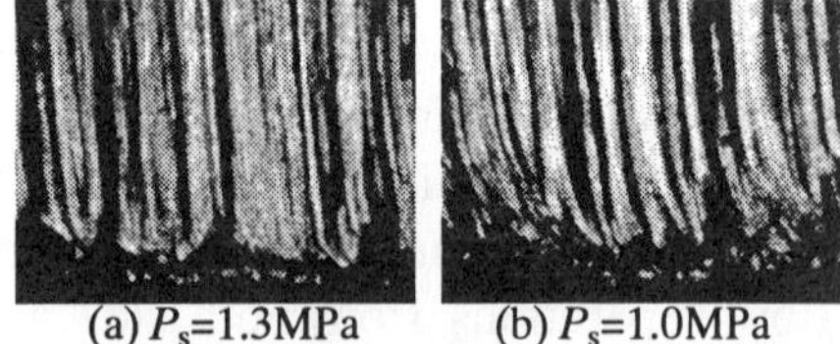

(a) $P_s=1.3$MPa   (b) $P_s=1.0$MPa

Figure 6. Photos of laser-cut surface (bottom edge) (Laser power: $P_L = 6$ kW (CW), SUS316L(t:10 mm), $Vc = 1.5$ m/sec, Gas jet: Nitrogen).

shown in Fig.5 for $Ps = 0.5$ MPa and $Ps = 1.0$ MPa. It can be seen that the shear force acting on a cutting front increases with x at a kerf inlet, and reaches maximum inside a kerf before an abrupt drop near the exit. Although little difference in the maximum values between cases for $Ps = 0.5$ MPa and

$Ps = 1.0$ MPa, a significant difference in the distribution and values was observed near the kerf exit, i.e., lower values for a lower Ps case. This must affect on the difference of cutting performance and also on the striations, or surface roughness, in cases with different gas jet pressure.

3. 3. Correlation between Gas Jet and Cutting Performances

Photos of laser-cut surfaces of lower edge, showing the effects of the increase in feed pressure of a gas jet Ps on cutting performance, are shown in Fig.6. As shown in these figures, striations occurring on a bottom edge of a surface cut with lower pressure are significantly curved and its surface roughness is larger compared to the higher pressure case. Although not given in Fig.6, little difference in striations of upper parts of the section between different pressure cases was observed. Corresponding to the relative shear force distribution, there is the significant difference in striations between cases of different pressures near the kerf exit, but little difference above those sections. At the bottom edge of a cutting front, a significant flow separation is occurring in the $Ps = 1.0$ MPa case, while such a flow separation is blown out of a kerf in the $Ps = 1.3$ MPa case. Therefore, it can be seen that the curve of the striations in Fig.6 (b) is mainly influenced by the flow pattern inside the kerf.

4. CONCLUSIONS

From the Schlieren observations and CFD results, a flow separation and an associated recirculation zone occurring at the bottom of a cutting front were found. From the cutting experiments, it was found that better cutting performances were achievable with the conditions in which wider and deeper penetration of the flow field inside the kerf was obtainable, i.e., with higher feed pressure of the gas jet.

REFERENCES

1. H. Zefferer, et al., *DVS* **135**, pp.210-214, 1991.
2. F. B. Thomassen, et al., *Lasers in Manufacturing – 1st International Conference Papers*, pp.169-180, 1983.
3. B. A. Ward, *Proc. ICALEO* **44**, pp.94-101, 1984.
4. J. Firet, et al., *SPIE Proc.* **801**, pp.243-250, 1987.
5. F. O. Olsen, *Proc. 4th Nordic Laser Material Processing Conference*, pp.77 – 88, 1993.
6. D. Leidinger, at al., *SPIE Proc.* **2207**, pp.469-479, 1994.
7. H. Horisawa, et al., *SPIE Proc.* **3888**, SPIE, pp.644-653, 2000.
8. B. G. Semiletenko, et al., *J. Eng. Phys.* **23**, pp.1122-1126, 1972.
9. G. T. Kalghatgi, et al., *The Aeronautical Quarterly* **27**, pp.169-185, 1976.
10. G. Broden, et al., *Welding and Cutting* **8**, pp.124-126, 1989.
11. V. Poncon, et al., *Welding in the World* **30**, pp.279-282, 1992.

A NEW METHOD FOR REDUCTION OF RESIDUAL STRESS OF WELDED JOINT USING ULTRASONIC VIBRATIONAL LOAD

Shigeru Aoki*, Seiji Hirai*, Tadashi Nishimura** and Tetsumaro Hiroi*

*   Department of Mechanical Engineering, Tokyo Metropolitan College of Technology
**Department of Production Systems Engineering, Tokyo Metropolitan College of Technology

Abstract

A new method for reduction of residual stress using ultrasonic vibrational load during welding is proposed. The proposed method is examined experimentally. First, in order to simulate butt-welding of thin plates, reduction of residual stress is examined by using a specimen with a groove and specimens supported on the supporting device. Next, in order to simulate repair welding of press die, reduction of residual stress is examined by using a thick plate with a groove. From these experiments, it is found that tensile residual stress near the bead is reduced when the specimen is shaken during welding.

Keywords

Welding, Ultrasonic vibration, Residual stress, X ray diffraction method, Butt-welding, Repair welding

1.INTRODUCTION

Welding is widely used for construction of many structures. Since welding is a process using locally given heat, residual stress is generated near the bead. Tensile residual stress degrades fatigue strength(ASM 1983, Donko 1983). Some reduction methods of residual stress are presented(Gnirss 1988, Hassen and Tholen 1978). For example, heat treatment and shot peening are practically used. However, those methods need special tools and are time consuming.

In this paper, a new method for reduction of residual stress using ultrasonic vibrational load during welding is proposed. The proposed method is examined experimentally for some conditions.

First, in order to simulate butt-welding of thin plates under simple condition of residual stress, reduction of residual stress is examined by using a specimen with a groove is used. The specimen is supported on the supporting device and welded along the groove using an automatic CO_2 gas shielded arc welding machine. The specimen is shaken by ultrasonic vibrational load during welding. Residual stress in the direction of the bead is measured by using a pararelled beam X-ray diffraction method. For comparison, residual stress of the specimen without ultrasonic vibrational load is also measured. Next, two thin plates are supported on the supporting device and butt-welded. Finally, in order to simulate repair welding of press die, reduction of residual stress examined by using a thick plate with groove. The specimen is welded along the groove. From these experiments, it is concluded that tensile residual stress near the bead can be reduced when ultrasonic vibrational load is used during welding.

2.EXPERIMENT OF BUTT-WELDING OF THIN PLATES

Reduction of residual stress of butt-welded thin plates is examined by using a specimen with a groove and specimens supported on the supporting device.

2.1 Preliminary Experiment

First, in order to simulate butt-welding of thin plates under simple condition of residual stress, reduction of residual stress is examined by using a specimen with a groove at the center. Size and shape are shown in Fig.1. In this case, residual stress in the direction of thickness is uniform. The specimen is supported on the supporting device and welded along the groove using an automatic CO_2 gas shielded arc welding machine. Fig.2 shows experimental set-up. Welding is completed through one pass. Material of specimen is rolled steel for general structure (JIS SS400). In order to eliminate residual stress induced by rolling, the specimen is annealed at 800°C for one hour and cooled in a furnace to 200°C. Diameter of the wire is 1.2mm. The applied voltage is 25V and current is 200A. The velocity of welding is 30cm/min. The specimen is shaken at the point 110mm from the center of the groove by ultrasonic vibrational load during welding. Frequency of vibrational load is 17.8kHz. The signal from the frequency synthesizer is amplified by a wide band amplifier. Ultrasonic vibrational load is applied by an exponential ultrasonic vibration horn excited by a ultrasonic vibrator.

Residual stress in the direction of the bead is measured since stress in this direction is the maximum. Residual stress is measured by using a pararelled beam X-ray diffractometer with scintillation counter after removing quenched scale chemically. Table 1 shows the conditions for X-ray measurment. For comparison, residual stress of the specimen without ultrasonic vibrational load is also measured. Fig.3 shows measuring points of residual stress. The point C is on the bead. The points B and D are 10mm from the bead. The points A and E are 50mm from the bead.

Fig.4 shows residual stress in the direction of the bead. Symbols ○ are results for specimen with ultrasonic vibrational load and symbols ● are results for specimen without ultrasonic vibrational load. Tensile residual stress on the bead, the point C, is significantly reduced when ultrasonic vibraional load is used during welding. At the point B and D, 10mm from the bead, tensile residual stresses are also reduced. At the point A and E, 50mm from the bead, compressive residual stresses are not changed greatly. Form Fig.4, tensile residual stress near the bead is reduced when the specimen is shaken during welding by ultrasonic vibrational load.

2.2 Butt-welding of Thin Plates

Next, two thin plates are butt-welded. Fig.5 shows size and shape of specimen. Two plates are hold on to the supporting devices and shaken by ultrasonic vibraitonal load at the point 110mm from the center of the groove. The groove is V-shaped. The root opening is 1mm and the groove angle is 30°. The welding conditions are same as the preliminary experiment. These measuring points are same as Fig.3.

Fig.6 shows residual stress in the direction of the bead. Symbols ○ are results for specimen with ultrasonic vibrational load and symbols ● are results for specimen without ultrasonic vibrational load. Tensile residual stress on the bead, the point C, is significantly reduced when ultrasonic vibraional load is used during welding. At the point B and D, 10mm from the bead, tensile residual stresses are also reduced. At the point A and E, 50mm from the bead, compressive residual stresses are not changed greatly. From Fig.6, tensile residual stress near the bead is reduced when the specimen is shaken during welding by ultrasonic vibrational load.

3.REPAIR WELDING OF BLOCK

For repair of press die, welding is used. Cracks are sometimes found at welded line caused by residual stress. In order to simulate repair welding of press die, reduction of residual stress is examined by using a thick plate with a groove. Fig.7 shows size and shape of the specimen. Fig.8 shows experimental set-up. The specimen is welded along the groove and shaken at the point 35mm from one edge of the groove. Fig.9 shows measuring points of residual stress. Residual stress in the direction of the bead is measured. The point C is on the bead. The points B and D are 10mm from the bead. The points A and E are 40mm from the bead.

Fig.10 shows residual stress in the direction of the bead. Symbols $\bigcirc$ are results for specimen with ultrasonic vibrational load and symbols ● are results for specimen without ultrasonic vibrational load. Tensile residual stress on the bead, the point C, is significantly reduced when ultrasonic vibraional load is used during welding. At the point B and D, 10mm from the bead, tensile residual stresses are also reduced. At the point A and E, 40mm from the bead, compressive residual stresses are not changed greatly. From Fig.10, tensile residual stress near the bead is reduced when the specimen is shaken during welding by ultrasonic vibrational load.

4.CONCLUSIONS

A method for reduction of residual stress using ultrasonic vibrational load during welding is proposed. The proposed method is examined experimentally. First, in order to simulate butt-welding of thin plates, reduction of residual stress is examined by using a specimen with a groove and specimens supported on the supporting device. Next, in order to simulate repair welding of press die, reduction of residual stress is examined by using a thick plate with a groove. From these experiments, it is found that tensile residual stress near the bead is reduced when the specimen is shaken by ultrasonic vibrational load during welding.

REFERENCES

American Society of Metals, *Metals Handbook*, 1983, 856-95

Donko, J.C., A Review of Weld Residual Stress in Austenitic Stainless Steel Piping. –Recent Trends in Welding Science and Technology-, Proceedings of the 2nd International Conference on Trends in Welding Research, ASM International, Materials Park, OH, 1989, 113-18

Gnirss, G., Vibration and Vibratory Stress Relief. Historical Development, Theory and Practical Application, Welding in the World, Vol.26, 1988, 4-8

Hassen, I. and Tholen, A., Plasticity due to Superimposed Macrosonic and Static Strains, Ultrasonics, March, 1978, 57-64

Table 1 Conditions of X-Ray Measurement

Characteristic X-rays	Cr-Kα
Diffraction plane	α-Fe(211)
Filter	Vanadium foil
Stress determination	$\sin^2\psi$ method
Irradiated area	$2\times4\text{mm}^2$
Tube voltage and current	30kV, 8mA
Scan condition of 2θ	Step scanning
Divergence angle shift	$1.0°$
Peak determination	Half value width method

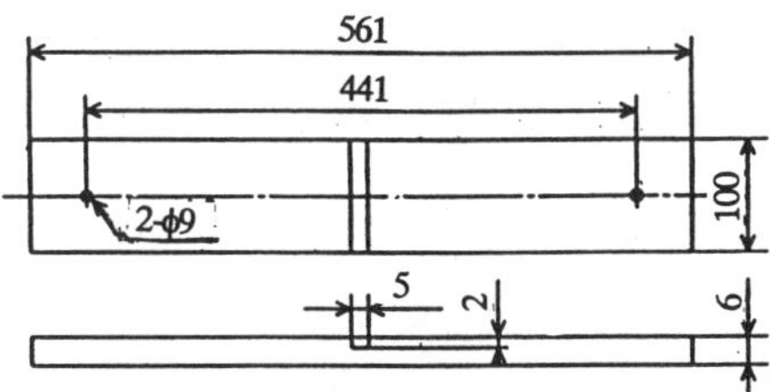

Fig.1 Size and shape of specimen of thin plate with a groove

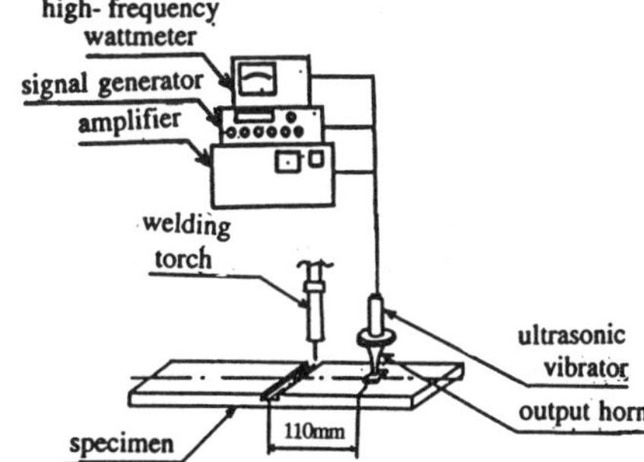

Fig.2 Experimental set-up for welding of thin plate

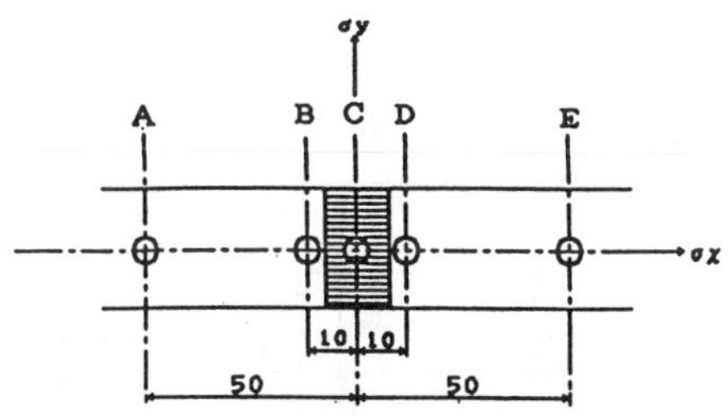

Fig.3 Measuring points of residual stress on thin plate

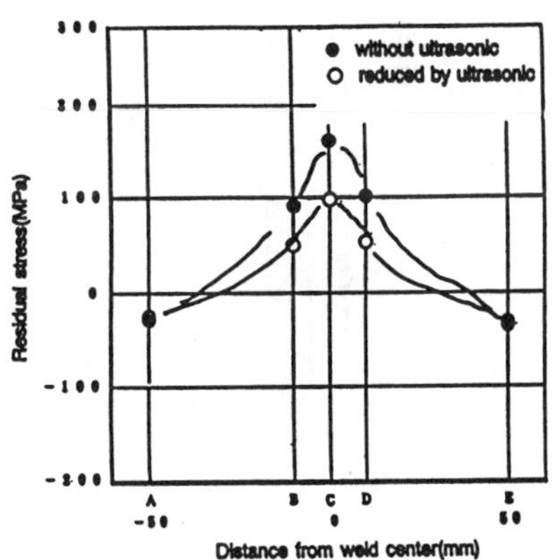

Fig.4 Residual stress on thin plate with a groove

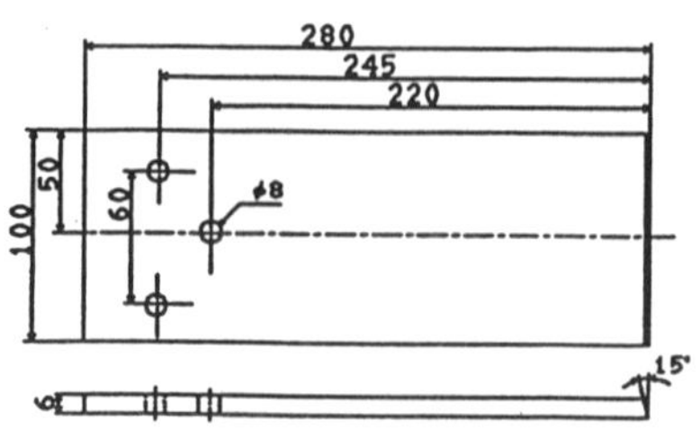

Fig.5 Size and shape of specimen of thin plate for butt-welding

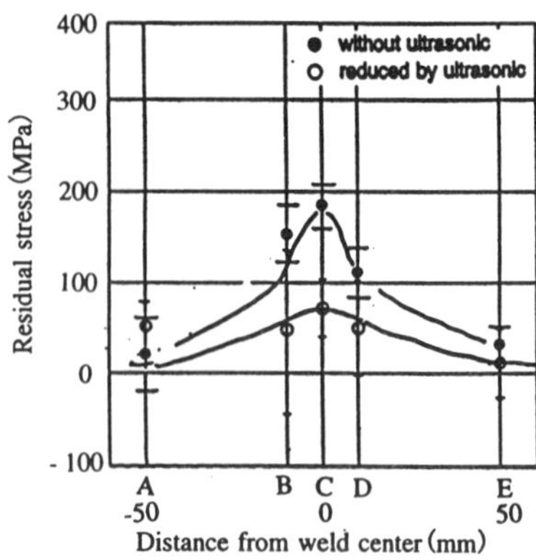

Fig.6 Residual stress on butt-welded thin plates

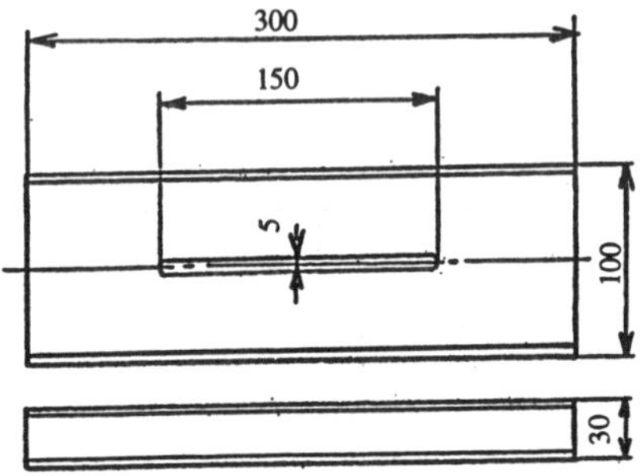

Fig.7 Size and shape of specimen of thick plate with a groove

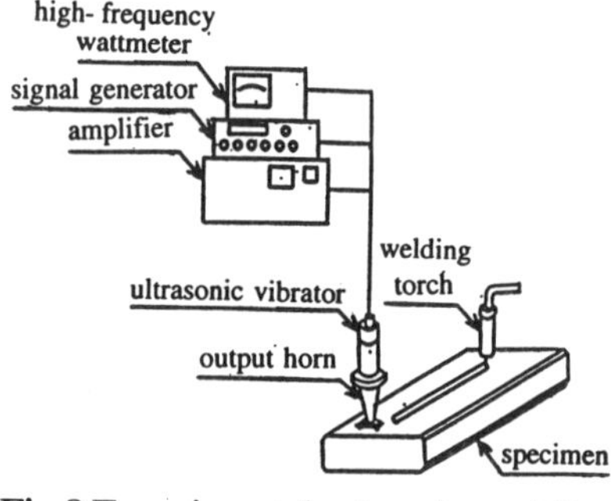

Fig.8 Experimental set-up for welding of thick plate

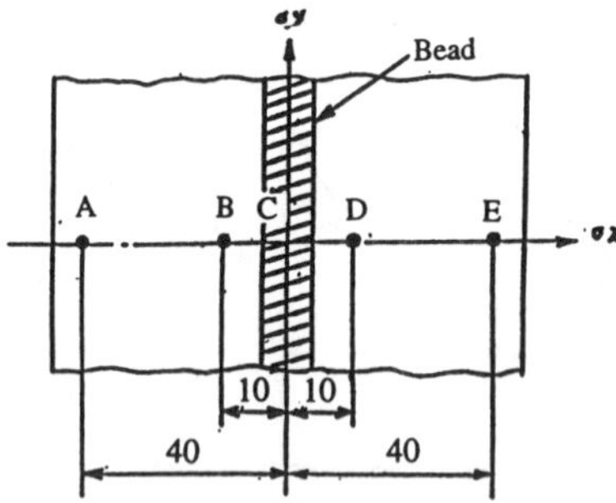

Fig.9 Measuring points of residual stress on thick plate

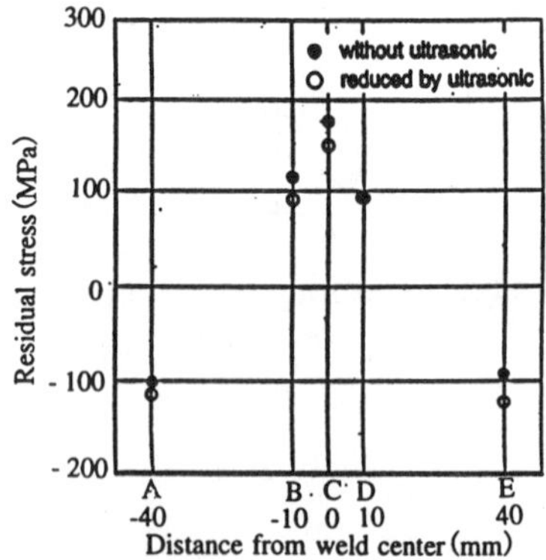

Fig.10 Residual stress on thick plate

348

Three-dimensional Microcasting

Hiroyuki Noguchi and Masao Murakawa

Nippon Institute of Technology

Abstract

Investigations on the methods of fabricating and casting microsized parts were carried out. Alumina powder with an average diameter of $0.7\,\mu$m was dispersed in a self-hardening liquid binder, and the powder density was increased by applying centrifugal force to fabricate a cast with good shape replicability and high strength. By setting the burning temperature below $800\,°C$, casting with self-destruction ability can be accomplished. The three-dimensional microshape of an ant could be fabricated.

Keywords

Microcasting, casting, three-dimensional micromachines, ant

1. Introduction

Significant progress has been made with regard to manufacturing methods of micromachines with two-dimensional configurations. The LIGA （an acronym from German words for lithography, electroplating, and molding） process is one such method. However, the development of a method whereby one can obtain three-dimensional micromachines has so far been very slow. It is thought that in the future, such micromachines will be manufactured in large quantities.

In this study, investigations on the methods of fabricating and casting microsized parts were carried out. In order to improve the accuracy of castings, it is necessary to use fine casting sand. Compared to general casting sand, fine casting sand has lower density and higher volume, making it very difficult to handle. To fabricate a casting mold with uniform density using fine powder, first, the fine powder was uniformly distributed in large amounts of binder, then the powder density was increased using centrifugal force, and the binder was solidified to fabricate a casting mold[1].

For microshaped parts, since it is difficult to remove cast parts from the casting mold, the casting mold must have sufficient handling strength and self-destruction ability after casting.

2. Fabrication of Casting molds

In the experiments, the samples shown in *Table 1* were used. The binder was made by diluting a mixture of soluble phenol resin organic self-hardening casting adhesive (Chem Rez 630 HSL) and hardener by twofold with water. An appropriate amount of alumina powder with average particle diameter of 0.7 μ m was added to the binder and stirred to make the slurry.

Fig. 1 shows the method of fabricating the casting mold. An appropriate amount of this slurry was poured into a test tube with an internal diameter of 11 mm. The burning-up model (an actual ant) was fixed to the tip of a metal needle (ϕ 0.7mm) by adhesive and set at the designated position (5mm from the base of the test tube). Centrifugal force was applied, using a centrifugal separator (TAITEC: VC-36N), to increase the density of the powder uniformly distributed in the binder and to separate the excess binder.

Table 1   Materials Used

Alumina Powder	ADMATECHS AO-502 Average diameter：0.7 μ m
Binder: Water-soluble phenol	Ashland Chemical Company, USA Chem Rez 630 HSL
Hardener	Ashland Chemical Company, USA Chem Rez 6028
Dilution	Water

The casting mold (green casting mold) was removed one hour later from the test tube, the needle was removed, pressure was decreased using a vacuum dryer (200℃ 6 Torr) and water was completely vaporized (dry state). Subsequently, the casting mold (dry state) was debinded and burned under controlled temperature, as shown in *Fig. 2*, in air in an electric furnace.

Rotation speed ：2000min^{-1}(max)

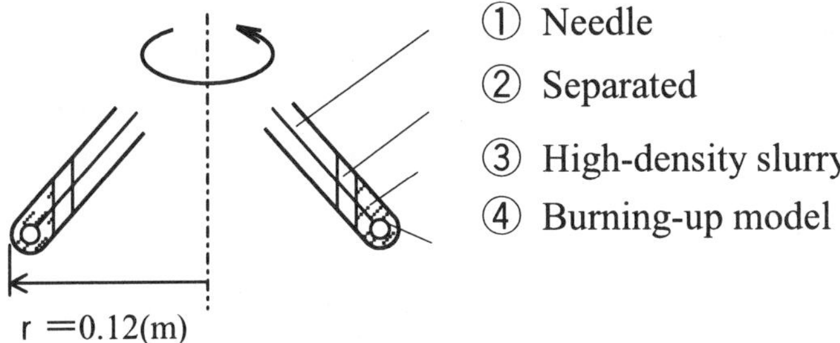

Fig. 1　Method of fabricating the casting mold

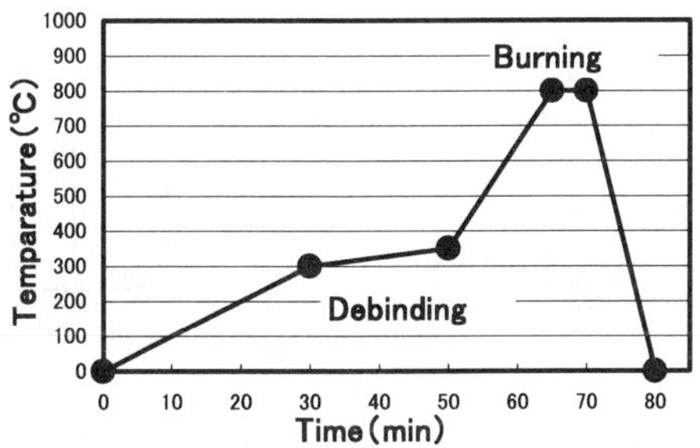

Fig. 2　Debinding and burning temperature of the mold

3. Characteristics of Casting mold

As the burned product acquires self-destruction strength, part of the alumina becomes sintered when the burning temperature is above 800℃, resulting in the loss of the self-destruction nature of the casting mold.
It was also confirmed that the casting mold burned at temperatures up to 800℃ self-destructs when placed in water after casting.

4. Casting Method

The most effective way of preventing solidification during casting is to heat the casting mold itself to a temperature above the fusion point of the cast metal.

Fig. 3 shows the present casting method. First a casting mold preheated to 200℃ was immersed in a metal (tin) that was heated and melted (300℃ in air) in an electric furnace. It was then pressurized from outside using the pressure of the melt. Vacuum suctioning was then performed in a vacuum furnace for five minutes at a maximum of 6 torr to heat the casting mold and deair the inside of the casting mold. After this, the vacuum furnace was left open in air for one minute and the casting mold was filled with melt using air pressure.

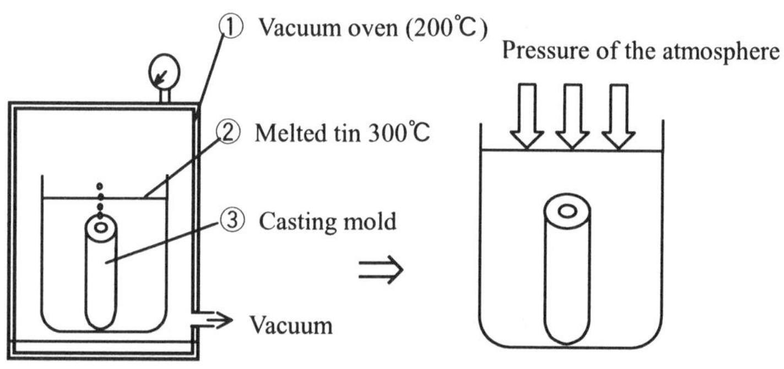

Fig. 3   Casting method

5. Experimental Results

Fig. 4 shows the SEM micrograph of the ant cast by this method. The three-dimensional microshape of the feelers and legs of the ant have been successfully cast. *Fig. 5* shows part of the compound eye of the cast ant. The cast product made using the casting mold fabricated of alumina powder with an average particle diameter of 0.7 μ m was confirmed to be able to replicate a shape of about 10 μ m. After casting, sufficient self-destruction can be achieved by placing the casting mold in water, and the cast microshape can be removed without damage.

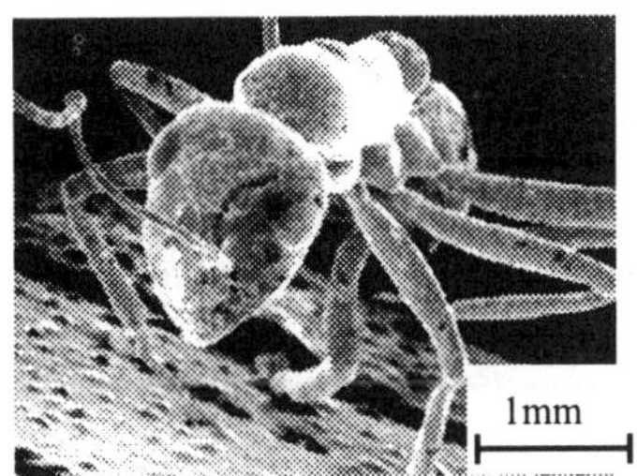

Fig. 4   SEM micrograph of a cast ant

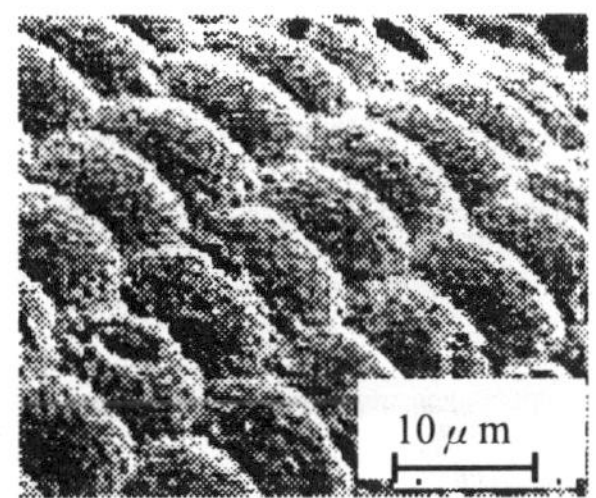

Fig. 5 Part of the compound eye of the cast ant

6. Conclusions

Alumina powder with an average particle diameter of $0.7\,\mu$m was dispersed in a self-hardening liquid binder, and the powder density was increased by applying centrifugal force to fabricate a cast with good shape replicability and high strength. By setting the burning temperature below 800℃, casting with self-destruction ability can be accomplished. It is thus possible to remove the mold with the cast completely intact. To achieve this a desirable feature, a suitable temperature at which there is no sintering of the fine powder should be used. Furthermore, by performing casting using a casting mold with self-destruction ability, the three-dimensional microshape of an ant could be fabricated.

In this study, at present the model is a concrete object. However, in the future it may become possible to make the three-dimensional model itself in large quantities, for example, by means of the rapid prototyping method.

Reference

1) H. Noguchi, M. Murakawa: Study on Microcasting (Report 1), The Japan Society for Precision Engineering 2000 Autumn Seminar Proceedings, p. 275.

IMPROVEMENT OF PROPERTIES OF SIC MIRROR SURFACE EPITAXIALLY GROWN ON SILICON SUBSTRATE

Akira KAKUTA, Katsumi HASHIMOTO,
Nobuyuki MORONUKI and Yuji FURUKAWA

Tokyo Metropolitan University, Graduate School of Mechanical Engineering

1-1 Minami-ohsawa, Hachioji-shi, TOKYO, JAPAN, 192-0397

E-mail: kakuta-akira@c.metro-u.ac.jp

Abstract

We are applying molecular beam epitaxy (MBE) to the surface finishing process, by which single crystal and atomically smooth surface can be obtained. In this paper, the proper conditions are discussed to improve the surface properties in hetero-epitaxy of silicon carbide on silicon substrate. The substrate temperature during the process, which strongly affects the epitaxial growth process, was changed between 1023K and 1373K. It is found that the proper substrate temperature, 1073-1173K in this case, improves the roughness up to 1.4 nm Ra while keeping good crystal structure.

Keywords

Atomically smooth surface, Silicon carbide, Molecular beam epitaxy

1. INTRODUCTION

Single crystal silicon carbide (SiC) is one of the potential materials for the precise structures in severe environment, such as X-ray mirrors. But it is difficult to achieve high accuracy by using conventional cutting and/or grinding processes because of its high hardness. We have been applying molecular beam epitaxy (MBE), which is one of the crystal growth techniques in vacuum, to the surface finishing process in atomic level. We are now extending this process from homo-epitaxy of silicon on silicon substrate (Furukawa et al., 1996) to hetero-epitaxy of SiC on silicon substrate (Moronuki et al., 2000). However, the properties of the obtained SiC surface, such as chemical composition and roughness, are not satisfactory. This paper aims to find the proper substrate temperature during the process to improve the surface roughness, chemical composition and hardness. The substrate temperature is one of the most important parameters, because it governs the energy to synthesize SiC from dissociated silicon and carbon molecules.

2. EXPERIMENTAL APPARATUS AND PROCEDURE

Figure 1 shows the setup for experiments. The silicon and carbon molecular beams were generated by an electron beam evaporator and dissociating pure acetylene gas, respectively. These beams were supplied to the substrate ((111) plane, ϕ100mm), that was heated up to the specified temperature. A film-thickness monitor measured the thickness of the grown layer. The crystal structure of the substrate surface was evaluated by reflection high-energy electron diffraction (RHEED) pattern in-situ.

Figure 2 shows the hetero-epitaxy process. After the cleaning, the substrate surface was carbonized by supplying the carbon molecular beam only. The main purpose of this process is to inhibit the desorption of silicon atoms from the surface to environment at high temperature (Zekentes et. al, 1995). Subsequently, SiC was grown by supplying both the silicon and carbon molecular beams, while the substrate was kept between 1023K and

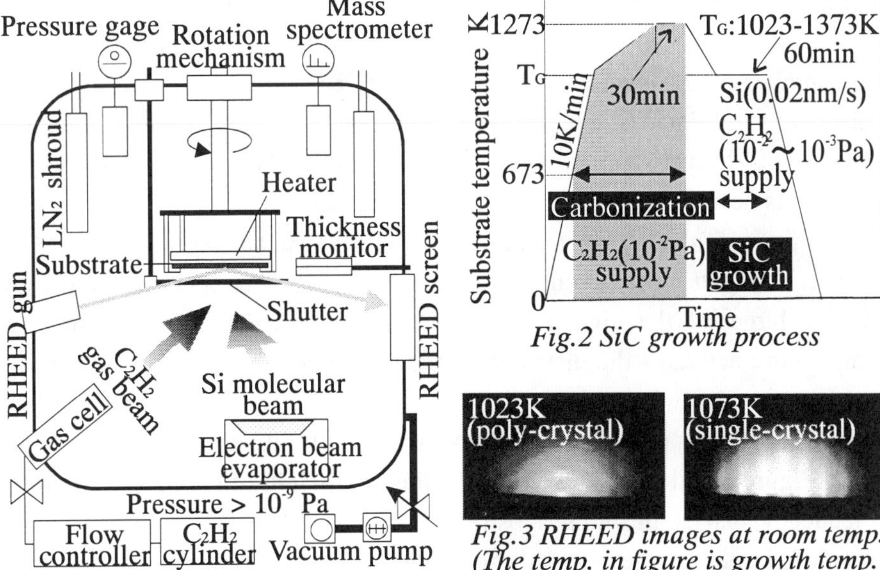

Fig.1 Schematic of experimental set-up

Fig.2 SiC growth process

Fig.3 RHEED images at room temp. (The temp. in figure is growth temp. The pressure of C_2H_2 was $10^{-2}Pa$)

1373K at every 50K. The partial pressure of acetylene gas was kept at 10^{-2} or 10^{-3}Pa to investigate its effect. The thickness of the grown layer was 180nm. The chemical composition of the layer was evaluated with Auger electron spectroscopy (AES). The surface roughness was observed with atomic force microscope (AFM). The hardness was measured with nano-indenter.

3. EXPERIMENTAL RESULTS

Figure 3 shows the typical RHEED images. The crystal structure processed at 1023K was poly-crystal SiC because the concentric circular (halo) pattern is observed. However, when the temperature is higher than 1023K, it became single-crystal because the pattern has become streaky. These results mean that there exists a threshold energy to drive the epitaxial growth. The equivalent temperature is considered as 1073K in this case.

Figure 4 shows the typical results of AFM observation. It is found from the microscopic observation (the left column in fig. 4) that the SiC surface is covered with nuclei of which diameter is less than 150nm and height is less than 10nm. The existence of the three dimensional nuclei suggests that the growth mode was Volmer-Weber mode. Also, it is found from the macroscopic observation (the right column in fig. 4) that there are many pits on the surface. The shape is triangular with specific orientation and all of the sidewalls are estimated as {111} crystal planes. It is considered that the silicon atoms of the substrate diffuse to the surface through the pits. The number of pits decreases when the acetylene gas pressure is high. It is considered that insufficient carbon supply resulted in insufficient coverage of carbonized layer over substrate. It is found that the higher acetylene gas pressure is necessary to obtain smooth SiC surface without pits.

Figure 5 shows the mean area of the pits analyzed from the AFM images. The pits became small at the low substrate temperature and they became larger at the high temperature. It is reasonable because the high temperature activates the diffusion of the silicon atoms and thus the surface becomes rough. It is considered that the lower substrate temperature, less than 1173K in this case, is proper to obtain smooth SiC surface.

Figure 6 and 7 show the mean radius and height of the nuclei. There is a tendency that at the high substrate temperature the radius of the nuclei becomes larger, while the height of the nuclei becomes scattered. The increase of the radius with the temperature can be related to the surface-migration-distance. A line in Fig.6 shows the theoretical migration distance of the silicon molecules adsorbed on ideal (111) SiC surface. The estimated migration distance corresponds with the actual radius. It is considered that the adsorbed atoms which receive large energy at high substrate temperature can migrate for long distance and coalesce each other to make large nucleus, thus, the surface becomes rough at high substrate temperature. The temperature less than 1223K in this case are considered to be proper to obtain smooth surface. The best roughness in this study was 1.4 nm (Ra) at 1073K.

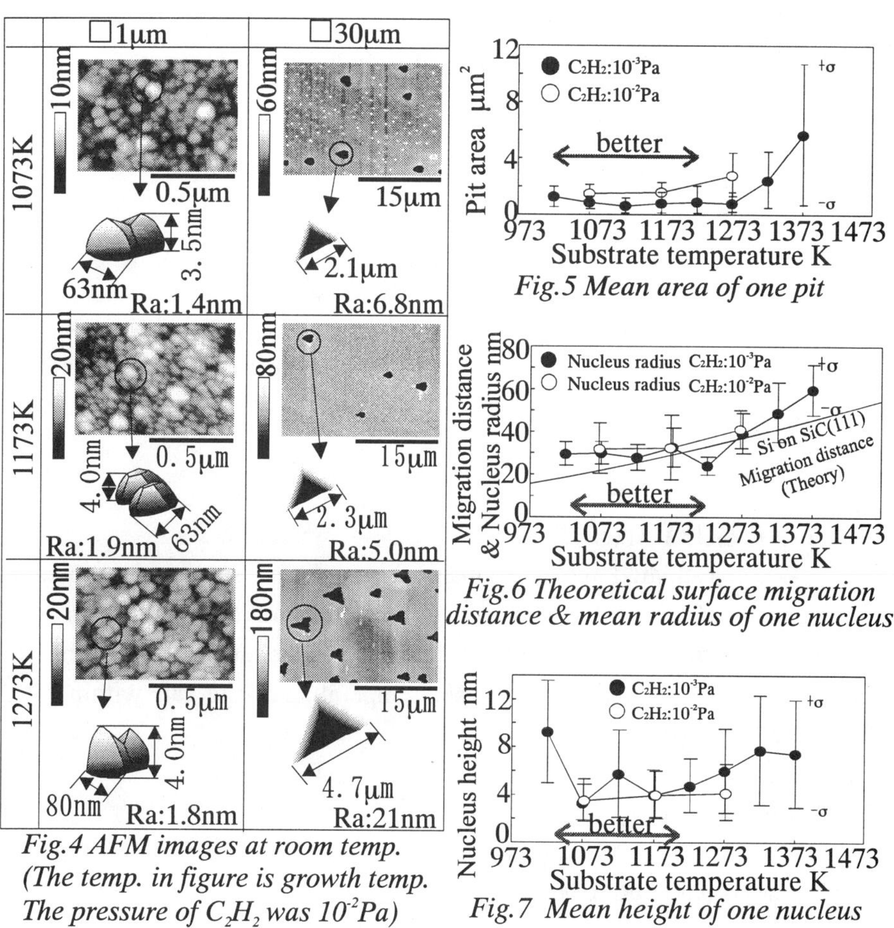

Fig.5 Mean area of one pit

Fig.6 Theoretical surface migration distance & mean radius of one nucleus

Fig.7 Mean height of one nucleus

Fig.4 AFM images at room temp.
(The temp. in figure is growth temp.
The pressure of C_2H_2 was $10^{-2}Pa$)

Figure 8 shows the ratio of silicon/carbon composition of the grown SiC layer. The ratio was about 1 both at 1073K and 1373K. However, in the region between 1123K and 1323K, the ratio has become low as 0.8. The

357

relationship between this ratio and the substrate temperature cannot be explained. It is suggested, based on these results, that the temperature at 1073K or 1373K in this case is proper from the viewpoint of the chemical composition.

Figure 9 shows the hardness of the SiC layer. When the substrate temperature is lower than 1173K, the hardness became lower than the other temperature region. It is considered that the synthesized material is not SiC but other chemical compound or SiC with other structure. The reason is estimated as the shortage of energy that the adsorbed atoms receive. Therefore, it is suggested that the substrate temperature region higher than 1173K in this case is proper from the viewpoint of hardness.

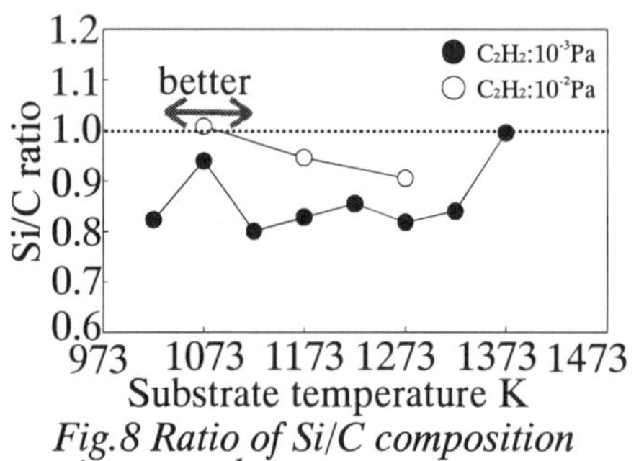

Fig.8 Ratio of Si/C composition in grown layer at room temp.

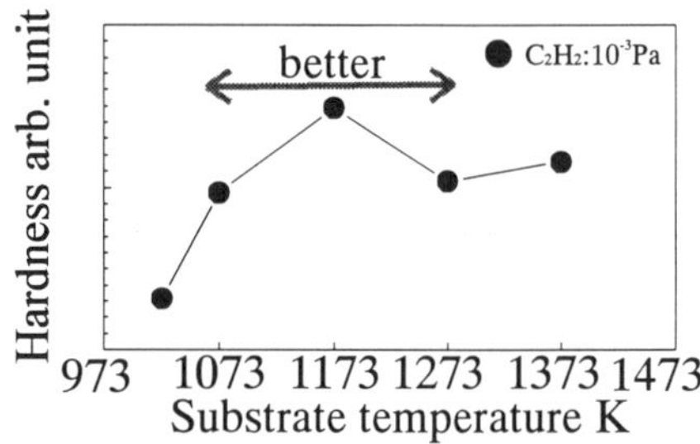

Fig.9 Hardness of grown layer at room temp.

4. CONCLUSIONS

We investigated the relationship between the substrate temperature during the SiC/silicon hetero-epitaxy process and the obtained surface properties from the viewpoint of roughness, chemical composition and hardness. It is found that the proper temperature range fall within 1073-1173K in this case.

REFERENCES

Y. Furukawa, A. Kakuta: Molecular Beam Epitaxy (MBE) as an Ultraprecision Machining Process: Annals of the CIRP 1996; 1:197-200

N. Moronuki, Y. Furukawa: An analysis of surface properties of hetero-epitaxially grown SiC surface on Si substrate: Annals of the CIRP 2000; 1:447-450

K. Zekentes, V. Papaioannou, B. Pecz, J. Stoemenos: Early stages of growth of B-SiC on Si by MBE: Journal of Crystal Growth 1995; 157:392-399

Part Ⅱ

Grinding / polishing / ultraprecision machining

PRECISION GRINDING OF MICRO ASPHERICAL SURFACE

H. Suzuki, O. Horiuchi, H.Shibutani
Toyohashi University of Technology, Tempaku, Toyohashi, Aichi, 441-8580, Japan

T. Higuchi
University of Tokyo, 7-3-1, Hongo, Bunkyo, Tokyo, 113-8656, Japan

Abstract

Recently, micro aspherical glass lenses are required for electric devices, optical devices and advanced optical fiber transmission equipments. The glass lenses are manufactured by glass molding method by using ceramics dies such as tungsten carbide. Therefore molding dies are most important and they must be ground with ultra-precision by diamond wheel. In this paper, our developed grinding methods/systems are discussed according to the workpiece shape.

Keywords

Micro aspherical grinding, diamond wheel, ceramics die, glass lens molding

1. INTRODUCTION

Recently, micro aspherical lenses are required for electric devices (CD, DVD), optical devices and advanced optical fiber transmission equipments. In those systems, glass lenses are more superior in optical characteristics than plastic lenses. The glass lasses are manufactured by glass molding method using ceramics dies such as WC (tungsten carbide). Therefore ultra-precision grinding technology is most important. There are many types of axi-symmetric aspherical lens shape, size and curvature. In this paper, two types of our developed grinding methods are discussed. [1][2].

2. INCLINED MICRO GRINDING

Micro axi-symmetric aspherical lenses are used in optical transmission devices. In those cases, the 45 degrees inclined micro grinding system is useful. The size of the smallest lens is about 0.25 mm in diameter and the approximate radius curvature as shown in *Figure 1*. In these systems, an axi-symmetric aspherical shape is commonly used and expressed as follows:

$$Z(Y) = C_v X^2 / \{1 + [1-(K+1)C_v^2 X^2]^{0.5}\} + \sum_{i=1}^{n} C_i X^i \qquad (1)$$

where, X is a radial direction, Z is an axial direction, and C_v, K, C_i ($i=1$ - n) are aspherical constants.

2.1 Grinding system

Figure 2 shows a schematic diagram of the micro aspherical grinding machine. The grinding wheel axis is 45 degrees inclined from workpiece rotational axis. The micro wheel was resinoid bonded diamond wheel as shown in *Figure 3* and the wheel diameter was very small. The wheel was formed to a precise symmetrical column by a cup type of GC micro wheel on the machine. The grinding spindle was an air bearing and the maximum rotational rate of grinding wheel was 150,000 rpm. The grinding head was actuated by the simultaneous 2-axes (X, Z) linear scale feedback system. The wheel height was adjusted with Y-axis table.

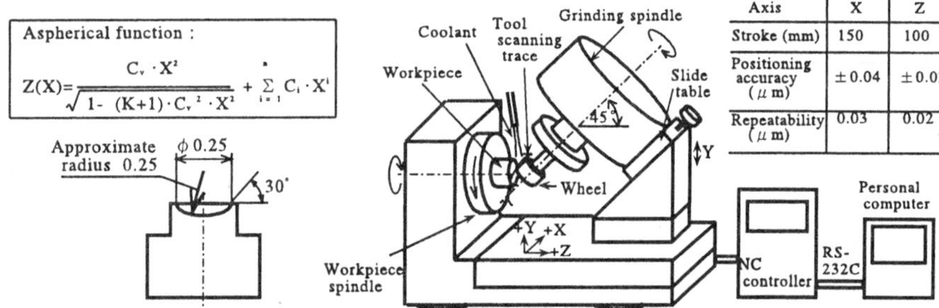

Axis	X	Z
Stroke (mm)	150	100
Positioning accuracy (μm)	±0.04	±0.02
Repeatability (μm)	0.03	0.02

Figure 1 Shape of a target aspherical molding die

Figure 2 Schematic diagram of micro aspherical grinding machine

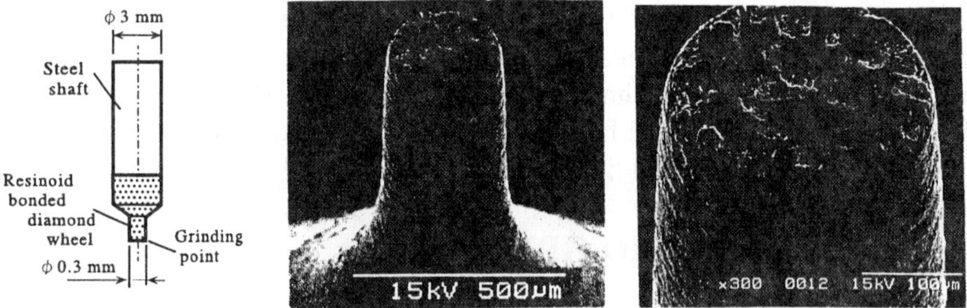

Figure 3 Shape of micro wheel and SEM photographs

2.2 Grinding method

The wheel center point (X_o, Y_o, Z_o) can be obtained from the coordinate of grinding point (X_g, Y_g, Z_g) and the normal vector $\vec{n}$ (a, b, c) at the grinding point as shown in *Figure 4* by the following formulae.

$$\left. \begin{array}{l} X_o = X_g + a \cdot r/l + a \cdot R/L \\ Y_o = Y_g + b \cdot r/l + (b+c) \cdot R/(2 \cdot L) \\ Z_o = Z_g + c \cdot r/l + (b+c) \cdot R/(2 \cdot L) \end{array} \right\} \quad (2)$$

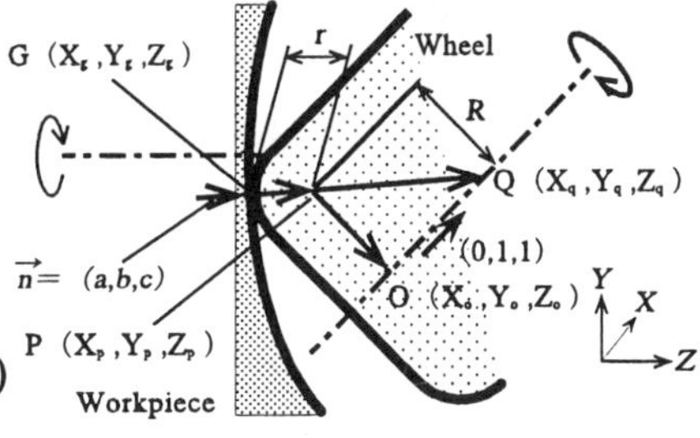

Fig.4 Calculation of tool path

where, $l=(a^2+b^2+c^2)^{0.5}$ and $L=\{a^2+(b+c)^2/2\}^{0.5}$.

Moreover, this machine is actuated by 2-axes (X, Z) drives, the wheel center point (X_o, Z_o) that realize $Y_{o\,(X_g=0)} = -R/2^{0.5}$ is calculated numerically by using Newton-Raphson method.

2.3. Grinding results

As a workpiece, the molding die of about 0.25 mm in radius curvature and radius was tested and the material was tungsten carbide. The grinding conditions are shown in *Table 1*. As a wheel, a resinoid bonded diamond wheel of #1200 was used. The wheel shape was columnar and the tip end was formed sharply. In *Eq.2*, the initial tip radius of the wheel r was left to be zero and it increased as the ware. *Figure 5* shows a SEM photograph of

Table 1 Grinding conditions

Grinding wheel	Resinoid bonded diamond
Grain size	#1200
Diameter	$\phi 250\ \mu$m
Rotational rate	
Wheel	150,000 rpm
Workpiece	1000 rpm
Feed rate	0.05 mm/min
Depth of cut	0.5 μm

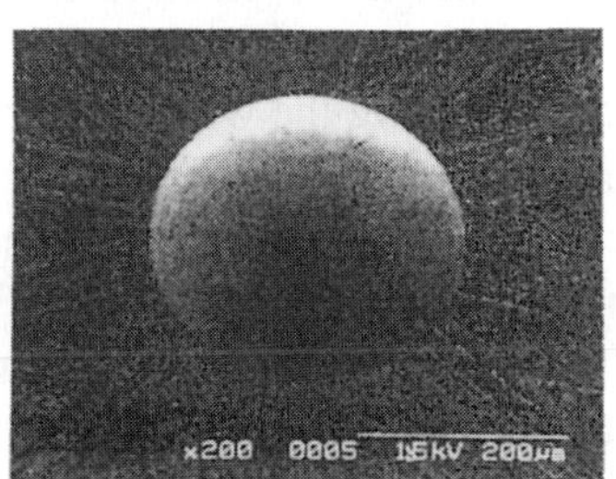

Figure 5 SEM photograph of ground workpiece

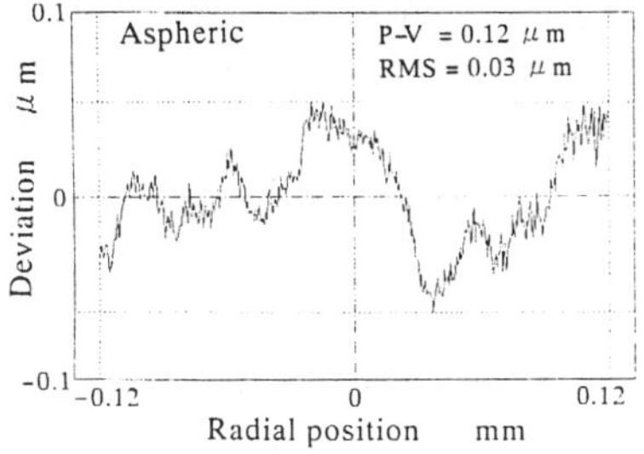

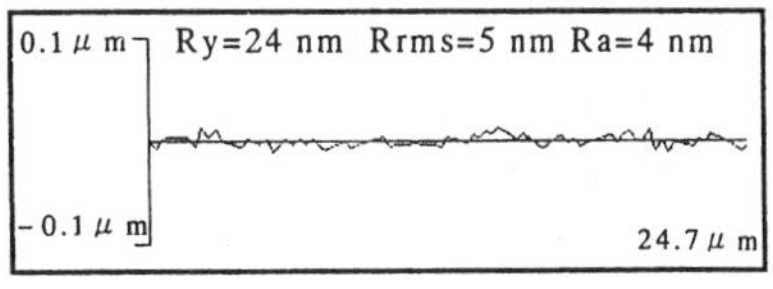

Figure 5 Form deviation and surface roughness profiles of ground surface

ground surface. *Figure 6* shows the deviation profile and surface profile after grinding. The form deviation profile was measured with Form Talysurf. The form accuracy of about 0.06 μmP-V and mirror surface were obtained.

3. FRESNEL SHAPE GRINDING

Micro fresnel glass lenses are useful for the next generation type of optical devices because of its thin structure and excellent optical characteristics. In the conventional machining method, the lens materials were restricted to plastics, because the cutting tool was a single crystal diamond and the molding die material was restricted to soft metals. Therefore ceramics dies must be ground ultra-precisely with diamond wheel. An axi-symmetric fresnel shape is expressed as the next equation:

$$Z(Y)=\mathrm{mod}\{g(Y),b\} \tag{3}$$

where, Y is a radial position, Z is an axial position, b is a depth of groove and $g(Y)$ is an aspherical function expressed in Eq.1.

3.1 Grinding method and system

A schematic diagram of newly developed micro fresnel grinding method is shown in *Figure 7*. As a wheel, a disk type of diamond wheel, whose edge was trued as a sharp knife-edge, was used and the wheel scanned along the workpiece radial position vertically. The wheel air spindle rotated vertical and the workpiece air spindle rotated horizontally. The feature of this grinding system was that the wheel rotated in the parallel with the workpiece rotational direction at the grinding point. With the NC-controlled sharp edged wheel, the axi-symmetric aspherical fresnel shaped workpiece will be generated. The grinding spindle was actuated in X, Y and Z-axes by the linear scale feedback system of 1 nm positioning resolutions. In this grinding experiments, the grinding spindle was simultaneous 2-axes (Y, Z) controlled. The wheel horizontal position was adjusted with X-axis table and was fixed in the aspherical grinding.

3.2 Grinding results

The workpiece diameter was about 4 mm. The material was WC. The wheel feed rate was varied between 0.025 - 0.4 mm/min in order to decrease the grinding force change. The 45 degrees sharpen wheel was trued on the machine by using a disk type of #200 diamond wheel. A 3-dimensional topography of ground workpiece is shown in *Figure 8*. The form profile was measured by non-contact type of laser scanning instrument (NH-3, Mitaka-

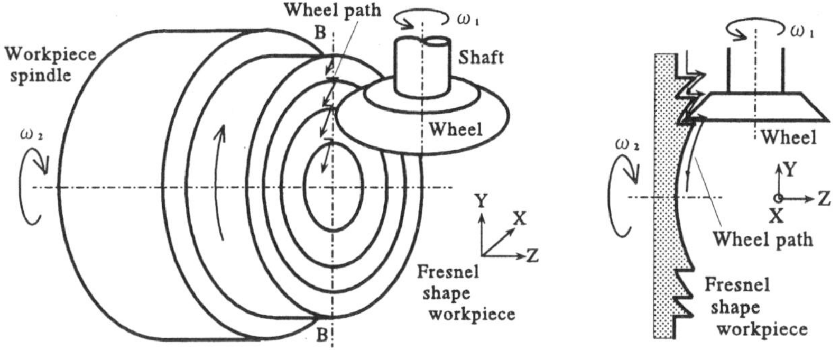

(a) Schematic illustration   (b) Cross-section BB

Figure 7 Schematic diagram of micro fresnel grinding system

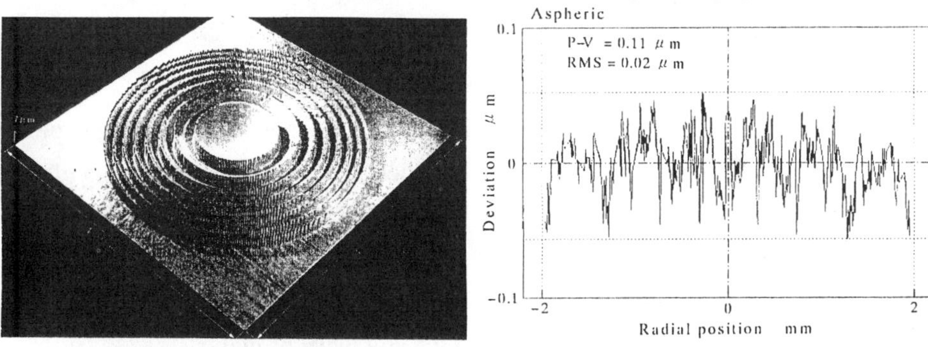

Figure 8 3-D topography   *Figure 9* Form deviation profile

koki). *Figure9* shows a form deviation profile in the radial direction. The form accuracy of about 0.1 μ mP-V was obtained. The profile was measured with contact type of surface measurement instrument (UA-3P, Panasonic) with a 2 μ m diamond stylus. A sharp edge was created.

4. CONCLUSIONS

This paper summarized as follows:

(1) 45 degrees inclined type of grinding method is useful for micro axi-symmetric aspherical surface.

(2) Vertical scanning type of grinding method with knife-edged wheel is useful for fresnel shape workpiece.

REFERENCES

1. Suzuki, H. et al., Proceedings of ASPE'98 Annual Meeting; 1998 Oct.25-30; 149-152
2. Suzuki, H. et al., Proceedings of euspen'99 International Conference; 1999 May 31-June 4; 318-321

MIRROR GRINDING OF SILICON WAFER WITH SILICA EPD PELLETS

Takashi FUKAZAWA [a], Naruto FUWA [a], Jun-ichi IKENO [b],
Hideo SHIBUTANI [c], Osamu HORIUCHI [c] and Hirofumi SUZUKI [c]

[a] *Graduate Student, Department of Production Systems Engineering, Toyohashi University of Technology*
[b] *Graduate School of Science and Engineering, Saitama University*
[c] *Department of Production Systems Engineering, Toyohashi University of Technology*

Abstract

This paper deals with mirror grinding of silicon wafers using EPD pellets of silica aggregate grains. The EPD pellets were fabricated by using electrophoretic deposition phenomenon. Mirror surfaces without scratch marks can be obtained by grinding with the silica EPD pellets. The grain size has a small influence on the grinding performance of silica EPD pellets. Grinding with the silica EPD pellets does not cause any warp of wafer.

Keywords

Silicon wafer, mirror grinding, silica aggregate grain, electrophoretic deposition phenomenon, EPD pellet.

1. INTRODUCTION

Semiconductor wafers and optical elements are usually finished by polishing with loose abrasives. However it takes a long time to obtain a smooth surface by polishing and there are great demands to shorten the processing time. On the other hands, grinding with bonded abrasives is generally superior to polishing in obtaining high form accuracy, high machining rate and low environmental contamination. Therefore the importance of mirror grinding is recently increasing in semiconductor and optics industries.

In mirror grinding of brittle materials such as silicon wafers, it is essential to accomplish ductile mode grinding using a grinding wheel of very fine abrasive grains, though it is difficult to make such a grinding wheel and to keep the working surface sound without loading.

One of the authors has developed a method to fabricate an ultra fine abrasive pellet using electrophoretic deposition phenomenon [1,2]. The pellet was named EPD pellet. It had a uniform distribution of ultra fine silica particles and the bonding strength was suitable for mirror grinding of brittle material.

Silica aggregate grains composed of ultra fine primary particles with about 10 nm diameter are considered suitable to use as abrasives for mirror grinding tool of brittle materials, because during grinding they may be collapsed into ultra fine primary particles.

Purpose of this study is to make new EPD pellets of silica aggregate grains and to investigate grinding performance of the new pellets.

2. FABRICATION OF EPD PELLET

The new EPD pellets of silica aggregate grains were fabricated by applying electrophoretic deposition phenomenon. The grain size were 0.55, 5, 10 and 20 µm in diameter. Sodium alginate, one of the hydrophilic electrolytes, was used as the bonding agent. The fabrication processes of the EPD pellets as shown in Figure 1 were as follows,

(1) Mix silica aggregate grains with sodium alginate solution uniformly in a vessel.

(2) Form the deposition layer with the above mixture applying electrophoretic deposition.

(3) Fabricate pellets by cutting the deposition layer.

(4) Dry the pellets in air.

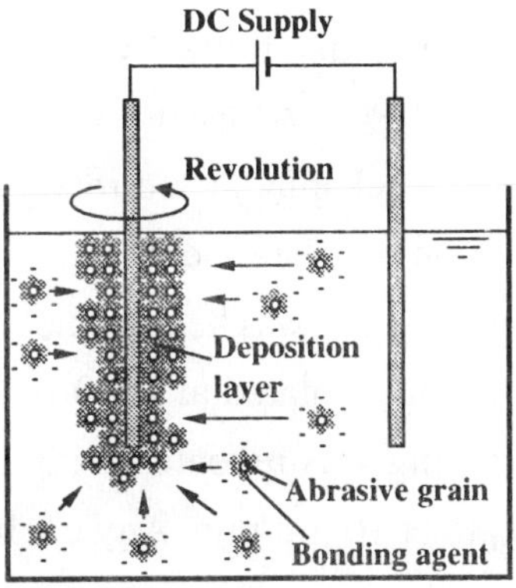

Figure 1. Fabrication of EPD pellet.

The fabrication conditions of EPD pellets were 20 % mass concentration of silica aggregate grains, 3 % mass concentration of sodium alginate, the applied voltage of 10 V and the processing time of 30 min. As shown in Figure 2, the fabricated EPD pellet has no cracks and the silica aggregate grains are

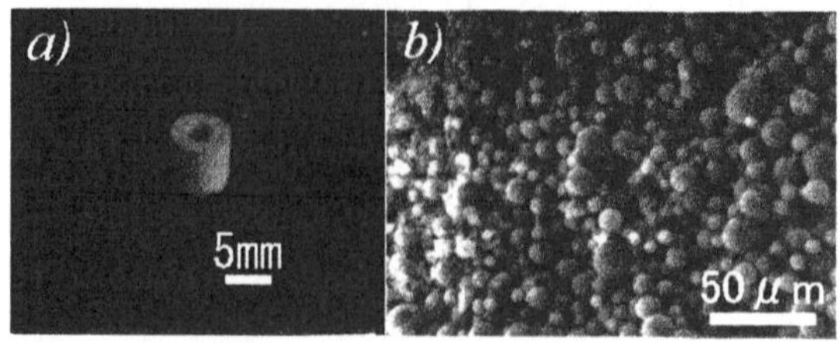

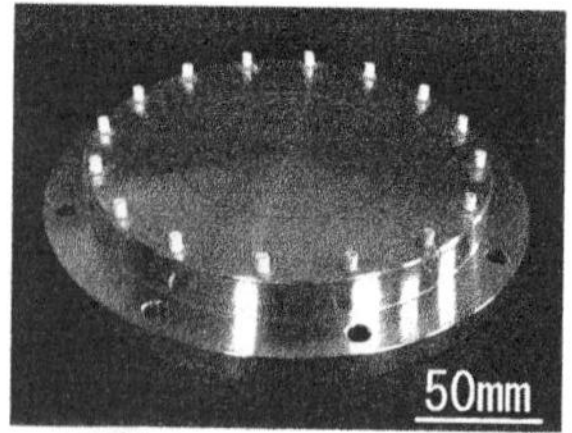

Figure 2. Fabricated EPD pellet, a) appearance, b) SEM image of the cross section

Figure 3. Appearance of grinding wheel

distributed uniformly. A cup type of grinding wheel was constructed by gluing 16 pellets on an aluminum disk of 210 mm diameter, as shown in Figure 3.

3. GRINDING EXPERIMENTS

Grinding experiments of silicon wafer with the silica EPD pellets was performed to investigate the grinding performances. Figure 4 shows the ultraprecision vertical surface grinder with a rotary table (Nissin Machine K.K.: VPG-1A) used for experiments.

Figure 4. Appearance of ultra pre-cision grinding machine

The wheel spindle and rotary table had a hydrostatic bearing. The axial feed motion of wheel head had a resolution of 10 nm. The workpiece of 8 inch silicon wafer was held on the ceramic vacuum chuck of rotary table. The grinding wheel rotated at 1500 rpm and the rotary table at 50 rpm. Face plunge grinding was repeated with a spark-out of 1 min after an infeed of 10 μm depth at 10 μm/min. Any coolant was not used.

4. EXPERIMENTAL RESULTS

In all the experiments, mirror surfaces without scratch marks were obtained as shown in Figure 5. The surface roughness obtained at total infeed of 900 μm measured by Zygo New View 100 were almost same and almost inde-

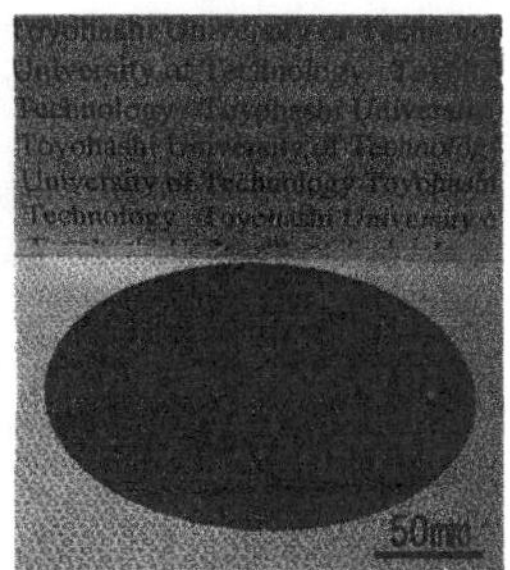

Figure 5. Surface condition of silicon wafer after grinding

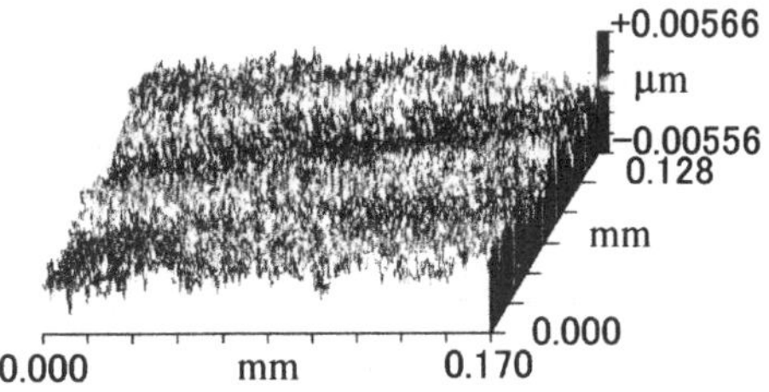

Figure 6. Surface roughness of silicon wafer after grinding

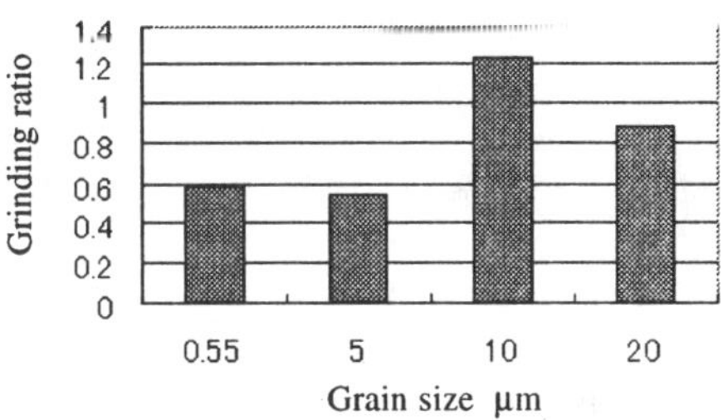

Figure 7. Grinding ratio of new EPD pellet

pendent of grain size. The surface roughness was about 10 nm P-V, as shown in Figure 6. The grinding ratios measured at total infeed of 900 μm were 0.6-1.2 and the influence of grain size on the grinding ratio seemed rather small, as shown in Figure 7.

Moreover the diamond grinding for preparation of specimen resulted in a warp of the specimen due to subsurface damage but the warp was disappeared by the successive grinding with the silica EPD pellets, as shown in Figure 8. It is considered that the residual strain by conventional diamond grinding was removed. These results may suggest that material removal were performed by not only mechanical action but also chemical reaction between grains

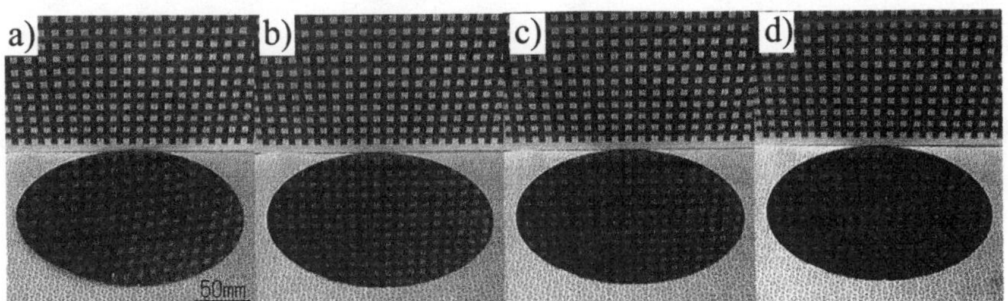

Figure 8. Warp of thin wafer after grinding at total infeed of a) 0 μm, b) 100 μm, c) 200 μm and d)300 μm

and work.

Figure 9 shows the result of SEM observation of EPD pellet after grinding. A worn grain of hemispherical shape is observed on the working surface of pellet. This result suggests that the silica aggregate grain was partially collapsed into a number of ultra fine primary particles, which played important roles of mechanical and chemical actions during the material removal.

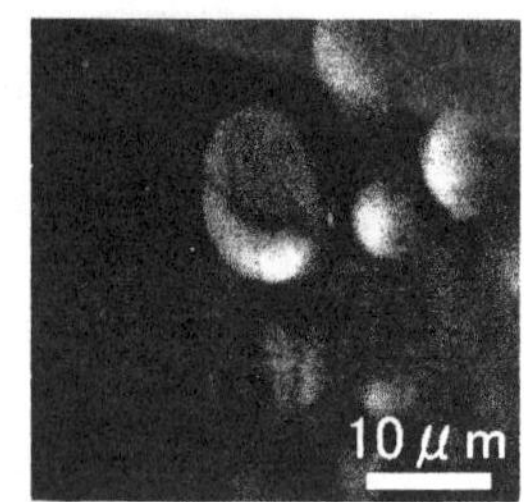

Figure 9. Working surface of EPD pellet after grinding

5. CONCLUSIONS

The paper represented the silica aggregate grain EPD pellets were manufactured by way of trial and investigated the grinding performance of the new grinding wheel. The following are conclusions of this study.

(1) New EPD pellets which consist of silica aggregate grains could be made.

(2) Mirror surface without scratch marks was obtained by grinding with the silica EPD pellets.

(3) The grain size has a small influence on the grinding performance.

(4) Grinding with the silica EPD pellets does not cause any warp of wafer.

(5) Worn grains were found on the working surface of the pellets.

(6) These results indicate that material removal were performed by not only mechanical action but also chemical reaction.

References

1. Ikeno J., Tani Y. and Sato H., Nanometer Grinding Using Ultrafine Abrasive Pellets -- Manufacture of Pellets Applying Electrophoretic Deposition, Annals of the CIRP, 1990; 39(1): 341-344.
2. Shibutani H., Ikeno J., Horiuchi O. and Yano K., Mirror Grinding of Silicon Wafer with EPD Pellet, Proceedings of 9th ICPE, Osaka, August 29 – September 1, 1999; 98-102.

NEW FABRICATION PROCESS OF DEEP PARABOLOIDAL MIRROR BY COMBINATION OF ELID-GRINDING AND ELECTRO-FORMING TECHNIQUES FOR LAMAN SPECTROSCOPY INSTRUMENT

Hitoshi OHMORI, Jianqiang GUO, Weimin LIN, Takashi MATSUZAWA, Shinya MORITA, Hidetoshi SATO and Hideo TASHIRO

The Institute of Physical and Chemical Research, Wako-shi 351-0198, Saitama, Japan

Abstract

In this paper, ELID grinding and electro-forming techniques were applied to fabricate a deep concave paraboloidal mirror which will be used in Laman spectroscopy instrument. First, an injection mold with paraboloid was fabricated by ELID grinding and polishing. After the injection mold was ground by a #4000 metal resin bonded diamond grinding wheel, Ra, Rz and Ry were 8nm, 36nm and 60nm respectively. In order to further reduce the surface roughness, the injection mold was polished by diamond paste with average grit size of $0.25\,\mu$ m. Half an hour later, the surface roughness was Ra2nm, Rz15nm and Ry32nm. Secondly, the injection mold was used to electro-form the deep concave paraboloidal mirror.

Keywords

ELID, grinding, mirror grinding, precision grinding, dressing, truing

1.INTRODUCTION

The Laman spectroscopy analysis is used more and more widely in medical field. However, the conventional light dispersed system that consists of a set of

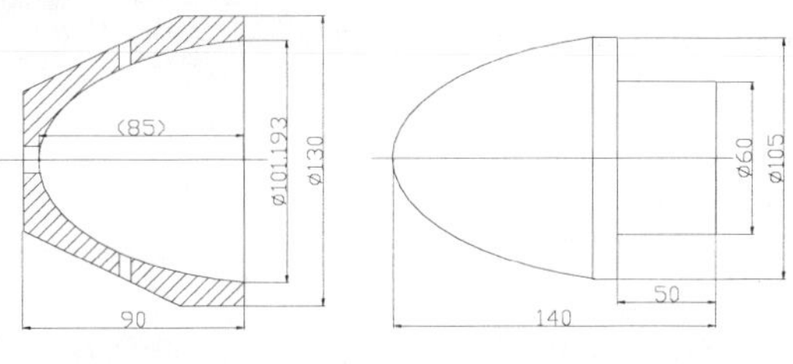

(a) Paraboloidal mirror   (b) Injection mold

Figure 1 Outline of paraboloidal mirror and injection mold

mirrors lost some of reflection light. In order to improve Laman spectroscopy instrument, a new light dispersed system that consists of a single paraboloidal mirror was developed by some researchers.

Figure 1(a) is the outline of the deep concave paraboloidal mirror, which will be used in the new light dispersed system above-mentioned. In this paper, the deep concave paraboloidal mirror was fabricated by the combination of ELID grinding and electro-forming. First, an injection mold with convex paraboloid showed in figure 1(b) was fabricated by ELID grinding and polishing. Then the injection mold was used to electro-form the deep concave paraboloidal mirror.

2.ELID GRINDING PRINCIPLE

To get high accuracy and low surface roughness of these materials with high hardness, grinding is always used as finishing process. The finer the grain is, the lower the surface roughness is. It is proved that grinding of these materials by metal bonded fine super-abrasive wheel is of obvious advantage. But when the grinding wheel is worn off, the dressing of abrasive wheel is very difficulty.

To solve the difficult dressing problem of metal bonded fine super-abrasive, ELID (electrolytic in-process dressing) grinding was proposed by one of the authors in 1987 [1]. ELID grinding provides enough space to hold grinding debris and proper protrusion of abrasive grain, thus realizes stable grinding of the metal bonded diamond grinding wheel over a long time.

It is easy to construct an ELID system on a conventional grinder. This system includes a metal bonded diamond grinding wheel, electrolytic power supply, electrode and electrolytic coolant (also used as grinding coolant). The metal bonded diamond grinding wheel is connected to the positive pole of the electrolytic power supply and the electrode is connected to the negative one. There is a gap of 0.1mm between the grinding wheel and electrode.

Electrolysis occurs in this gap under the existence of voltage and grinding coolant. As a result, the bonded material is electrolyzed and a stable oxide layer is formed on the surface of the grinding wheel.

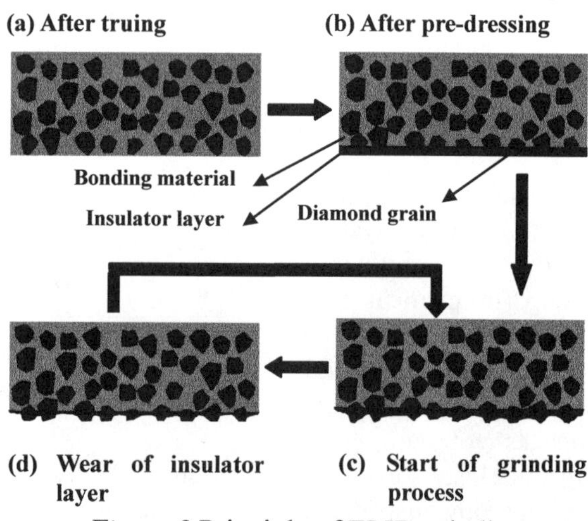

Figure 2 Principle of ELID grinding

When the grinding process starts, the thickness of insulator layer formed in the pre-dressed stage is reduced. The electric balance in ELID circuit cannot be hold. The current between the grinding wheel and electrode increases and the bonded material comes to be electrolyzed again. With the increase of insulator layer thickness, the current reduces and then the electric balance in ELID circuit reaches again. And thus there is always proper protrusion of abrasive grains on grinding wheel surface to provide the good grinding ability. Figure 2 shows the principle of ELID grinding.

ELID grinding consists of the following steps:

① Truing: Eliminates the eccentricity to enhance the axial straightness of the grinding wheel surface.

② ELID initial pre-dressing: An oxide layer is formed on the surface of grinding wheel in this process.

③ ELID grinding: Employs a metal bonded abrasive wheel and electrolysis to realize stable grinding over a long period of time [1-2].

Table 1   Conditions of ELID grinding experiment

Grinding machine	Machining center (MAZAK VQC-15/40)				
Grinding wheel	#600 CIBD wheel (size φ150×w10mm)				
	#2000 MRBD wheel (size φ150×w10mm)				
	#4000 MRBD wheel (size φ150×w10mm)				
Grinding fluid	AFG-M diluted to 2% with water				
ELID conditions	Open circuit voltage: Vp=100V Maximum current: Ip=12A				
	Pulse interval: τ_{on}=2μs, τ_{off}=2μs				
Truing	Truing wheel speed: 200rpm Speed of #600 wheel: 1061rpm Depth of cut: 1 μm/pass Speed of #2000 wheel: 1061rpm Feed speed: 50mm/min Speed of #4000 wheel: 849rpm				
Grinding	Wheel	Speed of wheel (rpm)	Speed of workpiece (rpm)	Feed rate (mm/min)	Depth of cut (μm /pass)
	#600	1061	500	100	5
	#2000	1061	500	100	2
	#4000	849	300	50	1
Measuring instrument	Surftest-701 (Mitutoyo Co., Ltd.)				

3.ELID GRINDING EXPERIMENT

3.1 Experimental Conditions

The diameter and length of the workpiece was 105mm and 140mm respectively. The shape of the workpiece was paraboloid. See figure 1. The material used was stainless steel with hardness above HRC58.

Table 1 shows the conditions of this

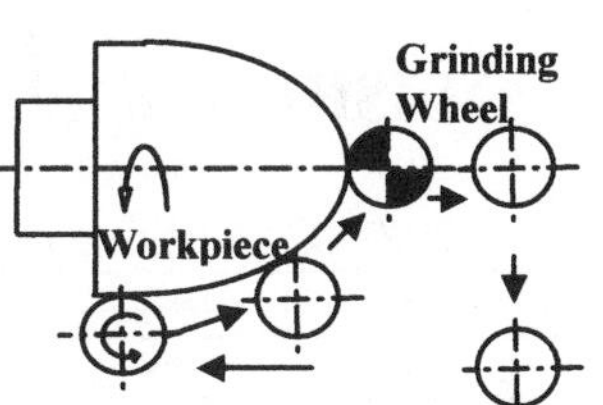

Figure 3 Moving path of grinding wheel

experiment. It was carried out on a vertical machining center. Grinding wheels with mesh size of #600, #2000 and #4000 were used respectively.

To obtain mirror surface, NC data was calculated according to the principle of arc interpolation. The maximum calculation error of NC data is 0.094μm. Figure 3 shows the moving path of the grinding wheel.

3.2 Experimental Procedure

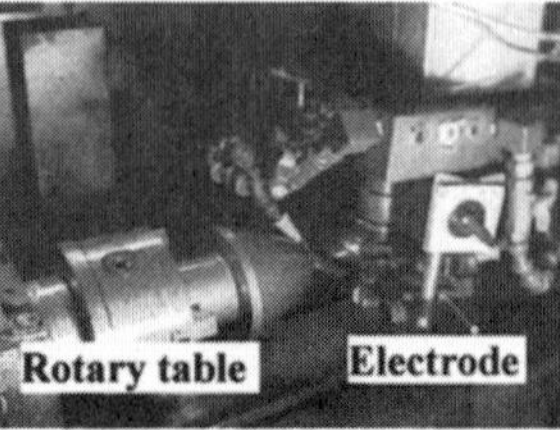

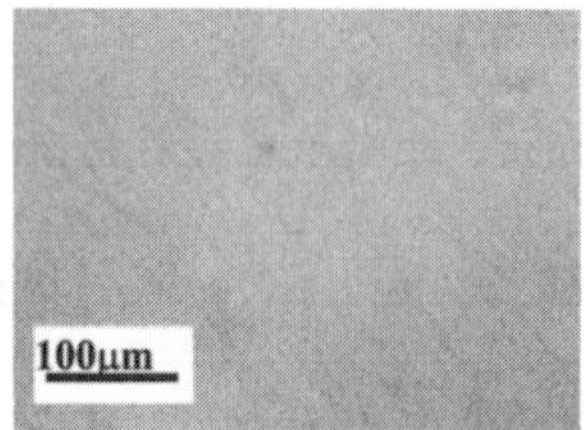

Figure 4 Setup of electric discharge truing

Figure 5 Setup of the ELID grinding

Figure 6 Micrograph of ground surface

First, each grinding wheel was trued by means of the electric discharge method when used. Figure 4 shows the setup of electric discharge truing of grinding wheel.

Secondly, the grinding wheel was pre-dressed by electrolytic method at the speed of 200rpm until the working voltage was close to the open circuit voltage 100V.

Finally, ELID grinding was carried out on a vertical machining center. The workpiece was fixed on a rotary table that rotated around the horizontal axis. Figure 5 shows the setup of this experiment. In this process, the workpiece was ground by #600, #2000 and #4000 diamond grinding wheel. The #600 abrasive wheel was used for rough grinding and #4000 for finish grinding.

After the grinding process, the roughness of the ground surface was measured with Mitutoyo Surftest-701. Ra, Rz and Ry were 8nm, 36nm and 60nm respectively. Figure 6 shows the micrograph of the ground surface.

4.POLISHING EXPERIMENT

Table 2 Conditions of polishing experiment

Machine	Polishing machine FP-0806（NAGASE)
Abrasive	Diamond paste with average grit size of 0.25μm
Rotary speed of polisher	300rpm
Rotary speed of workpiece	100rpm
Feed speed	300mm/min
Press	0.1MPa

In order to further reduce the surface roughness, the workpiece was polished by diamond paste with average grit size of 0.25 μ m on the Nagase flex polisher FP-0806. The workpiece was tilted at 60 degrees relative to the horizontal axis so that the whole surface could be polished in one pass. Figure 7 shows the polishing experimental setup. After polishing for only half an hour, the surface roughness was Ra2nm, Rz15nm and Ry32nm. Figure 8 is the micrograph of polished surface and figure 9 shows the finished workpiece.

Figure 7 Setup of polishing

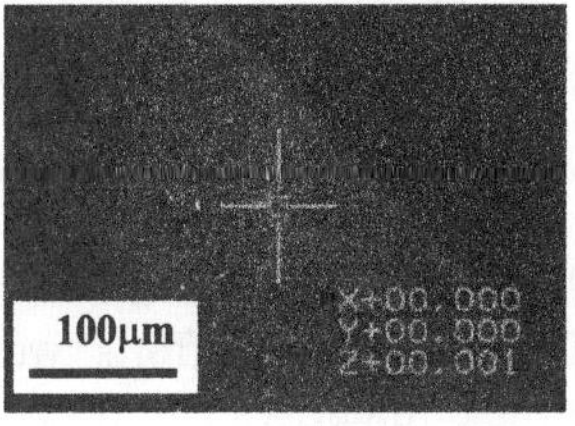

Figure 8 Micrograph of polished surface

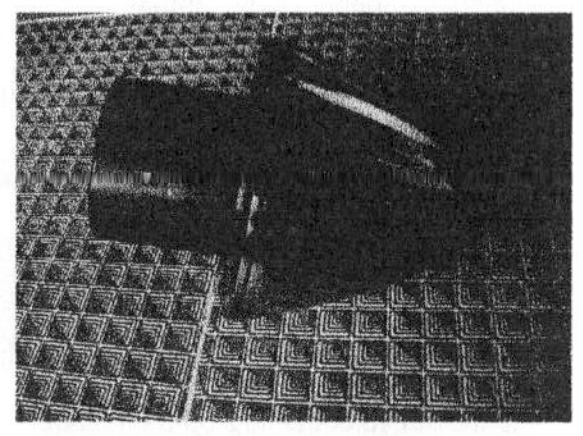

Figure 9 Photograph of workpiece

5.ELECTRO-FORMING

After polished, the injection mold of convex paraboloid was used to electro-form the deep concave paraboloidal mirror. Figure 10 shows the injection mold and the electroformed concave paraboloidal mirror.

Figure 10 Photograph of paraboloidal mirror

6.CONCLUSIONS

A deep concave paraboloidal mirror was fabricated by the combination of ELID grinding and electro-forming techniques in this paper. After the injection mold was ground by a #4000 metal resin bonded diamond wheel, surface roughness was Ra8nm, Rz36nm and Ry60nm respectively. Surface was polished by diamond paste with average grit size of 0.25 μ m, roughness was further reduced to Ra2nm, Rz15nm and Ry32nm. Finally, the deep concave paraboloidal mirror was made by the electro-forming method.

REFERENCE
1. H.Ohmori, T.Nakagawa, Mirror surface grinding of silicon wafer with electrolytic in-process dressing, Annals of CIRP, 1990, 39(1): 329-332
2. H.Ohmori, Electrolytic in-process dressing (ELID) grinding technique for ultraprecision mirror surface machining, International journal of the JSPE, 1992, 26(4): 273-278

A Study on the Surface Integrity of Single Crystal Silicon Ground by CIFB-diamond Wheels(ELID) and Resin-bonded Diamond Wheels

C. L. Chao[1], K. J. Ma[2], D.S. Liu[2], S.C. Sheu[2], Y.S. Lin[2] and H. Y. Lin[3]

[1]Department of Mechanical Engineering, E790, Tam-Kang University, No.151 Ying-Chuan Road, Tamsui, Taipei Hsien, Taiwan-251, R.O.C. (email: clchao@mail.tku.edu.tw, Tel: -886-922-043212 Fax: -886-3-3900681)

[2]Department of Mechanical Engineering, Chung-Cheng Institute of Technology, Ta-Hsi, Taoyuan, Taiwan-335, R.O.C.

[3]MEMS Division, MIRL, Industrial Technology Research Institute, Chu-Tung, Hsin-Chu, Taiwan-310, R.O.C.

Abstract

Both metal-bonded (ELID) and resin-bonded diamond wheels were used in this study to grind silicon (100) wafers and efforts were made to investigate the characteristics of obtained surface integrity. The results showed that, depending on the wheel and machining conditions, amorphous layer, nanocrystals, polycrystals, dislocations and micro-cracks were observed in both cases. It was found that the ELID ground surfaces were very sensitive to the ELID parameters and the optimized conditions were difficult to obtain and be maintained. In the case of resin bond wheel, except when feedrate is very low (say 2μm/min) or wheel is loaded, the amorphous layer generated were normally thinner than those generated by ELID grinding under the same grinding conditions.

1. Introduction

Single crystal silicon, having many advanced physical and mechanical properties, is now widely used in the semiconductor industry (account for more than 90% of the semiconductor devices). Owing to the increasing demands on brittle materials such as advanced ceramics, glasses and single crystal silicon, researches on ductile-mode grinding and related cutting theories, material removal mechanism have attracted many researchers' attention (Bifano et al 1988; Blake and Scattergood 1988; Chao et al 1989, 1997, 1999, 2000; Puttick et al 1989, 1990; Itoh et al 1998; Schinker and Doll 1983, 1987, Abe 2000). As a results, the traditional lapping, etching, polishing routine of making wafers is suggested by many researchers to be replaced by precision grinding and polishing if the requirements of flatness, TTV and roughness are to be fulfilled when producing 12"~16" wafers. In order to minimize the polishing works and the resulted deterioration of form accuracy, it is important to reduce the grinding induced surface/subsurface damage. It is well known that the grit size of abrasive on the grinding wheel has profound effect on the obtained surface roughness. Generally speaking, the smaller the grit size, the better surface finish and the less damaged layer could be achieved. However, when abrasive gets smaller the wheel has bigger chance to be loaded by swarf (chips). As a result, the wheel constantly needs to be redressed and the process becomes impractical to be employed in the real production. The ELID (electrolytic in-process dressing) technique, developed by Ohmori[1992, 1994,1995, 1999], offers a way of in-process monitoring/dressing the grinding wheel which enables the wheel of ultra-fine abrasives to be used. However, ELID grinding has its own difficulties. Problems like peak voltage/peak current to be used, gaps to be maintained, on/off duration of dressing, type of electrolyte, shape of electrode, wear of wheel and correlations between ELID parameters and grinding conditions are somehow making ELID grinding complicated to implement. That is apparently one of the reasons some people choose to remain in using resin-bonded wheel for their fine grinding works.

2. Experimental Setup

In order to clarify the underlying differences between ELID grinding and grinding by resin-bonded wheel, both ELID and resin-bonded wheel grinding were conducted in this study. A Nachi RGS20N ELID grinding machine was used in this study for carrying out grinding experiments. The power supply and CIFB(cast iron fiber bonded) diamond wheels were made by Fuji (ELIDer 630) and Noritake respectively. The (100) silicon wafer was placed on the work spindle which was set to rotate at a relatively low speed (100~400rpm). The grinding head was set to operate at the speed of 1000 to 3000rpm. The feed rates, peak voltages and peak currents ranged respectively from 2μm/min to 8μm/min, from 30V to 60V and from 2A to 10A. A pulse duration (Ton/off) of 2μSec and gap of 0.4mm were used in the study. An Okamoto Grind X VG-40 precision grinding system was used to carry out the grinding tests of resin bonded wheel. Apart from the parameters for ELID, the grinding conditions (speed, feed, total stock removed..) were kept as close as possible to those used in CIFB/ELID grinding.

3. Results and Discussions

Specimens ground under various wheels/conditions and grinding parameters were subsequently observed using SEM(scanning electron microscope), AFM(atomic force microscope) and HRTEM(high resolution transmission electron microscope) to analyze its surface and subsurface. The results showed that the obtained surfaces, depending on the wheel and machining conditions, were characterized by amorphous layer, poly-crystalline layer, occasionally micro-cracks/ distributed dislocation loops (~300nm into the substrate) and substrate of perfect single crystal (*Figure 1* and *Figure 2*).

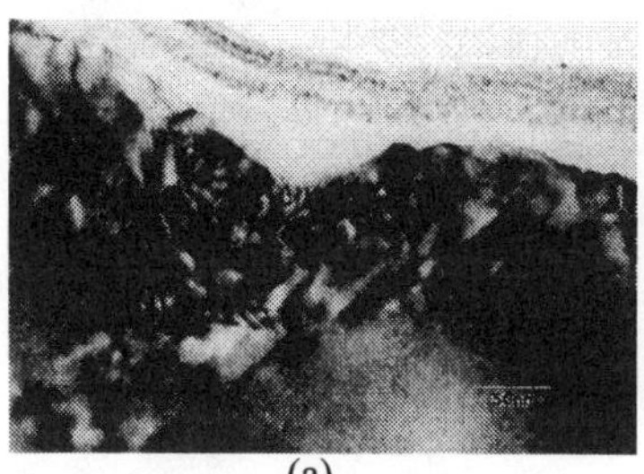
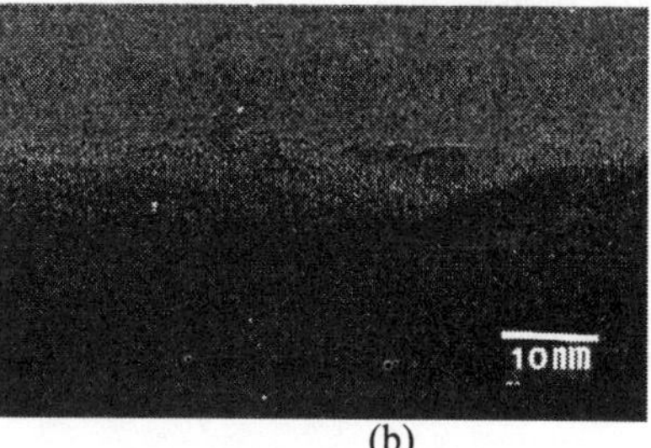

(a)	(b)

Figure 1 HRTEM micrographs of the ELID ground silicon surfaces using #6000 diamond wheel and (a) 60V, 10A, 3000/100rpm, 2μm/min (b) 60V, 10A, 3000/ 100rpm, 8μm/min, zone axis ∶ [110]

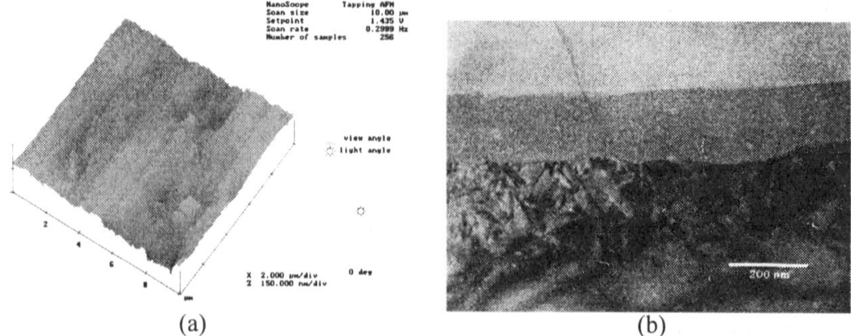

Figure 2 (a)AFM (b)HRTEM micrographs of the ELID ground silicon surfaces (#6000, 30V, 2A, 2000rpm, 400rpm, 2μm/min) zone axis : [110]

Generally speaking, the obtained surface integrity relied heavily on the conditions of abrasives. Apart from the machining conditions such as speed and feedrate, ways of dressing and types of wheels also played important roles in the conditions of abrasives. Having the relatively greater abrasives retaining force, metal bond wheels are good in achieving form accuracy but extra efforts have to be made in dressing if good surface roughness is to be persuited. In the case of ELID grinding, ELID parameters heavily involved not only in preventing wheel from loading but also in maintaining the cutting capability of abrasives. However, when grinding with a resin bond wheel the ability to self-sharpen became the major factor of keeping the new sharp abrasives to emerge to the surface.

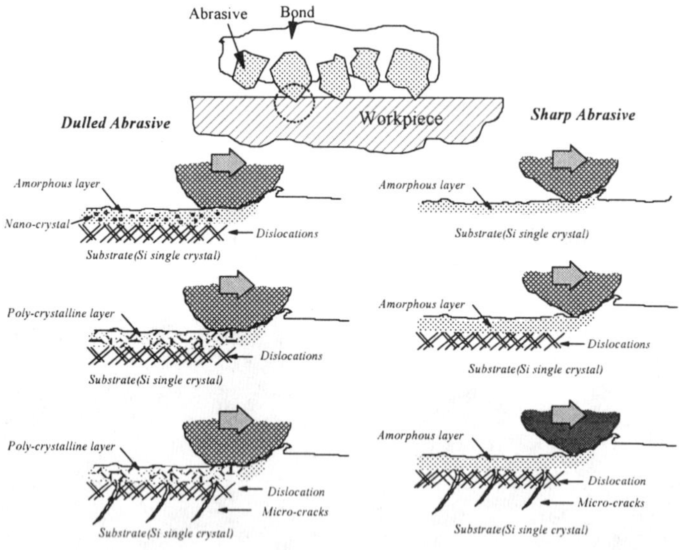

Figure 3 Schematical representation of the surfaces generated by dulled and sharp abrasives at various depth of cut (from top to bottom, shallow to deep cut)

Shown in *figure 3* are the schematical representations of the surfaces generated by dulled and sharp abrasives at various depth of cut (from top to bottom, shallow to deep cut). A dulled abrasive tends to generate relatively more heat during the cutting process than sharp ones. As a result, relatively thicker amorphous layer, scattered nanocrystalline silicon or polycrystalline silicon are normally the consequences. In the case of ELID grinding, thick oxide layer and high feedrate favors wheel to produce sharp new protruding grains so that a stable cutting condition can be reached. When thick oxide layer and low feedrate is used, there is still chances for some protruding grains getting excess attritious wear before it is pulled out. The worn grains will dull the wheel and generate much friction heat and subsurface damage.

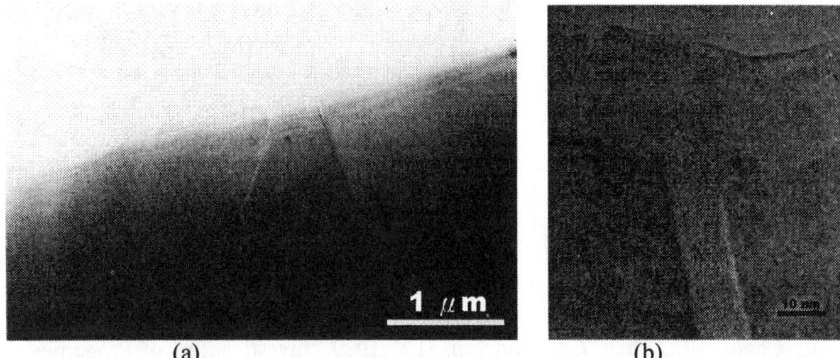

(a)   (b)

Figure 4 HRTEM micrographs of the silicon surfaces ground by resin bond wheel showing the (a) amorphous layer and microcracks (b)detailed view of the crack on the right-hand side (#2000, 2400rpm, 100rpm, 5μm/min) zone axis ∶ [110]

In the case of resin bond wheel, abrasives refresh in a relatively higher rate due to its low abrasive retaining force. This means that, except when feedrate is very low (say 2μm/min) or wheel is loaded, the wheel tends to cut with relatively sharp abrasives. This also reflected in the HRTEM observations that the amorphous layer generated by grinding with resin bond wheel were normally thinner than those generated in ELID grinding under the same grinding conditions.

4. Conclusions

1. Depending on the wheel and machining conditions, amorphous layer, nanocrystals, polycrystals, dislocations and micro-cracks were observed both on ELID ground and resin-bond wheel ground surfaces
2. It was found that the ELID ground surfaces were very sensitive to the ELID parameters and the optimized conditions were difficult to obtain and be maintained. In the case of grinding with resin bond wheel, the surface integrity was relatively insensitive to the machining conditions in comparison to ELID grinding.
3. In the case of ELID grinding, thick oxide layer and high feedrate favors wheel to produce sharp new protruding grains so that a stable cutting condition can be reached. When thick oxide layer and low feedrate is used, there is still chances for some protruding grains getting excess attritious wear before it is pulled out.

379

The worn grains will dull the wheel and generate much friction heat and subsurface damage.

4. In the case of resin bond wheel, except when feedrate is very low (say 2μm/min) or wheel is loaded, the wheel tends to cut with relatively sharp abrasives and the amorphous layer generated were normally thinner than those generated by ELID grinding under the same grinding conditions.

References

Abe, K. 2000 "Next Generation Manufacturing Technologies for Super-Large and Super-flat Silicon Wafer", Advances in Abrasive Machining, Proc. of ISAAT2000, pp21-28, Oct. 2000, Hawaii, U.S.A.

Bifano, T.G.; Dow, T.A. and Scattergood, R.O. 1988 "Ductile-regime Grinding of Brittle Materials", Proc. Conf. on Ultra-Precision in Manufacturing Engineering, Achen, Germany

Blake, P.N. and Scattergood, R.O. 1988 "Ductile Regime Turning of Ge and Si", Proc. 2nd Inter-society Sympos. of Ceramic Materials and Components, ASME Winter Ann. Mtng., Chicago, Nov. 30-Dec.2, p249

Chao, C.L. and Gee, A.E. 1989 " Investigation of Brittle/Ductile Transition in Single Point Machining Glassy Materials Using a Ruling Engine", 5th Int. Precision Engineering Seminar & Annual Meeting of Amer. Soc. for Precision Engineering, 18-22 Sept., Monterey, CA, USA

Chao, C.L.; Lin, H.Y.; Fang, C.M.; Shy, T.L. and Hung, Y.T. 1997 "The Machinability of Single Crystal Silicon", "Progress in Precision Engineering and Nanotechnology", Ed. By H.Kunzmann et al, Braunschweig, Germany, p578

Chao, .L.; Chen, G.L., Sheu, C.C. and Lin, H.Y. 1999 "Investigations of Properties of the Oxide Layer in ELID Grinding", Proc. of the 3rd Conf. On Nanotechnology and Micro-systems, May 4-5, ITRI, Hsin-Chu, Taiwan, pp2-43 ~ 2-50

Itoh, N.; Ohmori, H.; Kasai, T.; Karaki-Doy, T. and Horio, K. 1998 "Characteristics of ELID Surface Grinding by Fine Abrasive Metal-Resin Bonded Wheel", Int. J. Japan Soc. Pre. Eng., Vol.32, No.4, p.273

Ohmori, H. 1992 "Electrolytic In-Processing Dressing(ELID) Grinding Technique for Ultraprecision Mirror Surface Machining" Int. J. Japan Soc. Prec. Eng., Vol.26, N0.4, p.11

Ohmori, H. and Takahashi, I. 1994 "Efficient Grinding Technique Utilizing Electrolytic In-Process Dressing for Precision Machining of Hard Materials", Adv. Of Intelligent Production, Eiji Usui (editor), JSPE, Elsevier Sci. B.V.

Ohmori, H. and Nakagawa, T. 1995 "Analysis of Mirror Surface Generation of Hard and Brittle Materials by ELID Grinding with Superfine Grain Metakkic Bond Wheels", Annals of the CIRP, Vol.44/1, p.287

Ohmori, H.; Takahashi, I. And Nakagawa, T. 1990 "Mirror Surface Grinding of Silicon Wafers with Electrolytic In-Processing Dressing", Ann. CIRP, Vol.39, No.1, p.329

Puttick, K.E.; Rudman, M.R.; Smith, K.J.; Franks, A. and Lindsey, K. 1989 "Single-Point Diamond Machining of Glasses", *Proc. Royal Soc.* A.426, pp19-30

Puttick, K.E. and K.J.; Franks 1990 "The Physics of Ductile-brittle Machining Transitions: Single-point Theory and Experiment", *JSPE* Vol. 56, No.5, pp788-792

Schinker, M.G. and Doll, W. 1987 "Turning of Optical Glasses at Room Temperature", *SPIE* Vol. 802, pp70-80

Schinker, M.G. and Doll, W. 1983 "Basic Investigations into the High Speed Processing of Optical Glasses with Diamond Tools", *SPIE*, Vol. 381, pp32-38

A STUDY ON SCRATCH REDUCTION USING THE SONIC DISPERSION OF CMP SLURRY

Sung-Hwan CHO[a], H.J. KIM[a], H.Y. KIM[a], K.J. KIM[a], H.D. JEONG[b]

[a]*Graduate School of Precision Mechanical Engineering,*
Pusan National University
[b]*School of Mechanical Engineering, Pusan National University,*
San 30, Changjeon-Dong, Kumjeong-Ku, Pusan, 609-735, Korea
Tel : +82-51-510-3210, Fax : +82-51-518-8442,
E-mail : atommy@dreamwiz.com

Abstract

Huge particles agglomerated in CMP slurry cause scratches on the wafer in CMP. This study focuses on the reduction of the scratches in CMP with the aid of ultrasonic and megasonic energy. During ultrasonic irradiation into the CMP Slurry, the abrasives in the slurry are dispersed by ultrasonic cavitation. As for megasonic irradiation, the abrasives are also dispersed by enormous megasonic acceleration.

Keywords

CMP(Chemical mechanical polishing), Dispersion, ILD(Interlayer dielectric), Megasonic, pH, Ultrasonic

1. INTRODUCTION

There are a variety of defects on a wafer surface after CMP. The most significant problem in CMP is scratch generation on metal or dielectric layers which causes the malfunction of electric devices. There are various reasons for scratch generation in CMP. Especially, large particles in slurry chemicals give a dominant effect upon scratch generation[1].

In this study, to know whether ultrasonic waves and megasonic waves affect the dispersion of abrasives in slurry or not, the volume distributions of slurry particles are measured according to irradiation time[2][3]. And for effective dispersion, the pH and temperature of slurry are measured according to the irradiation time as well. Then the dispersion stability of slurry is checked. And TEOS wafers are polished to figure out the impact of the slurry by ultrasonic irradiation. Finally, by processing tungsten wafers with EPW 2000 slurry, we perceived the relation between the scratches on the wafer and large abrasives in slurry at metal CMP.

2. EXPERIMENTS

2.1 Experiments of Abrasive Dispersion with Ultrasonic and Megasonic Waves

2.1.1 Particle Dispersing Experiments Using Ultrasonic waves

Figure 1 shows the changes of the abrasive volume ratio (vol%) and abrasive diameters(D_{10}, D_{50}, D_{90}) in slurry according to ultrasonic irradiation time. Figure 1(a) shows, after ultrasonic irradiation for 1 minute, huge abrasives over $1\mu m$ which bring scratches during CMP are dispersed and reduced. Figure 1(b) shows in detail after the

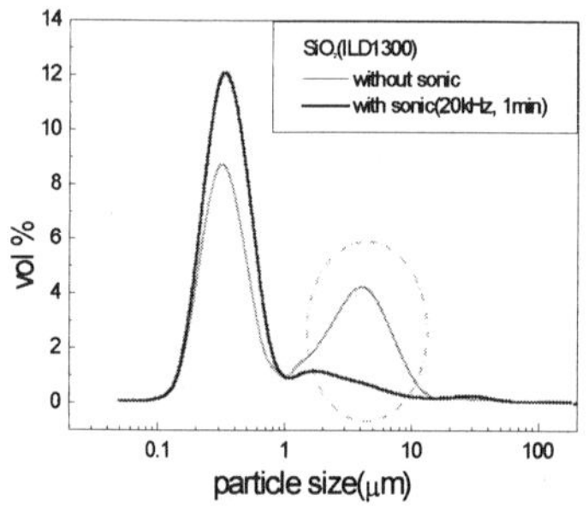

(a)The vol% for each time   (b)The particle size changes

Figure 1. The dispersion of ILD 1300 slurry using ultrasonic(20kHz)

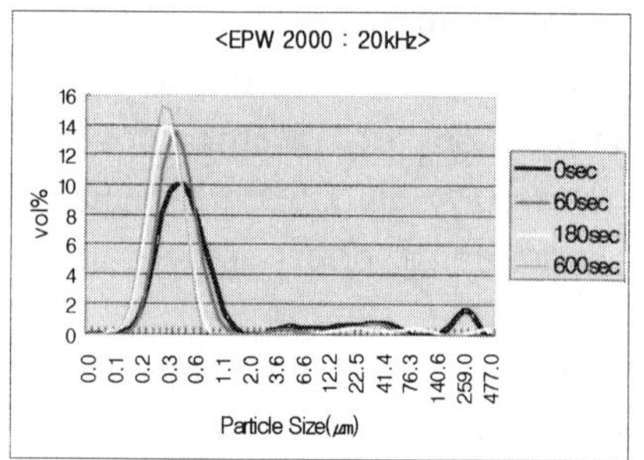

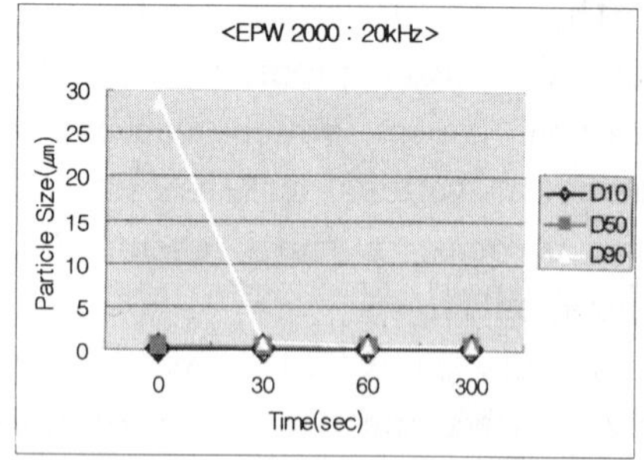

(a)The vol% for each time   (b)The particle size changes

Figure 2. The dispersion of EPW 2000 slurry using ultrasonic(20kHz)

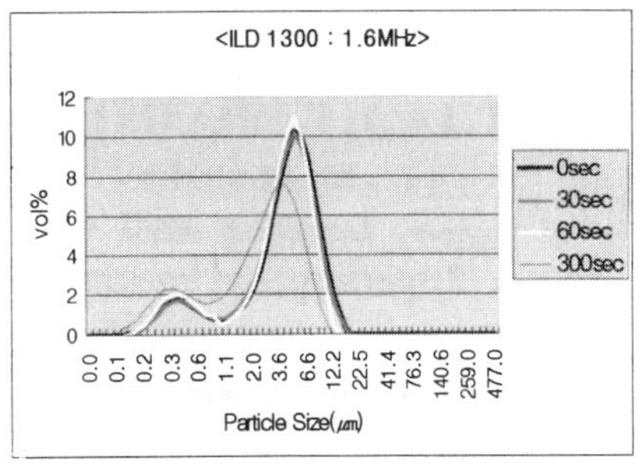

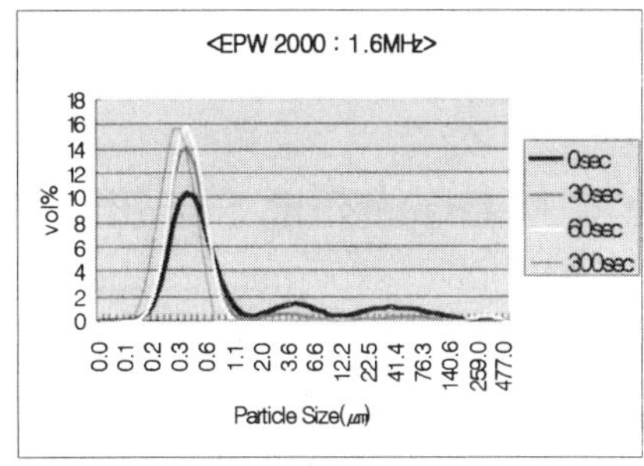

(a)The vol% of ILD 1300   (b)The vol% of EPW 2000

Figure 3. The partlcle dispersion using megasonic(1.6MHz)

ultrasonic irradiation, the number of huge abrasives is reduced drastically while the size of abrasives in the fine region is almost intact.

Figure 2 indicates that like in the ILD 1300 slurry case, when EPW 2000 CMP slurry is irradiated by ultrasonic wave, huge abrasives which cause scratches during the CMP process are also dispersed into the increasing number of fine abrasives.

2.1.2 Particle Dispersing Experiments Usin₂ Megasonic waves

Figure 3 reveals that megasonic irradiatiᴗɴ has the same effects of silica(ILD 1300 slurry) and alumina(EPW 2000 slurry) abrasives dispersion as those of ultrasonic.

2.2 The Measurement of Temperature and pH in the Processes of Ultrasonic and Megasonic Irradiation in Slurry

Figure 4 shows the temperature and pH changes of ILD 1300 slurry when the slurry is irradiated by ultrasonic and megasonic waves. It is clear that as the longer is the ultrasonic and the megasonic irradiation time the higher goes up temperature and the lower becomes pH. In addition, it seems the megasonic waves rather than the ultrasonic waves cause more drastic changes of temperature and pH even if output powers, which are 1200W for the ultrasonic and 30W for the megasonic, and slurry amounts, which are 2000㎖ for ultrasonic and 200㎖ for megasonic, are considered. Especially, when it comes to megasonic waves, pH shock has been detected at the first stage of irradiation. It is presumed that the pH shock affects the momentary agglomeration of abrasives in slurry as it is shown in figure 3(b).

2.3 A Dispersion Stability Experiment

For this experiment, abrasives in slurry dispersed by megasonic waves are observed to see what changes happen as time passes. After alumina abrasives dispersed in slurry are getting agglomerated again little by little as time passes and huge abrasives over $200\mu m$ are produced. After 5 days, as it is clear in figure 5, there are only two sorts of abrasives, one is the abrasives in fine region and the other is abrasives around $200\mu m$. This means abrasives dispersed into fine region by the megasonic waves were practically unchanged and the huge abrasives cohered by enormous megasonic acceleration keep agglomerating as time passes.

Therefore, to get a stable CMP process by megasonic irradiation it

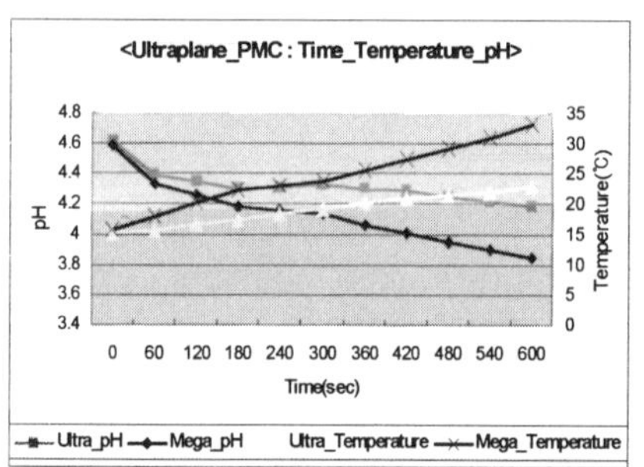

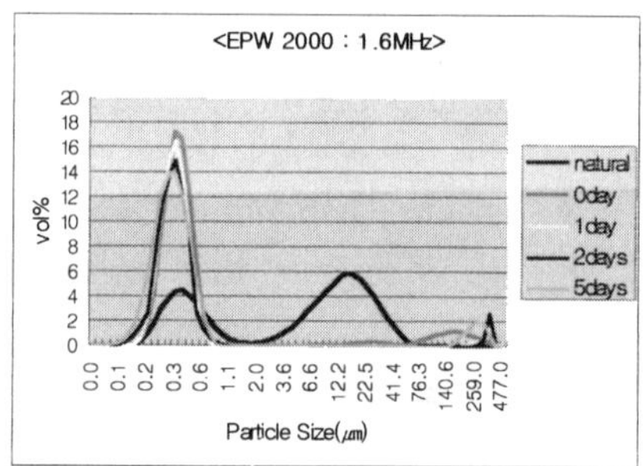

Figure 4. The temperature and pH change of ILD 1300

Figure 5. The stability experiment

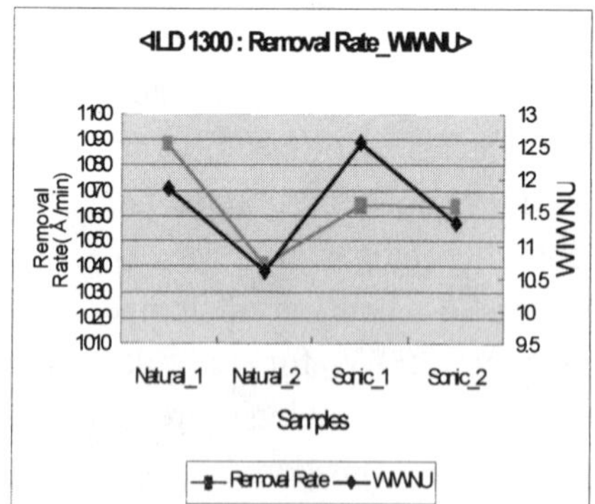

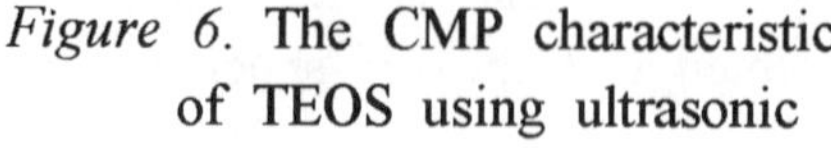

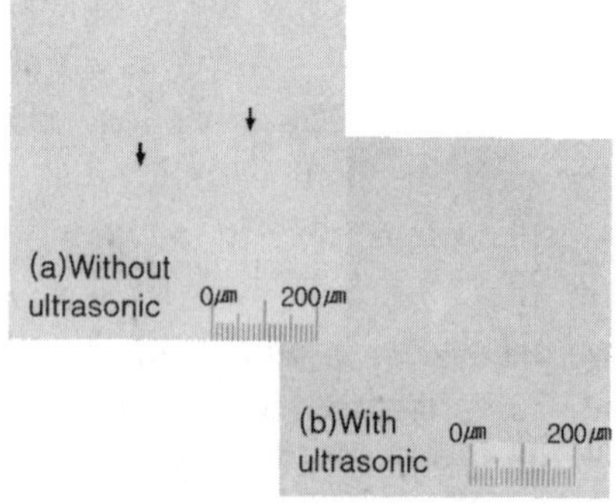

Figure 6. The CMP characteristic of TEOS using ultrasonic

Figure 7. The images on W wafer after CMP

is needed to prevent re-agglomeration into huge abrasives according to time, by completely dispersing huge abrasives into the range of process particles(about $0.1 \sim 1 \mu m$).

2.4 Experiments of TEOS wafer's CMP Characteristic and Scratch Reduction in Tungsten wafers by the ultrasonic

This experiment is designed to understand the effects that the property changes of slurry by the ultrasonic and megasonic waves give to CMP process. Figure 6 shows CMP characteristic doesn't change a lot whether the ultrasonic irradiation is given or not. So, the stable CMP process can be got with slurry irradiated by the ultrasonic waves.

Finally, a tungsten wafer CMP process is performed to make sure that the scratch reduction effect works in the real metal CMP process. Figure 7(a), the picture of the tungsten wafer surface which is processed by a natural EPW 2000 slurry, shows a couple of scratches. In contrast, in figure 7(b), we barely see scratches.

3. CONCLUSION

With the experiments, we assure that when huge abrasives in CMP slurry are dispersed by the particle dispersing effect of the ultrasonic and megasonic waves, scratches can be reduced in the CMP process. However, the characteristics of dispersion are different according to the sonic frequency, irradiating time, output power, the sort of slurry, the amount of slurry and the sort of abrasive in the slurry. Therefore with the fittest ultrasonic waves and megasonic waves, we can disperse huge abrasives in the slurry and then prevent the scratch generation in the CMP process.

4. REFERENCES

[1]S. H. Li, R. O. Miller, *Chemical Mechanical Polishing in Silicon Processing*. San Diego, 1997.
[2]S. I. Hatanaka, T. Taki, M. Kuwabara, M. Sano, S. Asai, Effect of Process Parameter on Ultrasonic Separation of Dispersed Particle in Liquid. Japanese Journal of Applied Physics, May, 1993; vol. 38, no. 5B, pp. 3096-3100.
[3]A. K. Hipp, G. Storti, M. Morbidelli, Particle Sizing in Colloidal Dispersions by Ultrasound. Model Calibration and Sensitivity Analysis, American Chemical society, 1999, pp. 2338-2345.

A STUDY ON THE CHEMICAL MECHANICAL MICRO MACHINING (C3M) OF SILICON

SangCheol Jeong and JunMin Park
Graduate School of Precision Mechanical Engineering, Pusan National University, Pusan 609-735, Korea

HaeDo Jeong
School of Mechanical Engineering, Pusan National University, Pusan 609-735, Korea

ABSTRACT

This paper proposes the concept and mechanism of chemical mechanical micro machining (C3M) for the silicon of semiconductor material. The experiment is focused on the form accuracy and efficiency comparison with pure mechanical and C3M. The hydrated layer of silicon reacted by the chemical solution is evaluated using SEM and XPS. Finally, it is confirmed that C3M is one of the powerful tools for micro machining or patterning with the aid of synergy effect of the chemical reaction

KEY WORDS

C3M(Chemical Mechanical Micro Machining), grooving, PCD(Poly Crystal Diamond), XPS(X-ray Photoelectron Spectroscopy)

1 INTRODUCTION
- Industrial needs and difficulty in micro machining

Milli-Structure means the system with the dimensional size of several millimeters. So it needs to fabricate the parts in size of several micrometers considering assembly.

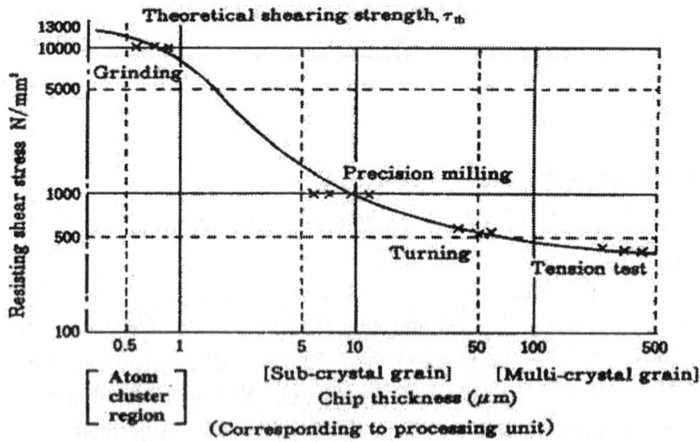

Fig.1 Example of radical increase for the resisting shear stress

It is required to organize the process with the high precision, high machining ability, and high form accuracy for applying for the production technology. But, as the resisting shear stress is radically increased in case of machining the microstructure according to the reduction of depth of cut, it is difficult to remove the diminutive material efficiently unlike the macro machining[1].

2. Research direction and basic principle

2.1 Research direction

This paper proposes the chemical mechanical micro machining (C3M) concept and applies on the silicon material and shows the hydrated layer by XPS. C3M is a noble method for micro machining with efficiency and accuracy. The mechanism of C3M for 2.5 dimensional features is now being developed for the broad applications, such as electronics, optics, information-communication and medical.

2.2 Basic principle

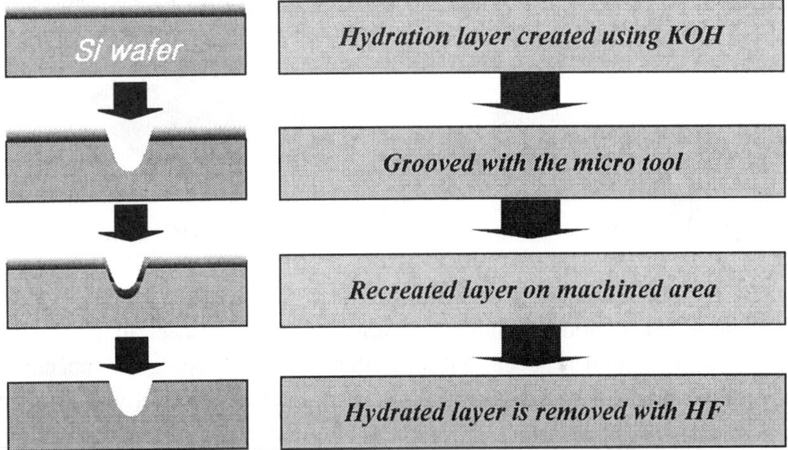

Fig. 2 Machining concept of C3M for Si wafer

A basic principle of C3M (Chemical Mechanical Micro Machining)[2] is followed. First, chemical solution reacts with the surface of material. Second, the reaction produces the chemically reacted layer, which might have other mechanical properties. Finally the reacted layer with/without substrate layer is machined with the mechanical method.

3. Experimental of silicon C3M

As one of the example for chemical mechanical micro machining, Si of the brittle nonmetallic material was chosen, widely consuming in semiconductor and MEMS processes. Surface patterning processes in semiconductor or MEMS have many problems with the processes of many numbers, the high cost system and system dependency using Si wafer in spite of high accuracy. However, C3M gives advantages in process flexibility and controllability, etc. without mask and photoresist in conventional lithography. In order to conquer these problems, C3M process can be applied.

3.1 Experimental setup

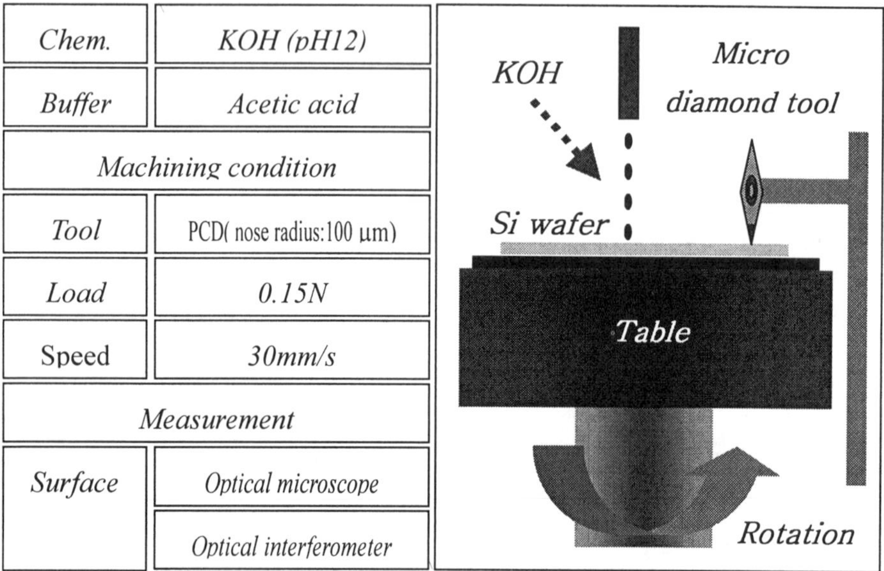

Fig. 3 Experimental setup for chemical mechanical micro machining

3.2 RESULTS AND DISCUSSES

While KOH solution with pH 12 was used with respect to the chemical reaction, the acetic acid was added to it for buffer chemical keeping the concentration change of OH⁻ small. In mechanical removal step, the tool material was PCD (poly crystal diamond) of 100 μm radius. The constant normal load of 0.15N was put on the surface of Si wafer. With the condition of pH 12, the bond of $\equiv$Si-OH(hydrated silica form) on the surface of Si was created and formed the $Si(OH)_4$ and broken by the strain induced by PCD micro tool.[3] Under the 0.15N load condition, the tool removes not only the upper $Si(OH)_4$ layer but also unreacted substrate.

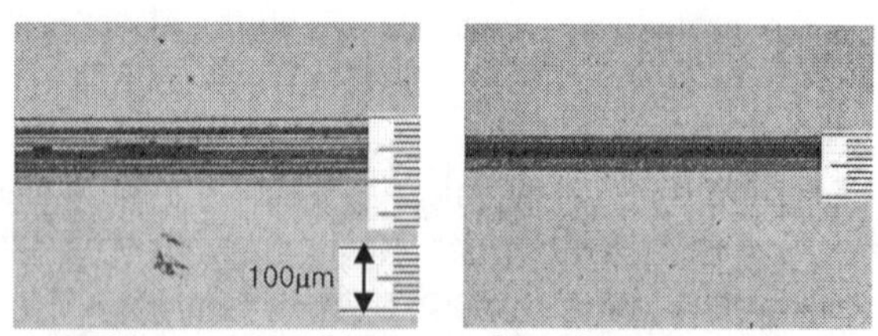

(a) Without reacted layer   (b) With reacted layer

Fig. 4 Pictures by optical microscope of grooved silicon wafer for 1 cycle

Fig. 4 by the optical microscope showed that the width of pattern with reacted surface was narrower than without reacted surface, 110 μm to 60 μm. Chipping which is found on the boundary of the removed area, is reduced to able to make neat profile on Si wafer in using C3M.

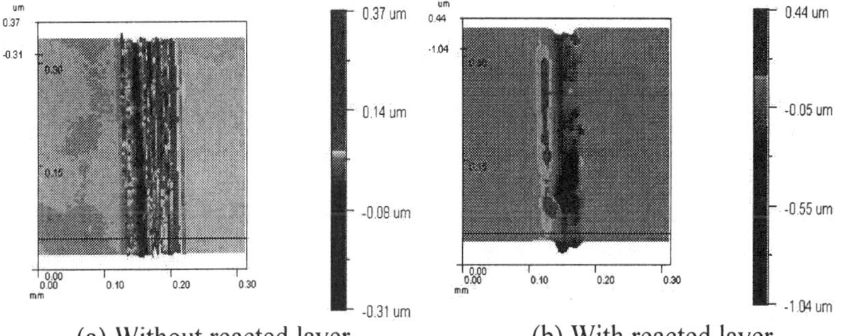

(a) Without reacted layer   (b) With reacted layer

Fig. 5 Pictures by optical interferometer of grooved silicon wafer for 1 cycle

The results from ACCURA, the optical interferometer represented that the depth of groove with reacted surface was over 2 times deeper than without reacted surface, 0.714 μm to 1.480 μm with same load of 0.15N.

4. SEM microgragh and the conformation by XPS of hydrated layer

Contrary to the general thin films, the hydrated layer created using the KOH solution wasn't able to be measured by the ellipsometry. To measure the layer, the hydrated surface of silicon wafer with the KOH solution was grounded and polished for the measurement of SEM. To measure thickness of the hydrated layer on silicon by the KOH solution, the specimen is prepared by the grinding and polishing and measured by SEM.

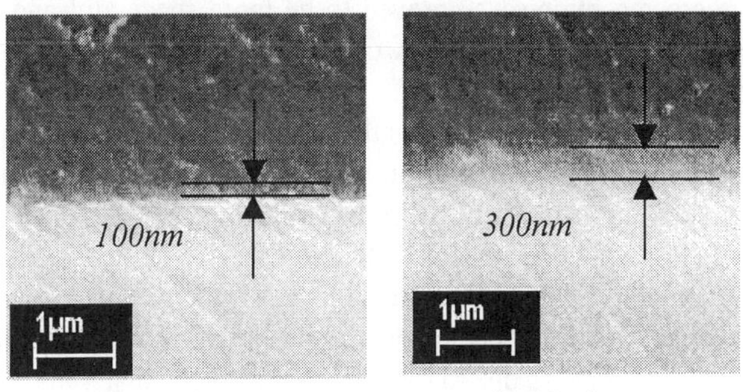

(a) 5min   (b) 10min

Fig. 6 Section view images of Noran Instrument SEM for hydrate layers

Fig. 6 by a Noran Instrument SEM shows how deep the hydrated layer by the KOH solution with pH12 is. A layer hydrated for 5 min was 100nm thick or so, and the other for 10 min was 300nm thick. Both layers had the gradation from light to dark from the silicon surface, which might mean the more active permeation of O^{2-} and OH^- by the KOH solution for some depth.

An EscaLab Model 250 X-ray Photoelectron Spectroscopy instrument presents the existence of oxygen on the hydrated layer with pH12, 0.1w% of the KOH solution for 5 min to 100nm in fig. 7.

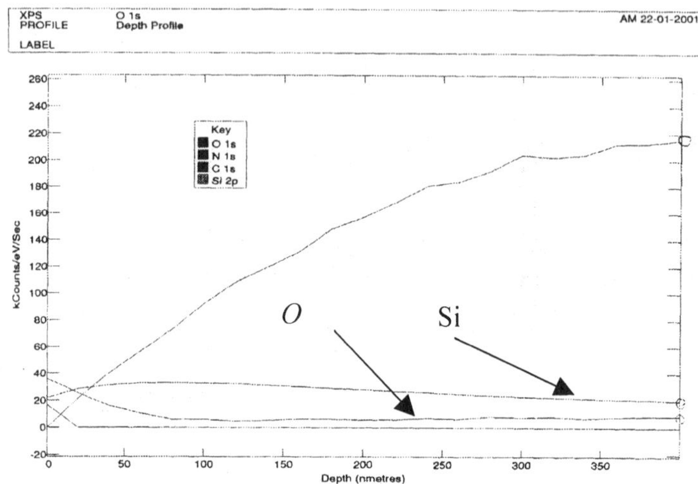

Fig.7 XPS depth profile for each atomic existence in hydrated silicon wafer

5 CONCLUSION

Chemically reacted surface and its mechanical machining gave an advantage over the groove machining with narrower and deeper grooves in comparison to mechanical only machining of Si wafer. Micro groove fabrication by chemical mechanical micro machining was proved to be more successful and more efficient in machining Si wafer. And the depth of chemically reacted layer is presented by SEM images and XPS depth profile.

This process will be applied on micro-groove needed parts of mold, requiring the high aspect ratio and high machining accuracy for example, micro channels, micro heat radiator, multi-fiber optical connector, etc.

REFERENCES

1   Norio Taniguchi, (1996). Nanotechnology. Fundamentals of nanotechnology, table 1.1.2 Distribution of defects in materials: failure due to movable dislocations in ductile materials and fracture due to micro cracks in brittle materials, p.16

2   JunMin Park, (2000), American Society for Precision Engineering, A Study on Fabrication and Application of Micro Tool by Using High Speed Chemical Etching.

3   Futoshi Katsuki, (2000) Journal of the Electrochemical Society, 147(6) p. 2328-2331, AFM studies on the Difference in Wear Behavior Between Si and SiO_2 in KOH Solution.

DEVELOPMENT OF A LAPPING FILM UTILIZING AGGLOMERATIVE SUPERFINE SILICA ABRASIVES FOR EDGE FINISHING OF A SILICON WAFER

Toshiyuki Enomoto [1], Yasuhiro Tani [2] and Kazuya Orii [3]

[1] Ricoh Company, Ltd., [2] The University of Tokyo, [3] Tokyo Magnetic Printing Co., Ltd.

Abstract

In the present manufacturing process of a silicon wafer, wafer edge finishing with loose abrasives is essential for preventing wafer breakage and particle contamination due to micro cracks. Loose-abrasive machining, however, has problems involving inefficiency and an unclean working environment. Therefore, fixed-abrasive machining such as film lapping has received attention as an alternative technology, but conventional lapping films have problems involving low finishing efficiency or low finished surface quality. In this study, agglomerative superfine silica abrasives are adopted in the lapping film to overcome these problems. The finishing experiments revealed that a high edge surface quality can be achieved with high efficiency.

Keywords

Lapping film, Agglomerative abrasive, Silica, Silicon wafer, Edge finishing

1. INTRODUCTION

Numerous micro cracks that cause wafer breakage and particle contamination exist on the edge of a silicon wafer after edge grinding. Thus, edge polishing with colloidal silica slurry is adopted after grinding [1]. Loose-abrasive machining, however, is inferior to fixed-abrasive machining in terms of finishing efficiency and working environment. Therefore, fixed-abrasive machining such as film lapping has received attention as a technology for replacing polishing with loose abrasives; however, the conventional lapping films have several problems during practical use.

In this paper, we identify these problems and propose the use of a new lapping film utilizing agglomerative superfine abrasives to overcome them.

2. PROBLEMS ASSOCIATED WITH A LAPPING FILM UTILIZING CONVENTIONAL ABRASIVES

2.1 Experimental Procedure

Three types of lapping films using conventional abrasives were prepared, as shown in Table 1. Diamond abrasive was adopted for achieving high finishing efficiency. Silica abrasive has a mechanical-chemical reaction with silicon [2], thus it was expected that a high surface quality can be obtained.

A finishing experiment on an edge of a 5" silicon wafer (thickness: 770 μm) was carried out with the lapping films. Figure 1 shows a photograph of the finishing apparatus. The finishing conditions are listed in Table 2. A lapping film was fed while oscillating in the orthogonal direction against the wafer revolution. The surface roughness of the wafer edge was adjusted to 0.3 μm Ry before finishing. The finishing cross hatch angle was set to be different from that of pre-finishing in order to evaluate the extent of material removed by observing the wafer edge microscopically.

Table 1. Symbols for lapping films using conventional abrasives

Symbol	Abrasive	Average diameter of abrasive
D	Diamond	1 μm
SL	Silica	4 μm
SS	Silica	20 nm
Abrasive content		40 vol%
Binding resin		Polyurethane
Backing material		Polyethylene terephthalate (thickness: 25 μm)

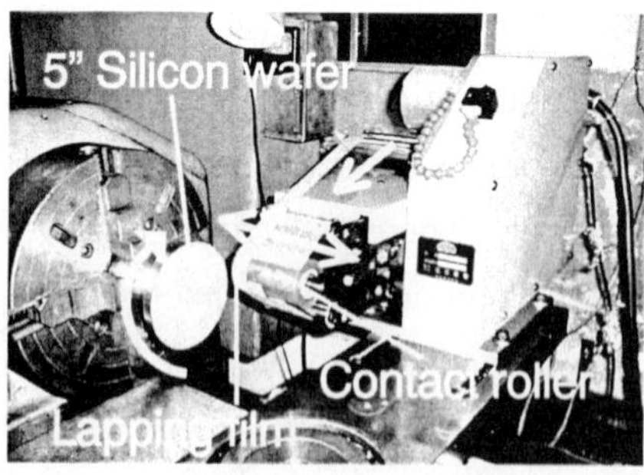

Figure 1. Overview of edge finishing setup

Table 2. Finishing conditions

Revolution of wafer	200 rpm
Film index speed	8 mm/min
Film oscillation width	4 mm
Film oscillation cycle	23 Hz
Contact pressure	1.34 MPa
Hardness of contact roller	Hs 50
Finishing time	10 min
Finishing fluid	None (dry)

2.2 Experimental Results and Discussion

Figure 2 shows the silicon wafer edge before and after finishing. A large amount of swarf was discharged when finishing with D and SL, and as a result, many cutting marks were generated (Figures 2 (b) and (c)). In the case using SS, a scratch-free surface was partly obtained, but the pre-finishing flaws remained (Figure 2 (d)).

The above results indicate the following:

- The surface of a lapping film with micron-size abrasive was so rough (7 μm Rt) that the effective cutting edge density was low and the force on the cutting edge was high. As a result, the finishing efficiency was high, but the finished surface became rough.

- The surface of a lapping film with superfine silica was so smooth (1 μm Rt) that the effective cutting edge density was high and the force on the cutting edge was low. As a result, a scratch-free surface could be achieved, but the efficiency was extremely low.

To overcome the above problems, that is, to achieve both high finishing efficiency and high surface finish, we introduced agglomerative superfine silica particles as abrasives.

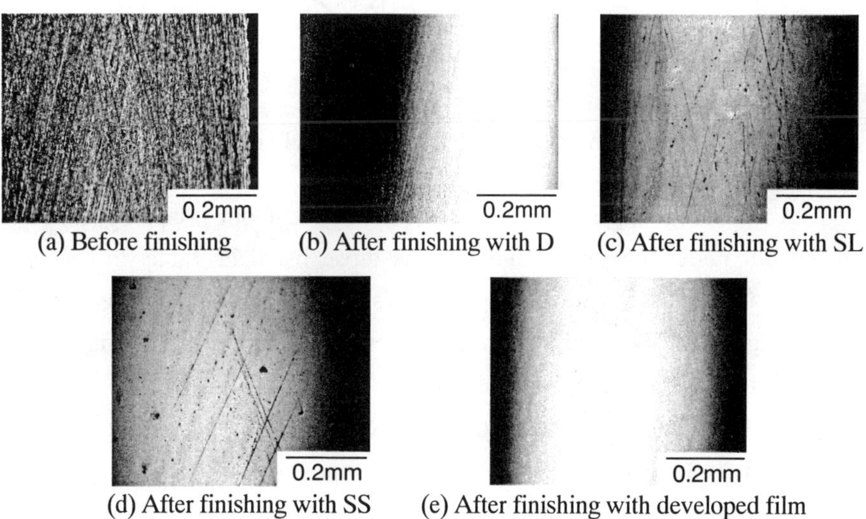

(a) Before finishing   (b) After finishing with D   (c) After finishing with SL

(d) After finishing with SS   (e) After finishing with developed film

Figure 2. Edge surface of silicon wafer before and after finishing

3. FINISHING PERFORMANCE OF A LAPPING FILM UTILIZING AGGLOMERATIVE SUPERFINE SILICA ABRASIVES

The powder (average diameter 7 μm), in which colloidal silica particles (average diameter 20 nm) were chemically agglomerated, was applied as an abrasive. The conceptual model of finishing with a newly developed lapping film is depicted in Figure 3. The finishing characteristics of the lapping film are considered to be as follows:

- At the start of finishing, the surface of the lapping film is so rough that the effective cutting edge density is low; this results in the achievement of high efficiency.
- In the process of finishing, agglomerative abrasives are worn flat uniformly, thus the grain depth of cut becomes extremely small; this results in the achievement of a high surface finish.

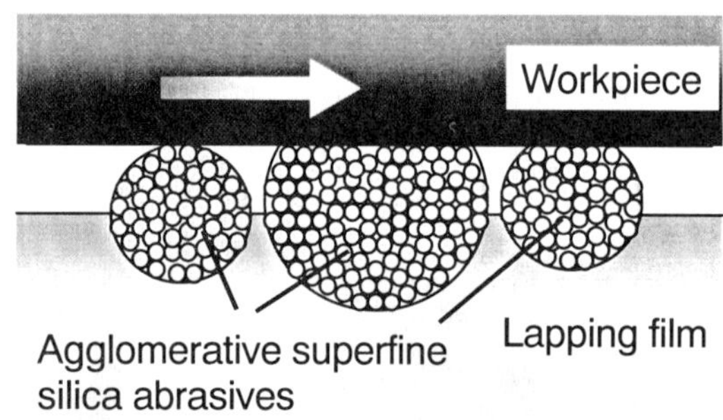

Figure 3. Conceptual model of finishing with newly developed lapping film

A finishing experiment with the newly developed lapping film was carried out under the same conditions as those listed in Table 2. The result reveals that an entirely scratch-free surface (shown in Figure 2(e)) was achieved. The results of the finished surface roughness are summarized in Figure 4. In conventional fixed-abrasive machining, the larger an abrasive, the larger the scattering of the injection heights of abrasives; therefore, the surface roughness machined by a tool with larger abrasives deteriorates. In contrast, it was found that the best surface roughness could be obtained by using the developed lapping film which has the largest abrasives among those tested.

Furthermore, to confirm that the good performance was a result of the mechanism mentioned above, observation of the lapping film was carried out. As shown in Figure 5, the film surface before finishing was rough (10 μm Rt) and agglomerative abrasives were truncated after finishing.

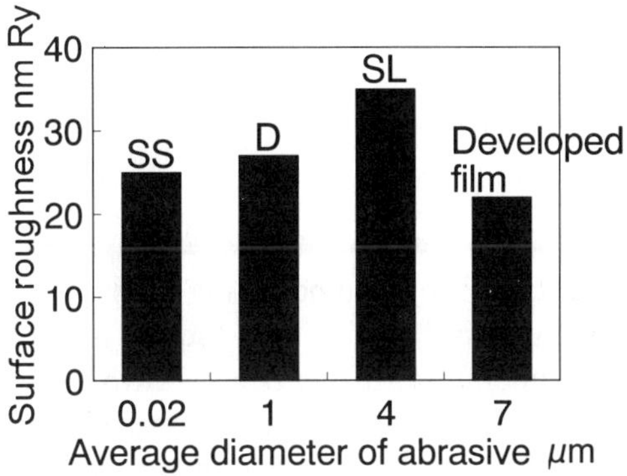

Figure 4. Comparison of surface roughness finished using different kinds of lapping films

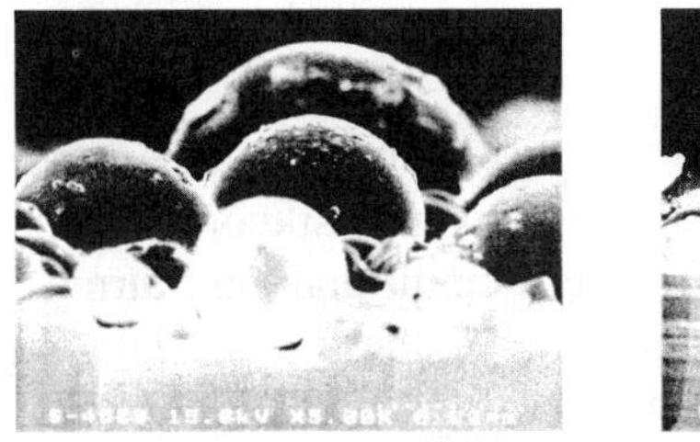

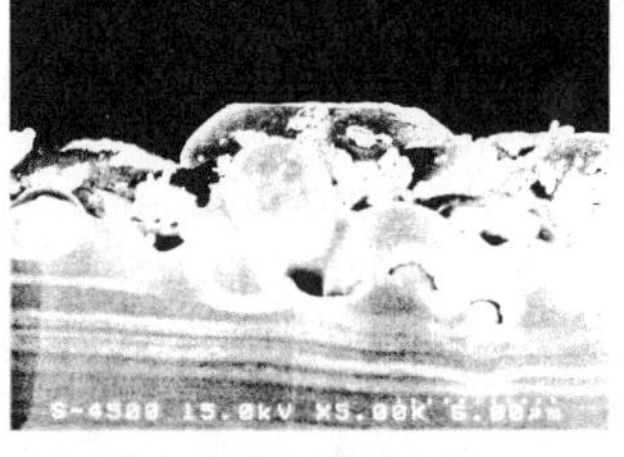

(a) Before finishing   (b) After finishing

Figure 5. Side view of a developed lapping film

4. CONCLUSION

A lapping film utilizing agglomerative superfine silica particles as abrasives was developed for finishing a silicon wafer edge. It was demonstrated that a scratch-free surface could be achieved with high efficiency using the developed lapping film.

REFERENCES

[1]   T. Abe: Silicon crystal growth and wafer manufacturing, Baifukan, Tokyo, (1994) 68.

[2]   H. Jeong, H. Noguchi and T. Nakagawa: Mirror grinding by using a stone with fine silica abrasives, Proceedings of Spring Conf. of JSPE, (1994) 513.

CRYOGENIC POLISHING METHOD
OF OPTICAL MATERIALS[*]

Feihu Zhang[1], Rongjiu Han[2], Yaowu Liu[2], Shu Pei[2]

(1.Harbin Institute of Technology, P. R. China)
(2.Academia Sinica, Changchun Institute of Optics and Fine Mechanics, P. R. China)

Abstract

This paper presents a cryogenic polishing method. The polishing liquid is frozen to ice polishing mold at first and then it is used to polish optical materials under low temperature. It has been shown by experiment that the polishing efficiency can be improved and super-smooth surface can be obtained. This is a new way to polish optical materials.

Keywords

Polishing, Cryogenic Polishing, Optical Material, Polishing liquid

1. INTRODUCTION

It is very difficult to get high form precision and super-smooth surface of hard and brittle materials, such as mono-crystal silicon and optical glass by ultra-precision machining. SPDT (single point diamond turning) that has just been getting mature has no ability to these materials. Traditional machining method, which is modified in a measure, is still the one to process these materials. In order to get high machining accuracy and efficiency of super-finishing, different physical and chemical processes are applied to machining process. For example, ultra-precision cryogenic SPDT was researched to cut ferrous metal [1].

All kinds of matter in cryogenic environment show different nature under different cooling temperature. Cryogenic polishing is to process material under $0\,^\circ\!C$. The essence of it is to make the work-piece and work environment remain the needed low temperature, so as to change the machining performance and improve the machining accuracy and efficiency.

2. CRYOGENIC POLISHING MOLD AND LIQUID

When applying cryogenic machining technology, it is very difficult to find a kind of coolant that has very good fluidity under $0\,^\circ\!C$. In order to get cryogenic grinding environment, K. Yokogawa developed a method by

[*] This project is supported by the National Science Foundation of China, No.59775072.

spraying high-pressure and low temperature air to the machining area[2]. H. Ohmori made the experiment of ice bond wheel[3]. In optical machining, water is the carrier of abrasive and coolant. So the best way to realize cryogenic polishing is to freeze the polishing fluid on the polishing pan and form an iced polishing mold. Through this, we can get a polishing mold that has the same accurate form as pitch polishing mold. It holds abrasive and acts like a fixed abrasive pan. When the iced mold contacts with work-piece and moves relatively, the polishing effect is made. Further, the hardness of the polishing mold can be adjusted by changing the cooling temperature, and the mechanical and chemical polishing effect can also be adjusted by adding some additives to the polishing liquid.

The polishing liquid used in cryogenic polishing is a suspension of water, abrasive and additives. This solution presents alkalescent or weak acid and the *pH* value of it can be adjusted so as not to erode work-piece. The abrasive used in the suspension should have dispersive capability and not aggregate, so the dispersant should be added. Freezing the suspension to solid needs a cryogenic cooling course which should be shortened as much as possible in order to make abrasive grains not aggregate in low layer. The abrasive whose grain size is more than μ m dimension makes uniform speed motion in dispersive medium. According to Stokes principle, the deposition speed v_0 is

$$v_0 = \frac{2(\rho - \rho_0)}{9\eta} gr^2 \qquad (1)$$

Where, r is the radius of abrasive grain, g is the gravitational acceleration, η is the viscosity of dispersive medium, ρ is the density of abrasive, ρ_0 is the density of dispersive medium. When abrasive grain is very small, the deposition speed is very low. For example, if the radius of an abrasive grain is $1\,\mu m$, the deposition speed is $3\text{-}5\,\mu m/s$ in different medium. So the deposition of abrasive will not happen, as freezing time is very short. Nanometer dimension abrasive is colloidal particle. It always suspends in liquid medium and doesn't deposit. So cryogenic polishing mold can be made whose abrasive disperses uniformly. In polishing experiment, SiO_2

abrasive is used, the grain size is 80nm and 20nm respectively.

3. CRYOGENIC POLISHING EXPERIMENT

The cryogenic polishing experiment is carried out on a four-axis polishing machine in normal temperature laboratory. Before polishing, the work-piece is pretreated under low temperature. The temperature of the ice mold is about -40 °C. In polishing process, CO_2 gas or appropriate quantity of dry ice is put to the polishing area, so as to retain the low temperature. The material used is mono-crystal silicon and Zerodur.

Before polishing, the surface roughness of silicon wafer is Ra 14.87nm. Then after 70 minutes cryogenic polishing, the surface roughness is Ra 1.29nm (Fig 1). In polishing process, the rotation speed of the main spindle is 250~300 rpm[4].

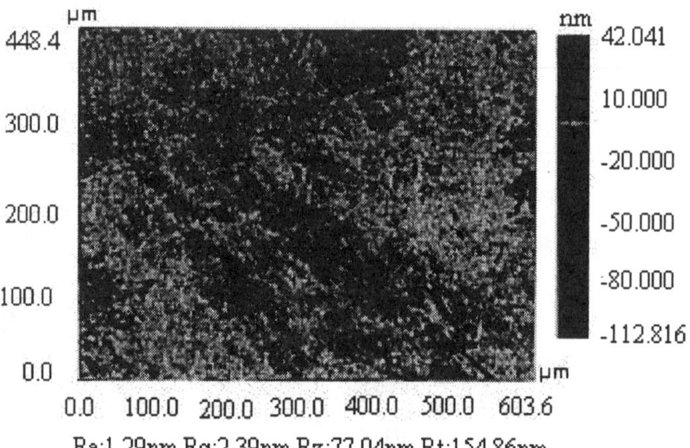

Ra:1.29nm Rq:2.39nm Rz:77.04nm Rt:154.86nm

Fig.1. Cryogenic polishing surface of mono-crystal silicon

Zerodur and K9 are representative optical materials. The experiment shows that the roughness of them can be less than nanometer dimension after cryogenic polishing. The polishing surface roughness of Zerodur is Ra 0.4nm(Fig.2) and form accuracy is over $\lambda/10$. The work-piece is a wafer whose diameter is 30mm.

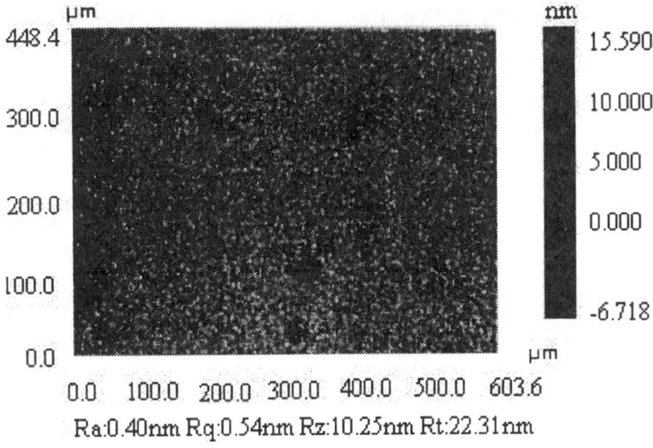

Fig.2 Cryogenic polishing surface of Zerodur

4. CONTRAST EXPERIMENT

In order to evaluate the effect of cryogenic polishing, a traditional asphalt polishing experiment is made with the same work-piece material and abrasive.

Before polishing the surface roughness of silicon wafer is Ra 12.04nm. After 16 hours polishing the surface roughness of it is Ra 3.03nm. The rotation speed of the main spindle is 60 rpm.

The surface roughness of Zerodur work-piece can reach the level of cryogenic polishing, but the polishing time is longer.

Comparing with the result of traditional asphalt polishing experiment, it can be seen that the surface quality and efficiency of cryogenic polishing is better. The reasons are as following:

1. In the cryogenic polishing process abrasive is fixed in ice mold, so the rotation speed of the main spindle can be increased to hundreds rpm. But the rotation speed of traditional polishing is largely limited, or the polishing effect will fall because the abrasive will be thrown out of the polishing area.

2. In cryogenic polishing process ice mold contacts with work-piece and makes relative motion, the material of work-piece surface is removed by the cutting function of abrasive. Meanwhile the heat produced between ice mold and work-piece will thaw a little ice and form a thin water film. Like traditional polishing, dispersive abrasive removes material by rolling. At the

same time, abrasive protruded on the ice surface cuts work-piece surface like fixed abrasive until it breaks away from ice mold, so the cutting effect of cryogenic polishing is better than traditional polishing.

3. In cryogenic polishing process hydrolysis function still acts like normal polishing in the thin water film between ice mold and work-piece mentioned above. It is very helpful for polishing surface quality. So cryogenic polishing is also a chemical and mechanical polishing process.

5. CONCLUSIONS

1. In cryogenic polishing process the ice-polishing mold has the effects of loose abrasive and fixed abrasive polishing, so the polishing efficiency can be improved and super-smooth surface can be obtained.

2. As the temperature of cryogenic polishing is not very low and ice-polishing mold can be made easily, no special equipment is needed; it can be applied to improve the traditional polishing process of optical materials.

ACKNOWLEDGEMENTS

The authors should thank Dr. Hitoshi Ohmori and Li Wei of Materials Fabrication Lab., Institute of Physical and Chemical Research(RIKEN) of Japan and Shi Weide of No.1 Chemical Reagents Factory of Tianjin. They made great efforts in helping to accomplish the experiment.

REFERENCES

[1] Jinnian Li, Ultra-precision cutting of ferrous metals and hard processing material, [doctor paper], Harbin University of Technology (1987).

[2] Kazuhiko Yokogowa, Substitute air condition for oil coolant to grind spherical surface with CBN, Mechanics and Tools, 12,8-9 (1993).

[3] Hitoshi Ohmori,Sei Moriyasu and Haedo Jeong, Iced bond wheel and its mirror surface grinding, Journal of the society of grinding engineers,Vol.44, No.12, 2000

[4] Rongjiu Han, etc., Cryogenic polishing technology of mono-crystal Silicon Wafer, Optical Fine Mechanics, 6(5), 104-109 (1998).

CHARACTERISTICS OF SMALL ROTARY TOOL IN POLISHING OF FUSED SILICA

Changling Liu, Toshio Kasai
Graduate School of Science and Engineering, Saitama University
Hitoshi Ohmori, Weimin Lin
Materials Fabrication Laboratory, The Institute of Physical and Chemical Research (RIKEN)

Abstract

In the finishing of high precision aspherical surfaces, computer controlled small tool polishing is usually used to maintain or improve the form accuracy. It is necessary to grasp the distribution of material removal of the small tool and the relationship between material removal rate and polishing conditions quantitatively. In our research, in order to investigate the characteristics of a small rotary tool, computer simulation based on Preston's law was conducted, and the result of it was compared to the experimental results obtained under various polishing conditions. The results show that the material removal volume is proportional to polishing pressure, rotation speed, inversely proportional to the scan rate of the small tool. The influences of inclination of workpiece and overhang of the small tool at the edge were also investigated.

Keywords

Small rotary tool, Polishing of fused silica, Distribution of material removal, Influence of polishing conditions

1. INTRODUCTION

Small tools are usually used in the finishing, especially computer-controlled finishing of aspherical surface since they can make a better fit with the workpiece and can easily realize controllable local material removal. Varied shapes and motions of small tools tend to generate different material removal rate and material removal distribution, and this influences the efficiency and form accuracy of finishing accordingly[1]. It is necessary to choose a proper tool in regard to the workpiece and make a full understanding of the characteristics of the small tool quantitatively. In this research, simulation of material removal distribution of a rotary tool was conducted based on the Preston's law. Further, The conformities with experimental results were inspected, under varied polishing conditions such as polishing load, rotation speed of polisher, scan rate, inclination angle of polisher axis, etc., in the polishing of fused silica with oxide cerium. The relationship between material removal and polishing conditions was also investigated.

2. SIMULATION OF MATERIAL REMOVAL DISTRIBUTION

Fig.1 shows the model for calculating material removal distribution. A polisher with a radius of R moves back and forth along a straight line in Y direction at a scan rate of v while rotating around the center of itself at a rotation speed of V. The polisher is pressed to the workpiece with constant load during polishing. Both the surface of polisher and the surface of workpiece are assumed planar, therefore the pressure under the polisher surface can be considered as uniformly distributed.

According to Preston's law, material removal is equal to the product of relative velocity, polishing time, polishing pressure and a coefficient. Because the last two parameters are of the same value for all points, Material removal distribution can be qualitatively described with the first two ones. For any position x in cross section, material removal volume can be calculated by integration of the hatched area using the following formula:

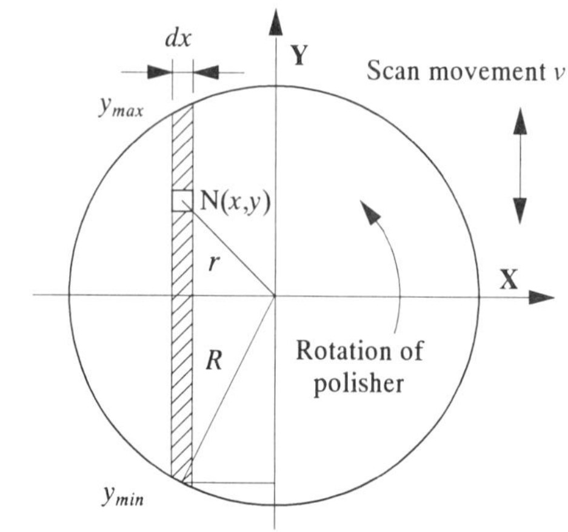

Fig.1 Relative movement of polisher and workpiece in simulation

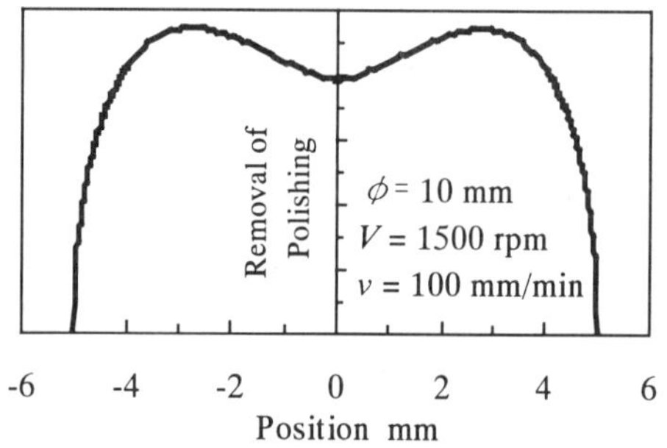

Fig.2 Simulation results

$$Removal(x) = \int_{y_{min}}^{y_{max}} \frac{2n\pi}{60v} \cdot \sqrt{x^2 + y^2} \cdot dy \qquad (1)$$

The simulation result is shown in Fig.2.

3. EXPERIMENT CONDITIONS AND METHOD

Experimental setup was shown in Fig.3. It operates in the same way as shown in Fig.1. The rotation movement of the polisher was driven by an electric motor and the rotation speed of it is from 100 to 2000 rpm. Polishing load can be preset by adjusting the position of a balance weight shown in the right side. Because polishing head (Polisher and motor) is mounted on an air bearing slider, it can move up and down freely to follow the workpiece at the desired polishing load accurately. The polishing pad is free to tilt to make a best fit to workpiece surface.

Fig.3 Experimental setup

Fused silica is used as material of workpiece, and CeO_2 with an average diameter of 3μ m is applied as abrasive. To avoid the influence of unevenness of the workpiece, surface profile was measured twice, before and after polishing, and polishing removal was obtained by subtracting the latter from the former. The measuring instrument was a non-contact optical instrument NH-3 (Mitaka kohki co.).

4. RESULTS AND DISCUSSION

Fig.4 shows the relationship between polishing removal and rotation speed of polisher. Reciprocal movement times were set inversely proportional to the rotation speed of the polishers. From Fig.4, we can see that the same removal depth and distribution was obtained for four conditions. This means that the polishing removal is proportional to the relative speed of polisher, which conforms to the Preston's law. However, distribution of them has no symmetry as the simulation results shown in Fig.2. This is caused by the inclination of the polisher axis.

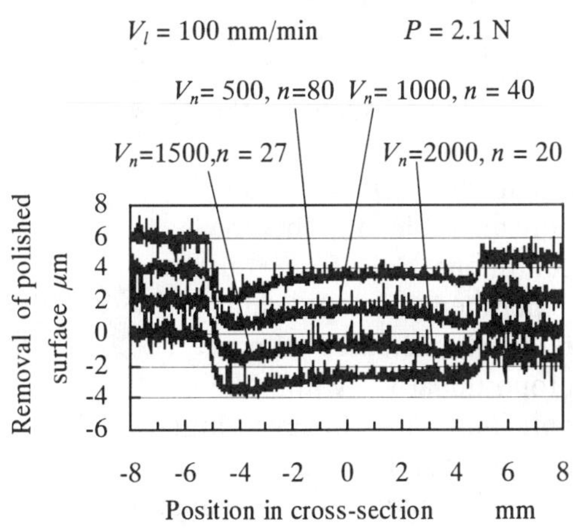

Fig.4 Relationship between polishing removal and rotation speed of polisher

By tilting the axis of the polisher to the opposite direction, bigger removal depth was obtained in right side than in left side. Though careful adjustment was carried out, it was found that polishing removal was very sensitive to the inclination; even a small inclination could cause great distortion in polishing removal distribution.

Fig.5 shows the relationship between polishing removals and scan rate. Reciprocal movement times were set proportionally to the scan rate. We can see that equal polishing removals were obtained for four conditions. Therefore, the conclusion that polishing removal is proportional to scan rate can also be drawn.

Fig.6 shows the relationship between polishing removal and polishing load. In the conditions of our experiment, polishing removal increases almost proportionally with the polishing load. However, bigger polishing load has the tendency to generate a bigger non-uniformity in polishing removal distribution.

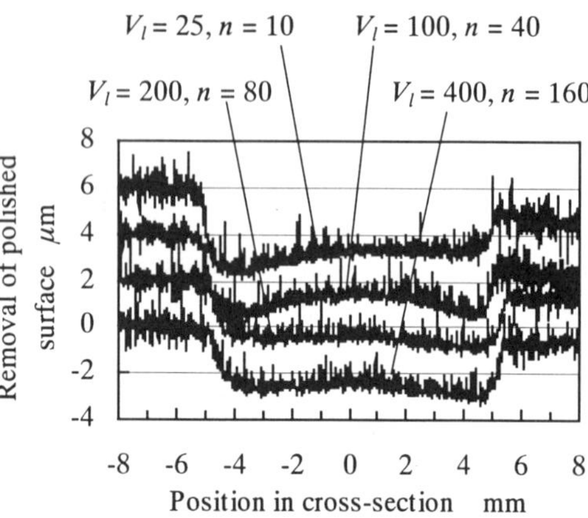

Fig5 Relation between polishing removal and scan rate

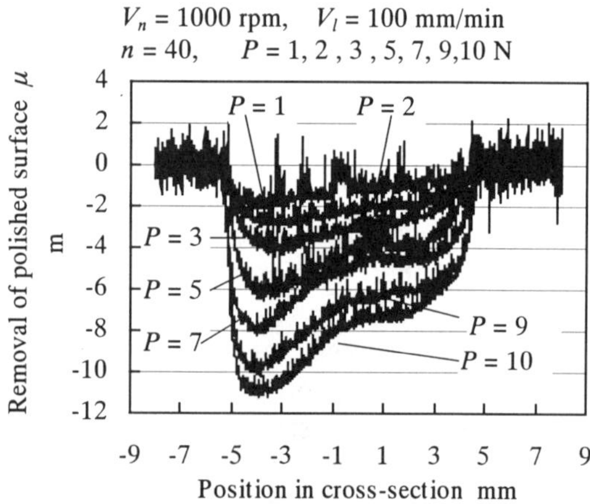

Fig.6 Relationship between polishing removal and polishing load

In small tool polishing, since the tool cannot move past the edge of workpiece, it is more likely to generate up edges. To decrease the width of up edge area, the small polisher is usually moved to hang over the edge. However, excessive overhang will cause unbalance of the small polisher. To clarify the behavior of the small polisher near the edge, polishing as shown in Fig.7 was made. Fig.8 shows the relationship between polishing removal and overhang of polisher. When the overhang is not bigger than 3mm, the up edge decreases with the increase of overhang; it demonstrates a good tendency to produce an expected edge profile. However, when the overhang was set to 4mm or bigger, the polisher could not contact the workpiece properly, and polishing removal occurred only at the very near edge area. This is considered a bad edge profile. The turning point of the two states of the polisher happens when the overhang is about one third of the diameter of the polisher.

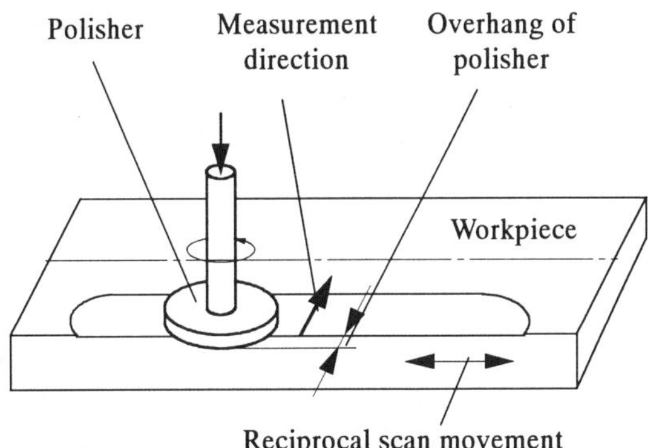

Fig.7 Polishing of edge area

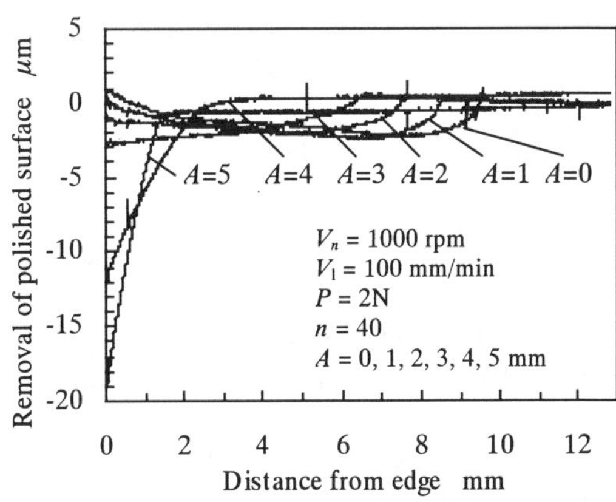

Fig8 Distribution of polishing removal affected by overhang of polisher

5. CONCLUTIONS

In this research, polishing removal volume and distribution in the polishing of fused silica with a small rotary tool were investigated. We drew the conclusion that experimental results conform to the Preston's law, i.e. material removal rate is proportional to the relative velocity, polishing pressure, and inverse to the scan rate. To get the best edge profile, the overhang of the polisher should be set to a value that slightly smaller than the one third of the polisher diameter. Symmetrical material removal distribution was difficult to obtain by the influence of inclination of the polisher axis relative to the workpiece surface.

REFERENCE

1. Robert A. jones. Computer-controlled optical surfacing with orbital tool motion, Optical Engineering. 1986; 6: 785-790.

DEVELOPMENT OF ELECTROSTRICTIVE POLYURETHANE(PU) FILMS AND ITS CHARACTERIZATION

Eun-Soo Jeong[a], Han-Soo Park[b], Hae-Do Jeong[c], Nam-Ju Jo[b]

[a]Graduate School of Precision Mechanical Engineering,
Pusan National University
[b]Department of Polymer Science & Engineering,
Pusan National University
[c]School of Mechanical Engineering, Pusan National University,
San 30, Changjeon-Dong, Kumjeong-Ku, Pusan, 609-735, Korea
Tel : +82-51-510-2463, Fax : +82-51-518-8442,
E-mail : jcobain@pusan.ac.kr

Abstract

This study was especially focused on the enhancement of electrostrictive characteristics of polyurethane films by the modification process of molecular position. The PU films were evaluated in terms of the ferro-electricity, capacitance, glass transition temperature(T_g) and thickness of films. Experimental results demonstrate that some of PU films have the ferro-electricity and high capacitance by the modification of molecular position.

Keywords

polyurethane(PU), electrostriction, piezoelectricity, micro actuator

1. INTRODUCTION

Recently, new synthetic methods and industrial needs lead to the development of functional polymer with piezoelectricity, pyroelectricity and etc. Generally, lead zirconate(PZT) and barium titanate($BaTiO_3$) are known as typical piezoelectric inorganic materials. Polyvinylidene fluoride(PVDF or PVF2: β-phase) and vinylidene fluoride copolymer can be also used as piezoelectric organic materials. Also, it is known that PU has electrostriction by the electrostatic interactions among molecules[1]~[3].

2. EXPERIMENTS

PU, electrostrictive material was synthesized by the

conventional prepolymer method. Generally, PU is composed of isocynate, chain extender and polyol. And morphologically PU is divided into hard segment by isocynate, chain extender and soft segment by polyol.

Isocynates were used 4,4'-diphenlymethanedissocyanate(MDI), etelenediamine(EDA) and chain extenders were 1,4'butanediol. Polyols were used polytetrametheleneadipatediol(PTAd), polycarbonate diol(PCD). In order to understand the effect of hard segment, diisocyanate and chain extender, the ratio of these components was controlled. PU was cast on the $1\mu m$ thick PVD Al film with Si substrate in the form of films of $10{\sim}100\mu m$ thick. In order to obtain the high capacitance, the modification process of molecular position was carried out by two methods shown in *Figure 1* and *Table 1*.

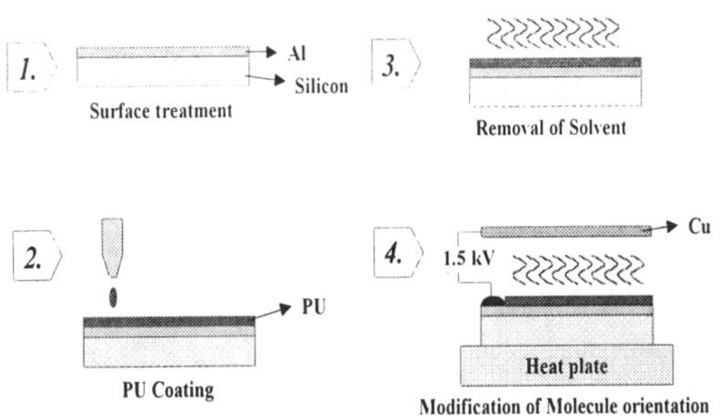

Figure 1. Process flow

Table 1. Process type

Process number	Process
1	1-2-3(raw)
2	1-2-4
3	1-2-3-4

One of the processes is an in-process modification method that solvent is slowly removed in electrical field at $55\pm5\,^{\circ}\text{C}$(in *Table 1* : process 2). Another is a sequential modification method that the voltage is applied to electrodes at glass transition temperature(T_g) of hard segment ($140{\sim}160\,^{\circ}\text{C}$) after complete removal of solvent(in *Table 1* : Process 3). And the third is a method of preparing a raw material(in *Table 1* : Process 1). Electrical field with $0.3{\sim}1.5$kV/cm intensity was charged between separated two electrodes. Cu and Al electrodes were located above and under the PU respectively. For the measurement of the electric characteristic, Au electrode layer was deposited on one side of PU film by thermal method. Electrostrictive materials used in the experiment are shown in *Table 2*.

3. RESULTS

The glass transition temperature(T_g) of hard segment and thickness of films were also evaluated by the differential scanning calorimeter(DSC) and the optical interferometer(ACCURA 1500M).

Table 2. Electrostrictive materials used in experiment

PU number	PU (HS=hard segment)
1	PCD,MDI,BD,PU (HS content=23wt%, Solid content=25wt%)
2	PTAd,EDA,BD,PUU (HS content=23wt%, Solid content=25wt%)
3	PTAd,MDI,BD,PU (HS content=23wt%, Solid content=25wt%)
4	PTAd,MDI,BD,PU (HS content=30wt%, Solid content=25wt%)
5	PTAd,MDI,BD,PU (HS content=40wt%, Solid content=25wt%)

The film thickness of PU is 10μm to 150μm. *Figure 2* shows the thermal changes of PU 3(PTAd, MDI, BD, PU, HS content=23wt%, solid content=25wt%). *Table 3* means the DSC results of each PU film. PU 1 has thermal changes at low temperature.

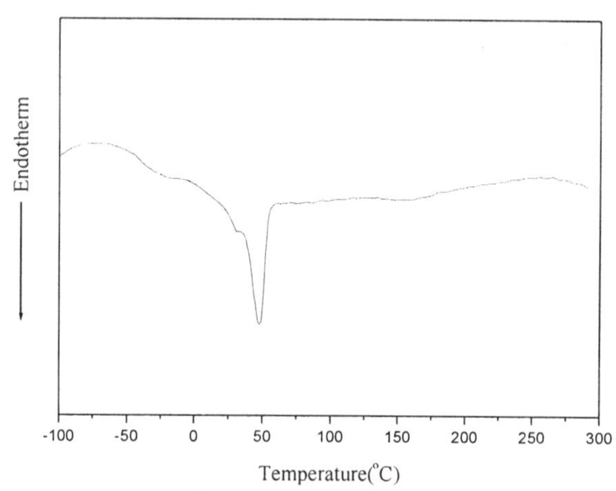

Figure 2. Thermal behavior of PU 3

Table 3. Results of DSC

material	solvent removing temperature(°C)	Tg temperature of hard segment (°C)	dissociation temperature(°C)
1	37.14 ~ 52.44	145.26 ~ 161.98	238.97
2	41.37 ~ 54.24	137.58 ~ 163.78	226.05
3	24.26 ~ 54.08	150.89 ~ 155.18	264.21
4	34.21 ~ 54.73	151.19 ~ 172.52	260.71
5	19.83 ~ 49.67	139.98 ~ 165.31	244.87

Electrical properties were measured using impedance/gain

phase analyser(HP4649A) and ferro-electrometer. But capacitance is somewhat different. Usually organic material has unstable characteristics at low frequency, *Figure 3* shows typical unstability of PU capacitance.

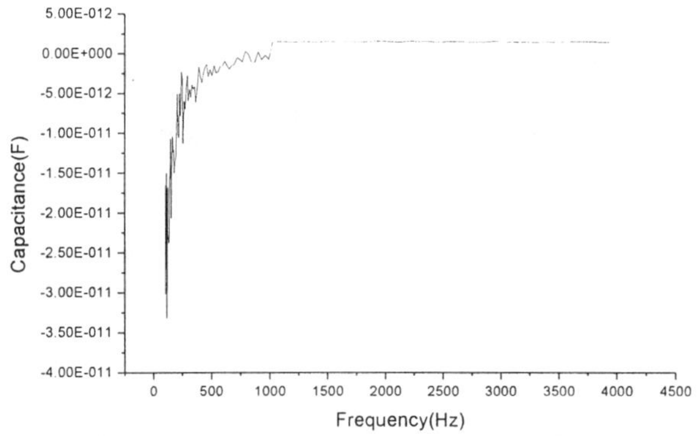

Figure 3. Capacitance of PU 1 at low frequency

The relative dielectric constant(RDC) of PU 5 fabricated by time controlled process 3 is shown in *Table 4.* And this shows effects of process time to RDC is small but strain and stress expect to be largely affected by process time. *Table 5* shows RDC of various type of PU and process. And PU films fabricated by in-process modification method have higher RDC than that by the sequential modification method. Namely process 2 is easier to handle molecule position than other processes.

Table 4. process-time-controlled PU RDC

process time (hour)	RDC
1/2	10.83587
1	12.24105
1 1/2	11.17349
3	9.708015
5	9.750249
7	11.43004
9	10.89547

Table 5. RDC of various type of PU and process

PU - Pro.	RDC
1 - 2	4.21452706
1 - 3	3.79330933
2 - 2	9.23970587
2 - 3	5.12503352
4 - 2	6.37317839
4 - 3	3.82596173
5 - 2	8.03694831
5 - 3	3.85854556

Figure 4 shows hysteresis loop of PU 2 fabricated by process

3. PU has lower remnant and spontaneous polarization value than other piezoelectric polymer like that of PVDF, but its strain expects to be larger than others.

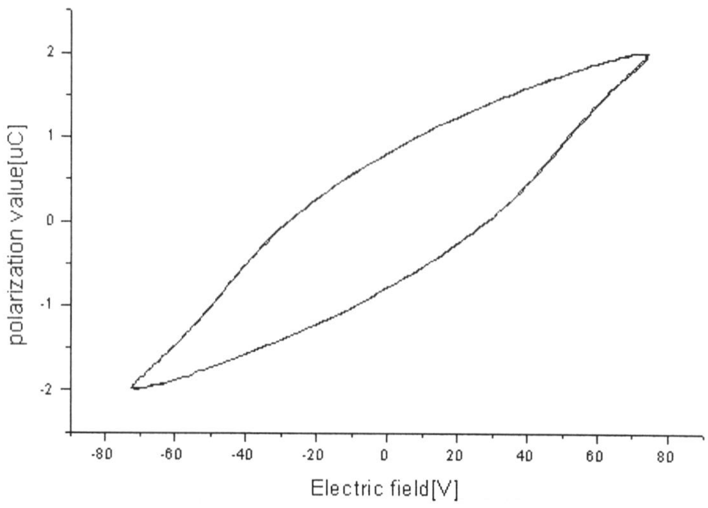

Figure 4. Hysteresis loop of PU 2

4. DISCUSSIONS

Finally, the electrostrictive characteristics of PU were investigated and then the process for the improvement of electrostrictive characteristics of PU was also developed. We found that some of them have ferro-electricity and the reason of ferro-electricity of PU is expected to changes of dipole orientation from electrical movement of relatively free elements. Electrostrictive PU has many advantages of convenience to handle, high elongation rate and low manufacturing cost, etc. Especially, it is expected that electrostrictive PU can be applied to micro-switch, micro-actuator and micro-pump for MEMS and etc.

[1] T. Ueda, et. al., R&D Div. of Nitta Co. : Polyurethane Elastomer Actuator. , Synthetic Metals 85, 1997; 1415-1416

[2] J. Kyokane, R&D Div . of Nitta Co. : ELECTRO-STRICTION EFFECT OF POLYURETHANE ELASTOMER(PUE) AND ITS APPLICATION TO ACTUATOR. , Synthetic Metals 103, 1999; 2366-2367

[3] Eiichi Fukada , Kobayasi Institute of Physical Research, Kokubunji, : Piezoelectricity in Polyurethane films induced by Electrostricion and Bias Electric Field. 10th International symposium on Electrets 1999; 655-658

PRECISION MACHINING OF FUTURE SILICON WAFERS

Prof. Dr.-Ing. Fritz Klocke
Dipl.-Ing. Dietmar Pähler

Fraunhofer Institute for Production Technology IPT
Steinbachstrasse 17, 52074 Aachen, Germany

Abstract

Ductile grinding of silicon wafers with superior surface quality and TTV is one way of precision machining. Most promising results are found with diamond grinding wheels using high grain concentrations and small D-size values. Innovative machining concepts employing the ductile grinding mode are presented promising advantages for future wafer generations to come.

Keywords
Rotational Grinding; Ductile Mode; Double-Side Grinding;
Large-Area Silicon Wafer;

1. ADJUSTING TO FUTURE INDUSTRY DEMANDS

Today high-purity silicon wafers are found at the starting point of virtually any microelectronic device. But, as prices of memory chips fall, manufacturing costs rise. One way to meet the industries' saving demands is to alternate the wafer production process. At the Fraunhofer Institute of Production Technology IPT a low damage grinding process is investigated for cost efficient yet quality ensuring production of silicon wafers. The process is designed to cause considerably reduced sub-surface damages (SSD) employing a ductile material removal mechanism. Among the ductile mode grinding research projects is the international Brite/EuRam III "BestWafer" project funded by the European Commission. Furthermore, improved processing chains for silicon wafers including innovative machine conceptions are in the focus of research at the Fraunhofer IPT as well.

The today initially as cylindrical ingot delivered silicon is sawn into very thin disc-like slices and subsequently processed by time-consuming and cost intensive lapping, etching and final polishing processes. Almost 40% of the costs of a standard 200 mm silicon wafer are accounting for the lapping and polishing processes. Present industrially applied silicon grinding processes are characterized by brittle material removal, with grains of the grinding surface causing sub-surface damages in the wafer upon penetration, ultimately spalling off silicon particles. The future process chain will consist of micro-rotation grinding for roughing, nano-rotation grinding for finishing and rotational polishing for mirror-like finishing of the wafers [Abe1991]. Costly conventional lapping and etching processes will be completely eliminated (*Figure 1*).

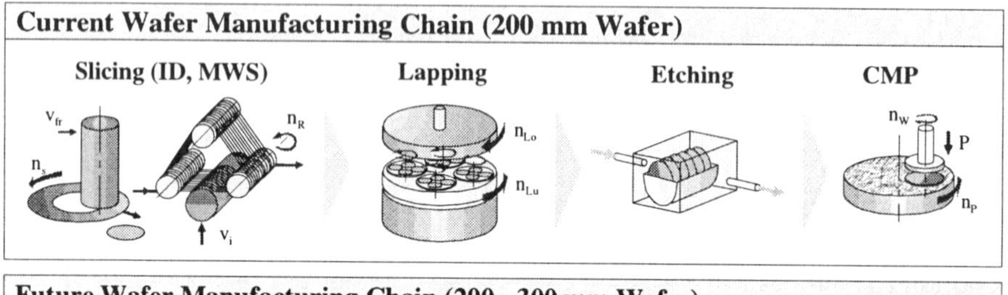

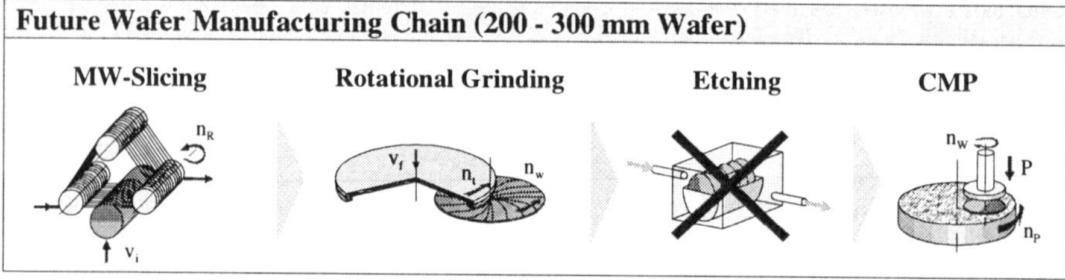

Figure 1. Comparing current and future wafer manufacturing chains the costly lapping and etching will be eliminated.

The International Technology Roadmap for Semiconductors (ITRS) latest 1999 edition already looks beyond the 300 mm wafers which are being installed as the new standard to the expected introduction of 450 mm in the next decade [ITRS1999]. According to the roadmap 450 mm wafers will be needed to manage the production costs of large chip devices such as Systems-On-A-Chip. Also a modification of the current process is needed since conventional lapping does not meet the requirements for wafers beyond 300 mm. The low damage grinding process is expected to reduce the costs of ownership due to higher yield along with improved quality, massive reductions in the use of agents and a more controllable mechanical processing of the wafers in general.

2. REDESIGN FOR IMPROVED WAFER QUALITY

Sub-surface damages caused by various faults such as cracks and crystallographic displacement are a major problem affecting the mechanical machining of silicon wafers. Yet the industry demands flawless wafers. Therefore, very gentle process stages have to be introduced in order to safely remove faulty layers from the wafers without inducing new SSDs. Unlike standard grinding tasks, wafer grinding is not just an ordinary surface finish. Next to surface quality, assessing the total thickness variation (TTV) with TTV < 1 µm is a very important factor. TTV is a measure of flatness and, along with the overall thickness range, it is one key factor in determining the quality of the finished wafer, with final values of ca. 800 µm.

The investigated wafer grinding method of rotational grinding with a vertically arranged diamond cup wheel is the most promising regarding form deviations and TTV values [Matsui1991]. The cup grinding wheel penetrates the rotating wafer in axial direction, covering half of the wafer (*Figure 2*) which is attached to the vacuum chuck in order to prevent the introduction of stress. The general advantage of this grinding method in comparison lies in the constant contact conditions in terms of contact length and area resulting in constant grinding forces. A symmetrical pattern subsequently appears on the surface of the wafer, characterized by arched grinding marks from the center towards the edges of the silicon wafer. For distinct tools in combination with fine tuned process parameters SSD-values down to 1 µm were achieved.

<table>
<tr>
<td>

Figure 2.
Kinematics of precision grinding of silicon wafers using a diamond cup wheel.

</td>
<td>

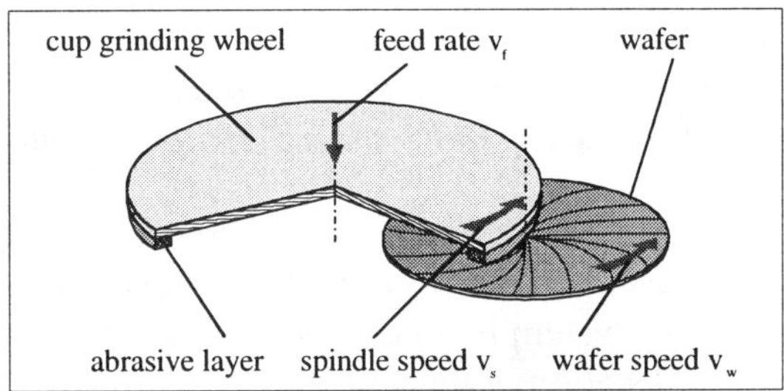

</td>
</tr>
</table>

Research objective is the optimization of the whole grinding system consisting of the machine tool, process layout and grinding wheel. To characterize the diamond grinding wheel it is necessary to determine the grain structure by means of shape distribution and actual grain size, since coarser grains might be responsible for a transition from ductile to brittle material removal due to the transgression of the critical depth of cut. Another describing component is the bonding system itself defining parameters such as thermal and chemical stability, the wheel's toughness and elasticity. The found system has to prove its reliability and wear resistance in the process.

3. DUCTILE ROTATIONAL GRINDING

The low damage grinding process is to be optimized regarding the necessary process stages to obtain virtually damage-free wafers at the last processing step. Therefore, the last step has to be a ductile machining operation (*Figure 3*). The new process design achieves the transition from brittle to ductile material removal by reducing penetration depth into the silicon wafer of the tool's grain. Ductile grinding reduces the material removal rate but results in virtually damage-free wafers with polished-like surface qualities. This process is successfully established with machining other brittle materials such as glass or advanced ceramics. Surface quality of finish-ground wafers is aimed to be far beyond $R_a < 0.01$ μm.

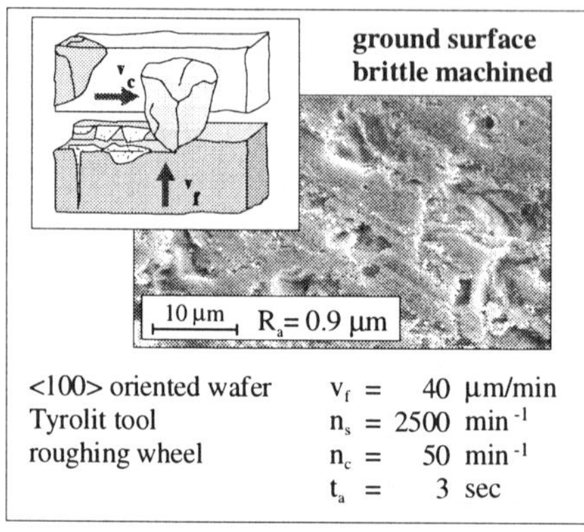

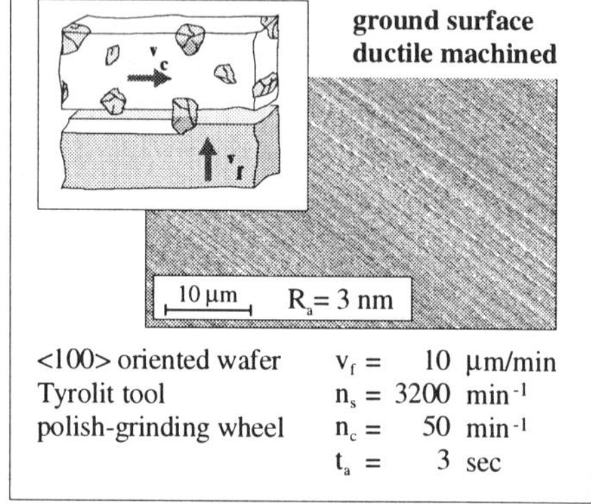

Figure 3. Employing ductile mode grinding results in superior surface qualities with up to $R_a = 3$ nm. Brittle machining does not meet the demands.

Momentarily extensive research is underway with variations of the so far found bonding system to achieve these goals. Developing a suitable bonding system is no easy task since the required properties - elasticity and thermal stability - do act controversially and distinct long-term properties are required as well. As the preferred bonding system a resin bond was introduced performing at much higher temperatures as formerly used systems. Furthermore, the concentration was set in favor of high values, promising positive effects on the surface quality due to the reduced depth of cut of each single grain.

It was found that the main influencing factor for attaining good surface qualities is grain size. For example, spindle speed affects surface quality only at lower speeds, just as feed rate influences surface quality only at low feed rates. Compared to grain size these influences can generally be ignored under normal machining conditions leaving grain size as the imperative quality parameter for silicon wafer grinding. For finish grinding polish-like qualities of $R_a = 3$ nm and $R_t \leq 30$ nm are projected using grinding wheels of D3, D6 and D9 grain qualities [Klocke2000]. A minimization of the grain size used in the process along with an optimization of machining parameters will be a successful road to travel on.

4. OUTLOOK: FUTURE MACHINING CONCEPTS

While further working on the grinding process's parameter optimization the eventual substitution of the costly stages of conventional lapping and etching can already be foreseen [Nishi2000]. Restructuring the wafer production operations in the area of plane-parallel machining is the key, giving the industry a competitive yet ecological tool. The projected savings for processing of wafers will even increase with the future introduction of silicon wafers of diameters of 300 and 450 mm and particularly with diameters of 675 mm already mentioned by the ITRS 1999.

As for the future innovative concepts are to take reduction of sawing induced waviness, increased throughput and superior geometrical and surface qualities into consideration. The chuck-based single side grinding concepts are to be optimized for superior TTV values and reduced SSD (*Figure 4, A*). The Double-Side Large Area Grinder is one approach with the wafer being supported between two coaxial mounted grinding wheels (*Figure 4, B*) with highly promising results regarding waviness. This significant improvement of the silicon wafer grinding process will reduce costs of ownership and serve beneficially to health and environment alike.

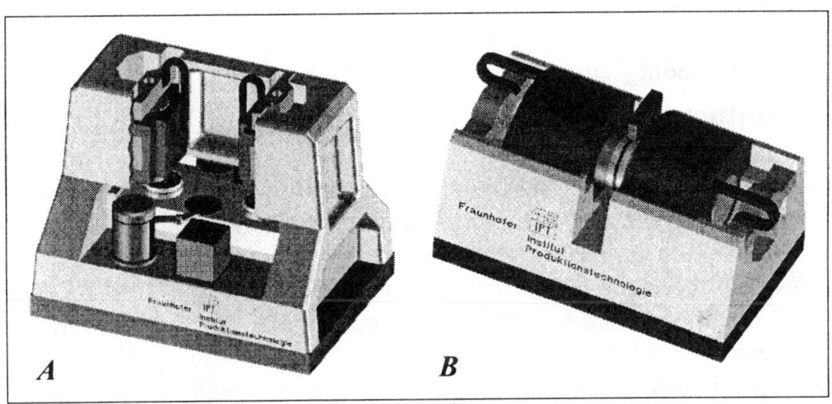

Figure 4. Design studies for Large Area Grinders by the Fraunhofer IPT, Aachen, for precision machining of silicon wafers greater than 300 mm.

References

Abe1991: Abe, T., A Future Technology for Silicon Wafer Processing for ULSI. Precision Engineering 1991; 4:251-55

Klocke2000: Klocke, F., Gerent, O., Pähler, D., Jakob, A. Flat rates on future silicon wafers: precision grinding. Industrial Diamond Review 2000; 2:149-56

ITRS1999: International Technology Roadmap for Semiconductors 1999. http://www.itrs.net

Matsui1991: Matsui, S., Parallelism Improvement of Ground Silicon Wafers. Transactions of ASME, Journal of Engineering for Industry 1991; 2:25

Nishi2000: Nishi, Y., Doering, R., *Handbook of Semiconductor Manufacturing Technology.* New York: Marcel Dekker, 2000.

High Speed Grinding Performance and Material Removal Mechanism of Silicon Nitride

Han Huang & Ling Yin

Gintic Institute of Manufacturing Technology, 71 Nanyang Drive, Singapore 638075

Abstract

This paper reports the high speed grinding performance of silicon nitride using resin bond diamond wheels. The investigation was focused on the effect of wheel speed on material removal mechanisms and ground surface quality. When brittle fracture was prevalent during grinding, surface quality was little affected by wheel speed. While using a fine grit wheel and a small depth of cut, the ground surface exhibited dominant "ductile" flow and the surface quality was influenced by wheel speed. The maximum chip thickness was used to interpret the associated material removal mechanisms. It was found that vibration due to high wheel speeds could limit the improvement of surface quality.

Keywords

High speed grinding, silicon nitride, removal mechanism, vibration analysis

1 Introduction

Silicon nitride has been extensively used to replace metals and alloys in industrial applications, such as bearings, pump seals and engine blades etc. However, processing difficulties and high costs (Inasaki, 1987, Malkin et al. 1996 & Tonshoff et al. 1999) associated with machining high quality surfaces have been the major barrier to its widespread use. To achieve the required roughness, shape and subsurface damage level, final grinding should be conducted in a "ductile" mode. Previous studies (e.g. Kovach et al. 1993) have indicated that increasing wheel speed could be an economic strategy for achieving "ductile" mode grinding. A concerted effort has been made toward high speed grinding of advanced ceramics (Klock et al. 1999, Hwang et al. 2000). Unfortunately, increasing the wheel speed may introduce factors, such as vibration and severe wheel wear (Hwang et al. 1999), into the process, which significantly influence grinding quality.

In this paper, efforts are made to study the high speed grinding performance of silicon nitride and the material removal mechanisms when using resin bond diamond wheels. The effects of wheel speed and other grinding parameters on surface quality are investigated. Material removal mechanisms associated with two resin bond diamond wheels of different grit sizes are revealed in terms of the maximum chip thickness. In addition, the effect of vibration caused by the unbalanced wheel at high speeds on surface quality is discussed.

2   Experimental Details

The mechanical properties of Si_3N_4 at ambient temperature are listed in Table 1. The specimens have dimensions of $40 \times 4 \times 3$ mm.

Table 1: The properties of Si_3N_4.

Material	Density [g/cm^3]	Elastic Modulus [GPa]	Micro Hardness [GPa]	Fracture Toughness [MPa.m$^{1/2}$]
Si_3N_4	3.2	290	17.5	5.5

Grinding experiments were conducted on an Okamoto high speed surface grinder. Resin bond diamond wheels of grit sizes of ~160 µm and ~20 µm were used. The wheels have a diameter of 200 mm and a width of 6 mm. Before grinding, the wheels were balanced at a speed of 40m/s. After balancing, the wheels were trued using a vitrified silicon carbide wheel (80M7V). The wheel was further dressed with a 400G alumina stick. Down grinding mode was employed. Detailed grinding conditions used are listed in Table 2. The wheels were re-dressed at the beginning of each grinding test.

Table 2: Parameters used in grinding of Si_3N_4..

Grit size	Depth of cut (µm)	Feed rate (mm/min)	Wheel speed (m/s)
160 µm	40	200, 500, 1000	40, 80, 120, 160
	10, 40, 100	500	40, 80, 120, 160
20 µm	0.5, 1, 2	200	40, 80, 120, 160

Three grinding passes were completed under each set of conditions. Grinding forces were measured using a piezoelectric dynamometer (Kistler 9257B). The roughness was measured using a profilometer (Taylor Hobson) perpendicular to the grinding direction. For each grinding test, three measurements were taken at different locations. Mean values and standard deviations were calculated. Ground surfaces were examined using a scanning electron microscope (SEM) and atomic force microscope (AFM).

3   Results

Grinding force and surface roughness

With the 160 µm grit size wheel the normal grinding force decreased with the increasing wheel speed, as expected. An increase in the wheel speed resulted in a smaller undeformed chip thickness, and thus a smaller grinding force. The effect of wheel speed on the normal grinding force was more significant when a larger depth of cut was used. With an increase in the feed rate from 200 mm/min to 1000 mm/min, the normal grinding force increased more than two times. When the largest feed rate 1000 mm/min. was used, the normal grinding force decreased more significantly with the increasing wheel speed. The surface roughness was improved when the depth of cut was reduced. However, increasing the grinding speed had little influence on the surface roughness, though the grinding force was decreased. Reducing feed rate from 1000 mm/min to 500 mm/min resulted in a large improvement on the surface roughness. However, little improvement was achieved when varying the feed rate from 500 mm/min to 200 mm/min.

Ground surface characterisation

SEM micrographs of ground surfaces with the 160 µm grit wheel are shown in Fig. 1. It is difficult to observe a significant quality improvement on the ground surfaces due to the increase in wheel speed. This implies that brittle fracture could be the dominant removal mode with the 160 µm grit size wheel. Usually four types of areas were observed, (1) ploughing area, (2) fractured area, (3) smeared area and (4) debris covered area. This is in an agreement with the findings in the work of Jahanmir et al (1999).

Fig. 1: SEM photos of ground surfaces (DOC 10µm, feed 500mm/min.).

4 Discussion

In high speed grinding, the increased wheel speed can substantially reduce the maximum undeformed chip thickness, h_m, resulting in a smaller grinding force. $h_m=(3/C\tan\theta)^{0.5}(v_w/v_s)^{0.5}(a_e/d_s)^{0.25}$, where C is the density of active cutting points, θ semi-included angle for the undeformed chip cross section, d_s wheel diameter, v_w workpiece feed rate, v_s wheel speed and a_e wheel depth of cut. The normal force is plotted as a function of h_m in Fig. 2 with C=20 & $\theta=60°$ for the 160 µm grit wheel. The data lie on one common curve, indicating that h_m substantially influences the force. When plotting roughness as a function of h_m in Fig. 3, the data seems to be concentrated into two groups. It is known that in a brittle mode grinding process, roughness is not significantly influenced by the h_m. However, if h_m surpasses a critical value, fracture will change to a larger scale mode, for instance, from micro-fracture to large grain dislodgement. As a result, the surface roughness value jumps to a higher level.

Bifano et al. 1991 indicated that in the machining of brittle materials a "ductile" mode material removal could be achieved when the specific material removal volume is sufficiently small. The critical un-deformed chip thickness for "ductile" grinding is estimated by $h_c=\beta(E/H)(T/H)^2$, where β is a constant related to the wheel topography, E the elastic modulus, H the hardness and T the fracture toughness. It was assumed that when h_m is smaller than h_c, "ductile" mode material removal should be achieved during grinding. To examine the effect of wheel speed on the brittle-ductile transition, a grinding test was designed for the 160 μm grit size wheel. In the experiment, a constant feed rate of 500 mm/min and depth of cut of 10 μm were used and the wheel speed was varied in order to achieve a h_m just below and above h_c. The calculated h_c was 0.25 μm (β=0.15). The h_m values were 0.36, 0.25, 0.21 and 0.18 μm for wheel speeds of 40, 80, 120 and 160 m/s, respectively. Unfortunately, the ground surfaces still exhibit characteristics of layers formed mainly via brittle fracture (Fig. 1). It was not possible to achieve a smooth surface dominated by "ductile" material removal with the 160 μm grit wheel. A possible reason is that the protrusion heights of diamond grains are widely distributed (compared with h_c), varying from 20 to 80 μm. Thus, the actual value of h_m for some active cutting points could be much larger than h_c.

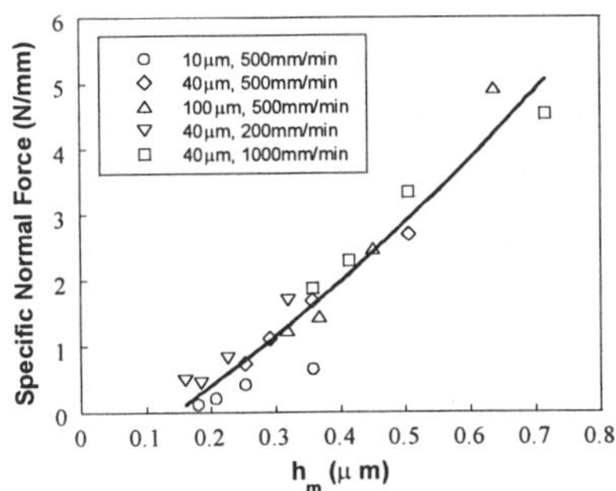

Fig. 2: Specific F_n Vs h_m

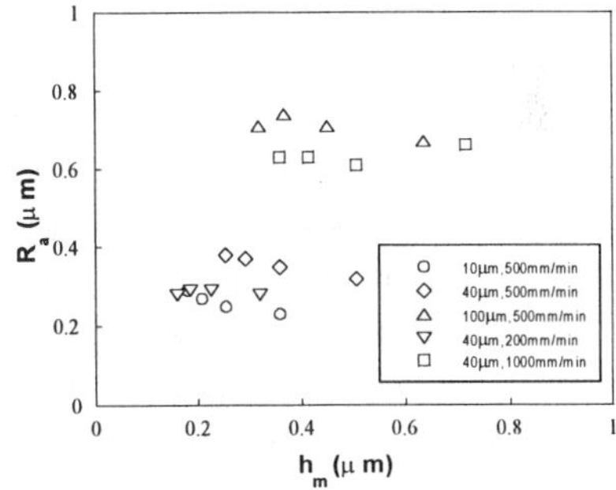

Fig. 3: R_a Vs h_m

To achieve a better surface quality, a wheel with a grit size of 20 μm was used. Under the conditions listed in Table 2, the maximum chip thickness was varied from 0.013 to 0.032 μm, much smaller than the h_c (0.25 μm). The surface roughness obtained using the fine grit wheel is shown in Fig. 4. It is apparent that the roughness is substantially reduced compared with those obtained using the coarse grit wheel. Comparison of the detailed surface textures formed using the two different wheels in Fig. 5 indicates that different removal mechanisms were prevailing during grinding, brittle fracture for the coarse grit wheel and ductile flow for the fine grit wheel.

In Fig. 4, it is found that roughness value is slightly decreased with increasing wheel speed when the wheel speed is below or equal to 80 m/s, except for the case of the depth of cut =2 μm. For all the depths of cut used further increasing the wheel speed leads to a significantly worsened surface. AFM examinations also confirm this result. This is attributed to vibration caused by the unbalanced wheel. The spindle vibration was measured. It is found that the vibration amplitude is extremely low at the balanced point (of

40m/s) as well as at the wheel speed of 80 m/s. The amplitude is almost doubled at 120 m/s and is four times higher at 160 m/s. The increased vibration at higher speeds limits the improvement of surface quality. The effect of wheel imbalance on the surface roughness becomes more influential at higher speeds.

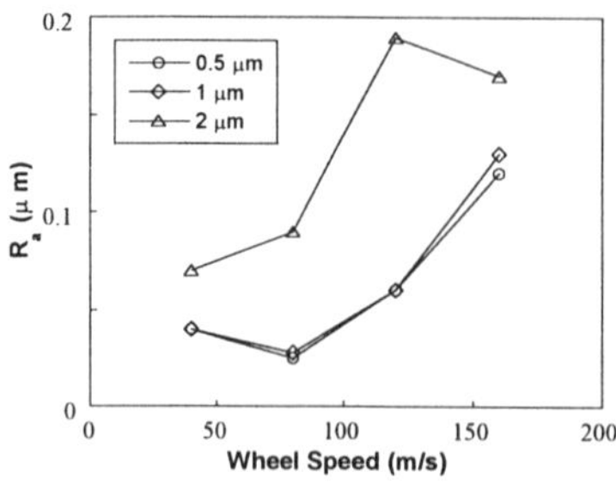

Fig. 4: R_a Vs Wheel speed.

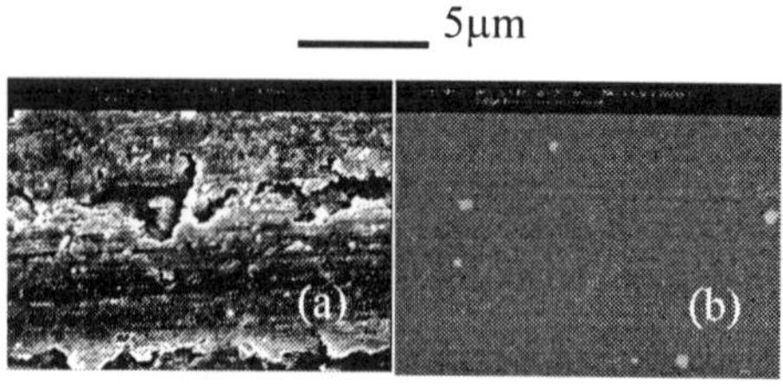

Fig. 5: SEM micrographs
(a) coarse grit wheel, $h_m=0.25\mu m$,
(b) fine grit wheel, $h_m=0.019\mu m$.

5 Conclusions

This study has revealed that the ground surface quality was not improved with increase in wheel speed when grinding in a region where material removal was dominated by brittle fracture. Increased wheel speed resulted in an improvement in ground surface quality when grinding was conducted within the region where ductile flow was prevalent. Vibration caused by wheel imbalance at higher speeds, however, limited the improvement and significantly worsened the surface quality.

References:

Bifano, T., T. Dow, & R.O. Scattergood, 1991, Ductile-regime grinding: a new technology for machining brittle material, ASME, Journal of Engineering for Industry, 113(5), p184-189.

Hwang, T.W., C.J. Evans & S. Malkin, 2000, An investigation of high speed grinding with electroplated diamond wheels, Annals of CIRP, 49/1, p245-248.

Hwang, T.W., C.J. Evans, E.P.Whitenton & S. Malkin, 1999, High speed grinding of silicon nitride with electroplated diamond wheels, I. Wear and wheel life, Manufacturing Science and Engineering, ASME, MED-Vol. 10, p431-442.

Inasaki, I., 1987, Grinding of hard and brittle materials, Annals of CIRP, 36/2, p463-471.

Jahanmir, S., H. Xu & L. Ives, 1999, Mechanisms of material removal in abrasive machining of ceramics, *Machining of Ceramics and Composites*, Marcel Dekker, Inc., p11-84.

Klocke, F., E. Brinksmeier, C. Evans, T. Howes, I. Inasaki, E. Minke, H. Tonshoff, J. Webster & D. Stuff, 1997, High speed grinding – fundamental and state of art in Europe, Japan and USA, Annals of CIRP, 46/2, p715-724.

Klocke, F., E. Verlemann & C. Schippers, 1999, High-speed grinding of ceramics, *Machining of Ceramics and Composites*, Marcel Dekker, Inc., p119-138.

Kovach, J.A. et al., 1993, A feasibility investigation of high speed low damage grinding of advanced ceramics, Proc. SME, 5th Int. Grinding Conf., Cincinnati, Ohio.

Malkin, S. & T. Hwang, 1996, Grinding mechanisms for ceramics, CIRP, 45/2, p569-580.

Tonshoff, H. K., T. Lierse & I. Inasaki, 1999, Grinding of advanced ceramics, *Machining of Ceramics and Composites*, Marcel Dekker, Inc., p855-118.

PRECISION MANUFACTURING OF MICRO MOULDING TOOLS OF GLASS AND SINTERED CARBIDE

H.-W. Hoffmeister, A. Wenda
Institute of Machine Tools and Production Technology, Technical University Braunschweig,
Langer Kamp 19 B, 38106 Braunschweig, Germany

Abstract

The production of miniaturized moulding tools or mould inserts of glass and sintered carbide is a field of research at the Institute of Machine Tools and Production Technology at the University of Braunschweig, Germany. The cost effective mass production of micro parts or micro structures by micro moulding is dependent upon the availability of high quality micro moulds. These moulding tools can be used for production of micro parts by hot embossing or micro injection moulding or micro forming processes.

Our challenge was to manufacture 3D structures in glass or sintered carbide with a smooth surface and as minimal edge chipping as possible. The technology of our choice was microgrinding, therefore this paper contributes to the understanding of a µ-grinding process technology, and its application for micro moulding tool production.

Keywords

micro moulding tools, mould inserts, grinding, glass, sintered carbide

1. STATE OF THE ART

The replication processes are based on plastic deformation or on solidification of liquids or pastes. The basic restriction of these processes is that only workpieces softer than the die or mold can be processed [1].

The micro moulding tools for hot embossing and micro injection moulding of polymers, mainly in use today, are normally based on nickel or brass masters made by LIGA or ultra precision milling (*figure 1*) [2]. Also silicon moulds, processed by lithography and etching technology, are available.

With the aim of extending the tool life of moulding tools and to mould other materials than plastics or polymers, harder materials for the tools, as well as, their machining technologies have been investigated.

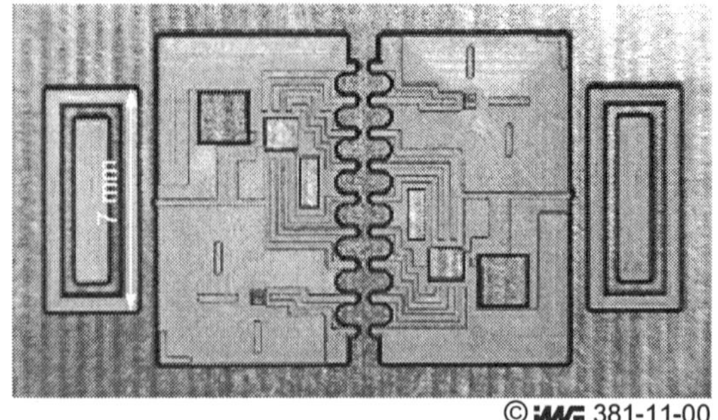

Figure 1: 3D-MID brass master structure for hot embossing or injection moulding manufactured by micromilling for HARTING EOB

1.1 Die/mould fabrication technologies for hard materials

Ultrasonic assisted diamond turning permits steel and glass parts to be manufactured in a turning process with the surface quality required of optical parts or even with micro-structures [3]. 3-dimensional micromachining of glass can be done by ultraprecision milling with a pseudo ball end mill, composed of a half-cut single crystal diamond and machining in the ductile mode cutting region [4]. For steel or sintered carbide micro moulding tools also micro-electro discharge machining is successfully applied [5]. Microgrinding allows machining of 3-dimensional structures in nearly every hard and brittle material [6].

2. NEW APPROACH

The advantage of moulding tools of glass or sintered carbide is the mechanical stability of the tool material which makes smaller structural dimensions possible and leads to a longer tool life. Our investigations on microgrinding proved that microgrinding allows the production of micro structures or micro components in theses materials (figure 2).

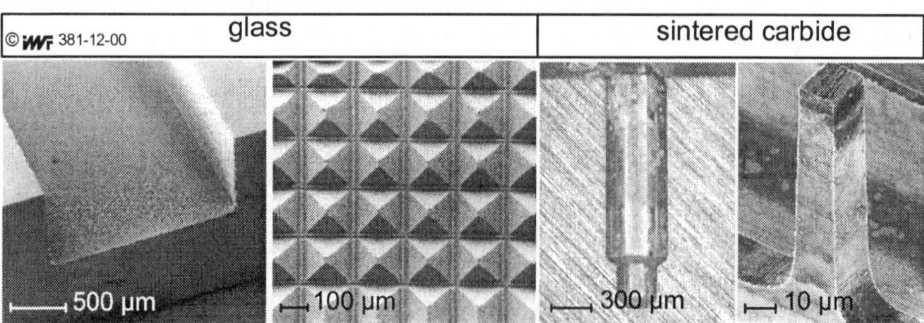

Figure 2: Microgrinding possibilities in glass and sintered carbide

2.1. Equipement

It seems to be obvious that a miniaturisation of products has to be combined with a miniaturisation of the form-, shape and surface roughness errors. For micromoulding parts with minimal structural dimensions of approx. 50 μm the desired errors for the shape and the geometry are in the range of 1 μm (2% of 50 μm). These tolerances are rather large compared with normal size precision cutting processes where a 2% error reaches 200 μm for a minimal structure size of 10 mm. Regarding the surface roughness it is convenient to desire a roughness at least 10 times lower than the geometrical tolerance, hence for a tolerance of 1 μm approx. 0.1 μm for R_{zDIN}.

To carry out our investigations, we used a CNC-controlled precision machine tool, equipped with measuring systems for the dynamic tool measurement (diameter and length) and a probing tool to determine the work piece position. The measured precision for the set-up was in the range of 2 to 4 μm for shape and geometry. These result shows the difficulties in reduction the tolerances in proportion to the size of the product.

3 MACHINING PROCESS

Modern grinding tools combined with a conventional technology gave us the advantage of a very flexible, fast and affordable moulding tool production.

For slots and straight grooves we use dicing blades, but for free form geometry the application of miniature abrasive pencils and core drills is necessary. The typical problems of the machined components are chipping at the edges (for glass) and insufficient surface roughness caused by the large diamond grains. Also the miniaturized abrasive pencils break easily due du the cutting forces.

For improvement in process stability and quality we successfully tested CVD-coated sintered carbide bodies as abrasive pencils. Smaller diamond cutting edges (approx. 2 to 10 μm), and the free tip shape allows the grinding of complex structures and an improvement in the surface quality.

On the example of a power beam splitter structure in glass, sintered carbide and brass, the prospects of micro grinding technology is demonstrated and compared to precision milling technology.

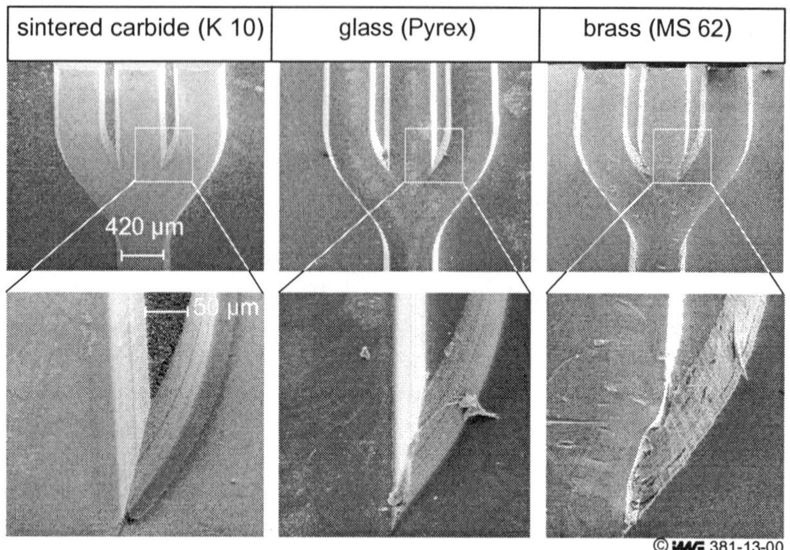

Figure 3: Power beam splitter moulding master

For the machining of glass it has to be mentioned, that chipping occurs rather often. But for grinding sintered carbides good machining results were obtained.

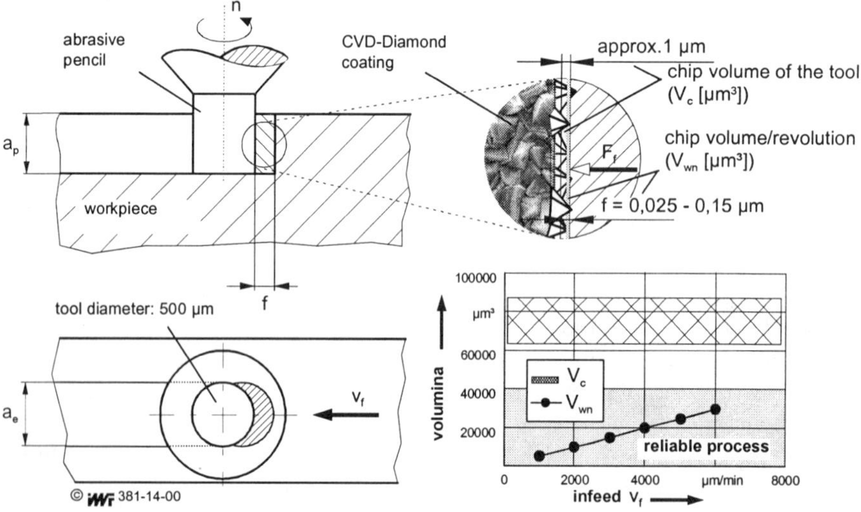

Figure 4: Grinding with minaturised CVD-Diamond coated abrasive pencils

The experiments were carried out in the ductile mode cutting region to obtain better surface quality. The typical process parameters were 45,000 rev/min for the spindle speed, a depth of cut of tool-diameter/3 and an infeed of 500 μm/min to 5000 μm/min. A coolant, for example emulsion or deionised water, is needed to remove the chips from the tool.

The used grinding process (figure 4) is characterized by very low cutting speeds of 0.2 to 2 m/s which is comparable to lapping or polishing, a stock removal rate of 800 μm³/(μm s) to 0.08 mm³/(mm s) and a specific cutting force in feed direction of approx. 5 to 15 N/mm², which is approx. the double of honing process forces. To avoid a load which leads to unacceptable tool-deformations or tool-breakage, the maximal chip load has to be considered using CVD-Diamond cutting tools and of course the infeed has to be adapted to the grindability of the workpiece material. *Figure 4* shows the available and the needed chip volume on the example of an abrasive pencil with CVD-Diamond coating and a diameter of 500 μm.

Summary and Outlook

Ground micro molding inserts of hard materials offers the extension of the moulding tool life and the possibility to mould other materials than plastics or polymers. Our next purpose is the investigation of the whole production chain from the mold making to the moulded product in collaboration with industrial partners to prove the advantages of the grinding technology for the mould fabrication.

REFERENCES

[1] Masuzawa, T.: *State of the Art in Micromachining.* Annals of the CIRP; Vol. 49/2, 2000, pp. 473-488

[2] Hoffmeister H.-W., Wenda A., Herrmann H.: The fabrication and application of micro milling cutters. Proceedings (Volume 1) of the MICRO.tec 2000, VDE World Microtechnologies Congress, September 25th/27th 2000, EXPO, Germany, pp. 125-129

[3] Klocke F., Rübenach O.: *Ultrasonic assisted diamond turning of steel and glass.* Proceedings of the international seminar on precision engineering and micro technology (EUSPEN); July 19th/20th 2000, Aachen, Germany, pp. 179-190

[4] Takeuchi Y.: *Developement of ultraprecision milling machine and its application to micromachining.* Proceedings of the international seminar on precision engineering and micro technology (EUSPEN), July 19th/20th 2000, Aachen, Germany, pp. 3-12

[5] Koch O., Wolf A., Ehrfeld W., Michel F., Gruber H.: *Micro-electro discharge machining for mould inserts generation-application and technology.* Proceedings (Volume 2) of the MICRO.tec 2000, VDE World Microtechnologies Congress, September 25th/27th 2000, Expo 2000, Hannover, Germany, pp.19-24

[6] Hoffmeister H.-W., Wenda A.: *Novel grinding tools for machining precision micro parts of hard and brittle materials.* Proceedings of the 15th annual meeting of the ASPE, October 22th/27th 2000, Scottsdale, Arizona, U.S.A., pp. 152-155

AN EXPERIMENTAL STUDY ON BORE HONING OPERATION
Influence of Unequal Over-run on Geometrical Error

Javad Akbari

Sharif University of Technology, Tehran, Iran
(Visiting Professor at Tokyo Institute of Technology)

Abstract

While horizontal honing is widely used in industry, most of research works have been performed on vertical honing machines. On the other hand, using honing fluid in a horizontal bore causes a different lubrication condition in this process, which has effect on the results. This study investigates the effect of tool over-run from two ends of the bore during horizontal honing. Experiments indicate that the amount of over-run has influence on shape accuracy and surface finish. Furthermore, unequal over-run causes a better condition for uniform lubrication, which increases the process quality.

Keywords

honing, geometrical tolerances, surface roughness, over-run

1. INTRODUCTION

High accuracy and tight tolerances of cylindrical holes in some engineering components such as valve bodies and auto industries are ever more demanding. *Honing* is an abrasive bore finishing technique which offers good geometrical tolerances. However, despite its widespread use, its process behavior is still under-explored. Some researchers such as Ueda et.al. (1982), Salje et.al. (1986) and Lee et.al. (1993) suggested some theoretical analysis beside their experiments studies on honing. On the other hand, many investigations have attempted to present empirical information for industrial applications without analysis, e g., Fischer (1986).

In most of the reported analytical works, vertical honing machine has been used, however in many industrial cases, a horizontal spindle honing machine should be used. For instance, in honing of long tubes, a vertical machine would have to be very tall and difficult to fit. Furthermore, many of manual honing machines have horizontal spindles which requires the operator to hold the part horizontally in hand while stroking the workpiece back and forth over the honing tool. Because of the nature of lubrication in vertical honing, which is partly based on the gravity of honing fluid, the mechanism of honing in horizontal spindle honing may have some differences with the vertical honing.

On any honing job, the abrasive stone has to work the entire length of the hole plus a little extra on each end, which is called the *over-run* or *over stroke,* (Figure 1). Experimental investigations in horizontal honing indicate that depending on the stone length, the bore geometrical accuracy is very sensitive to the amount of over-run (Akbari et.al., 2000). It was shown that when using longer stones, the bore geometrical tolerances are less sensitive to the over-run, although, they cause more geometrical error.

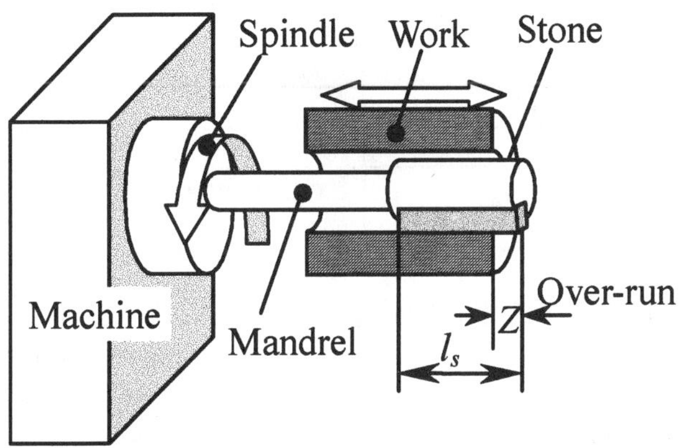

Figure 1. Schematic of honing process and over-run in operator side

In present investigation, experiments were performed on a horizontal-spindle honing machine to provide a description on the process specification such as degree of geometrical accuracy and surface topography inside the bore. Regarding the various distribution of surface roughness along the bore, the effect of shifting the center of stroke was investigated and discussed. It is shown that unequal over-run can be optimized for obtaining the best qualitative conditions in the bore finishing and accuracy.

2. TEST CONDITIONS

The honing tests were carried out on a power stroked SUNNEN model MBC-1800G horizontal-spindle honing machine under the conditions set out in Table 1. The coolant nozzles were arranged so that the honing oil enters the workpiece parallel to the honing tool. Two nozzles, one from each end were used. After the honing process, the workpieces were carefully cleaned and their cylindricity, roundness and straightness were measured on a Mitotoyo model KN810 CMM. Surface roughness on both ends of the bore was measured separately using Surftest 500 Mitutoyo roughness measuring machine.

Table 1. Honing conditions

Work material	Aluminum Alloy 7075
Work dimensions (mm)	I.D.=8, O.D.=15, L=32
Abrasive (SiC) grain size	#220
Stone length (mm)	32
Number of honing sticks	One
Honing peripheral speed (m/min)	25.1
Stroke/min (rpm)	80
Stroke length (mm)	29
Over-run ratio, λ	0.25, 0.375, 0.5, 0.625, 0.75, 0.813
Honing pressure (MPa)	0.2
Honing fluid	Mineral oil
Coolant supply system	External nozzles
Increase of radius (mm)	Max 0.1

3. EXPERIMENTAL RESULTS AND DISCUSSION

During honing process it is usual to set equal over-run at both ends of the work. Previous experiments (Akbari, et.al., 2000) indicates that the roughness is a little higher in the machine side of the bore. Measurement of work diameter in two ends of the bore also showed more stock removal at the machine side of the bore. Therefore, the idea of unequal over-run was proposed to increase the time of contact between the stone and the work in operator side. For this investigation, five different over-runs were examined as listed in Table 1. Figures 2 and 3 show the results of these experiments.

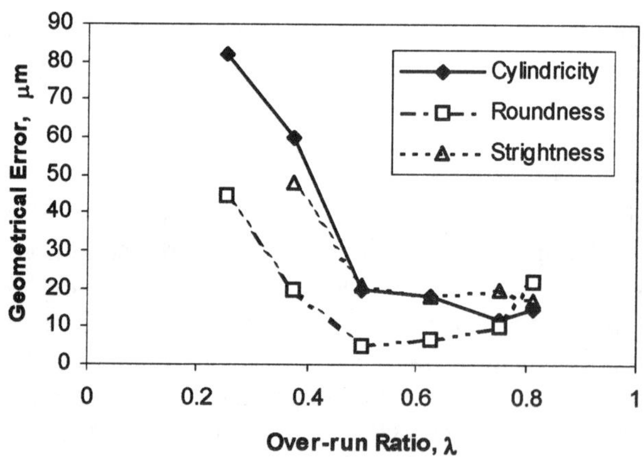

Figure 2. Variation of geometrical errors with over-run ratio

428

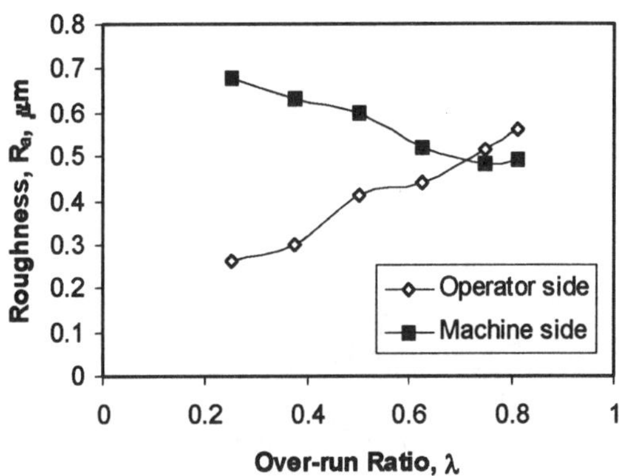

Figure 3. Surface roughness vs. over-run at two ends of the bore

Over-run ratio, λ, is defined as a non-dimensional parameter, which is proportional to the over-run in the operator side, Z (see Figure 1);

$$\lambda = Z_{operator\ side} / l_s \qquad (1)$$

in which l_s is the abrasive stone length.

With shifting the center of stroke toward operator (increasing λ), geometrical errors have a minimum when λ is about 0.6- 0.7, as Figure 2 shows. On the other hand, the relative motion between the work and the stone is a sinusoidal oscillation. With shifting the center of stroke toward operator, the abrading time in operator side increases and therefore, as shown in Figure 3, roughness increases in the operator side and decreases in the machine side.

The intersection between two curves is also about $\lambda = 0.7$. In this point, the difference between roughnesses along the bore is a minimum and the uniform distribution of surface roughness occurs.

For explaining the reason of roughness difference along the bore, especially for small diameter bores, it is necessary to notice to lubrication condition at two sides of the bore in horizontal honing. Figure 4 shows a cross section of honing tool while honing inside a bore. As can be seen in this figure, there is not enough space for honing fluid to enter the work from the machine side, while it can easily enter from the operator side. Better lubrication causes more sliding between the stone and work in the operator side but more abrading in the machine side. Shifting the stroke center toward the operator, gives more chance to honing fluid to enter the work from the machine side, which causes less geometrical error and more uniform distribution of roughness.

429

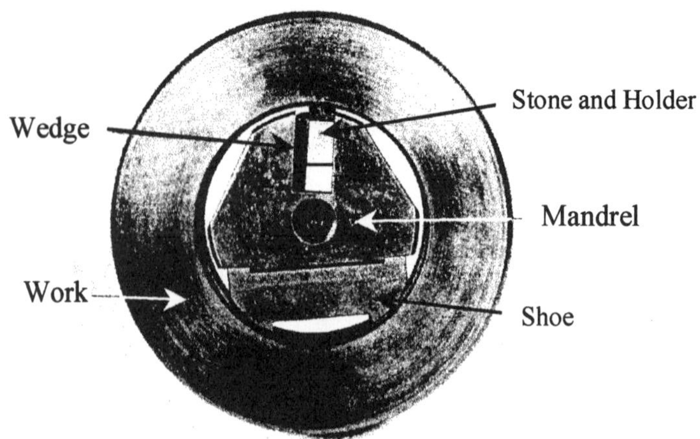

Figure 4. End view of honing tool inside the workpiece, after Fischer, 1986

4. CONCLUSION

Experimental study on honing process indicates that:
1. There is an optimum value of over-run for any specific work and tool length in which the geometrical errors minimize.
2. In horizontal honing, unequal over-run can increase the shape accuracy and cause uniform roughness distribution along the bore.

ACKNOWLEDGEMENT

The author completed this work at Professor Yoshio Saito's lab in Tokyo Institute of Technology. Mr. A. Reihani's valuable helps in collecting the experimental data is acknowledged.

REFERENCES

Akbari, J., Reihani, A., Study the Effects of Honing Parameters on Bore Quality, Proceeding 5th International Conference on Progress of Machining Technology, 2000 Sep. 16-20, Beijing, China, Aviation Industry Press, 2000; 317-21.

Fischer, H., Handbook of Modern Grinding Technology, Chapter 13, King, R.I. and Hahn, R.S. Eds.,: Chapman and Hall, New York, 1986.

Lee, J. and Malkin, S., Experimental Investigation of the Bore Honing Process, Trans. ASME, J. Engineering for Industry, 1993; 115: 406-14.

Salje, E., Mohlen, H., See, M.V., Comparison of Grinding And Honing Process, SME Manufacturing Technology Review, 1986: 649-53.

Ueda, T. and Yamamoto, A., An Analysis on Cutting Mechanism of Honing Operation, Seimitsukikai, 1982; 11: 1514-19 (In Japanese).

INVESTIGATIONS REGARDING THE OPERATING RANGE OF ULTRATHIN GRINDING WHEELS ON ALTIC

Hans H. Gatzen[1], Caspar Morsbach[1], and Gordon M. Jones[2]
[1]Institute for Microtechnology, Hanover University, Germany
[2]Seagate Technology, Minneapolis, USA

Abstract

Minimizing the street width, i.e. the distance required for the separation cut between thin-film elements on a wafer, is a key factor for an effective exploitation of a wafer's area. One potential approach to street width minimization is the use of ultrathin dicing wheels. This paper presents investigations of the operating limits in Al_2O_3-TiC wafer dicing with wheels of a width between 16 μm and 60 μm. Maximum wheel exposure limits for 16 μm to 40 μm wheels and feed rate limits for 40 μm to 60 μm wheels were identified. The results achieved exceed the established exposure and cut depth ratios by a factor of two.

Keywords

Grinding, dicing, ultrathin dicing wheels, street width, wheel exposure

1. INTRODUCTION

In the production of rigid disk read-write heads, a front-end process creates a batch of thin film transducers on an Al_2O_3-TiC (AlTiC) substrate. In a back-end process, the elements are separated into component parts by a combination of slicing and dicing processes. In order to maximize the areal yield of a wafer, the street width, i.e. the space between the elements reserved for the separation cut, has to be kept as small as possible. This requirement is of particular importance in the case of very small component parts such as thin film heads which occupy a surface area of only 1,010 μm x 300 μm. Development of thin dicing wheels creating 'ultra-thin' kerfs as a measure to reduce the street width results in increased die density and reduced component cost.

The process requirements for the separation of thin film heads tend to exceed the demands on semiconductor wafer dicing. For semiconductor applications the wafer material is mainly silicon. While the street width is to be minimized by a combination of small wheel and chip width [1], in general, there are no demands regarding the kerf side wall quality. Additionally, the ratio of the chip to the cut width is much lower on semiconductor applications due primarily to the material properties and the depth of cut (wafer thickness). The wafer material for thin film heads however is AlTiC and typical cut depths are over 1250 μm (compared to 300 μm on IC wafers). Besides minimizing the street width, there are stringent surface and edge quality requirements for the kerf side-wall, since the side-wall quality affects the read-write head functionality [2].

Ultrathin dicing wheels (i.e. wheels with a width smaller than 60 μm) utilize a metal binder for stability reasons. The grit size of the diamond grains used is limited due to the small wheel width. To ensure sufficient diamond embedding, as a rule of thumb, the minimal wheel width equals three times the diamond size. Therefore, the application of diamond is limited to a grain size of 3 μm to 8 μm.

From a production point of view, achieving the maximum wheel exposure allows for stable operating conditions and minimized process costs. As described by Ohbuchi [3], the wheel width significantly influences the wheel's stiffness and thus the tendency to axial deflect. When exceeding a certain aspect ratio (wheel exposure to wheel width), the wheels become extremely sensitive for wheel walk and axial deflection. If wheel walk occurs on ultrathin wheels during cutting, a substantial meandering of the kerf may be observed, causing an over-stretching and, ultimately, breakage of the wheel. This meandering cut must also be considered as part of the cut lane when calculating the density of the components on a wafer. When using ultrathin wheels the highest feasible wheel exposure aspect ratio is the limiting factor, if a minimum depth of cut is required.

2. EXPERIMENTAL PROCEDURE AND RESULTS

In order to investigate the performance of ultrathin dicing wheels with respect to wheel exposure and depth of cut, a series of tests was performed. The dicing saw used is equipped with a precision air-bearing spindle on which the dicing wheels are mounted, clamped in a pair of precision flanges. By using flanges with different diameters, the wheel exposure, i.e. the distance between outer flange diameter and outer wheel diameter, was varied. The wheels were hub-less type metal bonded dicing wheels. *Table 1* summarizes the parameters of the wheels under investigation. Due to the wheel exposure dependence on the wheel and flange diameter, the wheel exposure was chosen in incremental steps. As shown in *Table 1*, three wheel exposures ranging from 0.920 mm to 1.880 mm were chosen. Independently of the wheel exposure, the actual depth of cut was held at 500 μm for all tests and 1.4 mm for the tests with 30 μm to 60 μm thick wheels.

Table 1 Wheel Parameters

Width [μm]	Binder	Exposure [mm]	Grit [μm]	Depth of Cut [mm]
16	Metal	0.920	3-6	0.5
20	Metal	0.920, 1.700	4-8	0.5
30	Metal	0.920, 1.700, 1.880	4-8	0.5, 1.4
40, 50, 60	Metal	1.880	4-8	0.5, 1.4

AlTiC plates ("wafers") with a thickness of 1.25 mm and a length of 110 mm were used as the work piece. In the tests, a single pass down cut grinding mode was applied. Besides the feasible depth of cut, the edge quality is a crucial factor when evaluating the wheel performance. By cutting at various feed rates, the respective tendency for chipping was investigated. The edge quality was evaluated by measuring the average maximum chipping width over the entire cut length by optical microscopy in a top-side view of the wheel kerf.

Figure 1 (left) shows the chipping width of 20 μm and 30 μm wheels for a depth of cut of 0.5 mm and 0.920 mm exposure. The chipping remains constant over the entire range of the investigated feed rate (0.5 mm/s - 10 mm/s). It could be observed, that the 20 μm wheel yielded slightly smaller chipping. The SEM image (*Figure 1*, right) demonstrates the edge quality using a 20 μm wheel at a feed rate of 0.5 mm/s. A stable process with constant chipping width within narrow limits could be maintained over the entire cut length.

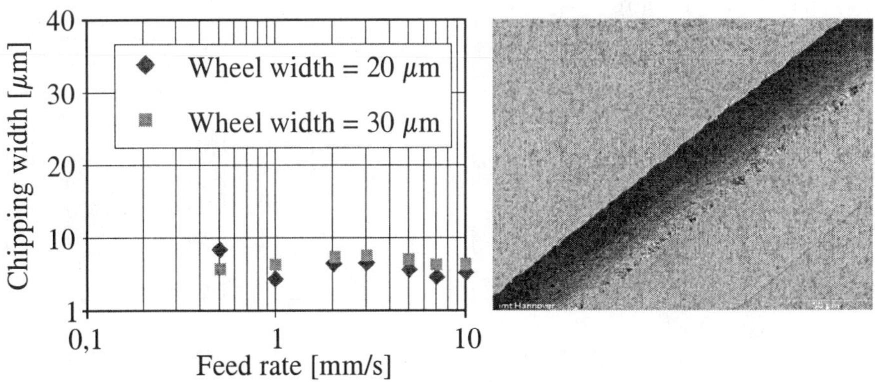

Figure 1 Chipping width for 20 μm and 30 μm wheels (left) and SEM image of a kerf cut with a 20 μm wheel at 0.5 mm/s (right)

With respect to the wheel performance limits, i.e. the maximum achievable wheel exposure and the highest feasible feed rates at a given depth of cut, the following results were achieved. An exposure of 1.88 mm resulted in a failure of both the 30 μm and 20 μm wheel and a substantial wheel walk could be observed. It resulted in a strong meandering of the kerf, indicating that the axial rigidity of the wheel was insufficient for maintaining a straight cutting direction. A wheel exposure of 1.88 mm therefore was found to be impractical for these wheel widths. On the other hand, a 40 μm wheel is capable of operating at 1.88 mm exposure and 0.5 mm depth of cut. In an additional test, the depth of cut was increased to 1.4 mm, which also proved feasible. In case of the 30 μm wheel, reducing the exposure lead to stable conditions. At an exposure of 1.70 mm, a depth of cut of 0.5 mm could be achieved with feed rates up to 10 mm/s. The 20 μm wide wheel required a further reduction of the wheel exposure to 0.920 mm to work in a stable regime. This is an exposure ratio of 46:1 and a cut ratio of 25:1. *Figure 2* (left) depicts the results of the wheel exposure investigations for these wheels. However, the 16 μm wheel failed even at an exposure of 0.920 mm and therefore could not be used for any of the test conditions.

Figure 2 (right) summarizes wheel exposure ratio and cut depth ratio respectively of the 20 μm to 40 μm wheels. The wheel exposure ratio exceeds 40:1 in all tests, in the case of the 40 μm wheel a cut depth ratio of 35:1 was achieved. Typical established exposure ratios are just under 30:1 and cut ratios are on the order of 14:1.

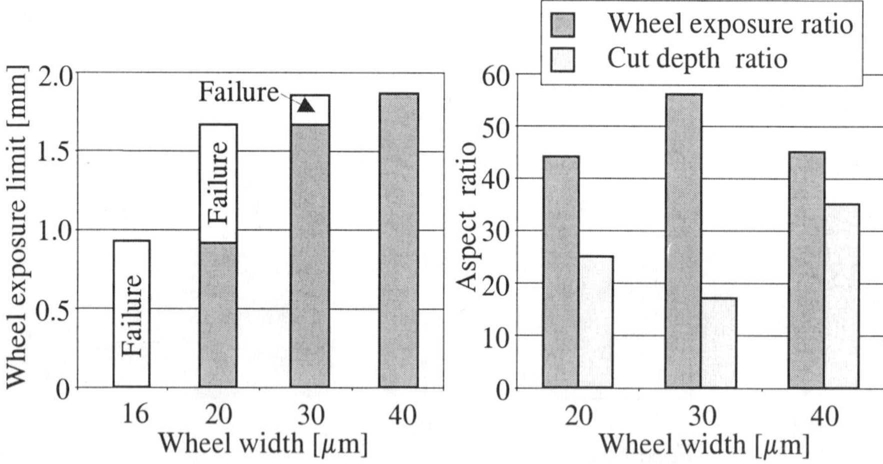

Figure 2 Wheel exposure limits (left) and wheel exposure ratio as well as cut depth ratio as a function of the wheel width (right)

In a second test series, the feed rate limitations for 1.4 mm depth of cut at 1.880 mm wheel exposure were investigated. The initial investigation

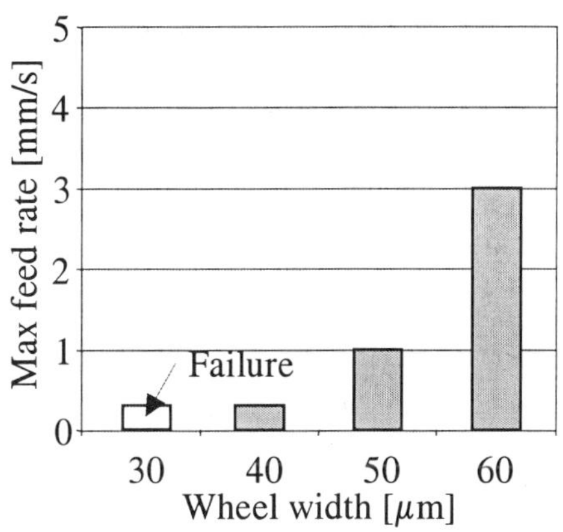

Figure 3 Maximum feed rate vs. wheel width at 1.4 mm cut depth

indicated the minimum feasible wheel for this cut depth. The results are shown in Figure 3 and indicate the limit to be 40 μm wheel thickness. This results in a wheel exposure ratio of 47:1 and a cut depth ratio of 35:1. As expected, the highest feasible feed rate rises with increasing wheel width. This is a result of higher rigidity of the wheel/flange setup due to a reduced wheel exposure ratio. Thus the thermo-mechanical load induced in the wheel during grinding by cutting forces and frictional forces is transferred into the flange without leading to a destructive deflection of the dicing wheel.

3. CONCLUSION

The investigations revealed operating limits of ultrathin dicing wheels. The 30 μm wheel was capable of working up to a wheel exposure of 1.7 mm, the 20 μm wheel up to 0.920 mm when grinding with a 0.5 mm depth of cut. The 16 μm wheel failed at 0.920 mm exposure and therefore could not be used at any of the test conditions. For wheel exposures found practical, feed rates up to 10 mm/s were achieved at a depth of cut of 500 μm without compromising the edge quality. A 1.4 mm depth of cut could only be achieved with wheels wider than 30 μm. Using 60 μm wheels a feed rate of up to 3 mm/s becomes feasible. The wheel exposure investigated substantially exceeds present industrial applications. Using ultrathin dicing wheels therefore offers the opportunity for a significant reduction in the dicing lanes for 'thick' (1250 μm) AlTiC wafers.

REFERENCES

[1] Inasaki, I.: Dicing Of Silicon Wafers. Abstracts MicroMat 2000, Berlin, Germany, Apr. 2000, p. 154

[2] Gatzen, H.H.; Morsbach, C.; Zeadan, J.: Advances in Dicing Wafers for Micro Electro-Mechanical Systems (MEMS), Proceedings of MICRO.tec 2000, Hanover, Germany, Sep. 2000, Vol. 2, pp. 621-625

[3] Ohbuchi, Y.; Matsuo, T.; Ueda, N.: Warp in High Precision Cut-Off Grinding of Al_2O_3-TiC Ceramic Thin Plate. Annals of the CIRP, Vol. 48, No. 1, 1999, pp. 285-288

ERRORS ANALYSIS OF BALL-HEADED WHEEL DISCHARGE DRESSING AND STUDY ON GRINDING EXPERIMENTS IN ULTRA-PRECISION ASPHERIC GRINDING SYSTEM*

Mingjun Chen, Shen Dong, Feihu Zhang, Dan Li

(Harbin Institute of Technology, P. R. China)

Abstract

In the paper, firstly dressing errors of ball-headed wheel was analyzed in dressing process theoretically. Afterwards author developed the dressing machine of the discharge principle, which have high profile accuracy of ball-headed wheel after dressing. Profile accuracy of the wheel is high after dressing. This dressing machine solved the problem of on-position dressing of the cast iron bonded diamond wheels and reduced the machine errors of aspheric surface part. At last the trued wheel was used in ultra-grinding machining aspheric part. Grinding result was shown that profile accuracy was 0.4μm, the surface roughness was less than 0.01μm.

Keywords

Ultra-precision grinding, Optical aspheric surfaces, wheel dressing

1. INTRODUCTION

With the development of science and technology, the optical aspheric surface parts play more and more important roles in many key instruments[1-2]. However, it is very difficult to machine the aspheric surfaces. During grinding aspheric surfaces, the shapes of wheels have heavy influence on the accuracy of machined surfaces[3]. Therefore, in order to improve the machining accuracy of the aspheric surface, dressing the wheels with high precision is the first important step. In this paper, the dressing error of the wheels was analyzed, then the dressing machine of the diamond wheel profiles was developed, finally grinding experiments of the aspheric surfaces were performed which show that the machine achieved the expected results.

2. THE DRESSING PRINCIPLE OF BALL-HEADED WHEEL

The principle of which is using the electrical corrosive phenomenon during discharge between cathode and anode to remove the extra material of the cast iron cohering material and to cater to the requirements of wheel size, shape and surface quality. The discharge dressing should be conducted under the following

---

*This project is supported by the National Science Foundation of China, No.59835180. and HIT.2000.61 Supported by the Scientific Research Foundation of Harbin Institute of Technology

condition: clearance between the tool electrode and the dressed surface of the wheel, the special electrical source for the discharge dressing, and the insulating liquid media etc.

The diagram of dressing the diamond wheel using cup-shaped electrodes is shown in Figure 1. The electrode and rotating axis of the wheel intersect at point of O, and the rotations of the electrode and wheel will achieve spherical surface. The relation between the radius R of the spherical surface of the wheel and the diameter d of electrode can be expressed by formula (1)

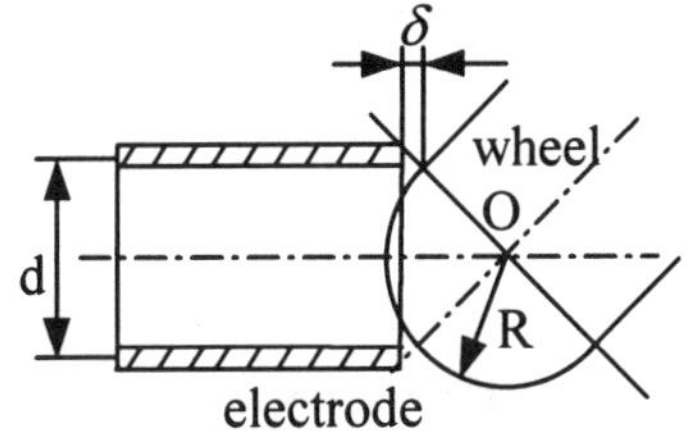

Figure 1. Principle tool electrode dressing ball-headed wheel

$$R = \frac{d}{2\sin\theta} \tag{1}$$

In the actual dressing, the electrode is mounted on the spindle of the machine tool; the angle θ between the axes of the electrode and the wheel is 45°. The diameter of the electrode can be computed using formula (1). Dressing wheels with various diameters need different electrode with various diameters. Therefore only by using different-diameter electrode, the dressing of ball head wheel with different diameters can be realized.

3. ERROR ANALYSIS OF DRESSING THE BALL-HEADED WHEEL

3.1 Effects of geometrical parameters on machining the spherical surface

If the decenter error δ of the electrode is ignored and only the angle error between the electrode and the wheel are considered, only the diameter error of the spherical surface is expected. This error can be achieved by the differential of formula (1)

$$dR = \frac{\sin\theta\,dr - r\cos\theta\,d\theta}{\sin^2\theta} \tag{2}$$

From the formula (2), we can know that, the error dr of the diameter of the electrode can result in the dr/sinθ error on the spherical surface; and the angle error dθ, will result in error $r\cos\theta d\theta/\sin^2\theta$ on the spherical surface.

3.2 Effects of decenter of the electrode on machining the spherical surfaces

If the error between the electrode and wheel to be dressed exists, shown in Figure 2, the equation of the rotating curve can be expressed

$$x^2 + y^2 = [r^2 - (r - \frac{R-z}{\sin\theta})^2] + (\frac{R-z}{\tan\theta} + \delta)^2 \tag{3}$$

Substituting (3) in (1), we can obtain formula (4)

$$x^2 + y^2 + (z + \delta \cot\theta)^2 = (R + \delta \cot\theta)^2 + \delta^2 \tag{4}$$

From (4), we can know that the achieved surface is still a part of spherical surface, the center is still the intersecting point of the two axes, and

the radius is $\sqrt{(R + \delta \cot\theta)^2 + \delta^2}$. Therefore, the error between the electrode and the wheel can only result in the radius error of the spherical and have no effect on the form accuracy of the spherical surface.

3.3 Incoplanarity of on machining the spherical surface

When the axes of the electrode and the wheel are not in the same plane, we call this incoplanarity. If the point of intersection of the axes of electrode and wheel has error δ in X direction, shown in Figure 3, then the rotating curve can be expressed

$$x^2 + y^2 = \left[\sqrt{r^2 - (r - \frac{R-z}{\sin\theta})^2} - \delta\right]^2 = (\frac{R-z}{\tan\theta})^2 \tag{5}$$

$$x^2 + y^2 + z^2 + \frac{2\delta}{\sin\theta}\sqrt{2R(R-z)\sin^2\theta - (R-z)^2} = R^2 + \delta^2 \tag{6}$$

From Figure 3, we can know that, whenever the error is in positive X direction or in negative X direction, only part of the curve involves the electrical machining. When $\delta \geq 0$, $x=0$, we can obtain from formula (6)

$$y^2 + z^2 + \frac{2R}{\sin\theta}\sqrt{2R(R-z)\sin^2\theta - (R-z)^2} = R^2 + \delta^2 \tag{7}$$

In formula, we can find that, in YOZ plane, the machining traces are no longer spherical surfaces, which means that incoplanarity of on machining the spherical surfaces can induce the profile error of the spherical surface.

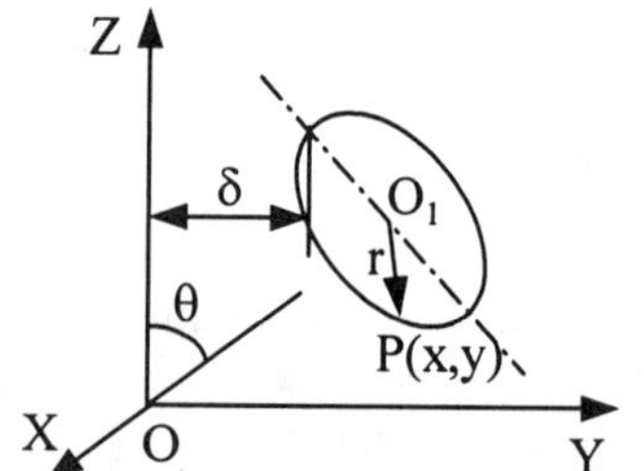

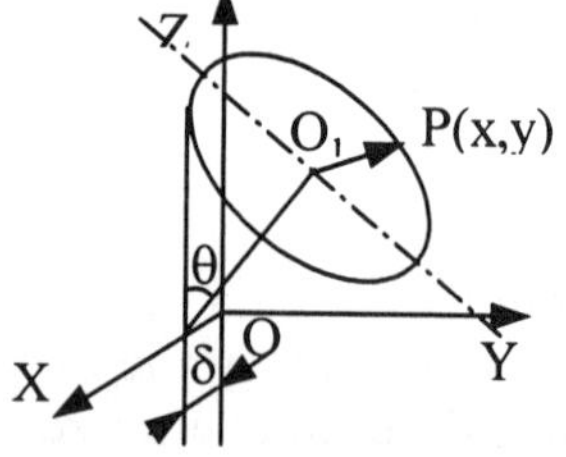

Figure 2. *Effect of spherical machining centererror of tool electrode*   **Figure 3.** *Effect of spherical machining of incoplanarity*

From what discussed above, we can see that in order to achieve high precision spherical surface, we should pay attention to the followings:

1. According to the diameters of the dressing wheels, the diameters of the electrodes should be accurately computed and machined;

2. When assembling the dressing machine, the proper angle between axes of the electrode and wheel should be ensured, and the center position of it with the

wheel to be dressed should be accurately adjusted;

    3. The two axes of electrode and wheel should be ensured coplanarity.

4. DESIGN OF THE DRESSING INSTRUMENT

The assembling error of wheels induced will affect the machining accuracy. So, in order to minimize the effect of the assembling error on the machining

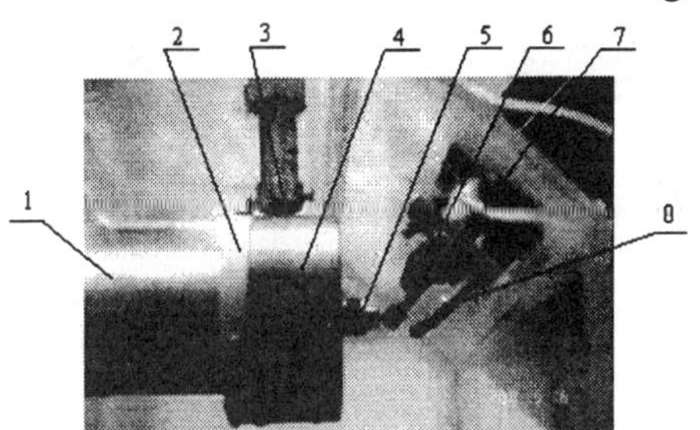

1 Transition tray, 2 dielectric, 3 brush(I), 4 axial sleeve, 5 electrode,
6 wheel, 7 brush(II), 8 coolant nozzle
Figure 4. *Wheel dressing machine*

accuracy, the wheels should be dressed. Therefore, in this paper, an on-position dressing machine of the wheels was designed, which is illustrated in Figure 4. The machine was directly mounted on the spindle of the machine tool, and the on-position dressing of the wheels can be realized without requirement to take

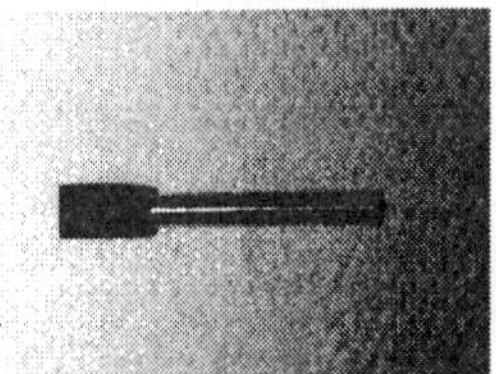

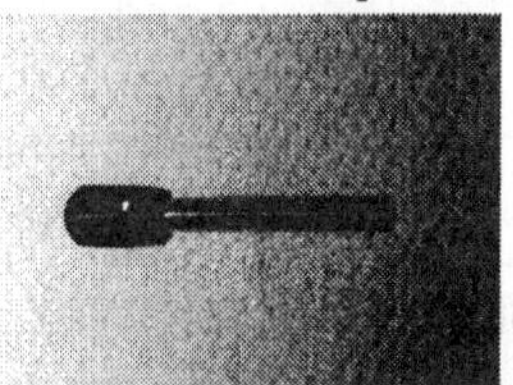

(a) Before dressing   *(b) After dressing*
Figure 5. *Before and after dressing wheel photo*

down the work-piece and the wheels. Furthermore, a special brush-assembling mechanism was designed to satisfy the requirements of conducting and rotating at the same time during the dressing process. Figure 5 is the photo of wheel before and after dressing.

5. GRINDING EXPERIMENTS USING DRESSED WHEEL

Optical glass (k9) was ground under the following condition: diamond wheel (the grit size w is 2.5µm), the speed of the wheel υ_s=30m/s, feed rate f=1µm/rev, grinding depth a_p=1µm. The ground surfaces were inspected using the NanoScope 3A developed by the American DI Company. Figure 6 is the inspected result, which shows that the surface roughness rms is 11.111nm and Ra is 8.096nm. The profile accuracy of the aspheric surface is 0.4µm.

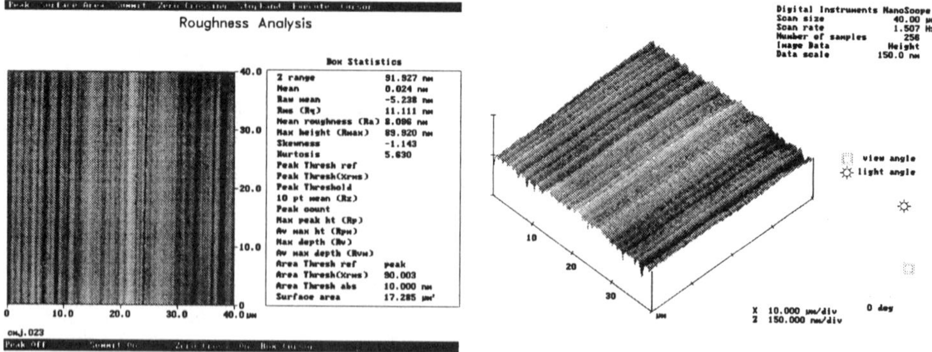

Figure 6. *AFM microscope graph of the grinding surface of K9 after wheel dressing (w=2.5μm, v_s=30m/s, f=1μm/rev, a_p=1μm)*

6. CONCLUSION

In this paper, after lots of theoretical analysis and experiments, the following conclusions can be reached:

1. The factors that affect the diameters of the wheels during discharge dressing were theoretically analyzed. The analyzing results show that geometrical parameters, the incoplanarity of the axes of electrode and wheel will influence the profile error of the ball-headed wheels; however the decenter error of axes of the electrode and wheel has no influence on the profile error of the ball-headed wheel;

2. The on-position dressing machine of the wheels was developed. Its structure is simple, dressing accuracy is high, and solve the problem of on-position dressing of the cast iron bonded diamond wheels;

3 For cast iron bonded diamond wheels of ultra-fine grits, high profile accuracy after dressing can be achieved, which can satisfy the requirement for the ultra-precision machining of the brittle materials such as optical glass etc., and the profile accuracy of the machined aspheric surface is 0.4μm, the surface roughness Ra is better than 0.01μm.

REFERENCE

[1] Yoshiharu NAMBA, Morihiko SAEKI, Takatomo SASAKI, 1994, Ultra-precision Grinding of KTP Crystals for Optical Surfaces, int. J. Japan Soc. Prec. Eng., Vol.28, No.1: 39-40.

[2] Katuo SYOJI, Tunemoto KURIYAGAWA, Libo ZHOU. Truing of Super-abrasive Wheels for From-grinding, int. J. Japan Soc. Prec. Eng., 1993, Vol.59, No3:485-490

[3] Mingjun CHEN, Shen DONG, Feihu ZHANG, Development of Ultra-precision Optical Aspheric Grinding System, China Mechanical Engineering, 2000, Vol. 11, No.8:849-851.

SIMULATION ON HIGH INTEGRITY SURFACE GENERATION OF ϕ300 Si WAFER WITH FIXED-ABRASIVE SOLUTION

Katsuo Sagawa[1], Hiroshi Eda[2], Libo Zhou[2], Jun Shimizu[2]

[1]*Graduate school, Ibaraki University*
[2]*Department of System Engineering, Ibaraki University*

Abstract

This study is done as one part of Regional Consortium Project of "Development of Transcendent Machine Tool and Core Technologies for ϕ300 Si Wafer", which aims to replace the conventional lapping, etching and polishing processes with fixed-abrasive solution. The newly developed grinding machine takes infeed-grinding form in order to keep the contact area unchanged and thereby to achieve the global flatness. The objective of this study is to simulate the achievable surface roughness by taking into account the contact stiffness between the abrasive and a point on the surface of workpiece, and to derive the principle of the grinding conditions and wheel specifications. Simulation results are as follows. The surface roughness at the center portion of the workpiece is better than that of at its fringe. Normal grinding force at the center portion of the workpiece is bigger than that of at its fringe. The processed wafer tends to be concave. The simulation results of surface roughness are found to agree with achieved experimental results.

Keywords

Simulation, Roughness, Grinding, Normal Force

1. INTRODUCTION

According to the roadmap of the semiconductor industry, the current ϕ200 mm (8") silicon wafer would be replaced by ϕ300 (12") by the year 2003. Present manufacturing process for silicon wafer included lapping, etching and polishing. We have developed a new grinding machine, which can achieve the final roughness equivalent to the polishing result by grinding only. This paper describes simulation model, analysis procedure, effects of grinding condition, the roughness and the flatness.

2. SIMULATION METHOD
2.1 Simulation Model

This simulation is executed in order to get surface roughness at infeed-grinding. Two kinds of elastic deformation have been taken into the consideration. One state is the contact deformation between abrasive grain and workpiece (u_c). The other state is the deformation of bond. Two kinds of elastic deformation are dependent on a normal force f_n. The relationship is shown in Figure 1. The total deformation u in Figure 1 is depth of geometrical interference between the abrasive grain and the workpiece. The t_c in Figure 1 is depth of scratch. Therefore, the relation of deformation is shown as Equation 1.

$$u = t_c + u_c + u_b \qquad (1)$$

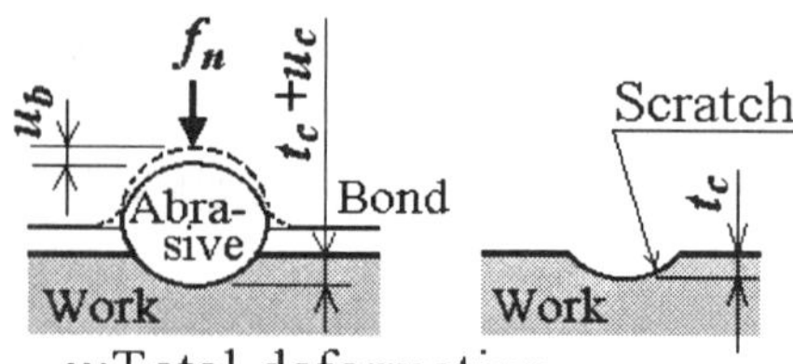

u:Total deformation
u_c:Elastic deformation
u_b:Elastic deformation of Bond
t_c:Depth of scratch
$\propto$ Prastic deformation

Fig.1 Simulation Model

2.2 Relationship u_b and f_n

Accordingly to the deformation of u_b.

$$u_b = G \cdot f_n \qquad (2)$$

Where G is the compliance of the bond. G is calculated by the Finite Element Method (FEM). The parameters used for calculation are shown Table 1.

As the results ;

$$u_b = 35 \cdot f_n \qquad (\phi 2 \mu m) \qquad (2.a)$$
$$u_b = 7.1 \cdot f_n \qquad (\phi 10 \mu m) \qquad (2.b)$$

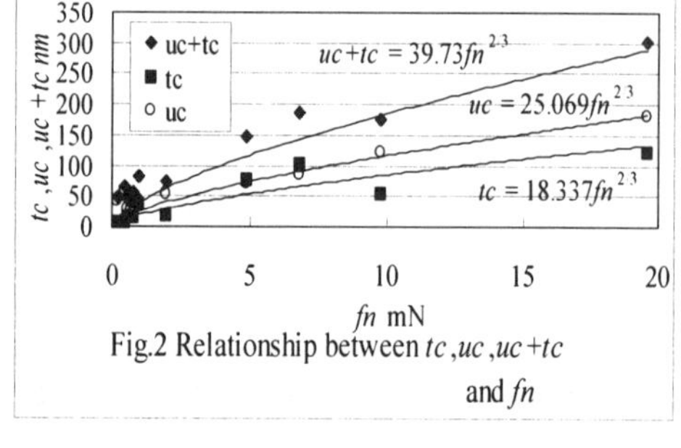

Fig.2 Relationship between tc, uc, $uc+tc$ and fn

The protrusion height of the abrasive grain from the bond surface is supposed to be $d_g/3$ in FEM.

Table 1 Parameters of FEM

Materials	Young's Modulus E(Gpa)	Poison's ratio ν	Radius of abrasive grain (μm)	Protrusion height (μm)
Diamond	1430	0.28	$\phi 2$	0.667
Resin Bond	7	0.3	$\phi 10$	3.33

2.3 Relationship between u_c, t_c and f_n

The u_c and t_c are derived on the basis of an indentation experiments. A diamond indenter with 1 μm radius was used to study the deformation taken place at the Si workpiece during the loading and unloading processes. The total deformation u_c+t_c at each different load f_n, the plastic deformation t_c after unloading and the elastic deformation u_c were shown in Figure 2. It was clear from the results that each deformation on the sphere contact, approximately agreed with Hertz's law ($\delta=25.1\cdot f_n^{2/3}$) and thereby was proportionally associated on to anther.

2.4 Extent t_c to depth of cut

At actual grinding, moreover, each abrasive underwent not only the normal force f_n but also the tangential force f_t. It was therefore necessary to transfer the indentation model into the grinding model by added in the generalized results from indentation tests. Therefore, this simulation has to consider f_t. The grinding model is shown Figure 3. The pressure distribution p_n by load f_n is determined by Hertz's law, on the other hand, vector s is determined by tangential load f_t, which is distributed equally at a unit area at the contact portion. And a vector p is determined by p_n and s. The position r of maximum $|p|$ relates to the f_n. The r in Figure 3 is the position where the maximum stress occurred. Figure 4 shows the relationship between f_n and r, in which r is proportionally associated Hertz's law. Based on the results of Figure 2 and Figure 4, we concluded that t_c as proportional to r. By fitting the relationship f_n and t_c to the already known result of scratching tests (f_n=3.1N, t_c=70nm)[1], Equation (3) is given as ;

$$t_c=7.19\cdot f_n^{2/3} \qquad (\phi 2\,\mu\text{m}) \qquad (3.\text{a})$$

$$t_c=4.23\cdot f_n^{2/3} \qquad (\phi 10\,\mu\text{m}) \qquad (3.\text{b})$$

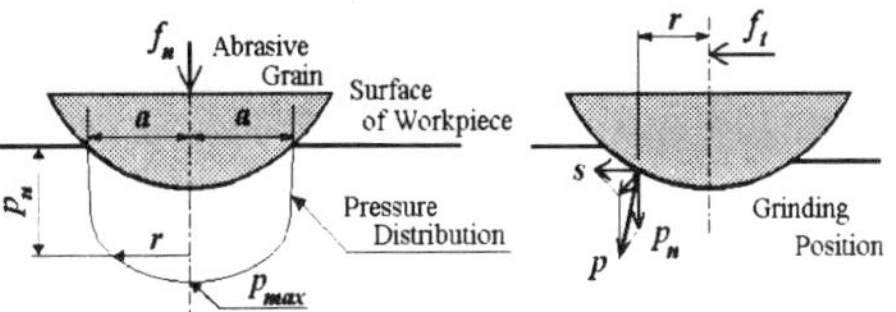

Fig.3 Grinding Model

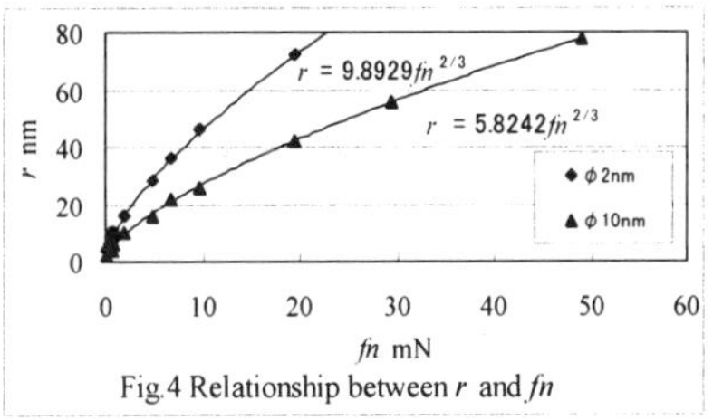

Fig.4 Relationship between r and f_n

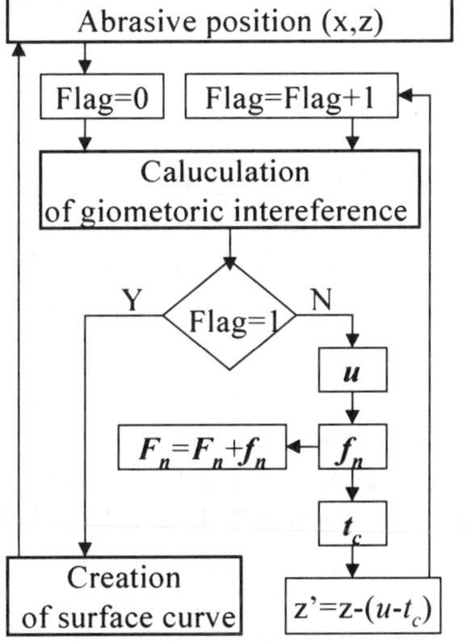

Fig.5 Simulation procedure

443

2.5 Total deformation u and Simulation

The total deformation u between abrasive grain and workpiece is calculated with Equation (4) using Hertz's law.

$$u=39.7 \cdot f_n^{2/3}+35 \cdot f_n \qquad (\phi 2 \mu m) \qquad (4.a)$$

$$u=23.2 \cdot f_n^{2/3}+7.1 \cdot f_n \qquad (\phi 10 \mu m) \qquad (4.b)$$

Table 2 Assumptions of Simulation

Abrasive grain	Sphere
The bond surface of grinding wheel	Flat
Maximum protrusion height	$d_g/3$
Wear of abrasive grain	Not considered

The surface roughness simulation is done at Figure 5. The assumptions of this simulation are shown in Table 2. Transcribing the geometrical shape of each abrasive grain, which behaves according to the above equations, generated the workpiece surface. The surface roughness is calculated from roughness curve.

3. Results of grinding simulation and evaluation

The simulation of grinding was executed at conditions shown in Table 3. This simulation is repeatedly done up 40 turns of workpiece. The surface roughness Ra, no more grinding force F_n in Figure 5 are calculated at every 5 rotations. This paper deals with that F_n is generated equally at each section by f_n. Figure 6 shows the relation of surface roughness Ra, F_n and d_c at 100mm away from the workpiece center. The Ra and F_n increase with d_c. In addition, Figure 6 shows that variations become smaller when abrasive grain is smaller. Figure 7 shows the relationship between Ra, F_n and R_w done at by d_c=10nm/rev, in which Ra at the fringe portion is large compared with that at.

Table 3 Conditions of simulation

Cup-type Grinding wheel	mm	$\phi 300$, 3t
Concentration		100
Diameter of abrasive grain	μm	$\phi 2$, $\phi 10$
Diameter of Si wafer	mm	$\phi 300$
Measured length	L μm	5, 20
Measured positions	R_w mm	50,100, 140
Wheel rotation speed	rpm	2000
Workpiece rotation speed	rpm	150
Depth of cut	d_c nm/rev	5, 10, 20

Also, F_n at the fringe is smaller compared with workpiece center. In addition, Figure 7 shows that the variations become smaller when abrasive grain is smaller. According to these figures are shown that $\phi 2 \mu$ m abrasive grain creates surface roughness of development aim (Ra=1nm). Figure 8 shows relationship between deviation of cutting rate and R_w, in which the cutting rate at fringe is large compared with center of it. Accordingly, the shape of workpiece is almost processed into concave lens shape.

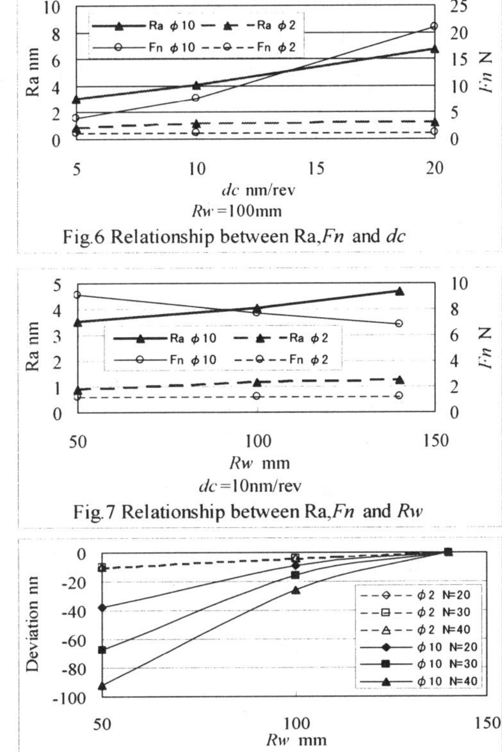

Fig.6 Relationship between Ra,Fn and dc

Fig.7 Relationship between Ra,Fn and Rw

Fig.8 Relationship between Deviation and Rw

4．CONCLUSION

This study executed a simulation to get a surface roughness of Ra$\leqq$1nm for ϕ 300 Si wafer. The results are shown in the following.

1) The surface roughness at the center portion of the workpiece is better than that of at its fringe.

2) The processed wafer tends to be concave.

3) Grinding wheel having $\phi 2 \mu$ m abrasive grain can get 1nm Ra at silicon wafer surface.

5. Acknowledgements

This research was sponsored by Regional Consortium from NEDO, the Grand-in-Aid for Exploratory Research (No.10875029) and the Regional Joint Research (No.11792007) from the Ministry of Education, Science and Culture of Japan. We extend sincere thanks to all project staff and researchers.

Reference

1) Yasuyuki Uemure,Yasuhiro Tani, Katuyoshi Satou, Dc and Fc value in ductility mode processing, Arts and science lecture article collection of the Society of Grinding Engineers (1995)51. (In Japanese)

EFFECT OF WHEEL REVOLUTIONAL SPEED ON STRIPED PATTERN ON SURFACES FINISHED BY HIGH-SPEED RECIPROCATION GRINDING

Nobuhito YOSHIHARA, Tsunemoto KURIYAGAWA and Katsuo SYOJI

Department of Mechatronics and Precision Engineering, Tohoku University
Aramaki Aoba 01, Aoba-ku, Sendai 980-8579, Japan

Abstract

Mirror-finished ground surfaces machined by high-speed reciprocation grinding contain striped patterns perpendicular to the worktable reciprocation direction, reducing the visual quality of the products. Simulations and grinding experiments have clarified that the striped patterns occur due to wheel vibration. It has also been demonstrated that small fluctuations in wheel revolutional speed produces significant changes in the striped patterns.

Keywords

Striped pattern, High-speed reciprocation grinding, Wheel revolutional speed

1. INTRODUCTION

The degree of integration required for modern ULSIs has resulted in a demand for narrower and more precise lead frame punches. Grinding machines that employ a high-speed reciprocating worktable or grinding wheel, a process known as high-speed reciprocation grinding, are commonly used for finishing lead frame punches. When the surface roughness is reduced to approach a mirror finish, striped patterns similar to chatter marks (called "vertical stripes" here) or crisscross patterns occur in the direction of worktable movement [1]. It is well-known that chatter marks are usually encountered in plunge grinding only, and represent a transfer of wheel imbalance to the workpiece surface. Accordingly, high-speed reciprocation grinding is more susceptible to chatter than creep-feed grinding where the table speed is extremely slow. As punch grinding involves cross-feed movement, grinding wheel vibration is transferred to the workpiece surface in different phases in each grinding path, and hence chatter marks should not appear.

The waviness of the ground surface with vertical stripes is not great, and there is no substantial damage to the flatness of the surface, although the marks can be seen with the naked eye, lowering the quality of the product. To develop a solution, it is important to clarify the cause of the vertical stripes in high-speed reciprocation grinding, and the mechanisms that lead to their appearance. This report proposes a method for simulating this phenomenon, and the validity of the results is shown through grinding experiments.

2. GRINDING EXPERIMENT

As illustrated in figure 1(a), the cross-section of the tungsten carbide lead frame punch is complex and extremely thin. The thickness of the thinnest section may be less than 0.1 mm. In this report however, for simplicity, a cantilever of thin flat plate was ground, as shown in figure 1(b). In the grinding experiment, a grinding machine with a high-speed reciprocation worktable was used. This type of worktable has a counterbalancing function (see figure 2), which effectively eliminates transience of the center of gravity of the grinding machine, thus reducing vibration during high-speed reciprocating motion.

The diamond wheel was trued with a cup truer (GC 400V) mounted on the machine to obtain a straight cross-sectional wheel profile. The wheel spindle was inclined 1 degree from grinding level, and grinding was performed using the front edge of the grinding wheel. Table 1 shows the grinding conditions.

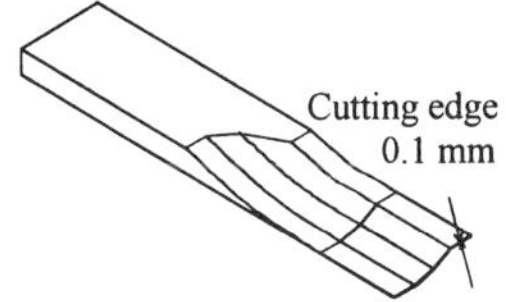

(a) Example of IC lead frame punch.

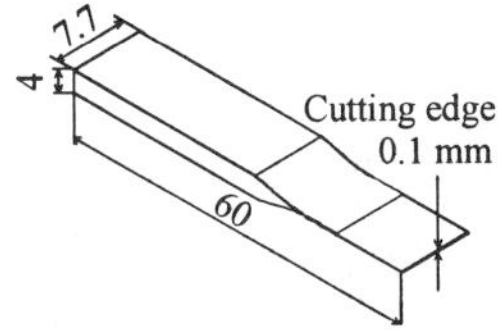

(b) Punch used in experiment.
Figure 1. Schematic of lead frame punch.

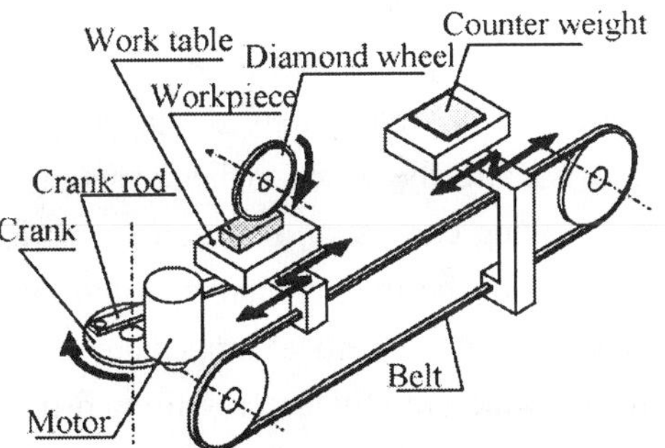

Figure 2. Mechanism of high-speed reciprocation worktable

The grinding experiment was carried out with a reciprocation frequency of 200 min^{-1}, a stroke length of 13 mm, a cross feed speed of 1 mm/min, and a wheel speed of 4778 rpm. The workpiece (7.7 W × 4 H × 60 L) was fixed on the work table. Figure 3 is a photograph of a workpiece surface, which has been ground down to a thickness of 0.5 mm. The surface roughness was

Table 1. Experiment Conditions

Grinding machine	High-speed reciprocation grinding machine (SHS80, Nagase Integrex. Co. Ltd)
Grinding wheel	SD400G100B ($\phi 125 \times 4.8\ W$)
Workpiece	Tungsten carbide KD20 (JIS) ($7.7\ W \times 4\ H \times 60\ L$)
Grinding parameters	Wheel depth of cut: Δ=10 μm Stroke length: L=13 mm Table reciprocating frequency: 200 mm^{-1} Wheel revolutional speed: 4778 rpm Cross feed rate: 1 mm/min

less than 0.05 μmRa, and clear vertical stripes were observed.

3. SIMULATION OF VERTICAL STRIPES

In the conditions of this experiment, where a cantilever-type flat punch was machined using high-speed reciprocation, it is conceivable that steady state vibrations and resonance of the worktable may be generated. From a previous experiment [1], however, it was

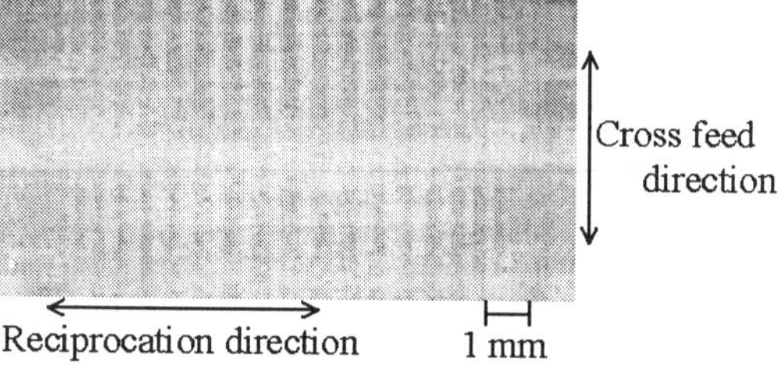

Figure 3. Example of vertical striped pattern on machined surface.

made clear that vibrations arising from the workpiece and worktable resonance were not directly related to the formation of vertical stripes. It was found that there is a striking correlation between the periodicity of the vertical stripes and wheel speed from the measurements of the cycles of vertical stripes in figure 3. This suggests that the vertical stripes are generated by interactions between grinding marks, which are themselves formed due to wheel vibration and have different phases.

We simulated the formation of vertical stripes on a workpiece surface ground by high-speed reciprocation grinding based on the assumption that the frequency of the vibration responsible for the vertical stripes is realized by applying the same frequency to the spindle. Figure 4 shows the simulation model. The grinding wheel vibrates   vertically at constant frequency, cross-fed at a constant rate, and reciprocated at a constant frequency. The surface with zero vibrational displacement in the simulation was defined as the ideal machined surface.

Downward deviation of the wheel from this ideal was marked black, as shown in the figure. Figure 5 shows the striped patterns obtained through simulation, applying the same conditions as those in the grinding experiment in figure 3. The results of the experiment and the simulation are largely in agreement, supporting the assumption that the vertical stripes are attributable to the transference of vibration from the wheel spindle to the workpiece.

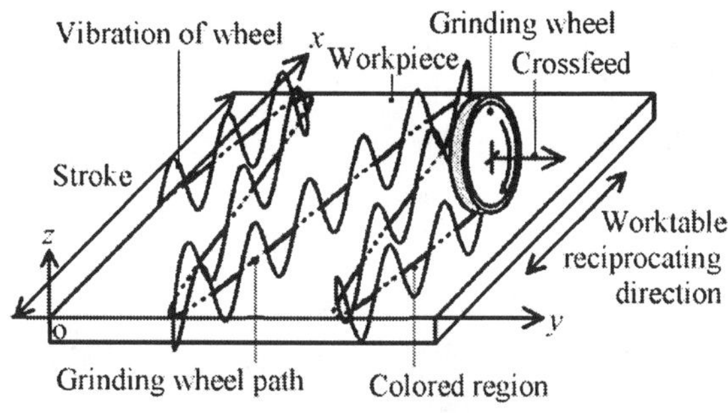

Figure 4. Simulation model

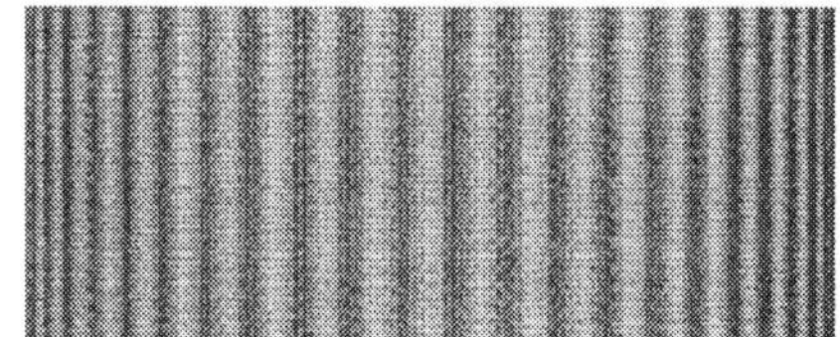

Figure 5. Results of simulation performed under same grinding conditions as in figure 3 (grinder spindle speed 4778 rpm)

The vertical stripes in these simulations are constituted of many small marks, defined with a shape parameter h/b, where h and b are the lengths of the sides of the rhomboid. The relationship between the shape parameter and grinding conditions is given by

$$\frac{h}{b} = \frac{2\pi f_1 \sqrt{xL - x^2}}{v_c}\left[1 - \frac{f_1}{f_2}\mathrm{int}\left(\frac{f_2}{f_1}\right)\right] \tag{1}$$

where f_1 is the frequency of worktable reciprocation, f_2 is wheel revolutional speed, L is the length of the stroke, v_c is the speed of cross-feed, and x represents the coordinate system, defined in the direction of reciprocation. It was found that the simulated pattern was crisscross marks in the case of $h/b\approx1$, horizontal stripes in the case of $h/b<<1$, and as vertical stripes in the case of $h/b>>1$. As the shape parameter h/b becomes extremely large, as is the case in high-speed reciprocation grinding, vertical stripes are likely to appear.

4. EFFECT OF WHEEL SPEED

In the grinding experiments it was unusual to observe clear vertical stripes, as shown in figure 3. The vertical stripes were distorted to some extent, and we even observed cases where the stripes changed to a completely different shape in the middle of the process. This is probably attributable to variations in wheel revolutional speed during grinding due to changes in the operating conditions of the spindle motor or grinding wheel. We therefore investigated changes in the vertical stripes in relation to changes in wheel revolutional speed.

Figure 6 shows the striped patterns produced at various wheel revolutional speeds, based on the conditions listed in table 1; +20 rpm and +40 rpm. As illustrated in figure 6(a), a 20 rpm increase induced changes from vertical stripes to an entirely different crisscross pattern, and as shown in figure 6(b), at +40 rpm, the vertical stripes were observed again. It thus became clear that even small fluctuations in wheel revolutional speed produce changes in the patterns on the ground surface.

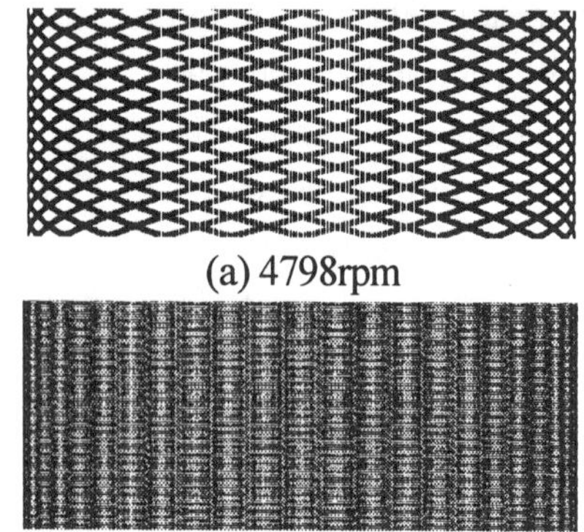

(a) 4798rpm

(b) 4818 rpm

Figure 6. Simulation results

5. CONCLUSION

Through simulations, the process by which vertical stripes sometimes appear on finished surfaces machined by tungsten carbide high-speed reciprocation grinding processes was analyzed, and the simulation results were verified in grinding experiments. The results can be summarized as follows:

(1)   In the reciprocation grinding process, vibration linked with the wheel speed produces vibrational displacement of the grinding machine, manifesting as striped patterns on the work surface.

(2)   When the grinding conditions are such that the shape parameters are sufficiently above 1, striped patterns that were vertical stripes were observed.

(3)   The low cross feed pitch in high-speed reciprocation grinding are conditions in which vertical stripes are likely to appear.

(4)   Wheel revolutional speed has a significant effect on vertical stripes.

References
[1] Nobuhito Y, Katsuo S, Tsunemoto K. Studies on High-reciprocating grinding. Proc. JSPE. Autumn Meeting; 1999; 309.   (in Japanese)

MECHANISMS FOR THE FORMATION OF GLOSSY WORKPIECE SURFACES IN GRINDING AND POLISHING OF NATURAL STONES

Hui Huang [1,2], Xipeng Xu [1,*], Hongjun Xu [2]

1. Provincial Key Lab of Stone Machining, Huaqiao University, 362011, P. R. China

2. School of Mechanical Engineering, NUAA, 210016, P. R. China

Tel: +86-595-2693567, Fax: +86-595-2686969, E-mail: xpxu@hqu.edu.cn

Abstract

An investigation is reported of the formation mechanisms for diamond machined stone surfaces by following their morphologies at each separated stage ranging from sawing to polishing. The temperatures at the machining zone were measured using a pair of thermocouple. The glossiness of the machined stone surfaces was found to be closely associated with the relative amount of ductile flowing on the surfaces and the temperatures was not high enough to cause chemical reactions in the machining processes.

Keywords

Natural stone, SEM, Temperature, Glossiness, Diamond

1. INTRODUCTION

Polished natural stones are popularly used as surface decorative materials of buildings due to their beautiful color and splendid glossiness, in which case the glossiness is one of the most important quality criterion and depends mainly on their polished surface finish. As precision machining processes that are widely used in the manufacture of components requiring fine tolerances and smooth surfaces, grinding and polishing are applied in the final stages to obtain high surface finish, which should be carried out under optimum conditions. A logical technological basis to achieve this objective requires a fundamental understanding of the formation mechanisms for stone surfaces in grinding and polishing processes.

---

* Presenter and the author for correspondence

Some past studies on microanalysis of stone surfaces machined by traditional abrasives indicated that a thin of micrometer scale "glossy film" was discovered on the polished stone surfaces [1] and high temperatures were assumed in the machining processes to account for the "glossy film" without providing any experimental evidence of the temperatures. Compared traditional abrasives, diamond tools are at present more popularly used to grind stones cost-effectively. Although some research on the grinding of stones using diamond tools have been reported [2], they focused mainly on the machining process itself and the optimization of operating parameters rather than material removal mechanisms. Therefore, the formation mechanisms of surface glossiness have not yet been understood clearly. The present research was undertaken to investigate the formation mechanisms by coupling temperature measurements with micro-observations on machined surfaces at separated machining stages using diamonds.

2. EXPERIMENTAL

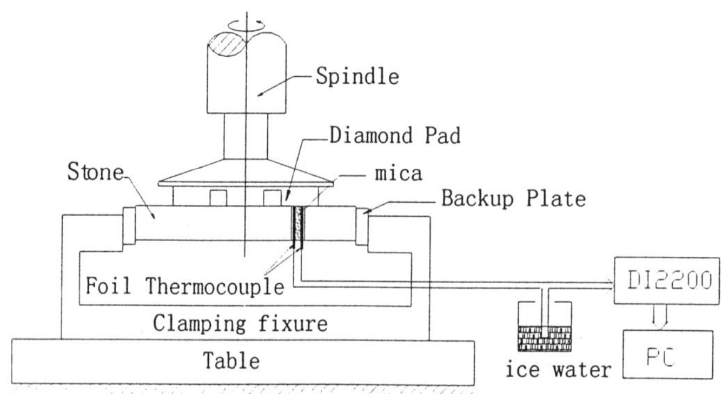

Figure 1. Illustration of the experimental set-up

Grinding and polishing experiments were carried out on a specially designed machine, where a diamond pad with 80 mm in diameter was fixed on a vertical spindle and a stone specimen was mounted on a horizontal table, as shown in Figure 1. The speed of the spindle was 4000rpm. The temperatures at the machining zone were measured by a kind of grindable thermocouple, which consists of a pair of constantan-iron foils. The outputs from the thermocouple were sampled by a DI-2200 recorder and then fed into a personal computer (Figure 1). Six kinds of rubber-bond diamond tools were applied at a range of grit sizes (coarse to fine) capable of processing stone from the roughest 'as sawn' stage to the smoothest finished stage. Tap water was used as coolant in the experiments. The workpiece materials are two kinds of natural red granite and black granite whose properties are shown in Table 1. A WGG60-S digital gloss meter was used to measure the gloss values of granite surfaces. The morphologies of the machined surfaces were investigated by a Hitachi S-520 scanning electron microscope (SEM).

Table 1. Compositions and properties of the workpieces

Workpieces	Shore's hardness	Compression strength （MPa）	Percentage of main minerals （%）		
			Quartz	Feldspar	Mica & others
Black granite	76	137	5	45	50
Red granite	114	125	25	70	5

3. RESULTS AND DISCUSSIONS
3.1 SEM observations on the machined surfaces

The glossiness of the surfaces machined using six kinds of diamond pads are shown in Figure 2, where the meshes of the diamond grit used for each stage are marked. Accordingly, Figure 3 shows the SEM pictures of the red granite surfaces machined using the six different diamond pads. The pictures indicate obvious

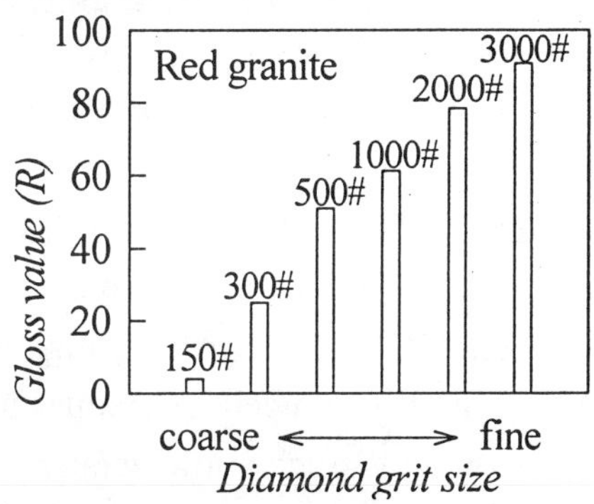

Figure 2. Gloss value versus grit size

traces of ductile flowing on the granite surfaces. The ductile flowing emerges in a small scale for #150 grinding, as shown in Figure 3-a. When the diamond grit size is decreased to #300 (Figure 3-b) and #500 (Figure 3-c), the ductile flowing spreads to a larger scale. The ground grooves shown in Figure 3-d become narrower and shallower and ductile flowing becomes more popular with the decreasing diamond grit size, as shown in Figures 3-e and 3-f. Coupled with Figures 2 and 3, it can be seen that the more amount of the ductile flowing, the higher the gloss value. The same phenomena were also found on the black granite surfaces.

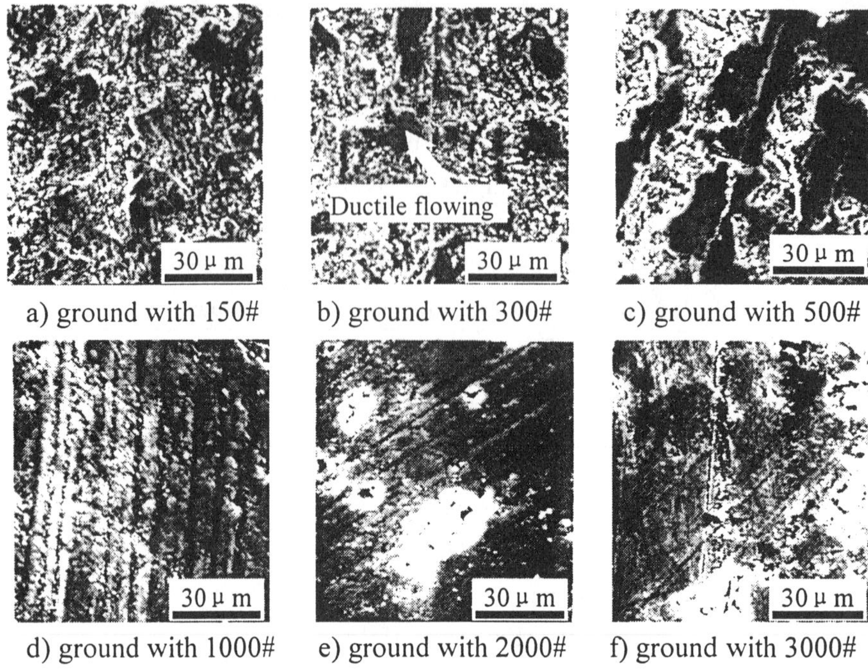

a) ground with 150#   b) ground with 300#   c) ground with 500#

d) ground with 1000#   e) ground with 2000#   f) ground with 3000#

Figure 3. SEM pictures of machined surfaces

3.2 Temperatures at the diamond-granite interface

As shown in Figure 4, the average temperatures at the machining zone are around 60℃ in wet machining, and 150℃ in dry machining, in which case the average temperature values are much lower than those assumed by past research [1]. Such a low temperature was also found in our recent study on surface grinding of granites with diamond grinding wheel [3] and might be impossible to cause chemical reactions as assumed by some of the pervious studies [1]. The result was also supported by a XRD detection of the machined stone surface in our recent study and might be attributed to the high thermal conductivity of diamond. Figure 5 shows that the average temperatures basically decrease with the reducing diamond grits size, which might be due to the increasing portion of heat conducted by diamond grits because of the increasing active cutting points of diamond grits at the pad-granite interface [3]. It should be mentioned that the temperatures at the diamond pad-granite interface are composed of two components, a "flash" temperature and a "background" temperature. The former is corresponding to the average temperatures at the pad-granite contact zone. Although "flash" temperatures may approach much higher as found in grinding of steels, they might not be responsible for the chemical reactions on the granite surface due to their extremely short duration. Due to possible contributions of "flash" temperatures to wear of diamond grains, a detailed analysis on tip temperatures will be discussed in future papers.

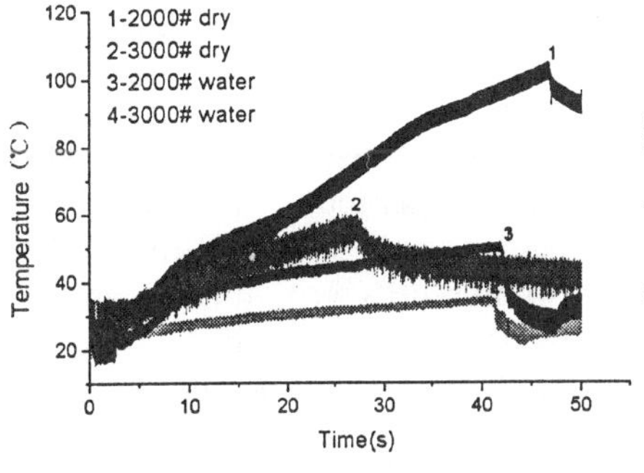

Figure 4. Temperature curves for wet and dry machining

Figure 5. Temperature curves for different diamond pads

4.CONCLUSIONS

The temperatures was found to be below 200 ℃ even in dry machining, which might be not high enough to cause chemical reactions on the machined stone surfaces. Coupled with the gloss values of the machined stone surfaces at separated stages, SEM observations indicated that there did not exist a stage after which the glossiness changed suddenly and the surface glossiness was mainly dependent on the amount of ductile plowing.

ACKNOWLEDGEMENT

The research was supported by Grant No.59775076 from the National Natural Science Foundation of the P.R. of China.

REFERENCES

[1] Sun D. Q., et al. A study on the grinding damage layer of the machined surface of rock.Chinese Journal of Mechanical Engineering 1993; 29(1): 1-5
[2] P. S. Midha, et al. An investigation into the optimization of stone processing using diamond tools. Proceedings of Intertech 2000; Canada, July. 2000: 1-11
[3] Xu X.P., Malkin S, Energy partition at the diamond-workpiece interface in grinding of granites. Journal of Tribology 2001; 21(1): 1-5

MECHANICAL PROPERTIES AND GRINDING CHARACTERISTICS OF HIGH PURITY POLYCRYSTALLINE CBN GRITS

Yoshio Ichida, Yoshitaka Morimoto, Ryunosuke Sato,
Masaki Saijo and Takayoshi Koinuma

Department of Mechanical Systems Engineering, Faculty of Engineering,
Utsunomiya University, 7-1-2 Yoto, Utsunomiya, 321-8585, Japan

Abstract

Some mechanical properties and grinding characteristics of newly developed high purity polycrystalline cBN grits have been investigated. This new cBN grit possesses an ultrafine crystal structure composed of sub-micron sized primary crystal grains. It also has a fracture strength 1.5 times higher than the conventional polycrystalline cBN grit. When grinding high-speed tool steel, the grinding ratio in using the new type cBN grits is 6 times higher than that in using the conventional polycrystalline cBN grits.

Keywords

High purity polycrystalline cBN grits, Friability test, Grinding characteristics

1. INTRODUCTION

Cubic boron nitride (cBN) has a high level of hardness next to the diamond and superior thermo-chemical stability to the diamond. Therefore, cBN abrasive grits have been widely used in various types of grinding operations, ranging from high efficiency grinding to high precision grinding. As the advanced machining systems such as ultra-high speed grinder, grinding center and so on are improved, the demand for the higher performance of cBN wheels is increasing (Brinksmeier et al., 1993). In order to enhance the grinding performance of cBN wheels, we have developed a new type of polycrystalline cBN grit by a direct conversion method from hexagonal boron nitride (hBN), which was produced by the chemical vapor deposition process (Ichida et al., 1996 and Ichida et al., 1997). The contents of impurities in this polycrystalline cBN are lower than that of the conventional polycrystalline cBN.

In this study, some mechanical properties and grinding characeristics of newly developed high purity polycrystalline cBN grits have been investigated and then compared with those of the conventional polycrystalline cBN grits.

2. EXPERIMENTAL PROCEDURE

A hBN disk made by the chemical vapor deposition process, whose dimension was 6 mm in diameter and 1 mm in thickness, was prepared as the starting material. It was set in a Ta capsule, and converted directly to polycrystalline cBN under an ultra high pressure of 7.7 GPa and at a

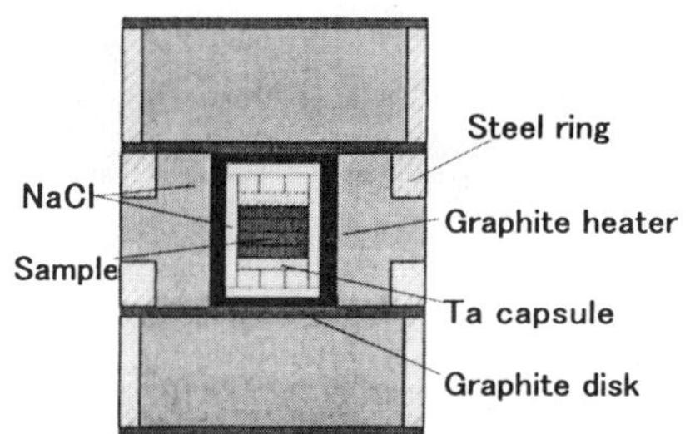

Figure1 The belt type-high pressure apparatus used in this study

Figure 2 A typical polycrystalline cBN sample synthesized by the direct conversion method

temperature higher than 2073 K using a modified belt-type high-pressure apparatus (*Figure 1*). The original one was reported by Yamaoka et al. (1992). *Figure 2* shows a typical polycrystalline cBN sample that was synthesized by this direct conversion method. The hardness of the cBN sample was measured by the indentation method using the Vickers indenter with a normal load of 500 g for 15 s. The contents of impurities in cBN sample were measured with an optical emission spectrometry (Ichida et al., 1996). Moreover, the cBN samples obtained were crashed into pieces using a roll crasher. The cBN abrasive grits with sizes of #60/80 and #80/100 were obtained by classifying the crashed cBN powder. The cBN grits with a size of #60/80 were used for a static compression friability test to measure the fracture strength (Yoshikawa et al., 1960 and Ichida et al., 1997) and an impulsive friability test to measure the fracture energy (Matsui et al., 1980 and Ichida et al., 1996). *Figure 3* shows a schematic view of the impulsive friability test and its conditions. On the other hand, a metal bonded grinding wheel with a concentration of 100 was made using the cBN grits with a grit size of #80/100 and was used for grinding experiments to evaluate the grinding performance of cBN grits.

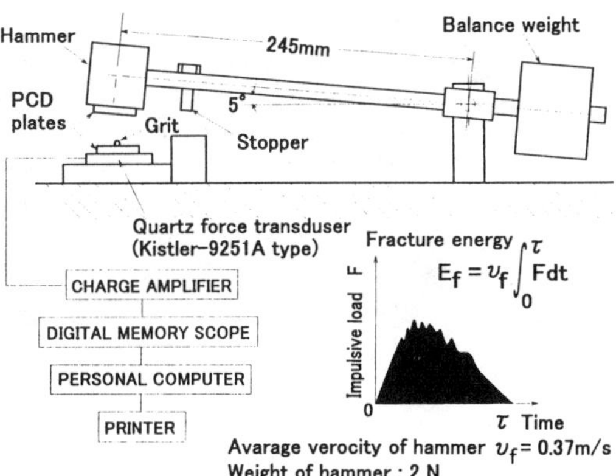

Figure 3 Equipment and conditions for impulsive friability test

3. MICROSTRUCTURE AND MECHANICAL PROPER-TIES OF HIGH PURITY POLYCRYSTALLINE CBN

Figure 4 shows the scanning electron microscope (SEM) images of the new polycrystalline cBN (hereafter denoted as cBN-U) and the conventional polycrystalline cBN (hereafter denoted as cBN-55) grits. The surface of

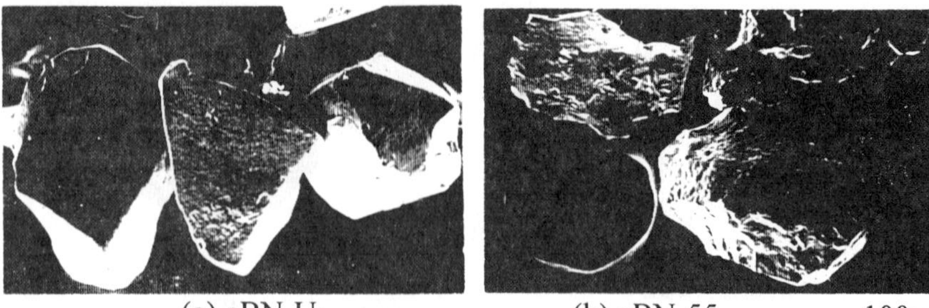

(a) cBN-U   (b) cBN-55   100μm

Figure 4 SEM images of high purity polycrystalline cBN (cBN-U) and conventional polycrystalline cBN (cBN-55) grits (Grit size: #60/80)

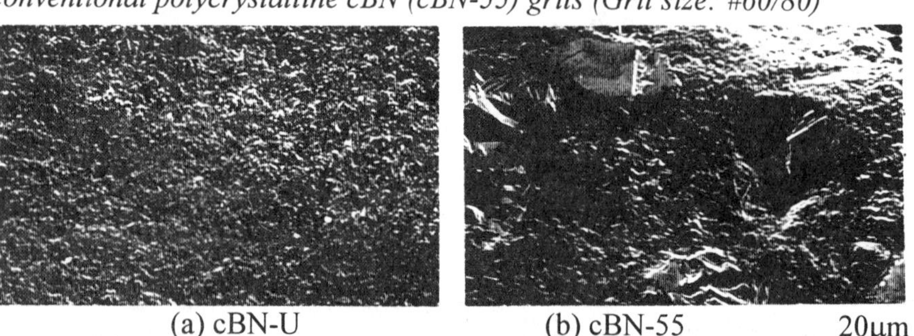

(a) cBN-U   (b) cBN-55   20μm

Figure 5 High magnification SEM images of cBN grit surfaces

cBN-U grit is smoother than that of cBN-55 grit. The higher magnificatin SEM images of these grit surfaces are shown in *Figure 5*. Some large primary crystal grains (Monocrystalline cBN) with a grain size of 10-30 μm are observed on the surface of cBN-55 grit. Namely, the conventional cBN-55 grit has a crystal structure composed of sub-micron sized primary crystal grains and coarse primary crystal grains that havs a size of 10-30 μm. On the other hand, the cBN-U grit has an ultrafine crystal structure composed of sub-micron sized primary crystal grains. The results of the impulsive friability test are presented in *Figure 6*. It shows that the fracture energy of cBN-U grits is around 1.47 times higher than that of the cBN-55 grits. Some properties measured for both cBN grits are summarized in *Table 1*. The tensile strength of cBN-U grits is about 1.52 times higher than that of cBN-55 grits (Load velocity: 1.67 N/s, Number of samples: 150). The optical emission spectrometry analysis of the grits found that the contents of metallic impurities in cBN-U grits are much lower than those in cBN-55 grits.

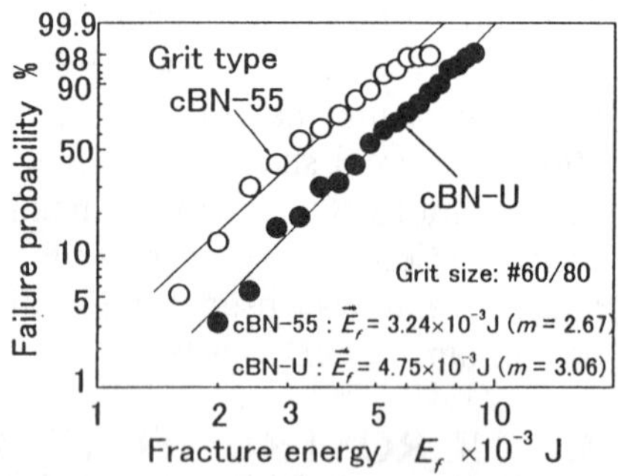

Figure 6 Results of impulsive friability test (m: Weibull modulus, Number of samples: 150)

458

*Table 1 Features of cBN abrasive grits (* Grit size: #60/80)*

cBN grit type (Symbol)		High purity polycrystalline cBN (cBN–U)	Conventional polycrystalline cBN (cBN–55)
Density	g/cm^3	3.4	3.4
cBN purity	wt %	> 99.9	> 99.9
Metallic impurities ppm	Al	—	10–100
	Ca	1	1–10
	Cu	< 10	100
	Fe	10–100	10–100
	Mg	< 1	1–10
	Mn	—	< 10
	Si	1	100
cBN primary grain size μm		< 1	< 20
Hardness HV GPa	R.T.	59.3	58.2
	1273 K	21	17
Tensile strength * σ_t GPa		3.55	2.34
Fracture energy * E_f J		4.75×10^{-3}	3.24×10^{-3}

4. GRINDING CHARACTERISTICS OF HIGH PURITY POLYCRYSTLLINE CBN GRITS

In order to make clear the grinding performance of the new type of cBN grits, grinding experiments for high-speed tool steel were conducted. The grinding conditions are listed in *Table 2*.

Figure 7 shows the variation of radial wheel wear with an increase of stock removal. The wear rate in grinding with cBN-U grits is lower than that of cBN-55 grits. The grinding ratio when using cBN-U grits is about 6 times higher than that when using cBN-55 grits at a stock removal of 20000 mm^3/mm. Thus, the cBN-U grits have a very high wear resistance as compared with the conventional cBN-55 grits. This property gives not only a high grinding ratio, but also a high stable grinding ability of the grit cutting edges during grinding.

Figure 8 shows the variation of surface roughness R_y with an increase of stock removal. It is confirmed that the increasing rate in surface roughness with an increase of stock removal when using cBN-U grits is lower than that in using cBN-55

Table 2   Grinding conditions

Grinding method	Surface grinding, Up cut
Grinding machine	Surface grinder (Okamoto/PSG–52DX) Main spindle motor power : 3.75 kW
Grinding wheel	cBN 80/100N100M Wheel diameter d_s = 200 mm Wheel width b_s = 10 mm
cBN grit type	High purity polycrystalline cBN (cBN–U) Conventional polycrystalline cBN (cBN–55)
Wheel speed	v_s = 30 m/s
Table speed	v_w = 150 mm/s
Depth of cut	a = 0.02 mm
Grinding width	b =5 mm
Coolant	Emulsion type (3.3 % dilution), 0.1 l/s
Workpiece	High speed tool steel (SKH51, HRC65) Dimensions : 100^l × 5^w × 30^h
Truing Dressing	Vertical rotary dresser (GC120 and WA120) Stick dresser (WA220G)

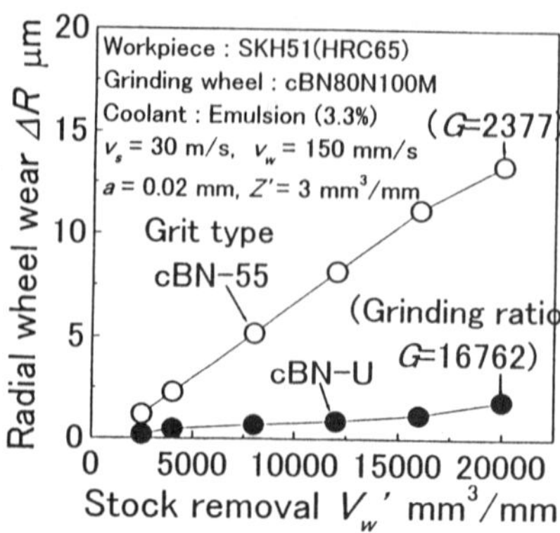

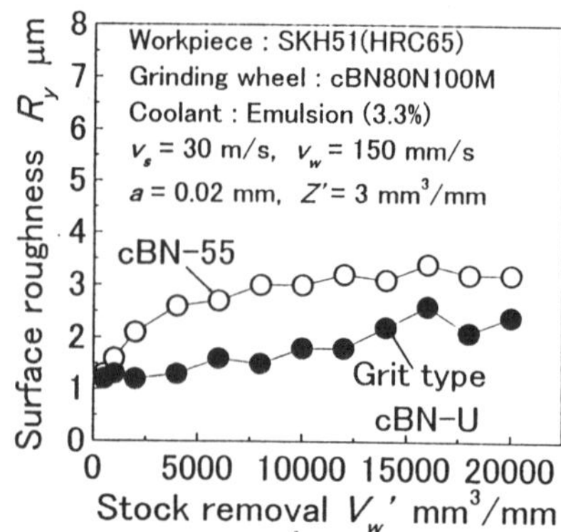

Figure 7 Relation between radial wheel wear and stock removal in grinding high-speed tool steel

Figure 8 Relation between surface roughness R_y and stock removal in grinding high-speed tool steel

grits. Thus, the estimated tool life of the wheel containing cBN-U grits, based on the roughness of the finished surface, is much longer than that of the wheel containing cBN-55 grits.

5. CONCLUSIONS

Some properties and grinding characteristics of newly developed high purity polycrystalline cBN (cBN-U) grits have been investigated. The main results obtained in this study are summarized as follows:

(1) The fracture energy of cBN-U grits is about 1.47 times higher than that of conventional polycrystalline cBN (cBN-55) grits.

(2) The cBN-U grits have a very high wear resistance as compared with the cBN-55 grits. Hence, the grinding ratio in grinding with cBN-U grits is about 6 times higher than that in grinding with cBN-55 grits.

(3) The estimated tool life of the wheel containing cBN-U grits, based on the roughness of the finished surface, is much longer than that of the wheel containing cBN-55 grits.

REFERENCES

Brinksmeier E., Minke, E., High-performance surface grinding-the influence of coolant on the abrasive process, Annals of the CIRP, 1993; 42(1): 367-370.

Ichida, Y., Kishi, K., Suzuki, M., Nikaido, T., Grinding characteristics of polycrystalline cBN abrasive grits having an ultrafine crystal structure, Trans. of the Japan Society of Mechanical Engineers, 1996; 62, 595: 1169-1175 (in Japanese).

Ichida, Y., Kishi, K., The development of nanocrystalline cBN for enhanced superalloy grinding performance, ASME J. of Manufacturing Science and Engineering, 1997; 119: 110-117.

Matsui, S., Syoji, K., Friability test of abrasive grains, J. of the Japan Society for Precision Engineering, 1980; 46, 11: 1416-1421 (in Japanese).

Yamaoka, S., Akashi, M., Kanda, H., Osawa, T., Taniguchi, T., Sei, H., Fukunaga, O., Development of belt type high pressure apparatus for material synthesis at 8 Gpa, J. of High Pressure Institute of Japan, 1992; 30: 249-258.

Yoshikawa, H., Sata, T., Fracture strength of abrasive grains, J. of the Japan Society for Precision Engineering, 1960; 26, 8: 476–481 (in Japanese).

INFLUENCE OF CUTTING EDGE WEAR ON DUCTILE-MODE GRINDING OF FINE CERAMICS WITH COARCE GRAIN SIZE DIAMOND WHEEL

Heiji YASUI, Yutaka HIRAKI, Masato SAKATA,
Michiko TSURUSAKI, Yasuyuki MURAYAMA
and Shigehiko SAKAMOTO

Dept. of Mechanical Engineering & Materials Science,
Kumamoto University, Kumamoto 860-8555, Japan

Abstract

The research aims to clear the influence of wheel surface conditions on the ductile-mode grinding of fine ceramics with coarse grain size diamond wheel. The wheel used is #140-mesh wheel. The increase of cutting edge ratio contributes to some extent to decrease the occurance of grinding cracks and to improve the surface roughness. The ductile-mode grinding without grinding cracks can be done at higher table speed for larger cutting edge ratio.

keyword

ductile-mode grinding, fine ceramics, cutting edge ratio

1. INTRODUCTION

Grinding operation is one of the most effective manners for high smoothness machining of fine ceramics. However, it is difficult to form crack-free high smoothness surface by ductile-mode grinding because of their mechanical properties of high brittleness. Then it is necessary for ductile-mode grinding that the suitable grinding conditions are selected. Some reports concerning ductile-mode grinding have been done with the wheel of finer grain size than #1500-mesh size. But the truing and dressing for the fine grain size wheel are difficult. In addition, the depth of cut and/or table speed are also limited to a certain small range because the active grain is thrown off easily from the wheel surface by small grinding force. From the point of view, in our pervious researches[1]-[3], the adaptation of the coarse #140-mesh grain size wheel to the ductile-mode grinding of hot-press silicon carbide (HPSC) ceramic was tried in the surface plunge grinding and found to be possible. This is one of a series of researches on the ductile-mode grinding of fine ceramics with coarse grain size diamond wheel. The research aims to clear the influence of wheel surface conditions on the ductile-mode grinding of fine ceramics with coarse grain size diamond wheel.

2. EXPERIMENTALS

The experiments are carried out by usual plunge grinding at a constant wheel depth of cut with conventional surface grinder. The microscope is set up on the wheel cover as shown in Fig.1 for measuring the cutting edge ratio by means of image processing. The cutting edge reatio is obtained by dividing the sum of cutting wear area by the observed wheel surface area. In addition, the rotary encoder and the position scale are assembled to the wheel spindle and wheel cover, so as to measure the same wheel surface.

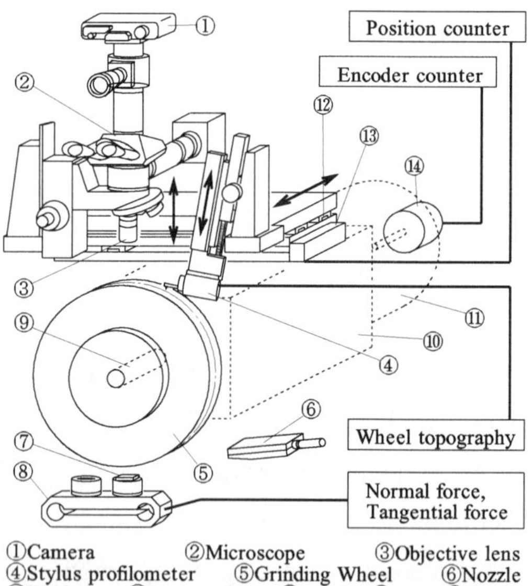

①Camera   ②Microscope   ③Objective lens
④Stylus profilometer   ⑤Grinding Wheel   ⑥Nozzle
⑦Workpiece ⑧Dynamometer ⑨Spindle ⑩Head ⑪Motor
⑫Cross roller guide ⑬Position scale ⑭Rotary encoder

Fig.1   Experimental arrangement

The grinding conditions are summarized in Table 1. The wheel and ceramic used are a metal-bonded diamond wheel of coarse grain size of #140-mesh and hot pressed silicon carbide （HPSC）

Table 1 Experimental conditions

Wheel	SD140Q50M
Workpiece	Hot pressed silicon carbide (**HPSC**)
Wheel speed	V_g = 20 m/s
Table speed	v_w = 0.05 - 5 mm/s
Depth of cut	t_t = 5 μm×5 passes
Coolant	Soluble (1/50) Flow rate : 12 L/min

ceramic. The ground workpiece surface is observed by the optical microscope and the scanning electric microscope （SEM）. The surface roughness is measured by the stylus profilometer, the three dimensional surface interferometer （WYKO TOPO-3D） and the atomic force microscope （AFM）.

3. RESULTS AND DISCUSSIONS

Figure 2 shows the microphotographs of the HPSC ceramic surface ground at the table speeds of v_w = 0.05, 0.5 and 5mm/s under the cutting edge ratio η of 0.43%（after truing） and 0.82%. In case of both cutting edge ratios, the grinding cracks decreases as table speed becomes slow. At v_w=0.05mm/s, the ground surface is almost the same as the crack-free surface.

Compared with the

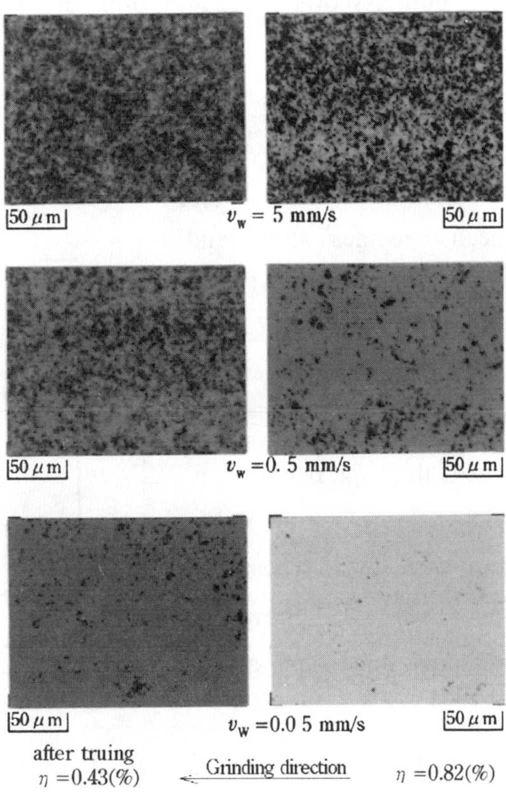

Fig.2 Relationship between surface roughness

grinding cracks under η =0.43% at the same table speed, however, those under η =0.82% are considerably few.

Comparison of the surface ground at v_w =0.05mm/s under η =0.82% with η =1.4%

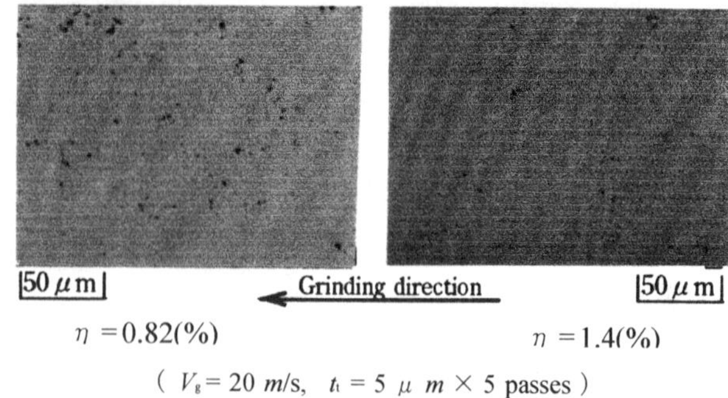

Fig.3 Influence of cutting edge ratio on ground surface

is made in Fig.3. It is found from the figure that the grinding cracks of η =1.4% is fewer than ones of η =0.82% though the difference is small.

Figure 4 shows the relationship between surface roughness, table speed and cutting edge ratio. The surface roughness parallel and normal to grinding direction which are denoted by (Rz)p and (Rz)n, are measured by WYKO TOP-3D and AFM, respectively. It is found from the figure that the (Rz)p decreases largely in all cases of cutting edge ratio as table speed becomes slow, and attains to about 20nm (P-V) at v_w=0.05mm/s under η =0.82% and η =1.4%. In the range of whole table speed, the (Rz)p under η =0.82% are much smoother than η =0.43%. The (Rz)p under η = 1.4%, however, is almost the same as η =0.82%.

Then it is considered that the change of cutting edge ratio in the range of

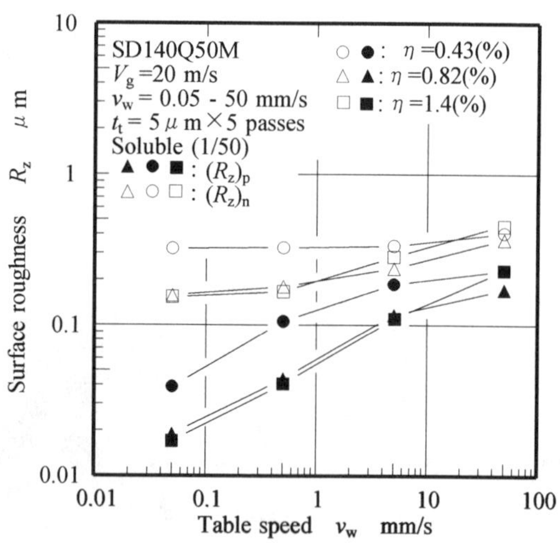

Fig.4 Comparison of HPSC ceramic surface roughness ground by the wheel of cutting edge ratio η = 0.43, 0.82, 1.4 %

below a critical value is closely related to the change of the (Rz)p. In the range of beyond the critical value, however, the cutting edge ratio has little influence on the (Rz)p. The (Rz)n as well as the (Rz)p decreases as table speed becomes slow. The ratio of decrease, however, is small with the decrease of table speed and the (Rz)n tends to converge to a given value.

4. SUMMARY

Influence of cutting edge ratio on ductile-mode grinding of silicon carbide ceramic with coarse #140-mesh grain size diamond wheel is examined. In all cases of cutting edge ratio, the grinding cracks decreases as table speed becomes slow. The increase of cutting edge ratio contributes to some extent to decrease the occurance of grinding cracks and to improve the surface roughness. The ductile-mode grinding without grinding cracks can be done at higher table speed for larger cutting edge ratio.

ACKNOWLEDGEMENT

The authors would like to thank Noritake Diamond Industries Co. Ltd., Yushiro Chemical Industry Co. Ltd. and Nihon Tungsten Co. Ltd., for grinding wheels, grinding fluids and workpiece materials. A part of this research is supported by Grant-In-Aid for Scientific Research 2000, in Japan.

REFERENCE

[1] Heiji YASUI et al: Influence of Sintering Method on High Smoothness Grinding of Fine Ceramics, Porc. 3th Int. Conf. on Progress of Cutting and Grinding, Vol.III(1996)530.

[2] Heiji YASUI et al: Ductile-mode High Smoothness Grinding of Fine Ceramics by Diamond Wheel of Coarse Grain Size (1st Report), J. JSPE, 63, 9(1997)1270.

[3] Heiji YASUI et al: Ultra-Smoothness Grinding of Fine Ceramics with #140-mesh Grain Size Diamond Wheel, Proc. 14th ASPE Annual Meeting (1999) 127.

STUDY ON INTERNAL GRINDING OF SMALL BORE

- Influence of Bearing Characteristics on Grinding Accuracy -

Katsuya YAMAUCHI[1] and Jun-ichiro TAKAGI[2]
[1]*ISUZU MOTORS Ltd.*   [2] *Yokohama National University*

Abstract

Internal grinding of small bore requires high-speed rotation of grinding spindle and also bearing stiffness and rotational accuracy to achieve very high grinding accuracy. As an attempt to apply air static bearings to internal grinding spindle, spindle performance and grinding accuracy of air static bearing which has high-speed rotation and high rotational accuracy but low bearing stiffness is examined and compared with the ball bearing spindle. As for the bearing stiffness, the difference becomes smaller as the rotational speed increases and L/d of wheel quill becomes larger. The non-repeatable run out of air bearing is approximately one fourth that of ball bearings, and is not influenced by the rotational speed. The grinding accuracy is remarkably influenced by elastic deformation of the wheel quill, however, scarcely affected by rotational accuracy of the bearing.

Keywords
Internal grinding, Air static bearing, Ball bearing, CBN wheel

1.INTRODUCTION

In order to improve grinding accuracy, the components of machine tool essentially need high accuracy and high stiffness. Though air static bearing has high rotational accuracy and high-speed rotation, it is rarely adopted to grinding machine due to low bearing stiffness. In this study, an internal grinding apparatus which equips both the grinding spindle with air static bearing and ball bearing is prototyped, and performance of bearing stiffness and rotational accuracy of grinding spindle which are expected to affect grinding accuracy are examined. And then, grinding experiments are carried out to verify the relation between the spindle characteristics and the grinding accuracy.

2. EXPERIMENTAL METHODS

The prototyped grinding apparatus equips both ball and air bearing spindles on the axial feed table unit. Three components dynamometer is installed on the radial feed table unit. An air static bearing spindle for the work and the rotary dresser is installed on the dynamometer.

In order to examine the radial stiffness of grinding spindle, static load is applied on the several setting part of the quill with a dead weight, and the displacement is measured by capacitance type non-contact displacement meter. When the grinding force is added, the angle of inclined spindle to the machine center line is measured. In order to estimate the variation of bearing stiffness during high-speed rotation, the outer race of ball bearing which is mounted on the quill is loaded with a dead weight, and the displacement of the outer race is measured.

Periodical run out and non-repeatable run out are sampled from the waveform measured by non-contact displacement meter when the spindle is rotating. A definite dynamic displacement is applied on the grinding spindle equipped with grinding wheel quill, then the eigenfrequency, damping ratio and damping time are measured. The eigenfrequency is analyzed from the waveform measured by non-contact displacement meter.

Finally, in order to investigate the influence of the characteristics of wheel spindle bearing on the grinding accuracy, surface roughness and roundness of ground works are measured. Plunge grinding without oscillation is adopted so that the characteristics of grinding wheel spindle reflect distinctly on the grinding accuracy.

3. EXPERIMENTAL RESULTS
3.1. Evaluation of radial stiffness of wheel spindle and quill

Figure 1 shows the spindle stiffness measured when the air pressure is changed while the spindle is at a stand still. As for the air static bearing, the bearing stiffness increases proportionally with the raise of supplied air pressure. Air static bearing has approximately one fourth of stiffness in comparison with that of ball bearing.

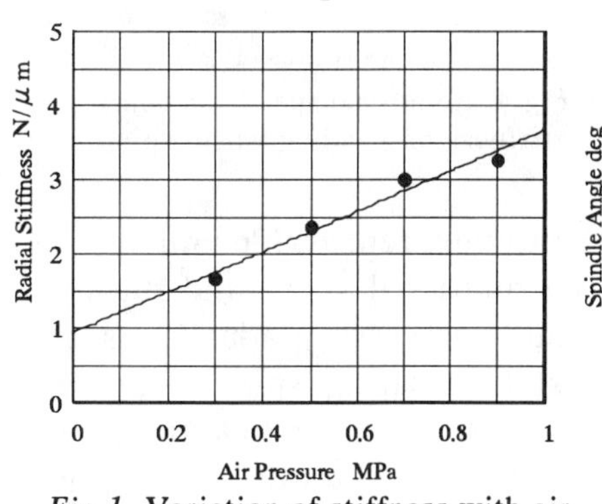

Fig.1. Variation of stiffness with air pressure

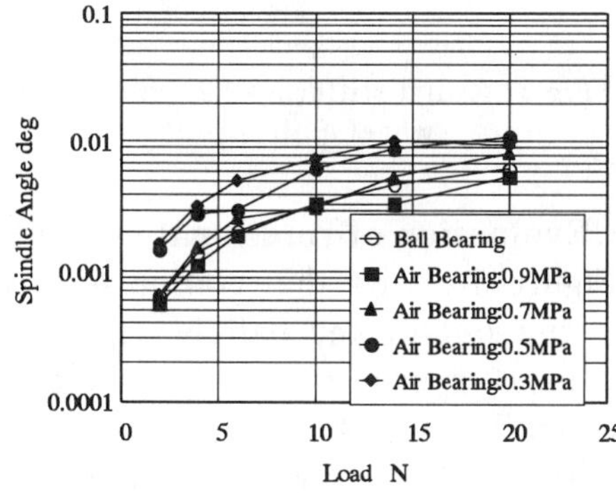

Fig.2. Relation between static load and spindle angle

467

Figure 2 shows the relation between the static load and the inclination angle of the spindle. Since the distance between front and tail bearings of air static spindle is longer than that of ball bearing spindle, the influence that the stiffness exerts on inclination angles of the spindle is decreased.

Figure 3 shows the radial stiffness measured at the several points from the spindle end. At the point close to the spindle end, the stiffness for the ball bearing spindle is high because of the high bearing stiffness itself. However, as L/d of wheel quill becomes larger, the difference between "ball" and "air" becomes smaller because the bending of wheel quill influences remarkably on the stiffness. It is expected that the stiffness of the bearing does not exert an influence on the displacement of grinding point in case of grinding small and deep bore.

Figure 4 shows the relation between rotational spindle speed and bearing stiffness. The stiffness of ball bearing spindle decreases slightly as the rotational speed increases. On the other hand, stiffness of air static bearing spindle increases gradually as the rotational speed increases. If it is possible to get higher rotational speed of wheel spindle, it is expected that the difference of spindle stiffness between "ball" and "air" will be smaller.

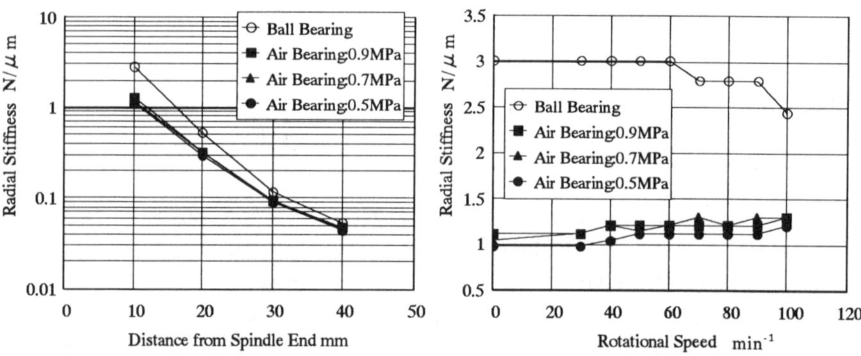

Fig.3. Radial stiffness of 2.7dia. wheel quill

Fig.4. Rotational speed and bearing stiffness of grinding wheel spindle

3.2.Evaluation of rotational run out and non-repeatable run out

Figure 5 shows the relation between rotational speed and run out. The run out of the ball bearing spindle increases gradually as the rotational speed increases. The air static bearing spindle has the resonance point between the rotational speed of 48000min[-1] and 60000min[-1], which increases proportionally with the raise of supplied air pressure. The run out decreases after exceeding the resonance point, and increases again after exceeding 80000min^{-1}.

Figure 6 shows the relation between rotational speed and non-repeatable run out of the wheel spindle. The value of air static bearing becomes smaller than that of ball bearing without relation to rotational speed and air pressure. Since the ball bearing supports the load by solid contact between balls and race, geometrical accuracy of these component parts has an effect on non-repeatable run out. On the other hand, as the air static bearing does not have any parts to be contacted while the spindle shaft is rotating, the rotational accuracy can be achieved beyond the geometrical accuracy of component parts.

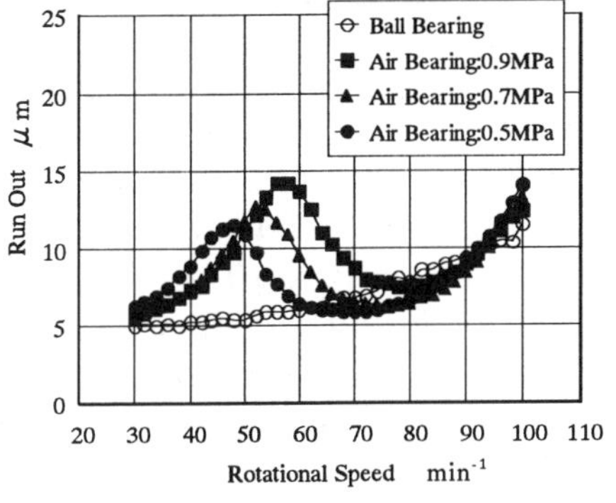

Fig.5. Relation between rotational speed and run out

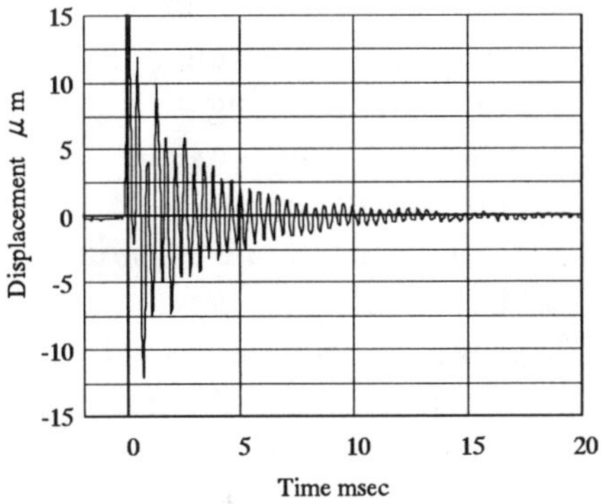

Fig.6. Relation between rotational speed and non-repeatable run out

3.3. Evaluation of damping characteristics of grinding spindle

Figure 7 shows damping waveform of the wheel spindle with ball bearing. The eigenfrequency is approximately 3.1kHz and damping ratio is approximately 0.015 .

Figure 8 shows damping waveform of the air static spindle. The

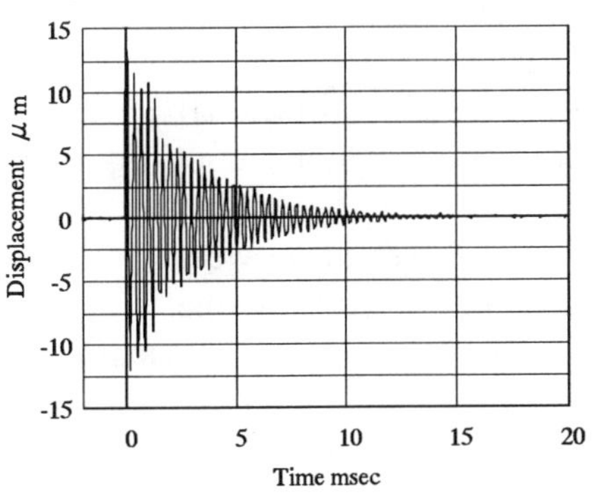

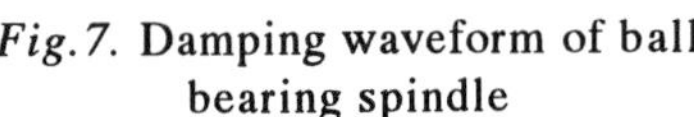

Fig.7. Damping waveform of ball bearing spindle

Fig.8. Damping waveform of air static bearing spindle

eigenfrequency is 2.4kHz, and damping ratio is approximately 0.02. Another eigenfrequency at 800~900Hz also exists.

3.4.Evaluation of grinding accuracy

Table 1 shows the radial stiffness of the wheel quill at grinding point and Table 2 shows grinding conditions respectively.

Table 1. Radial stiffness of grinding point

Quill	Grinding Spindle	Radial Stiffness N/μm
ϕ3 ×12	Air Bearing	0.7
ϕ3 ×12	Ball Bearing	1.4
ϕ3 ×4	Air Bearing	1.7
ϕ3 ×4	Ball Bearing	5.9

Table 2. Grinding condition

Plunge Grinding Condition	
Work Rotational Speed	1700 min^{-1}
Grinding wheel Rotational Speed	100000 min^{-1}
Stock Removal	0.1mm／R
Feed Speed	0.1mm/min
Spark Out	2sec

Table 3 shows results of grinding test. The surface roughness becomes gradually larger as the removals increases, however, it is not influenced by bearing characteristics and quill stiffness. The roundness advances for shortened quill which realizes higher stiffness. Bearing characteristics does not reflect on the grinding accuracy, because the stiffness at the grinding point is affected mainly by elastic deformation in case of low stiffness of wheel and quill.

Table 3. Result of grinding test

	Ball Bearing		Air Static Bearing	
Roughness Rz μm	0.63	0.71	0.66	0.65
Roundness μm	0.09~0.19	0.22~0.34	0.10~0.14	0.24~0.38

Left : Quill Length 3mm , Right :Quill Length12mm

4. CONCLUSIONS

1) In case of grinding small and deep bore, the stiffness at grinding point is influenced mainly by stiffness of grinding wheel quill, and the effect of bearing stiffness is small.

2) Air static bearing has high non-repeatable rotational accuracy, however, there exists resonance point which amplitudes the periodical run out.

3) Though the shape of spindle shaft is different between "air" and "ball", the damping characteristics are almost the same.

4) Grinding accuracy is influenced mainly by elastic deformation of wheel quill, and the difference of bearing characteristics scarcely reflect on the grinding accuracy.

A BASIC STUDY ON PROFILE GRINDING WITH ELID

Hisayoshi Shindo
Saitama Prefectural Industrial Technology Center
Hitoshi Ohmori
The Institute of Physical and Chemical Research
Toshio Kasai
Saitama University

Abstract

A profile grinding system by the Electrolytic In-Process Dressing (ELID) technique was developed for precision tool. In order to obtain precise surface for complicated shapes of workpieces, straight and v-face cast iron bond diamond grinding wheels were mounted on the machine along with an optical projection system. Using this system, basic experiment was made for processing stamping tools. To obtain higher surface roughness in the finish grinding, the rough and medium finish grinding conditions need to be further studied. This paper has reported on the results of the investigation into rough and medium finish grinding conditions.

Keywords

Profile grinding machine, ELID-grinding, Stamping tool, Cast iron bond diamond grinding wheel, Cemented carbide

1.INTRODUCTION

With the rapid growth of information technologies recently, new products (computer, portable phone, etc.) have to be developed one after another in a short time. Consequently, in stamping tool manufacturing, stamping tools with complicated shapes require especially high form accuracy and surface quality as well as short delivery time. Using a profile grinding machine, some kinds of stamping tools are fabricated. Electrolytic In-Process Dressing (ELID) grinding[1] is widely known to produce fine surface efficiently. The ELID grinding technique was applied to the profile grinding machine[2,3]. It is useful for processing stamping tools with fine surface roughness. In this study, it has to go through three more processes before it is finished. These are rough grinding, medium grinding and finish grinding. To obtain higher surface roughness in the finish grinding, the rough and medium finish grinding conditions need to be further studied. In this study, the rough and medium finish grinding conditions were investigated.

2.EXPERIMENT SYSTEMS

The profile grinding system is suitable for complicated tools. As the grinding wheel and workpiece are shown on the screen, the workpiece can be profiled by two numerically controlled axes or by hand-operation. Figure 1 shows the mechanism of the new profile grinder using the ELID-grinding technique. The electro-discharge profile truing process using v-face metallic bond wheels is also applied for the profile grinding system with ELID. The negative pole can be connected to the cathode or electrode. In case of electro-discharge truing, the negative pole can be connected to the cathode for electro-discharge truing. In the case of ELID, the negative pole can be connected to the brass electrode, which is 1/6 of the wheel periphery, with a width of 1mm more than the wheel rim's thickness. The positive pole is a metal bond grinding wheel, and is connected to the power source using a carbon brush. The gap between the grinding wheel and electrode is approximately 0.4mm. Grinding fluid flows between the grinding wheel and electrode.

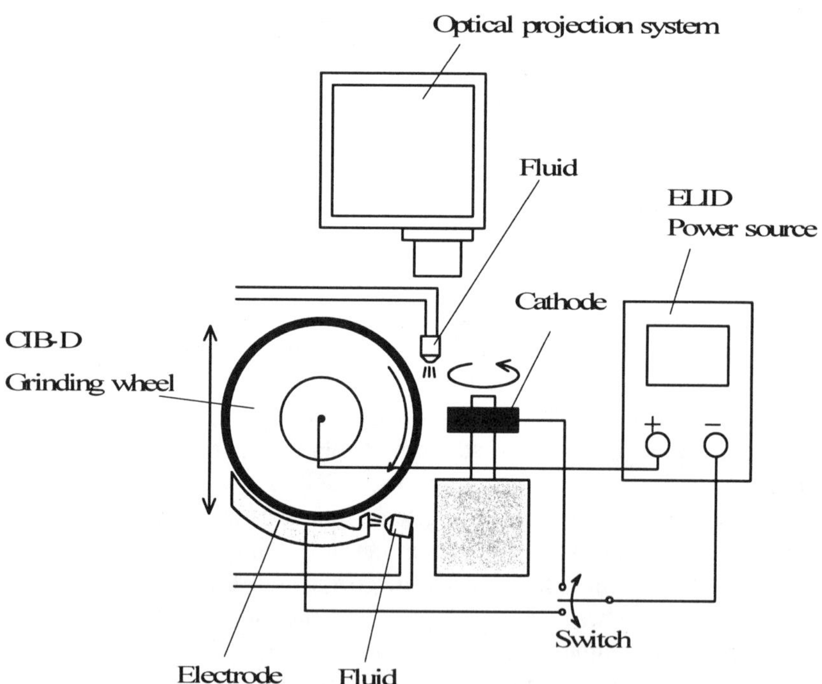

Figure 1. Mechanism of profile ELID-grinding system

3.EXPERIMENT ON ELID GRINDING

In this study, 200(straight), 600 and 4000(v-face) mesh size grinding wheels are mainly used for rough, medium finish and finish grinding respectively. To achieve fine surface roughness and form accuracy in finish grinding, it is important to investigate the conditions in rough and medium finish grinding. If the rough and medium finish grinding of the workpiece is precise enough, grinding wheel abrasion can be prevented in finish grinding. The grinding conditions were investigated with 200 mesh size and 600 mesh size grinding wheels. Feed rate, reciprocal frequency and wheel rotation speed were chosen from the grinding conditions in Table 1, and the condition of several grinding methods was attempted. The workpieces were constructed of cemented carbide (D60). Surface grinding was performed. The effect of this combination on the surface roughness was investigated.

4.RESULTS AND DISCUSSION

Table 1 shows the conditions of profile grinding with ELID. Figure 2 shows the relation between reciprocal frequency and surface roughness for each feed rate. At the reciprocal frequency of $120min^{-1}$, the surface roughness became the finest with the decrease in rate. A surface roughness of $1.45\mu mRy$ was obtained at a feed rate of 5mm/min, and reciprocal frequency of $120min^{-1}$.

Table 1. Conditions of profile grinding with ELID

	#200	#600
Grinding conditions		
Wheel rotation speed min^{-1}	3000~5500	3000~5500
Reciprocal frequency min^{-1}	100~160	120
Feed rate mm/min	5~60	0.4
Depth of cut μm	15	5
Spark out	1	1
Electrolytic conditions		
Open voltage V	60	30
Peak current A	12	1
On Time μs	2	2
Off Time μs	2	2

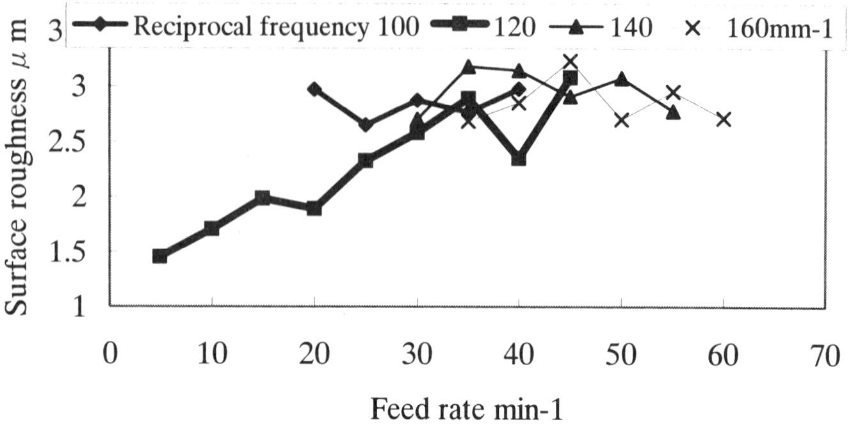

Figure 2. Relation between feed rate and surface roughness

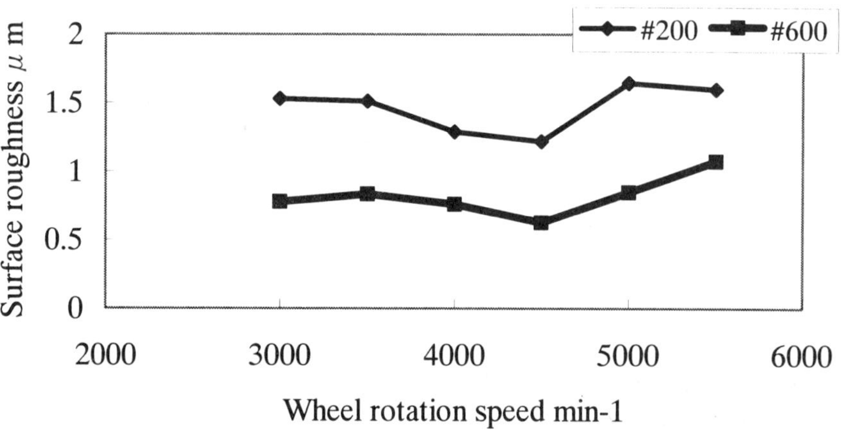

Figure 3. Relation between wheel rotation speed and surface roughness

Figure 3 shows the relation between wheel rotation speed and surface roughness. At 4500min⁻¹, surface roughness of 1.22μmRy was obtained. After optimizing conditions using a 200 mesh size grinding wheel, wheel rotation speed conditions were investigated. Using a 600 mesh size v-face grinding wheel, medium finish grinding was performed. Relation between wheel rotation speed and surface roughness is also shown in Figure 3. And a surface roughness of 0.63μmRy was obtained at 4500min⁻¹.

Figure 4. Grinding state   *Figure 5.* Stamping tool

At the optimum conditions, a stamping tool was processed on a trial basis. Figure 4 shows grinding state and Figure 5 shows stamping tool after grinding. A stamping tool with surface roughness of $0.67\,\mu$mRy was obtained.

5.CONCLUSION

This paper has reported on how to obtain high surface roughness in rough and medium finish grinding and through basic experiments on grinding conditions. Until now, workpieces with surface roughness of 0.63 μmRy were obtained. A stamping tool was processed on a trial basis. The stamping tool with surface roughness of 0.67μmRy was obtained. In order to improve the surface roughness, it is necessary to investigate other grinding conditions and finish grinding condition.

ACKNOWLEDGEMENT

The authors wish to thank Amada Mechanics, Inc. and Fuji Die Co., Ltd. for their cooperation.

REFERENCES

1. Hitoshi Ohmori. Electrolytic in-process dressing (ELID) grinding technique for ultra precision mirror surface machining. Journal of JSPE 1993; 9: 1451-1457
2. Hisayoshi Shindo, Hitoshi Ohmori. Micro-profile grinding system with ELID process for precision tool and die fabrication, Abrasive Technology, 1999 November 33 - November 24;Brisbane: World Scientific, 1999.
3. Hisayoshi Shindo, Hitoshi Ohmori, Toshio Kasai. Development of micro-profile ELID-grinding system for stamping tools. Advances in abrasive technology III; 2000 October 30 - November 2; Hawaii: Society of Grinding Engineering, 2000.

GRINDING CHARACTERISTICS OF CEMENTED CARBIDE CONCAVE MIRROR BY DESK-TOP TYPE 4-AXES MACHINE "TRIDER-X" WITH ELID SYSTEM

Yoshihiro Uehara[*1], Hitoshi Ohmori[*1], Yutaka Yamagata[*1], and Sei Moriyasu[*1]

*1 The Institute of Physical and Chemical Research (RIKEN)

Shin-Ya Morita[*2]

*2 University of Tokyo

Ken-Ichi Yoshikawa[*3], and Muneaki Asami[*3]

*3 The Nexsys Corp.

Takahiro Miura[*4]

*4 Ikegami Precision Tooling Co.,Ltd.

Abstract

In this study, we attempted to develop a desk-top 4-axes machine "TRIDER-X" mounting a ELID system. This paper describes concepts and history of the machine, and detailed investigations on the performance of the machine. A concave mirror was ground using the desk-top 4-axes machine "TRIDER-X" mounting an ELID system. Good form accuracy was achieved.

Keywords

ELID grinding, Micro fabrication, Plasma discharge truing, Cast iron bonded diamond wheel, Hard material, Concave mirror

1. INTRODUCTION

The miniaturization and reduction of the weight of portable acoustic equipment, portable picture equipment, and small size record instrument are being rapidly pursued and achieved in the recent years. This had led to growing demands for the miniaturization of high precision electronic parts, optical system parts, mechanism elements that compose those apparatus as well as high dimensional accuracy of these parts. Especially for optical system parts, configuration accuracy and surface roughness are demanded at the nanometer order. The improvement of surface roughness is increasing required for mechanism elements in addition to optical system parts in recent years. The ELID grinding method is able to satisfy the microfabrication requirement of these parts. This paper describes the grinding characteristics of concave mirror surfaces using a desk-top 4-axes machine "TRIDER-X" mounting an ELID [Electrolytic In-process Dressing] system [1]. Results of grinding experiments indicated that high efficient grinding with a good form accuracy could be achieved successfully.

2. DESK-TOP 4-AXES MACHINE

2-1. Concept of desk-top machine

The concepts of this desk-top machine are high efficient surface grinding, three dimensional form surface grinding, three dimensional form machining and driller, easy-to-carry through the reduction of machine weight, compact machine body, easy-to-operate using a PC based NC controller unit, high rigidity to design structure of machine.

2-2. Desk-top 4-axes machine "TRIDER-X"

Figure 1 shows the external view of the desk-top 4-axes machine. Table 1 shows the specifications of the desk-top 4-axes machine. This machine has four axes, X, Y, Z, and C. The three linear axes are supported by a roller slide mechanism on a cross roller guide, and are driven by a step motor. The X, Y and Z axes have a positioning resolution of 1 μ m with semiclosed-loop control, while the C axis has a positioning resolution of 0.002 $^\circ$ with semiclosed-loop control. The machine measures $580 \times 580 \times$

580 ㎜ in size, and weight 95 kg. The main body has a gate structure for high rigidity. The main spindle can be mounted in two directions for machining complicated shapes. The PC-based NC unit is used to facilitate operations.

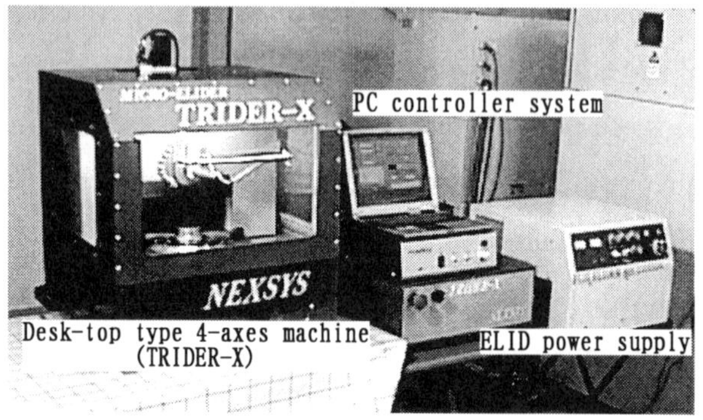

Figure 1 External view of desk-top 4-axes machine

Table1 Specifications of desk-top type 4-axes machine "TRIDER-X"

Stroke	X:100 ㎜, Y:100 ㎜, Z:100 ㎜
	C:±360°
Positioning Resolution	X, Y, Z:0.001 ㎜
	C:0.002°
Rotation speed of main spindle	500～40000rpm
Size of machine	580×580×580 ㎜
Weight of machine	95 kg
Input	AC100V

2-3. ELID grinding system for desk-top 4-axes machine

Figure 2 shows the ELID grinding system. Figure 3 shows the schematic of the ELID method. The ELID-grinding method is a complex grinding process adopting special electrolytic in-process dressing method.

The wheel serves as a positive Ve pole through smooth contact of a brush, while the electrode fixed below serves as a negative Ve pole. In the small clearance of approximately 0.1 ㎜ between the negative Ve and positive Ve poles , electrolysis occurs upon supply of the grinding fluid and electrical current.

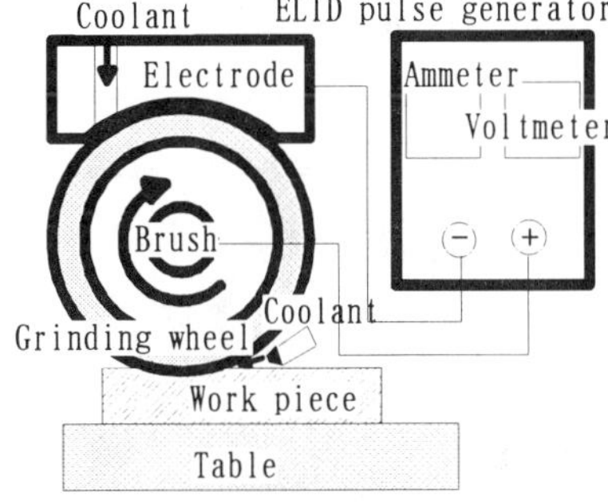

Figure 2 ELID grinding system   Figure 3 Schematic of ELID method

3. GRINDING EXPERIMENTS

The grinding conditions are summarized in Table 2. The desk-top 4-axes machine was employed for grinding. The grinding wheel was a #325 cast iron bonded diamond wheel, and the work piece was cemented carbide.

Table 2 Experimental Conditions

Grinding wheel	Cast iron bonded diamond wheel #325	
Work	Cemented carbide	
ELID conditions	Voltage (V)	60
	Maximum current (A)	10
	Pulse interval (μ sec)	2,2
Grinding conditions	The grinding wheel rotational speed （rpm）	4000
	Feed rate of Rotation table (㎜／min)	500
	Depth of cut (㎜／pass)	0.001

4. EXPERIMENTAL RESULTS

Figure 4 shows the concave mirror. The concave shape was measured using a non-contact three-dimensional shape measuring instrument NH-3 (Mitakakohki Co., Ltd), and the approximate curvature radius was calculated to be about 37.3 ㎜. Figure 5 shows the surface roughness profile. The surface roughness is 150nm in Ry and 20nm in Ra. The form error is about 20 μ m.

Figure 4 Concave mirror

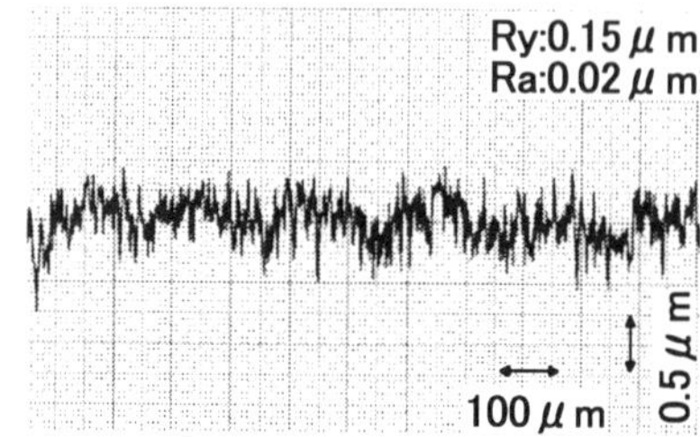

Figure 5 Surface roughness profile

5. CONCLUSIONS

The grinding characteristics of concave mirror surfaces were investigated using a desk-top 4-axes machine "TRIDER-X" with ELID system. The results of grinding experiments indicated that highly efficient grinding could be achieved successfully.

REFERENCES

1. Hitoshi Ohmori. Electrolytic In-Process Dressing (ELID) Grinding for Optical Parts Manufacturing. International Progress in Precision Eng. IPES, 7(1993),134-148

FINISH SURFACE GRINDING OF TITANIUM ALLOYS[*]

Zhejun Yuan, Bo Zhu, Zesheng Lu, Feihu Zhang
Harbin Institute of Technology, China

Abstract

In order to reduce wheel adhesion and improve the quality of ground surface, the ELID (electrolytic in-process dressing) grinding and cryogenic ELID grinding for grinding TC4 was developed. The various factors affecting the ground surface integrity were studied. The grinding forces were measured, grinding chips and the ground surface were studied under SEM. The grinding temperature was analyzed by using FEM. The results indicated that, by using ELID grinding and cryogenic cooling, the wheel adhesion, the grinding forces and grinding temperature were reduced and the ground surface integrity improved distinctly

Keywords

Titanium alloy (TC4), ELID, grinding, cryogenic cooling

1. INTRODUCTION

Titanium alloys, because of its excellent properties, is widely used in the aerospace and other industry [1], however its poor grindability is an unsolved problem. In grinding Ti-alloys, wheel adhesion is a particular phenomenon and it is the basic reason of its poor grindability[1][2].

In order to reduce the wheel adhesion and get good ground surface finish, ELID grinding and cryogenic ELID grinding were introduced for grinding Ti-alloys. The grinding forces, grinding temperature, mechanism of chip formation and surface integrity in ELID grinding and cryogenic ELID grinding of Ti-alloys were studied. Based on the results of our experiments, good ground surface finish TC4 was obtained in ELID grinding, especially cryogenic ELID grinding.

2. EXPERIMENTAL CONDITIONS

* Supported by Chinese National Foundation of Science

The experimental set-up for cryogenic ELID surface grinding is shown schematically in Fig.1. The lower end of TiC-alloy TC4 (Ti-6Al-4V) work piece was cooled by liquid nitrogen. Grinding force components Fn and Ft can be measured by the three-component KISTLER dynamometer respectively. The ground surfaces were studied by scanning electron microscope (SEM). Grinding conditions are given in Table 1.

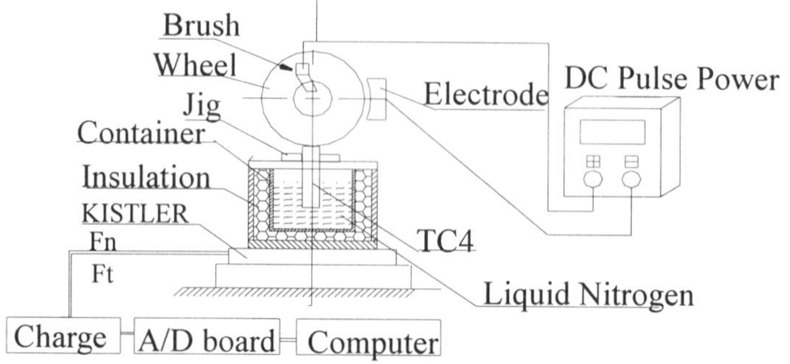

Fig.1 System for cryogenic ELID grinding

Table 1.   Experimental conditions

Items	Descriptions
Grinding wheel	metal bond CBN, ϕ200 mm $\times$ 15 mm, W10
Spindle speed	1500 rev/min
Table speed	6m/min
Down feed	1μm-5μm
Grinding condition	Wet grinding, ELID grinding, Cryo-ELID grinding

3. SEM STUDY ON CHIPS AND GROUND SURFACES

To study the mechanism of chip formation and the material removal in grinding under different grinding conditions, the grinding chips were collected and studied by SEM.

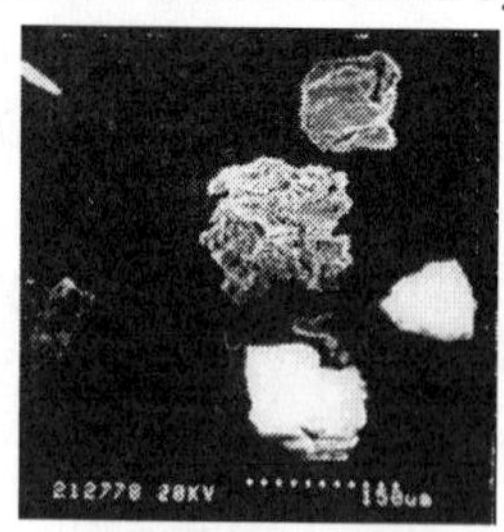

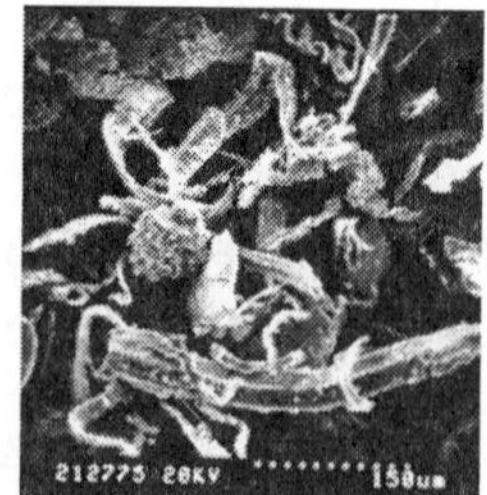

(a) Conventional   (b) ELID grinding   (c) Cryo-ELID grinding

Fig.2 Grinding chips of TC4

Fig.2 shows SEM photographs of the grinding chips obtained. In conventional grinding, the abrasive grits were dull due to the adhesion of Ti-alloy materials on the wheel surface, then, the grinding temperature was high and spherical shaped chips was formed, as shown in Fig.2 (a). In ELID grinding and Cryo-ELID grinding, the chips were predominantly cutting by sharp grits[3][4]. Therefore, ELID grinding gave fewer spherical chips and some fractured chips, and Cryo-ELID grinding gave predominantly lamellar and leafy chips, as shown in Fig.2 (b) and Fig.2 (c).

Fig.3 shows the SEM photography of ground surfaces of TC4. It indicates ploughing and plastic deformation to be the predominant mode of material removal in the conventional grinding. By Cryo-ELID grinding, the ground surface shows only few such defects, the area of the adhesive materials, side flow declined, indicating a lower temperature and less plastic deformation in the metal removal. The finish ground surface by Cryo-ELID grinding is better than by conventional and by ELID grinding.

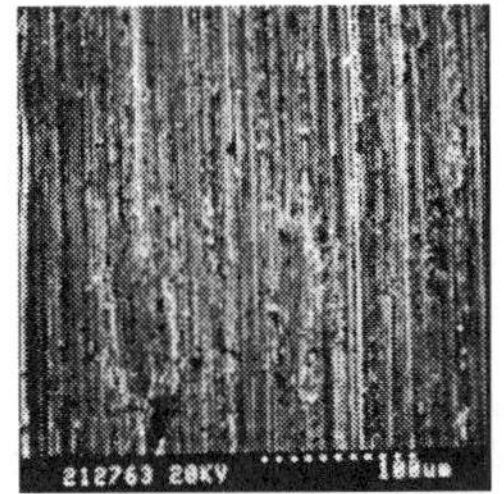
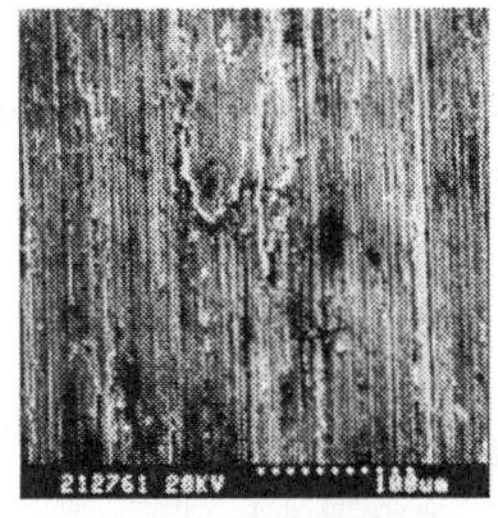
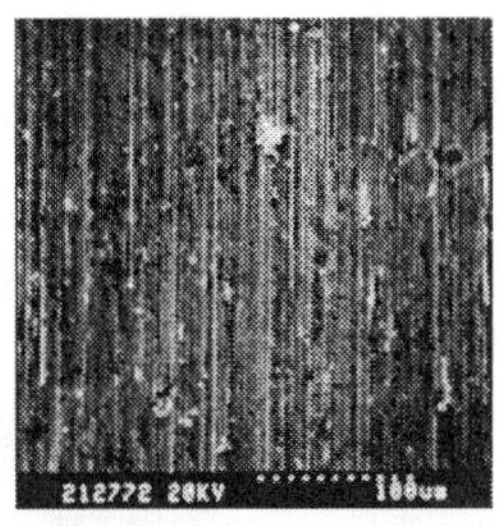

(a) Conventional grinding   (b) ELID grinding   (c) Cryo-ELID grinding

Fig.3 Ground surfaces of TC4

4.STUDY ON GRINDING FORCE AND TEMPERATURE

1) **Grinding forces:** Grinding forces are important parameters to judge the performance of any grinding process[5].

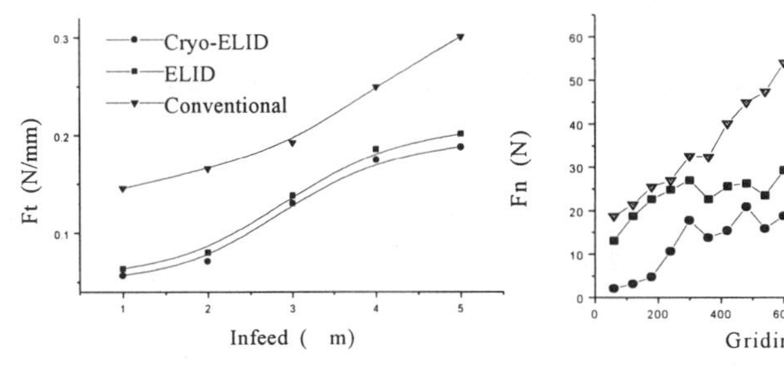

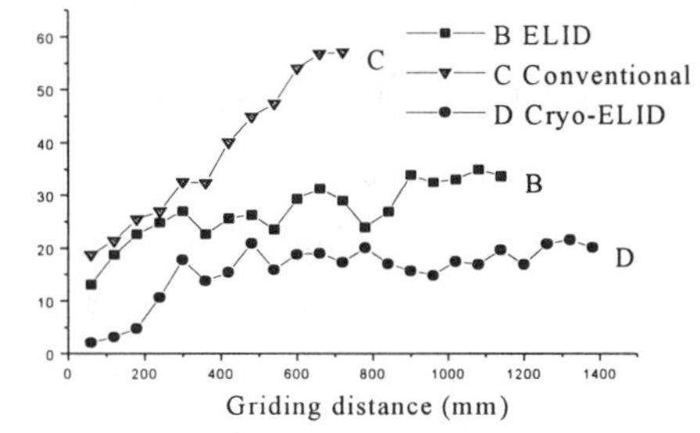

Fig.4 F_t variation with infeed   *Fig.5* F_n variation with grinding distance

Fig.4 shows that, in ELID grinding and Cryo-ELID grinding of TC4, the tangential (Ft) was reduced, in comparison with conventional grinding. Application of ELID and cryogenic cooling reduced the viscidity of TC4 , decreased the wheel adhesion and increased the chips capacity, hence reduced the grinding forces.

Fig.5 shows the variety of the grinding force Fn with the grinding distances. Because of strong adhesion of ground material to the wheel surface, the grinding force grows sharply with the increase of grinding distance in conventional grinding, besides, the ground surface finish is poor. In case of ELID and cryogenic ELID grinding, the grinding force grows only a little with the increase of grinding distance. Because ELID reduces the adhesion of grinding chips to the wheel, the abrasive grits keep sharp in the process of grinding, hence the grinding force is reduced. Cryogenic cooling reduces the viscidity of TC4, makes it more easily to be ground out, therefore the grinding force in cryogenic ELID grinding is the lowest.

2) Grinding temperature: The grinding temperature is an important factor effecting grinding performance. MSC finite element analysis (FEM) software was used to estimate the cryogenic effect on grinding temperature.

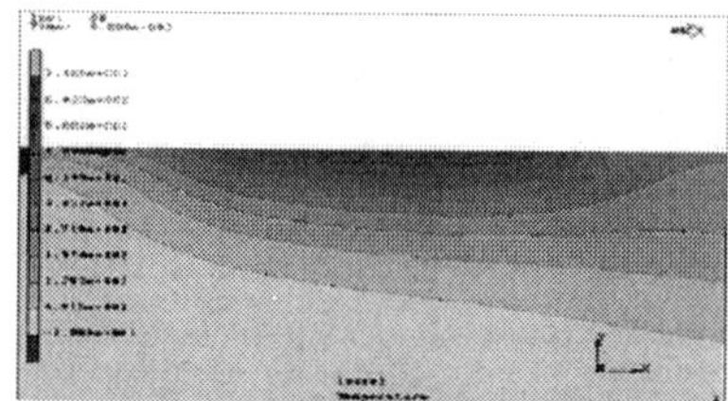

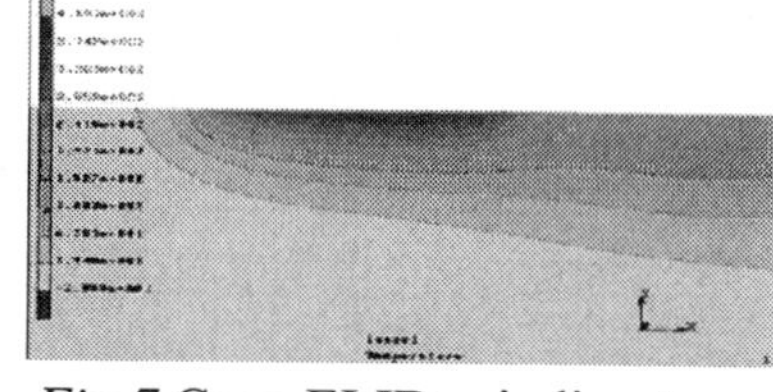

Fig.6 ELID grinding temp.   *Fig.7* Cryo-ELID grinding temp.

The Fig.6 and Fig.7 depicts grinding temp. FEM results. It shows that, in the grinding TC4, cryogenic cooling lowers grinding temperature from 747°C to 423°C (~40%) compared to conventional grinding. Low grinding temperature is favorable for finish grinding.

5. STUDY ON THE GROUND SURFACE INTEGRITY

1) **Surface roughness**: SEM photography of the ground surfaces under different grinding conditions (conventional grinding, ELID grinding and cryogenic ELID grinding) are given in Fig.3. Table 2 shows the ground surface roughness under these grinding conditions. With the increase of grinding distance in conventional grinding, the ground surface finish becomes poor, burned area and grinding cracks were often observed on the

ground surface.

Table 2.   Surface roughness

Grinding conditions	Surface roughness R_a (µm) in transverse direction
Conventional grinding	0.78
ELID grinding	0.47
Cryogenic ELID grinding	0.36

2) **Residual stresses**: the X-ray stress instrument was used to measure the residual stresses in the ground surface layer of TC4. In the case of ELID grinding the residual stress is +37.8MPa, and in cryogenic ELID grinding the residual stress is −1.9MPa. Cryogenic ELID grinding is favorable for finish grinding of Ti-alloys.

6. CONCLUSION

1) Studied the grinding chips by SEM shows that the chip deformation is largest in conventional grinding,. the next in ELID grinding, and least in cryogenic ELID grinding.

2) Application of ELID reduces the adhesion of chip to the wheel, therefore decreases grinding forces and produces good ground surface. This has been attributed to retention of grit sharpness and increase of chips capacity.

3) Cryogenic ELID grinding decreases the grinding temperature and reduce the viscidity of Ti-alloys, so it decreases the wheel adhesion, grinding forces, and produce good ground surface integrity.

7. REFERENCE

[1] Tarasov, L.P., How to grinding Titanium, American Machinist, Vol. 96 (1952), No. 23: pp 135-146.

[2] Kumar, K.V., Shaw, M.C., Metal Transfer and Wear in Fine Grinding, Wear, Vol. 82 (1982), No. 2: pp 256-270.

[3] A.B. Chattopadhyay, A.Boss and A.K. Chattopdhyay: Improvements in grinding steels by cryogenic cooling, Precision Engineering Vol.7 (1985): pp 94.

[4] S.Paul and A.B. Chattopadhyay, A Study of Effects of Cryo-cooling in Grinding, Machine Tools Manufact., Vol. 35 No. 1: pp 109-117.

[5] S. Paul and A.B. Chattopadhyay，The effect of cryogenic cooling on grinding forces, Machine Tools Manufact., Vol. 36.No.1: pp 64.

FLATTENING OF MICRO-FUNCTIONAL PARTS BY ELID LAP GRINDING

Nobuhide Itoh

Department of Mechanical Engineering, Ibaraki University

Hitoshi Ohmori, Yutaka Yamagata, Sei Moriyasu

The Inst. of Phys. and Chem. Res. (RIKEN)

Toshio Kasai

Department of Mechanical Engineering, Saitama University

Abstract

For the purpose of realizing the flat surfaces of composite parts made of components with different properties, the finishing of such parts by a lapping method based on the ELID technique (ELID-lap grinding) was investigated. The composite parts used in this experiment were fabricated from copper and resin by the film lamination method. After 15 minutes of grinding, surface bumps could be smoothly ground from 15 μ m to 0.4 μ m. The results confirmed that the method proposed is effective for smoothly grinding composite parts.

Keywords

ELID, metal-resin bonded diamond wheel, lap grinding, composite parts

1. INTRODUCTION

Since a research group in the U.S. developed and successfully manufactured a 100 to 300 μ m gear in 1987, the research and development of micro-functional parts have been actively carried out. Applications of these micro-structures for highly sensitive sensors and micro-machines are expected to grow rapidly in not only the electronic and optical fields, but in the medical, environment, and aeronautics fields as well. For this reason, various manufacturing and machining methods are being proposed for a wide variety of micro-structures, and studies are promoted to put them to practical application. The authors are also conducting studies on the manufacture of efficient micro-functional parts by the film lamination method[1]. With this method, first a glass board is given electrical conductivity by spattering and then laminated with an exposed photo-resistant film (about 50μ m). After this, unrequired portions are removed, and metal components are extracted from the remaining portions by electrical-casting to produce micro-structures. However, large bumps caused by the electrical casting are seen on the surface of the metal-structures, thus giving rise to the need to flatten these bumpy portions. This report describes the results of fundamental studies aiming to smoothly grind composite parts composed of resist and metal which have different machining properties by a lapping method applying the ELID technique (ELID lap grinding[2]).

2. PRINCIPLE OF ELID LAP GRINDING

Figure 1 shows the principle of the ELID lap grinding method which is a lapping method added with the ELID technique. ELID is a method in which a grinding solution with weak conductivity is supplied between the wheel which serves as the positive electrode and the negative electrode to promote electrolysis of the wheel surface. By applying this phenomenon, the bond material of the wheel surface is dissolved electrolytically. Thus, only the abrasives that are not affected electrically are protruded, achieving stable grinding.

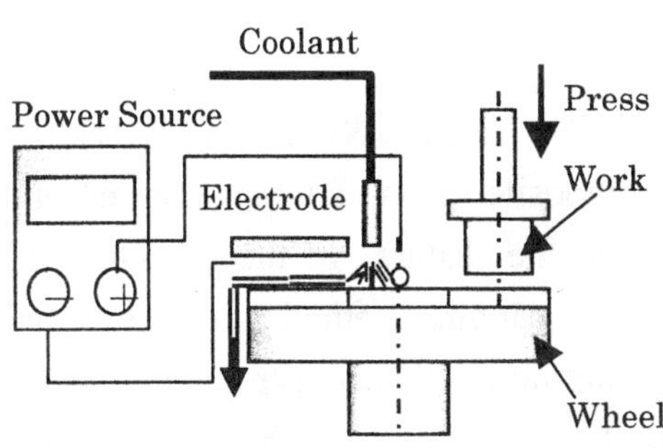

Figure 1 Principle of ELID-lap grinding

Figure 2 Close-up view of machine

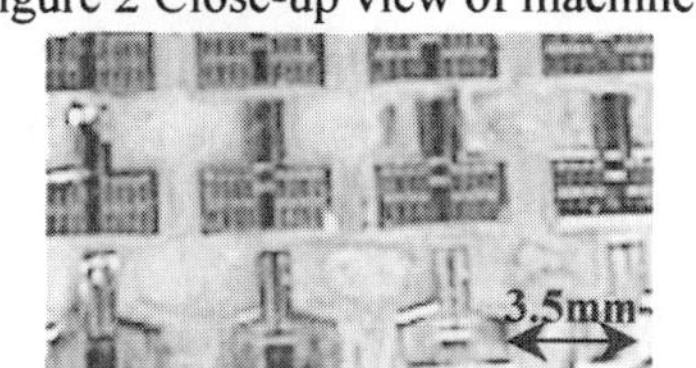

Figure 3 External view of workpiece

3.   EXPERIMENTAL SYSTEM

The experimental system used in this study consisted of;

(a) Grinding machine: Single-sided lapping machine (MG773-B) which has been remodeled (Figure 2) to apply the ELID technique.

(b) Wheel: Since a slight amount of grinding is required, a fine #20000 metal-resin bonded diamond wheel was used. The size of the wheel was 250 mm in diameter and 55 mm in width. The wheel shape was round and the rate of concentration was 100.

(c) Grinding solution and electrolysis power supply: General grinding solution with weak conductivity (AFG-M) diluted 50 times with tap water was used. The electrolysis power supply was a high frequency electrolysis power supply (ED630) for ELID.

(d) Work: The work used was a resin model from which copper has been extracted. (Figure 3).

4. EXPERIMENTAL METHOD

After truing the wheel with a #325 cobalt bonded cup wheel, initial electrolysis dressing was performed, after which the work was machined. In this experiment, since the amount of machining required was slight (about 30 to 40 μ m), to achieve the smooth surface finishing needed, a method which machines the work together with a harder material (simultaneous grinding method[3]) was applied using an ultrafine abrasive metal-resin bonded wheel. The grinding conditions were work and wheel rotation speed of 100rpm, and grinding pressure of 50 to 280kPa.

5. FEATURES OF SIMULTANEOUS GRINDING METHOD

Before the smoothing test of composite parts, silicon was ground together with cemented carbide to grasp the features of the simultaneous grinding method. Figure 4 shows a typical grinding method and the external view. Figure 5 shows the results. The grinding efficiency of the simultaneous grinding method for silicon was found to depend on the removal efficiency of the cemented carbide. The results confirmed that the grinding efficiency was lower than when   the silicon was ground alone. The surface roughness

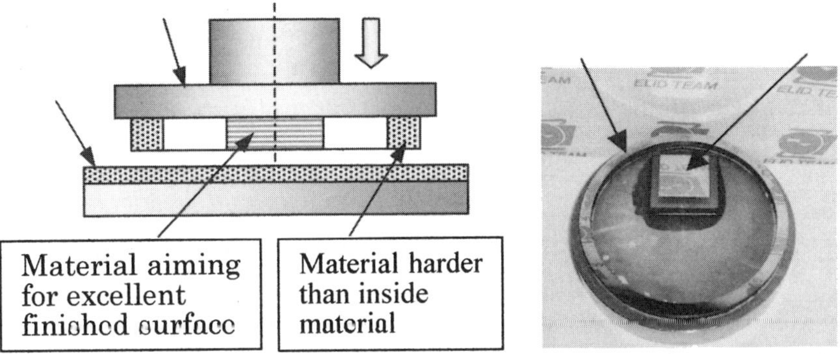

Figure 4 Schematic illustration of simultaneous grinding method

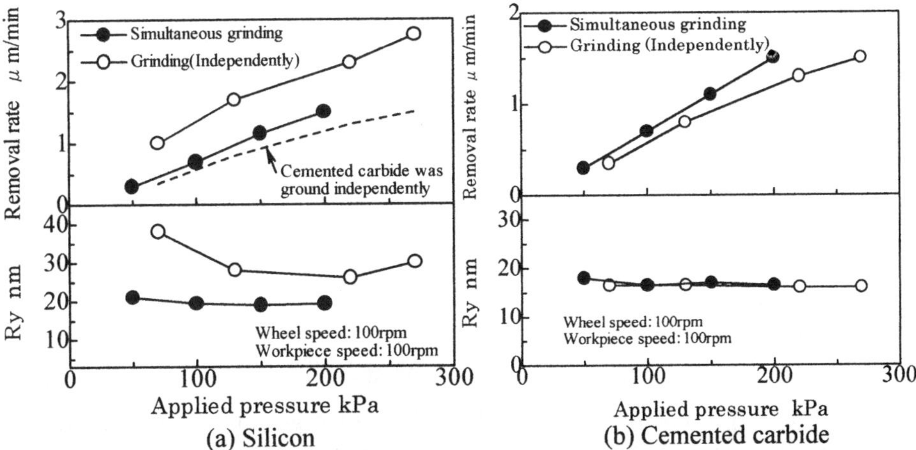

Figure 5 Results of grinding efficiency and surface roughness of silicon and cemented carbide in simultaneous grinding

of the silicon was also better.

6. SMOOTHING EXPERIMENTS FOR MICRO-PARTS

Experiments were also conducted on the smoothing of micro-parts (as shown in Figure 3) by the simultaneous grinding method described above. The grinding condition was work and wheel speed of 100rpm, and grinding pressure of 100kPa. Figure 6(a) shows the results of measuring the surface roughness of the workpiece before starting grinding. The bumps on the surface prior to grinding were 15μm at maximum. After 15 minutes of grinding, the surface was successfully ground smoothly (0.4 μ m) as shown in Figure 6 (b). Figure 7(a)(b) shows the differences in the surface properties before and after grinding. Microscopic observation also confirmed the smooth surface of the work. No abnormal electrolysis current, etc. was observed during the grinding process, indicating that stable grinding was successfully achieved.

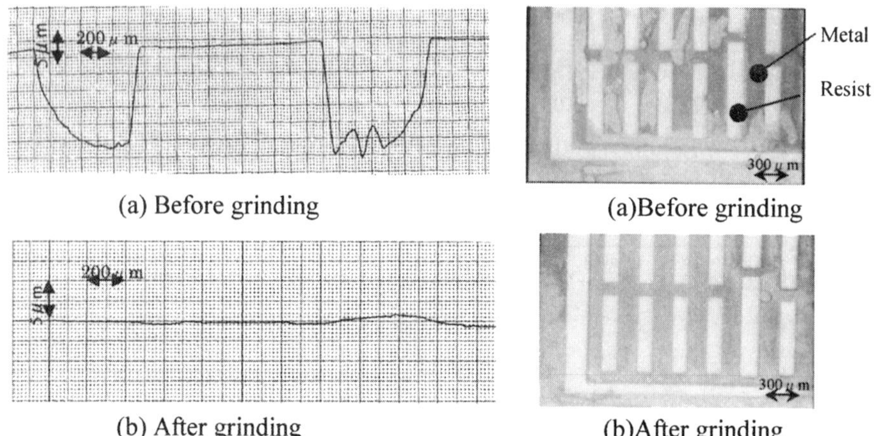

(a) Before grinding   (a)Before grinding

(b) After grinding   (b)After grinding

Figure 6 Differences in surface roughness before and after grinding

Figure 7 Differences in surface properties before and after grinding

7. CONCLUSION

The authors conducted fundamental studies to realize the efficient smoothing of micro-functional parts by ELID lap grinding using an ultrafine metal-resin bonded wheel. Experimental results confirmed that micro-functional parts can be ground smoothly using the grinding method developed.

8. REFERENCES

1. N.Itoh, H.Ohmori, S.Moriyasu and Y.Yamagata; Smoothing of micro-parts by ELID-lap grinding, Proceedings of JSPE Autumn Conference, 1998; 593

2. N.Itoh and H.Ohmori; Grinding Characteristics of Hard and Brittle Materials by Fine Grain Lapping Wheels with ELID, Journal of Materials Processing Technology, Vol.62, No.4, 1996; 315-320

3. N.Itoh, H.Ohmori, T.Kasai and T.Karaki-Doy; Study of Smooth Surface Finish by ELID-lap Grinding and Metal-Resin Bonded wheel, Abrasive Technology, Current Development and Application, JSGE 1999; 135-140

DEVELOPMENT OF SMALL TOOL BY MICRO FABRICATION SYSTEM APPLYING ELID GRINDING TECHNIQUE

Yoshihiro Uehara[*1], Hitoshi Ohmori[*1], Yutaka Yamagata[*1],
Sei Moriyasu[*1], and Weimin Lin[*1]

*1 The Institute of Physical and Chemical Research (RIKEN)

Ken-Ichi Kumakura[*2]

*2 Kumakura Co., ltd.

Shin-Ya Morita[*3]

*3 University of Tokyo

Tomoyuki Shimizu[*4], and Tetsuo Sasaki[*4]

*4 Nippon Institute of Technology

Abstract

In this study, we attempted to develop a cylindrical grinding machine
applying the ELID technique, for use as a microtool. We also carried
out a detailed investigation on the performance of the electrode turning
method of the ELID system; a microcylindrical shaft and microangular
shaft were ground using the cylindrical grinding machine applying
ELID technique. Good dimensional accuracy was achieved.

Keywords

ELID grinding, Microfabrication, Microtool, Plasma discharge truing,
Cast iron boned diamond wheel, Cylindrical grinder, Hard material

1. INTRODUCTION

In recent years, we have been witnessing the progress of high-speed
milling with high spindle rotation and high feed speed using microtools
capable of high accuracy and high quality for fabrication of detailed small
parts. Though these microtools have developed rapidly, it is difficult to apply
them in microfabrication technologies, because of the difficulties in
manufacturing and performing measurements. In this study, we developed a

cylindrical grinding machine applying the ELID [Electrolytic In-process Dressing] system [1] for microtools. We evaluated the performance of this ELID system with a newly developed rotational electrode. A microcylindrical shaft and a microangular shaft were ground, and good dimensional accuracy was achieved.

2. COMPACT CYLINDRICAL GRINDER

Figure 1 shows the external view of the compact cylindrical grinding machine. This machine has three linear axes, X, Y and Z. All linear axes are supported by a roller slide mechanism on a cross roller guide, and driven by step motor. The X and Y axes have a position resolution of 0.25 μm with semiclosed-loop control. The Z axis has a position resolution of 0.35 μm with semiclosed-loop control. A personal-computer-based NC unit is used.

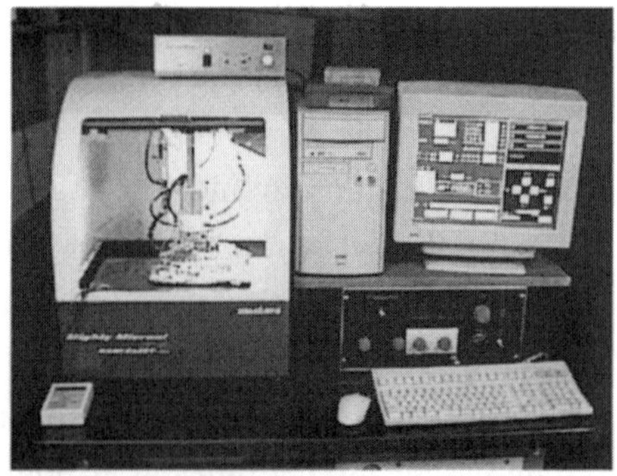

Figure 1 External view of compact cylindrical grinding machine

3. ELID GRINDING SYSTEM FOR COMPACT CYLINDRICAL GRINDING MACHINE

Figure 2 shows the ELID grinding system mounted to the compact cylindrical grinding machine. The ELID grinding system should be fixed to the electrode, but this is not possible in the present method. We therefore developed an ELID system with a rotating electrode that can move freely.

This system is composed of a plastic body, a bearing for the electrode to rotate, an insulating guide plate to prevent contact between the electrode and the grinding wheel.

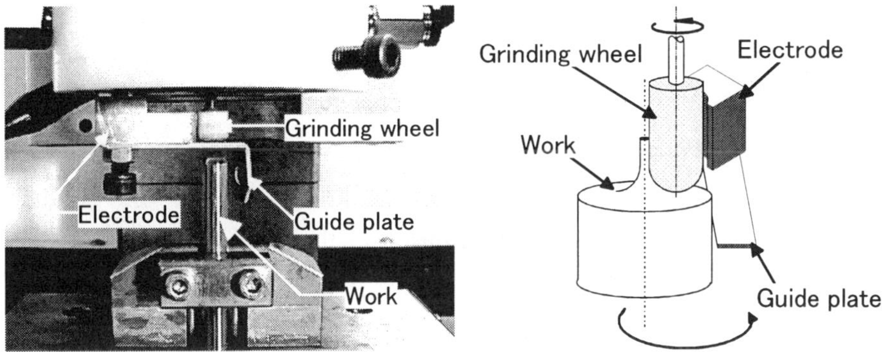

Figure 2 ELID grinding system

4. MACHINING EXPERIMENTS

Table 1 shows the experimental conditions.

Table 1 Experimental conditions

Wheel Mesh	#1200
Rotation speed (rpm)	20000
Depth of cut (μ m)	0.5
Peak current (A)	1
Open voltage (V)	30
On/Off time (μ s)	2 / 2

Figure 3 shows a schematic of the measurement point of the microangular shaft. Table 2 shows the dimensions of the microangular shaft along A and B. Figure 4 shows an overview of the microangular shaft. Figure 5 shows an overview of the microengraving characters.

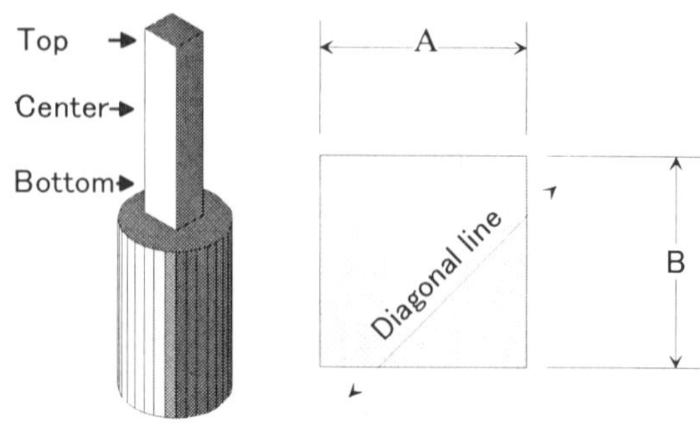

Figure 3 Schematic of measurement points of microangular shaft

Table 2 Diameters of microangular shaft along A and B

	A dimensions 71 μ m	B dimensions 71 μ m	Diagonal lines 100 μ m
Top	78 μ m	78 μ m	111 μ m
Center	78 μ m	79 μ m	111 μ m

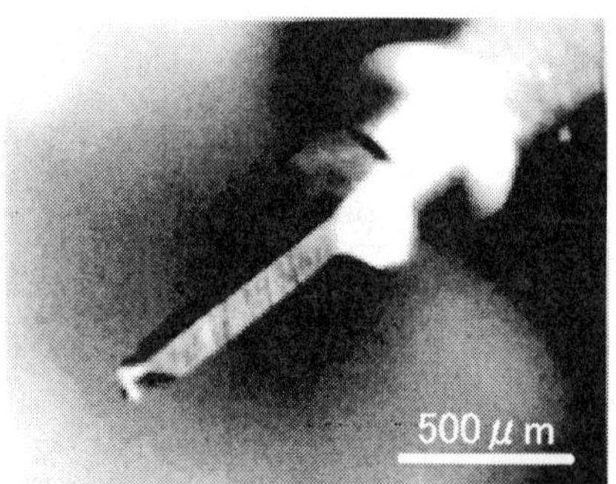

Figure 4 Overview of microangular shaft

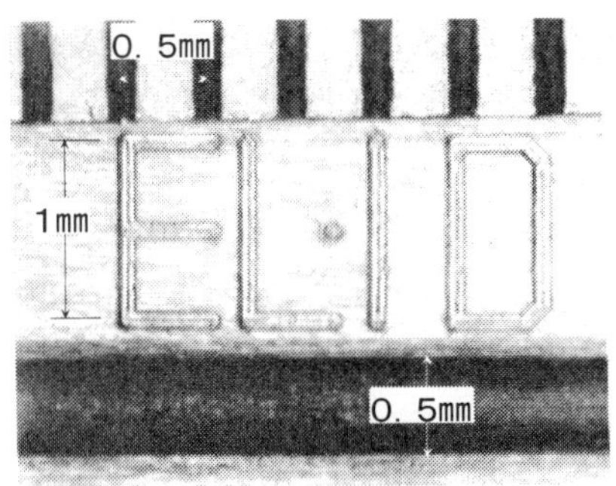

Figure 5 Overview of microengraving characters

5. CONCLUSIONS

The microtools developed were mounted with an angular shaft respectively ground using a compact cylindrical grinding machine applying the ELID technique. The results confirmed that dimensional accuracy could be achieved, and that the microtool with an angular shaft can be used for microengraving characters.

REFERENCE

1. Hitoshi Ohmori. Electrolytic In-Process Dressing (ELID) Grinding for Optical Parts Manufacturing. International Progress in Precision Eng. IPES, 7(1993),134-148

EFFECT OF POLISHING PADS IN FINISHING OF LARGE OPTICAL ELEMENTS

Weimin LIN, Hitoshi OHMORI, Yutaka YAMAGATA, Sei MORIYASU, and
Changling LIU

The Institute of Physical and Chemical Research, Wako-shi 351-0198, Saitama, Japan

Toshio KASAI

SAITAMA University, Urawa-shi 338-8570, Saitama, Japan

Abstract

In this research, a fused silica torus mirror with a surface roughness of 8.7 nm Ra were fabricated with highly efficient ELID (electrolytic in-process dressing) grinding, and then polished with a ϕ16 mm polishing pad. The polishing was implemented in two processes: pre-polishing and final polishing. In the pre-polishing process, polyurethane foam type sheet was used. As a result, surface roughness of about 1.2 nmRa was obtained in the measurement of an area of 0.169×0.126 mm^2 using an interferometric microscope. In the final polishing process, the polyethyleneterephthalate fiber felt sheet used produced very good surface roughness. Through the analysis of the polished surface texture, the factors influencing surface finish were also investigated.

Keywords

Polishing, polishing pad, X-ray mirror, fused silica, surface roughness, finishing

1. INTRODUCTION

In the SR (Synchrotron Radiation) Lithography System, many large aspheric mirrors (a long flat, torus or cylindrical surface) are being used. The optical components used in that system must have the ability to withstand damage caused by radiation and high heat load, which limits material selection. Fused silica, single crystal silicon, silicon carbide, etc. are the most commonly used materials [2]. It is very difficult to fabricate an optical surface with very low surface roughness and very high form accuracy, especially when the surface is applied in the optical system using short wave light source, such as X-ray. Usually lapping is implemented to obtain the required form accuracy, and then polishing is carried out to attain the desired surface roughness. But form accuracy is very difficult to maintain during the polishing process in which soft elastic polishing pad and loose abrasive are employed. In this study, a large cylindrical mirror, which has a length of 300 mm and was made of fused silica, ground by ELID (electrolytic in-process dressing)-grinding [1], [3] to high form accuracy with #4000 cast iron bonded diamond wheel, was polished with a rotational small tool.

The polishing was implemented in two processes: pre-polishing and final polishing. Different polishing pads were used. As a result, very good surface roughness was obtained.

2. POLISHING INSTRUMENT AND CONDITIONS

2.1 Polishing instrument

Figure 1 shows the large polishing machine used for polishing. It has three linear axes and one rotation axis, and the movements of the four axes can be controlled simultaneously using an NC system. An air cylinder was installed in the spindle of the Z-axis to provide the polishing load and an air turbine motor was used to drive the polishing tool, both the polishing load and rotation speed of the motor spindle can be regulated manually before polishing and are constant during the polishing process. The polishing tool was connected with the motor spindle with a ball-faced coupling, which permits the polishing tool to incline freely up to 60°, for the polisher to make good contact with the polished surface. It is possible to change the polishing pressure, head rotation speed, head scan type, and feed speed. The specifications of the machine are shown in Table 1.

Table1 Main specifications of polishing machine

Size of table	1200 mm × 600 mm
Max. weight of workpiece	2000 kg
Max. height of workpiece	400 mm
Travel	Table: 1400 mm X: 1400 mm, Y: 600 mm, Z: 400 mm, C axis: ± 200 °
Feed speed of axes	X, Y, Z: 2.5 - 50 mm/sec, C: 4 - 40 °/sec
Resolution of axes	X, Y, Z: 1 μm, C: 0.001°

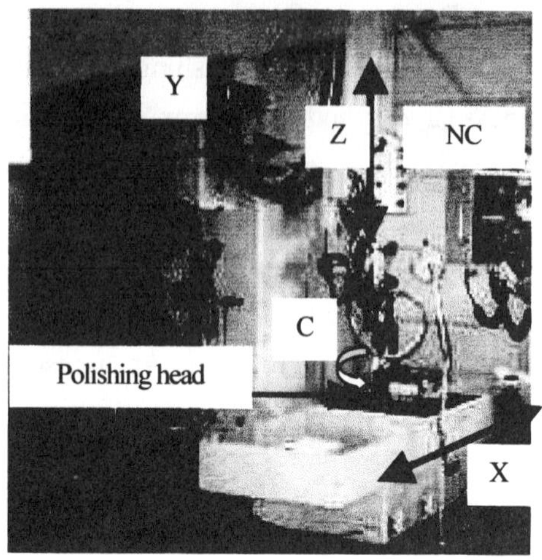

Figure 1 Photograph of polishing machine for large X-ray mirror polishing

2.2 Workpiece

The mirror is 300 mm long, 80 mm wide, 25 mm thick, and has a curvature radius of 45851 mm, 141.6 mm of the torus surface. Instead of lapping, which is usually used, ELID (electrolytic in-process dressing) grinding was implemented to generate the mirror figure. A #4000 (average grit size 4.06 μ m) cast iron bonded diamond wheel was used for the last pass of grinding. Figure 2 shows that very smooth surface of 8.7 nmRa was obtained in the measurement with Zygo New View100 surface profilers.

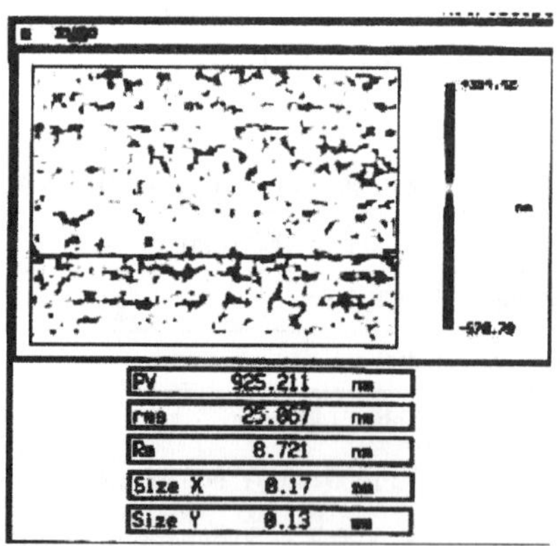

Figure 2 Surface roughness of ELID ground

Table2 Polishing conditions

	Preparing polishing	Final polishing
Machine	Large polishing machine	
Polishing pad	Polyurethane form type	Polyethyleneterephthalate fibers felt
Pad size	ϕ 16 mm	
Abrasives	Cerium oxide(average diameter of 1 μ m)	
Pressure	270 kPa	
Rotation speed	700 rpm	
Scanning speed	250 mm/min	
Scan pitch	1mm	
Scan type	Type 1 (Parallel X-axis)	
Polishing time	12 h	6 h

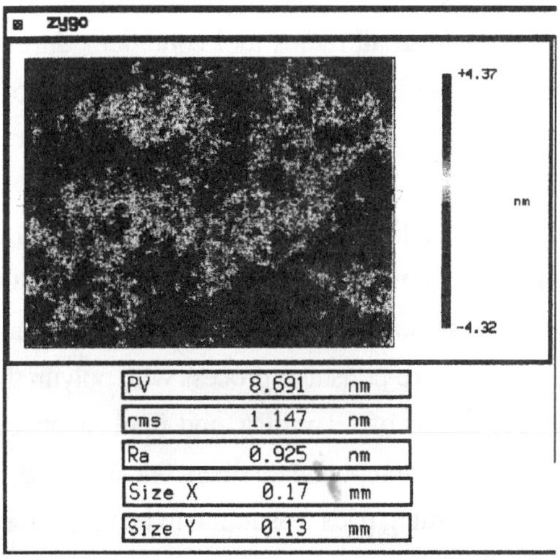

Figure 3 Surface roughness after final polishing

2.3 Polishing conditions

The workpiece polished with a vertical axis rotation tool was fixed on the table with two bolts at one side and four magnetic chucks at the other three sides. The polishing was implemented in two stages: pre-polishing and final polishing. In the polishing process, two types of polishing pad were used. In pre-polishing, polyurethane form type sheet with a Shore-A hardness of 67~82, density of 22 to

32, thickness of 1.5 mm was used. In the final polishing, a polyethyleneterephthalate fiber felt sheet with a SUBA400, and thickness of 1.27 was used. The other polishing conditions of the both process are shown in Table 2.

3. POLISHING RESULTS

Pre-polishing required a total time of 12 hours, which is considerably long for just smoothing the surface. It was performed until the grinding marks were completely removed. The surface roughness was found to be 1.1 nmRa when measured with a Zygo New view 100 interferometric microscopes over an area of 0.169×0.126 mm^2. Final polishing took about 6 hours. As a result, final 3D surface roughness of 0.9 nmRa was attained. Figure 3 shows the polished surface profiles. Very good surface roughness was obtained, indicating the effect of using different polishing pad. In this study, we used asphalt-polishing pellets to obtain super smooth surface, but none of the pellets could resist the heat that occurred in high rotation, high polishing pressure and small radius tool conditions during polishing. We will thus develop a new asphalt-polishing pellet to obtain super smooth surface.

4. CONCLUSIONS

In this study, a fused silica torus mirror ground by ELID grinding with a curvature radius of 45851 mm, 141.6 mm, and a length of 300 mm was polished with the local polishing method. The polishing tool had a diameter of 16 mm and vertical rotational axis. The pre-polishing process with polyurethane foam type sheet produced a surface roughness of 1.1 nmRa, and final roughness of 0.9 nmRa was obtained using a polyethyleneterephthalate fibers felt sheet and 1 μ m cerium oxide slurry. Use of different polishing pads was found to be useful for polishing.

REFERENCE

1.	Hitoshi Ohmori. Electrolytic in-process dressing (ELID) grinding technique for ultraprecision mirror surface machining. International Journal of the Japan Society for Precision Engineering, 26(4)(1992). 273-278.

2.	Jean Susini, Dieter Pauschinger, Roland Geyl, Jean-Jacques Ferme, Guy Vieux. Hard X-ray mirror fabrication capabilities in Europe. Optical Engineering, 34 (2) (1995), 388-395.

3.	Ohmori H., and Nakagawa T.: Mirror Surface Grinding of Silicon Wafer with Electrolytic In-process Dressing, ANNALS of the CIRP, Vol.39, No.1, pp.329-332 (1990).

EFFECTS OF CHANGES IN FLUID COMPOSITION ON MAGNETORHEOLOGICAL FINISHING (MRF) OF GLASSES AND CRYSTALS

Stephen D. Jacobs*†, Steven R. Arrasmith*, Irina A. Kozhinova*,
Sergei R. Gorodkin*, Leslie L. Gregg*, Henry J. Romanofsky*,
and Thomas D. Bishop II*
*Center for Optics Manufacturing and †Laboratory for Laser Energetics
University of Rochester, 240 East River Road, Rochester, NY 14623
Tel: 716-275-2478; Fax: 716-275-7225
email: sjac@lle.rochester.edu

Aric B. Shorey
Optical Coating Laboratory, Inc., 2789 Northpoint Parkway, Santa Rosa, CA, 95407

William I. Kordonski
QED Technologies, Inc., 1040 University Ave., Rochester, NY 14607

Abstract

Magnetorheological (MR) fluids for fine finishing of optical components consist of magnetic carbonyl iron (CI) particles and nonmagnetic polishing abrasives in a carrier liquid. With a solids loading of over 40 volume percent, these fluids incorporate surfactants and other additives to help stabilize them against sedimentation and corrosion during commercial finishing operations. Several experiments have been conducted with variations in the three primary ingredients that demonstrate the effectiveness of magnetorheological finishing (MRF) for a variety of glasses and crystals.

Key Words

deterministic polishing, magnetorheological finishing,
carbonyl iron, cerium oxide, nanodiamonds

1. INTRODUCTION

A deterministic polishing process called magnetorheological finishing (MRF) has been commercialized and is now being employed in the production of high precision optics in the United States, Europe, and Japan. The keys to MRF are a sub-aperture polishing tool with a constant removal rate, sophisticated yet user-friendly software for determining finishing solutions for figure correction, and CNC multiple axis machines with PC interfaces for rapid execution of the process [Golini, 1999].

In MRF, the magnetorheological polishing fluid is deposited by a nozzle on the rim of a rotating wheel, which transports the fluid to the workpiece surface. The wheel rim and the surface to be polished form a converging gap. A magnetic field is applied at the vicinity of the gap. The moving wall, which is the rim surface, generates a flow of magnetically stiffened MR polishing fluid through the converging gap, resulting in high shear stress over a portion of the workpiece surface and material removal. This area is designated as a polishing spot. The material removal is enhanced by nonmagnetic abrasive particles, constituents of the slurry, that are forced out to the polishing interface by a magnetic field gradient. Important benefits of this novel removal mechanism are as follows: a very clean, finished workpiece surface without embedded abrasive particles; no sub-surface damage; high material removal rates from 1 μm/min to 10 μm/min; and surface micro-roughness values of 0.5 nm rms.

Because the MR fluid forms a sub-aperture removal spot that conforms to the local curvature of the part, it is possible to finish flat or steeply curving surfaces. Industry is using the Q22-X MRF machine in three ways:

- to take conventional, pitch polished spherical surfaces from quarter wave to twentieth wave or better in surface shape accuracy in minutes;
- to put moderate aspheres with less than 20 μm departure onto spherical surfaces; and
- to polish out highly aspheric surfaces that have been deterministically micro-ground to within a few waves of the required final form.

Circular parts may vary in size from a few mm to 20 cm and larger. A new MRF machine, the Q22-Y, may be used in a rastering mode to finish prisms, square and rectangular flats, and cylinders. [Golini, 2000].

2. MR FLUIDS FOR FINISHING OF OPTICS

There are two magnetorheological (MR) fluids currently in widespread industrial use. One composition consists of cerium oxide in an aqueous suspension of magnetic carbonyl iron (CI) powder, and it has been found appropriate for almost all soft and hard optical glasses and low expansion glass-ceramics. Preparation of this "standard " MR fluid and its performance have been described in the literature [Jacobs et al., 1999]. The second composition uses nanodiamond powder as the polishing abrasive, and it is better suited to calcium fluoride, IR glasses, hard single crystals like silicon and sapphire, and very hard polycrystalline ceramics like silicon carbide [Jacobs et al., 1999; Arrasmith et al., 1999; Jacobs, 1995].

Considerations leading to a choice of nonmagnetic polishing abrasive are more complex than those encountered in conventional pitch or pad polishing. Not only do the hardness and chemistry of the abrasive grains need to be appropriate to the workpiece [Kaller, 1998], but the type of abrasive (median size, surface chemistry) can have a large or small effect on MR fluid flow in and out of the magnetic field. Fluid properties in an MRF machine circulation system must be held constant to realize constant rates of material removal during polishing. Variations in the corrosion resistance and sedimentation stability of the fluid have a significant impact, as was recently observed with some commercial cerium oxide products [Kozhinova et al., 1999].

The ability to readily change the MRF carrier fluid from water to a nonaqueous liquid opens up the possibility for processing soft, water-sensitive parts. Single crystals of potassium dihydrogen phosphate (KDP), an important electro-optic material for modulation and frequency conversion of laser radiation, have been successfully polished with an MR fluid based on a dicarboxylic acid ester. Laser damage testing at a pulse width of 1 ns shows that surface laser damage thresholds in the IR at $\lambda = 1054$ nm and in the UV at $\lambda = 351$ nm are equal to those achieved on diamond turned surfaces [Jacobs and Arrasmith, 1999]. Since this crystal has a Berkovich nano-hardness (1 mN load) of 1.5 ± 0.4 GPa compared to borosilicate BK7 glass at 7.7 ± 0.1 GPa. [Shorey et al., 2000], this finding offers convincing evidence that MRF-finished surfaces are extremely clean and contain no embedded particles.

3. ROLE OF WATER, CI, AND POLISHING ABRASIVE

Advances have recently been made toward understanding the mechanism of removal with MRF, based in part on the hardness of the CI particles [Shorey et al., 2000], the tribochemical interaction of cerium oxide or other abrasives with the workpiece surface [Shorey et al., 1999], and the type of slurry [Arrasmith et al., 1999 and Shorey et al., 2001]. Some preliminary results are summarized as follows for experiments with fused silica (FS) flats (nano-hardness 10 GPa) on a research test bed called the spot taking machine (no part rotation) [Arrasmith et al., 1999]:

Experiment I: Removal spots were taken with an MR fluid consisting of magnetic CI particles only in a *nonaqueous* carrier fluid. Removal rates were very low (~0.002 μm/min). Surface roughness for hard (>10 GPa) CI particles was high (>20 nm rms). This value dropped to ~2 nm for soft (<3 GPa) CI particles. In addition to impractical rates of removal, the dominant negative attribute of this fluid was found to be pitting of the FS surface by either the hard or soft CI.

Experiment II: Removal experiments were conducted using hard CI particles alone in an *aqueous* carrier fluid. While removal rates increased two orders of magnitude to ~0.2 μm/min, occasional pitting was still observed.

Experiment III: The additions of 0.1 to 1.0 vol. % concentrations of cerium oxide or diamond nano-abrasives to the MR fluid in experiment II were found to dramatically increase removal rates to ~5 μm/min. Pitting disappeared entirely, and surface roughness declined to <0.9 nm rms, the limit of the instrument used during the experiment. Parts examined with atomic force microscopy exhibited continuous grooves characteristic of excellent contact between the polishing abrasives and the work piece surface.

4. CONCLUDING REMARKS

Although material removal and surface polishing occur with just the magnetic CI particles in a nonaqueous carrier fluid, the MRF process is greatly enhanced by incorporating water and polishing abrasives (cerium oxide or nanodiamonds) into the MR fluid. Removal rates increase by over three orders of magnitude, surface pitting is eliminated, and roughness in the polishing zone decreases by over two orders of magnitude. We hypothesize that, upon coming into contact with the workpiece surface in the region of high magnetic field, the converging ribbon of MR fluid separates into layers. The top layer is composed of the carrier liquid and the polishing abrasives, while the bottom layer consists of compacted chains of magnetic particles attached to the rotating wheel. In future work we will attempt to verify and quantify the structural features of the MR fluid under the workpiece, in order to make further improvements to MRF.

5. ACKNOWLEDGEMENTS

The authors gratefully acknowledge the extensive contributions of the following individuals: co-workers at COM - C. Andrews, E. Fess, K. Grover, H. Pollicove, C. Pollicove, J. Ruckman, M. Stolberg; and colleagues at QED Technologies, Inc. - E. Cleaveland, P. Dumas, G. Forbes, D. Golini, S. Hogan, and. A. Sekeres. Financial support for this work was provided by QED Technologies, the U. S. Army Materiel Command, and DARPA.

6. REFERENCES

Arrasmith S. R., Jacobs S. D., Romanofsky H. J., Gregg L. L., Shorey A. B., Kozhinova I. A., Golini D., Kordonski W. I., Hogan S., and Dumas P., Studies of Material Removal in Magnetorheological Finishing (MRF) from Polishing Spots. Proceedings of the Finishing of

Advanced Ceramics and Glasses Symposium at the 101st Annual Meeting of the American Ceramic Society, pp. 201-212; 1999 April 25-28; Indianapolis, IN. The American Ceramic Society, 1999.

Golini D. Precision Optics Manufacturing Using Magnetorheological Finishing (MRF). Proceedings of the 9th International Conference on Production Engineering,: Precision Science and Technology for Perfect Surfaces; pp. 132-137; 1999 August 29-September 1; Osaka, Japan. The Japan Society for Precision Engineering (JSPE), 1999.

Golini D. Precision Optics Fabrication Using Magnetorheological Finishing (MRF), notes from The Center for Optics Manufacturing Summer School: New Machines, Tools, and Processes for Modern Optics Manufacturing, 2000 June 12-15; Rochester, NY. University of Rochester, 2000.

Jacobs S. D., Arrasmith S. R., Kozhinova I. A., Gregg L. L., Shorey A. B., Romanofsky H. J., Golini, D., Kordonski W. I., Dumas P., and Hogan S., An Overview of Magnetorheological Finishing (MRF) for Precision Optics Manufacturing. Proceedings of the Finishing of Advanced Ceramics and Glasses Symposium at the 101st Annual Meeting of the American Ceramic Society, pp. 185-199; 1999 April 25-28; Indianapolis, IN. The American Ceramic Society, 1999.

Jacobs S. D., and Arrasmith S. R., Development of New Magnetortheological Fluids for Polishing of CaF_2 and KDP. LLE Review 1999; 80: 213-219.

Jacobs, S. D. Nanodiamonds Enhance Removal in Magnetorheological Finishing. Finer Points 1995; 7: 47–54.

Kaller A. Properties of polishing media for precision optics. Glasstech Ber. Glass Sci. Technol. 1998; 71: 174-183.

Kozhinova I., Jacobs S., Arrasmith S., and Gregg L., Corrosion in Aqueous Cerium Oxide Magnetorheological Fluids. Technical Digest for the Optical Society of America Topical Meeting on Optical Fabrication and Testing, pp. 151-153; 2000 June 18-22; Quebec City, Canada. Optical Society of America (OSA), 2000.

Shorey A. B., Jacobs S. D., Kordonski W. I., and Gans, R. F., Experiments and Observations Regarding the Mechanisms of Glass Removal in Magnetorheological Finishing. Appl. Opt. 2001; 40: 20-33.

Shorey A. B., Kwong K. M., Johnson K. M., and Jacobs S. D., Nanoindentation Hardness of Particles used in Magnetorheological Finishing (MRF). Appl. Opt. 2000; 39: 5194-5204.

Shorey A. B., Gregg, L. L., Romanofsky H. J., Arrasmith S. R., Kozhinova, I. A., and Jacobs S. D., Material Removal During Magnetorheologial Finishing. Proceedings of the SPIE 3782: Optical Manufacturing and Testing III, pp. 101-113; 1999 July 20-23; Denver, CO. SPIE, 1999.

STUDY OF MAGNETIC ABRASIVE FINISHING PROCESS USING PULSED FINISHING PRESSURE AND ITS PROCESSING CHARACTERISTICS

Shaohui YIN, Takeo SHINMURA

Graduate School of Engineering, Utsunomiya University, 7-1-2 Yoto, Utsunomiya, Tochigi 321-8585, Japan

Abstract

New vibration-assisted magnetic abrasive finishing processes using pulsed finishing pressure in which the workpiece is vibrated in a vertical or elliptical direction to the machined surface are proposed in order to develop high-efficiency and high-precision finishing technology for 3-dimensional curve surfaces of micro components. The plane-finishing characteristics of the processes are described in this paper. The vertical vibration-assisted finishing process leads to a higher removal rates but a little rougher surface than conventional process. However, the elliptical vibration-assisted finishing process results in a good finish and high removal rates simultaneously.

Keywords

Magnetic abrasive finishing, Pulsed finishing pressure, Surface roughness, Stock removal, Vertical vibration-assisted, Elliptical vibration-assisted

1. INTRODUCTION

Conventional magnetic field-assisted finishing is one technique for generating a good finish and very little surface damage to the components due to the extremely low level of approximately constant finishing pressure [1, 2], Unfortunately, its finishing efficiency is low in machining 3-D complicated micro structures. It is necessary to improve the efficiency and machinability in finishing 3-D micro curve surface [3]. Changing the finishing pressure wave is a viable alternative. When the workpiece is vibrated in a vertical or elliptical direction to the machined surface, finishing pressure near the pulse shape is produced, and it is expected that the finishing performance will be considerably different from that in the case of constant finishing pressure. In this paper, we propose vertical vibration-assisted mode (Z mode) and elliptical vibration-assisted mode (ZX mode) using pulsed finishing pressure. The differences between plane finishing characteristics in the case of pulsed finishing pressure and constant finishing pressure were described; the effects of the elliptical vibration-assisted mode on finishing performance were examined.

2. PROCESSING PRINCIPLE AND EXPERIMENTAL SETUP

Figures 1 and 2 show schematic and external views of the experimental setup respectively. The finishing apparatus is installed on a table of a vertical milling machine comprising an electromagnetic coil, core, upper yoke, magnetic pole, workpiece, vibration table, X-Y stage, and lower yoke. The workpiece can be vibrated in vertical (Z), horizontal (X), or elliptical (ZX) directions by the use of two motorized cam systems. When mixed type magnetic abrasives are introduced into the gap between the tip of the magnetic pole and the workpiece, the abrasives are magnetically attracted by the magnetic pole and rotate with the magnetic pole. Relative motion is obtained consisting of vibration and rotation between the magnetic brushes and workpiece. The abrasives remove material acted by pulsed finishing pressure which is determined by magnetic force F_m, F_h, and vibrating force F_{vz}, F_{vx} acting on the abrasives against the machined surface.

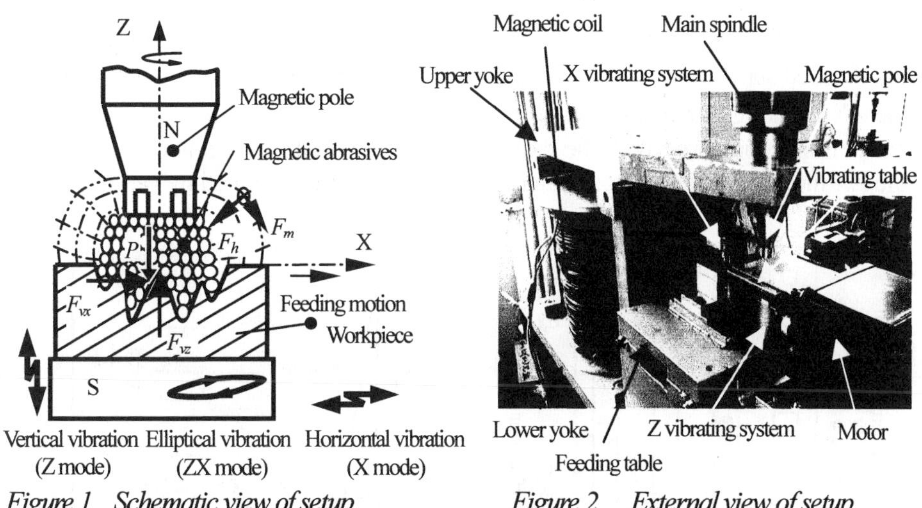

Vertical vibration   Elliptical vibration   Horizontal vibration
(Z mode)   (ZX mode)   (X mode)

Figure 1   Schematic view of setup   *Figure 2   External view of setup*

3. EXPERIMENTAL RESULTS AND DISCUSSIONS

3.1 FINISHING CHARACTERISTICS OF THREE KINDS OF VIBRATION MODES

Figure 3 shows the differences between surface roughness (Ry) and stock removal (M) in the case of pulsed finishing pressure (Z mode, ZX mode) and in the case of constant finishing pressure (X mode and non-vibration mode), in Figure 3, 'Non' 'ZX' 'Z' 'X' express non-vibration mode, ZX mode, Z mode, X mode respectively. It can be seen that stock removal in the case of X, Z and ZX modes are much higher than non-vibration mode. In the X mode, with X directional vibration-assisted, abrasive cutting edges scratch the material in many directions in the intermit cutting mode, do not trace the same path on the surface. This result in the crossing of the cutting marks and increase stock removal per working distance. In the Z mode, pulsed finishing pressure is produced, the mean value and the peak of finishing pressure increase remarkably. Owing to demonstrate effects of pulsed vibration cutting, and enhance behavior of deformation of the magnetic brushes, it results more higher stock removal than that in the X mode. In the ZX mode, the higher stock removal is attributed to the combined effect of pulsed finishing pressure from Z directional vibration and cross cutting effects from X directional vibration. Although it demonstrates cross cutting effects, the finishing pressure is lower than that in the Z mode. Therefore, there is not a great difference between stock removal in the ZX mode and that in the Z mode.

Without vibration assistance, the finished surface of 4.5 μ mRy before finishing gather to 0.47 μ mRy. In the Z mode, the finished surface gathered to 0.68 μ mRy, the machined surface becomes a rough surface somewhat. This is because cutting depth increase acted by higher finishing pressure. In the X mode, the finished surface was improved to 0.33 μ mRy, resulting cross cutting effects. In the ZX mode, the machined surface was improved near to 0.33 μ mRy, generate better finish compared with the case of the Z mode and non-vibration modes. In the ZX mode,   different cross-hatching patterns can be generated by varying the

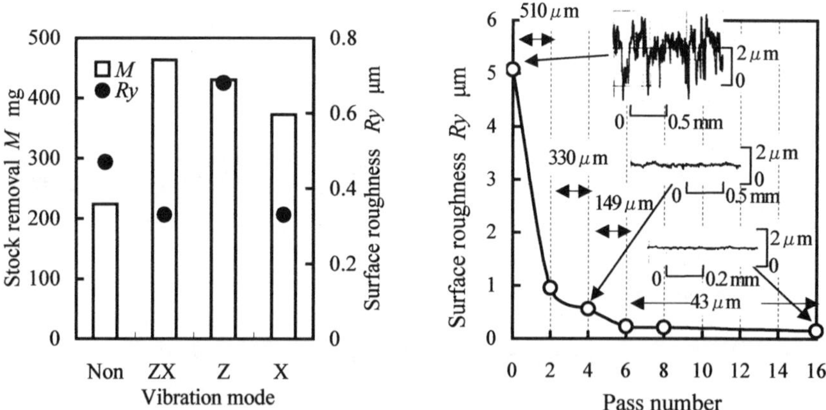

Figure 3   Effects of vibration mode   *Figure 4   Effects of many stages finishing*

[Conditions] Workpiece: Brass C2680, 38✕60 mm, thickness : 1.2 mm; Magnetic pole: SS400, 12 mm in dia., with six crossing grooves(width : 1mm, depth : 2 mm), Circumferential speed : 32 m/min; Clearance: 3 mm; Magnetic flux density: 0.7T; Pass number: 6; Stroke: 25 mm; Vibration amplitude: 3.5 mm in X direction, 1 mm in Z direction; Vibration frequency: 6 Hz in X and Z directions; Mixed type magnetic abrasives: Iron particles(4g supplied,330 μ m in mean dia.) and WA magnetic abrasives(1g supplied ,80 μ m in mean dia.); Feeding speed: 17mm/min; Lubricant: Straight oil grinding fluid(2 mL).

rotational speed and elliptical vibration. The effects of cross cutting pattern result in a reduction in surface roughness. The results of this experiment show that the ZX mode can overcome to some extent the shortcomings that lead to a rougher surface resulting from the Z mode and produce a better finish.

3.2 EFFECTS OF SOME PARAMETERS ON FINISHING PERFORMANCE IN THE ELLIPTICAL VIBRATION -ASSISTED PROCESS

In the elliptical vibration-assisted process, the stock removal and the finish generated depend on the rotational speed, elliptical vibration, magnetic abrasives and magnetic flux density. In the following, effects of some of the parameters on the surface roughness and stock removal will be discussed.

In order to test the finishing ability of the ZX mode, we carried out experiments using different kinds of iron particles during different stages. As shown in Figure 4, during the first 2 strokes, iron particles of $510\,\mu$m in mean diameter were used, the surface roughness decreased sharply to $0.96\,\mu$mRy, thereafter, $330\,\mu$m iron particles were used during the second 2 strokes, to $0.56\,\mu$mRy, $149\,\mu$m iron particles during the third 2 strokes, to $0.23\,\mu$mRy, thereafter, $43\,\mu$m iron particles were used, after the fourth 2 strokes, to $0.2\,\mu$mRy, thereafter, $0.15\,\mu$mRy (20 nmRa) fine surface was obtained. The results of this experiment show that this process has high finishing ability to obtain a mirror-like surface in a short time.

Figure 5 shows changes in surface roughness and stock removal with vibration frequency when the magnetic pole rotates at a low speed (3 m/min) and at a high speed (32 m/min). In the case of 3 m/min, stock removal is lowest with no vibration and increases remarkably with increasing vibration frequency. This is because cross cutting action mainly resulting from X directional vibration is enhanced with increase in vibration frequency. This phenomenon is difficult to be observed when the magnetic pole rotates at a high speed (32 m/min). The magnetic pole rotates at a high speed, enhance rotation cutting effect, reducing cross cutting effect resulting from X directional vibration. On the other hand, it can be seen that vibration frequency has little influence on surface roughness.

Figure 6 shows the influence of circumferential speed of the magnetic pole on stock removal and surface roughness. It can be seen that stock removal increases almost as a parabola with increase in circumferential speed when $V \geqq 3$ m/min. This is because the number of magnetic abrasives passing one point on the surface of the workpiece each unit time increases with increase in the circumferential speed. When $V \leqq 3$ m/min, stock removal is smaller than that in the case of $V = 0$. When the magnetic pole is static, finish action is generated as a result of X directional vibration. When $V \leqq 32$ m/min, stock removal increases

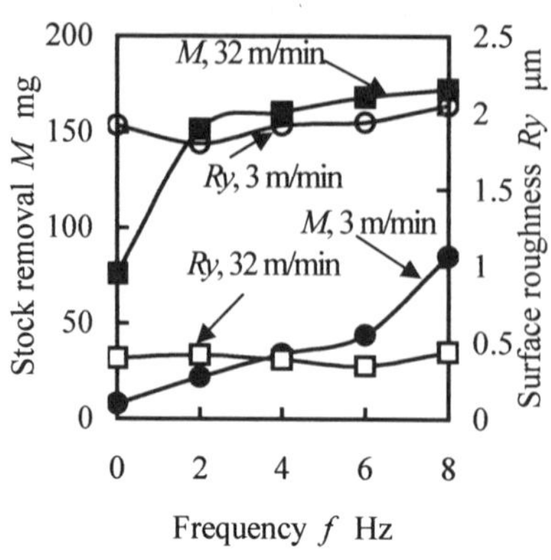

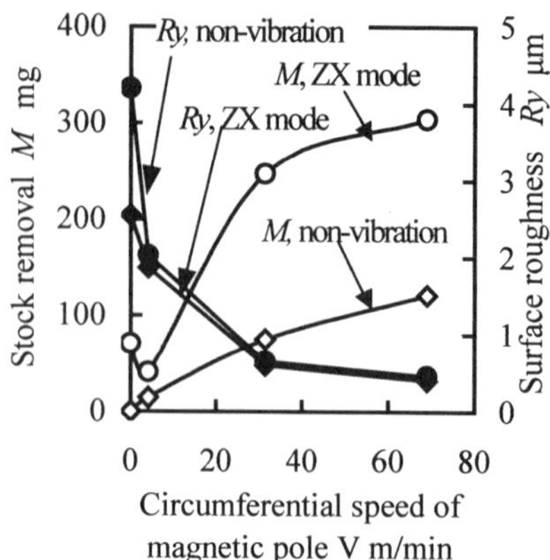

Figure 5   Effects of vibration frequency   *Figure 6   Effects of circumferential speed of magnetic pole*

[Conditions]   Pass number: 2; other conditions see Figures 3 and 4

more rapidly than that in the case of non-vibration. The value of surface roughness becomes smaller as the circumferential speed increases. Therefore, it is necessary to maintain a high-speed rotation to improve the finishing efficiency and obtain a good finish.

4. CONCLUSIONS

The conclusions of this report are summarized as follows.

1) The vertical vibration-assisted finishing process leads to higher finishing efficiency but a little rougher surface than conventional process.

2) The elliptical vibration-assisted finishing process combining some advantages of X and Z directional vibrations can obtain high finishing efficiency and a good finish simultaneously. The surface roughness (*Ra*) of brass plane can be finished to about 20 nm.

3) In the elliptical vibration-assisted finishing process, increasing circumferential speed of the magnetic pole was found to increase the rate of finishing as the best finish attainable. Increasing vibration frequency increase the rate of finishing when the magnetic pole rotate at a lower speed, but almost not increase in the case of high speed.

REFERENCES

[1] Fox M., Agrawal K., Shinmura T., Komanduri R.. Magnetic abrasive finishing of rollers. Annals of CIRP, 43/1/1994: 181-84.

[2] Shinmura T., Takazawa K., Hatano E.. Study of Magnetic abrasive finishing. Annals of CIRP, 39/1/1990: 325-28.

[3] Shinmura T., Takeshi Y.. Study of free form surface finishing by the application of magnetic abrasive machining. Proceeding of JSME annual meeting (IV) (in Japanese) 1997: 220-221

CHAIN FORMING PROCESS OF MAGNETIC BRUSH DEVELOPMENT SYSTEM USED IN LASER PRINTER

Nobuyuki Nakayama, Hiroyuki Kawamoto, Makoto Yamaguchi
and Janjomske Wiphut

Department of Mechanical Engineering, Waseda University

Abstract

Experimental, numerical, and theoretical investigations have been carried out on chain forming process of magnetic carrier beads in a magnetic brush development system of a laser printer. It was clarified that the chain length depends on both the density of magnetic carrier beads and the magnetic flux density and that the length is determined to minimize the total potential energy that consists of the gravitational and magnetic potential energy. These characteristics were also confirmed by the numerical calculation with the Distinct Element Method.

Keywords

Laser Printer, Magnetic Development, Electrophotography,
Distinct Element Method, Carrier Beads

1. INTRODUCTION

A magnetic brush development system in a laser printer is shown in Figure 1. Magnetic carrier beads with electrostatically attached toner particles are introduced into the vicinity of a rotatory sleeve with a stationary magnetic roller inside it (William, 1993). Carrier beads form chains on the sleeve by virtue of the magnetic field. Tips of chains touch the photoreceptor surface at the development area and toner particles on chains move to electrostatic latent images on the photoreceptor to form real images.

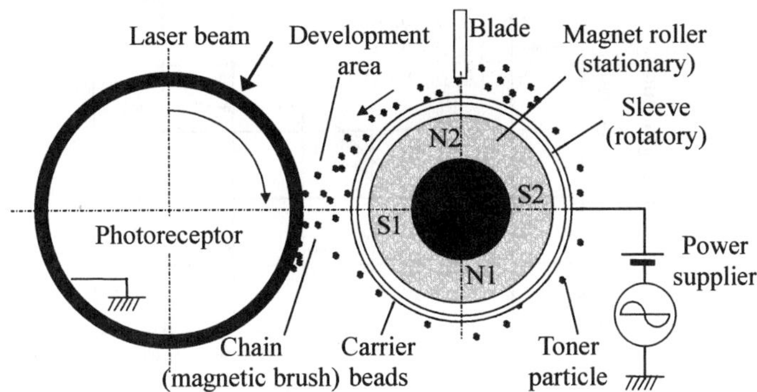

Figure 1. Magnetic brush development system in a laser printer.

Carrier chains play an important role in this development system. Therefore, in order to realize high quality imaging, it is necessary to clarify the relationship between kinetic characteristics of formed chains and design parameters, such as magnetic flux density and properties of carriers. In this study, experimental, numerical, and theoretical investigations have been carried out with respect to the chain forming process.

2. EXPERIMENT

2.1 Experimental Method

An experimental setup and an example of observed chains are shown in Figure 2. Carrier chains in the magnetic field created by a solenoid coil were observed by a digital camera and chain lengths were measured on the obtained image. Spherical carriers made by the polymerization method with 88.1 μm in diameter, 3550 kg/m^3 in volume density, and 4.34 in relative magnetic permeability (Toda Kogyo Corp. Hiroshima, Japan) were used for the experiment. Axial magnetic flux density B' along the center axis of the coil is approximated by $B'(z) = B_0(1-cz)$, where B_0 and c (= 66.87) are constants and z is the axial coordinate. B_0 is proportional to the coil current with a proportional constant 0.006156 T/A.

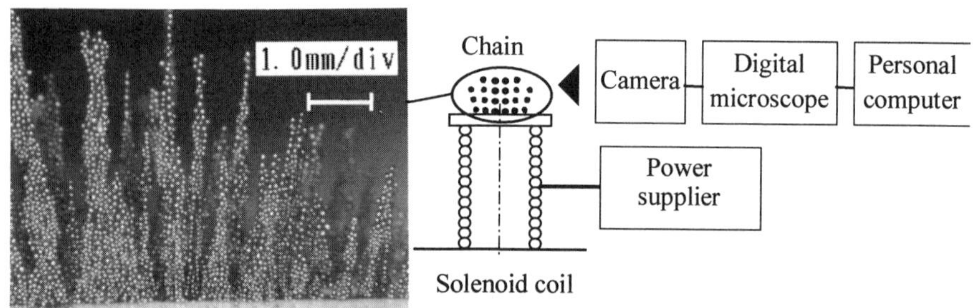

Figure 2. Experimental setup and carrier chains.

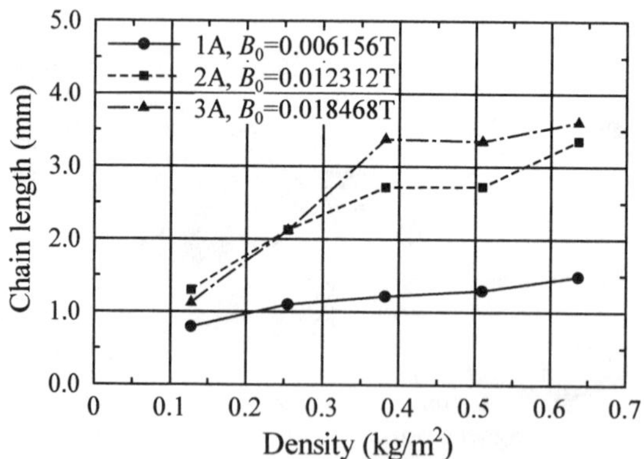

Figure 3. Relation between chain length and the density of carriers.

2.2 Experimental Results

Figure 3 shows relationships between the measured chain length at the center of the coil and the density of carriers, weight of carriers per unit area, on the coil for various magnetic flux densities. The chain length increases with the increase in the density of carriers and the magnetic flux density.

3. NUMERICAL SIMULATION
3.1 Numerical Method

Two-dimensional Distinct Element Method (DEM, Cundall, 1979) was used in the numerical simulation of chain forming process. In the calculation the momentum equations are solved with three degrees of freedom for each particle. In this study, mechanical interaction force, magnetic force, drag due to air flow, and gravity are considered in the equations. The effect of rolling friction and van der Waals force are neglected. The magnetic force F_j and rotational moment M_j of the j-th particle with the magnetic moment m_j are given by the following expressions under the assumption that each particle behaves as a magnetic dipole placed at the center of the magnetized particle (Paranjpe, 1986).

$$F_j = \left(m_j \cdot \nabla\right)B_j, \quad M_j = m_j \times B_j. \tag{1}$$

The magnetic moment m_j and magnetic flux density B_j at the position of the j-th particle are

$$B_j = B_j' + \sum_{\substack{k=1 \\ j \neq k}}^{N} \frac{\mu_0}{4\pi}\left(\frac{3m_k \cdot r_{kj}}{\left|r_{kj}\right|^5} r_{kj} - \frac{m_k}{\left|r_{kj}\right|^3}\right), \tag{2}$$

$$m_j = \frac{4\pi}{\mu_0}\frac{\mu-1}{\mu+2}\frac{a_j^{\,3}}{8}B_j, \tag{3}$$

where N is the number of the particles, μ_0 is the permeability of free space, μ is the relative permeability of particles, a_j is the diameter of the j-th particle and r_{kj} is the position vector from the k-th to the j-th particle. The first term in the right hand side of Equation (2) is the applied magnetic field by the coil and the second term is the field at the j-th particle due to dipoles of other particles.

3.2 Numerical Results

Simulated chain profile is shown in Figure 4. Initial condition of the calculation was that particles were placed at random positions in the area with 5 mm in width and 16 mm in height. Numbers of the particles were estimated from the density of carriers in the experiment. It is clearly recognized that the calculated profile was similar to that in the experimental observation shown in Figure 2.

Calculated dependency of maximum chain lengths on the number of particles and the magnetic flux density is plotted in Figure 5. Calculated results qualitatively agreed with the measured results in Figure 3. That is, the chain becomes longer with the higher density of carriers in the field of higher magnetic flux density. However, the simulated results give smaller values of the chain length than the experimental results. One of the possible reasons is that the particles are apt to fall off the chains in the simulation because of the neglected rolling friction. In addition to consideration of the effect of the rolling friction, three-dimensional DEM calculation is also necessary to improve the quantitative accuracy.

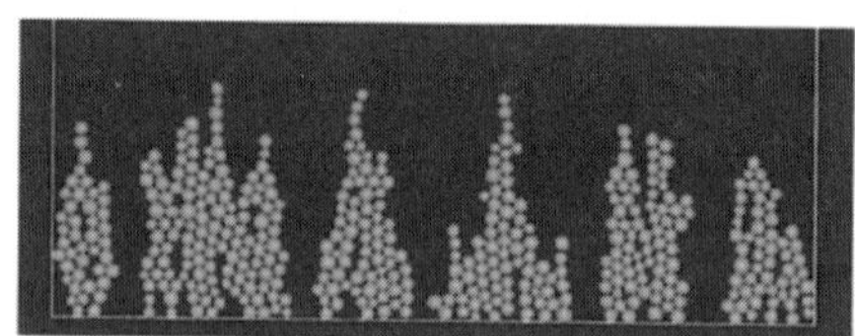

Figure 4. Numerically simulated profile of chains.

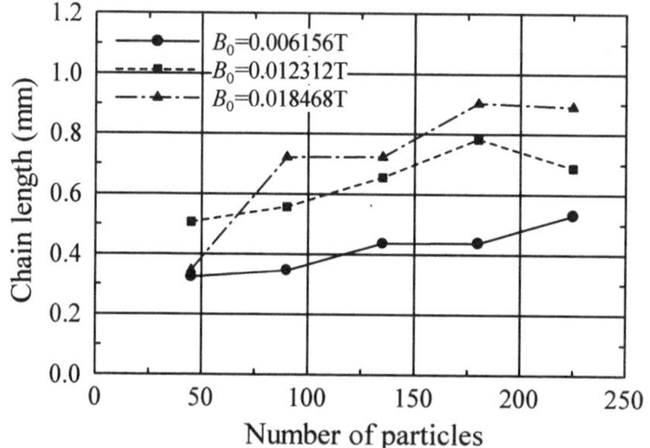

Figure 5. Calculated relation between chain length and number of particles.

4. THEORETICAL DISCUSSION

Chain lengths are to be determined to minimize its total potential energy. The total potential energy is given by the sum of magnetic energy U_m expressed by Equation (5) and gravitational energy.

$$U_m = -\frac{1}{2}\sum_{j=1}^{N} m_j \cdot B_j{}'. \tag{5}$$

It is assumed that the particles are connected along a straight line in vertical direction. On this assumption, the relation between the chain length and the energy was calculated to estimate two kinds of the most stable chain length. The length l_t is to minimize *total* energy and the length l_a is to minimize

514

average energy.

The relations between the chain lengths and coil current are plotted in Figure 6. The symbols show the experimental results and the lines are theoretically estimated results. Measured values are close to l_a with lower density of particles; on the other hand, the values are close to l_t with higher density of particles.

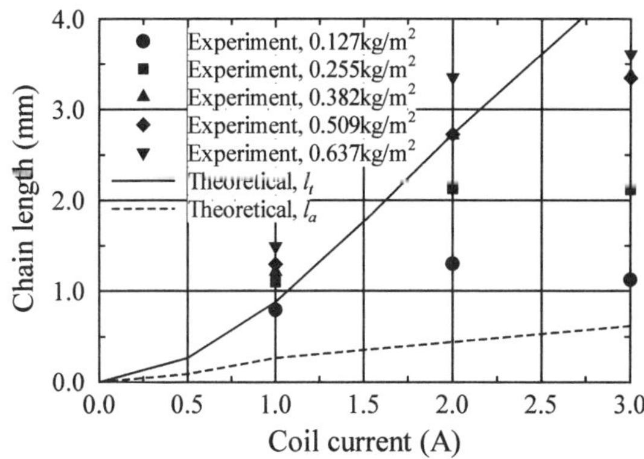

Figure 6. Calculated and measured chain length.

5. CONCLUSIONS

Experimental investigations, DEM simulations, and thermodynamic discussion were carried out on chain forming process of a magnetic brush development system in a laser printer. It was clarified that the chain lengths are determined to minimize the total potential energy that consists of the gravitational and magnetic potential energy and that the lengths depend on both the density of carriers and magnetic flux density. The process was simulated by the two-dimensional DEM simulation that gives qualitatively consistent result with the experimental one, but improvement of the model is necessary to predict quantitative characteristics of the process.

The authors would like to express their thanks to the Ministry of Education, Science and Culture of Japan for the financial support and Toda Kogyo Corp. for supplying carriers.

6. REFERENCES

Williams, Edgar M., *The Physics and Technology of Xerographic Processes*. Florida: Krieger Publishing, 1993.

Cundall, P.A. and Strack, O.D.L. A discrete numerical model for granular assemblies. Géotechnique 1979; 29-1: 47-65.

Paranjpe, R.S. and Elrod, H.G. Stability of chains of permeable spherical beads in an applied magnetic field. Journal of Applied Physics 1986; 60-1: 418-422.

Part III

Machine / element / measurement

PRECISION MACHINE TOOLS

Manfred Weck, Robert Hilbing, Christian Peschke

Abstract

Conventional cutting processes are highly qualified for the manufacturing of micro and precision parts. The paper describes the various parameters which have influence on the work piece quality and gives a review of different machine tools specially designed for high precision applications.

Keywords
Micro-structuring, micro-engineering, micro-machining

1 INTRODUCTION

Nowadays, micro-engineering is classified as an interdisciplinary key technology of the 21st century. The term micro-engineering covers an area of applications which extends from highly integrated mechanical and electrical systems with dimensions in the sub-micron range to macroscopic components which rely on their surface structure for their ability to perform special functions.

Both conventional metal cutting and micro-electronic manufacturing techniques can be applied successfully to manufacture micro-systems and micro-structured components. The great advantage of cutting operations, lies in the wide range of materials which can be machined and in the capacity for free, three-dimensional shaping. Consequently, these techniques are highly suitable for the manufacture of micro-injection mould and micro-embossing cavities or optical components.

The production machines, tools and working environment must have very special properties if the very high requirements in terms of form integrity and surface roughness as well as the extremely small dimensional deviations in micro-engineering, are to be achieved reliably.

2 DEMANDS OF MICROCUTTING

The manufacture of micro-components or micro-structures, imposes exacting requirements on each parameter which has any influence on the manufacturing process. In addition to giving due consideration to ambient condition, e.g. temperature constancy in the operating environment, it is vital to ensure that aspects such as the material quality of the workpiece, tool cutting edge geometry, machine tool and the manufacturing parameters are fully harmonised with one another and optimised. These factors are decisive in determining the smallest possible geometries of the parts to be manufactured (*Fig.1*).

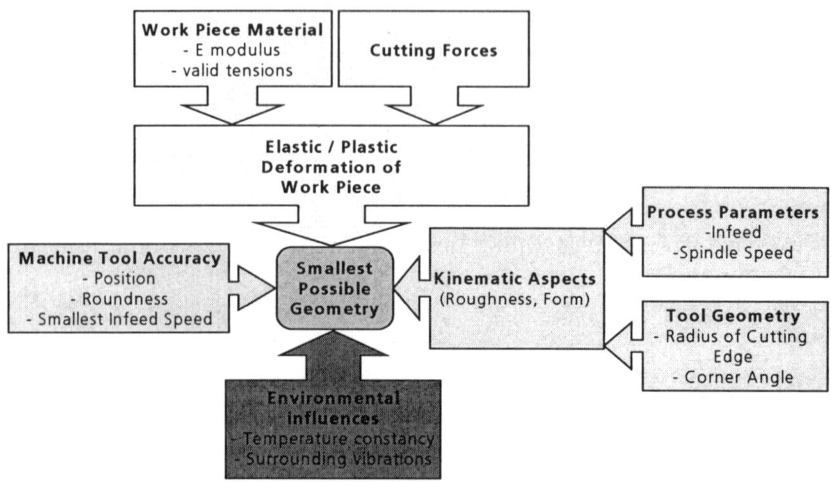

Fig. 1: Smallest possible geometry of manufacturing

The main problem in this context, is the accuracy of the machine tool and the quality of the tool. In micro-machining, it is absolutely essential to use precision machine tools with the highest levels of geometric and kinematic accuracy and thermal stability. It is also vital to ensure that the tool can be centred and positioned accurately, e.g. in a clamping device with integrated tool measuring facility. This applies equally to the clamping facility for the workpiece, which must be aligned with the tool with the utmost precision.

3 PRECISION MACHINE TOOLS

A clear development trend has emerged in the selection of the components largely responsible for the accuracy of precision machine tools. In the mean time, linear motors or ballscrews with high-precision DC motors, are used as feed drives in the vast majority of ultra precision machine tools. Due to the increased mechanical work involved, the use of friction drives or hydrostatic threaded spindles which also ensure excellent positioning, remains justifiable only in special cases. The trend in the selection of position measuring techniques, is currently moving away from the expensive laser interferometers to the more economical linear scales. The resolution of the linear scales has improved so much in the mean time that the measuring accuracy is close to that achieved by laser interferometers. Both the assembly work required and susceptibility to errors of the linear scales, are also considerably lower than those recorded when laser interferometers are used. In modern precision machines, the linear guides and the main spindles usually have hydrostatic bearings. This type of bearing system guarantees much greater rigidity than aerostatic bearing systems and, by virtue of its good absorbing characteristics, permits higher levels of process forces. This, in turn, permits operations such as grinding or hard fine machining to be conducted using CBN tools [FIS00].

When ultra-fine machining operations are performed using a geometrically defined cutting edge, the degree of manufacturing accuracy and the minimum structure size which can be achieved, depend primarily on the quality of the tool used. The high levels of hardness, wear resistance and thermal conductivity of mono-crystalline diamond, make this an optimum material for operations of this nature. The cutting edge rounding can be lapped to virtually atomic orders, enabling notch and groove free cutting edges to be manufactured.

The three-axis ultra precision machine UPM developed at the Fraunhofer IPT, is a good example of a micro processing machine tool (*Fig. 2*).

Fig.2: Ultra precision machine UPM (Fraunhofer IPT)

Micro-components and parts with structured surfaces can be manufactured on this machine either in conventional turning and milling or in fly-cutting and fast tool turning modes.

Following very precise tool measurement using the integrated optical measuring system, the tool cutting edge can be positioned in the working area of 250 mm^3 with sub-micron accuracy [TÖN01]. High-precision drive systems feed the air-bearing axes - the x-axis is equipped with a friction drive and the z-axis has a precision ballscrew. At the maximum rotational speed of 80,000 rpm, the air-bearing high speed spindle reaches a radial rigidity level of 30 N/μm and permits machining operations to be performed with a maximum radial error motion of less than 100 nm [WEC00]. Vibration-related and thermal problems are also minimised by the heavy, temperature-insensitive natural granite machine foundation in order to ensure the best possible manufacturing outcome. Pneumatic vibration absorbers are used, protecting the machine from ambient influences such as building vibration.

The Robonano Ui, manufactured by Fanuc Ltd. is a further example of an ultra precision machine tool (*Fig. 3*).

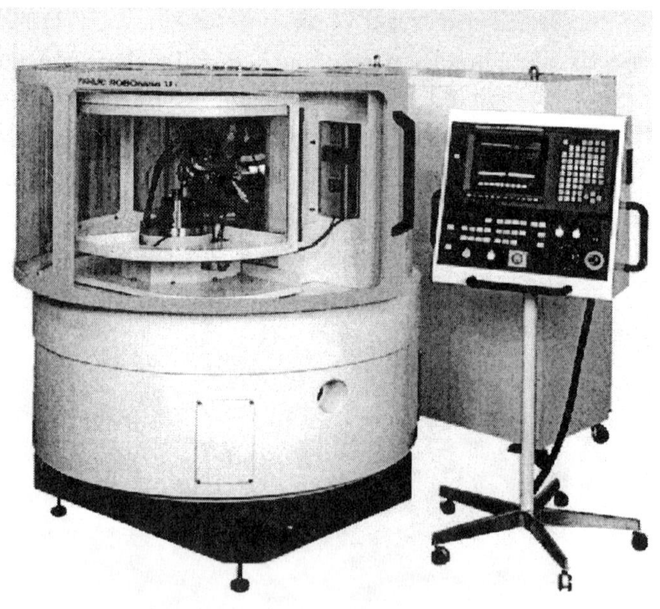

Fig. 3: Ultra precision machine tool Robnano Ui (Fanuc Ltd.)

Like the ultra precision machine tool UPM built by the Fraunhofer IPT, the Robonano Ui can be used for turning and milling operations. It is designed to manufacture three-dimensional shapes such as aspherical lenses and diffractive optics with the highest levels of precision. Surfaces or structures of this nature, are extremely difficult to manufacture using lithographic manufacturing techniques. Where highest precision is required in micro-milling operations, a mono-crystalline diamond is used as a tool on the air-bearing and compressed-air-driven high-speed spindle. The compressed air spindle permits rotational accuracy below 0.02 μm to be achieved at low rotational speeds. The rotational speed of this spindle can be set at values of between 50,000 rpm and 100,000 rpm.

The tool and the workpiece are positioned with a resolution of 1 nm using the linear axes which have completely air bearing linear axes. The x-axis and the z-axis have travel distances of 200 mm and 120 mm respectively.

4 SUMMARY

It is apparent that the machine concepts and the components selected for precision machine tools produced by various manufacturers, are very similar. This is attributable to the very highly specialised specifications which must be met when micro-components or micro-structured surfaces are produced. Certain solutions such as aerostatic or hydrostatic guidances have established themselves over the years.

This development process is unstoppable and it is likely that new technologies and measuring techniques will extend the machining limitations of precision machine tools still further.

5 REFERENCES

[FIS00]   Fischer, Stephan: *Fertigungssysteme zur spanenden Herstellung von Mikrostrukturen*. Aachen: Shaker, 2000

[TÖN01]   Tönshoff, H.K., Inasaki, I.: *Sensors Applications Volume 1: Sensors in Manufacturing*. Weinheim: Wiley-VCH, 2001

[WEC00]   Weck, Manfred, *Ultraprecision Machining of Microcomponents, in Proceedings / International Seminar on Precision Engineering and Micro Technology*. Voerde: Rhiem, 2000

EFFECTS OF BEARING SURFACE GEOMETRIES ON THE INCLINATION STIFFNESS OF AEROSTATIC THRUST BEARINGS

Susumu Ohishi*, Katsutoshi Tanaka**, Tetsuya Masada*

*Aoyama Gakuin University, Tokyo, Japan
**Toshiba Machine Co. Ltd., Numazu, Japan

Abstract

This paper describes the effects of bearing surface geometries on the inclination stiffness of aerostatic annular thrust bearings with inherently compensated restrictors (annular orifices). Guidelines are presented to improve the tilt stiffness by modification of the bearing clearance profile, and a desirable profile is proposed.

Keywords

Aerostatic thrust bearing, Inclination stiffness, Bearing surface geometry, Finite element analysis

1. INTRODUCTION

Aerostatic bearings or externally pressurized air bearings can be used today for many applications, even in high load conditions, due to the establishment of good design principles [Powell 1970, Shires 1968], the improvement of machining accuracy, and the development of various types of restrictors and compensation methods [Brzeski 1979]. For example, the development of groove compensation restrictors, which are shallow grooves machined on the bearing surface to act as restrictors, has led a considerable increase in the stiffness in terms of parallel movement. However the inclination or tilt stiffness is still considered to be low and remains to be studied in depth [Majumdar 1973].

The objective of this study is to clarify analytically and experimentally the effects of bearing surface geometries, i.e., bearing clearance profiles on the inclination stiffness of annular thrust bearings with annular orifices. These are drilled directly into the bearing surface and also called plain orifices or inherently compensated restrictors. The objective is then to suggest a desirable clearance profile contributing to higher stiffness without adding any costly compensators.

2. FINITE ELEMENT ANALYSIS

The following compressive Reynolds equation has been formulated and a computer program developed in order to obtain the pressure distribution of the air film in the bearing clearance space.

$$\frac{\partial}{\partial x}\left(\frac{h^3}{\mu}\frac{\partial p^2}{\partial x}\right) + \frac{\partial}{\partial y}\left(\frac{h^3}{\mu}\frac{\partial p^2}{\partial y}\right) = 12U\frac{\partial(\rho h)}{\partial x} + 12V\frac{\partial(\rho h)}{\partial y} \tag{1}$$

where $h(x,y)$=bearing clearance, ρ=density, μ=viscosity, p=pressure and U,V=sliding speed in x and y directions (U,V=0). The mass flow rate per unit width in bearings is given by (suffix a denotes ambient)

$$q = \vec{n}\cdot\left[\left(-\frac{h^3}{24}\frac{\rho_a}{p_a}\frac{\partial p^2}{\partial x} + \frac{Uh}{2}\frac{\rho_a}{p_a}p\right) + \left(-\frac{h^3}{24}\frac{\rho_a}{p_a}\frac{\partial p^2}{\partial y} + \frac{Vh}{2}\frac{\rho_a}{p_a}p\right)\right] \tag{2}$$

Full details of the theoretical treatment have been given by Ohishi [1997].

In the program, the total force on the element is given by the mean pressure multiplied by the area of the element. The moment on the element is then calculated by the force and the distance of the centroid of the element from the bearing center.

The developed program is not only capable of calculating the pressure, force, moment and flow rate for a given translational displacement and for a given inclination angle but also capable of obtaining, by iteration, the translational displacement and the inclination angle which balance the bearing force and moment against the applied force and moment.

It is important to note that the flow in the clearance space must be equated to the flow through the corresponding feed hole when the bearing surfaces tilt.

3. CONSIDERATION OF INCLINATION STIFFNESS

In aerostatic bearings, the air, as it flows from the source to the bearing exhaust, is subject to two flow resistances in series -, that of the restrictor and that of the space between the bearing surfaces. When a load is applied, the clearance between the bearing surfaces is reduced and the resistance of the air film increases relative to that of the restrictor. This results in an increase in pressure in the film which contributes to the restoring force. Thus, the bearing has stiffness.

Mass flow through a feed hole is given by the equation [Powell 1970]

$$m = C_D A \rho_s \sqrt{2RT_s} \cdot F\left(\gamma, \frac{P_d}{P_s}\right) \tag{3}$$

where γ is the specific heat ratio and suffices s and d denote at the inlet and at the outlet of the feed hole respectively. The coefficient of discharge C_D and the function F vary with pressure ratio. It should be noted that the area A for an annular orifice, which is a primary concern in this study, is a function of the bearing clearance.

Two thrust bearings are generally employed to support load in both directions. The inclination induces a restoring moment ($M_1 + M_2 - M_3 - M_4$)

as shown schematically in Figure 1. The moment is determined by the pressure profiles, which are very sensitive to the flow in the bearing clearance as can be seen from eq. (2). It is obvious from eqs. (2) and (3) that pressures are higher in smaller

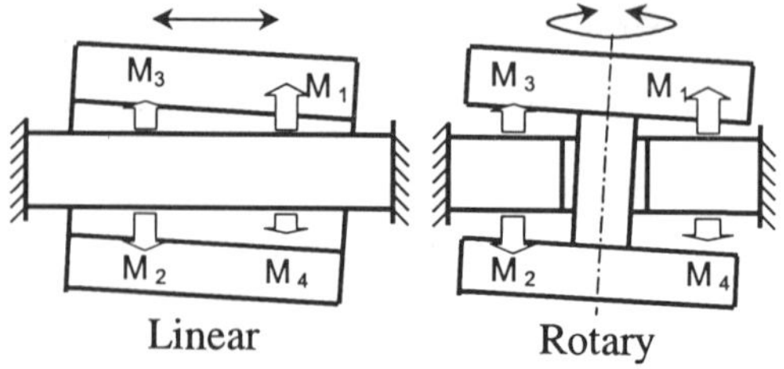

Figure 1. Combination of two thrust

clearances. However, when the bearing surfaces tilt, the higher pressures in regions with smaller clearance decrease because flows towards regions with larger clearance occur. This is clearly demonstrated in Figure 2, which is an

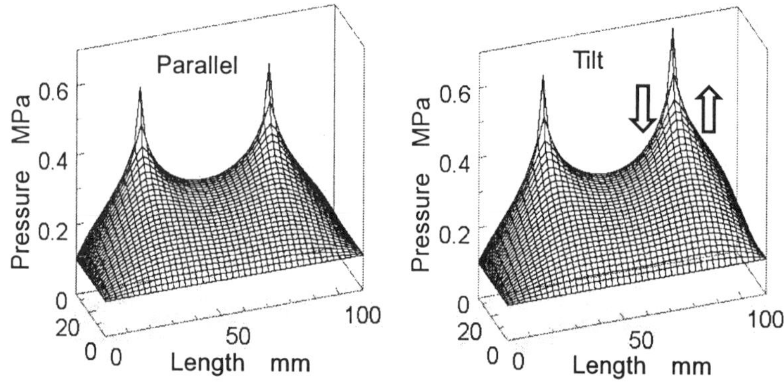

Figure 2. Pressure profiles by FEM

example of FEM calculations for a rectangular thrust bearing with two annular orifices. As the bearing surface tilts, the pressure increases in the region where the clearance becomes smaller towards the exhaust boundaries (upward arrow in Figure 2), but it reduces in the region where the clearance enlarges towards the exhaust boundaries (downward arrow). This is because the bearing area for each orifice is not independent. Experimental and analytical results for the rectangular slider bearings with 5.2 μm mean bearing clearance shown in Figure 3 clearly indicate that a bearing with groove compensation restrictor has higher translational stiffness than

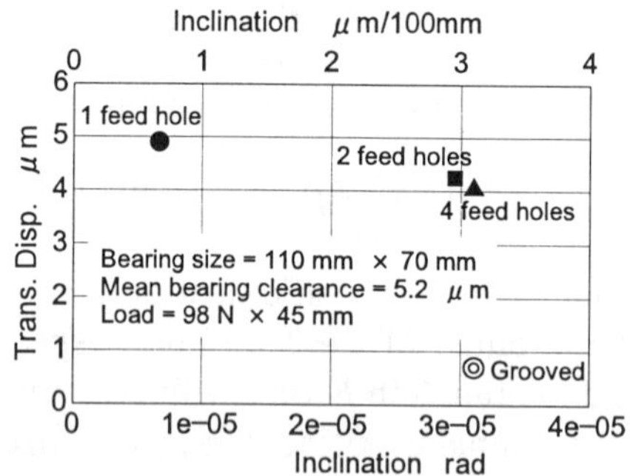

Figure 3. Stiffness comparison between orifice and grooved

other bearings with one, two and four feed holes (annular orifices). However, as far as the grooves are parallel to the tilt line, the inclination stiffness is almost the same as those with annular orifices.

From the above considerations, guidelines to improve the inclination stiffness with modifying the bearing surface geometry are:

(1) Keep the pressure around feed holes as high as possible.
(2) Avoid the influence of larger clearance on the flow in the smaller clearance.
(3) Keep the pressure difference between upper and lower bearings as high as possible.

It may generally be difficult to manufacture a desirable geometry on the rectangular bearing surface which will accommodate it to any direction of inclination. In the next section, therefore, these guidelines will be applied to annular thrust bearings for rotary motion, since the desirable surface geometry can be easily machined in this case because of their axisymmetry.

4. ANNULAR THRUST BEARINGS

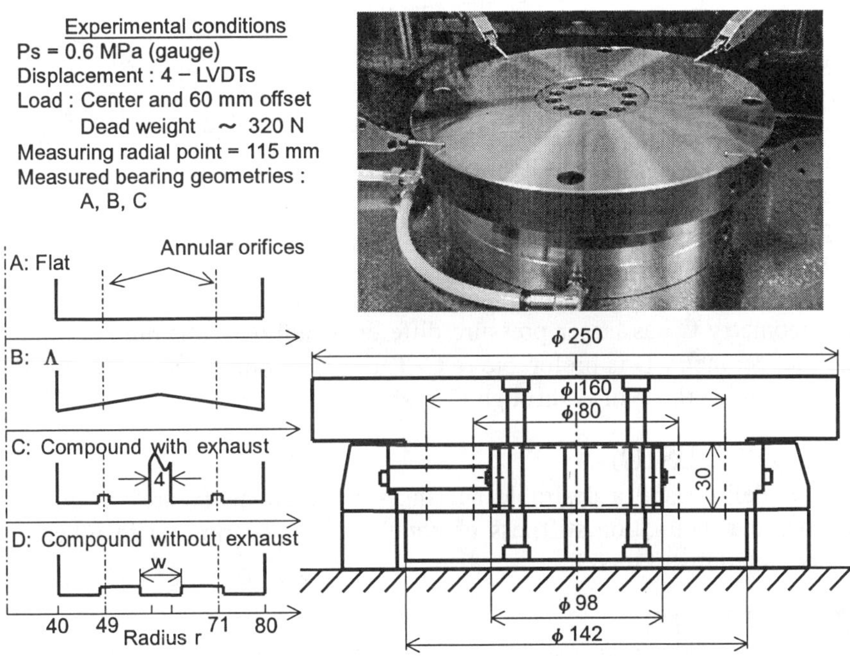

Figure 4. Surface geometries   *Figure 5*. Annular thrust bearing

Following the above guidelines, four different geometries of the bearing surface were examined as shown in Figure 4. Geometry B is expected to maintain higher pressure over a larger area. FEM analysis suggests that the height should be equal to the minimum clearance to obtain maximum

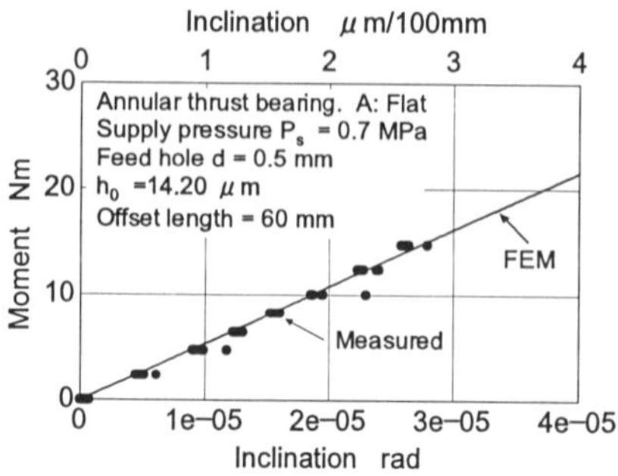

Figure 6. Inclination stiffness for A

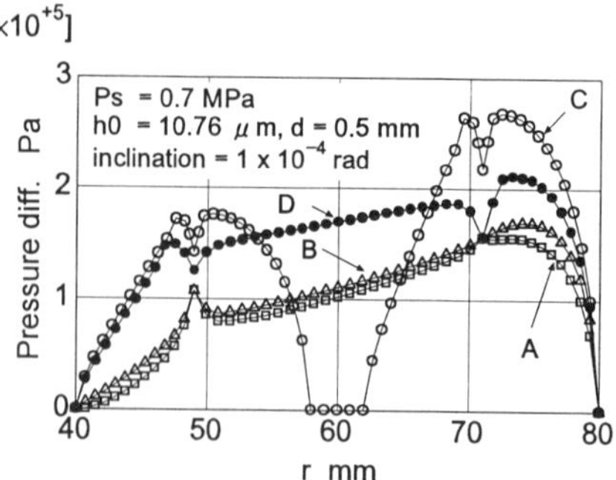

Figure 7. Pressure difference between upper and lower bearings

stiffness without instability. The bearing surface C has compound restrictors (0.5 mm annular orifices + 4 mm width $\times$ 5 μm depth groove compensations) and has an exhaust groove to avoid interference between adjacent feed holes. D is the same as C with the exception of the exhaust groove.

Figure 6 shows that the FEM is consistent with the measurements. The most important point with respect to the tilt stiffness is the pressure difference between upper and lower bearings. From the analytical results of Figure 7, it can be seen that geometry C has larger pressure difference, although the area is limited, and that geometry D is preferable to C. The stiffness of D can be expected to be about twice those of A through C as shown in Figure 8.

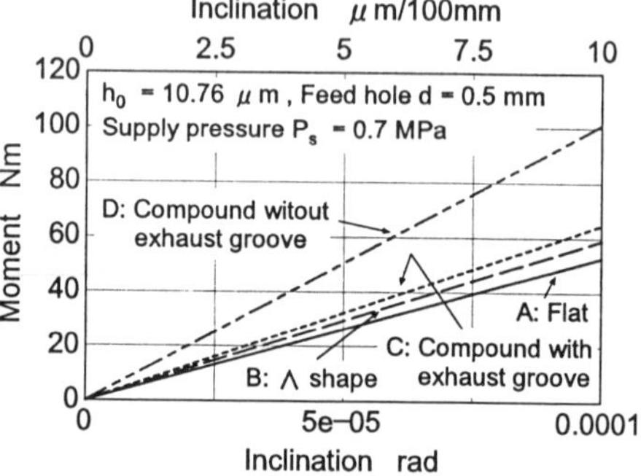

Figure 8. Comparison of bearing surface geometries

5. CONCLUSIONS

Guidelines and a desirable bearing surface geometry are presented to increase the inclination stiffness of annular thrust bearings with inherently compensated restrictors.

REFERENCES

Brzeski L., Kazimierski Z. High stiffness bearing. Trans. ASME 1979; 101:520-525.
Majumdar B. C., Singh K. C. Analysis of aerostatic thrust bearings with offset load. Int. J. Mach. Tool Des. Res. 1973; 13:65-76.
Ohishi S. Finite element analysis of air bearing characteristics. Proc. Int. Conf. Manufacturing Milestones toward the 21st Century MM21. JSME 1997:377-382.
Powell J. W. Design of aerostatic bearings. The Machinery Publishing, 1970.
Shires G. L. The design of pressurized gas bearings. TRIBOLOGY Nov. 1968:219-229.

DEVELOPMENT OF A THREE AXES TRAVELLING COLUMN ULTRAPRECISION MILLING MACHINE

Alberto Herrero, Ramón Bueno
Fundacion Tekniker
Avda. Otaola, 20.
20600 Eibar – Spain

Abstract

Precision machining is not a new field but the consequence of a top skill in existing conventional production techniques. The needs demanded to improve all products and the general tendency to miniaturisation is making all production systems' builders to invest in the R&D of the most precise manufacturing techniques.

Inside the machine tool field this research is driven towards the improvement of all manufacturing processes: grinding, turning, EDM-ing, milling, etc. The present document describes a research project in which the targeted processes are the diamond turning and the ultraprecision milling.

The project has been developed during the last 3 years and the assembled prototype can be configured as a diamond turning or a milling machine capable to reach a micron order and even sub-micrometer precision.

KeyWords
Advanced manufacturing systems, diamond turning machine, milling machine

1. INTRODUCTION

Precision machining is not a new field but the consequence of a top skill in conventional production techniques used in electronics, plastic and metal components production, etc. All these techniques, taken to the limit, constitute the group known as micro-technologies.

Considering the machining technologies, those parts that must comply with the tightest tolerances are only reachable using techniques like grinding, laser or EDM; apart from them, in the last years, the machine tool builders are developing "diamond turning lathes" and "ultraprecision milling machines" that reach micrometer and even sub-micrometer accuracy.

The present document presents the development of a 3-axes milling machine combining the experience in two close fields: machine tool design and metrology equipment design.

2. MACHINE DEVELOPMENT

The original idea of the project was to analyse several technologies in order to assemble a diamond turning machine designing the X and Y axes movement according to this concept (fig. 1).

Lathes are interesting for many applications and the diamond turning machines are offered as a commercial option by many companies. These machines are capable to produce a finished surface similar to those offered by grinders and the machining speed can be much higher. For all these reasons and thinking about business and commercial opportunities the project output was defined as mentioned.

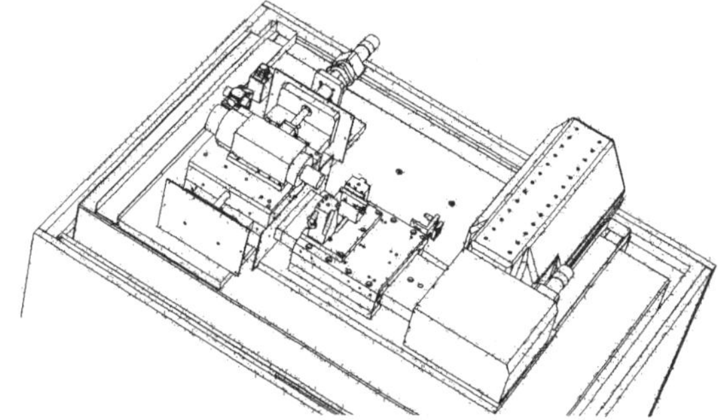

Figure 1.   Diamond turning configuration

As the project was being developed, in order to produce micro-parts, a 2-D machining system wasn't capable to produce complex pieces and the need of a 3rd axis was judged, and the project target was reconsidered: a modular machine that could be configurable as a diamond turning or as an ultraprecision milling machine.

Each configuration cannot be considered as an innovation because there are many companies and research centres that have produced such machines [Takeuchi, 2000; Weck 1997] although most of these models were conceived originally as diamond turning machines that latter were equipped with a fast-tool to provide them with a third axis. In these machines the workspace has got a strange shape very different to the cubic volume to which an operator is used.

The innovation that the project claimed was that the machine can be configured as a conventional lathe and also as a conventional (but very accurate) milling machine providing a cubic shaped working volume. The specifications that were imposed were the following:

- linear aerostatic bearings for the X, Y displacement
- laser interferometry for X, Y axes measurement
- small pieces machining (70 x 120 mm strokes are finally supplied)
- absolute precision and repeatability minor to 1 micrometer
- high speed spindle (10 000 rpm)

2.1 Components Selection

The air bearings were chosen due to the low friction, the cleanness and the enough stiffness for the negligible efforts that appear during the micromachining process.

The idea of incorporating laser interferometry into a conventional machine allows the control loop to provide better positioning and higher repeatability (2.5 nanometers resolution).

Concerning the lathe spindle, it must be capable to rotate with a very high precision avoiding any alignment error that could reduce the global accuracy. The equipped spindle can rotate up to 10 000 rpm and its run-out error is below 0.025 micrometers.

The switch to the milling configuration is performed removing the spindle of the lathe and disposing a module that consists of a vertical slider (Z axis) on which the headstock is assembled (fig. 2). The Z axis design is different to the other axes' design. It has been conceived as the vertical axis of a conventional milling machine and makes use of linear guides.

Analysing the measuring system for this axis, it was not possible to integrate a complete interferometric system (with the mass and thermal loads that it would suppose) and finally a high resolution linear scale was assembled in this axis (only 0.5 micrometers resolution).

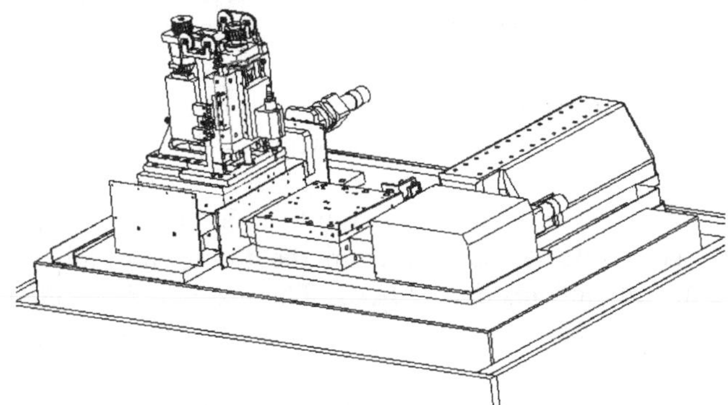

Figure 2.   Milling machine configuration

The micromachining of small chips should be performed at a very high cutting feed, and the options that were analysed for the spindle were the usage of an aerospindle or an electrospindle. The final election was performed taking into account that a minimum run-out should be reached. A high speed electrospindle capable to provide speeds between 5000 and 60 000 rpm was assembled. Its run-out error is slightly higher than that supplied by the lathe spindle.

Aiming to machine small components and microcomponents its workspace should be reduced in order to achieve a highest repeatability and precision (a 50 mm stroke was chosen). The machine strokes are enough to machine a watch metal case (70 x 120 x 50 mm).

Figure 3.   Z axis for milling machine configuration and floating nut detail

The vertical axis makes use of a conventional high precision ballscrew and the actuating force is applied using a floating nut (fig. 3) that allows the air to guide the slider in the transversal directions eliminating any torsion that could produce a pitch error. The floating nut solution is used in all axes.

The Z and Y axes (fig. 3) use precision ballscrews as actuators but the X axis has been provided with a frictional actuator. This actuator doesn't present the inconvenient of torque-to-linear-movement-conversion and, opposite to the ballscrew action, the pitch error when pushing the slider in any sense is avoided.

The whole assembly was mounted on a granite table in order to ensure the highest stability. The table is isolated from the floor vibration using dampers that remove all frequencies over 2 hertz. The machine is located in a temperature controlled and vibration-free room in which the temperature, humidity and dust are controlled.

The prototype presents a very good performance in X and Y axes, though making use of different actuators. The Z axis presents a lower resolution and that causes an unbalanced precision in this axis. In any case the absolute precision of the machine is very high: 0.03 micrometers of positioning precision in X and Y and 1 micrometer positioning precision in the Z axis. The repeatability is even lower in the three axes.

3. FIRST RESULTS

Currently the machine is still at the test stage although many pieces have already been machined. The main trouble that has been found during the tests resides on the tools' geometry and wear. In any case even the smallest commercial mills are too big to perform a micromachining process and special tailor made hardened steel 1-flute mills with a tip diameter minor to 20 micrometers are being used in current tests.

Just as an example of the machine ability, a 3D sculpture on a brass piece was machined using the ultraprecision milling machine. The sculpture was the face of Mozart in a very small size: 0.4 x 0.6 x 0.042 mm. The machining program was performed using a CAM software (fig. 4) and the machining lasted for more than 10 hours. The tool was a 20 micrometers tip diameter engraving tool and the feeds were as small as 0.001 mm/sec, the depth of cut was 1.5 micrometers and the spindle speed was fixed at 28000 rpm.

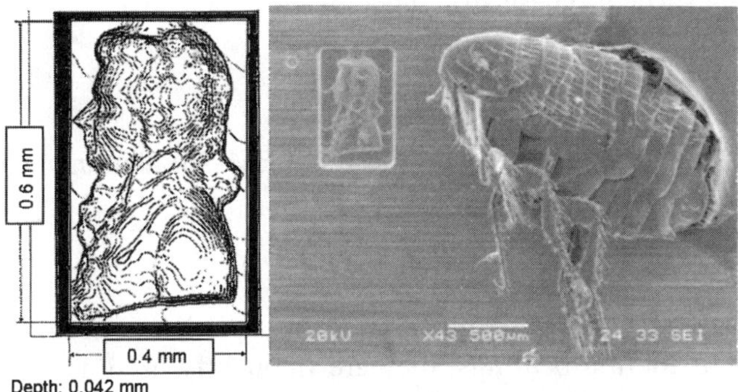

Figure 4.   Face of Mozart in comparison to a dog's flea

The small machining can be appreciated in comparison to a dog's flea using an optical microscope (fig. 4). The stairs that can be appreciated are due to the worst resolution in the Z-axis. The conventional cutting process can be optimised and is really powerful in order to produce microcomponents. The importance is higher when the part must be made of materials different from silicon (materials that cannot be machined with electronics' techniques). In those cases, the milling process, complemented by the EDM process in the machining of conductive materials, will be indispensable. The production of full-functional microproducts will need all these techniques in order to produce the different components.

4. REFERENCES

- Takeuchi, Y., Sawada, K. and Sata, T. Ultraprecision 3D Machining of Glass. Presented in the 46th General Assembly of CIRP/CIRP Annals; 1996. 1996 August 25 – 31; Como.Berne: Hallag Ltd., 1996; 45: 401-04.

- Weck M., Fischer, S. Development of an Ultraprecision Milling Machine. Proceedings of the 9th IPES/4th UME; 1997 May 26 – 30; Braunschweig. Braunschweig: H. Kunzmann et al. Eds. 1997; 379-81.

DEVELOPMENT OF HYDROSTATIC BEARINGS WITH GROOVE STRUCTURES

Manfred Weck, Jan Hennig, Markus Winterschladen

Abstract

Hydraulic linear motion bearings with integrated flow restrictors which eliminate the need for external compensation devices like capillaries or diaphragms, have been developed at the Fraunhofer IPT. This has been achieved by using of groove structures, only a few micrometers in depth, on the bearing surface. These structures produce their own, gap-dependent throttle effect.

Keywords
hydrostatic bearings, design, FEM-Simulation

1 INTRODUCTION

Because of the good geometrical characteristics of aerostatic and hydrostatic linear motion bearings, they are frequently used as force-carrying coupling elements between movable machine components in ultra-precision machines. Hydrostatic bearings are used when high machining forces rule out the use of aerostatic bearings [OEZ00]. In addition to high levels of static stiffness, hydrostatic bearings have excellent damping characteristics and work without slip-stick effects. The drawback of hydrostatic bearings is their considerable requirement for design and equipment work. The flow restrictors, in particular, which are essential for systems with one pressure source take a lot of space and have to be tuned specifically. There have been efforts to integrate these compensation devices into the bearing surface by using variable restrictors located on the bearing surface [SLO95]. This type of inherently compensated bearing provide load carrying capability only when at least two bearings are used and there are still two different elements: one restrictor area and one load carrying area.

At Fraunhofer IPT a self-compensating design has been developed, which combines the function of load carrying and self-compensation in the same structured area on the bearing surface. There is no connection between each individual bearing necessary besides a direct connection to the pressure source. Each bearing element is completely independent.

2 MECHANICAL PRINCIPLE

An analytical study of conventional hydrostatic bearings with a diameter of 30 mm shows that these behave unfavorably when load is applied eccentrically, since the pressure p_T, required to bear the load, is equally high at each point on the bearing. Therefore single conventional hydrostatic bearings do not provide any tilt stiffness. The effect of tilt stiffness is reached only when several bearings are combined [WEC97].

The idea underlying the structured bearings, is that the function of the restrictors can be integrated in the bearing surface in the form of small groove structures, thus replacing the restrictors [YOS98]. Whilst in the conventional bearing, the fluid can flow unhindered from the pressure source to the edge of the bearing (t >> h), the groove geometry is the decisive factor in determining flow distribution in the case of structured bearings, c.f. Fig. 1.

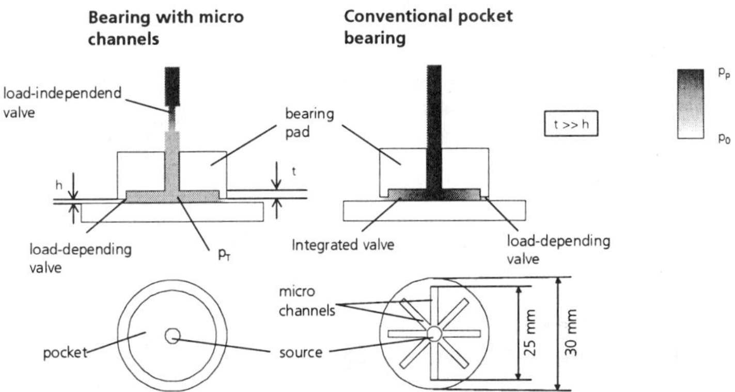

Fig. 1:   *Function principle of conventional recess bearings (a) and structured bearings (b)*

The general flow mechanical laws in the gap are described by the Navier-Stokes equations. It is extremely time-consuming to calculate these flow equations since in addition to the three velocity components, pressure and two physical values describing the flow turbulence must be taken into account.

3 NUMERICAL CALCULATION

The complex equation system for various structures of the small surface grooves mentioned above, was solved numerically using a FEM simulation program (ANSYS/FLOTRAN). Several models of bearings are solved by setting up a cluster of computers and carrying out a total of over 1,500 calculations. The geometric models were designed by taking account of symmetry conditions. In addition to using various gap heights to calculate bearing load and static stiffness, the groove widths and depths were varied to find the optimum geometry.

In a preliminary step, the groove profile was optimized on the basis of different star structures. In a second calculation phase, the results obtained in Phase 1, were used to calculate other structures with various groove profiles. The qualitative representation of the pressure distribution with various gap width, is shown by a structured bearing in form of a spiral structure in Fig. 2.

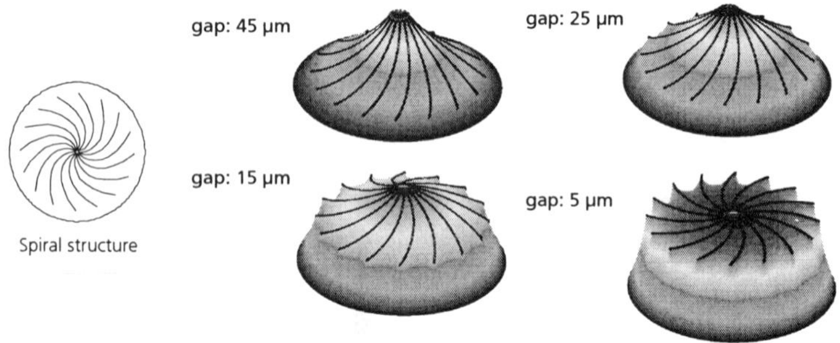

Fig. 2:   *Pressure distribution for various gaps*

In general terms, it has been observed that the maximum stiffness falls as the groove depth increases and the load capacity rises. This effect corresponds exactly to theory-based expectations since the pressure range is more evenly distributed when the groove profile is larger.

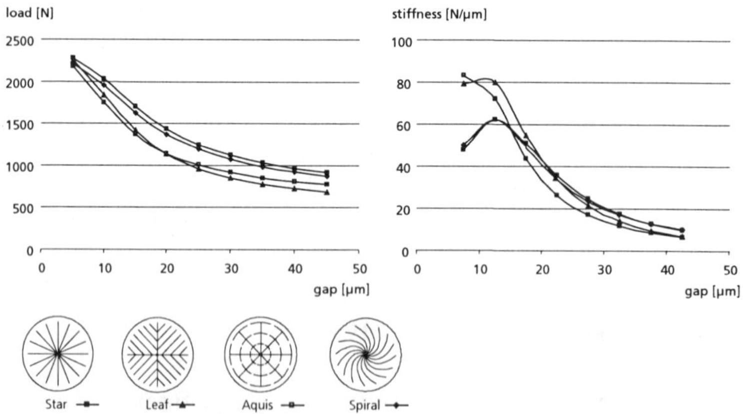

Fig. 3:   *Calculated load rates and static stiffness levels*

It can be concluded from the calculation that the groove structures can be divided into two distinct groups: Radial grooves, which transport the high pressure outwards, thus increasing the load capacity of the bearing by extending the pressure range outwards. Circumferencial grooves, which increase the static stiffness and the oil volume flow is more evenly distributed over the circumference. Combinations of radial and circumferencial elements demonstrate the best characteristics.

The tilt stiffness was also determined in numerical calculations and verified in experimental measurements, c.f. Fig. 4. The pressure maximum moves noticeably towards the side of the smaller gap. By tilting, the new structured bearing independently develops a moment which increases linear when the tilt angel is high.

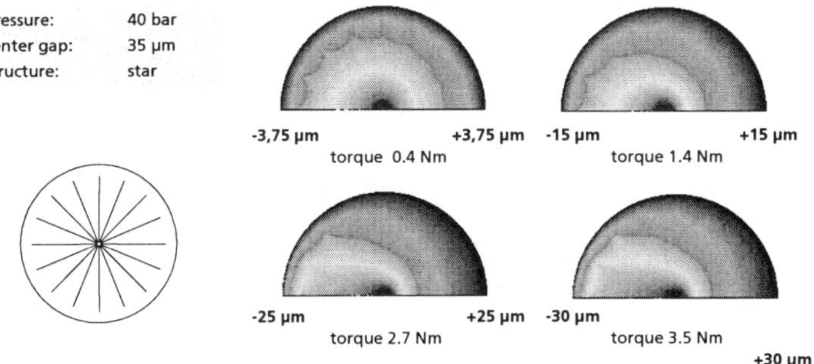

Fig. 4:   Displacement of the pressure maximum by tilting

4 EXPERIMENTAL VERIFICATION

In the next step, the results of the numerical calculation were tested in experimental investigations, c.f. Fig. 5.

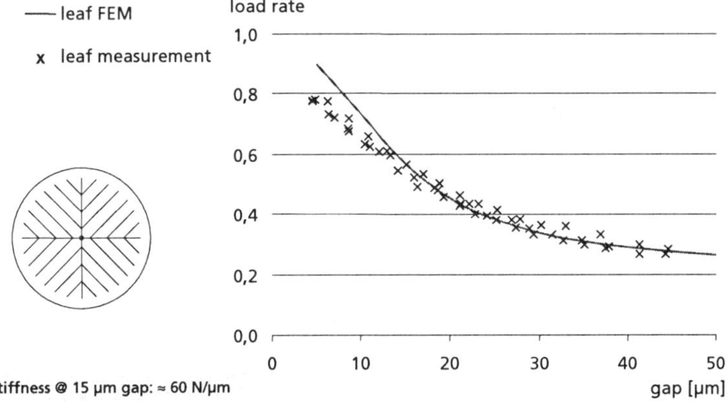

Fig. 5:   Measured and calculated results of the leaf structured bearing

A test bench, which used Strain Gauge force sensors to record the reaction in three-points and capacitance position sensors to record the gab variation, was specially designed and set up for this purpose (each type of a groove structure was tested). The small grooves were produced by milling on the diamond-machined bearing surface.

5 SUMMARY

The results of the research work outlined above, show that the structured bearings, achieve considerably higher levels of bearing force than those recorded by conventional bearings. It must be accepted that there will be a reduction of approx. 40 % in stiffness. In return, hydrostatic bearings are not further applied the need for flow restrictors. The leaf structure has emerged as a suitable starting point for further development. The individual levels of carrying force are higher than those of conventional recess bearings by a factor of approximately two. At 60..80 N/μm, the stiffness levels of the structured bearings show, in touch with the investigation by tilting the structured bearings, that these are suitable for industrial use, c.f. Table 1.

Bearing	*load F*	*stiffness k_{max}*	*tilt torque*
recess bearing	*1000 N*	*110 N/μm*	*0 Nm*
leaf structure	*1700..1800 N*	*60..80 N/μm*	*2..4 Nm*

Table 1:   Comparison between a conventional bearing and the leaf structured bearing

6 REFERENCES

[OEZ00]   Oezmeral, H., *Elektromagnetisches Fast Tool Servo System für den Einsatz in der Ultrapräzisionstechnik*. Aachen: Shaker, 2000.

[SLO95]   Slocum, A.H., Scagnetti P.A., Kane N.R., Brunner C., *Design of self-compensated, water-hydrostatic bearings*. Precision Engineering 1995; 17:173-185

[WEC97]   Weck, M., *Werkzeugmaschinen, Fertigungssysteme*. Berlin: Springer, 1997

[YOS98]   Yoshimoto S., Anno Y., Tamura M., Kakiuchi Y., *Axial load capacity of water-lubricated hydrostatic conical bearings with spiral grooves for high speed spindles*. Tribology International 1998; 31: 331-338

ASSESSMENT OF THERMOPHYSICAL PROPERTIES AT DESIGN STAGE OF MACHINE TOOL STRUCTURE WITH THERMAL SYMMETRICITY

Masayuki OKABE, Haruhisa SAKAMOTO and Shinji SHIMIZU
Precision Engineering Laboratory, Dept. of Mechanical Engineering,
Faculty of Science and Technology, SOPHIA University
7-1, Kioicho, Chiyoda-Ku, TOKYO, 102-8554 JAPAN

Abstract

This paper describes the relationship between thermal behavior and thermophysical properties of cross rail material of double column type machine tools. Assessment is performed to thermophysical properties of linear expansion coefficient, thermal conductivity and heat capacity. Cast iron is selected as a standard material and its thermophysical values are varied. By using FEM, numerical experiments are done with transient thermal stress analysis. The influence of thermophysical properties on the thermal deformation of cross rail is examined. It is found that linear expansion coefficient and thermal conductivity have greater influence on the maximum deformation. And then, thermal conductivity and heat capacity showed stronger influence on the time to the maximum deformation.

Keywords

Machine tool structure, Thermophysical analysis, Thermal deformation

1. INTRODUCTION

It can be said that the improvement of the machining accuracy of production parts is always a fight with the thermal behavior of machine tools. Especially, when higher accuracy must be kept for a long time, the appearance of thermal deformation or displacement is closed up as a serious problem [Bryan 1990], [Liangshen 1981], [Slocum 1987]. As the countermeasures for such a problem, up to now, preload regulation [Tu 1996], cooling of structural elements [Tanabe 1996], the compensation of thermal deformation [Ma 1999], [Ramesh 2000], [Weck 1998], etc. has been performed. It can be emphasized, however, that these treatments are subsidiary methods after casting, machining, and assembling main structures. In order to make the structural performance higher, it is necessary and essential to give the restraining capacity of thermal deformation to each structure. Therefore it is an important technology to resolve thermal problems at the design stage.

By the way, structures having thermal symmetricity are adopted commonly in machine tool design. For the most part, structural shapes are realized by using geometrically symmetric design. Even if an individual structure has such symmetricity, however, the resultant thermal deformation cannot be necessarily restrained owing to the deviation of heat source and/or the un-uniformity of temperature distribution. Moreover, when evaluating thermal deformation by the whole structure, the possibility of lower effect of partial symmetricity may occur.

Additional importance at design stage is the selection of proper materials.

Industrial materials have both elastic and thermophysical properties. However, there are no materials with the excellence for all properties and then trade-off problems always rise about the material adoption. And compared with elasticity, it is not so easy to capture the effect of thermophysical properties because they are related mutually. It can be said that the particular importance is to understand the contribution of thermophysical properties with structural shapes.

In this paper, the attention is paid to the relationship between structural shapes and the thermophysical properties of cross rail material of double column type machine tool. Thermal deformation caused by temperature gradient across the cross rail is assessed by FEM. The interrelation between thermophysical properties and the thermal behavior is examined and quantified.

2. ANALYTICAL MODEL AND CONDITIONS

Fig. 1 shows the mathematical model for a large double column type machine tool structure of a practical machine. The spindle head is installed in front of the cross rail and its movement along y-axis is in shorter stroke than other axes. While cutting for a long time, the spindle head position is maintained stationary on the cross rail. This means that the heat caused by spindle rotation brings heat source to cross rail. Then the thermal deformation of cross rail affects machining accuracy directly. From this reason, it is assumed that the spindle head remains at the middle of cross rail for a long time.

In the analysis, all structures are assumed to be midair boxes. Shell element was used as a finite element to represent the structural shape. Table 1 shows the type of shell element and the thickness of each structure. Cast iron was chosen as the standard material for all structures. Thermophysical values of cross rail were varied separately to examine the thermal effect on the resultant deformation.

As shown in Fig. 2, it was assumed that the heat source is located at slide way interface between spindle head and cross rail. The heat transfer boundary is set. The heat flows into the cross rail at the constant heat transfer coefficient of $150\text{W/m}^2\text{K}$ through the slide way interface. And then the heat flow through the lubricant was excluded. The temperature rise at spindle head was assumed to be $8\,^{\circ}\text{C}$, which was referred from practical experiments in a workshop. The temperatures

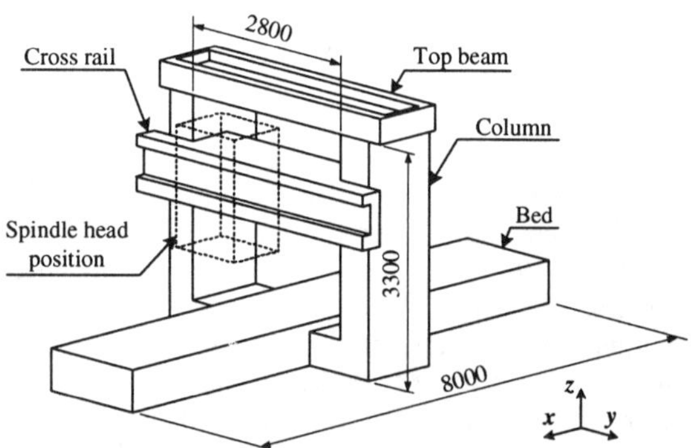

Fig. 1 Model of double column type machine tool

Table 1 Finite element type for main structures

Element type	Shell element with 4 nodes	
	Bed	30mm
Element thickness for main structures	Column	front face 40mm other face 20mm
	Cross rail	slide way 60mm other face 30mm
	Top beam	20mm

540

of heat source and atmosphere were 30 ℃ and 22 ℃ respectively. Other surfaces were kept as an insulated boundary and the initial temperatures were set to be 22℃.

Linear expansion coefficient α, thermal conductivity λ, and heat capacity ρc of the cross rail material were selected as thermophysical parameters. These values of cast iron have been changed within the range of other practical materials. Transient thermal stress analysis was done and thermal bending deformation of cross rail was assessed prior to temperature rise in the columns. FEM software MARC was used for analysis.

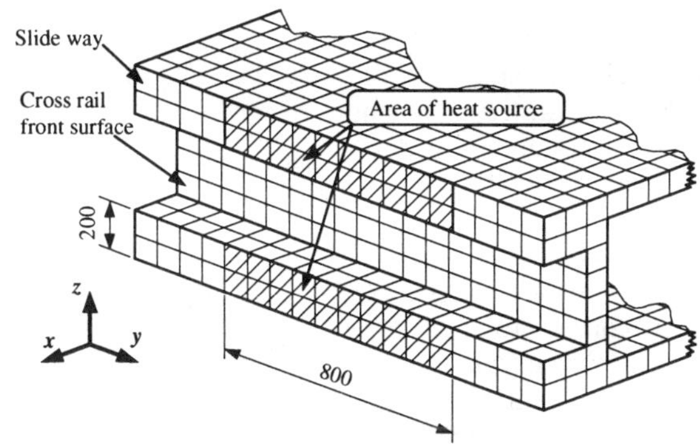

Fig.2 Heat source location on the cross rail

3. RESULTS

At first, it is important to recognize what kind of thermal deformation mode of cross rail dominates the spindle head deviation. To know this, basic mode of thermal deformation of cross rail was examined. As an example, Fig. 3 shows an analyzed result of temperature distribution and thermal deformation of cross rail after 6.5 hours. From the figure, it can be seen that the forward bending mode appears at cross rail. From the similar results, it was found that the dominant mode was the forward bending of cross rail induced by thermal stress. And then translation in x direction at the middle of cross rail showed the largest value in comparison with other translations and angular displacements. The attention was paid, therefore, to the cross rail translation in x because it has the greatest influence on the machining accuracy.

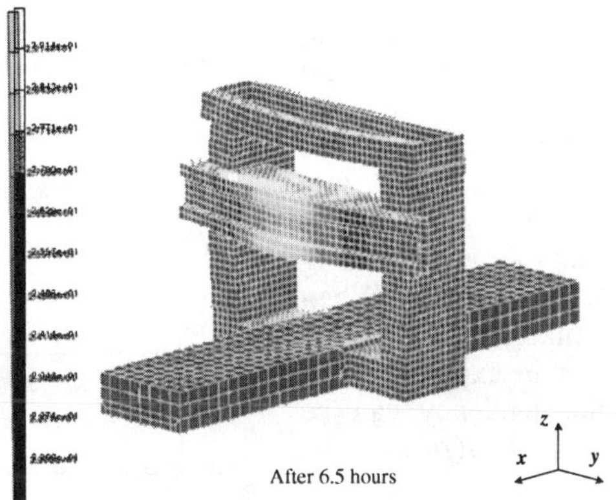

Fig. 3 Thermal deformation mode of cross rail

The evaluation point was chosen as shown in Fig. 4(a). The behavior of that point is shown in Fig. 4(b). Initially, the displacement increases with the time t. At time T, the displacement reaches the maximum dx but subsequently decreases gradually. Hereafter, the assessment was taken by using the maximum dx and the time T.

Fig. 5 shows the relationship between linear expansion coefficient α and the maximum displacement dx of cross rail. From the figure, it is found that α and dx are in the proportional relation, and dx becomes smaller with the smaller α.

541

Next, Fig. 6 shows the relationship between thermal diffusivity κ of cross rail and dx. When thermal conductivity λ is divided by heat capacity ρc, the quotient is defined as the thermal diffusivity κ. Therefore, two curves for dx are shown on the figure when λ was varied with constant $1/(\rho c)$ and when ρc was varied with constant λ. The maximum dx has the tendency to become smaller when λ is enlarged. On the other hand, when $1/(\rho c)$ is enlarged, dx hardly changes.

Fig. 7 shows the relationship between thermal diffusivity κ and the time T to the maximum dx. In the same way as Fig. 6, T was plotted for two cases of varying λ and varying $1/(\rho c)$ individually. When λ is enlarged, T becomes smaller in exponential. A similar tendency can be seen about $1/(\rho c)$ on the figure.

Based on above results, the quantification of the influence of α, λ, and $1/(\rho c)$ on dx was performed from Fig. 6. By using multiple regression analysis with normalization, the sensitivities for α, λ, and $1/(\rho c)$ became 74%, 25%, and 1% respectively. Therefore, it was found that α has a significant influence on the maximum displacement dx and then $1/(\rho c)$ has little effect on dx. Similar analysis was done for T from Fig. 7. As a result, normalized sensitivities of α, λ, and $1/(\rho c)$ were assessed as 1%, 53%, and 46% respectively. Then, it can be said that the effect

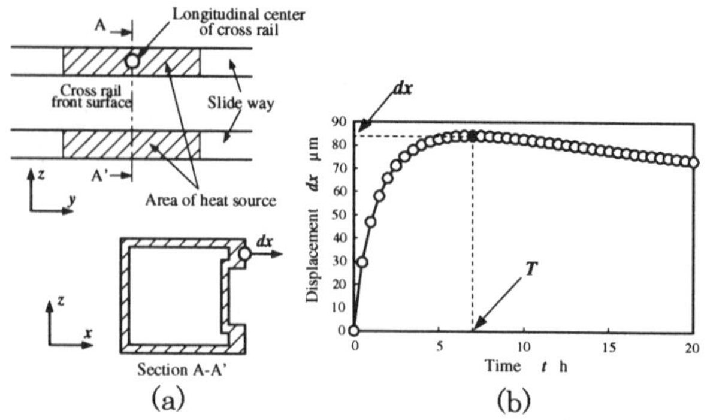

Fig. 4 Displacement of spindle head position

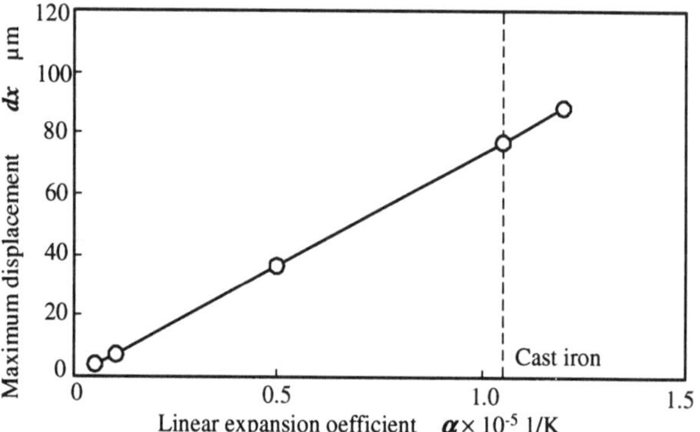

Fig. 5 Influence of expansion coefficient on dx

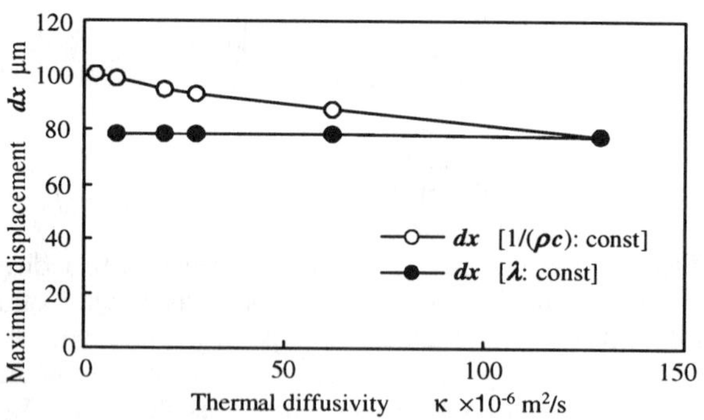

Fig. 6 Influence of thermal diffusivity on dx

of λ and $1/(\rho c)$ is predominant and α has little influence on T.

A quantitative evaluation mentioned above is one example of the discussion for the cross rail material of a double column type machine tool. However, it is expected that the thermal deformation mode like the cross rail of this research also appears in a structural element supported by both ends

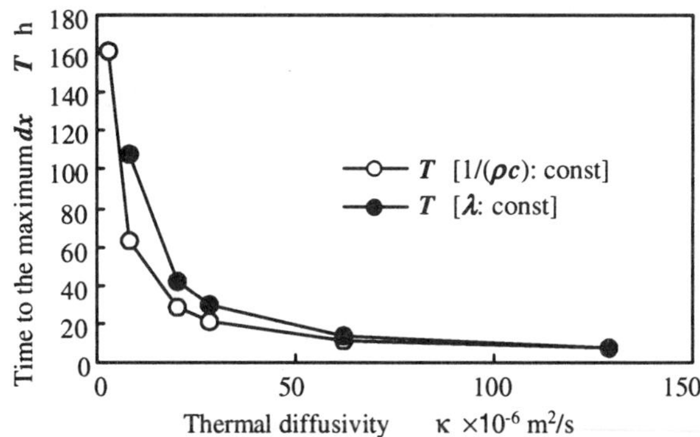

*Fig. 7 Influence of thermal diffusivity on **dx***

with heat source at the middle. In such a structure, the material with small α and with large λ should be selected to control the absolute value of translational displacement. Moreover, it can be said that the material with large diffusivity κ has greater effect to saturate the thermal displacement in a short time. This means that the ratio of λ and ρc of the material is large. Oppositely, if lower deformation rate is desired for a long time, the material with small κ, that is, with a small ratio of λ and ρc, should be selected.

4. CONCLUSION

Assessment was done for clarifying the influence of thermophysical properties on the thermal behavior of the cross rail in a double column type machine tool. Linear expansion coefficient, thermal conductivity, heat capacity were chosen as the thermophysical parameters. By using FEM, the contribution of thermophysical parameters were elucidated. It is found that linear expansion coefficient and thermal conductivity have greater influence on the maximum deformation. And then, thermal conductivity and heat capacity showed stronger influence on the time to the maximum deformation.

REFERENCES

1. Bryan J. International status of thermal error research. CIRP Ann. 1990; 39:645-56.
2. Liangshen Qu, Zhefong C., Zeng S. The thermal behavior of machine tool guideways. Manuf. Eng. Trans. 1981; 9:349-55.
3. Ma Y., Yuan J., NI J. A strategy for the sensor placement optimization for machine thermal error compensation. Trans ASME, J. of Manuf. Sci. Eng. 1999; 10:629-37.
4. Ramesh R., Mannan M.A., Poo A.N. Error compensation in machine tools - A review Part II, Thermal errors. Int. J. Mach. Tools Manuf. 2000; 40:1257-84.
5. Slocum A.H. Design to limit thermal effects on linear motion bearing performance. Int. J. Mach. Tools Manuf. 1987; 27:239-45.
6. Tanabe I., Yanagi K. Dual Cooling Jacket around Spindle Bearings with Feed-Forward Temperature Control System to Decrease Thermal Deformation. JSME Int. Journal, Ser. C 1996; 39:149-55.
7. Tu J.F., Stein J.L. Active Thermal Preload Regulation for Machine Tool Spindles With Rolling Element Bearings. Trans. ASME, J. Manuf. Sci. Eng. 1996; 118:499-505.
8. Weck M., Herbst U. Compensation of thermal errors in machine tools with a minimum number of temperature probes based on neural networks. Trans. ASME, J. of Dyn. Syst. Control 1998; 64:423-30.

CONCRETE-BASED CONSTRAINED LAYER DAMPING

Eberhard Bamberg and Alexander H. Slocum

Department of Mechanical Engineering,
Massachusetts Institute of Technology, Cambridge, USA

Abstract

This method allows a fabricated (welded) machine tool structure to be designed for minimum cost and maximum dynamic stiffness comparable to polymer concrete structures.

Keywords

Constrained-layer damping, replication, ShearDamper™, viscoelastic material, structural damping.

1. INTRODUCTION

Welded machine tool structures provide easy scalability in terms of size and outstanding flexibility in terms of rapid design and fabrication; however, damping of the structure is a very critical issue. Unlike cast iron or polymer concrete-based components, welded steel plates have virtually no internal damping. Filling the structure with concrete or sand can add damping but also a great deal of unwanted weight.

A better approach would be to use constrained layer damping (CLD) where a viscoelastic layer is sandwiched between the structure and one or more constraining layers. Kinetic energy from relative motion between the structure and the constraining layer as it occurs during bending or twisting gets dissipated by the viscoelastic layer. This mechanism introduces damping into the system, thereby limiting the structure's response to excitation frequencies near its modes [Marsh et al, 1996, Nayfeh et al, 1997].

2. CONCRETE CAST DAMPING DESIGN

The concrete-cast constrained layer damping design provides a novel solution to an existing patent, the ShearDamper™[1] [Slocum, 1994, 1995]. Instead of using expensive epoxy to fit steel constraining layers, that are made from a slotted steel tube, to the inside of a hollow structure (Figure 1), the new design makes use of a much simpler process. The constraining layers are replicated in-place by filling four "sausage-like" inserts, that are placed between the inside of the structure and the outside of a support tube, with expanding concrete. The inserts are made from 0.38 mm thick ISODAMP™ C-1002[2] with a perimeter according to:

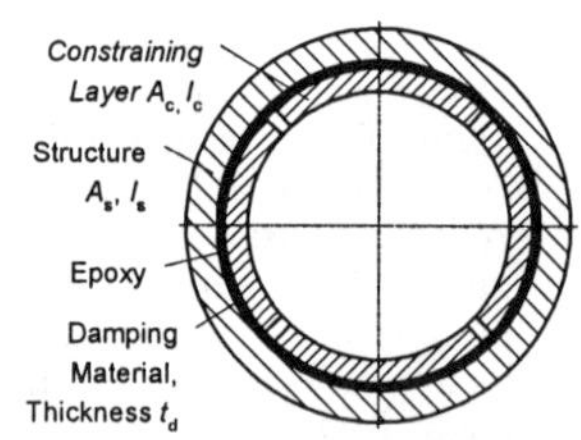

Figure 1 The ShearDamper™

---

1. ShearDamper™ is a registered trademark of AESOP, Inc.

$$P = (D_S - 2t_S)\left(\tfrac{\pi}{4} + 1\right) + D_{St}\left(\tfrac{\pi}{4} - 1\right) \qquad (2.1)$$

to ensure that the layers can form completely as illustrated in Figure 2. Eliminating voids is absolutely crucial, otherwise the concrete, as it expands during curing, cannot create the pressure required to keep the constraining layers firmly in place. Therefore, sufficient expansion of the concrete is an essential element of this design and is achieved by adding 1% Intraplast-N™[1] to the concrete mixture. This grout agent contains an aluminum powder that oxidizes during the curing of the concrete, thereby producing little hydrogen bubbles. The expanding gas counteracts the concrete's tendency to shrink and if dosed properly, actually causes its volume to increase.

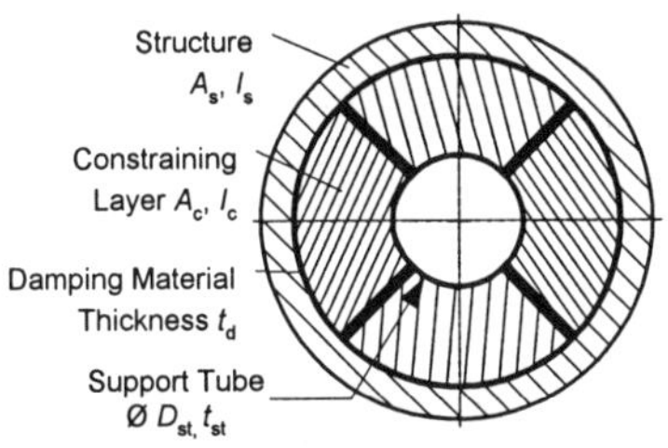

Figure 2 Concrete-cast design

2.1 Damping Performance

The performance of such a system is determined by the loss factor of the viscoelastic material and how well the stiffness of the constraining layers is tuned to the stiffness of the structure. For this purpose, Hale and Marsh introduced the stiffness ratio r, the ratio between the sum of the components' stiffness with respect to the system neutral axis and the sum of the components' stiffness with respect to their own neutral axis, respectively [Marsh et al, 1998]. The complex goal of maximizing damping is then reduced to maximizing the ratio r (2.2). From this simple equation we take that constraining layers should be as far away as possible from the system neutral axis, but this of course is limited by the fact, that as an internal system, dimensions cannot exceed that of the inside of the structure.

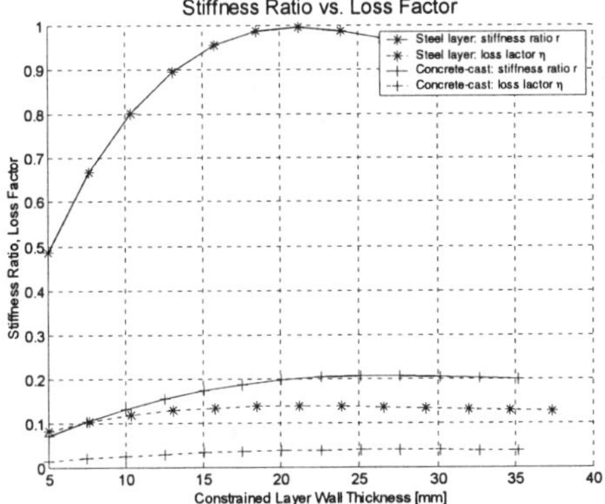

Figure 3 Stiffness ratio and modal loss factor in terms of CL thickness

$$r = \frac{EI_\infty - EI_0}{EI_0} = \frac{\sum_i E_i A_i (y_i - y_\infty)^2}{\sum_i E_i I_i} \qquad (2.2)$$

Figure 3 illustrates the development of the stiffness ratio r and the system loss factor η at resonance for an internal four quadrant system in terms of the thickness of the constraining layers. While the steel-based design offers

---

2. ISODAMP is a trademark licensed to AERO company
1. Intraplast-N™ is a registered trademark of Sika Corp.

better damping, its primary shortcoming also becomes apparent: the need to find an inner tube with the largest possible outer diameter to fit inside the structure AND the right wall thickness in order to maximize the stiffness ratio r.

The concrete-cast design, on the other hand, removes one of these constraints by using the inner tube as a simple support structure rather than as a constraining layer. Instead, only the outer dimension of the inner tube is of importance because it determines the thickness and with it the stiffness of the concrete constraining layers. The wall thickness and material of the inner tube are no longer design parameters, making this design simpler and less expensive. However, concrete has a lower modulus of elasticity, making the layers more compliant than layers made from steel. As a result, damping does not reach the level of performance of steel layers (see Figure 3).

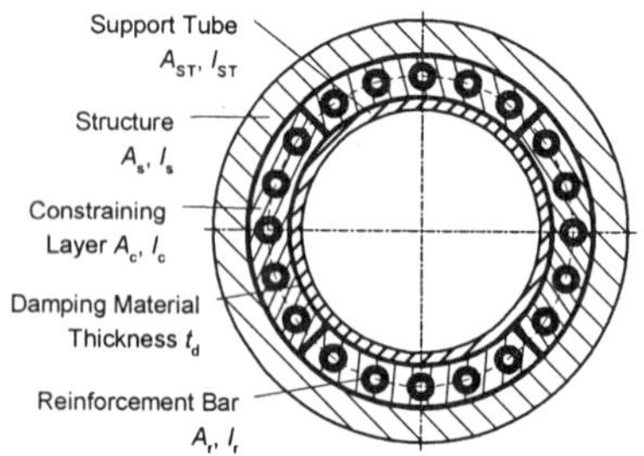

Figure 4 Reinforced concrete-cast constraining layer design

The effect of the lower Young's modulus of concrete can be partially offset by adding inexpensive rebars to the constraining layers. The rebars increase the stiffness of the constraining layers, thereby improving damping performance. Such a multiple reinforced design is illustrated in Figure 4 and the generic design equations are given in Table 1.

Table 1 Formulae for multiple reinforced constraining layers

$$y_c = \frac{\frac{2}{3}(R_c^3 - (R_c - t_c)^3)\sin\frac{\varphi}{2} - R_b\pi R_r^2 \sum_{n=0}^{N-1}\cos(\gamma_0 + n\gamma_n)}{\frac{\varphi}{2}\left(R_c^2 - (R_c - t_c)^2\right) - N\pi R_r^2}$$

$$y_r = \frac{1}{N}\sum_{n=0}^{N-1} R_b\cos(\gamma_0 + n\gamma_n)$$

$$A_c = \frac{\varphi}{2}\left(R_c^2 - (R_c - t_c)^2\right) - N\pi R_r^2$$

$$A_r = N\pi\left(R_r^2 - (R_r - t_r)^2\right)$$

$$I_{c,x} = \frac{1}{8}(\varphi + \sin\varphi)\left(R_c^4 - (R_c - t_c)^4\right) - \sum_{n=0}^{N-1}\left(\frac{\pi}{4}R_r^4 + \left(R_b\cos(\gamma_0 + n\gamma_n)\right)^2\pi R_r^2\right)$$

$$I_{r,x} = \sum_{n=0}^{N-1}\left(\frac{\pi}{4}\left(R_r^4 - (R_r - t_r)^4\right) + \left(R_b\cos(\gamma_0 + n\gamma_n)\right)^2\pi\left(R_r^2 - (R_r - t_r)^2\right)\right)$$

$$I_{c,0} = I_{c,x} - y_c^2 A_c$$

$$I_{r,0} = I_{r,x} - y_r^2 A_r$$

$$EI_\infty = 2E_c I_{c,x} + E_s(I_{r,x} + I_{St} + I_s)$$

$$EI_0 = 2E_c I_{c,0} + E_s(I_{r,0} + I_{St} + I_s)$$

2.2 Design Optimization

Using the approach laid out by Marsh and Hale, the damping of such systems can be estimated in terms of the modal loss factor, where higher values indicate better damping. The graphs shown in Figure 5 can be used to optimize the damping for a given structure. As expected, adding rebars to the concrete constraining layers increases the damping considerably. It is important to distinguish between optimum and actual damping performance, the former built around a damping layer with optimum thickness and the latter with a layer thickness that is actually available.

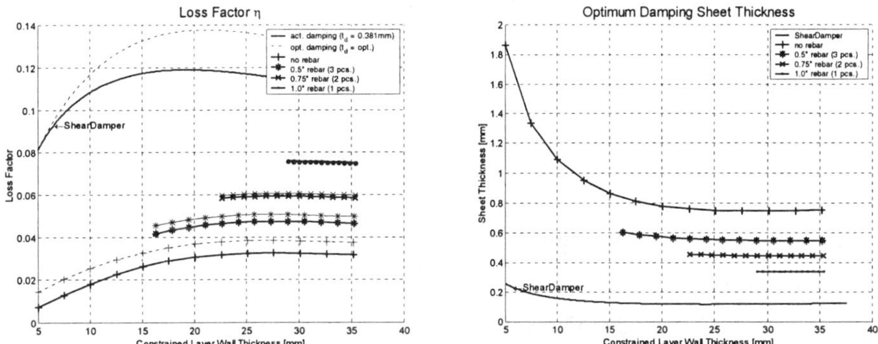

Figure 5 Loss factor and optimum damping layer thickness of CLD systems

3. EXPERIMENTAL RESULTS

3.1 Setup

The damping ratios were obtained by analyzing the transfer functions of a 3-D modal analysis using the Star System™[1] modal software. A 3-axis accelerometer was attached at 21 locations evenly spaced along the length and perimeter of the 48" long (1220 mm) prototype and the built-in Fast Fourier transformations (FFT) of a Hewlett-Packard 35670A frequency analyzer were used to transfer the transducer signals from the time into the frequency domain. In order to simulate a free-free boundary condition, the test tube was suspended at the first bending mode's nodal points at roughly 20% and 80% of the overall length using elastic cord. Fixed spacial excitation of the structure was accomplished by an impact hammer equipped with a Delrin tip.

3.2 Damping Results

The predicted and measured damping ratios for a 3.5" (88.9 mm) diameter and 48" (1220 mm) long structural tube equipped with a four-quadrant concrete-cast damping system and a 1.9" (48.3 mm) diameter support tube are given in Table 2. It is interesting to note how well theory and experiment agree on the first mode at 280Hz, while higher modes offer more damping than predicted. This unexpected performance plus might be attributed to the fact that concrete itself has some damping built-in that is highly frequency-dependent. Its influence, however, is not included in the predictions.

The frequency dependency in damping performance is consistent with experimental data from an earlier prototype that was shorter (24" = 610 mm long) and larger in diameter (5.5" = 139.7 mm). Loss factors measured at the

---

1. Star System™ is a registered trademark of Spectral Dynamics

first bending mode were considerably higher than calculated (Table 3).

Table 2 Loss factors for 48" x 3.5" prototype with concrete CastDamper

	1st Mode (283 Hz)	2nd Mode (740 Hz)	3rd Mode (1360 Hz)
theoretical	0.0260	0.0263	0.0265
measured	0.0180	0.0430	0.0640

Table 3 Loss factors for 24" long and 5.5" diameter prototypes (1st mode)

	ShearDamper™	CastDamper	CastDamper 3x0.5" rebars
theoretical	0.037	0.035	0.044
measured	0.0592 (1530 Hz)	0.1432 (1260 Hz)	0.3308 (1640 Hz)

4. CONCLUSIONS

The CastDamper design offers an excellent alternative to existing internal constrained layer damping designs. By eliminating the need for expensive epoxy, material cost for the prototypes were 70-75% lower compared to a ShearDamper design. Casting the constraining layers effectively decouples the shape of the support tube from the shape of the structure, making the CastDamper better suited for a wider range of structures.

In a ShearDamper design, the bottom needs to be sealed in order to prevent epoxy from leaking out. This can be rather challenging in cases where the bottom is not easily accessible. Such problems are non-existent with CastDampers because the damping inserts are closed at the bottom by default.

Damping performance of CastDampers is somewhat frequency-dependent, due to the internal damping of the concrete, and can be improved even further by adding inexpensive rebars to the constraining layers.

5. REFERENCES

1. Publications

Marsh, E.R., Hale, L.C., *Damping of Flexural Waves with Imbedded Viscoelastic Materials*, ASME Journal of Vibration and Acoustics, Vol. 120, No. 1, pp. 188-193, 1998.

Marsh, E.R., Slocum, A.H., *An Integrated Approach to Structural Damping*, Precision Engineering, Vol. 18, No. 2/3, pp. 103-109, 1996.

2. Conference Proceedings

Nayfeh, S., Slocum, A.H., *Flexural Vibration of a Viscoelastic Sandwich Beam in its Plane of Lamination*, ASME 16th Biennial Conference on Vibration and Noise, 1997.

3. Patents

Slocum, A.H., Marsh, E.R., Smith, D.H., *Replicated-In-Place Internal Viscous Shear Damper For Machine Structures And Components*, U.S. Patent # 5,799,924, AESOP, Inc., 1995.

Slocum, A.H, *Method and Apparatus for Damping Bending Vibrations While Achieving Temperature Control in Beams and Related*, U.S. Patent # 5,743,326, AESOP, Inc., 1994.

EFFECTS OF MANUFACTURING ERRORS ON THE ACCURACY FOR TRR-XY HYBRID PKM

Tsann-Huei Chang *, Shang-Liang Chen **, Min-Hsin Hsei **

*Industry Technology Research Institute, Taiwin, R.O.C.

**Institute of Manufacturing Eng., National Cheng-Kung University, Taiwin, R.O.C.

Abstract

A TRR-XY hybrid five DOF parallel link machine tool is built for this research to investigate the error model theory. The errors from the component machining and assembly are defined and considered into the inverse kinematic solution. The effects of the manufacturing errors on the accuracy of the machine tool are shown in this research.

Keywords

Parallel link, Machine, Manufacturing errors, Accuracy.

1. INTRODUCTION

Using parallel-link mechanism as the basic structure of the parallel-link machine tool is a new design concept and is one of the most important research fields and attracts many previous researchers [1-3]. However, the application of these machines as a machine tool has not proven itself in terms of accuracy enhancement over traditional machine tools. In general, the geometry of the machine tool (moving platform size relative to base size) has significant effects on the level of accuracy achievable. Therefore, an error model analysis for the parallel link machine tool to take the advantage of geometry to minimize the error and increase the accuracy is essential, interesting and is focused in this research. A TRR-XY hybrid five DOF parallel link machine tool is built for this research to investigate the error model theory (See Fig.1). The hybrid means that the machine tool is composed of a three DOF parallel link mechanism and a two DOF serial type XY-table. Here, "TRR" is the tool frame independent motion DOF of the upper parallel link mechanism. "T" stands for translation DOF and "R" stands for rotation DOF. From mechanism viewpoint, this system may also be named as PRS-PP. "P" stands for prismatic motion pair. "R" stands for rotation motion pair and "S" stands for spherical joint. Due to the fact that the XY-table is commonly used in the industry and the technology is well developed, this research focuses on the error model analysis of the three DOF parallel-link mechanism.

Fig.1 The hybrid parallel link machine tool developed for this research.

2. ERROR DEFINITIONS FOR THE MACHINE TOOL

The Denavit-Hartenberg notation method (D-H method) is adopted in this research to derive the inverse kinematic solution [4]. Two translation errors and three orientation errors are considered as the assembly errors of the machine tool structure. The error definitions are shown in Fig. 2 ~ Fig. 5.

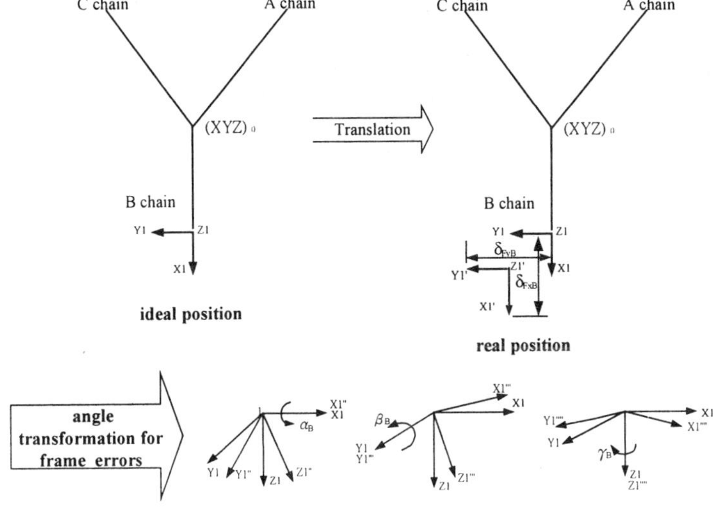

Fig. 2 Definition of coordinate transformation for frame errors

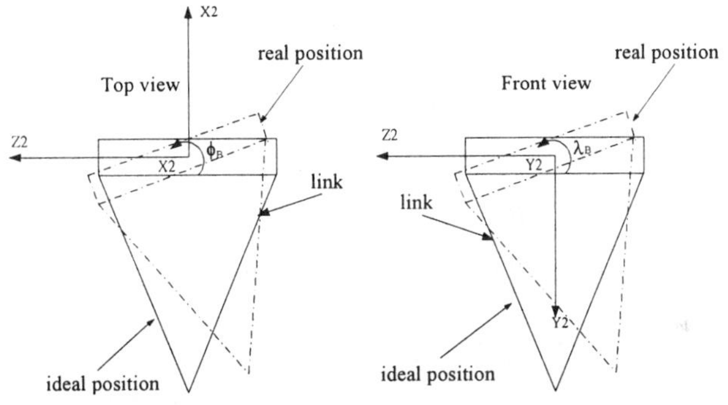

Fig. 3 Geometric definition for pin joint manufacturing errors

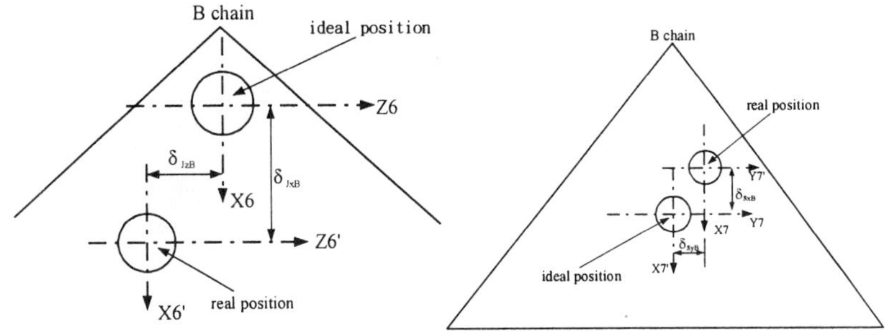

Fig. 4 Ball joint errors   Fig. 5 Spindle location errors

6. EFFECTS OF MANUFACTURING ERRORS ON THE POSITION ACCURACY

Fig. 6 shows the tool paths planning for the error model analysis. $\vec{i}, \vec{j}, \vec{k}$ are the components to represent the orientation of the tool axis.

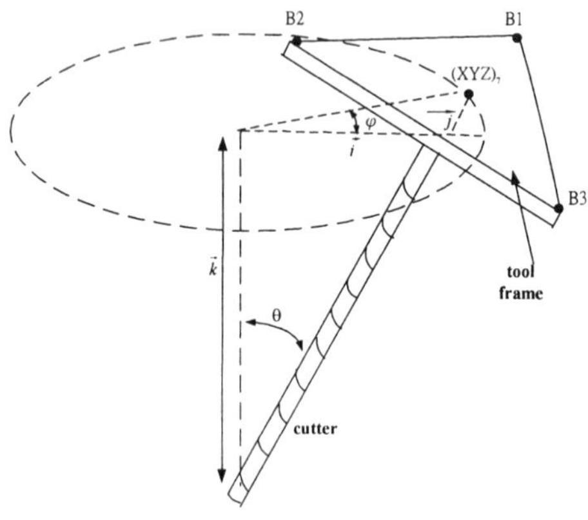

Fig. 6

Fig. 7 Shows the effects of the B-chain manufacturing errors on the variation of dS_A, dS_B, dS_C control position. ($\vec{k}$ =0.8) with (1)[Frame error]$_B$: α_B、β_B、γ_B=1^0，δ_{FxB}=δ_{FyB} =1mm；(2)[Pin error]$_B$: ϕ_B=1^0，λ_B=1^0；(3)[Ball error]$_B$: δ_{JxB}=1mm, δ_{JzB}=1mm；(4)[Spindle error]$_A$: δ_{SxB} =1mm, δ_{SyB} =1mm。

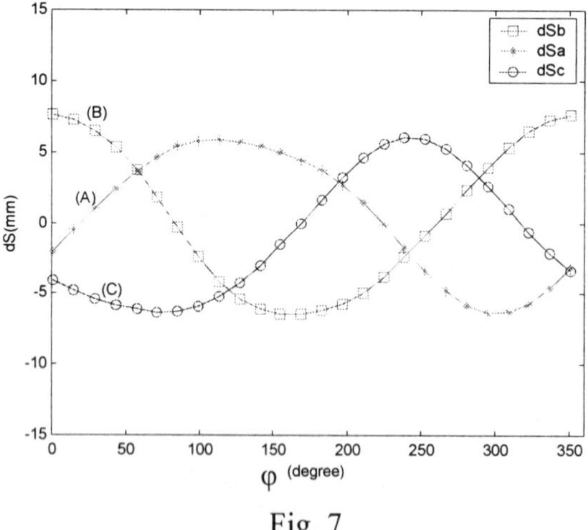

Fig. 7

Fig. 8 Shows the comparison on the effects of different type B-chain errors on the control position variation of dS_A, dS_B, dS_C. (k=0.8) with $\alpha_B=1^0$ ，$\beta_B=1^0$ ，$\phi_B=1^0$ 。originality

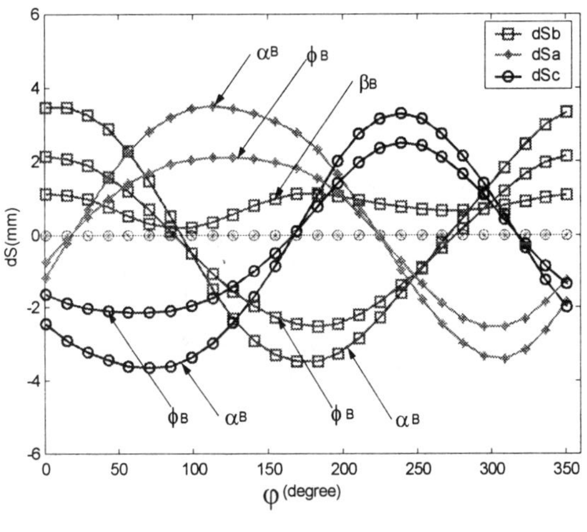

Fig. 8

Engineer is suggested to use small tool inclination angle in the real machining processes.   The effects of the r/R ratio on the position accuracy variation are found more significant than that for the L/R ratio.

9. ACKNOWLEDGEMENTS

Part of the research results was funded by I.T.R.I. of R.O.C. and NSC of R.O.C. (NSC 89-2212-E-006-023).   This financial support is gratefully acknowledged.

10. REFERENCES

1.   Giddings & Lewis Variax, (http://www.giddings.com/).

2.   Ingersoll, (http://www.mel.nist.gov/gallery/hex/hexph.htm).

3.   Toyoda,(http://www.toyoda-ouki.co.jp/_pub_html/sub_html/tmw/prodlines/paralink/paralink.jpg).

4.   Hartenberg, R., and Denavit, J., Kinematic Synthesis of Linkages. 1964, McGraw-Hill, New York.

EFFECT OF STATIC STIFFNESS OF GRINDING SYSTEMS ON A GENERATING MECHANISM OF WORKPIECE GEOMETRICAL ACCURACY

Hwa-Soo LEE* and Yutaka UCHIDA**

*Department of Mechanical Engineering, College of Science & Technology, Nihon University,
*Nikon Corporation

Abstract

In grinding operation, static stiffness between grinding wheel and workpiece affects on the generating process of workpiece geometrical forms. In this study, the effect of the stiffness on the center generating process of eccentric workpiece in cylindrical grinding is theoretically analyzed. And designing a grinding simulator which can set the stiffness of grinding system, theoretical analysis is experimentally confirmed. Depending on these results, a predicting method to estimate the grinding time required to complete the process is proposed.

Key Words

Cylindrical external grinding, Static stiffness of grinding system, Eccentric workpiece, Generation of workpiece geometry.

1. INTRODUCTION

One of the authors made clear that the static stiffness of grinding system affects on the generating process of workpiece size and on the machining efficiency [1][2]. Applying the similar procedure in above studies, the effect of static stiffness in grinding system on the center generating process is discussed theoretically and experimentally. In this study, a stiffness simulator which can set the static stiffness in grinding system reliably is designed, external grinding is carried out.

As the experimental results, it is clarified that this process depends on the stiffness of grinding system and on the grindability. Furthermore, depending on the results described above, an estimating method of the machining time to obtain the required roundness is proposed.

2. CENTER GENERATING PROCESS OF THE ECCENTRIC WORKPIECE

2.1. Foundation of Static Analysis

As shown in Figure 1, resultant stiffness between wheel and workpiece K_{res} can be represented as a serial connection among the stiffness of workpiece and wheel sub-systems K_w, K_s and contact stiffness K_{con}. As well known, grinding stiffness K_g, indicating the grindability, is defined as a ratio of true depth of cut a_a to normal grinding force F_n [1] [3]. Representing the ratio between K_{res} and K_g as;

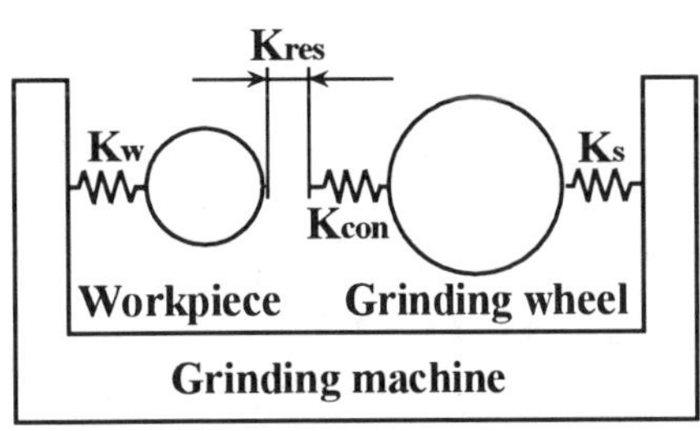

Figure 1 Static model of grinding system

$$\alpha = \frac{K_g}{K_{res}} \qquad (1)$$

it is a dimensionless parameter indicating the machinability of grinding system [1]. In this study, α is called as a ratio of grinding stiffness.

2.2. Center Generating Process

In order to make clear the plunge cut process of a workpiece with eccentricity E, this process can be considered that a workpiece with a geometrical shape $E \cdot sin(2\pi t / T)$ is ground linearly as shown in Figure 2. Where, T is a workpiece rotating interval and t is a grinding time. Elastic deformation, i.e. residual stock removal $y(t)$, due to the normal grinding force takes place and actual depth of cut decreases as shown with a dotted line.

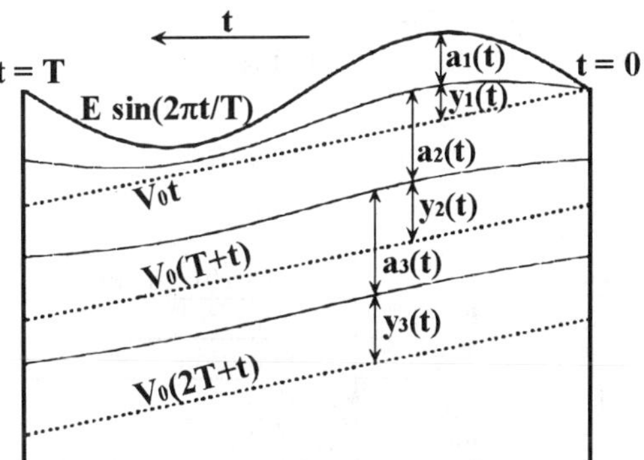

Figure 2 Schematic diagram of grinding process in eccentric workpiece

True wheel depth of cut in nth revolution $a_n(t)$ and residual stock removal $y_n(t)$ can be shown with α as follows.

$$a_n = \frac{1}{1+\alpha}\left(\frac{\alpha}{1+\alpha}\right)^{n-1} \cdot f(t) + \left\{1 - \left(\frac{\alpha}{1+\alpha}\right)^{n-1}\right\} \cdot V_0 \cdot t \qquad (2)$$

$$y_n(t) = \alpha \cdot a_n(t) = \left(\frac{\alpha}{1+\alpha}\right)^n f(t) + \alpha\left\{1 - \left(\frac{\alpha}{1+\alpha}\right)^{n-1}\right\} \cdot V_0 \cdot t \qquad (3)$$

From these equations, it can be seen that the depth of cut finally becomes $V_0 \cdot t$ and the residual stock removal becomes $\alpha V_0 t$. In case of spark-out process, they are represented as follows.

$$a_m(t) = \frac{1}{1+\alpha}\left(\frac{\alpha}{1+\alpha}\right)^{m-1} g(t) \qquad (4) \qquad\qquad y_m = \left(\frac{\alpha}{1+\alpha}\right)^m g(t) \qquad (5)$$

Where, m is a grinding time in spark-out process and $g(t)$ is a final true depth of cut at infeeding process.

3. EXPERIMENTAL SETUP AND METHOD

A grinding stiffness simulator is newly designed which can set K_w in certain amounts reliably. As shown in Figure 3, this simulator consists of a wheel head and two leaf springs supporting the wheel head above a machine table of ordinary external grinding machine. Machining condition is shown in Table 1.

Table 1. Experimental condition

Grinding wheel	WA60JV
Workpiece	S45C (HRC20)
Infeed speed: V_0	20 μ m/work rev.
Width of cut: b	16 mm
Wheel speed: V_s	1800 m/min
Work speed: V_w	5 m/min
Initial eccentricity: E	10, 100 μ m
Dressing	Single point diamond
Depth of cut	10 μ m
Lead	200 μ m

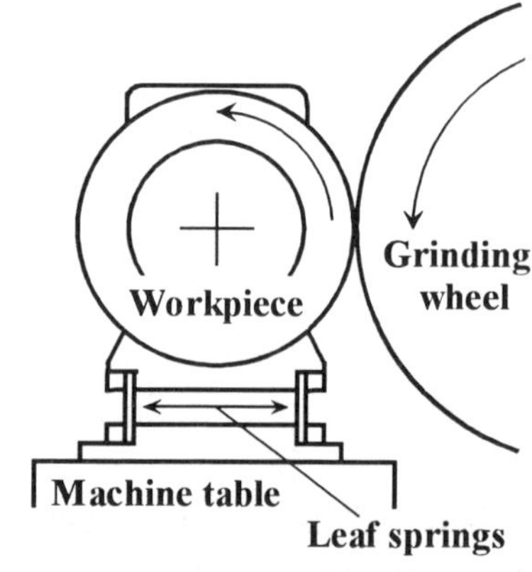

Figure 3 Construction of stiffness simulator for grinding machine

4. EXPERIMENTAL RESULTS AND DISCUSSION

Variation of workpiece size and normal grinding force F_n in infeeding process is shown in Figure 4. These results indicate a tendency that normal grinding force gradually increase with the grinding time t and converge to constant amounts. Simultaneously, preliminary set eccentricity of workpiece decreases with the grinding time and its converging time is short when α is small.

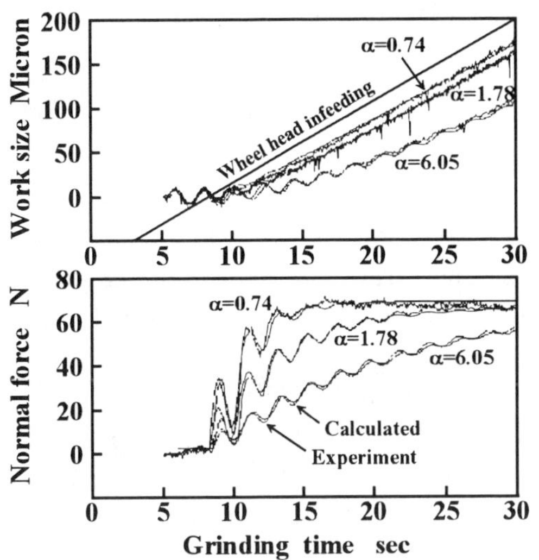

Figure 4 Center generating process in infeeding

In this figure, calculated results from Equations (2) and (3) are also shown by thin lines. Where, the contact stiffness between wheel and workpiece K_{con} has a hard spring characteristics depending on the normal grinding force F_n [4][5]. In this study, however, standing on the point to evaluate the Equations (2) and (3) experimentally, strict evaluation, in which K_{res} is non-linear, is not performed. K_{res} is assumed to be linear and an average K_{con} is used.

From the experimental results, it is confirmed that the true depth of cut $a_n(t)$ and residual stock removal $y_n(t)$ in infeeding process can be defined by Equations (2) and (3) respectively. In case of spark-out process, similar tendency has been confirmed experimentally.

5. GRINDING TIME TO OBTAIN ROUND WORKPIECES

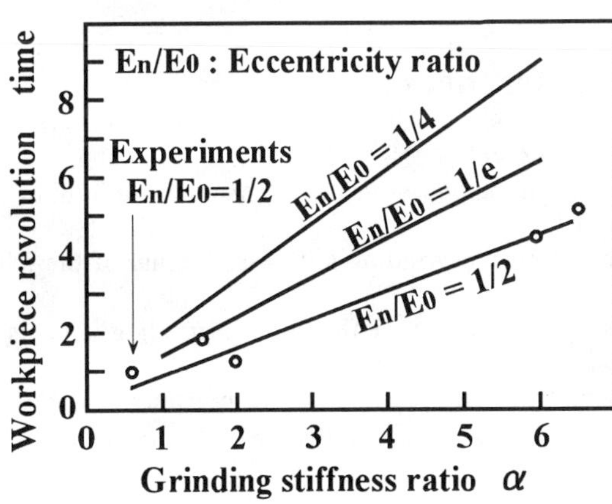

Figure 5 Relationship between grinding time and grinding stiffness ratio

Since the variation of depth of cut due to the workpiece eccentricity is defined by $f(t)$ in Equation (3), the eccentricity of workpiece is represented only by the first term in Equation (3). In case of spark-out, Equation (5) has to be considered. The amount of workpiece eccentricity after nth time grinding E_n in infeeding and/or spark-out process can be

557

represented as follows from Equations (3) and (5).

$$E_n = \left(\frac{\alpha}{1+\alpha}\right)^n E_0 \qquad (6)$$

Where, E_0 is an amount of initial eccentricity of workpiece. Consequently, decreasing process of workpiece eccentricity can be evaluated by the ratio of E_n/E_0, that is, $\{\alpha/(1+\alpha)\}^n$. Figure 5 shows calculated examples of the relationship between grinding stiffness ratio α and grinding time n. In this figure, experimental results under $E_n/E_0 = 1/2$ are plotted. From this calculated example, appropriate grinding time corresponding to required reducing rate of eccentricity can be selected easily.

4. CONCLUSIONS

The results obtained in this study can be summarized as follows.
1) Center generating process of the eccentric workpiece can be monitored in infeeding and spark-out process.
2) It is experimentally confirmed that the center generating process of the eccentric workpiece depends on the stiffness ratio between grinding stiffness and resultant static stiffness of grinding system.
3) Depending on the effect of the stiffness ratio on the grinding time to obtain the round workpiece shape, a method to estimate the required shortest spark-out time is proposed.

Acknowledgement

We are much grateful to Dr. Masumi Izumi and Mr. Yujiro Ito for having helped to do this experimental work.

References

1. Lee H, Furukawa Y. On the method to determine the stiffness of grinding machines. Bulletin of the JSPE; 1988 Vol.22, No.2; 127-132.
2. Lee H, Furukawa Y. Stiffness design method of grinding machine. Bulletin of the JSPE; 1990 Vol.24, No.2; 136-141.
3. Inasaki I, Yonetsu S. Experiments on the grinding stiffness. Journal of the JSPE 1970; Vol.36, No.3 207-211.
4. Brown R.H, Saito K, Shaw M.C. Local elastic deflections in grinding, Annals of the CIRP1971; Vol.XVIV 105-113.
5. Fukuda R, Tokiwa T. A study on the static stiffness between grinding wheel and workpiece, Journal of the JSPE; 1974 Vol.40 No.10 809~814.

DEVELOPMENT OF A NEW RESEATABLE MECHANISM WITH HIGH ACCURATE POSITIONING REPRODUCIBILITY

Satoshi Koga and Nobuhisa Nishioki
Mitutoyo Corporation

Abstract

A six-point-contact reseatable mechanism is most common as a simple and high reproducible one. To improve reproducibility of the reseatable mechanism, we have developed a High Reseatability Mechanism (HRM) which has the strain restoring effect, and have applied it to a touch trigger probe for a Coordinate Measuring Machine (CMM)[1],[2]. This paper presents the development of a new HRM with the improved strain restoring effect based on a dynamic simulation. Measurement results of the new HRM applied to the probe show that the fluctuation width of the relocated positions before the HRM action is 230 nm and 50 nm the after.

Keywords

reseatable mechanism, elastic deformation, strain restoring effect, Coordinate Measuring Machine, touch trigger probe

1. INTRODUCTION

In future manufacturing process for intricate and precise parts having dimensions of several mm, there would often be situations that a work being installed upon a detachable rigid tray has to be moved from one processing station to another, then returned to the previous processing station afterward. Therefore, a simple and high reproducible reseatable mechanism would be required between the tray and the processing station's receiver. Figure 1 shows an example of above reseatable mechanism which is called a six-point-contact reseatable mechanism.

Because a touch trigger probe for a Coordinate Measuring Machine (CMM) also needs a high reproducible reseatable mechanism, the six-point-contact reseatable mechanism is most commonly used. However, the reseatable mechanism brings a fluctuation of relocated positions caused by friction force and elastic deformations at the

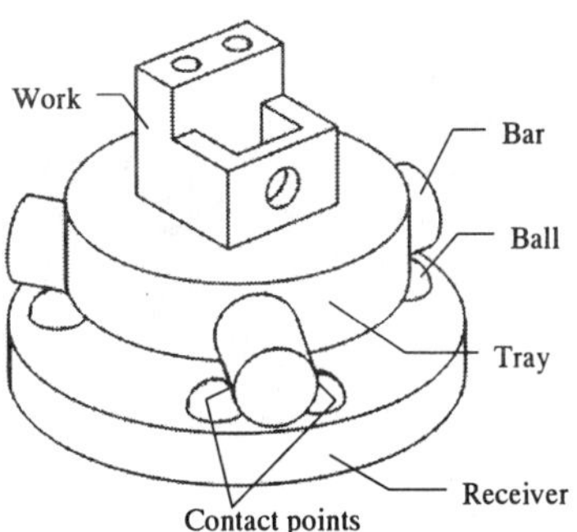

Figure 1. Example of reseatable mechanism

contact points between the tray and the receiver. The fluctuation is considered to be 0.1 μm order

To reduce the fluctuation, we have developed a High Reseatability Mechanism (HRM) which has the strain restoring effect and have applied it as the reseatable mechanism for the CMM's touch trigger probe[1],[2]. Furthermore, to drastically improve the strain restoring effect, a New HRM has been developed based on a dynamic simulation. A conspicuous improvement has been confirmed from experiments of the new HRM, and the contents are reported below.

2. NEW HIGH RESEATABILITY MECHANISM

2.1 Problem of usual HRM with straight bars

Relocated positions of the six-point-contact reseatable mechanism fluctuate depending on landing motions of the tray. It is impossible to remove the fluctuation of relocated positions as long as we can not make quite same landing motions each time.

An usual HRM is shown as Fig. 2. The technical point of the HRM is to reset strains at the contact points by making a relative displacement between three straight bars of the tray installing a stylus and six balls of the receiver. To make this relative displacement, the balls are moved to the radial directions by piezoelectric elements. We call this action "HRM action". After the HRM action, strains at the contact points are reset, so the tray is relocated to the designed position which does not depend on landing motions. However, it has been discovered from a result of HRM's dynamic simulation that the strains along axial direction are the only ones

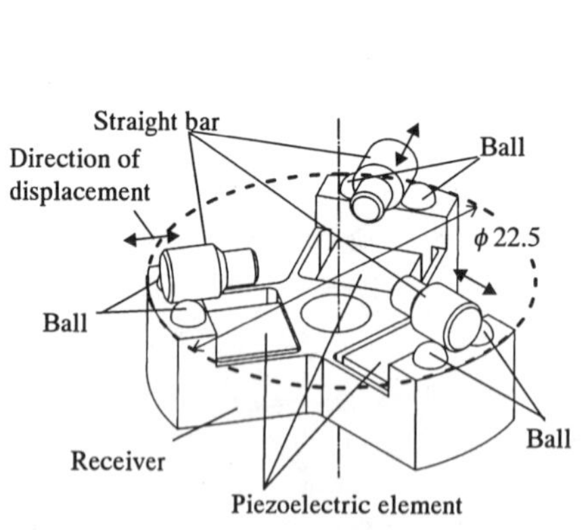

Figure 2. High Reseatability Mechanism (HRM)

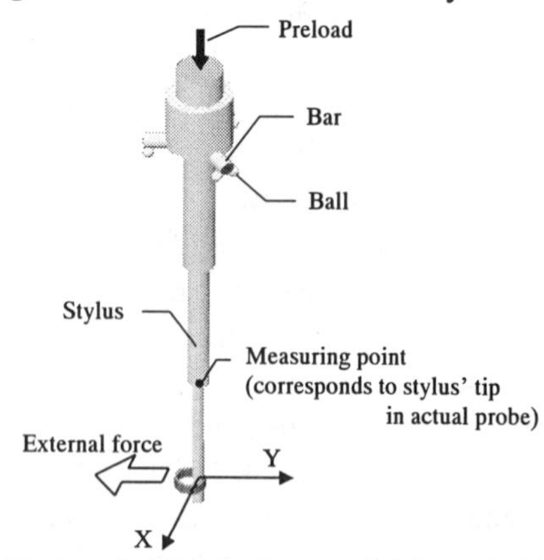

Figure 3. Simulation model for HRM

which are able to be reset because the HRM with straight bars makes one relative displacement along axial direction. Figure 3 is a dynamic simulation model of the HRM applied to the touch trigger probe, where ADAMS (mechanical dynamics simulation software) is used as a simulation tool.

2.2 New HRM with taper bars

As a solution for problem of the usual HRM, we propose to make the relative displacement along two directions crossing perpendicularly. If the relative displacement is made, the strains of all direction would be reset, and the fluctuation of relocated positions would be removed as well. In order to achieve the relative displacement simultaneously and easily, we have developed the new HRM with taper bars as shown in Fig. 4.

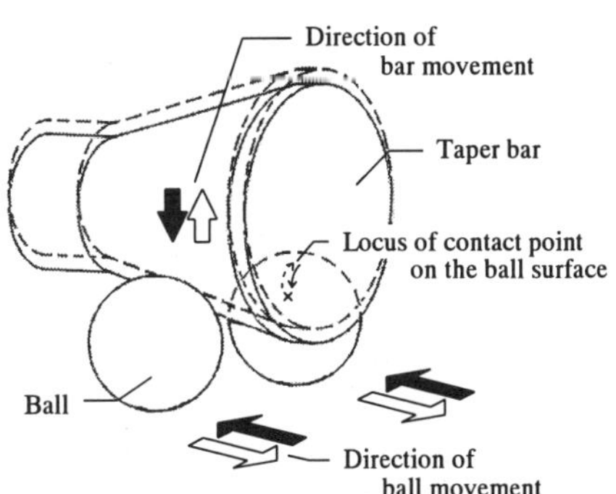

Figure 4. New HRM action with taper bar

When the new HRM is applied to the probe for a CMM, another effect is expected because the reproducibility of the stylus' tip becomes very important in the probe for a CMM. Figure 5 simply drawn in two dimensions shows the effect of the new HRM. The root of the stylus in the new HRM with taper bars moves in a direction which is opposite to the direction of the external force. Therefore, the difference between the

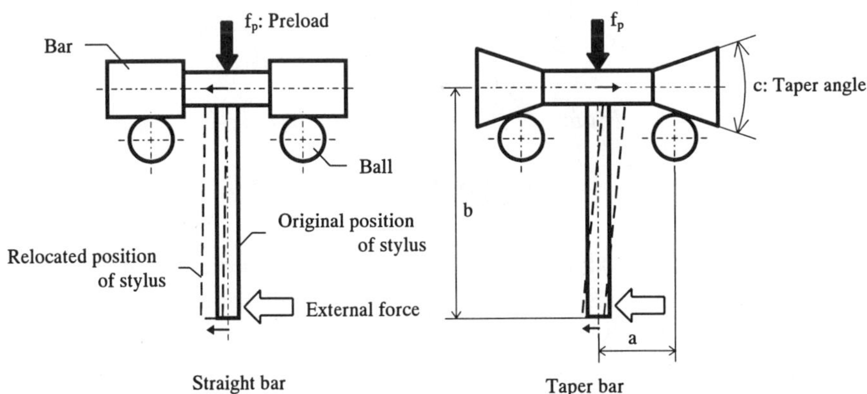

Figure 5. Difference of relocated positions

561

relocated position of the stylus' tip and the original position is smaller than the difference in the case of the straight bar. The effect is increased by selecting appropriate design parameters of the probe. Some design parameters which are used in the trial manufactured probe are shown as follows; a = 10 mm, b = 61.5 mm, c = 30° and f_p = 3.4 N.

3. PROBE ESTIMATION SYSTEM

Figure 6 shows a probe estimation system. This system has a image processing unit for measuring a position of stylus by detecting estimation-patterns on it. We use the geometrical center of the pattern as to representative positions. To decrease estimation errors, this system is installed in the room in which temperature is controlled and dynamic disturbance is reduced up to several "milligals" by using floating basement, moreover it is surrounded by heat insulating covers. A movement of the X-Y stage gives the external force to the stylus. The movement is equivalent to a measuring motion.

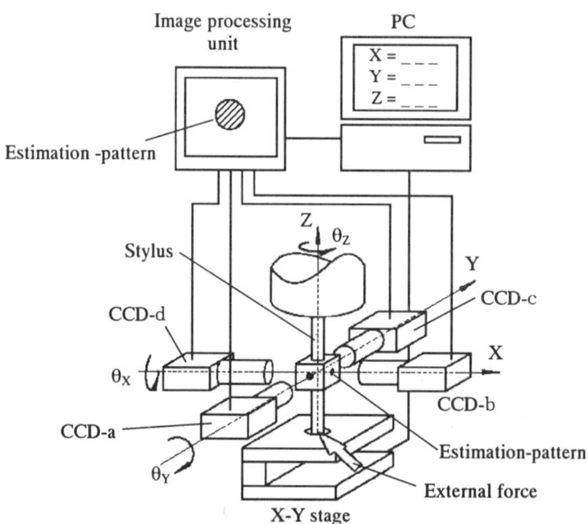

Figure 6. Probe estimation system

4. EXPERIMENTAL RESULTS

Figures 7 and 8 show the measured relocated positions of the stylus. The figure shows the relocated tip positions projected on X-Y plane being perpendicular to the stylus' axial direction. The external force is applied at 20 degrees interval, and the positions are measured thirty-six times (two laps). Circles shown in these figures are the minimum circles including all relocated positions in each case, and the diameters are evaluated as the fluctuation widths of relocated positions.   At first, we see from Fig. 7 that an epoch-making improvement has been made between two bars before HRM action since the fluctuation width of taper bar is 230 nm and the straight bar is 1600 nm.   Furthermore, after the HRM action the both fluctuation widths are 50 nm at the former and 200 nm at the latter.

From the result, just to change a straight bar to a taper bar, the strain restoring effect is greatly improved and the fluctuation widths of relocated positions are decreased.

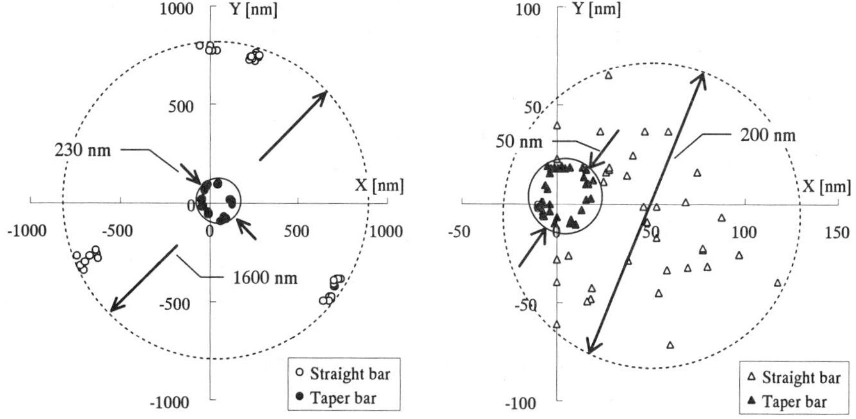

Figure 7. Relocated position before HRM action

Figure 8. Relocated position after HRM action

4. CONCLUSIONS

We have developed the new HRM with taper bars adopting the suitable design parameters, and have applied it to the touch trigger probe for a CMM.   Measurement results of the fluctuation width before the HRM action is 230 nm and 50 nm the after.   In conclusion, just to change a straight bar to a taper bar, the strain restoring effect is greatly improved and the fluctuation widths of relocated positions are decreased.

REFERENCES

(1) K.Hidaka, N.Ishikawa and K.Nishimura   High-Accuracy & High-Response Touch Trigger Probe for Coordinate Measuring Machine.   Proceedings of the 1st International Euspen Conference; 1999 May 31 – June 4; Bremen. Aachen: Shaker Verlag, 1999
(2) K.Hidaka and K.Nishimura   Development of Reseat Mechanism in Touch Trigger Probe for CMM.   Journal of the Japan Society for Precision Engineering 2000; 2: 298-303 (in Japanese)

COMPENSATION OF PERIODIC ERRORS BY GAIN TUNING IN MULTI-PROBE SENSORS

Ivan Godler and Tamotsu Ninomiya*

*Kitakyushu University, *Kyushu University*

Abstract

Periodic errors need to be compensated in precise shape measurements and in sensing of physical quantities. We propose a method to compensate any periodic error by using gain tuning technique. The tuning is performed at initial calibration of the sensor with multiple probes. The proposed method, needed number of probes, simulated and experimental results are shown in the paper.

Keywords

Periodic error, compensation, gain tuning, multi-probe sensor

1. INTRODUCTION

Accurate sensing of physical quantities is important in engineering practice. However, especially in the case of rotary types of sensors, periodic errors are frequently generated. To compensate the periodic error, two or more identical probes are positioned on opposite sides of a rotating shaft to generate error signals of opposite phases, which are mutually canceled. Needless to mention, probes are not always identical, positioning errors, and shape errors of the shaft and other influences contribute to non-symmetry, altogether producing non-ideal conditions for the error compensation.

A method to compensate eccentricity error in roundness measurements by using three or more probes was proposed by Gao [1], [2]. It was shown, that the eccentricity error could be removed from the measurement signal. However, in instrumentation practice, there is not only eccentricity that needs to be compensated. Other frequency components, e.g. roller bearings influence etc., need to be compensated too.

The authors developed a method to compensate periodic error in the case of torque sensing from a gear reducer (Godler [3]). To solve the same problem, a method by using Kalman filter was proposed by Taghirad [4], but it requires on-line calculations to compensate the periodic error, which makes it unsuitable for low-cost and fast real-time applications.

In this paper we present a method to compensate any arbitrary frequency component of an error by using three probes with tuned gains. Also, by applying more probes, multiple frequency components can be compensated.

The paper is organized as follows: in section 2 we derive a mathematical model of the periodic error, in section 3 we present the

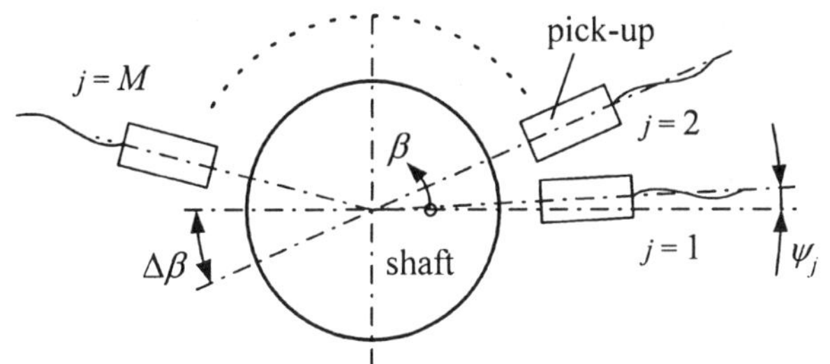

Figure 1. Model of a sensor

proposed method, in section 4 we derive needed number of probes, and in section 5 we show simulated and experimental results.

2. MATHEMATICAL MODEL OF A PERIODIC ERROR

A rotary type of a sensor with shaft's eccentricity can be used as a model to study periodic errors. An example of a sensor is schematically shown in Figure 1. Here, with an aim to produce a general model of a periodic error, we assume that M probes are arranged around a rotating shaft on equal angular distance $\Delta\beta$, with some angular positioning errors ψ_j at each of the probes. Using such a model, we can express the periodic error e_j from the j-th probe as a periodic function of the shaft's rotation angle β.

$$e_j = \sum_{i=1}^{N} a_{ij} \cos\left[f_i\left(\beta - (j-1)\Delta\beta - \psi_{ij}\right)\right] \tag{1}$$

In general, not only the shaft's eccentricity, but also other sources, including bearings influence, contributes to the periodic error. Therefore, in (1) it is assumed that N frequency components of a periodic error are present, each with its own frequency f_i, amplitude a_{ij}, and phase error ψ_{ij}. The parameters, contributing to the phase of a periodic error's frequency component can be collected into a phase angle ϕ_{ij}.

$$\phi_{ij} = f_i\left((j-1)\Delta\beta + \psi_{ij}\right) \tag{2}$$

A total output of a sensor is a sum of M outputs from M probes. Accordingly, the periodic errors from M probes are also added into an overall periodic error h. Considering this and applying (2), we obtain a mathematical model of the overall periodic error.

$$h = \sum_{j=1}^{M} e_j = \sum_{i=}^{N} \sum_{j=1}^{M} a_{ij} \cos\left(f_i\beta - \phi_{ij}\right) \tag{3}$$

3. PROPOSED PERIODIC ERROR COMPENSATION

The above-derived mathematical model of a periodic error is used to develop a method to compensate it. A condition to achieve zero periodic error is

$$h = \sum_{i=1}^{N} \sum_{j=1}^{M} a_{ij} \cos\left(f_i\beta - \phi_{ij}\right) = 0, \tag{4}$$

and is examined in the following.

A relation $\cos(f_i\beta - \phi_{ij}) = \cos\phi_{ij}\cos f_i\beta + \sin\phi_{ij}\sin f_i\beta$ is introduced into (4) to obtain the following equation.

$$h = \sum_{i=1}^{N} \sum_{j=1}^{M} a_{ij} \left(\cos \phi_{ij} \underline{\cos f_i \beta} + \sin \phi_{ij} \underline{\sin f_i \beta} \right) = 0 \tag{5}$$

Here the periodic error is decomposed into cosine and sine components of each frequency component (underlined in (5)). Notice that the amplitudes a_{ij} as well as the phases ϕ_{ij} contribute only to the amplitudes of the corresponding cosine and sine components of the periodic error. Recognizing the sum of cosine and sine components is zero regardless to the value of the argument only when amplitudes of both components are simultaneously zero and to achieve $h = 0$ in (5), the following homogenous system of equations must be satisfied.

$$\begin{aligned} \sum_{j=1}^{M} a_{ij} \cos \phi_{ij} &= 0 \\ \sum_{j=1}^{M} a_{ij} \sin \phi_{ij} &= 0 \end{aligned} \quad , \quad i = 1,2,\ldots,N \tag{6}$$

A solution of (6) can be found by adjustment of the phases ϕ_{ij} or by tuning the amplitudes a_{ij}. In general, after a sensor is manufactured, it is easier to tune the amplitudes than to adjust the phases. The method, which we propose is therefore to tune the amplitudes of separate probes. This can be realized by amplifier circuits or by software. The advantage of the method is that the gains can be tuned initially during the sensor's calibration and remain constant during the sensor's operation, with an assumption that the periodic errors do not change during the sensor's operation. In this case, on-line calculations and error estimations are not needed.

4. NUMBER OF PROBES

By gain tuning at each of the probes all frequency components of the error from one probe are amplified by the same factor. Having this in mind, we add a tuning factor k_j to each of the signals from the probes, affecting only the amplitudes a_{ij}. Equation (6) with gain tuning property added is therefore expressed in the following matrix form.

$$\begin{bmatrix} a_{ij} \cos \phi_{ij} \\ a_{ij} \sin \phi_{ij} \end{bmatrix}_{(N \times M)} \left\{ k_j \right\}_{(M)} = \mathbf{0} \tag{7}$$

A condition for nontrivial solution of k_j in (7) is related to the size of the matrix, and is expressed as $M \geq 2N + 1$. This gives a relation between the number of probes M and the number of frequency components N, which need to be compensated. A minimum number of probes M_{min} to compensate a given number of frequency components N is therefore

$$M_{min} = 2N + 1. \tag{8}$$

Notice that to perfectly compensate one frequency component ($N = 1$), minimum of three ($M_{min} = 3$) probes are needed. To perfectly compensate two frequency components ($N = 2$), minimum of five ($M_{min} = 5$) probes are

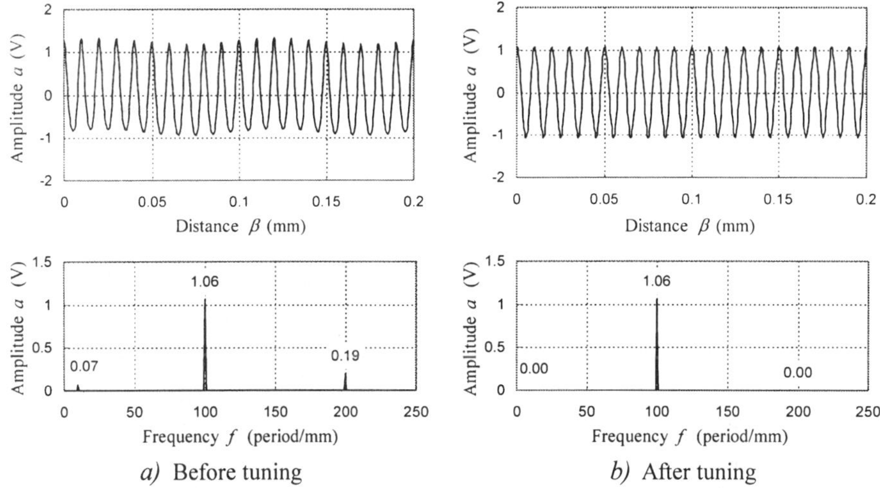

Figure 2. Simulated example of two frequency components simultaneously compensated by five probes

needed, and so on.

In the proposed method, it is not necessary to compensate the frequency components in order from the lowest one to the higher ones. Any frequency component can be freely selected for compensation. For example, we can select only the component that has the largest amplitude or any other frequency component, which is critical for a given application.

5. SIMULATION AND EXPERIMENT

A simple simulated example where two frequency components of the error signal are simultaneously compensated is shown in Figure 2. The frequency component 100 period/mm is used for distance sensing in a linear scale type sensor, while the other two frequency components are the sensing error (Figure 2a). By solving the system of equations (7), the two error frequency components are compensated (Figure 2b).

To verify the proposed method by an experiment we produced a rotary incremental encoder with five photo interrupters (probes) on the periphery of a slotted disc with 1000 slits per revolution. A frequency spectrum of the signal from one probe with a slotted disc constantly rotating is shown in Figure 3a. Additionally to the basic component of 1000 periods/revolution, there are higher harmonic components present in the signal. Amplitudes of the higher harmonics are very low, so that a bare eye cannot recognize their presence in the time signal.

Let us assume that we need to compensate the higher harmonic components of the signal to achieve more accurate sensor. A result of 2000 periods/revolution frequency component compensated by using three probes is shown in Figure 3b. The calculated gains are: −3.63, 8.38, and

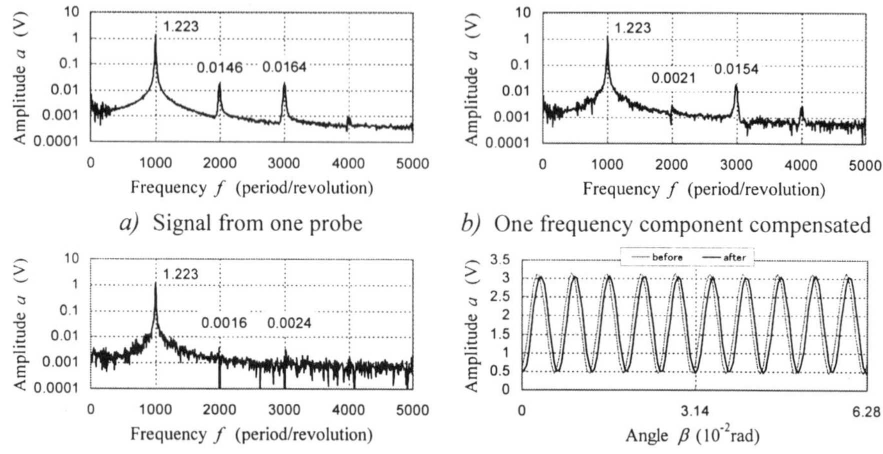

a) Signal from one probe

b) One frequency component compensated

c) Two frequency components compensated

d) Signal before and after tuning

Figure 3. Experimental results of one and two frequency components compensated by three and five probes respectively

−1.35. Another experimental result of two frequency components compensated is shown in Figure 3c. Here five probes were used with the gains: 2.19, 4.36, 3.44, 21.1, and −25.8. The frequency spectrum of the signal appears to be 'noisy', but understanding that the spectrum is not a time signal we realize that the tuning influenced amplitudes of various frequency components, appearing as a 'noisy' spectrum. However, the amplitudes do not differ significantly from the original spectrum, except the two higher harmonics, which we wanted to compensate. A time signal before and after compensation of the two frequency components is shown in Figure 3d, and exhibits no additional noise after tuning.

6. CONCLUSION

A method to compensate periodic errors was proposed in this paper and verified by simulation and experiments. The method is applicable to linear and rotary types of sensors with multiple probes. Equations to perform the gains tuning and to obtain the number of probes needed to compensate a given number of error's frequency components were derived.

REFERENCES

1. Gao W., Kiyono S., Nomura T. A new multi probe method of roundness measurements. Precision Engineering 1996; 1:37-45
2. Gao W., Kiyono S., Sugawara T. High-accuracy roundness measurement by a new error separation method. Precision Engineering 1997; 2/3:123-133
3. Godler I., Ninomiya T., Horiuchi M. Torque sensing from Harmonic Drives by using three strain gages, Proceedings of the 6th International Workshop on Advanced Motion Control; 2000 March 30 – April 1; Nagoya.
4. Taghirad H.D., Belanger P.R. Intelligent built-in torque sensor for harmonic drive systems. IEEE Transactions on Instrumentation and Measurement 1999; 6:1201-1207

PRECISION POSITIONING OF
A SURFACE MOTOR-DRIVEN XYΘ_Z STAGE
USING A SURFACE ENCODER

Wei Gao, Tomonori Nakada, and Satoshi Kiyono

Department of Mechatronics and Precision Engineering, Tohoku University

Aramaki Aza Aoba01, Sendai, 980-8579 JAPAN

Phone & Fax: +81-22-217-6951   E-mail: gaowei@cc.mech.tohoku.ac.jp

Abstract

This paper described a surface motor-driven $XY\Theta_Z$ stage equipped with a newly developed surface encoder as the position-detection sensor. The surface encoder consists of a two-dimensional angle sensor and a two-dimensional angle grid. The angle sensor is fixed on the stage base and the angle grid is mounted on the back the moving element of the stage. The multi-axis position of the moving element can be obtained from the angle sensor output. The surface encoder is placed inside the stage and the stage system is very compact in size. In this paper, experimental results in XY axes were presented. The travel ranges are 40 mm in XY axes the positioning resolution is better than 0.2 µm.

Keywords

metrology, multi-axis position detection, stage, surface motor, surface encoder

1. INTRODUCTION

Precision multi-axis (XY and/or $XY\Theta_Z$) stages are widely used in semiconductor manufacturing systems, precision machine tools and scanning probe measurement systems. These stages are required to have characteristics of high positioning accuracy, high speed, long travel range, and being compact in size.

Most of the multi-axis stages, however, are constructed by superposing one-axis stages. This results in an unbalanced and bulky structure, which reduces greatly the positioning accuracy and the speed of the multi-axis stage.

On the other hand, multi-axis planar motion stages driven by surface motors have being developed recently [1,2]. This kind of stages can easily generate translational motions in X and Y axes and rotational motions around Z axis with linear motors placed in the same plane. The single layer structure of the stage system is expected to overcome the problems of the conventional multi-axis stages.

However, it is a great challenge to measure the multi-degree-of-freedom position of such a stage. In this paper, we present a surface motor-driven $XY\Theta_z$ stage equipped with an accurate positioning system based on a newly developed surface encoder [3].

2. SURFACE ENCODER FOR MULTI-AXIS POSITION DETECTION

Figure 1(a) shows a schematic of the surface encoder which is used as the position-detecting sensor of the surface motor-driven stage. The surface encoder consists of a two dimensional angle sensor and a two-dimensional angle grid, on which two-dimensional sinusoidal waves are generated.

The height profile of the angle grid, which is a superposition of sinusoidal waves in the X and Y directions, can be expressed as:

$$f(x,y) = A_x \sin(\frac{2\pi}{\lambda_x}x) + A_x \sin(\frac{2\pi}{\lambda_x}y) \tag{1}$$

where, A_x, A_y are the amplitudes of the sine functions in the X direction and Y direction, respectively. λ_x and λ_y are the corresponding wavelengths.

The two-dimensional outputs $\alpha(x)$ and $\beta(y)$ of the angle sensor, which indicate the local slopes of the angle grid in the X direction and Y direction, can be obtained from the differentiation of f(x,y) ($\alpha(x) = \partial f / \partial x$, $\beta(x) = \partial f / \partial y$). The two-dimensional components (x and y) of the position can then be determined from the sensor output $\alpha(x)$ and $\beta(y)$ as:

$$x = \frac{\lambda_x}{2\pi}\cos^{-1}(\frac{\lambda_x}{2\pi A_x}\alpha(x)) \quad (2) \qquad y = \frac{\lambda_y}{2\pi}\cos^{-1}(\frac{\lambda_y}{2\pi A_y}\beta(x)) \quad (3)$$

Figure 1(b) shows a 3D view of a part of the angle grid generated by the

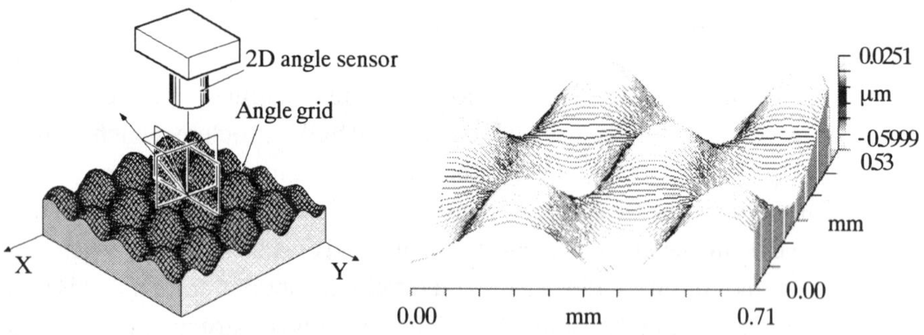

(a) Schematic of the surface encoder   (b) 3D view of a part of the generated angle grid

Figure 1 The surface encoder

technique of fast tool servo on a diamond turning machine[4]. The specifications of the angle grid used in the experiment are as follows:

Amplitude: $A_x=A_y=300\,\mu m$ Wavelength: $\lambda_x=\lambda_y=300\,nm$

3. SURFACE MOTOR-DRIVEN STAGE

Figure 2 shows the moving principle of the surface motor-driven stage. The stage basically consists of a moving element, a stage base and a surface motor. The surface motor is composed of four linear motors (two pairs in XY-axes), each has one permanent magnet array and a stator coil. The permanent magnet array is mounted on the back of the moving element, and the stator winding is placed on the stage base. The moving element can be moved in X-axis by the X-linear motors, and in Y-axis by the Y-linear motors as shown in Figures 2(a) and 2(b). The moving element can also be rotated around the Z-axis (Θ_z) if the X- or Y-linear motors generate opposite forces (Figure 2(c)).

Figure 3 shows the schematic and a photograph of the whole stage system. The angle sensor is fixed on the stage base and the angle grid is mounted on the back the moving element. The multi-axis position of the moving element can be obtained from the angle sensor output. The diameter of the angle grid was 55mm. The travel ranges of the stage are designed to be 40 mm in XY axes and approximately 10 degrees in Θ_z direction.

4. EXPERIMENTS

Experiments were carried out to investigate the basic performances of the stage system in XY axes. Figure 4 shows a result of testing the position-

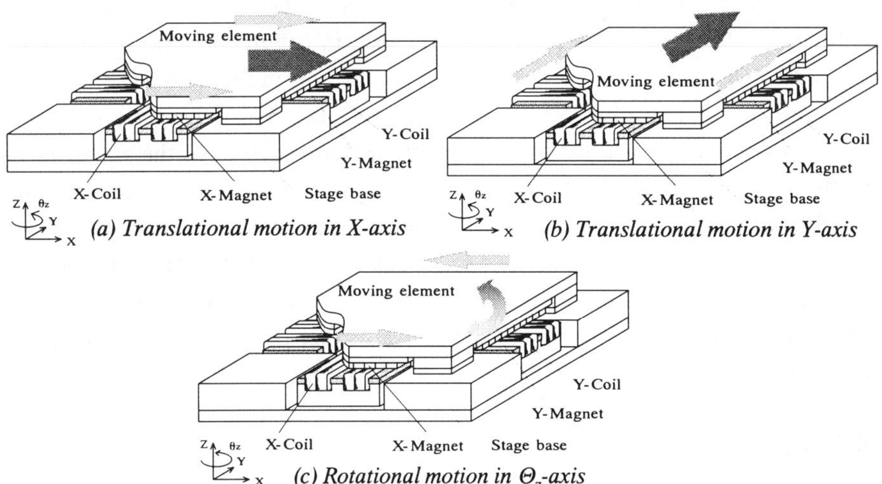

Figure 2 Moving principle of the surface motor-driven XYΘ_z stage

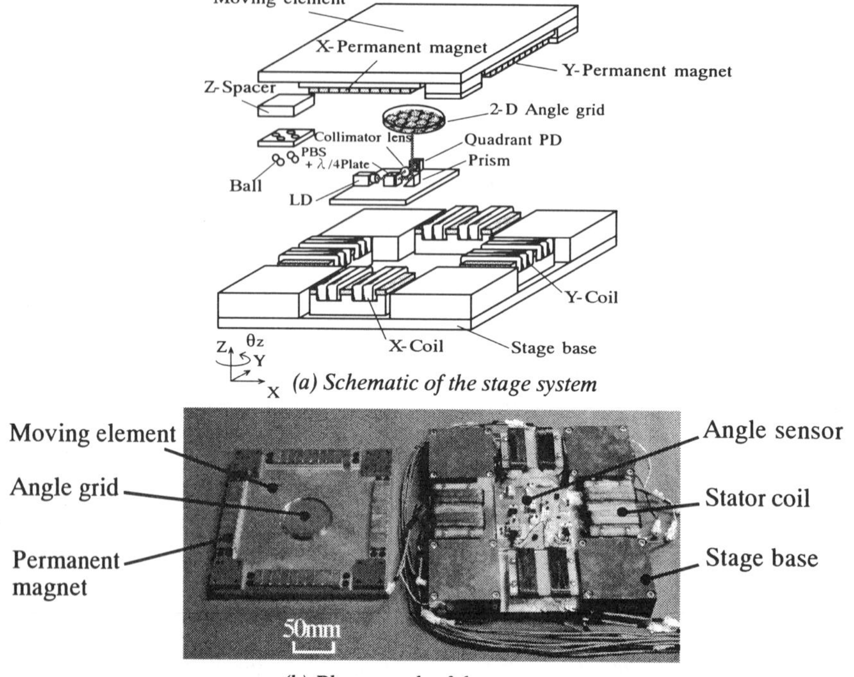

(a) Schematic of the stage system

(b) Photograph of the stage system

Figure 3 The surface motor-driven stage system with the surface encoder

detecting resolution of the surface encoder. The stage was driven along the X-axis with a 0.2 μm step and the movement was simultaneously measured by the surface encoder and an interferometer (HP 10706B, resolution 0.6 nm) which was placed outside of the stage. As can be seen in Figure 4, both the surface encoder and the interferometer responded to the step-movement of the stage, and the surface encoder has almost the same resolution as the interferometer.

Figure 5 shows a result of PTP (Point to Point) drives of the stage. The stage was driven along a L-shape route in XY-plane with closed-loop control using the output of the surface encoder. The step of each drive was 300 μm. A software PI controller was used for feedback control of the stage. It can be seen that the stage was driven fairly well in both X and Y axes. The positioning error can be reduced through calibration of the surface encoder, and this will be performed in our future work.

5. CONCLUSIONS

A multi-axis precision stage driven by a surface-motor has been proposed. A surface encoder placed inside the stage was used to detect the multi-axis

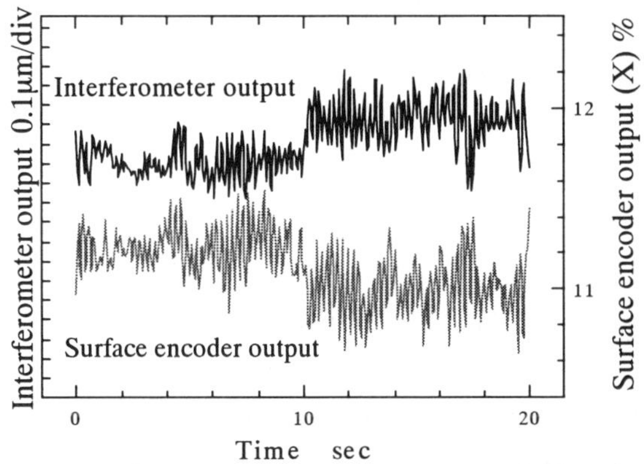

Figure 4 Surface encoder output at a small step drive

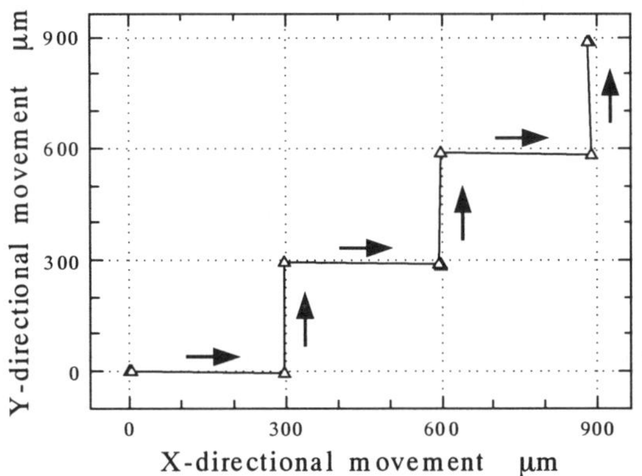

Figure 5 A closed-loop PTP drive result (L-shape drive with 300μm-step)

position of the stage. Basic performances of the stage as well as the surface encoder have been investigated by experiments.

ACKNOWLEDGMENTS

This work was supported by NEDO Industrial Technology Research Grant Program in '00 and a grant from Electro-Mechanic Technology Foundation. The work on the surface encoder was supported by a JSPS Grant-in-Aid for Scientific Research (No. 11305013). The authors appreciate Dr. R. J. Hocken of The Univ. of North Carolina at Charlotte, Dr. Y. Tomita of Sumitomo Heavy Industries Ltd., Mr. K. Yakura of NSK Ltd. for their advises and contributions.

REFERENCES

1. Y. Tomita, Y. Koyanagawa, and F. Satoh: A surface motor-driven precise positioning system, Prec. Eng., 16-3, (1994), 184-191

2. M. Holmes, R. Hocken, and D. Trumper: The long-range scanning stage: a novel platform for scanned-probe microscopy, Prec. Eng., 21, (2000), 191-209.

3. S. Kiyono, P. Cai, and W. Gao: An angle-based position detection method for precision machines, Int. J. of JSME, 42-1, (1999), 44-48.

4. W. Gao, Sudoh, S. Kiyono: Development of a machining system for generating 2D angle grid, Proc. of JSME General Assembly, III, (2000), 519-520.

FEEDFORWARD TRACKING CONTROLLER DESIGN BASED ON A LIMITED BANDWIDTH DESIRED MODEL

Lisong Wang, Feihu Zhang, Baoku Su

(Dept. of Control Theory and Engineering, Harbin Institute of Technology)

Abstract

To solve the high accuracy trace problem of precision machine tool, this paper presents a limited bandwidth desired model which utilizes future information. The model provides the transfer function with the following frequency characteristics. The gain is equal to one at given frequencies and zero at other frequency ranges. The phase is equal to zero for all frequencies. In this system, the feedback controller and feed-forward compensator were adopted to fulfil the model approximately. When the feed-forward and feedback controller were applied to the practical system, good trace result was obtained in experiment.

Keywords

Precision machine tool   Limited bandwidth desired model
Servo system   Future path information

1. INTRODUCTION

In high speed and high precision machine, tracking control has become more and more important. As future information on the desired output is available, preview control has drawn consideration in the past years. By using a few preview steps of desired output, the phase transfer frequency characteristics is zero for all frequencies and gain is equal to unity at zero frequency. But at high frequencies, the gain characteristic is not perfect. This paper proposed a new kind of model which can assure that gain is equal to one at given frequencies and zero at other frequencies. How to fulfill this model in controller design and verification are also expressed in this paper.

2. THE LIMITED BANDWIDTH DESIRED MODEL

The goal of designing a system is to make system output equal to input entirely. Suppose $G_B(z)$ is linear time invariant discrete system; $G_B(z)$ is asymptotically stable, and $G_c(z)$ is feedforward compensator. The closed loop transfer function $G(z)$ between output $x(t)$ and input $R(t)$ can be expressed as:

$$G(z) = \frac{x(z)}{R(z)} = G_C(z) \cdot G_B(z) \tag{1}$$

When $G_c(z)$ equals to $G_B(z)^{-1}$, the transfer function $G(z)$ has unity gain and zero phase lag . Supposed $G_B(z)$ is given by

$$G_B(z) = \frac{z^{-d} b^+(z^{-1}) b^-(z^{-1})}{a(z^{-1})} \tag{2}$$

where $\alpha(z^{-1})$ has all the poles in unit circle, and $b^+(z^{-1})$ contains all the zeros in unit circle and $b^-(z^{-1})$ contains all the zeros not included in unit circle. Thinking of the effect of uncancelled zeros $b^-(z^{-1})$, if $G_c(z)$ is given by:

$$G_C(z) = \frac{z^d a(z^{-1}) b^-(z)}{b^+(z^{-1})[b^-(1)]^2} \tag{3}$$

From equation (2) and (3), $G(z)$ becomes:

$$G(z) = \frac{b^-(z^{-1}) b^-(z)}{[b^-(1)]^2} \tag{4}$$

So, $G_c(z)$ is the inverse system of $G_B(z)$ based on pole-zero cancellation[1-2] . Let $z=e^{j\omega T}$, where T represent sample period. It is easy to verify that the phase lag between input and output is zero for all frequencies and the gain is one when ωT approach to zero. In practical application, the gain characteristic is of more importance. A gain characteristic which is equal to one at given frequencies and zero at other frequencies is often desired. In this paper, a limited-bandwidth-desired model was proposed as:

$$W(z,n) = \frac{1}{n^2}(1 + z^{-1} + z^{-2} + \cdots + z^{-n})(1 + z + z^2 + \cdots + z^n) \tag{5}$$

W(z,n) can assure not only the phase equals to zero for all frequencies, but also the gain characteristic is equal to one at any selection frequencies. The magnitude frequency characteristic curve is shown in Fig 1:

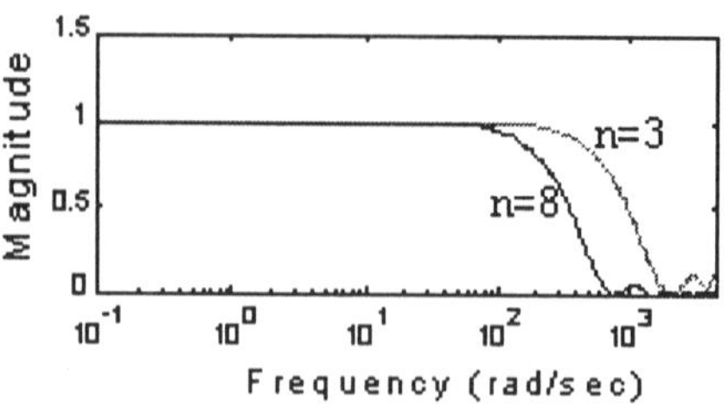

Fig 1. The magnitude frequency characteristic of W(z,n)

Where n represent the order of z. In this figure, n is set to 3 and 8 separately, and the desired model of different bandwidth is obtained.

It should be noted that both the inverse system design and the limited bandwidth desired model would use future information, which is feasible in CNC control.

3. SYSTEM CONFIGURATION

In an ultra-precision machine tool servo control system, an AC servomotor and a ball screw are used to implement the position servo feed. A two-frequency laser interferometer is used to detect the actual position. The system configuration is shown in Figure 2:

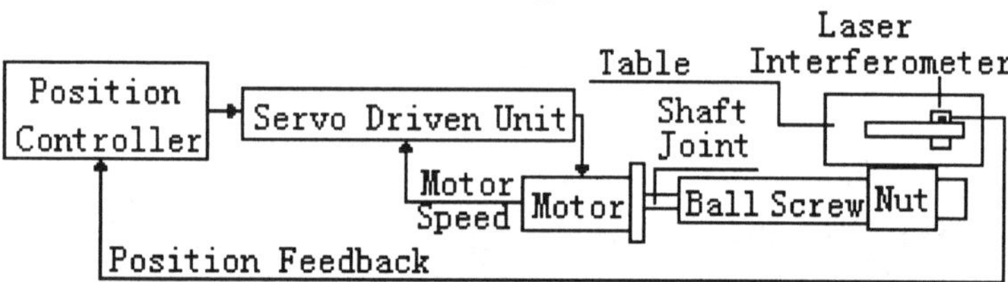

Fig 2. Diagram of the system configuration

By system identification, an z-plane open loop transfer function of the servo system is obtained as equation 6:

$$G_x(z) = \frac{0.4322 \times 10^{-4}(1+z^{-1})^2}{(1-z^{-1}) \cdot (1-0.7499z^{-1})}$$

(6)

4. CONTORLLER DESIGN

In this paper, a feedback plus feedforward controller is adopted, which is shown in Fig 3. The feedback controller D(z) is used to tranquilize plant and the feedforwand controller $G_F(z)$ is utilized to improve the system performance. As a type-I system, the goal of feedback controller design is to make the servo system

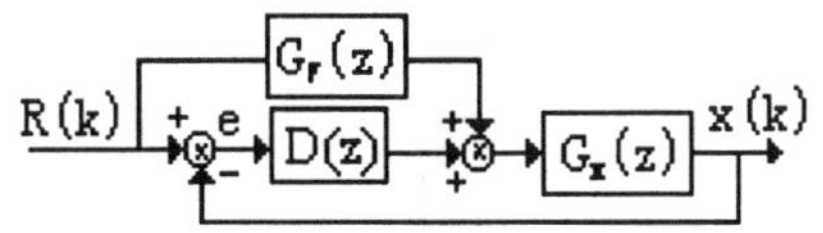

Fig 3. *The correction control system*

with broad bandwidth and high stiffness to get a fast response speed and anti-disturbance capability[3]. A conventional PI type feedback controller is designed for the system with bandwidth of 10Hz. The controller is as:

$$D(z) = \frac{91.0518 \cdot (z - 0.9969)}{z - 1} \tag{7}$$

The aim of this design to make the correction close-loop system frequency characteristic have the form of W(z,n), shown in equation 8:

$$W(z,n) = \frac{D(z)G_x(z) + G_F(z)G_x(z)}{1 + D(z)G_x(z)} \tag{8}$$

So long as $G_F(z)$ satisfy equation 8, the desired model W(z,n)can be acquired. To get $G_F(z)$, the inverse system of $G_x(z)$ should first be calculated. As uncancelled zeros exit in $G_x(z)$, an approximate stable inverse system of $G_x(z)$ is obtained by the method formulated in section 2, which is shown in $G_{ny}(z)$:

$$G_{ny}(z) = \frac{(1 - z^{-1}) \cdot (1 - 0.7499\ z^{-1}) \cdot (1 + z)^2}{0.4322 \cdot 10^{-4} \cdot 2^4} \tag{9}$$

Suppose $G_{F1}(z) = G_F(z)\ G_x(z)\ G_{ny}(z)$, from equation 8, G_{F1} is given as:

$$G_{F1}(z) = W(z,n) \cdot G_{ny}(z) + [W(z,n) - 1] \cdot D(z) \cdot G_x(z) \cdot G_{ny}(z) \tag{10}$$

When $G_{F1}(z)$ is replace of $G_F(z)$ of equation 8, the approximate limited bandwidth desired model is obtained. $G_{F1}(z)$ is the feedforwad controller used in our system. As $G_{ny}(z)$ needs two step advance information, so n plus 2 step advance information has to be known. In our system, n is set to be 8.

5. EXPERIMENT RESULTS

To testify the feasibility of the methods mentioned above, the feedback plus feed-forward controller were applied to the system of HCM-1 ultra-precision machine tool servo system, which has a detecting resolution of 5nm. The working max turning speed of the machine tool is about 50mm/mim, which corresponds to the motor rotating frequency of 1Hz. So we track a sinusoidal signal with magnitude of 2mm and frequency changed from 0.1Hz to 4Hz continuously within 1 second. The tracking error was shown in Fig 4, where curve 1 represent the situation only a feedback controller is used, and curve 2 is the situation both

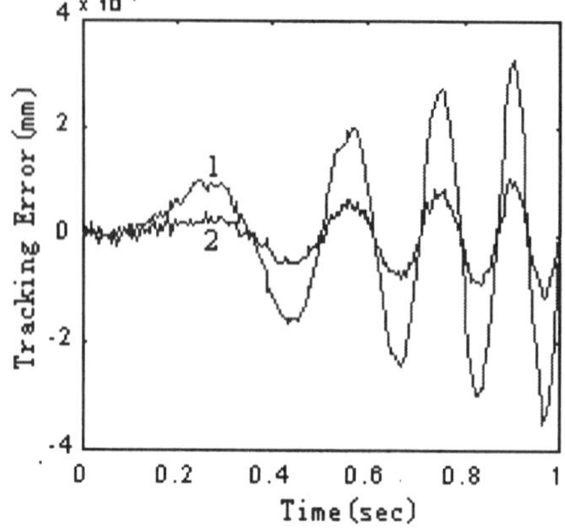

Fig 4. The tracking error curve

feedback and feeforward controller are used. It can be seen that the tracking error has been greatly reduced in the low frequency band by using our approximate desired model. The tracking error has been reduced from 100nm to 25nm at 1Hz.

6. CONCLUSION

In this paper, we proposed a limited bandwidth desired model, which was realized by using future path information. By selecting the order of the model, the bandwidth can be set at any selected range. Also the approximate implementation method is discussed in the paper. When the model is applied, good trace accuracy was obtained. It showed that the controller based on future path information has practical value in engineering application.

REFERENCES

[1] Hendrik Van Brussel. Accurate Motion Controller Design Based on an Extended Pole Placement Method and a Disturbance Observer. Annals of the CIRP. 1994,43(1):367-372

[2] Tomizuka.M. Zero Phase Error Tracking Algorithm for Digital Control. ASME Journal of Dynamic Systems, Measurement and Control.Mar.1987.Vol.109.pp.65-68

[3] Guangxiong, Control system design. Aerospace publishing house.1992

FUZZY-PID HYBRID VIBRATION CONTROL OF SLIDE CARRIAGE OF ULTRAPRECISION MACHINE TOOL

Dong Shen   Wang Jiachun   Li Dan

Precision Engineering Research Institute, Harbin Institute of Technology,

150001, Harbin, P. R. China

Abstract

Machine tool vibration is one of the key factors affecting the machining quality in precision and ultra-precision cutting. Hybrid vibration control of slide carriage is discussed in this paper. An active air-actuator based on PZT is developed to control the transverse vibration of slide carriage in the process of movement. Mathematical model of slide carriage vibration control is given. By using the merits of both fuzzy control and PID control, the composite fuzzy-PID control method is adopted. The experimental results show that this method can reduce the vibration of slide carriage of ultra-precision machine tool effectively.

Keywords

Slide carriage   hybrid vibration control   Fuzzy-PID control   air-actuator

1.INTRODUCTION

In precision and ultra-precision machining, the vibration of machine tool components is one of the key factors affecting the machining quality. For example, slide carriage vibration will worsen the machined surface of workpiece directly through tool post. How to eliminate vibration effectively is one of the critical problems in precision manufacturing. Some methods, such as aerostatic bearings, have been adopted in the design and manufacturing of precise slide carriage to reduce its vibration. By combining the active vibration control with passive vibration isolation, hybrid vibration control for slide carriage movement is discussed in this paper.

2.AIR-ACTUATOR

On the basis of aerostatic supporting, hybrid vibration control is obtained by adopting active vibration control technology to decrease the transverse vibration and swing motion around vertical axis of slide carriage. Because the transverse vibration control should perform in the process of slide carriage movement, conventional actuators are now unsuitable here. A non-contact active air-bearing actuator, as shown in Figure 1, is developed by the authors to reduce the carriage vibration.

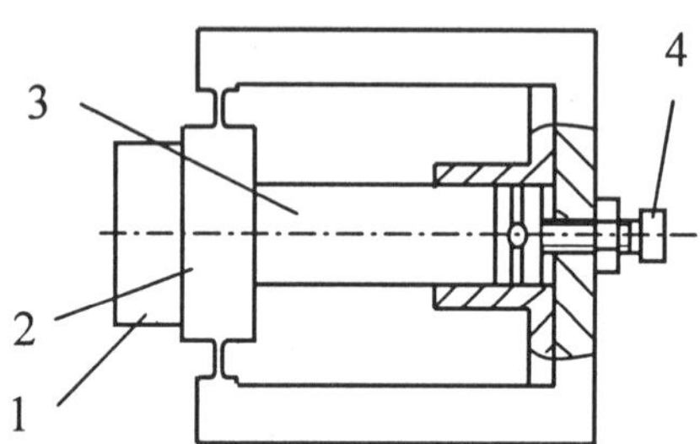

Fig.1 structural diagram of air-actuator
1-air-bearing pad; 2-elastic hinge;
3-piezoelectric ceramic; 4-adjusting screw

The amplified signal of slide carriage vibration signal drives the piezoelectric ceramics to change air film pressure by increasing or decreasing the thickness of air film between air bearing and the static guide. So the active controlling force changes. The elastic hinge enables the piezoelectric ceramics to stretch in time for good frequency response. Adjusting screw is used to preload the piezoelectric ceramics.

3.MATHEMATIC DESCRIPTION OF SLIDE CARRIAGE VIBRATION CONTROL SYSTEM

The slide carriage is supported by aerostatic film which performs passive damping. The active air-bearing actuator is fixed to the slide carriage. It provides active controlling force to the static guide plate through active air-bearing film. This forms a hybrid control with passive damping to lessen the slide carriage vibration. As the stiffness of slide carriage is sufficiently larger than that of the air film, the slide carriage can be simplified to be a concentrated mass pole.

If forming a coordinate with slide carriage mass point displacement $x_M(t)$ and turning angle $\theta_M(t)$, the slide carriage movement formula can be given as follows, taking slide carriage micro-vibration into account:

$$\begin{bmatrix} M+2m & ml_{BM}-ml_{CM} \\ ml_{BT}-ml_{CT} & I_M+ml_{CT}^2+ml_{BT}^2 \end{bmatrix}\begin{bmatrix} \ddot{x}_M \\ \ddot{\theta}_M \end{bmatrix}$$

$$+\begin{bmatrix} c_1+c_2+c_{10}+c_{20} & (c_1+c_{10})l_{BM}-(c_2+c_{20})l_{CM} \\ (c_1+c_{10})l_{BM}-(c_2+c_{20})l_{CM} & [(c_2+c_{20})l_{CM}^2+(c_1+c_{10})l_{MT}^2] \end{bmatrix}\begin{bmatrix} \dot{x}_M \\ \dot{\theta}_M \end{bmatrix}$$

$$+\begin{bmatrix} k_1+k_2+k_{10}+k_{20} & (k_1+k_{10})l_{BM}-(k_2+k_{20})l_{CM} \\ (k_1+k_{10})l_{BM}-(k_2+k_{20})l_{CM} & [(k_2+k_{20})l_{CM}^2+(k_1+k_{10})l_{BM}^2] \end{bmatrix}\begin{bmatrix} x_M \\ \theta_M \end{bmatrix}$$

$$=\begin{bmatrix} F \\ Fl_{ET} \end{bmatrix}-\begin{bmatrix} F_{BC} \\ F_{BCl} \end{bmatrix}$$

$$(1)$$

In this formula,

$$F_{BC}=m\ddot{w}_B(u_B)+c_{10}\dot{w}_B(u_B)+k_{10}w_B(u_B)$$
$$+m\ddot{w}_C(u_C)+c_{20}\dot{w}_C(u_C)+k_{20}w_C(u_C) \tag{2}$$

$$F_{BCl}=[m\ddot{w}_B(u_B)+c_{10}\dot{w}_B(u_B)+k_{10}w_B(u_B)]l_{BT}$$
$$-[m\ddot{w}_C(u_C)+c_{20}\dot{w}_C(u_C)+k_{20}w_C(u_C)]l_{CT} \tag{3}$$

the meanings of relative letters are: B and C are simplified force points of the forward and back air-buoyant block respectively, E is disturbing force point, M is slide carriage mass point, k_1 and c_1 are equivalent stiffness and damping of forward air-buoyant block respectively, k_2 and c_2 are equivalent stiffness and damping of back air-buoyant block respectively, k_{10}, c_{10}, k_{20} and c_{20} are air-film stiffness and damping of left and right actuator respectively, $F(t)$ is disturbing force, l_{IJ} is distance between point I and point J, $w_B(u_B)$ and $w_C(u_C)$ are the function of piezoelectric ceramics displacement over control voltage.

4.FUZZY-PID CONTROL

Fuzzy controller is a kind of newly developed controller. Its advantage is that mastering precise mathematical model of the controlled object is not needed. Two-dimension fuzzy controller is most widely used. But it has not satisfactory stable characteristics because of the lack of Fuzzy Integral control function. If it is combined with PID control, both good flexibility and adaptability of fuzzy control and good precision of PID control can be obtained.

In this paper the authors try to introduce composite Fuzzy-PID control to slide carriage vibration control system. Its control principle is shown in

Figure 2. This method applies Fuzzy control in large error field and PID control in small error field.

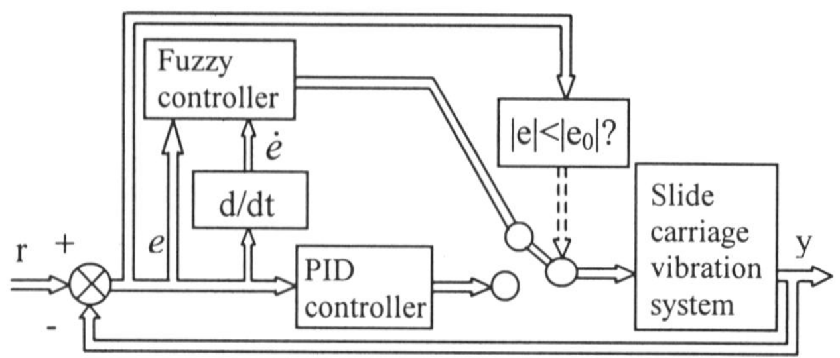

Fig.2 fuzzy-PID control principle

5.EXPERIMENTAL RESEARCH

Experiments are performed on the aerostatic guideway developed.

The Fuzzy control adopts two-dimension controller. Its control rules are shown in the following table.

Table　　　　　Fuzzy control rules （U）

E \ EC	-3	-2	-1	0	1	2	3
-3	3	3	2	2	2	1	1
-2	3	2	2	1	1	1	0
-1	2	2	1	1	0	-1	-2
0	2	1	1	0	-1	-1	-2
1	2	1	0	-1	-1	-2	-2
2	0	-1	-1	-1	-2	-2	-3
3	-1	-1	-2	-2	-2	-3	-3

In the experiment, the PID control parameters are determined as: K_p =12, K_i =3, K_d =2。 The conversion critical value of Fuzzy control and PID control is determined in experiments, as $|e_0|$=0.16μm。

Figure 3 shows the acceleration response characteristics of sliding carriage measured before control. Figure 4 and Figure 5 show the acceleration characteristics after PID control and fuzzy-PID control respectively. Comparison of the figures shows that the applications of PID control and fuzzy-PID control lessen the amplitude of acceleration by about 75% and 78% respectively. It is also measured that the displacement amplitude is lessened about 35% and the turning angle about 34% with fuzzy-PID control.

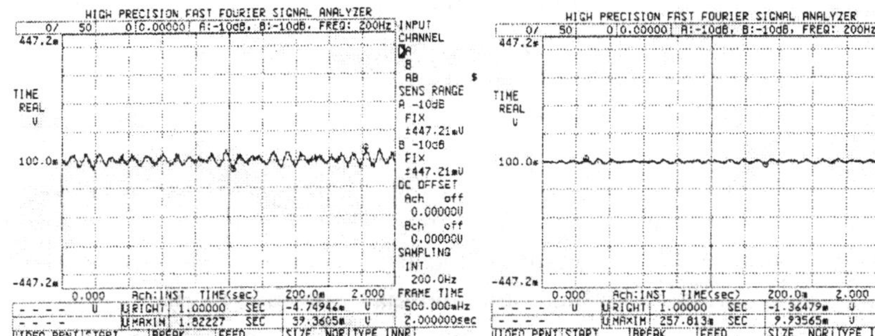

Fig.3 Dynamic response of slide carriage uncontrolled

Fig.4 Dynamic response of slide carriage with PID control

6.CONCLUSION

1. Hybrid vibration control system is set up by applying active control technology to the slide carriage of ultra-precision machine tool. It can improve the kinematic accuracy of slide carriage movement and machining quality of ultra-precision machine tool.

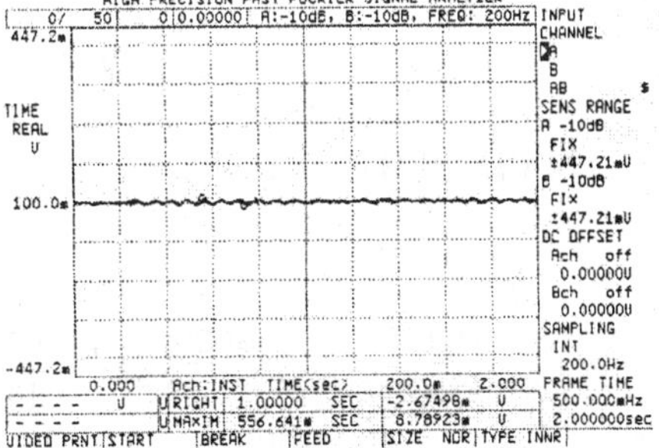

Fig.5 Dynamic response of slide carriage with fuzzy-PID control

2. A new type of active air-bearing actuator is developed in this paper to solve the control problem of transverse vibration during the longitudinal movement of slide carriage.

3. Fuzzy-PID control is used in this paper. Experiment results show that the amplitude of acceleration is lessened by about 78%. The controlling effect is a little better than that of PID control.

4.Experiments show that damping effect is good for frequencies lower than 80 Hz, but not good for frequencies above 80 Hz. This may be caused by the influence of air-bearing film.

REFERENCES

A.B.Palazzolo, et al. Hybrid active vibration control of rotor bearing systems using piezoelectric actuators. Journal of Vibration and Acoustics, 1993,115:111-119

Hansen, Colin. H. *Active Control of Noise and Vibration.* London: E & FN Spon, 1997

Ikawa N., Donaldson R. R., et al. Ultraprecision Metal Cutting——the Past, the Present and the Future. Annals of the CIRP. 1991, 40(2):586-589

NANOMETROLOGY - THE FRONTIER OF PRECISION

Robert J. Hocken
UNC Charlotte

Abstract

Nanotechnology, the engineering that deals with the production of systems whose tolerances lie in the domain of nanometers, is becoming increasingly important in many industrial fields. In order that nanotechnology create viable, commercial products, ultra-precision metrology will be required. We call this new metrology "nanometrology", and in this paper attempt to address some of the significant challenges faced by nanometrology in this new millenium.

Keywords

Nanotechnology; precision engineering; metrology; nanometrology.

1. INTRODUCTION

Metrology forms the backbone of modern manufacturing. Pioneers, such as Michelson [Michelson, 1889] and Rolt [Rolt, 1929], developed accurate measurements of length, displacement, angle, threads, surface texture, gears, flatness, straightness, etc. An equally important group of innovators directed their efforts toward production machinery [Schlesinger, 1978; Bryan, 1981]. Overlaying this hardware-based technology, researchers have developed tolerance representations [Wilhelm, 1992] and quality control systems [Taguchi, 1989]. Although metrology is essential in normal manufacturing, it is indispensable for "high-tech" products. These products, including optical components and fibers, VCRs, communications systems, computers, disk drives, as well as the machines that produce them, have critical dimensions, forms, and surfaces that have tolerances in the nanometer domain. Capabilities in materials modification [Hocken, 1992; Patten, 1996; Gao, 2000] and positioning [Holmes, 2000; Wang, 2000] have reached the near-atomic level, tolerances in production have been drastically reduced [Swyt, 1992], and surfaces and materials are being engineered at nano scales [Gonsalves, 2000]. New meso- and micro-scale products with nanoscale tolerances will be developed [NIST/NSF, 1999; Masuzawa, 2000] rapidly and will require assembly [Van Brussel, 2000] with nanometric precision. Further, new nanoscale products are being developed, and producing them will require nanometrology. For example, there are already 2273 U.S. patents (from 4298 inventors) for nanotechnology developments [Corbett, 2000], with many more to come.

2. NANOMETROLOGY CHALLENGES

Taniguchi discovered and elucidated a pattern that related achievable manufacturing tolerances with time and also coined the term "nanotechnology" [Taniguchi, 1974, 1995]. Modern usage has divided the field into two areas, so-called "molecular nanotechnology" and "engineering nanotechnology" [Corbett, 2000]. Molecular nanotechnology is based on "bottom-up molecular manufacturing" popularized by Drexler [Drexler, 1992]. Engineering technology is defined as "the study, development, and processing of materials, devices, and systems in which structure on a dimension of less than 100 nanometers is essential" [Corbett, 2000] and is more immediately practical to manufacturing. Nanometrology addresses the "... ultimate physical limitations of dimensional metrology" as they apply to ultra-precision manufacturing [Kunzmann, 1991]. Problems include "realizing a metric", referencing a metric, generating and measuring repeatable motion, and linking the test piece to the coordinate system [Teague, 1991; Hocken, 1994].

2.1 Realizing the Metric

Laser interferometry, based on heterodyne detection, is still the most viable method for realizing the metric [Badami, 1998]. Most common interferometers use a Michelson-like configuration, but commercial instruments have resolution limitations near a nanometer. A very promising technique for nanometrology is an x-ray interferometer suitable for practical measurement purposes [Miller, 1996; Smith, 2000]. Since x-ray interferometers have basic periodicity at the lattice spacing of silicon, resolution of these instruments could approach 1 picometer. The European Union already has a strong program in this area [Basile, 2000] but other investigators' concepts and new approaches are desperately needed.

2.2 Referencing the Metric

Relating the metric, the part, and the probe in the purest manner involves the creation of a stable metrology frame. Such frames need to be stable thermally, dimensionally, and dynamically. Required here is precision control of temperature, the dimensional stability of materials [Patterson,

1998], active vibration and acoustic isolation systems, and the detailed computer modeling of these systems at scales of accuracy heretofore not attempted. Only through the construction of actual instruments and very exacting experiments will metrology frames be constructed with nanometer and sub-nanometer stability.

2.3   Generating Repeatable Motion

The control of motion at nanometric levels involves ultra-precision drives and bearings. Much work has been done with linear motors and magnetic suspensions [Holmes, 1995, 2000; Hocken, 1995; Wang, 2000], but expanded efforts on flexural bearings, solid-contact bearings, air bearings, and magnetic bearings needs to be initiated. Test facilities for drive and motion control systems in order to reach sub-nanometric levels over macroscopic distances need to be built at labs around the world. Experience gained will help us control motion down to sub-atomic levels.

2.4   Linking to the Testpiece

One of the weakest areas in nanometrology is the area that has received much of the popular attention; that is, in probing systems that allow "imaging" at atomic levels. Here we refer to the scanning tunneling microscope, the atomic microscope, the near-field scanning microscope, and many other variants. All of these systems use highly non-linear actuators, PZTs, and therefore require extensive compensation to achieve non-distorted images. Also, they are inherently unstable over time and thus are unsuitable for long-term studies requiring the relocation of datums to nanometric precision. Currently high-precision confocal optical microscopes are still the best choice for tasks involving repeatability over long periods of time. It is therefore a major challenge in nanotechnology to develop highly-stable instruments capable of imaging atomic scale features in air.

3.   CONCLUSIONS

Nanotechnology, at this stage, reminds one of normal manufacturing at the end of the 19th Century in the sense that the tolerances required for the production of true nanotechnology products cannot really be achieved with available technologies. Major challenges exist in realizing the metric, referencing the metric, controlling motion, and linking the metrology system to the manufactured part or the test piece. Advances, however, have enabled us to see things at the atomic level; therefore, we are confident that we will be able to advance the state-of-the-art to reach the goals of sub-atomic, stable metrology needed for nanoscale production in the next decade.

REFERENCES

Badami, V., Patterson, S.   A Method for the Measurement of Nonlinearity in Heterodyne Interferometry. Proc. of the 13th Ann. Mtg. of the ASPE; 1998; St. Louis, MO.

Basile, G., et al. Combined Optical and X-ray Interferometry for High Precision Dimensional Metrology. Proc. Royal Soc.; 2000; London: A456.

Bryan, J. B.   A Simple Method for Testing Measuring Machines and Machine Tools: Parts I & II. Prec. Eng.; 1981; England.

Corbett, J., McKeown, P. A., Peggs, G. N., Whatmore, R.   Nanotechnology: International Developments and Emerging Products. Annals of the CIRP 2000; 49/2.

Drexler, K. E. *Engines of Creation: the Coming Era of Nanotechnology.*   New York: Doubleday, 1992.

Gao, W., Hocken, R., Patten, J. Experiments Using a Nano-Machining Instrument for Nano-cutting Brittle Materials. Annals of the CIRP; 2000; 40/1.

Gonsalves, K., Li, H., Perez, R., Santiago, P., Jose-Yacaman, M.   Synthesis of Nanostructured Metals and Metal Alloys from Organometallics.   Invited Paper, Coord. Chem. Revs. Special Issue: Organometallic Chem. at the Millennium; 2000; 206-207.

Hocken, R., Miller, J. Nanotechnology in Metrology. Proc. of the 5th Int'l. Symp. on Robotics and Mfg.; 1994; Maui, HI.

Hocken, R. J., Miller, J. A. Nanotechnology and Its Impact on Manufacturing. JAPAN/USA Symp. on Flex. Automation ; 1; ASME, 1992.

Hocken, R. J., et al. Research in Engineering Metrology - A Strategic Manufacturing Initiative. Proc. of the NSF Design and Mfg. Grantees Conf.; 1995; La Jolla, CA.

Holmes, M., Hocken, R., Trumper, D.   The long-range scanning stage: a novel platform for scanned-probe microscopy. Prec. Engrg. 2000.

Holmes, M., Trumper, D., Hocken, R.   Atomic-scale Precision Motion Control Stage (The Angstrom Stage). CIRP Annals 1995; 44/1.

Kunzmann, H. State-of-the-art and Ultimate Physical Limitations of Dimensional Metrology. Proc. ASPE; 1991; Santa Fe, NM.

Masuzawa, T. State of the Art of Micromachining. Annals of the CIRP 2000; 49/2.

Michelson, A. A., Morley, E. W. On the feasibility of establishing a light-wave as the ultimate standard of length. Am. J. of Sci. 1889; 37:225.

Miller, Jimmie A., Hocken, Robert, Smith, Stuart T., Harb, Salaam. X-ray calibrated tunneling system utilizing a dimensionally stable nanometer positioner. Prec. Engrg. 1996; 18:2/3.

NIST/NSF Workshop. Manufacturing Three-Dimensional Components and Devices at the Meso and Micro Scales. Proceedings of the Workshop; 1999; Gaithersburg, MD.

Patten, John, Hunsicker, Randal J., Ledford, Alton, Ferman, Cathie, Allen, Michael, Ellis, Clark. Automatic Vision Inspection and Measurement System for External Screw Threads. J. Mfg. Sys. 1994; 13:5.

Patterson, S. R. Interferometric Measurement of the Dimensional Stability of Superinvar. Lawrence Livermore National Laboratory Report UCRL-53787, 1988.

Patterson, S., Badami, V. U.S. Patent: Interferometry system having minimized cyclic errors. 1999.

Rolt, F. H., *Gauges and fine measurement.* 2 volumes; London: Macmillan, 1929.

Schlesinger, G. "Inspection Tests on Machine Tools (1927)." In *Testing Machine Tools*, 8th Ed., New York: Pergamon Press, 1978.

Smith, Stuart T., *Flexures – Elements of Elastic Mechanisms.* London: Gordon and Breach Science Publishers, 2000.

Swyt, D. Challenges to NIST in Dimensional Metrology: The Impact of Tightening Tolerances in the U.S. Discrete-Part Manufacturing Industry. NIST 1992; NISTIR 4757.

Taguchi, G., Elsayad, A., Hsiang, T., *Quality Engineering in Production Systems.* New York: McGraw Hill, 1989.

Taniguchi, N. On the Basic Concept of Nanotechnology. Proc. of ICPE; 1974; Tokyo.

Taniguchi, Norio. The state of the art of nanotechnology for processing of ultraprecision and ultrafine products. Prec. Eng. 1995.

Teague, E. C. Nanometrology. AIP. Conf. Proc. 241 Scanned Probe Microscopy; 1991; Santa Barbara, CA.

Van Brussel, H., et al. Assembly of microsystems. Annals of the CIRP 2000; 49/2.

Wang, C., Hocken, R. Interferometry with Angstrom Resolution. Proc. of the ASPE Spring Topical Meeting; 2000; Tucson AZ.

Wilhelm, R. G., Lu, S. C-Y. *Computer Methods for Tolerance Design.* New York: World Scientific Publishing Co., 1992.

DEVELOPMEMT OF ULTRAPRECISION 5-AXIS MACHINE TOOL EQUIPPED WITH ELLIPTICAL VIBRATION CUTTING DEVICE

Toshimichi Moriwaki[1], Eiji Shamoto[1], Katsutoshi Tanaka[2],

Makoto Matsuo[3] and Michio Osada[3]

[1]*Mechanical Engineering Department , Kobe University*

[2]*Toshiba Machine Co. Ltd.,* [3] *Towa Co. Ltd.*

Abstract

A new ultraprecision 5-axis machine tool was developed. The simultaneous 5-axis control consists of three linear motion control in X, Y and Z axes and two rotary motion control in C axis of the main spindle and B axis of the rotary table. The nominal resolutions of each linear motion and rotary motion are 1nm and 0.0001deg., respectively. One of the specific features of the machine is that it is equipped with an ultrasonic elliptical vibration cutting device, placed on the rotary table, to machine various kinds of hardened steel dies with complicated shapes, such as Fresnel lens and other free forms. The present paper describes the detail of the machine and presents some cutting test results.

Keywords

Ultraprecision machine tool, 5-axis control, elliptical vibration cutting

1. INTRODUCTION

Demands for ultraprecision dies and molds are much increasing in recent years, as mechanical, electric/electronic and optical parts are mass produced, which are typically used for most recent opto-electric or opto-mechanical devices, such as CD/DVD, printer, projector and other IT devices[1]. In order to meet such demands, the authors have developed a new ultraprecision cutting method of hardened die steels, which is named ultraprecision ultrasonic elliptical vibration cutting method[2]. The hardened steels have been successfully cut into mirror surfaces with carefully polished single crystal diamond cutting tool, which is oscillated ultrasonically in such a way that the vibration locus of the cutting edge forms an ellipse.

In order to manufacture ultraprecision dies and molds with complicated shapes, such as those for Fresnel lens and other free forms, a new ultraprecision 5-axis machine tool has been developed and fabricated jointly by the authors. The details of the machine developed are described here focusing some specific features of the machine. Some cutting test results are also introduced here.

2. CONFIGURATION AND SPECIFICATIONS OF MACHINE

The ultrasonic elliptical vibration cutting device employed here vibrates the single crystal diamond cutting tool elliptically in a plane including the main cutting force and the thrust force directions. It sits on the table of the machine, and hence the workpiece is fixed to the rotating main spindle. In order to turn the rotary workpiece into complicated forms it is necessary to have two linear motions in X and Z directions. The X motion is given to the tool table and the Z motion to the work spindle, respectively. It is also essential to add one rotational motion to the tool table so that the approach angle of the cutting tool can be controlled freely.

In order to index the rotary position of the workpiece, the rotary motion control is given to the work holding main spindle. Additional linear motion control in Y direction is given to the tool table so that planer cutting or grooving of the workpiece can be performed by combination of the three linear motions. Major specifications of the 5-axis control of the machine are summarized in Table 1.

Table 1. Major specifications of machine.

Axis	Stroke	Resolution
X: Tool post traverse	300 mm	0.001 μ m
Y: Tool post up/down	75 mm	0.001 μ m
Z: Work spindle	150 mm	0.001 μ m
B: Tool post rotation	$\pm$180 deg	0.0001 deg
C: Work rotation	360 deg	0.0001 deg

3. MACHINE AND ITS ELEMENTS

Figure 1 shows overview of the machine fabricated. The main spindle, to which the workpiece is attached, is 80mm in diameter, supported by hydrostatic air bearings, and driven by a synchronous motor. Its range of rotational speed is from 10 to 1500 min^{-1}. The runout of the main spindle measured is less than 0.05 μ m in both radial and axial directions.

The detail of the ultrasonic elliptical vibration cutting device is shown in Figure 2. Two directional bending mode of ultrasonic vibration is applied to the stepped elliptical vibration tool. The single crystal diamond cutting edge is attached to the end of the vibrator. The structure of the ultrasonic elliptical vibration cutting device is basically the same with the one developed in the past and reported elsewhere[2].

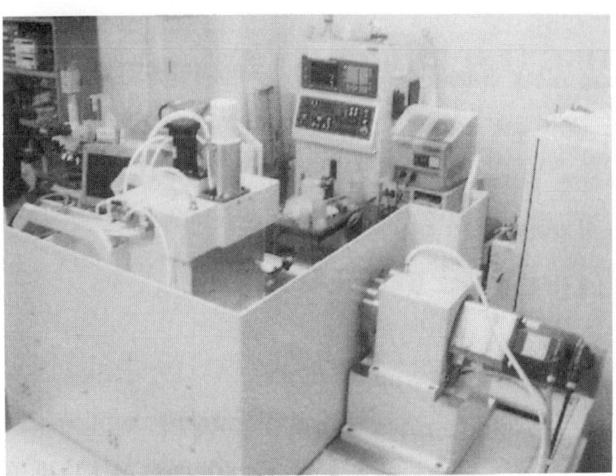

Figure 1. Overview of machine.

Figure 2. Detail of ultrasonic elliptical vibration cutting device.

Another specific feature of the machine is application of V-V roller guide ways for linear motion of the table, which guarantees high rigidity, high endurance stability and low friction in the moving direction as well as high precision. The measured accuracy in the straightness of the linear motion is $0.2\,\mu$ m along X-axis, $0.1\,\mu$ m along Y-axis and $0.1\,\mu$ m along Z-axis, respectively for the whole strokes of the individual motions. It is understood that the machine has good motion accuracy.

4. SOME CUTTING TEST RESULTS

Some cutting tests were carried out to machine hardened die steels with the single crystal diamond cutting tool. Figures 3 and 4 show SEM photographs of dies for micro Fresnel lens and light guide plate of LCD, respectively. The work materials are die steels (JIS:SUS440C) hardened to HRC55.

The focal length of the Fresnel lens is 30mm. The depth of the groove of the die is $20\,\mu$ m, and the pitch of the grooves varies from 120 to $350\,\mu$ m. The geometrical accuracy and the surface roughness are measured with an electron roughness analyzer (ELIONIX ERA8000). The measured roughness

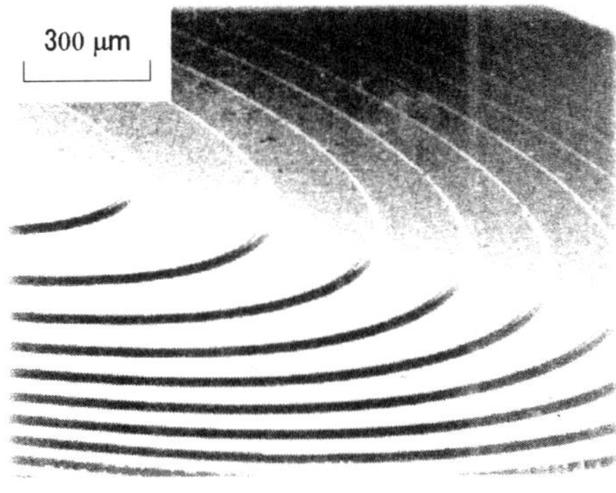

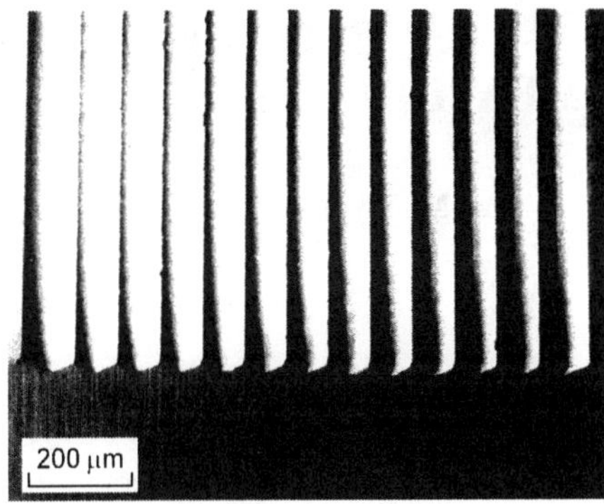

Figure 3. SEM photograph of die for Fresnel lens.

Figure 4. SEM photograph of die for light guide plate.

of the machined die surface is 0.08 μ mRy. The vertical angle and the depth of the groove of the die for the light guide plate are 117deg. and 30 μ m, respectively. The average pitch between grooves of the die measured is 99.963 μ m, which is in good agreement with the set value of 100 μ m. Good geometrical accuracy and good surface quality are obtained in both cases.

5. CONCLUSION

A unique ultraprecision 5-axis machine tool equipped with the ultrasonic elliptical vibration cutting devices has been developed and fabricated. Satisfactory results were obtained by some cutting tests of hardened dies for specific purposes. It is expected that the machine can be used to manufacture various kinds of ultraprecision dies.

REFERENCES

1. Moriwaki T., Shamoto E. Recent development in ultraprecision machining and machine tool technology. Proceedings of International Seminar on Precision Engineering and Micro Technology; 2000 July 19 – 20; Aachen: 161-170

2. Shamoto E., Moriwaki T. Ultraprecision diamond cutting of hardened steel by applying elliptical vibration cutting. Annals of the CIRP 1999; 48/1: 441-444

593

DEVELOPMENT OF ON-MACHINE PROFILE MEASURING SYSTEM WITH CONTACT-TYPE PROBE

Sei Moriyasu, Shin-ya Morita, Yutaka Yamagata, Hitoshi Ohmori,
Weimin Lin, Jun-ichi Kato and Ichiro Yamaguchi
The Institute of Physical and Chemical Research (RIKEN)

Abstract

To achieve both high precision machining and high efficiency machining, the on-machine measurement system is inevitable. But conventionally there are no desirable sensors for on-machine profile measurement in the point of measurement accuracy and size. Authors have developed an ultraprecision profile measurement probe for on-machine measurement, which has three key technologies: laser interferometric displacement sensor and ultra-low pressure air-slide are inside and real-time measurement is possible by transmitting the current position data directly from the machine scales to the PC simultaneously together with the displacement data from the probe. The newly developed probe has high linearity, repeatability and traceability.

Keywords

Probe, Profile measurement, On-machine measurement, Form error, Aspherical optics, Linearity, Repeatability, Traceability

1. INTRODUCTION

On-machine measurement, in which the measuring device is mounted on the machine, is widely known as the way which leads to the improvement of the machining accuracy and high productivity that the positioning error in attach and removal of the workpiece can be removed, the requiring time for the initial set of the workpiece and the air-cut time can be reduced. However, no sensor whose size is small enough to attach on the machine without the interference in machining and which is so accurate that the evaluated form of the aspherical optics can be guaranteed is commercially available[1, 2]. In this paper, a new small-size contact-type profile measuring probe with high accuracy was designed, and the characteristics of the developed sensor was evaluated.

2. DESIGN OF NEWLY DEVELOPED PROBE

Figure 1 shows the simple illustration of the newly developed profile measuring probe. A sapphire ball is fixed on the tip of the measuring probe, and the probe shaft is supported by the air slide. On the other side of the probe shaft, the reflection mirror is fixed, and it is possible to measure the form of the measured object which is constantly contact to the tip ball of the probe shaft by the measurement of the displacement between the endface of the optical-fiber-type laser interferometric displacement sensor and the reflection mirror. Measuring force can be changed from 0 to 500mgf by controlling the pressure of the input air to the thrust port with the electropneumatic regulator. The stiffness of the air slide can be kept high by keeping the pressure of the input air to the bearing port constant.

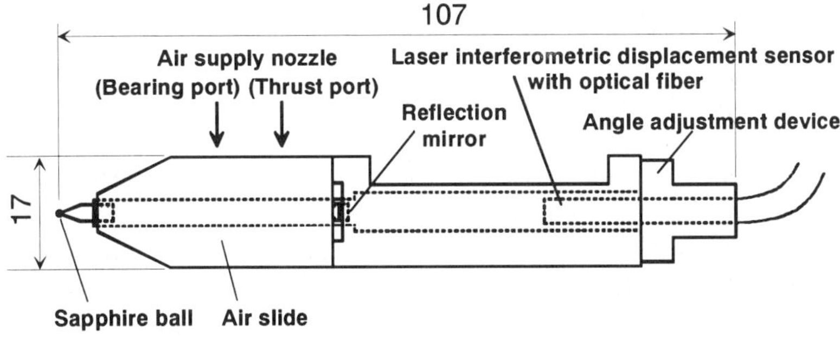

Fig.1 Illustration of new profile measuring probe

3. DESIGN OF DATA I/O INTERFACE

When the form measurement is conducted by the profile measuring probe located on the machine, it is the conventional way that NC data is sent from the computer to the NC controller, and after confirming that the machine reaches the directed point, the displacement in this point is obtained from the sensor to the computer. By using this method, it takes, however, much time to measure because the measurement is conducted after stopping the machine in each point. This paper proposes that the current position data of the machine is obtained directly from the scales of the machine to the computer, and at the same time the displacement data from the profile measuring probe is also obtained to the computer shown in **Figure 2**. Since the real-time measurement without stopping in each point was achieved by this method, the measurement with high-accuracy and high-efficiency can be achieved.

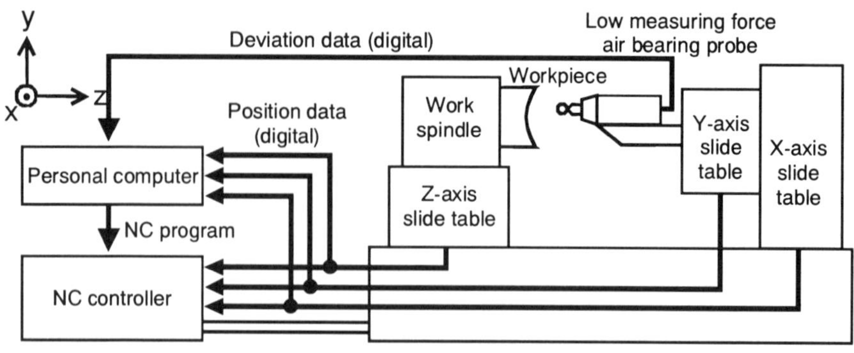

Fig.2 Real-time measuring system

4. PERFORMANCE EVALUATION

To evaluate of the performance of the constructed profile measuring probe, three kinds of measurement experiments were conducted.

To check the linearity of the probe signal, the displacement data of the probe and the machine axis which was parallel to the probe axis was obtained to move the machine axis within 0.75mm. The deviation of the obtained data was calculated by the linear fitting. The linearity of the probe signal to the machine scale was less than 0.1µm in $\pm 3\sigma$ (**Figure 3**).

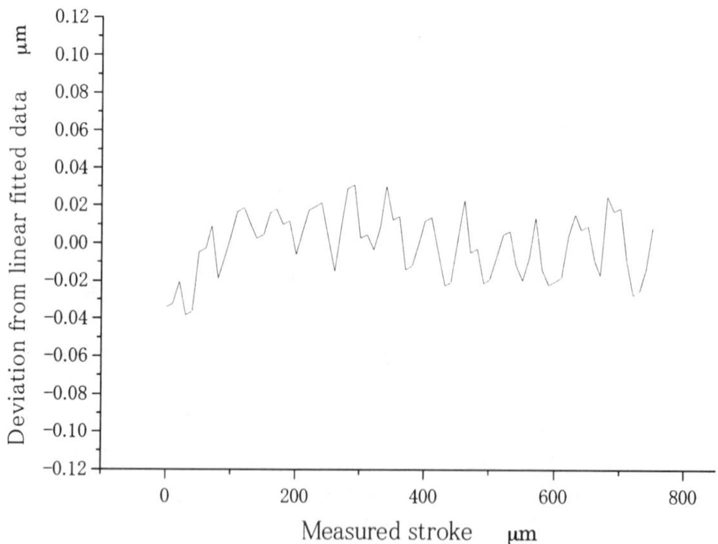

Fig.3 Linearity of newly developed probe ($\pm 3\sigma = 0.07$µm)

To check the repeatability of the measured data with the probe, the profile measurement of a sphere workpiece whose maximum slope angle was $60°$ was conducted three times per direction to four directions X+, X-, Y+, Y-, and the dispersion of the measured data from the average data curve was evaluated by the standard deviation σ. The measurement could be conducted relatively stable under the condition that the measuring force was 100-150mgf and the measuring speed is 50-100mm/min. The repeatability of the measured data was achieved to $\pm3\sigma < 0.1\mu m$ in all directions (**Figure 4**).

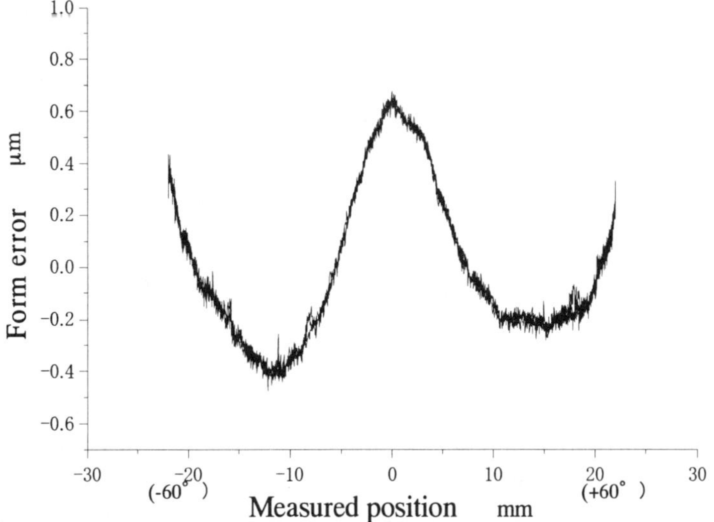

Fig. 4 Repeatability of measurement with the probe
(X: + $\rightarrow$ -, $\pm3\sigma = 0.068\mu m$)

To check the traceability of this measurement system, the profile measurement of a sphere master was conducted on the machine. The curvature radius of the sphere master was measured and evaluated by Form Talysurf (Taylor Hobson), and the sphereness was measured by the interferometer GPI-xp (Zygo) and guaranteed in $\lambda/10$ beforehand. The curvature radius of the tip ball of the probe was also measured and evaluated beforehand by UA3P (Panasonic). After the measurement, the form error was calculated by using the evaluated values of the curvature radius of the sphere master and the tip ball. The form error was $0.15\mu m$ ($\pm3\sigma$) in the measured area whose maximum slope angle was $\pm45°$ (**Figure 5**). This data shows the great traceability of this developed measurement system to the conventional ultraprecision profile measurement systems.

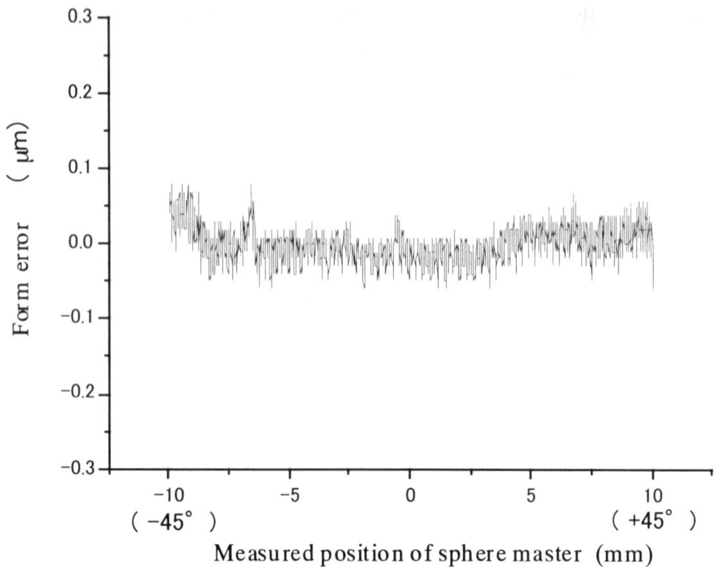

Fig.5 Form error in the measurement of sphere master with the probe

5. CONCLUSION

In this paper, a new ultraprecision profile measuring probe for on-machine measurement was developed. The size of this probe is small enough to install on the machine. The special air slide, by which the measuring force can be precisely controlled (about 0-500mgf) and the interferometric displacement sensor are included in the probe. High efficiency of measurement was achieved because of the real-time measurement, which becomes possible by obtaining the current position data of the machine directly from the machine scales. High linearity of $\pm 3\sigma < 0.1\mu m$ within 0.75mm stroke, high repeatability of $\pm 3\sigma < 0.1\mu m$ within $\pm 60°$ in 4 directions, and great traceability of $\pm 3\sigma < 0.15\mu m$ within $\pm 45°$ were successfully achieved.

References:

1. Negishi M., Matanabe K., Matsusita K., Kasahara T., Hosaka K. A High-Precision Coordinate Measurement Machine for Aspherical Optics. Precision Science and Technology for Perfect Surfaces. JSPE Publication Series No.3; 1999; 354-359.
2. Shiozawa H., Fukutomi Y. Development of Ultra-Precision 3D-CMM with the Repeatability of Nanometer Order. Precision Science and Technology for Perfect Surfaces. JSPE Publication Series No.3; 1999; 360-365.

AN INSTRUMENT TO MEASURE THE STIFFNESS OF MEMS MECHANISMS

Jin Qiu, Joachim Sihler, Jian Li, Victoria Sturgeon, Micah Smith, Alexander Slocum

Precision Engineering Research Group, Massachusetts Institute of Technology
77 Massachusetts Avenue Room 3-470, Cambridge, MA 02139, USA
Tel: +1-617-253-1953   Fax: +1-617-258-6427   Email: jqiu@mit.edu

Abstract

The Deep Reactive Ion Etching process (DRIE) has made it possible for an increasing number of MEMS devices to be made with flexures as their main components. However, the DRIE creates slightly tapered cross sections in addition to other fabrication variance or even failure, which affect the predicted stiffness. The flexures typically move in the plane of the wafer and it is often desirable to measure the force-displacement characteristics of such flexures after fabrication to close the design feedback loop. However, to our knowledge there is no commercially available instrument to perform this task. This paper introduces an instrument specifically designed to quickly record the force-displacement curve of MEMS flexures. The first prototype features a resolution of 10 nm for the displacement, and 100 μN for the force. It has been successfully used in practice [1].

Keywords

MEMS, DRIE, stiffness tester, flexure design, instrument design, compliant mechanism.

1. INSTRUMENT DESIGN

Fig 1: Side view of the instrument.

1.1 General Layout

The prototype is shown in Figure 1, with its major components comprised of flexure designs [2] cut from aluminum sheet stock by an OMAX 1632 Abrasive Waterjet Machining Center. The measurement method is based on mechanical contact between a needle probe and the MEMS flexure to be tested. For the tests, the wafer is placed onto the wafer holder. The x-, y-, and z-axes, realized by moving stages, are used to coarsely locate the needle probe tip in the vicinity of the MEMS flexure to be measured, and then the stages are locked in place. A micrometer moves the data acquisition unit forward, guided by two flexural bearings [3]. The data acquisition unit carries the probe assembly that has a built-in reference flexure. By measuring the movement of the data acquisition unit and the deflection of the reference flexure, the force and displacement of the MEMS flexure can be derived.

The data acquisition unit itself is also supported by flexural bearings, because they offer friction and backlash free motion over small distances. The length and thickness of the flexural bearings are chosen such that the maximum bending stress during movement is not exceeded. The two crab-leg shaped flexural bearings can be seen in Figure 1.

1.2 Measurement Principle

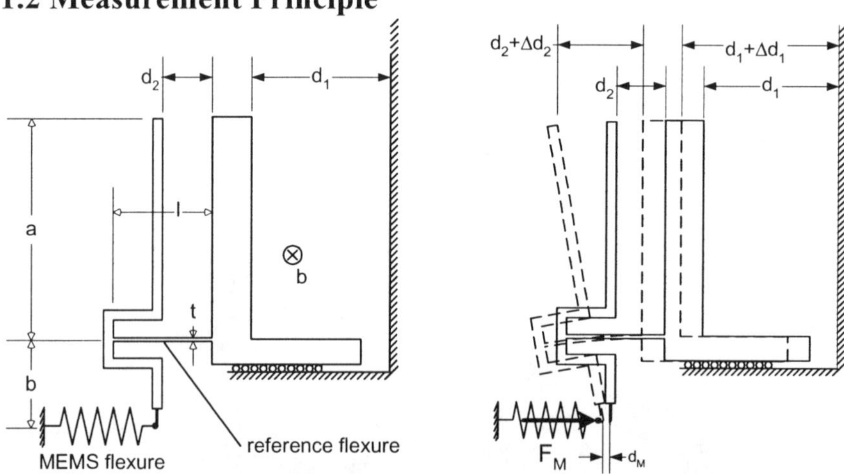

(a) Probe assembly and the reference flexure.   (b) Deflections during measurement
Fig 2: Measurement method.

The functionality of the instrument during a measurement cycle is shown in Figure 2. The probe tip is placed close to the MEMS flexure to be measured (Figure 2a). Then, the data acquisition unit moves forward, the MEMS flexure and the reference flexure are deflected. The movement of the data acquisition unit Δd_1 and the deflection of the probe tip are measured by displacement sensors. Note that the deflection of the reference flexure is amplified by the amplification lever and then picked up by the sensors as Δd_2 at the tip of the lever. The MEMS flexure exerts a force F_M on the probe tip which is directly related to the displacement Δd_2. d_M is the displacement of both the MEMS flexure and the probe tip.

The following two equations are used to layout the geometry of the probe assembly, while its actual behavior is recorded by calibration prior to the measurement, which is required to compensate for the compliance of the probe.

$$F_M = \Delta d_2\, EI/abl, \text{ where } I = b\,t^3/12 \tag{1}$$

$$d_M = \Delta d_1 - b/a\,\Delta d_2 \tag{2}$$

To enhance force sensitivity, an amplification lever is attached to the reference flexure, and it is designed such that it experiences a true rotation around a spatially fixed pivot point. It can be shown that if the centerline of the lever goes through the middle point of the relaxed reference flexure, a moment applied to the end of the reference flexure will cause the lever to rotate truly (Figure 3), i.e. there is no parasitic vertical error motion which would have an influence on the sensor readings. This phenomenon is governed by the following equation:

$$\varepsilon = \delta - \theta\, l/2 = M(l^2/2EI) - M(l/EI)(l/2) = 0 \tag{3}$$

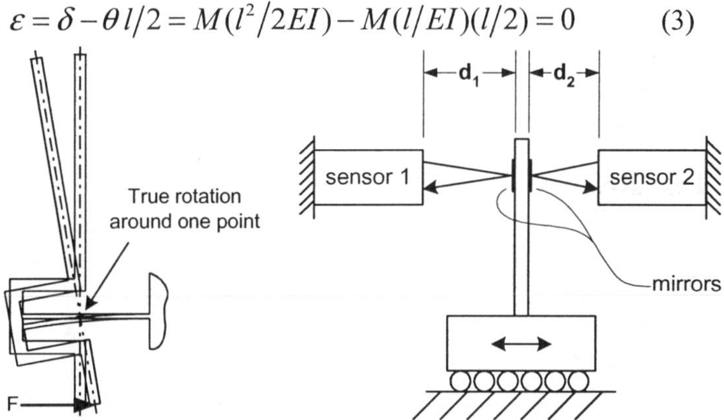

Fig 3: Deflection of the Amplification lever. Fig 4: The optical sensor setup

1.3 Displacement Sensors

Two different types of position sensors have been tested: LVDT (Linear Variable Differential Transformer) and HP1500 optical barcode readers in a differential configuration according to [4]. The optical sensors turn out to be two orders of magnitude lower in cost and have a resolution one order of magnitude better than the LVDT's. The working principle of the optical sensors is shown in Figure 4. A light source inside the optical sensor emits light that is reflected by a mirror attached to the moving structure. The returning light beam goes through a lens and reaches a photodiode inside the optical sensor. The distance between the mirror and the sensor determines the focal condition of the light received by the photodiode, which in turn emits a current dependent on that distance. The differential configuration of the optical sensors as shown in Figure 4 is used to achieve common noise rejection of the sensor signals.

2. CALIBRATIONS AND MEASUREMENT RESULT
2.1 Calibration

The calibration of the instrument is carried out in three steps. First, the displacement sensor for Δd_1 is calibrated using the micrometer reading. Next, the displacement sensors for the probe assembly (Δd_2) are calibrated by locking the probe tip to ground and turning the micrometer. Since now the relative movement of the needle tip with respect to the data acquisition unit is equal to the micrometer reading, a relation between the relative needle tip movement and the sensor readout for Δd_2 can be established. The last step is to find a relation between the displacement Δd_2 and the force exerted on it. A precision music wire with known mechanical properties is attached to ground and loaded in bending by the probe needle tip. Since the sensors are already calibrated by now, the displacement of the needle tip with respect to ground is known. Using bending beam equations for the music wire, the force F_M can be calculated and mapped to the sensor readings. Ultimately, the music wire could be precalibrated by the National Standard Institute.

2.2 Test Results

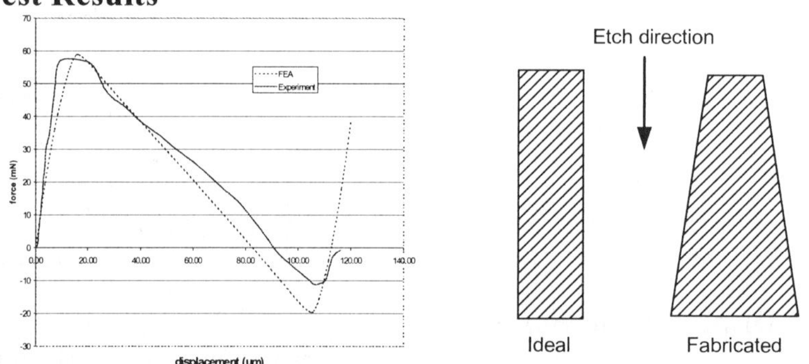

Fig 5: Test data of a MEMS flexure.   Fig 6: Tapering shape of DRIE cut

The MEMS mechanism that was first tested with the instrument is a novel centrally-clamped parallel-beam bistable mechanism [1] fabricated by DRIE. Figure 5 shows stiffness test results obtained by this instrument compared to an FEA simulation. Since a small deviation in the feature width has a third power impact on the stiffness of bending beams, the deviation is believed to result from the inaccuracy of the fabrication process. DRIE always creates somewhat tapered cross sections (Figure 6), in addition to other fabrication variance or even failure, which makes it hard to accurately predict the behavior of the final device. For proper design feedback it is necessary to experimentally measure the final product.

3. CONCLUSIONS AND FUTURE WORK

An instrument for the characterization of MEMS flexures that move in the wafer plane was introduced. The proper functionality was shown by measuring a MEMS device fabricated by one of the authors. The instrument has a high force resolution of 0.1mN and displacement resolution of 10nm, achieved with relatively low cost. A Labview program has been used to automate the calibration and the data acquisition cycle. The force displacement graph is shown on a computer screen in real time.

Due to the variance of the music wire diameter that was used for calibration, the measured force has an error of approximately %6. More precise ways of calibration will be explored. To make possible the batch testing of multiple devices on a wafer, a future design of the instrument is planned to involve stepper motors that drive all the axes to achieve full automation of the entire measurement process. To account for different force and displacement measurement ranges, it is also planned to have exchangeable reference flexures with different thickness or lengths. The wafer holder could also be mounted upright so that stiffness measurements can also be done in the direction perpendicular to the wafer surface (for example for membranes).

4. ACKNOWLEDGEMENTS

The authors would like to acknowledge the help of Prof. Jeffrey Lang, Roger Cortesi, James White, and Carolyn Phillips, and the sponsorship of MIT course 2.75 as well as ABB in the development of this instrument.

5. REFERENCES

[1] Qiu Jin, Lang Jeffrey and Slocum Alexander, "A Centrally-Clamped Parallel-Beam MEMS Bistable Mechanism," Proc. of the IEEE MEMS-01 Conference 353-357 2001 January, Interlaken, Switzerland.
[2] Slocum, Alexander H., <u>Precision Machine Design</u> 521-538, Prentice Hall, 1992.
[3] L. Saggera and Sridhar Kota, "A new design for suspension of linear microactuators," J. Dynamic Systems and Control vol 55-2, ASME 1994.
[4] Colin JH Brenan et al., "Characterization and use of a novel optical position sensor for microposition control of a linear motor," Rev. Sci. Instrum. 64(2), February 1993.

DEVELOPMENT OF X-RAY STEPPER WITH HIGH OVERLAY ACCURACY FOR 100-nm LSI LITHOGRAPHY

M. Fukuda, H. Morita, T. Haga, M. Suzuki, H. Tsuyuzaki,
A. Shibayama, S. Ishihara,
H. Aoyama*, S. Mitsui*, T. Taguchi*, Y. Matsui*

NTT Telecommunications Energy Laboratories
*ASET Super-fine SR Lithography Laboratory
3-1 Morinosato Wakamiya, Atsugi, Kanagawa 243-0198, Japan

Abstract

X-ray lithography using synchrotron radiation (SR) is a promising tool for fabricating large-scale integrated circuits (LSIs) with device feature sizes as small as 100 nm. The making of 100-nm devices requires both a high overlay accuracy and a printing resolution of less than 35 nm. We have developed an SR x-ray stepper meeting this requirement that utilizes air-bearing lead screws, an optical heterodyne detection system, and magnification correction by means of thermal stress control. Exposure results demonstrate that the exposure system achieves a total overlay accuracy of better than 30 nm, including stepper, mask and wafer errors.

Keywords

X-ray lithography, 100-nm devices, X-ray stepper, air-bearing lead screw, optical heterodyne detection, thermal stress control magnification correction, overlay accuracy

1. INTRODUCTION

The patterns of large-scale integrated circuits (LSIs) continue to become finer and denser. X-ray lithography using synchrotron radiation (SR) is a promising tool that can help maintain this trend down to the 100-nm regime. SR x-ray lithography uses a vertical stepper with a proximity gap of less than 30 μm for the one-to-one printing of mask patterns on a wafer [1]. Since proximity x-ray lithography uses a wavelength of about 0.7 nm, it has the potential to delineate lines and spaces with dimensions as small as 60 nm. But to make 100-nm devices, an overlay accuracy of less than 35 nm is essential [2]. An SR x-ray stepper (SS-3) that meets this requirement has recently been developed. It utilizes air-bearing lead screws, an optical heterodyne position detection system, and magnification correction by means of thermal stress control [3]. This paper describes the key technologies employed in the stepper and the results of exposure experiments using it.

2. STEPPER CONFIGURATION
2.1 Improvement of Stage Positioning

Extremely accurate positioning is essential to achieving a high overlay accuracy. This requires that non-linear elements be eliminated and that the stage be stiff enough. The SS-3 utilizes air-bearing lead screws (ABLS) in the horizontal (x) and vertical (y) directions to position a wafer. The advantage of an ABLS is that it allows the stage to move without friction [4]. To improve the stiffness, we have developed a tandem nut system in which two double nuts are mounted in series (Figure 1).

The vertical xy-stage is shown in Figure 2. The stage movement is constrained by an L-shaped air-bearing guide. The stroke of the stage is 280 mm x 230 mm. A wafer to be exposed is mounted from the back of the stage. A constant-tension spring cancels out the influence of gravity.

The performance of the stage was evaluated by positioning it in small 5-nm steps. Measurements were made using a capacitance displacement sensor with a low-pass filter with 1-kHz cutoff frequency. The results for the x- and y-directions in Figure 3 indicate that the stage has a positioning resolution of 5 nm and that there is no backlash in the steps. Furthermore, the vibration of the stage has a very small magnitude of about 10 nm.

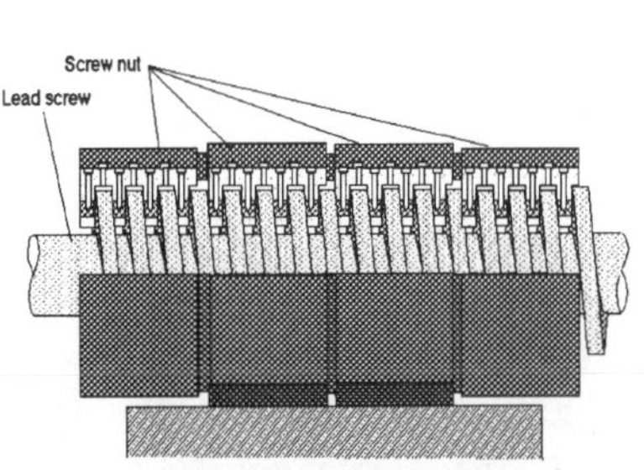

Figure 1. Structure of tandem nut system for high stiffness ABLS

Figure 2. Vertical xy- stage

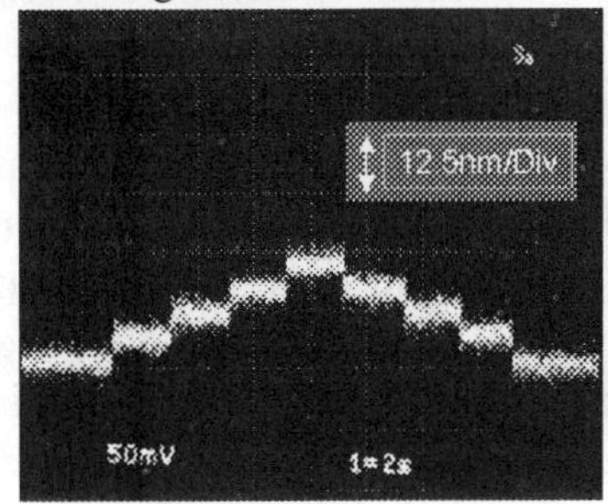

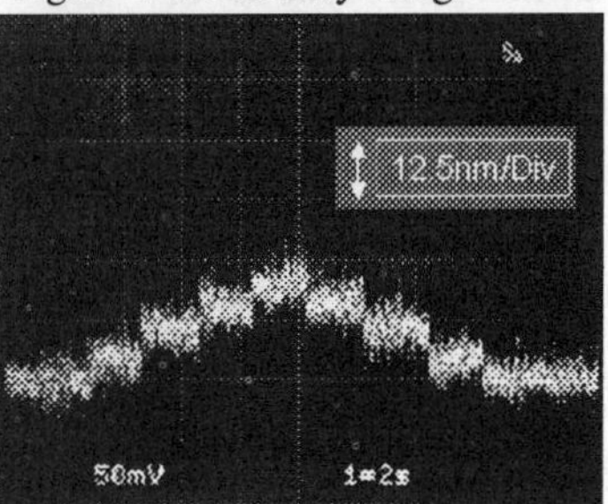

(a) Horizontal (x) axis   (b) Vertical (y) axis

Figure 3. 5-nm step psitioning

2.2 Two-wavelength Detection for Mask and Wafer Alignment

Accurate detection is also essential to ensure precise alignment between a mask and a wafer. An optical heterodyne detection system can detect the relative position between a mask and wafer with a very high resolution [5]. But the drawback of this system is that, since it uses a laser beam, the signal level varies with the thickness of the material on the wafer surface. We have developed a two-wavelength alignment system that ensures an adequate signal level regardless of the thickness of a resist and/or deposited film on a wafer. It employs two semiconductor lasers with wavelengths of 830 nm and 685 nm for the heterodyne system. The setup is shown in the Figure 4. The diffraction marks have a grating pitch of 5 μm, which provides a periodic phase difference in the pitch of 2.5 μm.

Figure 5 shows the phase difference between a mask and a wafer as measured with the stepper when the wafer is moved in the x- and y-directions. The phase difference varies linearly with wafer motion in a saw-tooth fashion. A good linearity of 10 nm over a distance of 2.5 μm was obtained.

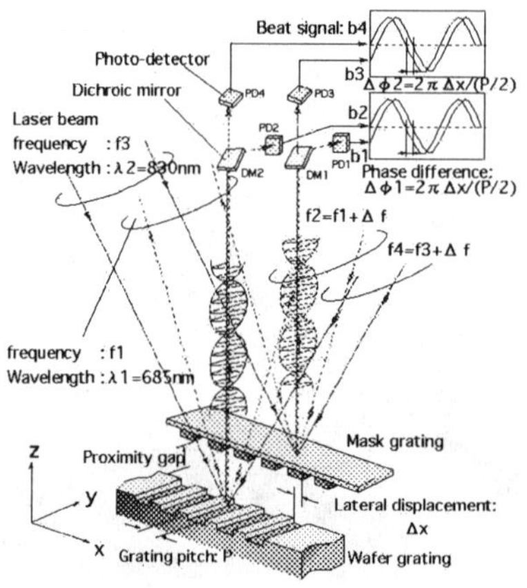

Figure 4. Optical heterodyne optics

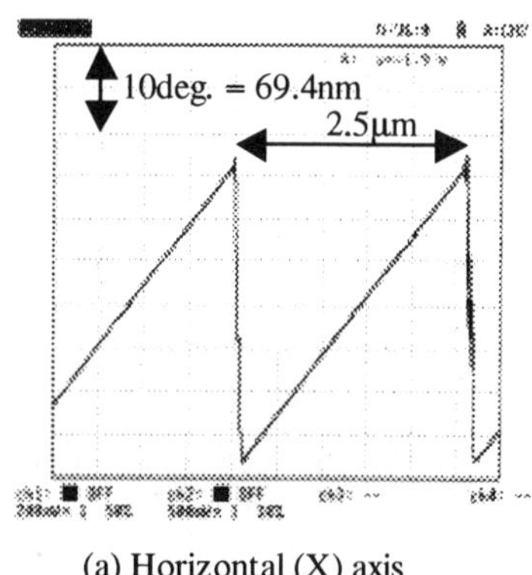

(a) Horizontal (X) axis

Figure 5. Optical heterodyne signal

2.3 Magnification Correction by Thermal Stress Control

As a wafer passes through various processes, the field size changes with the stress of the surface materials on it. This causes the placement of patterns to vary from layer to layer, leading to pattern deformation. Correcting the deformation is an effective way to improve the overlay accuracy. A thermal stress control method has been developed to do this [6]. Figure 6 shows how the size of a pattern is reduced. A wafer is heated up, clamped with a vacuum chuck, and then cooled down to the exposure temperature. At this time the wafer has tensile stress that keeps the wafer size large during exposure. After the wafer is released from the chuck, the exposed pattern shrinks. Figure 7 shows how much the chip size varies with the temperature of the wafer. The degree of magnification obtainable was investigated through measurements of exposed patterns at various temperatures. It was found that errors as large as 4 ppm of the wafer size can be corrected with this method.

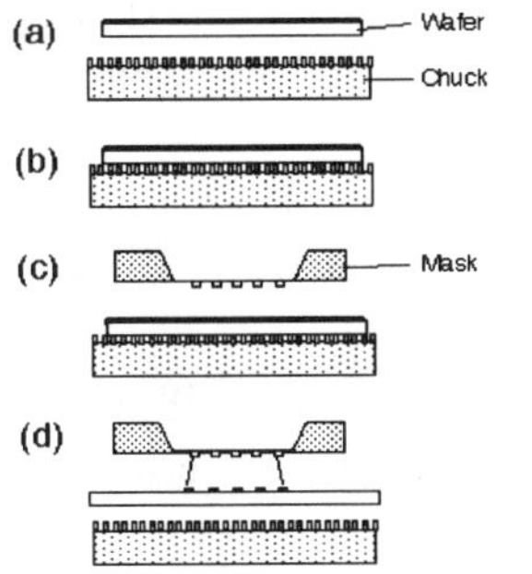

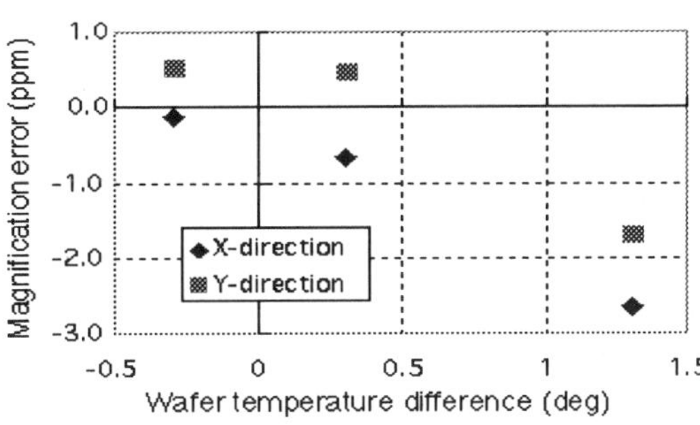

Figure 6. M agnification Correction by Thermal Stress Control: M ethod

Figure 7. M agnification Correction by Thermal Stress Control: R ange

3. MEASURED OVERLAY ACCURACY

The overlay performance of the stepper was examined by means of exposure experiments [7]. The error map and histogram in Figure 8 show the overlay accuracy between the gate and contact layers of 100-nm device patterns. The gate pattern was replicated on a wafer by exposure and etching; an isolation layer was deposited; and then contact patterns were replicated on top to make the contact layer. The error between the gate- and contact-layer patterns was measured with an overlay measurement machine; KLA500. The total overlay accuracy (mean + 3σ) was found to be better than 30 nm. According to the 1999 International Technology Roadmap, the target overlay accuracy for 100-nm devices is 35 nm.

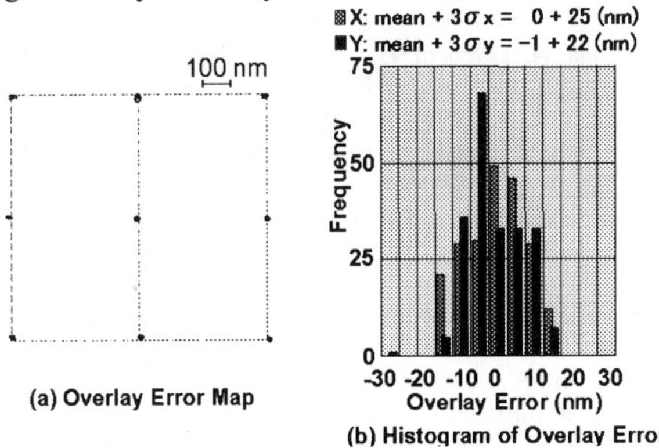

Figure 8. Overlay Error between Gate and Contact Layers

4. CONCLUSION

We have developed an x-ray stepper that employs air-bearing lead screws, optical heterodyne optics, and magnification correction by means of thermal stress control. The following results were obtained.

(1) The vertical xy-stage, which has a tandem nut system and ABLS, provides a positioning resolution of 5 nm.

(2) The optical heterodyne system uses two wavelengths and can detect the position between a mask and wafer with a resolution of 5 nm and a linearity of 10 nm over a distance of 2.5 μm.

(3) Magnification correction of the exposure field size can be carried out by thermal stress control during chucking. The results show that corrections up to a magnitude of 4 ppm can be made.

(4) Exposure experiments yielded a total overlay accuracy of 28 nm (3σ), which includes mask and wafer errors and the alignment error of the stepper.

These results demonstrate that this x-ray exposure system is capable of printing 100-nm device patterns.

ACKNOWLEDGEMENTS

The authors would like to thank Mr. T. Shimizu for his assistance in the exposure experiments. This work was a collaborative effort between ASET and NTT. ASET is supported by the New Energy and Industrial Technology Development Organization (NEDO).

REFERENCES

1. S. Ishihara, M. Suzuki, M. Kanai, and M. Fukuda: An advanced X-ray stepper for 1/5-um SR lithography, MicroElectronic Engineering, Vol. 17, No. 1-4, March 1992, pp. 141-144

2. International Technology Roadmap for Semiconductors: 1999 Edition, Edited by International SEMATECH, published by the Semiconductor Industry Association. USA (1999) p. 147

3. M. Fukuda, M. Suzuki, T. Haga, N. Takeuchi, H. Morita, K. Deguchi, T. Taguchi, H. Aoyama, S. Mitsui, and Y. Matsui: Performance of X-ray stepper for next-generation lithography; Jpn. J. Appl. Phys. Vol. 38 (1999), pp. 7059-7064

4. S. Ishihara, M. Kanai, A. Une, and M. Suzuki: A vertical stepper for X-ray lithography, J. Vac. Sci. Technol. B7(6) (1989), pp. 1652-1656

5. M. Suzuki and A. Une: An optical-heterodyne alignment technique for quarter micron X-ray lithography, J. Vac. Sci. Technol. B7(6) (1989), pp.1971-1974

6. H. Aoyama, S. Mitsui, T. Taguchi, Y. Tanaka, Y. Matsui, M. Fukuda, M. Suzuki, T. Haga, and H. Morita: Magnification correction by changing wafer temperature in proximity X-ray lithography, J. Vac. Sci. Technol. B17 (1999), pp. 3411-3415

7. H. Aoyama, T. Taguchi, Y. Matsui, M. Fukuda, K. Deguchi, H. Morita, M. Oda, and T. Matsuda: Overlay evaluation of proximity X-ray lithography for 100-nm device fabrication, , J. Vac. Sci. Technol. B18 (2000), pp. 2961-2965

WIDE RANGE CAPACITIVE GAP SENSOR WITH REPRODUCIBLE NANO-METER-ORDER RESOLUTION

Nobuhiro Ishikawa, Nobuhisa Nishioki

Mitutoyo Corporation

Abstract

A capacitive gap sensor with high-performance has been developed. The sensor is characterized by a sensor electrode having a guard electrode, a phase compensated AC-DC converter and a digital linearizer to accomplish the compatibility of high S/N with high speed of response.   A resolution of order 10nm, a high frequency response of about 10kHz at the phase-delay of $45°$ and a wide measurement range of 100μm with 0.03% linearity error are realized by the sensor with the above mentioned features.   It is estimated that the expanded uncertainty(k=2) of the sensor is 20.6nm.

Keyword

Capacitive gap sensor, Non-contact measurement, High resolution, High frequency response, Wide measurement range, High linearity

1. INTRODUCTION

One of non-contact and high-resoluble means to measure a gap between a conductive work surface and a sensor is a capacitive gap sensor, which is more effective than a optical interferometer, because the structure is very simple and the acquired data is absolute and reliable in an environment of air turbulence.   However, it has some shortages of the narrow measurement range , the low response and the difficult linearization.

These problems have been improved by ① raising S/N without reducing its frequency response with a sensor electrode and a AC-DC converter, and ② using a digital signal processing for an exact linearization in all of its measurement range.

2. SENSOR OUTLINE

The sensor consists of a sensor head unit and a control unit. A photograph of a trial manufactured sensor head unit is shown in Figure 1, and a block diagram of the sensor is shown in Figure 2. The sensor electrode of the sensor head unit is driven by 200kHz AC signal from the exciter of the control unit. The capacitance which corresponds to the gap between the sensor electrode and the work is detected as the voltage amplitude by an AC bridge type capacitance detector. In the control unit, the amplitude is converted and linearized, and then the linearized analog output in proportion to the gap is given.

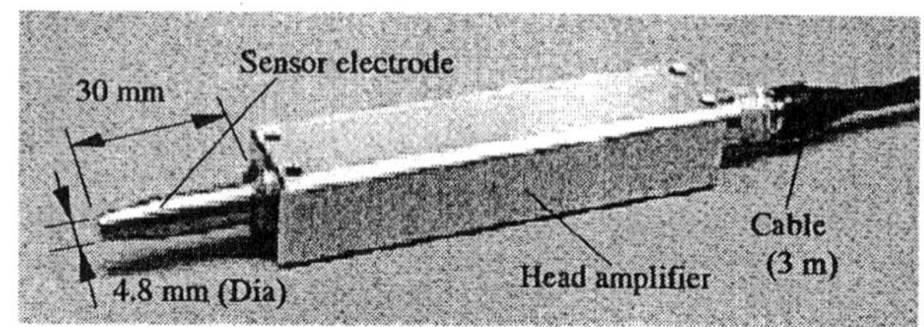

Figure 1. Sensor head unit

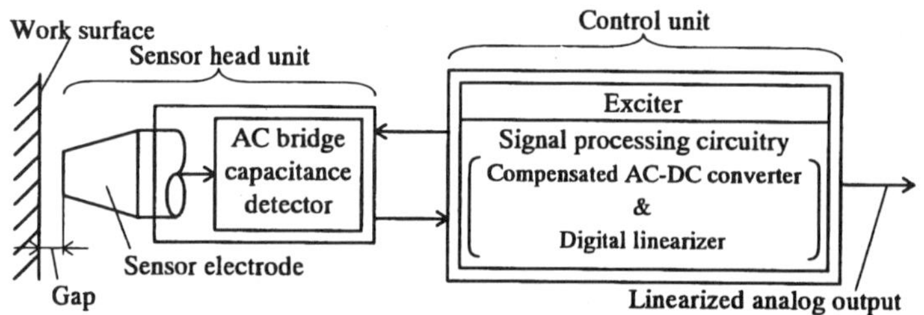

Figure 2. Block diagram of capacitive gap sensor

3. SENSOR ELECTRODE

The structure of the sensor electrode is shown in Figure 3. It comprises of an outer and center electrodes which detect the capacitance for measuring the gap, and the guard electrode to suppress the stray capacitance. The effect of the guard electrode is shown in Figure 4. As a result, the guard electrode can improve S/N to about one hundred times more compared with the case of not using the guard electrode.

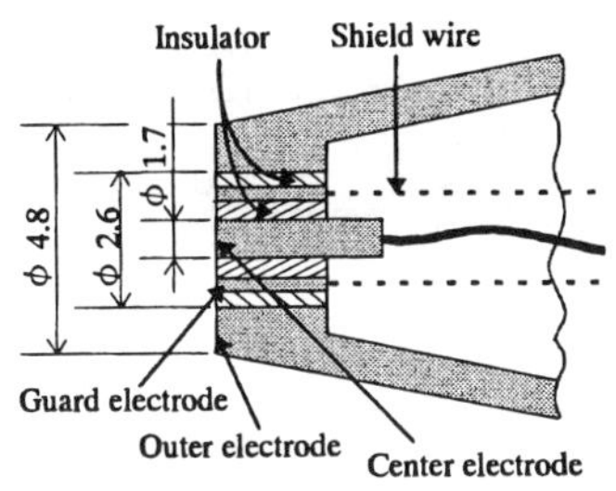

Figure 3. Structure of sensor electrode

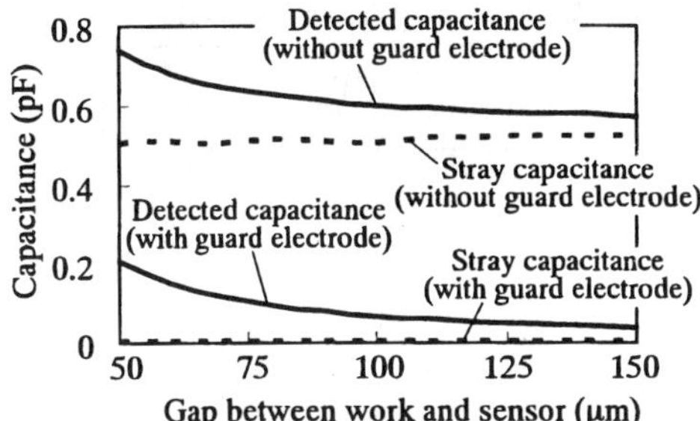

Figure 4. Effect of guard electrode

4. SIGNAL PROSSEING CIRCUITRY

The signal processing circuitry consists of a phase compensated AC-DC converter and a digital linearizer. A block diagram of the AC-DC converter is shown in Figure 5. The AC-DC converter samples the sensor head unit output of 200kHz at 400kHz by using two A/D converters alternately. The conversion speed is ten times as high as the conventional AC-DC converter which consists of a full wave rectifier and a low pass filter. Moreover, even if the frequency of the sensor head unit output changes slightly, it makes the accurate AC-DC conversion possible because the phase difference between Sps1 and Sps2, Sps3 and Sps4 is kept at 90° .

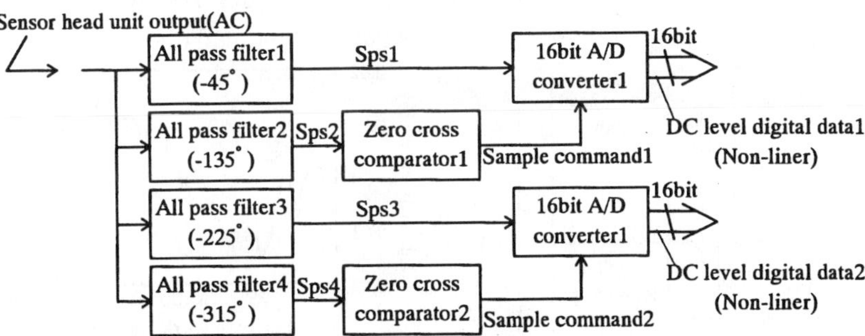

Figure 5. Block diagram of phase compensated AC-DC converter

A block diagram of the linearizer is shown in Figure 6. The linearizer makes the correction for the non-linearity of the A/D converters possible because of using two ROM tables correspond to two A/D converters respectively. Moreover, it has the very wide dynamic range more than 80dB, and the performance doesn't change for a long time.

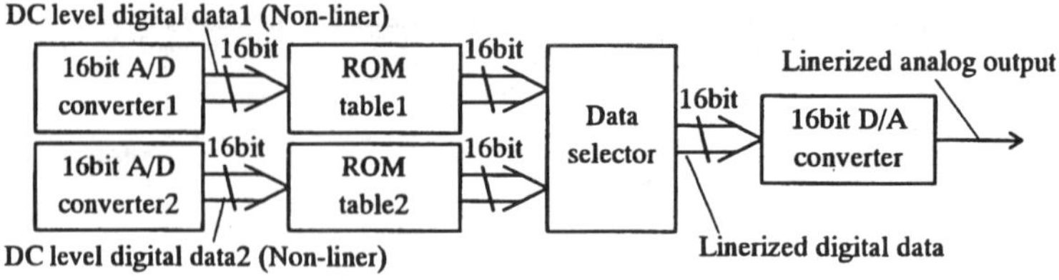

Figure 6. Block diagram of digital linearizer

5. EXPERIMENTAL RESULT

A measurement example of the stage motion in 50nm steps is shown in Figure 7. The figure shows that the sensor has a resolution of order 10nm. Here it contains a steady vibration of about 10nm from a building floor.

Figure 8 shows the calibration and estimation system for linearity of the sensor.

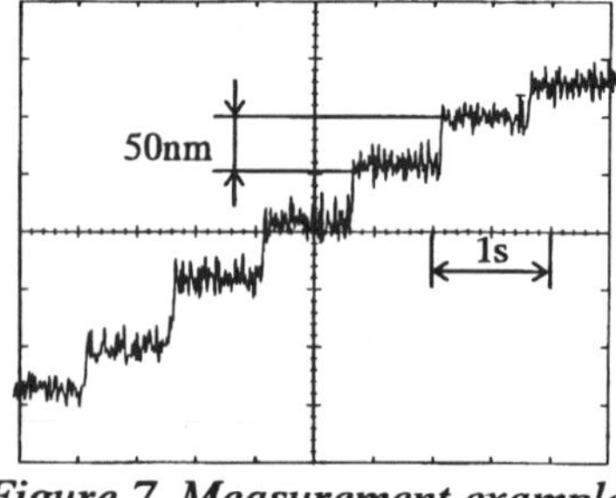

Figure 7. Measurement example

The calibration equipment[1] has the interferometer installing a variable length vacuum cell. The sensor is calibrated by the system. The linearity error which is estimated by the system is shown in Figure 9. The result shows that the sensor has a wide measurement range of 100μm with 0.03% linearity error. The measurement range is twice as wide as the conventional one.

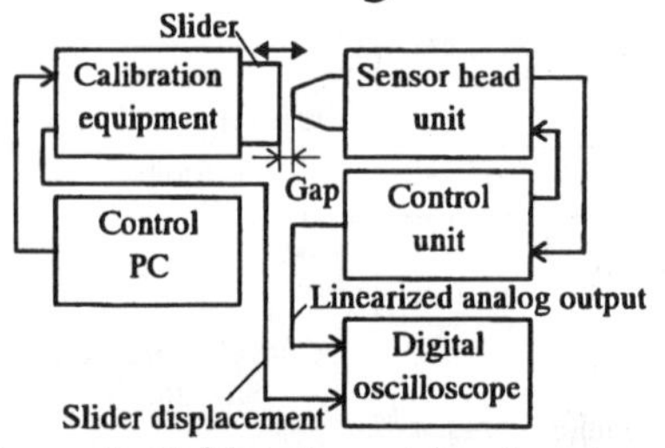

Figure 8. Calibration and estimation system for linearity

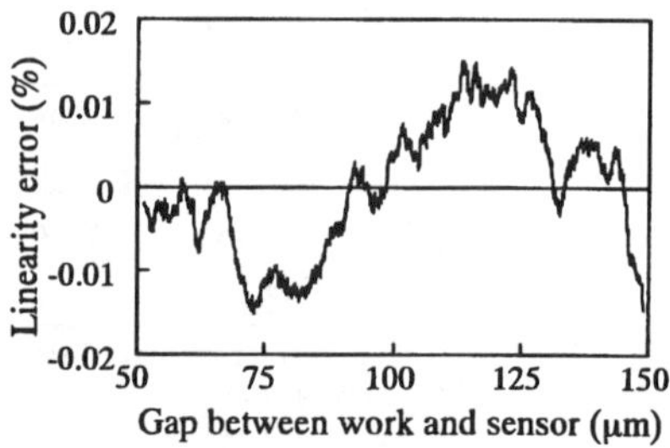

Figure 9. Linearity error

Figure 10 shows the estimation system for frequency response of the sensor. The frequency response of the system is over 10kHz. Figure 11 shows the frequency response of the sensor. From the result, the frequency response of the sensor is about 10kHz at the phase-delay of $45°$. That is 40% larger than the conventional one.

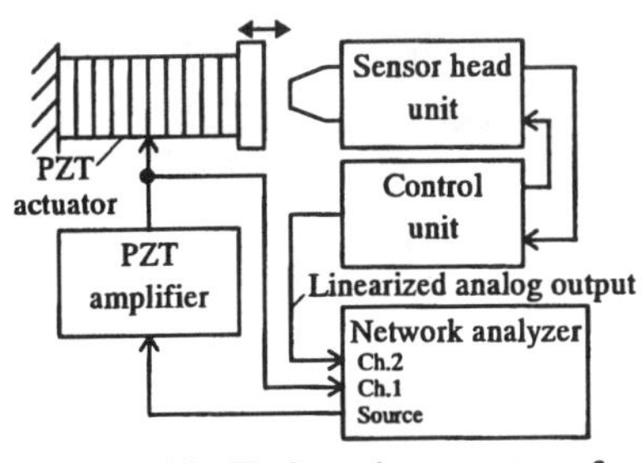

Figure 10. Estimation system for frequency response

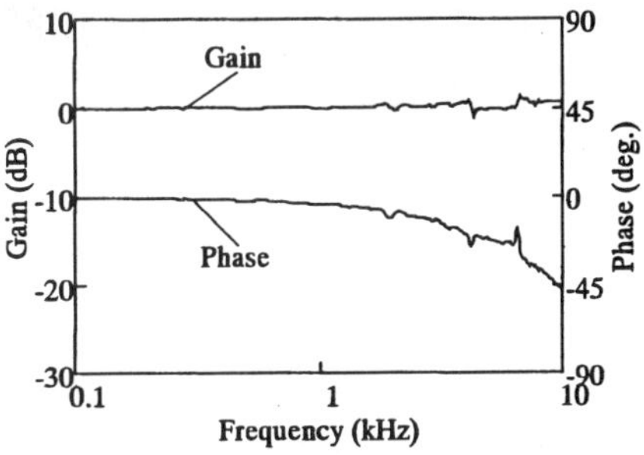

Figure 11. Frequency response

Table 1 shows the uncertainty budget for the sensor. We pick up major uncertainty factors and estimate the expanded uncertainty(k=2) to be 20.6nm.

Table 1. Uncertainty budget for the capacitive gap sensor

Uncertainty factor	Estimation condition	Standard uncertainty (k=1)
Relative permittivity of the air T=293.15K, H=50%RH A=1013hPa, d=100μm	ΔT=0.01K	0.180nm
	ΔH=2%RH	0.301nm
	ΔA=2hPa	0.061nm
Sensor's tilt	Tilt=20μrad d=100μm	0.001nm
Thermal gradient change of sensor at 5minutes	ΔT=0.01K Length=20mm Material:SUS303	2.43nm
Resolution	Reference of Figure7.	10.0nm
Combination standard uncertainty (k=1)		10.3nm
Expanded uncertainty (k=2)		20.6nm

where T:Temperature, H:Humidity, A:Atmospheric pressure, d:Gap between sensor and work

6. CONCLUSION

We have developed the capacitive gap sensor. Some experiments show that the sensor has a resolution of order 10nm, a high frequency response of 10kHz(-45°) and a wide measurement range of 100μm with 0.03% linearity error. It is estimated that the expanded uncertainty(k=2) of the sensor is 20.6nm.

REFERENCE

(1)H. Sakai, Y. Kuriyama, H. Masuda, M. Ogihara, H. Oozeki. Development of calibration equipment with sub-nano meter order resolution using laser interferometer combing a variable length vacuum cell: Metrologia-2000 international conference on advanced metrology; 2000 December 4-7

DESIGN AND DEVELOPMENT OF MULTI-FUNCTION SENSOR USING A GROUP OF MICRO CANTILEVERS

Kenji Uchiyama and Nobuyuki Moronuki

Graduate School of Engineering, Tokyo Metropolitan University
1-1 Minami-ohsawa, Hachioji, Tokyo 192-0397, Japan
E-mail: uchiyama-kenji@c.metro-u.ac.jp

Abstract

It is difficult to measure precisely physical quantities in micro domain since effect of plural quantities that are negligible in macro domain becomes large. This study proposes a multi-function sensor that consists of multiple micro-cantilevers, which utilize bimorph construction. A measurement principle for detection of the plural physical quantities was briefly explained and a design guide was shown for fabrication of the micro-cantilevers with different length. The experimental results show that temperature and flow velocity can be detected simultaneously and independently using present multi-function sensor.

Key words
Multi-Function sensor, Micro-Cantilever, Bimorph

1. INTRODUCTION

Many micro actuators and sensors have been developed through use of semiconductor manufacturing technologies. Particularly micro integrated sensors are necessary for measurement physical quantities because of small space. The micro-cantilevers have been applied as sensor element because its construction is simple and characteristic, such as natural frequency or displacement at the free end, is affected by the physical quantities. An air flow sensor uses large deflection at tip of the cantilevers (Ozaki, 2000). Accelerometer and angular velocity sensor utilize inertia force and Corioris force acting on mass of the cantilever. There is a thermal sensor using bimetal effect of the cantilever. Shift of natural frequency of cantilever can be applied as gas and pressure sensors.

These physical quantities are not dependent and often correlate with each other. In order to sense the physical quantities accurately, it is necessary to overcome the interference problem. Typical sensors have a compensation circuit to measure the specified physical quantities. However its construction becomes complicated.

A multi-function sensor that consists of a group of micro-cantilever is introduced in this paper. From measurement principle, plural quantities can be detected by using a group of the cantilevers with different length. The effectiveness of the sensor design was verified by the experiment about measurement of temperature variance and flow velocity.

2. MEASUREMENT PRINCIPLE

Figure 1 illustrates the concept for measurement of the physical quantities by using micro-cantilevers. If the dependency is formulated by a function of the cantilever size, the complex effect can be separated by a group of cantilevers with different size. Thus a group of bimorph cantilevers was used to detect the plural physical quantities. The proposed sensor is designed to measure temperature variance and flow velocity in this study. It was assumed that the drag force acting on the cantilevers could be regarded as uniformly distributed one since the displacement at the free end of cantilevers is sufficient small. The tip deflection of cantilever with respect to temperature variance ΔT and flow velocity U can be formulated as:

$$v_t = C_t l^2 \Delta T \tag{1}$$

$$v_f = C_f l^4 U^2 \tag{2}$$

where v_t and v_f denote the tip deflection due to temperature variance and flow velocity. Parameters C_t and C_f are the coefficients that are defined by material characteristic and geometric property of the cantilevers with length l. When the cantilevers are simultaneously influenced by the temperature variance and flow velocity, the tip deflection v is expressed by sum of v_f and v_t. We are concerned with the cantilevers with different length l_1 and l_2 to detect them separately. Then equations of the temperature variance ΔT and flow velocity U are obtained as follows:

$$\Delta T = (l_2^4 v_1 - l_1^4 v_2)/C_t l_1^2 l_2^2 (l_2^2 - l_1^2) \tag{3}$$

$$U = \sqrt{(l_2^2 v_1 - l_1^2 v_2)/C_f l_1^2 l_2^2 (l_2^2 - l_1^2)} \tag{4}$$

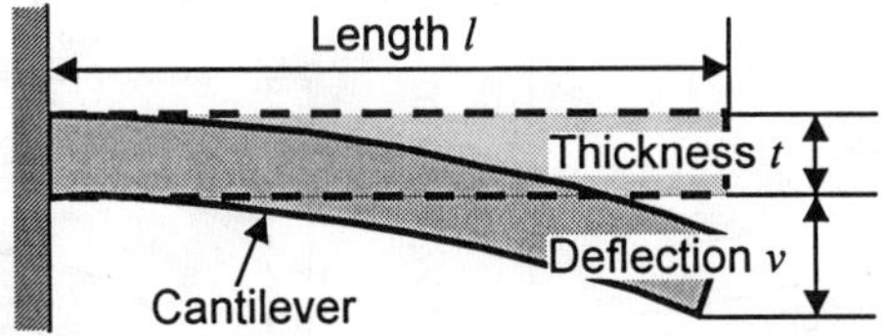

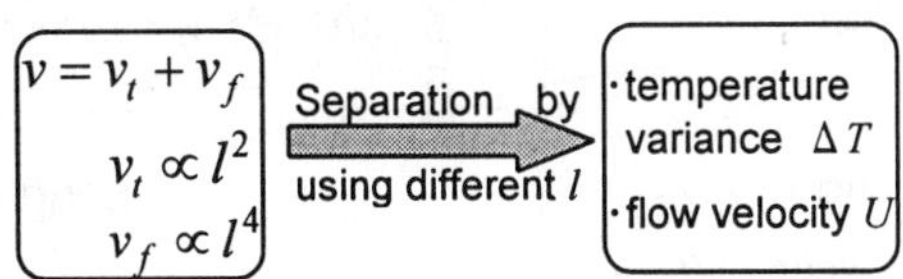

Fig.1 Schematic illustration of measurement principle

It is clear from these equations that temperature variance and flow velocity can be measured separately when each length is known and tip deflection is measured.

3. DESIGN AND FABRICATION OF CANTILEVER

3.1 Design Guidelines

It is necessary to establish a guideline for design of the micro-cantilevers in terms of its geometrical characteristics. Firstly measurement range of physical quantities should be determined. The range of temperature variance and flow velocity is set to $0\sim5$ degrees and $0\sim$ 2m/s in this study. The bimorph cantilevers consist of Ti layer (Young's modulus: 66.7GPa, thermal coefficient: 8.6×10^{-6}) and SiO_2 layer (Young's modulus: 116GPa, thermal coefficient: 0.50×10^{-6}). Semiconductor manufacturing technology readily fabricates SiO_2 cantilevers. The reason of selection of Ti is that it has high adhesive force to SiO_2 and there is large difference between two values of thermal coefficient.

Secondly thickness of the cantilevers is decided. Figure 2(a) shows relation between the thickness ratio of Ti and SiO_2 and the ratio of C_f and C_t. It is assumed for the calculation that ratio of U^2 and ΔT is equal to 1. The thickness value of SiO_2 layer t_{sio2} is fixed as 0.9μm by the restriction of fabrication process of cantilevers. Deflection due to the temperature variance is selected similarly to that due to the flow velocity. From Eqs.(1) and (2), the each deflection is function of coefficients C_f and C_t. Ratio between these coefficients is expected to be 1, and thickness ratio is better more than 0.1 in this case.

Figure 2(b) shows relation between the displacement and the length of the cantilevers when the thickness ratio is assumed to be 0.33. A design guideline was based on the derived formulation. Then the range of design of cantilevers' length is determined at intersection points of each curved line as

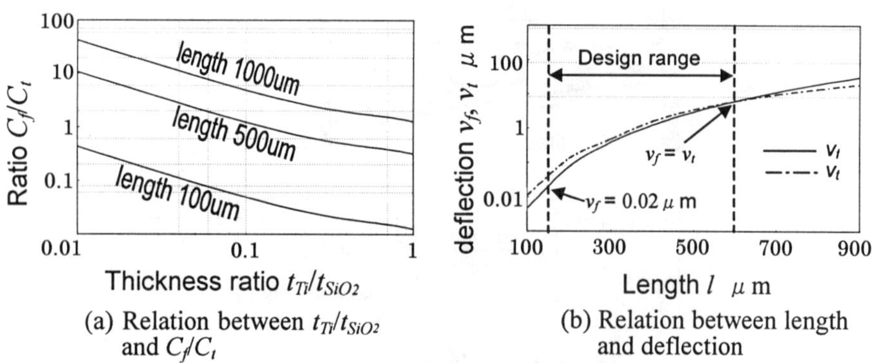

(a) Relation between t_{Ti}/t_{SiO2} and C_f/C_t

(b) Relation between length and deflection

Fig. 2 Design guide for fabrication of micro-cantilevers

shown in the figure. The upper limit of design range of the cantilever length is at the point of $v_f/v_i=1$. The lower limit is determined by the resolution of sensors such as a laser micrometer. According to these design guideline, the range of cantilever length can be chosen from $150\,\mu$ m to $600\,\mu$ m as shown in the figure.

3.2 Fabrication

The SiO_2 cantilevers were fabricated by applying anisotropic etching of single crystal silicon with SiO_2 layer, $1.0\,\mu$ m, and the KOH solution (35wt%, 333K) was used as etchant in this process. The vapor deposition is utilized to deposit Ti on the SiO_2 cantilevers. Figure 3(a) and 3(b) show the SEM photograph and side view of example of bimorph cantilevers. The dimensions of the typical cantilevers are about $500\,\mu$ m in length, $100\,\mu$ m in width, and $1.1\,\mu$ m in thickness. The thickness of Ti and SiO_2 layer is $0.2\,\mu$ m and $0.9\,\mu$ m, respectively.

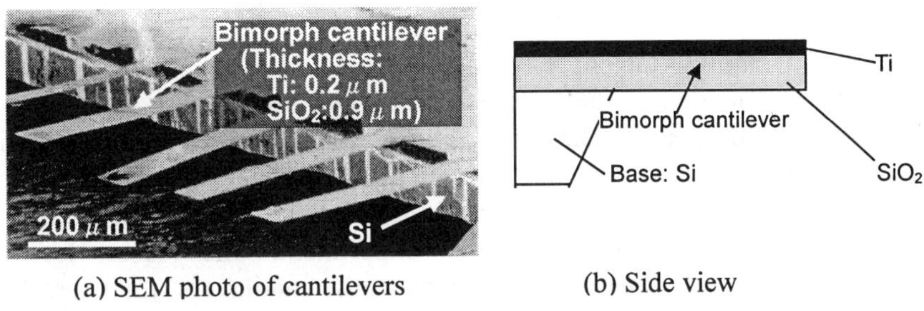

(a) SEM photo of cantilevers   (b) Side view

Fig.3 Bimorph cantilevers that consist of SiO₂ and Ti

4. EXPERIMENTAL RESULTS

The relation between the displacement of the cantilevers and the plural quantities are evaluated in a clean room in which temperature and relative humidity are controlled as 20 degrees and 50 %. A laser micrometer (resolution: $0.02\,\mu$ m) measures the displacement at the free end of the cantilever.

Figure 4 shows experimental results of evaluation of the proposed multi-function sensor. Relation between the tip deflection of the cantilever and the flow velocity is shown in Fig.4(a) when the effect of temperature variance on the cantilever is sufficient small, i.e. $\Delta T \approx 0$. The experimental values agree with the theoretical values. Figure 4(b) shows the relation between tip deflection and the temperature variance. The more the temperature variance increases, the more the error between the deflection and the temperature variance increases. It is considered that there may be the variance of thickness of Ti layer.

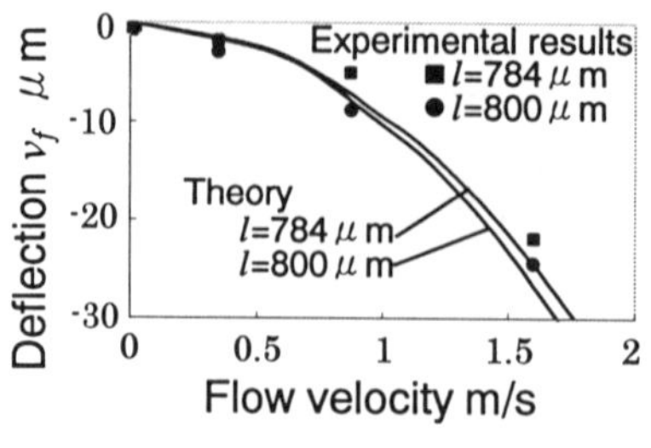

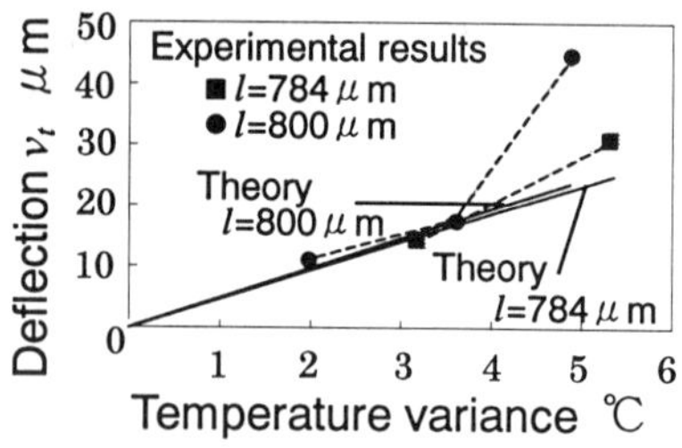

(a) Flow velocity (ΔT=0℃)　　　(b) Temperature variance (U=0)

Fig.4 Evaluation of multi-function sensor

Table 1　Experimental results

Other	U (m/s)	0.85		
sensors	ΔT(°C)	1.0	1.2	1.5
Proposed	U (m/s)	1.47	0.63	0.47
sensor	ΔT(°C)	1.60	2.89	5.47

Table 1 shows the experimental results of the detection of the temperature variance and the flow velocity. The flow velocity was given as constant value 0.85 m/s. The temperature variance and the flow velocity detected by the cantilevers coincided with the theoretical values qualitatively, but the experimental values were not same values in comparison with other sensors which are employed in this experiment. The reason for this difference is considered to be the variance of the cantilevers' thickness. The experimental results have proven the possibility that the temperature variance and the flow velocity are detected by using micro-cantilevers with different length.

5. CONCLUSIONS

The multi-function sensor that consists of the group of bimorph micro-cantilevers is proposed to detect the temperature variance and flow velocity in this paper. The results are summarized as follows:

(1) The design guideline is given of the thickness ratio and the length of the cantilevers.
(2) The proposed sensor actually detects the temperature variance and the flow velocity.

The cantilever that has less variance of its thickness should be fabricated to improve the accuracy.

REFERENCES
Ozaki Y. et al., An air flow sensor modeled on wind receptor hairs of insects, Proceedings of the Annual International Conference on Micro Electro Mechanical Systems; 2000 January 23-27; Miyazaki, Japan; 531-536.

A TOOL TO ESTIMATE THE ADHESIVE FORCES BETWEEN MICROCOMPONENTS AMONG EACH OTHER AND BETWEEN MICROCOMPONENTS AND GRIPPERS OR MAGAZINES

Prof.-Dr.-Ing. Dr. h.c. Jürgen Hesselbach, Dipl.-Ing. Christiane Graf
Institute of Machine Tools and Production Technology
Technical University Braunschweig, Germany

Abstract

The handling behaviour of microcomponents is characterized by adhesive forces. Since first approaches do not clearly describe these adhesive forces in order to support the process of designing handling tools for microassembly, a tool is developed to estimate the adhesive forces between microcomponents among each other and between microcomponents and grippers or magazines. Theoretical approaches and a test stand to measure the adhesive forces are the base of this software tool that facilitates the development or the selection of the most suitable gripper and magazine for a specific microcomponent.

Keywords

Adhesive force, handling of microcomponents, microassembly

1 INTRODUCTION

Grippers and magazines form the interface between handling tools and the components which have to be handled. Therefore they always have to be developed specifically fitting to the handled components. In order to facilitate the right selection of grippers and magazines the components can be characterized with the aid of different features. Through the help of this a restriction of suitable grippers and magazines can be carried out. These features are summarized for components with macroscopic lengths (> 1mm) in guideline VDI 3237 [VDI]. Based on the systematization of this guideline the handling behaviour of components can be estimated before developing the grippers or magazines. In this way possible problems can consequently be recognized and eliminated or at least reduced to a minimum by the selection of suitable grippers and magazines.

For microcomponents such a systematization does not exist yet. Up to now only approaches can be found in the preliminary design of standard DIN 32563 [DIN]. However, general features like geometry, weight and material are not sufficient to describe the handling behaviour of micro-components. Above all estimations of the probable handling behaviour are not made clear. But especially this handling behaviour is very important for the handling of microcomponents because it differs very much from the handling behaviour of components with macroscopic lengths.

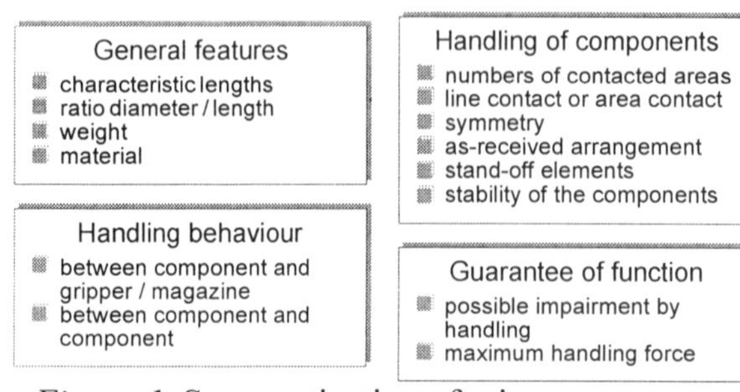

Figure 1. Systematization of microcomponents

Therefore the handling behaviour represents an important feature of the systematization to characterize microcomponents (Figure 1). This characterization of each microcomponent considerably facilitates the selection of suitable grippers and magazines.

2 THE HANDLING BEHAVIOUR OF MICROCOMPONENTS

Components can hook with each other or adhere to each other due to magnetism. These problems concern microcomponents as well as components with macroscopic lengths. In addition the handling behaviour of microcomponents is characterized by adhesive forces. On the one hand these adhesive forces impair the handling at the interface component – gripper or magazine because the components adhere to them. On the other hand microcomponents can sometimes be separated hardly because they adhere to each other on account of adhesive forces. First approaches do not clearly describe this behaviour in order to support the development of handling tools for microassembly. Therefore a tool was created by the authors to fill this gap.

2.1 Adhesive forces

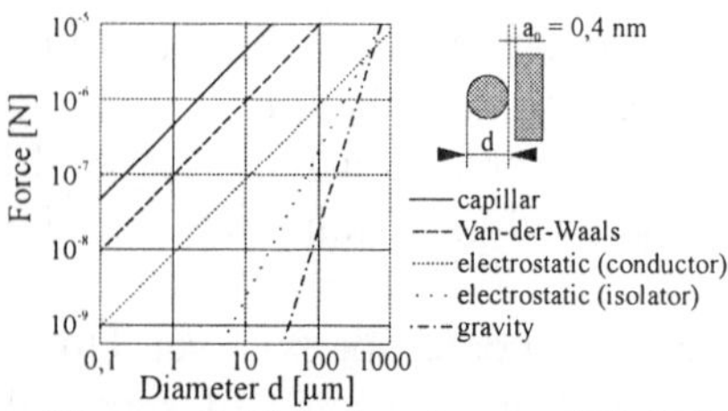

Figure 2. Existing forces [Oh98]

If two bodies get into touch with each other, physical interactions, among other things, cause adhesive forces. However, these adhesive forces do not have any effects on the handling behaviour of bigger components since in this case gravity forces are noticeably more dominant than adhesive forces. The smaller components are, the larger becomes the ratio adhesive force / gravity force (Figure 2).

Therefore microcomponents adhere to other microcomponents or to the gripper or magazine (Figure 3). Adhesive forces result from liquid bridges, Van-der-Waals-interactions and electrostatic interactions and on account of a combination of these three different interactions. They depend on the geometry of the components, the distance of the bodies from each other and on surface roughness.

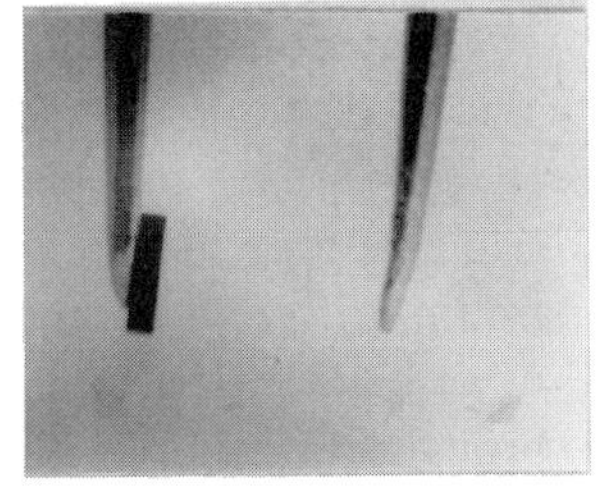

Figure 3. Gripper and silicium component [Oh98]

In several publications first approaches referring to these three different interactions can be found. However adhesive forces only are considered separately and independent from each other and also only for specific geometry pairs (plate/plate, sphere/plate, sphere/sphere) with a constant distance of $a_0 = 0{,}4$ nm [Löffler92], [Schubert90]. Rumpf describes the adhesive forces between sphere and plate with variable distance, but he does not continue this first approach [Rumpf75]. Therefore different diagrams based on these first theoretical approaches have been carried out.

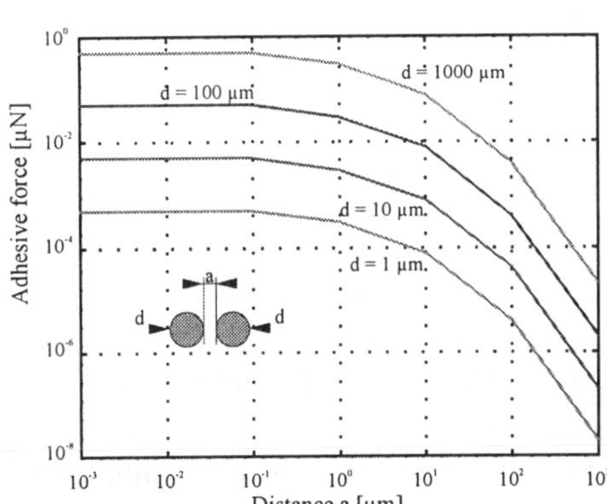

Figure 4. Van-der-Waals-force

These diagrams show the three types of adhesive forces for different pairs of geometry forms with variable length of components and with variable distances between them. Figure 4 shows an example of the Van-der-Waals-force between two spheres depending on variable diameters and variable distances between the spheres.

2.2 Simulation of adhesive forces between two microcomponents or between microcomponent and gripper / magazine

Since always a combination of the three adhesive forces will occur in reality, a test stand was assembled at the Institute of Machine Tools and Production Technology to get more information about adhesive forces and their effects on the handling behaviour of microcomponents (Figure 5). One component is fastened on the force sensor, the other on the piezo actuator. With the aid of the piezo actuator, the distance between both components can be reduced. The whole process can be observed by a camera.

On the one hand the handling behaviour of microcomponents be-tween each other can be simulated by bringing together and afterwards separating different components. On the other hand microcomponents can be brought together with a bigger plate and afterwards separated in order to simulate the handling behaviour while gripping and storing.

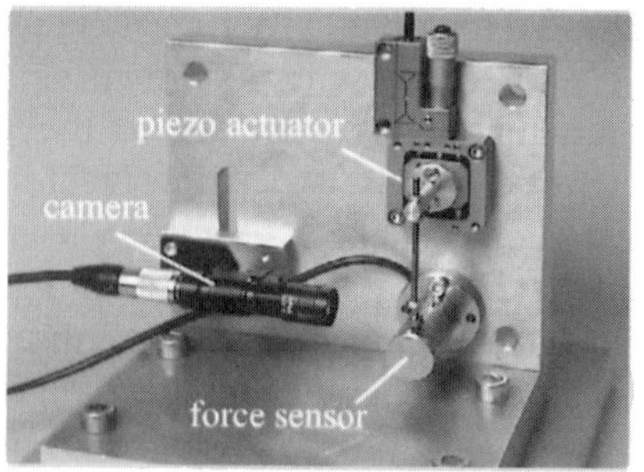

Figure 5. Test stand

Figure 6 and 7 show an example of adhesive forces measured by the force sensor for two spheres with the diameter d = 1000 μm.

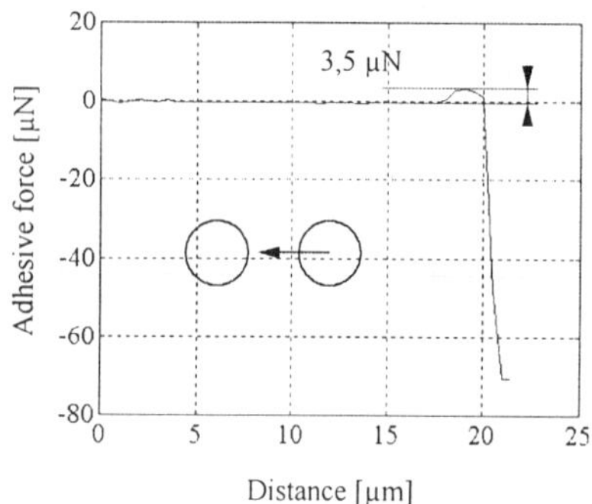

Figure 6. Adhesive force while bringing two microspheres together

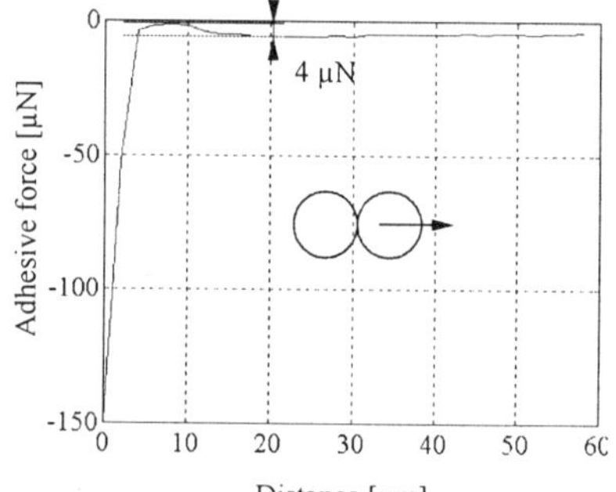

Figure 7. Adhesive force while separating two microspheres

The results of the theoretical approaches and of the tests can be sum-marized in the following way:

- Adhesive forces resulting from contacting two bodies are only signifi-cant for bodies with microscopic length, because in this case gravity forces do not have any influence.
- The bigger the difference between the length of both bodies, the big-ger is the tendency of the smaller body to adhere to the other one.
- If two components get in contact with eachother, the adhesive force is not so high as if one component has to be separated from another component or from the gripper or magazine. Furthermore the adhesive force occurs only during a smaller distance.
- With growing distance the adhesive force quickly gets smaller.
- The smoother the surface of the body, the bigger is the contact area that is significant for the adhesive forces.
- If a body with a rough surface is pressed towards a second body, the number of contacts grows. Therefore the adhesive force grows too.

3 A SOFTWARE TOOL TO ESTIMATE THE ADHESIVE FORCES OF MICROCOMPONENTS

With the aid of these test results fundamental guidelines for the development of grippers or magazines can be acquired. So on the one hand it can be taken advantage of the adhesive behaviour for example by developing electrostatic grippers and magazines. On the other hand the conclusions of the test results can be used to reduce the adhesive forces to guarantee an optimal gripping and storing process. Furthermore the theoretical approaches and the test results are used to get different diagrams about the handling behaviour of microcomponents. These diagrams are deposited in a data base of a software tool (Figure 8).

This program can be used to inform about the adhesive behaviour of microcomponents. After the input of the most important features of the micro-component, the program displays the classification of the microcomponent according to the systematization and an unambiguous codification of the micro-component.

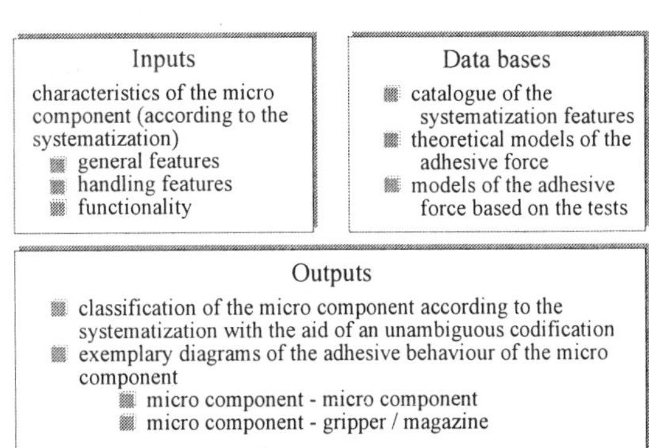

Figure 8. Software tool to estimate the adhesive forces

Furthermore it shows exemplary diagrams of the adhesive behaviour between the used microcomponents among each other and towards the grippers and magazines. As a result a software tool is given to the user to estimate the adhesive forces and their effect on the handling behaviour of microcomponents. It allows a development or a selection of the most suitable gripper and magazine for a specific microcomponent.

4 REFERENCES

[DIN]   DIN. Klassifizierung von Mikrobauteilen. DIN-Entwurf 32563, 2000.

[Löffler92] Löffler, F., Raasch, J., *Grundlagen der Mechanischen Verfahrenstechnik.* Vieweg, 1992.

[Oh98]   Oh, H.-S., *Elektrostatische Greifer für die Mikromontage.* Braunschweig: VDI-Fortschrittsberichte, 1998.

[Rumpf75]   Rumpf, H., *Mechanische Verfahrenstechnik.* Carl Hanser Verlag, 1975.

[Schubert90] Schubert, H., et.al., *Mechanische Verfahrenstechnik.* Deutscher Verlag für Grundstoffindustrie, 1990.

[VDI]   VDI. Fertigungsgerechte Werkstückgestaltung im Hinblick auf automatisches Zubringen., Fertigen und Montieren. VDI-Richtilinie 3237, 1967.

HIGH PRECISION COLLIMATION USING TALBOT INTERFEROMETRY

Shinya Haramaki*,Shunsuke Yokozeki**,Akihiro Hayashi**,Hiroshi Suzuki**

*Department of Mechanical Engineering, Ariake National College of Technology

**Department of Mechanical System Engineering, Faculty of Computer Science
and Systems Engineering Kyushu Institute of Technology

Abstract

This study concerns the high precision beam collimation method using the Talbot interferometry. When the collimation beam is obtained by placing point-source of light in focal position of the collimation lens, the defocus value is calculated from a moire fringe width in the Talbot interferometry using two grating. With the advance of collimation, the moire fringe is hard to be confirmed. Using the movement of moire fringe when one of grating shifts, the expanded moire fringe width is measured high-precisely from the phase difference of intensity change at two detecting points. In this paper two methods are proposed. (1).Fourier image is approximated in the sine wave, and the phase difference is obtained by four-frames method. (2).The correlation coefficient is calculated, and the phase difference is obtained from the peak position of it.

Keywords

collimation, Talbot interferometry, moire fringe, phase difference

1. INTRODUCTION

The high precision measurement system has to be realized for micro/ nano technology, some optical measurement systems[1] have been proposed and reported for it. In optical measurement system, collimation beam plays important roles in optical systems. The quality of beam collimation may affect the performance of measurement system. This paper deals with the beam collimation using the Talbot interferometry. The Talbot interferometry[2] is known by its simple configuration and then the low cost, and beam collimation is one of the useful applications.

2. THE TALBOT INTERFEROMETRY

The Talbot interferometry is a sort of shearing interferometer to which the Fourier self-imaging phenomenon called Talbot-effect and the moire fringe detection method are applied. The optical arrangement of the Talbot interferometer is shown in Fig.1. The interferometer simply consists of a point-source of light S_0, a collimation lens L, and two grating Gm and Gr for diffraction and detection.

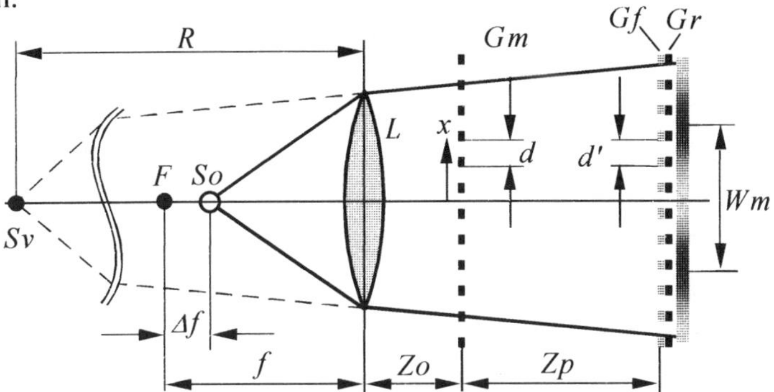

Fig. 1. Schematic diagram of the Talbot interferometry.

3. THE PRINCIPLE OF COLLIMATION

By the coherent beam passing through the diffraction grating Gm located just behind the collimation lens L, then the so-called Fourier images Gf of the diffraction grating Gm are appeared at the every distance nd^2/λ. Where n is an integer, d is the grating constant, and λ is the wavelength of the laser beam. This is Talbot-effect. The characteristic of Fourier images is influenced by the shape of wave-front of the beam. If a point-source of light is not at the focal point F of collimation lens L, the beam passing through the collimation lens L has a spherical wave-front. When the defocus value and focal length of collimation lens L are Δf and f respectively, the curvature R of spherical wave-front is

$$R = \frac{f^2}{\Delta f} - f .$$

(1)

If the detection grating Gr is put on the Fourier image Gf, moire fringes are observed at the observation plane. The width of fringes Wm is expressed by

625

$$Wm = \frac{R + Zo + Zp}{Zp} d \qquad (2)$$

Where Zp is the distance between gratings. The defocus value Δf can be computed by the eq. (1) and (2). The collimation can be done by adjusting the position of the lens L for the defocus value to be minimized. This is the basic principle of the collimation by using the Talbot interferometer.

4. HIGH PRECISION COLLIMATION METHOD

As the degree of the collimation improved, moire fringe could not be clearly observed within the observation area[3] because the width of moire fringe is extended over the width of observation area. This problem can be resolved by shifting the grating Gm. As shown in Fig.2, the intensities of moire pattern observed at the two detecting points A and B are changed by scanning the grating so as a periodic wave being out of phase. Then the width of moire fringe can be also obtained from a phase difference e between the intensities observed at the two detecting points by

$$Wm = \frac{d}{e} w \ . \qquad (3)$$

,where w is the distance between two detecting points. From eq. (1), (2) and (3) , the relationship between a phase difference and a defocus value is deduced as the following equation.

$$\Delta f = \frac{f^2}{wZp} e \qquad (4)$$

If the phase difference can be accurately detected, a more precise collimation may be realized by using eq. (4).

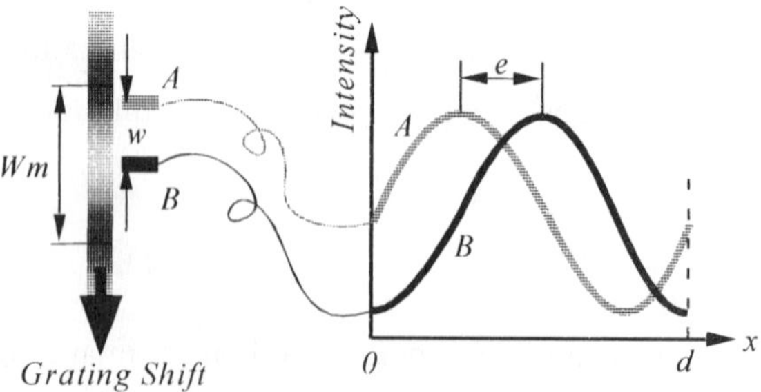

Fig. 2. Illustration of grating shift method.

5. THE DETECTION METHOD OF PHASE DIFFERENCE

Two methods have been applied to detect a phase difference from the changes of intensity observed at two detecting points.

5.1 The Four-Frames Method

Under the assumption that the intensity observed at two detecting points would be sine wave with the period of the grating constant, the phase difference of two intensity waves can be obtained approximately from the intensity values I_0, I_1, I_2, I_3 observed every 1/4 period of the grating constant. Here, it is a phase for the period of grating constant d.

$$p_A = \frac{d}{2\pi} tan^{-1}\left(\frac{I_3 - I_1}{I_0 - I_2}\right) \tag{5}$$

Therefor, the phase difference e for the period d can be obtained by $e = p_A - p_B$. To improve the accuracy of calculation, however the intensity distribution of moire fringe must be approximated to sine distribution. To satisfy this approximation, the distance of the grating between Gm and Gr is so adjusted that the intensity distribution of moire fringes becomes quasi-sinusoidal.

5.2 The Correlation Coefficient Method

Both changes of intensity observed at two detecting point would be similar wave. The correlation coefficient $R_{AB}[k]$ of two intensity waves data $A[i]$, $B[i]$ can be calculated by the following expression.

$$R_{AB}[k] = \sum_{m=0}^{N-1} A[m]B[k+m] \quad (k = 0,1,\cdots,N-1) \tag{6}$$

, where N is number of partitions for the period of grating constant. Therefor, a phase difference e for the period of grating constant can be obtained by finding a division point k to be maximum of the correlation coefficient and calculating by $e = kd/N$. To improve the accuracy of peak detecting, division points which has the correlation coefficient more than a threshold are averaged.

6. EXPERIMENTAL RESULTS

In order to examine the accuracy of phase difference detection, the intensity distribution at two detecting points put in the distance of 5mm is measured for the two methods mentioned in the above. The calculated phase difference and detection error from regression line are shown in Fig.3 and 4. The 3 times accuracy has been obtained by the correlation coefficient method for the four-frames method. Both results agree with the theoretical formula (4). The beam divergence is kept within 2.4 μrad under this experimental condition.

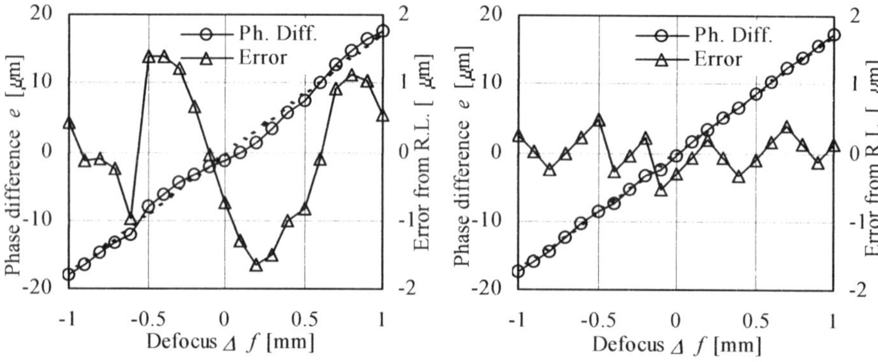

Fig.3. Relationship between phase difference and defocus by the four-frame method.

Fig.4. Relationship between phase difference and defocus by the correlation coefficient method.

7. CONCLUSION

To improve the accuracy of beam collimation, two methods to detect the defocus value have been discussed for the Talbot interferometer and the effectiveness of two methods has been confirmed. The accuracy of beam collimation can be easily improved by only setting the parameter of collimation system.

REFERENCES

1. Darlin JS, Kothiyal MP, Sirohi RS. Hybrid wedge plate-grating interferometer for collimation testing. Optical Engineering 1998; 37-5: 1593-1598.
2. Liu Q; Ohba R. A simple real-time method for checking parallelism between the two gratings in Talbot interferometry. Optics Communications 2000; 175-1-3: 19-26.
3. S.Yokozeki, K.Patorski and K.Ohnishi. Collimation Method using Fourier Imaging and Moire techniques Optics. Communications 1975; 14: 401-405.

A Novel Approach For Simultaneous Measurement Of The Linear Guideway Errors Of A Machine Tool With A Volumetric Optical Encoder

Ulrich Mueller *, Yoshihito Kagawa *, Kazuo Yamazaki *, Jan Braasch **

IMS-Mechatronics Laboratory
Dept. of Mechanical & Aero. Eng.
University of California
Room 1065-Bainer Hall
One Shields Avenue
Davis, CA. 95616-5294, U.S.A.
**

Email: mechatronics@ims.engr.ucdavis.edu
Tel/Fax: (530) 752 8253

Dr. Johannes Heidenhain GmbH
Postfach 1260
D-83292, Traunreut, GERMANY

E-mail: braasch@heidenhain.de
Fax: +49-8669-38609,

Abstract

Precision machine tools require in-process measurement of volumetric accuracy for machining and calibration. This paper describes the development and evaluation of the mounting of a previously presented volumetric linear encoder for measuring the on-axis volumetric motion accuracy. The encoder is based on three cross-grid gratings as measurement bodies and a sensor-head consisting of six optical sensing elements. In order to implement this device for machine tool testing, mounting flexures were used as mounting link. This allows repeatable measurements under varying environmental conditions.

Keywords

Measuring instrument, Optoelectronic sensor, Motion calibration,

INTRODUCTION

Goal of this research is to provide a calibration tool for machine tools with prismatic guideways. It can measure all six guideway related errors simultaneously therefore eliminating the need for changing setups to test for particular errors such as straightness or pitch, yaw, roll. It also reduces setup time and setup errors caused due to misalignment and abbé offset [1]. Moreover, it is applicable for dynamic as well as static measurement. In order to implement the encoder as in situ compensation device it is part of a compensation scheme shown in Figure 2 which includes error extraction, error synthesis and error compensation. Currently this algorithm is implemented on a PC for testing and evaluation. In order to run in real-time it will have to be implemented in hardware. This could be done with an application specific circuit (ASIC) or a digital signal-processing unit (DSP) (Figure1).

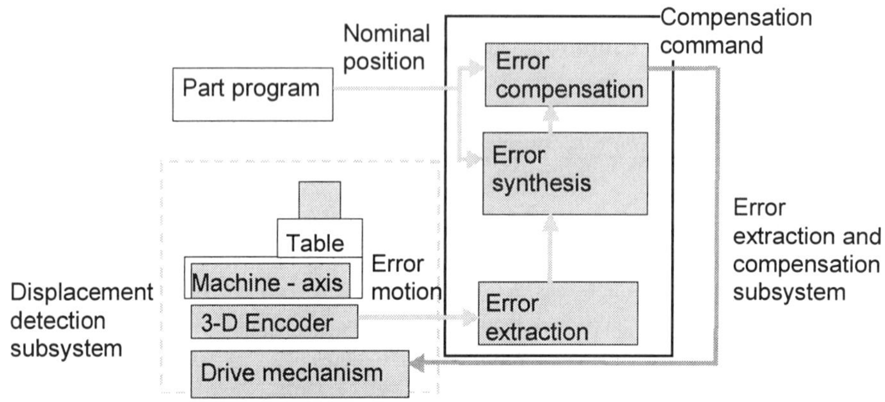

Figure 1. Encoder compensation scheme

ENCODER PROTOTYPE

The Measurement system itself consists of 3 cross grid diffraction gratings that are attached to a beam acting as base for the measurement normal. These gratings are the receivers for six optical displacement sensors that are moving along with the guideway carriage of the inspected guideway (Figure 1). The sensors have a maximum resolution of 5nm [2]. First, a prototype was built to evaluate the scanning method and the compensation scheme.

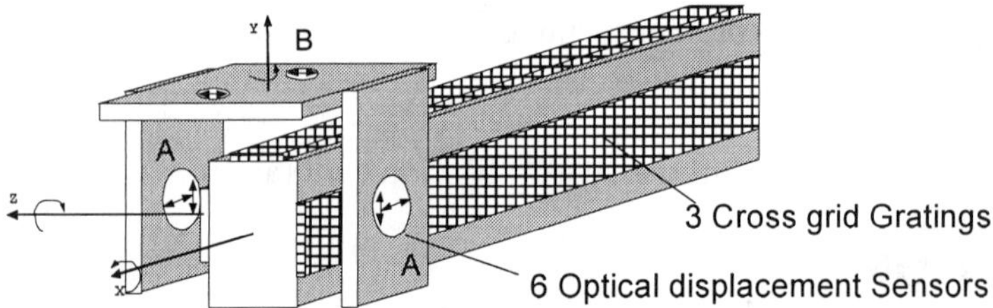

Figure 2. Encoder schematics

The device (Figure 3) was tested in comparison with a laser interferometer setup. The testing showed the feasibility of the device as long as mounting accuracy and machining accuracy of the sensor-carrying bracket was either high or included as measured data in a lookup table. It was also concluded that due to the varying environmental operating conditions a compliant mounting scheme should be used. In the prototype testing, these conditions were kept stable. The straightness measurement in comparison to a laser interferometer system is identical within 0.5µm [3]. For use in the machine environment, a mounting structure is proposed that allow the defined thermal expansion of the mounting beam. Therefore, a de-coupling mechanism from the machine structure has to be provided.

Figure 3. Encoder prototype for feasibility study

IMPLEMENTATION

The full-scale 1m long encoder consists of a stainless steel thin-wall beam in order to minimize static deflection and maximize its natural frequency. It has three diffraction gratings with a 4µm grating period. (Figure 4). The sensors are mounted on a quadratic bracket that has five roller bearings attached of which one is suspended with a spring in order to keep the unit aligned before or after use. These bearings can be removed after the mount is attached to the machine tool in order to allow non-touching measurement (Figure 5).

Figure 4. Encoder prototype (1m long)

The following criteria's had to be considered for the design of the device:

- The static deflection of the device should be small and de-coupled from the mounting surface
- Thermal elongation of the device has to be possible and should also not influence the shape and deflection of the thin wall beam nor the gratings
- A thermal and geometric reference has to be established

- The mounting structure has to provide slip stick free motion for a range of a few hundred mm

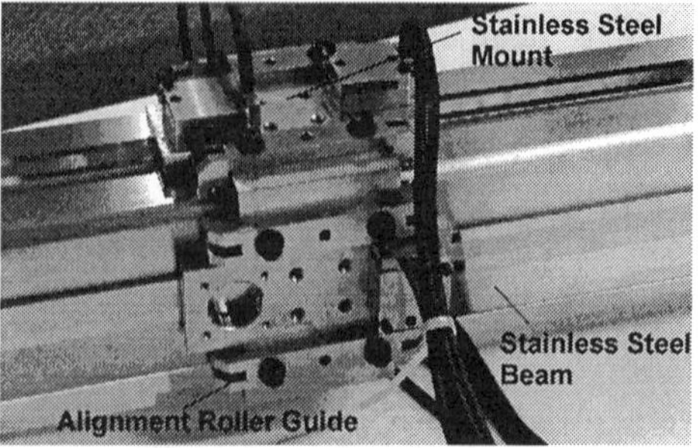

Figure 5. The sensor mounting bracket and its guidance bearings

To de-couple the measurement normal from the mounting surface unit, 5-Degrees of freedom (DOF) have to be provided. In order to provide a thermal reference for the device one end is kept axially rigid giving it only a bending DOF. The other end does not only allow bending but also torsion and axial elongation as shown in figure 6. In order to provide the necessary DOF while maintaining a slip stick free setup the use of flexures was proposed. Not only do they provide frictionless motion but they can also be designed to fulfill the rigidity requirement. Figure 7 shows such de-

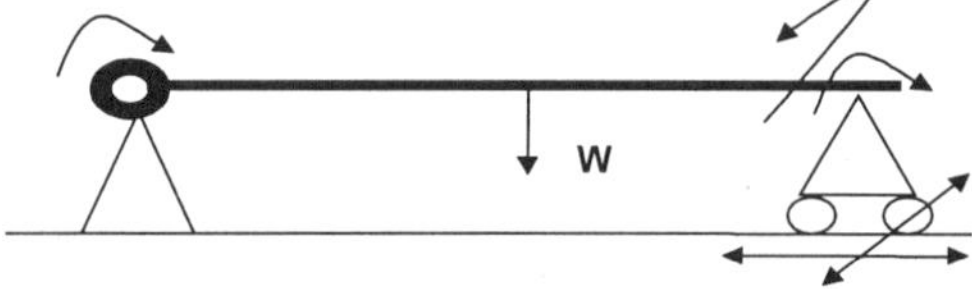

Figure 6. Decoupling model of the encoder

couplings. In order to provide the desired degrees of freedom while maintaining rigidity and providing a defined thermal and geometrical reference the mounting flexures have to be properly dimensioned.

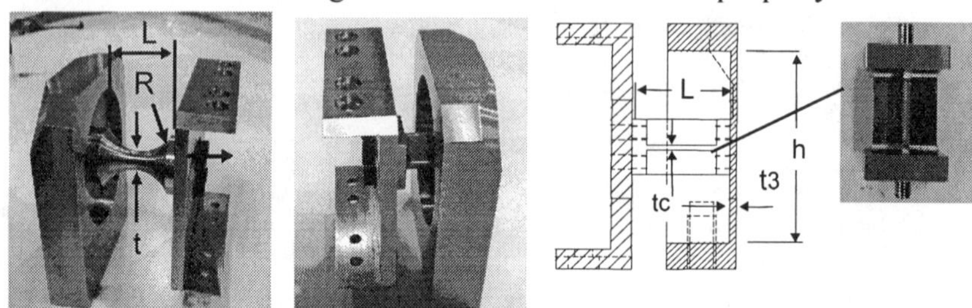

Figure 7.Two and three DOF mount.

The 3-DOF flexure consists of a cruciform type of flexure for toroidal motions and a disk type flexure for translational motion as well as bending. With the mathematical descriptions [4], one can create graphs for each flexures bending and torsion as well as axial stiffness depending on the individual geometric variables such as length, diameter and thickness. These graphs allow an intuitive way of deciding on the flexures parameters following the desired stiffness values.

TESTING AND MACHINE SETUP

The encoder has been mounted and tested on a long stroke-machining center in order to evaluate both the installation and the testing procedures for such a device. This device was initially aligned with measurement gages and then attached to the spindle. The result of a straightness measurement is shown in figure 7. Ongoing testing is carried out to compare the measurements to a laser interferometer setup-for varying environmental conditions.

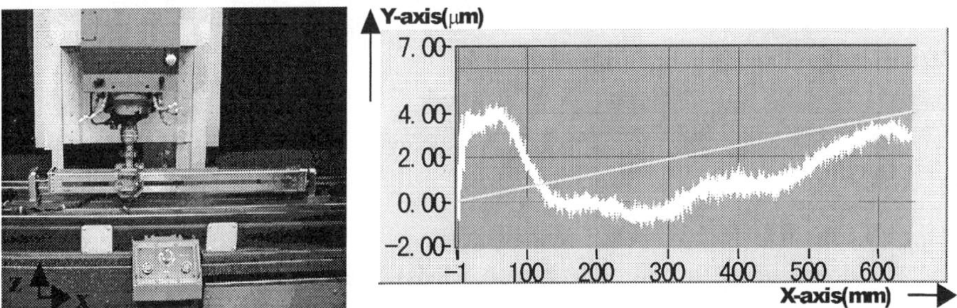

Figure .7 Machine-testing setup and straightness measurement

CONCLUSIONS AND OUTLOOK

The Volumetric encoder system was implemented for the use in the machine tool environment. The necessary mounting mechanisms were designed. The main contributors to the system errors of the measurement device were identified and where possible measured and included as compensation curves. Tests are currently carried out to simulate the thermal expansion in order to show the effectiveness of the flexure mount design.

REFERENCES

[1]   S. Sartori, G.X. Zhang 1995, Geometric Error Measurement and Compensation of Machines, Annals of the CIRP, 44/2:599-609.

[2]   A. Spies, 1997, Linear and Angular Encoders for the High-Resolution Range, Proceedings of the 9th International Precision Engineering Seminar.

[3]   Kazuo Yamazaki, Ulrich Mueller, Jiancheng Liu, and Jan Braasch 2000, A Study on the Development of a Three Dimensional Linear Encoder System for In-Process Motion Error Calibration and Compensation of Machine Tool Axes, Annals of the CIRP, 49/1/2000, p. 403.

[4]   S. T. Smith, 2000, Flexures: Elements of Elastic Mechanisms, Gordon & Breach Science Pub; ISBN: 9056992619

EVALUATION OF STAGES OF NANO-CMM

M. Fujiwara, A. Yamaguchi, K. Takamasu, S. Ozono

The University of Tokyo

Abstract

The Coordinate Measuring Machines (CMMs) are widely used for the three-dimensional measurements of workpieces. For solving the limits and the drawbacks of the traditional CMMs, we have started developing nano-CMM that measures three dimensional parts in nanometer resolution. In this article, we evaluate the repeatability and the straightness of stages of nano-CMM.

Keywords

CMM (coordinate measuring machine),nano meter measurement, friction drive

1. INTRODUCTION

Coordinate Measuring Machine (CMMs) have been developed and widely used to measure quickly and complex shapes with high accuracy as improving precision of industrial workpieces. The system and the key technology of traditional CMMs come to maturity in this 10 years. However, the limits and the drawbacks of the traditional CMMs are clearly such as the limit of accuracy, measuring range, measuring speed and so on.

Therefore, we have started developing novel systems and key technology as "nano-CMM project". in this project, our intention is developing the CMM with nanometer resolution to measure three dimensional position, orientations and parameters of three-dimensional features.

2. BASIC CONCEPT OF NANO-CMM

Figure 1 shows the basic construction of prototype nano-CMM. Almost all specifications of nano-CMM are 1/100 of the specifications of traditional CMMs. For developing nano-CMM, we established the specifications and key points of each factor, such as scales, actuators, tables and a probing system. Firstly, we decide that nano-CMM has simple and the symmetric constructions made of single material for the stability of measurements.

scale : an optical glass scale for abstract accuracy, large measuring range and high stability.

Actuator : A friction drive system for large moving range, high resolution and feedback control by scale.

Table : Symmetric construction of sliders with a scale and an actuator, and a double Vee groove with Teflon films for stability.

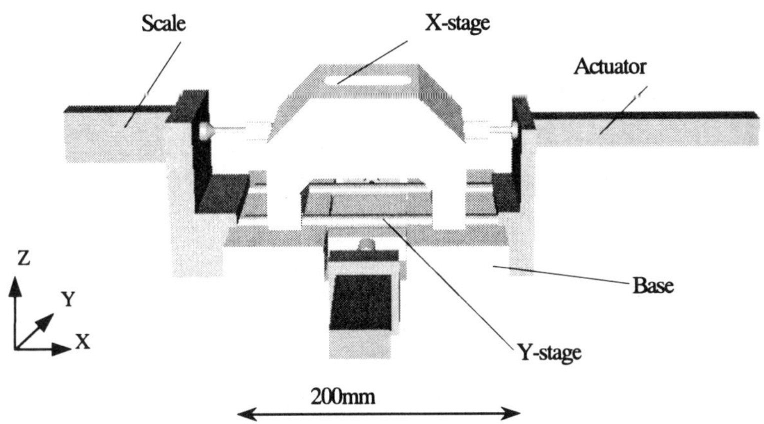

Figure 1.   Basic construction of nano-CMM

3. EVOLUATION USING ELECTROSTATIC CAPACITY DISPLACEMENT METER

The straightness and the repeatability of stages of the prototype are evaluated using the measurements on a surface of a gauge block. A optical glass scale has been used as the scale to measure a gauge block. However, the straightness of nano-CMM is so small ,and it is not enough to realize precise shape that only glass scale evaluates the straightness. Therefore electrostatic capacity displacement meter is used for measuring. Table 1 shows specifications of the optical glass scale and the electrostatic capacity displacement meter. Figure 2 (a) illustrates the straightness of X-stage measured with the optical glass scale and Figure 2 (b) illustrates the straightness of X-stage measured with the electrostatic capacity displacement meter.

From these evaluations, the straightness measured with a optical glass scale shows approximately the actual condition of the stages.

Table 1. Specifications of the optical glass scale and the electrostatic capacity displacement meter

	optical glass scale	electrostatic capacity displacement meter
accuracy	100 nm	4 nm~40 nm
resolution	10 nm	0.1 nm~1 nm
measuring renege	10 mm	±1μm~±10 μm
	in contact	without contact

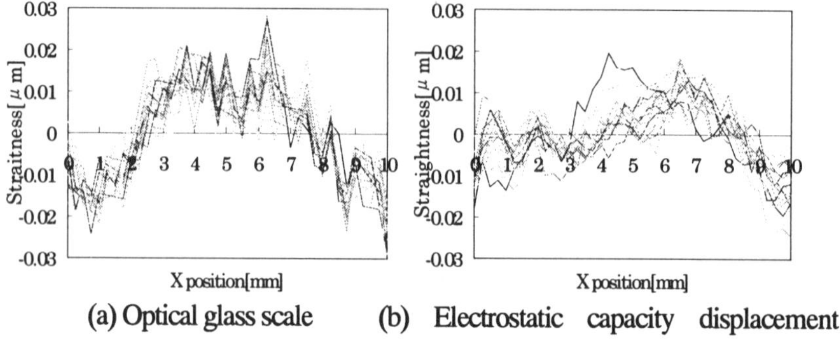

(a) Optical glass scale   (b) Electrostatic   capacity   displacement

Figure 2.   Straightness of X-stage

Table 2.   Straightness and repeatability measured with the optical glass scale and the electrostatic capacity displacement meter

	optical glass scale	electrostatic capacity displacement meter
straightness	30 ~ 40 nm	30 nm
repeatability	20 nm	16 nm

4. MEASURING A TILT ANGLE OF THE STAGE

We measured straightness in order to evaluate precision of stages till now. However, we realize only 1 dimension of length by measuring straightness. Therefore we measure a tilt angle of the stage in order to realize three-dimensional behavior of the stage. Two glass scales are used for this measurement. Figure 3 illustrates constitution of the experiment, and figure 4 illustrates the difference of measurement value of two glass scales and the tilt angle of X-stage.

From these evaluations, we recognize, when the stage changes direction,   the tilt angle changes in discontinuity. It is because the contact state between the stage and guide shifts.

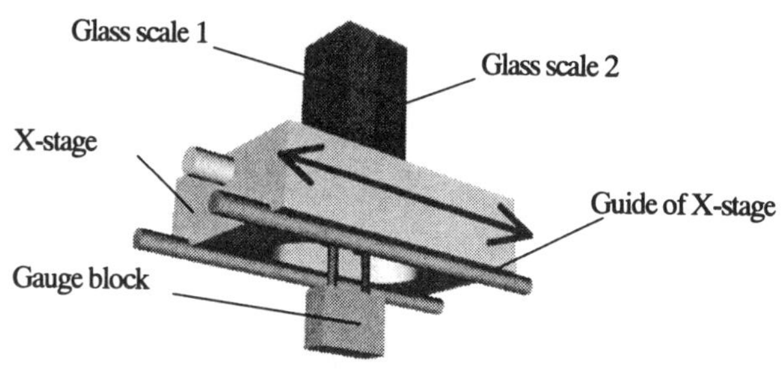

Figure 3. Constitution of the experiment to measure a tilt angle of X-stage

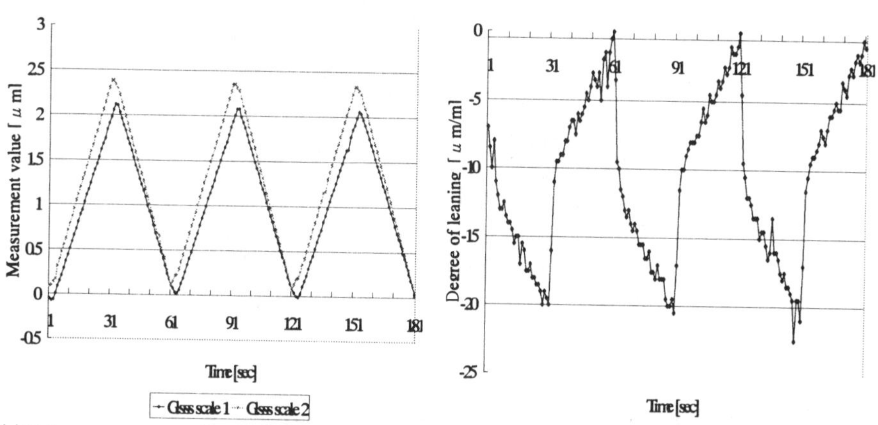

(a) Measurement values of glass scale 1 and 2   (b) Tilt angle of X-stage

Figure 4. Result of the experiment to measure a tilt angle of X-stage

5. DEVELOPMENT OF NEW STYLE STAGES OF NANO-CMM

we develop the new prototype to solve the problems of our prototype. The main improvements are listed as follows:

- material : low thermal expansion cast iron.
- shape : more simple and symmetric shape.
- actuator : adapt friction drive to Z-stage.

Figure 5 illustrates the construction of X-stage and Y-stage of new prototype. In this construction, X-stage is simplified and downsized and the center of gravity of X-stage move into the lower part. Figure 6 illustrates the overview of the new prototype.

637

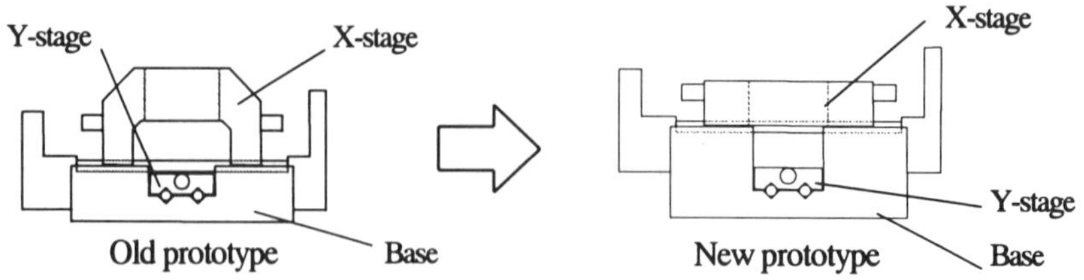

Figure 5. Construction of X-stage and Y-stage of new prototype and old prototype

Figure 6. Overview of new prototype

6. CONCLUSION

In this article, we introduced our developing projects "nano-CMM project" carried out in the University of Tokyo. We reached the following conclusion from the developments and the series of experiments of nano-CMM :

- We introduce the basic concept of nano-CMM.
- The prototype of nano-CMM are made and tested using a optical glass scale and a electrostatic capacity displacement meter.
- Straightness of the prototype of nano-CMM measured with a optical glass scale is similar to straightness with a electrostatic capacity displacement meter.
- We measure a tilt angle of the stage, and recognize, when the stage changes direction, the contact state between the stage and guide shifts.
- We develop the new prototype with low thermal expansion cast iron.

REFERENCES

[1] K. Takamasu, S. Ozawa, T. Asano, A. Suzuki, R. Furutani and S. Ozono, Basic Concepts of Nano-CMM(Coordinate Measuring Machine with Nanometer Resolution), *Jpn.-China Bilateral Symp. on Advanced Manufacturing Eng.*, 1996,155

EVALUATION METHOD OF THERMAL DISPLACEMENT OF MACHINE TOOLS

Shinji SHIMIZU and Noboru IMAI

Dept. of Mechanical Engineering, Faculty of Science and Technology, Sophia University

Abstract

Recently speeding-up in machine tools' spindles tends to increase greatly thermal displacement of the machine tools. Then the standardization of the measurement and evaluation methods of the thermal displacement is advancing in ISO[1]. However, the standard proposes only to use the maximum value of the displacement in the specified testing time. This is not enough to evaluate the characteristics of the thermal displacement of the machine tools. In this paper, we propose the universal evaluation methods for the machine tools with a high-speed spindle. The evaluation methods are constructed from two parts according to the running conditions: preparation and running conditions. The results from our proposed measurement and evaluation method clearly show their effectiveness.

Keywords

machine tool, high-speed spindle, thermal displacement, evaluation, ISO

1. INTRODUCTION

Speeding-up of spindles in machine tools has been advanced greatly to realize high-productivity with high accuracy in machining. This leads to increase of heat generation, then the thermal displacement of the machine tools becomes the serious problems. Therefore it makes necessary to measure and evaluate the thermal displacement, and then in ISO the standardization of its test method is progressing now. However, the standard proposes only to use the maximum value in the specified testing time, nevertheless the measured values have a lot of information, that is, transient behavior of the displacement in the five directions and temperature. In this study, we propose the universal methods to evaluate the characteristics of the thermal displacement of the machine tools with a high-speed spindle in detail and to make clear their effectiveness.

2. PROPOSITION OF EVALUATION METHOD OF THERMAL DISPLACEMENT

Table 1 shows the proposed evaluation method. The tests are constructed from two parts: the preparation condition for running and the running condition in machine tools. In the preparation condition, there are three conditions: temperature change of the environment, power-on of the machine and warming-up condition.

The environmental temperature (ET) is changed by the ground temperature,

Table 1. Proposed evaluation method

	Preparation condition for running	Running condition
Test condision	1. Temperature change of the enviroment 2. Power-on of the machine 3. Warming-up	1. Cold-start 2. Constant spindle speed 3. Patterned spindle speed
Contents of the evaluation	1. Influence of the temperature change 2. Correlation between the temp. of heat source and thermal displacement 3. Max. thermal displacement 4. Time constant of thermal disp. and temp. change	1. Max. thermal displacement 2. Time constant of thermal disp. and temp. change 3. Correlation between the temp. of heat source and thermal disp. 4. Relationship between the temp. of heat source and thermal disp. 5. Relationship between spindle speed and temperature 6. Relationship between spindle speed and time constant

sunshine, air convection, and heat from the tested objects. To evaluate the influence of ET change on the thermal displacement of the machine tools in a workshop, the ET is intentionally changed periodically in the temperature-controlled room.

The thermal displacement is caused also by power-on of the machine, even if the spindle stops, because of heat generation from motors and other electric devices, and temperature change of machine structures induced by coolant oil circulation. Therefore, the thermal displacements and the temperatures of elements generating the heat such as motors and a control box are measured for a period after power-on of the machine. Then the correlation between the both is examined. Limiting the power-on devices in this test make possible to evaluate which devices influence the thermal displacement. In addition, the influence of the warming-up condition recommended by manufacturers also should be evaluated to check the validity of the recommended time and pattern of the warming-up operation.

The test results from the above preparation condition can be used for predicting main factors of the thermal displacement during the test under the running condition and for the basic information to estimate the uncertainty in the other test results same as the ETVE (Environment Temperature Variation Error) test in ISO.

In the running condition, there are three conditions close to the real running operation: cold-start condition, constant spindle speed condition and patterned spindle speed condition. The cold start, that is, the operation without warming-up is strongly demanded for high productivity. Therefore, it becomes necessary to carry out the test to evaluate its characteristic. In addition to this test, the test under the constant spindle speed operation or the patterned spindle speed operation is necessary. The patterned spindle speed operation means that the spindle speed is increased and decreased with the stepwise pattern as we specified. Under these conditions, the thermal displacement and the temperature of the elements supposed to be heat source, their time constant and their correlation are evaluated.

In the following sections, some of the test results from the proposed methods will be shown.

3. MEASURING DEVICE FOR THERMAL DISPLACEMENT OF HIGH-SPEED MACHINE TOOLS

640

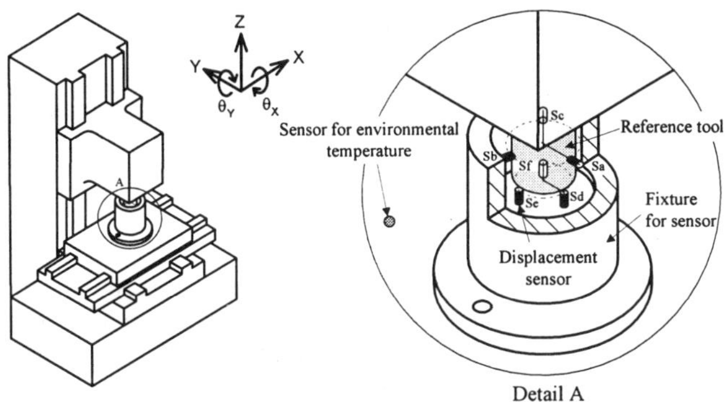

Figure 1. Set-up of the proposed measuring device for thermal displacement of the vertical spindle machining center

Fig.1 shows a measuring device to measure the thermal displacement of the vertical machining center at high-spindle speed[2]. This device consists of a sensor fixture mounted on the table and a reference tool inserted into the spindle nose. The sensor fixture has six non-contact type displacement sensors. Three of them are fixed in the radial direction and other three of them are fixed in the axial direction. Thermal displacement Δx, Δy can be obtained by the three points method with the three radial sensors without the effects of the thermal expansion and centrifugal expansion of the reference tool. Displacement Δz, $\Delta\theta x$ and $\Delta\theta y$ can be obtained by the three axial sensors. Adopting these principle, the length of the reference tool can be shorter than one of the conventional measurement method, so the principle can be applied to the machine tools with a high-speed spindle.

4 EVALUATION RESULTS AND DISCUSSION

4.1 Influence of environmental temperature change

Fig.2 shows the thermal displacements of the single column type machining center without the power supply. The cycle of the environmental temperature(ET) change is 70 minutes, and its amplitude is 1.6 °C. Δx, Δy and Δz have the same cycle as the temperature. The amplitude of Δx is 1.1 μm, Δy is 1.0 μm, Δz is 4.0 μm. Table 2 shows the influence coefficient K_E obtained from this test. K_E ($=\Delta X/\Delta T$) is defined

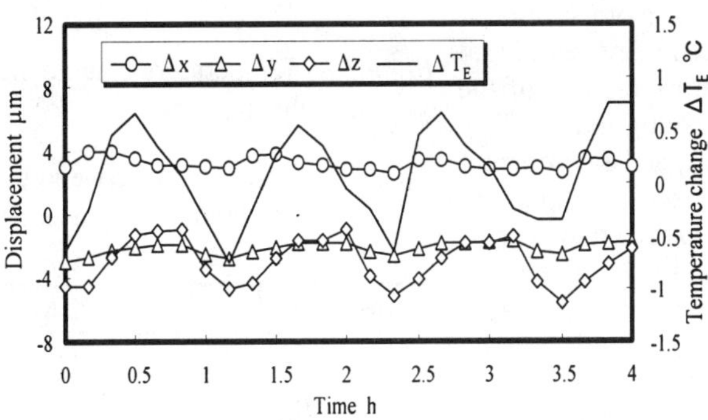

Figure 2. Thermal displacements with the change of environmental temperature

Table 2. Influence coefficients of the environmental temperature

Direction of thermal displacement	X	Y	Z
Influence coefficients K_E (μ m/°C)	0. 66	0. 78	2. 78

by the amplitude ratio of the thermal displacement change ΔX to the temperature change ΔT. It shows that K_E of X and Y direction are close, but Z direction is four times of them. This means that Z direction is more sensitive to the ET change.

As mentioned above, using this value, we can evaluate the extent of the influence of the ET change. Therefore, manufacturer can offer the value of the K_E for customers as a universal evaluation index. Moreover, the prediction of allowable temperature change become possible according to the required machining accuracy, and then the criterion can be obtained for the room temperature control.

4.2 Influence of power-on of the machine

Fig.3 shows the thermal displacements of the single column type machining center during four hours after power-on. Δx and Δy decrease slowly and then come to about -6 µm and saturate after four hours. Similarly $\Delta\theta y$ increase slightly and slowly, and then saturate. Δz and $\Delta\theta x$ have unstable characteristics, changing in the minus or plus direction for one hour after power-on, then reaching to maximum or minimum value and changing to the opposite direction. The heat sources to cause these characteristics seem to be motors for driving axis or electric devices. Fig.4 shows temperature change of the motor of the driving axis (ΔT_M), control box (ΔT_C) and environment (ΔT_E). The temperature of the motor and control box fixed to the column rise to about 9 °C. The environmental temperature increases slightly (0.7 °C) for four hours by the influence of these temperature. The correlations coefficient K_R between the thermal displacement and

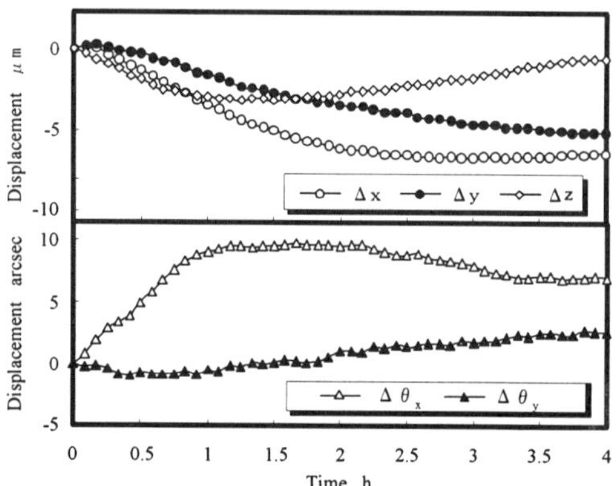

Figure 3. Thermal displacements after power-on of the machine

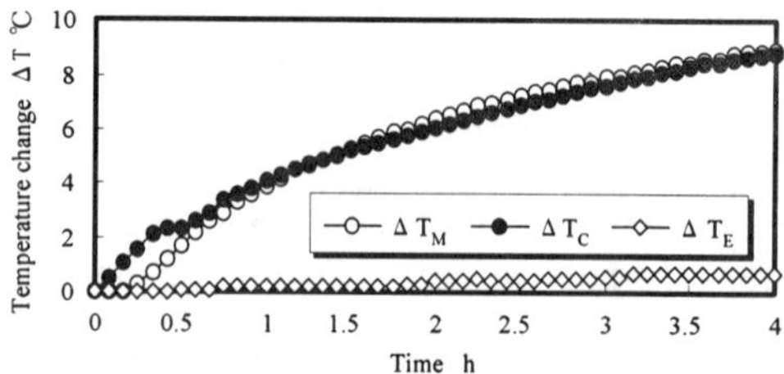

Figure 4. Temperature change after power-on of the machine

Table 3. Correlation between the temperature and thermal displacement after power-on of the machine

Temperature	Δx	Δy	Δz	$\Delta\theta_x$	$\Delta\theta_y$
Motor	-0.97	-0.99	0.16	0.53	0.91
Control box	-0.95	-0.99	-0.23	0.45	0.93
Environment	-0.84	-0.94	-0.44	0.25	0.95

the temperature change are shown in table 3. There are strong correlations in Δx, Δy and $\Delta\theta y$. But Δz do not correlate with any temperature changes. This seems that Δz is caused by other heat source.

As mentioned above, by evaluating of the thermal characteristics after power-on, the considerations become possible on the causes of the thermal displacement except spindle rotation.

4.3 Influence of cold start

Fig.5 shows thermal displacement after cold start of the double column type machining center. Each thermal displacement changes like a first order lag system. Table 3 shows the time constants and the displacements after 100 minutes when the values reach nearly constant. This machine has the thermal symmetry structure for the YZ plane at the center of the X axis, since Δx is smaller than Δy and Δz. This is a feature of the double column type machine. The time constants of $\Delta\theta x$, $\Delta\theta y$ and Δz, are nearly same value. Δz is mainly influenced by the thermal expansion of the spindle. Accordingly, it seems to be that $\Delta\theta x$ and $\Delta\theta y$ are influenced by the thermal displacement of the spindle. The time constant of Δy is larger than the other directions. $\Delta\theta x$ possibly causes Δy, but their time constants are not same. Therefore, it is reasonable to think that Δy is influenced by the thermal distortion of the structural elements far from the spindle, for example, the cross rail or the columns.

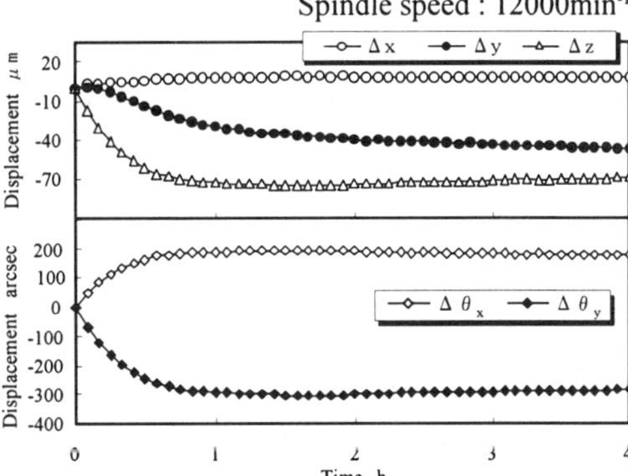

Figure 5. Thermal displacement of the double column type machining center

Table 3. Thermal displacements after 100 minutes in the case of cold start

Direction	Δx	Δy	Δz	$\Delta\theta_x$	$\Delta\theta_y$
Max. displacement (mm,arcsec)	9	−47	−75	−192	−302
Time constant (h)	0. 38	1. 03	0. 28	0. 28	0. 29

Spindle speed : 12000min⁻¹

5. CONCLUSION

We proposed the universal methods to evaluate the characteristics of the thermal displacement of the machine tools with a high-speed spindle. It was made clear from the results that the proposed methods are effective to evaluate the thermal characteristics of the machine tools and to consider the cause of the thermal displacement.

6. REFERENCES

[1] ISO/TC 39/SC 2, *Test Code for Machine Tools Part 3 Determination of Thermal Effects.* Geneva:ISO/FDIS 230-3, 2000.

[2] Qi X., Shimizu S., Imai N. Measuring method of thermal displacement for machine tools at high-speed spindle rotation. Journal of the Japan Society for Precision Engineering 1999; 65: 396-400

TAUT WIRE STRAIGHTEDGE REVERSAL ARTIFACT

James G. Salsbury* and Robert J. Hocken**

** Mitutoyo America Corporation, 958 Corporate Boulevard, Aurora, IL, 60504-9102 USA*
*** University of North Carolina at Charlotte, Charlotte, NC 28223-0001 USA*

Abstract

The design and test of an artifact for the measurement of machine straightness error motion is presented. The artifact is specifically designed for measuring straightness on machines equipped with video probe sensors. The artifact utilizes a taut wire and can be easily measured using reversal, or error separation, techniques. The reversal mathematics is presented along with an analysis of the uncertainty of the straightness measurement process. The experimental results indicate the artifact can be a useful and highly accurate metrology tool, particularly for video measuring machines.

Keywords

Metrology, straightness, self-calibration, video CMM, uncertainty

1. INTRODUCTION

Straightness error motion is one of the fundamental geometric error motions found in all linearly moving machine elements. For this reason, the measurement of straightness has been well studied over the years [Bryan, 1989; Campbell, 1995; Estler, 1985; Hocken, 1980], and standards have been written containing procedures for the measurement of machine straightness errors [ASME, 1998; ISO, 1996]. Some of the metrology tools available for straightness measurements include calibrated straightedges, laser interferometers, autocollimators, and taut wires.

Due to the physical nature of straightness measurements, one of the key metrology methods used for high precision straightness measurement is straightedge reversal, a type of self-calibration where the machine and artifact error can be uniquely separated [Evans, 1996]. Straightedge reversal therefore permits the use of uncalibrated artifacts for precision straightness measurements and allows for straightness measurements with an uncertainty on the order of the repeatability of the machine under test.

For measuring machines equipped with a video probe sensor, often called vision or video CMMs, the measurement of machine straightness parallel to the focal plane of the video probe, i.e. perpendicular to the optical axis of the probe, can be a challenge using traditional techniques. Physical space limitations often preclude the use of instrumentation such as laser interferometers, and the fixed orientation of the probe sensor makes reversal measurements on typical artifacts impossible. One artifact that works well with video systems is the taut wire; however, in searching the literature, no examples were found of using taut wires in combination with straightedge reversal to achieve high precision straightness measurement. In this research, therefore, an inexpensive novel taut wire artifact was developed with the goal of measuring straightness, via reversal, with an uncertainty around 100 nanometers on common industrial video CMMs.

2. ARTIFACT DESIGN

The taut wire straightedge artifact fabricated for this research is shown in Figure 1. With video probes, two-dimensional features can be measured with the highest accuracy; since true two-dimensional artifacts do not exist, the goal is therefore to develop a reversible three-dimensional artifact that best approaches the measurement properties of two-dimensional features. A thin wire appears to be almost two-dimensional in nature as seen by a video probe, depending on the video system magnification.

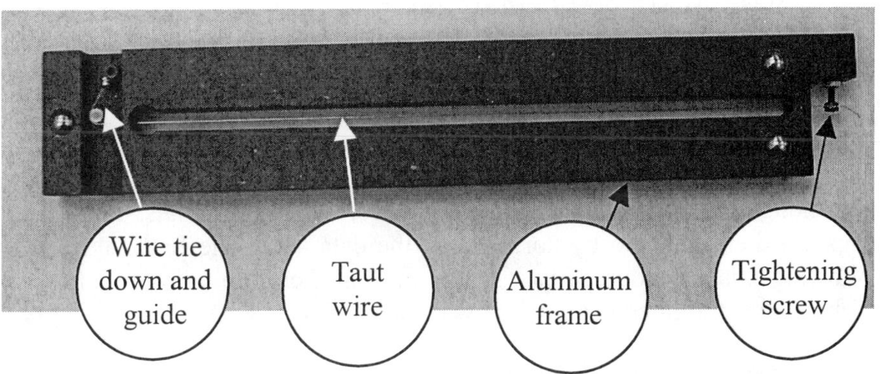

Figure 1. Taut wire straightedge, 200 mm long with 0.25 mm diameter wire.

Many of the design parameters for the artifact were determined by experimentally testing a variety of wire materials and sizes. In the final artifact shown, the length of the wire is about 200 mm and the wire diameter is 0.25 mm. The best diameter wire depends on the field of view, or magnification, of the video sensor. If the diameter is too small, then measurements are difficult to perform. If the diameter is too large, then edge effect problems can lower the quality of the results. A wire diameter of approximately 50% of the field of view worked well for all the cases tested.

Wires made from a variety of materials and from different manufacturers were tested with various levels of success. One of the problems was the quality of the wires, as many were found to have a dominant spiral form error with a periodicity of only a few millimeters, which was considered to be inadequate for reversal purposes. In the end, the wire that worked the best for the artifact was a quality steel guitar string.

3. STRAIGHTEDGE REVERSAL

Straightedge reversal using an artifact, including the taut wire straightedge, is done by measuring the straightness of the artifact in two different orientations. The second, or reverse, orientation is simply a 180° rotation of the artifact about its own axis relative to the first, or normal, orientation. In both cases, the measured straightness contains some superposition of the machine straightness error, M, the error in the straightedge itself, S, the error in the angular alignment of the straightedge to the machine axes, Ψ, and an offset error, Δ. For measurements made along the nominal machine X-axis, with straightness errors perpendicular to the axis of motion and in the focal plane of the probe, the measured straightness in the normal, N, and reverse, R, positions can be described by

$$N(x) = M(x) + S(x) + x[\Psi_1] + \Delta_1 \tag{1}$$
$$R(x) = M(x) - S(x) + x[\Psi_2] + \Delta_2 \tag{2}$$

Adding and subtracting these two equations results in

$$M(x) = \tfrac{1}{2}\,[N(x) + R(x)] - \tfrac{1}{2}\,[x(\Psi_1 + \Psi_2)+\Delta_1+\Delta_2] \tag{3}$$
$$S(x) = \tfrac{1}{2}\,[N(x) - R(x)] - \tfrac{1}{2}\,[x(\Psi_1 - \Psi_2)+\Delta_1 - \Delta_2] \tag{4}$$

The alignment terms are removed from Eqs. (3) and (4) by fitting regression lines though the final resulting data and reporting the residuals. The offset terms are then removed by normalizing the data set to the first data point. Following this procedure, Eqs. (3) and (4) can therefore be solved for the machine and straightedge straightness respectively.

4. EXPERIMENTAL RESULTS

The capability of the taut wire artifact for measuring machine straightness error motion was tested using two video CMMs. In one experiment, measurements were made in two independent positions, the first parallel to the X-axis of machine "A" and second parallel to the Y-axis of the machine. No changes to the artifact were made between the measurements. Using Eq. (4), the in-situ calibration of the wire straightness should therefore be the same for both measurements, within the uncertainty of the measurements. The measurement results are shown in Figure 2, with the error bars representing the estimated expanded uncertainty of the measurements (using a coverage factor k=2). Assuming the machine and artifact straightness errors change slowly spatially, then adequate fixturing will result in these sources of uncertainty being negligible. The only significant source of uncertainty in the reversal process is therefore the repeatability of the measurement process. A statistical analysis resulted in an estimated uncertainty of 140 nm for these measurements [Salsbury, 2000].

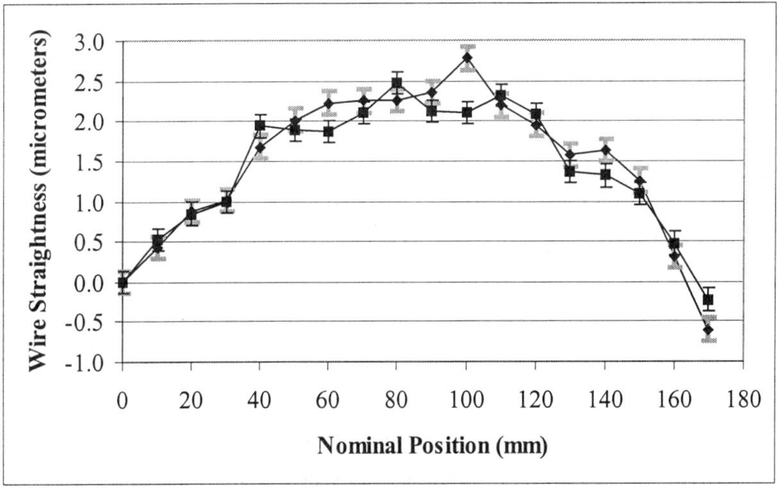

Figure 2. Calibration of taut wire straightness in two independent positions.

In another experiment, the machine straightness was measured on CMM "B" using a steel straightedge and the taut wire artifact, as the machine is equipped with both a contact and a video probe system. The resulting machine straightness results, as calculated using Eq. (3) for both methods, are shown in Figure 3. Even though the measured machine straightness is small, the two methods yielded comparable results within the estimated uncertainty of the measurements. The uncertainty of the measurements, shown as the error bars in Figure 4, is estimated at 480 nm for the contact probe and 120 nm for the video probe.

5. CONCLUSIONS

A taut wire straightedge reversal artifact is presented for use with measuring machine straightness error motion. The artifact is particularly useful for video measuring machines. Experimental results were shown to confirm the expectations of the artifact and demonstrated its capability when used with reversal procedures on common video CMMs. Straightness measurements with uncertainty near 100 nanometers were achieved. Since machine repeatability was the primary source of measurement uncertainty, and not the taut wire artifact, it is likely that straightness measurements with lower uncertainty would be possible for more accurate machines. That testing, along with extending the use of taut wires in the design of other artifacts, such as squares and length gages, is left for future work.

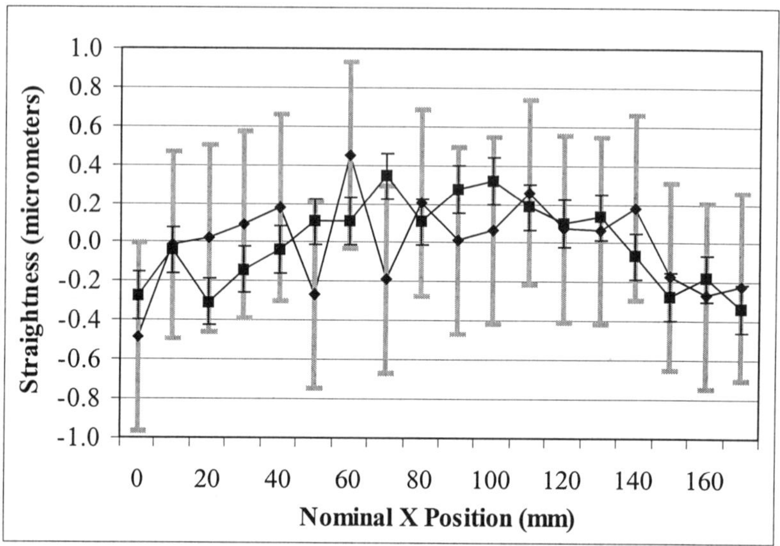

Figure 3. Machine straightness as measured with contact and video probe.

6. REFERENCES

ASME B5.54-1992 (R1998), *Methods for Performance Evaluation of Computer Numerically Controlled Machining Centers*, New York, NY, 1998.

Bryan, J. B., Carter, D. L., "How straight is straight," *Am Machinist*, 133(12), 1989.

Campbell, A., "Measurement of lathe Z-slide straightness and parallelism using a flat land," *Precision Engineering*, 17(3), 1995.

Estler, W. T., "Calibration and use of optical straightedges in the metrology of precision machines," *Optical Engineering*, 24(3), 1985.

Evans, C. J., Hocken, R. J., Estler, W. T., "Self-calibration: reversal, redundancy, error separation, and absolute testing," *CIRP Annals*, 45(2), 1996.

ISO 230, *Test Code for Machine Tools, Parts 1-4*, Geneva, Switzerland, 1996.

Salsbury, J., *Three-Dimensional Metrology of Video Coordinate Measuring Machines*, Ph.D. Thesis, University of North Carolina at Charlotte, 2000.

Hocken, R. J., ed., *Technology of Machine Tools, Vol. 5: Machine Tool Accuracy*, UCRL 52960-5, 1980.

AUTOMATED SYSTEM FOR THREE-DIMENSIONAL ROUGHNESS TESTING

Prof. J.Rudzitis, Assoc. prof. J.Krizbergs, Msc.Eng. M.Skurba
Department of Production Engineering, Riga Technical University, Riga, Latvia

Abstract
This paper presents theory for 3D surface roughness evaluation and application of automated system for 3D measurements.
Keywords
3D parameter automated system, original software, surface analysis.

1. INTRODUCTION

The present surface roughness research is based on machine part's surface cross section profile analysis. But in practical applications machine parts surface roughness behaves as a 3D object. That is why it is necessary to create a new theoretical and practical basis for machine parts surface assessment as a 3D quantity.

2. THEORETICAL RELATIONS FOR 3D ROUGHNESS ASSESSMENT

Let normal homogeneous random field $h_1 = h(x, y)$ describes rough surface. On the basis of its continuity we can specify commonly used roughness parameters:

Ra - mean arithmetic deviation of surface; η_u-relative bearing area ; **Nv** - density of summits; **Hm** – average peak height; $\mathbf{k_x}$, $\mathbf{k_y}$ –mean summits curvature along 'X' and 'Y' axes; $\mathbf{k_v}$-full summits curvature; ∇- slope module; **S** - surface area; **Vγ** - volume of irregularities.

We have obtained equations for mathematical expectation for mentioned parameters in the way similar to presented below for some parameters.

2.1. Mean arithmetic deviation of surface

Major parameter of surface roughness describing height properties of surface, is the average roughness.

$$Ra = \frac{1}{A} \iint_{\Omega} |h(x,y)| dx dy \, ,$$

where $h(x, y)$ - deviation of surface from least squares mean surface; A- size of an investigation area Ω. If to designate $g = |h(x,y)|$, mathematical expectations Ra for surface and profile are identical, as

$$E\{Ra\} = \frac{1}{A} \iint_{\Omega} dx dy \int_{0}^{\infty} g f(g) dg,$$

where $f(g)$ - density function of absolute value of the deviation of the surface from mean surface. We obtain

$$E\{Ra\} = \sqrt{\frac{2}{\pi}}\, \sigma .$$

2.2. Relative bearing area

A relative bearing area at level u we shall understand the ratio of the total platforms of cut of surface at level u over nominal area A considered:

$$\eta_u = \frac{1}{A} \iint_{\Omega} \xi(h,u) dx dy ,$$

where $\xi(h,u)$ - variable depending on surface height and level u measured from mean plane, as follows:

$$E\{\eta_u\} = \frac{1}{A} \iint_{\Omega} E\{\xi(h,u)\} dx dy = 1 - \phi\left(\frac{u}{\sigma}\right) .$$

2.3. Other parameters

Mathematical expectations for other parameters [Rudzitis J., 1992]:

$$E\{N_v\} = c \frac{2\pi}{3\sqrt{3}} E^2\{m_1\} ;$$

$$E\{k_v\} = \frac{1}{2}\left[E\{k_x\} + E\{k_y\}\right] = \frac{1}{2} \pi^2 \sigma E^2\{n_1(0)\}(1 + c^2) ;$$

$$E\{\nabla_h\} = \frac{\pi}{2} \sigma E\{n_1(0)\} \sqrt{\pi(1 + c^2)} ;$$

$$E\{S\} \approx 1 + \frac{\pi^2}{2} \sigma^2 E^2\{n_1(0)\}(1 + c^2) ;$$

$$E\{V_\gamma\} = \frac{1}{\sqrt{2\pi}} e^{-\gamma^2/2} - \gamma[1 - \phi(\gamma)],$$

where c- surface anisotropy ratio; m_1- profile peak density and n_1 –profile zero-crossings density in x direction respectively; λ- surface texture parameter; σ - root mean square deviation of a surface, γ - normalized surface section level, $\phi(\gamma)$ - function of the Laplace.

3. DESCRIPTION OF THE MEASUREMENT SYSTEM

For verification of theoretical equations the three-dimensional measurement system has been used.

The 3D system is composed of following three parts:

-The mechanical structure including two step motors, a support with setting and adjusting screws with an inductive sensor, stage (X-Y), gearbox, column stand;

-The electronic part including a transducer, a frequency filter, a digitization and amplification circuitry. This unit receives the output signal from the sensor and transforms it into a form necessary to the computer to receive it, as well as controls the step motors.

-The computer which is used to control all operating procedures, to calculate parameters and to display drawings and results; specially developed adapter passes the measurement results into the memory of PC.

The system is capable of dealing separately with roughness, waviness, summary surface and also showing them by means of graphical images. For this purpose, the user, in a dialogue mode, sets the filter size $\mathbf{F_x}$ and $\mathbf{F_y}$ along $\mathbf{x}$ and $\mathbf{y}$-axes. Then the surface roughness r_{ij} at point $\mathbf{h_{ij}}$ is determined as difference between the height measured at this point and the mean height of the nearest $\mathbf{ij}$ points:

$$r_{ij} = h_{ij} - \frac{\sum_{i=1}^{Fx}\sum_{j=1}^{Fy} h_{ij}}{Fx \cdot Fy}.$$

4. BORE SURFACE ANALYSIS

Specimens of air compressor cylinder surface were tested by above-mentioned system. Total scanned area is square 3.2x3.2 mm for all specimens. Tests were managed and test data processed by system software, developed in the Riga Technical University. Results are placed in Table1.

4.1. Parameter values

Table 1. Test results.

Specimen No	Surface roughness parameters					
	$R_a, \mu m$	$N_v, 1/\mu m^2$	$K_m, 1/\mu m$	$K_x, 1/\mu m$	$K_y, 1/\mu m$	$\eta_{10}, \%$
1	0.22	90.62	0.84	1.65	0.03	51
2	0.36	103.20	0.67	1.32	0.02	62
3	0.65	100.00	2.49	4.87	0.11	34
4	0.28	112.50	1.68	3.31	0.05	38
5	0.38	121.88	1.28	2.51	0.05	58

4.2. Graphical presentation of test results

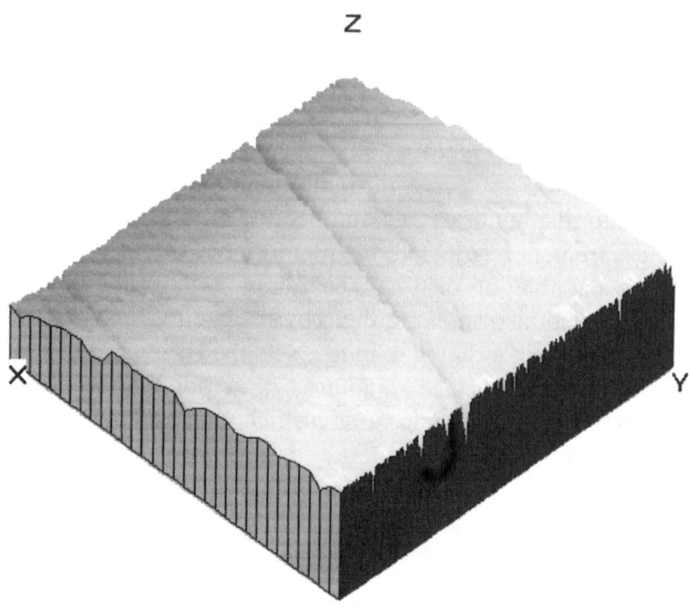

Figure 1. View of measured surface

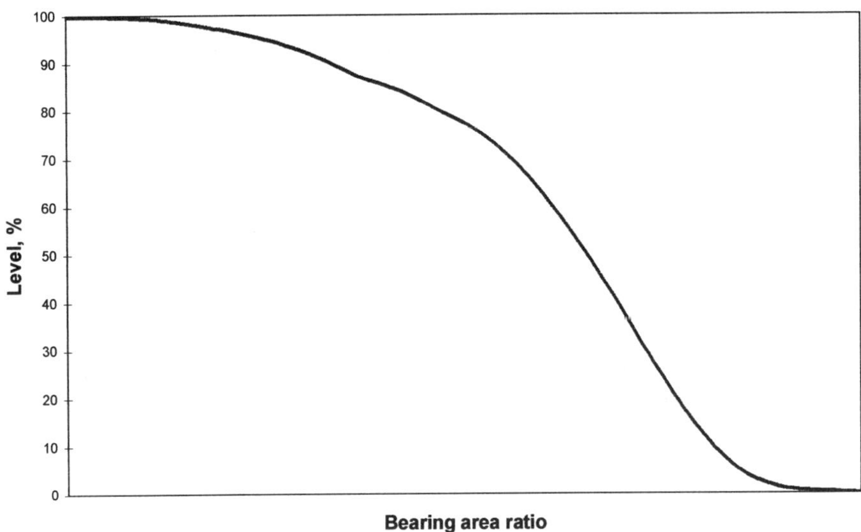

Figure 2. Surface bearing curve.

Surface view (Figure 1) allows qualitative assessment of machined surface, but bearing curve (Figure 2) can be used for quantitative evaluation of surface contact properties.

5. CONCLUSION

The difference between proposed 3D system and current profile analysis is based firstly on use of another datum – plane instead of line and secondly on another set of parameters, describing area features of asperities. The system can be used in any field where machine parts quality is an important issue. In industry it is surface microgeometry investigation (composite material, metal, plastic, wooden etc. items and machine parts), in medicine it could be dermatological investigations, in biology, for example, plant surface investigations, in criminology - dactiloscopy and other fields.

6. REFERENCES

Conference proceedings
Rudzitis J. Surface Roughness Topography Investigations. Proceedings of the VIII. Internationales Oberflächen Kolloquium, 1992, February 3-5, Chemnitz. TU Chemnitz, 1992.

DEVELOPMENT OF MICRO-ROUGHNESS MEASURING PROBE USING LONGITUDINAL TAPPING MODE BY ULTRASONIC VIBRATION SENSOR

Akinori Saitoh, Kazuhiko Hidaka, Takashi Yamagiwa, Kunitoshi Nishimura

Mitutoyo Corporation

Abstract

This paper presents a new micro-roughness measuring probe which uses an ultrasonic vibration sensor in a longitudinal tapping mode. The sensor is controlled by a measuring force control system including a fine and a coarse servo driving mechanism for non-destructive, accurate and rapid measurement. The measuring force is 1μN–20μN, and the frequency response is 0.1kHz–1kHz, depending upon the measuring force. Furthermore the two servo driving mechanisms enable a wide measuring range. The surface of a gauge block measured with nm resolution is shown as an example of the performance of the probe.

Keywords
micro-roughness measuring probe, force sensor,
measuring force control system, active balancing system.

1. INTRODUCTION

There has recently been a demand in mechanical and optical engineering, for instruments capable of an accurate and rapid non-destructive measurement of the form and roughness of fine surfaces. However, it is difficult for conventional tracers to measure the surfaces owing to the high measuring force. Micro-roughness measurements using a SPM (Scanning Probe Microscope) as an AFM (Atomic Force Microscope) are limited to a narrow measuring range and a low measuring speed.

Moreover, the lateral tapping mode of the AFM which utilizes the bending vibration of a stylus is robust for contamination, but usually deforms the image in various directions owing to the low bending rigidity of the stylus.

This paper presents a new micro-roughness measuring probe used in a longitudinal tapping mode[1]. The probe can non-destructively measure in a range approximately 100 times as wide as the conventional SPM and at a high response speed of 1kHz.

The features of this technique are : The longitudinal tapping mode is expected to produce an image independent of the measuring direction. The non-destructive measurement is made by keeping the measuring force low and constant. The expansion of the measurement range of the probe is realized by the tandem connection of a fine servo driving mechanism with a coarse servo driving mechanism with a wide movable range. The high response and resolution are realized by the use of a force sensor with an ultrasonic vibrating stylus and a fine servo mechanism with quick response PZT actuators.

2. PROBE   STRUCTURE

The structure of the probe is shown in *Fig. 1*, and the force sensor is shown in *Fig. 2*. The force sensor, the fine servo driving mechanism and the coarse servo driving mechanism are connected in tandem. The stylus of the force sensor is not vibrated laterally but longitudinally by the PZT actuators as shown in *Fig. 2*. The longitudinal resonant frequency is 450 kHz. When the vibrating tip ball comes into contact with the surface of a

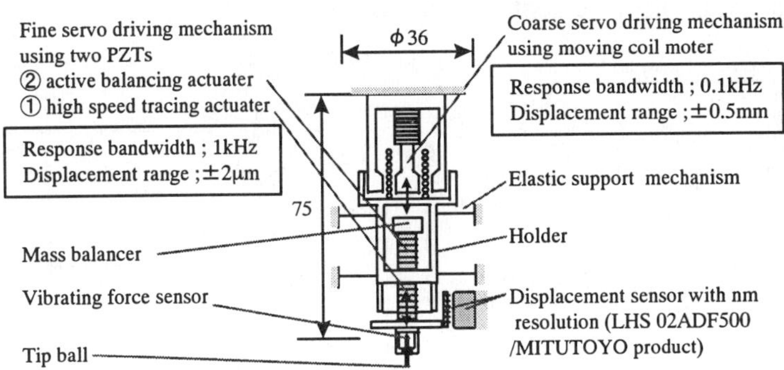

Figure 1. Probe structure

specimen, the vibration amplitude decreases due to the vibration restraint. The decreasing amplitude is detected as a signal by sensing electrodes on the PZT actuators and this is demodulated into a DC signal by a signal processing circuit.

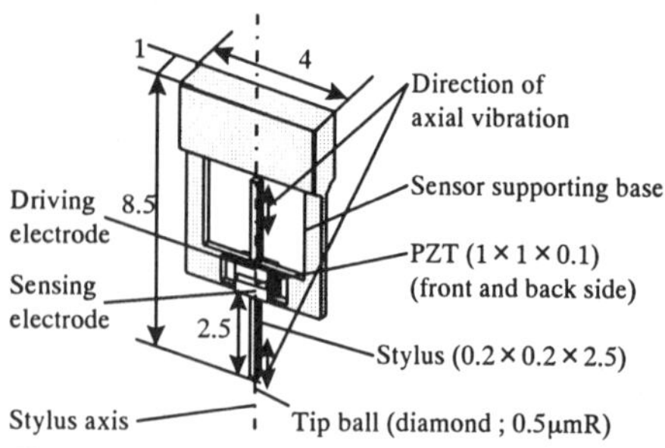

Figure 2.   Overview of force sensor

The fine and coarse servo driving mechanisms are controlled so that the DC signal amplitude equals the desired force value.   Also, the fine servo driving mechanism consists of a high speed tracing actuator (PZT ①), an active balancing actuator (PZT ②) and a mass balancer which has almost the same mass as the force sensor (see in *Fig. 1*).   These are arranged symmetrically above and below the supporting point of the holder. The PZT actuators are adjusted to cancel any reaction force between them. Thus, the movement of the fine servo driving mechanism does not affect the movement of the coarse servo driving mechanism.[3]   On the other hand, the holder is connected to a base with two flexure plates used as an elastic support mechanism and is moved linearly toward the top and the bottom by a moving coil motor.

3. PROBE   CHARACTERISTICS

An experimental system to measure the force sensitivity of the force sensor is shown in *Fig. 3*.   The measuring force of the sensor is calculated from the axial displacement of the sensor and the rigidity of the plate

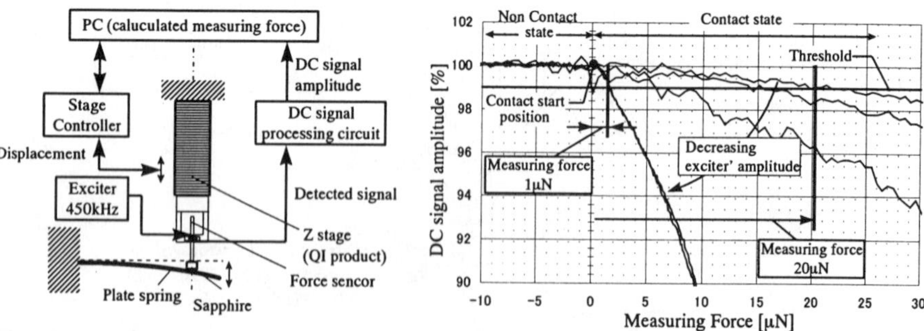

Figure 3. Experimental system   *Figure 4. Static force sensitivity of sensor*

spring[2]. *Fig. 4* shows four relationships between the measuring force and the DC signal amplitude of the force sensor under the condition of constant excitation amplitude. The figure shows that the static sensitivity changes with the excitation amplitude of the driving electrodes. We can select the measuring force of the force sensor in the range from less than 1μN to 20μN.

A schematic of the experimental system for the frequency response of a force control system including the force sensor and the fine servo driving mechanism is shown in *Fig. 5*. An example of the result is shown in *Fig. 6*, which shows the frequency response of 1kHz (at which the gain is -3dB and the phase is -130 degree) at a force of 20μN.

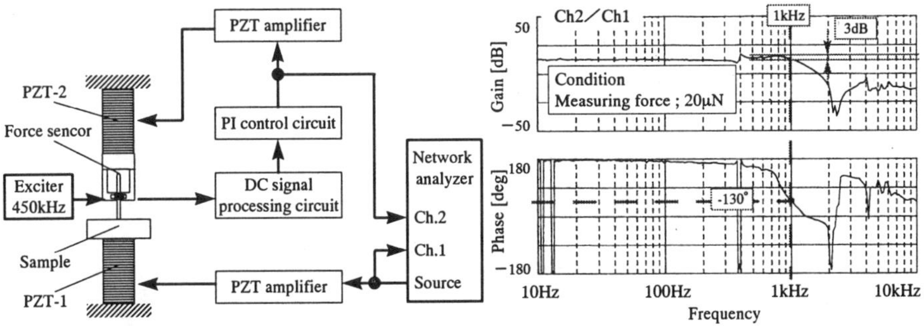

Figure 5. Schematic of the experimental system of the frequency response	Figure 6. Frequency response of the force control system

The effect of the active balance system is shown in *Fig. 7*. The dotted line shows the response characteristics without the active balancing and the solid line with the active balancing. This figure shows that resonance at the natural frequency of 56Hz of the coarse servo driving mechanism is substantially reduced.

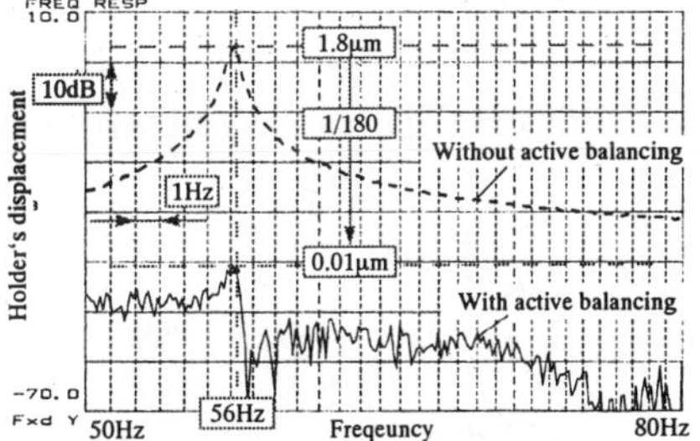

Figure 7. Effect of the active balancing mechanism

Finally, the surface form of a gauge block is measured with the probe with the above mentioned improvements. The result is shown in *Fig. 8*. The figure demonstrates that the probe can detect a surface roughness of 12nmPV with 1nm resolution .

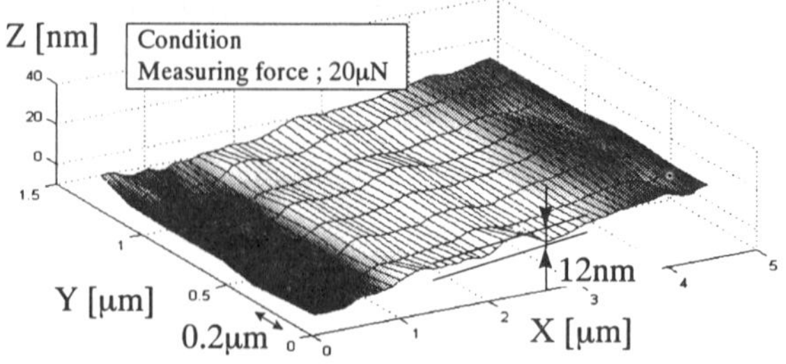

Figure 8. Example of the measurement of the surface form of a gauge block

4. SUMMARY

A micro-roughness measuring probe with a wide measuring range used in a longitudinal tapping mode has been described and its characteristics have been measured. It was confirmed experimentally that the measuring force was 1μN and the frequency response was 1kHz. In the near future, micro-roughness measurement with both a lower measuring force and a higher response frequency will be realized.

5. ACKNOWLEDGMENT

This work was supported by the Proposal-Based New Industry Creative Type Technology R&D Promotion Program from the New Energy and Industrial Technology Development Organization (NEDO) of Japan.

Reference

(1) Akinori Saitoh, Kaoru Matsuki, Kazuhiko Hidaka, Kunitoshi Nishimura, Development of Micro-Roughness Measuring Probe, Proceedings of annual autumn meeting of the Japan Society for Precision Engineering; 2000/10: p291 (in Japanese).

(2) Hideki Miyazaki, Takashi Kasaya, Koichi Kayano, Tomomasa Sato. Adhesive force acting on micro object, Transactions of the J.S.M.E, 1998, 4, 134-141.

(3) Kazuhiko Hidaka, Akinori Saitoh, Kunitoshi Nishimura. Study of Balance Mechanism in Micro-Positioning System, Proceedings of annual autumn meeting of the Japan Society for Precision Engineering, 2000/10: p337 (in Japanese).

EFFECTS OF SURFACE ROUGHNESS OF SI SUBSTRATE AND PT-C MULTILAYER COATED FILM ON X-RAY REFLECTIVITY

Yoshiharu Namba[*], Kazuhiro Zushi[*] and Taketoshi Tanaka[*]

Koujun Yamashita[†], Yuzuru Tawara[†] and Takashi Okajima[†]

[*] Department of Mechanical Engineering, Chubu University

[†] Department of Physics, Nagoya University

Abstract

The roughness effects of bare uncoated silicon substrates and Pt-C multilayer coatings on grazing-incidence X-ray reflectivity were studied at the range of less than 0.5nm rms in surface at the wavelength of 0.834 nm. The X-ray reflectivity depends upon the surface roughness of Si substrates and Pt-C multilayer coating even in the range of less than 0.5nm rms in surface roughness. The measured specular reflectivity on the float-polished Si surface shows 91.2%, which is 98% compared with that on the ideal surface. The maximum specular reflectivity on the Pt-C multilayer coating on the smoothest Si substrate was measured 83.1% which is 99% compared with that on the ideal Pt surface.

Keywords

X-ray reflectivity, Pt-C multilayer coating, surface roughness, float polishing

1. INTRODUCTION

There is a need for extremely smooth, low-scatter mirrors for grazing-incidence X-ray applications and normal incidence soft-X-ray mirrors as well as for longer wavelengths. Substrates used for high-reflectivity X-ray mirrors must be exceedingly smooth, because multilayer coatings will replicate the roughness of the spatial wavelengths on the substrates as well as add additional short-spatial-wavelength roughness. The best way to test X-ray mirrors is to measure specular reflectivity on mirrors, using an X-ray wavelength close to that for which the X-ray mirrors are designed.In this paper we are studying the roughness effects of bare uncoated substrates and Pt-C multilayer coatings on grazing-incidence X-ray reflectivity as well as the relationship between the surface roughness of multilayer coatings and bare substrates.

2. EXPERIMENTAL PROCEDURE

Silicon single crystal samples of (001) surface were polished with various methods including the float polishing. Pt-C multilayers were coated on the Si surfaces with DC magnetron sputter method. The number of layer pair is 20, and periodic distance is 4.1 nm. The top layer is Pt.

The surface roughness of Si substrates and Pt-C multilayer were measured with a SPA 270 + SPI 3800 scanning probe microscope system (also called an AFM) built by Seiko Instruments, which was used in the dynamic force microscope mode. With this technique the microcantilever of 44-N/m force constant in the z direction was oscillated near its resonant frequency of 344 kHz as it was scanned over the sample surface. The tip radius was stated by the manufacturer to be less than 0.01 μm, and the tip length was longer than 10 μm. Images sizes were from 5nm square to 10 μm square.

Grazing-incidence specular reflectivity was measured with a precision X-ray reflectometer with Al-Kα 0.834 nm (1.486-keV) radiation source. The reflectometer had a 10-m long beam line, which included an X-ray source and a sample chamber that were connected through vacuum ducts. The X-ray source was a windowless rotating anode X-ray generator capable of operating between 5 and 60 kV at a maximum beam current of 200 mA. The X-ray beam width was defined by a 0.2-mm-diameter pinhole and a 0.1-mm movable slit in front of the sample chamber. The 1-m diameter, 0.8-m-high sample chamber was evacuated to a pressure of 1×10^{-6} Torr by a turbo molecular pump.

3. EXPERIMENTAL RESULTS AND DISCUSSION

The X-ray reflectivity is the function of wavelength, incident angle, surface roughness and reflecting material. "Figure 1" shows the theoretical relation between the X-ray reflectivity and surface roughness on Si substrate at the

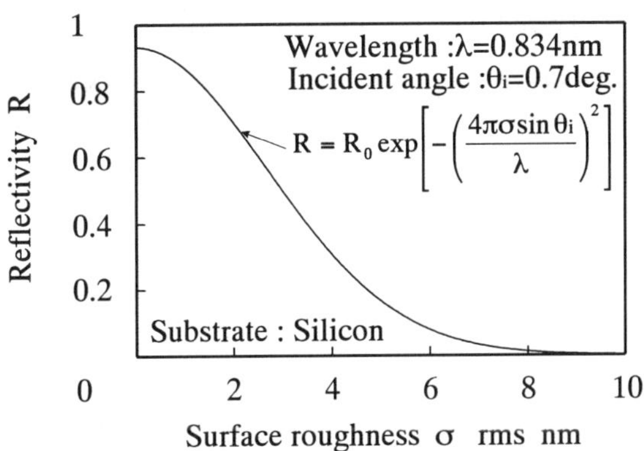

Fig. 1 *The theoretical relation between the X-ray reflectivity and surface roughness on Si single crystal substrate at the wavelength of 0.834nm at the grazing incidence angle of 0.7 degree.*

wide range of surface roughness. The smoother surface shows the higher reflectivity, so that we need the X-ray mirror surfaces smoother than 0.5nm rms in surface roughness.

"Figure 2" shows the relation between the specular reflectivity at the wavelength of 0.834 nm on polished Si substrates and surface roughness at the grazing incident angle of 0.7 degree. In "Figure 2" the solid line shows the

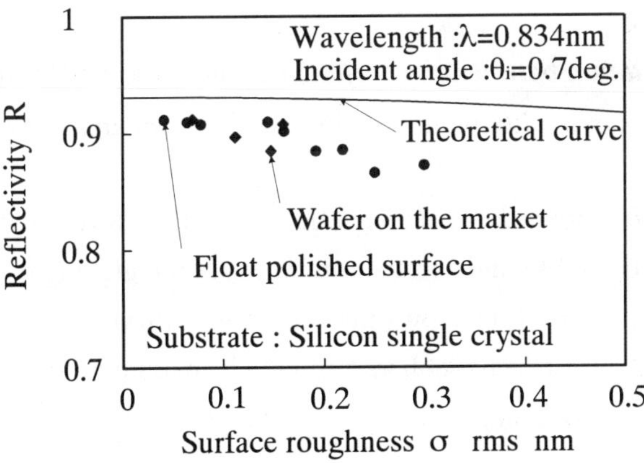

Fig. 2. *The relation between the X-ray reflectivity and surface roughness on Si single crystal substrate at the wavelength of 0.834nm at the grazing incidence angle of 0.7 degree.*

theoretical curve between the specular reflectivity and rms surface roughness on Si substrates. The maximum specular reflectivity is calculated at 93.1% on the zero surface roughness Si. The measured specular reflectivity on the float-polished Si surface shows 91.2%, which is 98% compared with that on the ideal surface.

"Figure 3" shows the relation between the surface roughness on Pt-C multilayer coatings and on Si substrates. The surface roughness of coated Pt is rougher than that on the substrates.

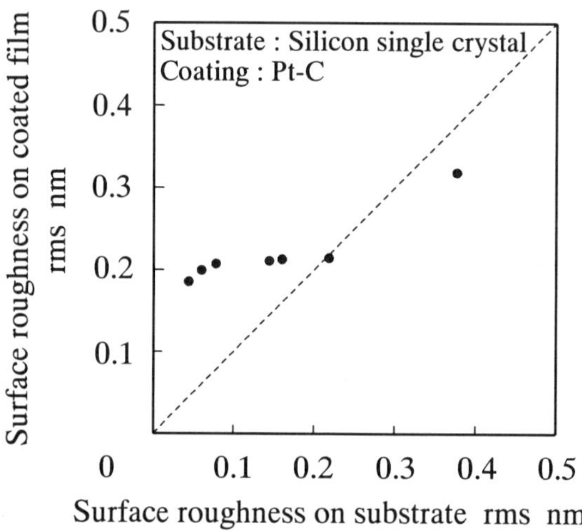

Fig. 3 The relation between the surface roughness of Pt-C multilayer and Si substrate, measured with an atomic force microscope in the 10μm square.

"Figure 4" shows the specular reflectivity at the wavelength of 0.834 nm on multilayer coated Pt and surface roughness at the grazing incident angle of 0.7 degree. The solid line shows the theoretical curve between the specular reflectivity and rms surface roughness on Pt. The maximum specular reflectivity is calculated at 84.0% on the zero surface roughness Pt. The measured maximum specular reflectivity on the multilayer coating shows 83.1% which is 99% compared with that on the ideal Pt surface.

These results may be used for making the mirrors for synchrotron radiation and X-ray lithography as well as X-ray telescopes.

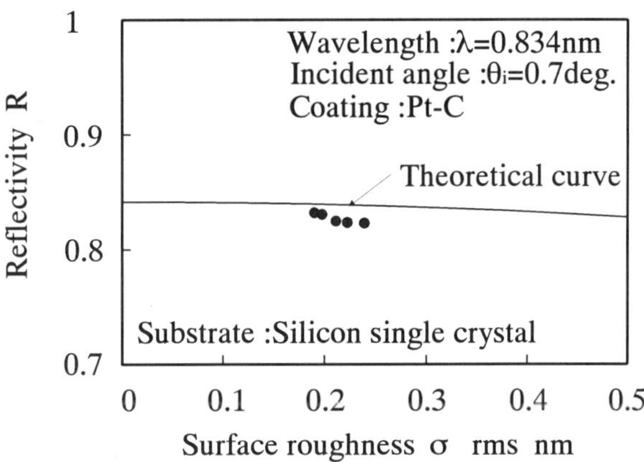

Fig. 4 The relation between the X-ray reflectivity and surface roughness of Pt-C multilayer at the wavelength of 0.834nm at the grazing incidence angle of 0.7 degree.

4. CONCLUSIONS

The roughness effects of bare uncoated substrates and Pt-C multilayer coatings on grazing-incidence X-ray reflectivity were measured at the wavelength of 0.834nm at the grazing incidence angle of 0.7 degree. The following conclusions may be made on the results of this study.

1.   The grazing-incidence X-ray reflectivity of bare uncoated substrates and surface roughness of Pt-C multilayer coatings depend upon the surface roughness of Si single crystal substrates even in the range of less than 0.5nm rms in surface roughness.

2.   The measured reflectivity on the float-polished Si surface shows 91.2%, which is 98% compared with that on the ideal surface. The measured maximum specular reflectivity on the Pt-C multilayer coating on the smoothest Si substrate shows 83.1% which is 99% compared with that on the ideal Pt surface.

5. ACKNOWLEDGMENTS

A part of the work was supported by a Grant-in-Aid for Specially Promoted Research No.07102007 of the Ministry of Education, Science, Sports and Culture, and also Grant-in-Aid for Scientific Research (B) No. 12450063 of Japan Society for the Promotion of Science.

DETECTION OF SURFACE DEFECTS ON STEEL BALL IN BEARING PRODUCTION PROCESS USING A CAPACITIVE SENSOR : *Performance of Prototype System*

Takashi Matsuda[a], **Motohiro Sato**[a], **Takehiro Yata**[a], **Akira Kakimoto**[b]

[a]Shizuoka Univ., 3-5-1 Johoku, Hamamatsu 432-8561, Japan
[b]Kakimoto Co. Ltd., 4-20-21 Sumiyoshi, Hamamatsu 432-0906, Japan

Abstract

The prototype ball inspection system has been developed to automatically discard defective balls in a ball bearing production line. This system has two unique features; (a) the only movable parts are balls moved in lapping oil by an electromagnet system, and (b) the instrument with a capacitive sensor is designed for work under lapping oil. In this paper, it is investigated experimentally whether this system is applicable to discard defective ball bearings with the surface defect which is larger than a hole defect of 0.01 mm in diameter and depth. As a result, it is suggested to be possible by improving the mechanical system.

Keywords

Ball bearing, Steel ball, Capacitive sensor, Surface detect

1. INTRODUCTION

It is important to detect surface defects on balls (indicated as defects hereafter) and discard defective ball bearings (indicated as balls hereafter) efficiently in the manufacturing line before they are assembled in a ball bearing assembly. Now, optical sensors are mainly adopted for inspecting a ball as proposed by Valliapan(1992). However, generally speaking, these systems have the following drawbacks: (a) residual oil on a ball must be removed before inspection and oil must be reapplied to prevent its oxidation after inspection, (b) exposing the entire ball surface to optical sensors, is accompanied by the large mechanism which brings the frictional wear to the ball and itself and many expensive optical sensors, and (c) residual oil film or fine dust on the ball surface will be perceived as a defect.

The developed system avoids the above drawbacks. Until now, it is clarified that the system is practicable for discarding the defective ball 4 mm in diameter with a hole defect 0.05 mm in diameter and 0.01 mm in depth by Kakimoto(1996). In this paper, as the first step for applying the system to discarding the defective ball with 0.01 mm in diameter and depth, it is examined what has an effect on detecting a defect in this system.

2. PROTOTYPE BALL INSPECTION SYSTEM

This system is composed of the following three segments, such as the electromagnet system for moving and positioninng a ball, the ring-shaped capacitive sensor with stabilizer and driving motor for detecting a defect, and the electric circuit for driving, detecting and sorting a ball. And the instrument for detecting a defect is an object of study, because it is ascertained the other has no problem to put it to practical use.

2.1 Whole Setup for Detecting a Defect

As shown in *Figure 1*, the instrument is composed of the upper and lower blocks of 8 electromagnets, the upper and lower ceramic plate, and the capacitive sensor with stabilizer placed between them. Each of the magnets is made by winding a formal coated copper wire around the ferrite bar attached between a ferrite pole piece and a base. The gap between the ceramic plates is close to the diameter of the ball, and the upper plate is slightly thinner than the lower one to adhere the ball to the upper one by weak net force.

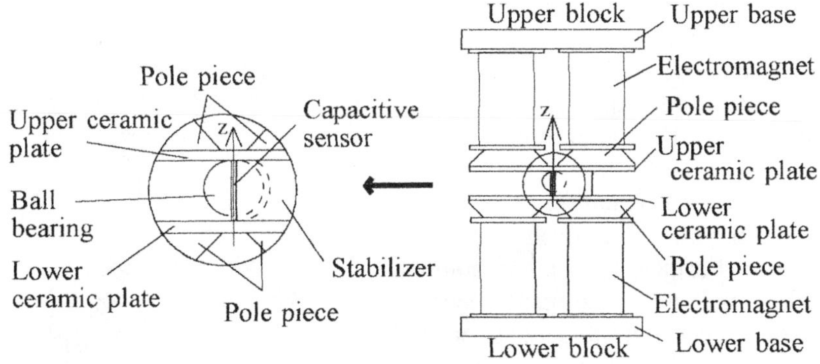

Figure 1.   Instrument for spinning a ball and detecting a defect on the ball

2.2 Driving Motor for Spinning the Ball to be Inspected

Figure 2 shows the upper block of 4 electromagnets in *Figure 1*. As shown in *Figure 2*, the voltage application produces the counter clockwise rotating magnetic field whose speed is 3000 rpm on axis z. Two opposing

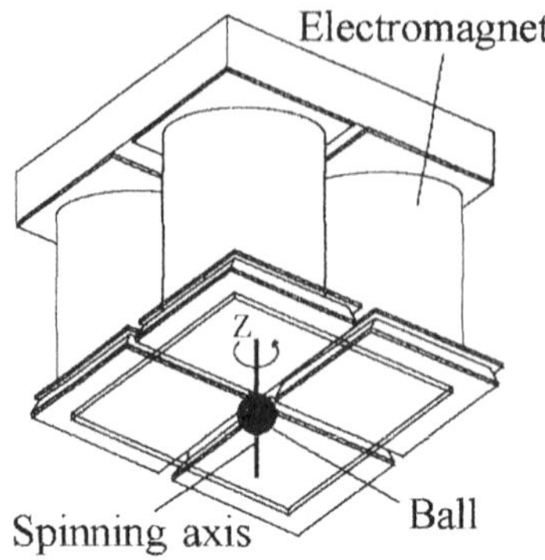

Figure 2. Drive motor for spinning a ball

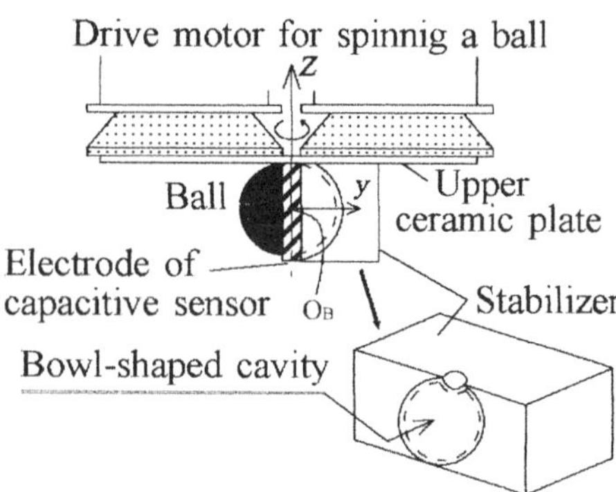

Figure 3. Stabilizing a spinning ball on oil film generated between ball and bowl-shaped cavity

groups of 4 electromagnets, doubles the torque on the ball.

2.3 Stabilizer and Ring-shaped Capacitive Sensor

As shown in *Figure 3*, a stabilizer with a bowl-shaped cavity concentric with a ring-shaped capacitive sensor, stabilizes a ball concentrically with them. The diameter of the cavity and the sensor are 4.2 mm on the assumption that oil-film is 0.1 mm in thickness between the ball bearing and the cavity. The center O_B of a spinning ball should be placed at the origin of the coordinate O-xyz fixed with the sensor. However, it is put out of order due to the fluctuation of magnetic field and oil-film thickness, manufacturing errors and etc. as shown in *Figure 4*. The coordinate z_B is designed to be zero, and x_B and y_B are varying. Then the fluctuation of x_B and y_B have an substantial effect on the detective performance. Further, the ball is spinned on another axis and examined, because the sensor cann't detect defects around the upper and lower poles near the rotational axis of the ball. The test is not carried out here.

3. EXPERIMENTAL RESULTS AND DISCUSSION

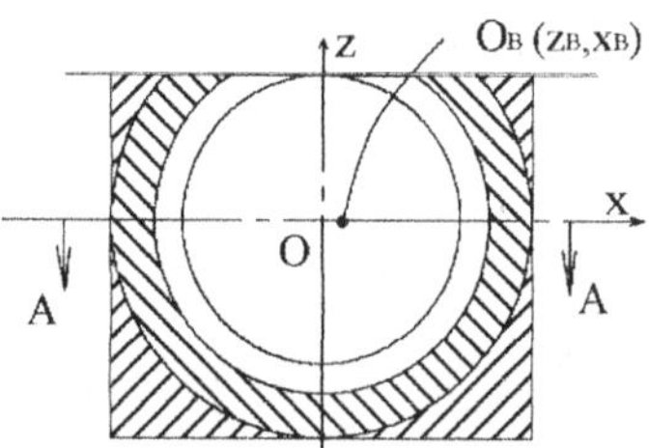

Figure 4. Relative position among a ball, capacitive sensor and bowl-shaped cavity

The tests were done using the reference balls with artificial hole-shaped defects drilled by a laser processing. Output signal, that is, the capacitance fluctuation is taken after the ball setup at random is spinned around the axis z for three seconds, and then the output signal is taken at one minute interval for four minutes. The test is repetited 4 times for each following ball; (1) ball with no defect (indicated as ball-0), (2) ball with a defect of 0.01 mm in diameter and depth (indicated as ball-1), (3) ball with a defect of 0.02 mm in diameter and 0.01 mm in depth (indicated as ball-2), and (4) ball with a defect of 0.05 mm in diameter and 0.01 mm in depth (indicated as ball-5).

3.1 Detectable Defect

Figure 5 shows the favorable output signals selected from the experimental results for detecting a defect and the movement of the ball. As shown in *Figure 5*, it seems obvious that the defect of ball-2 can be detected in this system. And, the defect of ball-1 can be detected by making the gap between the sensor and the ball smaller.

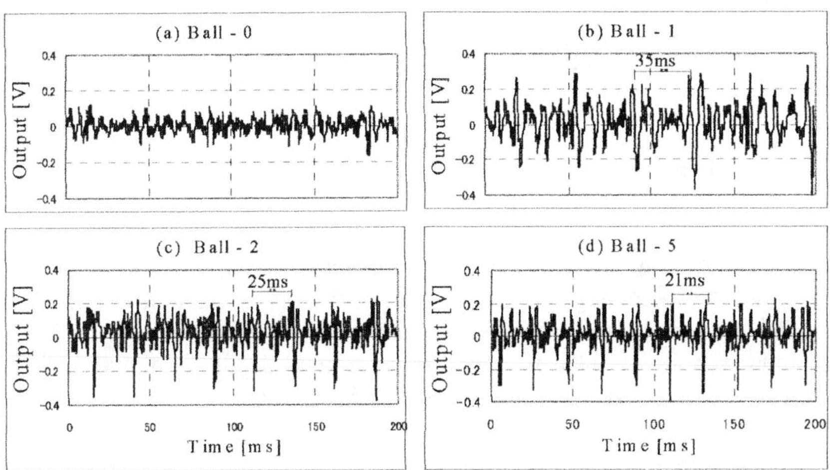

Figure 5. Output signal of high-pass filter for detecting a defect

3.2 Position of an Inspected Ball

As shown in *Figure 5* (c) and (d), a defect is detected once a rotation of the ball. And the ball moves periodically in the same period as the rotation of the ball as shown in *Figure 6* (b). It seems that these results from the center of the ball oscillated in either the area $x_B>0$ or $x_B<0$. It is necessary to heighten the concentricity between the sensor and the ball.

3.3 Movement of a Defect

A defect should be move toward the minimum energy condition of a spinning ball according to the principle of mechanics. However, there are no effects of the defect movement on the detection for 4 minutes' working, because it results from the defect being microscopic.

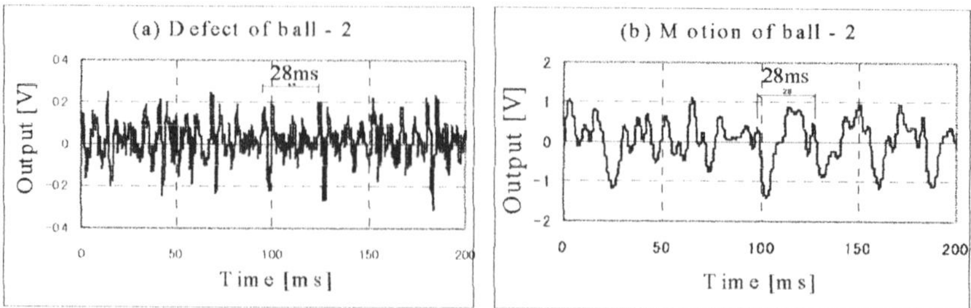

Figure 6. Output signal of high-pass filter for detecting a defect, and of law-pass filter for detecting the motion of a ball

3.4 Reproducibility of Detection

The probability of detecting a defect is around 50% in the random repetition tests. From the results, it seems to depend on the relative position of a defect to the sensor whether the defect can be detected or not. It is necessary to heighten the concentricity between the sensor and the ball.

4. CONCLUSION

The results in this paper suggest the following.

(1) The detecting circuits and the capacitive sensor can be put to practical use for detecting the hole defect 0.01 mm in diameter and depth.

(2) This system is applicable to discard defective balls whose defect is larger than a hole of 0.01 mm in diameter and depth in a ball bearing production line by improving the mechanical system for detecting a defect.

The authors wish to thank the Minebea Electronics Co. for financial support, and Humo Co. for production of this system. Thanks are also due to Syuichi Naito, Satoshi Nonaka, Akihiro Eto, Lillian Overman, Akihiro Hirao, Aiji Matsumoto, Tomohiko Atumi, Katumi Tamanuki, Yoichi Shinriki and Hiromi Yagi for useful contributions.

5. REFERENCES

[1] Kakimoto A. Detection of surface defects on steel ball bearing in production process using a capacitive sensor, Measurement 1996;17(1):51-7.
[2] Valliapan R, Lieu DK. Defect characterization of roller bearing surfaces with laser doppler vibrometry, Precision Engineering 1992;14(1)35-41

ADVANCED 3D-MEASURING TECHNIQUES FOR QUALITY CONTROL OF CUTTING INSERTS

Jürgen Leopold[1], Ichiro Inasaki[2]

1: Department for Calculation and Testing; Society for Production Process and Development
Lassallestr. 14; D – 09117 CHEMNITZ, Germany
2: Faculty of Science & Technology; Keio University
3-14-1 Hiyoshi, Kohoku-ku; YOKOHAMA-SHI, Japan

Abstract

In this paper, it will be pointed out, a new method in 3D- cutting insert inspection - using an effective optical method combined with Neural Nets. For the characterisation of the three dimensional topography, a backpropagation based neural network tool has been developed and applied for cutting insert inspection. The inspection system was applied to different cutting insert examples (geometry, substrate, coating).

Keywords

Cutting tool, moiré technique, neural nets, precision engineering

1. INTRODUCTION

The visual inspection of cutting tools and cutting inserts is one task within manufacturing that has been automated at a comparatively slow space. Many inspection tasks also require a substantial amount of reasoning capability to make an accept - reject decision or classify the type of defects. On reason is, that many inspections tasks also require flexibility. It is well known, that cutting tools and cutting inserts are difficult to analyse by two-dimensional image-processing methods. This is the reason for the developing of a three-dimensional testing method, based on the moiré technique.

A special software tool has been developed for a real - time comparison of the three-dimensional master - type with the produced - type. On the other hand, for an industrial acceptance, the inspection time must in the range of processing time - in general in the time area of some seconds.

To our knowledge, nearly all of the existing automated inspection systems - with the exception of experimental systems - have been designed to inspect a single object or part whose position is highly constrained. Flexibility is thus one of the important research issues in the international automated visual inspection world, although great advances in the flexibility of experimental and finally of commercial vision systems is probably more likely in long - rather than short - term. Due to in - creased international competition, the market for automated visual inspection applications has become global.

2. 3D-INSPECTION USING MOIRÉ TECHNIQUE

A system, which has the superposition of two periodic structures or intensity distribution, is called a moiré - system. In the projection moiré method are two gratings: the projection grating (Projection Unit) and the reference grating (Viewing System) [6]. The used inspection system for cutting inserts is given in Figure 1.

In this method the fringes formed by the viewing system of the deformed object fringe through the reference grating in front of the camera. The usual best resolution of this technique is 100 line pairs per millimetre. A variant of this is not use a master reference grating before the camera but to use the columns of the CCD camera array in the camera to provide the reference. Results of superposition of the two line types are moiré fringes . These lines are of high accuracy, compared with the original line systems. In the out-of-plane direction, the sensitivity is approximately 1/10000 in relation with the observed area of the cutting insert.

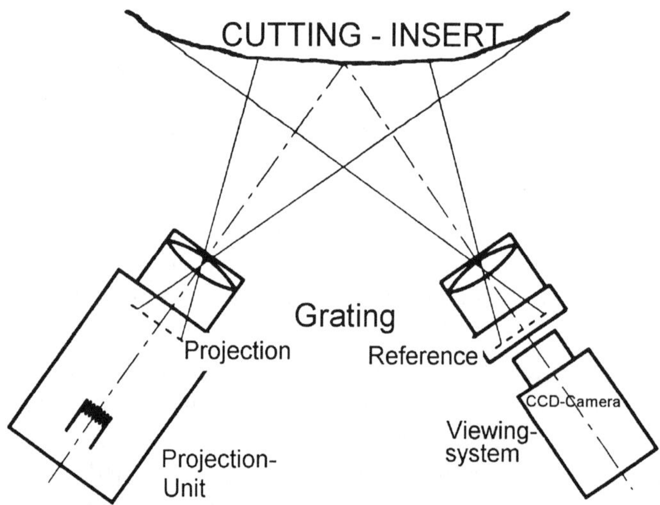

Figure 1: Projection Unit and Viewing System

3. NEURAL NET APPLICATION

Particularly Artificial Neural Networks with their special features are used for solving many problems in pattern recognition, optimisation, diagnostics, and forecasting.

For Quality Monitoring and Quality Control during the manufacturing as well as during the process, Artificial Neural Networks are applied in many branches [1], [2], [3], [4], [5].

As an out - of - process technique, a solution for cutting state recognition using digital image processing was presented. This system evaluates important features of the digitised image of the tool edge and permits in consideration of designated cutting parameters (speed, cutting depth, hardness etc.) a prognosis for tool life. A neural net identification of the surface quality of cutting inserts can be constructed by the method given in Figure 2.

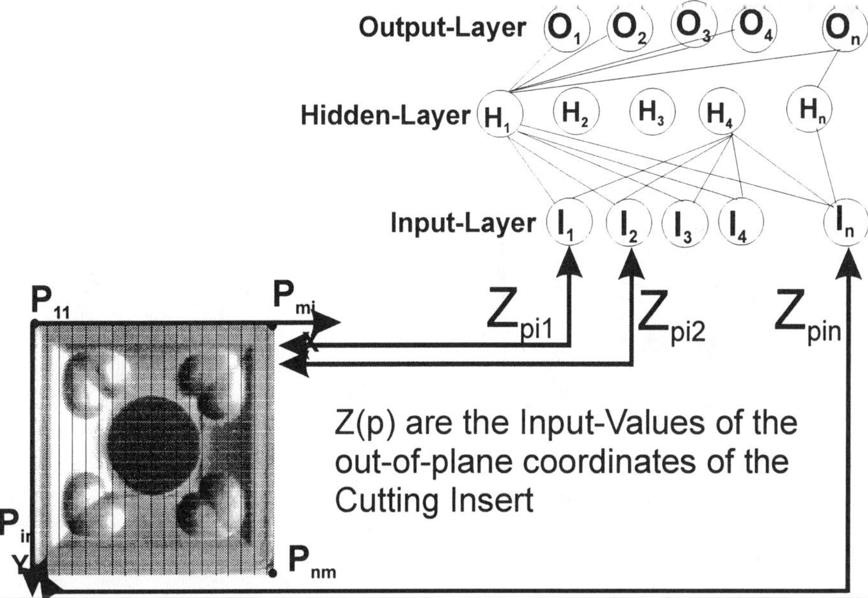

Figure 2: Neural Net Application for quality control of 3D-Cutting Inserts

4. APPLICATIONS FOR 3D-CUTTING INSERTS

The methods developed as mentioned above, has been applied to determine the 3D-Geometry of Cutting Inserts and for the Detection of 3D-Surface Errors. A special system has been developed for the comparison of the master Insert with a produced Insert. The main errors for the detection of the out-of-plane geometry are in the range of 1 Micron for a 10 mm x 10mm cutting inserts (Figures 3 - 6).

A special testing method for comparison of master- with produced inserts is also developed. Figure 7 gives the survey to the method - based on Moiré technique.

Figure 3: Grey-level Cutting Insert

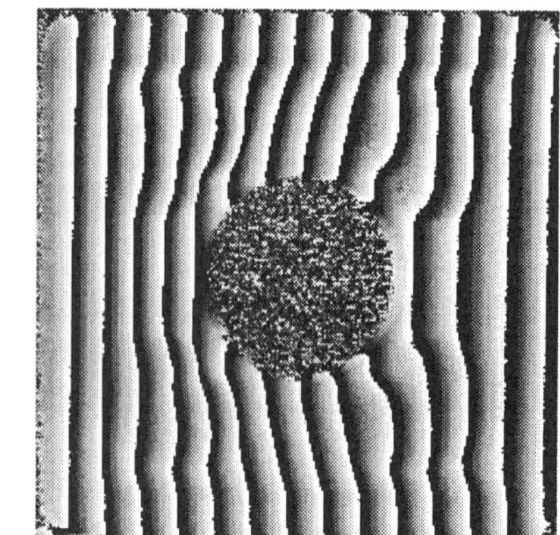

Figure 4: Phase map of the Cutting insert

Figure 5: Grey-coded 3D-geometry

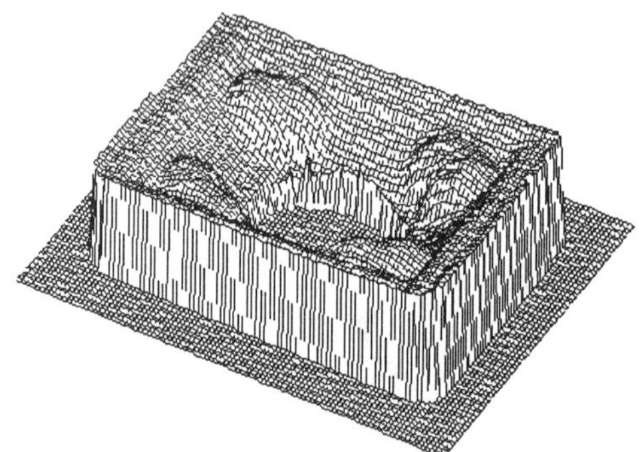

Figure 6: 3D-Cutting insert

5.   CONCLUSION

The combination of Moiré techniques with Neural Nets Application is a basis for the 3D-Quality Inspection of Cutting Inserts. The method, developed by the authors, has been applied to different types (geometry , substrate material and coatings) of cutting inserts. Recently a new inspection system based on a two-camera system was developed. With the new software tool, inspection and computing times is less than 10 seconds. The results of the 3D-shape measurement based on this optical method where compared with conventional CMM – methods and gives a high accuracy for the shape measurement. Based on these investigations, a real-time quality control of the 3D-topography of cutting inserts was carried out.

6.   REFERENCES

[1]   Inasaki, I.: Monitoring of Turning Process. Workshop on Tool Condition
        Monitoring (TCM). First Meeting of the CIRP Working Group on TCM. Paris,
        January   1993, S. 112-118

[2]   Leopold, J.: 3D-shape and pattern recognition using optical methods a neural nets
        DFG-Project proposal Le 746 (30.11.1994)

[3]   Leopold, J.: Image Processing and Neural Nets; Workshop „NATHAN", March
        31st, 1995; Lyon, France

[4]   Leopold, J.; Günther, H.: Quality Control of the Cutting Process with Neural Nets;
        27th CIRP International Seminar on Manufacturing Systems; May 21-23, 1995,
        Ann Arbor, Michigan, USA

[5]   Teshima,T.; Shihbasaka,T.; Takuma,M.;Yamamoto,A.:
        Estimation of Cutting Tool Life by Processing Tool Image Data with Neural
        Networks. Annals of the CIRP, 1993; 1: 59-62

[6]   Leopold, J.; Hertwig, M.; Günther, H.; Staeger, B.: Three-dimensional Measure-
        ment of Macro- and Microdomains Using Optical Methods; Optics and Lasers in
        Engineering 1997; 28:1-16

ACKNOWLEDGEMENTS

The authors wish to acknowledge the support of the Foundation for
Promotion of Advanced Automation Technology (PAAT)/Japan with the
work described in this paper.

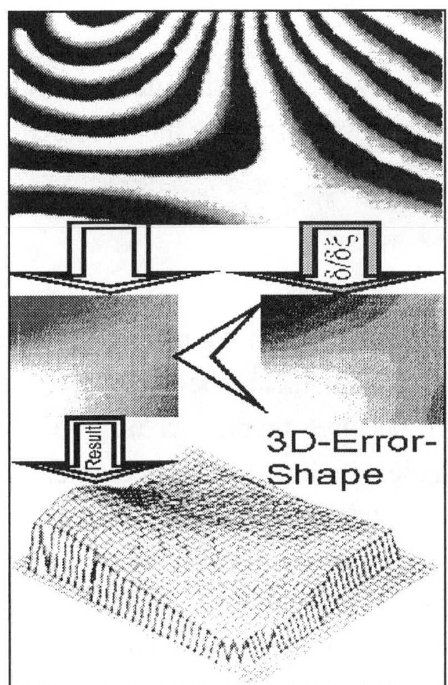

Figure7: Comparison of master - with manufactured cutting insert

AUTONOMOUS PROFILING MEASUREMENT OF TWO DIMENSIONAL SECTION FORMS USING SMALL-SIZED ULTRASONIC PROBE

Kaoru Matsuki, Kazuhiko Hidaka

Mitutoyo Corporation

Abstract

This paper presents a small-sized ultrasonic probe which is able to measure a micro-part surface form with sub-μN measuring force non-destructively. A force sensor installed in the probe has a Ni-Cr stylus of 3mm length which is axially vibrated by a bulk-PZT at longitudinal resonant frequency, and a tip ball of ϕ30 μm diameter made of glass at the end of the stylus. The stylus aspect ratio of 100 is in the highest level around the world. An algorithm for autonomous profiling measurement using output-signal of the ultrasonic probe and a measurement result of a small gear surface form as an example are shown.

Keyword

ultrasonic probe, force sensor, axial vibration, resonant frequency, PZT, autonomous profiling measurement.

1. INTRODUCTION

The conventional probes could not measure a micro-part surface form, such as an inner form of a chemical fiber nozzle and a micro-gear surface form of a dial gauge, because there was no probe which could fit in. Furthermore, a small hole of ϕ100 μm order diameter with a bottom and a hole with a wider inside diameter than the entrance one's could not been measured by any optical measurements (for example, projected light beam and CCD area sensor).

The probe which can measure the above mentioned artifact by a tapping mode has been recently developed[1]. As the tapping is realized by sensing of bending vibration restraint of a cantilever type system, the measurement performance seems to change depending on a difference between its approach direction and bending direction.

Because a new type probe does not use bending but uses axial vibration of a stylus, and the axial vibration of the tip ball realizes equal vibration restraint independent of the approach direction, the micro-part surface form can be precisely measured.

2. STUCTURE AND CHARACTERISTICS OF A FORCE SENSOR

A force sensor installed in a small-sized ultrasonic probe is shown in *Fig. 1*. A stylus of 3 mm length has a tip ball of ϕ30 μm diameter on the top. A surface of a bulk-PZT, on which a stylus is mounted, is divided into a driving and a sensing electrode. The stylus of the force sensor vibrates in a longitudinal resonant state when the driving electrode is excited by sinusoidal voltage with resonant frequency. A change of the amplitude due to vibration restraint caused by contact with a specimen is detected by the sensing electrode[2].

One of features of the force sensor is a miniaturized glass tip ball. This tip ball is formed to a spherical shape by utilizing surface tension in melting state. As shown in the photograph of *Fig. 2*, a melted glass ball is mounted like a water drop at the tip of Ni-Cr wire stylus. The surface tension makes ball. The assembly technology of the vacuum tube shows that the affinity of the glass for the Ni-Cr wire is extremely high[3].

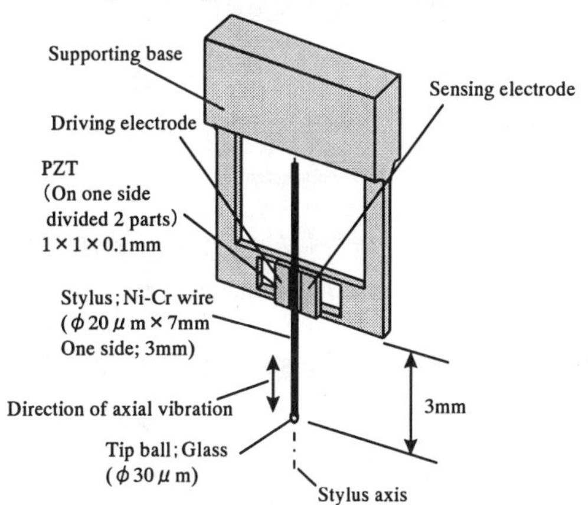

Figure 1. Structure of force sensor

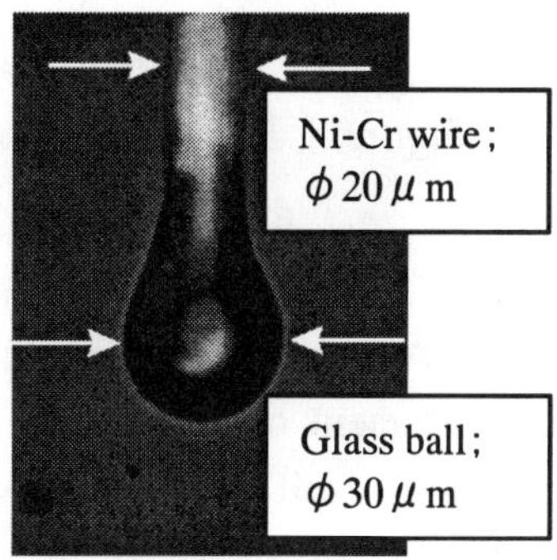

Figure 2. Magnified tip ball

Another feature is high S/N and high Q value. As a thin stylus of φ20 µm is directly glued on the surface of the PZT, so the sensor sensitivity is very high. The resonant frequency is approximately 349kHz and the Q value is approximately 200.

Fig. 3 shows a experiment system which evaluates the characteristics of the force sensor. The sinusoidal detecting signal is demodulated in the detecting circuit. The specimen is quasi-statically moved by a PZT. A static sensitive characteristics of the force sensor is evaluated by comparing the displacement sensor output of the specimen directly installed on the PZT with the detecting circuit output.

Fig. 4 shows an example of the static sensitive characteristics of the force sensor in case which the stylus tip contacts a specimen at a right angle. The vertical axis shows the demodulated output signal from the force sensor, and the horizontal axis shows the relative displacement between the force sensor and the work-piece. The relative displacement from the position which the force sensor output begins to change to the position of a commanded threshold level is defined as a deviation. As measuring force is calculated to be 0.15µN from the stylus flexural rigidity and the deviation, any damages to a specimen or wearinesses of the tip ball is considered to be very rare.

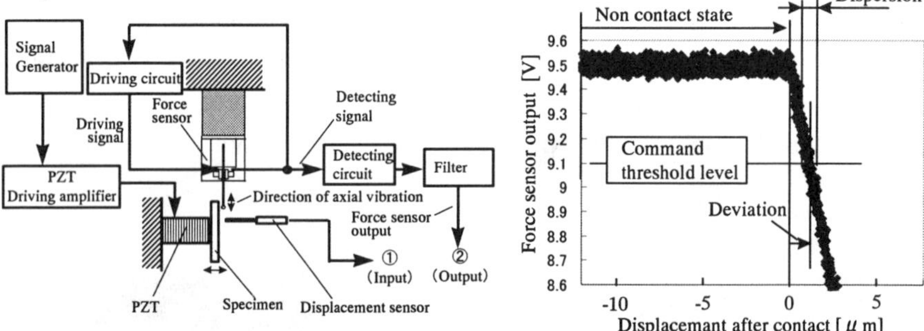

Figure 3. Estimation system of force sensor

Figure 4. Example of static sensitive characteristics

A dynamic characteristics is evaluated by sinusoidally changing the displacement of the PZT in the estimation system shown in *Fig. 3*. *Fig. 5* shows a result of frequency response. The response of the force sensor is 50Hz. (The response of the force sensor is defined as the frequency of the phase of -45degree.) This result means the force sensor system is able to follow undulations of 50 numbers in 1 second.

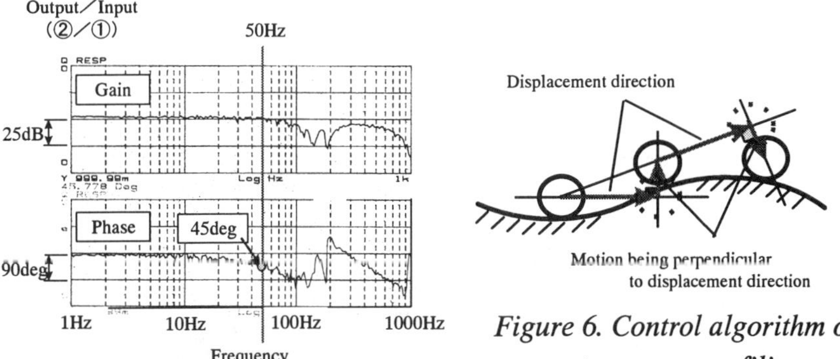

Figure 5. Frequency response of

force sensor

Displacement direction

Motion being perpendicular
to displacement direction

Figure 6. Control algorithm of

autonomous profiling

3. AUTONOMOUS PROFILING MEASUREMENT

An algorithm of autonomous profiling measurement is shown in *Fig. 6*. The displacement of the designated direction and amount is given to the probe with the tip ball contacted to the specimen, and then the force sensor output changes according to the vibration restraint due to relative displacement between the tip ball and the surface of the specimen. The following motion which is perpendicular to the direction of the last displacement is controlled according to the force sensor output.

The displacement direction of the next step is on the same straight line which is connected to the passed two points. The autonomous profiling measurement is realized under this process. *Fig. 7* shows a block diagram for autonomous profiling control system. The control computer receives the X-Y stage position signal and the force sensor output, determines the next displacement direction and amount,

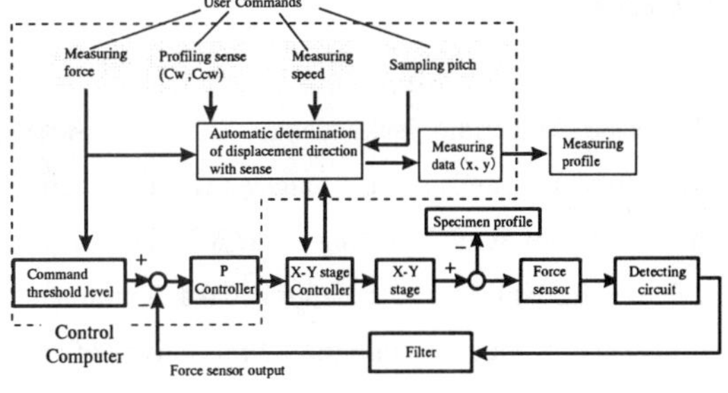

Figure 7. Block diagram for autonomous

profiling control system

and controls the measuring force to be constant.

A part form of gear has been measured by using the algorithm. The result is shown in *Fig. 8.* The figure shows that the measuring result which is recorded 30 times overlapped agrees with the form. The dispersion is 2 µm when the measuring speed is 30 µm/s.

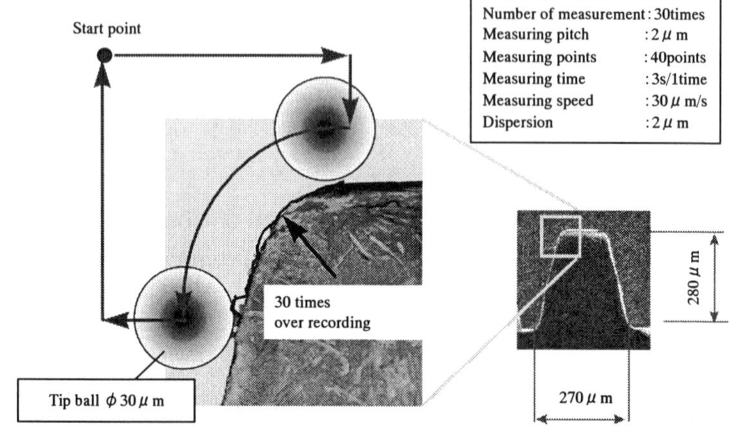

Figure 8. Measurement result example using autonomous profiling

4. SUMMARY

The small-sized ultrasonic probe which has a 3mm length stylus with a tip ball diameter of φ30µm and its aspect ratio 100 has been developed, and the two dimensional sectional form of micro-part has been measured with good reproducibility by autonomous profiling algorithm.

Reference

(1)Masaki Yamamoto, Isaku Kanno, Shinichiro Aoki, Profile measurement of high aspect ratio micro structures using a tungsten carbide micro cantilever coated with PZT thin films, Proceedings of annual meeting of the Micro Electro Mechanical Systems, p217, 2000
(2)Nobuhiro Ishikawa , Kazuhiko Hidaka, Kunitoshi Nishimura, High-speed touch signal processing for touch signal probe, Proceedings of annual autumn meeting of the Japan Society for Precision Engineering, M22,p531, 1998,(in Japanese).
(3)Seikiti Miyagi, Glass to metal seal, Corona Co., p37, 1942, (in Japanese).

Interferometry with Null Optics for Testing Aspherical Surfaces at 1nm Accuracy

T.Gemma[a], S.Nakayama[a], H.Ichikawa[b], T.Yamamoto[b], Y.Fukuda[b], T.Onuki[c], T.Umeda[c]

[a] *Optical Technology Development Dept., Core Technology Center, Nikon Corporation*
  1-6-3, Nishiohi, Shinagawa-ku, Tokyo 140-8601, JAPAN
[b] *Lens Engineering Development Dept., Precision Equipment Company, Nikon Corporation*
  1-10-1, Asamizodai, Sagamihara, Kanagawa 228-0828, JAPAN
[c] *Production Equipment Dept., Precision Equipment Company, Nikon Corporation*
  1-6-3, Nishiohi, Shinagawa-ku, Tokyo 140-8601, JAPAN

Abstract

This paper describes null interferometry at 1nm accuracy for testing aspherical surfaces of sub-mm deviation from the best fitting sphere. We have developed the two kinds of null compensators. The one was the null lens composed of almost "perfect" spherical surfaces and homogeneous glass. The other was the zone plate manufactured through the lithography process. The measurement results using these null compensators were compared with the results by the ultra-precision CMM (Coordinate Measuring Machine). These three measurements differed only by an amount of 1.6nm rms. This result proved the accuracy of our null interferometry was almost 1nm rms.

Keywords

aspherical surface, interferometry, null testing, zone plate, CMM

1. INTRODUCTION

Aspherical surfaces are widely used in many kinds of optical systems to improve their performance and to decrease volume of optics. DUV (248nm, 193nm) and VUV (157nm) lithography optics is not an exception. The shrinkage of pattern size in semiconductor devices requires high NA optics. But it is difficult to design small optics using only spherical surface elements. Aspheric testing is a key technology to realize the small and high quality DUV and VUV optics.

The point diffraction interferometry (PDI) is well known as a metrology at sub-nm accuracy for testing EUV aspherical mirrors which have a few

μ m deviation from the best fitting sphere [1-2]. PDI really has very high accuracy, but it cannot test DUV and VUV aspherical surfaces, because of too many fringes caused by a sub-mm deviation.

We have developed the technology to test large deviation surfaces at 1nm accuracy using a null optics.

2. NULL TEST USING COMPENSATOR

2.1 Sample Asphere

At first, some aspherical surfaces were designed for DUV lithography optics as samples. They have more than 0.1mm deviation from the best fitting sphere. And a few nm RMS accuracy is needed to test them.

2.2 Null Compensator

We chose the null interferometry [3] to test such large deviation aspheres at very high accuracy. Our approach is to develop the "perfect" compensators. Two kinds of null compensators were designed and manufactured for each asphere. They were designed as optics which change plane waves to aspherical waves impinging normal to the aspheres.

2.3 Null Lens

The first kind of compensator is a null lens system composed of spherical lenses. The error sources of the null lens are divided into two categories.

The errors of radius of curvature, refractive index, lens thickness and surface distance affect mainly low order wave aberration. This wave aberration can be minimized by redesigning null lens system using all the data of each lens elements.

The higher order errors of each element can be measured accurately with the absolute test method, but it is impossible to reduce the higher order aberration by redesigning. To avoid the higher order wave aberration, we try to reduce the higher order shape error and inhomogeneity of each lens, and also the numbers of lens elements. We decided to design null compensator using lens elements less than three. It is difficult to match the null wavefront perfectly to the asphere using small numbers of lens elements. The resultant aberration between the null wavefront and the asphere is corrected numerically using ray trace data.

2.4 Computer Generated Diffractive Optical Element

The second kind of compensator is a computer generated diffractive optical element (zone plate). A zone plate can produce the high order wavefront matching to the aspheric surface shape easier than null lens. The biggest error source of a zone plate is the patterning error. The best way to reduce the patterning error is making the zone plate pattern pitch large. So we designed a zone plate compensator composed of a zone plate and a spherical lens to make the pattern pitch more than $20\,\mu$ m.

3. MEASURING SYSTEM
3.1 Interferometer

Fig .1 illustrates the optical setup. It is a Fizeau type interferometer.

When testing with a null lens, the numerical correction of the resultant aberration is necessary. To correct it accurately, the relationship between the point on the CCD in the interferometer and the surface on the asphere must be known. So the distortion of the interferometer was designed less than 0.1%.

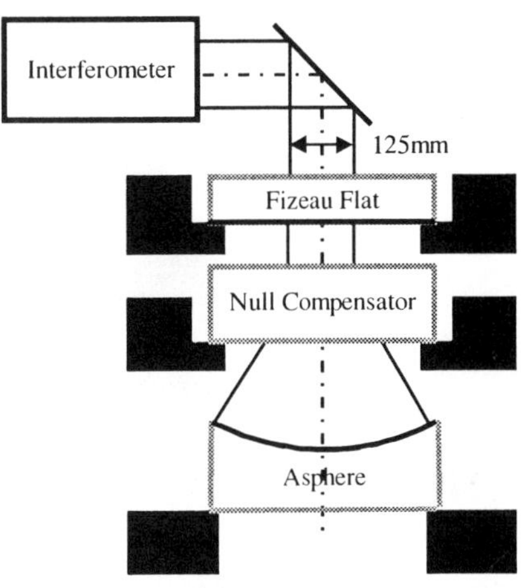

Fig.1 Optical Setup

To avoid the null wavefront changing, the temparature of the interferometer room was thermally controlled within 0.1degree. The asymmetric error of the null compensator was calibrated by the rotation test of the asphere.

All the technologies to fabricate the fine null compensator and the enviromental control together with some correction method bring the very accurate interferometric aspherical testing.

3.2 Ultra Precision 3D-CMM

An ultra-precision CMM [4] was used as a reference method. This machine has a 3D metrology frame, vacuumed laser passes and low force contact probe system. Its repeatability is 2.1nm.

4. MEASUREMENT RESULTS

4.1 Repeatability

The difference of the two successive interferometric measurements was smaller than 0.3nmRMS (Fig.2). Especially the difference of the radial profile was smaller than 0.06nmRMS.

This measurement was repeated periodically over several months. The radial profile changed only 1.0nmRMS. It proves that the compensators are very stable and reliable.

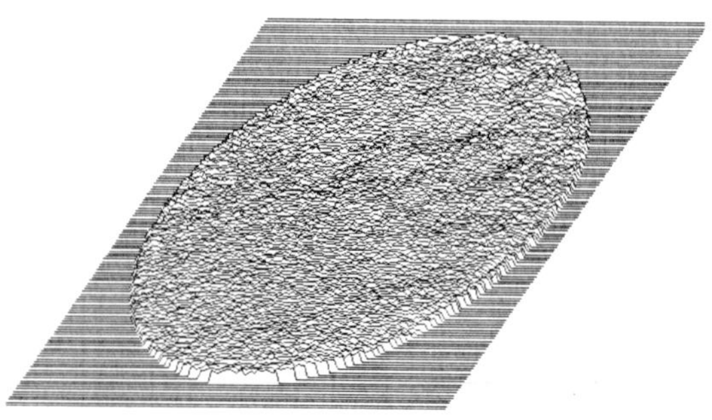

Fig.2 Repeatability of interferometric measurements

4.2 Accuracy

The comparison results of the radial profile are shown in Table 1. The differences of the RMS values between three diferent methods were smaller than 1.8nm (=12.7-10.9nm). The RMS of the surface difference from the average of the three method was smaller than 1.6nm. Especially RMS of the difference between two null tests were smaller than 1.3nm.

Fig.3 is the radial profile got from the three different measurements. It shows that each methods have the high accuracy even for the surface with 60nm shape error.

Table 1. Comparison of three measurement method

Test Method		Profile (RMS)	Difference of Profile (RMS)
Null Test	Null Lens	10.9nm	1.3nm
	Zone Plate	11.8nm	0.7nm
Tactile	3D-CMM	12.7nm	1.6nm

5. CONCLUSION

The interferometric null test at 1nm accuracy for testing aspherical surfaces of sub-mm deviation from the best fitting sphere was developed. The accuracy evaluated from the totally different three test methods was 1.6nm rms. This technology made the high performance and high-NA DUV and VUV lithography optics possible.

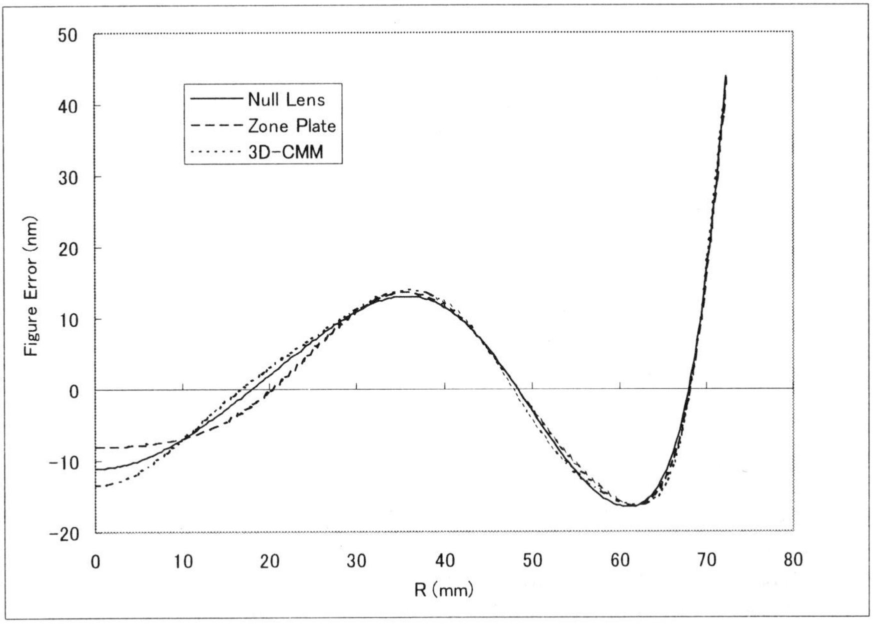

Fig.3   The comparison of the three different radial profile

References:

[1] Sommergren, G.E.: Phase shifting diffraction interferometry for measuring extreme ultraviolet optics, OSA Trends in Optics and Photonics Vol.4, Extreme Ultraviolet Lithography, Glenn D. Kubiak and Don R. Kania, Eds. (Optical Society of America, Washington DC 1996), pp.108-112

[2] Yamamoto, T., Fukuda Y., Otaki K., Ota K., Nishiyama I. and Okazaki S.: Advanced point diffraction interferometer for EUV aspherical mirrors, Proc.SPIE 4343 (in Press), (2001)

[3] Offner, A. and Malacara, D. "Null Tests Using Compensators" in *Optical Shop Testing*, Malacara, D.: John Wiley & Sons, 1992.

[4] Shiozawa, H., Fukutomi, Y., Ushioda, T. and Yoshimura, S.: Development of Ultra-Precision 3D-CMM Based on 3D Metrology Frame, Proc.ASPE 1998 Annual Meeting (1998)

ANALYSIS OF DEFECTS ON SiO$_2$ FILMED WAFER
- Evaluation of CMP Defect Detection Schemes
Using Computer Simulation (BEM) -

Taeho HA, Takashi MIYOSHI, Yasuhiro TAKAYA, Satoru TAKAHASHI

Department of Mechanical Engineering and Systems, Graduate School of Engineering,
Osaka University, 2-1 Yamadaoka, Suita, Osaka, 565-0871, JAPAN

Abstract

CMP defect detection schemes using computer simulation are proposed for analyzing defects on SiO$_2$ filmed wafer surface. Sizes of defects can be estimated by the defect classification map. Therefore, killer defects can be easily categorized. We also developed the scattered light analyzing system that can measure spatial distribution of the scattering light intensity. Basic experimental confirmation shows good agreement with the simulation results. It is expected that this simulation tool will extend capabilities of the scattered light analyzing system for defect measurement.

Keywords

defect, measurement, CMP, laser, scatter, thin film, simulation

1. INTRODUCTION

The technical trend of manufacturing ULSI circuit is pursuing wafer size increase and process geometries decrease at the same time. This requires many series of complex and time-consuming process steps. In many of these steps, the size and density of defects can be severe factor of yields as geometry line widths decrease. CMP (Chemical Mechanical Polishing) becomes a popular technique to meet the planarizaiton requirement for the current wafer fabrication technology. CMP processing introduces several types of polish-related defects, which are residual slurry and microscratches, pits and voids in the polished surface. If there exist these kinds of defects on the surface, the characteristic scatter resulting from the defect appears at a certain range of angle. The size and the type of microdefects are estimated from the defected intensity values by locating detectors at a proper position. In this paper, CMP defect detection scheme is investigated using computer simulation to detect and classify defects. Basic experiments are conducted to validate the simulation results.

2. NUMERICAL METHOD AND SIMULATION MODEL

Numerical analysis helps in understanding new phenomena and designing defect detection system. In this paper, electromagnetic scattering simulation tool is developed based on BEM (Boundary Element Method) for the defect detection scheme. BEM is a technique for solving complex integral equations by reducing them to a set of simpler linear equations. In case of BEM, expansion and weighting functions are defined only on a boundary surface. The equation solved by BEM is generally a form of the electric field integral equation or the magnetic field integral equation. Both of these equations can be derived form Maxwell's equations. *Figure 1* shows SiO_2 filmed wafer with defect, coordinate system and basic parameters of simulation. Incident angle and scattering angle are denoted as φ_i and φ_s. Parameters, such as optical configuration (wavelength, polarization, etc.), film thickness and defect size are changeable. Gaussian beam condensed by a lens is used as incident beam. 2D scratch models are used for computer simulation when D, W mean depth and width of the scratch, respectively. The cross sectional profiles of the defects are defined as half-ellipse shape (concave).

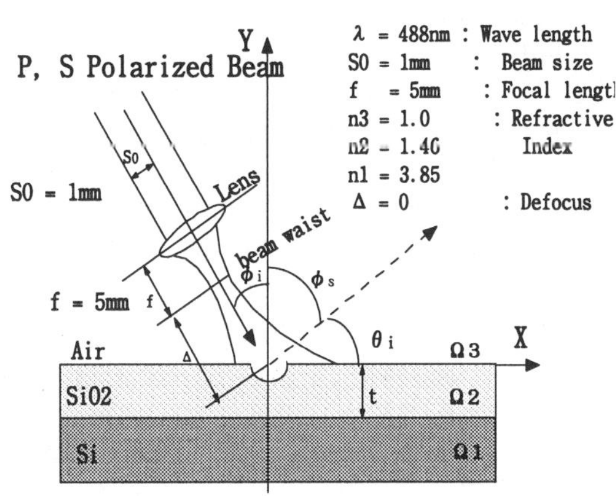

Figure 1 coordinate system and basic simulation parameters

3. EXPERIMENTS

3.1 Experimental Setup

Figure 2 indicates a schematic diagram of the scattered light analyzing system. This system utilizes Ar+ (488nm) laser for incident beam, PMT (photomultiplier) for detecting the scattered light intensity from the defect. There are three rotation stages and one XY stage in this system. R1 and R2 stage enable to rotate the detector unit mounting the PMT in both directions of φ_s and θ_s(*Figure 2(a)*). By rotating each stage, we can

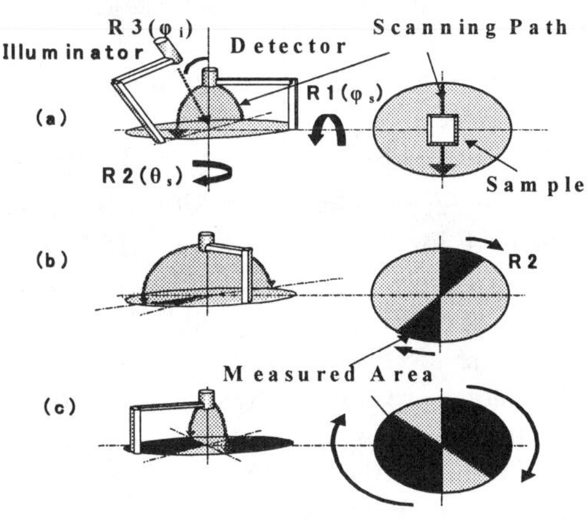

Figure 2 Schematic diagram of Scattered light analyzing system

measure the light intensity distribution spatially scattered from the defect. Firstly, the light intensity distribution in the scattering plane is detected by scanning the R1 stage(φ_s) to the sample from 0° to 180°(*Figure 2(a)*). Secondly, after this scanning, the R2 stage(θ_s) is changed by a slight angle in θ_s direction(*Figure 2(b)*) and the detection experiment is preformed in the same way as mentioned above. As a result, we get the 3D spatial (θ_s-φ_s) scattering light intensity distributions by this series of measurements (*Figure 2(c)*). While illuminator is rotated by the R3 stage (φ_i) to change incident angle from 0° to 90° (*Figure 2(a)*). The illumination position is controlled by the XY stage that mounts a sample. Parameters such as incident angle, polarization of beam, the range of acceptable scattering angle (N.A. : Numerical Aperture) for PMT and so forth are changeable in the scattered light analyzing system. The acceptable scattering angle (N.A.) means the range of angle that PMT can accept at a time.

3.2 Experimental Confirmation and Discussion

Several experiments are conducted to confirm simulation results. The incident angle 75°(p-polarized) is used for these experiments. The acceptable scattering angle (N.A.) is set at 20°. The defects sizes used in experiments are shown in *Table 1*.

Table 1 size of defect

	Depth (μm) Width (μm)
A	D 0.204 W 0.658
B	D 0.113 W 0.551
C	D 0.105 W 0.328
D	D 0.059 W 0.341

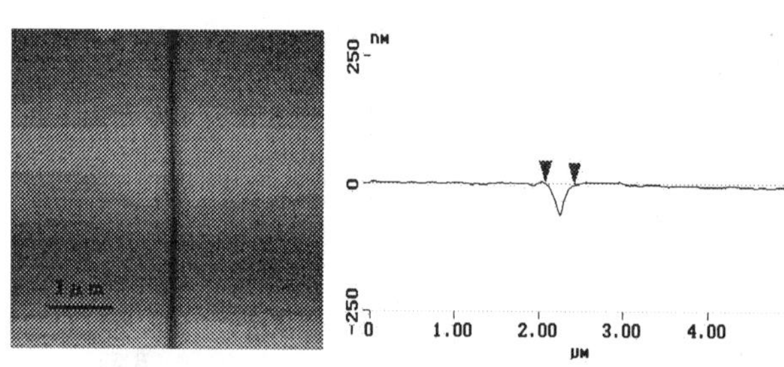

Figure 3 AFM image and profile

(D : Depth 0.059μm, Width 0.341μm)

These defects are made on SiO_2 film surface (film thickness : 0.5μm) by FIB (focused ion beam) process. AFM image and its cross-sectional profile (defect D) are indicated in *Figure 3*. *Figure 4(a)* shows the scattering light intensity distribution (φ_s : 40°~150°) for each defect in the incident plane when the incident angle is 75° (φ_s, p-polarized). *Figure 4(b)* represents the simulation results corresponding to the experiments. To make same conditions as experiments, the simulation value at each scattering angle is set for the sum of the scattering light intensities in the acceptable angle of 20° at each angle. It is shown that the experimental results agree well with the simulation results for each defect, namely both the light intensities of *Figure 4(a)* and *(b)* increase with the increase of depth and width of the defects

686

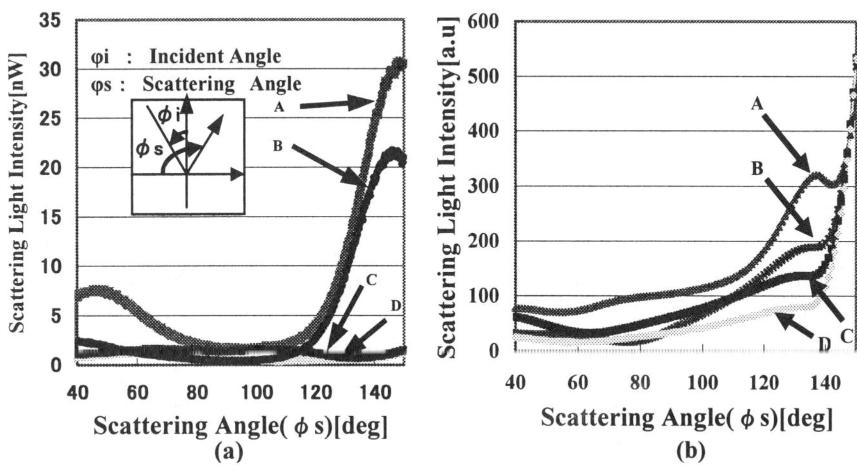

Figure 4 (a) Scattering light intensity($\varphi_i=75°$)   (b) simulation results

between 120° and 140°. Especially, fluctuation tendency for each defect shows good matches around 70° and 140°(φ_s).

4. NUMERICAL ANAYSIS AND PROPOSAL OF DEFECT CLASSIFICATION MAP

Defect model parameters in this simulation vary in depth 0.02 to 0.1µm, in width 0.1 to 0.8µm and in film thickness 0.5 to 1.0µm. The incident light (p-polarized) angle is restricted to normal incidence for this numerical analysis. *Figure 5* shows the defect classification map for various defects sizes when the film thickness is 0.6µm. In order to reduce influence of the

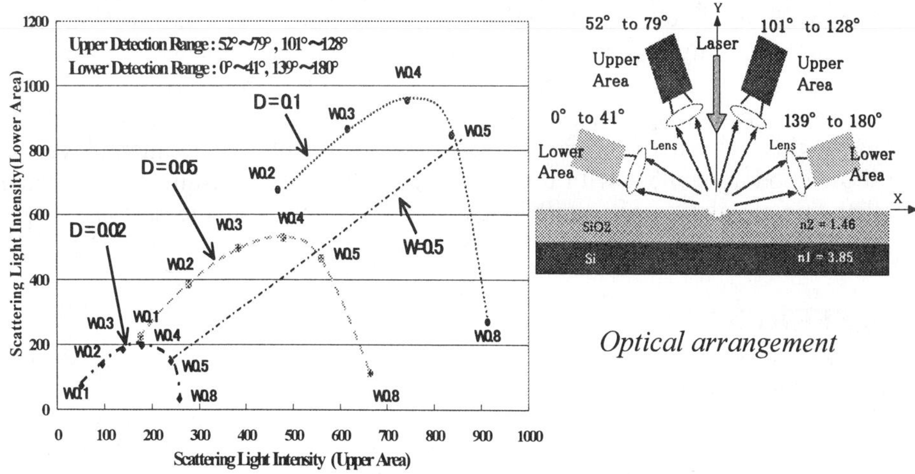

Figure 5 Defect classification map (film thickness 0.6µm)

systematic error, the difference between the scattering light intensities of the defects and flat surface is used as the scattering light intensity. The points plotted in *Figure 5* mean the scattering light intensities detected at upper area and lower area for the same defects as shown in the optical arrangement (*Figure 5*). The lateral axis is scattering light intensity at upper area ($52°\sim79°$, $101°\sim128°$) and the vertical axis is that at lower area ($0°\sim41°$, $139°\sim180°$). There are three parabolic plots in the map. Each parabolic plot consists of the points representing the scattering light intensities obtained from the same defect depth. The detected values at upper area become larger as the defect widths increase in the same parabolic plot. When the defect depth becomes deeper, the detected values at both upper and lower area increase. Therefore, the sizes of defects with respect to the depth D and width W are separately obtained using this map and the killer defects, which affect the yield enhancement of ULSI, can be detected.

The most suitable detection angle range for each film thickness depends on its reflective ratio occurred by interference between the SiO_2 film surface and Si wafer surface. Design of the proper detection angle range improves the detection sensitivity. Optimal detection angle range for film thickness 0.6μm is estimated to be $52°\sim79°$, $101°\sim128°$ for upper area and $0°\sim41°$, $139°\sim180°$ for lower area.

5. CONCLUSION

In this paper, CMP defect detection schemes by using numerical analysis based on BEM have been proposed. Depths and widths of defect are separately obtained by the scattered light intensity distribution. It is shown that the killer defects can be easily categorized in the defect classification map. The simulation results suggest that the proper setting of detection angle range is important for classification of defect on different film thickness. Also, we developed the scattered light analyzing system. This system has ability to measure the spatial scattering light intensity distribution for certain given conditions like incident angle, polarization of light, detection angle range and so on. Several basic experiments are carried out using this system to validate the simulation results. Experimental values show good agreement with the calculated values. As a result, it is expected that this simulation tool will extend the capabilities of the scattered light analyzing system for the defect measurement.

REFERENCES

1. Kumagaya, Nobuaki, electromagnetic wave and boundary element method. Morikita publisher, 1987 (in Japanese)
2. T. Ha, T. Miyoshi, Y. Takaya, S. Takahashi, Light scattering simulation for silicon wafer's CMP defects on SiO2 film, Proceedings of The Japanese society for precision engineering. Spring Meeting. 2000, pp.603 (in Japanese)

PALMTOP PANTOGRAPH MECHANISM WITH LARGE-DEFLECTIVE HINGES FOR MINIATURE SURFACE MOUNT SYSTEMS

Mikio Horie* ,Toru Uchida**, and Daiki Kamiya*

* Precision and Intelligence Laboratory (P & I Lab.), Tokyo Institute of Technology

4259 Nagatsuta-cho, Midori-ku, Yokohama 226-8503, Japan

** Graduate school, Tokyo Institute of Technology

Email: mhorie@pi.titech.ac.jp; Phone/Fax: +81-45-924-5048

Abstract

In the present paper, a new surface mount system with parallel arrangement miniature manipulators is proposed for use in system downsizing. The miniature manipulator consists of a molded palmtop pantograph mechanism, which is composed of large deflective hinges and links, both made of the same materials. In order to create such systems, first, durability of the pantograph mechanism has been confirmed by fatigue tests. Next, the input and output displacement characteristics of the pantograph mechanism have been experimentally discussed. Finally, propriety of the proposed system has been confirmed.

Keywords

Machine element, Palmtop surface mount system, Large-deflective hinge, Molded pantograph mechanism, Fatigue, High polymer materials

1. Introduction

The present surface mount system has been developed by mainly focusing on the improvement of speed functions, however resulting in enlargement of the devices. It has become almost impossible to improve the speed of the system. Therefore, we should consider a productivity increase that would be made possible by minimizing each device, while filling designated spaces to the maximum capacity.

In this paper, we have reserached having next three originalities. (1) we propose a new surface mount system which consists of groups of the manipulators that have been minimized by an integral molded pantograph mechanism with hinges and links. (2) In addition, durability against repeated input displacement is confirmed with large deflective hinges, even the small size of which can obtain large angular displacements. (3) Moreover, this paper discusses minimization possibilities of a new system by a model-devised surface mount system, and has clarified the input-output displacement characteristics of a model-devised integral molded pantograph mechanism in the experiments.

2. Suggestions on a Palmtop Surface Mount System

The present mount system needs high structural stiffness in order to speed up functions of mounting work, which has a tendency to increase its overall size. Therefore, in this study, the new palmtop surface mount system shown in Fig. 1, is proposed based on the idea that placing as many miniature manipulators as possible within a working space should make it possible to minimize the size of devices while maintaining the present productivity. An integral molded pantograph mechanism made of high polymer materials is used for minimizing these manipulators. Since this type of mechanism is integrally molded with both link and hinge parts in an injection molding method, it does not require assembling, therefore making it easy to be devised even in such cases when minimizing is required.

In this study, large deflective hinges, which can obtain large angular displacements by producing larger deformation than the elastic region, are model-devised. Moreover, the durability against repeated input displacements is tested in such cases where large deflective hinges are used as the hinge parts of an integral molded pantograph mechanism. An upper part in Fig. 2 shows a large-deflective elastic hinge[1] which is used the deformation produced on the hinge surface by operations within a limited elastic dimension. A lower part in Fig. 2 shows a large deflective hinge whose aspect ratio is small. The subjective working of manipulators is determined as the substrate surface mount of a miniature electric device. From this substrate size, the moving space of the mechanism is determined as 50mm×40mm. In this way, the size of the mechanism is determined where the relative angular displacements between links around the moving space area can be set at 45°. The displacement enlargement ratios of a pantograph mechanism are set to be four times more than the input in the X direction, and five times more than that of the Z direction in Fig. 3. The length of hinge parts is set at 200 µm, and the thickness can be adjusted between 30 µm and 300 µm using a metal mold. In addition, the thickness of link parts and the width of the mechanism are both set at 5 mm. The shape and size of the model-devised integral molded pantograph mechanism are shown in Fig. 3. A large-deflective hinge shown in lower part of Fig. 2 is similar to the

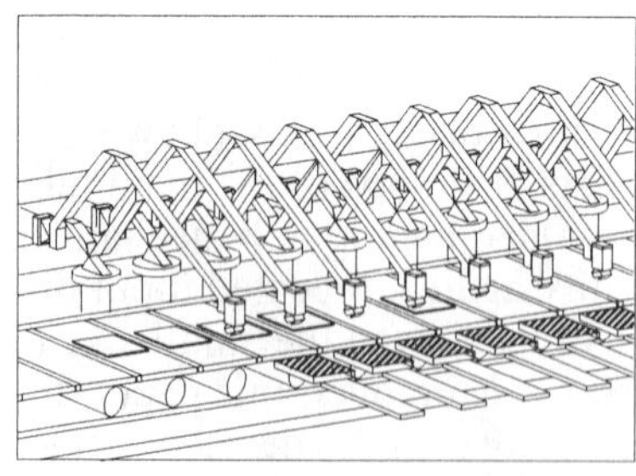

Figure 1. Proposed palmtop surface mount system

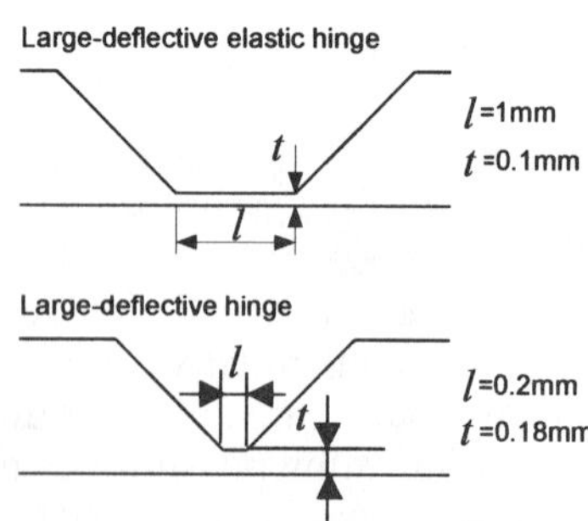

Figure 2. Dimension of large-deflective elastic hinge and large-deflective hinge

hinge at points A, B, C, D, F_1 and F_2 in Fig. 3. Polypropylene is used as it can both obtain a relatively larger deformation among high polymer materials and be easily molded.

3. Fatigue Tests of an Integral Molded Pantograph Mechanism with Large Deflective Hinges

Figure 4 shows the experimental apparatus with manufactured pantograph mechanism. The experimental apparatus consists of four parts: the vibrator based on the slider crank mechanism, the laser displacement sensors, the oscilloscope, and a manufactured pantograph mechanism. The test is conducted as follows: Using a vibrator, the sinusoidal displacement input at 7.5 mm vibration is given to the input part in the X direction of a pantograph mechanism in Fig. 3, and then the displacements of input-output ending points are measured by a laser displacement sensor. The pantograph mechanisms in this experiment were divided into three groups A, B and C according to their different hinge thickness $t<100$ μm, $100<t<150$ μm and $150<t$ μm respectively, as shown in Table 1. A hyphen mark in Table 1 shows that a pantograph mechanism does not fracture even after a one million-time repetition. The maximum input frequencies which does not fracture even after a one million-time repetition were determined as 15 Hz in Group A , 17.5 Hz in Group B and 20 Hz in Group C. Within the region used in this experiment, it is found out that the greater the thickness of the hinge, the more durable it can be when used at high speeds. Moreover, as shown in Table 1, it is obvious that the fracture of all the pantograph mechanisms except for that in Experiment Group B-No.13 is seen at hinge part D. Therefore, after microscopic observation of the hinge part D which is not broken in the fatigue test with one million number of cycles, we have next results. In case of hinge thickness more than 200 μm, a small crack was occured at the hinge, however, remaing 180 μm length of the hinge except the crack length. The hinge part has not a bending deformation state. It has a complex deformation state based on its small aspect ratio (Hinge length/ thickness) and its plastic material characteristic. Then, in this study, caliculating the forces acting at the hinge part D in the moving of the pantagraph mechanism based on the input frequency and hinge thickness in Table 1, we had the equevalent stress oftained as the above caliculating forces devide by the hinge section area (Hinge thickness t × Hinge width 5 mm). Figure 5 shows the relationship between the number of cycles and the equevalent stress. The thin line in Fig. 5 shows the line $\sigma = -0.78\log N + 5.97$ caliculated by use of the method of least squares and left ten data in Fig. 6. Sign $\longrightarrow$ in the figure shows that the pantograph mechanism did not fracture and also the conclusion time of the experiment.

4. Input-Output Displacement Characteristics of an Integral Molded Pantograph Mechanism

The accuracy of positioning repeatability at ending points E_1 and E_2 were measured after having the locus of the supposed surface mounting job moved back and forth between E_1 and E_2 on the experimental trajectry 40×20 mm in the working space 50×40 mm as shown in Fig. 6. Measuring the accuracy 15 times showed that the accuracy of positioning repeatability at E_1 was ±9μm, while that of E_2 was ±11μm. Next, no resonance in the X-input frequency range less than 100 Hz is confirmed dynamically in the experiments when X-input is ±7.5 mm and Z-input is zero, and no vibrations in the Z-direction at ending points E_1 and E_2 are confirmed.

In considering that the accuracy of positioning repeatability is quite high when this pantograph mechanism with 2-DOF of X- and Z-inputs in Fig. 3 is used as the surface mount system, and also considering that the accuracy of positioning repeatability of the present surface mount system is ±100 μm, the open loop control is therefore proven to be useful. Moreover, if the output ending point can hold some electric parts up to the maximum of 0.5 gf, the time of carrying the parts onto the substrate is then calculated to be 0.1 second based on the fatigue limit of the hinge parts. Furthermore, if each of the detecting, recognizing, maintaining, revolving, mounting of machine parts, as well as eliminating inferior parts requires 0.1 second, as does the present system, the mounting time for miniature manipulators will be 0.8 seconds. Since this enables

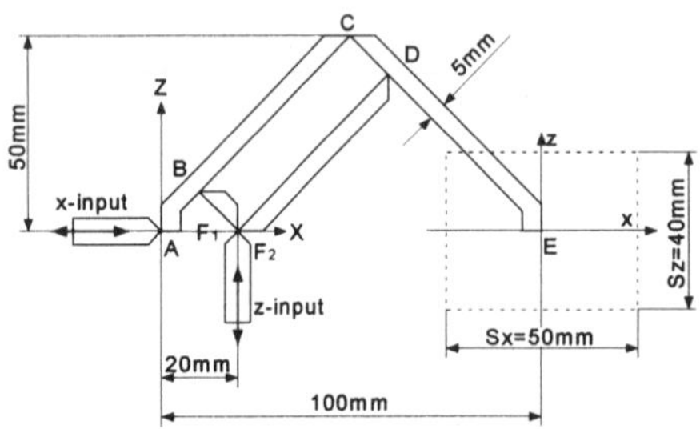

Figure 3. Dimension of pantograph mechanism

Figure 4. Manufactued pantograph mechanism

Table 1. Result of fatigue test

Group	A ($t{\leq}100$ μm)					B ($100<t<150$ μm)			
Exp. No.	1	2	3	4	5	6	7	8	9
Hinge thickness t [μm]	60	90	90	90	100	110	110	115	120
Input frequency [Hz]	19.4	15	15	20	22.2	5	20	18	15
Number of cycles N	100	–	153900	3140	100	–	3900	30600	–
Failed hinge	D	–	C, D	B,C,D	F_2, D	–	C, D	D	–

Group	B ($100<t<150$ μm)				C ($150{\leq}t$ μm)			
Exp. No.	10	11	12	13	14	15	16	17
Hinge thickness t [μm]	125	125	130	135	150	165	225	260
Input frequency [Hz]	17	20	26	20	26.8	17	10	20
Number of cycles N	–	6040	100	8600	100	–	–	–
Failed hinge	–	F_2, D	F_2, D	A	F_2, D	–	–	–

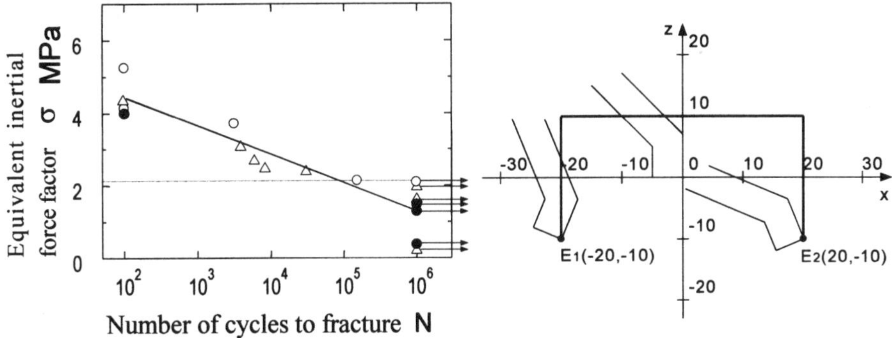

Figure 5. S-N curve(○: Group A, △: Group B, ●: Group C)

Figure 6. Experimental trajectory (Surface mount job is supposed)

over 8 manipulators to do the same work, a new system will be able to produce more machine parts than the presently used system. It is also found that a model-devised pantograph mechanism is quite small and only needs less than approximately 0.1 % of the space that rotary head parts of the presently used system require.These above-mentioned facts indicate that the presently used surface mounting system can be minimized even in cases where a great number of manipulators are placed in a certain working space with appropriate actuators.

5. Conclusion

In order to minimize the presently used system, a new surface mount system consisting of miniature manipulators was proposed, and integral molded pantograph mechanisms were then model-devised and researched. This research clarifies the durability as well as input-output displacement characteristics of the mechanisms. The results obtained from this research are as follows.
(1) An integral molded pantograph mechanism consisting of large deflective hinges with 200 μm of the length, 180 μm of the thickness and 5 mm of the width did not fracture in displacement input fatigue tests on the mechanism even after repeatedly used 1 million times. In this test, the maximum relative angular displacement between links was set at 45°.
(2) The accuracy of positioning repeatability of this integral molded pantograph mechanism was clarified to be ±11μm in the positioning test where the practical mount work had been estimated. Based on this, the mechanism was evaluated when used as a surface mount system. This indicates that there is some possibility that the proposed system can minimize the present surface mount system.

References

1.   M. Horie, M. Takayama, D. Kamiya and K. Ikegami, "Development of Pantograph Mechanisms Uniformly Molded with Large-Deflective Elastic Hinges and Links made from the Same Materials", *Proceeding of Movic'98* (The Fourth International Conference on Motion and Vibration Control), Vol.3, 913-918, 1998.

DYNAMICS OF PIN ELECTRODE IN PIN-TO-PLATE GAS DISCHARGE SYSTEM USED FOR OZONE-LESS CHARGER IN LASER PRINTER

H. Kawamoto, K. Takasaki, H. Yasuda and N. Kumagai

Department of Mechanical Engineering, Waseda University

Abstract

Dynamics of a pin electrode in a pin-to-plate discharge system have been investigated to realize a new ozone-less charger of a laser printer. Vertical coupled vibration of the pin electrode was observed when the spark discharge took place. The force applied to the pin electrode at the spark discharge was implicitly evaluated and it was elucidated that the force was attractive, in the order of 1 mN, and it became large in accordance with the increase of the applied voltage. The frequency of the vibration was smaller than the natural frequency of the pin electrode. The force at the positive discharge was larger than that at the negative discharge but the decrease of the vibration frequency at the positive discharge was smaller than that at the negative discharge.

Keywords

Spark Discharge, Electrophotography, Laser Printer, Electrostatic Force

1. INTRODUCTION

One of the most important issues of electrophotography technology is to reduce ozone emitted by charging devices, because not only does ozone damage photoreceptors and consequently causes deterioration of images, but it is also harmful for humans (Hansen 1986). A new charging system with pin electrodes is under development to realize extremely low ozone emission of a laser printer (Furukawa 1994). However, because the stiffness of the pin electrode is extremely low in this system, the electrode is deformed (Kawamoto 2000) and the force applied to the pin electrode induces abnormal vibration.

A series of investigation has been conducted on the kinetics of a wire-to-plate discharge system to clarify the mechanism of lateral oscillation observed in a corona charger used for a polyester film manufacturing machine (Ito 1996) but no systematic study has been performed for kinetics of the pin-to-plate system. In this study, the authors have investigated dynamics of the pin electrode in the pin-to-plate spark discharge system to realize the new ozone-less charger in electrophotography.

2. EXPERIMENTAL

Figure 1 shows an experimental set-up. A wire made of stainless steel was hung down perpendicular to a steel plate. The diameter of the wire used for experiment was 0.5 mm. The wire was connected to the free end of the cantilever plate made of stainless steel (stiffness of the cantilever, k, is 2.68 N/m, damping coefficient, c, is 0.00179, and the natural frequency, f_n, is 9.79 Hz). The displacement at the free end of the cantilever was measured by a laser displacement meter (Keyence Corp., Tokyo, LK-2000). Gap between the wire and the plate was adjusted using a mechanical stage attached at the back of the plate electrode. High voltage was applied to the gap by a DC power supply (Matsusada Precision Inc., Tokyo, HVR-10P (positive) and HVR-10N (negative), 0 ~ ±10 kV adjustable). Voltage was determined by a calibrated potentiometer of the power supply and current was measured by the voltage drop in a current-shunt resistor. Transducer signals of the vibration and the current were introduced to a digital oscilloscope and to an FFT analyzer. The surface of electrodes was frequently polished to prevent oxidation and chemical deposition on the tip of the discharge electrode due to gas discharge. Reproducibility of data was confirmed during experiments.

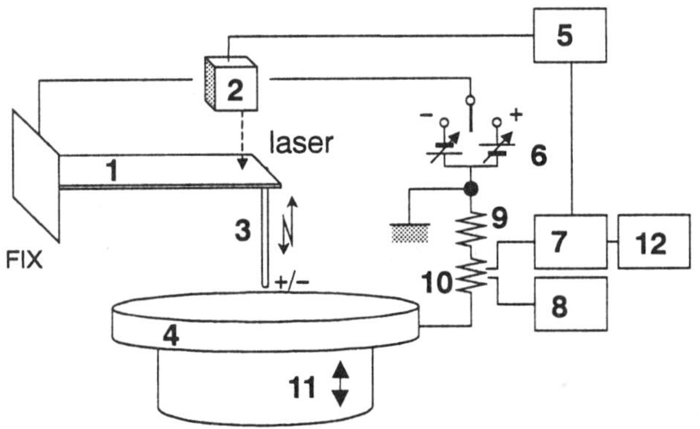

Figure 1. Experimental set-up. (**1**: stainless steel plate, cantilever, **2**: laser sensor, **3**: pin electrode, SUS304, **4**: plate electrode, steel, **5**: laser displacement meter, **6**: DC high voltage power suppliers, **7**: oscilloscope, **8**: DC volt meter, **9**: resistor, **10**: shunt resistor, **11**: mechanical stage, **12**: FFT Analyzer)

3. RESULTS AND DISCUSSION

When the voltage applied between the pin and the plate electrode was gradually increased, the following phenomena were observed: At the voltage lower than the threshold (about 3.5 kV), no substantial current flowed in the air gap and extremely small electrostatic pull force, in the order of 10 μN, was induced. Over the threshold voltage, the corona discharge took place and the corona current flowed. The order of the current was several 10 μA for the positive corona and 100 μA for the negative corona. At the same time, *repulsive* force, in the order 0.1 mN, was applied to the pin electrode both in the positive and negative corona due to the corona wind. If the voltage was more

695

increased, the corona discharge shifted to the intermittent spark discharge at 5 ~ 8 kV. At the same time, vertical vibration of the cantilever was induced. The vibration coupled with the occurrence of the spark discharge as shown in Figure 2. It is recognized that the spark discharge took place almost when the pin electrode approached to the plate electrode. The amplitude of the vibration was in the order of mm, which was about 10 times as large as the static displacement at the corona discharge. A pulse width of the spark discharge was 0.5~5 ms, which was short enough compared to a period of the free vibration. These experimental results suggested that the force at the spark discharge was attractive and its magnitude was much larger than the static repulsive force at the corona discharge.

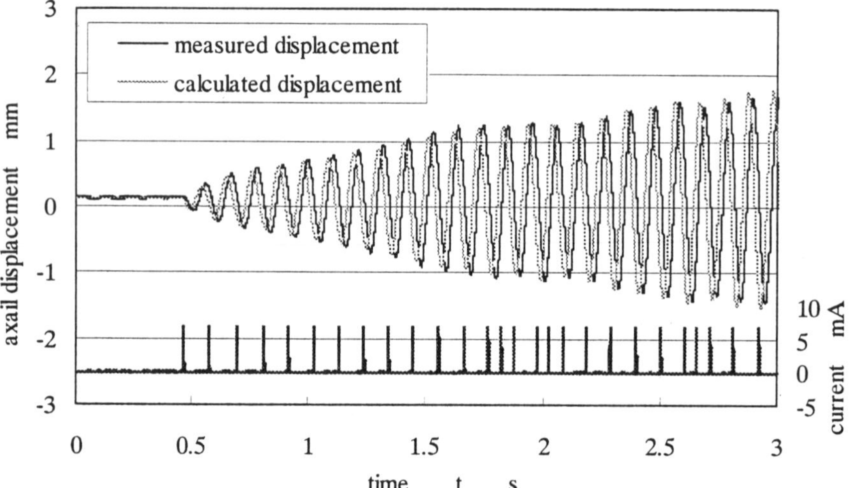

Figure 2. Vertical vibration of pin electrode and discharge current at the start of spark discharge. (air gap: 4 mm, applied voltage: 8.1 kV)

Figure 3 shows the frequency of the vibration. It is recognized that the frequency of the vibration at the corona discharge equals to the natural frequency, while the frequency at the spark discharge was a little smaller than the natural frequency. The decrease of the vibration frequency at the positive discharge was smaller than that at the negative discharge.

Figure 4 shows the force at the spark discharge. The force was derived by the method cited in Appendix. It was elucidated that the force was *attractive* and the order of the magnitude was 1 mN. The force at the positive discharge was larger than that at the negative discharge. The force became large in accordance with the increase of the applied voltage both in positive and negative discharge.

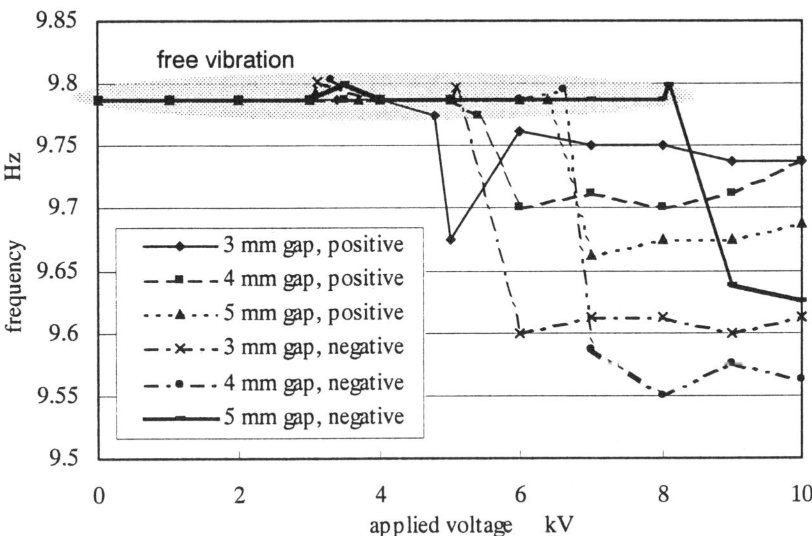

Figure 3. Frequency of free vibration and spark-discharge-coupled forced vibration. Because the force was static and vibration of the cantilever was not induced at the voltage lower than the onset of the spark discharge, the frequency of free vibration was measured and plotted in this figure in case of no spark discharge.

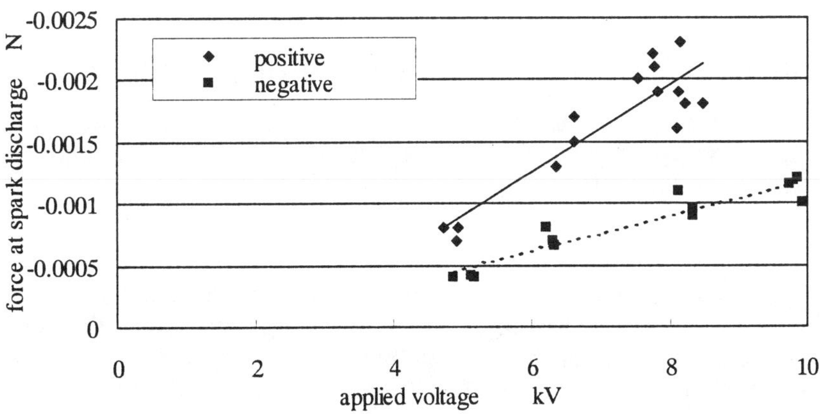

Figure 4. Electrostatic force applied to the pin electrode at spark discharge.

4. CONCLUDING REMARKS

Dynamics of the pin electrode in the pin-to-plate system have been investigated to realize a new ozone-less charger of the laser printer. The following is a summary of the investigation.

(1) Vertical vibration of the pin electrode was induced when the spark discharge took place. The vibration was coupled with the spark discharge. That is, a pulse current flowed almost when the pin electrode approached to the plate electrode.
(2) The frequency of the vibration was a little smaller than the natural frequency. The decrease of the vibration frequency at the positive discharge was smaller than that at the negative discharge.
(3) The force at the spark discharge was attractive, in the order of 1 mN, which was about ten times as large as the static repulsive force at the corona discharge. The force became larger in accordance with the increase of applied voltage. The force at positive discharge was larger than that at the negative discharge.

REFERENCES

Hansen T. B. and Andersen B., Ozone and Other Air Pollutants from Photocopying Machines, *Am. Ind. Hyg. Assoc. J* 1986; 47: 659-665.

Furukawa K., Ishii, H., Shiojima K. and Ishikawa T., Discharge Characteristics and OPC-Drum Charging Characteristics of Separated Saw-tooth Charging Device, *Proc. IS&T's Tenth Int. Congress on Advances in Non-Impact Printing Tech.*, New Orleans 1994: 34-38.

Kawamoto H., Takasaki K., Yasuda H. and Kumagai N., Statics of Pin Corona Charger in Electrophotography, *IS&T's NIP16: Int. Conf. on Digital Printing Tech.*, Vancouver 2000: 827-835.

Ito, Y., Dynamics of the Wire Electrode in a System of Wire and Plate Electrodes, Ph.D. Thesis, Keio University, 1996.

APPENDIX - method to derive force at spark discharge -

The vibration of the cantilever connected to the pin electrode is simplified as the single-degree-of-freedom with respect to the vertical displacement z (upper direction is designated to be positive).

$$m\ddot{z} + c\dot{z} + kz = F_z \qquad (1)$$

where m is an equivalent mass of the pin electrode and the cantilever. F_z is the vertical force acting on the pin electrode due to gas discharge. The force at corona discharge has already measured by a separate experiment but the force at spark discharge has been unclear. It is determined explicitly, i.e., the numerical calculation is repeated with a revised force from an initial guess until the calculated vibration response coincides with the measured. The calculation is conducted using the Runge-Kutta method on the lumped-force assumption that the force at the spark discharge and that at the corona discharge are constant at each discharge period.

EFFECTS OF PART FLEXIBILITY ON DYNAMIC BEHAVIORS OF MACHINE SYSTEMS WITH CLEARANCES IN THE JOINTS

Yu Wuyong   Ji Linhong   Jin Dewen

School of Mechanical Engineering, Tsinghua University, Beijing, China, P. R.

Abstract

Part flexibility and joint clearance are two of important factors that affect motion precision of machines. The fixed-boundary component modal synthesis (CMS) method and nonlinear impact model was used to study dynamic behaviors of mechanism with flexible parts and clearances at the joints. A test model of slider-crank mechanism was created and numerical simulation was made under several cases. After a coordinate transformation is applied for the fixed-boundary CMS method, the CMS method can be employed easily, improve the precision of computing and efficiency of CMS method and also present the effects of boundary constraints and impact. The coupling effects of flexible parts and clearances at the joints make the mechanism show particular dynamic behaviors that are different from the mechanism without clearances at joints or flexible parts.

Keywords

Clearance; Impact; Flexible multi-body dynamics; Dynamics of mechanical

1. INTRODUCTION

Very high precision and speed is often required by today's industry system, but many factors influence the dynamic performance of mechanisms. The element flexibility and joint clearance in the machine systems are those of most important factors that should be drawn attention to when designing and manufacturing a high technology system.

The flexibility of the elements and clearances in the joints may result in many such unexpectedly detrimental dynamic behaviors as increased vibration and noise, accelerate wear and fatigue, which result in the systems instability and premature failures, particularly lost of precision. The flexibility of parts might not be ignored when the speed of machines is high or the elements in the system have not enough stiffness. The presence of clearances in the joints of mechanism is inevitable. In the past, many researches on the dynamic responses of machine system with clearances in the joints have been reported [Deck J. F., Dubowsky S., 1993; Kakizaki T., 1993]. And many researches on the effects of flexible parts have been done [Shabana A.A., 1998]. But the combination effects of the flexibility of parts and clearances in the joints are very complex and need more research.

2. SYSTEM MODEL

Finite element method (FEM) is used to set up model of the total system, and component modal synthesis (CMS) is adopted to reduce the degree of freedom (DOF). First, each flexible part is taken apart from the system and established as a finite element model called component. It generally has very large number of DOF. Next, CMS method is employed to reduce the system DOFs. In order to consider the effects of force at the joints of component, Craig-Bampton method[Craig R.R., 1968] can be modified, which can efficiently reduce DOFs. Finally, all parts are assembled and constraints and load forces are applied.

In the CMS method, the DOFs of a component are [Craig R.R., 1968]

$$u = \begin{bmatrix} u_{\mathrm{I}} \\ u_{\mathrm{B}} \end{bmatrix} = \begin{bmatrix} \Psi_{\mathrm{N}} & \Psi_{\mathrm{C}} \end{bmatrix} \begin{bmatrix} q_{\mathrm{I}} \\ u_{\mathrm{B}} \end{bmatrix} = \Psi q \tag{1}$$

where:

u_{I}, u_{B}, u — interior, interface and system DOFs

$\Psi_{\mathrm{N}}, \Psi_{\mathrm{C}}, \Psi$ — normal, constraint and system transform modal

q_{I}, q — interface and system general DOFs

After the component equation is transformed using matrix Ψ, system DOFs are reduced. But in order to present the effects of interface at the joints and use CMS method easily, Craig-Bampton method must be modified. The component dynamics equation is

$$\begin{cases} \hat{M}\ddot{q} + \hat{K}q = \hat{F} \\ \hat{M} = \Psi^{\mathrm{T}}M\Psi \ , \quad \hat{K} = \Psi^{\mathrm{T}}K\Psi \ , \quad \hat{F} = \Psi^{\mathrm{T}}F \ . \end{cases} \tag{2}$$

Solve the general eigenvalue problem about equation (2) and can get a new eigenvector matrix H. Then

$$u = \Psi q = \Psi H \hat{q} = \hat{\Psi}\hat{q} \tag{3}$$

After the transformation using equation (3), component DOFs are reduced and CMS method can be applied into dynamics of mechanical easily. This method also can reflect the effects of constraints at the joints.

When clearances exist at the joints, the impact forces are very large and present nonlinear characteristics. The impact forces R can be model as follows

$$R = f(x, \dot{x}) = \boldsymbol{F}_{\mathrm{s}} + \boldsymbol{F}_{\mathrm{d}} \tag{4}$$

This is the parallel spring-damp model. x is relative displacement between two parts of a joint. $\boldsymbol{F}_{\mathrm{s}}$ is the spring force and $\boldsymbol{F}_{\mathrm{d}}$ is the damp force. They are both nonlinear forces. According to experiment or numerical simulation, they can be approximated as eaquation (5)

$$\begin{cases} \boldsymbol{F}_s = k\boldsymbol{x}^n \\ \boldsymbol{F}_d = c(\boldsymbol{x})\boldsymbol{x}\dot{\boldsymbol{x}} \end{cases} \tag{5}$$

k is a constant value and $c(x)$ is a quadratic polynomial. $\boldsymbol{R}$ is normal impact force. The tangential friction forces including two kinds of force: static friction and dynamic friction force. Usually when relative velocity becomes high, static friction force will change to dynamic friction force.

Using the CMS method and nonlinear impact clearance model, and employing Lagrange multiplier to introduce constraint conditions, the total system dynamics equation can be set up easily [Kakizaki T., 1993].

3. NUMERICAL EXPERIMENTAL RESULTS

A three-dimension slider crank mechanism (Figure 1) was numerically studied. The link was modeled as flexible body and the crank as rigid body. The material is steel, the Young's modulus E = 200 GPa, the Poisson ratio $\upsilon = 0.29$. There are clearances between slider and guide and the clearances are three-dimension. The contact force model is nonlinear including dynamic and static friction force. The crank was derived by a moment and an external force was applied on the slider.

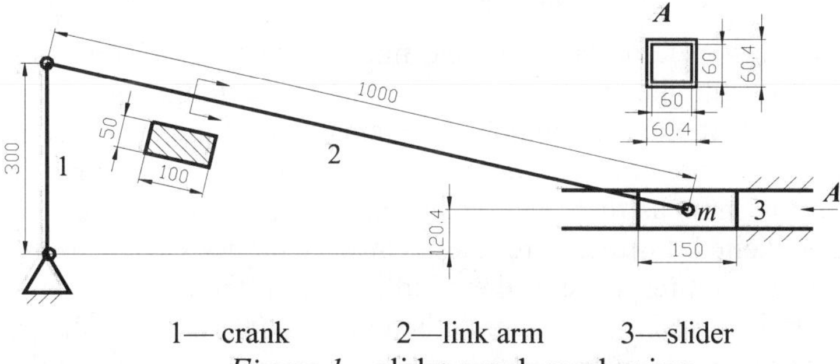

1— crank   2—link arm   3—slider

Figure 1   slider-crank mechanism

The link arm is a flexible part and is analyzed use CMS method. The two ends of the link are boundary points. The First 40 normal modals and 6 constraint modals are used to synthesize. We can get the transformation matrix $\widehat{\boldsymbol{\Psi}}$. From $\widehat{\boldsymbol{\Psi}}$, the first eight elasticity modals and six boundary modals are extracted. Numerical experiments show that other higher order modals have little effect on system and can be ignored.

At the joints with clearances, the impact forces usually are very great. The constraint modals can exactly express the impact effects. Figure 2 shows the results of numerical experiments. Under this case, the mass of slider is 400 kg and there is no clearance at the joints. The results with boundary modals are obviously different from those without boundary modals. When the mass of the slider is small, the forces at the joints are also small and have little effect on the system. When the mass of the slider is large or there are clearances at the joints, the force at the joints will significantly increase, such boundary condition as displacement, force and constraint must be considered carefully.

The impact forces on the slider are plotted on the Figure 3. The gravity, static and dynamic friction forces are taken into consideration. The impact forces mainly changed with the running of mechanism, but between two large peaks there are a lot of small peaks. The friction directly caused this crawl phenomenon. The crawl phenomenon will become serious while the gravity or service load is applied on the slider.

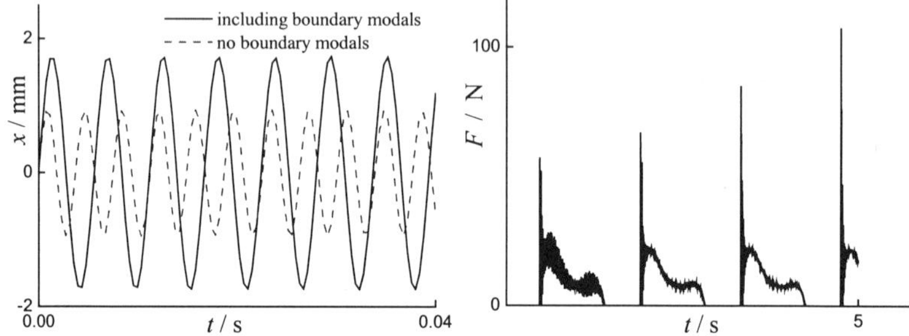

Figure 2 slider motions errors *x*   *Figure 3*   Impact force on the slider

The elastic deformation of the middle point of the link is show on figure 4. Owing to the clearances, the impact forces in the joints are impulse curves instead of flat and smooth functions. These impulse forces excited wide-band vibration of flexible bodies. Besides the vibrations of which periodicity is same as that of mechanism, a lot of other elastic vibration of the link is excited. Furthermore, these elastic vibrations interact. Figure 5 is the elastic spectral response of the middle point of the link. The Frequencies of the vibrations mainly concentrate between 0-200 Hz. Near the point of 1 Hz, the vibrations is mainly result of the running of the mechanism. Near 120 Hz, the link also vibrates greatly. All vibration is result of interaction between the movement of the mechanism and elastic deformation. Due to impact at the joints, some energy transfer to the flexible parts and excites vibration. If the flexible components have material damping, this energy will be consumed gradually.

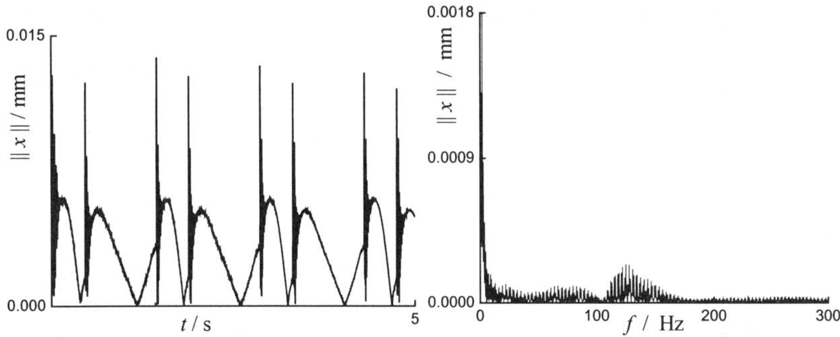

<table>
<tr><td>Figure 4 Elastic deformation of
the middle point of the link</td><td>Figure 5 Spectral response of
the middle point of the link</td></tr>
</table>

4. CONCLUSINOS

A test model of slider-crank mechanism was created and numerical simulation was made under several cases. After a coordinate transformation the CMS method can be employed easily, improve the precision of computing and efficiency of CMS method and also present the effects of boundary constraints and impact. The interaction between vibration of flexible body and impact force in the joint caused some unexpected behaviors. The flexible bodies will absorb a part of vibration energy, and the impact force in the joints decreased compared with under the case that there is no flexible body in the system. The magnitude and direction of load force also influence the performance of machine. In addition, the behaviors of this type of machine system show typical nonlinear characteristics.

References:

1. Craig R.R., Bampton M.C.C. Coupling of substructures fordynamics analyses. AIAA Journal, 1968, 6(7): 1313-1319
2. Deck J. F., Dubowsky S. On the Limitations of predictions of the dynamic response of machines with clearance connections. ASME J. Mechanical Design，1993，116（9）: 833-841
3. Kakizaki T., Deck J.F., Dubowsky S. Modeling the spatial dynamics of robotic manipulators with flexible links and joint clearances. Trans. ASME Journal of Mechanical Design, 1993, 115(4): 839-847
4. Shabana A.A., Hussien H.A., Escalona J.L. Application of the absolute nodal coordinate formulation to large rotation and large deformation problems. ASME J. Mechanical Design, 1998, 120(60):188-195

ANALYSIS OF FEED DRIVE SYSTEM OF X-Y TABLE CONSIDERING THE FRICTION

Qiang SUN[*], Xuesong MEI[**] and Masaomi TSUTSUMI[*]

*:Graduation School of Tokyo University of Agriculture and Technology

**:Xi'an Jiaotong University

Abstract

Friction plays a significant role in machine tools incorporating mechanical parts with relative motion, but it is generally an impediment for the servo control system. In this study, taking an X-Y table of NC machine tools, a mathematical model is investigated with considering the friction generated at both ball-screw and guide way. We introduce a two-dimensional frictional model into the mathematical model, which is a function of the sliding velocity and the instantaneous separation of the sliding bodies. Using the developed model, we analyze the behavior of feed drive system by conducting the circular motion tests as well as the simulation of the behaviors of X-axis, Y-axis and quadrant glitch. It is confirmed that the developed model can well express the actual behavior of the X-Y table driven by a ball-screw.

Keywords

Friction, feed drive system, quadrant glitch, X-Y table

1.Introduction

It is always required to improve the working accuracy of NC machine tools. However, effect of the friction generated in the feed drive mechanism is not negligible and the contouring accuracy of NC machine tools affected by the errors such as quadrant glitch generated by the friction force in the circular motion.

Many works have been carried out to reduce the deteriorate effect of friction. One general method is to raise the gain of the NC servo controller. However, too large gain will cause unstable conditions, so such technique is not enough to maintain the higher accuracy.

The other proposal is focused on predicting motion error accurately and to compensate motion error. In this case, the key point is to establish a friction model properly. Lots of applied friction models are commonly described friction as a single-valued, empirically determined function of the sliding velocity[1] such as Coulomb friction.

Friction is also affected by the lubrication condition, surface roughness, etc. Although those models are convenient for simulation and control, they are hardly to depict friction behavior exactly when motion conditions change in large range.

Recently, Polycarpou and Soom have proposed a two-dimensional frictional model at lubricated line contact condition, operating in boundary and mixed lubrication regimes and estimating the friction behavior well over large sliding condition[2)3)]. In this study, a mathematical model of feed drive system for NC machine tools with considering the friction is newly proposed, and its effectiveness is confirmed through experiments.

2. Experimental Equipment

The X-Y table of the NC machine tool is the object of this study. As shown in *Figure 1*, X-axis is mounted on Y-axis, the table is driven by two three-phase AC servomotors through the ball screw. The target position of the table is inputted on the computer installed with Matlab[4)]. The control device includes DSP board and servo amplifier with angular velocity, current and table position feedback as well as various compensations. The control device with sampling time of 300µs is independent to the ability of the personal computer. The experimental data shown in this paper are obtained based on the output of position and velocity of table measured by each linear scale.

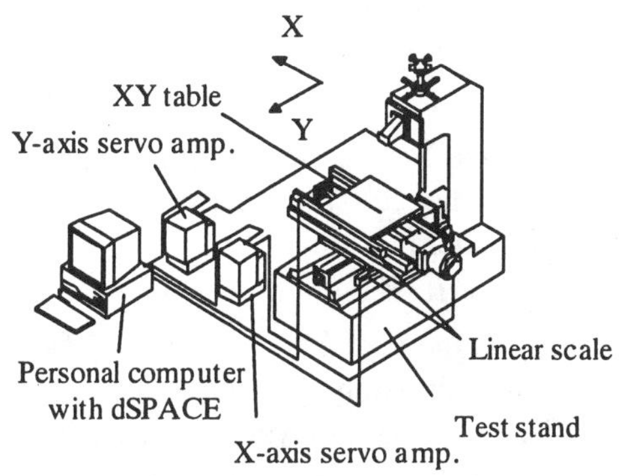

Figure1. Experimental equipment

3. System Dynamic Model

To express the behavior of feed drive system properly, considering the angular displacement of motor and ball screw and the linear displacement of nut and table, a dynamic model with four-degrees of freedom shown in *Figure 2* is adopted in this study. By adding the friction torque of ball crew and friction force of linear guide way, the model is used for the analysis of feed drive system. The motion equations are as follows:

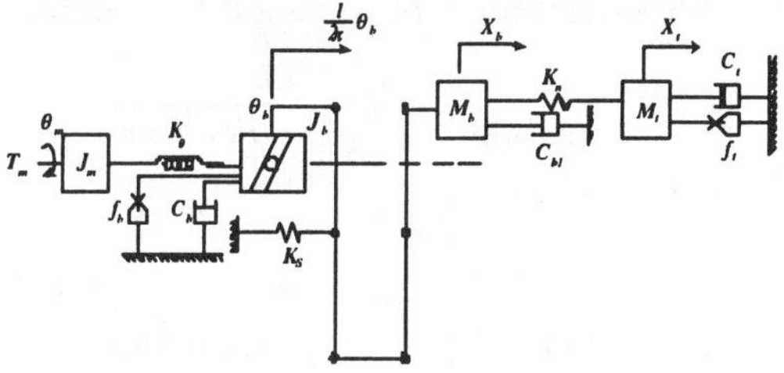

Figure 2. System dynamic model

$$
\left.\begin{aligned}
J_m\ddot{\theta}_m &= -K_\theta(\theta_m - \theta_b) - \frac{K_{cp}+K_e}{L+R_m+K_{cp}}K_t\theta_m + \frac{K_{cp}K_{vp}V_{rg}K_t}{L+R_m+K_{cp}}u \\
J_b\ddot{\theta}_b &= K_\theta(\theta_m - \theta_b) - \frac{l}{2\pi}K_s\left(\frac{l}{2\pi}\theta_b - x_b\right) - C_b\dot{\theta}_b - T_f \\
M_b\ddot{x}_b &= \frac{l}{2\pi}K_s\left(\frac{l}{2\pi}\theta_b - x_b\right) - K_n(x_b - x_t) - C_{b1}\dot{x}_b \\
M_t\ddot{x}_t &= K_n(x_b - x_t) - C_t\dot{x}_t - f_t
\end{aligned}\right\}
\tag{1}
$$

where, T_f is the friction torque of ball screw and f_t is the friction force of guide way.

4. Friction Model

In this study, based on the Polycarpou's friction model, friction coefficient yielded in the X-Y table is calculated from Equation (2),

$$
\mu = C_1\left(\frac{\eta_i}{\eta_o}\right)\left(\frac{V^*}{R^*}\right)^{0.28} \cdot \frac{e^{-C_2\sqrt{v^*}}}{1+C_3\left(\frac{a}{b}\right)^2\left(\frac{\eta_i}{\eta_o}\right)R^*} + \frac{C_4}{1+C_5\left(\frac{a}{b}\right)^2 R^*}
\tag{2}
$$

It shows that besides the sliding velocity, many lubricant parameters have a great influence on the friction coefficient. Based Equation (2), proposed friction model is

$$
f = \mu \cdot W
\tag{3}
$$

where, μ is the coefficient calculated from Equation (2) and W is the normal load. By adjusting the parameters, a suitable friction model can be established for the analysis of feed drive system of the X-Y table.

5. Simulation and Comparisons with Experiment

Combining the friction model with the system model shown in *Figure 2*, both simulation and experiment were conducted to analyze the behavior of the feed drive system. As the circular test for the NC machine tools is a standard method in the analysis of the motion accuracy, the simulations as well as experiments are focus on the circular motion to discuss the working

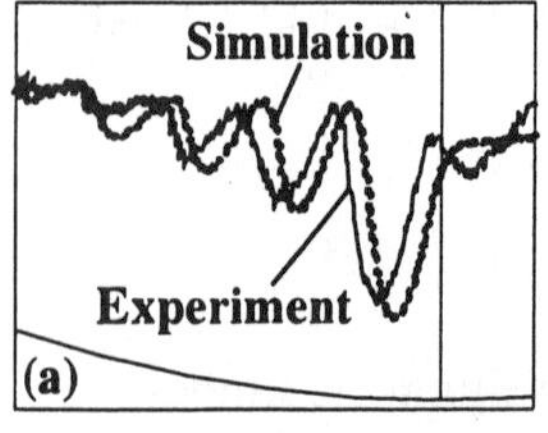

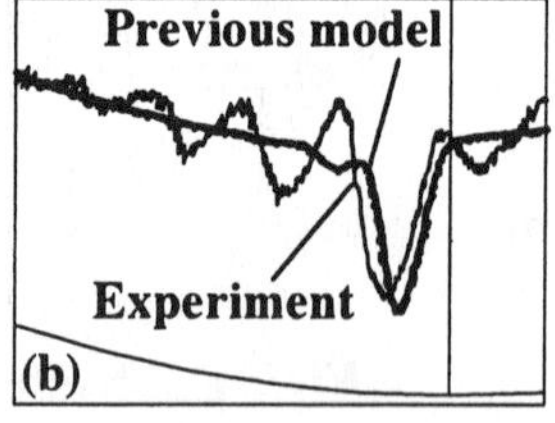

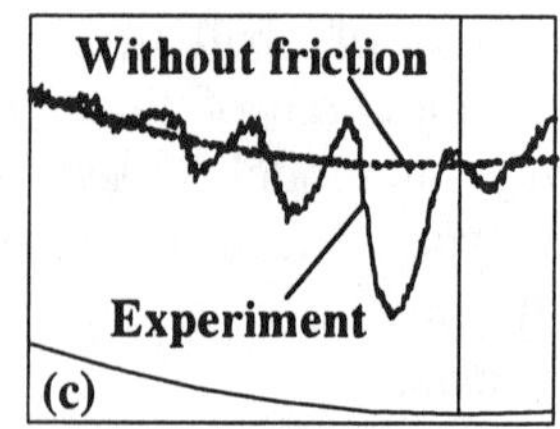

Figure 3. Contouring error

accuracy of the X-Y table.

The experiments were carried out changing the radius and velocity. *Fig. 3* shows the result of circular test with a feedrate of 4800mm/min and radius of 30mm as an example. For comparison, simulation without considering the friction and simulation using our previous friction model[1] were also performed. The enlarged view of around 270° of contouring error are shown in *Figures 3 (a)*, *(b)* and *(c)*. The velocity versus time of X-axis and that of Y-axis are shown in *Figure 4(a)* and the enlarged views around the velocity become zero are shown in *Figure 4(b)*, *(c)* and *(d)*. From these figures, the

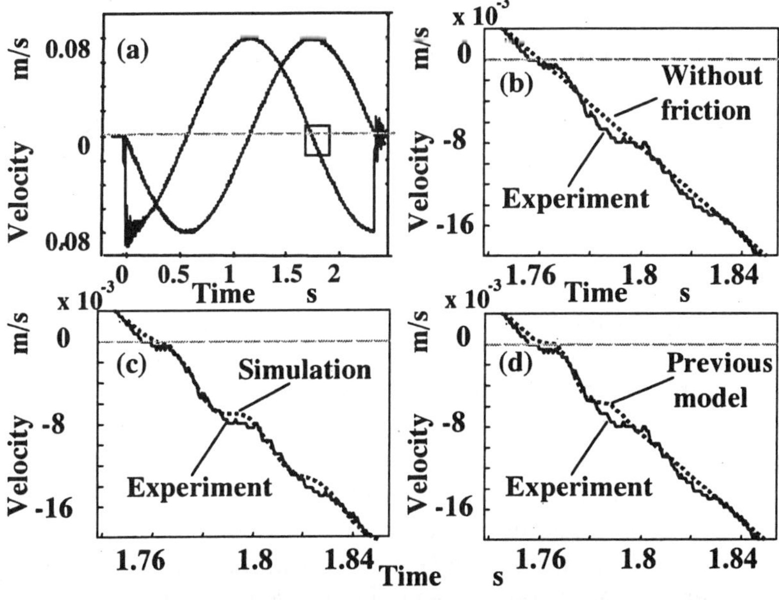

Figure 4. Velocity versus time

following conclusions are drawn.

1) The model developed in this study is effective because it depicts the behavior of feed drive system almost perfectly both in quadrant glitch and velocity of axes; while our previous model is difficult to express the behavior of the oscillation followed the quadrant.

2) When velocity becomes zero, the friction coefficient expressed by Equation (2) becomes large and the velocity has some changes in short interval. This is the reason of quadrant glitch generated at circular motion. The result of simulation without considering the friction cannot express the quadrant glitch.

The influence of feedrate on the quadrant glitch was investigated by changing the feedrate from 1200mm/min to 7200mm/min.

Figure 5 shows the relationship between the feedrate and the amplitude of quadrant glitch at $180°$ and $90°$. It can be seen from the experimental results that the amplitude of the quadrant glitch increases with the increasing of the federate and when the feedrate is over about 6000mm/min, the amplitude decreases. The simulation does not depict the trend completely that there is any shortage in Polycarpou's friction model.

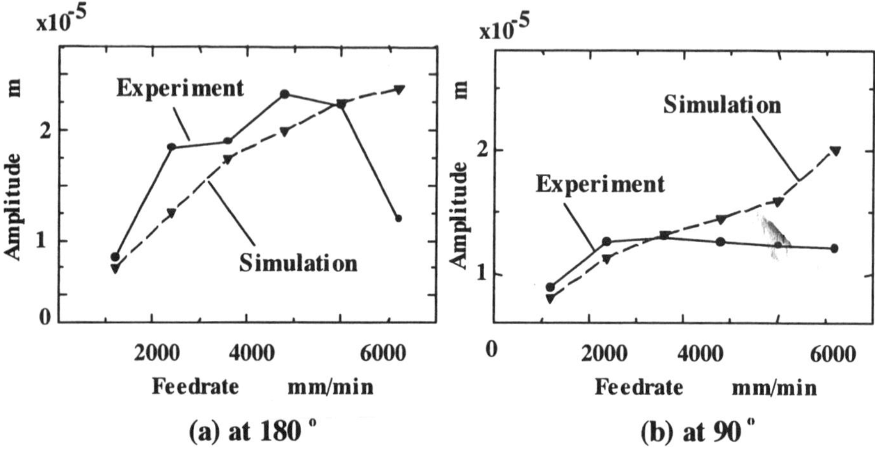

Figure 5. Relationship between feedrate and amplitude of quadrant glitch

6. Conclusion

In this study, a mathematical model with considering the friction yielded in ball screw and guide way for feed drive system of X-Y table was developed, and the behavior of circular motion was analyzed. From the fact that simulated results are consistent with those of experiment, it can be said that the developed model is effective, and the model may be used for compensating the friction of X-Y table driven by ball screws.

References

1) M.Tsutsumi, S.Ohtomo, Y.Okazaki, K.Sakai, K.Yamazaki, D.F.Ge, Mathematical model of feed drive mechanical system and friction for NC machine tools. JSPE, J. 61, 10, 1458-1462   (1995)

2) A.A. Polycarpou and A. Soom, Two-dimensional models of boundary and mixed friction at a line contact. Trans. ASME, J. Tribology, 117, (1995), 178-184

3) A.A. Polycarpou and A. Soom, A two-component mixed friction model for a lubricated line contact. Trans. ASME, J. Tribology, 118, (1996), 183-189

4) *Matlab; the language of technical computing*, the Math Works Inc. 1997.

ROTATIONAL ACCURACY AND POSITIONING RESOLUTION OF AN AIR-BEARING SPINDLE WITH ACTIVE INHERENT RESTRICTORS

Hiroshi MIZUMOTO,   Shiro ARII

Tottori University, Koyama 680-8552 Tottori, Japan

Makoto YABUYA

Nachi-Fujikoshi, Nameirkawa 936-0802 Toyama, Japan

Abstract

An active control system is incorporated into an air-bearing spindle designed for ultraprecision applications. The system uses the Active Inherent Restrictor (abbreviated AIR) for improving the rotational accuracy of the spindle. The AIR consists of a piezoelectric actuator having a hole, one end of which on the bearing surface acts as an inherent restrictor.   According to the spindle vibration detected, the AIR controls the airflow rate, and reduces the amplitude of the vibration to less than 10nm.   The control system also has the ability of positioning the spindle; one nanometer of step is clearly resolved.   Therefore, this air-bearing spindle with the AIR can be used as an ultraprecision positioning device.

Keywords

active control, air-bearing spindle, nano-technology, positioning, rotational accuracy

1. INTRODUCTION

The air-bearing spindle is key equipment for ultraprecision machine tools and measuring machines used in the field of nano-technology.   In the present paper, an air-bearing spindle having an active control system is proposed.   This control system controls the airflow to both radial and thrust bearings of the proposed air-bearing spindle by using the Active Inherent Restrictor (abbreviated AIR) invented by the authors.   It has been shown that the AIR is effective for improving the stiffness and the rotational accuracy of an aerostatic bearing [1]. The AIR was also incorporated into the aerostatic guideway of an ultraprecision machine tool to improve the straightness of the table motion [2].   The AIR consists of a piezoelectric actuator having a through hole, one end of which is small enough to function as an orifice when the actuator is embedded in the bearing surface.

Besides the active control system, the proposed air-bearing spindle has several features for improving the performance. First feature is that the thrust bearing is an opposed-pad type; the upper bearing is designed to have high stiffness, while the lower bearing has not restrictor and its bearing stiffness is zero. Second feature is that the radial bearing has a single-row of restrictors and its angular stiffness is zero. Experimental analyses for rotational accuracy and ultraprecision positioning of the air-bearing spindle are reported.

2. MECHANISM OF AIR-BEARING SPINDLE

Figure 1 shows the proposed air-bearing spindle with the AIR. The thrust bearing is an opposed-pad type; there are eight AIRs and displacement sensors on the upper bearing pad, while there is no restrictor on the lower bearing pad. Such structure reduces the influence of geometrical error of the lower thrust bearing on rotational accuracy because the stiffness of the lower bearing is zero. The radial bearing has a single-row of eight AIRs. Therefore, the radial stiffness of this bearing is enough for supporting the spindle, however, the angular stiffness is zero. Such structure reduces the interference between the thrust and radial bearings.

The mechanism of the AIR is shown schematically in Fig.2. Pressurized air is

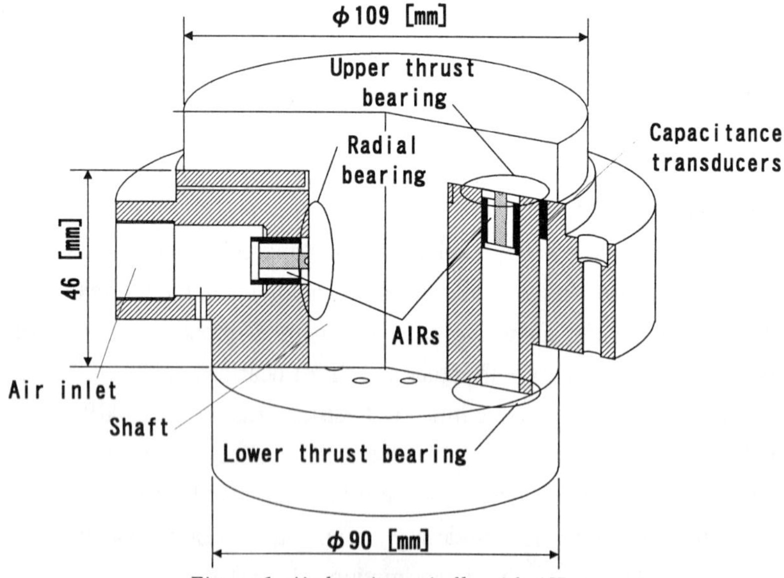

Figure 1. Air-bearing spindle with AIR

supplied through a piezoelectric actuator having a hole, one end of which on the bearing surface acts as an inherent restrictor. Orifice area of this restrictor is given by $\pi \cdot d \cdot h_d$, where d is the hole diameter and h_d is the restriction gap. According to the displacement of the spindle (change in the bearing gap h), the restriction gap h_d can be changed by

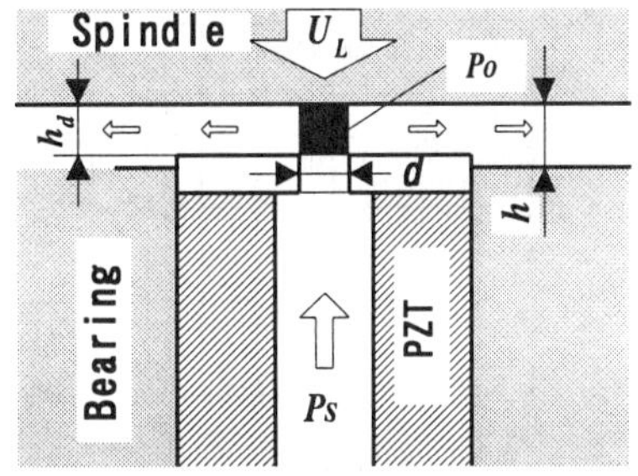

Figure 2. Mechanism of AIR

controlling the length of the piezoelectric actuator. When an additional load decreases h, h_d is increased to increase the bearing pressure Po, then the spindle returns to the original position and h is be kept constant. Consequently, the stiffness and the rotational accuracy of the spindle can be improved. The AIR also has the ability of positioning the spindle. By changing the length of the piezoelectric actuator, the bearing gap h can be changed and the spindle can be moved to a desired position. This means that the AIR moves the spindle by a distance proportional to the deformation of the piezoelectric actuator.

3. ROTATIONAL ACCURACY

The control system for the air-bearing spindle is shown in Fig. 3. In the thrust direction, vibration of the upper thrust plate is detected by a capacitance sensor embedded in the bearing surface. Detected vibration is A/D converted and input to a microcomputer. The computer calculates the control voltage for the piezoelectric actuator by using a PI algorithm. The calculated voltage is D/A converted and supplied to the actuator to reduce the vibration. The radial vibration is measured at a reference ball on the thrust plate. By using the same process as in the thrust direction, the radial vibration can be reduced.

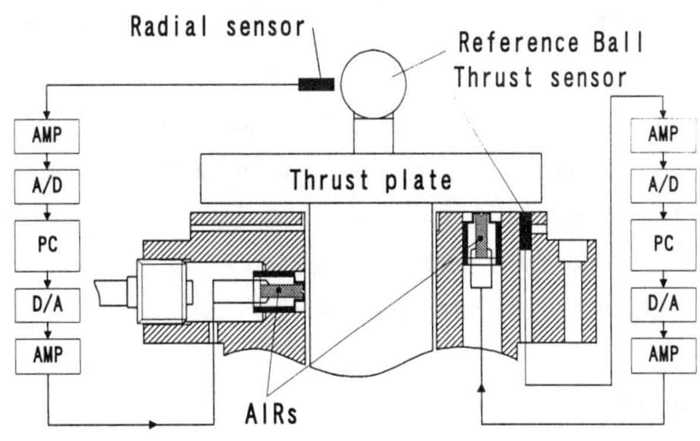

Figure 3. Control system for air-bearing spindle

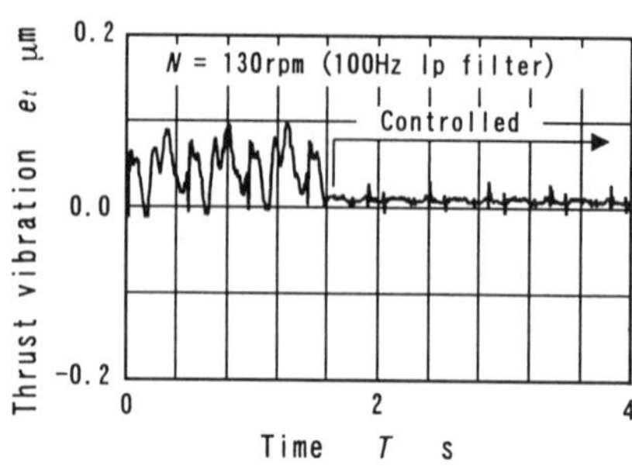

Figure 4. Effect of AIR on thrust vibration

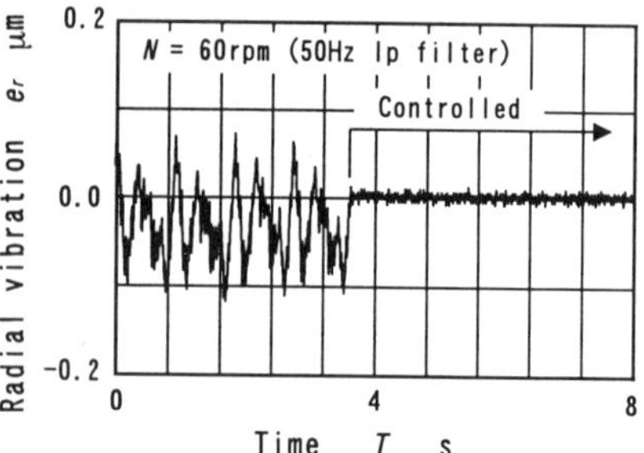

Figure 5. Effect of AIR on radial vibration

The effect of the AIR on the rotational accuracy in thrust direction is shown in Fig. 4, where N is the rotational speed and the low-pass filter is used.   Without active control, the amplitude of the spindle vibration measured on the thrust bearing surface is about 0.1µm.   After the active control starts, the amplitude of the vibration is reduced to less than 10nm.   Figure 5 shows the effect of the AIR on the radial vibration.   In the radial direction, the amplitude of the vibration under control is also less than 10nm.   Thus the rotational accuracy can be improved by the AIR.

Figures 4 and 5 show that the waveform of the vibration without the active control is not a simple sinusoidal curve but a complex curve having two peaks per one revolution of the spindle. Such complex vibration is caused by the interference between the conical motion of the spindle on the radial bearing and the inclination of the spindle on the thrust bearing.   Both conical motion and inclination show one cycle per one revolution of the spindle, however, there is a phase lag between these motions in general.   As a result of this phase lag, the vibration with two peaks occurs.   In the proposed spindle, the radial bearing has no angular stiffness because of its single-row of restrictors.   Therefore, if there is error in the perpendicularity between the radial and thrust bearing surfaces of the spindle, the inclination of the spindle is reduced by the AIR of the thrust bearing, however, the spindle rotates in the conical mode.   As shown in Fig. 3, the AIR of the radial bearing compensates the vibration at a cross section of the spindle where the reference ball is attached. Consequently, even if the spindle rotates in the conical mode, the rotational accuracy in the radial direction at the reference ball can be improved.

4. POSITIONING

As mentioned above, the control system shown in Fig. 3 can move the spindle to a desired position.   The control computer detects the difference between the current

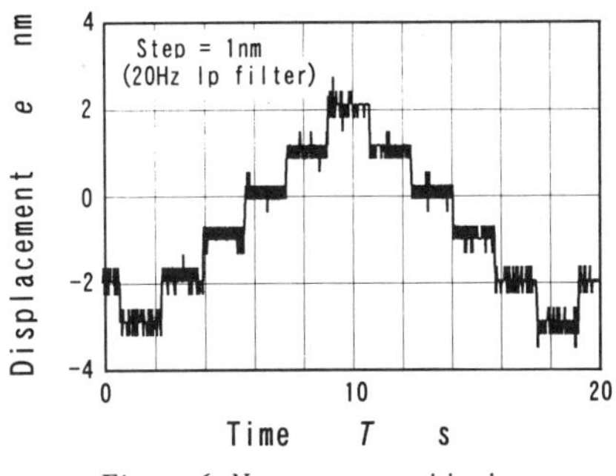

*Figure 6. Nanometer positioning
in thrust direction*

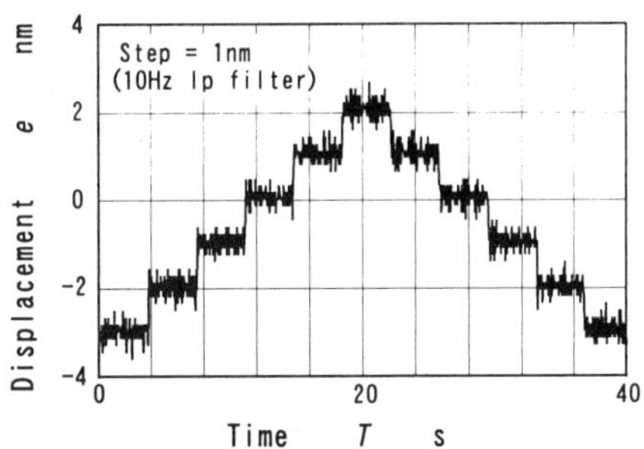

*Figure 7. Nanometer positioning
in radial direction*

and the desired positions of the spindle, then outputs control voltage proportional to the difference in positions to the actuator. Thus the positioning of the spindle can be realized. Figure 6 shows the result of step positioning in thrust direction, where the spindle is programmed to move by one nanometer and stop for a while. Such a step is repeated five times in upward, then in downward. It is shown in Fig. 6 that each step is clearly resolved and the positioning resolution is at least one nanometer. The result of positioning in the radial direction is shown in Fig. 7. In the radial direction, noise level is slightly higher than in the thrust direction, however, one nanometer of step is also clearly resolved. Such ability of positioning is useful when adjusting the spindle for micro machining or ultraprecision measurement.

5. CONCLUSIONS

The effect of active control using the Active Inherent Restrictor on the performance of the air-bearing spindle is analyzed and following results are obtained:

(1) Rotational accuracy of the spindle with the active control can be less than 10nm in both thrust and radial directions.

(2) This air-bearing spindle with the AIR can also be used as an ultraprecision positioning device. Positioning resolution is at least 1nm in both thrust and radial directions.

6. REFERENCES

1) H. Mizumoto, S. Arii, Y. Kami, K. Goto, T. Yamamoto and M. Kawamoto: Active Inherent Restrictor for Air Bearing Spindles, Precision Engineering, **19**, 2/3 (1996) 141.

2) Mizumoto H., S. Inoue, S. Arii, Y. Kami and M. Yabuya: An Ultraprecision Machine Tool with Active Control Aerostatic Guideways for Improving Quality of Machined Surface, Proc. of 9th ICPE (1999) 165.

PRECISION CONTROL OF
RADIAL MAGNETIC BEARING

Xiaoyou Zhang[†], Tadahiko Shinshi[††], Lichuan Li[††],
Kee-Bong Choi[††] and Akira Shimokohbe[††]
[†] *Interdisciplinary Graduate School of Science and Engineering,
Tokyo Institute of Technology*
[††] *Precision & intelligence Laboratory, Tokyo Institute of Technology*

Abstract

The aim of the research is to realize a radial magnetic bearing with the rotational accuracy of several nanometers. In this paper, a partial bias current method for driving electromagnets is introduced. The heat generation by the method is smaller than that by a conventional bias current method. Furthermore, the method has higher force-slew-rate than a zero bias current one at the zero force point. Then a repetitive controller is applied to eliminate the periodic runout caused by the unbalance and the profile error of the spindle. The experimental results show that the partial bias current method has the merits of precision positioning capability, low heat generation and low rotational electromagnetic damping. The repetitive controller can eliminate the periodic runout and achieve the rotational accuracy of 15.1nm at 1500min^{-1}.

Keywords

Magnetic bearing, Electromagnet, Bias current, Rotational accuracy, Repetitive control, Rotational loss

1. INTRODUCTION

Magnetic bearings can compensate spindle motion errors caused by unbalance and disturbance forces because of their active motion control capability. Moreover the magnetic bearings are free from friction and can be used in a vacuum so that they can be candidates for high accurate and high-speed bearings in the future. However few magnetic bearings of nanometer rotational accuracy have been reported [Shinshi et al., 2000].

The purpose of this research is to realize the precision magnetic bearing. In this paper, first, an experimental radial magnetic bearing mechanism is introduced in which roundness of the spindle and internal surfaces of the electromagnets are accurate. Secondary, a partial bias current method for driving electromagnets is introduced. The heat generation by the method is smaller than that a conventional bias method. Furthermore, the method has higher force-slew-rate than a zero bias one at the zero force point. The rotational accuracy, rotational electromagnetic damping and heat generation of the partial bias current method are experimentally compared with those of conventional bias and zero bias methods. Then, the repetitive controller is applied to eliminate the periodic runout and improve the rotational accuracy.

2. EXPERIMENTAL RADIAL MAGNETIC BEARING

An experimental radial magnetic bearing is shown in Figure 1. The spindle motions in two radial directions are controlled by four pairs of electromagnets. The axial and two tilting motions are constrained by the thrust air bearings. The maximum diameter of the spindle is 80mm and its mass is 3.29kg. The radial displacements are measured by two capacitance displacement sensors (ADE, Microsense II 5130, 0.2mm measuring range).

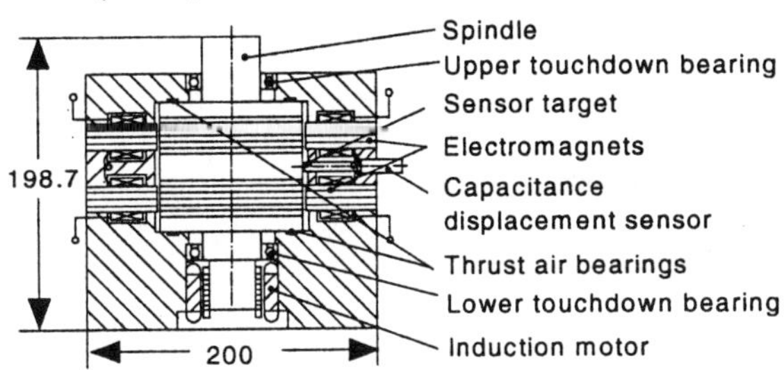

Figure 1 Scheme of experimental magnetic bearing

3. CONTROL SYSTEM

Usually electromagnets of the ordinary magnetic bearings are driven with a bias current, as shown in Figure 2(a). The copper loss caused by the bias current results in the heat generation which reduces mechanical accuracy of the magnetic bearing.

One simple solution is to drive the electromagnets with a zero bias current, as shown in Figure 2(b). The heat generation of the electromagnets can be minimized. However, the drawback is the zero-force slew rate at zero-force point [Maslen et al., 1996]. To overcome the zero-force slew rate and reduce the heat generation of the electromagnets, the partial bias current method is proposed, as shown in Figure 2(c).

The three bias current methods with an optimal state feedback controller are applied to the magnetic bearing, as shown in Figure 3. The nonlinear compensator is applied to compensate the nonlinear relationships of the

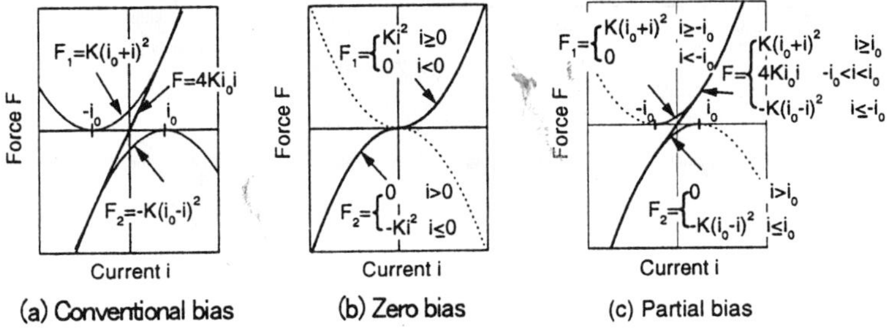

Figure 2 Current-force characteristics

715

force/current/airgap of the zero bias and the partial bias methods. The gains from the input V of the nonlinear compensator to the output F of the electromagnet are the same on the three bias current methods. The optimal controller works so as to make the output of the displacement sensor constant.

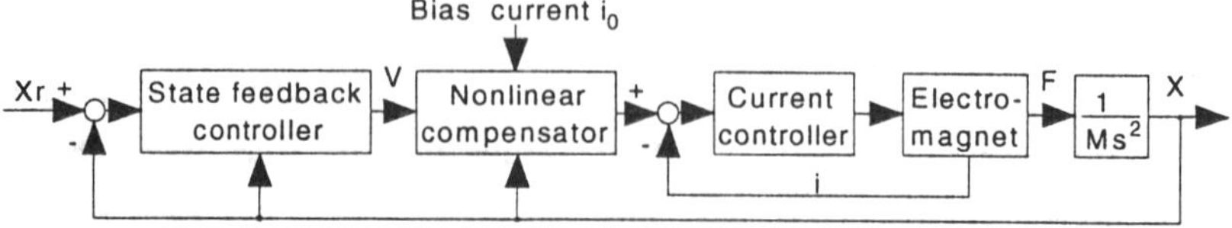

Figure 3 Block diagram of the magnetic bearing system

4. PERFORMANCE TEST

4.1 Rotational accuracy

The rotational accuracies using the conventional bias (i_0=4.0A), the zero bias and the partial bias currents (i_0=1.0A) are experimentally compared. Figure 4 shows the outputs of the displacement sensor in X direction. The outputs in Y direction are similar to those in X direction. The rotational accuracy is expressed by 3σ of the runout $e=(x^2+y^2)^{1/2}$, where x and y are the outputs of the displacement sensors. The rotational accuracies using the conventional bias and the partial bias are approximately 10nm at 500 min^{-1}, which are higher than that using the zero bias. The rotational accuracies decrease due to the unbalance forces as the rotational speed increases.

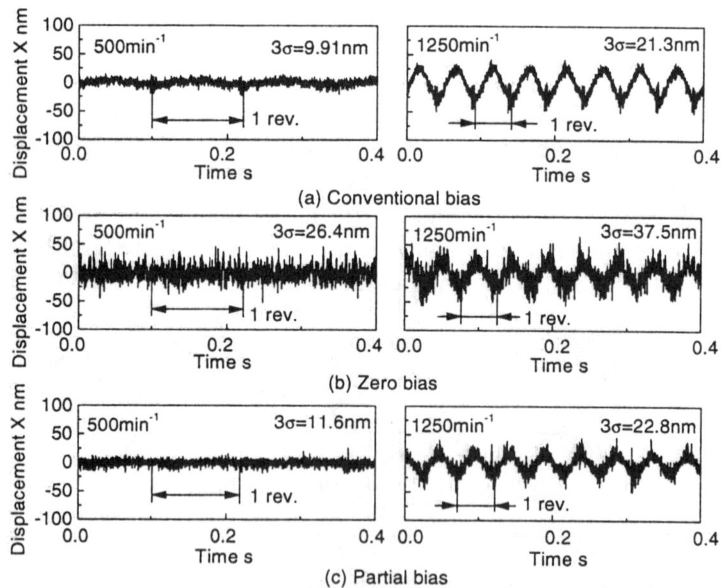

Figure 4 Displacements of the spindle in X direction

716

4.2 Rotational loss

In order to measure the rotational electromagnetic damping of the spindle, the rundown characteristics of the spindle are measured, as shown in Figure 5. The dynamics of the spindle in the rotational direction is given by Equation (1):

$$J\dot{\omega} + C\omega = 0 \qquad (1)$$

Where ω is the rotational speed, J is the moment of inertia, and C is the damping coefficient. Using the rundown characteristics and Equation (1), the damping coefficients of three bias current methods are identified. The damping coefficient using the conventional bias is twice as much as those using the zero bias and the partial bias.

Moreover, the temperature of the coils is measured to evaluate the heat generation of the electromagnets. Using the conventional bias, the temperature of the coils rises by 5.4°C within 8 minutes. On the contrary, the temperature using the zero bias and the partial bias rises by 0.5°C within 2 hours.

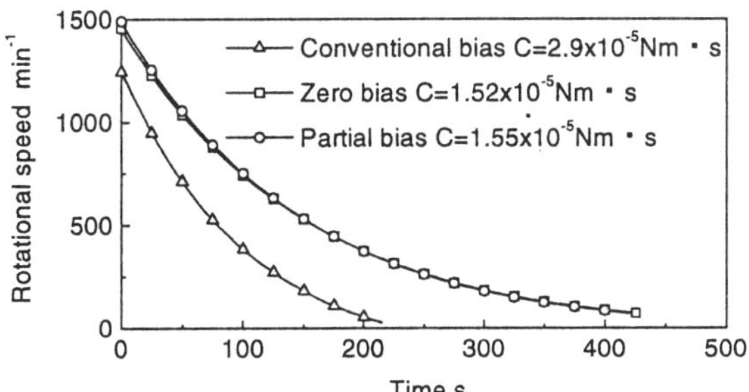

Figure 5 Rundown characteristics

5. IMPROVEMENT OF ROTATIONAL ACCURACY

5.1 Repetitive controller

The periodic runout is caused by the unbalance of the spindle as the rotational speed increases, as shown in Figure 4, which can be eliminated by repetitive control. The spindle does not rotate at a constant speed due to air turbine driving. Therefore, a repetitive controller depending on the rotational angle [Horikawa and Shimokohbe, 1992] is applied to the magnetic bearing. The block diagram is shown in Figure 6. λ is a delay operator(=z^{-1}). The delay time of λ depends on the rotational speed. $F_n(\lambda)$ is a linear phase filter, which stabilizes the system.

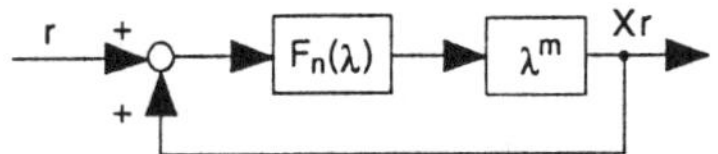

Figure 6 Block diagram of the repetitive controller

717

5.2 Rotational accuracy using repetitive controller

The outputs of the displacement sensor at 1500min^{-1} are shown in Figure 7. The periodic runout is eliminated, and the rotational accuracy is improved from 28.8nm to 15.1nm. Figure 8 shows the rotational accuracy from 0min^{-1} to 1500min^{-1}. The rotational accuracies over 500min^{-1} are improved by the repetitive control.

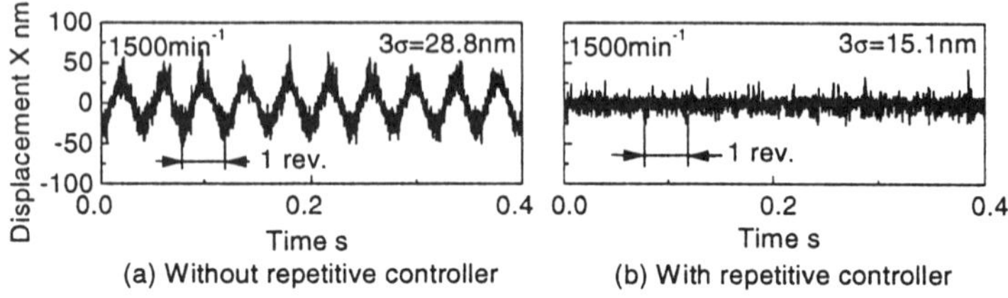

Figure 7 Displacements of the spindle in X direction with the partial bias

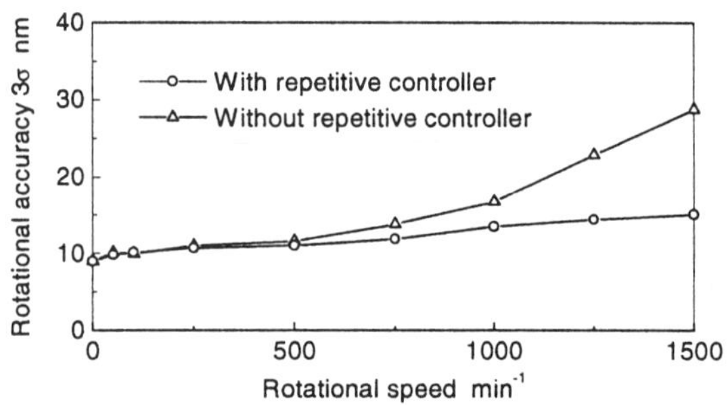

Figure 8 Rotational accuracies with the partial bias

6. CONCLUSIONS

To realize a precision magnetic bearing, the partial bias current method having the merits of precision positioning capability, low heat generation of electromagnets and low rotational electromagnetic damping, was proposed and tested. The experimental radial magnetic bearing using the partial bias current method and the repetitive controller attained the rotational accuracy of 15.1nm at a speed of 1500min^{-1}. This research is supported in part by Grant-in-Aid for Scientific Research No. 11555049 from JSPS.

REFERENCES

Horikawa O., Sato K., Shimokohbe A. An active air journal bearing, Nanotechnology 1992; 3: 84-90

Maslen E. H., Allaire P. E., Noh M. D., Sortore C. K. Magnetic bearing design for reduced power consumption. Trans. of the ASME, Journal of Tribology 1996; 10: 839-846

Shinshi T., Iijima C., Zhang X., Choi K., Sato K. Shimokohbe A. Precision radial magnetic bearing. Proceedings of the ASPE Annual Meeting; 2000 October 22-27; Scottsdale, Arizona.

THE EFFECTS OF JOINT CLEARANCE ON THE MOTION PRECISION OF THE MACHINES

Jia Xiaohong   Ji Linhong   Jin Dewen   Zhang Jichuan
School of Mechanical Engineering, Tsinghua University, Beijing, China, P. R.

Abstract

Joint clearance is one of important factors which affect motion precision of machines. In this paper the mathematical model of a new-type joint, tripod-ball sliding joint, is established, and the new concepts of basal system and active system are put forward. Based on the mode-change criterion established here, the consistent equations of motion in full-scale are derived by using Kane method. The dynamic behavior of the crank-slider mechanism with clearance in the tripod-ball sliding joint is investigated. The results including motion error and impact acceleration are obtained and the advantage of this new joint is shown.

Keywords

Clearance, dynamics, motion precision

1. INTRODUCTION

Achieving high accuracy at low cost is one of the greatest challenges in the manufacturing industry. The most important factors affecting that accuracy are the static and dynamic stiffness of the machine, which significantly depend on precision of the machine itself. Joint clearance plays an important role in the analysis of machining precision [Wang Jinsong, 1999].

Small joint clearances in the machines are necessary for assembly and mobility, and it will become larger with wear or thermal deformation. When clearance size is not suitable, not only additional displacement but also impact force will be caused, and as a result, system precision will be reduced [Jungkeun R., 1996].

The tripod-ball sliding joint (TBS joint), shown in Figure 1, is a new-type prismatic joint, which has an attractive friction coefficient as small as 0.003. Owing to its high transmission efficiency, long life and low wear, it shows good prospects in application in high precision machines. However, the joint clearance is one of fundamental obstructions of the improvement of motion accuracy [Jia Xiaohong, 1999].

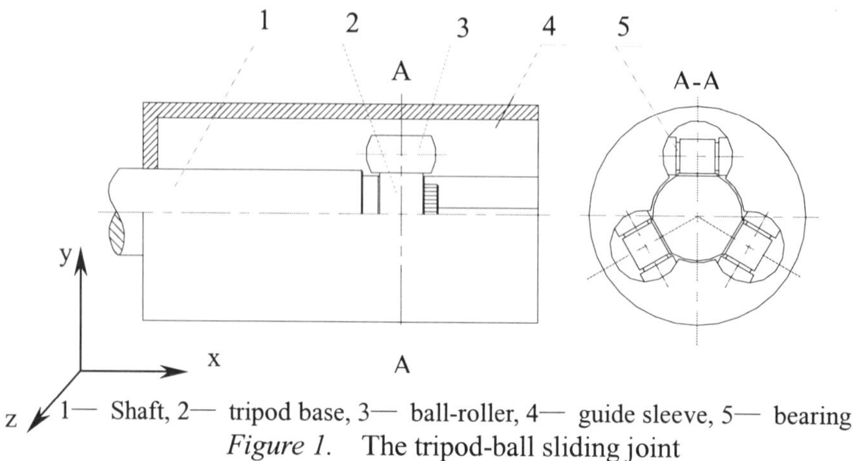

1— Shaft, 2— tripod base, 3— ball-roller, 4— guide sleeve, 5— bearing

Figure 1. The tripod-ball sliding joint

2. MATHEMATICAL MODEL

The discontinuous contact model is used to describe the dynamics of the joint with clearance, which consists of free-flight mode and following mode. The contact force between the two joint elements can be obtained as follows.

$$p = \begin{cases} 0 & \text{(free - flight mode)} \\ \dfrac{2}{3(k_1 + k_2)} \cdot (\dfrac{\pi m_a}{k(e)})^{3/2} \cdot (\dfrac{1}{\Sigma \rho})^{1/2} \delta^{3/2} + \mu \delta \dot{\delta} & \text{(following mode)} \end{cases} \quad (1)$$

Where:

δ, $\dot{\delta}$ — the deflection of the contact surface and its one-order derivative

$k(e)$, m_a $\Sigma \rho$ — some parameters obtained from the geometry of the contacting joint elements

$k_i = 1 - v_i^2 / E_i$, E_i — the Young's modulus, v_i — Poisson ratio

μ — a material property, independent of surface deformation

When any group of ball-roller of the TBS joint doesn't contact with the groove, the system is defined as basal system, otherwise it is the active system. Due to the contact force p_i between the joint elements i-th, a set of extra constraint equations can be established as follows, and be written as equation (2).

$$\Phi(\dot{u}_r, u_r, q_r, t) = 0 \quad (2)$$

Taking a crank-slider as a mechanism being investigated. In fact, when the crank rotates continuously, contact mode of the TBS joint will be changing frequently and unpredictably. The basal system remains invariable unless the constraint equations take effect, and then the active one will be aroused. In order to distinguish whether the constraint equations is living or not, a special criterion λ was defined as equation (3), which is determined by the kinetic parameters and external forces of joint elements.

$$\lambda_i = \begin{cases} 1 & (\delta_i \geq 0) \\ 0 & (\delta_i < 0) \end{cases} \qquad \lambda = \Sigma\, \lambda_i \qquad (3)$$

With the help of the mode-change criterion λ, a set of consistent equations of motion for both free-flight and following mode can be obtained, rather than two different sets of equations in other publications [Stammers C. W., ,1991 and Miedema B., 1976], as shown in following:

$$M(\dot{u}_r, u_r, \ q_r, \ t) + \lambda\Phi(\dot{u}_r, u_r, \ q_r, \ t) = 0 \qquad (4)$$

3. NUMERICAL AND EXPERIMENTAL RESULTS

The implicit variable-step Runge-Kutta method was employed to solve the motion equations (4), the initial time step was typically 10^{-5} s. The initial values were obtained from the nominal mechanism which has no clearance under the same conditions.

No numerical problems were encountered as the program switched from one mode to another mode, by virtue of two reasons, one is the small time step, and the other is using a nonlinear spring-dashpot model while impact happens which improve the stability of the mode-change effectively.

The effects of the clearance on the motion precision were studied comprehensively. The results of motions errors around z-axis and the effect of clearance size is shown in Figure 2 and Figure 3. The theoretical response shows good agreement with the experimental response to some extent.

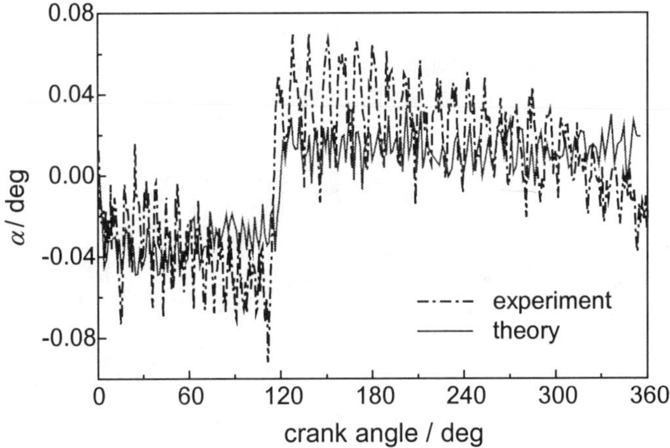

Figure 2   Motions errors

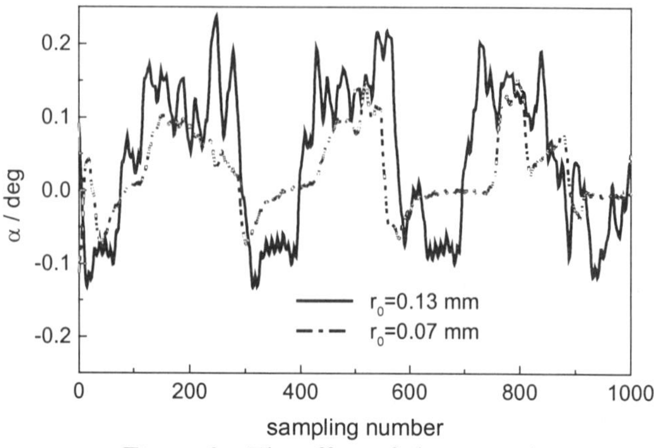

Figure 3   The effect of clearance size

Moreover, the results about joint force gotten here have much difference from dynamic behavior of other normal joint [Bahgat, 1979]. Figure 4 shows the resultant force F_y of the TBS joint along y-axis and Figure 5 shows the contact force F_a of joint elements of group a-th. Obviously, even though there is remarkable variation of F_a when sudden shock happens in motion error in Figure 2, F_y has no great impact. This is very useful for motion stability. The TBS joint has an advantage to improve the dynamic behavior of the mechanism. It is because the geometric structure of the TBS joint improves the distribution of forces of the system and impacts caused by clearance are reduced.

According to the analysis of the contact state of each group of joint elements in one cycle of crank rotation, it can also be found that there are at least two groups in contact at any angle of crank rotation, and the complete breakaway between joint elements never occurs. As a result, no large impact will take place. This agrees with the result above.

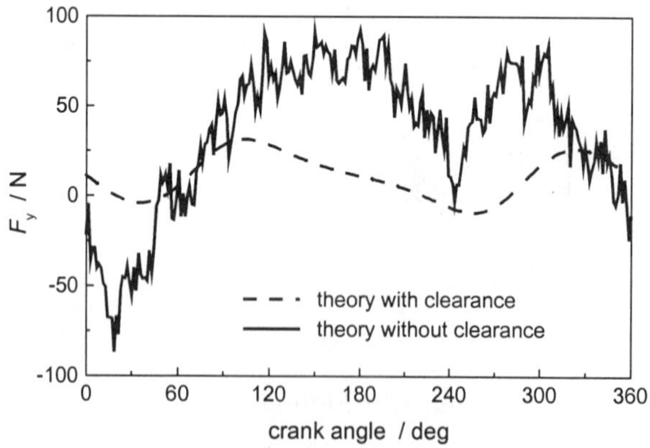

Figure 4   The resultant force F_y

722

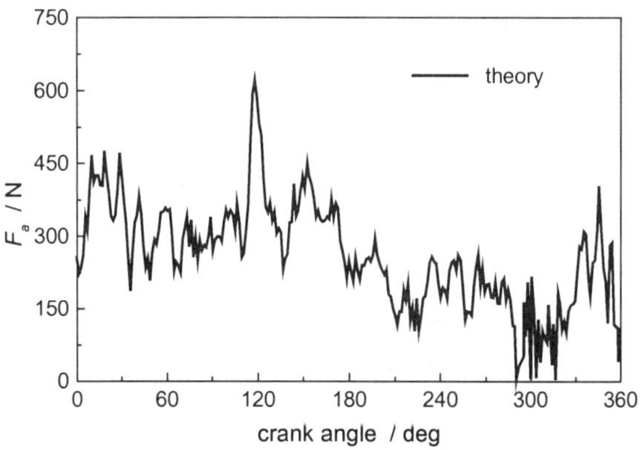

Figure 5   The contact force F_a

The mechanism with clearance is a typical nonlinear system, and there will be chaotic response over a special range of parameters, details can be obtained in Ref. [Jia Xiaohong, 1999].

4. CONCLUSINOS AND PROSPECTS

(1)   The influence on motion precision of the joint clearance was investigated, and some quantitative results were obtained.

(2)   The mathematic model of the new-type tripod-ball sliding joint with clearance established in this paper is effective to study its kinematic and dynamic effects. The results verify that the TBS joint has the advantage to meet the command of high precision.

(3)   The consistent equations for both free-flight and following mode are presented by defining a mode-change criterion. The method may be extended for mechanisms with multiple clearance joints or other complicated systems, and thus it has universal significance.

References:

1. Bahgat B. M., Osman M., Sankar T. S. On the effect of bearing clearances in the dynamic analysis of planar mechanisms. J. Mech. Engng. Science 1979; 21(6): 429-437
2. Jia Xiaohong. Investigation on the dynamics of the universal transmission system of high-speed locomotive and tripod-ball sliding joint with clearance. PhD Dissertation at Tsinghua University 1999
3. Jungkeun R., Adnan A. Dynamic response of a revolute joint with clearance. Mech. Mach. Theory 1996; 31(1): 121-134
4. Miedema B., Mansour W. M. Mechanical joints with clearance: a three-mode model. ASME J. Engng. Ind. 1976; 11: 1319-1323
5. Stammers C. W., Ghazavi M. A theoretical and experimental study of the dynamics of a four-bar chain with bearing clearance: pin motion, contact loss and impact. Journal of Sound and Vibration 1991; 150(2): 301-315
6. Wang Jinsong, Huang Tian, Yang Xiangdong. Design theory and methodology of 6-THS parallel machine tools. Journal of Tsinghua University 1999; 8: 25-29

TOOL TRAJECTORY ERROR FOR TABLE-TILTING MACHINES

Yean-Ren Hwang and Ming-Che Ho

Dept. of M.E., National Central University

Chung-Li, Taiwan, R.O.C

Abstract

Five-axis machining is the primary method for generating complex surfaces, such as the surfaces of turbine blades. These surfaces are machined by end-mill cutters along specified tool paths, which are generated only based on the geometry of the sculptured surface regardless the structural parameters. The step length of the tool paths directly affects the accuracy of the manufactured surface and machining time. Our research shows the trajectory of the cutters (or cut error) varies with different structures of machines, rotation angles, and rotation radii. Compared to traditional algorithms, simulation results show that the cut errors are reduced using our algorithm.

Keywords

Computer aided manufacturing, five axis NC, cut error, step length

1. INTORDUCTION

Five-axis machining has several advantages over three-axis machining including better surface finishing, higher removal rate and less setup time

[1-2]. Current CAM algorithms have three major stages to generate NC machine codes, i.e. generating CC points, generating cutter location data and post-processing. For a specified cut tolerance, the main concern of tool path generation is to have as large a step length as possible to reduce the machining time. With different 5-axis structures, most CAM algorithms generate the same CC, CL points and CO angles at first two stages. Hwang and Liang [3] pointed that the cut error will be different even for the same CC points for different structures of five-axis machines.

For three-axis machining, the cut trajectory is a straight line passing through the two consecutive CC points, and the curve between two CC points can be approximated as an arc with the radius R. [4-5]. For a specified tolerance, e_{max}, the step length s_f is estimated as $\sqrt{8Re_{max}}$. For five-axis NC machines, the cut trajectory is not a straight line although the tool holder movement is a straight line. Under the same machining conditions (i.e. same arc curvature, step length, tilt and lead angles), Figure 1 shows that the cut error for spindle-tilting type machines is independent of the position of the machined part [3]. However, Figure 2 shows that, the cut error is larger for the curve far away from the rotational axis.

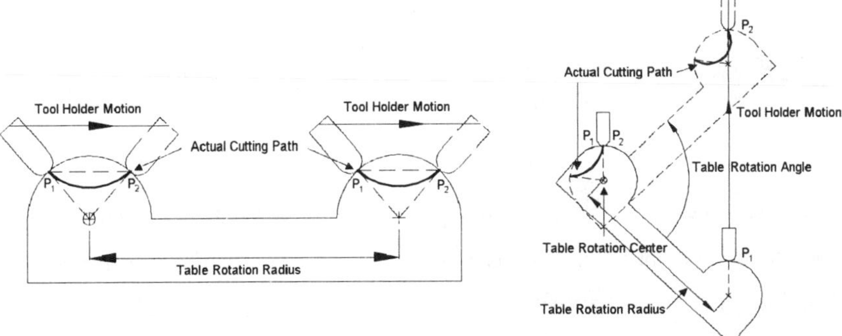

Figure 1. Tool movement of spindle-tilting type NC machining	*Figure 2.* Tool movement of table-tilting type NC machining

2. STEP LENGTH ESTIMATION

To obtain the desired tool angles and the cut error for the table tilting machining, we offer several steps:

(1) Calculate the roll, pitch, and yaw angles of tool position vector at each CC points (based on machine coordinate system).

(2) Since the tool axis of machine is fixed in Z-axis, the machined part needs to be rotated in order to obtain desired tool angle.

(3) Approximate the actual cut trajectory between two consecutive CC points.

(4) Calculate the maximum cut error and maximum allowable step length.

2.1 Variable Lead Angles Machining

For many manufacturing process, the tilt angles are kept zero and the tools are kept with specified lead angles to normal of the machined surface. Therefore, the step length estimation result for variable lead angle machining is as follows:

$$s \approx R_c \Delta\alpha + \sqrt{\frac{R_c^{2}\left[8e_{max} + (R_c + \rho)\Delta\alpha\left|\sin 2\alpha_1\right|\right]}{R_0\left|\sin(\hat{\beta} + \alpha_1)\right|}} \text{, Where } \hat{\beta} = -\theta\!\!\big/\!\!_2 - \xi - \alpha_1 + \beta \quad (1)$$

From the above equation, the step length depends on e_{max}, R_c, R_0, ρ, α_1, $\Delta\alpha$, and $\hat{\beta}$. We say that R_0 is the rotational distance between the rotational axis and the approximated arc, and $\hat{\beta}$ is the relationship of rotational angles. The tool's corner radius, ρ, can be considered fixed value although it may have small deviation due to tool wear. The radius of the approximated arc between two consecutive CC points, R_c, is the reciprocal of the curvature along the cut trajectory. The detail definitions are shown in Figure 1.

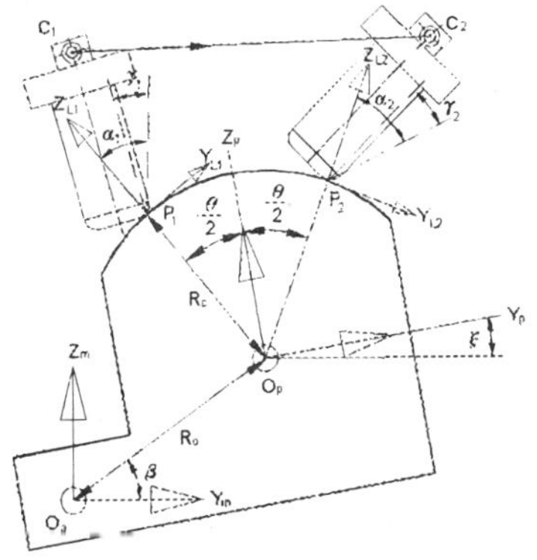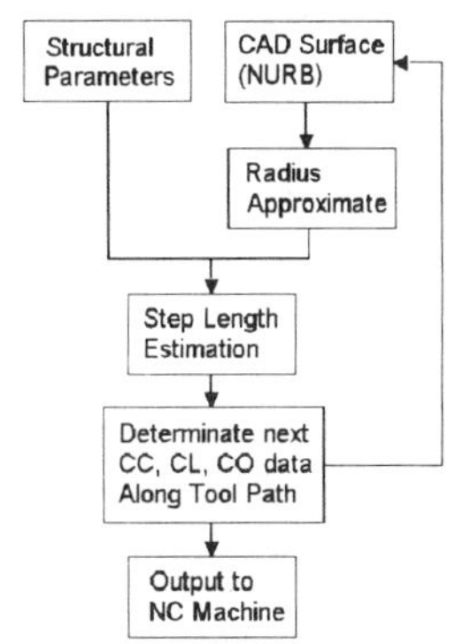

Figure 3. Tool position   *Figure 4.* Function flow chart

As long as the first CC point is selected, R_c, R_0, and β are determined. However, α_1 depends on not only the CC point's position but also the selected tool orientation method. Since the second CC point depends on the step length, α_2, $\Delta\alpha$, θ, and ξ are also dependent on the step length. Most current CAM architectures only consider R_c. However, as shown in Equation 1, the step length should depend on R_c, R_0 and $\hat{\beta}$. Figure 4 is a function flow chart shows how our algorithm is integrated with the CAD system. Figure 5 shows the tool paths of three arcs with different locations. The CC points are shown as the '+' sign.

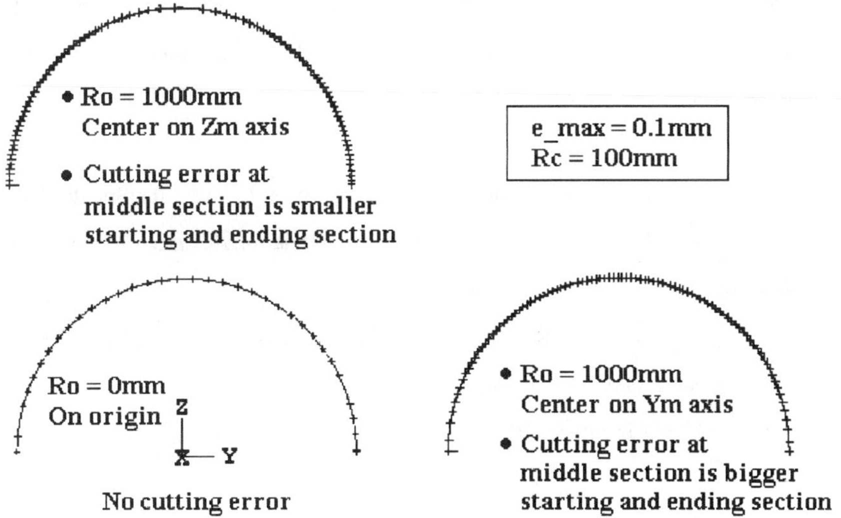

Figure 5. Tool path for the arc at different locations

3. CONCLUSION

Current CAM algorithms generate tool path only base on the part geometry and independent of the five-axis machine structures. Only the post processors consider the structure. The advantage of separating general processor and post processors is to make CAM general processors simply. Since the actual cut trajectory is no longer a straight passing through two consecutive points where there exists table rotation. Once the CC, and CL-CO data are determined, the post processors may not have enough information to re-generate the tool path. Therefore, the structure parameters should be considered in the tool path generator processes.

ACKNOWLEDGE

We gratefully acknowledge the financial support from National Science Council grant NSC89-2212-E-008-035.

REFERENCE

1. Vickers, G.W. and Quan, K.W., "Ball-Mills versus End-Mills for Curved Surface Machining", ASME Journal of Engineering for Industry, pp. 22-26, vol.111, Feb. 1989.

2. Cho, H.D., et. Al., "Five-axis CNC milling for effective machining of sculptured surfaces. ", International Journal of Production Research, Nov. 1993, vol.31, (no.11): 2559-73.

3. Hwang, Yean-Ren and Liang, Cha-Shong "Cutting Error Analysis for Spindle-Tilting Type five-Axis NC Machines", The International Journal of Advanced Manufacturing Technology, vol.14, pp 399-405, 1998.

4. Sarma, R. and Dutta, D., "The Geometry and Generation of NC Tool Paths", Journal of Mechanical Design, pp253-258, June 1997.

5. Zeid, I., *CAD/CAM Theory and Practice*. New York: McGraw-Hill, Inc., 1991.

Automated Calibration for Assembly Device Installation Based on Plug & Produce Concept

Haruka KIKUCHI, Yusuke MAEDA, Masao SUGI, and Tamio ARAI

Arai Lab., Dept. of Precision Engineering, School of Engineering, The University of Tokyo

7-3-1 Hongo, Bunkyo-ku, Tokyo 113-8656, JAPAN

TEL +81-3-5841-6486   FAX +81-3-5841-6487

kikuchi@prince.pe.u-tokyo.ac.jp

Abstract

Reconfigurability is essential to complete highly flexible and agile assembly systems. Easier installation of assembly devices is indispensable for easier reconfiguration. In this paper, we propose an automated calibration method based on "Plug & Produce" concept to support device installation for parts handling with two cameras. We also present experimental results on accuracy for calibration between two robot manipulators.

Keywords

Calibration, Plug & Produce, DLT (Direct Liner Transformation) Method

1. INTRODUCTION

Key issues of manufacturing systems have been changed to flexibility or agility in this decade. Arai et al. proposed "Plug & Produce" concept [1], which is a methodology to introduce a new assembly device into a manufacturing system easily and quickly.

In this paper, we propose a method of automated device-to-device calibration for Plug & Produce shown in Fig. 1. Using two uncalibrated CCD cameras and markers attached to each device, we obtain relative relationship between two neighboring devices.

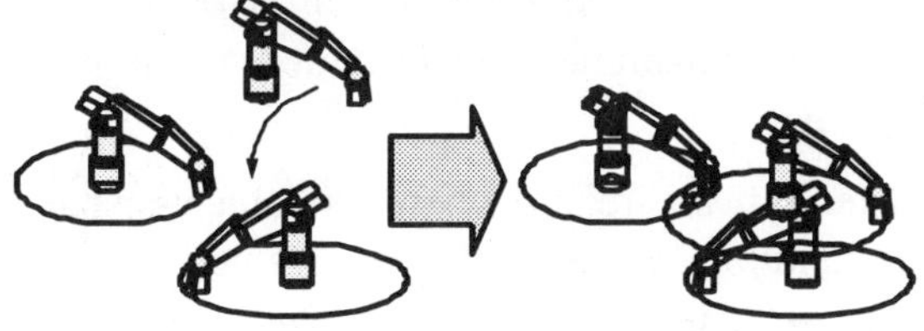

Fig. 1 Plug-in of Assembly Device

2. CALIBRATION FOR PLUG & PRODUCE

In this section, we propose a method of automated calibration. Our proposal has the following characteristics:

- The relative positional relationship between two devices can be obtained from two uncalibrated CCD cameras.
- Calibration is carried out automatically. Two cameras observe markers attached on each device while the device is moving.
- The two cameras can be set anywhere as far as the marker motion is visible.

Note that we aim not to localize the new device in the world reference frame, but only to know relative positional relationship between two neighboring devices around the relay point. So calibration is executed around only their relay point.

2.1 Procedures of Calibration

Actual procedures of calibration are as follows:

1) Set two cameras at appropriate positions.
2) Drive the existing device and take n samples of marker's coordinates in the image plane of each camera. Simultaneously, calculate the marker position in the local reference frame of the existing device with its internal sensors.
3) Calibrate the cameras using the DLT (Direct Linear Transformation) method [2]. The concrete way of calculation is in section 2.2.
4) Drive the new device and take m samples of marker's coordinates with the cameras. The positions of the marker are measured in the local reference frame of the existing device. Simultaneously calculate the marker position in the local reference frame of the new device with its internal sensors.
5) Compare the coordinates of the markers in the local reference frame of the existing device (measured in the previous step) and those in the local reference frame of the new device (calculated by its internal sensors), and calculate a homogenous transformation matrix between them. The actual way of calculating homogenous transformation matrix is in section 2.3.

2.2 DLT Method

In the DLT Method, the relation between object-space reference frame (the XYZ system) and image-plane reference frame (the UV system) are:

$$(u \quad v) = \left(\frac{B_1 x + B_2 y + B_3 z + B_4}{B_5 x + B_6 y + B_7 z + 1} \quad \frac{B_8 x + B_9 y + B_{10} z + B_{11}}{B_5 x + B_6 y + B_7 z + 1} \right) \tag{1}$$

To identify DLT parameters for camera j ($j=1,2$), $\mathbf{B}_j = \left(B_{j1} \quad \cdots \quad B_{j11} \right)^T$, we need a group of points whose x, y & z coordinates are already known. They are called control points. The ith control point can be given by the drive of existing device as $\mathbf{x}_i = (x_i \quad y_i \quad z_i)^T$, and the uv at camera j as $\mathbf{u}_{ji} = (u_{ji} \quad v_{ji})^T$. Rearranging (1) with respect to DLT parameters:

$$
\begin{pmatrix}
x_1 & y_1 & z_1 & 1 & -u_{j1}x_1 & -u_{j1}y_1 & -u_{j1}z_1 & 0 & 0 & 0 & 0 \\
0 & 0 & 0 & 0 & -v_{j1}x_1 & -v_{j1}y_1 & -v_{j1}z_1 & x_1 & y_1 & z_1 & 1 \\
& & & & & \vdots & & & & & \\
x_n & y_n & z_n & 1 & -u_{jn}x_n & -u_{jn}y_n & -u_{jn}z_n & 0 & 0 & 0 & 0 \\
0 & 0 & 0 & 0 & -v_{jn}x_n & -v_{jn}y_n & v_{jn}z_n & x_n & y_n & z_n & 1
\end{pmatrix}
\mathbf{B}_j =
\begin{pmatrix}
u_{j1} \\ v_{j1} \\ \vdots \\ u_{jn} \\ v_{jn}
\end{pmatrix}
\tag{2}
$$

11 DLT parameters are calculated by least-squares method. To identify DLT parameters, at least 6 control points are necessary $(n \geq 6)$.

After identifying DLT parameters, we can calculate 3D coordinates from two UV coordinates. Rearranging (1) with respect to the unknown point $\mathbf{x} = (x \quad y \quad z)^T$:

$$
\begin{pmatrix}
B_{1,1} - B_{1,5}u_1 & B_{1,2} - B_{1,6}u_1 & B_{1,3} - B_{1,7}u_1 \\
B_{1,8} - B_{1,5}v_1 & B_{1,9} - B_{1,6}v_1 & B_{1,10} - B_{1,7}v_1 \\
B_{2,1} - B_{2,5}u_2 & B_{2,2} - B_{2,6}u_2 & B_{2,3} - B_{2,7}u_2 \\
B_{2,8} - B_{2,5}v_2 & B_{2,9} - B_{2,6}v_2 & B_{2,10} - B_{2,7}v_2
\end{pmatrix}
\mathbf{x} =
\begin{pmatrix}
u_1 - B_{1,4} \\
v_1 - B_{1,11} \\
u_2 - B_{2,4} \\
v_2 - B_{2,11}
\end{pmatrix}
\tag{3}
$$

$\mathbf{x}$ is calculated by least-squares method, too.

2.3 Calculate Homogenous Transformation Matrix

Homogenous transformation matrix between two reference frames is:

$$
\begin{pmatrix}
x_{ei} \\ y_{ei} \\ z_{ei} \\ 1
\end{pmatrix}
=
\begin{pmatrix}
T_1 & T_2 & T_3 & T_4 \\
T_5 & T_6 & T_7 & T_8 \\
T_9 & T_{10} & T_{11} & T_{12} \\
0 & 0 & 0 & 1
\end{pmatrix}
\begin{pmatrix}
x_{ni} \\ y_{ni} \\ z_{ni} \\ 1
\end{pmatrix}
\tag{4}
$$

where $\mathbf{x}_{ei} = (x_{ei} \quad y_{ei} \quad z_{ei} \quad 1)^T$ are the ith samples of coordinates measured with the existing device, and $\mathbf{x}_{ni} = (x_{ni} \quad y_{ni} \quad z_{ni} \quad 1)^T$ are those with the new one. Rearranging (4) to use least-squares method:

$$
\begin{pmatrix}
x_{n1} & y_{n1} & z_{n1} & 1 & 0 & 0 & 0 & 0 & 0 & 0 & 0 & 0 \\
0 & 0 & 0 & 0 & x_{n1} & y_{n1} & z_{n1} & 1 & 0 & 0 & 0 & 0 \\
0 & 0 & 0 & 0 & 0 & 0 & 0 & 0 & x_{n1} & y_{n1} & z_{n1} & 1 \\
& & & & & & \vdots & & & & &
\end{pmatrix}
\mathbf{T} =
\begin{pmatrix}
x_{e1} \\ y_{e1} \\ z_{e1} \\ \vdots
\end{pmatrix}
\tag{5}
$$

To identify $\mathbf{T} = (T_1 \quad \cdots \quad T_{12})^T$, at least 4 control points are necessary $(m \geq 4)$.

3. CALIBRATION EXPERIMENTS

We experimented on the accuracy of calibration for plug-in of a new manipulator. Fig. 2 shows our experimental setup. We used two 6-degree-of-freedom manipulators with ±0.03mm repeatability and two monochrome CCD cameras with 640x416 resolution. Each manipulator has an LED marker on its endpoint.

Fig.   2   Experimental Setup

We measured rms error and maximum calibration error between two local reference frames. The control points for calculating both DLT parameters and homogenous transformation matrix is located on from 2x2x2 to 5x5x5 grid (8-125 points), which covers a space of 200x200x200mm. Actual calibration errors are measured at 5x5x5 grid that is located on the relay point. We obeyed the following attentions to reduce errors.

- For effective use of camera resolution, two cameras shoot marker motion as large as possible in the camera frame.
- To reduce errors in the direction of optical axis, two cameras should keep a certain distance each other.

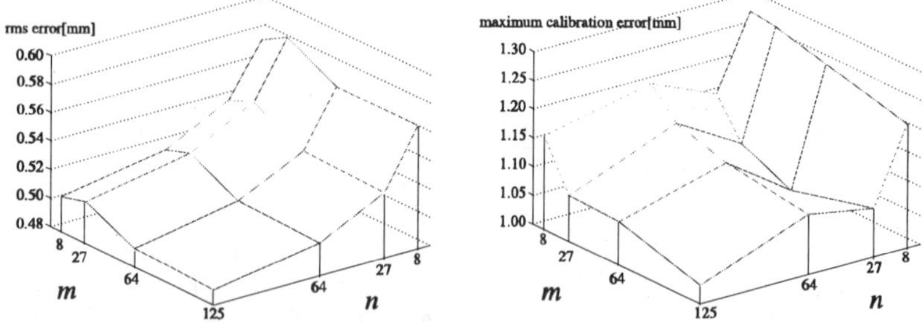

Fig. 3   Relation Between Calibration Errors and n, m

Fig. 3 shows relation between the errors and the number of control points, *n* (see section 2.2) and *m* (see section 2.3). For example, when we used 125 points for calibrating both the existing device and the new one, the relation between two local reference frames is:

$$\begin{pmatrix} x_e \\ y_e \\ z_e \\ 1 \end{pmatrix} = \begin{pmatrix} 1.002 & 0.017 & 0.008 & -618.39 \\ -0.016 & 1.003 & -0.007 & -6.92 \\ 0.001 & 0.010 & 1.001 & -24.14 \\ 0 & 0 & 0 & 1 \end{pmatrix} \begin{pmatrix} x_n \\ y_n \\ z_n \\ 1 \end{pmatrix} \qquad (6)$$

In this case, the maximum calibration error was about 1.03mm and the rms error was 0.49mm. The achieved accuracy was at the same level as the result of manipulator localization by using 3D coordinate measuring machine [3].

Required time for calibration is within 3 minutes when the number of control point is 8. It takes about 10 minutes for the most precise calibration with 125 control points. For any sake, most of the procedures are automated except for setting two cameras.

4. CONCLUSION

This paper is summarized as follows:
- We proposed a method of automated calibration between an existing device and a new one based on Plug & Produce concept for the installation of a new device in assembly systems.
- Calibration between two robot manipulators was demonstrated. The accuracy of 0.49mm rms error was achieved. Identification of internal parameters of the CCD cameras and the manipulators is required for more accuracy.

ACKNOWLEDGMENTS

This research is supported by IMS/HMS (Holonic Manufacturing Systems) project.

REFERENCES

[1] Arai, T. et al., "Agile Assembly System by "Plug & Produce"", Annals of the CIRP, Vol. 49, No. 1, pp. 1-4, 2000.
[2] Shapiro, R., "Direct Linear Transformation Method for Three-Dimensional Cinematography", Research Quarterly, Vol. 49, pp.197-205, 1978
[3] Zhuang, H. et al., "A New Method for Pose Fitting from Two 3D Point Sets and Its Application to Robot Localization," Proc. of IEEE Int. Conf. on Robotics and Automation, pp. 655-660, 1996.

CALIBRATION OF 2-DOF PARALLEL MECHANISM

O. Sato, M. Hiraki, K. Takamasu and S. Ozono

Department of Precision Engineering, The University of Tokyo,
7-3-1 Hongo, Bunkyo-ku, Tokyo, 113-8656, Japan
Tel: +81-3-5841-6472, FAX: +81-3-5841-8556
e-mail: ri@nano.pe.u-tokyo.ac.jp

Abstract

In order to avoid some drawbacks of traditional Coordinate measuring machine (CMM) based on serial mechanism, we are developing Parallel-CMM based on a parallel mechanism. We have already built the prototype of 3-DOF Parallel CMM and now we are researching about the calibration of our Parallel CMM. When we use a CMM to measure the objects, we need to calibrate the geometrical parameters of the CMM to evaluate the uncertainty of measurement. In this paper, we discuss the details of calibration for 2-DOF parallel mechanism instead of 3-DOF parallel mechanism.

Keywords

Parallel mechanism, Calibration, Coordinate Measuring Machine

1. INTRODUCTION

We are developing Parallel-CMM (Coordinate Measuring Machine) based on a parallel mechanism [HIRAKI 1997]. In parallel mechanism, the base unit and end-effector are connected by many links in parallel. The advantages of parallel mechanism are its robustness against external force and error accumulation. We have already built the prototype of 3-DOF Parallel CMM shown in Figure 1 [HIRAKI 1999], and now we are researching about the calibration of our Parallel CMM.

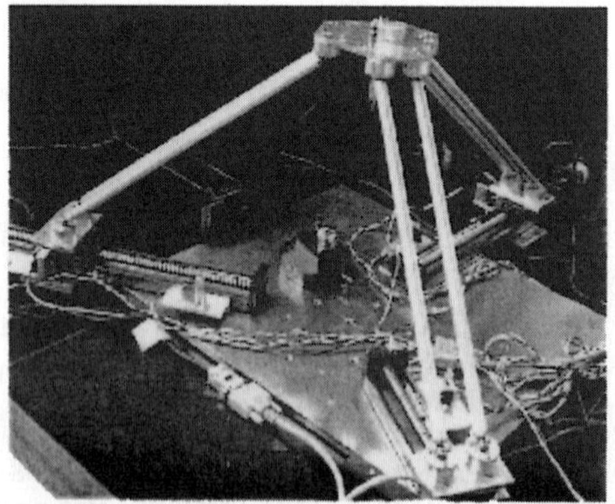

Figure 1   Prototype of Parallel CMM

If the end-effector move in parallel to the base unit, each pair of the rods can be replaced by one rod and three virtual rods are regarded as sides of trigonal pyramid, which upper vertex is the center of end-effector. Figure 2 shows the virtual link model. When we use a CMM to measure objects, we need to calibrate geometrical parameters of the CMM to evaluate the uncertainty of measurement. For this prototype, we need to identify parameters shown in Fig.2.

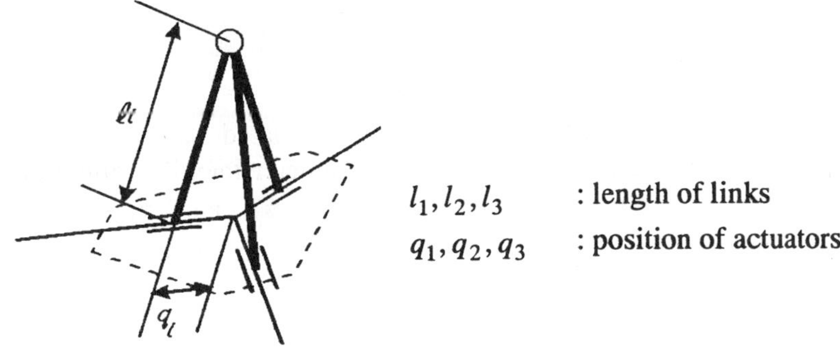

Figure 2   Geometrical model of Parallel CMM

There are some methods for identifying both the geometrical and kinematical parameters. One is calculating the parameters by measuring artifacts whose size is known. To identify the parameters efficiently with high accuracy, what artifact we should measure i.e. what the sets of measured points we should get? Now we think about the calibration of 2-DOF parallel mechanism like Figure 3 to discuss about the above questions. This mechanism has two actuators (q_1, q_2), two links (l_1, l_2), and one end-effector (P).

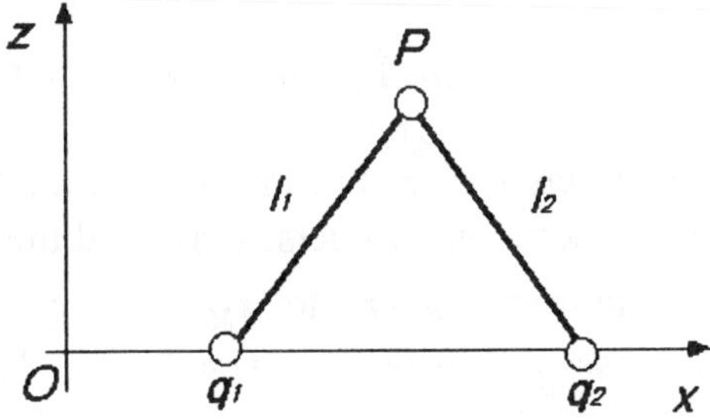

Figure 3   2-DOF parallel mechanism

Through the simulation to calibrate the geometrical parameters for the parallel mechanism, we got the basic knowledge about the way to calibrate the Parallel CMM.

2. KINEMATICS AND PARAMETER IDENTIFICATION

Forward kinematics of the model in Figure 3 can be solved analytically as follow equations.

$$x = \frac{q_2{}^2 - q_1{}^2 - l_2{}^2 + l_1{}^2}{2(q_2 - q_1)} = \frac{q_1 + q_2}{2} - \frac{2l\Delta l}{q_2 - q_1}$$

$$z = \frac{1}{2}\sqrt{-\frac{16l^2\Delta l^2 - 4(l^2 + \Delta l^2)(q_2 - q_1)^2 + (q_2 - q_1)^4}{(q2 - q1)^2}} \tag{1}$$

$$(l_1 = l + \Delta l, l_2 = l + \Delta l)$$

Here, the kinematical parameters are calculated using the result of measure by other measuring system. The parameter identification model is shown in Figure 4. Two coordinate planes $x_M - z_M$ and $x_P - z_P$ are the measuring machine's and the parallel mechanism's respectively.

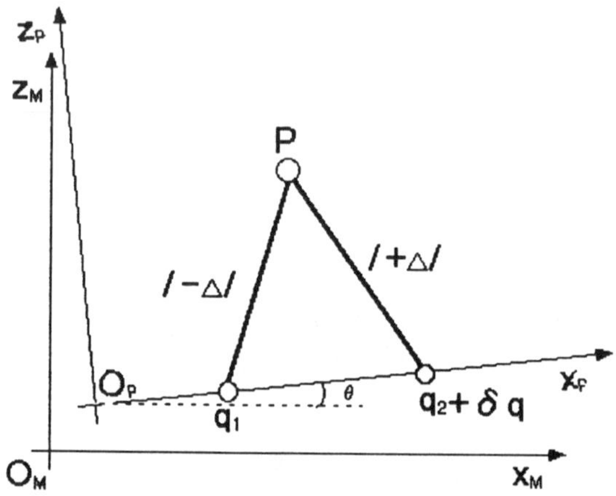

Figure 4   Parameter identification model for 2-DOF mechanism

Kinematical parameters we need to identify are the length of the links (l_1, l_2), the offset value between encoders (δq), and the parameters by the coordinates transforming from $x_P - z_P$ to $x_M - z_M$ (x_m, z_m and θ).

We identified these parameters in simulation. The condition was follows.

The number of measured points	:20
The error of measuring machine	:0.01 mm
The error of encoder	:0.02 mm

Under this condition, we arranged the measured points under some rules. The good way to identify parameters with high accuracy is arranging as $q_1 + q_2$ is constants (Figure 5).

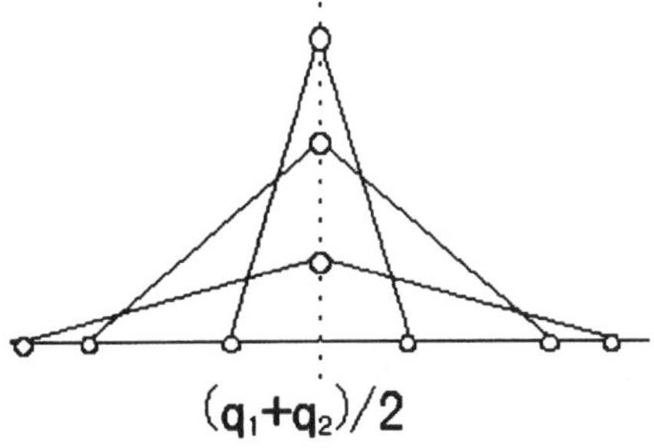

Figure 5   Measured points arranged as $q_1 + q_2$ is constants.

We got worse result of simulations under other rules by arranging measured points than above rule. The reason $q_1 + q_2$ constant arrangement gives better result than others is it gives smaller condition number of the pseudo-inverse matrix of Jacobian than other arrangings give.

Table 1 shows the identified parameters when the measured points were arranged as $q_1 + q_2$ is constants. It shows kinematic parameters are identified with high accuracy.

Table 1   Result of simulation: The measured points were arranged as $q_1 + q_2$ is constants. The measuring error is regarded as 20 μ m.

	Set values	Identified value
l	100.5550	100.5709
Δl	0.6130	0.6090
δq	0.9480	0.9568
x_m	-0.1720	-0.1511
z_m	-1.1130	-1.1314
θ	-0.00822	-0.00770

(unit : mm and rad)

After the calibration of the kinematic parameters, we evaluated uncertainty of the position of end-effector. Figure 6 and Table 2 show estimated error of positions of the end-effector. The error of positions after calibration is at most 10 μ m. The minimum error can be achieved when this system uses the sets of measured points arranged as $q_1 + q_2$ is constants.

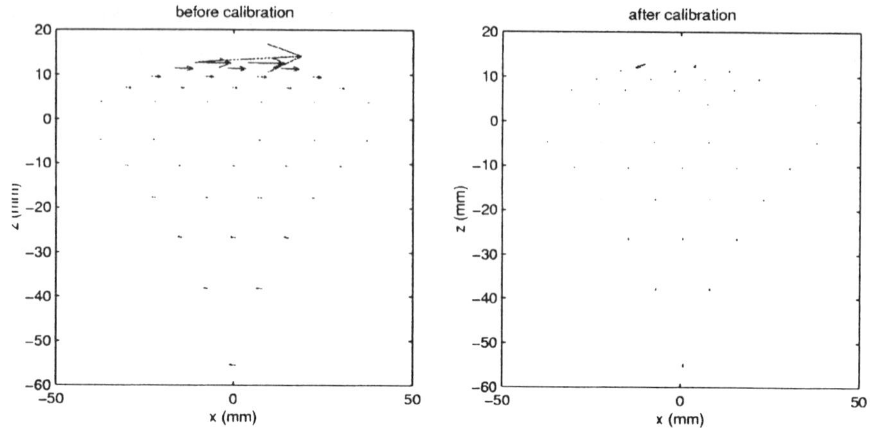

Figure 6   Uncertainty of the position of end-effector

Table 2   Error of positions: The maximum error of positions estimated before calibration and after calibration.

Direction	Before calibration	After calibration
x	0.574	0.016
z	0.466	0.006

(unit : mm)

3. CONCLUSION AND FURTHER STYDY

Through the simulation of calibration of 2-DOF parallel mechanism, we got the knowledge about what the sets of measured points should be got to identify kinematic parameters with high accuracy. The best one is to get the measured points arranged as $q_1 + q_2$ is constants, in other words, make the condition number of the pseudo-inverse matrix of Jacobian small.

Next, we need to try the identification of the kinematical parameters of 3-DOF Parallel CMM with the sets of measured points that gives small condition number of the pseudo-inverse matrix of Jacobian.

4. REFERENCES

[HIRAKI 1997] M. Hiraki, N. Yoshikawa, K. Takamasu and S. Ozono, The Development of Parallel-CMM: Parallel-Coordinate Measuring Machine, Proceedings of the 9th International Precision Engineering Seminar; 1997 May; Brawnschweig, Germany

[HIRAKI 1999] M. Hiraki, N. Yoshikawa, K. Takamasu and S. Ozono, Development of 3 DOF Parallel-CMM, Proceedings of IMEKO-XV World Congress; 1999 June; Wien, Austria.

738

DEVELOPMENT OF A CALIBRATION EQUIPMENT USING LASER INTERFEROMETER COMBINING A VARIABLE LENGTH VACUUM CELL
Structure and Measurement

Hiroki Masuda, Yutaka Kuriyama, Hisayoshi Sakai, Morimasa Ueda

Mitutoyo Corporation

Abstract

According to a requirement concerning measuring instruments that have to be certified their traceability, a new calibration equipment for displacement sensors has been developed. In order to estimate measurement uncertainties correctly, it is effective to make the traceability chain shorten. Thus, a 633nm commercial stabilized He-Ne laser wavelength with relative measurement uncertainty of 10^{-9} order and a variable length vacuum cell for laser pass are applied for the laser interferometer system in this equipment. Moreover, in order to keep its mirror's posture constant for the laser beam, a friction-drive unit and air bearing guide are employed. The expanded uncertainty (k=2) of this equipment is estimated to be 2.80nm, in the stroke of 50mm and the resolution of 0.08nm.

Keywords

Laser interferometer, Variable length vacuum cell, Friction-drive, Air bearing guide

1. INTRODUCTION

As for recent trends concerning measuring instruments, the concept of traceability is becoming widespread, and "ISO/IEC 17025" requires to show measurement uncertainties in order to certify traceability of measuring instruments. Generally, estimating uncertainty of a measuring instrument is to combine all uncertainties related to national standards with an unbroken traceability chain. Therefore, when the traceability chain is long, estimating uncertainties sometimes brings on complicated tasks and underestimation.[3]

On the other hand, many measuring instruments, such as roundness measuring machines and surface roughness testers, have probes as displacement sensors. Recently, some of displacement sensors become to have nanometer order resolution and uncertainty, so conventional calibration equipment can not always determine their uncertainties.

In order to avoid unexpected matters as above, we have developed a new calibration equipment using a 633nm commercial stabilized He-Ne laser. In this paper, we describe the configuration and features of the calibration equipment, and estimate its measurement uncertainty.

2. BASIC CONCEPT

At the beginning of this development, following items are taken into consideration as basic concept.

(1) The calibration objects of the equipment are displacement sensors from the above-mentioned point.

(2) We arrange a measurement laser beam in a vacuum. It is well known that relative measurement uncertainties of laser interferometers operating in air are worse than 10^{-6} or even 10^{-7} order because of influence of air turbulence.[1]

(3) We make arrangement to the calibration equipment as standing on Abbe's principle to minimize errors owing to the motion mechanism. So the measurement laser beam axis, the axis of guide-way, and the displacement direction of probes are coincided in a straight line.

3. CONFIGURATION AND FEATURES

Figure 1. shows a schematic drawing of the calibration equipment based on above concepts.

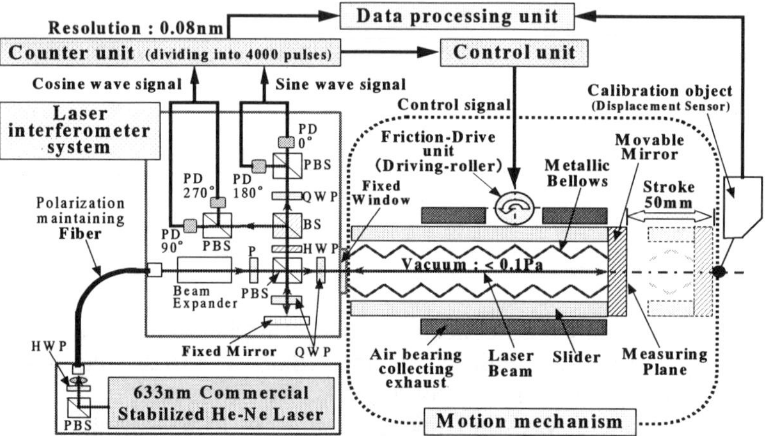

Figure 1. Schematic drawing of the Calibration Equipment

BS: beam splitter, HWP: half wave plate, QWP: quarter wave plate, P: polarizer,
PBS: polarizing beam splitter, PD: photodetector

This equipment is composed of a laser interferometer system, motion mechanism, control unit, counter unit, and data processing unit. The laser beam from the light source is led to the laser interferometer through the polarization-maintaining fiber.

In the laser interferometer system, the movable mirror is mounted on the end face of the slider, also fixed to the end of the metallic bellows inside of the slider. Thus, a variable length vacuum cell for optical pass is formed. In order to keep this mirror's posture constant for the laser beam, the slider is supported by air bearings and driven by the friction-drive unit.

3.1 Laser Interferometer System

This laser interferometer system is composed of the single-pass Michelson type. The laser beam from the interferometer is divided into four beams which are $0°$,$90°$,$180°$, and $270°$ phase sifted by wave plates and polarizing beam splitters and others. Then they are photoelectric-converted into electric signals by photodetectors. And further, every half wave-length (316nm) is divided into 4000pulses by processing the electric signals at the counter unit. As a result, the minimum resolution of 0.08nm is obtained.

3.2 Motion Mechanism

It is said that a friction drive has a desirable mechanism to reduce mechanical vibration. Also, the number of contact area of a slider must be reduced to get highly accurate motion of the slider without deformation or hysteresis. From these points of view, following schemes are applied.

In a conventional friction drive unit, a backup-roller is placed on opposite side of a slider to support the contact force of a driving-roller, but in this equipment, an air bearing supports the contact force. Then, the slider with a square cross section is guided by air bearings supporting from four directions with non-contact. Moreover, the air bearings have a function to collect exhaust not to cause thermal influence on the equipment. It is confirmed that the collecting efficiency of exhaust from the air bearings is more than 96 %. In the means described above, the mechanical contact of the slider is made only at the one area.

Figure 2 shows the kinematics accuracy of the movable mirror, that is measured with an autocollimator. As this experimental result, it's confirmed that the mirror's angular errors are less than 4μ rad in the stroke of 50mm.

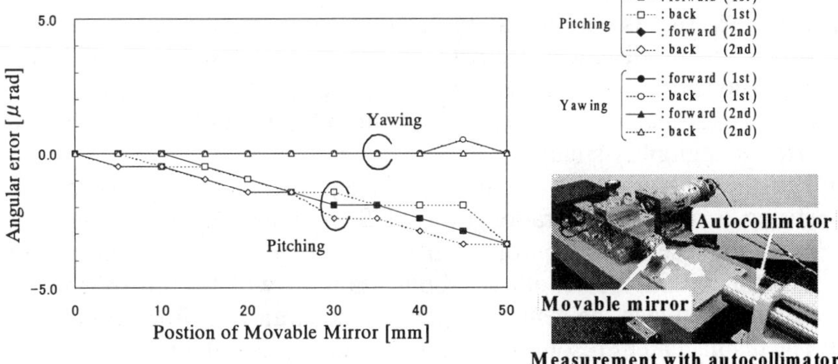

Figure 2. Kinematics accuracy of the movable mirror

4. ESTIMATION OF UNCERTAINTY

Table 1 shows the uncertainty budget of the calibration equipment. We pick up four major uncertainty factors, then estimate each standard uncertainty and expanded uncertainty.

Table 1. Uncertainty budget for the calibration equipment

Uncertainty factor	Estimation Condition	Standard uncertainty (k=1)	Standard uncertainty(k=1) Stroke:L=50 [mm]
Wavelength of laser	633nm commercial stabilized He-Ne laser	$1.0 \times 10^{-9} \times L$	0.05nm
Kinematics accuracy of mirror's posture angular error : 0.07μ rad/mm	Sine error by offset (0.05mm) and angular error	$4.81 \times 10^{-9} \times L$	0.24nm
	Cosine error by angular error	$2.45 \times 10^{-15} \times L^3$	0.31×10^{-3} nm
Thermal gradient change of mirror and base ± 0.005K at calibration time 5 min.	Mirror's material : Synthesis quartz	0.01nm	0.01nm
	Base's material : Low thermal expansive cast	1.2nm $+ 3.5 \times 10^{-9} \times L$	1.38nm
Resolution	Resolution : 0.08nm	0.03nm	0.03nm
Combined standard uncertainty (k=1)			1.40nm
Expanded uncertainty (k=2)			2.80nm

Some explanations on the each standard uncertainty are given as follows.

(1) The wavelength of the 633nm commercial stabilized He-Ne laser is calibrated by comparing with wavelength of an iodine stabilized He-Ne laser (stability: 10^{-11} order) which is traceable to the national length standard. This wavelength calibration causes an uncertainty of about 0.05nm.

(2) The sine error is combined the mirror's angular error and Abbe's offset which means distances from the measurement laser beam axis to the center axis of guide-way and to the displacement axis of sensor. The sine error is estimated to be 0.24nm in the 50mm stroke, while the cosine error is negligibly small.

(3) The thermal gradient change in this equipment is at most ± 0.005K during a normal calibration time of 5 minutes. On the above thermal condition, the expansion of the base causes measurement uncertainty of 1.38nm within the length of about 400mm, which corresponds to the distance between the center of interferometer and a calibration object.

(4) The displacement resolution of this equipment is 0.08nm. The dispersion between two digital value is regarded as triangular distribution, when a quantization error is transformed into a standard uncertainty.[4]

5. CONCLUSIONS

(1) A practical calibration equipment for displacement sensors using a laser interferometer combining a vacuum cell, a friction drive unit, and air bearing guides has been developed (Figure 3).

(2) Estimation of a measurement uncertainty owing to electrical interpolation error in detail is our subject to be analyzed in near future, at least in every interval of half wavelength (316nm), the expanded uncertainty is estimated to be 2.80nm.

(3) It is recognized that the uncertainty factor concerning temperature is most influential factor. As for the next subject, the research of design concept about sub-nanometer order compensation system for thermal influences might be mentioned in future.

(4) Using the variable length vacuum cell for laser pass is so effective not only to exclude the uncertainty of laser wavelength compensation which includes uncertainties of measurement temperature, atmospheric pressure, humidity, CO_2 concentration, and compensation equation error, but to eliminate influence of air turbulence with the motion of the slider. If the interferometer system is used in the air, the uncertainty of laser wavelength compensation would be estimated to be more than 5nm (k=1) in the stroke of 50mm[4], moreover the air turbulence would bring on influence of about 20nm (k=1).[2]

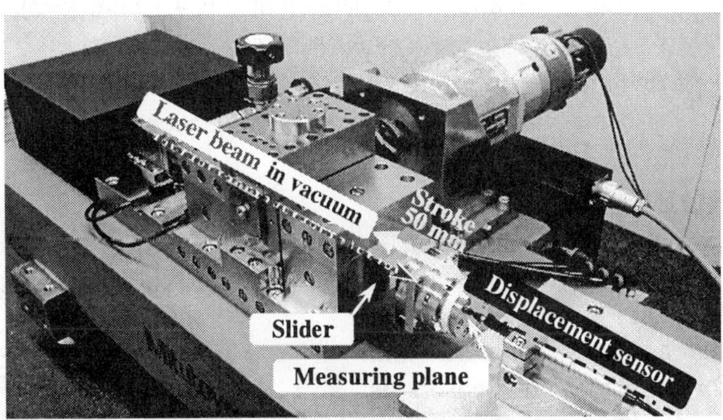

Figure 3.   Calibration equipment

References

(1) H. Kunzmann : Scales vs Laser Interferometers Performance and Comparison of Two Measuring System, CIRP Annals. 1993; Vol.42/2 :753-767

(2) H. Shiozawa : Deveropment of Ultra-Precision 3D-CMM Based on 3D metrology Frame, Proceedings of ASPE 1998 Annual Meeting ; 15-18.

(3) K. Tanaka : Traceability and Uncertainty in Measurement, Journal of Japan Society for Precision Engineering, 1999; Vol.65/7 : 945-952 (in Japanese)

(4) Y. Yamaryo : Measurement Uncertainty for Calibration of Standard Scale, Journal of Japan Society for Precision Engineering, 1999; Vol.65/7 : 953-957 (in Japanese)

DEVELOPMENT OF CALIBRATION EQUIPMENT USING LASER INTERFEROMETER   COMBINING A VARIABLE LENGTH VACUUM CELL

Control And Application

Hidekazu Oozeki , Shingo Kiyotani , Motonori Ogihara

Mitutoyo Corporation

Abstract

According to demand to make traceability chain of length measurement shorter, the calibration equipment which has nano-meter level positioning accuracy and 50 mm strokes for displacement sensors is developed. The problems in the high-precise positioning control system are resonance characteristics of controlled object and non-linear characteristics of a slight area in the neighborhood of positioning point [2]. In order to solve these problems, the original band-elimination filter and the gain scheduling method are applied. As the result, ±0.38 nm(±3 σ) which is positioning accuracy is achieved. The capacitive gap sensor is calibrated by using this calibration equipment. As the result, it is recognized that the non-linear errors of the capacitive gap sensor which is ±40 nm within the range of the measurement of ±20 μ m and the dispersion of it is 4.19 nm(k=2).

Keywords
Vacuum laser interferometer, DSP, Gain scheduling,
band-elimination filter, positioning accuracy

1. INTRODUCTION

According to a requirement concerning measuring instruments that have to be certified their traceability ,a practical calibration equipment which has at nano-meter level positioning accuracy and 50 mm strokes for displacement sensors is developed [1]. In order to estimate measurement uncertainties, it's effective make the traceability chain shorten.

Figure 1 shows the appearance of the calibration equipment. In the calibration equipment, a 633 nm commercial stabilized He-Ne laser with relative measurement

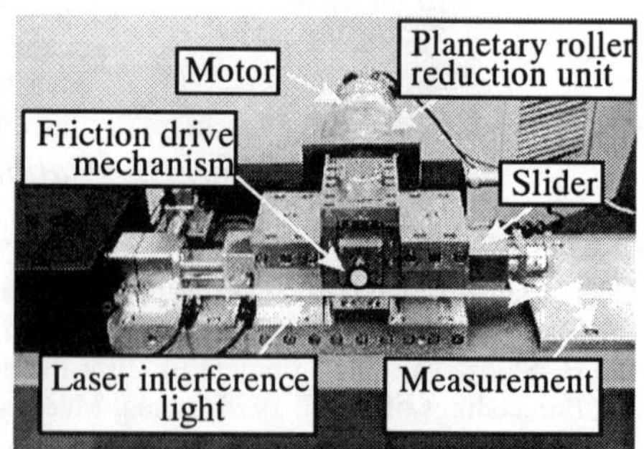

Figure 1.
Appearance of calibration equipment

uncertainty 10^{-9} order and a variable length vacuum cell for laser pass are applied to the laser interferometer system. This calibration equipment is to arrange the direction of the measurement axis capital and the vacuum laser interference light axis on the same and these axis to be suitable for the abbe's principle. The slider is supported by air bearings and driven by friction-drive mechanism .

The expanded uncertainty(k=2) of this calibration equipment is estimated to be 2.80 nm when thermal gradient change is at ±0.005 K.

2. OUTLINE OF CONTROL SYSTEM

Figure 2 shows the scheme of the control system. The control system is consists of the position, the speed, and the current control loop. The position and the speed control loop are basically the PI control by the digital signal processor (DSP), and the control cycle is 300 μ sec. However, the position command with the acceleration and deceleration profile (S-character) of velocity and the equal velocity profile is generated by the position control system every each time when feeding.

The position detecting resolution is achieved 0.08 nm by the dividing the half laser wave length （316 nm） into 4000. It is possible that the number of division of 4000 or 400 is chosen. In order to confirm positioning performance of this system, the number of division of 4000 is chosen in this paper.

The band-elimination filter with an analog circuit is added as positive amends element of the speed control in order to attenuate the resonance intensity. In order to reduce noise of amplifier, the current control loop become a linear current amplifier.

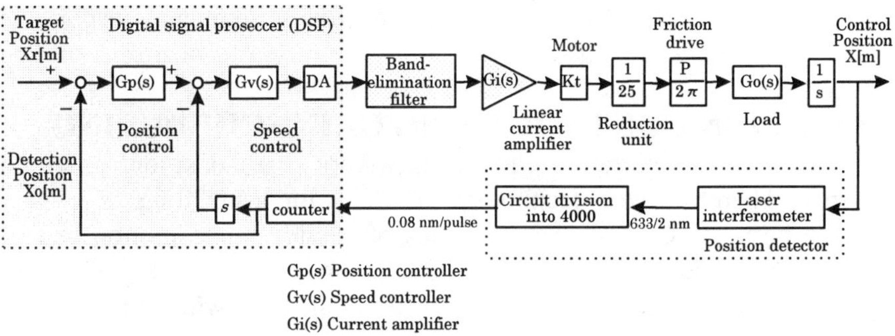

Gp(s) Position controller
Gv(s) Speed controller
Gi(s) Current amplifier

*Figure 2.*Schematic diagram of control system

3.BAND-ELIMINATION FILTER

The one of problems in the high-precise positioning control system is resonance characteristics of controlled object. In this system, the gain of the control system cannot increase by the resonance characteristic. In order to interception the resonance characteristic, the original band-elimination filter is added to the control system.

The structure of the band-elimination filter which uses the tandem phaseshifter is shown in Figure 3, and the frequency response of the filter are shown in Figure 4. This filter is robust to the error of the circuit constants by small variations because it is composed by the tandem phaseshifter and the adder; therefore, when the filter is composed of an analog circuit, the filter with a very steep interception characteristic can be made stably. The proportional gain of the speed loop is increased 10 times by the effect of band-elimination filter.

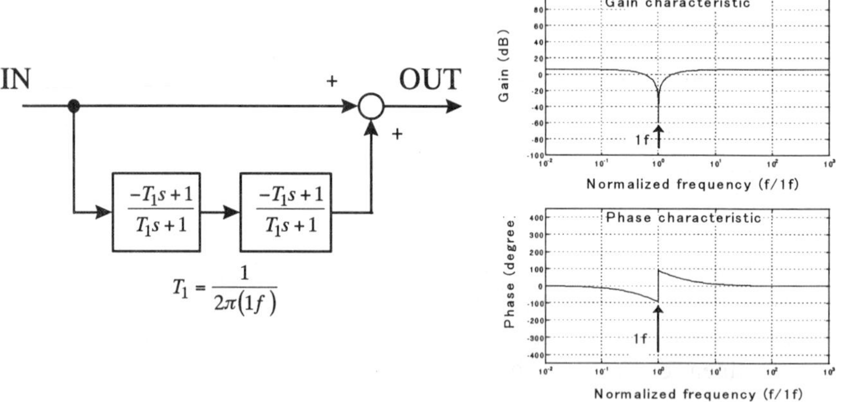

Figure 3. Structure of band-elimination filter

Figure 4. Frequency response of band-elimination filter

4. POSITION CONTROL USING GAIN SCHEDULING

As for this calibration equipment, a very small positioning error is required. Therefore, the technology of the gain scheduling is used to improve both the control characteristics of a wide range motion and the control characteristic of a slight area positioning. Concretely, the proportion gain of the position loop is increased 100 times after the wide range motion , and the positioning error is limited in a slight area.

5.RESULT OF POSITIONING IN A SLIGHT AREA

This calibration equipment usually moves in a wide range according to the acceleration and deceleration profile (S character) of velocity and the equal velocity profile. However, after the wide range motion positioning in a slight area begins. In a slight area , small step commands were added up wise and down wise.

Figure 5 shows the response of motion control system at the 2 nm step inputs, and the 1nm step inputs. The response of position follows to 1 nm step input by high accuracy, and there is not lost motion when the direction of sending reverses. The positioning error of this calibration equipment is ±0.38 nm(±3 σ) at 10 sec samplings with 3 msec intervals.

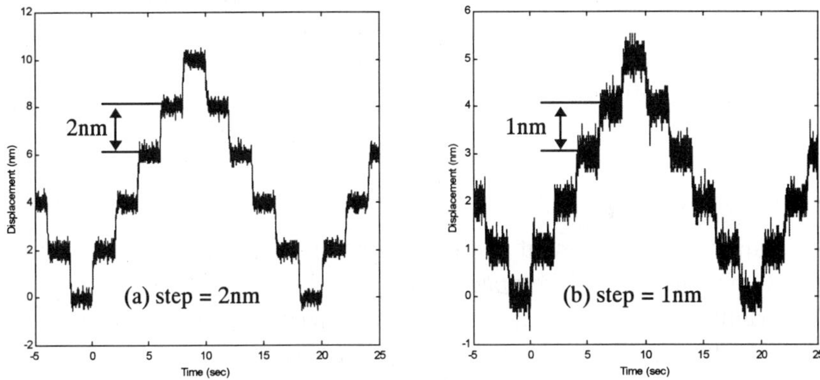

Figure 5. Response of position at stepwise inputs

6.CALIBLATION EXAMPLE OF A CAPACITIVE GAP SENSOR

As an example, the capacitive gap sensor is calibrated by using the calibration equipment. As a calibration method, the direction of the capacitive gap sensor axis and the slider axis is to arrange on the same. The gap between the sensor and the slider is changed three shuttling (position instruction which changed like the sine wave) within the range of ±20 μ m, and the output signal of the capacitive gap sensor and the laser interferometer is measured in every 75 msec. The shuttling measurements of 3 times takes 300 sec.

A result of calibration of the capacitive gap sensor is shown in Figure 6, which shows the non-linear fluctuating errors. The linearity error which is made by the alignment error between the slider and the sensor is excepted by method of least squares after the measurement data is acquired.

The high frequency component which is mainly electric noise was removed with the low pass filter. The thermal gradient change in the measurement is less than ±0.005 K. As the result, non-linear errors of the capacitive gap sensor which is ±40 nm within the range of the measurement of ±20 μ m is recognized.

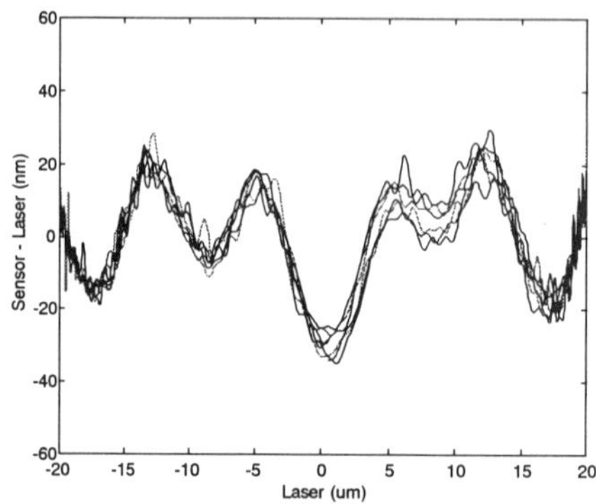

*Figure 6.*The non-linear error of a capacitive gap sensor （with filter）

Moreover, it is calculated by removing the non-linear error of the capacitive gap sensor that the dispersion of the capacitive gap sensor is 4.19 nm(k=2). The dispersion includes both the electric noise and the data processing error such as analog to digital conversion .

7. CONCLUSION

(1)The calibration equipment with nano-meter level positioning accuracy and 50 mm strokes is developed.

(2)The original band-elimination filter which used the phaseshifter and the gain scheduling method is applied to the control system of the calibration equipment. As the result, ±0.38 nm(±3 σ) which is the positioning accuracy of this calibration equipment is achieved .

(3)The capacitive gap sensor is calibrated by using this calibration equipment. As the result, non-linear errors of the capacitive gap sensor which is ±40 nm within the range of the measurement of ±20 μ m is recognized. Moreover, it is calculated by removing the non-linear error of the capacitive gap sensor that the dispersion of the capacitive gap sensor is 4.19 nm(k=2). The dispersion includes both the electric noise and the data processing error such as analog to digital conversion .These factors will be improved and decrease in future.

REFERRENCES

(1)H.Sakai : Development of Calibration Equipment with Sub-nanometer order Resolution Using Laser Interferometer Combining a Variable length Vacuum Cell,Metrologia-2000,pp464-470

(2)M.Takahashi,J.Otsuka : Precise Positioning Using Friction/Traction Drive（1st～6th reports）,JSPE 1988～1993

A MEASUREMENT FORCE ESTIMATION METHOD OF TAPPING STYLUS

TAKAHASHI Ken, HAYASE Masanori and HATSUZAWA Takeshi

Tokyo Institute of Technology
4259 Nagatsuta-cho Midoriku, Yokohama 226-8503, Japan
Tel & Fax   +81-45-924-5036
E-mail: ktakahas@pi.titech.ac.jp

Abstruct

A new estimation method for measurement force caused by a topographic measurement system using a microfork (*tapping stylus*) is proposed. Although Hertz model has been used to estimate the repulsive force in AC-mode AFM, it is not clear that the model can be applied to the tapping phenomenon at high frequencies. In this paper, experimental results of the near field characteristics and a numerical simulation for the motion of the microfork are compared. The tapping force can be calcurated using contants obtained by experimental results.

Keywords

tapping stylus, tapping force, nearfield characteristic, numerical simulation

1. INTRODUCTION

To image surface topography of microstructures, a measurement system giving less damages to samples with a fine vertical resolution is necessary. A new measurement system[1], whose resolution is greater than stylus profilers but less than scanning probe microscopes, using a metallic microfork(Figure 1) has been developed. During measurement, the tungsten probe, fabricated using a DC electropolishing[2], fixed at an end of the leg is vertically vibrated at the resonant frequency of the fork and it taps the sample surface. Consequently, the system gives less damages on samples than conventional stylus profilers because the scratching force is decreased.

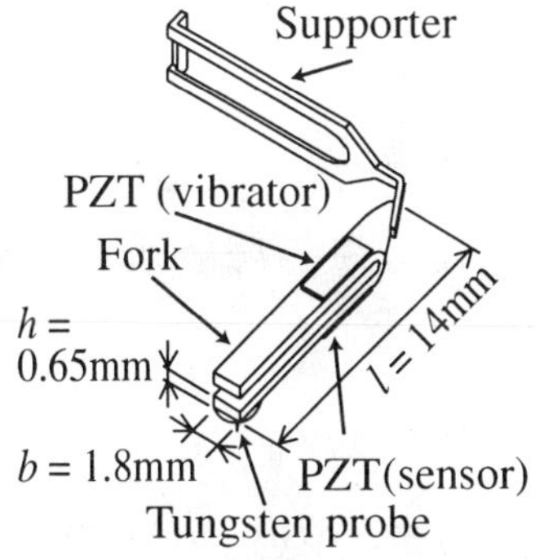

$\rho = 8.15 \times 10^3 \text{kg/m}^3$
$E = 1.8 \times 10^{11} \text{N/m}^2, Q = 2000$
Resonant frequency = 2965Hz

Figure 1. A sketch of the piezo-driven microfork

In the AC mode AFM, Hertz model has sometimes been used to estimate repulsive force[3]. However, it is not clear that the model can be applied to the tapping phenomenon at high frequencies.

In this paper, a novel estimation method of measurement force for the tapping stylu is described. Comparing the simulation with experimental results of the amlitude dependence on the distance from the surface(*near field characteristic*), the tapping force can be obtained.

2. MODELS
2.1 Principle

When the vibrating microfork approaches to the sample, the amplitude becomes smaller because of a mechanical interaction. As it varies by the properties of sample surface, characteristic of the specimen is obtained when the near field characteristic of the sample is measured.

To begin with, the motion of the microfork during tapping is simulated. For the sake of shorten the calculation time,

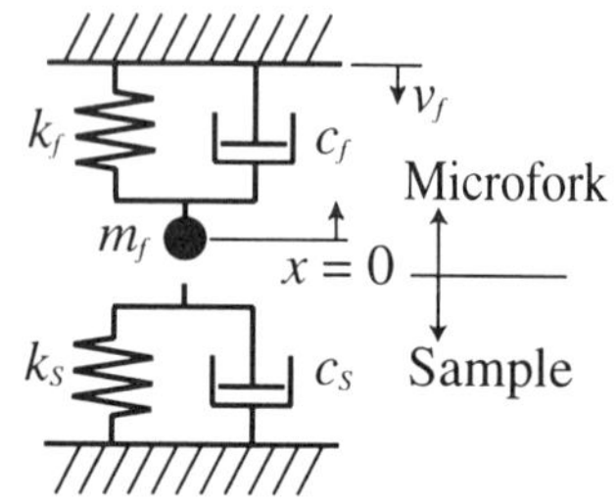

Figure 2. A numerical simulation model of the fork and a sample

the microfork is assumed to be a simple beam. Then, the beam is modeled as a spring(k_f)-dumper(c_s) system with an effective mass(m_f) shown in Figure 2. On the other hand, a spring(k_s) and a dumper(f_s) are used to make a model of the specimen.

2.2 Equation of motion

The equation of motion of the microfork including tapping force(F_s) is shown as follows:

$$x''(t) + 2\gamma\omega_0 x'(t) + \omega_0^2 x(t) = x_{st}\omega_0^2 \sin(\omega t) + F_s(t) \tag{1}$$

where x_{st}, 2γ and ω_0 are values set to $x_{st} = \frac{F_0}{k_f} = \frac{A_0}{Q}$, $2\gamma = \frac{1}{Q} = \frac{c_f}{\sqrt{m_f k_f}}$, $\omega_0 = \sqrt{\frac{k_f}{m_f}}$ using A_0(initial amplitude) and Q(quality factor), respectively.

$$\omega_0 = \left(\frac{\lambda}{l}\right)\sqrt{\frac{EI}{\rho b h}} \tag{2}$$

where $\lambda = 1.8751$, E(Young's modulus), I(geometric moment of inertia) and b, h, l, ρ(width, height, length and density) shown in Figure 1.

The probe can be assumed to be a rigid body. Under this assumption, F_s can be expressed as follows.

$$F_s(t) = \begin{cases} -k_s\{x(t) - (-A_0)\} - c_s x(t) & x < -A_0 \\ 0 & x \geq -A_0 \end{cases} \tag{3}$$

3. CURVES OF NEAR FIELD CHARACTERISTICS
3.1 Experiment

To measure near field characteristics, the vibrating microfork is fed to the specimen at a speed of v_f=0.02μm/s by a piezo actuator.

Table 1.   Experimental results of the near field characteristics

Sample	Amplitude m	α	d_s m	z m
Rubber	5.04×10^{-7}	0.344	2.05×10^{-7}	2.55×10^{-7}
	5.92×10^{-7}	0.343	2.33×10^{-7}	2.92×10^{-7}
	6.99×10^{-7}	0.339	2.77×10^{-7}	3.47×10^{-7}
	7.95×10^{-7}	0.337	3.38×10^{-7}	4.07×10^{-7}
Epoxy	5.06×10^{-7}	0.706	7.10×10^{-8}	1.22×10^{-7}
	5.96×10^{-7}	0.696	8.08×10^{-8}	1.40×10^{-7}
	7.04×10^{-7}	0.719	1.02×10^{-7}	1.73×10^{-7}
	8.01×10^{-7}	0.708	1.13×10^{-7}	1.93×10^{-7}
Duralumin	4.91×10^{-7}	0.801	2.36×10^{-8}	7.27×10^{-8}
	6.01×10^{-7}	0.817	2.55×10^{-8}	8.55×10^{-8}
	7.03×10^{-7}	0.780	2.43×10^{-8}	9.47×10^{-8}
	8.01×10^{-7}	0.808	2.59×10^{-8}	1.06×10^{-7}

Rubber, epoxy and duralumin are selected as specimens. Near field characteristics are measured five times for each samples and each initial amplitudes of 0.5μm, 0.6μm, 0.7μm and 0.8μm.

3.2 Experimental results

Some examples of experimental results are shown in Figure 3. The inclination of the curves(α) and the depth(d_s) at the point 90% of the initial amplitude is obtained from those results. Distance z indicates the feeding distance of the stage from the point of begining tapping to 90% of the initial amplitude. Table 1 shows the average values of α, d_s and z for each samples and the initial amplitudes.

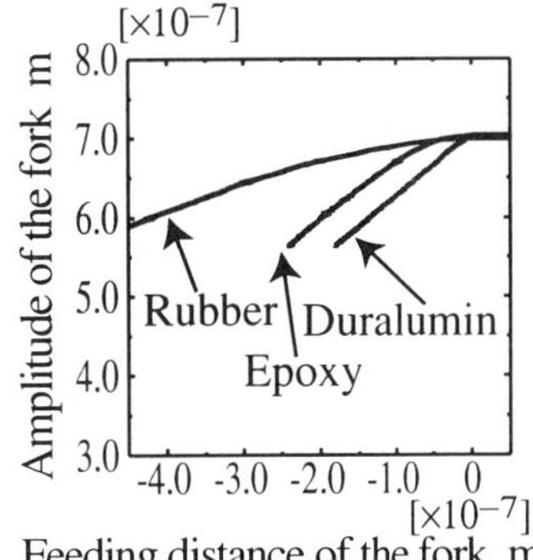

Figure 3. Experimental results of near field characeristics

4. SIMULATION
4.1 Simulation for near field characteristics

Using equation(1), motion of the microfork during tapping is simulated. Spring constant k_s and dumping coefficient c_s are given as follows:

$$k_s = 1.0 \times 10^5 \sim 2.0 \times 2.0^7 \quad \text{N/m} \tag{4}$$

$$c_s = \begin{cases} 0, 1, 3, 5, 7 \ \text{Ns/m} & \text{(rubber)} \\ 0, 5, 10, 15, 20 \ \text{Ns/m} & \text{(epoxy)} \\ 0, 50, 100, 150, 200 \ \text{Ns/m} & \text{(duralumin)} \end{cases} \tag{5}$$

Figure 4 shows an example of the simulation result for epoxy. The horizontal axis indicates given k_s and the vertical axis of lower figure shows inclinations at the point of 90% of initial amplitude.

751

The upper figure shows depths obtained by the simulation when the stage is fed by a distance of z after the probe begins to touch the sample. The dotted lines in the figure indicates the experimental values.

4.2 Identification of k_s and c_s

By combinating the values at intersection points of simulations and experiments in Figure 4, the results of simulations and experiment on d_s or α can be equalized. (k_s, c_s) combinations are obtained each from the depth and the inclination figures. But only one set of (k_s, c_s) exists for each specimen and initial amplitude because the dotted lines is obtained from one sample. When the both sets of (k_s, c_s) is plotted on one figure whose x-asix is k_s and y-axis is c_s like Figure 5, the intersection point is the required combination of (k_{s0}, c_{s0}). The (k_{s0}, c_{s0}) of the other samples are obtained in same ways and the results are listed in Table 2.

Concerning duralumin, two lines does not cross in case of big initial amplitude. Moreover, also in case of smaller initial amplitude, the values of c_s result in fairly large. As a pulsed strong force is applied to the probe during tapping on duralumin, vibration mode may change resulting in the difference between the model and the real microfork has been occured.

4.3 Derivation of tapping force

Tapping force is calculated by substituting obtained values of (k_{s0}, c_{s0}) for equation(1). Figure 5 shows the simulation results for epoxy. The horizontal and the vertical axis indicates

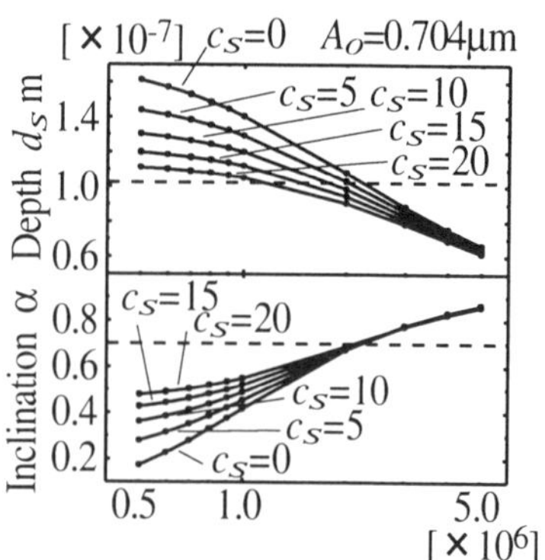

Figure 4. Simulation result for exoxy

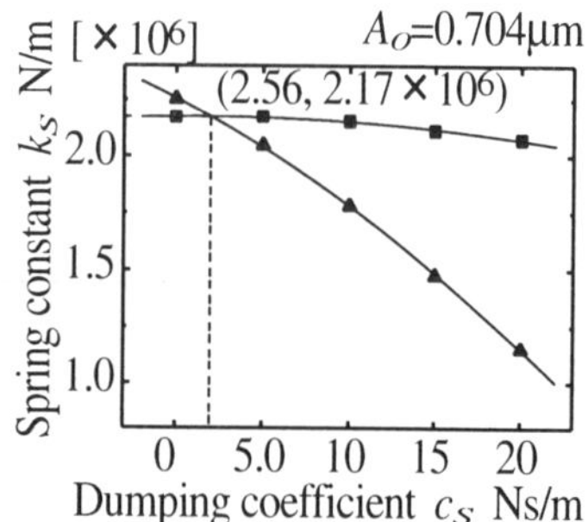

Figure 5. (k_s, c_s) dependence on epoxy

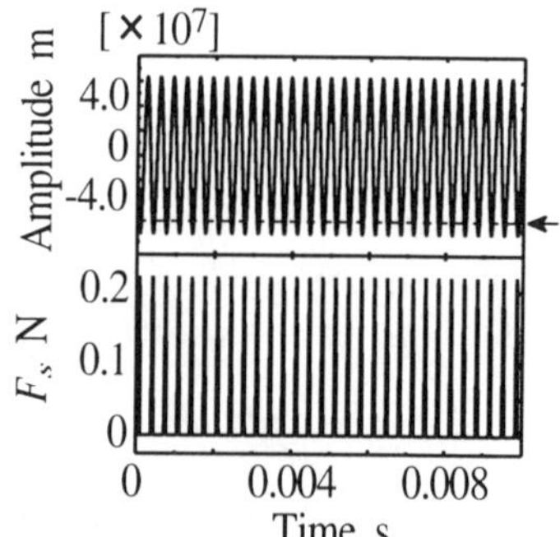

Figure 6. Tip motion and tapping force for epoxy

time and the tip amplitude(upper) and tapping force(lower), respectively. The arrow in the upper figure means the surface of the sample. In Table 3, maximum values(f_{tap}) and mean values($= \frac{1}{T}\int_0^T F_{tap}dt$) are

Table 2. k_{s0} and c_{s0} obtained from k_s-c_s curve

Sample	Amplitude m	k_{s0} N/m	c_{s0} Ns/m
Rubber	5.04×10^{-7}	3.13×10^5	3.74
	5.92×10^{-7}	3.05×10^5	4.21
	6.99×10^{-7}	2.96×10^5	4.21
	7.95×10^{-7}	3.02×10^5	3.66
Epoxy	5.06×10^{-7}	2.27×10^6	2.90
	5.96×10^{-7}	2.23×10^6	6.53
	7.04×10^{-7}	2.17×10^6	2.56
	8.01×10^{-7}	2.28×10^6	1.61
Duralumin	4.91×10^{-7}	4.64×10^6	1.28×10^5
	6.01×10^{-7}	5.15×10^6	1.54×10^5
	7.03×10^{-7}	—	—
	8.01×10^{-7}	—	—

Table 3. Maximum and mean values of the tapping force

Sample	Amplitude m	F_{max} N	Mean value N
Rubber	5.04×10^{-7}	6.47×10^{-2}	1.36×10^{-2}
	5.92×10^{-7}	7.19×10^{-2}	1.49×10^{-2}
	6.99×10^{-7}	8.31×10^{-2}	1.73×10^{-2}
	7.95×10^{-7}	1.00×10^{-1}	2.13×10^{-2}
Epoxy	5.06×10^{-7}	1.60×10^{-1}	1.92×10^{-2}
	5.96×10^{-7}	1.79×10^{-1}	2.12×10^{-2}
	7.04×10^{-7}	2.20×10^{-1}	2.70×10^{-2}
	8.01×10^{-7}	2.56×10^{-1}	3.10×10^{-2}

shown for each samples and tapping amplitudes.

From these experiments, tapping force becomes greater and the contact time gets shorter in case of solid sample.

6. CONCLUSION

To estimate tapping force of tapping stylus, experimental results of near field characteristics are compared with numetical simulation results. Using this novel method, it is obtained that few hundreds mN of tapping force is given to the specimen. Despite of such a great magnitude of tapping force, damages on the sample are small. Tapping measurement method is proved useful for measurement of soft materials.

REFERENCE

1. Ken TAKAHASHI and Takeshi HATSUZAWA. Surface topographic measurement system using a microfork. International conference Mechatronics 2000; 388

2. Ken TAKAHASHI, Masanori HAYASE and Takashi HATSUZAWA. Optimum Shape Fablication of Tungusten Probe using DC Electropolishing. The 10th International Conference on Solid-State Sensors and Actuators; 1999;552-555

3. J. Chen, R. K. Workman, D. Sarid and R. Hoper. Numerical Simulations of a Scanning Force Microscope with a Large-amplitude Vibration Cantilever, Nanotechnology, 1994; 5,199.

Investigation of the Plastic Deformation and Fracture of Carbon Nanotubes

K. J. Ma[1], C. L. Chao[2], C. W. Tang[1], K. H. Chen[3] and L. C. Chen[4]

[1]*Department of Mechanical Engineering, Chung-Cheng Institute of Technology, Taoyuan, Taiwan, R.O.C.*

[2]*Dept. of Mechanical Engineering, E790, Tam-Kang University, No.151 Ying-Chuan Rd, Tamsui, Taipei Hsien, Taiwan-251, R.O.C (email: clchao@mail.tku.edu.tw, Tel: -886-922-043212 Fax: -886-3-3900681)*

[3]*Institute of Atomic and Molecular Science, Academia Sinica, Taipei,Taiwan,R.O.C.*

[4]*Center for Condensed Matter Science, National Taiwan University, Taipei,Taiwan R.O.C.*

Abstract

Carbon nanotubes have demonstrated fascinating mechanical properties. The tensile strength as well as plastic deformation and fracture behaviour of carbon nanotubes, probably more important for material applications, are still lacking experimentally. It is too small to be manipulated with conventional tensile test facilities. We have developed a technique with which deformation and fracture behaviour of carbon nanotubes under a tensile load can be examined. High resolution TEM image shows that bonding breakage prefers to initiate at outmost graphene layer where a higher tension tends to build up. It is believed that slippage between the graphene layers is responsible for the strain relaxation and plastic deformation of multi-walled carbon nanotubes.

1. Introduction

Carbon nanotubes have demonstrated fascinating mechanical properties. The extremely high stiffness and excellent flexibility during bending and compression has been observed experimentally and studied theoretically. The tensile strength as well as plastic deformation and fracture behaviour of carbon nanotubes, probably more important for material applications, are still lacking experimentally. It is too small to be manipulated with conventional tensile test facilities. Two approaches were suggested to resolve this problem- - to grow longer nanotubes, or to design proper instruments for a conclusive test. Recently, longest carbon nanotube were produced by CVD process. Pan *et al* (1999) have directly measured the Young's modulus and tensile strength of multiwalled carbon nanotubes by pulling very long (2mm) aligned nanotube rope with a specially designed stress-strain puller. The Young's modulus and tensile strength are lower than those calculated previously. The contacting between mounting grips and nanotube ropes only limited to the outer layers of the nanotubes was one of the factors leading to a lower strength. More recently, Yu *et al.* (2000)

have developed a manipulate tool with a nano-stressing stage located within a scanning electron microscope. Individual MCNTs were picked up and then attached at each end of a section length onto the opposing tips of AFM cantilever probes. Each nanotube section was then stress loaded and then observed in situ in the SEM. Firm nanotube attachment on to the AFM tips is critical in the tensile test. The bonding strength at interface may also influence the measured tensile strength and plastic deformation behaviour of carbon nanotubes. In this study, we reported simple techniques without contact problems, with which deformation and fracture behaviour of carbon nanotubes under a tensile/torsion forces can be examined.

2. Experimental Setup

Two approaches were used to induce the deformation and fracture of carbon nanotubes:

(1) Well aligned multi-walled carbon nanotubes (MWNTs) which were wrapped by the closely contact graphitic carbon layer were synthesized by CVD process (*Figure 1*). Delicate breaking the amorphous carbon layer by diamond indentor allows the wrapped amorphous carbon layer to crack and separate from the enclosed carbon nanotubes. This provides a tensile stress on the carbon nanotubes and results in plastic extension of carbon nanotubes.

(2) Amorphous carbon layer was deposited on the multiwalled carbon nanotubes which were produced by MWCVD method. The bonding strength between amorphous carbon and carbon nanotubes is excellent. Amorphous carbon/ nanotubes layer was scratched by the diamond indentor. Fracture and separation of amorphous layer provides a tensile stress on the carbon nanotubes and leads to the plastic extension of carbon nanotubes.

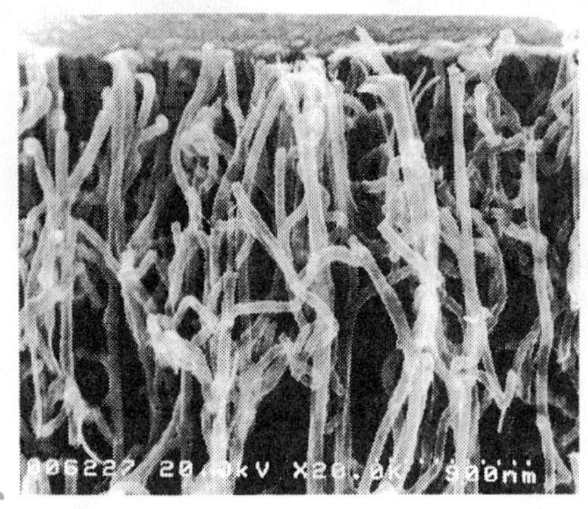

Figure1 SEM image of aligned MWNTs.

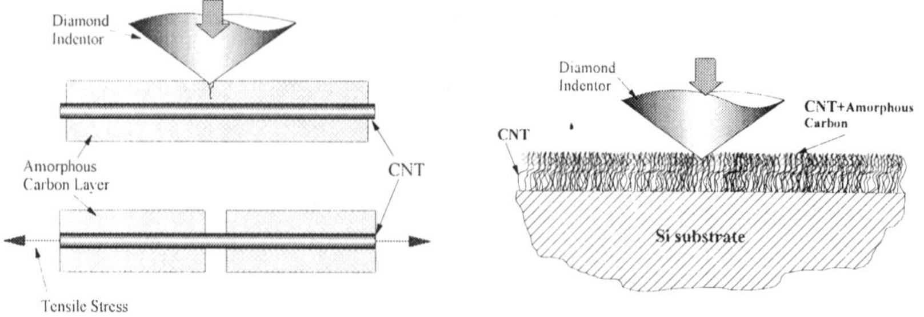

Figure 2 Plastic extension of CNT

Figure 3 Amorphous carbon layer was deposited on the MWCNTs

3. Results and Discussions

High resolution image (*Figure 4*) showing the structure of multi-walled carbon naotubes with a diameter between 20~ 50 nm which were produced by MWCVD process. Figure 5 shows the SEM image of initiating plastic deformation of carbon nanotube under tensile force. It is worth to note that the necking may occur in different locations simultaneously. Further increasing the tensile force leads to the nonuniform plastic extension and fracture of carbon nanotubes, as shown in *Figure 6*. Due to shear force and instability involved in the deformation process, twists and necking can be observed on the heavily deformed carbon nanotubes.

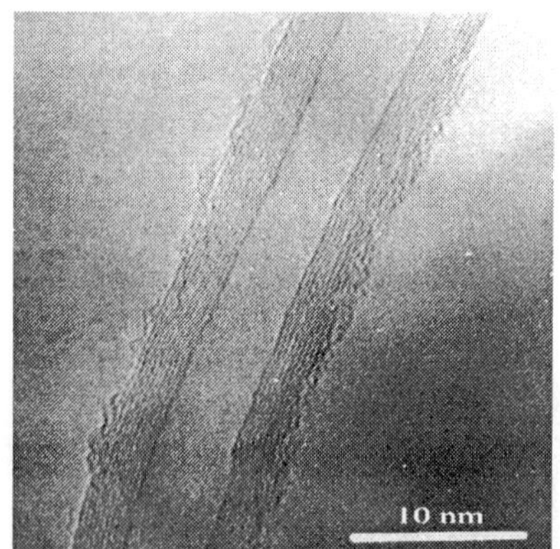

Figure 4 TEM image of MWNTs

Figure 5   SEM image showing the initiating plastic deformation of CNT under tensile force

756

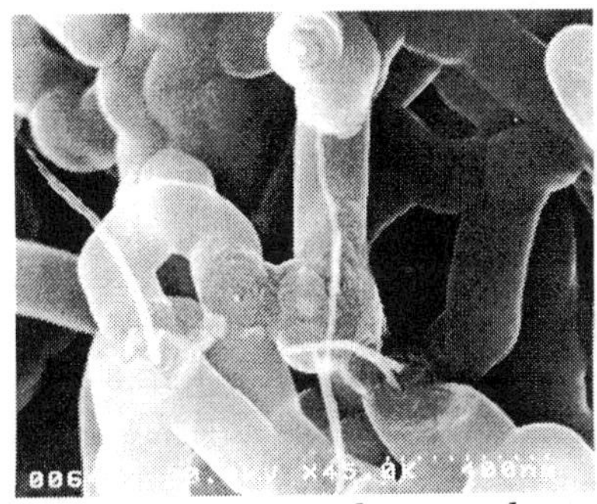

Figure 6   SEM image showing the plastic deformation and fracture of CNT

Figure 7 SEM image showing helical appearance of CNT after breakage of wrapped amorphous carbon layer

In the case of amorphous carbon wrapped multi-walled carbon nanotubes (synthesized by CVD process), the Helical appearance of MWNTs can be observe after the breakage of wrapped amorphous carbon layer (*Figure 7*).

For ideal MWNTs, the layer interact between each other only through van der Waal forces, slippage between one another is possible and which may provide a mechanism for the strain relaxation and plastic deformation. Similar results were also reported by other researchers,and which is explained by the sword-in-sheath type fracture mechanism.   High resolution TEM image(*Figure 8*) shows that bonding breakage prefers to initiate at outmost graphene layer where a higher tension tends to build up. Accordingly, some covalent bonds within the inner graphene layers are possibly to be broken locally and simultaneously. The resulted hole in MWNTs becomes a precursor of fracture.   Heavily distorted and merged graphine layers can be observed near the necking/breaking area.

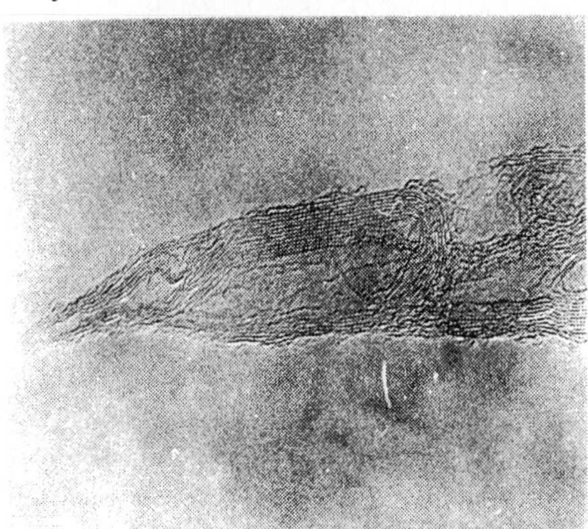

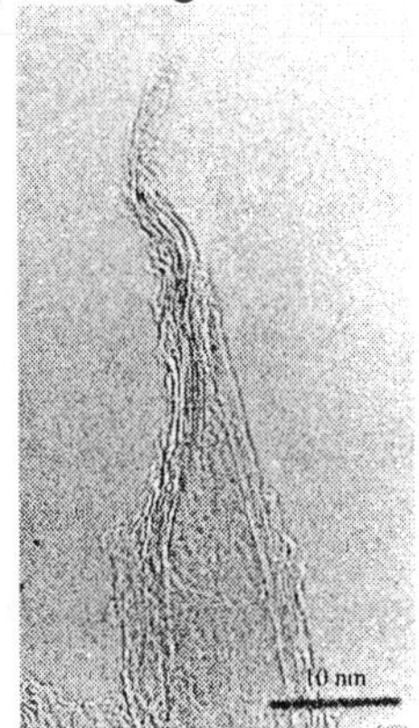

Figure 8 HRTEM image of fractured CNT

Figure 9 HRTEM image of fractured CNT

It is suggested that vigorous motion of tubes results in collisions and slidings between the graphene layers which leads to the merging of graphene layers (*Figure 9*).   Yakobson and his colleagues(1996).suggested that after the nucleation of a first (5-7-7-5) defect in the hexagonal network, either brittle cleavage or plastic flow were possible, depending on tube symmetry, applied tension, and temperature.   They also stressed that under high strain and low temperature conditions, all tubes are brittle.   We have showed that the significant plastic extension can occur in multi-walled carbon nanotubes even under high strain and low temperature conditions.   Further work must be undertaken to interpret the mechanisms of deformation and fracture of carbon nanotubes.

4. Conclusions

We have developed a technique with which deformation and fracture behaviour of carbon nanotubes under a tensile load can be examined.   It is believed that slippage between the graphene layers is responsible for the strain relaxation and plastic deformation of multi-walled carbon nanotubes.

5. References

M. Buogiorno Nardelli, B. I. Yakobson and J. Bernholc,   "Brittle and Ductile Behaviour in Carbon Nanotubes", Physics Review Letters, 81(21), 4656-4659(1998).

Z M. R. Falvo, G. Clary, R. M. Talor II, V. Chi, F. P. Brooks, Jr., S. Washburn and R. Superfine,   "Bending and Buckling of Carbon Nanotubes under Large Strain", Nature, 389, 582-586 (1997).

. W. Pan, S. S. Xie, B. H. Chang, C. Y. Wang, W. Liu, W. Y. Zhou, and W. Z. Li, "Very Long Carbon Nanotubes", Nature, 394(13), 631-632 (1998).

Z. W. Pan, S. S. Xie, B. H. Chang, L. F. Sun, C. Y. Wang, W. Y. Zhou, and G. Wang, and D. L. Zhang, "Tensile Tests of Very Long Aligned Multiwall Carbon Nanotubes", Applied Physics Letters, 74(21), 3152-3154 (1999).

B. I. Yakobson, C. J. Brabec and J. Bernholc,   "Nanomechanics of Carbon Tubes: Instabilities Beyond Linear Response", Physics Review Letters, 76, 2511-2514(1996).

M. F. Yu, O. Lourie, M. J. Dyer, K. Moloni, T. F. Kelly, and R. S. Ruoff, "Strength and Breaking Mechanism of Multiwalled Carbon Nanotubes Under Tensile Load", Science, 287(28), 637-640 (2000).

Molecular Dynamics Simulation on Dependence of Atomic-Scale Stick-Slip Phenomenon upon Probe Tip Shape

Jun Shimizu   Hiroshi Eda   Libo Zhou

Ibaraki University, 4-12-1 Nakanarusawa, Hitachi 316-8511

Abstract

This study intends to clarify the generation mechanism of the atomic-scale stick-slip phenomenon just like observed through the surface observation process using the atomic force microscope (AFM).   This paper reports on the correspondence between probe tip shape and generation form of the atomic-scale stick-slip phenomenon, it has been analyzed by use of the molecular dynamics (MD) simulation.   A 3-dimensional model is proposed where the specimen and the probe are assumed to consist of mono-crystalline copper and rigid diamond, respectively, and the effect of the cantilever stiffness is taken into consideration.   From the simulation results, it is clarified that the generation form of atomic-scale stick-slip phenomenon differed, when the probe has single atom tip or plane tip consisted of several atoms.

Keywords

Atomic force microscope, Stick-slip, Molecular dynamics, Cantilever

1. INTRODUCTION

The atomic force microscope (AFM) is one of the microscopes widely used in the fields of advanced technology, because it is applicable to any material and possesses atomic-scale resolution even in air.   Atomic-scale stick-slip phenomenon was discovered by Mate et al.[1] thorough the AFM surface measurement for the first time.   It is a unique phenomenon in such friction process at an atomic level.   Recently, many researchers have made efforts to clarify the atomic-scale friction mechanism experimentally by using the AFM [2] and analytically by using the molecular dynamics (MD) method [3] or something else [4], and some worthwhile insights have been reported.   However, there remain unknown problems, that is, whether a surface image obtained through the AFM surface observation is expressed by each specimen surface atom or periodicity of the specimen surface, and the behavior of the specimen atoms at contact point, and so on.

This study aims to clarify the generation mechanism of atomic-scale stick-slip phenomenon as observed through the surface observation process by the AFM [1,2].   This paper reports on the difference of the generation form of atomic-scale stick-slip phenomenon when the probe has different type tip shapes.   It has been analyzed by use of the MD simulation.

2. SIMULATION MODEL

The specimen and the probe are assumed to consist of mono-crystalline copper and rigid diamond, respectively. The scanning direction of the probe is taken as x axis and direction perpendicular to the sliding direction is taken as y axis. The x-y plane corresponds to specimen surface. The direction perpendicular to the specimen

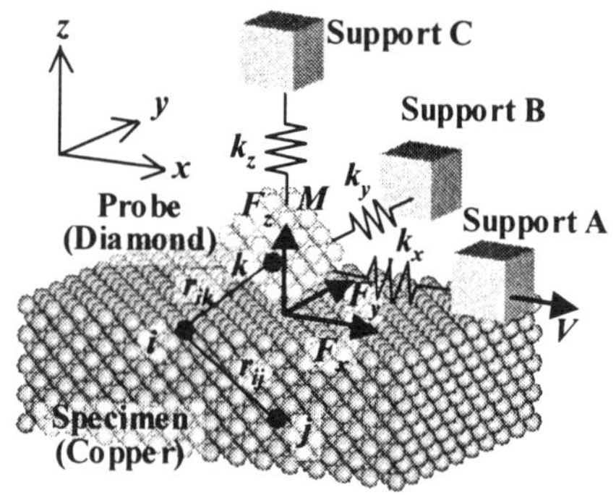

Figure 1. MD simulation model

surface is taken as z direction. They are chosen according to a right hand coordinate system. A Morse potential [5] existed is applied for the description of interaction between a pair of copper atoms. Because the actual potential between a pair of a copper and a carbon atom has not been clarified, a Morse potential proposed by Inamura et al. [6] is applied. 723 Cu atoms are subject to the simulations, here. In order to consider the influence of spring constant of AFM cantilever, an analytical model as shown in Figure 1 is proposed. The probe tip and the supports A, B and C are connected by means of springs in x, y and z directions, their spring constants are k_x, k_y and k_z, respectively. Support A moves at the constant sliding speed V and support B and C move at the same sliding speed as that of the probe tip with keeping the constant distance between the specimen surface. In the friction process, spring forces exert on the probe tip. Now, by taking the mass of the probe tip to be M, its position in the x direction at time t to be $x(t)$, its relative displacement to the support in the y and z directions to be $y(t)$ and $z(t)$, respectively, and the force exerted on the probe tip in each direction to be $F_x(t)$, $F_y(t)$ and $F_z(t)$, respectively, the following equations of motion are obtained.

$$M\ddot{x} = F_x(t) + k_x\{Vt - x(t)\} \tag{1}$$

$$M\ddot{y} = F_y(t) - k_y y(t) \tag{2}$$

$$M\ddot{z} = F_z(t) - k_z z(t) \tag{3}$$

where, $F_x(t)$, $F_y(t)$ and $F_z(t)$ are obtained through the MD calculations. Leap flog method is applied to the integral calculations of Eq.(1)-(3) and Newton's motion-equations (MD calculations), here.

3. RESULT AND DISCUSSION

The x, y and z axes correspond to [100], [010] and [001] of copper, respectively. Cu(001) corresponds to the specimen surface of well-defined in an atomic level. The sliding positions or directions of the probe are set as shown in Figure 2. In this case, the probe slides almost just on top of the surface atoms (shifted $a/1000$ in the positive direction of the y axis). In order to investigate the influence of the difference of probe tip shape on friction mechanism, a carbon atom and a diamond lattice are selected as the probes as shown in Figure 3. Table 1 shows simulation parameters, respectively. To reduce the noise due to the lattice vibration on the reactive forces, initial temperature is set at absolute zero. The cantilever stiffness is set at 5 N/m in both x, y and z directions referring to that of actual AFM cantilever.

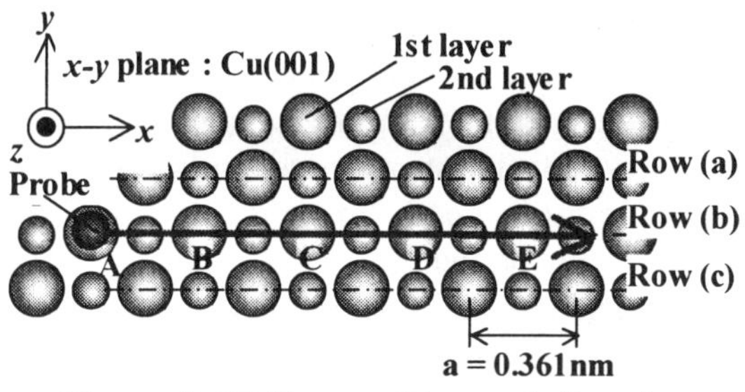

Figure 2. Sliding position and direction

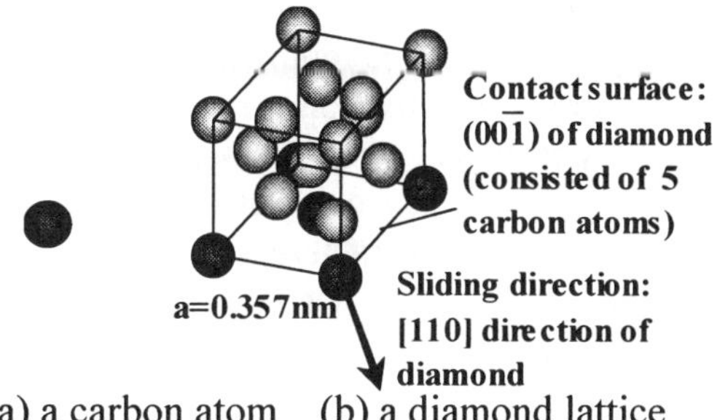

(a) a carbon atom   (b) a diamond lattice

Figure 3. Shapes of probe tip

Table 1. Simulation parameters

Probe tip	(1) a carbon atom (C)
	(2) a diamond lattice (C)
Specimen	Copper (Cu)
Initial temperature K	0
Mass of probe M kg	(1) 0.19945×10^{-26} (an atom)
	(2) 3.5901×10^{-26} (18 atoms)
Sliding speed V m/s	5
Spring constant N/m	5 ($k_x = k_y = k_z$)

3.1 By carbon atom probe

Figure 4 shows simulation results of relationship between (a) friction (x), (b) lateral(y), (c) normal(z) spring forces, (d) friction coefficient and (e) velocity of prove in x direction, and position of probe tip in x direction, respectively, by using a carbon atom probe tip.   Where, position of probe tip means that of its center.   Figure 5 shows snapshots of sliding process, where x_1 to x_4 are as shown in Figure 4(a).

In Figure 4(a), four stick-slip waveforms are confirmed and a state where the probe tip slides on top of four specimen surface atoms (B,C,D,E) are understood from friction force.   The probe tip does not rest and it slips comparatively at low speed during the stick period.   The reason why is caused by stick-slip behavior of specimen surface atom as shown in Figure 5.   By comparing friction with lateral forces, a behavior in which the probe tip wraparound specimen surface atoms (B,C,D,E) when it slides on top of the specimen surface atoms is observed from lateral force, and generation of the 2-dimensional friction phenomenon even in one-dimensional sliding process is confirmed. From normal force, a state where the probe slides over atomic irregularities of specimen surface (B,C,D,E) is confirmed. By comparing friction force, friction coefficient with velocity of probe, it is proven that the negative friction characteristics is established, as well as macroscopic stick-slip phenomenon.

3.2 By diamond lattice probe

Figure 6 shows simulation results of relationship between spring forces, friction coefficient and velocity of prove in x direction, and position of probe tip in x direction, respectively, by using a diamond lattice probe.

The stick-slip waveforms in friction force are different from those in the case of single-atom probe (see Figure 4). Because stick-slip phenomenon in this case is affected by the surface atoms of the row (a)-(c) as shown in Figure 2. 2-dimensional friction process in lateral force is also weakened comparing with Figure 4. In the change of normal force, the state where the probe tip slides over atomic irregularities of specimen surface is hardly recognized

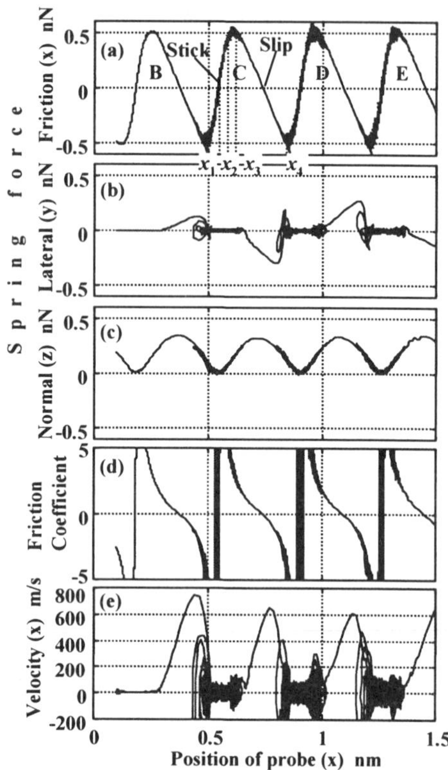

Figure 4. Relationship between spring forces, friction coefficient and velocity of prove, and position of probe tip (single atom probe).

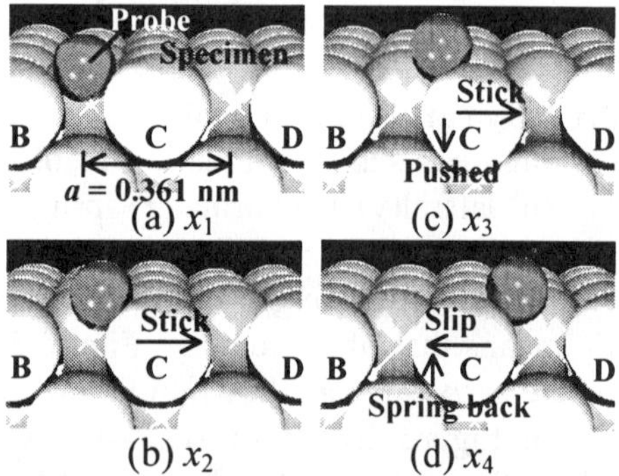

Figure 5. Snapshots of sliding process (single atom probe)

because the specimen surface atoms are remarkably pushed to the depth direction when the probe slides over them.

4. CONCLUSIONS

In order to clarify the generation form of atomic-scale stick-slip phenomenon, which depends on probe tip shape in AFM surface observation, 3-dimensional MD simulations have been performed. Conclusions obtained are summarized as follows:

(1) The atomic-scale stick-slip phenomenon can be observed, when the probe with the plane tip consisted of several atoms is utilized, even though the stick-slip waveform is different from that of single atom tip.

(2) The state where the probe slides over atomic irregularities of specimen surface is hardly recognized from normal force when the plane tip probe is utilized, because the specimen surface atoms are remarkably pushed to the depth direction when the probe slides over them.

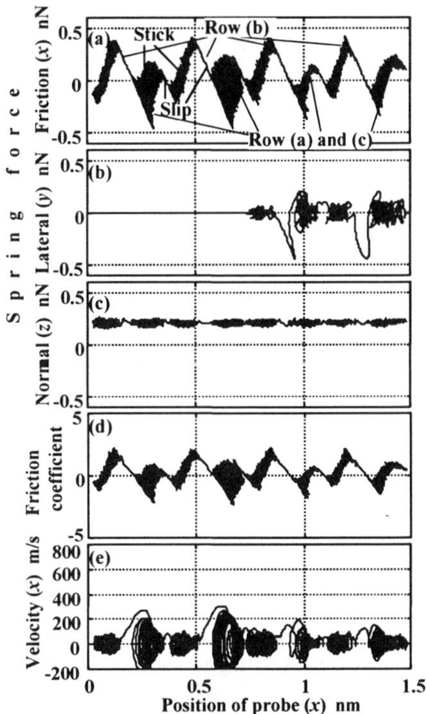

Figure 6. Relationship between spring forces, friction coefficient and velocity of prove, and position of probe tip (a lattice probe).

ACKNOWLEDGEMENT

This research was partially sponsored by the Grant-in-Aids for Scientific Research from the Japan Society for the Promotion of Science.

REFERENCES

[1] Mate C.M., McClelland G.M., Erlandsson R., Chiang S. Atomic-scale friction of a tungsten tip on a graphite surface. *Phys. Rev. Lett.* 1987; 59:1942-45.

[2] Morita S., Fujisawa S., and Y. Sugawara, Spatially quantized friction with a lattice periodicity. *Surface Science Reports* 1996; 23:1-41.

[3] e.g. Landman U., Luedtke W.D. Ribrarsky M.W. Stractural and dynamical consequences of interactions in interfacial systems. *J. Vac. Sci. Technol.* A 1989; 7:2829-39.

[4] Sasaki N., Tsukada M., Fujisawa S., Sugawara Y., Morita S., Kobayashi K. Analysis of frictional-force image patterns of a graphite surface. *J. Vac. Sci. Technol* B 1997; 15:1479-82.

[5] Girifalco L.A., Weizer V.G. Application of the Morse potential function to cubic metals. *Phys. Rev.* 1959; 114:687-90.

[6] Inamura T., Takezawa N. Atomic-scale cutting in a computer using crystal models of copper and diamond. *Annals of CIRP* 1992; 41:21-4.

TOOL WEAR MONITORING IN MILLING PROCESS WITH LASER SCAN MICROMETER

Takashi MATSUMURA, Takahiro MURAYAMA, and Eiji USUI
Tokyo Denki University

Abstract

An approach is presented to monitor flank wear in milling process with laser scan micrometer. Flank wear is monitored with measuring the change of cutter diameter and height. Being performed during air cutting time, the presented approach predicts tool wear process based on monitored data. Monitoring intervals are optimized with estimating prediction error and time loss for monitoring.

Keywords

Monitoring, Milling, Tool wear, Laser scan micrometer

1. INTRODUCTION

Tool wear monitoring is an important issue in long time operations such as mold machining. Monitoring approaches are generally classified into direct method which observes tool wear with optical instrument [1] and indirect method which estimates tool wear size with detecting the change of cutting process signals such as cutting force and AE [2]. As for direct monitoring, the recent advance of laser measuring technique brings about the possibility of reliable direct monitoring. This paper presents an approach for tool wear monitoring with laser scan micrometer. The presented approach measures the change of edge shape during air cutting time. Sequential prediction of wear process and sequential determination of monitoring intervals are discussed to apply the monitoring approach to practical use.

2. TOOL WEAR MONITORING

This paper discusses monitoring of flank wear on side cutting edge and end cutting edge in Fig. 1. Flank wear land on side cutting edge can be monitored with measuring the change of cutter diameter during air cutting time. Laser scan micrometer is placed on machine tool table so that scanning plane is perpendicular to tool rotation axis as shown in Fig. 2. Fig. 3 shows the cross section of tool in scanning plane. Laser beam, which is scanned from Edge A to Edge B, cannot reach to light requirement device while tool

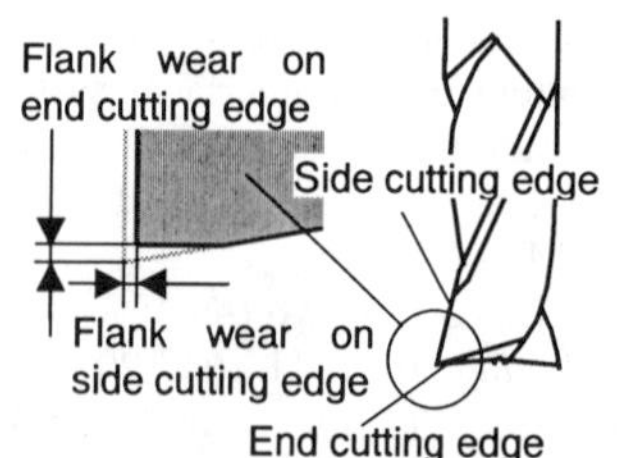

Fig. 1 Flank wear of milling tool

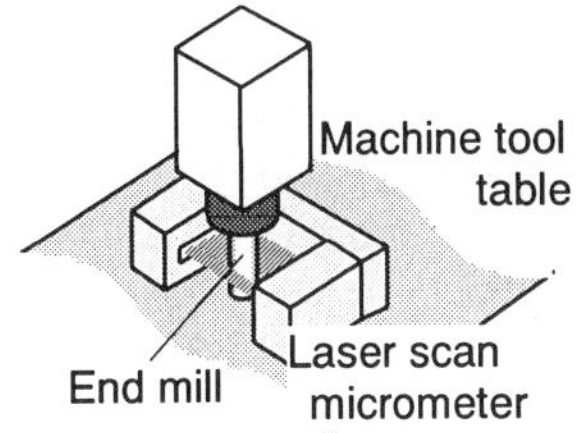

Fig. 2 Measurement of cutter diameter

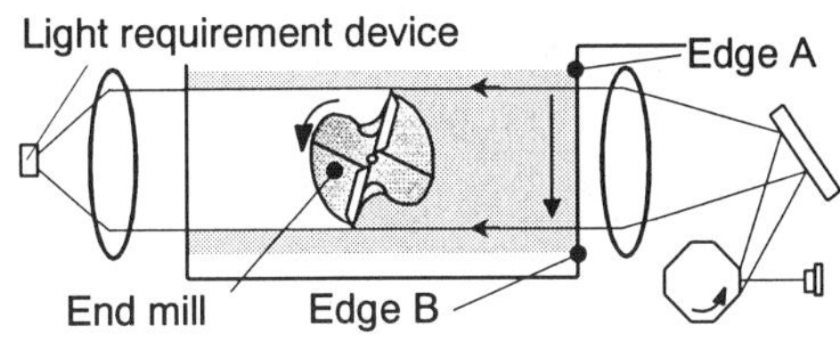

Fig. 3 Measurement of cutter diameter with laser scan micrometer

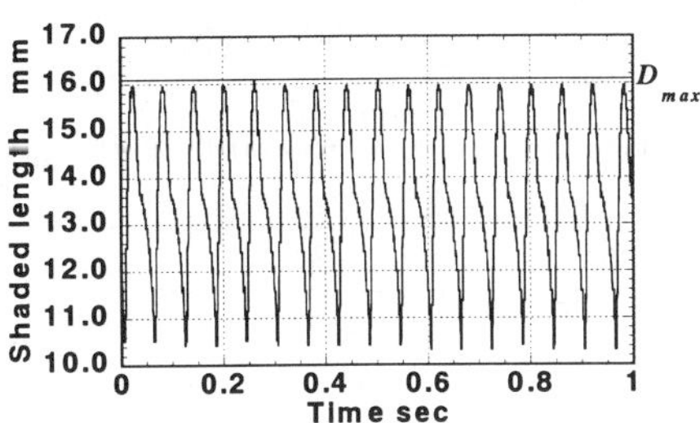

Fig. 4 Shaded length during rotating cutter

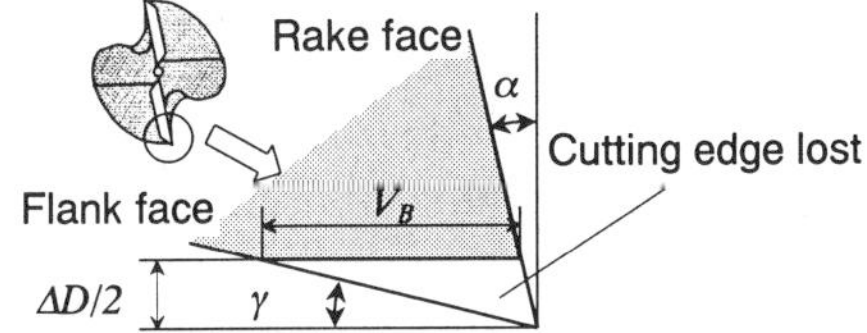

Fig. 5 Flank wear on side cutting edge

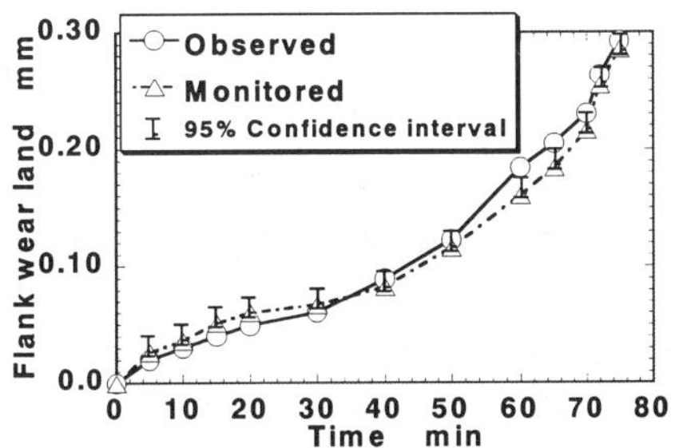

Fig. 6 Flank wear process on side cutting edge
Cutting conditions: workpiece, 0.45% Carbon steel; cutting speed, 50m/min; feed rate, 100mm/min; axial depth of cut 8mm; radial depth of cut, 0.2mm; tool, HSS, $\phi16mm$, and 2edge; lubrication, dry.

intercepts laser beam. Laser scan micrometer can measure the length intercepted by tool. The length shaded by rotating cutter changes as shown in Fig. 4. Because cutter rotates during laser scanning, cutter diameter D can be given as follows:

$$D = \frac{D_{max}}{\cos\left(\dfrac{\omega D_{max}}{2v}\right)} \quad (1)$$

where ω and v are angular velocity of tool and laser beam scanning rate respectively. D_{max} is the maximum value in Fig. 4. Cutter diameter decreases with tool wear as shown in Fig. 5. Flank wear V_B can be calculated by:

$$V_B = \left(\frac{1}{\tan\gamma} - \tan\alpha\right)\frac{\Delta D}{2} \quad (2)$$

where α and γ are rake angle and relief angle of milling cutter; and ΔD is the decrease of diameter. Coolant is supplied before measurement to clean and cool tool edge. Fig. 6 shows flank wear process monitored with 95% confidence intervals and the process observed by microscope. It is verified that the presented approach can monitor flank wear process accurately.

As for monitoring of flank wear on end cutting edge, laser scan micrometer is placed on machine tool table so that cutter axis is in the scanning plane as shown in Fig. 7. Because laser scan micrometer can

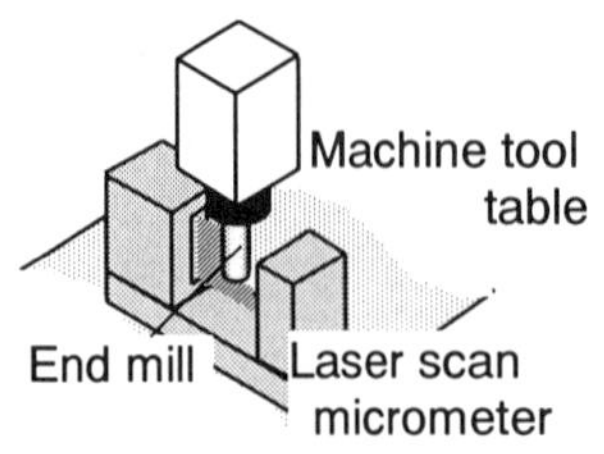

Fig. 7 Measurement of cutter height

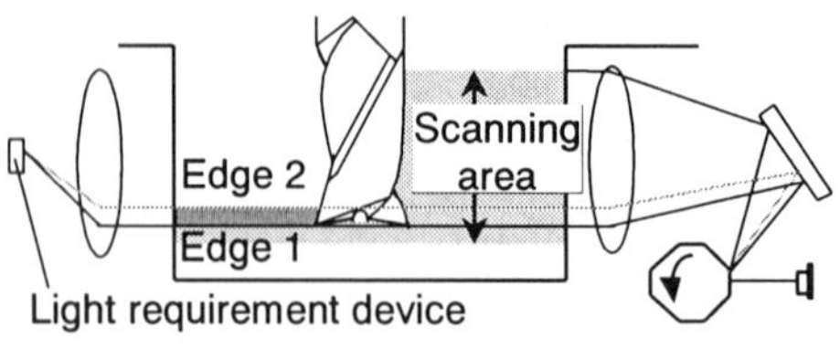

Fig.8 Measurement of cutter height with laser scan micrometer

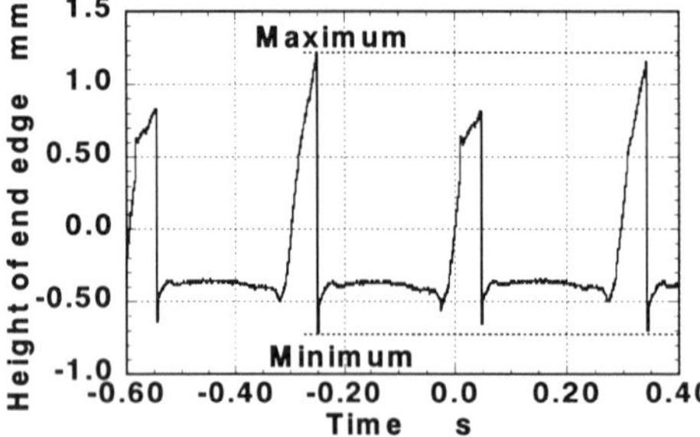

Fig. 9 Height of end edge

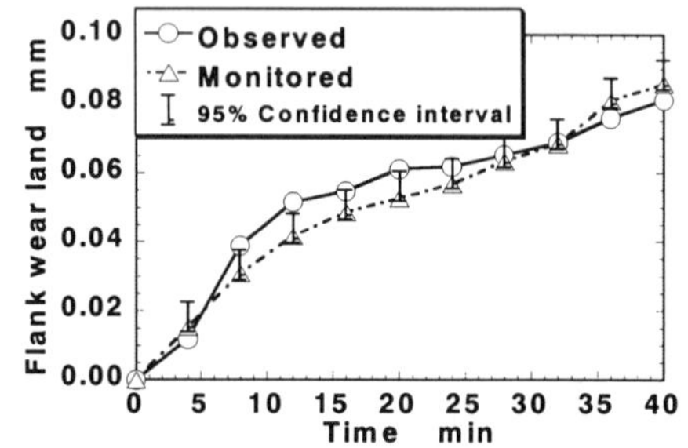

Fig. 10 Flank wear on end cutting edge

Fig. 11 Flank wear process on end cutting edge
Cutting conditions: workpiece, 0.45% Carbon steel; cutting speed, 50m/min; feed rate, 100mm/min; axial depth of cut 0.5mm; radial depth of cut, 16mm; tool, HSS, φ16mm, and 2edge; lubrication, dry.

measure the distance from the end of shaded area by tool to end of scanning area in Fig. 8, the distance varies with cutter rotation as shown in Fig. 9. The maximum value indicates the height at center of cutter (Edge 2); and the minimum one is the height at the cutting edge (Edge 1). Flank wear on end cutting edge is monitored with measuring the distance between Edge 1 and Edge 2. Flank wear length V_{BL} in Fig. 10 is given as follows:

$$V_{BL} = \frac{\Delta h}{\tan \beta} \tag{3}$$

where Δh is the decreases of the distance between Edge1 and Edge 2 from initial one; and β is inclination angle. It is proved that the presented approach can monitor flank wear length accurately as shown in Fig. 11.

The presented approach has the following features:

(1) The monitoring requires only the original tool shape. This approach does not need experiments to get the relation between the cutting process signals and tool wear, which are usually performed in indirect methods[2].

(2) Flank wear monitoring is independent of workpiece, cutting conditions, and machine tool because of monitoring in air cutting process. Cutting processes do not affect the monitoring performance.

3. SEQUENTIAL WEAR PREDICTION

Performed only during air cutting, the presented approach cannot monitor tool wear all the time in cutting process. Therefore tool wear until the next monitoring can be predicted by extrapolation with linear regression. Linear regression at time T_i is sequentially made using n recent data monitored from T_{i-n+1} to T_i. Flank wear at T_{i+1} can be predicted by extrapolating on the regression line. Fig. 12 shows an example of prediction,

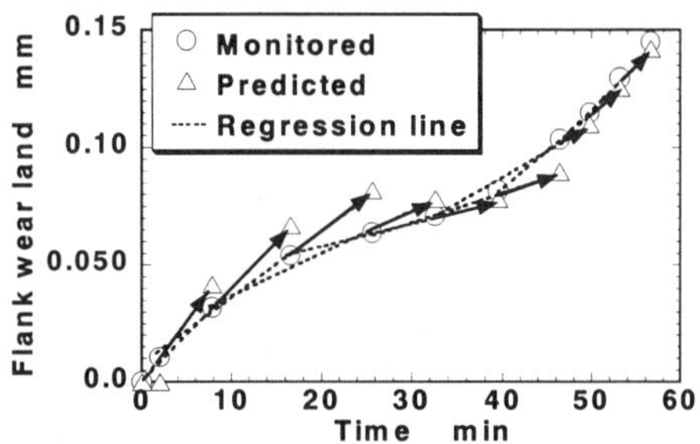

Fig. 12 Sequential prediction of flank wear with optimizing monitoring intervals
Prediction order, n=3; machining time, 60min; maximum flank wear land, 0.15mm.

where broken line shows regression line. Flank wear at each monitoring can be predicted by extrapolating on the solid line with arrow.

4. SEQUENTIAL DETERMINATION OF MONITERING INTERVAL

Because this approach has time loss for monitoring, monitoring intervals are an important issue. If monitoring interval is short, total time loss in an operation increases with the number of measurements though error of wear prediction decreases due to frequent monitoring. On the other hand, if monitoring interval is long, time loss decreases. However discrepancy of prediction increases with monitoring interval. Therefore proper monitoring intervals should be determined with considering expected prediction error and time loss for measurement. Error rate $(dEr/dt)_j$ at T_j can be calculated by:

$$\left(\frac{dE_r}{dt}\right)_j = \sqrt{\left(\frac{\left(V_{B,pred}\right)_j - \left(V_{B,mon}\right)_j}{T_j - T_{j-1}}\right)^2} \tag{4}$$

$(V_{B,pred})_j$ is flank wear at T_j, which is predicted at T_{j-1}; and $(V_{B,mon})_j$ is monitored one at T_j. Error rates from T_{i-n} to T_i are averaged as follows so that more recent error rate has a larger influence on the calculation of intervals:

$$\left(\frac{dE_r}{dt}\right)_{ave} = \int_{T_{i-n}}^{T_i} \frac{t - T_{i-n}}{T_i - T_{i-n}} \left(\frac{dE_r}{dt}\right)_i dt \left/ \int_{T_{i-n}}^{T_i} \frac{t - T_{i-n}}{T_i - T_{i-n}} dt \right. \tag{5}$$

Prediction error for ΔT at time T_i is $(dEr/dt)_{ave}\Delta T$. Expected prediction error $f(\Delta T)$ is given by normalizing $(dEr/dt)_{ave}\Delta T$ on the scale of tolerable maximum flank wear V_{Bmax}:

$$f(\Delta T) = \left(\frac{dE_r}{dt}\right)_{ave} \Delta T \left/ V_{B\,max} \right. \tag{6}$$

On the other hand, total time loss $T_{mon,all}$ at time T_i can be evaluated by the following equation on the assumption that monitoring in the rest of

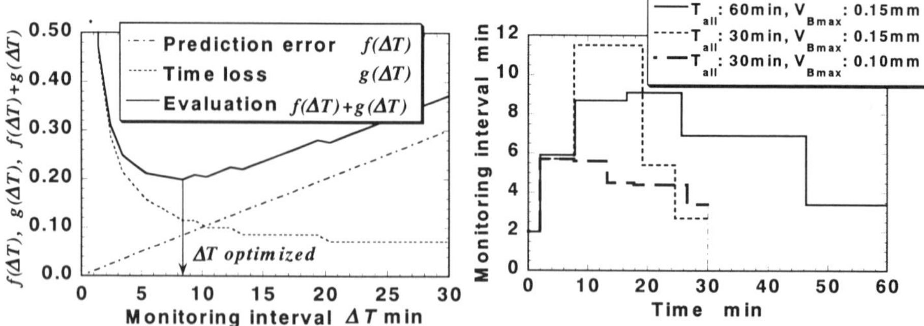

Fig. 13 Optimization of monitoring intervals

Fig. 14 Effect of T_{all} and V_{Bmax} on monitoring intervals

machining time T_{res} is performed at the interval ΔT:

$$T_{mon,all} = \begin{cases} T_{mon}\{(\text{int})(T_{res}/\Delta T)+i-1\} & (T_{res}=k\Delta T) \\ T_{mon}\{(\text{int})(T_{res}/\Delta T)+i\} & (T_{res}\neq k\Delta T) \end{cases} \qquad (7)$$

where k is integer; and T_{mon} is monitoring time. Expected time loss $g(\Delta T)$ can be given by normalizing $T_{mon,all}$ on the scale of total machining time T_{all}:

$$g(\Delta T)=T_{mon,all}/T_{all} \qquad (8)$$

Fig. 13 shows the change of $f(\Delta T)$, $g(\Delta T)$, and $f(\Delta T)+g(\Delta T)$ with ΔT. $f(\Delta T)$ increases monotonously with ΔT; while $g(\Delta T)$ decreases stepwise with increasing ΔT. Consequently $f(\Delta T)+g(\Delta T)$ has the minimum value for ΔT. Therefore monitoring interval ΔT can be given to minimize $f(\Delta T)+g(\Delta T)$. Monitoring intervals in Fig. 12 are determined according to the above procedure. Monitoring intervals change with prediction error. Monitoring intervals are short after large prediction error at 25min and 46 min; and the interval is long after small error at 39min. Fig. 14 shows the effect of machining time T_{all} and V_{Bmax} on monitoring intervals. It is shown that short machining time makes monitoring intervals long; and small tolerable maximum flank wear V_{Bmax} makes the intervals short.

5. CONCLUSION

Flank wear in milling process is monitored by measuring the change of cutter diameter and height with laser scan micrometer. Wear process can be predicted by extrapolation with linear regression of monitored data. Monitoring intervals are optimized with estimating prediction error and time loss for monitoring.

REFERENCE

1. Jeon, J.U. and Kim, S.W. Optimal Flank Wear Monitoring of Cutting Tool by Image Processing. Wear 1988; 127: 207-217.
2. Byrne, G., Dornfeld, D., Inasaki, I., Ketteler, G., Konig, W., and Teti, R. Tool Condition Monitoring (TCM): The Status of Research and Indusrial Application. Annals of the CIRP 1995; 44: 541-567.

STUDY ON A MODIFIED CONCAVE FILTER FOR MULTI-OSCILLATING MODALITIES COMPENSATION IN ULTRA-PRECISION MACHINE TOOL

Baoku Su[1], Lisong Wang[1], Shen Dong[2]

(1.Dept.of Control Theory and Engineering, Harbin Institute of Technology)

(2.Dept.of Mechanical Engineering and Automation, Harbin Institute of Technology)

Abstract

The existence of several oscillating modalities is the main factor that affects the performance of the ultra-precision machine tool control system. To compensate for this effect, a single Butterworth concave filter was added at different place and in order to improve the amplitude margin, a modified Butterworth concave filter was also studied. By applying the modified Butterworth concave filter to the practical system, the profile accuracy had been increased from 0.6μm to 0.3μm.

Keywords

Precision machine tool   Concave filter   Oscillating modality
Butterworth filter

1. INTRODUCTION

Precision machine tool is one of the most important instruments to produce metal optics. Generally, due to some mechanical structure restrictions of the machine tool, there are multi-oscillating modalities in the servo system, such as the distortion resonance of the shaft-joint, oscillation produced by aerostatics guide when the stiffness of its gas film is low. These vibrations will then be translated into the surface of the part as surface finish error. So, these resonance modalities are the decisive factors that affect the positioning accuracy of machine tool servo system greatly. Generally Butterworth concave filters are used to compensate the effects of multi-oscillating modalities, but its compensation effect is limited. In this paper, a modified Butterworth filter was studied to improve the control accuracy of the precison machine tool.

2. SYSTEM CONFIGURATION

In the servo system of the ultra-precision machine tool, a servomotor driven by an AC servo driven unit and a ball screw are used to implement the position servo feed. A two-frequency laser interferometer is used to detect the actual position. An aerostatics pressure guide and air-floating table are used to reduce the effect of friction generated during the transmission process. In the servo driven unit, a closed loop of speed regulation structure is used, in which a proportional regulator is adopted. The dynamic model of the position servo system is shown in figure 1:

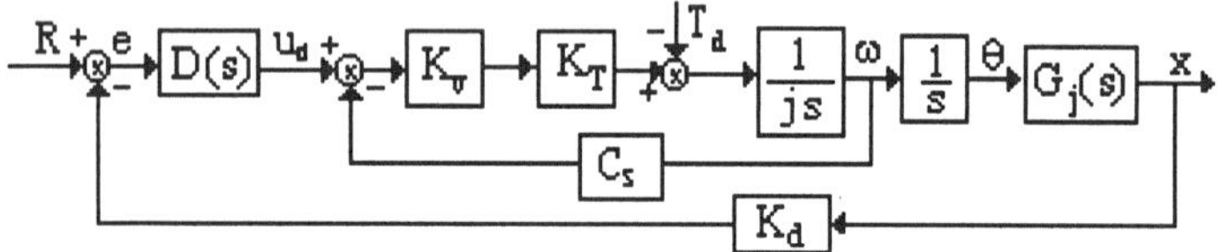

Fig1. Dynamic model of the position servo system

where K_T is the motor moment constant equal to 16N •m/v; D(s) is a position controller. The proportionality coefficient of the tachometer C_S equals to 0.796V/Rad/s. K_d is the circuit enlargement factor of the laser interferometer, which equals to 1. The detection resolution of the laser interferometer is 5nm. T_d is the disturbance moment of the system mainly produced by the friction and motor's ripple torque. $G_j(s)$ is a mechanical transmission link. The proportionality coefficient of the rate fixer k_v equals to 32 s^{-1}. R is the input signal and x is the position output of the table. By system identification, an open loop transfer function of the servo system is obtained as equation 1; it can be seen that the resonance appears at the frequency of 43Hz, 65Hz and 83Hz.

$$G(s) = \frac{0.6912 \cdot (\frac{1}{307^2}s^2 + \frac{0.28}{307}s + 1) \cdot (\frac{1}{415^2}s^2 + \frac{0.9}{415}s + 1) \cdot (\frac{1}{584^2}s^2 + \frac{1}{584}s + 1)}{s \cdot (\frac{1}{285.5}s + 1) \cdot (\frac{1}{264^2}s^2 + \frac{0.32}{264}s + 1) \cdot (\frac{1}{333^2}s^2 + \frac{0.14}{333}s + 1) \cdot (\frac{1}{490^2}s^2 + \frac{0.26}{490}s + 1)} \tag{1}$$

3. CONTORLLER DESIGN

As a type-I system, the goal of its design is to make the servo system with broad bandwidth and high stiffness to get a fast response speed and anti-disturbance capability[1]. Firstly, a conventional PI type controller is designed for the system with the bandwidth of 10Hz. So, the controller is:

$$D(s) = \frac{285.599 \cdot (0.3183s + 1)}{s} \tag{2}$$

As mentioned above, two harmonic peaks appear at the frequency of 246rad/sec and 408rad/sec, which bring about sinusoidal vibration error in system, So concave filters should be used to restrain the effect of the two oscillating modalities. As the Butterworth filter [3] has the characteristic of the monotony amplitude response of a flat pass band, so a two-level Butterworth filter H(s) was adopted. The dynamic equation of H(s) is expressed in equation 3:

$$H(s) = \frac{s^2 + \omega_0^2}{s^2 + Bs + \omega_0^2} \tag{3}$$

where ω_0 is the center of the inhibited band with a bandwidth of B. As only a single filter is used to compensate for multi-oscillation, the filter frequency ω_0 has to be chosen carefully. A concave filter added at different frequency in the system is studied. The frequency characteristic curve of the open-loop system is shown in figure 2, where curve 1, and curve 2 correspond to a concave filter added at the frequency of 264rad/sec and 522rad/sec respectively.

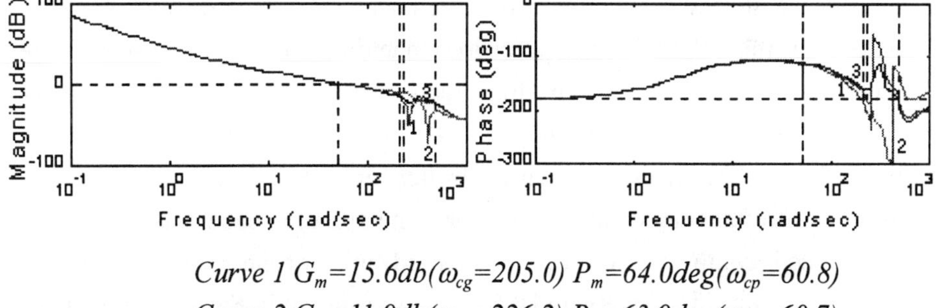

Curve 1 G_m=15.6db(ω_{cg}=205.0) P_m=64.0deg(ω_{cp}=60.8)
Curve 2 G_m=11.9db(ω_{cg}=226.2) P_m=63.9deg(ω_{cp}=60.7)
Curve 3 G_m=21.6db(ω_{cg}=462.4) P_m=67.3deg(ω_{cp}=60.9)
Fig 2. Open-loop frequency characteristic curve of the system with different concave filter methods

As shown in figure 2 that a single concave filter can increase the amplitude margin of the system with a proper value of parameter B. And when the concave frequency is set at the first oscillating frequency of 264rad/sec, the system has the better relatively stable performance. But the amplitude margin of the system is not enough. In order to improve the amplitude margin further, a modified concave filter is studied, which is expressed by the equation 4.

$$H^{'}(s) = \frac{s^2 + B^{'}s + \omega_0^2}{s^2 + Bs + \omega_0^2} \tag{4}$$

By introducing an item of B's to the numerator, the improved concave filter can make the change of phase angle at the center frequency ω_0 become gentle. Thus, when it is added to the system, there will be a phase displacement of -180° at the higher frequency, and the amplitude margin is high. The open-loop frequency characteristic of the system with improved concave filter having the same value of B and concave frequency of curve1 is shown as curve 3 in figure 2, which has the best relatively stable performance compared to the curve1 and curve2.

4. EXPERIMENT RESULTS

To testify the feasibility of the methods mentioned above, the controller with and without the modified Butterworth concave filter was applied to the HCM-1 ultra-precision machine tool servo system. The machine tool is a two freedom ultra-precision lathe, the servo system of y-axis has the similar characteristic as that of x-axis which was mentioned above. So we designed the similar controller for y-axis. Two aspheric optical surfaces were machined for comparison. The spindle speed was 1440r/min and the feed length was 2μm/rev. The turned finish of the parts was detected by STM produced by American DI Co and was shown in Fig3. It can be seen that the surface roughness Ra is 14.976nm and rms is 21.904nm when the controller had no concave filters, and Ra is 4.547nm and rms is 14.928nm when the controller adopted the concave filter. The profile accuracy was also measured, which had been increased from 0.6μm to 0.3μm when the modified concave filter was applied.

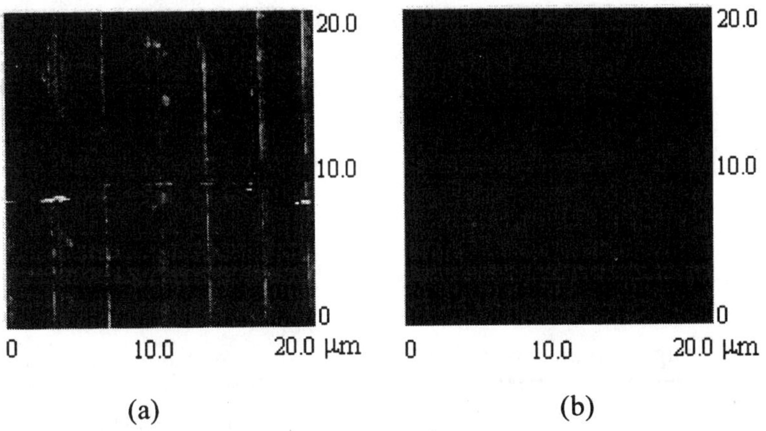

(a) Ra: 14.976nm   rms: 21.904nm without modified concave filters
(b) Ra: 4.547nm   rms: 14.928nm with a modified concave filter
Fig 3.The comparison figure of roughness analysis

5. CONCLUSION

Digital concave filters can be utilized to compensate for several oscillating modalities in the ultra-precision machine tool servo system. This paper studied the method of adding concave filters to compensate for the effect of oscillating modalities. First we studied the method of using a single Butterworth concave filter to compensate for several oscillating modalities. Then an improved concave filter was proposed, which has the better relatively stable performance. In order to test the effect of the improved filter to compensate for the oscillating modalities, comparison experiment was performed, which shows that it is an effective method to use digit concave filter to increase the profile accuracy of the parts in ultra-precision maching.

REFERENCES

[1] Guangxiong, Control system design. Aerospace publishing house.1992

[2] Wei Kexin, MATLAB language and automatic control system design. Mechanical industry publishing house.1997.

[3] D.E Johnson, The speedy utility design of active filter. People's postage publishing house.1980.

NEW PRACTICAL CONTROL OF POINT-TO-POINT POSITIONING SYSTEMS : ROBUSTNESS EVALUATION

Wahyudi [a], K. Sato [a] and A. Shimokohbe [b]

[a] Interdisciplinary Graduate School of Science and Engineering,
[b] Precision and Intelligence Laboratory,
Tokyo Institute of Technology
4259 Nagatsuta, Midori-ku, Yokohama 226-8503, Japan

Abstract

This paper presents robustness evaluation of the new practical control method for point-to-point (PTP) positioning systems. The proposed controller consists of a nominal characteristic trajectory element and PI elements. The trajectory element is determined experimentally based on open-loop step responses of the position and velocity of the plant. So it includes the non-linearity effects of the plant such as friction and saturation. The trajectory is also used to determine the PI coefficients. Therefore the proposed controller does not required an exact model and parameters of the plant. It is proved experimentally that the proposed controller is better than the conventional PID ones in robustness to parameter variation.

Keywords

PTP, Positioning system, nominal characteristic trajectory, PI element, robustness.

1. INTRODUCTION

Positioning systems play important role in the industrial equipment such as machine tools and robotics. They are required to have not only fast response with little or no overshoot, but also robustness. In order to satisfy the requirements, many types of controllers such as PID (Chang et al., 1997), optimal (Workman et al., 1987), sliding mode (Sankaranarayanan and Khorrami, 1997) and robust controllers (Endo et al., 1996) have been proposed. However those controllers require the plant model and its parameters which are difficult to be identified. This paper describes a new practical control method for point-to-point (PTP) positioning systems which do not require an exact model and its parameters. Robustness of the proposed controller is evaluated and compared experimentally with a conventional PID controller.

2. PROPOSED CONTROLLER DESCRIPTION

2.1 Proposed Controller Concept

The structure of the proposed controller is shown in Fig. 1. Here, the objective of the controller is to make the plant motion follow the nominal characteristic trajectory and to end the plant motion at the origin of the phase plane $(e, \dot{e})$. PI elements are used to reduce signal u_p which represents the difference between the actual error rate $\dot{e}$ and that of the trajectory. The PI elements are designed by using only information from the nominal characteristic trajectory.

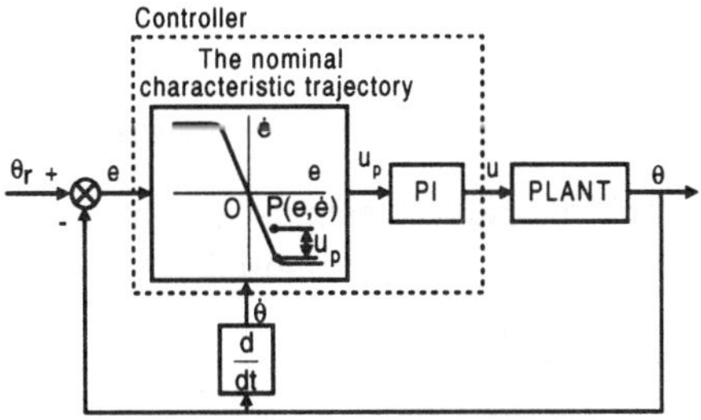

Figure 1. Proposed Controller Structure

2.2 The Nominal Characteristic Trajectory Determination

Fig. 2 shows the stepwise input adopted in this paper, and the velocity and position responses due to the input. The velocity and position responses are used to make the nominal characteristic trajectory on the phase plane. As PTP motion control is to stop an object at a certain position, the deceleration response indicated by b and d in Fig. 2(a) is used. The rated capacity of the actuator is indicated by the maximum velocity (a and c in Fig. 2(a)) when the rated input u_r is applied to the actuator. From a, b, c and d in Fig. 2(a), the trajectory a, b, c and d in Fig. 2(b) is decided.

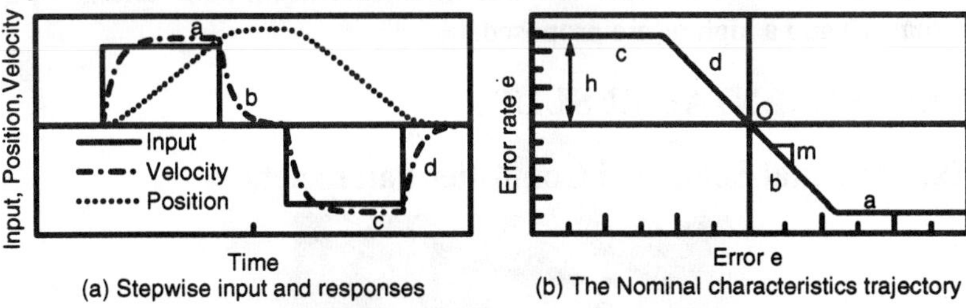

(a) Stepwise input and responses   (b) The Nominal characteristics trajectory

Figure 2. The Nominal Characteristic Trajectory Construction

2.3 PI Coefficients Design

The proposed controller does not require an exact model and parameters of the plant for controller design. Here, a theoretical background justifying the derivation is given. First, it is assumed that the plant is a following second order system :

$$\frac{\Theta(s)}{U(s)} = K \frac{\alpha}{s(s+\alpha)} \tag{1}$$

where $\Theta(s)$ represents the position of the object, $U(s)$, an input to the actuator, and K and α, positive constants. The simplified plant parameters can be related with the slope m and maximum error rate h of the nominal characteristic trajectory (Fig. 2) as follows :

$$\alpha = -m \tag{2.a}$$

$$K = -\frac{h}{u_r} \tag{2.b}$$

Based on the simplified plant dynamics at small e and $\dot{e}$ (near origin), the closed-loop transfer function of the proposed control system is :

$$\frac{\Theta(s)}{\Theta_r(s)} = \frac{\alpha}{s+\alpha} \frac{2\zeta\omega_n s + \omega_n^2}{s^2 + 2\zeta\omega_n s + \omega_n^2} \tag{3}$$

where $2\zeta\omega_n = \alpha K K_p$ and $\omega_n^2 = \alpha K K_i$. By using ζ and ω_n as design parameters, the PI coefficients can be expressed as follows :

$$K_p = \frac{2\zeta\omega_n}{\alpha K} \tag{4.a}$$

$$K_i = \frac{\omega_n^2}{\alpha K} \tag{4.b}$$

where K and α are calculated using Eqn (2.a) and Eqn (2.b). In order to realize a high response system, the design parameters should be selected so that the closed-loop dominant pole is α. So the poles and zeros influenced by ζ and ω_n should be insignificant. In order to make the insignificant poles and zero, ζ larger than 10 and a high ω_n are proposed.

3. EXPERIMENTAL RESULTS

3.1 Experimental Setup and Controller Parameters

Figure 3. An Experimental Rotary Positioning System
An experimental rotary positioning system as shown in Fig. 3 is used for

robustness evaluation. Its nominal parameters are shown in Table 1. The system consists of an AC servo motor, a servo driver and mechanical elements. The angular position is measured with an optical encoder of a $2\pi \times 10^{-4}$ rad resolution. The angular velocity used in the experiment is obtained by using a backward difference algorithm.

The proposed controller is applied to the system with nominal parameter (System 1). It is includes the nominal characteristic trajectory shown in Fig. 4. The PI coefficients designed using parameters $\zeta = 13$ and $\omega_n = 29$ rad/s. Two PID controllers are also applied to System 1. First PID controller (PID 1) is tuned using Ziegler-Nichols rule, and second PID controller (PID 2) is tuned in order to have similar bandwidth with the proposed controller. For robustness comparison, the controllers are also applied to the system with increased inertia (System 2, Inertia load of 10 x J_n) and increased friction (System 3, Friction torque of 6 x $\tau_{f\,max}$).

Table 1. Nominal Parameter of The Rotary Positioning System

Parameter	Value	Unit
Inertia load, $\mathbf{J_n}$	1.17×10^{-3}	Kgm^2
Motor resistance, $\mathbf{R}$	1.2	Ω
Motor inductance, $\mathbf{L}$	8.7	mH
Torque constant of the motor, $\mathbf{K_t}$	0.57	Nm/A
Motor voltage constant, $\mathbf{K_b}$	0.57	Vs/rad
Frictional torque, $\boldsymbol{\tau_{f\,max}}$	0.215	Nm
Viscous friction, $\mathbf{C}$	1.67×10^{-3}	Nms/rad
Proportional current controller, $\mathbf{K_{cp}}$	26.2	V/A
Integral current controller, $\mathbf{K_{ci}}$	3.619×10^{3}	V/As
Proportional velocity controller, $\mathbf{K_{sp}}$	8.6×10^{-2}	As/rad
Velocity constant, $\mathbf{K_v}$	42	Rad/Vs
Slew rate of the amplifier, S_R	0.1	V/μs

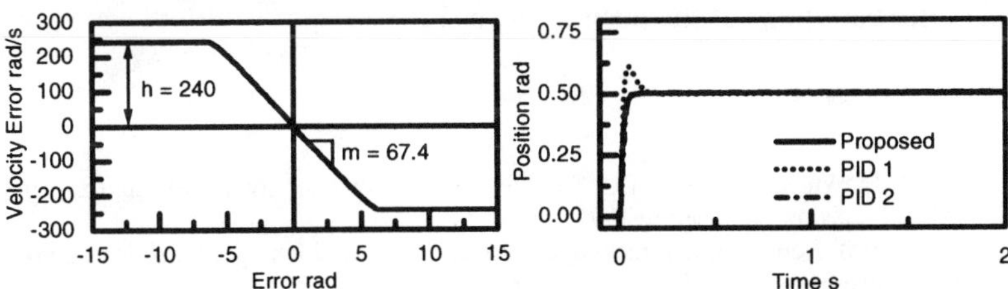

<table>
<tr><td>

Figure 4. The Nominal Characteristic Trajectory

</td><td>

Figure 5. Step Responses to a 0.5 rad Step Input, System 1

</td></tr>
</table>

3.2 Positioning Performance and Robustness Comparison

Fig. 5, 6 and 7 show responses to a 0.5 rad step input. In this paper, the positioning accuracy is defined as the positioning error at 2 s. In System 1, the positioning accuracy of the proposed controller and PID 1 and PID 2 are 1.42 x 10^{-4}, 1.42×10^{-4} and 7.66 x 10^{-4} rad respectively. Moreover the proposed controller gives a smaller overshoot than PID 1. In System 2, the positioning accuracy of the proposed controller and PID 2 are 1.42 x 10^{-4} and 2.91 x 10^{-3} rad respectively. Moreover the proposed controller give a smaller overshoot than that of PID 2. On the other hand, PID 1 results in unstable system. In System 3, the positioning accuracy of the proposed controller, PID 1 and PID 2 are 1.42 x 10^{-4}, 1.42 x 10^{-4} and 1.48 x 10^{-3} rad respectively. Hence, the proposed controller has better performance than the conventional PID controllers in terms of robustness to parameter variations of the system.

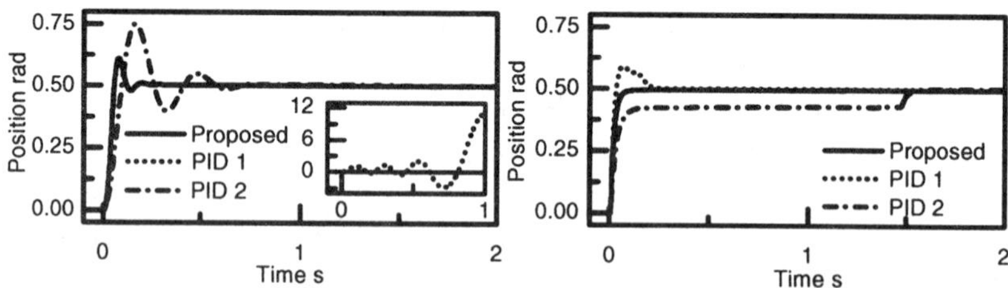

Figure 6. Step Responses to a 0.5 rad Step Input, System 2

Figure 7. Step Responses to a 0.5 rad Step Input, System 3

4. CONCLUSIONS

A new control method for point-to-point (PTP) positioning systems has been proposed and discussed. The proposed control method does not require an exact model of the plant and its parameters. The robustness of the proposed method is evaluated experimentally. The experimental results prove that the proposed controller is better than the conventional PID ones in robustness to parameter variations of the positioning system.

REFERENCES

Chang S.B, Wu S.H. and Hu Y.C. Submicrometer overshoot control of rapid and precise positioning. Precision Engineering 1997;20:161-170.

Endo S., et al. Robust digital tracking controller design for high-speed positioning systems. Control Engineering Practice 1996; 4:527-536.

Workman M.L., Kosut R.L. and Franklin G.F. Adaptive proximate time-optimal servomechanisms : continuous time case. Proceedings of the American Control Conference, 1987 June 10-12; Minneapolis, USA.

Sankaranarayanan S., Khorrami F. Adaptive variable structure control and applications to friction compensations. Proceedings of the 36th IEEE Conference on Decision & Control, 1997 December 10-12; San Diego, USA.

THE RESEARCH OF AN INTELLIGENT OPEN CNC SYSTEM

Jing Zhang, Lisong Wang, Sheng Li, Feihu Zhang

Harbin Institute of Technology, China

Abstract

Nowadays, manufacturing enterprises have come to an agreement that open architecture controllers should be set up. Demands for increased control capabilities require the open computer numeric control (CNC) to comprise intelligent building elements. This paper presented HCM-I, an intelligent open CNC system, in which some intelligent elements are developed. The graphic platform of the system make it convenient to build and (re)configure a system by choosing icons. And with the intelligent programming module of the system, users can use natural language to describe the machining parts. It can translate the natural language, such as English or Chinese, into the standard NC language. These intelligent elements bring the system more flexibility and convenience to use.

Keywords

Open CNC system   Graphic platform   Intelligent programming

1. INTRODUCTION

Open architecture CNC controls has become very popular for years. The hardware and software components of an open system can be (re)used. And so, the open system is more efficient and economic. Several well-known research projects, such as OSACA, OMAC, OSE, have been conducting research on it and gained some progress[4].

An HCM-I open CNC system is introduced here. Its software and hardware are easy to be (re)used and augmented among different platforms. By choosing and combining these components, end users can configure a satisfying system according to their practical situation and specific needs. In addition, a graphic platform module and an intelligent programming module are developed, which make it convenient for the users to control the system.

2. SYSTEM CONFIGURATION

Many of the popular used operating system, such as Windows9X/NT, can not satisfy the requirement of real-time control. So, it is more practical to build an open-architecture, PC-based controller using a readily available motion control processor[1,3]. The two major computing hardware components of our system are the operator workstation and the motion control processor.

The motion control processor can perform various time-critical functions including servo control, path planning, and programmable logic control. The operator workstation running a Microsoft Windows 9X/NT operating system can deal with the other functions, such as operator interface, simulation, diagnostics and network functions. Figure 1 illustrates the configuration of the system.

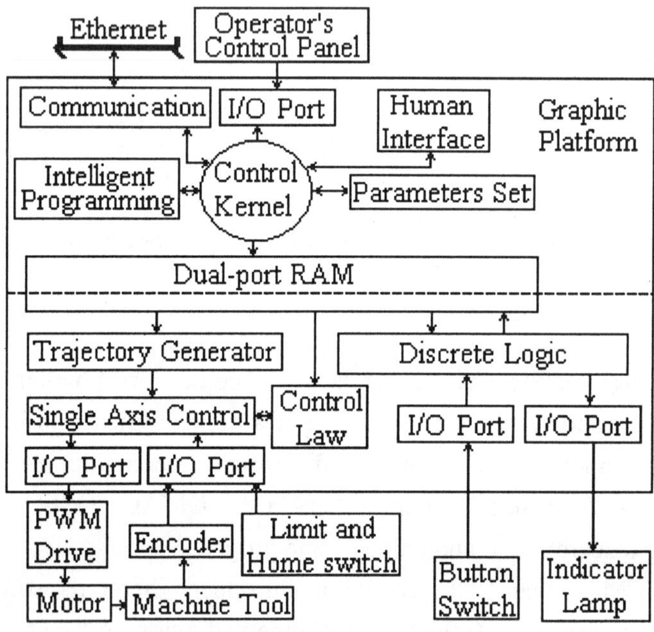

Fig1. System Configuration

Users can build and configure their CNC systems by the interface provided by the graphic platform. The control kernel of the platform can transmit data among software and hardware modules to run the whole system. Via the operator's control panel and the human interface, users can set the parameters of the motion control processor. These parameters include amplifier gains, backlash compensation, acceleration/deceleration rates, error limits, homing parameters for each axis and so on. The system can also get the part program (e.g. RS274) by the net communication module or the intelligent programming module. The Dual-port RAM is used to transmit necessary information between the platform and the motion control processor, such as part program, some parameters and the running status. The discrete logic module supports all the logic operations. The I/O ports are used to read or write to the process sensors and actuators.

3. GRAPHIC PLATFORM

Graphic platform is developed with the compiler of Visual C++. It is used to connect the motion control processor and the machine tool with the user. There are several software modules in the platform, including simulation module, inspection module, communication module and so on. Different modules fulfil different functions and the implementation of each one does not restrict that of another. This architecture makes the system robust because the responsibilities of the functional blocks are less likely to conflicts. Also, the system is quite flexible to integrate different modules with the functional requirements. Finally, new modules can be added to the system to fulfil new function without a reworking of the entire system.

The graphic platform provides a simple way for the users to build their own system. When the platform is open, an editing window is open automatically for the user to edit his system form. By opening the libraries, each of which consists of several functional modules represented by different icons, the user can choose what he need with the mouse. By pulling the icons to the edit window and connecting the arrows between different icons, the user will build his system.

With the dual-port RAM, the data are exchanged among the platform and the other components of the system. For example, the parameters of the motion control processor can be configured, the motion files can be transmitted from the platform to the motion control processor, the servo data and the running status of the machine tool are collected by the platform.

The platform can also fulfil a lot of other functions. With the data collected, an inspection module[2] can be used to show users the performance status of machine tool in the form of graphics and files. With detailed directions provided by the platform, users will find it convenient to add new developed functional modules to the platform. In addition, the technique of "client/server" is introduced to our communication module to enable the users to control the machine tool in a network environment on this platform.

4. INTELLIGENT PROGRAMMING MODULE

Intelligence is an outstanding property for an open CNC system. The traditional manual programming requires that the programmers are skilled in programming and familiar with the knowledge of part design and manufacturing. What's more, for parts with a complex structure, some errors might happen with the manual programming. But with the intelligent programming module presented here, users can use natural language to describe the machining parts. It can translate the natural language, such as English or Chinese, into the standard NC language (e.g. RS274). For example, if the sentences of "An optical parabola surface is to be machined. The equation is $y=x^2$ and the caliber is $\phi30$. ..." are input, the appropriate G code would be automatically generated.

Figure 2 is the flowchart of this module. The input natural language is mainly used to describe the actions of the machine tool, such as the machining process route of the part, the feed trajectory, the offset and some cutting parameters.

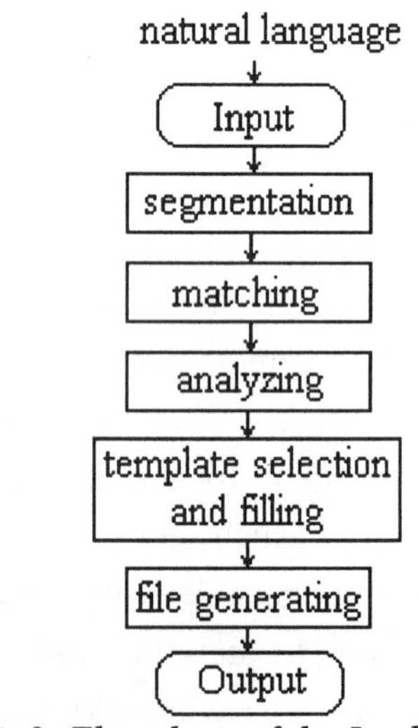

Fig2. Flowchart of the Intelligent Programming Module

First of all, a lexicon of the key words in the field of machining, such as offset, cutting, feed, part, ellipse, line and so on, should be constructed and stored in the computer. A set of templates of NC language is set too, which is a frame of the NC codes with some parameters. When a paragraph of natural language was input, it was segmented into a set of phrases and words using the principle of maximum length match, which is a kind of artificial intelligence technique. After searching and matching the key word listed in the lexicon, the meaning of the natural language is analyzed. Then, the machining process and the corresponding parameters are determined, which would be used to select and fill into the templates. By combining these templates, a file written in standard NC language is generated and output at last.

5. CONCLUSION

In conclusion, an intelligent CNC system with an open-architecture, PC-based controller is introduced in this paper. As the hardware and software components of the system are made up of some reusable modules, it is convenient for the user to configure it to serve different applications. In our application, the system was configured to control an aspherical optical curve precision grinding machine tool by choosing the appropriate icons on the graphic platform. Although the final evaluation of the new controller must await the results of grinding tests, the intelligent elements of the system have brought great convenience and flexibility to the system.

REFERENCES

[1] Howard K. McCue, Open Architecture Controller Activities in Technology Enabling Agile Manufacturing (TEAM). Proceedings of SPIE; 1996 November 20-21; Boston, Massachusetts: published by SPIE 2912. 1997.

[2] Jeniece May, Intelligent Inspection System. Proceedings of SPIE; 1996 November 20-21; Boston, Massachusetts: published by SPIE 2912. 1997.

[3] Kang Lee, Lisa Fronczek, Robert Gavin, Richard Schneeman, PC-based Machine-tool Controller Implemented as Testbed for Smart Transducer Interface. Proceedings of SPIE; 1996 November 20-21; Boston, Massachusetts: published by SPIE 2912. 1997.

[4] Yoram Koren, Zbigniew J.Pasek, A.Galip Ulsoy, Uri Benchetrit, Real-Time Open Control Architectures and System Performance. Annals of the CIRP. Vol. 45. 1996:377-380.

ROBOT INTELLIGENCE AUGMENTED BY INFORMATION TECHNOLOGY

Shinsuke Sakakibara

FANUC LTD

Abstract

Intelligent robots make it possible to automate advanced assembly tasks by using simple peripheral devices alone, thus achieving high automation ratio and payability. Also they are capable of considerably lowering expenses per hour and heighten competitiveness at factories as they extend the operating time of facilities by enabling unmanned operation during overtime and/or on holidays by means of sensor-based multiple checking.

Furthermore, the level of intelligent functions that robots are provided can be further augmented by making use of information technology (IT).

Keywords

intelligent robot, vision, force, sensor, offline programming, remote monitor, IT

1. INTRODUCTION

Companies must survive intense international competition. That makes the requirements on manufacturing systems all the more severe, and the role required of robots as components of such systems more important. In particular, the robots that have been provided intelligent functions (simply called "intelligent robots" from here on) are of great attention as they can automate advanced assembly tasks by simple peripheral devices alone, thus achieving high automation ratio and payability. Furthermore, intelligent robots are capable of considerably lowering expenses per hour and heighten competitiveness at factories as they extend the operating time of facilities by enabling unmanned operation during overtime and/or on holidays. The tasks performed by intelligent robots can be reliably executed in an unmanned manner by checking the performance of them in different ways with multiple sensors.

On the other hand, the communications functions of robot controllers have rapidly evolved, and the latest controllers now support 100Base-TX high-speed Ethernet. Intelligent robots are linked directly and indirectly to personal computers or EWS in the factory or offices by high-speed networks, and as a result support can now be received more easily from these numerous information processing devices located at the back of the network. The level of intelligent functions that robots are provided can be further augmented by making use of IT. Actual examples will be introduced below.

2. PROVIDING ROBOTS WITH INTELLIGENT FUNCTIONS

Due mainly to the rapid progress of sensor technology, intelligent robots have succeeded in attaining the automation of advanced assembly tasks such as automobile transmissions, that have so far been difficult for robots, and in considerably simplifying peripheral devices that have been a major factor in increased facilities costs.

Figure 1 shows an example where a force sensor is used to automatically assemble a multistage clutch unit used in the transmission of automatic vehicles. In this example, the robot fits the hub into the respective plates in the housing while matching their center positions and phases. The force sensor is used to keep the pressing force constant and perform the matching operation of the center positions and phases of the hub and plates. This task is performed reliably because the force sensor can detect whether the task is performed successfully or not in the end.

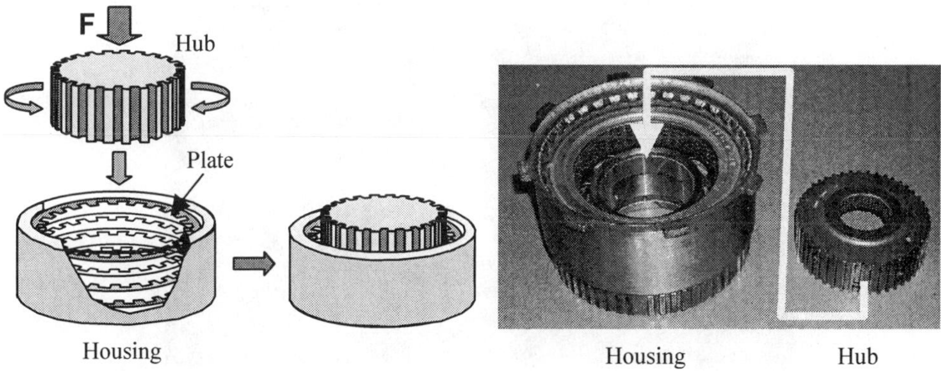

Figure 1   Automatic assembly of clutch unit by force sensor

Figure 2 shows an example where an intelligent robot uses a vision sensor to handle cast parts roughly piled in a box. This obviates the need for dedicated part supply equipment.

By the introduction of intelligent robots in production lines, expenses per hour at factories can be considerably reduced as peripheral devices such as dedicated part supply equipment are almost no longer required as described above.

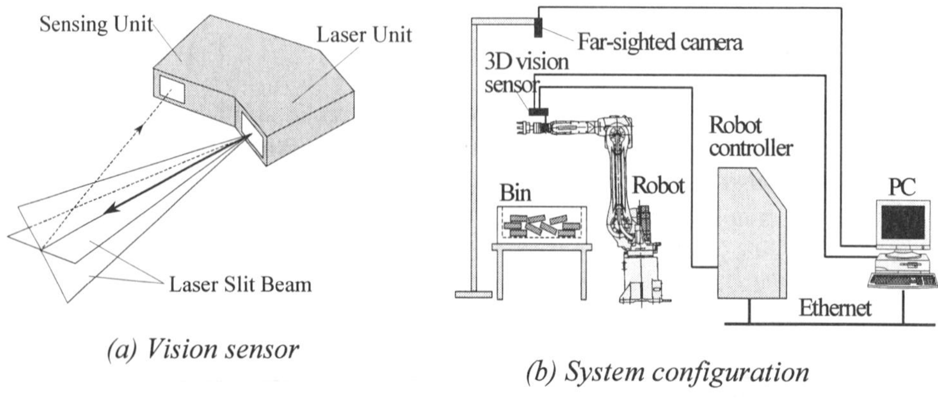

(a) Vision sensor

(b) System configuration

(c) Handling example

Figure 2   Handling of roughly piled parts by vision sensor

3. UTILIZATION OF INFORMATION TECHNOLIGY

The latest high performance of robot controllers, as well as sensor technology mentioned above, play a considerable role in providing robots with intelligent functions. In addition to advanced control functions such as a high-sensitivity collision detection and synchronous operation of multiple robots linked by Ethernet, some robot controllers lately utilize 100Base-TX, thus enabling high-speed communications with computers. The following describes the utilization of IT from a viewpoint of improving the level of intelligent functions that robots are provided.

3.1 Teaching

Although many man-hours are required for teaching operations to robot, offline programming system can be effectively utilized to reduce man-hours. This system has various geometric, kinetic and control models that enable the accurate simulation of robot cycle times. Also, 3D CAD product design data can be loaded in as it is, thus enabling the setup of work site environment models and facilitating interference checks between the robot and other robots, parts and peripheral devices. By improving absolute positioning accuracy, data output from this system can be directly utilized on the robot. *Figure 3* shows an example where vision sensor and offline programming have been combined to facilitate teaching of arc welding tasks. The robot motion program is automatically generated as the vision sensor tracks the weld lines after designating the start point, end point, and via points on screen. The generated motion path can be checked on screen before the robot is actually operated.

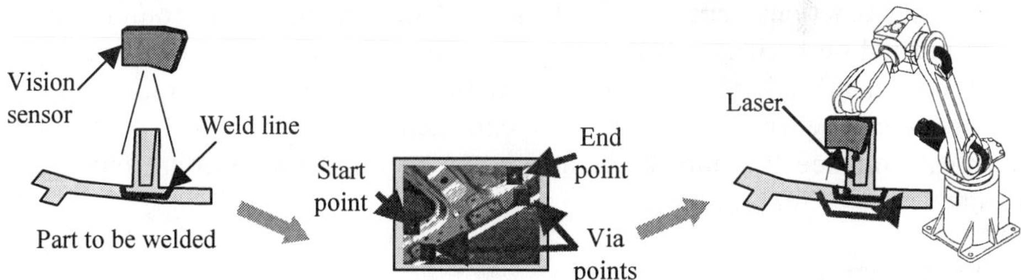

Figure 3   Sensor teaching with offline programming system

3.2 Remote Monitoring

Manufacturing systems are strongly required to respond quickly in the event of a malfunction. In particular, during automatic operation such as in night shifts or on holidays where operators are not in attendance at the factory, it is often required that the operating status of robots and other FA equipment is monitored at an office located away from the factory so that the appropriate response can be made quickly. Also as described above, unmanned operation during overtime and on holidays is effective in reducing expenses per hour at factories. Consequently, the importance of multiple sensor-based checking and remote monitoring functions is on the increase.

The following functions have been achieved for remote monitoring of robots:

- Display of robot operating status
- Display of line operating status
- Recording of mass monitored data in logs
- Display of alarm occurrence state and analysis of states (I/O, sensors, etc.)
- Management of multiple robot data
- Batch backup of robot motion programs and settings, etc.

4. CONCLUSION

Thanks to rapid progress in sensor technology, intelligent robots can raise the automation ratio by automating such advanced tasks as were so far difficult to be automated and make payable systems by obviating any dedicated peripheral devices. Also, sensor-based multiple checking of task and remote monitoring functions supported by IT enable safe unmanned operation during overtime and on holidays. This enables considerable reduction of expenses per hour on manufacturing systems incorporating intelligent robots, and aids increased competitiveness. It is also expected that the support of IT will drastically reduce the number of man-hours required for teaching intelligent robots.

REFERENCES

Shinsuke Sakakibara, et al. "AN INNOVATIVE AUTOMATIC ASSEMBLY SYSTEM WHERE A TWO-ARMED INTELLIGENT ROBOT BUILDS ROBOTS." 27th International Symposium on Industrial Robot, pp937-942, 1996.

STANDARD API FOR OPEN-ARCHITECTURE CNC AND IT'S APPLICATION TO HMI AND OPERATION MONITORING

Shigeru UENO* , Mamoru MITSUISHI, Kazuo MUTO***, and Shozo TAKATA****,**

*Technical Research Institute of JSPMI, ** The University of Tokyo, *** Polytechnic University, **** Waseda University

Abstract

JOP (Japan Open System Promotion Group) has been conducted the CNC API standardization for three years to provide the environment to implement custom HMI and communication tools with upper controllers. The specification of the API set: PAPI (Principal Application Program Interface) has defined and becomes new Japanese standard. This API has developed for open CNC systems that involve three types of open controllers including proprietary NC systems. PAPI consists of two group API sets such as Basic API and Extended API. This paper describes its development concept, specification and example results that could simultaneously control two different manufacturers controller by the same HMI. This development could accelerate the spread of open NC systems in real manufacturing field.

Keywords

API, Controller, CSV, HMI, NC, Open Controller, Operation Monitoring, PC, Standardization

1. INTRODUCTION

Although an open architecture controller system has been told that it would become next generation controller system in FA area 1)2), there still are few manufacturers who introduce it in machining system. The reason is fully depending upon the lack of its standard systems and standard API. JOP (Japan Open System Promotion Group) has been conducted the CNC API standardization for three years. The specification of the API set PAPI

(Principal Application Program Interface) has defined and becomes new Japanese standard.   In this paper, its basic concept and an application result to CNC Human-Machine-Interface (HMI) is reported.

2. BASIC CONCEPT

It is a critical point   to specify the meaning of an open architecture NC system. Before starting its detail developments, the development group discussed its classifications.   The results were; this API should be developed for three types of open controllers such as 1) PC +proprietary NC type, 2) PC + NC Board type and 3) Software NC on PC+I/O Boards type, as proprietary NC systems still keep its strong position in the machine tool manufactures field.

In the type 1), the controller is the existing CNC, which consists of the motion controller, and PLC. Setting from the API includes those processed directly by the motion controller and those passed to the PLC. Commands passed to the PLC are mediated with input information from the panel, etc. Type 2) consists of specially designed NC board, which includes PLC functions. API processing methods are similar to the type 1).   Type 3) is the software PC based controller. The controller is made according to an open agreement. In general, almost all basic processing is realized in the software on the PC.

PAPI is classified into three categories with their functional characteristics. Category one is called as CNC system APIs that include Display, Input, Resource manager and Communication. Category two is called CNC Devices APIs that consists of Interpolation, Acceleration, Deceleration, Servo control, Spindle control and Sequence control. The final one, category three is called CNC Application APIs that include Shared data and Preprocessing (NC programming).

Figure1 explains the relations between the categories and target CNC systems. The horizontal axis indicates three possible types of CNC systems. Also it indicates the real time processing speed that required in PC system. In case of PC +proprietary NC type, processing speed in the PC does not required so much. On the contrary, fully software based CNC may be requested high data processing rate. The vertical column explains classified API categories that can be applied to the all kind of CNC systems. PAPI covers major CNC system APIs, also support some of CNC application and CNC device control APIs . Remained areas are provided for the future extension.

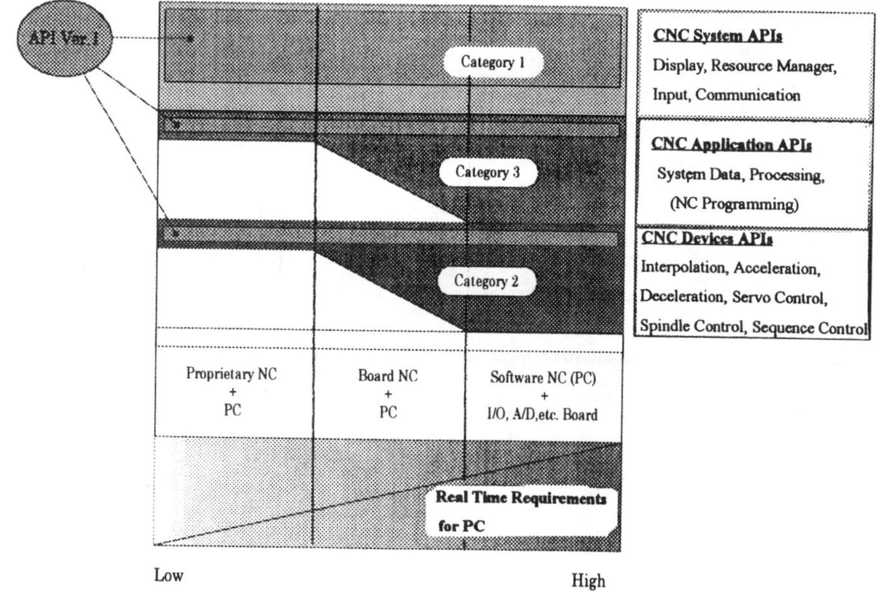

Figure 1. Category of CNC system and its APIs

3. PERFORMANCE TEST OF THE NEW API

To verify its performance, the developed API set has implemented into two existing different kinds of CNC systems that are manufactured by MITSUBISHI Electronic Corp., and FANUC LTD., Basic control HMI has developed by the students of Tokyo University Mitsuishi Lab.. All system has installed in Windows NT system. Figure 2 shows the total configurations demonstrated at Mechatro-tech Japan in Nagoya, Oct.13-16, 1999. The aims of demonstration were to show that PAPI is actually implementable in commercial CNCs, to show how PAPI is effective to realize a custom HMI, and to show that PAPI is also useful for the communication with upper level controllers. Management data items for operation monitoring were transferred to upper control PCs in terms of a standard format defined in JOP using Comma Separated Value (CSV) . The contents of the operation information are Main program name, Machine name, Alarm, Power Condition, Mode position and Program status.

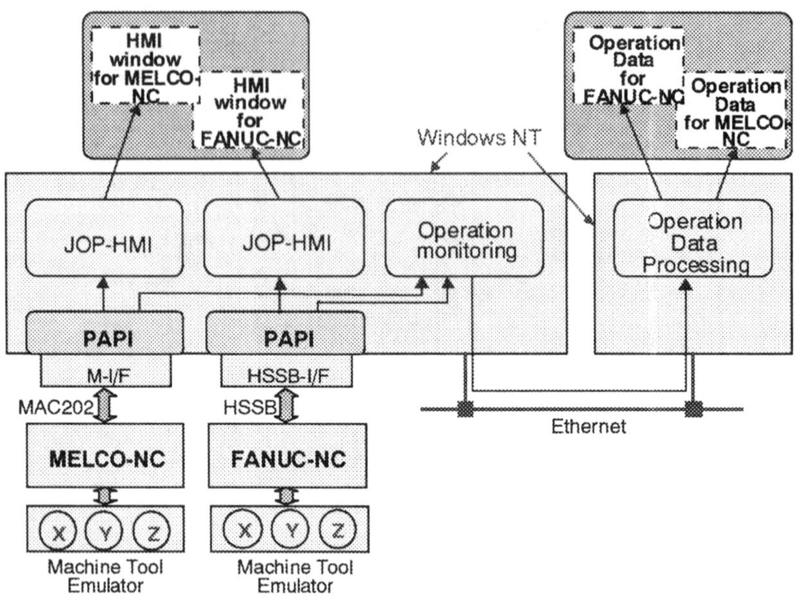

Figure 2.   Demonstration Configuration in Mechtro-tech Japan 99

These items are illustrated in the figure 3. The demonstration results were quite successful, that connected two NC system could operate by the same HMI simultaneously, and also could communicate with the higher level controller that provides production-scheduling software.

4. Conclusion

Standard API for open-architecture CNC has developed and its performance was examined through the demonstration operation. Followings are the conclusion of this development.

(1) Standard API for Open-architecture CNC has developed with two categories. Developed API(PAPI) could well cover existing basic open architecture CNC performances.

(2) Existing CNC system could operate by the same HMI with PAPI.

(3) Communication with higher level controller could be managed by the PAPI and production-scheduling software.

These results indicate that developed PAPI has a good ability and affinities with existing CNC systems.

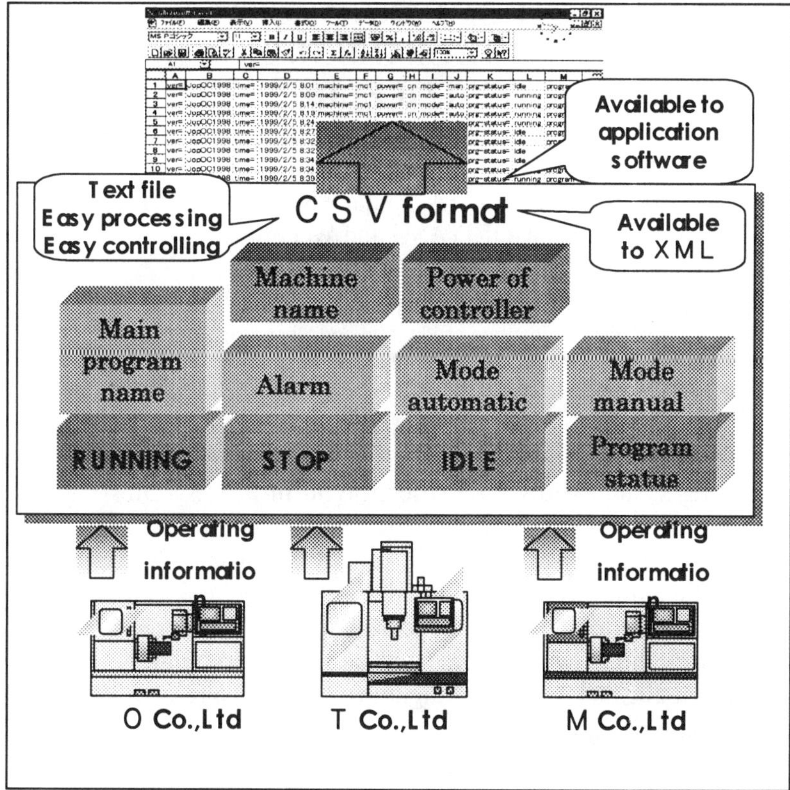

Figure 3. Standardization for Communication with Upper level Controller

Acknowledgement

This development was supported by MSTC (Manufacturing Science and Technology Center) JOP program. The authors would like to express our thanks to MSTC for their kind supports during this project.

Reference

1) Chrysler, Ford, GM, Requirements of Open, Modular Architecture Controllers for Applications in the Automotive Industry, Version 1.1,(1994)

2) Bradley N. Damazo et al. Open Architecture Controls for Precision Machine Tools, Proceedings of ASPE Spring Topical Meeting on Mechanisms and Controls for Ultra precision Motion, 1994 April 6-8, Tucson, AZ:26-31.

DEVELOPMENT OF LARGE ULTRAPRECISION MIRROR SURFACE GRINDING SYSTEM WITH ELID

Kazutoshi KATAHIRA, Hitoshi OHMORI, Masahiro ANZAI,
Yutaka YAMAGATA, Akitake MAKINOUCHI,
Sei MORIYASU and Weimin LIN

RIKEN(The Institute of Physical and Chemical Research)
2-1 Hirosawa Wako-shi Saitama Japan

Abstract

This paper introduces a new ultraprecision mirror surface machining system with ELID (Electrolytic In-Process Dressing). The machine has three linear axes, which can be controlled at a feeding resolution of 10nm under full closed feedback. 3-axes double hydrostatic guideway are used for sliding the machine, and hydrostatic bearing is used for the grinding wheel spindle. Ultraprecision mirror surface grinding, which achieved a surface roughness of 0.007 microns in Ra and surface straightness of 0.25 microns for a metal mold steel, was stably realized.

Keywords

ELID grinding technique, ultraprecision mirror surface machining, double hydrostatic guideway, submicron feeding control

1. INTRODUCTION

ELID (Electrolytic In-Process Dressing) grinding technique is expected to contribute greatly to precision grinding, particularly in the field of mirror finishing hard and brittle materials. ELID-grinding can produce ultraprecision grinding performances especially under specific machine such as hydrostatic bearing and submicron feeding control. The authors thus proposed a new ultraprecision mirror surface machining system with ELID.

The machine has three linear axes, which can be controlled at a feeding resolution of 10nm under full closed feedback. 3-axes double hydrostatic guideway are used for sliding the machine, and hydrostatic bearing is used for the grinding wheel spindle. This paper introduces the concept of the developed ultraprecision ELID-grinding system, the specifications, and trial use in metal mold steel.

2. CONCEPT OF LARGE ULTRAPRECISION MIRROR SURFACE GRINDING SYSTEM

To grind metal mold steel effectively, the concept design of the large ultraprecision mirror surface machining system was proposed as shown in *Figure 1*. The machine has three linear axes, which can be controlled at a feeding resolution of 10nm under full closed feedback. The step-response of Z-axis movement by feed resolution of 10nm is shown in *Figure 2* as measured using Microsense. A unique double hydrostatic guideway are used for 3-axes sliding the machine. That is composed of main-table and sub-table that is shown in *Figure 3*. It makes sliding straight travel very smooth, high accuracy, high damping and dynamic stiffness. And also, hydrostatic bearing is used for the grinding wheel spindle. The specifications of the developed machine are shown in *Table 1*. The maximum workpiece size is 1200mm*500mm along the X and Y axes. The X-axis can be driven at the 9m/min in maximum feed-speed.

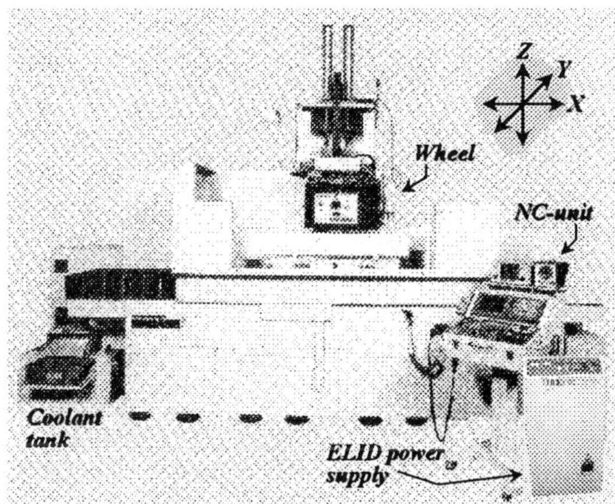

Figure 1   Ultraprecision mirror surface machining system with ELID

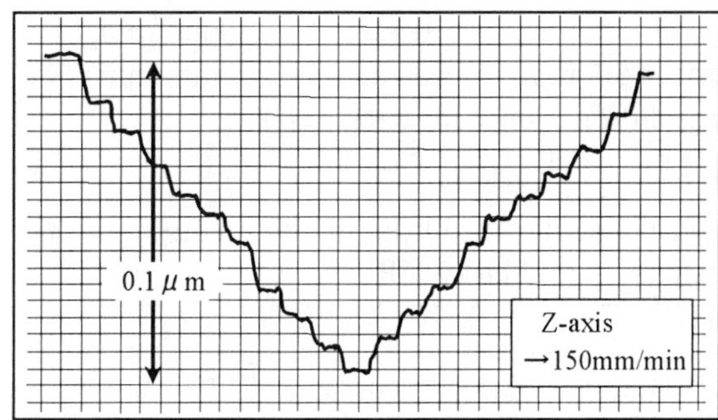

Figure 2   Step-response of Z-axis movement by feed resolution of 10nm

3. CONFIGURATION OF SYSTEM

The developed machine is mainly composed of the high stiffness machine frame, NC controller, ELID-system, and supporting devices such as oil tanks and temperature control units. *Figure 4* shows the external view of the grinding wheel with the ELID electrodes. The chemical solution type coolant is supplied to both the grinding wheel for ELID and cooling, cleaning and lubrication of the grinding points. The electrode can be adjusted by using NC to a specific gap.

The ELID power supply has a capacity open-circuit voltage of 90V and a peak-current of 20A, generating square pulse undone of 1-10 msec during the ON-OFF time at a maximum duty ratio of 80%. The power supply can be switched automatically by the M-code in the NC program. The NC unit used can be connected with personal computer through high-speed data transfer interface, and can display information including position data, controlling parameters, and NC-programs, etc. on the PC console.

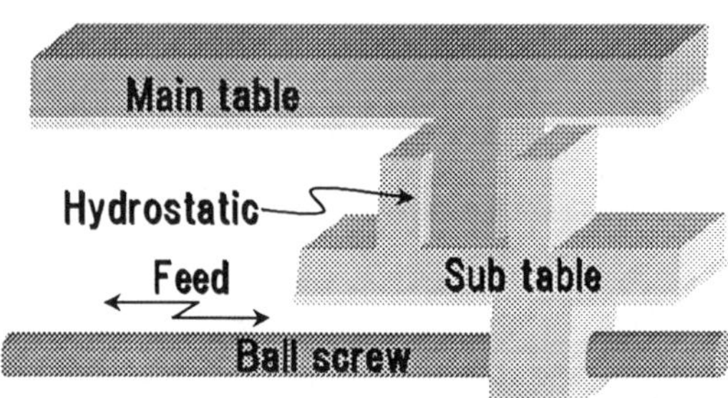

Figure 3   Schematic illustration of double hydrostatic guideway

Table 1   Specifications of developed grinding system

Linear Axis: Double hydrostatic guideways
X-axis: stroke=1400mm, feed-rate=1-9000mm/min, Y-axis: stroke=550mm,   feed-rate=1-1000mm/min, Z-axis: stroke=500mm,   feed-rate=1-1000mm/min (Feeding resolution: 10nm for each axis)
Rotational Axis: Hydrostatic bearing
Grinding wheel spindle: 1800rpm in maximum Grinding wheel motor: 11 kw 4P
ELID-capacity
90V(open voltage), 20A(mean current), 40A(peak current), Duty ratio: 80% in maximum, Square wave pulse generator

4. ULTRAPRECISION GRINDING OF METAL MOLD STEEL

ELID-grinding of a metal mold steel was conducted. The workpiece was a metal mold steel (SKD11) measuring 250mm in length and 200mm in width, and 50mm in height. Cast iron-cobalt hybrid bonded #325, #1200,and #4000 diamond wheels was trued on the machine by the plasma discharge method. The truing wheel used was the metallic bond diamond type. *Table 2* summarizes the grinding conditions. Relatively high efficiency could be achieved under these conditions to profile and finish the mirror surface by ELID. The finished surface roughness was approximately 0.057 microns in Ry, 0.007 microns in Ra by the #4000 cast iron-cobalt hybrid diamond wheel. *Figure 5* shows an example of surface profile. A very accurate surface with straightness of 0.25 microns per 250mm length was attained.

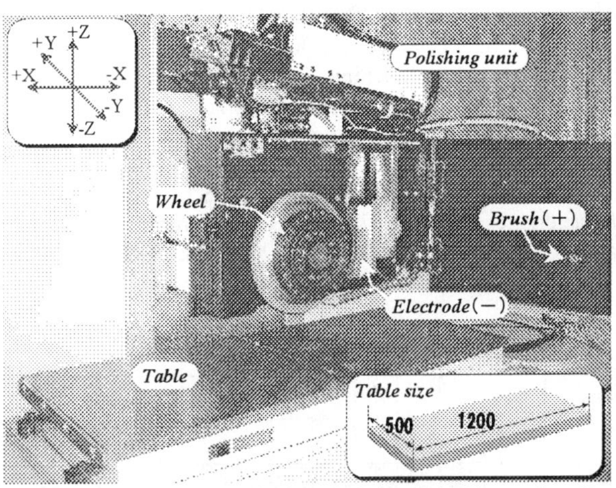

Figure 4   External view of the grinding wheel
with the ELID electrodes

Table 2   ELID grinding conditions (#4000)

Grinding conditions
Wheel speed : 4m/min
Depth of cut : $1.0\,\mu$m(coarse) $0.5\,\mu$m(finish)
Feed-pitch (Y-axis): 1.0mm(coarse) 0.5mm(finish)
Table feed-rate (X-axis): 4m/min

ELID conditions
Open voltage (Eo):90V, Peak current (Ip):20A
Pulse timing (τ on/ τ off):4μs
Pulse wave: rectangle

5. CONCLUSION

This paper proposed and introduced the concept of a large ultraprecision mirror surface machining system with ELID. The system configuration, features, and specifications were described. Especially, that has the double hydrostatic guideways with resolution of 10nm using full-closed feedback. Ultraprecision mirror surface grinding was stably realized in metal mold steel.

The authors would like to express their heartfelt thanks to Nagase Co.Ltd. for their kind cooperation.

REFERENCES

1. H.Ohmori and T.Nakagawa, Mirror Surface Grinding of Silicon Wafers with Electrolytic In-Process Dressing, *Annals of the CIRP* 1990; 39: 320.
2. H.Ohmori, Electrolytic In-Process Dressing (ELID) Grinding Technique for Ultra-Precision Mirror Surface Machining, *Int.Journal of JSPE* 1992; 26: 273.
3. H.Ohmori, Utilization of conditions in precision grinding with ELID (Electrolytic In-Process Dressing)for fabrication of hard material components, *Ann. CIRP* 1997; 46: 261-264
4. K.Katahira, H.Ohmori, M.Anzai, A.Makinouchi, S.Moriyasu, Y.Yamagata and W.Lin, Grinding Characteristics of Large Ultraprecision Mirror Surface Grinding System with ELID, Advances in Abrasive Technology 2000; The Society of Grinding Engineers; 125-128

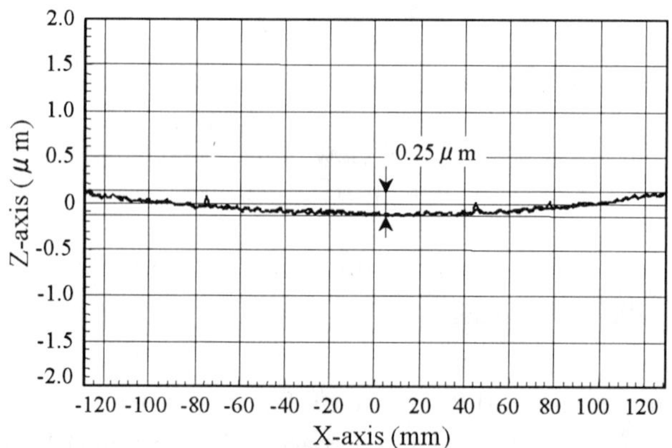

Figure 5 Straightness of workpiece finished with ELID-grinding

Development of a new measurement system for straightness error by a heterodyne interferometer with a grating

Shin Asano, Takayuki Goto, Hisashi Tanimura*, Kimiyuki Mitsui**

Advanced Technology Research Center, Mitsubishi Heavy Industries, LTD

**IC Equipment Division, Nikon, Co*

***Keio University*

Abstract

A new straightness error measurement optical system has been developed. This system consists of three components: (i) A He-Ne Zeeman laser, (ii) A grating secured onto a slide table and (iii) A heterodyne interferometer where a beat signal, which includes a phase shift $\delta\phi$ depending on an in-plane displacement δx of the grating, is generated. Straightness error can be measured, because $\delta\phi$ is proportional to δx. In this study, using this system, straightness error of a slide table is measured experimentally. Straightness error measured by this system is good agreement with one measured by a commercial laser interferometer system. The effectiveness of this system is confirmed by this result.

Keywords

Straightness, Grating, Heterodyne Interferometer, Machine tools, Accuracy

1. Introduction

Increasing demands for higher accuracy of machine tools and measuring machines have led to a great need to measure and characterize straightness error of slide-ways of these machines [1][2][3]. The final target of our study is to develop an in-process compensation system for straightness error of a slide table of precision machines. As the first step in our study, we propose a new straightness error measurement optical system in resolution of $0.02\,\mu$m or less. In this paper, the principle of this system is described in detail, then the basic characteristics of this system and experimental results of straightness error measurement are shown.

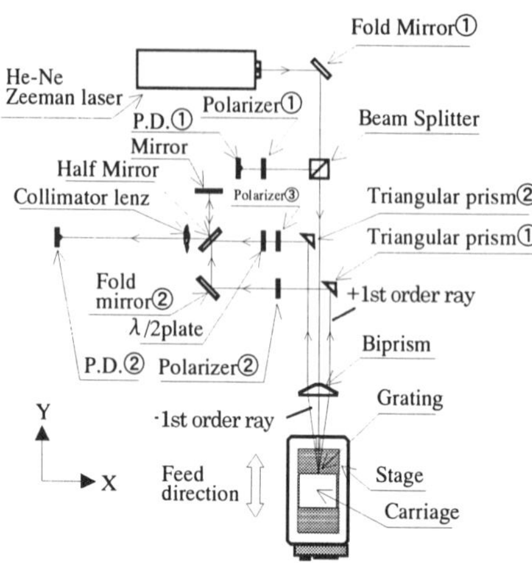

Fig 1 Proposed Optical system

2. Measurement Principle

The schematic configuration of the proposed measurement system is shown in Fig.1. A combined beam from a laser source contains two partial beams of slightly different frequencies that are in orthogonal linear polarization. The combined beam travels from the laser source to a beam splitter and is divided into transmitted and reflected beam. The reflected beam passes through polarizer① (at $45°$ to the polarizing direction of the combined beam), and falls on PD (Photo-Detector) ①. Partial beams of the reflected beam are superimposed to generate a beat signal at PD①. Since the phase of this signal isn't affected with an influence of a motion of the carriage of the stage, it can serve as a reference .

The transmitted beam enters to a grating that is secured onto the carriage. This beam is mainly diffracted in three directions of the order -1, 0, +1 by the grating. Among these, diffracted angles of $\pm$1st order are expressed by equation (1).

$$\sin \theta_{\pm 1} = \pm \frac{\lambda}{d} \qquad (1)$$

λ : *the wave length of the diffracted beams*
d : *the period of the grating*

A biprism makes the $\pm$1st order beams parallel. The partial beam of the +1st order, of which the polarization direction is horizontal and the frequency is $\omega 1$, pass through polarizer②, and continues toward the half mirror. Half of this beam, after passing through the half mirror, arrives at the mirror. Then this beam is reflected and directs toward the half mirror again. The half mirror reflects this beam so that half of it can be transmitted to PD②.

On the other hand, the partial beam of the -1st order, of which the polarization direction is vertical and the frequency is $\omega 2$, pass through polarizer③, and travels to the half wave plate. By passing through the half wave plate, the polarization direction of this beam changes from vertical to horizontal. Half of this beam pass through the half mirror and is transmitted to PD②.

In result, both partial beams of $\pm$1st order are superimposed to generate a beat signal at PD②. As shown in Fig.2, when the grating moves δx in the lateral direction, each amplitude of the $\pm$1st order diffracted is expressed by equation (2) and (3).

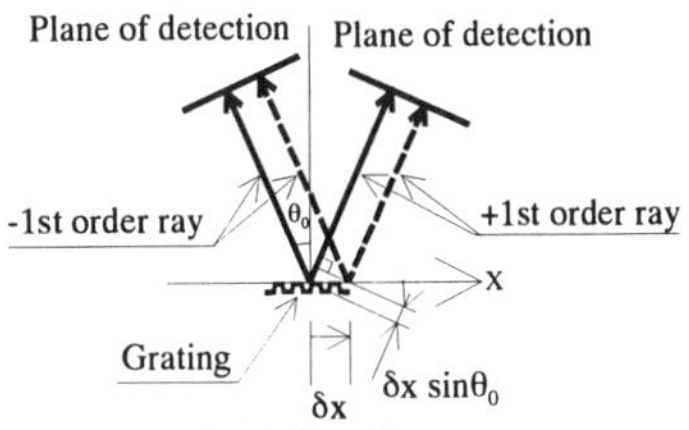

Fig.2 Principle of measurement

$$u_1 = a_1 \exp j\left(\omega_1 t - \phi - \frac{2\pi}{\lambda_1}\sin\theta_1\right) \qquad (2)$$

$$u_2 = a_2 \exp j\left(\omega_2 t - \phi + \frac{2\pi}{\lambda_2}\sin\theta_1\right) \qquad (3)$$

Therefore, the intensity of the beat signal at PD② is expressed by equation (4).

$$I \approx a_1^{\,2} + a_2^{\,2} + 2a_1 a_2 \cos(\delta\omega t - \delta\phi)$$

$$\delta\phi = \frac{2\pi}{\lambda_0} 2\delta \sin\theta_1 = 2\pi\frac{2\delta x}{d} \qquad (4)$$

Since phase shift $\delta\phi$ is measured by comparing the phase of the reference at PD① and the beat signal at PD②, it represents the change of the optical path length difference caused by the grating displacement δx.

3. Experimental results

3.1 Basic characteristics

In this experiment, we used a micro-positioning stage that uses a piezoelectric device as a drive actuator, and a grating was secured onto the stage. In-plane displacement is caused by the stage. The amount of in-plane displacement was measured by this system and simultaneously measured by a capacitance gauge attached parallel to in-plane direction of the grating as a reference. The sensitivity and resolution was evaluated based on a comparison of the two measurement.

At first, the sensitivity was evaluated. As shown in Fig.3, it is made clear that the grating displacement of 3.4 μm corresponds to the phase shift of 2π rad. This result shows the sensitivity of this system is 1.816 (rad/μm).

Next, we measured the resolution by translating the grating in displacement of 0.02 μm. As shown in Fig.4, it is clear that this system has a resolution of 0.02 μm or less.

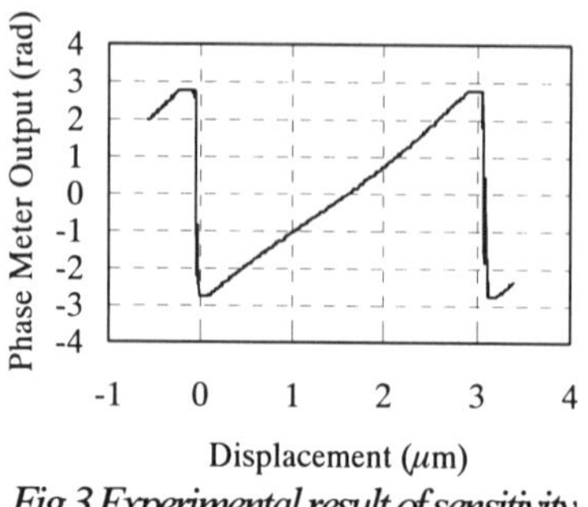

Fig 3 Experimental result of sensitivity

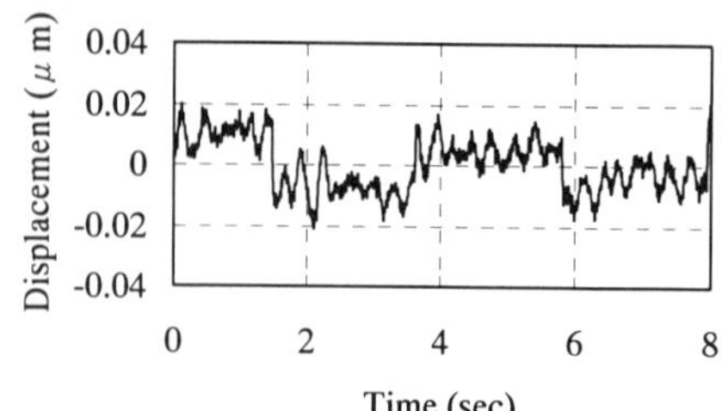

Fig 4 Experimental result of resolution

3.2 Straightness error measurement

The straightness error of the actual feed stage was evaluated using this system. Straightness error that occurred with the stage was simultaneously measured by a commercial laser interferometer system; then the resulting straightness error curve was compared to that obtained this system.

Fig.5 shows the schematic configuration of the evaluation testing apparatus. The stage moved at a speed that was proportional to the command voltage that was input to the motor drive amp from PC via DAC. Both outputs from this system and a commercial laser interferometer system were retrieved by PC via ADC and counter. Stage positioning information obtained from a linear encoder attached to the stage was retrieved by PC via counter. As for the straightness error data obtained by the laser interferometer system, the displacement was measured at two locations of the L-shaped reference mirror mounted on the carriage, and the average of these values was found.

The straightness error measured by our system includes translational error motion due to yawing of the carriage. In this evaluation, by geometrically estimating the amount of this error at the position of the grating and adding this amount to the straightness error data of the laser interferometer system, we obtained a compensated curve with the laser interferometer.

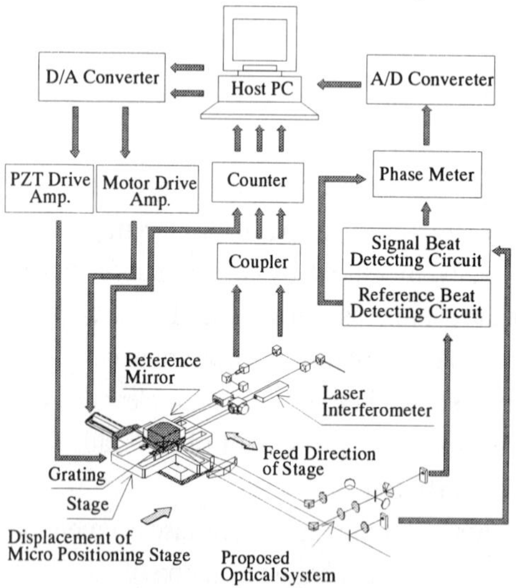

Fig 5 Evaluation testing apparatus

Fig.6 shows a result of straightness error curve in this experiment. From this result, it is made clear that there was extremely close agreement between this system and the laser interferometer system, and this system is an effective means to measure straightness errors.

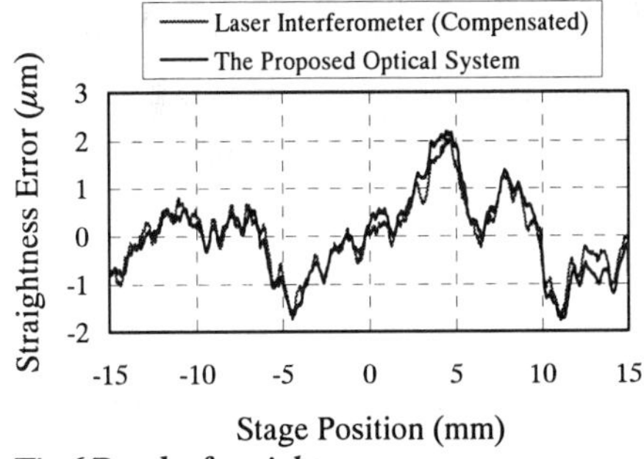

Fig 6 Result of straightness error measurement

4. Yawing measurement system

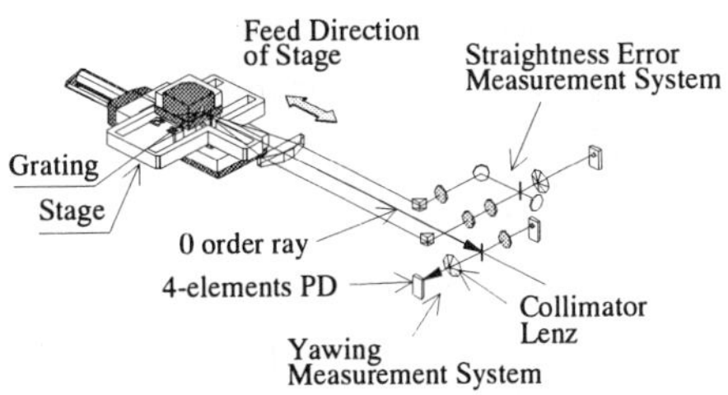

Fig 7 Yawing measurement system

In order to measure translational and rotational straightness error motions simultaneously and separately, we developed the yawing measurement system, included in the straightness error measurement system that was discussed in advance. Fig.7 shows the schematic configuration of the yawing measurement system. This system is consisted of a collimator lenz and a 4-elemnts PD. With this constitution, a deflection of the 0 order diffracted ray, which occurs due to yawing, can be easily detected. As a result for evaluating a correlation between the deflection and yawing, both correlate closely to each other. And yawing can be measured in a resolution of 0.5sec or less.

5. Conclusion

The final purpose of this study was to develop a system that, in process, could measure and control straightness errors for precision machines. As the first step, a new straightness error measurement system was developed, and it was shown that this system is an effective means to measure straightness errors in the experiments.

References

(1) W. Tyler Estler, Calibration and use of Optical Straightness in the Metrology of Precision Machines, Optical Engineering, 24-3 (1985) 372.

(2) Zongto Ge and Satoshi Kiyono, Measurement of 2-D Surface Profile and Motion Errors by Using Software Datum, JSME. Int. J.C., Vol.41, No.2 (1989) 199.

(3) N. Ikawa, S. Shimada and H. Morooka. Laser Beam as a Straight Datum and its Application to Straightness Measurement at Nanometer Level, Annals of the CIRP, Vol.37/1 (1988) 523.

DEVELOPMENT OF MEASURING SYSTEM EQUIPPED WITH WIRE PROBE TO MEASURE FINE CONTOUR SHAPE OF PENETRATED SPECIMEN

Haruki OBARA*, Masaki TSUJI*, Tsuyoshi OHSUMI*
and Masatoshi HATANO*

*Toyama University, Japan

Abstract

We describe a system developed to measure the fine contour shape of a penetrated specimen. The system uses a fine wire as a probe having a partially circular convex shape. The wire probe moves along the programmed contour path, and approaches the surface of the specimen in a cross section at the desired height. The contact of the wire probe to the specimen is detected based on the displacement of the wire probe on the outside of the specimen. The displacement of the wire is measured with an optical detector we have developed. The measured results and the possibility of using the system for submicron measurement are described.

Key words

Measuring system, wire probe, penetrated specimen, contour shape

1. INTRODUCTION

It is necessary to measure the accuracy of parts machined by wire EDM or other penetrated parts, which have a fine contour shape on the order of 0.1 mm and a height of 50mm or more. Conventional three-dimensional measuring apparatus are not suitable for this measurement, because their probes are not sufficiently fine and long to measure these parts. Consequently the accuracy of these parts is measured with a projector, however a projector cannot be used to check the error of the shape in a cross section at any desired height, except at the top and bottom surfaces of the parts. Masuzawa et al. developed the vibro-scanning method [1] to measure the accuracy of fine holes, in which a fine vibrating cantilever is used. It is not applied in our case because it is difficult to make a sufficiently long lever. We have developed a new system which uses a wire as a probe. In principle this method has no restriction on the thickness of the specimen, and it is possible to measure a fine contour shape. In the developed system, a table is moved in 1 μ m steps, but it is possible to realize submicron measurement using an extrapolation method explained below. The principle of this system and results are described.

2. PRINCIPLE OF MEASUREMENT

Figure 1 shows a schematic of this method. The wire probe comprises a fine wire and has a circular convex portion. It moves along a contour path and approaches to the surface of the specimen. The contact of the convex portion of the wire probe on the surface of the specimen is detected with a detector we have developed, and the contact position is entered in a computer, to calculate the error of the specimen. The straightness of the walls of the specimen should be better than the value of the difference between the radius of the convex portion and the radius of the wire.

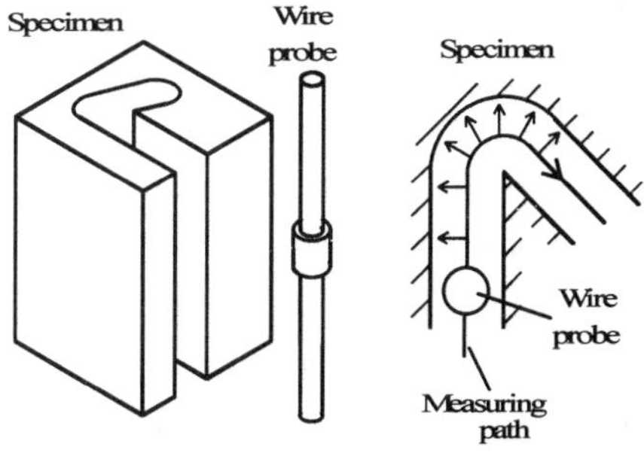

Figure 1   Schematic of measurement

3. MEASURING SYSTEM

The most important point in measurement is how to detect the contact of the wire probe on the surface of the specimen. In order to identify this contact, we use a method where it is detected based on the displacement of the wire probe on the outside of the specimen. Figure 2 shows a schematic of the developed optical detector. The detector is composed of slits of 0.3 mm width, optical fibers and light/detect amplifiers for each x and y axis. The light is partially interrupted by the wire, and the wire displacement is detected from the light intensity. The output of the detector is compared with a reference voltage to detect the contact. The detector is set beneath the top wire guide.

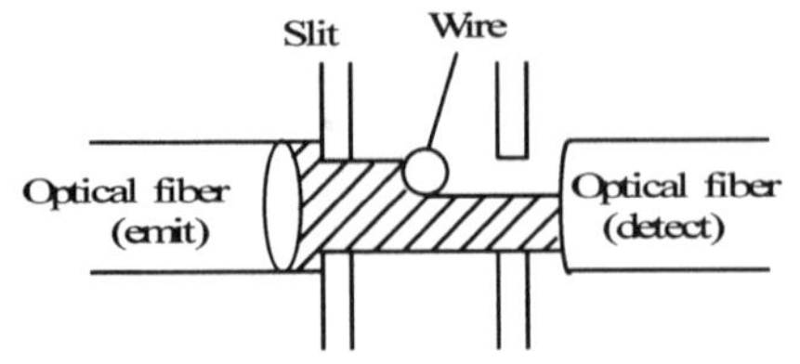

Figure 2   Schematic of detector

When the wire probe moves 1 μ m toward the specimen after the contact, the wire probe on the outside of the specimen moves less than 1 μ m, and the displacement is detected by the detector.

Figure 3 shows the developed measuring apparatus. The wire probe is made of a tungsten wire 0.1 mm in diameter, and the convex portion is a pipe 0.24 mm in diameter. The wire probe is suspended between two guides at

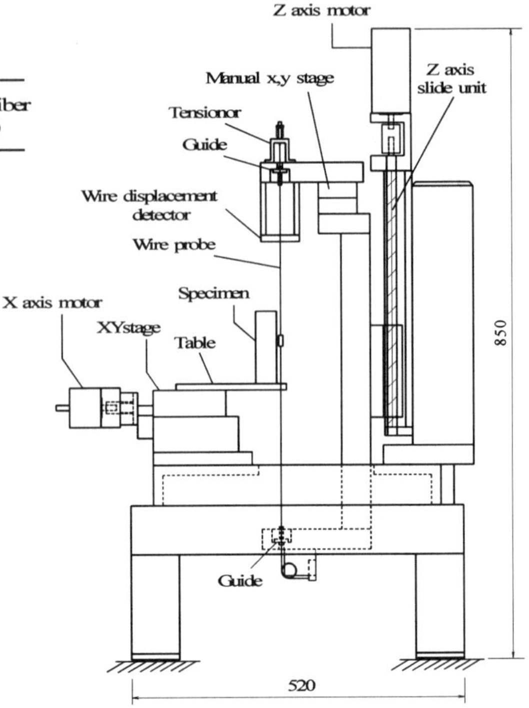

Figure 3   Schematic of measuring apparatus

the top and bottom of an arm. A tension of 9.8 N is supplied to the wire probe. The arm moves in the z direction, and the convex portion of the wire probe is set at the desired height of the specimen. The specimen is placed on an x, y table. The table is moved by a personal computer in 1 μ m steps. The wire probe moves along a contour path and approaches to the wall of the specimen from a point on the contour path. The contour path is programmed with G codes used in NC machines, and measuring points on the contour path are preset. When the wire probe contacts the surface of the specimen, the contact position is entered into the personal computer, and the wire probe returns to the contour path. The personal computer calculates the contact point on the surface of the specimen and the error of the shape of the specimen.

4. TEST RESULTS

Tests were carried out for specimens machined by wire EDM. Figure 4 shows an example of the measurement. The specimen is 2 mm thick, and has a roughness of 3 μ m Ry. Another test was carried out for a 40 mm thick specimen, but the result was similar to that in Figure 4, and is not presented here. Figure 5 shows a magnified shape of Figure4. Measurements were taken four times; it took 1 hour for each measurement. $\bigcirc$ shows a position of contact on the surface of the specimen, and $+$ shows the error of the shape magnified 10 times. In Figure 5, plot points are scattered around 5 μ m. This is caused by the roughness of the specimen. It appears that the specimen has an error of 9 μ m in the x direction, but this is not true. It is caused by the pitch error of the table. From these figures, we conclude that this system functions well.

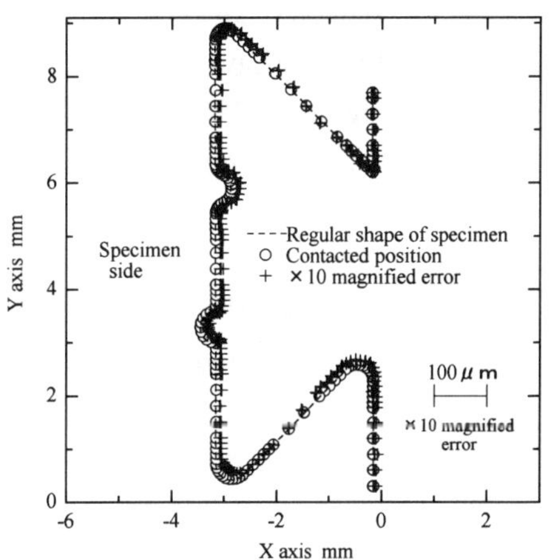

Figure 4   Measured result of specimen

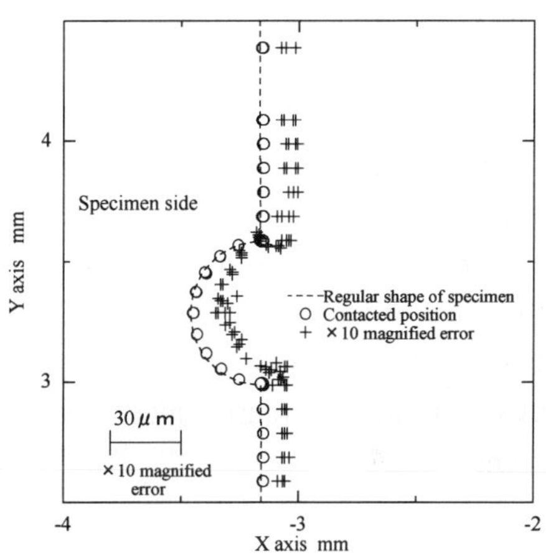

Figure 5   Magnified shape

5. POSSIBILITY OF SUBMICRON MEASUREMENT

The table is moved in 1 μ m steps. But when the positioning accuracy of the table is on the submicron order, it is possible to use this method for submicron measurement. Figure 6 shows the schematic. The wire probe moves toward the specimen, after contact is made on the wall, and thus the wire probe hangs over the specimen. The positions of the table and the output of the detector are memorized at each position. Subsequently the contact position, where the detected level is 0, is obtained by extrapolation as shown in Figure 6. We call this the extrapolation method, and we call the previous method the threshold method.

Figure 7 shows the repeatability of the two methods using a cylindrical bar as a specimen. The tests were carried out 50 times. The repeatability of the extrapolation method is about 0.3 μ m; the result of the threshold method appears more stable than the result of the extrapolation method. However, the accuracy of the threshold method is theoretically restricted to ± 1 μ m and thus the precision of the extrapolation method is better than that of the threshold method. Consequently this method is available for submicron measurement.

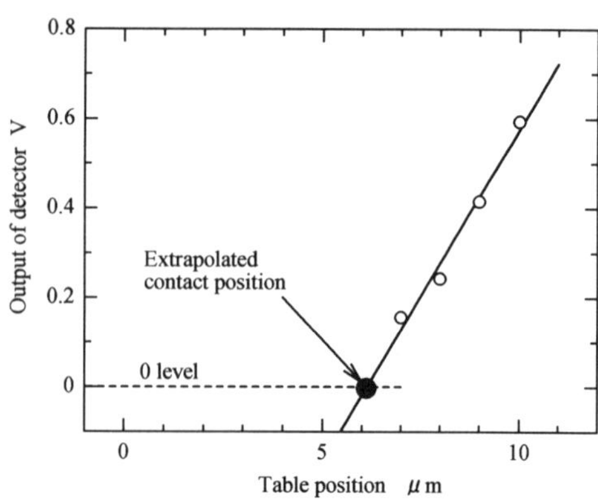

Figure 6   Schematic of extrapolation method

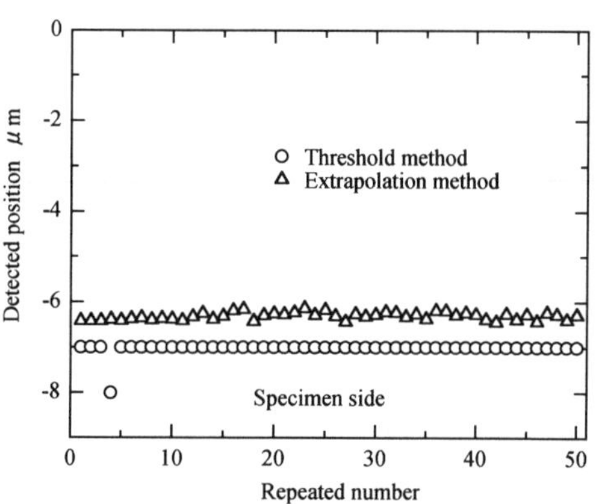

Figure 7 Accuracy of extrapolation method and threshold method

6 . CONCLUSIONS

In order to measure the shape of a penetrated specimen at any desired height, a new measuring system using a wire probe was developed. This system functions well, and can be used for submicron measurement.

7 . REFERENCES

[1] T. Masuzawa, Y. Yamasaki and M. Fujino: Vibroscanning Method for Nondestructive Measurement of Small Holes, Ann. CIRP, 42, (1993)

DEVELOPMENT OF ULTRA PRECISION LONG END STANDARD INSTALLING VACUUMED OPTICAL PATH LASER INTERFEROMETER

Yutaka Kuriyama*   Jun Ishikawa**
Morimasa Ueda*   Yuuichirou Yokoyama*
MITUTOYO Corporation   *National Research Laboratory of Metrology**

Abstract

Reducing uncertainty of the standard is one of the most effective factors to realize the high-accuracy length measurement. However, scales (e.g. Gage Blocks, Standard Scales) have uncertainty of $10^{-6} \sim 10^{-7}$ at most. Therefore, to supply practical length standard that has uncertainty of 10^{-8} or less, we have developed an ultra precision long end standard installing a vacuumed optical path laser interferometer. It has been confirmed that the length change owing to the installation posture change of the standard is able to be kept within 1nm(p-p) in the total length of 500mm.

In this paper, the basic experiments and the results that have obtained by the trial manufacture are reported.

Keywords

End standard, Wavelength in vacuum, Uncertainty, Laser interferometer, Temperature control, Posture control

1. INTRODUCTION

Nowadays, the standard of length is the wavelength of I_2 stabilized He-Ne laser in vacuum, and its uncertainty is very small, $5 \times 10^{-11}(k=2)$. However, scales (e.g. Gage Blocks, Standard Scales) widely used in industry as practical standard of length in air have uncertainty of $10^{-6} \sim 10^{-7}$ at most (Figure 1), and it is in about 10^4 times inferior

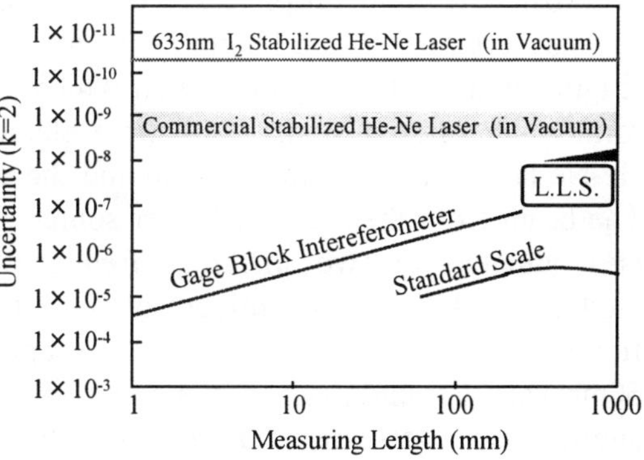

Figure 1. Uncertainty of length calibration systems

to the wavelength's. Also, it is necessary and important to have countermeasures to keep the enviromental conditions constant, to isolate the dynamic disturbances and to maintain the designed posture constant for the high-accuracy measurement of the above-mentioned scales. Therefore, we have developed an ultra precision long end standard called Laser Lock Scale (L.L.S.) shown in Figure 2. It is robust for the change of the above-mentioned operating conditions, and has less uncertainty of 10^{-1} than the previous products'. The basic experiments and the results that have obtained by the trial manufacture (500mm in total length) are reported here.

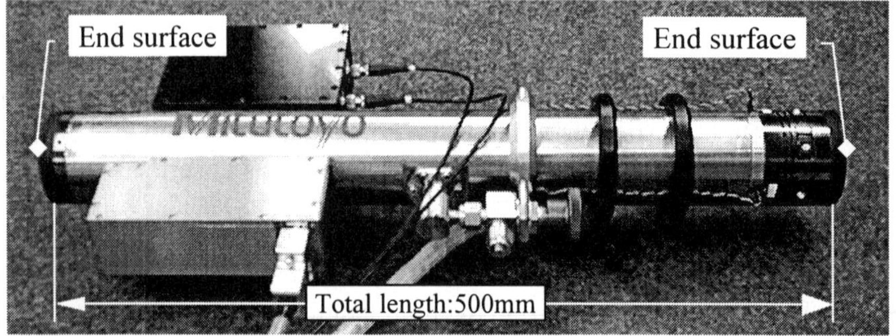

Figure 2. Trial manufacture of L.L.S.

2. SYSTEM STRUCTURE

The L.L.S. is long end standard of the total length 500mm and 1000mm. The aim is to realize the uncertainty $1 \times 10^{-8}(k=2)$ in spite of installing posture change, under the environment of the operating temperature change: $\pm 1K$ in the range of 18~23 °C and the dynamic disturbance of 30mGal or less. The two major demanded functions of end standard are "keeping a length constant" and "having a calibrated absolute length", and the L.L.S. is a system which is especially paid attention to the former. The system consists of three elements, which are the I_2 stabilized He-Ne laser, the scale mechanism and the controller, as shown in Figure 3. The beam from the stabilized laser source (which is the standard of length) is modulated to two different frequency beams by an acousto-optics modulator (AOM), and they are directed to the scale mechanism by using fibres. The scale body is a cylinder-shaped chamber (outside diameter is 60mm) and has end plates at the both ends. The body is made of SuperInvar, and the end plates are made of a low thermal expansion glass to suppress their thermal expansion along optical path. In order to exclude the influence of refractive index of air on the optical path, the inside of the scale body is evacuated. The surfaces of end plates face each other. They function as mirrors of a heterodyne interferometer. The length and posture between both plates can be kept constant anytime by detecting their changes simultaneously with an interferometer and by controlling them with PZTs installed at one end.

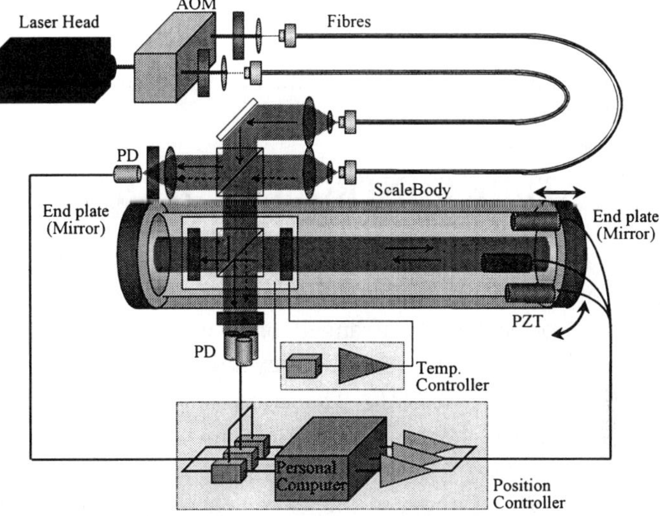

Figure 3. System block diagram

3. ESTIMATION OF THE UNCERTAINTY

The uncertainty factors of the system are extracted, analyzed, and the operating conditions are decided based on the result.[1]

Table 1. The uncertainty budget for L.L.S. in control

Factor	Target Condition	Standard Uncertainty	Std.Uncertainty L=500mm/1000mm
1) Wavelength accuracy of the laser	633nm I_2 Stabilized He-Ne laser	$2.5 \times 10^{-11} \times L$	0.0125nm / 0.025nm
2) Refractive index of air on the optical path	vacuum level $< 5.0 \times 10^{-2}$Pa	$3.6 \times 10^{-10} \times L$	0.18nm / 0.36nm
3) Optical path-length change	Temp. change of the optical parts ± 0.025K	1.5nm	1.5nm
4) Thermal expansion and contracton of the end plates	Temp. change of the environment ± 1K	0.6nm	0.6nm
5) Controlled accuracy of the end plate	In spite of the installing posture	1.1nm	1.1nm
Combined Std. Uncertainty (k=1)			2.0nm
Expanded Uncertainty (k=2)			4.0nm

There are two groups of uncertainty factors when they are divided roughly (A) the factors in control and (B) the factors due to the absolute length calibration. The expanded uncertainty in total length L=500mm / 1000mm is 5nm / 10nm(=L$\times 10^{-8}$, k=2) by the combination of (A) and (B), and the both can be expected same 4.0nm(k=2) like showing expanded uncertainty by the (A) factor in Table 1. Therefore, it is necessary for the target accuracy achievement to calibrate the value of the (B) factor with 3.0nm / 9.1nm(k=2).

4. TEMPERATURE CONTROLLED OF OPTICAL PARTS

As discussed previously, one of the major uncertainty factors is the optical path-length change due to the thermal expansion and the refractive index change of optical parts arranged on a vacuumed optical path. It is necessary for the target accuracy achievement to suppress temperature change of them within $\pm$0.025K. Therefore, the optical parts are installed in a holder with a foil heater and a platinum resistance thermometer to control their temperature. Moreover, the holder is designed to suppress the thermal transfer to/from the outside as much as possible. Presently, the temperature change of the holder is controlled to be within $\pm$0.01K(p-p) under the environment of external operating temperature 20$\pm$0.5 $^{\circ}$C in the system.

5. POSTURE CONTROLLED AND EVALUATION OF ENDPLATE

In order to precisely control the posture of an end surface, it is necessary for the optical path to be kept in vacuum, for the end plate to be tilted and for the body to be expanded. Therefore, the end plate is actuated by the micro-motion-mechanism with the elastic hinges, the metallic bellows and three PZTs. Also, to detect the displacement, the beam (Dia. = 8mm) of the interferometer that is divided into three signals is received with PDs. The length is kept constant with one of the interference signals, and the posture is kept constant with the rest two signals and the first one. The PZTs are driven to equal the phase difference between the rest two signals and the first one.

To confirm the effect of these mechanisms on controlling the posture of the movable end plate, the change in length is measured by using the internal interferometer. The change is caused by the installation orientation change of the scale mechanism from the horizontal to the vertical.

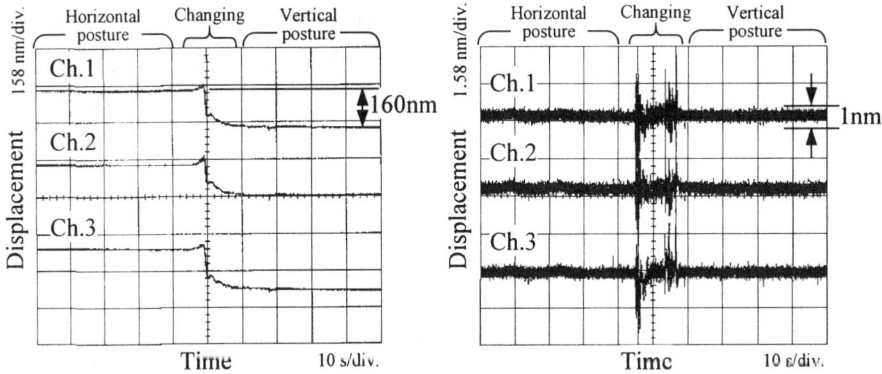

Figure 4(a). Without end plate control   *Figure 4(b)*. With end plate control

The results of the measurements are shown in Figure 4(a) and 4(b). In case of not using end plate control is shown in Figure 4(a). The displacements of about 160nm implies that the scale body is compressed by its own weight and the shortage is not compensated. Figure 4(b) shows in case of using end plate control. It has been confirmed to be able to keep the length and the posture change within 1nm (p-p) even if installing orientation of the scale mechanism is changed in the horizontal to/from the vertical.

6. CONCLUTIONS

The ultra precision end standard installing vacuumed optical path laser interferometer has developed, and the following results are obtained.

1) The Uncertainty is evaluated and their amounts are assigned under operating conditions.
2) A trial manufacture is developed, and the temperature change of optical parts has been confirmed to achieve $\pm0.01K$ by giving the temperature control to a target $\pm0.025K$.
3) It has been confirmed that the length change owing to the installation posture change of the scale mechanism is able to be kept within 1nm(p-p) for the target of 1.1nm.

The calibration technology of absolute length, the length stability under various operating temperature and the dynamical disturbances will be practiced.

REFERENCES

1) Morimasa Ueda, Jun Ishikawa, Error analysis of high precision block gage with vacuum space installed laser interferometer, Proceedings of the annual autumn meeting of the JSPE (1999), p431

FLATNESS MEASUREMENT SYSTEM USING FIZEAU TYPE SOLID INTERFEROMETER WITH HIGH SPEED ACQUIRING MULTI-PHASE INTERFEROGRAMS AND HIGHLY EFFECTIVE ERROR DECREASING ALGORITHM.

Kazuhiko Kawasaki, Yasushi Ueshima, Naoki Mitsutani, Hiroshi Haino

Mitutoyo Corporation

Abstract

For the purpose of achieving high accurate and high speed flatness measurement, a new interferometer acquiring Multi-Phase interferograms using Two orthogonally polarized lights has been developed (hereafter referred to as MP2 interferometer). In this interferometer, the characteristic errors occur because three interferograms are acquired by dividing a measuring light into three. Therefore, we propose a new algorithm by which the native errors of this interferometer are surprisingly decreased.

Keyword

Interferometer, Bias, Amplitude, Phase shift, Polarization, Fizeau, Reflectance

1. INTRODUCTION

In the conventional phase shifting method using one camera for a flatness measurement, more than three phase shifted interferograms must be acquired at each position where a surface of reference flat is intermittently moved along optical axis. For that reason, the measuring time becomes long and the repeatability of the measured flatness becomes lower. In order to solve these problems, we have developed a new high-speed solid interferometer acquiring three phase shifted interferograms simultaneously using three cameras without mechanical movement [1],[2]. However, because three interferograms are acquired in three coordinates respectively in our interferometer three biases and three amplitudes which compose each interferogram may be different from each other in each pixel unit. An actual phase shifts added to each interferogram in the interferometer are different from designed one. Furthermore, the bias and the amplitude change depending upon a surface under test.

So we propose a new algorithm for decreasing error caused by the difference among three interferograms and for taking the measuring result independent of the characteristics of the surface under test. The algorithm utilizes the biases, the amplitudes and the phase shifts that are measured on the surface under calibration S for the surface under test T by defining the reflectance ratio γ of both surfaces.

2. OUTLINE OF MP2 INTERFEROMETER AND ALGORITHM

The system of MP2 interferometer is shown in Fig. 1[1],[2]. The objective quarter waveplate ⑦ that placed in between the surface of reference flat ⑥ and the surface under test ⑧ of a Fizeau interferometer, which brings two orthogonally polarized lights. Then

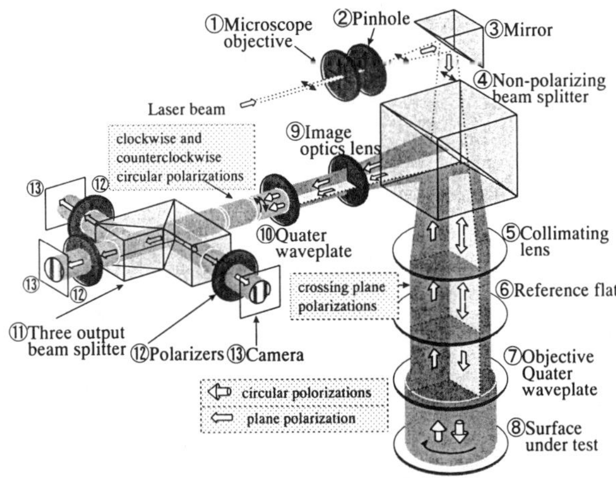

Figure1 Schematic diagram of MP2 interferometer

these polarized lights are converted into clockwise and counterclockwise circular polarized lights by a quarter waveplate ⑩. After the lights are divided into three parts by the three output beam splitter ⑪ and each part passes through each polarizer ⑫, three phase shifted interferograms are obtained simultaneously where the interferograms are focused on the CCDs by an image optics lens ⑨.

Three images intensity $I_i(x,y)$ are expressed in Eq. (1-1),(1-2),(1-3).

$$
\begin{cases}
I_1(x,y) = B_1(x,y) + A_1(x,y)\cos\left[\Phi_1(x,y)\right] = B_1(x,y) + A_1(x,y)\cos\left[\phi(x,y)\right] & (1-1) \\
I_2(x,y) = B_2(x,y) + A_2(x,y)\cos\left[\Phi_2(x,y)\right] = B_2(x,y) + A_2(x,y)\cos\left[\phi(x,y)+\alpha(x,y)\right] & (1-2) \\
I_3(x,y) = B_3(x,y) + A_3(x,y)\cos\left[\Phi_3(x,y)\right] = B_3(x,y) + A_3(x,y)\cos\left[\phi(x,y)+\beta(x,y)\right] & (1-3)
\end{cases}
$$

The values $B_i(x,y)$ and $A_i(x,y)$ show biases and amplitudes respectively. The subscript i indicates the cameras' number (i=1,2,3). Also, x and y are the coordinates after positional adjustment by the software. The $\phi(x,y)$ is the phase amount between lights reflected the both surfaces. $\alpha(x,y)$, $\beta(x,y)$ are the phase differences among interferograms, which are decided by the setting angles of the polarizers in front of each CCD camera, which are called the phase shifts.

A data acquiring system for biases, amplitudes and the phase shifts is shown in Fig.2. The surface under test is shifted for n times by h_j interval along the shifting direction just like the conventional systems. Biases and amplitudes are calculated by Eq.(2) and Eq.(3)[3] using several interferograms acquired by each of three cameras of the MP2 interferometer.

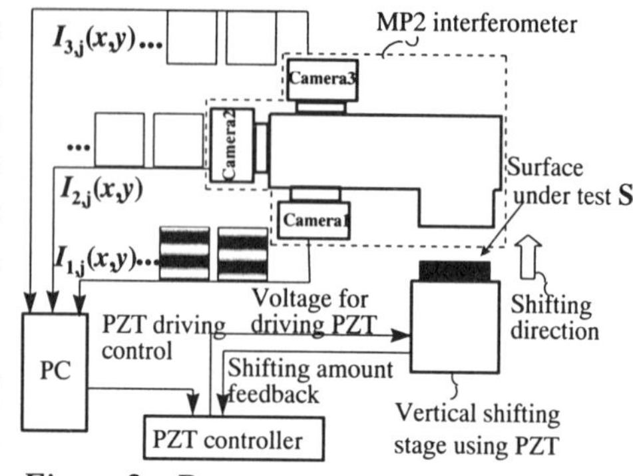

Figure2 Data acquiring system for biases, amplitudes and phase shifts

$$B_i(x,y) = \frac{1}{n}\sum_j^n I_{i,j}(x,y) \qquad (2)$$

$$A_i(x,y) = \frac{2}{n}\left\{\left[\sum_j^n I_{i,j}(x,y)\cdot\cos(\delta_j)\right]^2 + \left[\sum_j^n I_{i,j}(x,y)\cdot\sin(\delta_j)\right]^2\right\}^{1/2} \qquad (3)$$

$$\delta_j = \frac{4\pi}{\lambda}\cdot h_j = \frac{2\pi}{n}\cdot j; \quad j = 1,...n$$

Three phases $\Phi_i(x,y)$ expressed in Eq.(1-1), (1-2), (1-3) are calculated using Eq.(4)[3] from acquired those images. Then, by subtracting $\Phi_1(x,y)$ from $\Phi_2(x,y)$ and $\Phi_3(x,y)$, it is possible to get the actual phase shifts $\alpha(x,y)$ and $\beta(x,y)$ which are optically realized by the interferometer in each pixel respectively.

$$\Phi_i = \tan^{-1}\left[-\sum_j^n I_{i,j}(x,y)\sin(\delta_j)\bigg/\sum_j^n I_{i,j}(x,y)\cos(\delta_j)\right] \quad i = 1,2,3 \quad (4)$$

If we can substitute simultaneously acquired three interferograms $I_i(x,y), B_i(x,y), A_i(x,y), \alpha(x,y)$ and $\beta(x,y)$ into Eq.(5) and Eq.(6), real flatness without analysis errors caused by differences among biases and amplitudes, and differences between designed and actual phase shift is obtained.

$$I_i{'}(x,y) = \frac{I_i(x,y) - B_i(x,y)}{A_i(x,y)} \quad (i = 1,2,3) \quad (5) \qquad \phi = \tan^{-1}\left\{\frac{CI_1{'} + DI_2{'} + EI_3{'}}{FI_1{'} + GI_2{'} + HI_3{'}}\right\} \quad (6)$$

$$C = \sin(\alpha)\cos(\alpha) + \sin(\beta)\cos(\beta), \quad D = -\sin(\alpha) - \cos(\beta)\sin(\alpha - \beta), \quad E = -\sin(\beta) + \cos(\alpha)\sin(\alpha - \beta)$$

$$F = \sin^2(\alpha) + \sin^2(\beta), \quad G = -\sin(\beta)\sin(\alpha - \beta), \quad H = \sin(\alpha)\sin(\alpha - \beta)$$

Fig.3 shows a result of a flatness measurement inside 20mm square at the surface under test. Its PV value is assured less than 20nm. (a) and (b) show the results without and with error decreasing algorithm respectively.

Comparing (a) with (b), the PV value is decreased from (a) 71nm to

816

(b) 29nm. The PV value of (b) is equal within nm order to the result of PV value calculated from some phase shifted interferograms which are acquired using only one camera in MP2 interferometer; which is so called conventional phase shifting method.

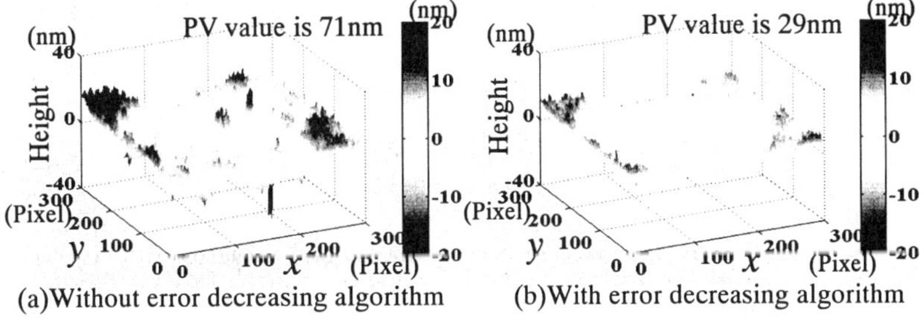

(a)Without error decreasing algorithm   (b)With error decreasing algorithm

Figure 3 Examples of measuring surface under test

3. ALGORITHM FOR REFLECTANCE RATIO γ

The bias and the amplitude depend on characteristics of a surface under test because an intensity of light reflected from the surface under test varies. When the reflectance ratio of surface under calibration S and surface under test T is defined γ, the models of intensities of reflected lights are shown in Fig.4(a) and (b).

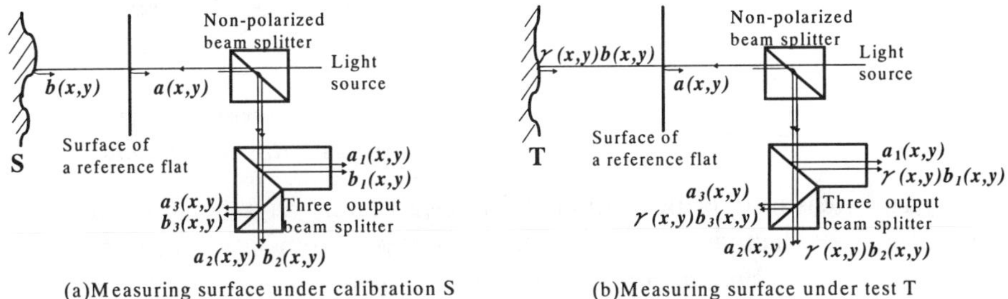

(a)Measuring surface under calibration S   (b)Measuring surface under test T

Figure4 Models of reflected light intensities

In case γ is not equal to 1, phase ϕ calculated in Eq.(6) includes phase error ϕ_{error} shown in Eq.(7). The phase error ϕ_{error} corresponding to reciprocal of an integer period of phase ϕ.

$$\phi_{error} = \phi' - \phi$$

$$\approx -\frac{1}{2}\left(\frac{\gamma-1}{\sqrt{\gamma}}\right)^2 \left[L_1^2 + M_1^2\right]\cos(V-W) + \frac{\gamma-1}{\sqrt{\gamma}}\sqrt{L_1^2 + M_1^2}\cos(\phi-W) - \frac{1}{2}\left(\frac{\gamma-1}{\sqrt{\gamma}}\right)^2 \left[L_1^2 + M_1^2\right]\cos(2\phi - V - W) + \cdots \quad (7)$$

ϕ';Calculated phase value(when $\gamma \neq 1$) ϕ;Ideal phase value(when $\gamma = 1$)

$$L_1 = \frac{1}{2}\sqrt{\frac{b_2}{a_2}}, \quad M_1 = \frac{1}{2}\left(\sqrt{\frac{b_1}{a_1}} - \sqrt{\frac{b_3}{a_3}}\right), \quad \tan(V) = \frac{L_1}{M_1}, \quad \tan(W) = \frac{-M_1}{L_1}, \quad \left(\text{in case of } \alpha = \frac{\pi}{2}, \beta = \pi\right)$$

817

However, using the intensities a_i acquired without the surface under test and calculating phase ϕ in Eq.(8), the result of phase ϕ is acquired independent of γ.

$$\phi = \tan^{-1}\left\{\frac{C\cdot(I_1-a_1)+D\cdot(I_2-a_2)+E\cdot(I_3-a_3)}{F\cdot(I_1-a_1)+G\cdot(I_2-a_2)+H\cdot(I_3-a_3)}\right\} \tag{8}$$

$$C=(B_2-a_2)A_3\cos(\beta)-(B_3-a_3)A_2\cos(\alpha),\quad D=-(B_1-a_1)A_3\cos(\beta)+(B_3-a_3)A_1,\quad D=(B_1-a_1)A_2\cos(\alpha)-(B_2-a_2)A_1$$

$$E=(B_2-a_2)A_3\sin(\beta)-(B_3-a_3)A_2\sin(\alpha),\quad F=-(B_1-a_1)A_3\sin(\beta),\quad G=(B_1-a_1)A_2\sin(\alpha)$$

The results simulated on assumption that the measuring ideal plane is tilted so that two fringes appear in one observation area are shown in Fig.5. The simulated conditions are shown in this figure. Fig.5(a) shows the result from Eq.(4) and Eq.(5), and Fig.5(b) shows the result calculated in Eq.(8) in which the reflectance ratio γ is considered. An undulated shape corresponding to a period of interferograms' phase ϕ shown in Eq.(8) appears in Fig.5(a). On the other hand, the undulated shape is remarkably decreased in Fig.5(b). These results mean that in case surface under test T is different from calibration S, substituting values which are acquired using the calibration system shown in Fig.2 into in Eq.(8), errors caused by dividing measuring light into three are decreased.

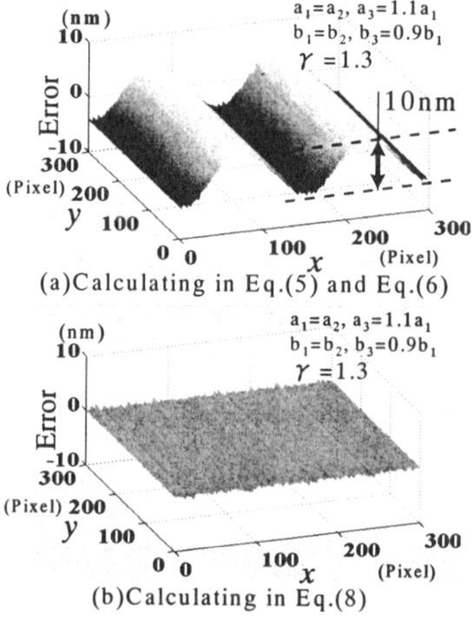

Figure5 Results of simulations

4. CONCLUSION

The measuring error of the MP2 interferometer have been remarkably decreased by effective error reducing algorithm. From now on, our target will be to develop an algorithm decreasing the residual error such as system error.

Reference

[1]Kazuhiko Kawasaki, Naoki Mitsutani, Hiroshi Haino, Profile measurement using interferometer simultaneously taking phase shifted fringes. Proceedings of annual spring meeting of the Japan Society for Precision Engineering; 2000 March 22 - 24; Tokyo. Japan,   (in Japanese).

[2]Kazuhiko Kawasaki, Yasushi Ueshima, Naoki Mitsutani, Hiroshi Haino, Profile measurement using interferometer simultaneously taking phase shifted fringes. Proceedings of annual spring meeting of the Japan Society for Precision Engineering; 2000 October 7 - 9; Nagoya. Japan, (in Japanese).

[3]J. E.. Greivenkamp and J. H. Bruning. "Phase Shifting Interferometry" In Optical Shop Testing 2ed. , Daniel Malacara, ed: John Wiley & Sons, Inc. , 1992

PROPOSAL OF ABSOLUTE LENGTH MEASURING MACHINE BY COMBINING CRYSTALLINE LATTICE SCALE AND LASER INTERFEROMETRY

P. Rerkkumsup, M. Aketagawa, K. Takada

T. Takagi, T. Watanabe, S. Sadakata

Department of Mechanical Engineering, Nagaoka University of Technology
1603-1 Kamitomioka, Nagaoka, Niigata, 940-2188 Japan

Abstract

A new method for absolute length measurement with a resolution of sub-nanometer and a measuring range of a few millimeters by combining a "crystalline lattice scale" and laser interferometry is proposed. The crystalline lattice scale is realized using a regular crystalline surface as a reference and a scanning tunneling microscope (STM) as a detector. The crystalline lattice scale and laser interferometer are utilized as fine and coarse reference scales, where the lattice spacing and optical wavelength are basic references.

Keywords

Absolute length measurement, crystalline lattice scale, sub-nanometer, scanning tunneling microscope, atomic force microscope, laser interferometer, phase modulation.

1. INTRODUCTION

Since precision technology, e.g., ultra-large-scale integration (ULSI) circuits and nanotechnology has progressed rapidly, new methods for length measurement applicable to the millimeter range with sub-nanometer resolution are required. Currently, the standard method for length measurement is laser interferometry. A heterodyne interferometer with a Zeeman laser [1] or a homodyne interferometer using the bi-fringes counting method [2] is widely used in the industry. However, it is difficult to determine arbitrary length with the sub-nanometer accuracy using the interferometers, because they have the non-linearity problem of the fringe interpolation [3]. A phase modulation homodyne interferometer, that can determine the optical path difference of wavelength times integer ($\lambda \times m$) with the accuracy of 10 pm or less, is proposed [4].

The lattice spacing of approximately 0.2 nm for some regular crystalline lattices is uniform and stable over a long range, when the crystals are stress free. These crystals can be used as reference scale with a sub-nanometer resolution instead of laser interferometry. X-ray interferometry (XRI) using silicon crystal has been developed to determine the lattice spacing of silicon at National Metrological Laboratories [5]. Moreover, the combined optical and x-ray interferometer (COXI) [6] has been developed for absolute length measurement with sub-nanometer accuracy at European metrological laboratories. However, XRI is very delicate for an adjustment to obtain x-ray fringe, and not commonly used in the industry. On the other hand, a scanning tunneling microscope (STM) [7] and an atomic force microscope (AFM) [7] are becoming a powerful and popular tool in surface engineering fields and can be used to obtain "images of atoms" on a regular crystalline surface. Therefore, such crystalline lattice can be used as a "crystalline lattice scale" with sub-nanometer resolution by combining them with STM/AFM. We have shown the feasibility of the crystalline lattice scale using a graphite crystal (highly oriented pyrolytic graphite: HOPG) as the reference scale and a dual-tunneling-unit scanning tunneling microscope as the detector [8].

2. PRINCIPLE OF THE ABSOLUTE LENGTH MEASURING MACHINE

Figure 1 and **2** illustrate the schematic diagrams of the absolute length measuring machine (ALMM). The sample of interest and reference crystal (HOPG) are set on piezo-driven stages A and B, respectively, which are just beneath AFM and STM heads with tips and YZ scanners. Both stages travel along 1-dimensional motion axes X_A and X_B. (A coarse motion mode with millimeter travel may also be added in stage A of figure 2.) The stage A and B are independently controlled and their motion axes X_A and X_B are aligned to be parallel with the two arms of the Michelson interferometer. The interferometer can measure motion (= displacement) difference between the X_A and X_B axes. 3-dimensional images of the both samples can be obtained using the combination motions of the stages A/B and AFM/STM heads. The compact thermo-stabilized vacuum cell with a temperature stability of 1 mK and a vacuum of 10^{-4} torr is employed to remove thermal deformation and fluctuation of refractive index of air.

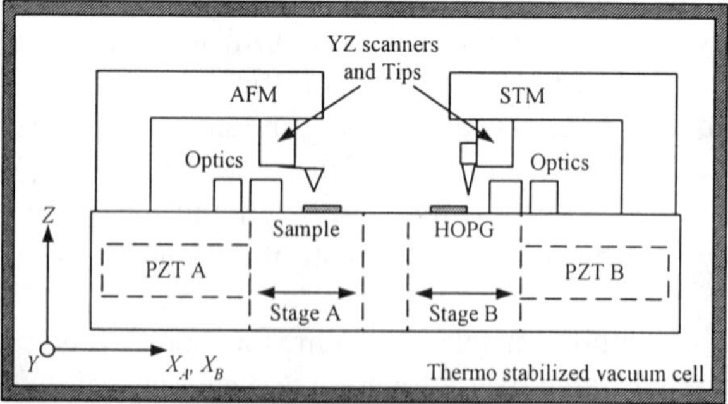

Figure 1. Side view of the absolute length measuring machine.

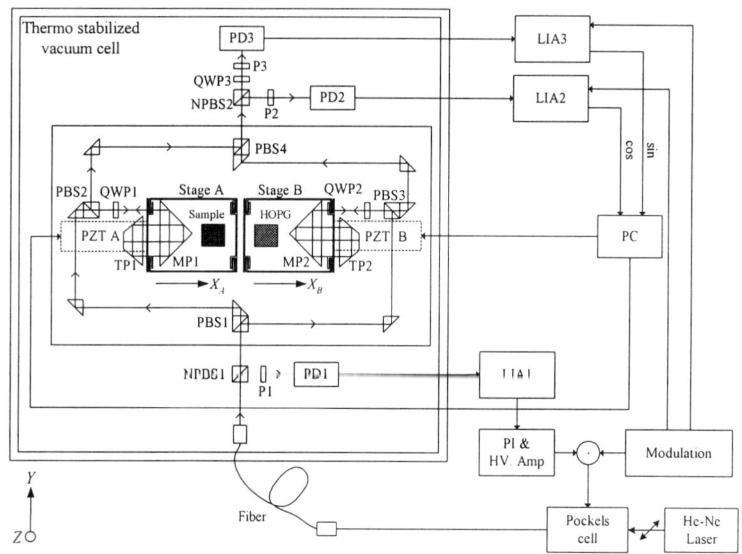

Figure 2. Top view of the absolute length measuring machine.

The operation of the interferometer is shown in Figure 2. The linear polarized beam, whose polarization angle is 45° from the vertical, emitted from the frequency-stabilized He-Ne laser passes through the pockels cell and the single-mode polarization preserving fiber. The pockels cell is used to supply phase modulation and a phase bias shift between the two polarization components p and s. Removing the instability of beam pointing and a heat source of the He-Ne laser can be attained using the fiber. Then the beam passes through the non-polarizing beam splitter 1 (NPBS1), and a part is split into the photodiode 1 (PD1) through polarizer 1 (P1) for the reference phase detection with the lock-in amplifier 1 (LIA1). The beam incident on the polarized beam splitter 1 (PBS1) is divided into two beams, s and p components. The two beams go along the two arms of the interferometer via PBS2 and 3, quarter wave plates 1 and 2 (QWP1/2), moving prism 1 and 2 (MP1/2) on the stages and trapezoid prism 1 and 2 (TP1/2). The beams travel back and fourth four times between MP1/2 and TP1/2. The two beams re-passed through QWP1/2 and the PBS2/3 are recombined at the PBS4. The recombined beam is divided at NPBS2 into two parts, one goes to PD2 via P2 and the other to PD3 via QWP3 and P3, respectively. Output signals form the PD2 and 3 are fed into the LIA2 and 3 to determine the phase shift corresponding to the motion difference between the stage A and B using phase modulation and lock-in detection. The output signal of the LIA1 is used to control the bias phase shift of the pockels cell so that amplitudes of the s and p components at the PBS1 should be the same. The output signals of LIA2 and 3 are the cosine and sine function of the optical path difference.

The motion difference can be measured and controlled at the dark fringes points (= null points) with picometer resolution. In our design, the motion difference between the stages A and B should be $\lambda/4 \times m$. The following steps represent the measurement procedure of the ALMM.

(1) *Initial to zero calibration point*: At the starting position x_{A0} of scan for stage A, displacement of stage B x_B is controlled and locked so that the fringe of the interferometer should be dark (null point). The stage B is then at position x_{B0}.

(2) *Scanning of the stage A for the AFM*: The stage A is then scanned for the AFM imaging to the ending position x_{A1}. The moving translation ($x_{A0}-x_{A1}$) will consist of integer part of optical fringe $n \times \lambda/4$ and fractional part f of an optical fringe $\lambda/4$. n is then determined from the output of the LIA2 and LIA3.

(3) *Scanning of the stage B for the STM*: The stage B is then scanned for the STM atomic imaging to position x_{B1} to shift the optical fringe to the next dark (null point). The translation ($x_{B0}-x_{B1}$) equals to the fractional part of the optical fringe and it can be determined by counting the number m of atoms in the STM image, in which lattice spacing is d.

(4) *Determination of total translation length*: The total translation equals to $n \times \lambda/4 + md$.

3. PRE-EXPERIMENT

In order to realize the ALMM project, a lattice spacing on crystalline surface must be determined. The lattice spacing of a bulk state may not be same as the one on a surface state. To resolve the problem, we preliminary made a measuring machine for lattice spacing on crystalline surface [9], which consist of a flexure piezo-driven X axis-stage, a STM head with tip and a YZ-scanner and multi-path differential interferometer. **Figure 3** shows the outline of the machine, which are almost same as the ALMM, except a AFM head and another flexure stage. (Unfortunately, the interferometer is under construction.) The stage is fabricated from low linear expansion cast iron whose linear expansion coefficient is less than 0.8×10^{-6} K^{-1}. A first resonance frequency of the stage is about 2 kHz. A maximum travel of the stage is about 10 μm (applied voltage to the piezo

Figure 3. Photograph of the measuring machine for lattice spacing on crystalline surface

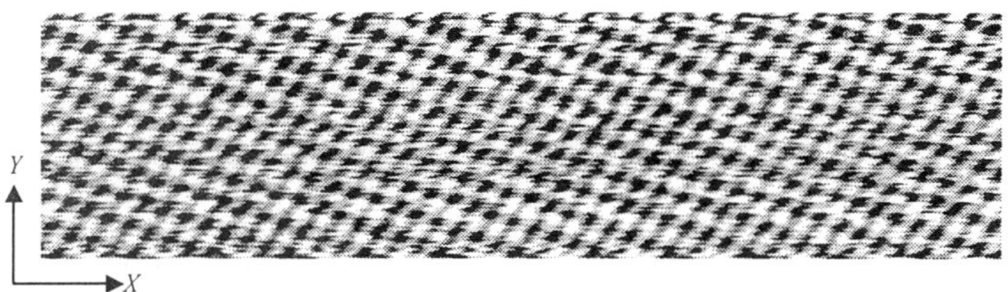

Figure 4. Part of a long atomic image of HOPG. The X-axis was selected as the measurement and the fast scamming axis. The scanned area was 1 μm (X) × 2 nm (Y). Tip speed was 2 μm/s. The scanned time for one line along the fast scanning axis and the scanning period of the complete image were 0.5 and 256 s, respectively.

is 1 kV). Pitch and yaw motion are less than 1.5 μrad for maximum travel. **Figure 4** shows a part of atomic image of HOPG obtained using the machine, whose length is 1 μm. We obtained a good quality image of HOPG for counting the number of atoms in the image. We will report determination process for lattice spacing on HOPG crystalline surface in near future.

4. CONCLUSION AND FUTURE SCOPES

We proposed an absolute length measuring machine (ALMM) by combining a crystalline lattice scale and a laser interferometry. In order to realize the ALMM, we preliminary made a measuring machine for lattice spacing on crystalline surface. 1 μm long atomic STM image of HOPG was obtained with the machine.

In near future, we can report 1) performance of a phase modulation homodyne laser interferometer, 2) determination process of lattice spacing on HOPG crystalline surface, and 3) a prototype of the ALMM.

REFERENCES

1) eg., Agilent Technologies, Model 10897B; 2001.
2) eg., S. Gonda *et al.* Real-time, interferometrically measuring atomic force microscope for direct calibration of standards. Review of Scientific Instruments 1999; 70, 8: 3362-8.
3) R. Thalmann, W. Hou. Limitations of interpolation accuracy in heterodyne interferometry. Proc. of the 7th International Precision Engineering Seminar; May 1993; Kobe, Japan: 11-23.
4) G. Basile, A. Bergamin, G. Cavagnero, G. Mana. Phase modulation in high-resolution optical interferometry. Metrologia 1991-2; 28: 455-61.
5) eg., NIST, PTB, NPL, IMGC, NRLM.
6) G. Basile, P. Becker, A. Bergamin *et al.* Combined optical and X-ray interferometry for high-precision dimensional metrology. Proc. Royal Society Lond. A 2000; 456: 701-29
7) G. Binnig, H. Rohrer, Ch. Gerber, E. Weibel. Tunneling through a controllable vacuum gap. Applied Physics Letters 1982; 40, 2: 178-9
8) M. Aketagawa, K. Takada *et al.* Long atomic imaging over a 5-μm-long region using an ultralow thermally drifted DTU-STM. Review of Scientific Instruments 1999; 70, 1: 133-6.
9) T. Hashida *et al.*, Proc. Spring meeting of JSPE; 2000 March 22-24; Tokyo: 589.

Evaluation of acceleration and deceleration profile of velocity and a new design method using Gaussian distribution curve

Motonori Ogihara , Yuwu Zhang , Hidekazu Oozeki , Yutaka Kuriyama

Mitutoyo Corporation

Abstract

In this paper, we propose a new evaluation index of smoothness, and an originally developed design method of the profile using the Gaussian distribution curve (GD curve). Moreover, the smoothness of GD curve is compared with the conventional quadratic curve and the cycloid curve by using the frequency analysis. These profiles of velocity are applied to the fine measuring machine, and the validity of the evaluation index is confirmed.

Keywords

acceleration , velocity , Gaussian , evaluation , frequency expansion

1. INTRODUCTION

In recent years, numerical control of machining tools and measuring machine require higher speed and accuracy. A fine measuring machine which has nano-meter order accuracy and hundreds of mm strokes requires a smoother and shorter time acceleration and deceleration profile of velocity.

On the other hand, there was no mathematical index which quantitatively evaluated the characteristic of such a profile of velocity.[2] Therefore, we proposed a new evaluation index of smoothness which is based on uncertainty relation.[1] This numerical index is possible to evaluate a profile of velocity of space and frequency region.

2. A NEW EVALUATION INDEX OF THE PROFILE OF VELOCITY

In general, the profile is called the S shaped acceleration and deceleration profile in Figure 1, where vibrations due to acceleration and deceleration are induced corresponding to several modes of resonance. It is important that this curve has localization and smoothness in both of space and frequency region.

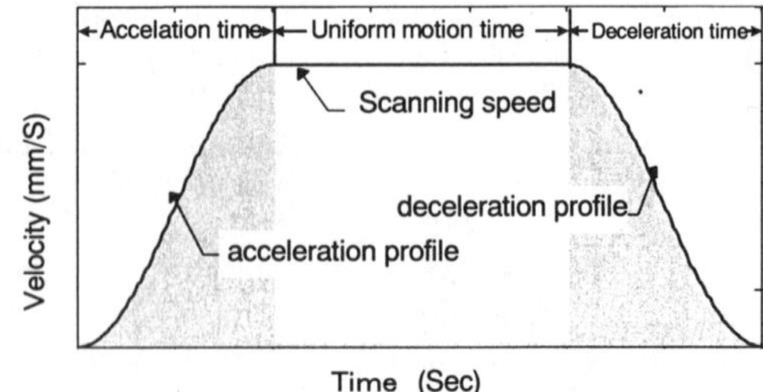

Figure1. Total profile of velocity

The smoothness P is shown by the product of $\Delta\omega$, which indicates a frequency expansion of a filter, and Δt , which shows a spatial expansion of a impulse response , defined by Eqs.(1) and (2), where, Δ means standard deviation. Index P which shows the localization of the filter is shown by Eq.(3). The smoothness S is defined to show the new index of the profile of velocity's smoothness, and S is shown by Eq.(4). In order to evaluate the acceleration and deceleration profile of velocity , frequency extension ω is replaced by spatial extension t using Parseval equality (Eq.(5)); if the acceleration and deceleration curve is to be $A(t)$, the smoothness S can be shown as Eq.(6) .

However, The smoothness S is able to evaluate a continuous function. Because, the profile of velocity should be a continuous function not only in a specified section but also before and behind that. If the profile of velocity is discontinuous, it is converted in the Fourier series which has enough accuracy.

Uncertainty relation

$$(\Delta\omega)^2 = \frac{\int_{-\infty}^{\infty}\omega^2 \cdot F^2(\omega)d\omega}{\int_{-\infty}^{\infty}F^2(\omega)d\omega} \qquad (1)$$

$$(\Delta t)^2 = \frac{\int_{-\infty}^{\infty}t^2 \cdot f^2(t)dt}{\int_{-\infty}^{\infty}f^2(t)dt} \qquad (2)$$

$$P = \Delta t \cdot \Delta\omega \quad \geq 0.5 \qquad (3)$$

$$S = \frac{1}{4\cdot P^2} \quad \leq 1 \qquad (4)$$

The Parseval equality

$$\int_{-\infty}^{\infty}|F(\omega)|^2 d\omega = 2\pi\int_{-\infty}^{\infty}f^2(t)dt \qquad (5a)$$

$$\int_{-\infty}^{\infty}\omega^2|F(\omega)|^2 d\omega = 2\pi\int_{-\infty}^{\infty}\{f'(t)\}^2 dt \qquad (5b)$$

Evaluation index of profile of velocity

$$S = \frac{1}{4}\cdot\frac{\left\{\int_{0}^{\infty}A^2(t)dt\right\}^2}{\int_{0}^{\infty}(t-\bar{t})^2 A^2(t)dt \cdot \int_{0}^{\infty}\{A'(t)\}^2 dt} \qquad (6a)$$

$$\bar{t} = \frac{\int_{0}^{\infty}t \cdot A^2(t)dt}{\int_{0}^{\infty}A^2(t)dt} \qquad (6b)$$

3. DESIGN METHOD OF A NEW PROFILE OF VELOCITY USING GD CURVE

When acceleration and deceleration curve $A(t)$ is assumed to be a GD curve which extends to $\pm\infty$, it becomes the smoothest acceleration and deceleration curve as the smoothness $S = 1$. However, it is impossible to use it as the acceleration and deceleration curve, because addition-subtraction speed time will become infinity. Then, Figure 2 shows a part of the GD curve which is a part of the total GD curve. It is converted into the normalized area of the normalized time of $t=0$ to 1, and this is used as an acceleration and deceleration profile of velocity . This partial curve is shown as Eq.(7). Figure 3 shows the relation of the smoothness and the ratio of the maximum acceleration v.s. average acceleration to range cut out ,n. The smoothness is saturated in the area of n=3 or more, and an acceleration monotonously increases with n. Therefore, the GD curve has to be used in the area of n= 2 to 3.

$$f(x) = \frac{1}{\sqrt{2 \cdot \pi}} \cdot e^{-\frac{x^2}{2}} \qquad (7a)$$

$$A(t) = \frac{1}{k} \cdot e^{-\frac{x^2}{2}} - \left(\frac{1}{k} - 2\right) \qquad (7b)$$

$$k = \frac{1}{2 \cdot n} \cdot \int_{-n}^{n} e^{-\frac{x^2}{2}} dx - e^{-\frac{n^2}{2}} \qquad (7c)$$

$$x = (t - 0.5) \cdot n \cdot 2 \qquad n = 2.28 \qquad (7d)$$

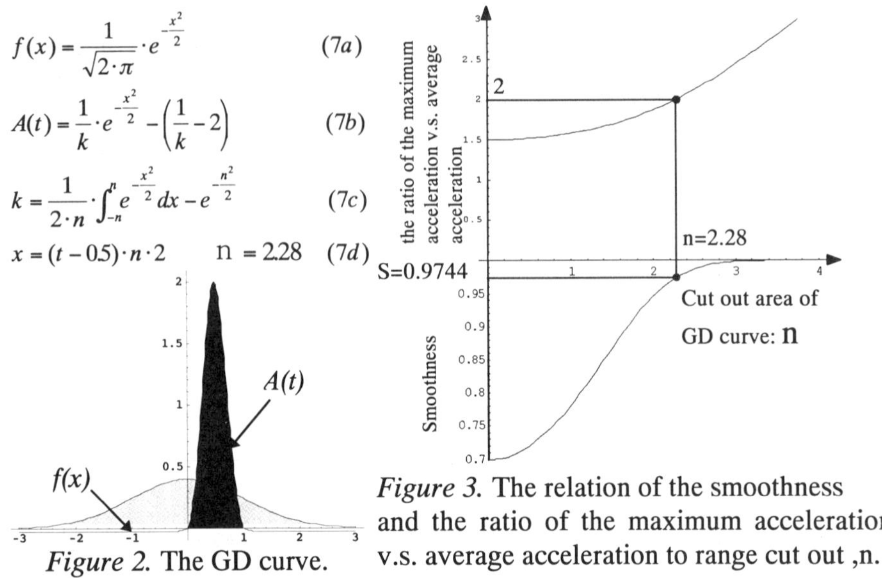

Figure 2. The GD curve.

Figure 3. The relation of the smoothness and the ratio of the maximum acceleration v.s. average acceleration to range cut out ,n.

4. SMOOTHNESS OF THE PROFILE OF VELOCITY

The smoothness S of the profile of velocity are shown in Table 1. 5 types of profile of velocity are evaluated as follows; ① what is applied to the quadratic curve(Eq.(8)) , ② to the sine curve (Eq.(9))[3], ③ to the quartic curve (Eq.(10)), ④ to the cycloid curve (Eq.(11)), and ⑤ to the GD curve(Eq.(7)). All the profile of velocity are normalized so that the ratio of the maximum acceleration v.s. average acceleration becomes 2. However, the addition-subtraction speed time of these curves (①,②,③) is shortened by their own characteristics.

Table 1 smoothness comparison

No	Name	Smoothness
①	The quadratic curve	0.7000
②	The sine curve	0.7753
③	The quartic curve	0.9167
④	The cycloid curve	0.9495
⑤	The GD curve	0.9744

$$A(t) = -128/9 \cdot x^2 + 2 \qquad (8a)$$

$$x = (t - 0.5) \qquad (8b)$$

$$A(t) = 2 \cdot \sin(4 \cdot x - 2 + \pi/2) \qquad (9)$$

$$A(t) = \frac{1}{k} \cdot (30 \cdot x^4 - 60 \cdot x^3 + 30 \cdot x^2) \qquad (10a)$$

$$x = t/k - (1-k)/2 \quad k = 0.9375 \qquad (10b)$$

$$A(t) = 1 - \cos(2 \cdot \pi \cdot t) \qquad (11)$$

5. COMPARISON OF THE PROFILE OF VELOCITY

The GD curve is compared with the quadratic curve which is the most simply function, and the cycloid curve which has so far been considered the best acceleration curve.[3] These velocity curve are shown Figure 4. Likewise, these acceleration curves are shown in Figure 5. In particular, there is little difference in the profile of velocity.

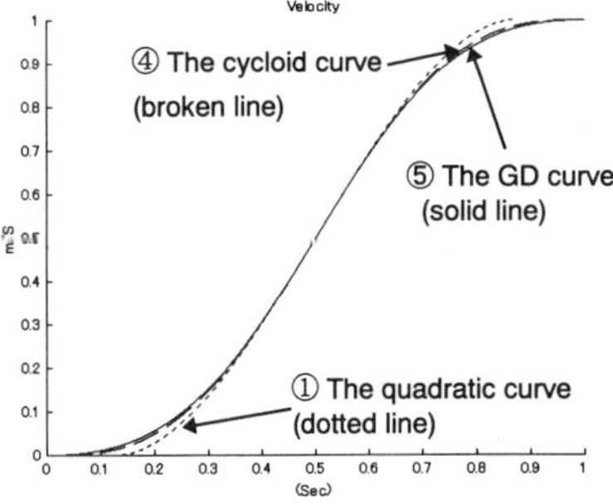

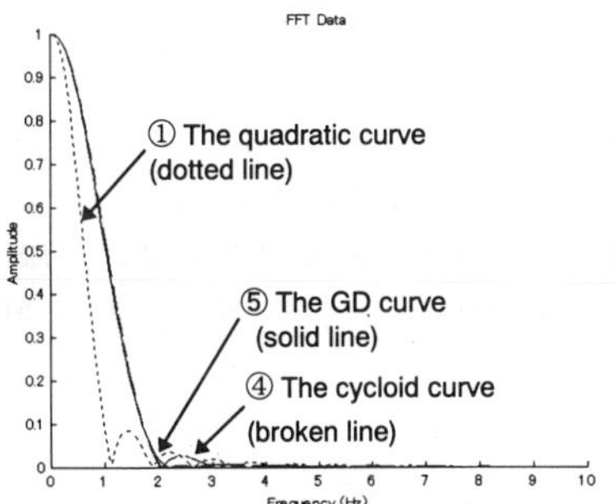

Figure 4. The profile of velocity   *Figure 5.* The acceleration curves

The results of the frequency analysis for these acceleration curves are shown in Figure 6 and 7. The high order frequency component of the curve is considerably different among these acceleration curves. It is confirmed that the high order frequency component of the GD curve is less than other acceleration curves'. By this result, the validity of smoothness S which is the index of the smoothness proposed in this paper has been confirmed.

Figure 6. The frequency analysis (extended figure)   *Figure 7.* The frequency analysis

6. APPLICATION OF PRECISE TABLE FOR MEASURING MACHINE

The profile of velocity is applied to a precise table of measuring machine. This table is driven by AC servo motor, and its resolution is determined at 3.125nm by a linear encoder. This full closed positioning system is controlled by PI control.

The velocity tracking error is used as an index of the evaluation figure. Because, it is very important in the measuring machine which is measured by continuous measurement at uniform velocity, for example a roughness tester. If there are large velocity tracking error, the measuring machine could not measure accurately. Moreover, there are linear relationships between the acceleration and the velocity tracking error.

Figure 8 shows the velocity tracking error at the profile of velocity. (the command velocity :5mm/sec , the maximum acceleration :50mm/sec^2). The frequency analysis are shown in Figure 9. It is similar to the frequency analysis of acceleration curves (Figure 7). The GD curve is better than other curves, and other curves have more unnecessary vibration. The above results, as well as analysis result, confirm the validity of the smoothness S .

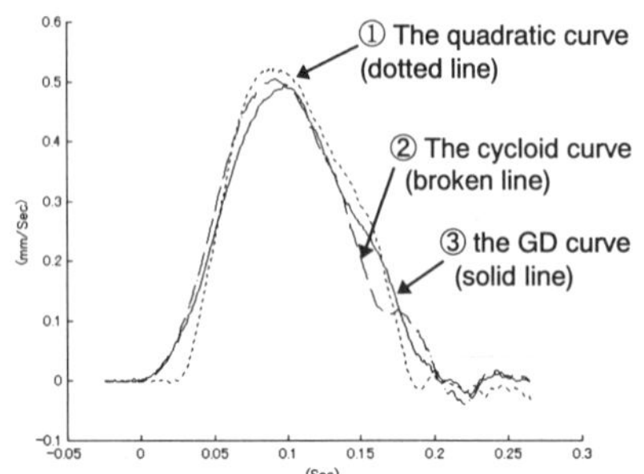

Figure 8. The velocity tracking error   Figure 9. The frequency analysis of the velocity tracking error

7.CONCLUSION

According to the smoothness S which we propose, the curve is more smooth in the order of the following; GD curve, cycloid curve, quartic curve, sine curve, and quadratic curve. It is confirmed that the GD curve brings less unnecessary vibrations by the frequency analysis of the tracking error at the measuring machine table.

REFERENCES

(1) Motonori Ogihara, Yutaka Kuriyama: Development of precise table for measurement with friction drive mechanism, Regular Meeting proceedings of Ultra-precision positioning committee,the japan Society for Precision Engineering,No99-5,1/21/2000,pp9-20,(in japanese).

(2) Takanori Yamazazaki,Masaki Seto and Masaomi Tsutsumi: Desing of Acceleration and Deceleration Commands for NC MachineTools,Journal of the japan Society for Precision Engineering,ol66,No8,2000 ,pp1260-1264 ,(in Japanese).

(3) Tosio Ogiso, Fujio Tajima, Akinobu Takemoto, Nobumoto Kezuka: Optimization of the Driving Pattern for Assembly Robots, JSME Robotics and Mechatronics,1990-B,pp151-154,(in Japanese).

Thermal and mechanical FEA (Finite Element Analysis) of wafer heating for EB stepper

Satoshi Takahashi, Yuichiro Miki,

Kenji Morita*, Noriyuki Hirayanagi*, Tomoharu Fujiwara*,

Motoharu Tateishi**, Kazuo Maeno***

Technical Systems Department, Core Technology Center, Nikon Corporation, 1-6-3 Nishi-Ohi, Shinagawa-ku, Tokyo 140-8601, Japan
**2nd Development Department, Development Management Department, IC Equipment Division, Precision Equipment Company, Nikon Corporation*
***Technical Division, Japan Operations, MSC JAPAN LTD., 2-39, Akasaka 5-chome, Minato-ku, Tokyo 107-0052, Japan*
**** Urban Environment Systems Department, Faculty of Engineering, Chiba University, 1-33 Yayoi-cho, Inage-ku, Chiba 263-8522, Japan*

Abstract

In this study the wafer heating problem for EB stepper is simulated numerically. The two modeling methods of averaged heat loading and of dynamic meshing are proposed. By using these methods, it is possible essentially to reduce the computational time with high accuracy.

Keywords

Heat Conduction, Finite Element Method, Adaptive-h Method, EB

1. INTRODUCTION

Electron Beam Projection Lithography (EBPL) systems are developed as the next generation lithography systems. One of the typical EBPL systems is SCALPEL ® [1] of Lucent Technologies and the other is EB stepper[2] of Nikon and IBM. These EBPL systems require a higher power EB and larger one-shot-image size than in conventional EB lithography. In these arrangement thermal deformation is a significant key issue[3].

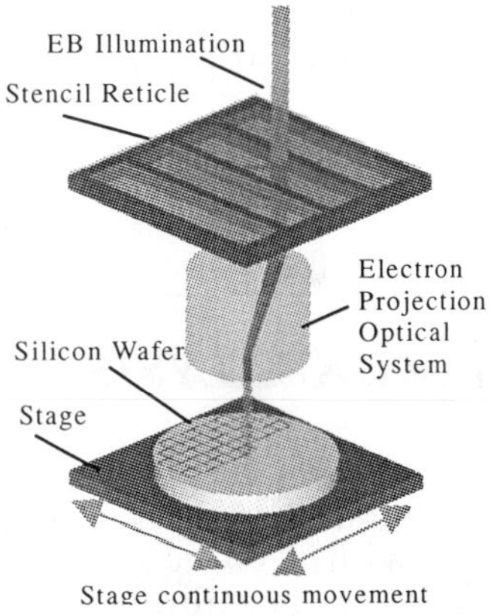

Figure 1 Schematic of EB Stepper

Figure1 shows a schematic of EB stepper. One shot size (sub-field) on the silicon wafer is $250\,\mu\text{m} \times 250\,\mu\text{m}$, and the complete field of a die (i.e. $10\text{mm} \times 10\text{mm}$) is exposed by means of the stitching of sub-fields using EB deflection and stage-scanning motion. The dies are created on entire wafer of $\phi\,300\text{mm}$. The high-power electron beam with EB stepper causes a significant and dynamic heat deformation of wafer during high-throughput operation, which turns into a Pattern Placement Error (PPE) in plane shown Figure 2.

In SCALPEL ®, this expansion- induced PPE is simulated about sub-field scale response, and the correction strategy for PPE is investigated[4] This PPE simulation requires excessive computational time, as the finite element (FE) model consists of a lot of fine elements of sub-fields scale. The goal of our work is to make highly accurate modeling with realistic computational time. In this paper, two modelings are proposed for reduction of computational time.

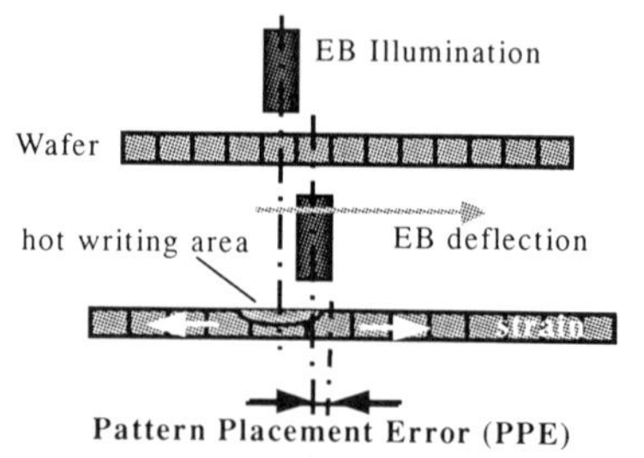

Figure 2 Beam heating of wafer

2. FE MODEL AND BOUNDARY CONDITIONS

Modeling and analysis of FE transient thermal deformation are performed using standard/customized MSC.MARC®. As for the global boundary conditions, one-shot is power input of 2.5W for a sub-field, corresponding to the EB current of 25 μA and acceleration voltage of 100kV, under adiabatic conditions. Pulse duration is 62.5 μsec and the dwell time between each shots is 25 μsec. The mechanical boundary conditions are assumed without chucking for wafer. The material properties of silicon wafer are shown in Table 1.

Table 1. Material Properties of Silicon Wafer

Modulus of elasticity:1.07E+11 [Pa], Poissons ratio : 0.2
Mass density : 2.34E+3 [Kg/m^3], Thermal conductivity : 1.68E+2 [W/(m·K)]
Coefficient of thermal expansion : 2.4E-6 [1/K], Specific heat : 7.60E+2 [J/(kg·K)]

3. LOCAL ANALYSIS

The FE model picks up one local area of the silicon wafer for computational efficiency. The purpose of this local analysis is to get PPE data with accuracy for the three dimensional (3D) calculation, and to confirm possibility of 2D plane modeling. The local analysis model and its scanning procedure are shown Figure 3. Heat load for 3D model is imposed as surface heat flux on the top wafer, while homogeneous and volumetric heat generation is given for 2D plane

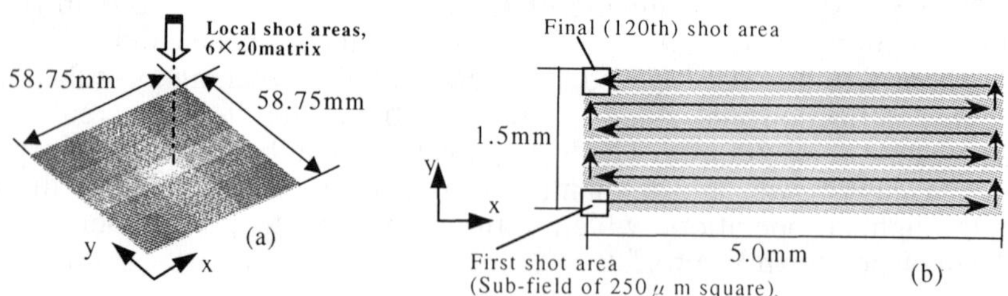

Figure 3 FE model (a), Local shot areas [6rows×20columns] (b)

830

model through the thickness (0.775mm). The results of both 3D and 2D analysis are plotted in Figure 4. The PPE are normalized by the maximum PPE of 3D analysis. The PPE of 2D analysis are one third of 3D analysis results, while the shape of the profiles are similar in both X and Y directions. Though the 3D results reflect the real situation, 2D analysis can be also used to evaluate the rough tendency of global wafer.

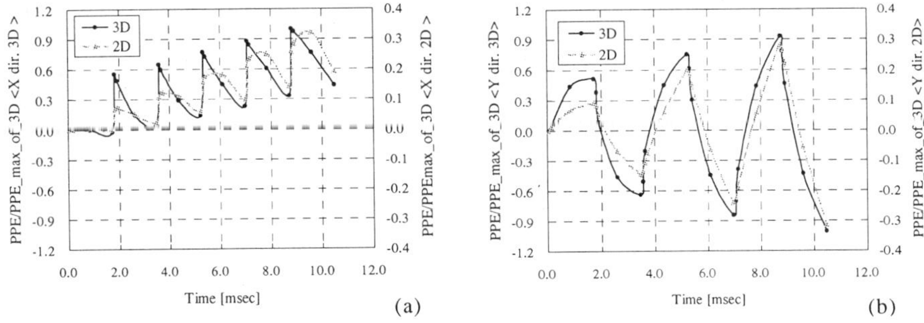

Figure 4 Comparison of PPE between 3D and 2D analysis, the x direction (a) and the y direction (b).

4. GLOBAL ANALYSIS

4.1 AVERAGING TECHNIQUE FOR HEAT LOADING

The exposure areas on the wafer are made up a large number of dies. In fact, a die currently being exposed is influenced by the large areas exposed before. We focus on one die and attempt to represent correctly for the PPE occurring in that die. The averaging

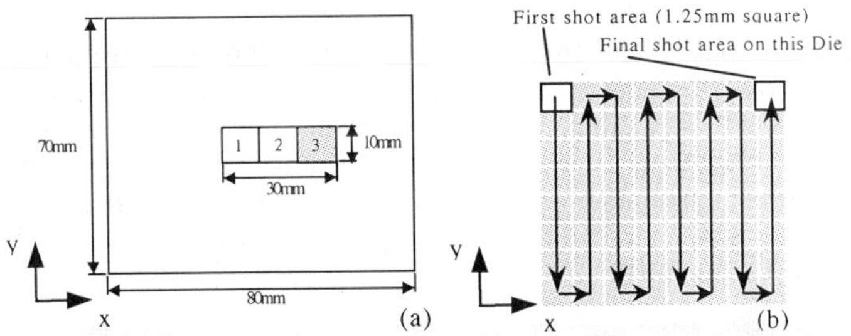

Figure 5 Schematic of the beam exposure areas for the global analysis model and scanning direction from 1 to 3 for the die (a),   and scanning procedure in the die of 8×8 matrix (b).

Table2.Scanning procedure for global analysis

	1st	2nd	3rd	Die
Case1 Three actual shots				
Case2 One actual and two averaged shots				
Case3 Two actual and one averaged shot				
Case4 Same to Case 3 (Reduced mesh model)				

technique is to deposit the averaged input energy for areas being exposed before, except the focused die. The 2D plane model of die scale is shown in Figure 5. A die is made up of 8×8 matrix, and one cell is 1.25mm square. Four scanning procedure models are generated as in Table 2. Case1 represents actual scanning. The focused die (No.3) is made actual scanning at all cases. Moreover, the FE models of Case 1, 2 and 3 are identical. In Case4 the mesh number of FE is approximately one tenth of the other cases. Figure 6 shows the PPE of all cases. Both qualitative and quantitative agreements are good for all cases except Case2. Furthermore, Case1 required thirty times as many as computational time of Case4. These results show that the necessary scanning procedure for FE analysis is to shot the die just before the focused die, i.e. the 2nd die in this case. Then the mesh number around the focused die and formerly exposed dies can be reduced greatly.

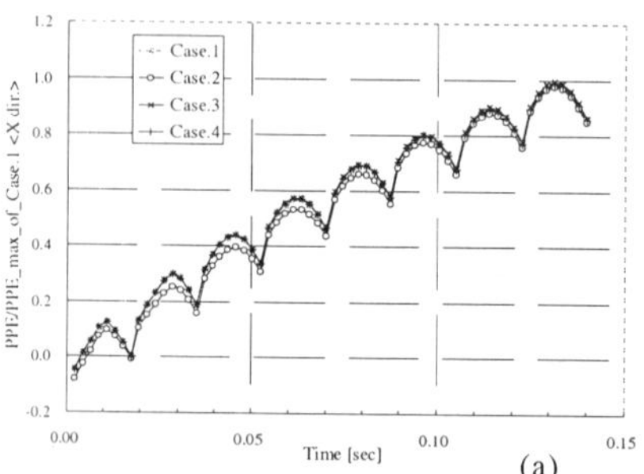
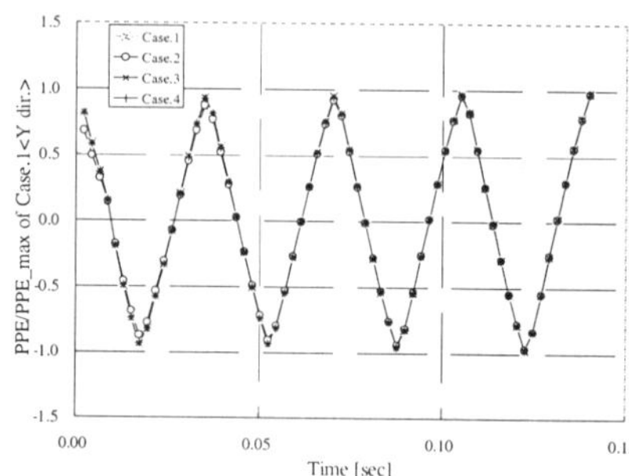

Figure 6 Comparison of the PPE from Case 1 to Case 4 on the 3rd die, the x direction (a) and the y direction (b).

4.2 DYNAMIC MESHING APPROACH

On the next trial, for the reduction of mesh number of entire FE model, we developed a dynamic mesh generation procedure applying adaptive-h method. This procedure performs repeatedly fine or coarse re-meshing in a dynamic way and adapts the mesh for heat load position. This dynamic meshing method is applied to the model shown in Figure 5, and scanning procedure of Case3/4 (i.e. Case5) is used. The results are shown in Figures 7 and 8. Figure 7

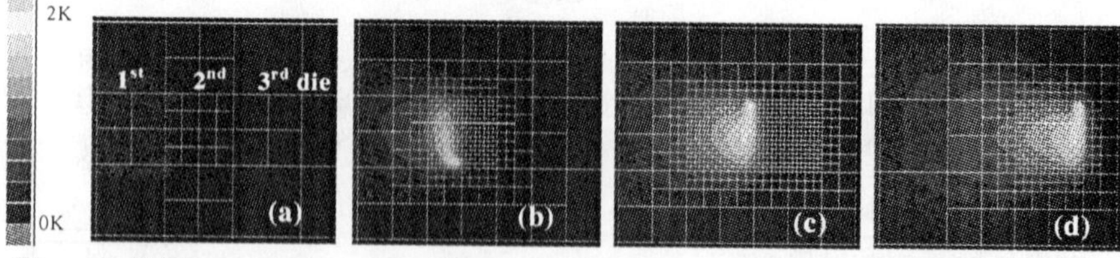

Figure 7 Temperature distribution of the wafer, after averaged shots on rough meshed 1st die (a), on the 2nd die and fine re-meshed that die (b), final shot of 2nd die and fine re-meshed 3rd die (c), final shot of 3rd die and coarse re-meshed 2nd die.

shows transient temperature distribution of the wafer. The PPE on each directions are shown in Figure 8. The agreement between Case1 and Case5 is good. The computational time of Case5 is one fifteenth of Case1. This dynamic meshing approach is to calculate in the limited mesh number. Therefore, the more focused die number increases, the more relative computational time is essentially reduced with high accuracy.

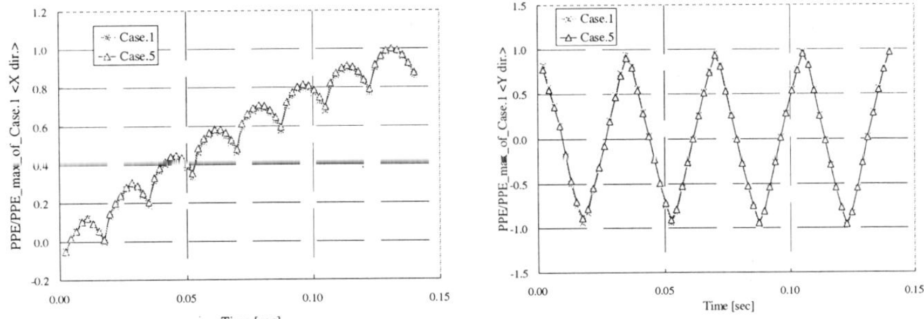

Figure 8. Comparison of the PPE between Case1 and Case5 on the 3rd die, the x direction (a) and the y direction (b).

5. CONCLUSION

Through finite element analysis of the wafer heating, the following has been concluded:
1) From the numerical simulations the accurate pattern placement error (PPE) is obtained by 3D local analysis, while 2D plane analysis is also able to evaluate roughly for this local transient thermal deformation.
2) As the trial case, averaged heat loading and dynamic meshing methods are applied for global analysis of 2D plane model. Applying these methods, the computational time is reduced below one fifteenth of actual scanning model. The expansion to full 3D model of these methods is feasible sufficiently keeping such effect as high accuracy and reduction of computational time on 2D analysis.

ACKNOWLEDGMENT
The authors would like to express their thanks to Mr. Masaya Miyazaki and Mr. Kazuaki Suzuki of Nikon, Mr. Kazuhiro Maeda of the former Nippon MARC for their encouragement, helpful suggestions and discussions.

REFERENCES
1) S.D.Beger and J.M.Gibson, New approach to projection-electron lithography with demonstrated 0.1μm linewidth, Appl. Phys. Lett. 57, 153-155, 1990.
2) K.Suzuki, et al., Nikon EB stepper: its system concept and countermeasures for critical issues, SPIE 3997, 214-224, 2000.
3) K.Morita, et al., (1999) Thermal Characteristics of Scattering Stencil reticle for Electron Beam Stepper Jpn. J. Appl. Phys., 38-7027, 1999.
4) B. Kim, et al., Finite element analysis of SCALPEL wafer heating, J. Vac. Sci. Technol. B17(6), Nov/Dec, 2883-2887, 1999.

Part IV

Manufacturing / CAD / CAM

MODELING OF MACHINING PROCESS PLANNING BASED ON INVERSE FORM-SHAPING FUNCTION

Fumiki Tanaka, Naosuke Toyama, Takeshi Kishinami

Graduate School of Engineering, Hokkaido University

Abstract

This paper deals with a model of process planning based on an inverse form-shaping function. A machining process model, defined as the form-shaping function, has been proposed and analyzed by many researchers. If the process planning is considered as an inverse process of the machining process, the process planning model can be easily established by inverting the form-shaping function. To accomplish this, the parametric form-shaping function as the machining process model and inverse form-shaping function as machining process planning model are proposed.

Keywords

Machining process planning, Form-shaping function,
Machining process model

1. INTRODUCTION

In a machining process, input information is an NC program and output is a machined shape. On the other hand, in a process planning, input information is the machining feature and output information is the NC program. Therefore, output information of the machining process is input information of the process planning, and vice versa. If the machining process is considered as the inverse process of the process planning as shown in Figure 1, we can easily establish the process planning model by inverting the machining process. The model of machining process has been proposed using a form-shaping function in a previous report (Tanaka et. al. 2000). In this paper, a model of process planning based on the inverse form-shaping function is proposed.

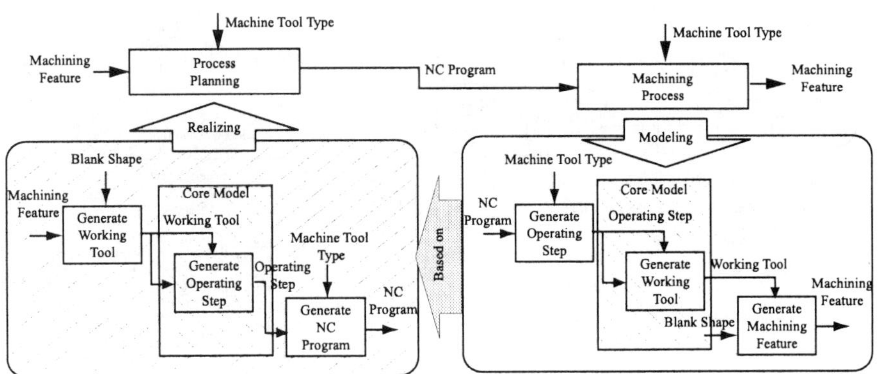

Figure 1. Modeling methodology of process planning.

2. MACHINING PROCESS MODEL

One of the mathematical models of machining process is the form-shaping function proposed by V.T. Portman (Reshetov and Portman 1988). Using the machining motion matrices A and the cutting edge vector r_T, we can obtain the form-shaping function (equation 1) that represents the cutting points r_0 on the workpiece coordinate frame.

$$r_0 = A(q_1, q_2, \cdots, q_l)\, r_T(q_{l+1}, \cdots, q_{l+m})$$

$$\text{with} \begin{cases} f_1, \cdots, f_{Lf} : \textit{Functional Constraints} \\ g_1, \cdots, g_{Le} : \textit{Enveloping Constraints} \end{cases} \tag{1}$$

where, q_i is an i-th relative motion variable or a parameter of the cutting edge, l is the number of the relative motion variables of the machine tool, and m is the number of parameters of the cutting edge vector. However, we cannot obtain the process planning model directly by inverting the form-shaping function. Therefore, the machining process needs to be modeled in consideration of the actual phenomena of the process for determining the functional constraints and the cutting edge shape from the required shape in the process planning.

The machining process can be considered as consisting of the activities shown in Figure 2. In this figure, an operating step is the machining information related to a cutting edge and a working tool that is a functional model of the cutting edge. The operating step consists of the controlled table motion, the enveloping constraints and the macro tool that is generated by rotating the cutting edge around the spindle axis. The working tool consists of the swept macro tool, the envelope of the swept macro tool and the working edge which is the contact portion between the macro tool and the required shape. In the machining process activity, the activity which generates the working tool from the operating step is the core model of the machining process, because the operating step represents the basic machining operation, and the working tool directly corresponds to the machining feature.

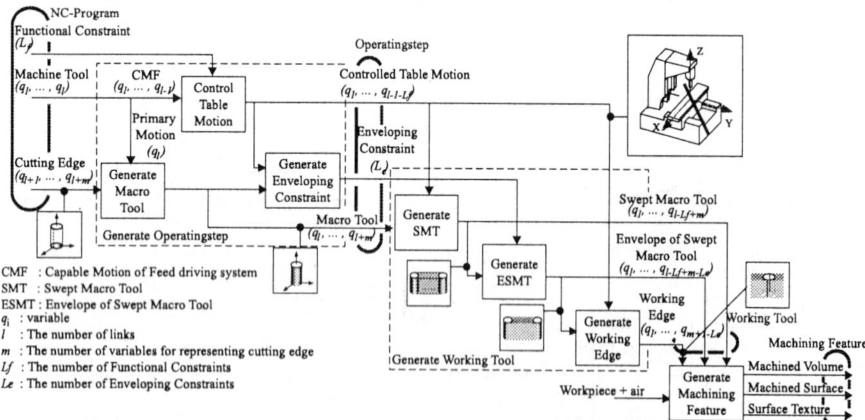

Figure 2. Model of machining process.

3. PARAMETRIC FORM-SHAPING FUNCTION

According to the model of machining process, as mentioned above, we propose an abstract core model of the machining process in order to capture its general and fundamental characteristics.

The characteristics of the operating step and the working tool are represented by the number of variables. As shown in Figure 2, the numbers of the variables of the operating step are l-1-Lf, $m+1$, $-Le$ and the numbers of the variables of the working tool are $\gamma=l$-$Lf+m$, $\alpha=l$-$Lf+m$-Le, $\lambda=m+1$-Le. Therefore, the type of operating step and working tool are represented by triple numbers (l-1-Lf, $m+1$, $-Le$) and (γ, α, λ) where, $m+1$ is the dimension of the macro tool, l-1-Lf is the degree of freedom of the controlled table motion, Le is the number of equations of the enveloping constraints, γ is the number of variables of the swept macro tool which is not less than the dimension of the machined volume, α is the number of the variable of the envelope of the swept macro tool which is not less than the dimension of the machined surface, λ is the dimension of the working edge which is the same as the dimension of the surface texture as shown in Figure 3.

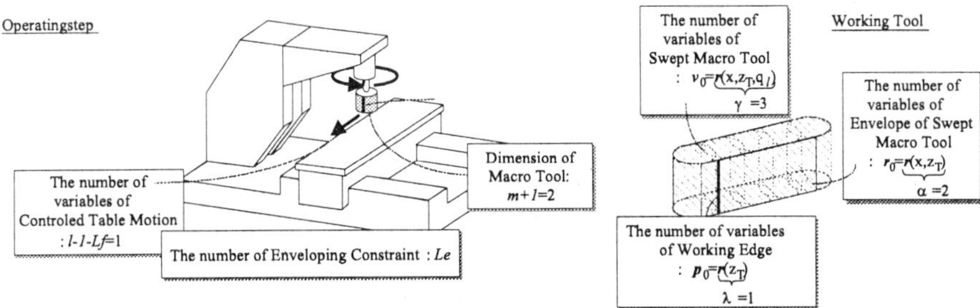

Figure 3. Parameters of working tool and operating step.

The relationship between the numbers of the variables of the working tool and operating step is given by equation 2. Equation 2 is considered as a parametric form-shaping function of the machining process which gives a mapping of the type of operating step to the type of working tool.

$$\begin{bmatrix} \gamma \\ \alpha \\ \lambda \end{bmatrix} = \begin{bmatrix} 1 & 1 & 0 \\ 1 & 1 & 1 \\ 0 & 1 & 1 \end{bmatrix} \begin{bmatrix} l-1-L_f \\ m+1 \\ -Le \end{bmatrix} \tag{2}$$

4. PROCESS PLANNING BY INVERSE FORM-SHAPING FUNCTION

The process planning model is considered as a inverse model of the machining process as mentioned above. The abstract core model of the machining process, defined as the parametric form-shaping function, is represented by equation 2 in the linear equation. Therefore, we propose an abstract core model of process planning, defined as a parametric inverse form-shaping function, which could represent the linear equations related to the numbers of the variables in the working tool and operating step represented by equation 3.

$$\begin{bmatrix} l-1-L_f \\ m+1 \\ -Le \end{bmatrix} = \begin{bmatrix} 0 & 1 & -1 \\ 1 & -1 & 1 \\ -1 & 1 & 0 \end{bmatrix} \begin{bmatrix} \gamma \\ \alpha \\ \lambda \end{bmatrix} \qquad (3)$$

Using the parametric inverse form-shaping function, the process planning is performed as shown in Figure 4. At first, from the machining feature, we determine the type of working tool as specifying γ, α and λ. Then, we derive the type of operating step from equation 3. Consequently, we determine the number of functional constraints L_f and the dimension of cutting edge m by specifying the number of the relative motion variables of the machine tool l. To derive the complete information of the machining process, we need the geometrical information of the machining feature.

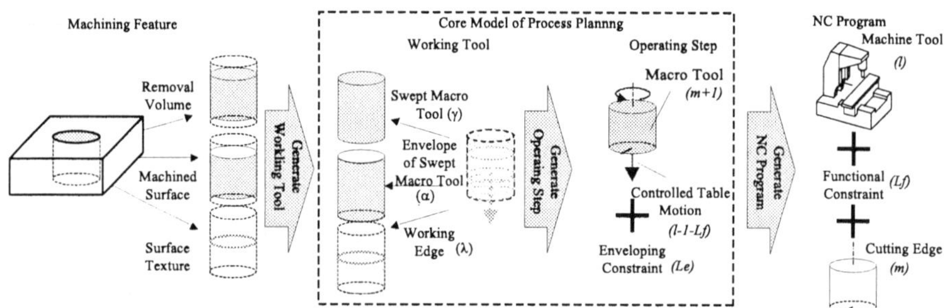

Figure 4. The process planning based on parametric inverse form-shaping function.

5. EXAMPLE

We show some examples to demonstrate the advantages of the proposed model. Figure 5 shows the examples of process planning for a cylindrical depression shape using a proposed parametric inverse form shaping function.

For example, in the case of the removal volume as the volume and the surface texture as a circle, we determine the type of working tool as specifying $\gamma=3$, $\alpha=2$ and $\lambda=1$. Next, we determine the type of operating step ($Le=1$, $l-1-L_f=1$, $m+1=2$) from equation 3. If we use the three axis-milling machine ($l=4$), we determine the number of functional constraints ($L_f=2$) and the dimension of the cutting edge ($m=1$).

As shown in this figure, not only the geometrical information of machined surface, but also the parametric inverse form-shaping function is necessary to determine the machining process unambiguously.

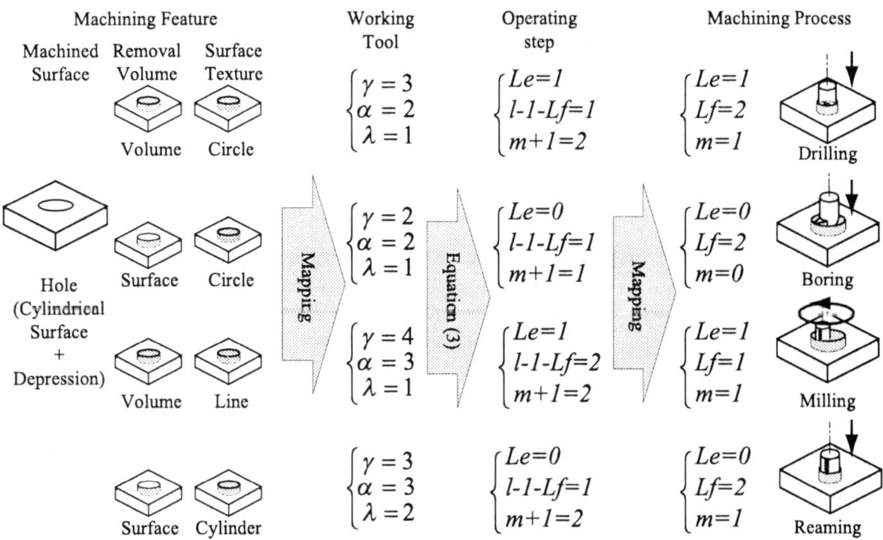

Figure 5. Examples of process planning for cylindrical depression shape.

6. CONCLUSIONS

In this paper we proposed a model of process planning based on an inverse form-shaping function. The conclusions are as follows:

(1) We proposed the parametric form-shaping function which represents the relation between the type of working tool and operating step as the abstract core model of the machining process.

(2) We proposed the inverse parametric form-shaping function by inversing the parametric form-shaping function as the abstract core model of process planning.

(3) We showed examples of the process planning using inverse form-shaping function to demonstrate the advantages of the proposed model.

REFERENCES

Gupta, S.K., Kramer, T.R., Nau, D.S., Regli, W.C. and Zhang, G. (1994) Building MRSEV models for CAM applications, Advances in Engineering Software, 20, 121-139.

Portman V.T., Reshetov D.N., Accuracy of Machine Tools. New York: ASME Press, 1988

Shirur, A. and Shah, J.J. (1996) Machining Algebra for Mapping Volumes to Machining Operations, Proc. of the 1996 ASME Design Engineering Technical Conferences and Computers in Engineering Conference, 96-DETC-DFM-1303.

Tanaka F, Toyama N, Kishinami T., Feature models of machining operation and machined shape for single cutting edge machining, Proceedings of 2000 JAPAN-USA FA Symposium, 2000

EASY CONSTRUCTION OF 3D MODEL
FROM 2D INFORMATION

Yoshinori Urabe, Hideki Aoyama

Department of System Design Engineering, Keio University

Abstract

In the early stage of style design processes, designers start by expressing their ideas of a product on sketches. This study aimed to develop a system for constructing three-dimensional models from two-dimensional sketches to support the style designs of automobiles. The developed system allows the construction of models without obstructing the aesthetic expression of the designer. This paper describes an automatic identification method of a sketch as well as a simple three-dimensional model construction system that builds 3-D models from the 2-D information of sketches mainly for automobile body.

Keywords

Sketch, Three-dimensional Model, CAD, Aesthetic Design, Style Design, Sensitivity Expression, Automobile

1. INTRODUCTION

Recently, CAD systems are being increasingly applied to design activities of industrial products. Although 3-D CAD systems are suitable for function designs, they are very difficult to use for constructing complex three-dimensional shapes for product designs, because the currently used 3-D CAD systems are not fully efficient for supporting designer's activities based on aesthetic sense. [1] Consequently, the Reverse-Engineering Technology is still being used mainly in the style design. The Reverse Engineering Technology is a method of constructing three-dimensional models from physical models made of clay or wood which are easy for expressing the aesthetic sense of designers. Though physical models are not required in the final process, considerable time and costs are needed to build these physical models.   This calls for some improvements in the style designing process.

On the other hand, in the early processes of style designing, designers express their ideas of products on sketches.   This means that if three-dimensional CAD models can be built directly from sketches, the aesthetic expressions of the designer can be faithfully reproduced in the design, the lead time can be shortened, and costs will be lower. [2] [3]

This study is thus conducted with the aim of developing a style design support system which is able to construct three-dimensional models easily from two-dimensional sketches drawn by designers.   This system is targeting for use in the design process of industrial products, especially the complex curved surfaces of automobiles, etc.

2.   PROCEDURE FOR CONSTRUCTING 3-D MODEL

The objective of this study is to develop a system that supports the construction of three-dimensional models easily from two-dimensional sketches, targeting for use in the design process of industrial products such as automobiles, etc.   After construction of a three-dimensional model, the model shape will be improved with other computer-aided systems.

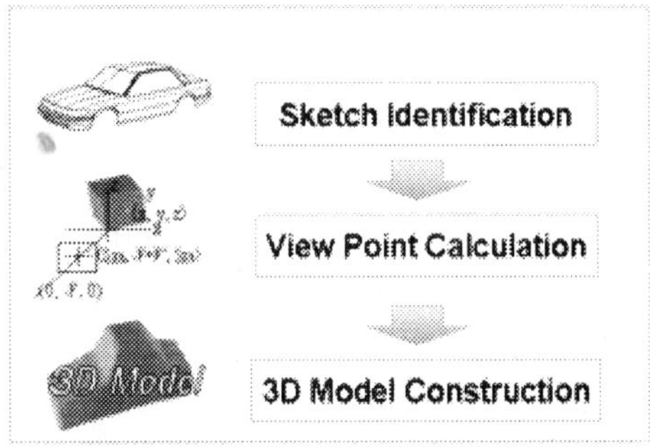

Fig. 1 Procedures for constructing 3-D

As shown in Fig.1, three-dimensional models are roughly built in the following procedures:

(1) The information of the two-dimensional sketch drawn by designer is fed into a computer by a scanner.
(2) The lines of the input sketch are identified by the system.
(3) The viewpoint direction is recognized according to a prior condition of the left-right symmetry of a car body.
(4) The rough three-dimensional model will be constructed with the identified sketch-line information according to the condition of the left-right symmetry of a car.

3. IDENTIFICATION OF LINES ON SKETCH

In order to construct a 3-D model, lines have to be extracted from sketches. Sketches always contain lines with noise so that automatic extraction and identification of lines are required with elimination of the noise. The following describes the line identification method applying the peculiar algorism with circle operations used in the system proposed.

(1) A sketch is fed into a computer by using a scanner, and pixel data are constructed by binarization of the PGM data made from the sketch data.

(2) When the binarization data is scanned with a circle having the specified radius, the number of the black pixel included in the circle is counted. The first pixel point on which the circle has the specified number of the black pixel is extracted as the start point P_0 for identifying the lines as shown in Fig 2.

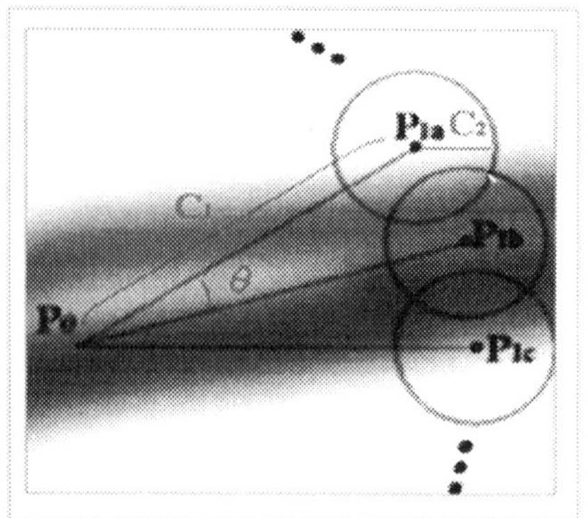

Fig. 2 Circle operation algorithm

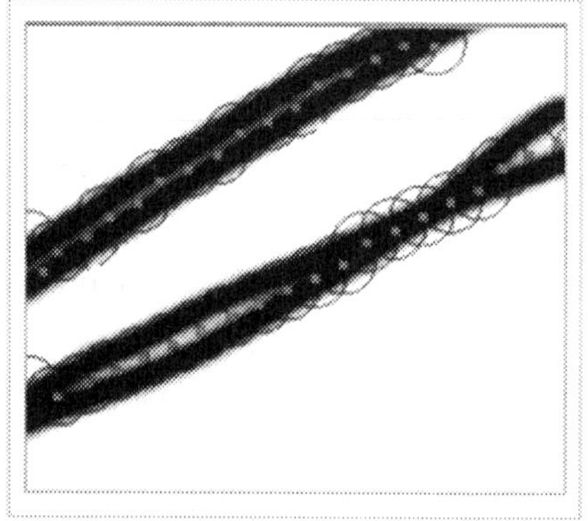

Fig. 3 Identification process of lines

(3) Several circles with the specified radius as shown in Fig. 2 are placed around the point P_0 with the constant distance C1 and the constant interval angle θ for the points P_0. As shown in Fig. 2, the center point P_{1b} of the circle which has most abundant black pixel number is determined as the point on the line connecting the points P_0.

(4) The lines on the sketch are identified by repeating the above procedure (3). Figure 3 shows one of the result of line identification.

4. DETERMINATION OF VIEW POINT

The relationship between the viewpoint, the screen coordinates, and the model coordinates are needed to build three-dimensional model. The left front point of the car body is located at the origin point of the world coordinates, and the model coordinates are turned by α around the Z-axis and by β around the X-axis for the world coordinate. In this case, the viewpoint must be located on the Y-axis of the world coordinate.

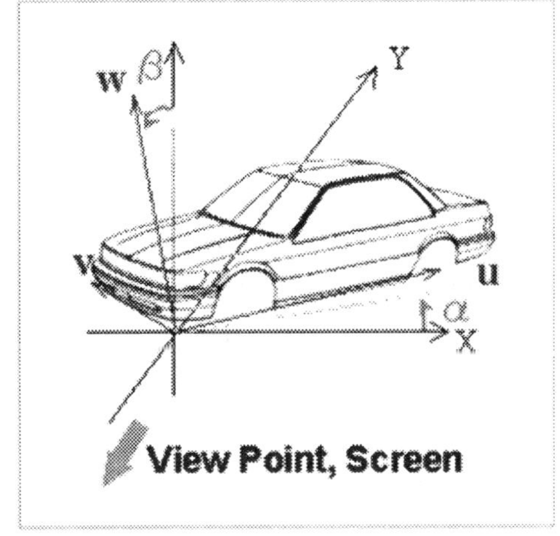

Fig. 4 World & model coordinate

The angles α and β can be estimated from the projected vectors of the vectors **u** (car length) and **v** (car width) shown in Fig. 4. The distances F from the viewpoint to the car and F' from the viewpoint to the screen are also estimated from the two the vectors **u** and **v**. These parameters: rotation angles α and β, and distances F and F' are used in the process of building a three-dimensional model as described below.

5. MODEL CONSTRUCTION

The symmetry of the car body is used for modeling of a three-dimensional car. The center point of a symmetrical pair points always exists on the center plane of the car body. The followings are the procedures to built a 3-D model:

(1) The inclination and continuity of the line are taken into consideration to search for pair points: right side and left side points of the car.

(2) Assuming the condition that the center point is always on the center plane of a car, the three-dimensional coordinate value of this center point is calculated.

(3) Since the height of these three points: a pair of the left-right points and

the center point, is always the same, other three-dimensional coordinate values of these points are estimated respectively by calculating the height w of the center point.

(4) By repeating these processes (1)-(3) for all curves along the width of the automobile, the three-dimensional model can be built from a two-dimensional sketch as shown in Figure 5.

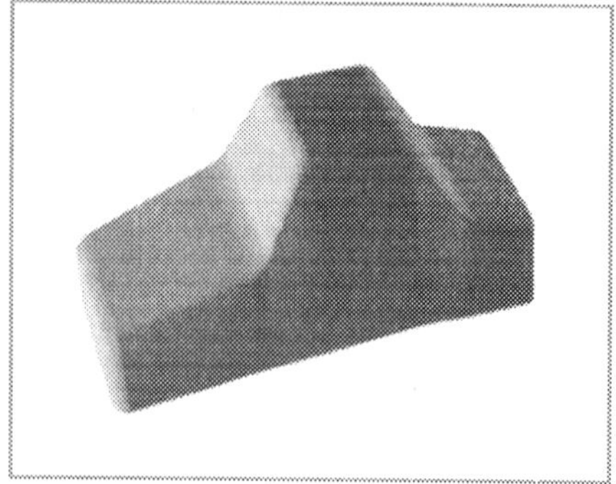

Fig 5. 3D Model

6. CONCLUSIONS

In this study, a system for supporting the style design process of industrial products with complicated shapes, especially automobile body form was developed. With this system, three-dimensional models were easily constructed from these sketches on which designers can express their ideas. In developing the system, methods of extracting lines directly from two-dimensional sketches and building three-dimensional models easily from the sketches were proposed.

REFERENCES

[1] Susannah Ravden and Graham Johnson: "Evaluating Usability of Human-Computer Interfaces", Ellis Horwood Limited, (1989), p.5-8, p.15-19.

[2] Robert C.Zeleznik, Kenneth P. Herndon, John F. Hughes: "SKETCH: An Interface for Sketching 3D Scenes", Computer Graphics Proceedings, Annual Conference , (1996), p.163-170.

[3] Takeo Igarashi, Satoshi Matsuoka, Hidehiko Tanaka: "Teddy: A Sketching Interface for 3D Freeform Design", ACM SIGGRAPH'99, Los Angels, 1999, pp.409-416.

VIRTUAL REALITY FOR NC-PROGRAMMING

H.K. Tönshoff, V. Böß, N. Rackow

Institute of Production Engineering and Machine Tools, University of Hannover

Abstract

NC-programming systems for three- to five-axis machining need a large amount of interactivity in order to create tool paths of an appropriate quality. A new user interface using virtual reality technologies is presented in this paper. The main advantage arising from the use of VR-devices is a substantial relief in tool positioning and error correction. A high-precision force feedback produces a realistic feeling of the tools' movement on the workpiece surface (Figure 1). Additional assistance is given to the user by stereoscopic visualization. Apart from increasing comfort and intuition in programming difficult part operations, it will produce NC-code in strongly reduced programming time as well.

Keywords

Advanced Manufacturing Systems, CAM, NC-programming,
Virtual Reality, Five-axis machining

1. INTRODUCTION

The number and performance of CAX systems, i.e. computer aided tools in design and manufacturing is still rising today. In CAD, the migration from computer aided 2D drafting to 3D modeling is done by nearly all major systems. This implicates further developments in NC-programming systems, e.g. the generation of 3D tool paths. Advanced CAM systems, either as an additional module to a CAD system or as a stand alone program, usually import 3D models via a neutral geometry interface (e.g. IGES or STEP). Afterwards they offer a more or less large number of functions to the NC-programmer in order to produce suitable tool paths interactively. These functions range from trac-

58/29332 © IFW

Figure 1. NC-programming using force feedback devices

ing 2D lines (e.g. for sheet metal cuts) up to generating complex five-axis motion sequences for milling sculptured surfaces. In contrast to the former task, which can be done automatically, the latter job requires an important amount of interactivity. In this case, users have to decide whether to optimize the milling operations by time-intensive procedures (mainly characterized by the creation of auxiliary geometry) based on their own skills and knowledge or to accept the inadequacies of the tool paths produced automatically. While programming of up to three-axis operations is possible with nearly all known software systems, five-axis programming support is still found rarely on the market today. The main reason stated frequently for this fact, is the difficult programming of tool paths that was mentioned above and can only be described insufficiently by conventional procedures, which were developed for one- or two-dimensional geometry. Today's techniques and user interfaces are not flexible enough or too complex in their application for the description of spatial tool positions and orientations. Hence NC-programmers avoid simultaneous five-axis machining wherever possible.

2. BETTER TOOL PATHS IN LESS TIME

However, five-axis machining has substantial advantages compared with conventional 2 ½- or 3-axis milling, particularly within the area of machining sculptured surfaces by reducing manual finishing effort. Obviously, demand arises for innovative procedures which provide faster and better programming operations for five-axis machining. In order to supply this, solutions have to be found for the following tasks:

- The creation of error free and technically optimized tool paths with a reasonable amount of time must be supported. Most important for this are an improved visualization technique, which enables a real 3D view, and also specialized input devices for 3D pointing.

- The user should not be restricted in any way regarding tool positioning. Moreover suitable means are necessary which support direct input of five-axis motions. An input device which allows the user to manipulate the movement of the tool directly along the workpiece surface will increase intuition.

- The system should be able to determine as much information as possible about the produced machining conditions and to convert them both into optical as well as haptic feedback.

A suitable approach to meet the mentioned requirements is the use of virtual reality. Due to the rapid development of the arithmetic performance and storage capacities, modern IT-systems are able to execute the necessary complex calculations in real-time.

While methods of virtual reality are already used in some CAD applications for design (Fujiki 1999, Ma 1999), tele-operation (Mitsuishi 1999), simulation of kinematics and assembly processes (Steffan 1998), NC-programming systems still rely on conventional methods and simulations. A large step towards user-friendly and intuitive NC-programming is possible by an integration of tool path programming and 3D simulation. The tool paths can be created, corrected and/or optimized simultaneously. A close-to-reality representation, navigation and interaction facilitates direct feedback of the current machining situation.

In order to be as close to reality as mentioned above, a combination of 3D visualization and 6-DOF (Degree Of Freedom) input device with 3-axis force feedback was chosen at the Institute of Production Engineering and Machine Tools (IFW) at the University of Hannover. The five-axis milling operation was taken as an exemplary task, since this process is known to cause many complications during NC-programming. The VR-device enables the following new programming and work techniques:

- The 3D-visualization unit creates a close-to-reality feeling for the user who gets direct information about the actual position of the tool in reference to the workpiece. Ambiguities and problems with the recognition of depth are avoided.

- Beyond visualization, the user is able to feel the workpiece in its present processing status. Very precise force feedback of all linear axis of the input device enables the orientation of the tool on the workpiece by really feeling the surface boundaries.

- In interaction with software capable of computing the intermediate processing status, real-time cutting force calculation may be performed and transmitted to the user. Therefore, critical loads can be felt immediately and corrected intuitively.

3. HOW VR CAN SUPPORT NC-PROGRAMMING

Although VR will not replace all well established techniques known in NC-programming, it offers important means to simplify the large amount of manual data input needed for high quality NC-programs. A haptic three dimensional pointer supports the programmer in the use of standard strategies implemented in CAM software. All established automatic process strategies can be applied to different areas but choosing edges, faces or solids can be done in a much more comfortable way with even feeling them through the force-feedback device. Three dimensional coordinates can be put in directly by pointing to them. This will not only save time but will increase the comfort of work as well. The implementation of named functionality should be a comparatively small expense for suppliers of CAM software.

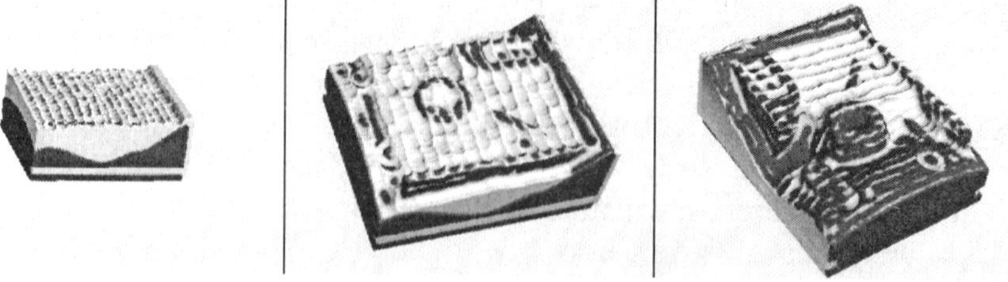

Figure 2. Example of manually generated tool paths with final part (dark) and raw material (light)

A more difficult but very effective way how VR can be used creating NC-programs is given by a completely new approach. Usually the person who intends to remove a specific volume of a workpiece has in mind a certain tool-path and –orientation. The easiest way to describe this path to the system would be to take a virtual tool of the same shape as the real one and move it the way the real tool should do. By using an input device with at least five degrees of freedom, haptic feedback and stereoscopic output, the programmer can teach in the tool path the way he would guide the real machine tool including all natural boundaries like workspace or collisions depending on different tool shapes and sizes and their active part. Furthermore he can be guided by the system using artificial force fields to keep the tool inside predefinable boundaries.

However, this approach requires some new tasks to be solved by the computer: The accuracy of a manual input is of course not as precise as a conventionally calculated path, so the coarse input data has to be processed into a straight, technically reasonable path (Figure 2). Parts of the information generated by this can be directly transformed into feedback creating a "helping hand" that guides the programmers hand into, for example, a continuos removal rate or parallel sub paths. Further correction can be done after the manual user input, e.g. on the feedrate or by smoothing lines.

Both of these two ways how VR can be implemented into CAM software complement each other. Direct input only makes sense when supported by strategies for the major part of material removal. On the other hand practice shows that there will always remain single sequences that are not covered by conventional strategies.

MATERIAL REMOVAL AND COLLISION DETECTION

The conventional way of verifying NC-programs is to simulate material removal in a separate software module that directly uses the z-buffer of the system's graphics card or a faceted model of the part and tool with interference detection. Usually this process offers a visual check for collisions and rest material but it does not provide means for online corrections of the errors detected during simulation.

The approach pursued at IFW integrates the simulation into the toolpath generation procedure. All geometry involved in the process like tool, raw stock, final part and machine environment is mapped into the virtual scene. Hence, the programmer gets direct visual and haptic feedback of the current situation with the possibility to test alternative paths simultaneously.

4. SUMMARY

Using VR, a programming environment can be supplied, which goes far beyond today's systems' capabilities. Users can use their technical know-how intuitively and in a more convenient way. The first step to implement both approaches will be to expand a conventional CAM system with the capabilities of VR in- and output. This will supply a three dimensional pointing device that can be used to get coarse data for direct input of complete tool paths. A lot of work has to be done to optimize this data considering all constraints and wishes of the NC-programmer. The final goal is an intelligent teach-in system that is able to identify the intention of the user and to derive optimized technically reasonable tool paths.

REFERENCES

Chen E. Six Degree-of-Freedom Haptic System for Desktop Virtual Prototyping Applications. Proceedings of the First International Workshop on Virtual Reality and Prototyping; June 1999, Laval France; 97-106.

Fujiki H., Aoyama H. Development of Virtual Clay Modeling System. Proceedings of the 32nd CIRP International Seminar on Manufacturing Systems; 1999 May 24 – 26; Leuven.

Ma W., Zhong Y., Tso S.-K. Intuitive and Precise Solid Modeling in a Virtual Environment through Constraint-Based Manipulations. Proceedings of the 32nd CIRP International Seminar on Manufacturing Systems; 1999 May 24 – 26; Leuven.

Mitsuishi M., Tanaka K., Yokokohji Y, Nagao, T. Remote Rapid Manufacturing with „Action Media" as an advanced User Interface, Proceedings of the 1999 IEEE International Conference on Robotics & Automation, 1999 May 10 - 15; Detroit, Michigan; 1782-87

Steffan R., Schull U., Kuhlen T. Integration of virtual reality based assembly simulation into CAD/CAM environments. Proceedings of the 24th Annual Conference of the IEEE Industrial Electronics Society, 1998 August 31 – September 4; Aachen, Germany; 4:2535-37.

Tönshoff H.K., Rackow N., Böß V. Virtual Reality in der NC-Programmierung: Einsatz virtueller Realität zur Programmierung fünfachsiger Fräsbearbeitungsfolgen. wt Werkstatttechnik 2000; 90:297-301

DEVELOPMENT AND ESTIMATION OF CASE RETRIEVAL MODEL FOR MACHINING KNOWLEDGE SUPPORT SYSTEM ON MACHINE DESIGN

Yoshio FUKUSHIMA
Takasaki University of Health and Welfare

Shigeru NAGASAWA
Nagaoka University of Technology

Abstract: This paper deals with machine design support system for a novice engineer. By use of this system, it is possible to retrieve the similar past case to new parts from database, which consists of practical cases. In order to handle the machine design case in computer system, the classification method of machine design case items is described.   Also, it is recognized that the retrieval result has the characteristics to judge the validity of new parts.

Keywords: Machine Design, Support System, Novice Engineer, Case Retrieval

1.   INTRODUCTION

In machine design, it is necessary to investigate a new design process from many points of view such as strength calculation, production cost etc. Even for expert engineers, not making mistake in a design work is unusual. Getting machined parts into production is a continuous process of trial and error. But compared with a novice engineer, an expert engineer has the ability to solve a problem before machining.   Because they have empirical knowledge about machine design, they can predict that there's something wrong with the design. However, a novice engineer is not able to predict what the result will be.   Therefore, the modification cycle of trial and error takes more time, and profit loss tends to go up.   Recently there have been some investigations about this design process method and these investigations are being watched. [1]

Also, a lot of companies recognize that there is a problem in shortening the education period of the designer and in maintaining the ability to develop. These problems need to be solved. There are many investigations [2] [3] from the point of view of a design support system.   There are many references for an expert engineer but there are few references for a novice. Accordingly, the development of a support system for a novice is very important. Therefore, this study deals with the development of a support system for a novice. "The novice " in this paper is an engineer, who has basic knowledge of drawing and basic knowledge of engineering.

The purpose of the support system is: for a novice to be able to use past cases to create new designs and to judge the validity of a new design.   This paper is mainly about the following items :

(1) For the retrieval of similar machine parts, the classification method of machine design category is described, and the trial system is programmed.

(2) By using retrieval result and the characteristic which the results possess, the judgment method of "the validity of the part design's value" compared with past cases is described.

2.   THE CLASSIFICATION METHOD OF DESIGN CACE

2.1 Retrieval Attributes and Associate Attributes

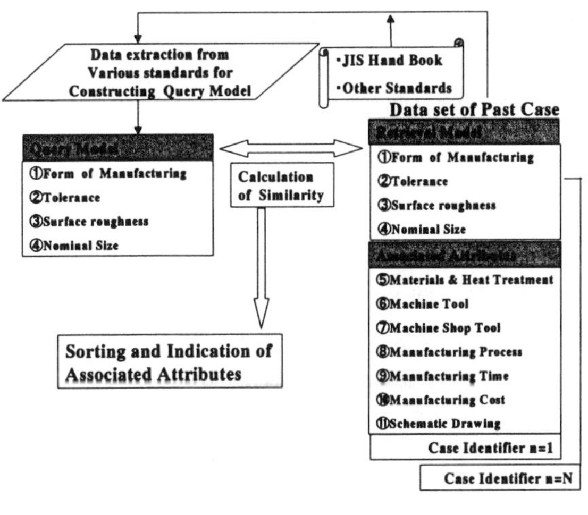

Figure 1 Attributes of Query Model and Past Case

The items that should be written in the drawing are "Form of manufacturing, Nominal size, Material, Surface roughness, Tolerance etc.". Here, the machine design classification method is described. First of all, as Retrieval attributes, "Form of manufacturing, Tolerance, Surface roughness, Nominal size" were selected. Next, as Associated attributes, the items that are shown in Fig.1(⑤～⑪) were selected. Associated attributes are shown as part of the retrieval result and they are important information for the user. That is, a past case consists of 11 attributes. (Fig.1) Assuming a set of retrieval attributes in case to be "Retrieval Model", and assuming a set of retrieval attributes selected by the user to be "Query Model". However, these two models consist of the same attributes, each generated process is different as follows.

"Retrieval Model" ⋯ Extract attributes (①～④) from a past case(①～⑪).

"Query Model"⋯ The attributes selected by user in input interface from various standards.

The support system calculates the similarity (resemblance distance: dc) between these two models, and indicates similar cases for Query model as the result.

2.2 Exclusive Selection Attributes and Quantization Attributes

Retrieval attributes are classified into two categories "A. Exclusive selection attributes", "B. Quantization attributes", in consideration of the characteristics of each attribute.   "Form of manufacturing" corresponds to "A", and other 3 attributes correspond to "B". The sample items in each retrieval attributes are shown in Table1. $Y[X[i] , j]$ indicates the j th sub-attribute or the value of attribute which i th retrieval attribute can take. ("j" is the position number.) The items in X[1] should be selected exclusively because

Table1 Example Contents of Retrieval Model

Retrieval Attribute X [i]	Sub Attribute or Value Y [X [i], j]
X[1] : Form of Manufacturing	Screw, Pitch of Holes, Inside Diameter, Outer Dia., Depth etc
X[2] : Tolerance	0.05,0.1,0.3,0.5, 0.7, 1.2, 1.6 (j=1~7)
X[3] : Surface Roughness	6.3S, 25S, 100S, None (j=1~4)
X[4] : Norninal size	0,3,5,7,10,30,50,80,120,160 (j=1~10)

these are independent concepts. On the other hand, the items in X[2],X[3],X[4] are continuous value, but these are quantized. By quantizing within a certain continuous range, we can treat them as finite numbers. Also, quantizing is convenient, because we can restrict an input value, which does not exist.   Next, the selection standard of sub-attribute and value of attribute for this system is described. "General Standard (ex. JIS hand book)" or" Standard of the companies concerned" are available. Therefore, we can achieve practicable system, which has customary items in an organization concerned.

3.   RETRIEVAL METHOD BY RESEMBLANCE DISTANCE

The items in retrieval attribute put into cord vector in retrieval calculation. Equation (1) indicates the method to calculate the resemblance distance (dc) between Q and C (Fig.2). This "dc" is calculated by use of inner product of cord vector.[4)5)]This method is applied to this study shown in Fig.3. Here, W [i, j, k] is frequency weight (FW). The "j" of Y[X[i], j] of X[i] which is selected by user in Query Model is assumed to be "k".

$$Q = (q_1 , q_2) , C = (c_1 , c_2)$$
$$dc = (\|Q\| \|C\| / Q \cdot C) - 1 \quad \cdots (1)$$
$$\|Q\| = sqr(q_1^2 + q_2^2), \quad \|C\| = sqr(C_1^2 + C_2^2)$$
$$Q \cdot C = q_1 C_1 + q_2 C_2$$

Figure2 Resemblance Distance Calculation

This "k" is made a reference value and W [i, j, k] is decided. However C is a past case in database, element of value which each case possesses is assumed to be "1", and other elements are assumed to be "0". This method has generality because it does not use the attribute value directly. "k " and "W[i,j,k]" are necessary to this method. FW is applied to the attribute X[2] $\sim$ X[4].

Example:

Query Model   : Q= (Xq[2], Xq[3], Xq[4])

Retrieval Model : C= (Xc[2], Xc[3], Xc[4])

Xq[2]= (W[2,1,k], W[2,2,k],…, W[2,7,k])   Xc[2]= (0,1,0,0,0,0,0)

Xq[3]= (W[3,1,k], W[3,2,k], W[3,3,k], W[3,4,k])   Xc[3]= (0,0,0,1)

Xq[4]= (W[4,1,k], W[4,2,k],…, W[4,10,k])   Xc[4]= (0,0,0,0,0,0,0,0,0,1)

Figure3 Application of "dc" calculation for this study

In this study, equation (2) is used for the determination of FW. ("m" is the distribution coefficient.)

$$W[i,j,k] = \begin{cases} (1-m\,|k-j|)^2 & \text{Where } |k-j| \leq (1-0.1)/m \\ 0.01 & \text{Where } |k-j| > (1-0.1)/m \end{cases} \quad \cdots (2)$$

4.   RETRIEVAL PERFORMANCE

The database of past case and retrieval system were programmed by use of spreadsheet application. In this section, some experiments of performance of this system is described.

4.1 Distribution Coefficient

An experiment is necessary to decide the suitable retrieval condition. For that, coefficient m is changed in this case, and the influence on the result is examined. The experimental condition is shown as below.

Query Model (QM1)$\rightarrow$ <u>Form of manufacturing [X[1],2]</u> = "pitch of holes"

<u>Tolerance Y[X[2],4]=0.15;</u>

<u>Surface roughness Y[X[3],4]=None;</u>

<u>Nominal size Y[X[4],5]=10;</u>

Distribution coefficient$\rightarrow$<u>Condition A</u>: 0.3 ; <u>B</u>: 0.45 ; <u>D</u>: 0.15 ; <u>E</u>: 0.01;

<u>C</u>: value which is selected is assumed to be "1", and other elements are assumed to be "0".

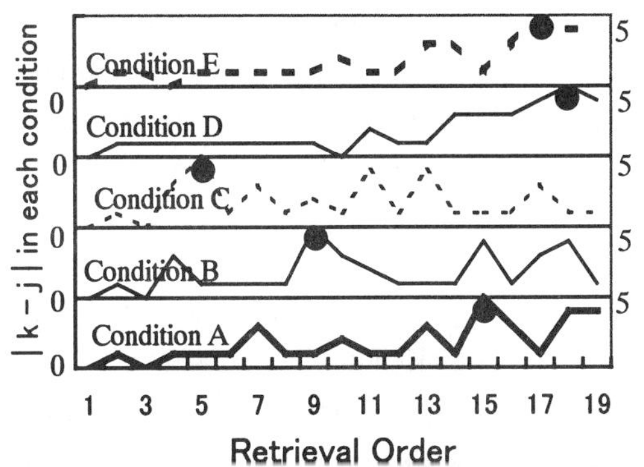

A "●" sign in figure indicates the max. | k-j | in each condition .

Figure 4 Relation bet. Retrieval Order and |k-j|

X[2]	X[3]	X[4]
1st: 0. 15(4),	None(4),	10(5)
2nd: 0.15(4),	None(4),	7(4)
2nd: 0.1(3),	None(4),	7(4)

Figure5 Retrieval result of QM1

Fig.4 shows the result of this experiment. Here, the nominal size attribute in the result is paid to attention. The x-axis indicates the retrieval order. There are 19 "pitch of holes" cases in database. The y-axis indicates the absolute value of difference | k -j | between value "k" of nominal size attribute of QM1 and value "j" of nominal size attribute of retrieved cases. The retrieved case with larger | k-j | value does not posses an appropriate value for QM1. Judging from the shape of the line graph and the order of retrieved case which possesses the maximum | k-j | value ("5"), it seems that condition D (m=0.15) is the optimum condition in this experiment. By using condition D, suitable result was retrieved (Fig.5). Though a lot of other methods are devised, we could check the suitable result by use of this condition.

Query Model: QM2

X[2] = 0.15 or 0.5 (Qj= 4 or 7)

X[3] = 30   (j= 6)

X[4] =12S   (j= 4)

Rj: " j " of tolerance X[2]

   in higher ranking attributes

   (Within the 5th place)_

Qj: " j " of tolerance X[2]

   in Query Model

Qj	Rj in each place	|Qj−Rj|
4	1st → 4	0
	2nd → 4	0
	3rd → 4	0
	4th → 3	1
	5th → 3	1
7	1st → 7	0
	2nd → 5	2
	3rd → 4	3
	4th → 5	2
	5th → 5	2

Figure 6 Result of Retrieval
Characteristic Experiment (1)

4.2 Design Value Judgment by use of Characteristic of Retrieval Result

It is possible to judge the validity of the design's value by using retrieval result. For example, X[2] (Tolerance) in the Query Model(QM2) is changed, and the influence on the result is examined. (Fig.6) Here, regards "j" of result as "Rj", also "j" of QM2 as "Qj" for convenience. When Qj= 4 or 7 ,the Rj which results (1st~5th place) possess are paid to attention. |Qj−Rj| is 0 or 1 when Qj=4. That is, these results are suitable for Qj. On the other hand,   the maximum |Qj−Rj| is 3 (3rd place)when Qj=7. Some inappropriate cases were retrieved for the reason that there is not suitable case in database. Because Qj=7 is not suitable for this QM2. In other words, if the database consists of practical cases, the changed attribute with increasing |Qj −Rj| is not suitable any longer for other attributes. And it indicates the changed attribute goes away from

855

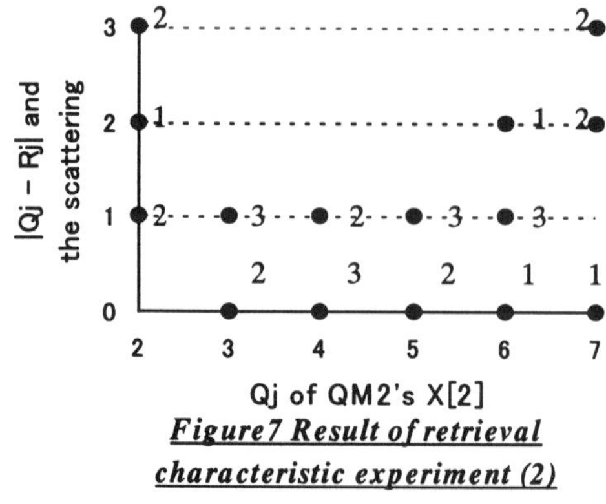

Figure7 Result of retrieval characteristic experiment (2)

database (standard of the companies concerned). Fig.7 shows the relationship between changed Qj(2~7) and "|Qj−Rj| and the scattering ". The figures fixed to each point in fig.7 show the number of retrieved cases. As already shown in Fig.6, when Qj=4, |Qj−Rj| is 1 or 0 and "3 cases(1st, 2nd and 3rd) with |Qj−Rj| =0", " 2 cases(4th,5th) with |Qj−Rj| =1" were retrieved. A certain Qj value is assumed to be a reference value, and |Qj−Rj| and the scattering are increasing. In fig.7 the vicinity of Qj=4 is a reference value. Thus, the user can investigate a new design from two viewpoints of "similar retrieved cases" and "|Qj−Rj| and the scattering". Therefore, it is possible to check the adjustment of a new design for past case by using this system. And it is able to supplement beginner's sense.

5. CONCLUSION

Early education of the designer is a matter of importance.   In this study, we investigated the machine design support system for a novice.   The results of this study are as follows;

 (1) There are two categories of attributes, "A. Exclusive selection attributes" and "B. Quantization attributes", in consideration of the characteristics of machine design case. Also there are suitable retrieval methods for each category. When using " Quantization attributes",   only a certain range of attributes can be handled.   The method to calculate the distribution coefficient according to quadratic function is effective in retrieving the Quantization attributes.

(2)By checking the results (|Qj−Rj| and the scattering) , we can judge the validity of the composition of the Query Model.

This study deals with the support method of machine design for novice designers. As a result, a design in which practical knowledge is used can be created.

6. REFERENCES

1) Conference Proceedings; S.Ahmed et al: The Relationships between Data Information and Knowledge Based on a Preliminary Study of Engineering Designers , ASME DETC99, (1999)

2) Journal Article; Akagi: Artificial Intelligence in Engineering Design, Trans. JSME ,55−516,C(1989-8)

3) Journal Article; Tomitsu : A Contribution to the Development of an Expert System , JSME, 56-521, C(1990-1)

4) Journal Article; Nagasawa, at al; Estimation of Similary for Retrieve Data of Finite Element Modeling, Simulation, 14-13

5) Book; David Ellis : New Horizons in Information Retrieval (Japanese Edition), Maruzen Co., Ltd., 1994

INTEGRATED PROCESS PLANNING AND PRODUCTION CONTROL
A Flexible Approach Using Co-operative Agent Systems

H.K. Tönshoff, P.-O. Woelk
Institute of Production Engineering and Machine Tools, University of Hannover

O. Herzog, I.J. Timm
Center for Computing Technologies, University of Bremen

Abstract
Nowadays, strong borderlines exist in industrial production between process planning, production control and scheduling systems caused by an extreme specialisation and independent historical paths of system evolution. Thus, a gap exists between the involved systems, which implies loss of time, information and in consequence loss of quality and prolonged time-to-market. In order to bridge this gap there is a strong need for new ideas integrating these worlds. Innovative fundamental concepts and methods for management and control of integrated information logistics, of production scheduling and process planning are necessary. An architecture of a suitable, agent-based approach is presented in this paper.
Keywords
Planning and Scheduling for Production, Intelligent Manufacturing, Process Planning, Distributed Artificial Intelligence, Multiagent Systems

1 INTRODUCTION
Many enterprises face radical changes within global production. They have to offer highly customised and innovative products to meet customers' requirements and to discover new markets. Thus, companies have to solve the change towards flexible, demand driven production. New and more information have to be handled and a considerable speed-up of the development and manufacturing processes is needed. However, strong borderlines exist in information logistics. The traditional approach of separating planning activities (e.g. process planning) from implementing activities (e.g. production control and scheduling) results in a gap between the involved systems, which implies loss of time and information. Thus, innovative fundamental concepts and methods for management and control of integrated information logistics, of production scheduling and of process planning are necessary. These concepts have to take several topics into consideration, which characterise the disadvantages of the current situation.

Conventional static process plans take mainly technological aspects into consideration. However, economical aspects (e.g. capacity and current load of resources) remain disregarded. Furthermore, process plan modifications, which may become necessary due to unexpected events (e.g. machine breakdown), carried out by a centralised process planning group are very time-consuming. Thus, modifications take place on shop floor level, which will lead to feasible, but not to optimal results. Additionally, the complexity of manufacturing processes and the knowledge, which is necessary for process planning, increases due to new manufacturing technologies. Despite the fact, that this knowledge is available at shop floor level (e.g. by well trained machine operators), it often takes a long time until this knowledge is available at a centralised process planning group. Thus, the advantages of the application of innovative manufacturing technologies, which are often more suitable regarding ecological aspects, remain unused.

Therefore, the interest in agent technology and agent related topics has risen enormously in the last decade (Jennings et al., 1998). Co-operative agents can act autonomously, communicate with other agents, are goal-oriented (pro-active) and are using explicit knowledge (Weiss, 1999). Especially in the naturally distributed production engineering domain they seem to be a promising approach (e.g. Kaihara et al., 1997). An adequate application of co-operative agents should meet the following three criteria (Müller, 1997): A *natural distribution* of the participating entities (e.g. resources at the shop floor), a *dynamic environment* where structures and conditions are continually changing, and *complex interaction* between the individual entities. Since most industrial environments meet these requirements, the use of an agent-based approach with autonomously, co-operatively and purposefully acting intelligent software units seems to be suitable to make possible short term and flexible reaction in manufacturing.

2 THE "INTAPS" APPROACH

The agent-based integration of process planning and production control in this very flexible and distributed way is the main topic of the current research project "IntaPS". The approach of co-operative agents for this purpose is based on two substantial components, which link together information systems of earlier stages of product development and the resources on the shop floor (see Fig. 1). This link is realised by decentralised planning on shop floor level and by rough level process planning.

2.1 Decentralised Planning Unit on Shop-Floor Level

A co-operative multiagent system implements decentralised planning on shop-floor level. Within this architecture three different types of agents are used, resource, order, and service agents.

Each relevant resource of the production system and its environment is represented by one **resource agent**. Consequently, agents exist for e.g. machines, assembly workplaces, transportation devices as well as staff and virtual resources like business information systems, CAM systems for NC code generation or legacy information systems. Resource agents provide local knowledge bases of the associated resources. The entirety of all resource agents represents the shop floor model of the whole production system (Tönshoff et al., 1999).

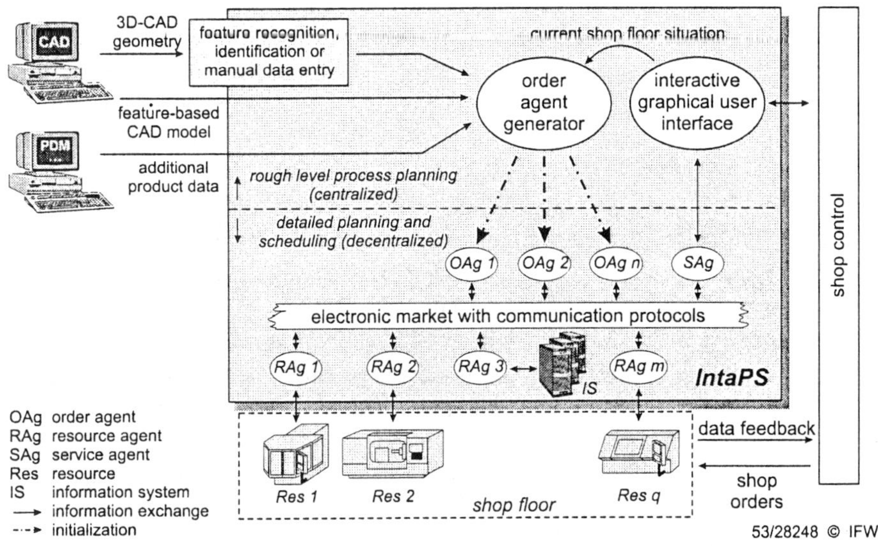

Figure 1. Architecture of the "IntaPS" approach

Order agents are representing orders which have to be manufactured. They pursue the goal of market-based optimisation and endeavour to optimise utility functions respective individual goals. For example, an order agent for a "rush order" will rate the goal "finished on due date" higher than the goal "using cost-saving manufacturing processes". Thus, individual utility functions with respect to order priority are used to evaluate possible action alternatives. Due to its autonomy and pro-activity, an order agent is able to recognise internal and external disturbances and to react appropriately.

Service agents are used for human interaction, transparency and maintenance purposes. Thus, human users like process planners are enabled to track actions performed by agents and to interact with the system. Processes hidden inside the agent system are made visible and comprehensible. Some service agents perform system monitoring tasks and will request user interaction under particular conditions, e.g. if the knowledge of some agents is insufficient or if the agent communication comes to a "deadlock situation".

The detailed process planning and scheduling takes place co-operatively within an **electronic marketplace**. Order agents and resource agents are interacting according to a "three-phase-model". Starting with a "negotiation phase" required manufacturing skills and due dates as well as capabilities and capacities are communicated. Suitable sequences of manufacturing operations are resulting from auctions between appropriate partners. The optimal sequence of manufacturing operations is accepted as detailed plan. The second phase is called "verification phase" and ensures the feasibility of the detailed plan. The order agent examines continuously whether its detailed plan is executable under the current conditions. As mentioned before, changing conditions (e.g. by breakdown of resources) are recognised. This causes order agents to analyse the consequences and to identify those parts of their detailed plans which are affected. If necessary, the order agent enters a "re-negotiation phase" and tenders parts of the detailed plan for a new auction. The re-negotiation phase leads to an improved alternative detailed plan which substitutes the previous plan. Afterwards the verification phase is resumed and lasts until the order is finished. To ensure a global optimisation within the three-phase-model open, adaptive communication protocols are used as introduced in (Timm et al., 2001). In contrast to "classical" multiagent design the so-called "emergent behaviour" should be reached without extensive analysis and design of communication protocols. The open, adaptive communication protocols can be generated, refined, or adapted autonomously.

2.2 Centralised Rough-Level Process Planning

Centralised rough level planning pre-processes incoming data and generates a rough process plan. These data are geometrical or technological information about the workpiece (e.g. from CAD systems), further organisational information related to products and orders (e.g. from PDM or ERP systems) as well as information concerning the current shop floor situation. Order agents are synthesised with respect to these information and initialised with a rough level process plan. This plan contains information only, which the agent is not able to recognise from the environment by itself (e.g. constraints between manufacturing operations). Thus, order agents obtain a maximum scope for the allocation of suitable resources and time slots for manufacturing. Furthermore, the centralised part of the "IntaPS" architecture provides a graphical user interface for interaction with the system.

3 OUTLOOK AND SUMMARY

In this paper a system architecture based on the application of co-operative agents is proposed for improvement of information logistics in the area of process planning and production control. Current research activities deal with the creation of a structured application domain model which is a representative sample of a real-world production system as well as with open, adaptive communication protocols. A prototype system is under development and will be used for evaluation of the approach in early 2002.

Due to the approach of a wide integration of process planning and production control, capacity information and due dates will be taken into consideration for early stages of process planning as well as process planning knowledge will be used for short term scheduling decisions at the shop floor. Therefore, problems will be eliminated which result from time-delayed return of manufacturing knowledge and capacity data and/or from other lacks of information flows e.g. from the use of static process plans. Thus, enterprises are able to react to changing requirements of their environment in a more flexible way and will face the challenges of international competition successfully.

ACKNOWLEDGEMENT

The presented work is being funded by the "Deutsche Forschungsgemeinschaft" (DFG) within the projects He 989/5-1 and To 56/149-1 as part of the Priority Research Program 1083 "Intelligent Agents and Realistic Commercial Application Scenarios". Further information about the "IntaPS" project is available in the internet (http://www.intaps.org).

REFERENCES

Jennings N.R., Wooldridge M.J., *Agent Technology: Foundation, Applications, and Markets*. New York: Springer, 1998.

Kaihara T., Fujii S. A self-organization scheduling paradigm using coordinated agents. Proceedings of the 8th International Conference on Production Engineering (ICPE); 1997 August 10 – 20; Sapporo. London: Chapman & Hall, 1997.

Müller H.-J. Towards Agent Systems Engineering. International Journal on Data and Knowledge Engineering, Special Issue on Distributed Expertise 1997; 23:217-245

Timm I.J., Tönshoff H.K., Herzog O., Woelk P.-O. Synthesis and Adaptation of Multiagent Communication Protocols in the Production Engineering Domain. Proceedings of the 3rd International Workshop on Emergent Synthesis (IWES); 2001 March 12th – 13th; Bled.

Tönshoff H.K., D'Agostino N., Ehrmann M. Dynamic modelling of process planning knowledge based on a shop-floor model. Proceedings of the 32nd CIRP International Seminar on Manufacturing Systems; 1999 May 24 – 26; Leuven.

Weiss G., *Multiagent Systems - A Modern Approach to Distributed Artificial Intelligence*. Cambridge, Massachusetts: The MIT Press, 1999.

DYNAMIC JOB SHOP SCHEDULING USING A NEURAL NETWORK
Multi-stage Training and Selection of Input Information

Toru EGUCHI, Fuminori OBA, Satoru TOYOOKA

Faculty of Engineering, Hiroshima University

Abstract
This paper proposes a scheduling method using a multi-layer feed-forward neural network to minimize tardiness cost in a dynamic job shop environment. The proposed neural network is trained by a two stage training method to outperform the best dispatching rules in various levels of due-date tightness and job arrival rate.
Keywords
Scheduling, Job shop, Tardiness cost, Neural network, Mutual information

1. INTRODUCTION

In this research we deal with a dynamic scheduling in job shop type production systems that can process a low volume and high variety of part types. In dynamic scheduling environments, dispatching rules are the most practical method to realize real-time scheduling. Over the years, dispatching rules have been developed on the basis of the experience of a vast amount of numerical experiments. However, the performance of the dispatching rules varies depending on scheduling conditions and no rule has found to perform best under various different shop conditions.

In this paper we propose a neural network approach to dynamic job shop scheduling. A multi-layer feed-forward neural network (NN) is used to calculate the priority of job to be processed next on a machine to minimize tardiness cost. In this approach, we have already proposed the training method that performs well in the medium due date tightness and medium load level (Eguchi et al. 1999). This paper proposes a multi stage training method of the NN to outperform any dispatching rules in various levels of due-date tightness and job arrival rate. We also try to prune the insignificant input units. Simulation results show the effectiveness of the proposed method.

2. DYNAMIC JOB SHOP SCHEDULING PROBLEM

The problem studied in this paper is described as follows. Consider the job shop composed of M machines which process jobs that enter the shop continuously in time. Each job J_i has to be processed n_i operations $\{o_{ij} \ (j = 1, 2,\ldots, n_i)\}$ in numerical order of j. The processing time of the operation o_{ij} is p_{ij}. The aim of scheduling is to minimize tardiness cost TC:

$$TC = \sum_{i=1}^{N} v_i \times \max(\,0, C_i - d_i\,)\,/\,N$$

where C_i, d_i, v_i are the completion time, due date, and delay penalty per unit time of job J_i respectively, and N is the total number of jobs under consideration. Jobs arrive at the shop randomly according to a Poisson process that has arrival rate λ. The due date of each job J_i is determined as follows:

$$d_i = r_i + k_i \times \sum_{j=1}^{n_j} p_{ij}$$

where r_i, k_i are the arrival time and due date tightness factor of job J_i respectively.

3. SCHEDULING USING A NEURAL NETWORK
3.1 Structure of the Neural Network

The proposed NN is used to calculate the priority of each waiting job in the buffer of machine and the job with the highest priority is selected as the next job to be processed. A three-layer feedforward neural network (see Fig.1) is designed for this purpose. Each unit has a sigmoid function that outputs a value between 0 and 1. The input information of the NN is the status of shop floor and waiting job such as the percentage of tardy jobs in the current shop, the queue length of the machine buffer, the imminent processing time of job, the current slack value of job, delay penalty v_i and so on. The number of input units is 36.

3.2 Two Stage Training of the Neural Network

The NN can be trained by using training examples. However, it should be noted that no direct training data of each job selection in dynamic scheduling environment is given beforehand because it is impossible to tell the influence of each job selection on the final scheduling performance before the whole schedule completes. Therefore the NN is trained by a random optimization method using a set of whole scheduling problems instead of the input-output pairs of each job selection examples.

The status of shop floor and waiting job

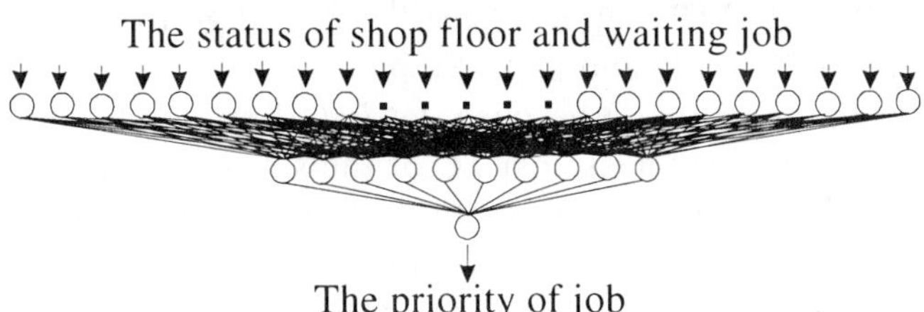

The priority of job

Figure 1 Structure of the neural network

863

This approach causes the difficulties of searching the optimal weight data. First of all, it is very time consuming because the scheduling problems for training should be fairly large size, say 1000 jobs, in order to simulate the steady state condition of dynamic scheduling. Furthermore, random optimization methods without derivative information are generally inefficient. We adopt a simulated annealing method, which is designed to avoid converging a poor local optimum, as a random optimization method. However, it is not always easy to find a good solution in a reasonable amount of time if the solution space is complicated. It is very time consuming and risky to try to find the optimal weight data for all scheduling conditions by only one time trial starting from arbitrary initial weights data. In order to overcome this problem, we propose a two stage training method as given below.

3.2.1 The First Stage Training

In the first stage, the total solution space is divided into several partial spaces according to the parameters that represent scheduling conditions such as machine utilization rate and the tightness of due date. Table 1

Table 1 Problem division

		The due date tightness factor k_i		
		2~3	3~4	4~5
Average job arrival interval $1/\lambda$	45	P_{11}	P_{12}	P_{13}
	50	P_{21}	P_{22}	P_{23}
	55	P_{31}	P_{32}	P_{33}

shows an example of the space division. The aim of this approach is to increase the certainty of finding good weight data by limiting the search space. For each partial problem, a set of scheduling problems for training is prepared and a NN is trained by the random optimization using the simulated annealing. The algorithm is as follows:

Step1. Set the total training number L. Let $l=1$.
   Set the initial weights vector $w^{(1)}$ randomly.

Step2. Generate a random noise vector ξ.
   Calculate $\Delta S = S(w^{(l)} + \xi) - S(w^{(l)})$
   If $\Delta S < 0$ then $w^{(l+1)} = w^{(l)} + \xi$ and go to Step4.
   else go to Step3.

Step3. Calculate the acceptance probability $P = \exp(-\Delta S/T(l))$,
   where $T(l) = c / \ln(1+l)$, c is the cooling parameter.
   If $\mathrm{Rnd}[0,1] < P$ then $w^{(l+1)} = w^{(l)} + \xi$ and go to Step4.
   else $w^{(l+1)} = w^{(l)}$ and go to Step4.

Step4. If $l=L$ then stop. Otherwise, let $l=l+1$ and go to Step2.

$S(w)$ is the average tardiness cost TC of the set of scheduling problems solved by the NN with weight vector w. $\mathrm{Rnd}[0,1]$ represents a random value determined by the uniform distribution between 0 and 1.

Training should be carried out several times with different values of cooling parameter c and initial weights data in order to search the globally optimal solution. The weights data that can achieve the best scheduling performance is selected as the final weights data.

3.2.2 The Second Stage Training

In the second stage, the input-output pairs for job selection are collected by applying the trained NN to the set of scheduling problems of each scheduling condition in the dynamic scheduling environment.

Next, we try to reduce the number of input units of the NN. Reducing the number of input units decreases the cost of collecting the information of shop status and has the possibility to increase the generalization ability of NN. The mutual information MI (Battiti 1994) is used to evaluate the significance of the relation between each input information and the priority of each job using the input-output pairs.

$$MI = \sum_{c=1}^{Nc} \sum_{f=1}^{Nf} P(c,f) \log_2 \frac{P(c,f)}{P(c)p(f)}$$

where, Nc is the number of divided intervals of output range and Nf is that of input range. $P(\ )$ is the probability function for the divided interval. $P(c,f)$ is the joint probability function. The input units that have high value of mutual information are assumed to be important. On the Contrary, the input units that have low value of mutual information are eliminated because they are assumed to be insignificant.

After eliminating commonly insignificant input information for all scheduling conditions, a new NN is trained by a back-propagation algorithm using the input-output pairs of all scheduling conditions. The training of this stage is comparatively fast because back-propagation methods can utilize the derivative information of search space. Figure 2 shows the overview of the training method.

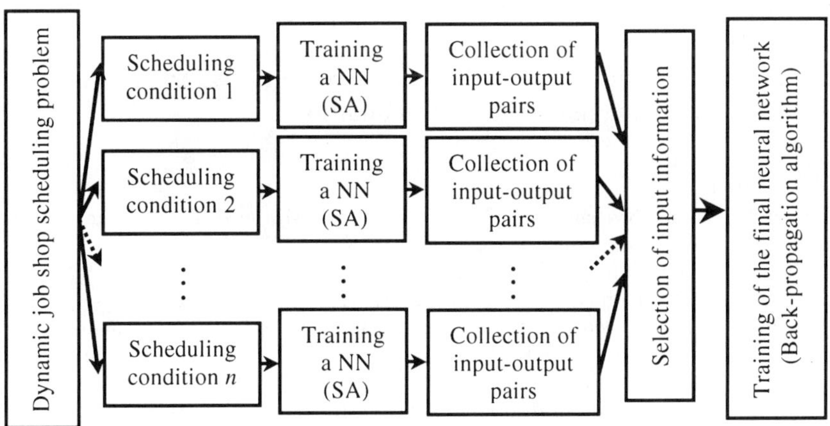

Figure2 Overview of the multi-stage training method

Table2 The best dispatching rule and the improvement rate
using the NN for each scheduling condition

		Due date tightness factor k_i		
		2.0~3.0	3.0~4.0	4.0~5.0
Average job arrival interval $1/\lambda$	45	1.3% (*W_S/RPT+SPT*)	6.4% (*W_S/RPT+SPT*)	24.1% (*W_CR+SPT*)
	50	2.7% (*W_CR+SPT*)	23.8% (*ATC*)	42.1% (*ATC*)
	55	4.7% (*W_CR+SPT*)	20.5% (*ATC*)	24.2% (*ATC*)

4. NUMERICAL EXPERIMENT

We prepared nine scheduling conditions in terms of three levels of due date tightness and three job arrival rates (see Table1). First, NNs are trained for the selected five conditions (P_{11}, P_{13}, P_{22}, P_{31}, P_{33}) using the simulated annealing. The scheduling problems for training are ten problems of $N=1000$, $M=4$, $n_i=2\sim4$. The final NN is trained by a back-propagation algorithm after eliminating nine input units that are commonly insignificant for all scheduling conditions, and adding two input units, which are average machine utilization rate and the average due date tightness factor k_i of the jobs in the current shop. A hundred problems of $N=500$, $M=16$, $n_i=8\sim 16$ were newly generated for each scheduling condition. Table2 shows the percent differences between the average tardiness cost using the NN and that using the best dispatching rules (Vepsalainen and Morton 1987, Anderson and Nyirenda 1990). It is confirmed that the trained NN performs best under all scheduling conditions. As for the input units pruning, the NN performed as good as or better than the NN without pruning input units.

5. CONCLUSION

In this paper a dynamic scheduling method using a NN was studied. The advantage of the proposed training method is the ability to build up a robust NN step by step. The weights data obtained in the first training stage, which is the most time consuming, can be reused if we try to make a more robust NN by adding new scheduling conditions.

Reference

Anderson,E.J. and Nyirenda,J.C., Two New Rules to Minimize Tardiness in a Job Shop, International Journal of Production Research, 1990;28, 2277-2292.

Battiti, R: Using Mutual Information for Selecting Features in Supervised Neural Net Learning, IEEE Trans. on NEURAL NETWORK, 1994; 5:4:537-550.

Eguchi, T., Oba, F. and Hirai, T.: A neural network approach to dynamic job shop scheduling, In *GLOBAL PRODUCTION MANAGEMENT*, Kai Meritins, Oliver Krause, Burkhard Schalloc, ed., 152-159:Kluwer Academic Publishers, 1999.

Vepsalainen,A.P.J. and Morton,T.E., Priority Rules for Job Shops with Weighted Tardiness Costs, Management Science, 1987; 33, 1035-1047.

AN ORGANIZATIONAL CONCEPT FOR MANUFACTURING OF HIGH PRECISION PRODUCTS

W. Eversheim, I. Fricker, M. Koschig, N. Michalas

Laboratory for machine tools and production engineering (WZL)
Chair for production engineering
Aachen University of Technology (RWTH), Aachen, Germany 52056
Tel: ++49-241-807379, Fax: ++49-241-8888293, email: n.michalas@wzl.rwth-aachen.de

Abstract

Manufacturing of products of high quality and high precision requires the use of high end Machine Tools. The use of such machines for manufacturing a wide range of products, often demanding on various manufacturing processes, results in a complex organizational concept. The information and material flow is enormously complex, not able to be covered by traditional organizational concepts.

'Autonomous production cells' is a concept that allows the production of high quality and precise products, based on the capability to accomplish autonomously complex processes without external intervention. The basic characteristics of 'Autonomous Production Cells' are presented in this paper. Several concepts for configuration of the cells are analyzed. The special requirements of precision engineering are taken into consideration. A concept for scheduling orders will be also presented.

Keywords
Decentralized Production, Scheduling, Manufacturing Systems

1. INTRODUCTION

Improvement of quality is a topic that becomes a crucial factor for compliance of customer demands, fulfillment of ecological restrictions, and thus economical success. Precision engineering can be seen as a resulting development to meet these increasing demands [McKeown, 1991].

The aim of precision engineering is the economical production of high quality parts. According to a specific field of application there can be different aspects of quality, which are relevant for a certain part in order to guarantee its function. Therefore, precision engineering deals with all possible kinds of quality such as high precision of dimensional accuracy, accuracy of shape as well as high quality of the surface finish.

Another aspect of the rising demands is the requirement of process knowledge and the need for qualified staff, as the production processes in precision engineering become increasingly complex and sophisticated. As a consequence, skilled personnel is a basic element for precision engineering.

In addition quality of produced parts is not only depending on the single process as such but also on the sequence of processes respectively. In particular the sequence of fixturing on different tool machines, e.g. re-clamping of parts, can lead to a loss of quality due to a possible loss of the initial reference system.

Besides efficiency, flexibility gets more important in manufacturing enterprises. This has led to new organisational approaches like Autonomous Production Cells (APC). APCs have been developed with the objective of carrying out manufacturing processes for a long period without external interferences. Such cellular systems are able to react independently to disturbances of the production process, enabling long lasting production cycles without any external intervention [Eversheim, 1996].

APC is a concept for precision engineering, as high performance machine tools are used. Furthermore skilled staff is supposed to run the production process with planning and manufacturing integrated in one single organizational unit for minimum interference. Moreover quality data is fed back for increasing manufacturing excellence. Suchlike APC fits demands for precision engineering.

2. AUTONOMOUS PRODUCTION CELLS (APC)

The concept for APC deals with the development of a new type of distributive manufacturing system. Autonomy in this case refers to the ability of carrying out manufacturing processes for a long period without external interference. To reach this objective APCs need the functionality to avoid problems and disturbances wherever possible and to react internally when problems and disturbances take place.

An autonomous production system is being designed with integrated procedures together with milling and laser beam welding processes, and CAD/CAPP/CAM process chain. The correspondent integrated procedures just include initial- and re-planning, control, man-machine interfaces, gripping and clamping, and disturbance management. Here, one of the main approaches to enlarge the autonomy of production units is the integration of planning functions, which not only allows event oriented action on planning level but also extends the possibilities to react on disturbances and modifications of requirements.

The ability to run alternative planning procedures is identified as an important characteristic of an APC. In addition, a powerful information management to allow the integration of planning and re-planning functions is the prerequisite for consistent data exchange. A feature-based approach based on the unified product data model was chosen in this case.

3. CONFIGURATION OF THE CELLS

An important issue for the framework of the autonomous production cells is the configuration of the cells. In industrial practise, there are five general approaches, namely: the job shop, project shop, cellular system, flow line and continuous systems [Chryssolouris, 1992]. Taking into consideration the basic characteristics of the products of APCs and of precision engineering (high quality, small to medium batches, small size products) the project shop method (products with great size and/or weight), continuous systems (production of liquids, gases, powders) and flow lines (suitable for use in automotive industry) seem to be inappropriate for the concept of the APCs. A short description will be provided for the rest of the methods.

In a job shop, machines with the same or similar material processing capabilities are grouped together. In this structure, the part or lot of parts moves through the system visiting the different work centres according to the parts' process plan. The material and information flow resulting from the typical products of APCs, make the system extremely complex and inappropriate for the APC concept.

In manufacturing systems organized according to the cellular plan the equipment is grouped corresponding to the process combinations that occur in families of parts. Each cell contains machines that can produce a certain family of parts. A version of this method is the Flexible Manufacturing Systems (FMS). In this approach, each cell is reconfigured according to the needs of each new product.

A comparison of the described approaches is presented in Figure 1.

	Shop floor	FMS	APC
Material flow	○	●	●
Information flow	○	●	●
Flexibility	●	●	◕
Reconfiguration simplicity	●	○	●
Adaptation on disturbances	●	◑	●
Specialization on technologies within the cell	●	◔	●
Specialization on products within the cell	○	◑	●

Measure of fit: ● High ○ Low

Figure 1. Comparison of approaches

The cellular system is the most suitable approach for the APC concept. It allows the specialization of each cell to a specific part family, guaranteeing maximum quality and accuracy of the processes and maximum utilization of knowledge of the operators. Moreover, the system guarantees a simplified material and information flow, allowing cell-internal adaptations on occurring disturbances.

A precondition for use of the method in the praxis is the analysis of the product spectrum and the formation of product families. It is essential for the success of the implementation of the concept that 70-80% of the manufactured products can be completely manufactured within one cell.

4. SCHEDULING

Main task of the production scheduling is to allocate tasks corresponding to jobs, requested by customers' orders to the appropriate facility. For the APC concept executing most of the planning functions within the cell is necessary, in order to have control and to react autonomously in case of disturbances.

In order to provide the highest degree of flexibility on scheduling within the cell, while retaining control of the whole factory, it is necessary to split the scheduling procedure into two levels (Figure 2).

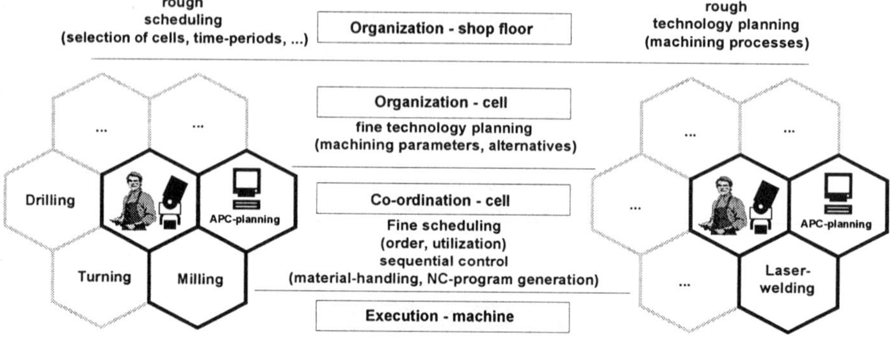

Figure 2. Scheduling in APC: Multi-Machine Concept

4.1 Rough scheduling

At the first level, central control of the factory takes place. A first analysis of the orders takes place in order to identify which cell will execute the order. The selection of the cell is based on two aspects. First, with a rough process planning it is identified which are the requested processes for the manufacturing of the product. Taking into consideration the abilities of each cell and the processes that can be carried out, the possible cells that can execute the order are identified. After the identification of the possible cells, a rough scheduling must take place, in order to identify which cell can execute the order, catching the specified deadlines. Actual workload data must be taken into consideration for this purpose. The free capacity of each suitable cell is provided in form of a rough Gantt chart [Li and Willis, 1992]. After performing the rough-cut capacity planning the cell is identified that can execute the order within the specified period. In case that more cells are identified as suitable, criteria like cost, utilization, or earlier execution are taken into consideration. When the appropriate cell is identified, the available data of the order are forwarded to this cell.

4.2 Decentral Scheduling

At the second level, the scheduling problem is regarded as a typical decentralized production control. Within each cell, the orders allocated to the cell have to be scheduled in order to meet the production goals. For the scheduling at this level, existing methods of scheduling are used. The main points which make the concept to differ and guarantee the success are:

• Bound of the parameters of the scheduling problem. In each cell, a limited number of orders must be scheduled, having also a limited number of available resources.

• Integration of process planning and scheduling. The procedures of process planning and scheduling are executed by the same person within the cell. Suitable planning tools allow the planner to take into consideration the actual workload of the cell and generate alternative process-sequences, achieving a better resource utilization.

• The simulation tools that are developed within the APC concept allow a detailed simulation of all the processes needed for the manufacturing of the product. The simulations provide a very accurate calculation of the needed operational times, so that the fine scheduling of the tasks can be easy and accurate.

5. CONCLUSIONS

The presented concept of the APCs is a suitable organizational concept for product manufacturing within the area of precision engineering. The objectives of the APC, for manufacturing of high quality products in small to medium batches with a high grade of autonomy, define the organizational environment for precision engineering.

The partition of the scheduling procedure into two levels allows a better control of the resources. It allows furthermore the better reaction on disturbances. The APC project is financed by the Deutsche Forschungs-gemeinschaft (DFG).

6. REFERENCES

Chryssolouris, G. *Manufacturing Systems – Theory and Practice*. New York: Springer-Verlag, 1992.

Eversheim, W. et. al. Adaptive process planning in autonomous production cells. Annals of the German Academic Society of Production Engineering 1996.

Li, K.Y., Willis, R.J. An interactive scheduling technique for resource-constrained project scheduling. European Journal of Operational Research 1992; 56:370-379.

McKeown, Pat. Progress in Precision Engineering. Proceedings of the 2nd International Conference on Ultraprecision in Manufacturing Engineering in Braunschweig; Berlin, Heidelberg, New York, 1991.

CONSIDERING FEATURE INTERACTIONS FOR THE DEVELOPMENT OF A PROCESS PLANNING SYSTEM

Muljadi Hendry, Ando Koichi, Ogawa Makoto, Sakurai Takehiko
Shibaura Institute of Technology

Abstract

This paper presents the role of feature interactions in the development of our proposed process planning system. Features that have interactions require special consideration when determining a machining sequence. This paper describes the machining sequence determination for interacting features in a setup. Precedence constraint rules for the machining sequencing of each type of feature interaction are determined. And a method to determine the machining sequence of interacting features in a setup is proposed.

Keywords

Features, Feature Interaction Type, Machining Sequence, Process Planning

1. INTRODUCTION

Our research puts its goal in the generation of a flexible and integrated Computer Aided Process Planning System. The word `flexible` implies that the system can manage the dynamic changing that may occur in the stage of production planning, such as increased production, machine breakdowns, layout change etc. The word `integrated` implies that the system can integrate design, manufacturing and scheduling activities. Figure 1 shows the overview of our proposed CAPP System.

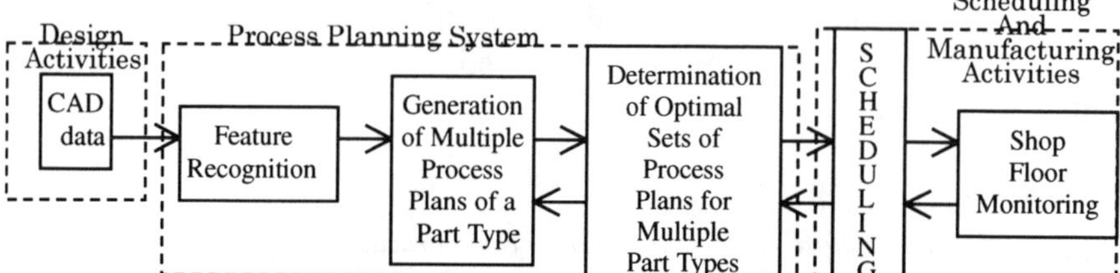

Figure 1. Overview of a Flexible and Integrated CAPP System

The system starts with the interpretation of product design data (CAD data) that leads to feature recognition. Generation of multiple process plans of a part type is considered to provide alternative plans in the determination of optimal sets of process plans for all the part types in the shop, and provide flexibility in shop floor scheduling. In order to bring this CAPP system to realization, we implemented feature concept as the basis for the CAPP

System. Based on the feature concept, we develop a system called Feature-Based Flexible Process Planning System (FBFPPS) to bring the flexibility to realization. FBFPSS is developed to make the generation of multiple process plans for a part type possible. FBFPPS is divided into 3 phases: 1) feature recognition, 2) setup generation and 3) machining sequencing. The methods for feature recognition phase and setup generation phase have been proposed (Hendry, 2000).

After features have been categorized into setups, we need to determine the machining sequence of the whole features in a part. We divide this machining sequencing phase into 3 steps: 1) machining sequencing of interacting features in a setup, 2) machining sequencing of all features in a setup, and 3) setup sequencing.

In this paper, we focus our topic in the first step of the machining sequencing phase. Interacting features are features that touch or overlap other features. Features that have interactions require special consideration when determining a machining sequence since precedence constraints exist in the machining sequencing of interacting features. We will present feature interaction types and determine the precedence constraint rules for the machining sequencing of each type of feature interaction in section 2. In section 3, we present our proposed method, called Interacting Feature Sequencing (IFS) Method to determine the machining sequence of interacting features in a setup.

2. FEATURE INTERACTIONS

2.1 Interacting Features in a Setup

In a setup, there may be interacting and non-interacting features. Figure 2 shows a part that has 3 interacting features (feature f_1, feature f_2 and feature f_3 interact each other) and 1 non-interacting features (feature f_4) in a setup.

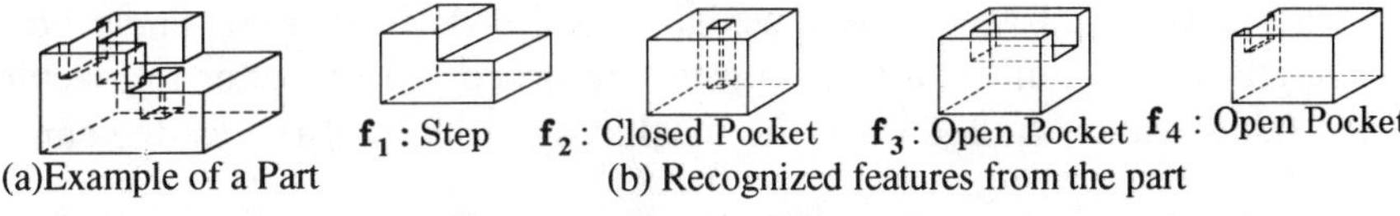

(a)Example of a Part   (b) Recognized features from the part

Figure 2. Example of a Part with interacting and non-interacting features

2.2 Approach-faces and Feature Interaction Types

In our research, we consider the relation between the occupied area of the approach face of features to define the feature interaction types. An approach-face of a feature is the face of the feature which, when projected, would intersect the exterior surface of the machined part, and constitute the plane of which a cutting tool must pass to create that feature. Figure 3 shows the approach face of each features in figure 2.

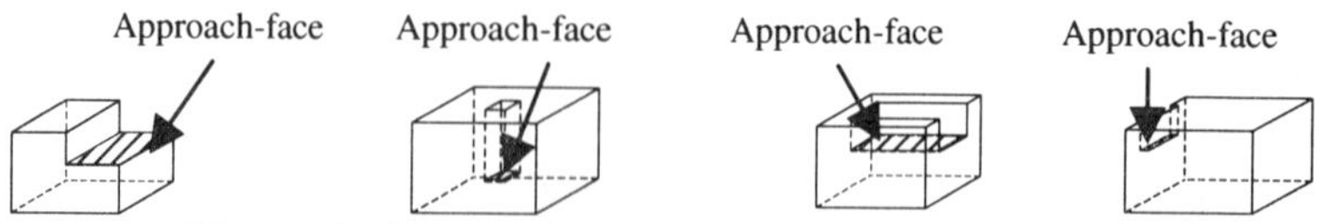

Figure 3. Approach-faces of features in figure 2

By considering the relation between the occupied area of the approach face of features, 4 types of feature interaction: 1) nested inside interaction type, 2) partially overlapping interaction type, 3) adjoining interaction type, 4) no interaction type, can be identified. Figure 4 shows feature interaction types.

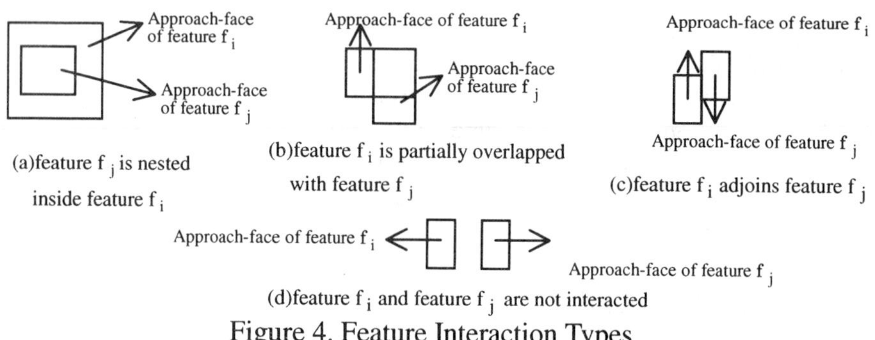

Figure 4. Feature Interaction Types

2.3 Precedence Constraint Rule

For the machining sequence of interacting features, we propose the precedence constraint rules for each type of interaction.

1. For nested inside interaction type, machine feature with bigger surface first. The exception of this rule would be for the case of a feature nested inside a wedge. To machine a wedge, a ball endmill is used. The process using a ball-endmill is a time consuming process. It is better to machine the other feature first, so that the machining volume of the wedge can be decreased.
2. For adjoining interaction type or the partially overlapping interaction type, machine feature with bigger surface first.

3. INTERACTING FEATURES SEQUENCING

To determine the machining sequence of interacting features in a setup, we develop a method called Interacting Features Sequencing (IFS) Method. The procedure of this method is as follows.

Step 1: prepare the feature ranking table. List all features in a setup as shown in table 1.

Table 1. Feature Ranking Table

	f_1 .... f_n		$\sum_{i,j=1}^{n} X_{i,j}$	Rank
f_1	$X_{1,1}$	$X_{1,n}$		
f_n	$X_{n,1}$	$X_{n,n}$		

$$X_{i,j} = \begin{cases} 1, & \text{if the machining sequence is feature } f_i \rightarrow \text{ feature } f_j \\ 0, & \text{otherwise,} \end{cases}$$

and is obtained from the algorithm below.

Let $A(ff_i)$ be the occupied area of approach face of feature f_i and $A(ff_j)$ be the occupied area of approach face of feature f_j.

For i=j, $X_{i,j} = 0$, $X_{j,i} = 0$

If $A(ff_i) \cap A(ff_j) = \phi$,

 $X_{i,j} = 0$, $X_{j,i} = 0$

Else if $A(ff_i) \cap A(ff_j) = A(ff_i)$

 If feature f_j is wedge then $X_{i,j} = 1$, $X_{j,i} = 0$

Else $X_{i,j} = 0$, $X_{j,i} = 1$

If $A(ff_i) \cap A(ff_j) = A(ff_j)$

 If feature f_i is wedge then $X_{i,j} = 0$, $X_{j,i} = 1$

Else $X_{ij} = 1$, $X_{ji} = 0$

If $A(ff_i) \cup A(ff_j) = \cup A(ff_i, ff_j)$

 If $A(ff_i) \geq A(ff_j)$ then $X_{i,j} = 1$, $X_{j,i} = 0$

 Else $X_{i,j} = 0$, $X_{j,i} = 1$

Else

 If $A(ff_i) \geq A(ff_j)$ then $X_{i,j} = 1$, $X_{j,i} = 0$

 Else $X_{i,j} = 0$, $X_{j,i} = 1$

End

Table 2 shows the feature ranking table of features in figure 2.

Table 2. Feature Ranking Table of Part in figure 2

	f_1	f_2	f_3	f_4	$\sum\limits_{i,j=1}^{n} X_{i,j}$	Rank
f_1	0	1	1	0	2	1
f_2	0	0	0	0	0	3
f_3	0	1	0	0	1	2
f_4	0	0	0	0	0	4

Step 2: draw oriented graphs. From table 2, take rank 1 and draw graphs as shown in figure 5(a). Then take rank 2 and draw graphs as shown in figure 5(b). The edge from node f_1 to node f_2 can be eliminated, since a node shall not be an end-vertex of two edges. Continue the process until all the features are taken. Figure 5(c) shows the machining sequence graph of interacting features in figure 2. The machining sequence for interacting features of part is feature $f_1 \rightarrow$ feature $f_3 \rightarrow$ feature f_2. Also, one non-interacting feature (f_4) is identified.

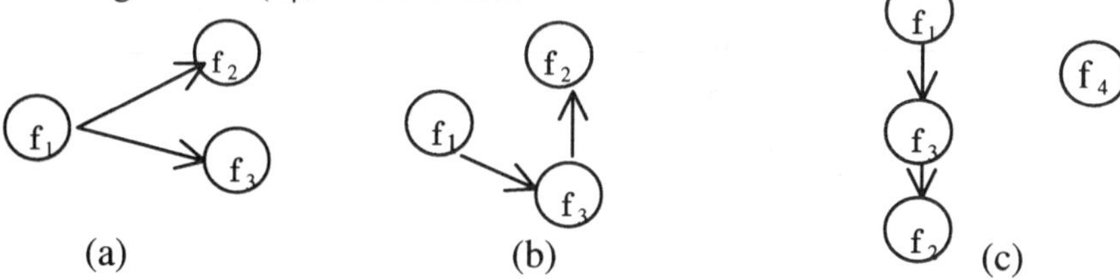

(a)   (b)   (c)

Figure 5. Machining Sequence Graph of Interacting Features in Figure 2

4. CONCLUSION

In the development of Feature-Based Flexible Process Planning System, we consider feature interactions as a crucial factor in the determination of the machining sequence since precedence constraints exist in the machining sequencing of interacting features. Our proposed method has the ability to identify interacting and non-interacting feature, and is effective for the machining sequencing of interacting features. Thus, it shows that this method is a useful basis for the development of the FBFPPS.

5. REFERENCE

Hendry Muljadi, Koichi Ando, Takehiko Sakurai, Makoto Ogawa. Setup Generation in a Feature-Based Flexible Process Planning System. Proceedings of the Fifth International Conference on Progress of Machining Technology; 2000 Sept 16-20; Beijing, Beijing: Aviation Industry Press, 2000

Strategic Product Development Decision-making Based on Optimization Incorporating the QFD Method

Masataka Yoshimura and Kazuhiro Izui

Department of Precision Engineering
Graduate School of Engineering
Kyoto University

Abstract

During the conceptual design stage of an optimization problem, optimization of the product quality, its function, and development time, as well as other factors, must be considered in addition to details concerning product structure. To support such complex concurrent design endeavors, we propose a method for deciding the directions of product design development using QFD methods and additional optimization techniques. In order to optimize design parameters and alternatives contained in various HOQ stages, the relationships of design parameters are translated from the initial set of matrices into hierarchical genotypes. HGAs can solve this type of problem and a Pareto optimum solution set displays the outcome graphically. This output can be employed by a decision-maker to chart the most effective development strategy for a given product.

Keywords

Quality function deployment, Development strategy, Customer satisfaction, Development time, Hierarchical genetic algorithms

1. Introduction

During the optimization of the design and production of a product, the ultimate success of the product often depends on the sequence of decision-making taken during its development. Upstream decisions made during the various stages of design and development as they proceed sequentially can have a profound impact on the final result. Therefore, it is crucially important to determine the best strategy for the design problem's frame structure and development policy at the earliest possible stage. If this can be done successfully, the efficiency of the design and development processes will be improved, and the final product will be more likely to embody to an optimal design.

To manage product information at the start of the planning stage, in order to determine the best strategy for the product's design and development, so-called Quality Function Deployment (QFD) methods (Akao 1990) have attracted considerable attention. QFD methods assist product development by analyzing market trends and translating customer demands into specific product quality characteristics and product structures, and by deploying the relationships between such demands and resulting product characteristics in a systematic and easily understood manner. Information that can be readily applied to strategic decision-making from an early stage can be collected using QFD methods. However, one shortcoming of conventional QFD methods is that they are not designed to support reaching final decisions based on the collected information, and points of fundamental importance must be entrusted to the judgment of decision-makers. A sophisticated support system to obtain the optimal development strategy is needed, since conventional QFD methods do not address the evaluation of the product development system as a whole, in terms of the harmony and conflict among its parts. This paper discusses improved QFD methods that can be applied to large-scale systems, and that can derive optimal product development and production strategies.

This paper proposes a method for applying optimization techniques to the decision-making required when a product's development strategy is created. First, an optimization problem model is constructed using QFD techniques and then, methods for applying Hierarchical Genetic Algorithms (HGAs), which can efficiently optimize compound problems and can find solutions for contained hierarchical deployment problems (Yoshimura and Izui 1999), to these QFD constructions are discussed.

2 Hierarchical Genetic Algorithms

When a product is analyzed using two or more hierarchical matrices, various hierarchical alternative solutions must be evaluated when searching for an optimal optimization solution (Malmqvist and Svensson 1999). However, it is difficult to evaluate such hierarchical alternative solutions using the simple matrices produced by QFD techniques, since the arrangement of the design information and alternative solutions is inflexible. Therefore, although QFD techniques are effective for evaluating detailed performances when the frame structure of the product has been decided, conventional QFD techniques are ineffective for the evaluation of hierarchical alternative solutions. For example, changes in the quality characteristics matrix are not reflected in either the alternative solution or detailed parts matrices, because the representation used in these matrices is inflexible and unsuitable for dealing with the hierarchical alternatives. Yoshimura and Izui have developed Hierarchical Genetic Algorithms (HGAs) to make possible the optimization of such hierarchical problems. HGAs can solve large-scale optimization problems that use hierarchical genotypes to represent hierarchical alternatives.

3. Product Development Decision-making
3.1 Translation of QFD matrices into hierarchical genotype representation

When various House Of Quality (HOQ) stages are analyzed, one or more decision-makers may propose a number of improvement plans, which must be considered as to whether or not they should be adopted into the new product design. During such consideration, optimization techniques are required to obtain optimal alternatives. Therefore, these improvement plans are

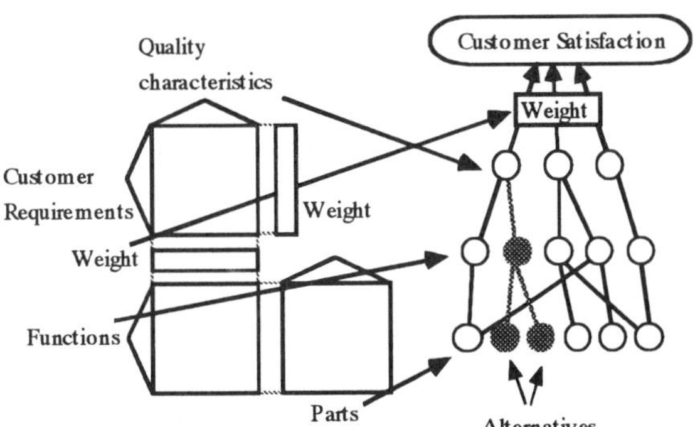

Fig.1　A example of translation from HOQs into hierarchical genotypes

treated as decision variables in the main optimization procedure and to optimize them, we translate the design parameters relationships from the initial set of planar matrices into HGA genotypes, as shown in Fig.1. If design parameter features in lower layer matrices are controlled by design parameter values in upper layer matrices, lower design parameters genes wind up positioned below upper genes, which then assume control. Thus, obtained the HGA genotypes contain n layers when n stages of HOQ are used in the QFD analysis.

3.2　Configuration of the objective functions

The objective of design and development strategy optimization here is to produce products that are as attractive to customers as possible. Therefore, in this paper, customers' satisfaction is adopted as one of the objective functions. The levels of customers' requirements are represented as weighted values in conventional QFD methods. Using quality matrices, these weighted importance levels can be transformed into relative weighted importance levels of quality characteristics, as shown in equation (1).

$$\mathbf{w}_{QC} = \mathbf{Q}_1 \mathbf{w}_{CR} \qquad (1)$$

Here, $\mathbf{Q}_1$ shows a quality matrix for customer requirements and product quality characteristics. The matrix represents the relationships using discrete symbols such as ◎, ○ and △. Using the weighted importance levels of the quality characteristics, the customer satisfaction S can be calculated from achieved quality characteristics, as shown in equation (2).

$$S = \sum_i w_{QC_i} s_i \rightarrow \max \qquad (2)$$

where a normalized quality characteristics s_i $(i = 1, \cdots, n)$ represent how well S is achieved. The optimization problem is formulated as a multi-objective problem that includes the estimated production cost and production developmenttime, as explained below.

3.3　Consideration of production developmenttime

When optimization problems are solved using HGAs, a number of alternatives are proposed via various QFD procedures and then evaluated, and an optimal solution is obtained from among the alternatives. In most cases, additional plans that include the possibility of new production techniques or radical design improvements are also proposed through the QFD procedure, to attempt to improve customer satisfaction levels. Thus, it is necessary that such alternative plans be included in the optimization problem, regardless of whether or not they are ultimately adopted. Modern market realities make it imperative that product development time be as short as possible but also that when newly developed technologies become available, that these be effectively employed to make a superior product. In order to include the product development time in the optimization formulations, the time required to integrate new technologies is estimated. The additional development plans and production technology levels are included in the formulation of the optimization problem. Furthermore, the development time is also included as one of the objective functions.

4. Case study

As a case study to illustrate the methods proposed above, a design, development and production problem for a 2 DOF robot arm is presented. Here, customer requirements, quality characteristics and parts details are analyzed in two HOQ stages. First, HOQs for this problem are constructed as matrices, shown in Fig.2. From this matrix, the weighted parameters for the quality characteristics used in equation (1) can be calculated.

Two kinds of robot arm mechanisms were considered in this case study. The first is a direct-drive type where the second arm is operated by a motor installed at the flexing joint, and the second is a linked-drive type, where two motors are at the base of the first arm and the second arm is operated via additional links. These two mechanisms offer fundamental alternatives

Quality characteristics

Customer requirements	Movement precision	Power consumption	Deflection	Size	Weight	Movement range	Weight
Speed	○	△					0.30
Precision	◎		△	○		△	0.30
Durability		△	◎				0.15
Versatility	○		◎	○		◎	0.08
Running Cost		◎					0.13
Installability		△		◎	○		0.04
Weight	0.29	0.13	0.16	0.15	0.01	0.08	

Direct-drive type							Weight
Arm1 material	○	△	◎		○		0.10
Arm1 length	△	△	△	◎	◎	◎	0.10
Arm1 shape	◎	△	◎		○		0.13
Motor1	○	△			◎		0.06
Arm2 material	○	△	◎		○		0.10
Arm2 length	△	△	△	◎	○	◎	0.10
Arm2 shape	△	△	◎		○		0.07
Motor2	◎	○	○		◎		0.13
Controler1	○	◎					0.08
Controler2	◎	◎					0.12

Fig.2 HOQs for the robot arm problem

requiring an intelligent decision as to which mechanism should be adopted. Therefore, two matrices are required for the second stage HOQ that represents relationships between the quality characteristics and the parts details for the two arm types. Since hierarchical genotype representation can deftly manage the complex interrelationships among multiple matrices, the entire set of HOQ matrices are translated into hierarchical genotypes.

The undecided parts details at the lowest hierarchical level are also treated as decision variables. In this example, the arm lengths, arm materials, shapes of structural members and motor specifications, as well as parameters for PD control are all decision variables. Furthermore, additional development plans for alternative structural shapes are proposed during all stages of the HOQ analysis and are included as decision variables. Estimates of the development time and expected performance impact for each of these plans are included the optimization problem.

This design and development was solved using HGAs. The obtained Pareto optimum solutions set for the three objective functions, namely customer satisfaction, cost and development time, are graphically displayed in Fig.3. Since this figure clarifies the exact relationships between customer satisfaction levels S, cost C and development time T, the decision-maker can effectively determine an optimal development plan for this product design problem.

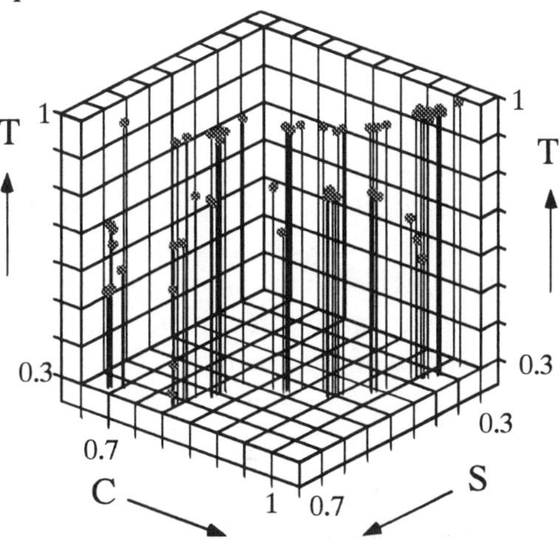

Fig.3 Obtained Pareto solutions set

References

Akao, Y., *Quality Function Deployment*, QFD –Integrating Customer Requirements into Product Design,Productivity Press, Portland, 1990

Yoshimura, M. and Izui, K., Smart Optimization of Machine Systems Using Hierarchical Genotype Representations, Proceedings of the ASME 1999 Design Engineering Technical Conferences, Las Vegas, Nevada, 1999.

Malmqvist, J. and Svensson, D., A Design Theory Based Approach Towards Including QFD DATA in Product Models, Proceedings of the ASME 1999 Design EngineeringTechnical Conferences, Las Vegas, Nevada, 1999.

DEVELOPMENT OF A LOGISTIC OPERATING CURVE FOR AN ENTIRE MANUFACTURING DEPARTMENT
- Logistic Process Operating Curve (LPOC) –

Hans-Peter Wiendahl, Michael Schneider

Institute of Production Systems, University of Hannover
Callinstraße 36, D-30167 Hannover / Germany
Phone: 00 49 5 11 / 7 62 – 34 94
Fax: 00 49 5 11 / 7 62 – 38 14
E-Mail: schneider@ifa.uni-hannover.de

Abstract

This paper describes the derivation, calculation and fields of application of the Logistic Process Operating Curve (LPOC) that is currently being developed by the Institute of Production Systems at the University of Hannover. The LPOC aims at depicting the correlation between mean performance, mean throughput time and mean WIP (work in process) level of a network of work centers. Thus it is possible to adequately describe the behavior of an entire manufacturing department due to the relevant interactions between the single work centers.

Keywords

PPC, Production Controlling, Logistic Operating Curve (LOC).

1   INTRODUCTION

In order to control the 'Dilemma of Operation Planning' (see figure 1) (Gutenberg, 1972) by using an objective-oriented manufacturing process organization, it is essential to analyze and quantify the interdependencies between the logistic key-factors throughput time, schedule reliability, WIP level and machine utilization. Moreover, it is important to understand how the logistic key-factors can be adjusted to target values. For this purpose, the Institute of Production Systems at the University of Hannover developed the theory of Logistic Operating Curves (Wiendahl, 1995). The LOCs describe how mean performance and mean throughput time of a work center change with the WIP level. Figure 1 shows an example of the LOCs of a work center. The calculation of LOCs is extensively covered in (Wiendahl,

Nyhuis, 1999). The LOCs have found many applications, both, in industrial use and in the development of new methods for PPC.

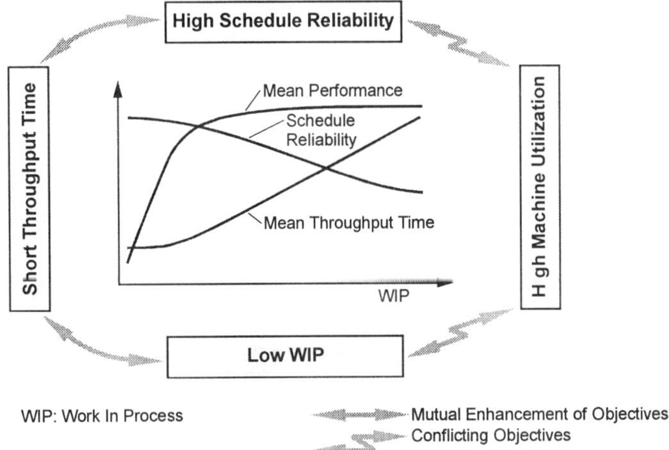

Figure 1: Logistic key-factors and the Dilemma of Operation Planning

However, so far the LOCs can only be calculated for single work centers. Consequently, the interdependencies between the work centers can be very complex as well. In particular, the relevant interactions between the characteristic behavior of single work centers cannot be described adequately with the LOCs for single work centers. For example it cannot be analyzed yet how capacity constraints of a single work center affect the total output of an entire manufacturing department. Apart from that, Production Managers often do not have much time for a detailed analysis of every single work center.

2 LOGISTIC PROCESS OPERATING CURVES FOR A MANUFACTURING DEPARTMENT

For this reason it was necessary to develop a new method to describe the behavior of a network of work centers when the mean WIP changes. The chapter will describe the derivation of a new kind of LOC for manufacturing departments: *Logistic Process Operating Curve (LPOC)*. The LPOC is based on the LOC of a single work center. LPOCs depict the correlation between mean performance, mean throughput time and mean WIP level of a network of work centers. Thus it is possible to assess and to organize single processes as well as entire manufacturing process chains or manufacturing departments.

2.1 Derivation

LPOCs can be calculated on the basis of individual LOCs according to a bottom-up-concept (see figure 2).

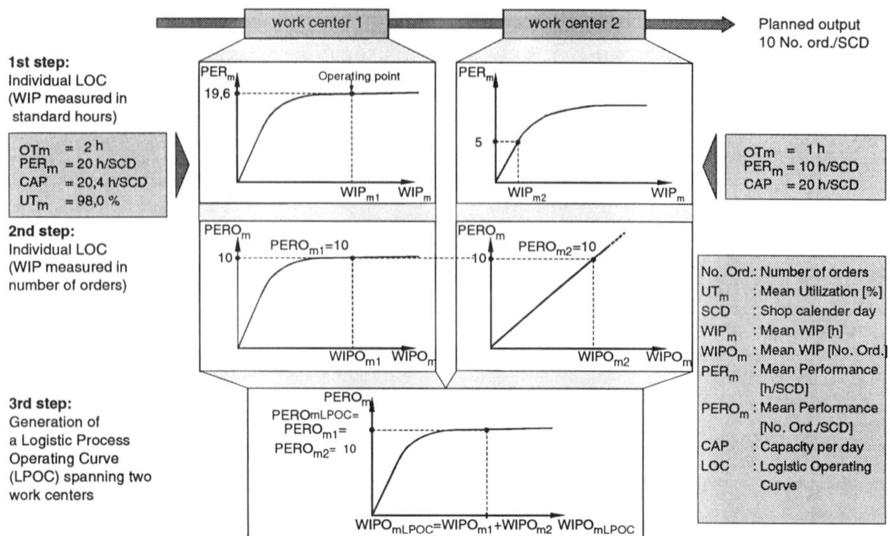

Figure 2: Development of a LOC for an entire manufacturing department (Logistic Process Operating Curve, LPOC)

The major step for generating the LPOC is to describe the connection between the individual LOCs of the single work centers in the material flow of the manufacturing department by measuring a flow parameter at every work center. It has been found useful to outline the linkage of various work systems by the material flow with the number of incoming and outgoing orders. In order to calculate the mean performance of a single work center the mean performance of the entire manufacturing department is multiplied with the material flow coefficient which describes the percentage of orders that are being processed at a single work center (Wiendahl, Nyhuis, 1999):

$$PER_{m,i}(t) = PERO_{m,i}(t) \cdot TO_{m,i} = \{PER_{m,Dep}(t) \cdot MFC_i\} \cdot TO_{m,i} \qquad \textbf{(1)}$$

$PER_{m,i}(t)$: Mean performance of work center i [h/SCD]

$PERO_{m,i}(t)$: Mean performance of work center i [No. ord./SCD]

$TO_{m,i}$: Mean order time of work center i [h]

$PERO_{m,Dep}(t)$: Mean performance of the manufacturing department [No. ord./SCD]; SCD = Shop calendar day

MFC_i : Material flow coefficient of the work center i [-]

t : Running parameter ($0 \leq t \leq 1$)

This information can be taken from the material flow matrix, which can be calculated using feedback data from the shop floor. To calculate the material flow coefficient it is necessary to normalize the material flow in relation to the total output of the entire manufacturing department. The mean WIP in number of orders can be calculated from the mean WIP in hours (Wiendahl, Nyhuis, 1999):

$$WIPO_{m,i}(t) = \frac{WIP_{m,i}(t)}{TO_{m,i}} - \frac{UT_{m,i}(t) \cdot TO_{v,i}^2}{100} \qquad (2)$$

$WIPO_{m,i}(t)$: Mean WIP of the work center i [No. ord.]
$WIP_{m,i}(t)$: Mean WIP of the work center i [h]
$UT_{m,i}(t)$: Mean utilization of the work center i [%]
$TO_{m,i}$: Mean order time of work center i [h]
$TO_{v,i}$: Variation coefficient of mean order time [-]

After having calculated the individual LOCs the shape of the LPOC can be quite easily calculated from the WIP and output values of the individual LOCs. The WIP level of the LPOC and therefore the entire manufacturing department can be calculated by summing up the WIP levels taken from the individual LOC of every single work center:

$$WIPO_{m,LPOC}(t) = \sum_{i=1}^{n} WIPO_{m,i}(t) \qquad (3)$$

$WIPO_{m,LPOC}(t)$: Mean WIP of the LPOC [No. ord.]
$WIPO_{m,i}(t)$: Mean WIP of the work center i [No. ord.]
n : Number of work centers in the manufacturing department

The maximum output of the manufacturing department can also be derived from the individual LOC of the single work centers. Therefore the bottleneck of the manufacturing department has to be identified. The bottleneck determines manufacturing output and thus LPOC output:

$$PERO_{m,LPOC}(t) = PERO_{m,BN}(t) / MFC_{BN} \qquad (4)$$

$PERO_{m,LPOC}(t)$: Mean performance of the LPOC [No. ord./SCD]
$PERO_{m,BN}(t)$: Mean performance of the bottleneck [No. ord./SCD]

On the basis of the maximum output of the bottleneck and the sum of the WIP taken from the individual LOCs the LPOC can be determined. In order to evaluate LPOC's basic concept, the LPOCs were tested in several simulation runs. The results of the simulations proved correctness of LPOC's concept. The shape of the calculated LPOC traced the shape of the simulated LPOC.

2.2 Scope of Application

When applying the LPOC one has to bear in mind that relevant information can be drawn from the individual LOCs of the single work centers as well as from the LPOC itself. It is one of the LPOC's most significant advantages that it condenses the information of various individual LOCs in one single characteristic curve. According to a top-down-concept the Production Manager can quickly analyze the logistic performance in the manufacturing department by means of the LPOC without checking each of the individual LOCs of every single work center. For a further analysis the individual LOCs of the single work centers provide detailed information. One can divide the scope of application of the LPOC into three main parts: Process design and evaluation, Production Planning and Control and Production Controlling (see table 1).

	Detailed application
Process Design	Operations Sequence
Production Planning& Control	Target WIP
	Target Throughput Times
	Lot-sizing
	Capacity Planning
Production Controlling	Actual WIP
	Actual Throughput Times
	Logistical objective trade-off

Table 1: Fields of application of the LPOC

LPOCs can be compared with each other, providing a means of assessing the effects of different process designs on the logistic key-factors. As a result, the need for modified process sequences or for new manufacturing processes can be identified. Also, the consequences of the interaction between different orders or processes which are competing to be processed at a work center can be evaluated. Within an application scope of PPC the determination of lot-sizes is supported. For the purpose of Production Controlling the LPOC presents the relevant measures mean performance, mean throughput time and mean WIP of the manufacturing department in one chart.

All in all the Logistic Process Operating Curve represents a comprehensive method for modeling and assessing complex networks of work centers such as entire manufacturing departments by using a systematic approach based on the connection of individual LOCs.

3   REFERENCES

Gutenberg, E, *Grundlagen der Betriebswirtschaftslehre. Band 1: Die Produktion.* Berlin, Heidelberg, New York, 1973.

Wiendahl, Hans-Peter, *Load-oriented Manufacturing Control.* Berlin, Heidelberg: Springer-Verlag, 1995.

Wiendahl, Hans-Peter; Nyhuis, Peter, *Logistische Kennlinien.* Berlin, Heidelberg: Springer-Verlag, 1999.

FIXTURING FEATURE
FOR DETERMINING OPTIMUM CLAMPING POINTS:

Concept of Fixturing Feature and Its Application
to 2-D and 3-D Workpiece Models

Ikuhiro Zaitsu, Hideki Aoyama and Tojiro Aoyama
School of Integrated Design Engineering, Keio University
3-14-1 Hiyoshi, Kohoku-ku, Yokohama, 223-8522 Japan

Abstract

In manufacturing process, engineers often estimate the deformation of clamped workpieces from their features, and select clamping points with minimum deformation. In this paper, the fixturing features which composed of the form features, the compliance features and deformation features are proposed in order to determine the optimum points for clamping a workpiece. And the effects of the method of deciding the best clamping points applying this concept are confirmed for 2-D and 3-D workpieces.

Keywords

fixturing, feature, automation, CAD, CAM

1. INTRODUCTION

The process planning including fixturing planning in the manufacturing is an important factor influencing machining accuracy and productivity. Fixturing planning is the most time-consuming process and it depends on the heuristic knowledge and experience of skilled workers. Therefore demands for the automation of fixturing have been growing(1,2).

One of the key factors to determine clamping points is to minimize the deformation of a clamped workpiece. Such optimization of clamping points becomes possible by estimating the deformation of clamped workpiece. In this paper, the characteristic shapes of workpieces relevant to deformation due to fixturing are considered as the fixturing features. Fixturing becomes to be planned easily by a feature based CAD which includes fixturing features as shown in Figure 1.

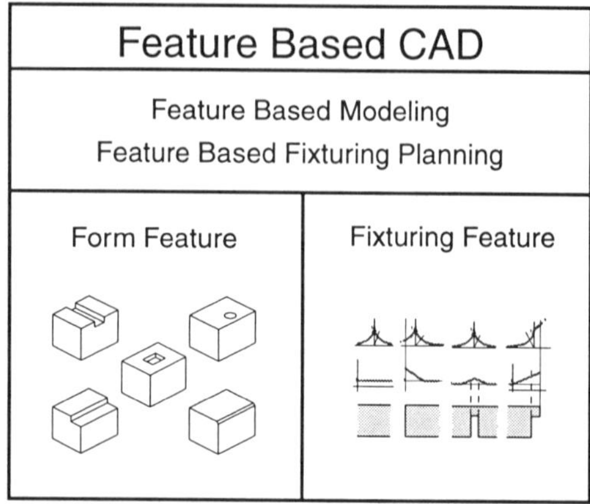

Figure 1. A new concept of feature based CAD

2. FIXTURING FEATURE

The deformation of the clamped workpiece extends to not only the clamping point but also surrounding. The deformation at the clamping point and the ratio of the deformation in the surrounding to the deformation at the clamping point characterize the deformation of the clamped workpiece. The compliance of the workpiece, the deformation at the clamping point, is called compliance feature. The ratio of the deformation in the surrounding to the deformation at the clamping point is called deformation feature. These features are classified in the form feature of the clamping part and they are called fixturing feature. In this study, four fixturing features are defined by simplifying the deformation of the clamped workpiece as shown in Figure 2.

3. APPLICATION TO A 2-D WORKPIECE MODEL

A 2-D workpiece model shown in Figure 3 contains two flat features, two edge features and a bridge feature as fixturing features. It is assumed that the workpiece is placed on a base plate and its top surface is clamped. The compliance of this workpiece is obtained by superimposing the compliance features of contained fixturing features. At the each boundary between two features, the larger value of two features is used. Assuming that this workpiece

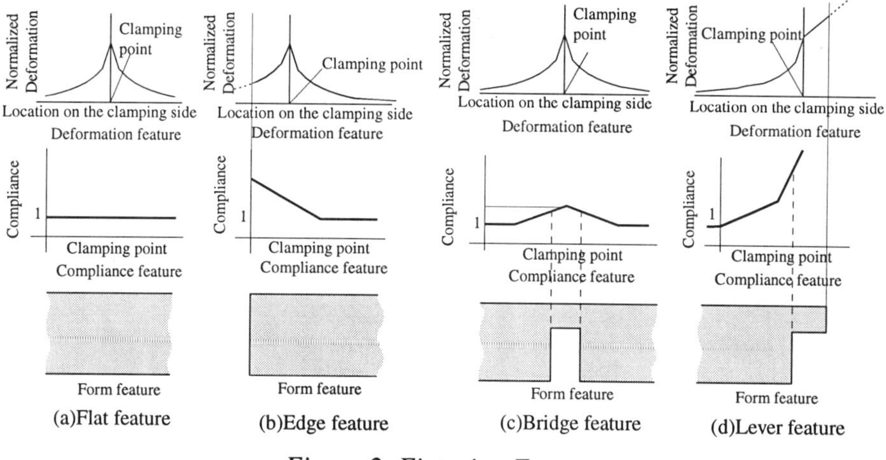

Figure 2. Fixturing Feature

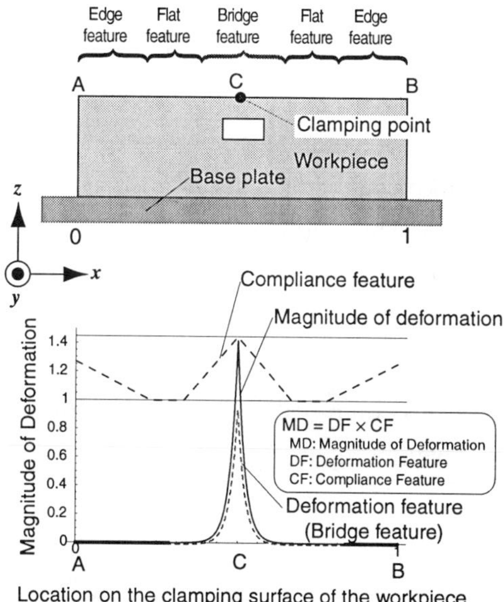

Figure 3. Applying fixturing feature to a 2-D

is clamped at the point C, the compliance value at the point C is 1.45 from the compliance feature of this workpiece. The magnitude of deformation is obtained by applying the deformation feature multiplied by the compliance value (1.45). It is possible to obtain the deformation by superposing the deformations caused by respective clamping when a workpiece is clamped at two or more clamping points. The optimum clamping points are shown in Figure 4 that is obtained by applying fixturing feature. In this study, the optimum clamping points are determined as the clamping points that make the maximum deformation minimum.

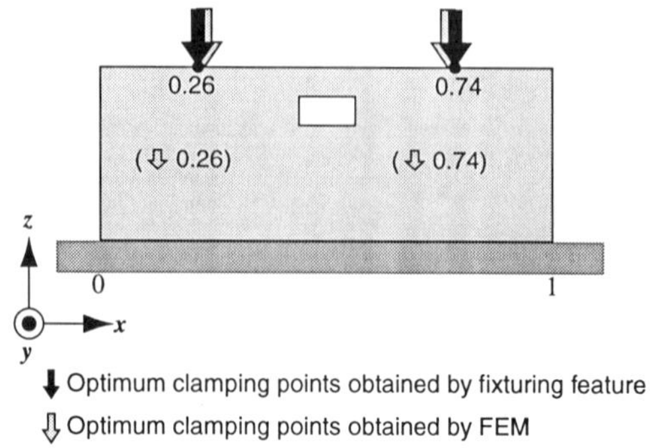

Figure 4. Optimum clamping points (2-D model)

4. APPLICATION TO A 3-D WORKPIECE MODEL

Figure 5(b) shows the compliance feature obtained about the sectional shape when a 3-D workpiece model shown in Figure 5(a) is sectioned in the direction of the arrow A. The compliance feature of the workpiece shown in Figure 6 can be obtained by multiplying the value of the compliance features in the direction A and B. The magnitude of deformation can be obtained by multiplying the deformation feature by the compliance value given in Figure 6. When the workpiece is clamped at three clamping points, the optimum clamping points are given in Figure 7.

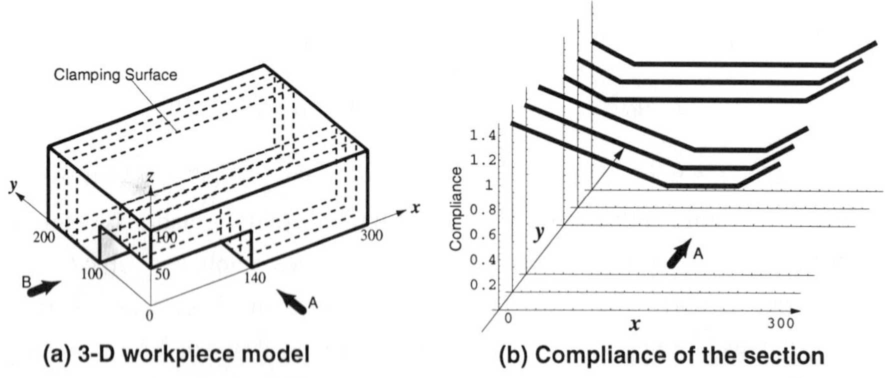

(a) 3-D workpiece model **(b) Compliance of the section**

Figure 5. Fixturing feature for a 3-D workpiece model

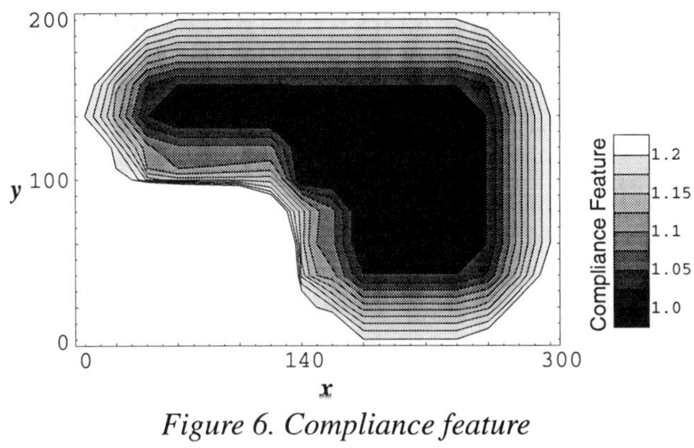

Figure 6. Compliance feature

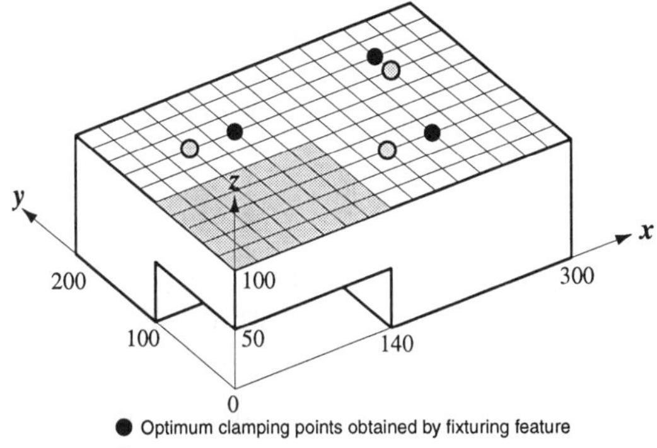

Figure 7. Optimum clamping points (3-D model)

5. CONCLUSIONS

(1) Workpiece deformation due to clamping force is characterized by fixturing features, deformation features and compliance features.

(2) Applying the fixturing features, the optimum clamping points of a workpiece can be determined.

(3) The effectiveness of the proposed method was confirmed by the FEM analysis.

6. REFERENCES

(1) Soh-Khim, Ong and Andrew Yeh-Ching, Nee, 1995, "A systematic approach for analyzing the fixturability of parts for machining," MED-Vol. 2-1/MH-Vol. 3-1, Manufacturing Science and Engineering, ASME, pp.747-761.

(2) Ikezoe, K. and Aoyama, T., 1994, "Development of automatic fixturing design system," Transactions of Japan Society of Mechanical Engineers, Vol. 62, No.595, pp.203-205.

FEATURE-BASED FIXTURE PLANNING

FOR WORKPIECES

Toru Inoue and Ichiro Inasaki

School of Integrated Design Engineering, Keio University

3-14-1 Hiyoshi, Kohoku-ku, Yokohama, 223-8522 Japan

+81-45-563-1141

E-mail: inoue@ina.sd.keio.ac.jp

Abstract

For decades, manufacturers have envisioned a fully integrated environment for CAD, CAPP, and CAM. Nevertheless, the development in fixturing planning, the critical link from product design to manufacturing, fell short to achieve a seamless integration into computer-supported manufacturing system. In this paper, a computer supported procedure to determine the fixturing plan of a workpiece by applying feature technology and utilization of a fixturing knowledge base is introduced. Also, the application of the concept is demonstrated by a prototype system.

Keywords

feature , fixturing, CAD, CAPP

1. INTRODUCTION

In process planning, the design of fixture to hold a workpiece for machining operation is considered to be one of the most critical steps in product manufacturing. Very often, it takes years of training and experience for a skilled worker to be able to perform the fixturing planning task adequately. The challenge stems from the fact that fixturing planning process involves a great deal of human knowledge and various rule-of-thumb that are very difficult to be automated in Computer-Aided Process Planning (CAPP) [1]. To overcome the obstacle, several research efforts have been spent in finding alternative ways to capture the fixturing-related knowledge in fixturing planning systems. Our approach is to adapt the feature technology to represent the workpiece data model and to be utilized in conjunction with a fixturing knowledge base. Fixturing features are used to describe about the workpiece and the fixturing knowledge base is used to model the reasoning process in fixture design.

2. FEATURE-BASED FIXTURING PLANNING

The concept of fixturing features has been built around the feature methodology. Fixturing features are defined as the entities with attributes that represent the engineering meanings to be utilized in the area of fixturing planning. As shown in Fig. 1, the fixturing features of a workpiece could be categorized into geometrical, mechanical, functional, and operational attributes. The geometrical attributes are simply the translations of the workpiece's CAD data and presented in the form of feature primitives (Fig. 2). In addition to the physical properties, the mechanical attributes indicate the material composition and surface properties of the workpiece. Specific to the fixturing planning process, the functional and operational attributes both carry the information on how the surfaces on a workpiece are to be handled.

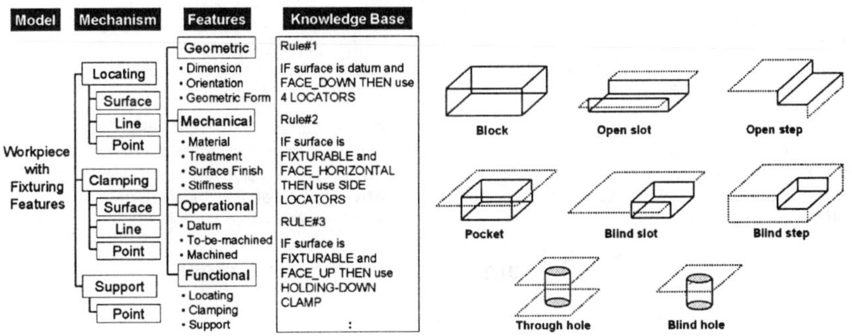

Fig. 1 Fixturing Features and Inference Process   Fig. 2 Examples of Feature Primitives

3. KNOWLEDGE BASE AND INFERENCE PROCESS

The knowledge base represents the core database for the fixturing planning system. It houses the collection of production rules related to fixturing. Some examples of rules are shown in Fig. 3. The implementation of the knowledge base tries to resemble itself to represent the "know-how" of skilled workers in the design of fixtures. In conjunction with the fixturing mechanisms, as shown in Fig. 1, the inference process utilizes sub-tasks such as surface locating, point clamping, and so on to search through the domain of fixturing in the knowledge base. The inference technique takes on the approach of forward chaining to navigate through the rule content. The search continues until it reaches the pre-stated goal. In this case, the final goal is to secure the workpiece with fixturing elements of locators, clamps, or support in a 3-dimentional space. Similar to any human decision-making process, the inference may result in none or more than one solutions, or fixture configurations, for a designated workpiece. In an effort to avoid possible ambiguities, which may arise during the process, the knowledge base is limited to fixturing rules that are readily applicable in the design of fixtures. Instead of declaring a comprehensive rule base in order to capture vary details about fixturing, the knowledge base features only rules that are essential in succeeding feasible fixture configurations.

IF	surface *is* DATUM		**IF**	surface *is* FIXTURABLE
AND	surface *face* DOWN		**AND**	surface *face* HORIZONTAL
THEN	*place* 4 vertical_locators		**THEN**	*place* 2 side_locators
AND	*place at* 15% of the lengths from either edges		**AND**	*place at* 15% of the lengths from either edges
AND	*locate* 1st side		**AND**	*locate* 2nd side
IF	surface *is* FIXTURABLE		**IF**	surface *is* FIXTURABLE
AND	surface *face* UP		**AND**	surface *face* SIDEWAY
THEN	*place* 4 vertical_clamps		**THEN**	*place* 2 side_clamps
AND	*place at* vertical_locators' trajectory points		**AND**	*place at* side_locators' trajectory points
AND	*reject* 6 DoF		**AND**	*reject* 4 DoF

Fig. 3 Examples of Production Rules]

4. SYSTEM DESIGN AND IMPLEMENTATION

To experiment with the feasibility of the feature-based fixturing system with the support of a knowledge base, software program is developed for the implementation of fixturing features and the integration of the knowledge base. Fig. 4 shows a schematic of the program architecture. The program is written in Windows programming environment and has a familiar Graphical User Interface (GUI). The GUI provides the visual representation of the workpiece, as well as fixture configurations recommended by the fixturing planning system. Hidden from the GUI, there is the control console that manages the inquiry of several databases. The console interprets a user's request for a feature-based data model and refers it to the inference engine. The inference engine then searches through the knowledge base and the fixel database and returns with fixture configurations that can be adapted for the workpiece. Currently, the knowledge base is structured in a way to accommodate the conventional 3-2-1 fixturing method. Due to the complexity of the fixturing planning process, the developed system undertakes solely prismatic parts. That is, no rotational parts are being investigated for now. In order to accommodate the addition or modification of the information, a rule editor and a fixel editor are also in place to update the respective databases. Through the GUI, a relatively simple feature editor is built-in to allow the user to make minor changes to the feature-based data model. Note that the feature editor is not meant to replace the functions of computer-aided design.

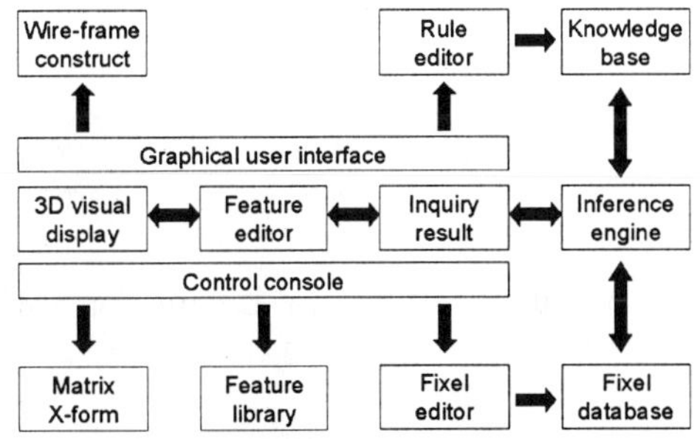

Fig. 4 Program Architecture

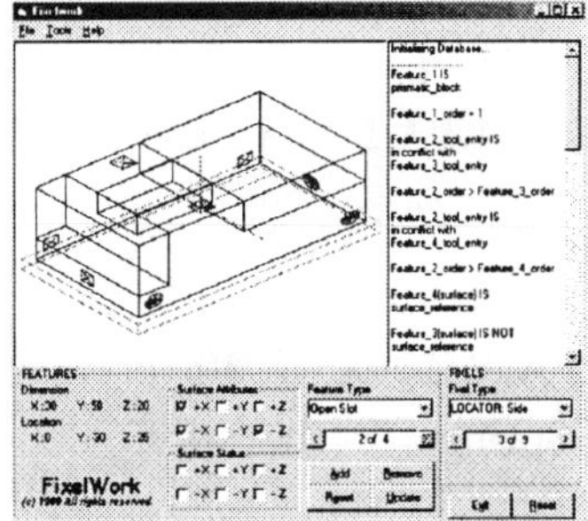

Fig. 5 Graphical User Interface

5. RESULT

A sample workpiece is tested on the fixturing planning system. The workpiece is consisted of four feature primitives; namely a prismatic block, an open slot, and two open steps. The result is shown in Fig. 5. The evaluation of the coplanar relations between features and the reference surface yields a logical machining sequence of having the larger open step machined first, followed by the second open step, and finally the open slot. Since all the features are oriented in the same face with respect to the prismatic block and the machining of all features still leaves room for the placement of fixture elements, a single fixture configuration is sufficient for this particular workpiece does not provide enough space for placing four vertical clamps, as the 3-2-1 locating principle would have required, only two vertical clamps are applied.

6. CONCLUSION

While the example with prismatic part is relatively simple, it illustrates that a feature-based data model is able to provide the fixturing planning system with pieces of information to be applied in referring to the fixturing knowledge base. The collection of fixturing rules in the knowledge base also demonstrates that it can serve as highly functional means to generate feasible fixture configurations.

7. REFERENCES

[1] A.Y.C.Nee, K.Whybrew, A.S.Kumar. Advanced Fixture Design for FMS, Springer-Verlag, 1995

SIMULATION OF MANUFACTURING SYSTEM DESIGN FOR UNCERTAIN MARKET CONDITIONS

Hideo Fujimoto, Alauddin Ahmed

Nagoya Institute of Technology, Japan

Abstract

This paper describes a model for technology acquisition to decide the technology and the size of investment in the face of uncertainties in product and technology evolution, and products' life cycles. Uncertainty in product mixes results from the stochastic product stream and overlapping of the life cycles, whereas uncertainty in the acceptable choices of technologies results from uncertainty in product mix and technological evolution. The level of technology is measured by the system's flexibility. A decision model based on Markov process has been developed and implemented on a moderate size problem.

Keywords

Product Evolution, Technological Evolution, Diversity and Flexibility

1. INTRODUCTION

This paper addresses the investment problem in uncertain market conditions where uncertainty exists in product evolution, technological evolution, and products' life cycles. The key decision variables are the timing and the size of investment and the level of flexibility associated with the facility, where the flexibility is measured in terms of the ability to change from one product to another. This study is related to the works by Fine and Li (1988), Monahan and Smunt (1989), and Rajagopalan at el. (1998). Fine and Li (1988) has the limitation of handling only two products in their model, while Monahan and Smunt (1989) does not include product evolution. Rajagopalan at el. (1998) gives a comprehensive mathematical model for investment in uncertain technological breakthroughs; however, it handles only single product with growing demand and does not include product evolution in the model. This paper tries to overcome those limitations, describes a model for investment decision considering the above uncertainties, develops a solution method by using Dynamic Programming, and implements the model for solving a finite planning range problem. The model is much closer to reality compared to other works available in literature in the sense that it considers overlapping of the lifecycles and there is no restriction on the life of the products. Diversity of the environment outside the firm has been used to justify the flexibility.

2. MODEL DESCRIPTION

Assumptions. Only one product is introduced newly to the market or withdrawn from the market in one period. Minimum product life of any product is one period. Scrapped products do not reappear in the market.

Cost Functions. Three cost functions are considered in this model: fixed cost I, variable cost O and switching cost S.

Fixed cost includes the cost of equipment, tools, and other set up per unit product produced in the period and is expressed as a function of the level of flexibility and the capacity required to meet the demand.

The variable cost represents per unit variable production cost in a given scenario and includes two multiplying cost factors: operational cost and holding cost. Cost of operation is again obtained from two multiplying functions related to the environments inside and outside the firm.

Switching costs will mean cost of changes in level of flexibility. It is a function of the existing level of flexibility and the upgraded level of flexibility.

Problem ($\mathcal{P}$) Formulation: To decide on the size C_t and the level of flexibility f_t of the facility to satisfy the demand until period j, given the current product mix Y_0, technology mix T_0, the probability distribution of new product evolution $w(t)$ and the probability distribution of new technology evolution $y(t)$.

Let the demand and the capacity of the product mix be expressed by vectors $D_t = (d_{t,1}, d_{t,2}, \ldots\ldots d_{t,N})$ and $C_t = (c_{t,1}, c_{t,2}, \ldots\ldots c_{t,N})$ respectively. In each notation, subscript t refers to the period t. $d_{t,i}$ is the demand of product i and x is the total number of products in the mix. $c_{t,i}$ is the total capacity of the plant if only i were produced.

Cn_t represents the total capacity lost due to all the set up changes for producing the products in the mix in the required quantity.

The same technology may need different lead-times to produce different products. A multiplying factor $\omega_{i,r}$ is used to convert demand of any product i to an equivalent of the representative product r. The firm must meet demand of all products at all time within the planning range. Therefore,

$$\sum_i \omega_{ir} d_{t,i} \le c_{t,r} - Cn_t \tag{1}$$

A state is characterized by product mix Y_t, demand vector D_t, capacity vector C_t, flexibility index representing the current facility f_{t-1}, and flexibility index representing the facility level of the new facility f_t. The evolution process of new products and technology can be modeled as Markov processes. $P_{m,n}$ denotes the probability of introducing product n after m. $Q_{\alpha\beta}$ denotes the probability of development of technology β after α. The transition process is controlled by (m_t, q_t, r_t, s_t) where m_t is the most recent product appeared on or before period t, q_t is the most recent technology appeared on or before period t, r_t is the period product m_t appeared first and s_t is the period technology q_t appeared. The probability of occurrence of i-th transition, $\phi_i(\cdot)$ is obtained from, $P_{m,n}, Q_{\alpha\beta}$, $w(t)$ and $y(t)$.

Let A_t be the set of all permissible alternatives of (f_t, C_t) at time t.

Also let $F(Y_t,C_t,D_t,A_t,f_{t-1},f_t)$ denote the expected total cost incurred from the beginning of period t through the end of the planning horizon. $L(Y_t,C_t,D_t,A_t,f_{t-1},f_t)$ is the long term total expected cost through the end of the planning period, subject to the condition that all future decisions are taken optimally. Then we have,

$$F(Y_t,D_t,C_t,f_{t-1},f_t) = \min_{(f_t,C_t)\in A_t}\left[L(Y_t,D_t,C_t,f_{t-1},f_t)\right] \qquad (2)$$

$$L(Y_t,D_t,C_t,f_{t-1},f_t) = I(D_t,C_t,f_t) + O(Y_t,D_t,C_t,f_t) + S(C_t,f_{t-1},f_t) + \\ \sum_i \phi_i(m_t,q_t,r_t,s_t) \times F(Y_{t+1},D_{t+1},C_{t+1},f_t,f_{t+1}) \qquad (3)$$

3. ALGORITHM

An efficient algorithm based on Dynamic Programming has been developed for solving the equations (2) and (3) subject to the constraint (1). As usually done, the long running cost at the last period of the planning range is taken to be zero. Four specific cases for the possible combinations of the states in the subsequent period are considered. These are cases of introduction of a new product, withdrawal of an existing product, simultaneous introduction and withdrawal, and finally of no change in the product mix. The transition probabilities for all the possible combinations are derived. For space constraints detail of the algorithm is omitted here.

4. COMPUTATIONAL STUDY

Simulation has been carried out for a finite planning horizon ranging from 6 to 12 periods, with unpredicted product evolution, demand and associated technological evolution. Initially the firm starts with products a, b and c and has the possibility of further introducing products d, e, and g. the firm has an option of choosing from a range of technologies. δ_i, the drop in the flexibility for the introduction of product i is selected to be 0.1. Initially the mean of inter-arrival period for products is taken to be 3 with variance 1, and the life of the products to be 3 with variance 1.

For the given problem, the long run cost can be minimized by procuring a flexible system of flexibility level 0.7 to satisfy the demand outcomes up to period 3.

5. CONCLUSION

In this paper, a simulation has been studied for addressing the capacity and technology acquisition with replacement in the face of uncertainty in the product mix and technological evolution. A solution method based on Dynamic Programming has been proposed for the model. It introduces measurement of diversity of the environment outside the firm to justify flexibility of the facility and it shows that the firm's choice of investment and technology depends on the diversity of the environment it has to operate in. Finally the model has been implemented for a moderate size problem that is very much similar to that of a real life one. In simulation, it was observed that for the same products-market-life, the time between products evolutions have greater effect on the result than the time between technological evolutions. It was also found that after a particular length of planning range there is no significant effect on the first period decision of investment.

REFERENCES

Fine, C.H. and Li, L.: "Technology Choice, Product Life Cycles, and Flexible Automation", *J. Manuf. and Oper. Manage.*, 1988; 1: 372-399

Gaimon C. and Singhal V.: Flexibility and the Choice of Manufacturing Facilities under Short Product Life Cycles, *Eur. J. Opl. Res.*, 1992; 60: 211-223

Monahan and Smunt: "Optimal Acquisition of Automated Flexible Manufacturing Processes", *Oper. Res.*, 1989; 37/ 2: 288-300

Rajagopalan S.: "Capacity Expansion and Equipment Replacement: A Unified Approach, *Oper. Res.*, 1998; 46/ 6: 846-857

Rajagopalan, Singh and Morton: "Capacity Expansion and Replacement in Growing Markets with Uncertain Technological Breakthroughs", *Manage. Sci.*, 1998; 44/1: 12-30

CONSTRUCTION OF MASTER MODEL FROM 3D SCANNED DATA USING RECURSIVE SUBDIVISION SCHEME*

[1]Bing-Yin, REN [2]Ichiro, HAGIWARA [3]Qing-Xin, MENG

[1]*Advanced Manufacturing Technology Center, Harbin Institute of Technology*
92, West Da-Zhi Street, Nan-Gang District, Harbin, 150001, China
[2]*Dept of Mechanical Science & Engineering, Tokyo Institute of Technology*
1-12-1. O-okayama, Meguro-Ku, Tokyo, 152-8552, Japan
[3]*School of Mechanical and Electrical Engineering, Harbin Engineering University*
41, Wenmiao Street, Nan-Gang District, Harbin, 150001, China

Abstract

Construction of master model of a product from 3D scanned data has become a more and more important topic in the field of CAD/CAM. By means of the recursive interpolating subdivision scheme on an initial mesh of 3D scanned data, a polygonal master model that consists of smaller planar patches may be directly obtained. The principle of constructing the initial mesh is presented. Recursive subdivision scheme for triangular mesh is introduced. A quadtree data structure is designed to store all the necessary information of interpolation subdivision.

Keywords

Master model, Recursive interpolation, Subdivision, CAD/CAM

1 INTRODUCTION

Constructing the master model of complex object from 3D scanned data is one of the most important tasks to the scientists and engineers in the fields of CAD/CAM. In last decades, parametric method of tensor-product Non-Uniform Rational B-Spline (NURBS) plays an important role in constructing the geometrical models from 3D scanned data. For example, Zhao and Okubo et al. [1996] presented fitting measure points of clay model of a car body to a parametric surface of the master model of car body. Yau and Chen [1997] discussed the construction of complex geometry in reverse engineering using rational B-Spline. In fact, the complicated surface of arbitrary topology must be decomposed into a set of trimmed NURBS patches. Besides, the smooth parametric surface of the master models must be dispersed to be **polygonal models** in the actual process of FEM analysis,

Rapid Prototyping (RP) and NC machining of the product.

On contrary, the recursive subdivision scheme generalizes classical parametric approaches of surface modeling to arbitrary topology and can generate a smooth surface in the limit by repeated application of a fixed set of refinement rule on initial control mesh. Besides, surface of subdivision is a typical **polygonal** surface. Recursive subdivision scheme overcomes all the shortcomings of parametric approaches of surface modeling. DeRose and Kass et al. [1998] have achieved great success in applying subdivision schemes to geometric modeling in computer animation. Subdivision technique demonstrates greater potential not only in a wide range of solid modeling, computer graphics, but also in CAD, FEM, CAM and RP.

The mesh produced by subdivision scheme can be **triangle** or **quadrilateral**. Since the triangular mesh is more flexible for satisfying the arbitrary topological shape, this paper will only discuss the interpolation subdivision scheme on triangular meshes, including the construction of initial mesh, data structure and the interpolating coefficients of subdivision scheme.

2 CONSTRUCTION OF INITIAL MESH

The initial mesh is the basis of the subdivision scheme. Recursive subdivision scheme can produce a smooth surface in limit whose shape is controlled by the initial mesh. For any 3D scanned datasets, different algorithms may result in different initial meshes, and different initial meshes may lead to different recursive times for certain geometric error.

When the discrete points of 3D scanned data are connected to be triangular initial mesh, two points on the boundary are firstly connected to be an edge of a triangle. The third point of the triangle is defined by the ratio of length between the longest edge and the shortest edge of each possible triangle.

As shown in *Figure 1*(left), let P_i ($i=1,2,...,n$) be the neighboring vertex at the same side of the edge $P_A P_B$ of a triangle, calculate the ratio of length between the longest and the shortest edge of each triangle $\Delta P_A P_B P_i$:

$$C_i = \max\{l_{AB}, l_{iA}, l_{iB}\} / \min\{l_{AB}, l_{iA}, l_{iB}\} \qquad (i = 1,2,...,n)$$

If P_{i_0} is the vertex corresponding to $C_{i_0} = \min\{C_i$, and there are no vertices inside triangle $\Delta P_A P_B P_{i_0}$, the triangle $\Delta P_A P_B P_{i_0}$ will be part of the initial mesh. Otherwise, select a vertex P_{j_0} from those vertices of P_j ($j=1,2,...,m$, $m<n$) inside the triangle $\Delta P_A P_B P_{i_0}$ by the previous principle, the triangle $\Delta P_A P_B P_{j_0}$ will be part of the initial mesh. By means of such a

principle, a better initial mesh may be obtained as *Figure* 1 (right).

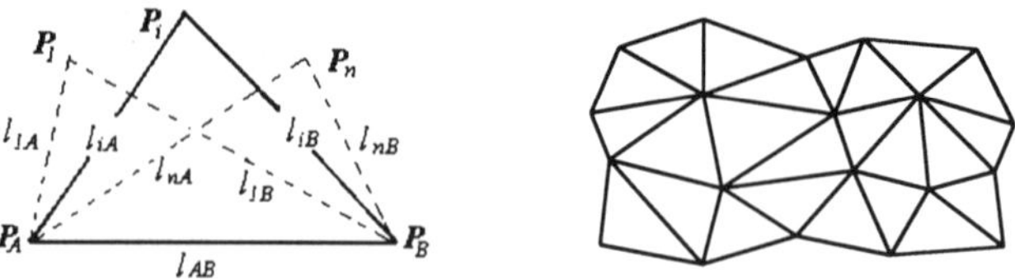

Figure 1 Construction of initial triangular mesh

3 INTERPOLATION SCHEME OF SUBDIVISION

For each subdivision scheme defined on arbitrary topological initial mesh, new vertices are inserted by means of different local affine combinations of neighboring vertices. The most well-known interpolating subdivision scheme for triangular meshes is the "butterfly" scheme proposed by Dyn et al. [1990]. It needs a topologically regular setting of the initial control mesh in order to obtain a smooth C^1 surface in limit. Zorin et al. [1996] developed an improved interpolating subdivision scheme that results in much smoother surfaces even from irregular initial meshes.

When calculate a new vertex between two regular vertices of valence 6, as shown in *Figure 2* (left), 10 vertices stencil is used, the weight coefficients corresponding to different neighboring vertex are as follows:

$$a = 1/2 - w; \quad b = 1/8 + 2w; \quad c = -1/16 - w; \quad d = w; \quad 0 \le w \le 1/16$$

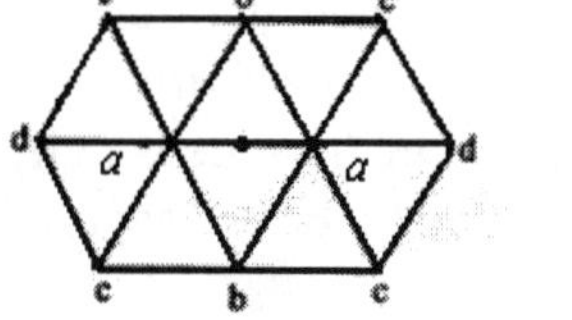
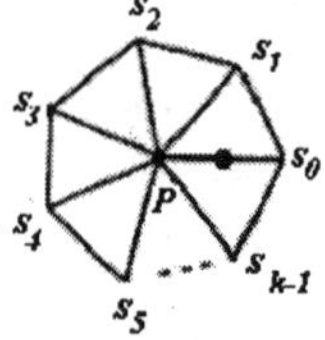

Figure 2 Masks for vertex interpolation

When calculate a new vertex between a regular vertex of valence 6 and a extraordinary vertex of valence k ($k \ne 6$), only the neighboring vertices of the extraordinary vertex are used in the stencil, the weight coefficients corresponding to different neighboring vertex (*Figure 2* right) are given by

$$S_0 = \frac{5}{12}; \quad S_1 = S_2 = -\frac{1}{12}; \qquad for \quad k = 3;$$

$$S_0 = \frac{3}{8}; \quad S_1 = S_3 = 0; \quad S_2 = \frac{1}{8}; \qquad for \quad k = 4;$$

$$S_i = \frac{1}{k}\left(\frac{1}{4} + \cos\frac{2i\pi}{k} + \frac{1}{2}\cos\frac{4i\pi}{k}\right), \quad i = 0,1,\ldots k-1; k \geq 5, k \neq 6.$$

When calculate a new vertex between two extraordinary vertices of valence k ($k \neq 6$), the average of the two new vertices defined by the previous scheme is used.

When calculate a new vertex between two vertices on the boundary, 6 vertices are used to obtain a boundary curve of C^2 in limit.

$$\begin{cases} P_{2i}^{(k+1)} = P_i^{(k)} & \left(0 \leq i \leq 2^k n\right) \\ P_{2i+1}^{(k+1)} = \left(\frac{9}{16} + 2\alpha\right)\left(P_i^{(k)} + P_{i+1}^{(k)}\right) - \left(\frac{1}{16} + 3\alpha\right)\left(P_{i-1}^{(k)} + P_{i+2}^{(k)}\right) + \alpha\left(P_{i-2}^{(k)} + P_{i+3}^{(k)}\right), 0 \leq i \leq 2^k n - 1 \end{cases}$$

Where $P_i^{(k)}$ are vertices on the boundary; $0 \leq \alpha < 0.02$. The above scheme will become as a 4-point interpolation scheme when $\alpha = 0$.

Figure 3 shows an example for constructing an initial mesh from 3D discrete points, a polygonal surface with finer triangular mesh is obtained by applying interpolating subdivision scheme on the initial mesh.

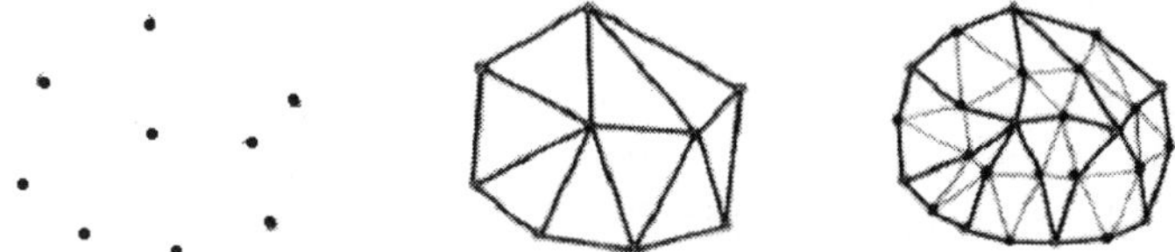

Figure 3 Example of vertex interpolation subdivision scheme

4 DATA STRUCTURE

For the interpolating scheme of **uniform subdivision**, each edge of a triangular mesh is split into two, each triangle will be recursively replaced by four sub-triangles. As for **non-uniform subdivision**, one or two edges of the triangle are split into two, the triangle will be replaced by two or three sub-triangles, as shown in *Figure 4* (left and middle).

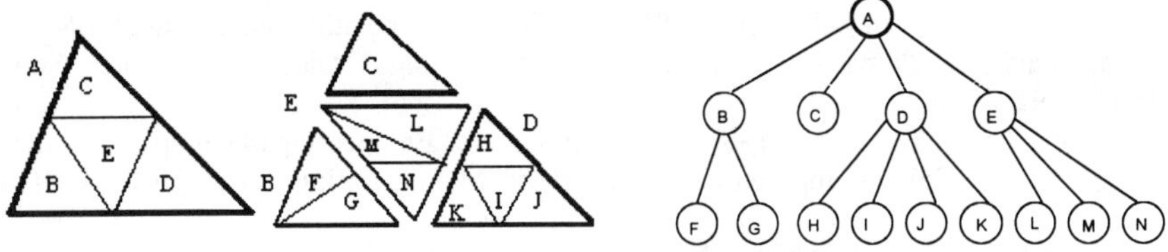

Figure 4 Quadtree representation of non-uniform subdivision of triangle

All the information of vertices and triangular mesh can be recorded by **quadtree typed data structure**. The triangle A of initial mesh is the root of the tree, each sub-triangle (e.g. B, C,..., N) is a node of the tree, those sub-triangles with the highest level of subdivision are the leaves of the tree.

The whole initial mesh of a can be treated as a special **forest of triangle quadtrees**

By such a data structure, neighbor relations with a single quadtree can be resolved in the standard way by ascending the tree to the least common parent when attempting to find the neighbor across a given edge. Neighbor relations between adjacent trees are resolved explicitly at the level of a collection of roots. Sub-triangles rooted at any level can be created by assembling a new sub-trees with some set of triangles as roots of their child quadtrees. Each non-leaf triangle (a node of the tree) contains a pointer to a block of N ($N \leq 4$) sub-triangles.

5 CONCLUSIONS

The principle for constructing the initial control mesh is presented. A quadtree typed data structure is designed to store all the information of recursive subdivision scheme. By means of vertex interpolating scheme, a polygonal master model of complex object for practical use may be obtained from 3D scanned data.

Recursive subdivision scheme can handle arbitrary topological surface with ease, only a few number of neighboring vertices are used in computing the new vertex. According to the demands on admission error and the geometric size of the mesh, the levels of subdivision may be defined adaptively.

Note: Part of the work was finished when the first author worked as a visiting researcher at the Department of Mechanical Science & Engineering, Tokyo Institute of Technology, 2-12-1 Ookayama, Meguro, Tokyo, 152-8552, Japan

REFERENCES

Bingyan Zhao, Shingenori Okubo and Houjun Tang et al. Fitting Measure Points to a Parametric Surface. Proceeding of JSME(Part C), 1996,63(616):4378-4384

Denis Zorin, Peter Schroder and Wim Sweldens. Interpolating Subdivision for Meshes with Arbitrary Topology. Computer Graphics Proceedings, Annual Conference Series, 1996:189-192

Hong-Tzong Yau and Jenq-Shyong Chen. Reverse Engineering of Complex Geometry Using Rational B-Splines. Int. Journal of Advanced Manufacturing Technology. 1997(13): 548-555

Nira Dyn, David Levin and John A. Gregory. A Butterfly Subdivision Scheme for Surface Interpolation with Tension Control. ACM Transactions on Graphics, 1990,(2):160-169

Tony DeRose, Michael Kass and Tien Truong. Subdivision Surfaces in Character Animation, Computer Graphics Proceedings, Annual Conference Series, 1998. SIGGRAPH'98: 85-94

A STUDY ON KNOWLEDGE ENGINEERING FOR EQUIPMENT DESIGN AND PRODUCTION

Seiji Takeo*   Norio Kurochi**   Shigeo Tsuchiya*   Hitoya Nakamura*

*NEC Corporation   **Hokkaido Institute of Technology

Abstract

In order to verify and realize new ideas quickly in manufacturing enterprises, the activation of knowledge is indispensable.   However, worldwide expansion of manufacturing process outsourcing and the remarkable increase in the generation gap among engineers makes knowledge activation more difficult. Hence, a new systematic aid for engineering expertise and experienced skills management should be taken into account.   In this paper, we will propose a new concept for engineering knowledge support systems, composed of categorically layered Knowledge Base and Knowledge Driver.

Keywords

Knowledge transfer, knowledge engineering, enterprise management, knowledge base, knowledge driver, product model, CAD, intranet

1. INTRODUCTION

In recent years, knowledge transfer has become one of the most important subjects in the manufacturing industry.   According to a typical case investigation, about 40 percent of the design problems discovered in the development process are caused by knowledge transfer deficiencies. On-the-job development with skilled engineers is critical in promoting knowledge transfer.   Nevertheless, the aging of skilled engineers and a shortage of younger generation engineers make O.J.D. more difficult. Furthermore, because of manufacturing process outsourcing, it becomes more difficult for engineers to acquire and learn information concerning production engineering.   In order to solve the problem of knowledge transfer, authors have successfully proposed engineering knowledge support systems (Hino et al., 1993, Aoyagi et al., 1997, Tsuchiya et al., 1998, Aoyagi et al., 1998, Sakamoto et al., 1999, Takeo et al., 2001).

In the second chapter, a workflow model, which combines supply chain and knowledge data in manufacturing enterprise management, is defined.   In the model, knowledge data is defined as a kind of enterprise management information.   In the third chapter, we clarify that knowledge data has a layered structure, classified into Case, Manual and Formula layers in the Knowledge Base.   A newly developed Knowledge Driver acts as a design processor for Knowledge Base and performs agent-like functions between the engineers and the Knowledge Base.

2. WORKFLOW MODEL

A workflow model on manufacturing enterprise activities is defined in this chapter.   As shown in Fig.1, four kinds of product models are assigned, corresponding to each process of the supply chain from sales to maintenance. The concept model is the idea for the design engineer, the electronic model represents CAD data, the physical model is itself a product and the resource model is the parts or materials of the product. Design, manufacturing, maintenance and recycling production processors, where the product models are converted from the concept model to the resource model, are also defined.

The concept model is converted to an electronic model through the design processor. The electronic model is then converted to a physical model through the manufacturing processor. The physical model is also maintained through the maintenance processor for operating continuously. Finally, the physical model is converted to a resource model through the recycling processor, which finishes operating and is either reused or recycled. Thus, in this conversion process, the knowledge data to which the four production processors refer is defined as enterprise management information.

As described above, the workflow model on manufacturing enterprise activities can be defined as a series of product model conversion processes. In short, the product model is defined as input and output, the production processor is used as a model converter and the enterprise management information is referred to as knowledge data for the conversion with these processes.

Since our division assumes responsibility for the structural design of telecommunication equipment, this paper discusses the structural design processor and related knowledge data (hatched portions in Fig.1).

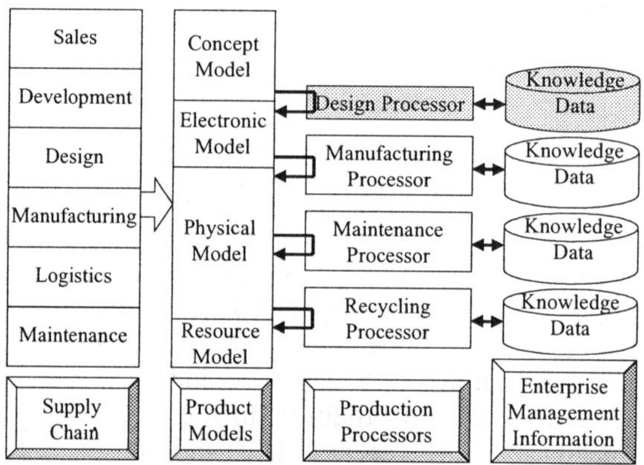

Fig.1 Workflow model of manufacturing enterprise activities

3. SYSTEM DEVELOPMENT
3.1 Classification of Explicit Knowledge

Knowledge can be divided into tacit knowledge that is not documented and explicit knowledge that is documented (Nonaka & Takeuchi, 1995). Knowledge data is defined as a subset of tacit knowledge. The authors investigated and classified knowledge data into Reports, Standards and Algorithms. Figure.2 illustrates the results. Since the reports are the knowledge data in its most basic documented forms, they occupy the largest ratio. The standards are considered as a rearrangement and integration of the reports. Further, the algorithms can be considered as the essence and logical formulation of the standards.

3.2 Layered Knowledge and Knowledge Base

From the above considerations, the three forms of the knowledge data classify not only every descriptive form, but also every combination level. Hence, it is appropriate that the three forms of the knowledge data have layered structures according to their combination levels instead of their descriptive forms. Now that we know this property, let us formally call the three forms of the knowledge data Cases, Manuals, and Formulas instead of Reports, Standards and Algorithms, respectively. Knowledge Base (KB) is defined as a database layered in the order of Cases, Manuals and Formulas from the tacit knowledge side as shown in Fig.3.

Cases contain what we call reports on design, quality, experiment and investigation. These are also the most basic explicit knowledge related to the tacit knowledge. Since the assistance of engineers who wrote the reports is regularly needed for referencing, a case-based interpretation is useful.

Manuals indicate standards, and are explicit knowledge where combination processing, such as merging or overlapping elimination, was applied to Cases. Thus, browsing convenience is an important issue.

Formulas are the most refined explicit knowledge where combination processing, such as logical formulation, was applied to Manuals. Since we have several algorithms based on structural parts geometry, integration through geometric processing on 3D-CAD is desired.

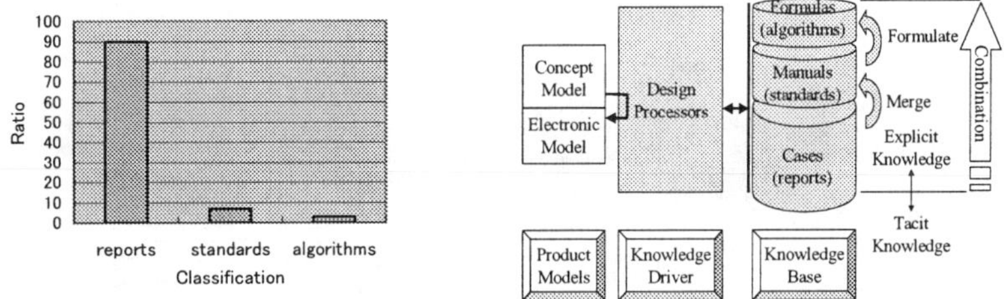

Fig.2 Results of knowledge data ratio   Fig.3 Three-layered Knowledge Base

3.3 Knowledge Driver

Since each layer of KB differs in combination levels and descriptive forms, the optimum design processor is considered essential for every respective layer in order to use KB effectively. As shown in Fig.3, let us call such a design processor Knowledge Driver (KD), where we will discuss the optimum KD for each layer.

Since we have several algorithms based on structural parts geometry for the formula, a geometric processing engine should be installed on 3D-CAD by using an Application Programming Interface. Cost estimating procedures and manufacturing rules for structural parts are installed on 3D-CAD. This engine automatically examines the features of the parts geometry on 3D-CAD, calculates the machining time and material cost, and checks its manufacturability.

Normally, since the referential relationship among manuals has a complicated structure, such as N generations or loops, manual browsing is troublesome for engineers. Hence, for Manuals, the manual browsing should be simplified. An environment is developed whereby Manuals are obtained as databases in a HTML format over the intranet and KB can be browsed via an information retrieval engine. This engine activates simultaneously with the geometric processing engine described above. For Case, attention was paid to improving thermal and electromagnetic compatibility designs that play important roles in structural design. Since both design processes are performed by using past design cases and past experiments, a case reasoning engine is installed for selecting the optimum design parameter from past design cases. ID3 was installed for the thermal design, and Design of Experiment was applied for the electromagnetic compatibility design.

3.4 System Configuration

Figure.4 shows system configuration. This system is composed of KB, KD and Interface. As described before, KB is composed of three-layered knowledge data. KD also has three engines corresponding to each layer of KB. Each engine has an interface between engineers and KD, the 3D-CAD, Browser and Application. Therefore, engineers can be supported from KB by using KD through Interface.

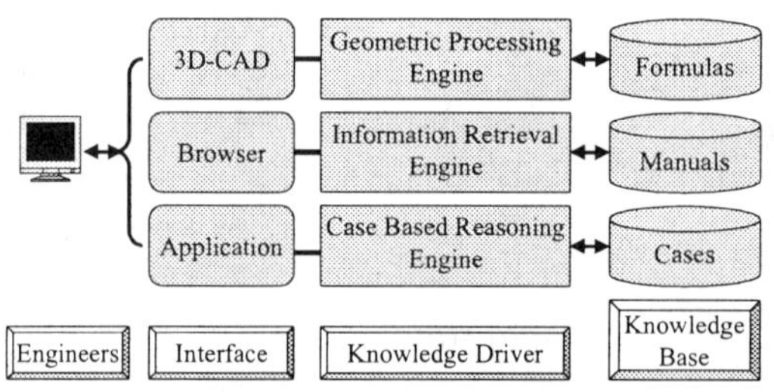

Fig.4 System configuration

4. EXAMPLES

Figure.5 shows an example of Manuals browsing through the information retrieval engine. Figure.6 shows an example of sheet metal design application. In this instance, if an unskilled designer is working against the production engineering rules described in Formula, KD is activated onto 3D-CAD, geometric features of the design along information in KB are reviewed, design problems are detected and then the geometry is indicated. Manuals are also activated through Browser that shows the designer how to correct them.

Fig.5 Example of Manuals retrieval

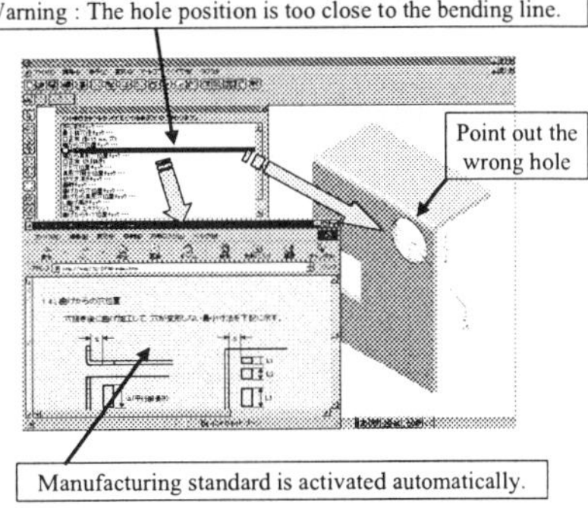

Fig.6 Example of sheet metal design

5. CONCLUSIONS

Engineering knowledge transactions among variously experienced engineers, including skills, expertise and sophisticated engineering technologies have been performed too rarely.   The overall achievements in this study can be summarized as follows.

(1) A workflow model in manufacturing enterprise activities is proposed as a series of product model conversion processes which refer to knowledge data.   With this model, the role of knowledge data in manufacturing enterprise is clearly defined.

(2) We discovered from this investigation that knowledge data is classified into three categories according to its descriptive form.   Since each form is equivalent to the knowledge combination level, a triple-layered Knowledge Base can be defined.

(3) By assigning a Knowledge Driver to each layer of the Knowledge Base, the newly developed knowledge support system was integrated into each design work.

References

Aoyagi, T., Hino, Y. & Takeo, S. (1997). Development of cost simulation system for sheet metal parts assembling. Proceedings of JSPE Spring Conference, 905.

Aoyagi, T., Hino, Y., Tsuchiya, S., Kitajima, Y. & Nakamura, H. (1998). Development of DFM (Design For Market/Manufacturability) 2nd Report –Development of design support system by 3D-CAD–. Proceedings of JSPE Spring Conference, 55.

Hino, Y., Nakamura, H., Tsuchiya, S. & Takeo, S. (1993). Development of cost simulation system for sheet metal parts. Proceedings of JSPE Fall Conference, 19.

Nonaka, I. & Takeuchi H. (1995). The Knowledge-Creating Company. Oxford University Press, Inc.

Sakamoto, N., Nakamura, H., Aoyagi, T., Jin, D. & Teshima, S. (1999). Development of DFM (Design For Market/Manufacturability) 3rd Report –Development of thermal design support system by ID3–. Proceedings of JSPE Spring Conference, 279.

Tsuchiya, S., Otsuka, T., Tokinaga, T., Asari, H., Kozai, H., Tanabe, Y. & Takeo, S. (1998). Development of DFM (Design For Market/Manufacturability) 1st Report –Sharing and transference of design and production engineering for manufacturability–. Proceedings of JSPE Spring Conference, 54.

Takeo, S., Tsuchiya, S., Nakamura, H., Hino, Y., Aoyagi, T. & Miyoshi, T. (2001). Development of Knowledge Driven Digital Design System for Telecommunication Equipment. Journal of JSPE. 3, 515.

LIFE CYCLE SUPPORT OF MECHANICAL PRODUCTS USING NETWORK AGENT

M. Suzuki, K. Sakaguchi and H. Hiraoka,

Dept. of Precision Mechanics, Chuo University

Abstract

A framework for the system of network agents that correspond to every parts of a mechanical product is proposed to support its life cycle activities. Each Part agent accompanies a real part through its life cycle via network. Part agent that accumulates the maintenance information on the part could know its residual life and take appropriate maintenance actions. To investigate the effectiveness of the scheme, simulation is carried out for a prototype system on workstations in a network. The prototype has a function to simulate the deterioration of the product and to represent assemblies. Initial results of the simulation are shown.

Keywords

Network agents, life cycle support, deterioration analysis, maintenance, assembly.

1. INTRODUCTION

To realize sustainable development, products are required to be under appropriate control throughout their life cycle. Parts should be monitored if some trouble is happening and should be disassembled when the failure is detected or anticipated, and should be repaired or recycled. For that purpose, every part would better have intelligence and autonomy with which it detects the effects from the environment, evaluates its damage and decides its own action.

As a scheme to realize this property, we propose "Part agent" system [Sakaguchi 99, 00]. Nowadays products are always connected to computers throughout their life, such as, CAD when it is designed, CAM when it is actually produced and control system when it is used. And as these computers are connected via network, we can expect in the near future, that we assign a network agent to a particular part and program it to follow their real counter-part wherever it goes. We call this agent a "Part agent".

To realize this scheme, in addition to the network development, it is required, firstly, small intelligent chips that can be attached to parts and can communicate with the network, and secondly, small sensors that can detect detailed real world properties. Lastly, network agent system framework should be necessary in which Part agents behave properly. This paper describes our proposal of Part agent system framework with an assumption that intelligent chips and sensors will also be realized in near future. Some results are shown on the prototype software we have developed and on its simulation without including real parts.

2. NETWORK AGENTS FOR INDIVIDUAL PART

In our scheme, each part has a Part agent that follows it through the network. Part agents gather the information to support the activity of the part by using appropriate application, e.g., activating sensors at the site and asking remote database for necessary information. Two kinds of application interface are provided for Part agents as shown in Figure 1. One is to activate the local applications at a site and to communicate with them. The other interface is with lower level communication agents that support the

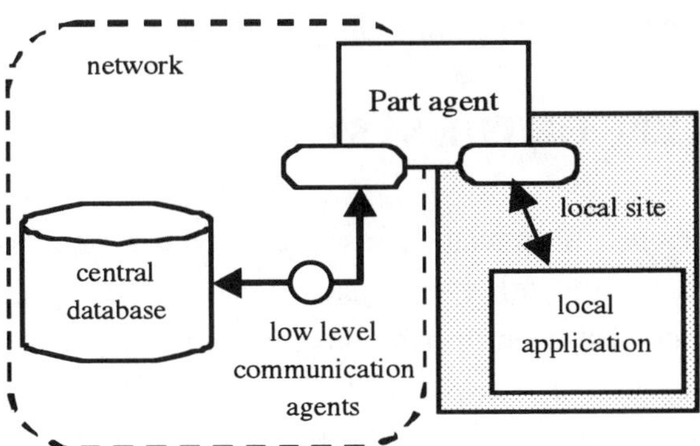

Figure 1. Configuration of the network agents for the part

communication with databases in the network, as we consider more practical that all the information about the product is stored in databases in the network. With this mechanism Part agents can be programmed to decide the action appropriate for the part based on the acquired information.

3. LIFE CYCLE SUPPORT USING NETWORK AGENTS

A part will change its position according to its life cycle stages. We assume several life cycle stages and the accessibility from each part to the network in each life cycle stage. To support activities of parts, Part agents would have the following functions.

(1) Moving through network: Part agent moves to the site corresponding to the destination of actual part based on the information of its transfer.

(2) Detecting the wear of parts: Using available local sensors, Part agent monitors the conditions of part to detect its deterioration. Part agent will keep in the database historical record of part about its state, the applied stress and its environmental conditions. Part agent can use the record if necessary, to analyze and predict possible deterioration.

(3) Representing assembly: Most products are assemblies. Maintenance requires replacing only faulty parts in an assembly to a new one. Healthy parts extracted from damaged assembly would be re-used for other assemblies. To represent assemblies, a Part agent that corresponds to an assembly should have the information on the Part agents that correspond to its components. 'Assembly' Part agents should have functions to manage the behavior of whole assembly including their 'component' Part agents.

4. PROTOTYPE OF THE SYSTEM

We developed a simulation-based prototype system to evaluate the Part agent scheme. Network agent system Bee-gent [Bee-gent] is used as a base system. Bee-gent agent can be programmed to move via network and call local application as necessary. For lower level communication with central database we use agents of CORBA-based system Voyager [Voyager]. To represent life cycle stages, the system is installed in four workstations in a network, representing four local sites, i.e., 'production', 'assembly', 'use', and 'maintenance'. Part agents are generated at 'production' site, assembled in products at 'assembly' site, as shown in Figure 2. At 'use' site, random stress data is added to the state data of Part agent. With the stress data, Part agent calls deterioration analysis application and

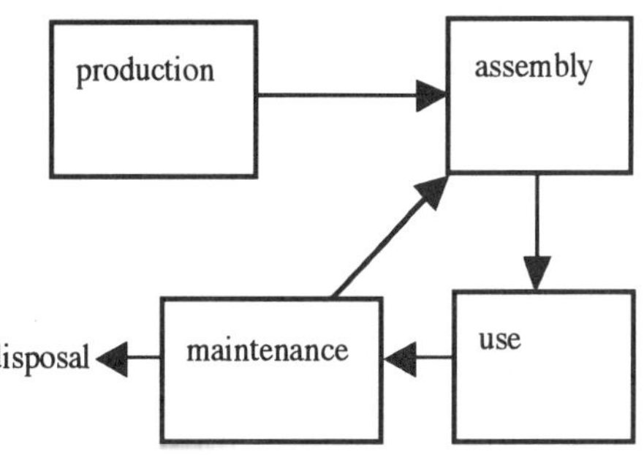

Figure 2. Stages for Part agents in the Experimental System

checks if the corresponding part is worn. Deterioration of these parts is analyzed in function of time and load. When one or more Part agents conforming a 'product' have the results that exceeds wear level, all the Part

Figure 3. Display of the Experimental Prototype

915

agents of the 'product' go to 'maintenance' site and are disassembled. Part agents that have worn data are discarded and those with healthy data go back to 'assembly' site for re-using. As an example we implemented three types of Part agent corresponding to a belt, a motor and a head cartridge carrier of a printer. Display of a UNIX workstation is shown in Figure 3 where preliminary simulation is carried out with 10 'product's. Each window shows the activity of agents in one of four life cycle stages.

Each Part agent acts based on the stress it endures. This results in more effective use of parts than time-based maintenance. The effectiveness of the system is more apparent in an additional simulation where a Part agent selects based on its load history one of multiple 'use' stages with different load. These early results of the simulation show a good performance as expected.

5. SUMMARY

In this paper we propose a framework of life cycle support system for mechanical parts using network agents representing each part. The simulation-based prototype system called Part agent system is implemented with good results, although more research is necessary for details and implementation.

To realize this schema, research is required on the communication between Part agent and actual part. Small chips are also required that includes high performance small computer with large memory, wireless network device, and high performance sensors.

REFERENCES

[Beegent] Bee-gent homepage. Toshiba Co., http://www2.toshiba.co.jp/beegent/index_j.htm
[Sakaguchi99] Sakaguchi, K., Hiraoka, H., et al. Re-use of Parts Using Agents. Proceedings. of EcoDesign'99 Japan Symposium; 1991 December 10-11; Tokyo: 1999 (in Japanese).
[Sakaguchi00] Sakaguchi, K., Hiraoka, H., et al. Using Network Agents for the Maintenance of Mechanical Parts through their Life Cycle. Proceedings of EcoDesign2000 Japan Symposium, 2000 December 13-15; Tokyo: 2000 (in Japanese).
[Voyager] Voyager homepage. Tomen Information Systems Co., http://www.tomen.co.jp/tisco/Voyager/

AESTHETIC DESIGN
BASED ON "KANSEI LANGUAGE"

Manabu Ota and Hideki Aoyama
Department of System Design Engineering, Keio University

Abstract

The appearance of the industrial products serves as important factor in determining the value of products. However, CAD systems are not effective in the aesthetic design due to insufficient functions for expressing the designer's aesthetic ideas. The objective of this study is to develop the functions for aesthetic modeling based on designer's Kansei. From the idea that spoken language serves as a means of expressing human Kansei, in this study, attempts are made to develop a system to transform spoken language into aesthetic models.

Keywords

CAD, Modeling, Aesthetic design, Kansei language, Kansei

1. INTRODUCTION

The appearance of industrial products now serves as important factor for determining the value of the products. Free-form surfaces are therefore being used extensively for complex forms to differentiate products. It is difficult for designers to design industrial products using free-form surfaces on the CAD system, thus requiring them to cooperate with CAD operators in their design efforts. However, in this case, CAD operators often find it difficult to understand the Kansei of the designers because evidently the Kansei of each person differs. Since form design is a creative effort of the designer, which depends on the creativity and aesthetic taste of the designer to a great extent, it is difficult to support the Kansei of the designer on a computer. For this reason, the form-modeling environment of present design support systems using computers is often considered considerably different from the designer's sense[1].

Figure 1 shows a computer-supported aesthetic design processes proposed by the authors. The processes are carried out through two steps. In the first step, the ideas of the designer are input to the computer so that the

3-D model is constructed in the computer. In the second step: form modification process, insufficient forms are modified to precise forms. The aim of this study is to develop the form-modification-process using Kansei language. This system enables even inexperienced designers and users of present CAD systems to construct the desired aesthetic model easily using Kansei language.

2. KANSEI LANGUAGE

With increasing demands for high quality products, Kansei engineering is becoming increasingly important in aesthetic design. Kansei engineering is applied for identification between the Kansei of a person and physical form. Kansei language is considered as one of the means representing human sense: Kansei.

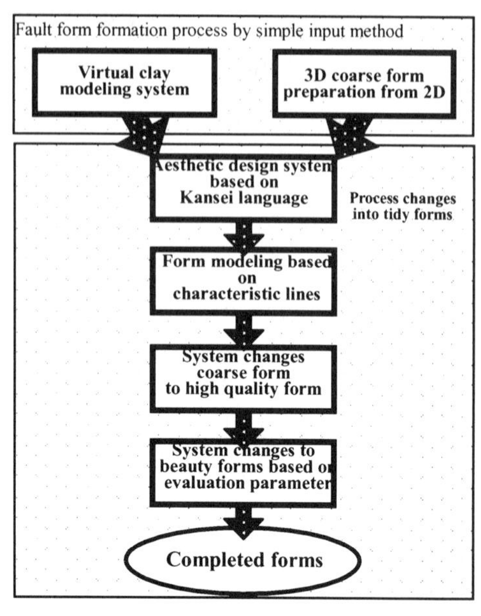

Figure 1 Aesthetic design processes

Table 1 Kansei language

Form	sharp, round, float, speed
Degree	terrible, a little, more, less

Kansei language registered in the system has two types: form and degree. Table 1 shows examples of the Kansei language for expressing form and degree. The form can be automatically modified by operation using the Kansei language.

3. OUTLINE OF SYSTEM
3.1 Method of Modifying Whole Form

In a previous study, Norihiko Mori extracted design elements forming the exterior design of cars from collected 77 kinds of cars, and the dimensions of each component of the exterior design of each car were measured[2]. The components were classified into eight independent design elements by the principal component analysis. Mori analyzed the relationships between the design elements and the Kansei of consumers. Figure 2 shows an example "degree of trapezoid" of eight form elements. Mori proposed design elements as parameters for modification of product design. A form can be modified based on the Kansei of the designer by controlling the degree of change of element parameters. In this system, this method is applied for modification of the whole form.

3. 2 Method of Modifying Local Form

(a) Modeling method

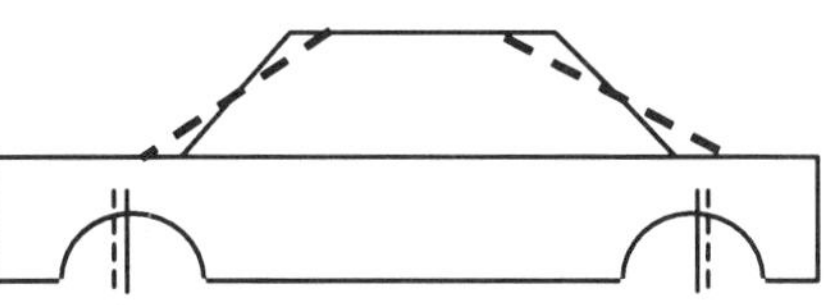

Figure 2 Degree of trapezoid

The basic model surfaces are expressed using Bezier surfaces. The flexibility for constructing surfaces connecting other surfaces with curvature continuity is limited because the arrangements of the control points of the Bezier surfaces to obtain curvature continuity are restricted. Consequently, as shown in Figure 3, new surfaces defined using control elements[3] between Bezier surfaces were constructed. The new surface defined using control elements can automatically produce curvature continuity between Bezier surfaces.

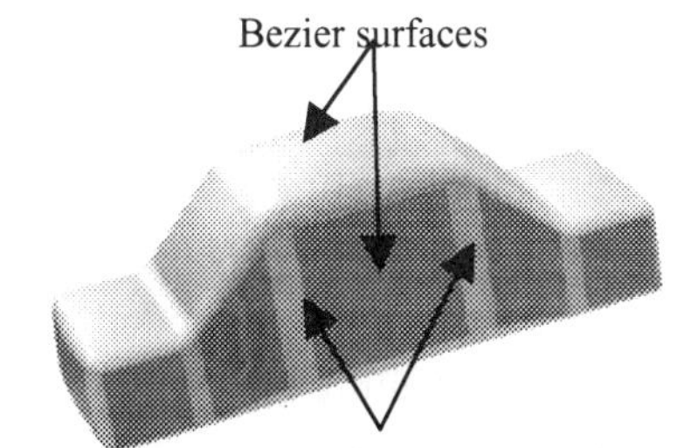

Figure3 Surfaces consist of model

As shown in Figure 4, when two Bezier curves pass through points Q_{i-3}, Q_i and Q_{i+3}, the curves are connected to satisfy the condition of curvature continuity at point Q_i.

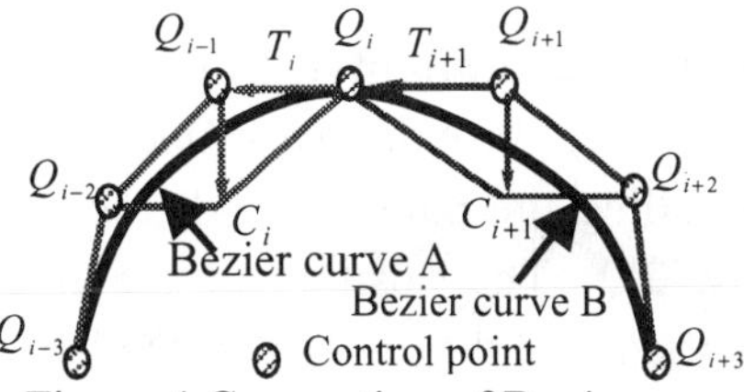

Figure 4 Connection of Bezier curve

The second differential vectors C_i and C_{i+1} of each curve at point Q_i must satisfy the condition of $C_i = C_{i+1}$ and the first differential vectors T_i and T_{i+1} of each curve at point Q_i must satisfy the condition of $T_i = T_{i+1}$ for connection with curvature continuity at point Q_i. The control elements satisfy these conditions so that the Bezier curves defined using the control elements have curvature continuity at the connecting points.

The Bezier surfaces are connected with the surfaces defined by the control elements for the flexible modification of the local form according to the following procedures:

(1) As shown in Figure 5, the connecting Bezier surfaces are divided into two surfaces, respectively. The divided lines are defied as "Div. 1".

(2) As shown in Figure 6, the divided surfaces are re-divided. The divided lines are defined as "Div. 2".

(3) The control points defining the subdivided surfaces are determined as the control elements of the surface between the subdivided surfaces, allowing the surface form to be constructed with high flexibility by controlling the division of the surface.

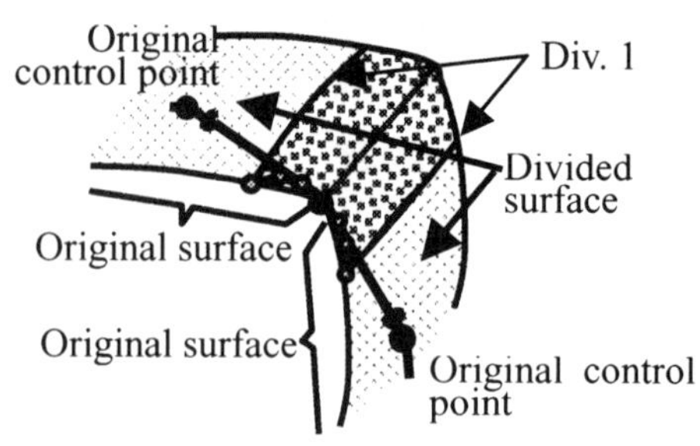

Figure 5 Division of original surface

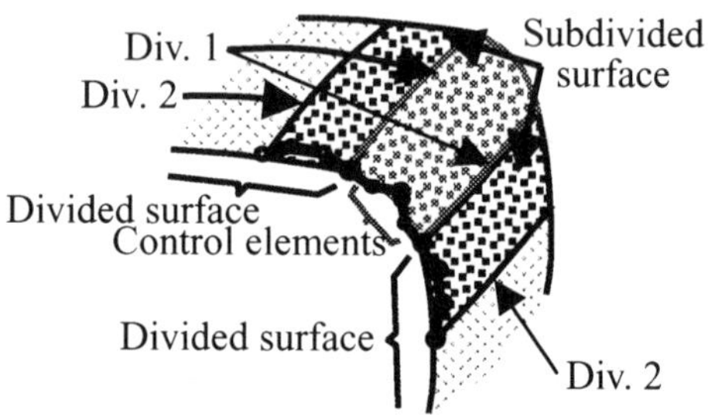

Figure 6 Control elements using subdivided Bezier surface

(b) Method of modifying local form

The local form is easily modified according to the Kansei of designers by controlling the parameters determining positions of the lines "Div. 1" and "Div. 2". The position of the line "Div. 1" determines the area to be modified, and the position of the line "Div. 2" decides the local form.

3.3 Learning System

The Kansei elements based on Mori are related to design elements through questionnaire for a lot of people so that the Kansei means the average sense. Therefore, since individual Kansei of designers are a little bit different from each other, a system based on Kansei should be fit for the designer sense using the system.

The system developed can recognize the individual sense of the designer using the system through the repeating utilization of the system. The designer controls the degree of modification to obtain the desired form by inputting Kansei language into the system by the repeating use.

4. EVALUATION OF SYSTEM

A computer system based on the above method was developed, and the effectiveness of the function was evaluated through basic experiments. Figure 7 shows the result of the experiment. "Speed" was input as "Kansei language which expresses form" for the original model shown in Figure 7(a). Figure 7(b) shows the results, which confirms that forms can be modified as a whole using Kansei language. "Round" was also input as "Kansei language for expressing form", while "little" was input as "Kansei language for expressing degree" for the model shown in Figure 7(b). Figure 7(c) shows the results, which confirm that forms can be partially modified using Kansei language.

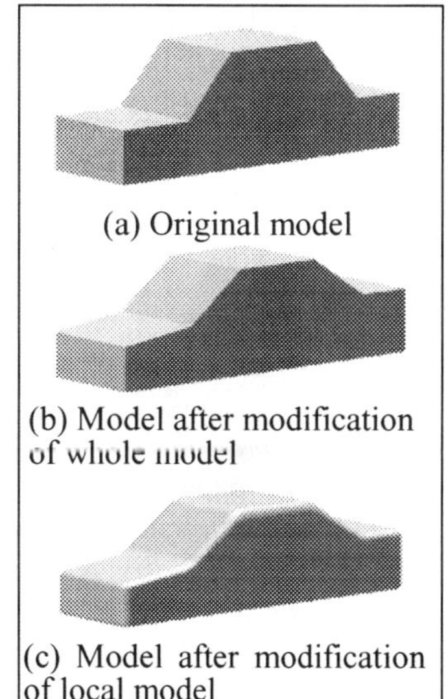

Figure 7. Results of experiments

This system enables designers to build the desired form easily and quickly.

5. CONCLUSIONS

This study can be summarized as follows:

- A method to modify the form of 3D models wholly and partially was proposed.
- A system to modify the form of a basic model as a whole or partially using Kansei language was developed.
- The learning function was developed to the system.
- The validity tests were carried out for the developed system.

References

[1] Mdes M T. Sketch mapping: Concept representation for stylists. Proceedings of International Symposium on Automotive Technology & Automation, 1998, 341-348.

[2] Norihiko Mori: *Engineering for design, Design planning of soft system*, Asakura Books, 1991, 137-142. (In Japanese).

[3] H.Aoyama, I.Inasaki, T.Kishinami, K.Yamazaki: A New Method for Constructing a Software Model of Sculptured Surfaces with C^2 Continuity from a Physical Model, JSPE, 1994, 7.

A DESIGN REVIEW AND REDESIGN SYSTEM FOR NURBS MODELS WITH A HAPTIC DEVICE

Hidetomo TAKAHASHI[*1]

[*1] Department of Mechanical Sciences and Engineering, Tokyo Institute of Technology
2-12-1, Ookayama, Meguro-ku, Tokyo 152-8552, JAPAN
TEL: +81-3-5734-2166,   FAX: +81-3-5734-2893
taka@mech.titech.ac.jp

Abstract

In this paper, a design review and redesign system of NURBS models with a haptic device is described. A haptic device permits operators to directly and intuitively review and redesign the models. First the overview of the system is described, then the direct haptic rendering algorithm for NURBS models with the homogeneous geometric Newton-Raphson method is described. Finally, the system is experimentally estimated for the accuracy of the direct haptic rendering for complex NURBS models, and for the effect of design review process.

Keywords

Design review, Haptic device, NURBS surface, Direct haptic rendering

1.INTRODUCTION

For CAD/CAM applications, it is required that systems are able to directly treat freeform surface models not to reduce the quality of the models such as continuity of the surface. And NURBS surface is the most popular expression in CAD/CAM systems. Some direct haptic rendering systems for NURBS have been proposed [1,2]. These systems approximately calculate the closest points, whose algorithm estimates a new surface parameter by a projection to the pre-calculated tangent plane. If the probe is moved rapidly or it is moved on the part where the curvature of the surface is large, this method is a failure. Fast and stable methods are required to calculate the closest point independent of the probe speed and the shape of the model.

The goal of this research is to realize stable haptic rendering of NURBS model of sufficient size for practical use, and to show the effect of the haptic device for design review.

2.DESIGN REVIEW SYSTEM WITH A HAPTIC DEVICE.

Figure 1 shows the hardware and software configurations of the system[3].

The haptic device control PC checks the collision between the tool and the model. It calculates the reaction from the model to the tool, then it controls the haptic display by force feedback. The model management PC displays the image of the freeform surface model. Each PC is connected to the Internet. The system is able to receive CAD/CAM data as IGES[4] files. IGES is the most popular format of 3D CAD/CAM systems. The functions to treat the elements related to freeform surfaces and trimmed surfaces for the IGES format, are implemented in this system. The haptic device has a 3DOF orthogonal mechanism consisting of three linear servo motors[5].

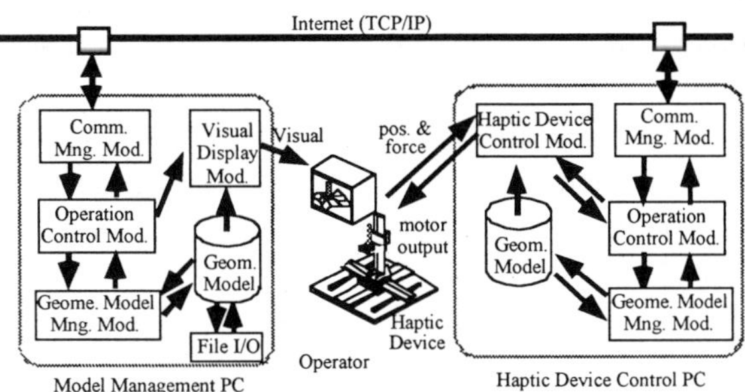

Figure 1   Design review system with a haptic device.

3 .DIRECT HAPTIC RENDERING ALGORITHM FOR NURBS MODELS.

In order to generate stable haptic sensation for the operator, the control frequency of the system requires more than 1kHz[6]. So the stable and rapid direct haptic rendering algorithm is needed to realize such a system. The direct haptic rendering algorithm consists of two stages: rough checking and the homogeneous geometric Newton-Raphson method.

3 . 1 . Rough Checking Algorithm

First using the bounding box, which covers all control points of the surface, the possibility of the collision between the tool and the surface is checked. Then the possibilities for each control mesh on a surface are checked by using local bounding box. One polyhedron on the mesh is selected and checked by using God-Object method[7]. The polyhedron, which consists of four control points, is divided by two triangles. One triangle is selected, and the God-Object is calculated for it. If the God-object is on that triangle, the surface parameter is estimated by linear interpolation of each vertex. If it is not on the triangle, one of the adjacent triangles is selected according to the area coordinates calculated in the God-object method. This process is repeated until the triangle with the God-object is found, or all neighborhoods of the triangle have been searched. Each edge has the pointer of the right edge connected to the end vertex and the left edge connected to the start edge the same as the radial edge structure, and with the two polygons contacted with the edge, the system is able to easily select the triangles.

3.2. Homogeneous Geometric Newton-Raphson Method.

In order to make the virtual probe to trace the surface of the geometric model, the closest point $S(u,v)$ that gives the minimum distance between the tool point and the surface, are required. The point is given as follows;

$$F(u,v,t) = S(u,v) - \{P + t\,n(u,v)\} = 0 \tag{1}$$

Where, u,v are parameters of the surface, $n(u,v)$ means a unit normal vector, t means the depth of the collision, and P means the tool point. $S(u,v)$ means the NURBS surface represented by Equation (2), $N_{i,k}$ means B-spline basis function of order k-th, w_{ij} means weight of control points and Q_{ij} means the position vector of a control point.

$$S(u,v) = \sum_{i=0}^{n}\sum_{j=0}^{m} N_{i,k}(u)\,N_{j,l}(v)w_{ij}Q_{ij} \left/ \sum_{i=0}^{n}\sum_{j=0}^{m} N_{i,k}(u)\,N_{j,l}(v)w_{ij} \right. \tag{2}$$

Equation (2) is rational form, so there are numerical instability. In order to stably calculate it, the homogeneous geometric Newton-Raphson method is proposed[8], and direct haptic rendering system for one surface was proposed. By using homogeneous coordinates, the equation is expressed as follows:

$$^{h}F(u,v,t,w) = {}^{h}S(u,v) - w\left(\begin{bmatrix} P \\ 1 \end{bmatrix} + t\begin{bmatrix} n \\ 0 \end{bmatrix}\right) = 0 \tag{3}$$

Where the script super on the right side means that the vector is represented in homogeneous coordinate.

$$^{h}S(u,v) = \begin{bmatrix} \sum_{i=0}^{n}\sum_{j=0}^{m} N_{i,k}(u)\,N_{j,l}(v)w_{ij}Q_{ij} \\ \sum_{i=0}^{n}\sum_{j=0}^{m} N_{i,k}(u)\,N_{j,l}(v)w_{ij} \end{bmatrix} \tag{4}$$

The Newton-Raphson method is calculated as follows:

Step-1. If $\left|{}^{h}F(u,v,t,w)\right| \le \varepsilon,$ (5)

then the surface parameters (u,v) satisfy the Equation (3) with the allowance ε. Go to end.

Step-2. The correction terms for the Newton-Raphson method, $\Delta u, \Delta v, \Delta t, \Delta w$ are calculated using Equation (6).

$$\begin{bmatrix} \Delta u & \Delta v & \Delta t & \Delta w \end{bmatrix}^{t} = -\begin{bmatrix} \dfrac{\partial^{h}F}{\partial u} & \dfrac{\partial^{h}F}{\partial v} & \dfrac{\partial^{h}F}{\partial t} & \dfrac{\partial^{h}F}{\partial w} \end{bmatrix}^{-1} {}^{h}F \qquad (6)$$

Step-3. Update the parameters as Equation (7).
$$u \leftarrow u + \Delta u, \ \ v \leftarrow v + \Delta v, \ \ t \leftarrow t + \Delta t, \ \ w \leftarrow w + \Delta w \qquad (7)$$

Step-4. Repeat Equations (5)-(7).

4 .EXPERIMENTAL RESULTS OF THE SYSTEM
4 . 1 . Accuracy of the NURBS Surface Model Tracing

First, how accurately the developed system can render the complex NURBS model, is checked. The example of the model is shown in Figure 2. The bulge on the center of the model is smoothly connected by minute surfaces to the surfaces on both sides and the model consists of ten surfaces. The haptic device control PC is Pentium ODP180MHz, and the control period is 1ms. The average height error of the tracing in one minute was 537μm.

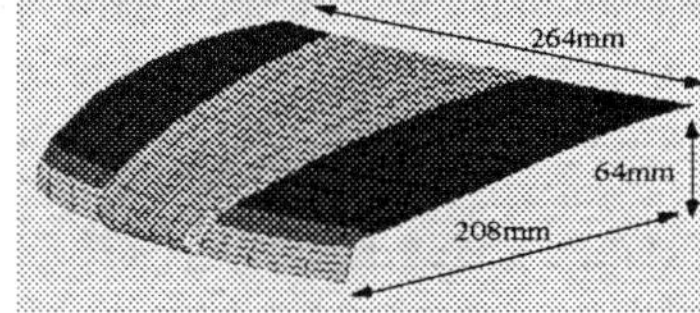

Figure 2　Model 1.

4 . 2 . Perception of the Defect in the Model.

In the design review, it is important to check the quality of the surface. The test model shown in Figure 3 is created by two surfaces connected in G1 continuity by lifting one surface up in a vertical direction. The heights of the gap Δh are set as 0,1,20,20,40, 60,80,100μm. After tracing the model with the defect; Δh=100μm which is perceived without fail, a subject tries to find the defect of the model, and each model is checked totally three times for one subject. There are five subjects, male, twenties. The result is shown in Figure 4. The gap is not perceived until the height is less than 20μm. If the gap is larger than 50μm, then the gap is almost perceivable. The motion of the haptic device is changed by the control parameters.

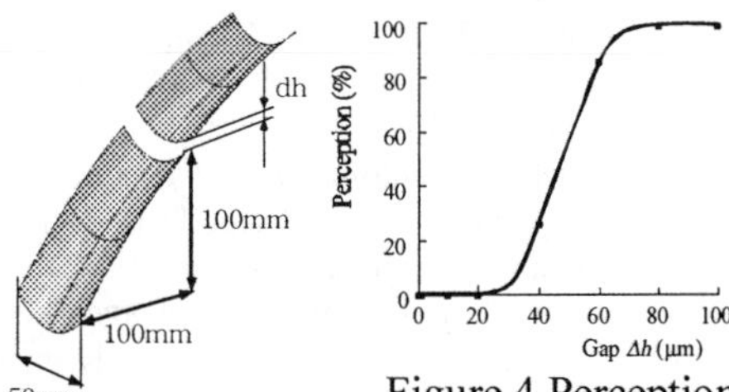

Figure 3 model 2　Figure 4 Perception of defect.

4 . 3 . Design Review

In this section, a visual design review and design review with this system are compared. Figure 2 is used. In visual review, by using a commercial CAD system (Rhinoceos, Robert McNeel & Associates), the operator can change the scale, position and orientation of the model by a keyboard.

By visual design review, most subjects reported that they could recognize the part where the outline clearly appeared (shown in Figure 5, part C). And they could not understand where the minute surface existed (part A to C). On the other hand, by using the haptic display, they could recognize that part A was smoothly connected by the minute surface. But they could not perceive that part B was smoothly connected.

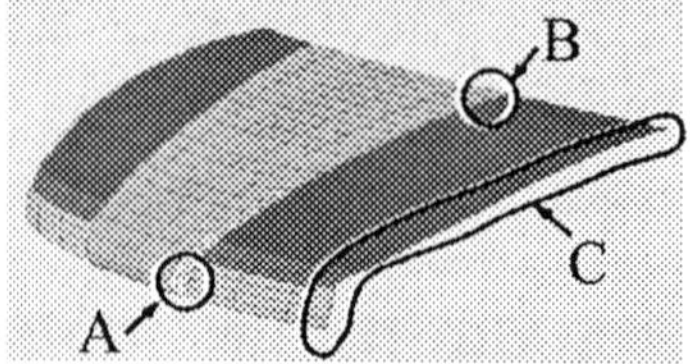

Figure 5   Design review.

5 .CONCLUSIONS

This paper described the design review system that enabled the operator to examine the model with practical size of NURBS surfaces. It was experimentally shown that the proposed system was able to display the minute gap that was not visually perceived by the operators.

ACKNOWLEDGEMENTS

The author would like to thank Prof. Hiroaki Funabashi of this institute for his precious advices, and Mr. Akio Suzuki of Digicrafter for his offer of a CAD model. Mr. Toshiyuki Ozawa, an undergraduate student of this institute cooperated in the study. This research is supported by Grants in Aid for Science and Research (11750199) from Japan Society for the Promotion of Science.

References

[1] Thompson II, T. V., Johnson, D.E., and Cohen, E.: Direct Haptic Rendering Of Sculptured Models, Proc. of Symposium on Interactive 3D Graphics, Providence,TI, pp.167, 176(1997)

[2] Thompson II, T. V. and Cohen, E.: DIRECT HAPTIC RENDERING OF COMPLEX TRIMMED NURBS MODELS, Proc. of ASME, Dynamic Systems and Control Division, pp.109, 116 (1999)

[3] Kanai S., Takahashi, H. and Kishinami T.: Networked haptic interfaces for distant redesign and review of free-form surfaces, Proc. of DETC'97 1997 ASME Design Engineering Technical Conferences, DETC97/DFM4366 (1997)

[4] US PRO, Initial Graphics Exchange Specification (IGES) 5.3

[5] Takahashi H. and Kanai S.: A study on virtual reality of NC programming for free-form surfaces, Int. J. Japan Soc. Precision Eng., 30, 3, pp.177, 182(1996)

[6] Stewart P., Buttolo P., Chen Y.: CAD DATA REPRESENTATIONS FOR HAPTIC VIRTUAL PROTOTYPING, Proc. of ASME DETC'97, DETC97/CIE4307

[7] Zilles C. B. and Salisbury J. K.: A constraint-based god-object method for haptic displays, Int. conf. on Intelligent robotics and systems, pp.146, 151(1995)

[8] Yamada A., Matsumura K. and Yamaguchi F.: Interference checking of Rational Curves and Surfaces Using a Homogeneous Geometric Newton Method, Journal of Japan Society for Precision Engineering, 61, 2, pp. 203,207(1995) (in Japanese)

Free-form Surface Manufacture for TRR-XY Hybrid Parallel Link Machine Tool

Tsann-Huei Chang[1], Shang-Liang Chen[2], Chao-An Kang[3], Kuan-Wen Chen[4]

[1,3,4] Taichung Technology Service Center, Mechanical Industry Research Laboratories, ITRI World Trade Center Taichung 2F,60 ,Tienpao Street ,Taichung 407,Taiwan

[2] Institute of manufacture engineering, National Cheng-Kung University, Tainan, Taiwan

Abstract

TRR-XY hybrid parallel link machine tool comprise of a 3 DOF parallel link mechanism and a conventional X-Y table. This work presents a surface interpolation theory for this machine tool. The CAD surface model and the machining parameters are also applied to this surface interpolation theory. After related calculations, the relative strokes of each driving axis at every sampling time are generated. Moreover, NURBS surfaces are adopted to examine the surface interpolation theory.

Keywords: Parallel machine, surface interpolation.

Introduction

Stewart (1965) developed the parallel link mechanism of Stewart-Platform, in which the parallel link mechanism is characterized by desired features such as a high stiffness, high payload, and absence of cumulative error. Owing to this trend, TRR-XY hybrid parallel link machine tool was produced, with the advantage of the parallel link mechanism, in which XY table increases workspaces. Figure 1 illustrates the machine structure. In contrast to the conventional machine tool, directions of the driving axes of this machine tool are not orthogonal and the motions of the driving axes are coupled to each other, implying that it must depend on the cutter tool's tracking path to calculate the path interpolation.

Owing to the demand for better appearance and quality, free-form surface applications has become increasingly important. In related machining plans, machines must allow the path of cutter machining to meet the demand path more precision. To solve related problems, TRR-XY hybrid parallel link machine tool must utilize more precise machining path interpolation to minimize the path tolerance. This study applies the surface interpolation theory to construct TRR-XY hybrid parallel link machine tool's surface machining interpolation calculation structure. More precision path interpolation can hopefully be used to minimize path error while machining to enhance the quality of machining free-form surface.

Surface Interpolation

In the conventional post-process, CAM system initially schedules the Cutter-contact(CC) paths on the parametric surface. Then, according to the defined tolerance and other relative parameters to calculate the Cutter-location(CL) path, straight line(G01), arc(G02, G03) or curve parameters are used to approach CL path and output approach results as NC codes. However, owing to the process of approaching tool path, the machining product incurs a slight error. Surface interpolation method has received increasing attention in recent years. In this method, scheduling tool paths and calculating inverse kinematics are incorporated into the interpolation device. Therefore, a machining device can directly accept the data of parametric surface, without going through the approach of CAD/CAM post process. In a related investigation, Lin and Koren[2] designed a surface interpolator of a ruled surface's five axial machining. In the interpolator, they designed ruled surface's defining parameters as input data. In the same method, Koren and Lin[3] designed five-axis interpolation system with a cubic surface. Lo[4] designed a three-axis surface interpolator with parametric surfaces.

Figure 2 presents the machining relationship of the surface interpolation with TRR-XY hybrid parallel link machine tool. The interpolator's input data are surface defining parameters and surface machining parameters. The calculation sequences for the interpolator is as follows: CC path is scheduled first, the CC path is then interpolated; and then the cutter tool position and inverse kinematics are calculated. Finally, the interpolator sends the appropriate commands of reference strokes to the control system at every sampling time.

Tool path planning

This study presents novel iso-parametric machining and iso-scallop machining theories, respectively. The IGES[5] format is also used to transform the parametric surface to obtain the parameters of the parametric surface designed by CAD software. Then, after analyzing the IGES file and the surface theory, the surface functions can be obtained as follows: (where u and v denote the surface parameters)

$$\mathbf{S} = \begin{bmatrix} x(u,v) \\ y(u,v) \\ z(u,v) \end{bmatrix} \quad u_{min} \leq u \leq u_{max}, v_{min} \leq v \leq v_{max} \tag{1}$$

Tool path planning also heavily emphasizes cutter angle planning. Herein, the cutter angle in the machining process is according to the normal vector and the tangent vector. Aiming on ball-end miller (Fig. 3), the original cutter axial direction is defined by lead angle λ and yaw angle ω.

Further more, in the layout of the end miller, only the layout of the lead angle of the cutter (Fig. 4).

In the machining process, the contacting position of the cutter and the surface allow us the define the cutter position, in which the ball-end miller's cutter position is as follows:

$$O_{cutter} = O_c + r \cdot \mathbf{N} \tag{2}$$

The end miller's cutter position is as follows:

$$O_{cutter} = O_c + \frac{r}{2} \cdot \sin(\lambda) \cdot \mathbf{N} - \frac{r}{2} \cdot \cos(\lambda) \cdot \mathbf{T} \tag{3}$$

Where O_{cutter} denote the cutter position and O_c refers to where the cutter contacts with the surface.

The cutter direction in the machining process can be obtained by the relationship between the normal and tangent vectors. For example, in terms of the end miller, the direction of the cutter can be determined as follows:

$$\mathbf{T}_{tool} = \begin{bmatrix} 1 & 0 & 0 & 0 \\ 0 & \cos(-\Phi) & -\sin(-\Phi) & 0 \\ 0 & \sin(-\Phi) & \cos(-\Phi) & 0 \\ 0 & 0 & 0 & 1 \end{bmatrix} \cdot \begin{bmatrix} \cos(-\Psi) & 0 & \sin(-\Psi) & 0 \\ 0 & 1 & 0 & 0 \\ -\sin(-\Psi) & 0 & \cos(-\Psi) & 0 \\ 0 & 0 & 0 & 1 \end{bmatrix} \cdot \begin{bmatrix} \cos(\lambda) & -\sin(\lambda) & 0 & 0 \\ \sin(\lambda) & \cos(\lambda) & 0 & 0 \\ 0 & 0 & 1 & 0 \\ 0 & 0 & 0 & 1 \end{bmatrix}$$

$$\cdot \begin{bmatrix} \cos(\Psi) & 0 & \sin(\Psi) & 0 \\ 0 & 1 & 0 & 0 \\ -\sin(\Psi) & 0 & \cos(\Psi) & 0 \\ 0 & 0 & 0 & 1 \end{bmatrix} \cdot \begin{bmatrix} 1 & 0 & 0 & 0 \\ 0 & \cos(\Phi) & -\sin(\Phi) & 0 \\ 0 & \sin(\Phi) & \cos(\Phi) & 0 \\ 0 & 0 & 0 & 1 \end{bmatrix} \cdot \begin{bmatrix} N_x \\ N_y \\ N_z \\ 0 \end{bmatrix} \tag{4}$$

Where $\mathbf{T}_{tool}$ denote the direction of the cutter, and

$$\sin(\Phi) = \frac{B_y}{\sqrt{B_y^2 + B_z^2}} , \quad \sin(\Psi) = \frac{B_x}{\sqrt{B_x^2 + B_y^2 + B_z^2}} , \quad \cos(\Phi) = \frac{B_z}{\sqrt{B_y^2 + B_z^2}} ,$$

$$\cos(\Phi) = \frac{\sqrt{B_y^2 + B_z^2}}{\sqrt{B_x^2 + B_y^2 + B_z^2}} ,$$

Where (B_x, B_y, B_z) is the bi-normal vector, (N_x, N_y, N_z) is the normal vector. In the same method, the direction of the ball-end miller repeats its swinging angle conversion.

Path Interpolation

Utilizing related machining path planning allows us to obtain the tool paths on the parametric surface as follows:

$$\mathbf{r}(t) = \{X(t), Y(t), Z(t)\} \qquad t_{min} \le t \le t_{max} \tag{5}$$

Where X, Y, Z denote the coordinates of the tool path and t represents the parameter of the tool path.

During interpolation of the tool path, each tool position corresponding to the sampling time can be defined by the machining feed rate. Based on work of tool path interpolation by Huang[6], Yang[7] , the feed rate(V_{feed}) can be obtained as follows:

$$V_{feed} = |\mathbf{V}(t)| = \left|\frac{d\mathbf{r}(t)}{dt}\right|\frac{dt}{dT} = \sqrt{\left(\frac{dX}{dt}\right)^2 + \left(\frac{dY}{dt}\right)^2 + \left(\frac{dZ}{dt}\right)^2} \cdot \frac{dt}{dT} \qquad (6)$$

Where T denotes time and $\mathbf{V}(t)$ represents the machining speed on the tool path.   After rearranging, the relation of parameter t and time T can be obtained.   By setting the tool path parameter at time T_i as t_i and the tool path parameter at the next sampling time T_{i+1} as t_{i+1}, the relation between tool path parameter and time can represented using Taylor's expansion:

$$t_{i+1} = t_i + \left.\frac{dt}{dT}\right|_{T=T_i} \cdot (T_{i+1} - T_i) \qquad (7)$$

By using eq(7), the relation between the tool path parameter and time can be defined.   Additionally, inserting the parameter into eq(5) allows us to obtai the positions of a tool path at each sampling time.   According to eq(2), eq(3) and eq(4), the cutter position and cutter direction can be interpolated.

Inverse Kinematics

Herein the geometric relations of each component and constraints on the pin joint are used to derive the inverse kinematics.   The related results was represented in Chang[8].

Simulation and Discussion

The relationship in inverse kinematics can be defined precisely by inserting actual dimensions of the machine tool to the related equations. The machine tool parameters are listed below:R_B=334 mm, L=1107 mm, R_S=200 mm, L_T =270 mm

In the following, a NURBS surface is used to simulate the surface interpolator. The machining parameters are defined as follows: λ = 5deg, ω = 0deg, cutter radius = 3mm, scallop height limit = 0.1mm, sampling time = 20ms, V_{feed} = 4 m/min.   when using the ball-end miller, the interpolation results for iso-parametric and iso-scallop are presented in Fig. 5 and Fig. 6, respectively.   When using the end miller, the interpolation results for iso-parametric and iso-scallop are displayed in Fig. 7 and 8, respectively. Because the surface interpolator is calculated directly on the desired machining surface, approaching tolerance does not need to be defined. Moreover, the machining results are more precise when the sampling time is decreased to several ms.

Summary

This work presents a novel surface interpolation theory for the TRR-XY hybrid parallel link machine tool.   The machining process of surface interpolation allows the cutter to achieve a uniform machining rate. According to the machine-dependent inverse kinematics, users are no need to deal with the complex post-processes. In contrast to the conventionally used NC program, interpolating with the sampling time of the control system minimizes the approach's tolerance of the tool path.

Reference

1. Stewart, D. A Platform with Six Degrees of Freedom. Proceedings of the Institution of Mechanical Engineering (London). 1965; 180:371-386.
2. Lin, R. S., and Koren, Y. Real-time five-axis interpolator for machining ruled surfaces. ASME Winter Annual Meeting, DSC, 1994; 55-2:951-960.
3. Koren, Y., and Lin, R. S. Five-Axis Surface interpolators. Annals of the CIRP, 1995; 44:379-382.
4. Lo, C. C. CNC machine tool surface interpolator for ball-end milling of free-form surfaces. International Journal of Machine Tools and Manufacture, 2000; 40:307-326.
5. Smith, B., Rinaudot, G. R., Reed, K. A., Wright, T., *Initial graphics exchange specification (IGES) Version 4.0*. Society of Automotive Engineers Inc, 1988.
6. Huang, J. T. and Yang, D. C. H. A generalized interpolator for command generation of parametric curves in computer-controlled machines. ASME 1992 Japan/USA Symposium on Flexible Automation, San Francisco, USA, 1992; 1:393-399.
7. Yang, D. C. H., and Kong, T. Parametric interpolator versus linear interpolator for precision CNC machining. Computer-Aided Design, 1994; 26-3:225-233.
8. Chang, T. H., Inasaki, I., Morihara K., Hsu, J. J. The Development of a Parallel Mechanism of 5-D.O.F. Hybrid Machine Tool. Parallel Kinematic Machines International Conference, Second European-American PKM Forum, september 2000.

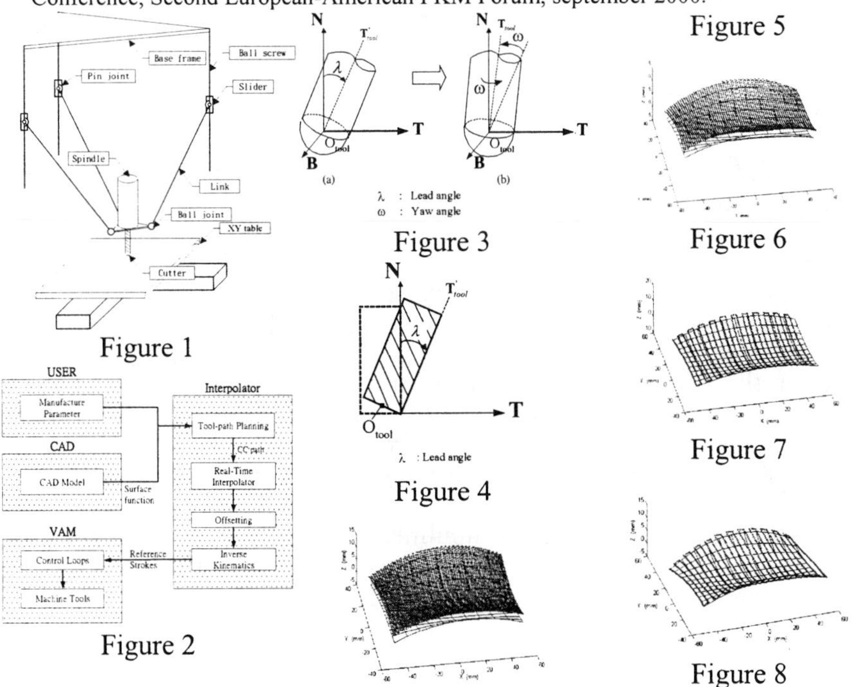

Figure 1

Figure 2

Figure 3

Figure 4

Figure 5

Figure 6

Figure 7

Figure 8

MANUFACTURING PROCESS CONTROL BASED ON THE DELTA OPERATOR

Tatsu Aoki

*Department of Electrical Engineering, Tokyo Metropolitan College of Technology
1-10-40 Higashi-Ohi, Shinagawa-ku, Tokyo 140-0011, Japan*

Abstract

This paper considers an improvement of the production rate in a manufacturing cell. The industrial robots are utilized to load and unload raw work pieces, semifinished parts, and finished parts. Because conventional robot controllers are usually designed by the help of a linear control theory, the actuator operates within a linear range. As a result, a high-power actuator is necessary to achieve a high production rate. In this paper a new nonlinear control methodology based on the delta operator is proposed to utilized the full performance of a given actuator. The simulated results show the effectiveness of the proposed algorithm. Furthermore, the production rate is considered.

Keywords

Microprocessor control, Delta operator, Nonlinear control systems

1. INTRODUCTION

In order to achieve the high production rate in a manufacturing cell, many constraints must be considered simultaneously such as actuator saturation of a machine, the cutting speed of the slowest machine,etc. Because linear control systems cannot work effectively in such systems, nonlinear control strategy is introduced to improve the performance.

On the other hand, it is necessary to make the sampling period be shortened at high-speed digital control. However, a control algorithm based on the z-transform or the shift form becomes unstable numerically in high-speed sampling. As a result, whole control systems become unstable. It is well known that a control algorithm based on the delta operator is useful in order to avoid numerical ill-conditions(Goodwin *et al.*, 1992).

In this paper a basic idea of nonlinear controller based on the delta operator for high-speed motion is proposed. Also, the simulated results are presented to show the effectiveness of the proposed algorithm.

2. PROPOSAL OF A NONLINEAR CONTROLLER

2.1 Control Algorithm Based on the Delta Operator

Stability boundary is represented as a unit circle on the z-plane in discrete-time control systems. Then stable discrete-time systems have all poles in the circle. Generally, it is necessary to make the sampling period T smaller to achieve a good control performance as well as the system's stability. Recently T can be set as small as 0.1ms by using DSPs. However, when T approaches zero, all poles converge to the stability boundary. In such situation, a control algorithm becomes numerically unstable, as shown in Figure 1. The difference between calculation results becomes extremely small and the word length in computation becomes relatively short to present such small value.

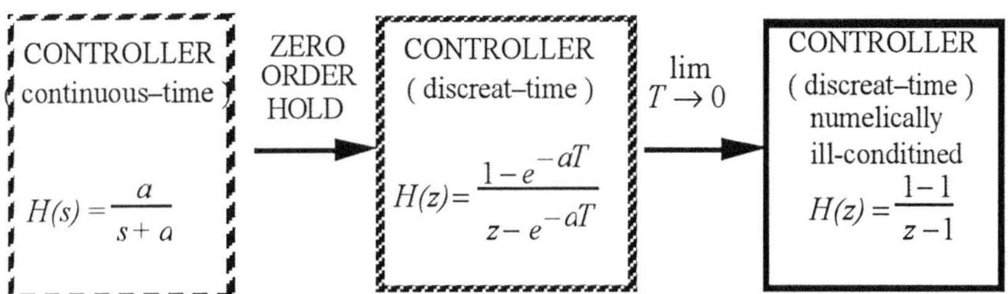

Figure 1. Digital controller based on shift operator z

On the other hand, the delta operator δ maps a unit circle in the z-plane into a larger circle of radius $1/T$ and center $-1/T$ on the δ-plane. When T approaches to zero, the larger circle on the δ-plane becomes the left-half plane. That is the stability boundary for the continuous-time systems. This means that the delta model converges to the continuous-time one, when T approaches to zero, as shown in Figure 2.

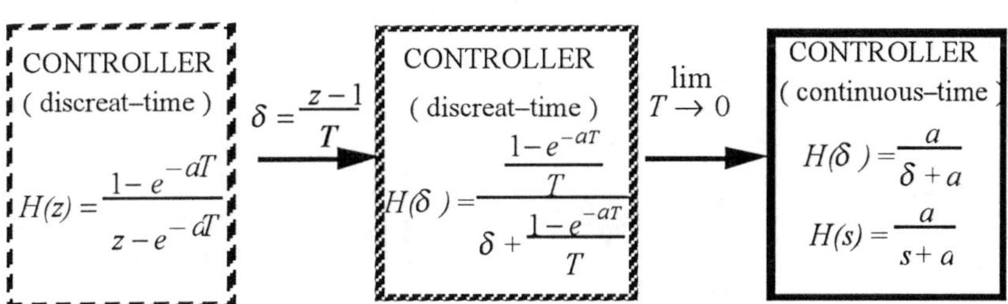

Figure 2. Digital controller based on the delta operator δ

In order to achieve high-speed sampling, a calculation is executed at fixed-point arithmetic for industrial robots. However, in fixed-point arithmetic the delta operation becomes numerically unstable by FWL.

2.2 Control Algorithm Based on the Modified Delta Operator

According above, the modified delta operator δ' was proposed for fixed-point arithmetic(Aoki and Furukawa, 1996). That is defined as subtracting unity from the shift operator z and then dividing by arbitrary constant T_δ. In order to extend the concept to nth-order systems, T_δ is set to unity, and the design parameters T_1-T_n were introduced(Aoki and Furukawa, 1996). Also, a new error reduction method named Variable Modulation Method (V. M. M.) was proposed to prevent overflow(Aoki and Furukawa, 1996). The principle is based on PWM. The dynamic range of the calculation becomes twice. The schematic diagram of the controller based on V.M.M. is shown in Figure 3.

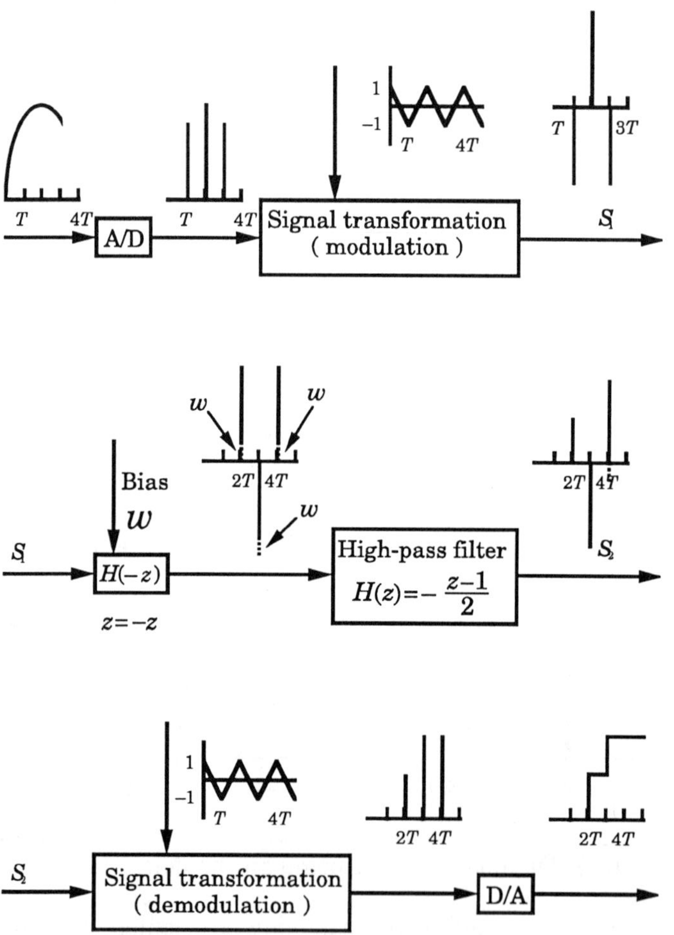

Figure 3. Schematic diagram of controller based on V.M.M.

2.3 Control Algorithm Based on the Extended V.M.M.

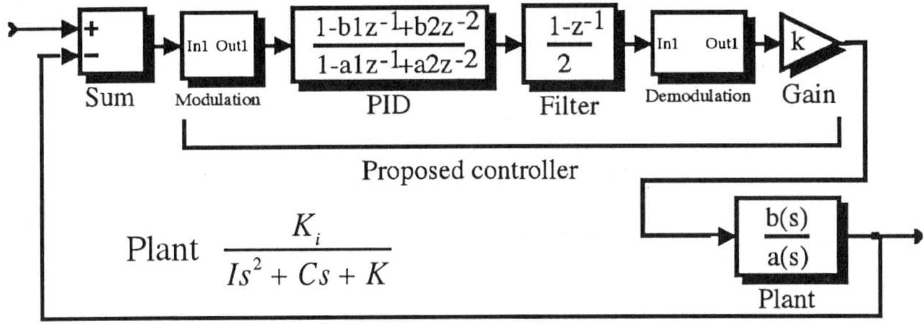

Figure 4. Block diagram of PID control system

As an illustration, the PID position control of a lightly damped robot with actuator saturation is considered, as shown in Figure 4. The control algorithms based on the shift and delta operator become the second-order systems. Step responses are shown in Figure 5.

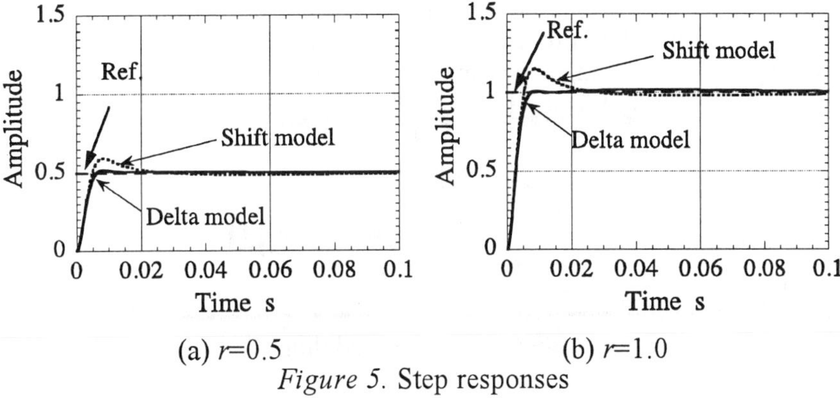

(a) r=0.5   (b) r=1.0

Figure 5. Step responses

Settling time is abut 0.02s and 0.01s, respectively. Thus, robots can load and unload parts in about half period than a conventional control in this case. Because the robot is lightly damped for high-speed motion, the amount of overshoot is large on the shift form. We cannot reduce the amount of overshoot by a linear control algorithm. On the other hand, the amount of overshoot on the delta form becomes very small in fixed-point arithmetic. This is because overflow occurs at the filter when the controller input is large. Because overflow introduces a step-like output in a conventional control algorithm, a control system becomes unstable in a conventional systems. In case of the extended V.M.M., overflow does not affect the output seriously due to signed modulation in a calculation. This algorithm positively utilizes overflow.

2.4 Improvement of the Production Rate in a Cell

As an illustration, a manufacturing cell consisting of a robot with a double gripper and two machines is shown in Figure 6. Thus, the robot can load a new part and unload a finished part without leaving machine tool bed. If the the robot's loading and unloading times t_1, t_3, t_5 are longer than the cutting times t_2 and t_4, the production rate becomes $t_1 + t_2 + t_3 + t_4 + t_5$ / part due to a serial processing.

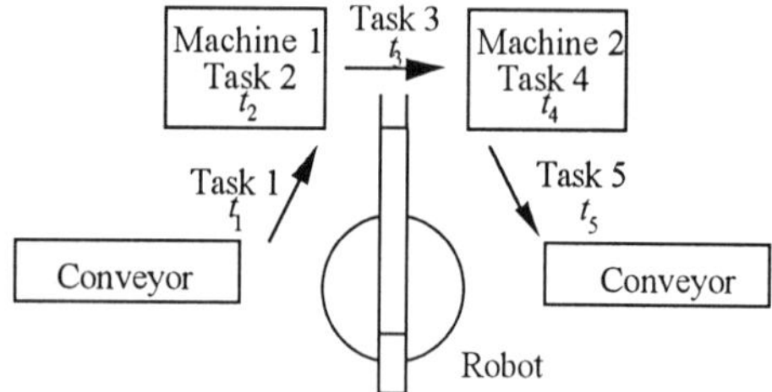

Figure 6. Manufacturing cell

However, the robot's loading and unloading times can be reduced by the extended V.M.M. because the methodology can reduce the settling time. If the conditions $t_1 < t_2$ and $t_3 < t_2$ and $t_3 < t_4$ and $t_5 < t_4$ can be realized for a parallel processing by using the extended V.M.M., the following task scheduling is possible, as shown in Figure 7.

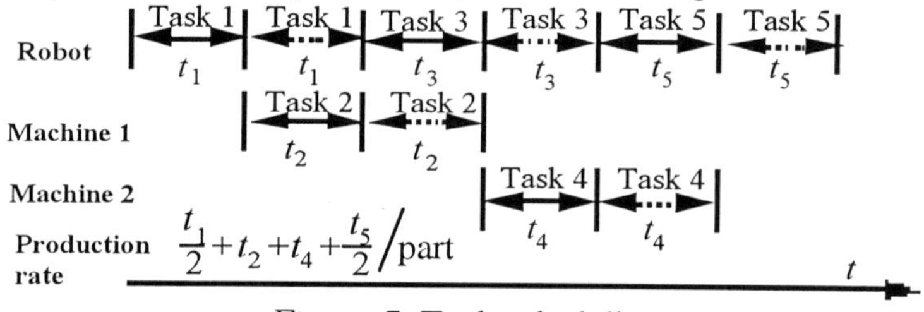

Figure 7. Task scheduling

3. CONCLUSIONS

In order to improve the production rate in a manufacturing cell, basic idea of nonlinear control methodology has been proposed by extending V.M.M. Simulation results show that its availability has been made clear.

REFERENCES

Aoki T., Furukwa Y. Proposal of Modified Delta Operation with V. M. M. and its Application to Controlling Algorithm in Fixed-point Arithmetic. Proceedings of Fourth International. Conference on Control, Automation, Robotics and Vision; 1996 Dec. 3 - Dec. 6; IEEE, 1996.

Goodwin G.C., Middleton R.H., Poor H.V. High-Speed Digital Signal Processing and Control: Proceedings of the IEEE. IEEE 1992; 80(2):240-59

ASSEMBLY RELIABILITY EVALUATION METHOD (AREM)

Tatsuya Suzuki[1], Toshijiro Ohashi[1], Masaaki Asano[2]

1) Hitachi Ltd., Production Engineering Research Laboratory, 292 Yoshida-cho, Totsuka-ku, Yokohama 244-0817, Japan

2) Hitachi Lighting Ltd., Ryugasaki Works, 69 Wakashiba-cho, Ryugasaki, 301-0041, Japan

Abstract

This paper introduces an effective new design tool called AREM that visualizes assembly fault potential. Its evaluation formula is constructed on a fault occurrence model and enables quantitative evaluation of the fault generation potential of both product and assembly shop.

The distinctive features of AREM are: (1) visualization of parts and operations having a high fault probability; (2) easy data input by symbol selection; and (3) quantitative evaluation of shop reliability, as well as product reliability. The efficacies of applying this method are: (1) improvement of product reliability; (2) reduction of the design and development period; and (3) improvement of shop operation reliability.

Keywords

Assembly, reliability, assembly fault, design, evaluation

1. INTRODUCTION

In recent years, occurrence of assembly faults in the assembly shop has been gaining more and more attention. Assembly faults are induced by a number of factors including: development of more sophisticated products; adoption of advanced and highly-developed technologies; insufficient production design caused by shorter development time; weakening ties between design and manufacturing due to a more complex and insecure supply chain (SC); and sudden and drastic changes in products and also manufacturing conditions. Many companies have experienced lost business opportunities as a result of production confusion due to a large number of fault occurrences at the start-up phase of new products. In addition, many companies have also lost the faith of their existing customers due to trivial product faults. Other factors that increase assembly faults are globalized production and outsourcing. Both of these factors result in reduced assembly skill, which leads to a considerable increase in the assembly fault rate.

In order to solve these problems, a method for evaluating design improvement and production improvement is required. The method should be able to quantitatively evaluate in advance the reliability of assembly and operations, in particular, those that have lower reliability and higher assembly fault probability. Although the problem of product quality must be handled primarily in the design department, assistance from other departments, such as quality assurance, manufacturing, and service departments is actually required. Therefore, a method of visualization and quantitative evaluation of the probability of operation before production is important.

Consequently, for the purpose of reducing the assembly fault rate and early establishment of mass production, the Assembly Reliability Evaluation Method (AREM) was developed in order to evaluate the reliability of assembly operation, i.e. evaluate the assembly fault potential, in advance.

2. DISTINCTIVE FEATURES OF AREM

AREM is a unique design evaluation tool that can distinctly separate the responsibilities of the structure of a product and the assembly shop conditions, and as a result can be used to systematically improve both product design and shop operation. The distinctive features of AREM are:

(1) Visualization of parts and operations having a high fault probability
  - Evaluation result index: Estimated assembly fault rate and assembly fault structure coefficient for operation/part/subassembly/product.
  - Prediction of fault phenomena: Output of a predicted fault phenomenon. Supports identification of fault cause and planning of countermeasures for improvement.
  - Basic shop fault rate (BSFR): Reflects the shop reliability level. Attaching parts that have a significant potential fault rate when produced overseas or in a different shop can be identified in advance.

(2) Easy data input by symbol selection
  - Attaching operations and characteristics of part: Movement and attributes of a part are specified using analysis symbols. The analysis symbols can easily be extracted from conceptual drawings, part drawings, samples, etc. of evaluated parts. The analysis symbols can then be simply input into a PC, and evaluation can be performed.
  - Assemblability evaluation is also possible concurrently: The data input for AREM can also be used for the Assemblability Evaluation Method.

(3) Quantitative evaluation of Shop reliability
  - The basic shop fault rate (BSFR) indicates the reliability level of the evaluated assembly shop. BSFR is calculated while estimating a product in production. It can also be estimated by the shop evaluation system prior

to production.

3. THEORY OF THE EVALUATION METHOD

AREM utilizes an assembly fault evaluation theory that is based on an assembly fault occurrence model, which was constructed by observing fault phenomena. The basic theory of the product assembly fault evaluation is outlined below (see Fig. 1).

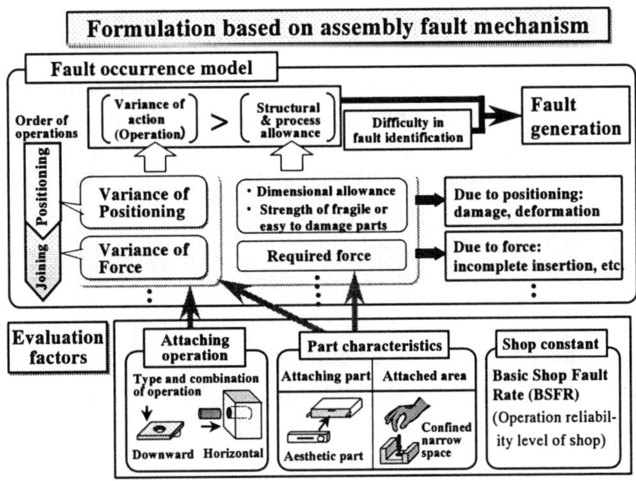

Figure 1. Assembly fault occurrence mechanism and evaluation theory

(1) Assembly faults are caused by operational inaccuracy or mistakes induced by the product's structure.

When operational variance exceeds various allowances dictated by the product's structure and/or process, an assembly fault occurs. Operational variance includes variance of positioning, locus, joining, etc. Allowance is influenced by such factors as dimensional allowance, fragile or easy to damage parts, and strength of press fitting force. Fault type and occurrence rate depend on operational variance, structural and process allowance, and difficulty of fault identification. For example, positioning variance causes damage and deformation.

(2) Factors for determining the operational variance and structural and process allowance

- In general, operational variance depends on the type, combination and number of assembly operations.
- The fault generation rate due to operational variance is weighted using the characteristics of the attaching part and the receiving part or area.

939

(3) Fault occurrence level in individual shops

The degree of fault generation in individual shops depends heavily on the shop conditions, including such factors as the operators, facilities and operating environment. To calculate the assembly fault rate, AREM uses a shop constant called the basic shop fault rate (BSFR).

The data input procedure for calculating the assembly fault rate is as follows.

- Attaching operation: An assembly operation is described through the selection and combination of analysis symbols provided beforehand. A fault generation potential coefficient is provided for each operation element.
- Part characteristics: Supplementary elements which remarkably increase or decrease the fault occurrence rate are defined. For each individual part, the corresponding elements are selected and input from among these elements. A coefficient that is used to weight the fault occurrence rate in respect of the individual part characteristics is provided.
- BSFR: The BSFR reflects the fault generation potential of the shop based on the capability of the workers, the assembly shop conditions, the facilities of the shop, etc. If the product is evaluated in advance using actual fault data, then the BSFR can be calculated.

The value of each coefficient and constant is decided based on actual fault data.

4. DETECTION CAPABILITIES OF AREM

Figure 2 shows the results obtained by applying AREM to an actual product. In the correlation diagram shown in Fig. 2, the vertical axis indicates the fault rate estimated by AREM and the horizontal axis indicates the actual fault rate. The closer the plotted points are to the 45-degree line, the higher the accuracy of the estimation.

During the application of AREM, a warning line was set at a certain fault rate, in this case 1 (index). The plotted points that lie above the warning line, that is, in the warning zone are detected faults and the design or way of operation of these parts must be reviewed. If any points lie in the Detection Failure Zone then this indicates that AREM failed to detect dangerous parts. All points that lie above the warning line but for which the actual fault rate is zero are false faults. That is, AREM indicates that these parts are dangerous but after actual production these parts did not show a fault.

It is important that AREM has a high detection rate/low false fault rate and, at the same time, a minimal detection failure rate. Overall, for this application, AREM detected most (97%) of the faults with a relatively low (less than 50%) false fault rate without failing to detect any major faults.

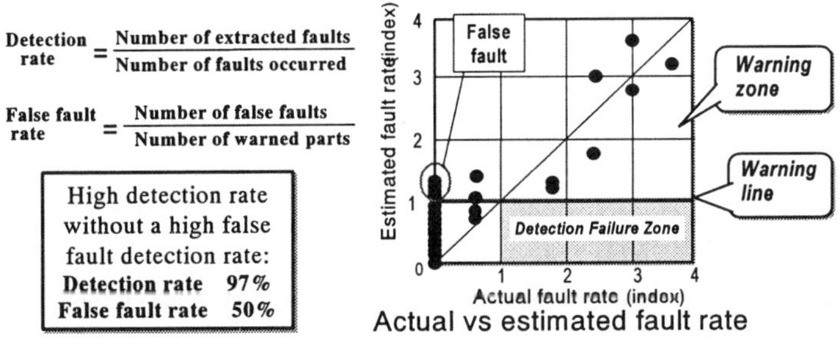

Figure 2. Fault detection capability of AREM (applied to FAX machines)

5. EFFECTIVENESS OF AREM APPLICATION

Hitachi Ltd.'s Tokai Works (Image & Information Media Systems Division) applied AREM to their VCR final assembly. As a result, they achieved quick start-up. After two design improvements and six manufacturing technology improvements based on the AREM evaluation results, the VCR was put into mass production. The assembly fault rate decreased to approximately 1/3 that of the previous model at the initial production stage, and mass production start-up went smoothly.

6. CONCLUSION

AREM is a new design tool that was developed to evaluate assembly reliability. The effectiveness of the method has been confirmed through actual applications. The distinctive features of AREM are as follows:
(1) Visualization of parts and operations having a high fault probability detection capability
(2) Easy data input by symbol selection as well as simultaneous assemblability evaluation
(3) Quantitative evaluation of Shop reliability and product reliability.

REFERENCE

Boothroyd, G., Dewhurst P., Knight W., *Product Design for Manufacture and Assembly.* Marcel Dekker, 1994.

MAINTENANCE SUPPORT SYSTEM EQUIPPED WITH MONITORING FUNCTION

Toshitake TATENO and Masanori IGOSHI

Department of Mechanical Engineering, Graduate School of Engineering, Tokyo Metropolitan University, 1-1 Minami-Ohsawa, Hachioji, Tokyo 192-0397, JAPAN

Abstract

Computer support for maintenance workers is discussed. First, the difference between manufacturing work and maintenance work is considered, and the functional requirements of a computer support system are discussed. In order to meet these requirements, a work support system equipped with a monitoring function is proposed. The system is structured with three-layered subsumption architecture. As an application of the proposed system, a safety management system is introduced. Through developing the system, suitable computer models for realizing the support system are introduced and the advantage of the monitoring function is shown.

Keywords

Work support, Monitoring, Maintenance, Safety management system

1. INTRODUCTION

The trends of recent products toward larger scale, complexity, diversity and a short-term life cycle lead to the problem that maintenance workers must perform many kinds of tasks in a short time. While the requirements for maintenance workers are increasing, most workers cannot have enough training time because of short budgets. Therefore, a maintenance worker often cannot solve the problem alone at a maintenance site. In order to support the work, a computer support is considered. In case of a scheduled maintenance, presentation of manual documents is one type of support. In practice, in airplane maintenance, some wearable presentation devices have been developed for workers to operate in order to refer to some manual documents. However, in the case of corrective maintenance including diagnosing and troubleshooting, more intelligent support should be provided by the system. Ordinal automatic diagnosing systems are limited to use in simple cases because the combination of diagnosing and presentation are not enough. In this paper, a work support system equipped with a function to monitor a worker's motion is proposed and the effectiveness of monitoring is discussed.

2. MAINTENANCE WORK

2.1 Characteristics of Maintenance Work

The difference between maintenance work and manufacturing work is considered. First, a maintenance worker must deal with more kinds of products than a manufacturing worker because maintenance must be given to all existing machines. As well, the order of maintenance is random. Second, a maintenance worker often must plan the work at the maintenance site because the operation task is not predictable, particularly in corrective maintenance. In the manufacturing process, it is a rare case when a manufacturing worker needs to decide something. Third, a maintenance worker must speedily adapt to new work conditions because maintenance work frequently changes due to newly revealed problems. Manufacturing work has the somewhat similar process of "kaizen" activity, but that is not so frequent.

2.2 Maintenance Support

For the above reasons, it is clear that maintenance support is a more serious problem than manufacturing support. The root of this problem is that maintenance workers must make many decisions on their own in a limited time. For that reason, computer support is applied for helping with decisions and for training in decision ability. The applications can be arranged into three categories, task management, safety management and worker management. Task management is an activity to improve task efficiency and reliability. Safety management is an activity to reduce incidents of accidental injury. Worker management is an activity to educate a worker on how to be an expert worker.

3. SUPPORT SYSTEM EQUIPPED WITH MONITORING FUNCTION

3.1 System Design

We have studied about a work support system for maintenance workers and have stressed that the system should have a monitoring function [Tateno and Igoshi,1999]. A monitoring function has the possibility of making rapid progress of the system ability. In the following sections, the system architecture is introduced.

The proposed system is similar to the associate system [Bushman et.al.,1993]. The support system monitors human motion and presents appropriate help based on analysis results of the motion data. Figure 1 shows the relation among human (worker), machine (support system) and work environment (work object). The support system may give not only intellectual help but also physical help. The intellectual help is given to the worker, and the physical action is given to the work object. Furthermore, the intellectual support to the worker is structured as three-layered support, and the each layer can make independent actions. This is one type of the subsumption architecture. We named it a three-layered monitor-action loop structure. Figure 2 shows the architecture as a three-layered structure.

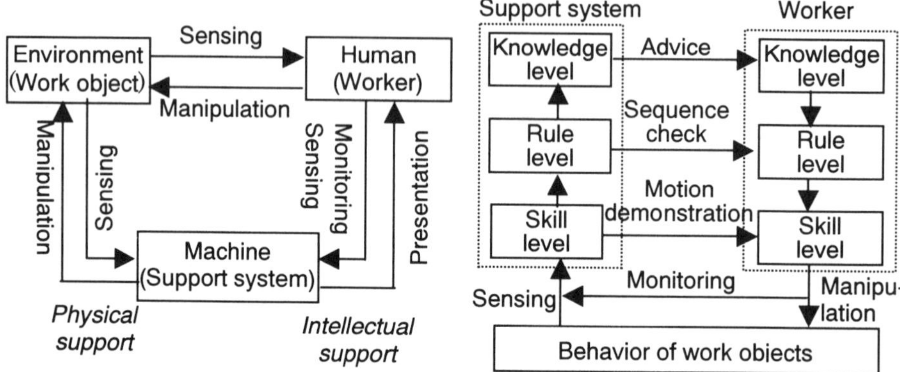

Figure 1 Relation between the human, machine and work environment

Figure 2 Three-layered monitor-action loop structure

Hashimoto [Hashimoto and Buss,1992] has proposed a style of three-layered structure for an intelligent work support system. However, the characteristic of the monitor-action loop has not been commented on. Here, we stress that the support system should include this structure, which is realized by the monitoring function. In each level, the following actions can be realized. The skill level layer includes an action that is a reactive action such as an alarm for when a worker's hand enters into a dangerous area. The rule level layer includes an action that is determined by predicted rules, such as a task sequence check. The knowledge level layer includes an action that is derived by some experiential models, which compiles many work examples or records worker experience, such as advice based on analysis results of historical data or a student model.

3.2 Expected Applications

We mentioned that there are three categories of the work support application. For each category, three levels of support can be considered. Table 1 shows samples of expected applications.

Table 1   Expected applications

Support level	Task management	Safety management	Worker management
Knowledge level	Diagnosis support	Human reliability analysis	Task scheduling support
Rule level	Task recording	Sequence check	Access to experience data
Skill level	Work demonstration	Emergency stop	Skill evaluation

4. PROTOTYPE

4.1 Safety Management System

As an application of the work support system, a prototype of the safety management system (Figure 3) was developed. The system is simple but has all the levels of the monitor-action loop structure. As a work object, we choose replacing the batteries of an electric car (Figure 4). The task is dangerous because the connector may generate a spark if the worker misses a connecting point. This prototype system checks the work sequence and presents a sign if the worker is in a dangerous situation.

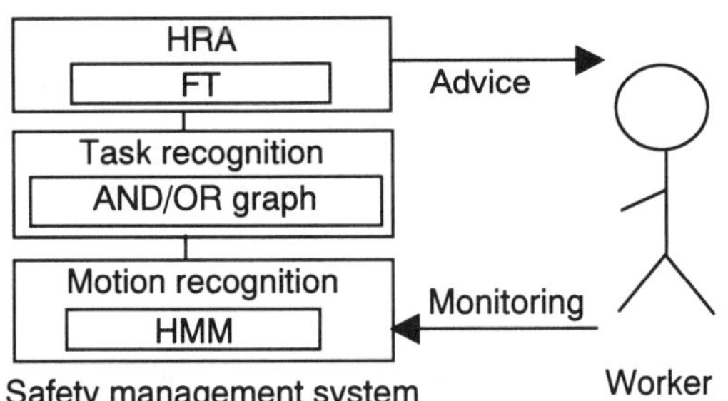

Figure 3 Safety management system

Figure 4 Battery replace work wearing data-glove.

4.2 Task and Motion Model

Not only for the safety management system but also for all applications, a task and motion model is used to monitor the worker. In the prototype system, a Hidden Markov Model (HMM) approach is used for motion recognition, and an AND/OR graph is used for task recognition. We adopted these models because it is relatively easy to add or delete part of model. These characteristics are useful for representing maintenance tasks because the task is usually limited in a product area. In the following, these models are introduced briefly.

HMM is a probability model. Each HMM for a task is made by abstracting some motion examples. The abstracting process is called HMM learning. The system must prepare HMMs for all tasks. After HMMs are prepared, the system can proceed to the recognition process. In the recognition process, the monitoring system computes the probability under the condition that an observed motion can appears in the learned HMMs. The HMM which gets the best probability is the recognized motion.

An AND/OR graph is used for representing work sequences. There are many ways to represent work sequence. However, it is known that the AND/OR graph is the best way for representing maintenance work [Homen de Mello and Sanderson,1991]. In the AND/OR graph, each node represents a subassembly of a product, and each arch represents action to build the subassembly. We can select an area that relates to the current task and choose the best way from a selected part of the model.

4.3 Human Reliability Analysis (HRA)

HRA is used for finding out the failure process. Fault Tree (FT) is a model for representing the process. Fault Tree Analysis (FTA) is a basic method for evaluating dangerous work. The FT is the combination of items that represent the states of the work. In this system, a simple combination is defined. When the condition in figure 5 is found in the task, the system presents a warning message.

A set of	(1) Work point P1 is finished	and
	(2) Workers hand is not in the work point P2	and
	(3) P2 is the next work point after P1	

Figure 5 Warning condition

4.4 Experiments and Discussion

A worker was tested doing work while wearing a data glove for sensing finger shape and position. The system worked perfectly, but the sensing device sometimes disturbed the task. Thus compactness of the device is required. Also, this test worker accidentally encountered a spark by contacting a connector with a wrench when he removed a stay bar on a battery. This fact shows the importance of the safely management system as well as a future problem in that the system should check not only human hands but also tools and work objects.

5   CONCLUSIONS

(1) The characteristics of maintenance work are considered different from manufacturing work.
(2) The structure of a work support system equipped with monitoring function is proposed and shown to match the requirements.
(3) As an application of the proposed system, a safety management system is introduced.

6   REFERENCES

Bushman J.B. et.al., Ally: An Operator's Associate System for Cooperative Supervisory Control, IEEE Trans. Sys. Man Cyber, 1993, Vol.23, pp.111-128.

Hashimoto H. and Buss M., Skill Acquisition for the Intelligent Assisting System Using Virtual Reality Simulator, The 2nd Int. Conf. On Artificial Reality and Tele-Existence ICAT'92, 1992, pp.37-46.

Homen de Mello L.S. and Sanderson A.C., And / or Graph Representation of Assembly Plans, IEEE Trans. on Robotics and Automation,1992 ,Vol.7, No.2, pp.211-227.

Tateno T. and Igoshi M., Monitoring and Guiding Human Operation in Maintenance Work: Towards Intelligent Associate System for Maintenance Worker, Proc. of Int. Denube Adria Association for Automation & Manufacturing Sympo., 1999, pp.545-546.

DEVELOPMENT OF SOFTWARE PLATFORM FOR REALIZING THE DYNAMIC RECONSTRUCTION OF SOFTWARE APPLICATIONS OF MANUFACTURING SYSTEMS

Toshifumi SATAKE*, Akihiro HAYASHI** and Guangming ZHANG***

*Numazu National College of Technology, Faculty of Control and Computer Engineering,
3600 Ohoka, Numazu, Shizuoka 410-8501, Japan, satake@cce.numazu-ct.ac.jp
**Kyushu Institute of Technology, Faculty of Computer Science and Systems Engineering,
680-4 Kawazu, Iizuka, Fukuoka 820-8502, Japan, hayashi@mse.kyutech.ac.jp
***University of Maryland, Faculty of Mechanical Engineering, A.V.Williams Building,
University of Meryland, College Park MD20742, U.S.A, zhang@eng.umd.edu

Abstract

This paper describes the development of a software platform for realizing the easy construction/reconstruction of software applications for manufacturing systems based on the distributed object approach. To maintain the productive capacity of manufacturing systems according to changes in production plans, it is required that software applications for constructing a manufacturing software system can be modified easily. This study aims to develop a software platform for providing that easy construction/reconstruction capability which can be further modified easily. This paper describes the concept of our software platform, technical development details and its functions.

Keywords

Manufacturing System, Reconstruction, Distributed Object Approach, CNC

1. INTRODUCTION

In the life cycle of a manufacturing system, a reconstruction is often forced to maintain the productive capacity or adapt to new products in maintenance phases which are needed to newly install/modify various types of software applications and machines. Whether software applications for constructing the manufacturing system are easy to modify is one of the important capabilities. The required operating time influences the total maintenance cost which has to be added to the product cost. The fact that software applications are fixed in computers after its installation obstructs the expansion of both the flexibility and the adaptability of manufacturing software systems. Especially, computers such as CNC and PLC, which have embedded software applications, are hard to modify because they both need to exchange ROM when their software applications are installed and modified.

To achieve the easy reconstruction of software applications for manufacturing systems, a software platform which can install and modify a software application by transporting and linking necessary functional modules according to demand is developed in this study. This paper describes the concept of the software platform and its technical summary. Finally, a simple system constructed by using this software platform is indicated to demonstrate effectiveness.

2. THE CONCEPT OF THE SOFTWARE PLATFORM

In this study, to install and modify software applications easily, all functional modules for constructing software applications are depicted as objects which can be relocated to any computer in the network computation environment formed from all computers of a manufacturing system connected to each other. All software applications are constructed dynamically by gathering necessary functional modules to specific computers from the network computation environment according to demand. In order to achieve this idea, the software platform is required to have some capabilities as follows:
    (1) management of locations of functional modules,
    (2) functional module transportation,
    (3) necessary gathering of functional modules.
(1) is the function to manage physical locations of each functional module for constructing a software application. (2) is to transmit the source codes or binary codes of one functional module and then install them into a target computer automatically after compiling and linking the transmitted functional module. (3) is used to collect necessary functional modules from the network computation environment to build a software application. This function consists of processes which are used to collect functional modules from the network computation environment, compile them and link their symbols to each other.

In order for our software platform to have the capabilities mentioned above, a communication environment between computers for constructing the network computation environment is needed to exchange various types of data with each other. CORBA is the standard for communicating messages between software applications when constructing a distributed system. Many CORBA-based software platforms have been developed to achieve a distributed computation environment and many commercial application systems have been developed.[3][4]   Our software platform should be developed on the message communication environment.   Our research group has developed a message communication environment for achieving functional module location management and functional module transportation.[1][2]   In this study, a prototype system of the software platform is developed to confirm the effectiveness of our idea which is to achieve the easy reconstruction/construction of software applications for manufacturing systems.

3. THE SOFTWARE PLATFORM

3.1 The basic components

Figure 1 shows the system configuration of the software platform proposed in this study.   The software platform consists of two software systems which are Broker and Binder.   These are installed into the computers forming the network computation environment in a manufacturing system.   Broker is installed in each computer with both IP address and port number and is connected to all the other Brokers.   All functional modules are provided logical names to identify with each other.   Broker manages the locations of functional modules by sharing the logical names of functional modules among all Brokers.   Broker manages the execution of Binder and relay messages to Binder via inter-process communication.   Messages are exchanged between functional modules via Binder and Broker.   A functional module is linked to Binder for changing to an executable state.   Binder has data of functional modules linked to itself, which are logical names, linking points on the memory and external symbols of those functional modules.   The functional module is executed in

a thread process provided by the OS. Binder is a message-driven type system and behaves like a server for offering execution results of functional modules. However, an executed functional module by Binder behaves like a client program which can use results of a remote functional module.

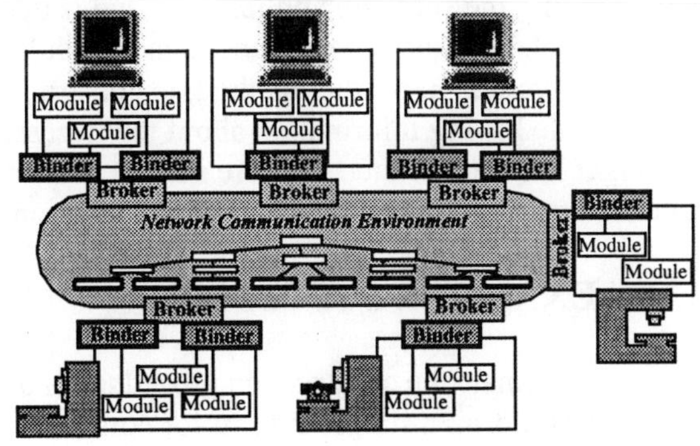

Figure 1   System Configuration

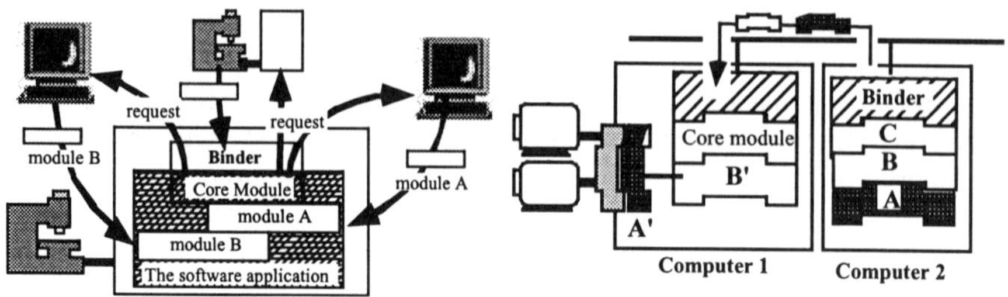

Figure 2   The application Building.   *Figure 3   Example system.*

3.2 The functions for building a software application

Binder provides three important functions for building a software application into specific computers in the network computation environment, as follows:

(1) Sending a copy of itself,

(2) Gathering necessary functions, and

(3) Binding external symbols.

(1) is used to send the source code of itself to another computer to build a software application.   (2) is used to request a functional module to send a copy of another functional module.   This function is used to expand first functional module's capabilities.   In these two functions, the copy of the functional module is linked to the destination's Binder.   When new functional modules are linked to the sender, a new logical name is given to distinguish the original module and its copy module.   The external symbols included into the sent functional modules are linked to each other by using function (3).   These functions provide two types of methods for building and installing the software application. One is that an operator or a program sends unilaterally necessary functional modules to target Binders.   As shown in figure 2, in another method, to construct the software application, a core module having information about the necessary modules first is sent to the target Binder and then the core module automatically gathers necessary functional modules from the network computation environment.

4. AN EXAMPLE SYSTEM

To estimate the construction procedure of a software application in the software platform, a simple system for controlling motors is constructed on the software platform developed on a Real-time Linux Operating System as the first approach to this research.   Figure 3 shows the software configuration of simple motor control software application.   This software application consists of a real-time module (Module A) coded to control motor revolutions according to instructions and a functional module (Module B) for sending instruction data to Module A.   First, the core module is sent to the Binder of computer 1.   Next, the core module sends messages for requesting Module B and Module A.   When the Binder of computer 2 receives the messages, its Binder sends the copy of Module A and Module B to the core module.   Module B' which is the copy of Module B is linked to the Binder of computer 1 after being compiled.   Module A' which is the copy of Module A is installed to the real-time process of the operating system after being compiled as a real-time module.   Finally, Module A' and Module B' open the communication line to each other and then Module B' sends instruction data to Module A'.   The functional modules for constructing this application can be exchanged with others and new functions can be appended to this system whenever this application is not being executed.

5. CONCLUSION

To provide the easy reconstruction capability to manufacturing software systems, we have developed the software platform on Real-time Linux Operating System as the first approach to this research.   This paper has described the concept and technical summary of the software platform and a simple system has been constructed to estimate the performance of the software platform.   By using this software platform, a software application can be constructed by gathering automatically necessary functional module from the network computation environment and can be reconstructed by replacing or appending functional modules according to demand via the network.   As the next step of this study, it is considered that our software platform should correspond to CORBA.

6. REFERENCES

1. Akihiro HAYASHI, Toshifumi SATAKE and Hiroshi SUZUKI, Development of Platform and Functional Tool Kit for Constructing/Executing a Decentralized Autonomous System, JSPE, Vol.64, No.4, 1998, 531-535
2. Akihiro HAYASHI, Toshifumi SATAKE and Hiroshi SUZUKI, Development of a Framework and a Software Tool for Constructing an Autonomous Distributed System, FAIM, May, 1996, 712-730.
3. B.C.Johnson, Distributed Computing Environment Framework: An OSF Perspective, http://www.osf.org, 1991
4. R.Otte, P.Patrick and M.Roy, UNDERSTANDING CORBA, Prentice hall, 1996.

Mechatronic Service and Investigation on the Example of a Forming and Milling Machine

Dr.Ing. J. Hamann, Dr.Ing. H.-P. Tröndle, A&D MC E55,
SIEMENS AG Erlangen,Germany

Abstracts

The dynamic performance of production machines is among others limited by the first mechanical resonance due to displacement of motor (locked-motor-frequency). By using Finite-Element-Methods we get the transfer-functions of the machine and thus these few resonances that restrict the capabilities of control. On the other hand FEM gives us the inertias and springs involved in these critical modes. By changing these mechanical elements we are able to improve the machine in the desired sense. In addition we can distinguish these modes that can be damped by active control and those that must be attenuated by means of construction. These informations are vital for avoiding a built-up machine that lacks the performance sold.

Keywords
Finite-Element-Method, Mechanical transfer function,Control,Eigenmodes

1. Mechatronic Service and Investigation

Mechatronics, electronics and software will work together more closely in modern devices, machines and industrial systems. Development of all these components together is called mechatronics. Therefore, mechatronics is the logic continuation of singular construction methods. Common simulation and verification of the mentioned components is very important, as the contruction engineer knows at an early stage in which way his machine will respond later on. The results of these investigations are more closely related to future reality as have done former separate approaches.

Mechatronic service as provided by SIEMENS offers several advantages to the machine manufacturer :

- Shorter developing time
- Creative products
- Higher safety in development

- Fewer prototypes
- Faster market entry
- Highly productive machines

When to use :

So far, mechatronic service has only been employed when finished machines did not work in the proper way. However it is more effective to implant an overall consideration at the very beginning of the engineering process.

Already after pre-development, possible weak points can be recognized and removed at low costs. As proven in the projects carried out so far, an early use of mechatronic service saves money and now and then even the survival of a company.

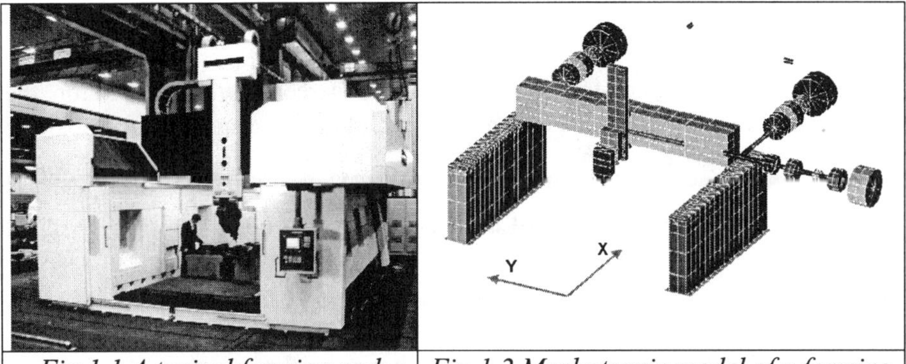

| Fig.1.1 A typical forming and milling machine | Fig.1.2 Mechatronic model of a forming and milling machine |

2. General Limits of Machine-Dynamics

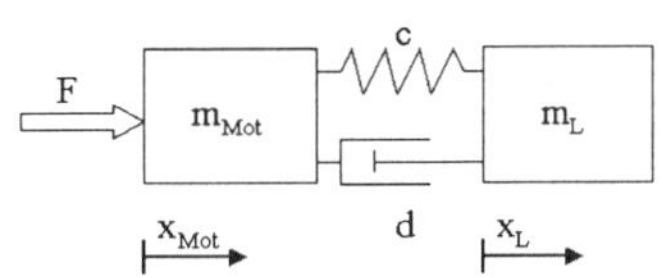

Fig. 2.1 Simple Model of a Production Machine

In a producing machine the movement of a load - carrying a tool - must be controlled by a motor (*Fig.2.1*). The machine builder is only responsible for his mechanical part. The features of this pure mechanical system are completely described by the transfer-function between the motion of load and

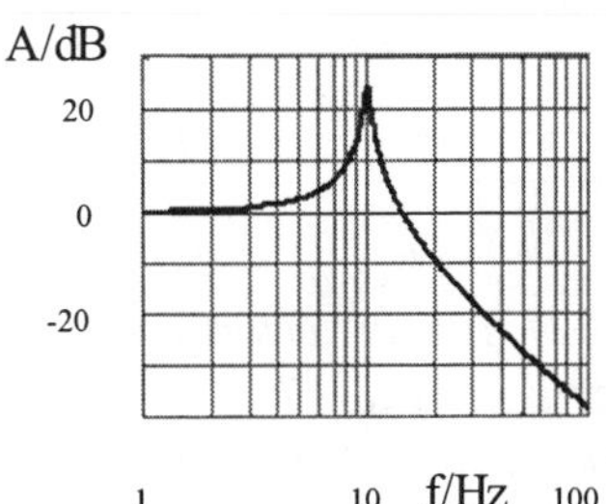

Fig.2.2 Transfer-Function between the Movements of Load and Motor : X_{Load} / X_{Motor}

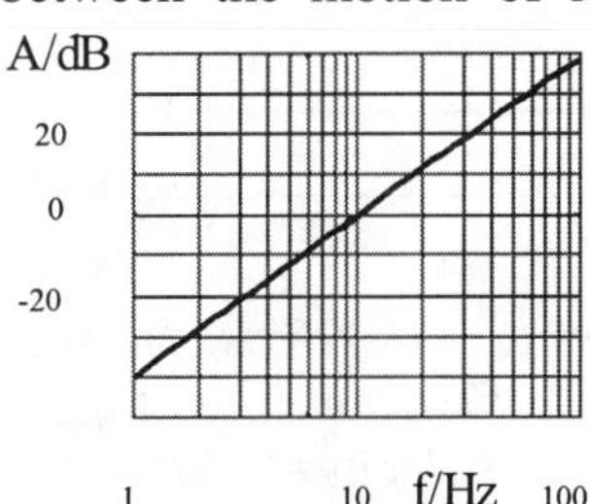

Fig.2.3 Transfer-Function between Deformation of Spring and Movement of Load : $\Delta X_{Spring} / X_{Load}$

that point where motor will be attached (*Fig.2.2*). This mechanical system - with a prescribed motion – has the same eigenvalues as if the motor would be locked. So its eigenfrequencies are called „locked-motor"-frequencies.

Beyond the first locked-motor frequency (in this example 10Hz) the load is dynamically rapidly decoupling of the motor (*Fig.2.2*). At resonance-point (10Hz) the deformation of spring equals the movement of load (*Fig.2.3*). Whereas at 30Hz the motor has to move 1000% of what is required at the load. So the first locked-motor-frequency is a strong dynamical limit of the mechanical system that cannot be easily exceeded without destroying the spring.

The first locked-motor-frequency is in the same way decisive for the attainable control dynamics [1]. The equivalent time constant Tex of position control will be greater than

$$\text{Tex} > 1.41/(2\pi*\{\text{First Locked-Motor-Frequency}\}) \qquad (2.1)$$

3. Example of Mechatronic Engineering

The principle of mechatronic analysis is shown on the example of a forming and milling machine (*Fig.1.1*). A simple FEM model of the machine is developed out of drawings or - in best case – with the help of the technical designer (*Fig.1.2*). In this example the X-axis is mechanically less critical than the Y-axis. So we concentrate on this latter axis.

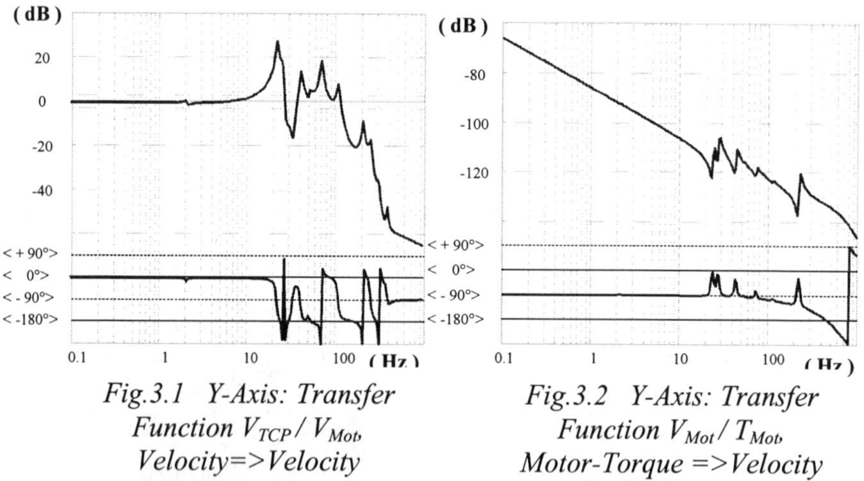

Fig.3.1   Y-Axis: Transfer Function V_{TCP}/V_{Mot} Velocity=>Velocity

Fig.3.2   Y-Axis: Transfer Function V_{Mot}/T_{Mot} Motor-Torque =>Velocity

Fig.3.1 shows the transfer-function of the mechanical part. A remarkable feature of this axis is that between 20 and 30 Hz two resonances turn up very close together. The dynamic behaviour of the machine will be limited by the resonance at 23Hz (*Eq.2.1*). .This mechanical mode appears as pole in *Fig.3.1* and as zero in *Fig.3.2*. Comparing eigenmodes and transfer-function of the mechanical system (*Fig.3.1,3.3,3.4,3.5*) we identify the springs and inertias that are actively involved with specific resonances :

- 23, 26Hz, pitching oscillations of support, can hardly be measured by the used linear scale. Seen from outside the shape of these two modes is nearly identical
- 43Hz, resonance of the supporting side panels
- 74Hz, oscillation in the structure of tool carrier

By this knowledge the machine may be improved, if necessary.

The 23Hz-mode of the Y-axis is attenuated by effect of control but not the neighbouring 26Hz-mode. In time response of velocities (*Fig.3.6*) the 26Hz-mode is revealing dominant in the motion of support.

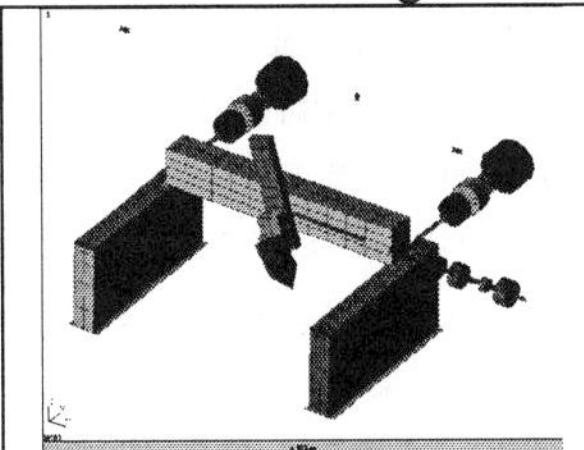

Fig.3.3 Y-Axis Natural Frequency: 23,26Hz Pitching Oscillation

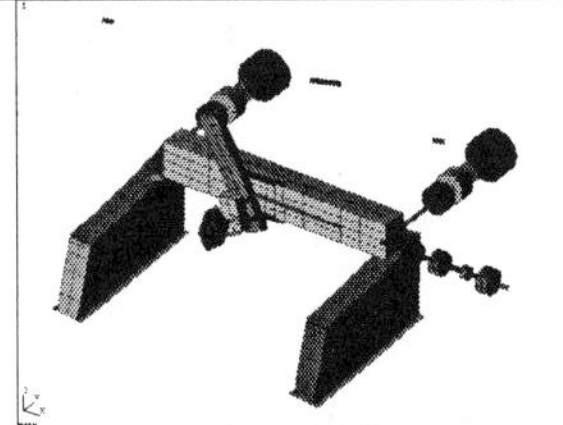

Fig.3.4 Y-Axis Natural Frequency: 43Hz Oscillation of Beam and Side Panels

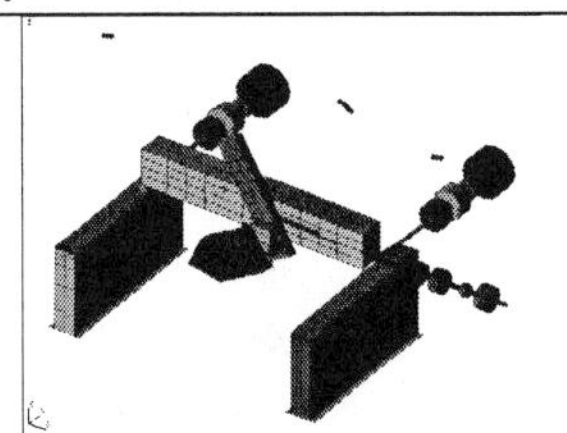

Fig.3.5 Y-Axis Natural Frequency: 74Hz Oscillation of Strukture of Tool Carrier

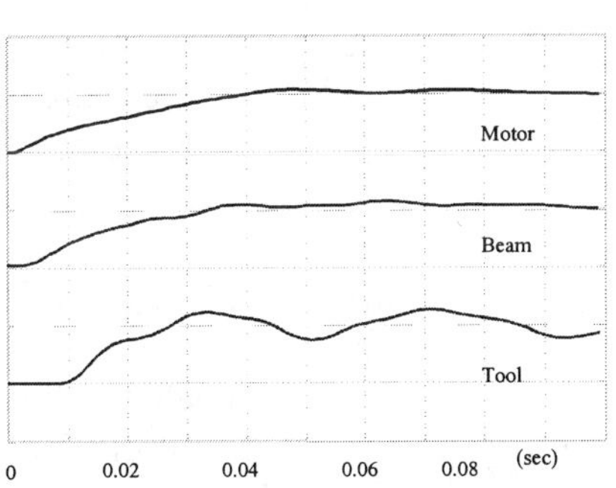

Fig.3.6 Y-Axis: Velocity-control, step in setpoint.
Depicted (from above) : motor, locus of position measurement,

The oscillation of the side panels (*Fig.3.4*) cannot be detected in contrast to the highly frequent structural oscillation of tool carrier (*Fig.3.5*). Due to their specific features at least 3 modes cannot be eliminated by means of control. This problem only can be solved by changes in mechanical construction. The attainable dynamic performance of control is given by the 23Hz of pitching/translational motion $KV < 2.3\text{m/min/mm}$ (Limit according *Eq.2.1*).

According to the compliance transfer response (*Fig.3.7*) the statical stiffness at TCP amounts to $2\text{N/}\mu\text{m}$. This relatively poor value is mainly caused by the quadratic effect of the long lever arm between TCP and the guidings at the beam.

A machine must be examined due to its behaviour when cutting Aluminium (~300Hz), steel (~80Hz) or titanium (~20Hz). According to *Fig.3.7* only light alloys can be cut on this machine. When cutting steel a chatter pattern will be produced on the surface due to the nearby resonance (74Hz) of the tool carrier. And because of the uncontrollable 26Hz pitching-resonance there is no question of working with titanium. This machine can be used for welding, laser-cutting or working with Aluminium. For a laser-system the dynamic performance of the machine is relatively poor.

One characteristic of performance of the controlled machine is the circle-test (*Fig.3.8*). In this case the outline produced by the Y-axis is about 50µm worse than that of the X-axis. Hence the Y-axis is the weakest axis and the pitching motion is strongly excited (*Fig.3.6*) by a step in setpoint. As to be seen in the eigenmode (*Fig.3.3*), the pitching motion could be weakened by enlarging the size of the cross-slide. Only by using a twin-drive above and below this mode can be attenuated by means of control. The first solution reduces the available work-space, the second one raises strongly the expense for drives....

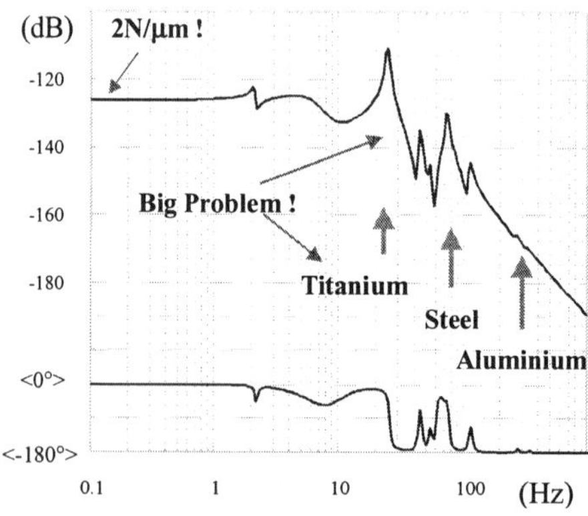

Fig.3.7 Compliance-Frequency-Response $\Delta Y_{TCP} / \Delta F_{TCP}$

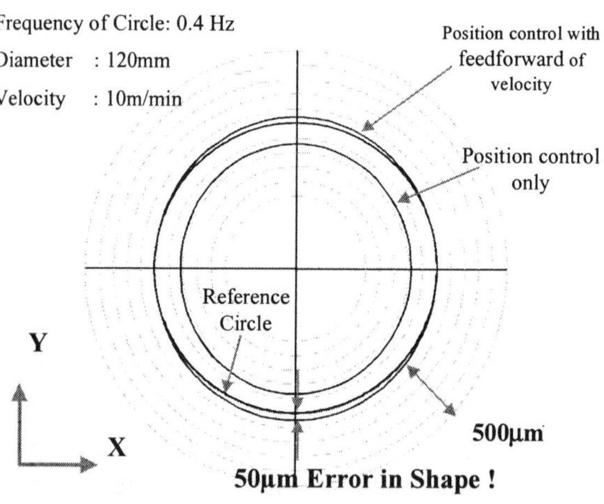

Fig.3.8 Shape of Circle, Central Position, Circle : Ø120mm, Velocity :10m/min

And another iteration of engineering starts up!

[1]   Hamann J.,Troendle H.P. The Sound Barrier in Controlling Elastomechanical Systems. 3.Magdeburger Maschinenbau-Tage 11.-13.September 1997, Tagunungsband 2, S.193-203, Logos Verlag Berlin

Gaussian-Based Free Form Deformation and its Application to Fashion Design

Norimasa Yoshida, Masaru Usui, and Katsuhiro Kitajima

Department of Computer Science
Tokyo University of Agriculture and Technology
2-24-16 Naka-cho, Koganei-shi, Tokyo, 184-8588 Japan
tel. +81-42-388-7453   fax +81-42-385-9747
E-mail: {norimasa, uma, kitajima}@cc.tuat.ac.jp

Abstract

This paper introduces an interactive and intuitive fashion design system using Gaussian-Based Free Form Deformations (GFFDs). GFFDs have several advantages over conventional Free Form Deformations. In GFFDs, Control points need not be positioned parallelpipedically and the locality of deformation can be easily controlled. Thus, GFFDs give designers more freedom in designing clothes. The theory of GFFDs, as well as some examples of clothes designed by our method, is presented.

Keywords

geometric modeling, Gaussian-Based free form deformations, fashion design

1. Introduction

Creating an interactive and intuitive fashion design system that arouses designer's creativity is a challenging problem. To provide intractiveness and intuitiveness for designing clothes, we present Gaussian-Based Free Form Deformations (GFFDs) and its application to fashion design. In GFFDs, control points can be placed more freely than conventional free form deformations[1][4] and the locality of deformations can be changed through modifying the standard deviations of Gaussian functions. We present the theory of Gaussian-Based Free Form Deformations and show some examples of clothes designed by our method.

2. Three Dimensional Fashion Design System

The manufacturing process of clothes goes through (1) design, (2) pattern making, (3) grading, (4) marking, (5) cutting, and (6) sewing. Conventional Apparel CAD systems[2][3] are mainly related to (2)-(5), and they did not aid the design process. We are developing a three-dimensional fashion design system, which is related to (1) and provides designers with interactiveness and intuitiveness.

To design clothes, various kinds of things, such as the geometry, material, and textile pattern of clothes, need to be taken into consideration. Since the geometry of clothes is very important for people in forming their impression, we mainly focus on the method that creates (modifies) the geometry of clothes.

3. Free-Form Deformations Based on Gaussian Functions

Free-Form Deformations(FFDs) based on Gaussian Functions are defined similarly to conventional FFDs that are based on Bernstein or B-spline functions[1][4], except that the basis functions are replaced by Gaussian functions. The GFFD of a point whose coordinates are (s,t,u) is performed by

$$\mathbf{P}(s,t,u) = \sum_{i=1}^{n} \mathbf{V}_i g_i(s,t,u) \tag{1}$$

where

$$g_i(s,t,u) = \frac{w_i G_i(s,t,u)}{\sum_{j=1}^{n} w_j G_j(s,t,u)} \tag{2}$$

is the i th basis function, and

$$G_i(s,t,u) = \exp(-(s-s_i)^2/2\sigma_{i,s}^2 \\ -(t-t_i)^2/2\sigma_{i,t}^2 - (u-u_i)^2/2\sigma_{ui,s}^2) \tag{3}$$

is a 3D Gaussian function of height 1 centered at (s_i,t_i,u_i). $\mathbf{V}_i$ is the i th control point, and $\sigma_{i,s},\sigma_{i,t},\sigma_{i,u}$ are the i th standard deviations in the direction of s,t,u, respectively. Since the 3D Gaussian functions are defined for arbitrary point coordinates in R^3, Eq.(1) defines a non-linear mapping from R^3 to R^3. Different from conventional FFDs, control points can be placed arbitrarily (not necessary to be placed parallelpipedically), and the effect of moving a control point on a deforming object can be modified through changing the standard deviations. To enable direct manipulation of arbitrary points on the surface, as well as any other points, we introduce manipulation points. By performing the inverse transformation of Eq.(1), control points are implicitly computed from manipulation points. The direct manipulation of GFFDs is performed by moving manipulation points.

4. Examples of Designed Clothes using GFFDs

In the design process, the designer should be able to freely modify the geometry of clothes(Our experimental system can also change the topology). Since GFFDs do not have restrictions of the positions and the number of manipulation points, the designer can freely modify the geometry of clothes.

The design proceeds by deforming the basic models prepared in our system. Fig. 1 shows some of the models.

The geometric modification of clothes is performed by placing and moving manipulation points. Fig. 2(a) shows original clothes, and (b) is the designed one from (a) by several GFFDs. The deformation process is drawn in Fig. 2(c). The designer manually places the manipulation points on the surface of the clothes, and moves them controlling the locality of the deformation.

To prevent the designer from placing many manipulation points and deciding their influence, the high-level user interface is prepared. The designer can place several groups of manipulation points and move each group separately. Fig.3 shows an example of designing a turtleneck by selecting and moving several groups of manipulation points.

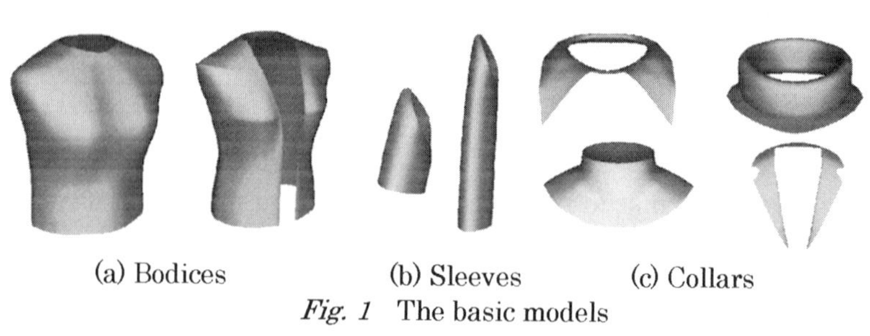

(a) Bodices (b) Sleeves (c) Collars

Fig. 1 The basic models

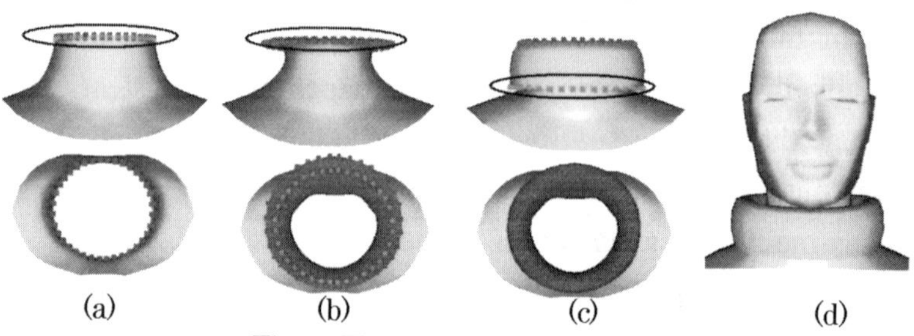

(a) original (b) designed (c) the process

Fig. 2 The design process by manual operations

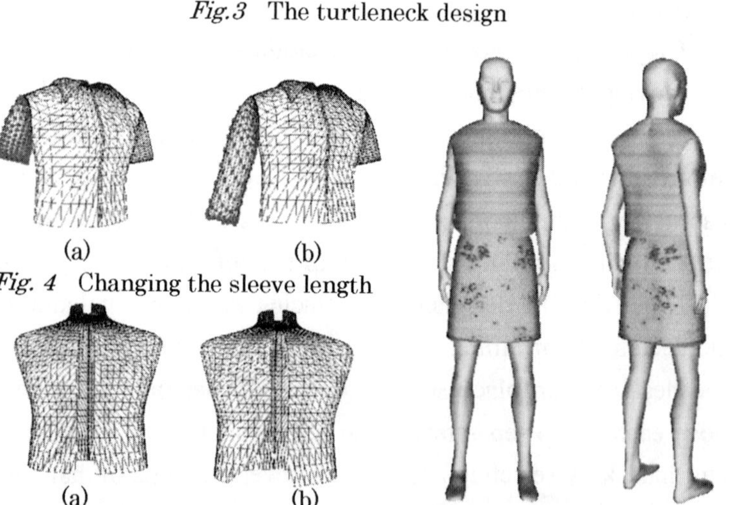

(a) (b) (c) (d)

Fig.3 The turtleneck design

(a) (b)

Fig. 4 Changing the sleeve length

(a) (b)

Fig. 5 Cutting by a curve

Fig. 6 Designed clothes worn by a virtual human

Since a GFFD is based on Gaussian functions, the displacement of a manipulation point influences the overall space. The influence gets exponentially small as it gets far away from the manipulation point. To remove small influence of displacing manipulation points at the regions the designer does not want to deform, the deformation region can be specified. Fig. 4 shows an example of changing the length of the sleeve by restricting the region in which a GFFD is performed by giving a constraint that the sleeve attaches to the bodice.

Several topological operations, such as cutting by a plane or a curve, are also prepared. Fig. 5 shows a topological modification of cutting the bodice by a curve.

Fig. 6 shows the designed clothes worn by a virtual human.

5. Conclusions

We have presented a Gaussian-Based Free Form deformation and its application to fashion design. In comparison with other FFDs, GFFDs do not have a topological restriction of control points and can control the locality of deformation. Due to these advantages, we have confirmed that GFFD is effective in designing clothes.

Acknowledgement

We express our gratefulness to Mr. Takashi Kondo, a fashion designer, in giving valuable comments.

References

[1] D. Bechmann: "Multidimensional free-form deformation tools," State of the Art Reports (STARs), Eurographics'98, 105-112, Sep. 1998.

[2] H. Okabe, H. Imaoka, T. Tomiha, and H. Niwaya: "Three dimensional apparel CAD system," Computer Graphics, 26(2), pp.105-110, 1992.

[3] T. Noma, K. Terai, and T. L. Kunii, "VIRGO A computer-aided apparel pattern-making system," Advanced Computer Graphics (Proceedings of Computer Graphics Tokyo'86), pp.379-401, Springer-Verlag, 1986.

[4] T. W. Sederberg and S. R. Parry: "Free-Form Deformation of Solid Geometric Models," Computer Graphics, 20(4), pp.151-160, 1986.

A Proposal of Assembly Model Framework Specialized for Unified Parametrics

Sho SANAMI, Norimasa YOSHIDA and Katsuhiro KITAJIMA

Tokyo University of Agriculture and Technology Kitajima laboratory

2-24-16 Naka-cho, Koganei-shi, Tokyo, 184-8588 Japan

tel. +81-42-388-7151   fax. +81-42-387-4619

E-mail: {sana, norimasa, kitajima}@cc.tuat.ac.jp

Abstract

Almost all of the previous papers discussing parametrics problems have a common premise that the necessary information solving parametrics is given by a human (not transformed from CAD data) and much attention has been paid in the methods of solving them.   Recently, the word 'assembly model' has spread widely.   We have already proposed in another paper a method of unified parametrics including not only conventional parametrics problems for dimensions of a single part, but kinematics problems.   In this paper, we discuss the relationship between our method of unified parametrics and the assembly model.

Keywords

Assembly model, Unified parametrics, Constraint equation

1. INTRODUCTION

The objective of this research is to propose an assembly model framework that can be applied to unified parametrics.   Since an assembly model is a core model for 3-dimensional CAD systems (3D-CAD), it should have various kinds of information that can be used in each phase such as design, analysis, manufacturing, etc.   However, we focus here on an assembly model framework specialized for the unified parametrics, which will be one of most important applications in 3D-CAD systems in the future.

Unified parametrics for an assembly model involves parametrics for not only a single part but also multiple parts.   Moreover, it involves kinematics problems.   In another paper[1], we proposed a method of automatically generating non-linear simultaneous equations expressing the latent conditions of assembly structure, which can be applicable to the above all problems. In our method, all the problems are solved by a unified method.

In this paper, we focus on an assembly model framework specialized for the unified parametrics and mention how to input/manipulate various kinds of necessary constraints to/in the assembly model.

2. ASSEMBLY MODEL FRAMEWORK
2.1 Outline

An assembly is composed of multiple parts. Some parts construct a subassembly hierarchically based on the mechanical functions of the parts. See Fig. 1.

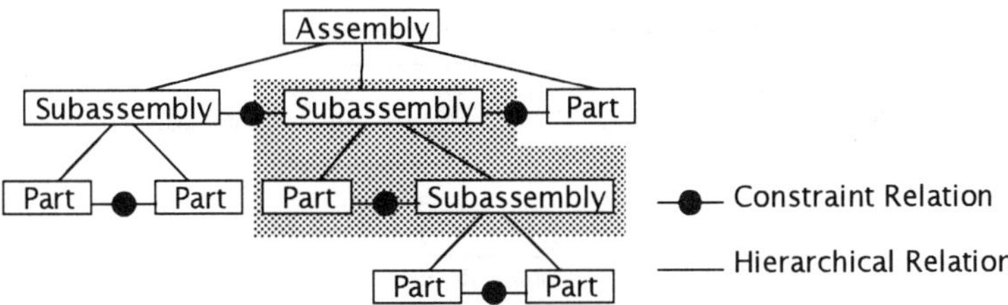

Figure 1.   Hierarchical structure of an assembly model

We call 'assembly', 'subassembly' and 'part' as 'component'. Each component has assembly features through which two components are related to each other. There are two types of relations, constraint relations and hierarchical relations, which are binomial. See Fig. 2.

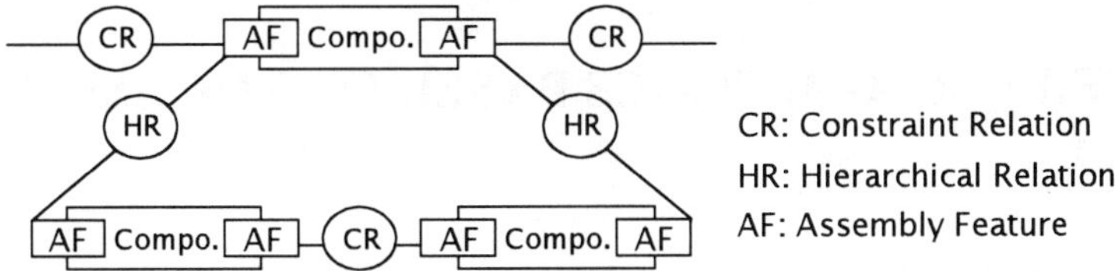

Figure 2.   A model representing components and relations

2.2 Data Structure

An assembly model is composed of components, constraint relations and assembly features.   See Fig. 3.

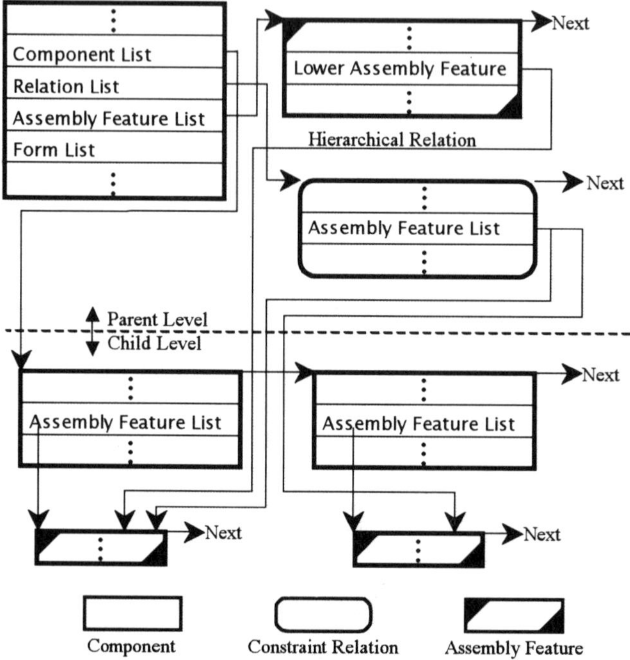

Figure 3.   Data structure of an assembly model

The information represented in the assembly model specialized for unified parametrics is roughly classified into two types: geometric information of a part and the constraint relations between components.   The former is stored in the Form List shown in Fig.3.   The latter is stored in the constraint relation in Fig.3 as transformation matrices.

3. UNIFIED PARAMETRICS BASED ON THE ASSEMBLY MODEL

We briefly show the method of automatically generating constraint equations by extracting the necessary information from the above assembly model.   The constraint equations for an assembly are classified into the following three types.   See Fig. 4.

1. RA: Equations based on a loop of constraint relations between child components.

2. RB: Equations based on constraint relations that do not belong to any loops.

3. RC: Equations based on hierarchical relations.

In RA. and RB., each constraint relation is extracted from the *relation list* in the data structure (Fig.3), and in RC, each hierarchical relation from the *lower assembly feature* (Fig. 3).

Thus, all the information in the data structure of the assembly model is used to generate the constraint equations.

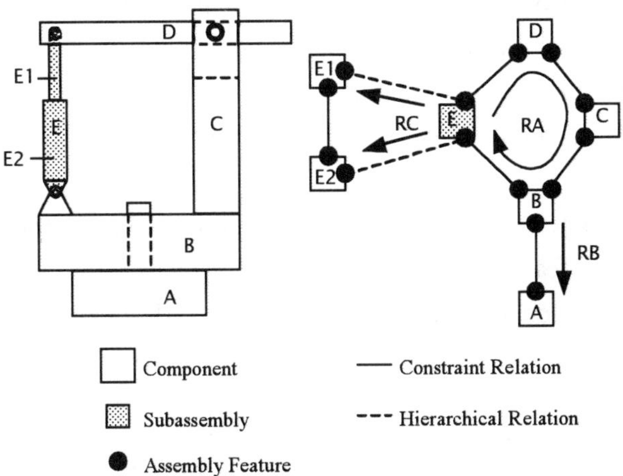

Figure 4. An example showing constraint/hierarchical relations

4. OPERATIONS TO DEFINE THE NECESSARY INFORMATION FOR UNIFIED PARAMETRICS

Here, we focus on how to define the necessary information in the data structure of the assembly model. User can define it through various kinds of operations, which are represented as a combination of the following three types of elementary operations:

1. Operations for defining a hierarchical structure

a. Add/delete a component to/from a component of parent level

The deletion of a component deletes its lower level components including their assembly features and constraint/hierarchical relations.

2. Operations for defining relations to the structure defined above

b.   Add/delete an assembly feature to/from a component

In the case of deleting an assembly feature, the constraint/hierarchical relations related to it are also deleted.

c.   Add/delete a constraint/hierarchical relation to/from a component

The deletion of a constraint/hierarchical relation does not have an effect on other information.

3. Operations for defining the initial condition of constraint equations or other equations such as those given by dimensions

d.   Set/delete a value for each parameter

The assembly model specialized for parametrics involves various parameters, such as the width, height, length, angle, etc.   This sets/deletes constant values for these parameters.

e.   Set/delete an equation describing the relations of dimensions, motion curves, etc.

Each parameter is given a unique name including a component name (and an assembly feature name).

Each operation has its inverse operation.   Using inverse operations, 'undo' functions can be realized.

5. CONCLUSION

In this paper, we proposed the assembly model framework specialized for unified parametrics and elementary operations to define the necessary information.   We have verified that constraint equations, which play an important role in unified parametrics, are efficiently generated by extracting the necessary information from not only hierarchical structures but also loop structures represented by using the framework.

REFERENCES

[1] S. Sanami, K. Kitajima.   A Study on Dimension Matching for Mechanical Assemblies. Proc. the 4th JAPAN-FRANCE CONGRESS & 2nd ASIA-EUROPE CONGRESS on MECHATRONICS; 1998:787-792

PRECISION MACHINING OF THIN-WALLED WORKPIECE BY NC COMPENSATING DEFLECTION

Ning He,   Kai Wu,   Zhigang Wang,   Chengyu Jiang   Dunwen Zuo

The College of Mechanical and Electrical Engineering,

Nanjing University of Aeronautics and Astronautics, Nanjing, 210016,China

Tel: (86) 25 489 2551, Fax: (86) 25 489 1501, E mail: drnhe@nuaa.edu.cn

Abstract

Machining deformation is one of critical problems in machining of thin-walled workpiece, which will result in poor size precision of the wall. This paper proposes a NC compensation method to control the machining quality of the thin-walled workpiece. In the method the finite element method (FEM) was applied to analyze the machining deformation quantity of the thin-wall structure. Then, we can feed the tool to cut excessive material that is left due to machining deflection according to the FEM results by NC compensation. Therefore, we can machine the thin-walled workpieces with high precision and high efficiency.

Keywords

FEM, thin-walled workpiece, machining deflection, NC compensation.

1. INTRODUCTION

Because of their poor stiffness, thin-walled workpieces are very easy to deflect under the cutting force during the process of cutting. It is difficult to machine the side surface of thin-walled workpiece with precision using an end mill. Even in CNC milling, in which tools are controlled exactly according to the contour of the thin-walled workpiece, the wall will be thicker on the top and thinner on the root. The most commonly used method in factories is spark-out finishing machining. This will increase working time and surface roughness, however, some parts will be over-tolerant yet.

Many researches have devoted their efforts to investigate the analysis and prediction of cutting force and deflection in end milling of thin-walled workpiece. Many researchers[1,2,5,7] use theoretical analysis and Finite Element Method (FEM) to calculate the deflection of workpiece and helical end mill during machining. The machining error of the side surface is obtained from the relative displacement of the workpiece and tool, and the calculated values of machining error almost coincide with the experimental values. These results are very useful to predict the machining accuracy and simulate the end milling process. As for the problem of high efficiency and high precision machining of thin-walled workpieces, beside the new approach of ultra-high speed machining, the NC compensation method can be applied by using of the FEM analyzed results.

Based on the FEM analysis of workpiece and tool deflections, this paper put forward a method of NC compensation to machine the thin-walled workpieces. In the method, the machining deflections will be analyzed by FEM, then according to the quantity of the deflection, we can feed the tool to cut excessive material that is left due to the deflection by NC compensation. In this way, the thin-walled workpiece can be machined with high precision and high efficiency on common 5 axis CNC machining center.

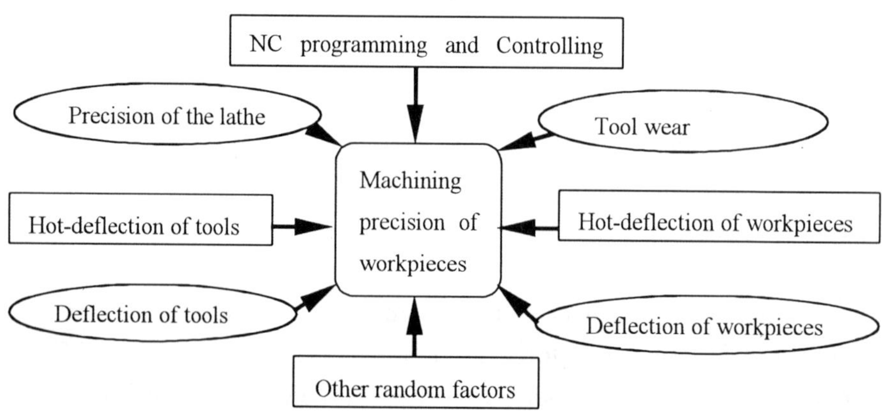

Figure 1 Factors affecting machining precision of the workpiece

2. MACHINING ERROR OF THIN-WALLED WORKPIECES

The factors affecting the machining precision of thin-walled workpieces are shown in Figure1. The precise NC machines are usually applied to mill thin-walled components in airspace industry. Because of thinner walls of the workpieces, their stiffness is so poor that they deform very seriously. Therefore, among the factors in Figure 1, the deflection of workpieces is the most important factor.

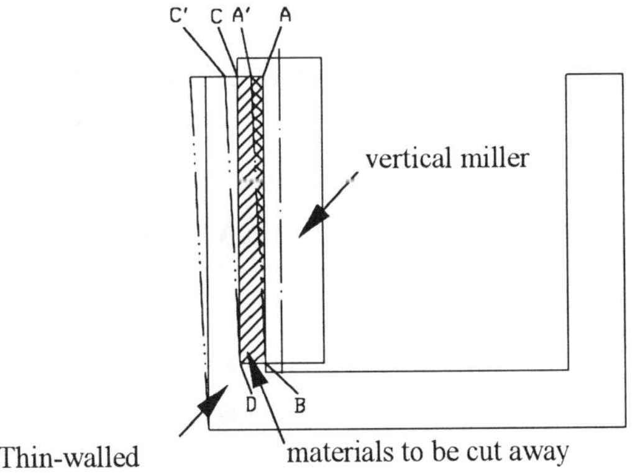

Figure 2 Machining sketch map of the typical

Figure 2 is a machining sketch map to illustrates the deflection of the thin-walled workpieces in machining process. The material ABDC need be cut away ideally. However, under the acting of milling force, point C is moved to Point C', so does the Point A. Therefore, only material A'BDC is cut away in practical machining process due to the cutting force. After the miller moves away from the milling surface, the wall recovers elastically, and material CDC' that should be cut away remains unremoved. It is the most common problem in the machining of thin-walled workpieces.

Take a thin-walled box of titanium alloy as an example. the cutting force distribution along miller axial is approximated to be ladder-shaped by theoretical analysis and the machining deflections are calculated by a ANSYS 5.4 FEM software under different cutting conditions.

The Figure 3 shows the deflections of the horizontal section at the middle of the wall height. It indicates that the deflection is in inverse proportion to the wall thickness. The maximum deflection takes place at the center of the wall. The Figure 4 shows the deflections of the vertical section. The higher part of the workpiece has larger deflection, while the lower part of it deforms little. The shape of remnants in vertical section is similar a del and this makes the wall thickness over-tolerence.

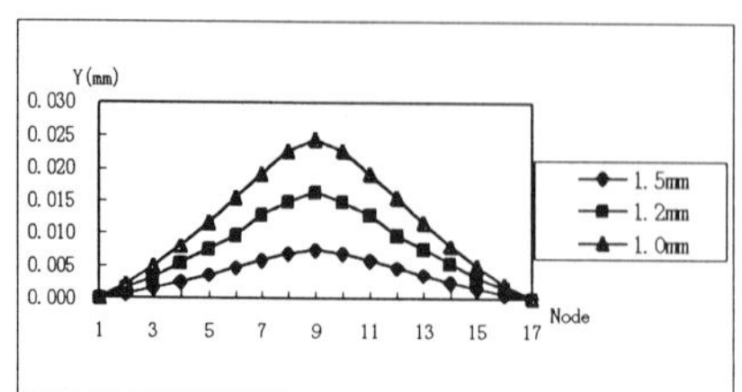

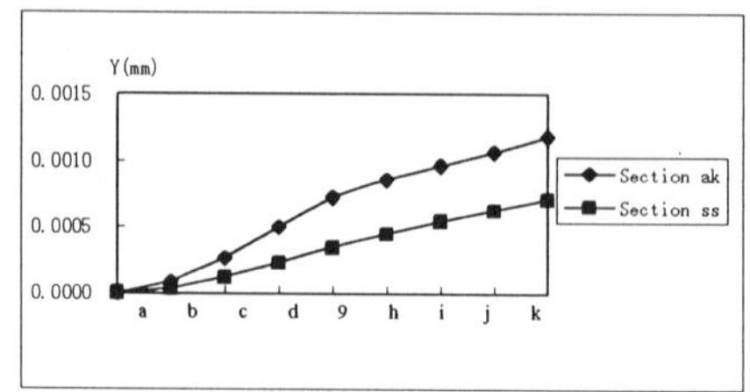

Figure 3 The deflections of
horizontal section

Figure 4 The deflections of
vertical Section

3. THE METHOD OF NC COMPENSATION

If we know the quantity of the deflection, the remnants can be cut away by NC compensation of tilting miller. Thus after machining, the wall recovers elastically to form desired shape and keep the dimension of the workpieces in tolerance.

We proposed a NC compensation method as showing in Figure5. The NC compensation system consists of cutting force model, machining deflection model and NC compensation model. The cutting force model, which can be set up by

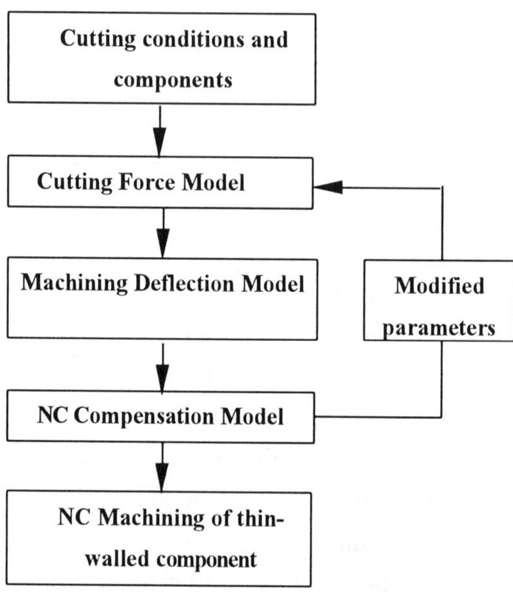

Figure 5 The NC compensation system for machining thin-walled workpiece

theoretical analysis or cutting force experiment, describes the distribution of the cutting force on miller under certain cutting conditions. The machining deflection model is based on FEM and it can simulate the deflection of the thin wall under the cutting force. The NC compensation model can send a compensation command to NC program according to the deflection given by machining deflection model. One of the compensations is to tilt miller. Because the NC compensation changes cutting parameters, the force and deflection will change too. Thus, cycle analysis is required to modify compensation until satisfied precision is reached. All the three models are operated prior machining on machining center. Therefore, if the three models can be established, the thin-walled workpiece can be machined on a common 5 axial CNC machining center with high precision and high efficiency.

4. CONCLUSIONS

The NC compensation provides a new approach to machine thin-walled workpiece with high precision and high efficiency on common CNC machining center. However, because of the complexity of workpieces, there are many works to be done to establish the models for different types of thin-walled structure, and some factors that were neglected in this paper should be taken into account.

REFERENCES

1. Budak E. and Altintas,Y. Modeling and avoidance of static form errors in peripheral milling of plates. Int.J.Mach.Tools Manuf.1995;35(3):459-476.
2. Iwabe H. Study on machining accuracy of thin wall workpiece by end mill, Trans. Of the JSME（C）1997-1;63(605):239-246 (In Japanese)
3. Lechniak Z. Methodology of off-line software compensation for errors in the Koji Teramoto, Coordinative Generation of Machining an fixturing plans by a modularized problem solver, CIRP Annals 1998; 47(1):437-440
4. Tlustry J. Techniques for the use of long slender end mills in high-speed milling, CIRP Annals 1996; 45(1):396-399
5. Tsai J.S. et al. Finite-element modeling of static surface errors in the peripheral milling of thin-walled workpieces. J. of Mater. Proc. Tech. 1999; 94:235-246
6. Zhang, Youzhen. *Principles of Metal Cutting*. Beijing:Aviation Industry Publisher, 1988(in Chinese)
7. Zhen L., Shun Y., Liang S. Y. Three dimensional cutting force analysis in end milling. Int. Journal of Mechanical Sciences 1996; 38(3):259-269.

Index of Contributors

Ahmed, A.   897
Akbari, J.   426
Aketagawa, M.   819
Alam, M.   324
Ali, M.Y.   142, 147
Ando, K.   872
Andrae, P.   97
Anzai, M.   794
Aoki, S.   344
Aoki, T.   932
Aoyama, H.   304
Aoyama, H.   604
Aoyama,H. 42, 887, 917
Aoyama, T.   887
Arai, T.   729
Arii, S.   709
Arrasmith, S.R.   501
Asami, M.   476
Asano, M.   937
Asano, S.   799

Bamberg, E.   544
Baumgarten, J.   27
BishopII, T.D.   501
Bo, Y.Z.   209
Boess, V. 847
Braasch, J.   629
Brinksmeier,E.   3, 174
Bueno, R.   529

Chang,T.H.   549, 927
Chao, C.L.   376, 754
Chen, K.H.   754
Chen, K.W.   927
Chen, L.C.   754
Chen, M.   436
Chen, S.L.   549, 927
Cheng, K.   294
Cho, S.H.   381

Choi, K.B.   714
Choy, C.H.   334
Choy, C.M.   42

Dai, Y.   244
Dan, L.   579
Dewen, J.   699, 719
Doege, E.   27
Dong, S.   294, 436, 769
Dornfeld, D.A.   47, 169

Eda, H.   441, 759
Edhi, E.   164
Eguchi, T.   862
Eisseler, R.   157
Enggalhardjo, M. 42, 334
Enomoto, T.   391
Eversheim, W.   867

Fricker, I.   867
Friemuth, T.   97
Fujii, M.   122
Fujimoto, H.   897
Fujiwara, M.   634
Fujiwara, T.   829
Fukazawa, T.   366
Fukuda, M.   604
Fukuda, Y.   679
Fukushima,Y.   852
Furudate, C.   209
Furukawa, Y.   354
Furusawa, T.   289
Fuwa, N. 366

Gao, W.   569
Gatzen, H.H.   431
Gemma, T.   679
Glaebe, R.   174
Godler, I.   564

Gomei, Y.   67
Gorodkin, S.R.   501
Goto, T.   799
Graf, C.   619
Gregg, L.L.   501
Guo, J.   371

Ha, T.   684
Haga, T.   604
Hagiwara, I.   902
Haino, H.   814
Hamann, J.   952
Hamann, J.C.   299
Han, F.   319
Han, R.   396
Hara, S.   194, 204
Haramaki, S.   624
Hashimoto, K.   354
Hata, S.   37
Hatano, M.   87, 804
Hatsuzawa, T. 137, 749
Hayakawa, S.   72
Hayase, M.   137, 749
Hayashi, A.   624, 947
He, N.   967
Heisel, U.   157
Hennig, J.   534
Herrero, A.   529
Herzog, O.   857
Hesselbach, J.   619
Hidai, H.   229
Hidaka, K.   654, 674
Higuchi, M.   179
Higuchi, T.   361
Hilbing, R.   519
Hino, H.   289
Hirai, S.   344
Hiraki, M.   734
Hiraki, Y.   461

Hirano, T.   219
Hiraoka, H.   912
Hirayanagi, N.   829
Hiroi, T.   344
Hirosawa, M.   137
Hirt, G.   32
Ho, M.C.   724
Hocken, R.J. 584, 644
Hoffmeister, H.W.   421
Horie, M.   689
Horisawa,H.   224, 339
Horiuchi, O.   361, 366
Hoshi, T.   164
Hsei, M.H.   549
Hu, B.H.   334
Huang, H   416
Huang, H.   451
Hung, W.N.P.   142, 147
Hwang, Y.R. 724

Ichida, Y. 456
Ichikawa, A.   289
Ichikawa, H.   679
Igoshi, M.   942
Iida, S.   117
Ikawa, N.   179
Ikeno, J.   366
Ikua, B.W.   189
Imai, N.   639
Imai, Y.   319
Inasaki, I.   669, 892
Inoue, T.   892
Ishida, T.   214
Ishihara, S.   604
Ishikawa, J.   809
Ishikawa, N.   609
Isogimi, K.   259
Itoh, N.   486
Itoh, T.   314
Itoigawa, F.   72, 269
Izui, K.   877

Jacobs, S.D.   501
Jeong, E.S.   406

Jeong, H.D.   381
Jeong, H.D.   386, 406
Jeong, S.C.   386
Jiachun, W.   579
Jiang, C.Y.   967
Jichuan, Z.   719
Jin, M.   102, 112
Jo, N.J.   406
Jones, G.M. 431

Kagawa, Y.   629
Kakimoto, A.   664
Kakuta, A.   354
Kamiya, D.   689
Kamlage, G.   62
Kaneeda, T.   107
Kang, C.A.   927
Kanzaki, S.   57
Kasai, T. 401, 471, 486,
            496
Katahira, K.   794
Kataoka, K.   314
Katayama, T.   264
Kathuria, Y.P.   329
Kato, J.   594
Kato, T.   37
Katsuki, A.   264
Kawada, M.   234
Kawai, T. 52, 77
Kawamoto, H.   511, 694
Kawasaki,K   814
Kawata, K.   269
Kikuchi, H.   729
Kim, H.J.   381
Kim, H.Y.   381
Kim, J.   169
Kim, K.J.   381
Kishinami, T.   837
Kitajima, K.   957, 962
Kiyono, S.   569
Kiyotani, S.   744
Klocke, F.   152, 411
Kobayashi, K.G.   122
Kobayashi, T.   67

Koerber, K.   62
Koga, S.   204, 559
Koinuma, T.   456
Kondo, S.   234
Kordonski, W.I.   501
Koroyasu, S.   289
Koschig, M.   867
Koshimizu, S.   184
Kozhinova, I.A.   501
Krizbergs, J.   649
Kulik, C.   62
Kumagai, N.   694
Kumakura, K.   491
Kumar, A.S.   309
Kunieda, M.   209,319
Kuriyagawa,T. 254,446
Kuriyama,Y. 739, 809,
              824
Kurochi, N.   907

Lapp, C.   97
Lee, C.W.   142
Lee, H.S.   554
Lee, K.   47
Lee, K.S.   324
Leopold, J.   669
Li, D.   294,436
Li, J.   599
Li, L.   714
Li, S.   244
Li, S.   779
Li, Y.D.   324
Liang, Y.   294
Lim, S.C. 309
Lin, H.Y.   376
Lin, W. 371,401, 491,
          496, 594,794
Linhong, J.   699, 719
Liu, C.   401
Liu, C.   496
Liu, D.S.   376
Liu, M.   239
Liu, Y.   37
Liu, Y.   396

Lu, Z.   481
Lun, Y.S.   376
Luo, X.   294

Ma, K.J.   376, 754
MACTEST   299
Maeda, Y.   729
Maekawa, K.   117
Maeno, K.   829
Makinouchi, A.   794
Masada, T.   524
Masuda, H.   739
Matsubara, T. 72, 269
Matsuda, T.   664
Matsui, Y.   604
Matsuki, K.   674
Matsumura, T.   764
Matsuo, M.   589
Matsuzawa, T.   371
Mei, X.   704
Meng, Q.X.   902
Meslin, F.   299
Michalas, N.   867
Miki, Y.   829
Min, S.   169
Mitsui, K.   799
Mitsui, S.   604
Mitsuishi, M.   789
Mitsutani, N.   814
Miura, T. 476
Miyamoto, S.   304
Miyoshi, T.   684
Mizumoto, H.   709
Morimoto, Y.   456
Morishige, K.   249
Morita, H.   604
Morita, K.   829
Morita, S.   371, 594
Morita, S.   476,491
Moriwaki, T.   589
Moriyasu, S. 794, 476,
    486, 491, 496, 594
Moromuki, N.   614
Moronuki, N. 132, 354

Morsbach, C.   431
Mueller, U.   629
Muljadi, H.   872
Murakami, H.   264
Murakawa, M.   102,112,
    349
Muramatsu, T.   42
Murayama, T.   764
Murayama, Y.   461
Muto, K.   789

Nagasawa, S.   852
Nakada, T.   569
Nakamura, H.   907
Nakamura, T.   72, 269
Nakanishi, E.   259
Nakao, Y.   82
Nakayama, N.   511
Nakayama, S.   679
Namba, Y.   659
Neumaier, T.   27
Ngothai, Y.   142
Ninomiya, E.   87
Ninomiya, T.   564
Nishi, D.   132
Nishi, K.   279
Nishii, T.   117
Nishimura, K.   654
Nishimura, T.   344
Nishioki,N.   194, 204,
    559, 609
Nitta. T   279
Noethe, T.   152
Noguchi, H.   349

Oba, F.   862
Obara, H.   87, 804
Obata, F.   189
Obata, K.   179
Obikawa, T.   127
Ogawa, M.   872
Ogihara, M.   744, 824
Oguchi, T.   137
Ohashi, T.   937

Ohishi, S.   524
Ohmori, H. 794, 371,
    401, 471, 476,
    486, 491, 496,
    594
Ohsumi, T.   87, 804
Okabe, M.   539
Okada, A.   219
Okajima, T.   659
Okamoto, Y.   219
Onikura, H.   264
Onuki, T.   679
Oozeki, H.   744, 824
Ori, R.I.   72
Orii, K.   391
Osada, M.   589
Ostendorf, A.   62
Ota, M.   917
Otsuka, J.   184
Ozono, S.   634,734

Paehler, D.   411
Park, H.S.   406
Park, J.M.   386
Pei, S.   396
Peschke, C.   519
Prabhuram, P.D.   199
Prakash, J.R.S.   309
Prinz, F.B.   122

Qiu, J.   599

Rackow, N.   847
Rahman, M. 309, 324
Rajurkar, K.P.   199
Rattay, B.   32
Ren, B.Y.   902
Rentsch, R.G.   274
Rerkkumsup, P.   819
Riemer, O.   3
Romanofsky,H.J. 501
Rudzitis, J.   649
Ruebenach, O.   152

Sadakata, S.   819
Sagawa, K.   441
Saijo, M.   456
Saito, A.   92
Saitoh, A.   654
Sajima, T.   264
Sakaguchi, K.   912
Sakai, H.   739
Sakaki, H.   12
Sakakibara, S.   784
Sakamoto, H.   539
Sakamoto, S.   189
Sakamoto, S. 234, 461
Sakata, M.   461
Sakurai, T.   872
Salsbury, J.G.   644
Sanami, S.   962
Sankaran, K.S.   324
Sasahara, H.   279
Sasaki, T.   491
Sata, T.   77
Satake, T.   947
Sato, H.   371
Sato, K.   774
Sato, M.   664
Sato, O.   734
Sato, R.   456
Sawada, K.   52, 77
Schneider, M.   882
Sendai, T.   319
Shamoto, E.   589
Shen, D.   579
Sheu, S.C.   376
Shibata, N.   67
Shibayama, A.   604
Shibutani, H. 361, 366
Shimada, S.   179
Shimizu, J.   441, 759
Shimizu, S.   539, 639
Shimizu, T.   491
Shimokohbe, A.   37, 714,
          774
Shindo, H.   471
Shinmura, T.   506

Shinozuka, J.   127
Shinshi, T.   714
Shorey, A.B.   501
Sihler, J.   599
Skurba, M.   649
Slocum, A.H. 18, 544,
          599
Smith, M.   599
Sornsuwit, N.   77
Stern, R.   3
Stirn, B.   47
Storchak, M.   157
Sturgeon, V.   599
Su, B.   574,769
Suga, T.   314
Sugi, M.   729
Sugisaki, K.   67
Sun, Q.   704
Suzuki, H.   624
Suzuki, H. 361, 366
Suzuki, M.   604
Suzuki, M.   912
Suzuki, T.   117
Suzuki, T.   937
Syoji, K.   254,446

Taguchi, T.   604
Takada, K.   819
Takagi, J.   239, 466
Takagi, T.   819
Takahashi, H.   922
Takahashi, K.   749
Takahashi, S.   684
Takahashi, S.   829
Takamasu, K. 634, 734
Takasaki, K.   694
Takata, S.   789
Takaya, Y.   684
Takeo, S.   907
Takeuchi, Y.   52, 77,
          214, 249
Takino, H.   67
Tamura, S.   224
Tanaka, F.   837

Tanaka, H.   179
Tanaka, H.   189
Tanaka, K.   524, 589
Tanaka, T.   659
Tang, C.W.   754
Tani, Y.   391
Tanimura, H.   799
Tashiro, H.   371
Tateishi, M.   829
Tateno, T.   942
Tawara, Y.   659
Timm, I.J.   857
Toenshoff, H.K. 62, 97,
          847,857
Tokura, H.   229
Tong, K.K.   42
Toyama, N.   837
Toyooka, S.   862
Troendle, H.P.   952
Tsuchiya, S.   907
Tsuji, M.   804
Tsuji, S.   289
Tsurusaki, M.   461
Tsutsumi, M.   92, 704
Tsuyuzaki, H.   604

Uchida, T.   689
Uchida, Y.   554
Uchiyama, K.   132
Uchiyama, K.   614
Ueda, M.   739, 809
Uehara, Y.   476, 491
Ueno, S.   789
Ueshima, Y.   814
Umeda, T.   679
Uno, Y.   219
Urabe, Y.   842
Usui, E.   764
Usui, M   957

Wahyudi   774
Wakuda, M.   57
Wang, L.574, 769, 779
Wang, Z.G.   967

Watanabe, R.   304
Watanabe, T.   87
Watanabe, T.   819
Weck, M.   519, 534
Wenda, A.   421
Wiendahl, H.P.   882
Winterschladen, M. 534
Wiphut, J.   511
Woelk, P.O.   857
Wu, K.   967
Wuyong, Y.   699

Xiaohong, J.   719
Xie, X.   244
Xu, H.   451
Xu, X.   451

Yabuya, M.   709
Yamada, H.   209
Yamagata, Y. 476, 486,
            491, 496,
            594, 794
Yamagiwa, T. 654
Yamaguchi, A.   634
Yamaguchi, I.   594
Yamaguchi, M.   511
Yamaguchi, S.   224
Yamamoto, T.   67, 679
Yamashita, K.   659
Yamauchi, K.   466
Yamauchi, Y.   57
Yamazaki, K.   629
Yan, J.   254
Yao, Y.   284
Yasuda, H.   694
Yasui, H.   234, 461
Yasunaga, N.   224
Yata, T.   664
Yin, L.   416
Yin, S.   506
Yokoyama, Y.   809
Yokozeki, S.   624
Yong, M.S.   334
Yoshida, N. 957, 962

Yoshihara, N.   446
Yoshikawa, K.   476
Yoshimura, M.   877
Yoshino, M.   127
Yu, Z.   199
Yuan, S.   142, 147
Yuan, Z.   284, 481
Yuge, Y.   264

Zaitsu, I.   887
Zhang, F.   396, 436,
            481, 574, 779
Zhang, G.   947
Zhang, H.   284
Zhang, J.   779
Zhang, M.   244
Zhang, S.X.   42
Zhang, X.   714
Zhang, Y.   824
Zhang, Y.F.   324
Zhao, X.   92
Zhou, L.   441, 759
Zhu, B.   481
Zuo, D.   967
Zushi, K.   659